Engineering Mathematics-II

ENGINEERING MATHEMATICS-I

In Accordance With Prescribed Syllabus of AMIETE–ET/CS/IT (New Scheme)

Code: AE–51 / AC–51 / AT–51

Also for engineering, technology and applied sciences examinations of various universities and engineering colleges of India and abroad

(Including Over 600 Problems With Solutions completely solved in detail)

BRIEF CONTENTS

1. Partial Derivatives
2. Applications of Partial Derivatives
3. Elements of Curve Tracing and Standard Curves
4. Multiple Integrals
5. Linear Algebra : Matrices, Vectors, Determinants, Linear Systems of Equations
6. Linear Algebra : Matrix Eigenvalue Problems
7. Ordinary Differential Equations of First Order
8. Applications of Differential Equations of First Order
9. Linear Differential Equations of Second and Higher Order
10. Series Solution of Differential Equations
11. Gamma, Beta and other Special Functions
12. Bessel's Differential Equation, Bessel's Functions, Legendre Differential Equations and Legendre Polynomials
13. Numerical Methods

Engineering Mathematics-II

In Accordance With Prescribed Syllabus of AMIETE–ET/CS/IT (New Scheme) Code: AE–56 / AC–56 / AT–56

Also for engineering, technology and applied sciences examinations of various universities and engineering colleges of India and abroad

(Including Over 500 Problems With Solutions completely solved in detail)

I.J.S. SARNA
B.E. Tech. (Mech. Engg.)
Delhi College of Engineering
Delhi University, Delhi

CBS PUBLISHERS & DISTRIBUTORS PVT. LTD.
New Delhi • Bengaluru • Pune • Kochi • Chennai

ISBN : 978-81-239-2371-0

First Edition : 2014

Published by Satish Kumar Jain and produced by V.K. Jain for
CBS Publishers & Distributors Pvt. Ltd.,
CBS Plaza, 4819/XI Prahlad Street, 24 Ansari Road, Daryaganj,
New Delhi - 110002, India. • Website: www.cbspd.com
e-mail: delhi@cbspd.com, cbspubs@airtelmail.in
Ph.: 23289259, 23266861, 23266867 • Fax: 011-23243014

Branches:

- ***Bengaluru:*** Seema House, 2975, 17th Cross, K.R. Road,
  Bansankari 2nd Stage, Bangalore - 560070 Ph.: 26771678/79
  Fax: 080-26771680 • e-mail: bangalore@cbspd.com
- ***Pune:*** Bhuruk Prestige, Sr. No. 52/12/2+1+3/2,
  Narhe, Haveli (Near Kátraj-Dehu Road by Pass), Pune-411051
  Ph.: +91-20-32404169 • Fax: 020-24464059
  e-mail: pune@cbspd.com
- ***Kochi:*** 36/14, Kalluvilakam, Lissie Hospital Road,
  Cochin - 682018, Kerala • e-mail: cochin@cbspd.com
  Ph.: 0484-4059061-65 • Fax: 0484-4059065
- ***Chennai:*** 20, West Park Road, Shenoy Nagar, Chennai - 600030
  e-mail: chennai@cbspd.com Ph.: 044-26260666-26202620
  Fax: 044-45530020

Printed at :
J.S. Offset Printers, Delhi

Dedicated to the memory of

My Respected Parents

Late Saran Kaur and Late Rawel Singh Sarna

Who are responsible for what I am today

and

My Esteemed Professor Late G.S. Sharma

Who taught Mathematics to Engineering Students at Delhi College of Engineering (Delhi University), Delhi for over four decades

Preface

I have the pleasure in presenting to my readers the second of a two-volume set written to meet the requirements of degree students of engineering, technology, applied sciences and equivalent professional examinations of various universities and engineering colleges of India and abroad. The text of this volume has been divided into seven chapters. The text presented deals with advanced topics in engineering mathematics usually included in the syllabi of degree courses of engineering technology, applied sciences and equivalent professional examinations of most of the universities.

While writing this book, I have made constant endeavour to include theory and large number of typical solved problems along with illustrations. The book also contain a number of supplementary problems with answers (hints have been provided wherever necessary).

A remarkable thing about this book is that it includes solved and unsolved problems collected from the past years examination papers of various universities and engineering colleges in order to familiarize the students with the types of problems asked in these examinations. So far as unsolved problems are concerned, vital hints have been provided wherever necessary so as to enable the readers to arrive at correct answer without any difficulty. Some major features of this book are :

- The text has been presented in such a form that it can be followed by the students with minimum guidance from others.
- Over 500 typical solved problems, from easy to difficult, supplemented with neat sketches wherever necessary have been provided.
- In each chapter maximum possible emphasis has been laid on theory.
- The notations used are as per latest standards.

It is hoped that the book will adequately meet the requirements and expectations of the readers.

Most sincere gratitude to various esteemed authors whose books I have consulted while accomplishing the task of writing this book. I am grateful to my publishers for their hard work and enthusiasm in publishing this book within a short period of time.

For further improvement of this book, I welcome suggestions from my readers.

New Delhi **I.J.S. Sarna**

Syllabus

Syllabus for AMIETE – ET/CS/IT (New Scheme)
Code : AE 56 / AC56 / AT 56

ENGINEERING MATHEMATICS – II

UNIT I — COMPLEX ANALYSIS

Introduction; Function of complex variable $w = f(z)$; Limit of a complex function; Continuity of $w - f(z)$; Derivative of $f(z)$; Analytic function; Cauchy Riemann equations (both in Cartesian and polar form); Harmonic functions; Application to flow problems; Construction of analytic functions using Milne Thomson method, Geometric representation of $w = f(z)$, standard transformation. Bilinear transformation; Conformal transformation; Special conformal transformations.

UNIT II — COMPLEX INTEGRATION

Line integral of $w = f(z)$; Cauchy's theorem; Cauchy's integral formula; Morera's theorem; Series of complex terms; Taylor's and Laurent's series; singularities of analytic function; Types of singularities; Residues; Residue theorem; Calculation of residues.

UNIT III — VECTOR CALCULUS

Introduction to vectors; Differentiation of vectors; Curves in space; velocity and acceleration; Scalar and vector point functions; vector operator del; Del applied to scalar point functions; Gradient; Del applied to vector point functions; Divergence and curl; Physical interpretation of Divergence and Curl; Del applied twice to point functions and products of point functions; vector identities.

UNIT IV — VECTOR INTEGRATION

Integration of vectors; Line integral; Surfaces; Surface integral; Green's theorem in a plane (only statement and problems); Stoke's theorem (only statement and problems); Volume integral; Gauss divergence theorem (only statement and problems).

UNIT V — NUMERICAL METHODS

Forward difference operator; Backward difference operator; Central difference operator; Shifting operator E (only the definition); Interpolation: Newton Gregory forward and backward interpolation for equal intervals; Lagrange's formula for unequal intervals; Divided differences; Newton's divided difference formula; Inverse interpolation; Numerical differentiation using Newton Gregory forward and backward interpolation formula; Numerical integration; Newton-Cote's quadrature formula; Trapezoidal rule; Simpson's $1/3^{rd}$ and Simpson's $3/8^{th}$ rule; Weddle's rule.

UNIT VI — PARTIAL DIFFERENTIAL EQUATION

Introduction; Formulation of partial differential equations; Solutions of a partial differential equations; Equations solvable by direct integration; Lagrange's linear partial differential equation of first order; Solutions of non linear partial differential equations by Charpit's method; Solution of homogenous partial differential equation by the method of separation of variables.

UNIT VII — THEORY OF PROBABILITY

Introduction; Basic terms and definitions; Probability and set notation; Theorem of total probability; Independent events; Theorem of compound probability; Conditional probability; Baye's theorem.

UNIT VIII — RANDOM VARIABLES

Random Variable; Discrete and continuous random variables; Discrete and continuous probability distribution; Probability mass and density function; Mean and variance of discrete and continuous probability distribution; Theoretical distributions; Binomial distribution; Constants of the Binomal distribution; Binomial frequency distribution; Applications of Binomial distribution; Poission's distribution; Constants of the Poisson distribution; Applications of Poisson distribution; Normal distribution.

Note: Students have to answer **Five** full questions out of **Eight** questions to be set from each unit carrying 16 marks.

Contents

Preface vii
Syllabus ix

1 Vector Calculus and Vector Integration 1.1–1.84

Introduction — Parametric Representation of Curves — Derivative of a Vector Function — Derivatives of Sums and Products of Vectors — Derivative of a Scalar Product — Geometric Interpretation of $r'(t)$ — Integrals of Vector Functions — Motion of a Body or Particle on a Curve — The Vector Differential Operator Del — Gradient of a Scalar Field — Geometrical Interpretation of Gradient — Directional Derivative — Properties of Gradient — Divergence and Curl of a Vector Field — Physical Interpretation of Divergence — Physical Interpretation of Curl — Vector Identities — Line Integrals — Applications of Line Integrals — Line Integrals Independent of the Path — Green's Theorem in the Plane (Transformation between double integrals and line integrals) — Surface Integrals — Divergence Theorem of Gauss (Transformation between Volume Integrals and Surface Integrals) — Green's Identities (Formulas) — Stokes's Theorem (Transformation between surface integrals and line integrals).

2 Partial Differential Equations 2.1–2.42

Introduction — Formation of Partial Differential Equations — Equations Solvable by Direct Integration — Linear Partial Differential Equations of the First Order — Non-linear Equations of the First Order — Charpit's Method — Linear Homogeneous Equations with Constant Coefficients — Rules for Finding the Complementary Function (C.F.) — Rule to write the Complementary Function — Particular Integral (P.I.) (Special Methods) — General Method for Finding P.I. — Non-homogeneous Linear Equations with Constant Coefficients — Non-linear Differential Equations of the Second Order.

3 Applications of Partial Differential Equations 3.1–3.38

Introduction — Initial and Boundary Conditions — Method of Separation of Variables (Product Method) — Classification of Linear Second Order Partial Differential Equations — Equation of Vibrating String — Boundary Conditions — One-Dimensional Heat Flow: $\frac{\partial u}{\partial t} = c^2 \frac{\partial^2 u}{\partial x^2}$ — Two Dimensional Heat Flow, $\frac{\partial^2 u}{\partial x^2} + \frac{\partial^2 u}{\partial y^2} = a^2 \frac{\partial u}{\partial t}$ — Solution of Transmission Line Equations Problems by the Method of Separation of Variables.

4 Complex Analysis and Complex Integration 4.1–4.126

Introduction — **Functions of a Complex Variable: Analytic Functions** — Limits — Continuity of a function —Complex Differentiation — Analytic Functions — Polar form of Cauchy-Riemann Equations —Derivative of $w = f(z) = u + iv$ in Polar Form — Properties of Analytic Functions — Method of Construction of the Conjugate Function — Complex Potential Function — **Integration of Complex Functions** — Contour Integrals (Line Integrals in the Complex Plane) — Cauchy's Integral Theorem — Simple Closed Curve, Simply and Multiply Connected Domain — Cauchy's-Goursat Theorem — Cauchy's Integral Formula — Cauchy's Integral Formula for Derivatives — Morera's Theorem — Cauchy's Inequality — **Power Series, Taylor and Laurent Series** — Power Series —Taylor Series — Taylor's Theorem — Laurent Series — **Zeros, Singularities and Residues** —Introduction — Zeros and Singularities of Complex Functions —Evaluation of Contour Integrals Using Residues — Residue Theorem — Evaluation of Real Integrals Using Residues — **Bilinear Transformations and Conformal Mapping** — Introduction — Inverse Transformation — Bilinear Transformation — Conformal Mapping.

5 Numerical Methods 5.1–5.94

Introduction — Differences; Forward Differences, Backward Differences, Central Differences, — Difference Operators, Δ, ∇, δ, E, μ — Relations between the Operators — Factorial Polynomial; Methods of Expressing an Algebraic Polynomial in Factorial Notation — Newton's Formulae For Interpolation; Newton's Forward and Backward Differences Formula for equal Intervals —Central Difference Interpolation Formulae; Gauss's Forward Formula, Gauss's Backward Formula, Stirling Formula, Bessel's Formula, Choice of Application of an Interpolation Formula — Interpolation with Unevenly Spaced Points; Lagrange's Interpolation Formula, Divided Differences, Newton's Formula for Divided Differences — Inverse Interpolation; Lagrange's Method, Method of Successive Approximations or the Method of Iteration, By Reversion of Series — Error in Newton's Interpolation Formulae — Numerical Differentiation — Introduction — Derivative using Newton's Forward and Backward Difference Formula — Derivative using Central Differences — Stirling's Formula — Maximum and Minimum value of a Tabulated Function — Numerical Integration, Introduction — Newton-Cotes quadrature or integration formula; Trapezoidal Rule, Simpson's 1/3 Rule, Simpson's 3/8th Rule, Weddle's Rule — Errors in Computation.

6 Statistical Methods 6.1–6.75

Introduction — Frequency Distribution — Graphical Representation of a Frequency Distribution — Frequency Curve — Measures of Central Tendency — Arithmetic mean (AM) — Short methods for Computing the Arithmetic Mean — Two Important Properties of Arithmetic Mean — Median (Med.) — Mode — Empirical Relation Between Mean, Median and Mode — The Geometric Mean (G.M.) and Harmonic Mean (H.M.) — Relation Between Arithmetic, Geometric and Harmonic Means — Partition Values — Quartiles, Deciles and Percentiles — Dispersion of Variation — Range — Quartile Deviation or Semi-inter-quartile Range — Mean Deviation — Variance and Standard Deviation — Variance of the Combined Series — Relation Between Quartile Deviation, Mean Deviation and Standard Deviation — Coefficient of Variation (C.V.) — Moments—Introduction — Moments about an Arbitrary Point '*a*' — Moments about the Mean — Relation Between the Moments about the Mean and Moments about any Point — Relation Between μ_r and μ'_r — Karl Pearson's Beta (β) and Gamma (γ) Coefficients based on Moments — Skewness

— Measures of Skewness — Kurtosis — Measure of Kurtosis — Curve Fitting — The Lines of Best Fit — Correlation — Scatter Diagram or Dot Diagram — Karl Pearson's (Product Moment) of Correlation Coefficient — Short-Computational Formulae — Correlation Table — Rank Correlation — Spearman's Modified Formula for Tied Ranks — Lines of Regression — Correlation and Regression.

7 Theory of Probability and Random Variables 7.1–7.70

Introduction — Random Experiments and Events — Definition of Probability — Theorems on Probability — Conditional Probability — Independent Events — Theorem of Total Probability — Inverse Probability—Baye's Theorem — Random Variables; Discrete Random Variable, Distribution Function, Continuous Random Variables and Distributions, Cumulative Distribution Function (cdf) — Theoretical Probability Distributions, Binomial Distribution, Poisson Distribution, Normal Distribution — Some Other Distributions, Uniform (or Rectangular) Distribution, Exponential Distribution, Beta Distribution.

Appendices A.1–A.15

Appendix-1 Some Constants, Some Important Conversion Factors, Greek Alphabet

Appendix-2 Mathematical Formulas

Appendix-3 Some Derivative and Integral Formulas

Appendix-4 Table I: Normal Distribution

Table II: Values of e^{-m}

Appendix-5 Solved Examination Paper AMIETE (New scheme) June 2013

CHAPTER

Vector Calculus and Vector Integration

1.1 INTRODUCTION

In this chapter, we shall study the vector differential and integral calculus. We often call this study as vector analysis. We first introduce few concepts.

Point Functions

A variable quantity the value of which at any point in a region of space depends on the position of the point, is called a *point function* or *function of position.*

There are two types of point functions: (*a*) *Scalar point function* (*b*) *Vector point function.*

Scalar Point Function

If to each point P in a region of space R, there corresponds a scalar $f(P)$, then f is called a *scalar function of position* or *scalar point function* and we say that scalar field f has been defined in R. Some of the examples of scalar fields are:

(*i*) The distance $f(P)$ of any point P from a fixed point P_0 in space is a scalar field. This is certainly independent of the choice of the coordinate system.

(*ii*) The temperature T at any point on the earth's surface at a certain time.

(*iii*) The density at any point of a certain body occupying given region.

(*iv*) $\phi(x, y, z) = 3xy^2z + x^2y$ defines a scalar field.

Vector Point Function

If to each point P in a region R of space there corresponds a vector $\mathbf{v}(P)$, then $\mathbf{v}$ is called a *vector function of position* or *vector point function* and we say that a vector field $\mathbf{v}$ has been defined in R. Some examples of vector fields are:

(*i*) velocity at each point of a fluid in motion.

(*ii*) tangent vector to a curve in space.

(*iii*) normal vector to a surface.

(*iv*) $\mathbf{v} = \mathbf{v}(P) = v_1\mathbf{i} + v_2\mathbf{j} + v_3\mathbf{k}$ defined at each point P.

(*v*) gravitational and electromagnetic force fields etc.

If we introduce cartesian coordinates x, y, z then the value of the scalar point function f at a point $P(x, y, z)$ may be written as $f(x, y, z)$ but we should bear in mind that this value is independent of particular choice of axes of co-ordinates and depends only on the position of P. In order to indicate this fact we sometimes write $f(P)$ instead $f(x, y, z)$.

Similarly at a point $P(x, y, z)$ the vector point function $\mathbf{v}$ may be written in the form:

$\mathbf{v}(x, y, z) = v_1(x, y, z)\mathbf{i} + v_2(x, y, z)\mathbf{j} + v_3(x, y, z)\mathbf{k}$ bearing in mind that $\mathbf{v}$ depends only on the position of P and defines the same vector for every choice of co-ordinate system. This fact is indicated symbolically by writing $\mathbf{v}(P)$ in place of $\mathbf{v}(x, y, z)$.

If the scalar and vector fields depend on time also, then we denote them as $f(P, t)$ and $\mathbf{v}(P, t)$ respectively. Both the fields are independent of the choice of the coordinate systems.

Differentiation of these scalar and vector point functions follow the same rules as those of ordinary calculus.

Thus if $f(x, y, z)$ and $\mathbf{v}(x, y, z)$ are respectively scalar and vector point functions, then

$$df = \frac{\partial f}{\partial x}dx + \frac{\partial f}{\partial y}dy + \frac{\partial f}{\partial z}dz, \qquad \frac{df}{dt} = \frac{\partial f}{\partial x}\frac{dx}{dt} + \frac{\partial f}{\partial y}\frac{dy}{dt} + \frac{\partial f}{\partial z}\frac{dz}{dt},$$

and

$$d\mathbf{v} = \frac{\partial \mathbf{v}}{\partial x}dx + \frac{\partial \mathbf{v}}{\partial y}dy + \frac{\partial \mathbf{v}}{\partial z}dz, \qquad \frac{d\mathbf{v}}{dt} = \frac{\partial \mathbf{v}}{\partial x}\frac{dx}{dt} + \frac{\partial \mathbf{v}}{\partial y}\frac{dy}{dt} + \frac{\partial \mathbf{v}}{\partial z}\frac{dz}{dt}.$$

Level Surfaces

If a surface $f(x, y, z) = c$ = constant, is drawn through any point P of the region such that at each point on the surface, the function has the same value as at P then such a surface is known as *level surface* through P. For different values of c, we obtain a family of non-intersecting surfaces. For example, if $f(x, y, z)$ represents temperature in a medium, then $f(x, y, z) = c$ represents a surface on which the temperature is a constant c. Such surfaces are called *isothermal* surfaces.

1.2 PARAMETRIC REPRESENTATION OF CURVES

A curve C in the two dimensional xy-plane can be parametrised by $x = x(t), y = y(t), a \le t \le b$. Then, the position vector of a point P on the curve C can be written as

$$\mathbf{r}(t) = x(t)\mathbf{i} + y(t)\mathbf{j}, \qquad \text{...(1.1)}$$

where t is called the parameter of the representation. The parameter t may be time or something else. Therefore, the position vector of a point on a curve defines a vector function, (Fig. 1.1). Similarly, a three-dimensional curve or a space curve C can be parametrised as

$$\mathbf{r}(t) = x(t)\mathbf{i} + y(t)\mathbf{j} + z(t)\mathbf{k}, \quad a \le t \le b. \qquad \text{...(1.2)}$$

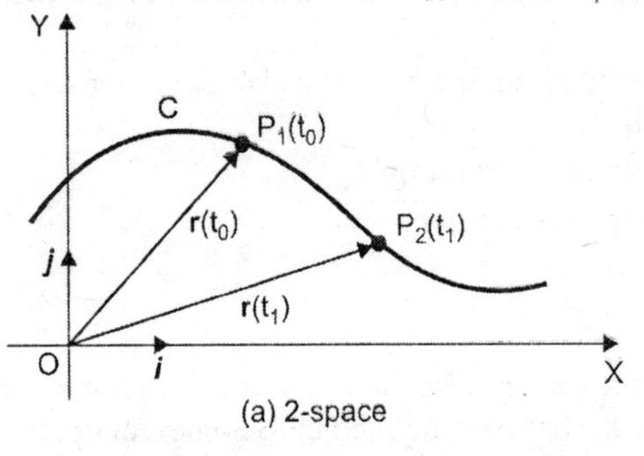

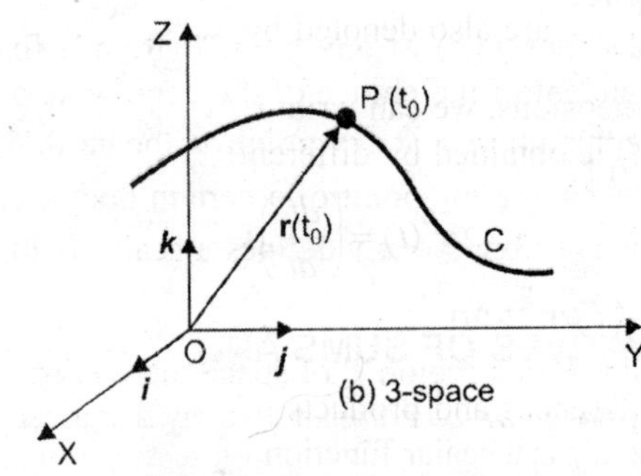

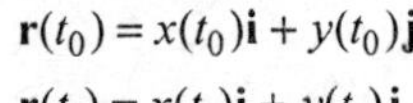

$$\mathbf{r}(t_0) = x(t_0)\mathbf{i} + y(t_0)\mathbf{j}$$

$$\mathbf{r}(t_1) = x(t_1)\mathbf{i} + y(t_1)\mathbf{j}$$

$$\mathbf{r}(t_0) = x(t_0)\mathbf{i} + y(t_0)\mathbf{j} + z(t_0)\mathbf{k}$$

Fig. 1.1: Position vector of a point.

The following are the parametric forms of some of the curves:

- **Straight line:** A straight line L which passes through the point A with position vector $\mathbf{a}$ and has the direction of a constant vector $\mathbf{b}$ can be represented in the parametric form as:

$$\mathbf{r}(t) = \mathbf{a} + t\mathbf{b} = (a_1 + tb_1)\mathbf{i} + (a_2 + tb_2)\mathbf{j} + (a_3 + tb_3)\mathbf{k}. \qquad \text{...(1.3)}$$

- **Circle:** The parametric form of the circle $x^2 + y^2 = a^2$ in the xy plane can be written as $\mathbf{r}(t) = a\cos t\,\mathbf{i} + a\sin t\,\mathbf{j}, \quad 0 \le t \le 2\pi.$ ...(1.4)
- **Ellipse:** The parametric form of the ellipse $\frac{x^2}{a^2} + \frac{y^2}{b^2} = 1, \quad z = 0$ can be written as

$$\mathbf{r}(t) = a\cos t\,\mathbf{i} + b\sin t\,\mathbf{j}, \quad 0 \le t \le 2\pi. \qquad \text{...(1.5)}$$

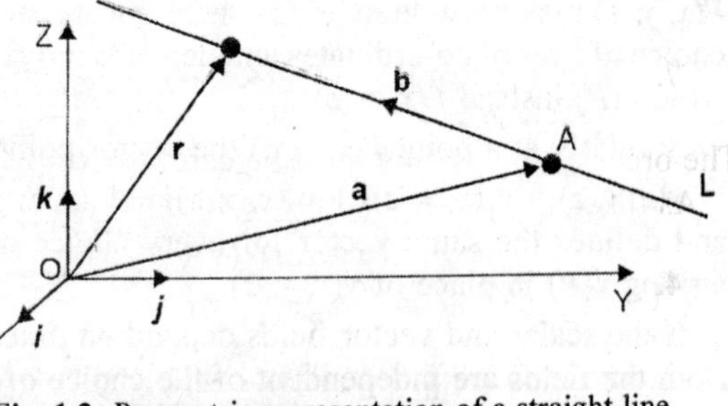

Fig. 1.2: Parametric representation of a straight line.

- **Parabola:** Consider the parabola as $y^2 = 4ax$. Its parametric equations are $x = at^2$, $y = 2at$. The vector function $\mathbf{r}(t) = at^2\,\mathbf{i} + 2at\mathbf{j}$ ...(1.6) represent a parabola in the xy-plane in standard form.
- **Circle in a plane in 3-dimensions:** Consider the circle having centre at origin and radius 'a', which lies in the plane $z = d$. The parametric form can be written as $\mathbf{r}(t) = a\cos t\,\mathbf{i} + a\sin t\,\mathbf{j} + d\,\mathbf{k}$. ....(1.7)

1.3 DERIVATIVE OF A VECTOR FUNCTION

If to each value of a scalar variable t there corresponds a value of a vector **r**, then **r** is called a *vector function* of t and is denoted by $\mathbf{r}(t)$.

Let the variable t be given a small increment Δt and let $\Delta\mathbf{r}$ be the corresponding change in **r** so that

$$\Delta\mathbf{r} = \mathbf{r}(t + \Delta t) - \mathbf{r}(t).$$

Then the derivative of a vector function $\mathbf{r}(t)$ with respect to t is defined by,

$$\frac{d\mathbf{r}}{dt} = \lim_{\Delta t\to 0}\frac{\Delta\mathbf{r}}{\Delta t} = \lim_{\Delta t\to 0}\frac{1}{\Delta t}[\mathbf{r}(t+\Delta t) - \mathbf{r}(t)], \quad ...(1.8)$$

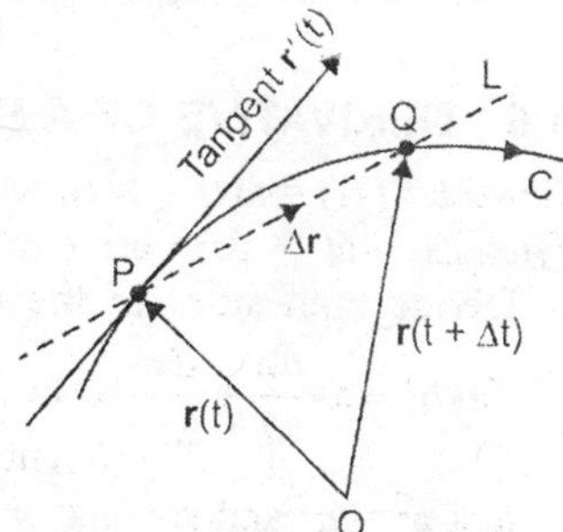

Fig. 1.3: Derivative of a vector function.

for all t for which the limit exists. This derivative is also denoted by $\mathbf{r}'(t)$, since division by the scalar Δt does not change the vectorial properties of $\Delta\mathbf{r}$, therefore the derivative $d\mathbf{r}/dt$ is also a vector. See Fig. 1.3. We call $\mathbf{r}'(t)$ a *tangent vector* of C at P because it has the direction of the tangent.

The derivative of $d\mathbf{r}/dt$ with respect to t is defined exactly in same way and is denoted by $\dfrac{d\mathbf{r}'(t)}{dt}$ or $\dfrac{d^2\mathbf{r}}{dt^2}$ and is called the **second derivative** of **r**. Similarly, higher order derivatives of any kind can be defined. The derivatives $\dfrac{d\mathbf{r}}{dt}, \dfrac{d^2\mathbf{r}}{dt^2}, \dfrac{d^3\mathbf{r}}{dt^3}, \ldots$ are also denoted by $\dot{\mathbf{r}}, \ddot{\mathbf{r}}, \dddot{\mathbf{r}}, \ldots$.

In three dimensions, we can write $\mathbf{r}(t) = f(t)\mathbf{i} + g(t)\mathbf{j} + h(t)\mathbf{k}$, f, g and h are differentiable functions of t. The derivative $\mathbf{r}'(t)$ is obtained by differentiating each component separately, that is,

$$\mathbf{r}'(t) = \left(\frac{df}{dt}\right)\mathbf{i} + \left(\frac{dg}{dt}\right)\mathbf{j} + \left(\frac{dh}{dt}\right)\mathbf{k} = f'(t)\,\mathbf{i} + g'(t)\,\mathbf{j} + h'(t)\,\mathbf{k}. \quad ...(1.9)$$

1.4 DERIVATIVES OF SUMS AND PRODUCTS OF VECTORS

The sums, differences and products of vectors can be differentiated by formulas similar to those of ordinary calculus. Thus if f is a scalar function of t and **u, v** and **w** are differentiable vector functions of a scalar t, then

1. $\dfrac{d}{dt}(c\mathbf{u}) = c\dfrac{d\mathbf{u}}{dt}$ (c constant)
2. $\dfrac{d}{dt}(\mathbf{u} \pm \mathbf{v}) = \dfrac{d\mathbf{u}}{dt} \pm \dfrac{d\mathbf{v}}{dt}$
3. $\dfrac{d}{dt}(f\mathbf{u}) = f\dfrac{d\mathbf{u}}{dt} + \dfrac{df}{dt}\mathbf{u}$
4. $\dfrac{d}{dt}(\mathbf{u}\cdot\mathbf{v}) = \mathbf{u}\cdot\dfrac{d\mathbf{v}}{dt} + \dfrac{d\mathbf{u}}{dt}\cdot\mathbf{v}$
5. $\dfrac{d}{dt}(\mathbf{u}\times\mathbf{v}) = \mathbf{u}\times\dfrac{d\mathbf{v}}{dt} + \dfrac{d\mathbf{u}}{dt}\times\mathbf{v}$
6. $\dfrac{d}{dt}(\mathbf{u}\cdot\mathbf{v}\times\mathbf{w}) = \dfrac{d\mathbf{u}}{dt}\cdot\mathbf{v}\times\mathbf{w} + \mathbf{u}\cdot\dfrac{d\mathbf{v}}{dt}\times\mathbf{w} + \mathbf{u}\cdot\mathbf{v}\times\dfrac{d\mathbf{w}}{dt}$
7. $\dfrac{d}{dt}[\mathbf{u}\times(\mathbf{v}\times\mathbf{w})] = \dfrac{d\mathbf{u}}{dt}\times(\mathbf{v}\times\mathbf{w}) + \mathbf{u}\times\left(\dfrac{d\mathbf{v}}{dt}\times\mathbf{w}\right) + \mathbf{u}\times\left(\mathbf{v}\times\dfrac{d\mathbf{w}}{dt}\right)$.

The order of the vectors must be carefully observed because cross multiplication is not commutative.

As illustrations we shall prove formulae (4), (5) and (6) only. Others can be proved on the same lines.

4. $\dfrac{d}{dt}(\mathbf{u}\cdot\mathbf{v}) = \lim\limits_{\Delta t\to 0}\dfrac{(\mathbf{u}+\Delta\mathbf{u})\cdot(\mathbf{v}+\Delta\mathbf{v}) - \mathbf{u}\cdot\mathbf{v}}{\Delta t} = \lim\limits_{\Delta t\to 0}\dfrac{\mathbf{u}\cdot\Delta\mathbf{v} + \Delta\mathbf{u}\cdot\mathbf{v} + \Delta\mathbf{u}\cdot\Delta\mathbf{v}}{\Delta t}$

$$= \lim_{\Delta t\to 0}\left(\mathbf{u}\cdot\frac{\Delta\mathbf{v}}{\Delta t} + \frac{\Delta\mathbf{u}}{\Delta t}\cdot\mathbf{v} + \frac{\Delta\mathbf{u}}{\Delta t}\cdot\Delta\mathbf{v}\right) = \mathbf{u}\cdot\frac{d\mathbf{v}}{dt} + \frac{d\mathbf{u}}{dt}\cdot\mathbf{v}.$$

5. $$\frac{d}{dt}(\mathbf{u}\times\mathbf{v}) = \lim_{\Delta t\to 0}\frac{(\mathbf{u}+\Delta\mathbf{u})\times(\mathbf{v}+\Delta\mathbf{v})-\mathbf{u}\times\mathbf{v}}{\Delta t} = \lim_{\Delta t\to 0}\frac{\mathbf{u}\times\Delta\mathbf{v}+\Delta\mathbf{u}\times\mathbf{v}+\Delta\mathbf{u}\times\Delta\mathbf{v}}{\Delta t}$$

$$= \lim_{\Delta t\to 0}\left(\mathbf{u}\times\frac{\Delta\mathbf{v}}{\Delta t}+\frac{\Delta\mathbf{u}}{\Delta t}\times\mathbf{v}+\frac{\Delta\mathbf{u}}{\Delta t}\times\Delta\mathbf{v}\right) = \mathbf{u}\times\frac{d\mathbf{v}}{dt}+\frac{d\mathbf{u}}{dt}\times\mathbf{v}.$$

6. $$\frac{d}{dt}(\mathbf{u}\cdot\mathbf{v}\times\mathbf{w}) = \frac{d}{dt}\mathbf{u}\cdot(\mathbf{v}\times\mathbf{w}) = \mathbf{u}\cdot\frac{d}{dt}(\mathbf{v}\times\mathbf{w})+\frac{d\mathbf{u}}{dt}\cdot(\mathbf{v}\times\mathbf{w})$$ [by formula (4)]

$$= \mathbf{u}\cdot\left[\left(\mathbf{v}\times\frac{d\mathbf{w}}{dt}\right)+\left(\frac{d\mathbf{v}}{dt}\times\mathbf{w}\right)\right]+\frac{d\mathbf{u}}{dt}\cdot(\mathbf{v}\times\mathbf{w})$$ [by formula (4)]

$$= \mathbf{u}\cdot\mathbf{v}\times\frac{d\mathbf{w}}{dt}+\mathbf{u}\cdot\frac{d\mathbf{v}}{dt}\times\mathbf{w}+\frac{d\mathbf{u}}{dt}\cdot\mathbf{v}\times\mathbf{w}.$$

1.5 DERIVATIVE OF A SCALAR PRODUCT

Consider $f(t) = \mathbf{a}(t)\cdot\mathbf{b}(t)$, where **a** and **b** are vector functions of the scalar t. Then we have $f(t+\Delta t) = \mathbf{a}(t+\Delta t)\cdot\mathbf{b}(t+\Delta t)$ or $f(t+\Delta t)-f(t) = \mathbf{a}(t+\Delta t)\cdot\{\mathbf{b}(t+\Delta t)-\mathbf{b}(t)\}+\{\mathbf{a}(t+\Delta t)-\mathbf{a}(t)\}\cdot\mathbf{b}(t)$.

Divide both sides of the identity by Δt and proceed to the limit as Δt tends to zero. It follows that $\frac{d}{dt}(\mathbf{a}\cdot\mathbf{b}) = \mathbf{a}\cdot\frac{d\mathbf{b}}{dt}+\frac{d\mathbf{a}}{dt}\cdot\mathbf{b}$. In particular, we have $\frac{d}{dt}\mathbf{a}^2 = \frac{d}{dt}(\mathbf{a}\cdot\mathbf{a}) = 2\mathbf{a}\cdot\frac{d\mathbf{a}}{dt}$.

But $\mathbf{a}^2 = a^2$ and so $\frac{d}{dt}\mathbf{a}^2 = \frac{2a\,da}{dt}$. Hence we have $\mathbf{a}\cdot\frac{d\mathbf{a}}{dt} = a\frac{da}{dt}$.

Further, if **a** is a vector of *constant magnitude*, $da/dt = 0$ and so $\mathbf{a}\cdot d\mathbf{a}/dt = 0$.

That is, the derivative of a vector of constant magnitude is either orthogonal to it or zero.

Remark 1

A vector changes if either its magnitude or direction changes or both magnitude and direction changes. The necessary and sufficient condition for a vector function to remain constant in magnitude or in direction or in both is given below:

(a) For the vector $\mathbf{v}(t)$ to be constant, $d\mathbf{v}/dt = \mathbf{0}$. That is, the derivative of a constant vector is the zero vector.

(b) For the vector $\mathbf{v}(t)$ to have a constant magnitude, $\mathbf{v}\cdot\frac{d\mathbf{v}}{dt} = 0$.

(c) For the vector $\mathbf{v}(t)$ to have a constant direction, $\mathbf{v}\times\frac{d\mathbf{v}}{dt} = \mathbf{0}$.

EXAMPLE 1.1 (*i*) Differentiate $\frac{t\,\mathbf{i}+t^2\mathbf{j}}{|t\,\mathbf{i}+t^2\mathbf{j}|}$ with respect to t.

(*ii*) If $\frac{d\mathbf{u}}{dt} = \mathbf{w}\times\mathbf{u}$ and $\frac{d\mathbf{v}}{dt} = \mathbf{w}\times\mathbf{v}$, show that $\frac{d}{dt}(\mathbf{u}\times\mathbf{v}) = \mathbf{w}\times(\mathbf{u}\times\mathbf{v})$.

Solution (*i*) Since $|t\,\mathbf{i}+t^2\mathbf{j}| = \sqrt{t^2+t^4}$.

Therefore, $$\frac{d}{dt}\left\{\frac{t\,\mathbf{i}+t^2\mathbf{j}}{|t\,\mathbf{i}+t^2\mathbf{j}|}\right\} = \frac{d}{dt}\left[\frac{t\,\mathbf{i}+t^2\mathbf{j}}{\sqrt{t^2+t^4}}\right] = \frac{d}{dt}\left[\frac{\mathbf{i}+t\,\mathbf{j}}{\sqrt{1+t^2}}\right]$$

$$= \frac{\sqrt{1+t^2}\,(\mathbf{j})-(\mathbf{i}+t\mathbf{j})\frac{t}{\sqrt{1+t^2}}}{1+t^2} = \frac{(1+t^2)\mathbf{j}-t\,\mathbf{i}-t^2\mathbf{j}}{(1+t^2)^{3/2}} = \frac{\mathbf{j}-t\,\mathbf{i}}{(1+t^2)^{3/2}}.$$

(*ii*) Now $$\frac{d}{dt}(\mathbf{u}\times\mathbf{v}) = \frac{d\mathbf{u}}{dt}\times\mathbf{v}+\mathbf{u}\times\frac{d\mathbf{v}}{dt} = (\mathbf{w}\times\mathbf{u})\times\mathbf{v}+\mathbf{u}\times(\mathbf{w}\times\mathbf{v}) = (\mathbf{w}\cdot\mathbf{v})\mathbf{u}-(\mathbf{u}\cdot\mathbf{v})\mathbf{w}+(\mathbf{u}\cdot\mathbf{v})\mathbf{w}-(\mathbf{u}\cdot\mathbf{w})\mathbf{v}$$

[$\because \mathbf{A}\times(\mathbf{B}\times\mathbf{C}) = (\mathbf{A}\cdot\mathbf{C})\mathbf{B}-(\mathbf{A}\cdot\mathbf{B})\mathbf{C}$]

$$= (\mathbf{w}\cdot\mathbf{v})\mathbf{u}-(\mathbf{u}\cdot\mathbf{w})\mathbf{v} = \mathbf{w}\times(\mathbf{u}\times\mathbf{v}).$$

EXAMPLE 1.2 Find $\mathbf{v}'(t)$ in each of the following cases

(*i*) $\mathbf{v}(t) = (\sin t)\,(5t^2\mathbf{i} + t\,\mathbf{j} + t^3\mathbf{k})$,

(*ii*) $\mathbf{v}(t) = (\sin 2t\,\mathbf{i} - \cos 2t\,\mathbf{j} + t\,\mathbf{k}) \cdot (\cos 2t\,\mathbf{i} - \sin 2t\,\mathbf{j} + t^2\mathbf{k})$,

(*iii*) $\mathbf{v}(t) = (6t^2\mathbf{i} - t\,\mathbf{j} + 3t^2\mathbf{k}) \times (t\,\mathbf{i} + t^2\mathbf{j} + 2t\,\mathbf{k})$.

Solution (*i*) Using formula 3, Article 1.4, we obtain

$$\mathbf{v}'(t) = (\sin t)\,(5t^2\mathbf{i} + t\,\mathbf{j} + t^3\mathbf{k})' + (\sin t)'\,(5t^2\mathbf{i} + t\,\mathbf{j} + t^3\mathbf{k})$$

$$= \sin t\,(10t\mathbf{i} + \mathbf{j} + 3t^2\mathbf{k}) + (\cos t)\,(5t^2\mathbf{i} + t\,\mathbf{j} + t^3\mathbf{k})$$

$$= (10t \sin t + 5t^2 \cos t)\mathbf{i} + (\sin t + t \cos t)\mathbf{j} + (3t^2 \sin t + t^3 \cos t)\mathbf{k}.$$

(*ii*) Using formula 4, Article 1.4, we obtain

$$\mathbf{v}'(t) = (\sin 2t\,\mathbf{i} - \cos 2t\,\mathbf{j} + t\,\mathbf{k}) \cdot (\cos 2t\,\mathbf{i} - \sin 2t\,\mathbf{j} + t^2\mathbf{k})' + (\sin 2t\,\mathbf{i} - \cos 2t\,\mathbf{j} + t\,\mathbf{k})' \cdot (\cos 2t\,\mathbf{i} - \sin 2t\,\mathbf{j} + t^2\mathbf{k})$$

$$= (\sin 2t\,\mathbf{i} - \cos 2t\,\mathbf{j} + t\,\mathbf{k}) \cdot (-2\sin 2t\,\mathbf{i} - 2\cos 2t\,\mathbf{j} + 2t\,\mathbf{k}) + (2\cos 2t\,\mathbf{i} + 2\sin 2t\,\mathbf{j} + \mathbf{k}) \cdot (\cos 2t\,\mathbf{i} - \sin 2t\,\mathbf{j} + t^2\mathbf{k})$$

$$= (-2\sin^2 2t + 2\cos^2 2t + 2t^2) + (2\cos^2 2t - 2\sin^2 2t + t^2) = 4(\cos^2 2t - \sin^2 2t) + 3t^2 = 4\cos 4t + 3t^2.$$

(*iii*) Using formula 5, Article 1.4, we obtain

$$\mathbf{v}'(t) = (6t^2\mathbf{i} - t\,\mathbf{j} + 3t^2\mathbf{k}) \times (t\,\mathbf{i} + t^2\mathbf{j} + 2t\,\mathbf{k})' + (6t^2\mathbf{i} - t\,\mathbf{j} + 3t^2\mathbf{k})' \times (t\,\mathbf{i} + t^2\mathbf{j} + 2t\,\mathbf{k})$$

$$= (6t^2\mathbf{i} - t\,\mathbf{j} + 3t^2\mathbf{k}) \times (\mathbf{i} + 2t\,\mathbf{j} + 2\mathbf{k}) + (12t\,\mathbf{i} - \mathbf{j} + 6t\,\mathbf{k}) \times (t\,\mathbf{i} + t^2\mathbf{j} + 2t\,\mathbf{k}) = \begin{vmatrix} \mathbf{i} & \mathbf{j} & \mathbf{k} \\ 6t^2 & -t & 3t^2 \\ 1 & 2t & 2 \end{vmatrix} + \begin{vmatrix} \mathbf{i} & \mathbf{j} & \mathbf{k} \\ 12t & -1 & 6t \\ t & t^2 & 2t \end{vmatrix}$$

$$= \mathbf{i}(-2t - 6t^3) - \mathbf{j}(12t^2 - 3t^2) + \mathbf{k}(12t^3 + t) + \mathbf{i}(-2t - 6t^3) - \mathbf{j}(24t^2 - 6t^2) + \mathbf{k}(12t^3 + t)$$

$$= -4t(1 + 3t^2)\mathbf{i} - 27t^2\mathbf{j} + 2t(1 + 12t^2)\mathbf{k}.$$

1.6 GEOMETRIC INTERPRETATION OF r′(t)

Let the vector $\mathbf{r}'(t) \neq \mathbf{0}$ at the point P, then we may draw tangent to the curve C, at P. The vectors $\Delta\mathbf{r} = \mathbf{r}(t + \Delta t) - \mathbf{r}(t)$ and $\dfrac{\Delta\mathbf{r}}{\Delta t} = \dfrac{1}{\Delta t}[\mathbf{r}(t + \Delta t) - \mathbf{r}(t)]$ are parallel. (See Fig. 1.4) Now, let $\Delta t \to 0$, the vectors $\mathbf{r}(t)$ and $\mathbf{r}(t + \Delta t)$ become close. If

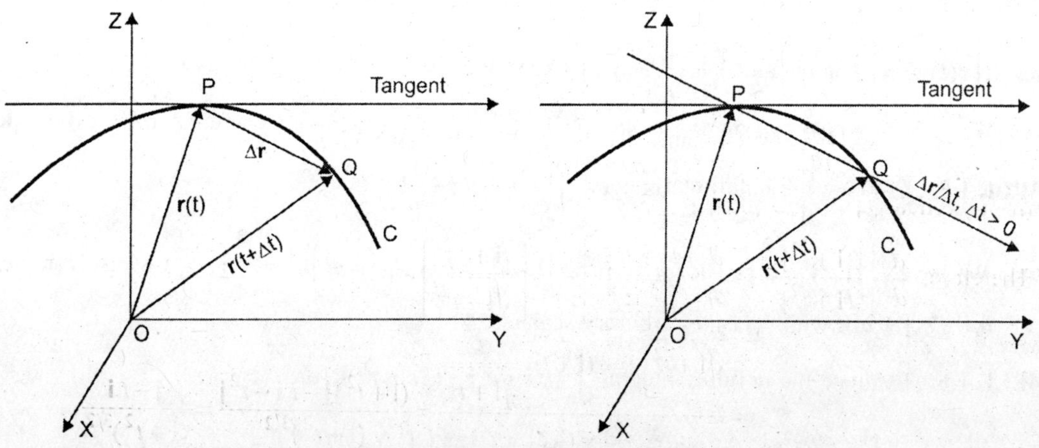

Fig. 1.4: Tangent to a curve

$\lim\limits_{\Delta t \to 0} \dfrac{\Delta\mathbf{r}}{\Delta t}$ exists, then the limiting position of the vector $\dfrac{\Delta\mathbf{r}}{\Delta t}$ is the tangent line to the curve C at the point P.

Thus $\dfrac{d\mathbf{r}}{dt} = \mathbf{r}'(t)$ ...(1.10) represents the tangent vector to the curve C.

EXAMPLE 1.3 Find the tangent vector to the curve whose parametric representation is $x = t^2 - 1$, $y = t + 1$, $z = t/(t + 1)$, at $t = 2$. Hence, find the unit tangent vector.

Solution The vector function that gives the position of a point P on the curve is

$\mathbf{r}(t) = (t^2 - 1)\mathbf{i} + (t + 1)\mathbf{j} + \left(\dfrac{t}{t+1}\right)\mathbf{k}$. Therefore, a tangent vector to the curve at any point is

$$\mathbf{r}'(t) = \frac{d\mathbf{r}}{dt} = \frac{d}{dt}\left[(t^2 - 1)\,\mathbf{i} + (t + 1)\,\mathbf{j} + \left(\frac{t}{t+1}\right)\mathbf{k}\right] = 2t\,\mathbf{i} + \mathbf{j} + [1/(1+t)^2\,]\mathbf{k}$$

and $\mathbf{r}'(2) = 4\,\mathbf{i} + \mathbf{j} + [1/9]\mathbf{k}$. Thus, $|\,\mathbf{r}'(2)\,| = \sqrt{16 + 1 + \dfrac{1}{81}} = \dfrac{\sqrt{1378}}{9}$.

Hence, the unit tangent vector at $t = 2$ is $\hat{\mathbf{r}}'(2) = \dfrac{\mathbf{r}'(2)}{|\,\mathbf{r}'(2)\,|} = [36\mathbf{i} + 9\mathbf{j} + \mathbf{k}]/\sqrt{1378}$.

1.7 INTEGRALS OF VECTOR FUNCTIONS

Let $\mathbf{r}(t) = x(t)\mathbf{i} + y(t)\mathbf{j} + z(t)\mathbf{k}$, where its component functions $x(t)$, $y(t)$, and $z(t)$ are each continuous in a specified interval I, then

$$\int \mathbf{r}(t)dt = \left(\int x(t)dt\right)\mathbf{i} + \left(\int y(t)dt\right)\mathbf{j} + \left(\int z(t)dt\right)\mathbf{k} \quad \text{...(1.11)}$$

is called an indefinite integral of $\mathbf{r}(t)$. If there exists a vector $\mathbf{R}(t)$ such that $\mathbf{r}(t) = \dfrac{d}{dt}\mathbf{R}(t)$ then

$$\int \mathbf{r}(t)dt = \int \frac{d}{dt}\mathbf{R}(t)dt = \mathbf{R}(t) + \mathbf{c},$$

where $\mathbf{c} = c_1\mathbf{i} + c_2\mathbf{j} + c_3\mathbf{k}$ with $c_1 + c_2 + c_3$ arbitrary scalars.

The **definite integral** between the limits $t = a$ and $t = b$ in such cases is given by

$$\int_a^b \mathbf{r}(t)dt = \mathbf{R}(b) - \mathbf{R}(a). \quad \text{...(1.12)}$$

EXAMPLE 1.4 If $\mathbf{r}(t) = 6t^2\mathbf{i} + 4e^{-2t}\mathbf{j} + 8\cos 4t\,\mathbf{k}$, evaluate the indefinite integral $\int \mathbf{r}(t)\,dt$.

Solution $\int \mathbf{r}(t)\,dt = \left[\int 6t^2 dt\right]\mathbf{i} + \left[\int 4e^{-2t}dt\right]\mathbf{j} + \left[\int 8\cos 4t\,dt\right]\mathbf{k}$

$= (2t^3 + c_1)\mathbf{i} + (-2e^{-2t} + c_2)\mathbf{j} + (2\sin 4t + c_3)\mathbf{k} = 2t^3\mathbf{i} - 2e^{-2t}\mathbf{j} + 2\sin 4t\,\mathbf{k} + \mathbf{c}$, where $c = c_1\mathbf{i} + c_2\mathbf{j} + c_3\mathbf{k}$.

EXAMPLE 1.5 Evaluate the indefinite integral $\int\left(t^2\mathbf{i} - 2t\,\mathbf{j} + \dfrac{1}{t}\mathbf{k}\right)dt$.

Solution We have $\int\left(t^2\mathbf{i} - 2t\,\mathbf{j} + \dfrac{1}{t}\mathbf{k}\right)dt = \left(\int t^2 dt\right)\mathbf{i} - \left(\int 2t dt\right)\mathbf{j} + \left(\int \dfrac{1}{t}dt\right)\mathbf{k} = \dfrac{t^3}{3}\mathbf{i} - t^2\mathbf{j} + \log_e|\,t\,|\mathbf{k} + \mathbf{c}$,

where $\mathbf{c} = c_1\mathbf{i} + c_2\mathbf{j} + c_3\mathbf{k}$ with c_1, c_2, c_3 arbitrary scalars.

EXAMPLE 1.6 Evaluate the definite integral $\int_0^1 \{t^2\mathbf{i} + e^t\mathbf{j} - (2\cos\pi t)\mathbf{k}\}\,dt$.

Solution We have $\int_0^1 \{t^2\mathbf{i} + e^t\mathbf{j} - (2\cos\pi t)\mathbf{k}\}\,dt = \left(\int_0^1 t^2 dt\right)\mathbf{i} + \left(\int_0^1 e^t dt\right)\mathbf{j} - \left(\int_0^1 2\cos\pi t dt\right)\mathbf{k}$

$$= \left[\frac{t^3}{3}\right]_0^1 \mathbf{i} + [e^t]_0^1\mathbf{j} - \left[\frac{2}{\pi}\sin\pi t\right]_0^1 \mathbf{k} = \frac{1}{3}\mathbf{i} + (e - 1)\mathbf{j}.$$

EXAMPLE 1.7 Solve the vector initial-value problem for $\mathbf{y}(t)$ by integrating and using the initial conditions to find the constants of integration. $\mathbf{y}''(t) = \mathbf{i} + e^t\mathbf{j}$, $\mathbf{y}(0) = 2\mathbf{i}$, $\mathbf{y}'(0) = \mathbf{j}$.

Solution We have $\mathbf{y}''(t) = \mathbf{i} + e^t\mathbf{j}$. Integrating, both sides of the differential equation with respect to t gives

$\mathbf{y}'(t) = t\,\mathbf{i} + e^t\,\mathbf{j} + \mathbf{A}$. ...(*i*) As $\mathbf{y}'(t) = \mathbf{j}$ for $t = 0$, we get $\mathbf{A} = \mathbf{0}$.

Therefore (*i*) takes the form $\mathbf{y}'(t) = t\,\mathbf{i} + e^t\mathbf{j}$. Integrating again, we have $\mathbf{y}(t) = \dfrac{t^2}{2}\mathbf{i} + e^t\,\mathbf{j} + \mathbf{B}$. ....(*ii*)

Use initial condition. As $\mathbf{y}(t) = 2\mathbf{i}$, when $t = 0$, we get $\mathbf{B} = 2\mathbf{i} - \mathbf{j}$.

Therefore (*ii*) takes the form $\mathbf{y}(t) = \left(\dfrac{t^2}{2} + 2\right)\mathbf{i} + (e^t - 1)\mathbf{j}$, the required solution of the given differential equation.

EXAMPLE 1.8 Given that $\mathbf{r}(t) = \begin{cases} 2\mathbf{i} - \mathbf{j} + 2\mathbf{k} & \text{when } t = 2, \\ 4\mathbf{i} - 2\mathbf{j} + 3\mathbf{k} & \text{when } t = 3. \end{cases}$ Show that $\displaystyle\int_2^3 \mathbf{r} \bullet \frac{d\mathbf{r}}{dt}\,dt = 10$.

Solution Since $\mathbf{r} \bullet \mathbf{r} = r^2$. Therefore, $\dfrac{d}{dt}(\mathbf{r} \bullet \mathbf{r}) = \dfrac{d}{dt}(r^2)$

$\Rightarrow\quad 2\mathbf{r} \bullet \dfrac{d\mathbf{r}}{dt} = \dfrac{d}{dt}(r^2)$ or $\mathbf{r} \bullet \dfrac{d\mathbf{r}}{dt} = \dfrac{1}{2}\dfrac{d}{dt}(r^2)$. Hence $\displaystyle\int_2^3 \mathbf{r} \bullet \frac{d\mathbf{r}}{dt}\,dt = \frac{1}{2}[r^2]_2^3$.

Now $r^2 = |\mathbf{r}|^2 = (4)^2 + (-2)^2 + (3)^2 = 29$, when $t = 3$ and $r^2 = (2)^2 + (-1)^2 + (2)^2 = 9$, when $t = 2$.

Therefore, $\displaystyle\int_2^3 \mathbf{r} \bullet \frac{d\mathbf{r}}{dt}\,dt = \frac{1}{2}[29 - 9] = 10$.

1.8 MOTION OF A BODY OR PARTICLE ON A CURVE

If $\mathbf{r(t)}$ is the position vector of a body or particle moving along a smooth curve in space, then $\mathbf{v}(t) = \dfrac{d\mathbf{r}}{dt} = \mathbf{r}'(t)$ is the particle's **velocity vector**, tangent to the curve. At any time t, the direction of $\mathbf{v}$ is the **direction of motion**, the magnitude of $\mathbf{v}(t)$ is the particle's **speed**, and the derivative, $d\mathbf{v}/dt$, when it exists, is the particle's **acceleration vector** $\mathbf{a}(t)$. In short

1. Velocity is the derivative of position: $\mathbf{v} = \dfrac{d\mathbf{r}}{dt}$.
2. Speed is the magnitude of velocity: speed $= |\mathbf{v}(t)|$.
3. Acceleration is the derivative of velocity: $\mathbf{a}(t) = \dfrac{d\mathbf{v}}{dt} = \dfrac{d^2\mathbf{r}}{dt^2}$,
4. The vector $\mathbf{v}/|\mathbf{v}|$ is the direction of motion at time t. The velocity of a moving particle can be expressed as the product of its speed and direction.

$$\text{Velocity } = |\mathbf{v}|\left(\frac{\mathbf{v}}{|\mathbf{v}|}\right) = \text{(speed) (direction)}.$$

EXAMPLE 1.9 The vector $\mathbf{r}(t) = \{2\log(t+1)\}\mathbf{i} + t^2\mathbf{j} + (t^2/2)\mathbf{k}$ gives the position of a particle in space at time t. Find the particle's velocity and acceleration vectors. Then find the particle's speed and direction of motion when $t = 1$. Write the particle's velocity at $t = 1$ as the product of its speed and direction.

Solution We have $\mathbf{r}(t) = \{2\log(t+1)\}\mathbf{i} + t^2\mathbf{j} + (t^2/2)\mathbf{k}$.

Therefore, $\mathbf{v}(t) = \dfrac{d\mathbf{r}}{dt} = \left(\dfrac{2}{t+1}\right)\mathbf{i} + 2t\,\mathbf{j} + t\,\mathbf{k}$, and $\mathbf{a}(t) = \dfrac{d\mathbf{v}}{dt} = \dfrac{d^2\mathbf{r}}{dt^2} = \left(\dfrac{-2}{(t+1)^2}\right)\mathbf{i} + 2\mathbf{j} + \mathbf{k}$.

At $t = 1$, the particle's speed and direction are: Speed: $|\mathbf{v}(1)| = |\mathbf{i} + 2\mathbf{j} + \mathbf{k}| = \sqrt{1+4+1} = \sqrt{6}$;

Direction: $\dfrac{\mathbf{v}(1)}{|\mathbf{v}(1)|} = \dfrac{1}{\sqrt{6}}\mathbf{i} + \dfrac{2}{\sqrt{6}}\mathbf{j} + \dfrac{1}{\sqrt{6}}\mathbf{k}$. Therefore, $\mathbf{v}(1) = |\mathbf{v}(1)|\left(\dfrac{\mathbf{v}(1)}{|\mathbf{v}(1)|}\right) = \sqrt{6}\left(\dfrac{1}{\sqrt{6}}\mathbf{i} + \dfrac{2}{\sqrt{6}}\mathbf{j} + \dfrac{1}{\sqrt{6}}\mathbf{k}\right)$.

EXAMPLE 1.10 If $\mathbf{r} = \mathbf{a} \cos \omega t + \mathbf{b} \sin \omega t$, where **a** and **b** are any constant non-collinear vectors and ω is a constant scalar, prove that $\mathbf{r} \times \frac{d\mathbf{r}}{dt} = \omega\,(\mathbf{a} \times \mathbf{b})$ and $\frac{d^2\mathbf{r}}{dt^2} + \omega^2\mathbf{r} = \mathbf{0}$.

Solution We have $\mathbf{r} = \mathbf{a} \cos \omega t + \mathbf{b} \sin \omega t$. ...(*i*)

Therefore, $\frac{d\mathbf{r}}{dt} = -\,\mathbf{a}\,\omega \sin \omega t + \mathbf{b}\,\omega \cos \omega t$. ...(*ii*)

$$\mathbf{r} \times \frac{d\mathbf{r}}{dt} = (\mathbf{a} \cos \omega t + \mathbf{b} \sin \omega t) \times (-\mathbf{a}\omega \sin \omega t + \mathbf{b}\omega \cos \omega t)$$

$$= (\mathbf{a} \times \mathbf{b})\, \omega \cos^2 \omega t - (\mathbf{b} \times \mathbf{a})\, \omega \sin^2 \omega t \qquad (\because \mathbf{a} \times \mathbf{a} = \mathbf{0} \text{ etc.})$$

$$= (\omega \cos^2 \omega t + \omega \sin^2 \omega t)\, \mathbf{a} \times \mathbf{b} \qquad (\because -\mathbf{b} \times \mathbf{a} = \mathbf{a} \times \mathbf{b})$$

$$= \omega\,(\mathbf{a} \times \mathbf{b}).$$

Differentiating (*ii*) with respect to *t*, we obtain

$$\frac{d^2\mathbf{r}}{dt^2} = -\,\mathbf{a}\omega^2 \cos \omega t - \mathbf{b}\omega^2 \sin \omega t = -\,\omega^2\,(\mathbf{a} \cos \omega t + \mathbf{b} \sin \omega t) = -\,\omega^2\mathbf{r} \quad \text{or} \quad \frac{d^2\mathbf{r}}{dt^2} + \omega^2\mathbf{r} = \mathbf{0}.$$

EXAMPLE 1.11 Find the angle between the tangents to the curve $\mathbf{r}(t) = t^2\,\mathbf{i} - 2t\,\mathbf{j} + t^3\,\mathbf{k}$ at points $t = 1$ and $t = 2$.

Solution We have $d\mathbf{r}/dt = 2t\,\mathbf{i} - 2\mathbf{j} + 3t^2\mathbf{k}$ a vector along the tangent at any point '*t*'.

If $\mathbf{T}_1$ and $\mathbf{T}_2$ are vectors along the tangents at $t = 1$ and $t = 2$ respectively then

$$\mathbf{T}_1 = 2\mathbf{i} - 2\mathbf{j} + 3\mathbf{k}, \text{ and } \mathbf{T}_2 = 4\mathbf{i} - 2\mathbf{j} + 12\mathbf{k}.$$

If θ is the angle between the tangents, we have $\mathbf{T}_1 \cdot \mathbf{T}_2 = |\mathbf{T}_1||\mathbf{T}_2| \cos\theta$

or $(2)(4) + (-2)(-2) + (3)(12) = \sqrt{4+4+9}\,\sqrt{16+4+144}\,\cos\theta$ or $48 = \sqrt{17}\,\sqrt{164}\,\cos\theta$

or $\cos\theta = \frac{48}{\sqrt{17}\sqrt{164}}$ or $\theta = \cos^{-1}\left(\frac{24}{\sqrt{697}}\right)$.

EXAMPLE 1.12 A particle moves along the curve $x = 2t^2$, $y = t^2 - 4t$, $z = 3t - 5$, where *t* is the time. Find the components of its velocity and acceleration at time $t = 1$ in the direction $\mathbf{i} - 3\mathbf{j} + 2\mathbf{k}$.

Solution The position vector of the particle at time '*t*' is $\mathbf{r}(t) = 2t^2\mathbf{i} + (t^2 - 4t)\mathbf{j} + (3t - 5)\mathbf{k}$.

Velocity, $\mathbf{v}(t) = \frac{d\mathbf{r}}{dt} = \frac{d}{dt}[2t^2\mathbf{i} + (t^2 - 4t)\mathbf{j} + (3t - 5)\mathbf{k}] = 4t\,\mathbf{i} + (2t - 4)\mathbf{j} + 3\mathbf{k}$.

Acceleration, $\mathbf{a}(t) = \frac{d^2\mathbf{r}}{dt^2} = \frac{d}{dt}\left(\frac{d\mathbf{r}}{dt}\right) = \frac{d}{dt}[4t\,\mathbf{i} + (2t - 4)\mathbf{j} + 3\mathbf{k}] = 4\mathbf{i} + 2\mathbf{j}$.

When $t = 1$, $\mathbf{v}(1) = 4\mathbf{i} - 2\mathbf{j} + 3\mathbf{k}$, and $\mathbf{a}(1) = 4\mathbf{i} + 2\mathbf{j}$. Unit vector in the direction $\mathbf{i} - 3\mathbf{j} + 2\mathbf{k}$ is

$$\frac{\mathbf{i} - 3\mathbf{j} + 2\mathbf{k}}{\sqrt{1^2 + (-3)^2 + 2^2}} = \frac{\mathbf{i} - 3\mathbf{j} + 2\mathbf{k}}{\sqrt{14}}.$$

The component of the velocity in the given direction is

$$= (4\mathbf{i} - 2\mathbf{j} + 3\mathbf{k}).\frac{\mathbf{i} - 3\mathbf{j} + 2\mathbf{k}}{\sqrt{14}} = \frac{(4)(1) + (-2)(-3) + (3)(2)}{\sqrt{14}} = \frac{16}{\sqrt{14}} = \frac{8\sqrt{14}}{7}.$$

The component of acceleration in the given direction is

$$= (4\mathbf{i} + 2\mathbf{j}) \cdot \frac{\mathbf{i} - 3\mathbf{j} + 2\mathbf{k}}{\sqrt{14}} = \frac{(4)(1) + (2)(-3) + (0)(2)}{\sqrt{14}} = \frac{-2}{\sqrt{14}} = \frac{-\sqrt{14}}{7}.$$

EXAMPLE 1.13 Find the tangent vector to the curve whose parametric representation is $x = t$, $y = 2t^2$, $z = 3t^3$ at $t = 2$. Hence, find the parametric representation of the tangent vector.

Solution The vector function that gives the position of a point P on the given curve is given by

$$\mathbf{r}(t) = t\,\mathbf{i} + 2t^2\mathbf{j} + 3t^3\mathbf{k}.$$

Therefore, the tangent vector is $\mathbf{r}'(t) = \mathbf{i} + 4t\,\mathbf{j} + 9t^2\mathbf{k}$, and so $\mathbf{r}'(2) = \mathbf{i} + 8\,\mathbf{j} + 36\,\mathbf{k}$ which is tangent to curve C at the point whose position vector is $\mathbf{r}(2) = 2\mathbf{i} + 8\,\mathbf{j} + 24\,\mathbf{k}$. The point is, P (2, 8, 24).

Thus the parametric equations of the tangent line are

$$\frac{x-2}{1} = \frac{y-8}{8} = \frac{z-24}{36} = t(\text{say}), \text{ that is, } x = 2 + t, \quad y = 8 + 8t, \quad z = 24 + 36t.$$

EXAMPLE 1.14 A particle moves along the curve $\mathbf{r}(t) = (t^3 - 4t)\mathbf{i} + (t^2 + 4t)\mathbf{j} + (8t^2 - 3t^3)\mathbf{k}$, where t is the time. Find the magnitudes of the tangential and normal components of its acceleration where $t = 2$.

(VTU, 2003S)

Solution Velocity, $\mathbf{v}(t) = \dfrac{d\mathbf{r}}{dt} = (3t^2 - 4)\mathbf{i} + (2t + 4)\mathbf{j} + (16t - 9t^2)\mathbf{k}.$

Acceleration, $\mathbf{a}(t) = \dfrac{d^2\mathbf{r}}{dt^2} = 6t\,\mathbf{i} + 2\mathbf{j} + (16 - 18t)\mathbf{k}.$

$\mathbf{v}(t)$, the velocity at $t = 2$ is $\mathbf{v}(2) = 8\mathbf{i} + 8\mathbf{j} - 4\mathbf{k}.$

$\mathbf{a}(t)$, the acceleration at $t = 2$ is $\mathbf{a}(2) = 12\mathbf{i} + 2\mathbf{j} - 20\mathbf{k}.$

Since the direction of velocity is along the tangent to the curve, therefore the tangential component of acc., is, $|\overrightarrow{PD}|$. Normal component of acc. is $\overrightarrow{PE}$. (Refer Fig. 1.5)

Fig. 1.5

Tangential component of acc.

$$= \mathbf{a}(2) \cdot \frac{\mathbf{v}(2)}{|\mathbf{v}(2)|} = (12\mathbf{i} + 2\mathbf{j} - 20\mathbf{k}) \cdot \frac{8\mathbf{i} + 8\mathbf{j} - 4\mathbf{k}}{\sqrt{64 + 64 + 16}}$$

$$= \frac{(12)(8) + (2)(8) + (-20)(-4)}{12} = \frac{192}{12} = 16.$$

$$|\overrightarrow{PF}| = |\mathbf{a}| = \sqrt{144 + 4 + 400} = \sqrt{548} = 2\sqrt{137}.$$

Therefore, $\overrightarrow{DF} = \sqrt{\overrightarrow{PF}^2 - \overrightarrow{PD}^2} = \sqrt{548 - 256} = \sqrt{292} = 2\sqrt{73}.$

$\therefore$ Normal component of acc., $\mathbf{a}(2) = |\overrightarrow{PE}| = |\overrightarrow{DF}| = 2\sqrt{73}.$

EXAMPLE 1.15 A particle P (position vector $\mathbf{r}$) is moving along a circle of radius r with angular speed ω. Show by vector methods that the acceleration is equal to $-\omega^2\mathbf{r}$.

Solution Refer Fig 1.6. P be the position of the particle moving along a circle of radius r with constant angular speed $\omega = d\theta/dt$.

The position vector of P is $\mathbf{r}(\theta) = (r\cos\theta)\mathbf{i} + (r\sin\theta)\mathbf{j}.$

$$\text{Velocity, } \mathbf{v}(t) = \frac{d\mathbf{r}}{dt} = [(-r\sin\theta)\,\mathbf{i} + (r\cos\theta)\,\mathbf{j}]\frac{d\theta}{dt}$$

$$= [(-r\sin\theta)\,\mathbf{i} + (r\cos\theta)\,\mathbf{j}]\,\omega.$$

$$\text{Acceleration, } \mathbf{a}(t) = \frac{d^2\mathbf{r}}{dt^2} = [(-r\cos\theta)\mathbf{i} - (r\sin\theta)\,\mathbf{j}]\,\omega\left(\frac{d\theta}{dt}\right)$$

$$= -[(r\cos\theta)\,\mathbf{i} + (r\sin\theta)\,\mathbf{j}]\,\omega^2 = -\omega^2\mathbf{r}.$$

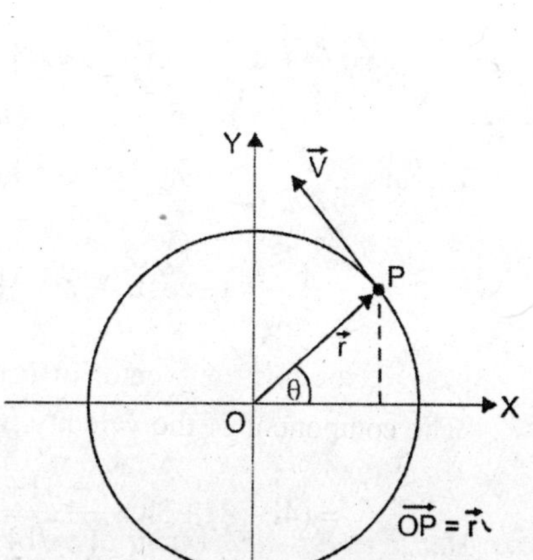

Fig. 1.6

Alternatively: $\mathbf{r} = (r\cos\omega t)\mathbf{i} + (r\sin\omega t)\mathbf{j}$. ...(*i*) Therefore, $\mathbf{v}(t) = \dfrac{d\mathbf{r}}{dt} = (-r\omega\sin\omega t)\mathbf{i} + (r\omega\cos\omega t)\mathbf{j}$,

and acc., $\mathbf{a}(t) = \dfrac{d^2\mathbf{r}}{dt^2} = (-r\omega^2\cos\omega t)\mathbf{i} - (r\omega^2\sin\omega t)\mathbf{j} = -\omega^2[(r\cos\omega t)\mathbf{i} + (r\sin\omega t)\mathbf{j}] = -\omega^2\mathbf{r}$, [from (*i*)].

EXAMPLE 1.16 Find by vector methods, the tangential and normal components of acceleration of a particle moving in a plane curve.

Solution Let $\hat{\mathbf{T}}$ and $\hat{\mathbf{N}}$ respectively denote unit vectors along tangent and normal at any point on the curve at any time *t*.

If **v** denotes velocity at *P* along the tangent then $\mathbf{v} = v\hat{\mathbf{T}}$ where $v = |\mathbf{v}|$.

$$\text{Acceleration along tangent } = \frac{d\mathbf{v}}{dt} = \frac{dv}{dt}\hat{\mathbf{T}} + v\frac{d\hat{\mathbf{T}}}{dt} = \frac{dv}{dt}\hat{\mathbf{T}} + v\frac{d\hat{\mathbf{T}}}{d\psi}\cdot\frac{d\psi}{ds}\cdot\frac{ds}{dt}$$

$$= \frac{dv}{dt}\hat{\mathbf{T}} + v^2\,\frac{1}{\rho}\,\frac{d\hat{\mathbf{T}}}{d\psi} \qquad \text{....(1)}$$

$$\left(\because \frac{d\psi}{ds} = \frac{1}{\rho}, \text{ where } \rho \text{ is radius of circle and } \frac{ds}{dt} = v.\right)$$

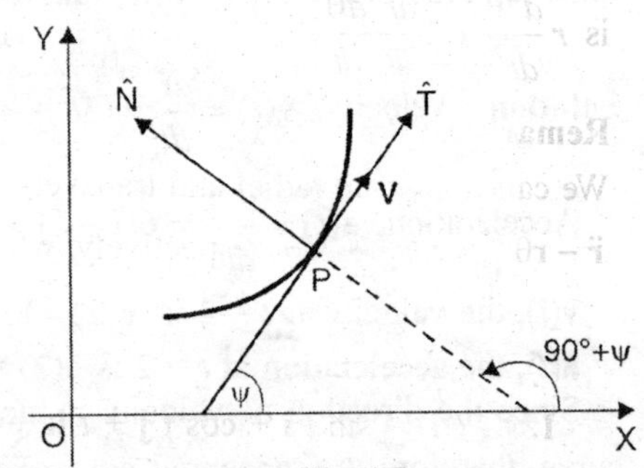

Fig. 1.7

Now $\hat{\mathbf{T}} = \cos\psi\mathbf{i} + \sin\psi\mathbf{j}$, and $\hat{\mathbf{N}} = -\sin\psi\mathbf{i} + \cos\psi\mathbf{j}$.

$$\therefore \quad \frac{d\hat{\mathbf{T}}}{d\psi} = -\sin\psi\mathbf{i} + \cos\psi\mathbf{j} = \hat{\mathbf{N}}.$$

Hence from (1) we have $\dfrac{d\mathbf{v}}{dt} = \dfrac{dv}{dt}\hat{\mathbf{T}} + \dfrac{v^2}{\rho}\hat{\mathbf{N}}$.

Thus tangential and normal components of acceleration respectively are $\dfrac{dv}{dt}$ and $\dfrac{v^2}{\rho}$.

EXAMPLE 1.17 Find the radial and transverse components of velocity and acceleration of a particle describing a plane curve. *(Kurukshetra, 2006; Rajasthan, 2006)*

Solution Let $P(r, \theta)$ be the position of a moving particle at any time *t* along a curve $r = f(\theta)$. Let $\hat{\mathbf{R}}$ and $\hat{\mathbf{S}}$ respectively denote unit vectors in the radial and transverse directions, that is, along and perpendicular to OP.

Then $\hat{\mathbf{R}} = \cos\theta\mathbf{i} + \sin\theta\mathbf{j}$,

and $\hat{\mathbf{S}} = \cos\left(\dfrac{\pi}{2} + \theta\right)\mathbf{i} + \sin\left(\dfrac{\pi}{2} + \theta\right)\mathbf{j} = -\sin\theta\mathbf{i} + \cos\theta\mathbf{j}$.

$$\text{Therefore, } \frac{d\hat{\mathbf{R}}}{dt} = (-\sin\theta\mathbf{i} + \cos\theta\mathbf{j})\frac{d\theta}{dt} = \hat{\mathbf{S}}\frac{d\theta}{dt} \qquad ...(i)$$

$$\text{and } \quad \frac{d\hat{\mathbf{S}}}{dt} = (-\cos\theta\mathbf{i} - \sin\theta\mathbf{j})\frac{d\theta}{dt} = -\hat{\mathbf{R}}\frac{d\theta}{dt}. \qquad ...(ii)$$

Fig. 1.8

Now **R**, the position vector of *P* is given by $\mathbf{R} = r\hat{\mathbf{R}}$ [where $r = |\mathbf{R}|$].

$$\text{Therefore, } \quad \frac{d\mathbf{R}}{dt} = \frac{dr}{dt}\hat{\mathbf{R}} + r\frac{d\hat{\mathbf{R}}}{dt} = \frac{dr}{dt}\hat{\mathbf{R}} + r\frac{d\theta}{dt}\hat{\mathbf{S}}. \qquad \text{...[By (i)]}$$

Hence *radial component* of velocity is dr/dt and *transverse component* is $r(d\theta/dt)$.

As $\dfrac{d\mathbf{R}}{dt} = \dfrac{dr}{dt}\hat{\mathbf{R}} + r\dfrac{d\theta}{dt}\hat{\mathbf{S}}$.

Therefore, $\dfrac{d^2\mathbf{R}}{dt^2}=\dfrac{d^2r}{dt^2}\hat{\mathbf{R}}+\left(\dfrac{dr}{dt}\right)\dfrac{d\hat{\mathbf{R}}}{dt}+\dfrac{dr}{dt}\left(\dfrac{d\theta}{dt}\hat{\mathbf{S}}\right)+r\dfrac{d^2\theta}{dt^2}\hat{\mathbf{S}}+r\left(\dfrac{d\theta}{dt}\right)\left(\dfrac{d\hat{\mathbf{S}}}{dt}\right)$

$$=\frac{d^2r}{dt^2}\hat{\mathbf{R}}+\frac{dr}{dt}\left(\hat{\mathbf{S}}\frac{d\theta}{dt}\right)+\frac{dr}{dt}\left(\frac{d\theta}{dt}\hat{\mathbf{S}}\right)+r\frac{d^2\theta}{dt^2}\hat{\mathbf{S}}+r\frac{d\theta}{dt}\left(-\hat{\mathbf{R}}\frac{d\theta}{dt}\right)$$

$$=\left[\frac{d^2r}{dt^2}-r\left(\frac{d\theta}{dt}\right)^2\right]\hat{\mathbf{R}}+\left[r\frac{d^2\theta}{dt^2}+2\frac{dr}{dt}\frac{d\theta}{dt}\right]\hat{\mathbf{S}}.$$

Hence the *radial component of acceleration* is $\dfrac{d^2r}{dt^2}-r\left(\dfrac{d\theta}{dt}\right)^2$, and *transverse component of acceleration* is $r\dfrac{d^2\theta}{dt^2}+2\dfrac{dr}{dt}\dfrac{d\theta}{dt}$.

Remark 2

We can also write radial and transverse components of acceleration in the form:
$\ddot{\mathbf{r}}-\mathbf{r}\dot{\theta}^2$ and $\mathbf{r}\ddot{\theta}+2\dot{\mathbf{r}}\dot{\theta}$ respectively, where dots denote derivative w.r.t. '*t*', that is $\dot{r}=dr/dt$ etc.

EXERCISE 1.1

1. If $\mathbf{r}(t) = \sin t\,\mathbf{i} + \cos t\,\mathbf{j} + t\,\mathbf{k}$, evaluate (*a*) $d\mathbf{r}/dt$, (*b*) $d^2\mathbf{r}/dt^2$, (*c*) $|d\mathbf{r}/dt|$, (*d*) $|d^2\mathbf{r}/dt^2|$.

   Ans. (*a*) $\cos t\,\mathbf{i} - \sin t\,\mathbf{j} + \mathbf{k}$, (*b*) $-\sin t\,\mathbf{i} - \cos t\,\mathbf{j}$, (*c*) $\sqrt{2}$, (*d*) 1.

2. If $\mathbf{r}=e^{-t}\mathbf{i}+\log(t^2+1)\mathbf{j}-\tan t\,\mathbf{k}$, find (*a*) $d\mathbf{r}/dt$, (*b*) $d^2\mathbf{r}/dt^2$, (*c*) $|d\mathbf{r}/dt|$, (*d*) $|d^2\mathbf{r}/dt^2|$ at $t = 0$.

   Ans. (*a*) $-\mathbf{i} - \mathbf{k}$, (*b*) $\mathbf{i} + 2\mathbf{j}$, (*c*) $\sqrt{2}$, (*d*) $\sqrt{5}$.

3. If $\mathbf{r}(t)=(t^3+2t)\mathbf{i}-3e^{-2t}\mathbf{j}+2\sin 5t\,\mathbf{k}$, find (*a*) $\dfrac{d\mathbf{r}}{dt}$, (*b*) $\left|\dfrac{d\mathbf{r}}{dt}\right|$, (*c*) $\dfrac{d^2\mathbf{r}}{dt^2}$, (*d*) $\left|\dfrac{d^2\mathbf{r}}{dt^2}\right|$ at $t = 0$ and give a possible physical significance.

   Ans. (*a*) $2\mathbf{i}+6\mathbf{j}+10\mathbf{k}$; (*b*) $2\sqrt{35}$; (*c*) $-12\mathbf{j}$; (*d*) 12.

   If *t* represents time, these represent respectively the velocity, magnitude of the velocity, acceleration and magnitude of the acceleration at $t = 0$ of a particle moving along a space curve whose parametric equations are: $x = t^3 + 2t$, $y = -3e^{-2t}$, $z = 2\sin 5t$.

4. Differentiate the following with respect to *t* (**a** and **b** are constant vectors):

   (*a*) $\dfrac{\mathbf{r}}{r^2}+\dfrac{r\mathbf{b}}{\mathbf{a}\cdot\mathbf{r}}$, (*b*) $\dfrac{\mathbf{r}\times\mathbf{a}}{\mathbf{r}\cdot\mathbf{a}}$, (*c*) $r^2\mathbf{r} + (\mathbf{a}\cdot\mathbf{r})\mathbf{b}$, (*d*) $\dfrac{\mathbf{r}+\mathbf{a}}{\mathbf{r}^2+\mathbf{a}^2}$.

   Ans. (*a*) $\dfrac{(d\mathbf{r}/dt)}{r^2}-2\dfrac{(dr/dt)\mathbf{r}}{r^3}+\dfrac{\mathbf{b}(dr/dt)}{\mathbf{a}\cdot\mathbf{r}}-r\dfrac{[\mathbf{a}\cdot(d\mathbf{r}/dt)]\mathbf{b}}{(\mathbf{a}\cdot r)^2}$, (*b*) $\dfrac{(d\mathbf{r}/dt)\times\mathbf{a}}{\mathbf{r}\cdot\mathbf{a}}-\dfrac{[(d\mathbf{r}/dt)\cdot\mathbf{a}](\mathbf{r}\times\mathbf{a})}{(\mathbf{r}\cdot\mathbf{a})^2}$,

   (*c*) $r^2\dfrac{d\mathbf{r}}{dt}+2r\dfrac{dr}{dt}\mathbf{r}+\left(\mathbf{a}\cdot\dfrac{d\mathbf{r}}{dt}\right)\mathbf{b}$, (*d*) $\dfrac{d\mathbf{r}/dt}{\mathbf{r}^2+\mathbf{a}^2}-\dfrac{2\mathbf{r}\cdot d\mathbf{r}/dt}{(\mathbf{r}^2+\mathbf{a}^2)^2}(\mathbf{r}+\mathbf{a})$.

5. If $\mathbf{A} = t^3\,\mathbf{i} + 2t\,\mathbf{j} + e^t\,\mathbf{k}$ and $\mathbf{B} = \sin t\,\mathbf{i} - \cos t\,\mathbf{j} + t\,\mathbf{k}$, find (*a*) $\dfrac{d}{dt}(\mathbf{A}\cdot\mathbf{B})$, (*b*) $\dfrac{d}{dt}(\mathbf{A}\times\mathbf{B})$.

   Ans. (*a*) $(t^3-2)\cos t+(3t+2)t\sin t+(t+1)e^t$,

   (*b*) $(4t+e^t\cos t-e^t\sin t)\mathbf{i}+(e^t\sin t+e^t\cos t-4t^3)\mathbf{j}+(t^3\sin t-3t^2\cos t-2t\cos t-2\sin t)\mathbf{k}$.

6. Evaluate the following:

(a) $\dfrac{d}{dt}\left(\mathbf{r}\times\dfrac{d\mathbf{r}}{dt}\right)$, (b) $\dfrac{d}{dt}\left(\mathbf{r}\cdot\dfrac{d\mathbf{r}}{dt}\times\dfrac{d^2\mathbf{r}}{dt^2}\right)$, (c) $\dfrac{d}{dt}\left(\dfrac{\mathbf{F}(t)}{|\mathbf{F}(t)|}\right)$.

Ans. (a) $\mathbf{r}\times\dfrac{d^2\mathbf{r}}{dt^2}$, (b) $\mathbf{r}\cdot\dfrac{d\mathbf{r}}{dt}\times\dfrac{d^3\mathbf{r}}{dt^3}$, (c) $[(\mathbf{F}\cdot\mathbf{F})\mathbf{F}'-\mathbf{F}(\mathbf{F}\cdot\mathbf{F}')]/(\mathbf{F}\cdot\mathbf{F})^{3/2}$.

[**Hint:** (c) $\dfrac{d}{dt}\left[\dfrac{\mathbf{F}}{|\mathbf{F}|}\right]=\dfrac{d}{dt}\left[\dfrac{\mathbf{F}}{(\mathbf{F}\cdot\mathbf{F})^{1/2}}\right]$.

7. If $\mathbf{u}=5t^2\mathbf{i}+t\,\mathbf{j}-t^3\mathbf{k}$, $\mathbf{v}=\sin t\,\mathbf{i}-\cos t\,\mathbf{j}$. Verify the formulae

(a) $\dfrac{d}{dt}(\mathbf{u}\cdot\mathbf{v})=\dfrac{d\mathbf{u}}{dt}\cdot\mathbf{v}+\mathbf{v}\cdot\dfrac{d\mathbf{v}}{dt}$. (b) $\dfrac{d}{dt}(\mathbf{u}\times\mathbf{v})=\mathbf{u}\times\dfrac{d\mathbf{v}}{dt}+\dfrac{d\mathbf{u}}{dt}\times\mathbf{v}$.

[**Hint:** $\mathbf{u}\cdot\mathbf{v}=5t^2\sin t-t\cos t$;
$\mathbf{u}\times\mathbf{v}=(t^3\sin t-3t\cos t)\mathbf{i}-(t^3\cos t+3t^2\sin t)\mathbf{j}+(5t^2\sin t-11t\cos t-\sin t)\mathbf{k}$].

8. If $\mathbf{r}(t)=(a\cos t)\mathbf{i}+(a\sin t)\mathbf{j}+(a\,t\tan\alpha)\mathbf{k}$, show that

(a) $\left|\dfrac{d\mathbf{r}}{dt}\times\dfrac{d^2\mathbf{r}}{dt^2}\right|=a^2\sec\alpha$, (b) $\dfrac{d\mathbf{r}}{dt}\cdot\dfrac{d^2\mathbf{r}}{dt^2}\times\dfrac{d^3\mathbf{r}}{dt^3}=a^3\tan\alpha$

9. If $\mathbf{r}(t)=5t^2\,\mathbf{i}+t\,\mathbf{j}-t^3\,\mathbf{k}$, find $\displaystyle\int_1^2\mathbf{r}\times\dfrac{d^2\mathbf{r}}{dt^2}\,dt$. **Ans.** $-14\mathbf{i}+75\mathbf{j}-15\mathbf{k}$.

10. If $\mathbf{r}=(\cos nt)\mathbf{i}+(\sin nt)\,\mathbf{j}$, where n is constant. Show that $\mathbf{r}\times\dfrac{d\mathbf{r}}{dt}=n\mathbf{k}$ and $\mathbf{r}\cdot\dfrac{d\mathbf{r}}{dt}=0$.

11. If $\mathbf{r}=a\cos t\,\mathbf{i}+a\sin t\,\mathbf{j}+bt\,\mathbf{k}$, prove that

(a) $\left[\dfrac{d\mathbf{r}}{dt}\;\dfrac{d^2\mathbf{r}}{dt^2}\;\dfrac{d^3\mathbf{r}}{dt^3}\right]=a^2b$, (b) $\left(\dfrac{d\mathbf{r}}{dt}\right)^2=a^2+b^2$, (c) $\left(\dfrac{d\mathbf{r}}{dt}\times\dfrac{d^2\mathbf{r}}{dt^2}\right)^2=a^2(a^2+b^2)$.

12. Given $\mathbf{v}(t)=t\,\mathbf{i}+(t^2-2t)\mathbf{j}+(3t^2+t^3)\mathbf{k}$, find $\displaystyle\int_0^1\mathbf{v}(t)\,dt$. **Ans.** $\frac{1}{2}\mathbf{i}-\frac{2}{3}\mathbf{j}+\frac{5}{4}\mathbf{k}$.

13. Find $\displaystyle\int_a^b\mathbf{v}(t)\,dt$, given $\mathbf{v}(t)=u^3\mathbf{i}+(3u^2-2u)\,\mathbf{j}+3\mathbf{k}$; $a=0$, $b=2$. **Ans.** $4\mathbf{i}+4\mathbf{j}+6\mathbf{k}$.

14. Find the unit tangent vector on the curve $x=2t+1$, $y=t^2-t$, $z=t^3$ at the point $t=2$.

Ans. $(2\mathbf{i}+3\mathbf{j}+12\mathbf{k})/157$.

15. Find the angle between the tangents to the curve $y=x^2$, $z=x^3$ at (1, 1, 1) and (–1, 1, –1). **Ans.** $\cos^{-1}(3/7)$
[**Hint:** The parametric equations of the curve are $x=t$, $y=t^2$, $z=t^3$.
The points (1, 1, 1) and (–1, 1, –1) correspond to $t=1$ and $t=-1$ etc.]

16. A particle moves along a curve where parametric representations are: $x=e^{-t}$, $y=2\cos 3t$, $z=2\sin 3t$, where t is the time. Find the magnitudes of the velocity and acceleration at $t=0$.

(PTU, 2003; VTU 2003S) **Ans.** $\sqrt{37}$; $5\sqrt{13}$.

17. If $\mathbf{r}=t^2\mathbf{i}-t^3\mathbf{j}+t^4\mathbf{k}$ is the position vector of a moving particle at time t. What is the component of its velocity in the direction of the vector $8\mathbf{i}-\mathbf{j}+4\mathbf{k}$ at $t=1$?
What is the acceleration of the particle at $t=1$? What are the tangential and normal components of its acceleration? **Ans.** $35/9$, $2\mathbf{i}-6\mathbf{j}+12\mathbf{k}$, $70/\sqrt{29}$, $\sqrt{(436/29)}$.

18. The acceleration of a particle at any time $t\geq 0$ is given by $\mathbf{a}=\dfrac{d\mathbf{v}}{dt}=12\cos 2t\,\mathbf{i}-8\sin 2t\,\mathbf{j}+16t\,\mathbf{k}$.
If the velocity $\mathbf{v}$ and displacement $\mathbf{r}$ are zero at $t=0$. Find $\mathbf{v}$ and $\mathbf{r}$ at any time t.

Ans. $\mathbf{v}(t)=6\sin 2t\,\mathbf{i}+4(\cos 2t-1)\mathbf{j}+8t^2\mathbf{k}$; $\mathbf{r}(t)=3(1-\cos 2t)\mathbf{i}+2(\sin 2t-2t)\,\mathbf{j}+8t^3\mathbf{k}/3$.

19. A particle moves so that its position vector is given by $\mathbf{r}(t) = (\cos \omega t)\,\mathbf{i} + (\sin \omega t)\,\mathbf{j}$. Show that the velocity $\mathbf{r}'(t)$ of the particle is perpendicular to $\mathbf{r}(t)$ and $\mathbf{r}(t) \times \mathbf{r}'(t)$ is a constant vector.

20. A particle moves along the curve $x = 3t^2$, $y = t^2 - 2t$, $z = t^3$, where t is time. Find (*a*) the magnitudes of its velocity and acceleration at time $t = 1$; (*b*) the components of velocity and acceleration at time $t = 1$ in the direction $\mathbf{b} = 4\mathbf{i} - 2\mathbf{j} + 4\mathbf{k}$. **Ans.** (*a*) $|\mathbf{v}(1)| = 3\sqrt{5}$, $|\mathbf{a}(1)| = 2\sqrt{19}$; (*b*) 6, 22/3.

21. The position vector of a particle at time t is $\mathbf{r}(t) = \cos(t-1)\,\mathbf{i} + \sinh(t-1)\,\mathbf{j} + \alpha t^3\mathbf{k}$. Find the condition imposed on α by requiring that at $t = 1$, the acceleration is normal to the position vector. **Ans.** $\alpha = \pm 1/\sqrt{6}$.

22. The acceleration of a particle at any time $t \geq 0$ is given by $\mathbf{a}(t) = d\mathbf{v}/dt = e^t\,\mathbf{i} + e^{2t}\,\mathbf{j} + \mathbf{k}$. If at $t = 0$, the displacement is $\mathbf{r}(0) = \mathbf{0}$ and the velocity is $\mathbf{v}(0) = \mathbf{i} + \mathbf{j}$, find $\mathbf{r}(t)$ and $\mathbf{v}(t)$ at any time t.

Ans. $\mathbf{r}(t) = (e^t - 1)\mathbf{i} + \left(\frac{1}{4}e^{2t} + \frac{1}{2}t - \frac{1}{4}\right)\mathbf{j} + \frac{1}{2}t^2\mathbf{k}$; $\mathbf{v}(t) = e^t\mathbf{i} + \frac{1}{2}(e^{2t} + 1)\mathbf{j} + t\,\mathbf{k}$.

23. The acceleration of a particle at any time t is given by $\mathbf{a} = d\mathbf{v}/dt = (t+1)\,\mathbf{i} + t^2\,\mathbf{j} + (t^2 - 2)\,\mathbf{k}$. If at $t = 0$, the displacement is $\mathbf{r}(0) = \mathbf{0}$ and the velocity is $\mathbf{v}(0) = \mathbf{i} - \mathbf{k}$, find $\mathbf{v}$ and $\mathbf{r}$ at any time t.

Ans. $\mathbf{v}(t) = \left(\frac{1}{2}t^2 + t + 1\right)\mathbf{i} + \frac{1}{3}t^3\mathbf{j} + \left(\frac{1}{3}t^3 - 2t - 1\right)\mathbf{k}$; $\mathbf{r}(t) = \left(\frac{1}{6}t^3 + \frac{1}{2}t^2 + t\right)\mathbf{i} + \frac{1}{12}t^4\mathbf{j} + \left(\frac{1}{12}t^4 - t^2 - t\right)\mathbf{k}$.

1.9 THE VECTOR DIFFERENTIAL OPERATOR DEL

The vector differential operator Del, written ∇ is defined in two and three dimensions by

$$\nabla = \mathbf{i}\frac{\partial}{\partial x} + \mathbf{j}\frac{\partial}{\partial y} \text{ and } \nabla = \mathbf{i}\frac{\partial}{\partial x} + \mathbf{j}\frac{\partial}{\partial y} + \mathbf{k}\frac{\partial}{\partial z} \equiv \nabla = \frac{\partial}{\partial x}\mathbf{i} + \frac{\partial}{\partial y}\mathbf{j} \text{ and } \nabla = \frac{\partial}{\partial x}\mathbf{i} + \frac{\partial}{\partial y}\mathbf{j} + \frac{\partial}{\partial z}\mathbf{k}.$$

The del operator is analogous to the derivative operator d/dx, which when applied to $f(x)$ produces the derivative $f'(x)$. The operator ∇, an inverted capital Greek delta, is also known as *nabla*.

1.10 GRADIENT OF A SCALAR FIELD

When the vector differential operator ∇ is applied to a differentiable function $z = f(x, y)$ or $w = F(x, y, z)$, defining a scalar field, we say that the vectors

$$\nabla f(x, y) = \left(\frac{\partial}{\partial x}\mathbf{i} + \frac{\partial}{\partial y}\mathbf{j}\right) f = \frac{\partial f}{\partial x}\mathbf{i} + \frac{\partial f}{\partial y}\mathbf{j} \qquad \text{...(1.13)}$$

and

$$\nabla F(x, y, z) = \left(\frac{\partial}{\partial x}\mathbf{i} + \frac{\partial}{\partial y}\mathbf{j} + \frac{\partial}{\partial z}\mathbf{k}\right) F = \frac{\partial F}{\partial x}\mathbf{i} + \frac{\partial F}{\partial y}\mathbf{j} + \frac{\partial F}{\partial z}\mathbf{k} \qquad \text{...(1.14)}$$

are the gradients of the respective functions or scalar fields.

Note that the del operator ∇ which when operated on a 'scalar field' produces a vector field. The vector ∇f is usually read "grad f" as well as "gradient of f" and "del f".

For instance, if $F(x, y, z) = xy^2 + 3x^2 - z^3$, then

$$\text{grad } F = \nabla F(x, y, z) = (y^2 + 6x)\,\mathbf{i} + 2xy\,\mathbf{j} - 3z^2\,\mathbf{k} \text{ and}$$

$$\nabla F(x, y, z) \text{ at } (2, -1, 4), \text{ that is, } \nabla F(2, -1, 4) = 13\,\mathbf{i} - 4\,\mathbf{j} - 48\,\mathbf{k}.$$

EXAMPLE 1.18 Find the gradient of the following scalar fields:

(*i*) $f(x, y) = x^3 - 3x^2y^2 + y^3$ at (1, 2),

(*ii*) $f(x, y, z) = 3x^2y - y^3z^2$ at (1, –2, –1), (*AMIETE, June 2011, 2009*)

(*iii*) $f(x, y, z) = xy^2 + 3x^2 - z^3$ at (2, –1, 4).

Solution (*i*) We have $\nabla f(x, y) = \left(\mathbf{i}\frac{\partial}{\partial x} + \mathbf{j}\frac{\partial}{\partial y}\right)(x^3 - 3x^2y^2 + y^3)$

$$= \mathbf{i}\frac{\partial}{\partial x}(x^3 - 3x^2y^2 + y^3) + \mathbf{j}\frac{\partial}{\partial y}(x^3 - 3x^2y^2 + y^3)$$

$$= (3x^2 - 6xy^2)\mathbf{i} + (-6x^2y + 3y^2)\mathbf{j}. \text{ Therefore, } \nabla f(1, 2) = -21\mathbf{i}.$$

(*ii*) We have $\nabla f(x, y, z) = \left(\mathbf{i}\frac{\partial}{\partial x} + \mathbf{j}\frac{\partial}{\partial y} + \mathbf{k}\frac{\partial}{\partial z}\right)(3x^2y - y^3z^2) = 6xy\,\mathbf{i} + (3x^2 - 3y^2z^2)\,\mathbf{j} - 2y^3z\mathbf{k}.$

The gradient of f at (1, –2, –1) is $\nabla f(1, -2, -1) = [6\,xy\,\mathbf{i} + (3x^2 - 3y^2z^2)\,\mathbf{j} - 2y^3z\,\mathbf{k}]_{(1,-2,-1)}$
$= -(12\,\mathbf{i} + 9\,\mathbf{j} + 16\,\mathbf{k}).$

(*iii*) We have $\nabla f(x, y, z) = \left(\mathbf{i}\frac{\partial}{\partial x} + \mathbf{j}\frac{\partial}{\partial y} + \mathbf{k}\frac{\partial}{\partial z}\right)(xy^2 + 3x^2 - z^3) = (y^2 + 6x)\,\mathbf{i} + 2\,xy\,\mathbf{j} - 3z^2\mathbf{k}.$

The gradient of f at (2, –1, 4) is $\nabla f = 13\,\mathbf{i} - 4\,\mathbf{j} - 48\,\mathbf{k}.$

EXAMPLE 1.19 *If* $\mathbf{R} = x\mathbf{i} + y\mathbf{j} + z\mathbf{k}$, $R = |r|$ and $\hat{\mathbf{R}} = \mathbf{R}/r$, then show that

(*i*) grad $r = \hat{\mathbf{R}}$, (*ii*) grad $(1/r) = -\hat{\mathbf{R}}/r^2 = -\mathbf{R}/r^3$. *(PTU 2003; UPTU 2002)*

Also find $\nabla\phi$ if $\phi = \log|\mathbf{R}|$. *(AMIE, W-2008)*

Solution (*i*) We have $r^2 = x^2 + y^2 + z^2$. Therefore grad r, that is,

$$\nabla r = \mathbf{i}\frac{\partial r}{\partial x} + \mathbf{j}\frac{\partial r}{\partial y} + \mathbf{k}\frac{\partial r}{\partial z} = \mathbf{i}\frac{x}{r} + \mathbf{j}\frac{y}{r} + \mathbf{k}\frac{z}{r} = \frac{x\mathbf{i} + y\mathbf{j} + z\mathbf{k}}{r} = \frac{\mathbf{R}}{r} = \hat{\mathbf{R}}.$$

(*ii*) $$\text{grad}\,\frac{1}{r} = \left(\mathbf{i}\frac{\partial}{\partial x} + \mathbf{j}\frac{\partial}{\partial y} + \mathbf{k}\frac{\partial}{\partial z}\right)\left(\frac{1}{r}\right) = \mathbf{i}\left(-\frac{1}{r^2}\frac{\partial r}{\partial x}\right) + \mathbf{j}\left(-\frac{1}{r^2}\frac{\partial r}{\partial y}\right) + \mathbf{k}\left(-\frac{1}{r^2}\frac{\partial r}{\partial z}\right)$$

$$= -\frac{1}{r^2}\left(\frac{x}{r}\mathbf{i} + \frac{y}{r}\mathbf{j} + \frac{z}{r}\mathbf{k}\right) = -\frac{1}{r^2}\left(\frac{\mathbf{R}}{r}\right) = -\frac{\mathbf{R}}{r^3} = -\frac{\hat{\mathbf{R}}}{r^2}.$$

We have $r = |\mathbf{R}| = \sqrt{x^2 + y^2 + z^2}$ and $\phi = \log|\mathbf{R}| = \frac{1}{2}\log(x^2 + y^2 + z^2).$

$$\therefore\ \nabla\phi = \frac{1}{2}\nabla\log(x^2 + y^2 + z^2) = \frac{1}{2}\left\{\mathbf{i}\frac{\partial}{\partial x}\log(x^2 + y^2 + z^2) + \mathbf{j}\frac{\partial}{\partial y}\log(x^2 + y^2 + z^2) + \mathbf{k}\frac{\partial}{\partial z}\log(x^2 + y^2 + z^2)\right\}$$

$$= \frac{1}{2}\left\{\mathbf{i}\frac{2x}{x^2 + y^2 + z^2} + \mathbf{j}\frac{2y}{x^2 + y^2 + z^2} + \mathbf{k}\frac{2z}{x^2 + y^2 + z^2}\right\} = \frac{x\mathbf{i} + y\mathbf{i} + z\mathbf{k}}{x^2 + y^2 + z^2} = \frac{\mathbf{R}}{r^2}.$$

EXAMPLE 1.20 Prove that:

(*i*) $\nabla f(r) = \frac{f'(r)}{|\mathbf{R}|}\mathbf{R}$, where $\mathbf{R} = x\mathbf{i} + y\mathbf{j} + z\mathbf{k}$ and $r = |\mathbf{R}|$, *(AMIETE, June 2009)*

(*ii*) $\nabla r^n = nr^{n-2}\,\mathbf{R}$.

Solution (*i*) We have $\nabla f(r) = \left(\mathbf{i}\frac{\partial}{\partial x} + \mathbf{j}\frac{\partial}{\partial y} + \mathbf{k}\frac{\partial}{\partial z}\right)f(r) = \mathbf{i}\frac{\partial}{\partial x}f(r) + \mathbf{j}\frac{\partial}{\partial y}f(r) + \mathbf{k}\frac{\partial}{\partial z}f(r)$

$$= \mathbf{i}f'(r)\frac{\partial r}{\partial x} + \mathbf{j}f'(r)\frac{\partial r}{\partial y} + \mathbf{k}f'(r)\frac{\partial r}{\partial z}, \text{ where } f'(r) = \frac{d}{dr}f(r).$$

Since $r = |\mathbf{R}| = \sqrt{x^2 + y^2 + z^2}$. Therefore, $\frac{\partial r}{\partial x} = \frac{\partial}{\partial x}(x^2 + y^2 + z^2)^{1/2} = \frac{x}{\sqrt{x^2 + y^2 + z^2}} = \frac{x}{|\mathbf{R}|}.$

Similarly $\frac{\partial r}{\partial y} = \frac{y}{|\mathbf{R}|}$ and $\frac{\partial r}{\partial z} = \frac{z}{|\mathbf{R}|}.$

Therefore, $$\nabla f(r) = \mathbf{i}f'(r)\frac{x}{|\mathbf{R}|} + \mathbf{j}f'(r)\frac{y}{|\mathbf{R}|} + \mathbf{k}f'(r)\frac{z}{|\mathbf{R}|} = \frac{f'(r)}{|\mathbf{R}|}(x\mathbf{i} + y\mathbf{j} + z\mathbf{k}) = \frac{f'(r)}{|\mathbf{R}|}\mathbf{R}.$$

(*ii*) Using the result of part (*i*), we have $f(r) = r^n$. Therefore $f'(r) = nr^{n-1}$.

Hence, $$\nabla r^n = \frac{nr^{n-1}}{|\mathbf{R}|}\mathbf{R} = nr^{n-2}\mathbf{R}.$$

Alternative Solution $$\nabla r^n = \mathbf{i}\frac{\partial}{\partial x}r^n + \mathbf{j}\frac{\partial}{\partial y}r^n + \mathbf{k}\frac{\partial}{\partial z}r^n.$$

Now $$\frac{\partial}{\partial x}r^n = \frac{\partial}{\partial x}(x^2+y^2+z^2)^{n/2} = \frac{n}{2}(x^2+y^2+z^2)^{(n-2)/2}\cdot 2x = nx(x^2+y^2+z^2)^{(n-2)/2} = nxr^{n-2}.$$

Similarly $\frac{\partial}{\partial y}r^n = nyr^{n-2}$ and $\frac{\partial}{\partial z}r^n = nzr^{n-2}$.

Therefore, $$\nabla r^n = nxr^{n-2}\mathbf{i} + nyr^{n-2}\mathbf{j} + nzr^{n-2}\mathbf{k} = nr^{n-2}(x\mathbf{i}+y\mathbf{j}+z\mathbf{k}) = nr^{n-2}\mathbf{R}.$$

EXAMPLE 1.21 (*i*) If $\mathbf{R} = x\mathbf{i} + y\mathbf{j} + z\mathbf{k}$ and $r = |\mathbf{R}|$, then find $\nabla|\mathbf{R}|^3$,

(*ii*) Find ∇f if $f(x, y, z) = (x^2+y^2+z^2)e^{-\sqrt{(x^2+y^2+z^2)}}$.

Solution (*i*) We have $|\mathbf{R}| = \sqrt{(x^2+y^2+z^2)}$, therefore $|\mathbf{R}|^3 = (x^2+y^2+z^2)^{3/2}$ and

$$\nabla|\mathbf{R}|^3 = \left(\mathbf{i}\frac{\partial}{\partial x}+\mathbf{j}\frac{\partial}{\partial y}+\mathbf{k}\frac{\partial}{\partial z}\right)(x^2+y^2+z^2)^{3/2}$$

$$= \mathbf{i}\frac{\partial}{\partial x}(x^2+y^2+z^2)^{3/2} + \mathbf{j}\frac{\partial}{\partial y}(x^2+y^2+z^2)^{3/2} + \mathbf{k}\frac{\partial}{\partial z}(x^2+y^2+z^2)^{3/2}$$

$$= \mathbf{i}\left[\frac{3}{2}(x^2+y^2+z^2)^{1/2}\cdot 2x\right] + \mathbf{j}\left[\frac{3}{2}(x^2+y^2+z^2)^{1/2}\cdot 2y\right] + \mathbf{k}\left[\frac{3}{2}(x^2+y^2+z^2)^{1/2}\cdot 2z\right]$$

$$= 3(x^2+y^2+z^2)^{1/2}[x\mathbf{i}+y\mathbf{j}+z\mathbf{k}] = 3r\,\mathbf{R}.$$

(*ii*) $\mathbf{R} = x\mathbf{i}+y\mathbf{j}+z\mathbf{k},\ r = |\mathbf{R}| = \sqrt{(x^2+y^2+z^2)}$.

$$\therefore\ \nabla f = \left(\mathbf{i}\frac{\partial}{\partial x}+\mathbf{j}\frac{\partial}{\partial y}+\mathbf{k}\frac{\partial}{\partial z}\right)(x^2+y^2+z^2)e^{-\sqrt{(x^2+y^2+z^2)}} = \sum \mathbf{i}\frac{\partial}{\partial x}\left[(x^2+y^2+z^2)e^{-\sqrt{(x^2+y^2+z^2)}}\right]$$

$$= \sum \mathbf{i}\left[2xe^{-\sqrt{(x^2+y^2+z^2)}} - (x^2+y^2+z^2)e^{-\sqrt{(x^2+y^2+z^2)}}\cdot\frac{1}{2}(x^2+y^2+z^2)^{-1/2}\cdot 2x\right]$$

$$= \sum \mathbf{i}\left[e^{-\sqrt{(x^2+y^2+z^2)}}\cdot x\left\{2-\sqrt{(x^2+y^2+z^2)}\right\}\right]$$

$$= e^{-\sqrt{(x^2+y^2+z^2)}}\left\{2-\sqrt{(x^2+y^2+z^2)}\right\}(x\mathbf{i}+y\mathbf{j}+z\mathbf{k}) = e^{-r}(2-r)\mathbf{R}.$$

1.11 GEOMETRICAL INTERPRETATION OF GRADIENT

Let $f(P) = f(x, y, z)$ be a differential scalar function. Let $f(x, y, z) = c$ = constant be the level surface S in space.

Let $x = x(t)$, $y = y(t)$, $z = z(t)$ be the parametric representation of the smooth curve C on the surface passing through a point P on the surface. A curve C in the space may be represented by a vector function $\mathbf{r}(t) = x(t)\mathbf{i} + y(t)\,\mathbf{j} + z(t)\mathbf{k}$. Since the curve C lie on the surface, S, then $f(x(t), y(t), z(t)) = c$. Differentiating both sides of this equation with respect to t and using the chain rule, we obtain

$$\frac{d}{dt}f(x(t), y(t), z(t)) = \frac{d}{dt}(c) \quad \text{or} \quad \frac{\partial f}{\partial x}\frac{dx}{dt} + \frac{\partial f}{\partial y}\frac{dy}{dt} + \frac{\partial f}{\partial z}\frac{dz}{dt} = 0 \qquad ...(1.15)$$

$$\text{or} \quad \left(\mathbf{i}\frac{\partial}{\partial x} + \mathbf{j}\frac{\partial}{\partial y} + \mathbf{k}\frac{\partial}{\partial z}\right)f \cdot \left(\mathbf{i}\frac{dx}{dt} + \mathbf{j}\frac{dy}{dt} + \mathbf{k}\frac{dz}{dt}\right) = 0 \text{ or } (\text{grad } f) \cdot \frac{d\mathbf{r}}{dt} = 0 \text{ or } \nabla f \cdot \mathbf{r}'(t) = 0. \qquad ...(1.16)$$

Refer Fig. 1.9. $\mathbf{r}'(t)$ is a tangent vector to curve C at the point P and lies in the tangent plane to the surface S at P. Hence $\nabla f(p)$ is orthogonal to every tangent vector at P. Its direction is away from the surface and is called the *outward normal* to S. Thus $\nabla f(P)$ is the vector normal to the surface $f(x, y, z) = c$ at the point P.

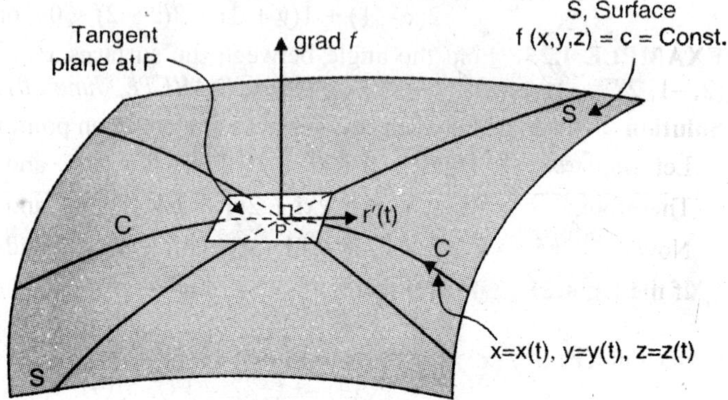

Fig. 1.9: Gradient as surface normal vector.

EXAMPLE 1.22 Find the normal vector and the equation of tangent plane to the surface $z = \sqrt{x^2 + y^2}$ at the point (3, 4, 5).

(AMIE- W-2010; AMIETE, Dec 2005)

Solution Let $f(x, y, z) = z - \sqrt{x^2 + y^2} = 0$ be the surface. Then, the normal vector is given by

$$\nabla f = \left(\mathbf{i}\frac{\partial}{\partial x} + \mathbf{j}\frac{\partial}{\partial y} + \mathbf{k}\frac{\partial}{\partial z}\right)(z - \sqrt{x^2 + y^2}) = \frac{-x}{\sqrt{x^2 + y^2}}\mathbf{i} - \frac{-y}{\sqrt{x^2 + y^2}}\mathbf{j} + \mathbf{k} = -\frac{x}{z}\mathbf{i} - \frac{y}{z}\mathbf{j} + \mathbf{k}.$$

The normal vector of f at (3, 4, 5) is given by

$$\nabla f(3, 4, 5) = -\frac{3}{5}\mathbf{i} - \frac{4}{5}\mathbf{j} + \mathbf{k}.$$

The tangent plane at the point (3, 4, 5) is given by

$$-\frac{3}{5}(x-3) - \frac{4}{5}(y-4) + 1(z-5) = 0 \quad \text{or} \quad 3x + 4y - 5z = 0.$$

EXAMPLE 1.23 Find a unit normal vector to the surface $x^2y + 2xz = 4$ at the point (2, –2, 3).

(AMIETE, June 2013, Dec 2007; AMIE, W-2005)

Solution Let $f(x, y, z) = x^2y + 2xz - 4 = 0$. Then $\partial f/\partial x = 2xy + 2z$, $\partial f/\partial y = x^2$ and $\partial f/\partial z = 2x$.

The normal vector is given by $\nabla f = \mathbf{i}\frac{\partial f}{\partial x} + \mathbf{j}\frac{\partial f}{\partial y} + \mathbf{k}\frac{\partial f}{\partial z} = (2xy + 2z)\mathbf{i} + x^2\mathbf{j} + 2x\mathbf{k}.$

At (2, –2, 3), we obtain the normal vector as $\nabla f(2, -2, 3) = -2\mathbf{i} + 4\mathbf{j} + 4\mathbf{k}$.
The unit normal vector is given by

$$\hat{\mathbf{n}} = \frac{\nabla f}{|\nabla f|} = \frac{(-2\mathbf{i} + 4\mathbf{j} + 4\mathbf{k})}{\sqrt{4 + 16 + 16}} = -\frac{1}{3}\mathbf{i} + \frac{2}{3}\mathbf{j} + \frac{2}{3}\mathbf{k}.$$

Another unit normal vector is $\frac{1}{3}\mathbf{i} - \frac{2}{3}\mathbf{j} - \frac{2}{3}\mathbf{k}$ having direction opposite to that above.

EXAMPLE 1.24 Find the normal vector and the equation of the tangent plane to the surface $xz^2 + x^2y - z + 1 = 0$ at the point (1, –3, 2).

Solution Let $f(x, y, z) = xz^2 + x^2y - z + 1 = 0$ be the surface. Then, the normal vector is given by

$$\nabla f = \left(\mathbf{i}\frac{\partial}{\partial x} + \mathbf{j}\frac{\partial}{\partial y} + \mathbf{k}\frac{\partial}{\partial z}\right)(xz^2 + x^2y - z + 1) = (z^2 + 2xy)\mathbf{i} + x^2\mathbf{j} + (2xz - 1)\mathbf{k}.$$

At (1, –3, 2), the normal vector is given by $\nabla f(1, -3, 2) = -2\mathbf{i} + \mathbf{j} + 3\mathbf{k}$.

The equation of the *tangent plane* at any point P_0 (x_0, y_0, z_0) on the surface $f(x, y, z) = c$ is

$$(x - x_0)\frac{\partial f}{\partial x}(P_0) + (y - y_0)\frac{\partial f}{\partial y}(P_0) + (z - z_0)\frac{\partial f}{\partial z}(P_0) = 0.$$

Therefore, the tangent plane at the point (1, –3, 2) is given by

$$-2(x - 1) + 1(y + 3) + 3(z - 2) = 0 \quad \text{or} \quad 2x - y - 3z + 1 = 0.$$

EXAMPLE 1.25 Find the angle between the surfaces $x^2 + y^2 + z^2 = 9$ and $z + 3 = x^2 + y^2$ at the point (2, –1, 2). *(AMIETE, June 2013, Kottayam, 2005; UPTU 2003; MDU, 2001)*

Solution The angle between two surfaces at a common point is the angle between their normals at that point.

Let $f_1(x, y, z) = x^2 + y^2 + z^2 - 9 = 0$ and $f_2(x, y, z) = x^2 + y^2 - z - 3 = 0$.

Therefore, $\nabla f_1(x, y, z) = 2x\mathbf{i} + 2y\mathbf{j} + 2z\mathbf{k}$ and $\nabla f_2(x, y, z) = 2x\mathbf{i} + 2y\mathbf{j} - \mathbf{k}$.

Now $\nabla f_1(2, -1, 2) = 4\mathbf{i} - 2\mathbf{j} + 4\mathbf{k} = \mathbf{n}_1$ and $\nabla f_2(2, -1, 2) = 4\mathbf{i} - 2\mathbf{j} - \mathbf{k} = \mathbf{n}_2$.

If the required angle is θ then

$$\cos\theta = \frac{\mathbf{n}_1 \cdot \mathbf{n}_2}{|\mathbf{n}_1||\mathbf{n}_2|} = \frac{(4\mathbf{i} - 2\mathbf{j} + 4\mathbf{k}) \cdot (4\mathbf{i} - 2\mathbf{j} - \mathbf{k})}{\sqrt{(4)^2 + (-2)^2 + (4)^2}\sqrt{(4)^2 + (-2)^2 + (-1)^2}} = \frac{(4)(4) + (-2)(-2) + (4)(-1)}{\sqrt{36}\sqrt{21}}$$

$$= \frac{16 + 4 - 4}{6\sqrt{21}} = \frac{8\sqrt{21}}{63}.$$ Thus the acute angle is $\theta = \cos^{-1}(8\sqrt{21}/63) = 54°25'$.

EXAMPLE 1.26 Find the values of l and m so that the surfaces $lx^2y + mz^3 = 4$ may cut the surface $5x^2 = 2yz + 9x$ orthogonally at (1, –1, 2). *(AMIETE, Dec. 2011)*

Solution Let $f_1(x, y, z) = lx^2y + mz^3 - 4 = 0$, $f_2(x, y, z) = 5x^2 - 2yz - 9x = 0$

and $\mathbf{n}_1$, $\mathbf{n}_2$ be the normal vectors to the given surfaces at (1, –1, 2).

Then $\nabla f_1(x, y, z) = 2lxy\,\mathbf{i} + lx^2\,\mathbf{j} + 3mz^2\,\mathbf{k}$, $\nabla f_1(1, -1, 2) = -2l\,\mathbf{i} + l\,\mathbf{j} + 12m\,\mathbf{k} = \mathbf{n}_1$.

and $\nabla f_2(x, y, z) = (10x - 9)\,\mathbf{i} - 2z\,\mathbf{j} - 2y\,\mathbf{k}$, $\nabla f_2(1, -1, 2) = \mathbf{i} - 4\,\mathbf{j} + 2\,\mathbf{k} = \mathbf{n}_2$.

If the given surfaces are orthogonal at (1, –1, 2) then the angle between $\mathbf{n}_1$ and $\mathbf{n}_2$ must be 90°, that is,

$$\mathbf{n}_1 \cdot \mathbf{n}_2 = 0$$

or $(-2l\,\mathbf{i} + l\,\mathbf{j} + 12m\,\mathbf{k}) \cdot (\mathbf{i} - 4\,\mathbf{j} + 2\,\mathbf{k}) = 0$ or $l - 4m = 0$. ...(*i*)

The point (1, –1, 2) lies on the surfaces $lx^2y + mz^3 = 4$ so $l - 8m + 4 = 0$. ...(*ii*)

On solving (*i*) and (*ii*), we obtain $l = 4$, $m = 1$.

EXAMPLE 1.27 If $\nabla f = (x + 2y + 4z)\,\mathbf{i} + (2x - 3y - z)\mathbf{j} + (4x - y + 2z)\,\mathbf{k}$, find the scalar function f.

Solution Since $\nabla f = \frac{\partial f}{\partial x}\mathbf{i} + \frac{\partial f}{\partial y}\mathbf{j} + \frac{\partial f}{\partial z}\mathbf{k}$. Comparing, we obtain

$$\frac{\partial f}{\partial x} = x + 2y + 4z \;\cdots(i), \quad \frac{\partial f}{\partial y} = 2x - 3y - z \;\cdots(ii), \quad \frac{\partial f}{\partial z} = 4x - y + 2z \;\cdots(iii).$$

Integrating the first equation, we obtain $f(x, y, z) = \frac{x^2}{2} + 2xy + 4zx + g(y, z)$. ...(A)

Substituting in the second equation, we get $\frac{\partial f}{\partial y} = 2x - 3y - z = 2x + \frac{\partial g}{\partial y}$, or $\frac{\partial g}{\partial y} = -3y - z$.

Therefore, $g(y, z) = -3\frac{y^2}{2} - yz + g(z)$.

Now Eq. (A) become $f(x, y, z) = (x^2/2) + 2xy + 4zx + (-3/2)y^2 - yz + g(z)$. ...(B)

Substituting in the third equation, we get

$$\frac{\partial f}{\partial z} = 4x - y + 2z = 4x - y + \frac{dg}{dz} \quad \text{or} \quad \frac{dg}{dz} = 2z.$$ Integrating, we get $g = z^2 + k$, constant.

Thus from Eq. (B), we obtain

$$f(x, y, z) = (x^2/2) + 2xy + 4xz - yz + (-3/2)y^2 + z^2 + k.$$

Remark 3

A vector field **v** is said to be conservative if the vector function can be written as the gradient of a scalar function *f*, that is $\mathbf{v} = \nabla f$. It may be noted that not every vector field is conservative.

EXAMPLE 1.28 Show that the vector field defined by the vector function $\mathbf{v} = xyz\,(yz\mathbf{i} + zx\mathbf{j} + xy\mathbf{k})$ is conservative. *(AMIETE, June 2011, 2008; AMIE-W-2010)*

Solution If the given vector field is conservative, then it can be expressed as the gradient of a scalar function $f(x, y, z)$. Therefore

$$\nabla f = \left(\mathbf{i}\frac{\partial}{\partial x} + \mathbf{j}\frac{\partial}{\partial y} + \mathbf{k}\frac{\partial}{\partial z}\right) f = \left(\mathbf{i}\frac{\partial f}{\partial x} + \mathbf{j}\frac{\partial f}{\partial y} + \mathbf{k}\frac{\partial f}{\partial z}\right) = \mathbf{v} = xyz\,(yz\mathbf{i} + zx\mathbf{j} + xy\mathbf{k}).$$

Comparing, we obtain

$$\frac{\partial f}{\partial x} = xy^2z^2 \quad \ldots(i), \qquad \frac{\partial f}{\partial y} = x^2yz^2 \quad \ldots(ii), \qquad \frac{\partial f}{\partial z} = x^2y^2z \quad \ldots(iii).$$

Integrating the first equation, we obtain

$f(x, y, z) = (1/2)\, x^2y^2z^2 + g(y, z)$. ...(A) Substituting in the second equation, we get

$$\frac{\partial f}{\partial y} = x^2yz^2 = x^2yz^2 + \frac{\partial g}{\partial y} \text{ or } \frac{\partial g}{\partial y} = 0 \Rightarrow g(y, z) = g(z).$$

Now Eq. (A) become $f(x, y, z) = (1/2)\, x^2y^2z^2 + g(z)$. ...(B)

Substituting this in the third equation, we get

$$\frac{\partial f}{\partial z} = x^2y^2z = x^2y^2z + \frac{dg}{dz} \text{ or } \frac{dg}{dz} = 0 \Rightarrow g = c, \text{ constant.}$$

Hence, $f(x, y, z) = \frac{1}{2}x^2y^2z^2 + c$. Therefore, there exists a scalar function $f(x, y, z)$ such that $\nabla f = \mathbf{v}$ and the vector field **v** is conservative.

1.12 DIRECTIONAL DERIVATIVE

Let $f(P) = f(x, y, z)$ be a differential scalar field. Then $\partial f/\partial x$, $\partial f/\partial y$, $\partial f/\partial z$ denote the rates of change of *f* in the directions of *x*, *y* and *z* axis, respectively. Let $\hat{\mathbf{b}} = b_1\mathbf{i} + b_2\mathbf{j} + b_3\mathbf{k}$ be any unit vector. Let P_o *be any point, its position vector is* $\mathbf{a} = a_1\,\mathbf{i} + a_2\,\mathbf{j} + a_3\,\mathbf{k}$. Refer Fig. 1.10. Let P_oC is parallel to $\hat{\mathbf{b}}$ and Q is a variable point on the directed line segment P_oC. Then the position vector of point Q is given by

$$\mathbf{r}(s) = \mathbf{a} + s\hat{\mathbf{b}}, \quad s \ge 0 = (a_1 + b_1 s)\mathbf{i} + (a_2 + b_2 s)\mathbf{j} + (a_3 + b_3 s)\mathbf{k} = x(s)\mathbf{i} + y(s)\mathbf{j} + z(s)\mathbf{k}.$$

The vector $\overline{P_oQ}$ is given by $s\,\hat{\mathbf{b}}$. Since $|\hat{\mathbf{b}}| = 1$, so the distance from P_0 to Q is s.

Then $$\frac{\partial f}{\partial s} = \lim_{s \to 0} \frac{f(Q) - f(P_0)}{s}$$

if it exists, is called the *directional derivative* of *f* at the point P_0 in the direction of $\hat{\mathbf{b}}$. It is the rate of change of *f* in the direction of **b**.

Therefore, $\frac{\partial}{\partial s} f(x(s), y(s), z(s))$ is the rate of change of *f* with respect to the distance *s*.

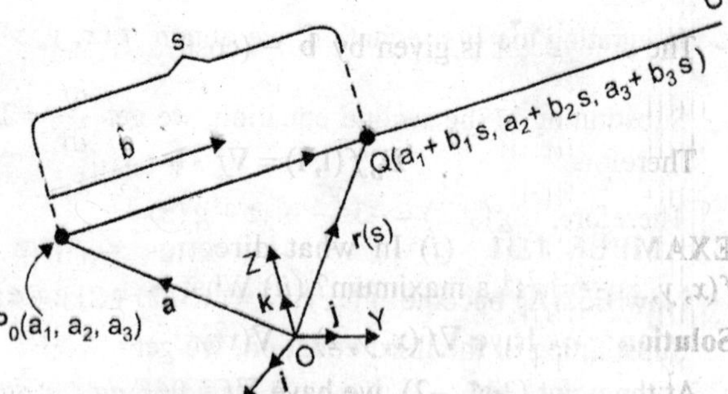

Fig. 1.10: Directional derivative

We have by the chain rule

$$\frac{\partial f}{\partial s} = \frac{\partial f}{\partial x}\frac{dx}{ds} + \frac{\partial f}{\partial y}\frac{dy}{ds} + \frac{\partial f}{\partial z}\frac{dz}{ds} \qquad \ldots(i)$$

where dx/ds, dy/ds, dz/ds are calculated at the point P_0 (where $s = 0$). We write (*i*) as

$$\frac{\partial f}{\partial s}=\left(\mathbf{i}\frac{\partial}{\partial x}+\mathbf{j}\frac{\partial}{\partial y}+\mathbf{k}\frac{\partial}{\partial z}\right)f\cdot\left(\mathbf{i}\frac{dx}{ds}+\mathbf{j}\frac{dy}{ds}+\mathbf{k}\frac{dz}{ds}\right)=\nabla f\cdot\frac{d\mathbf{r}}{ds}.$$

But $d\mathbf{r}/ds=\hat{\mathbf{b}}$ (a unit vector). Therefore, the directional derivative of f in the direction of $\hat{\mathbf{b}}$, denoted by $D_{\mathbf{b}}(f)$, is given by

$$D_{\mathbf{b}}(f)=\nabla f\cdot\hat{\mathbf{b}}=grad(f)\cdot\hat{\mathbf{b}}. \qquad ...(1.17)$$

Remark 4

If the direction is specified by a vector **u**, then we have $\hat{\mathbf{b}}=\mathbf{u}/|\mathbf{u}|$ and $D_{\mathbf{b}}(f)=\nabla f\cdot\mathbf{u}/|\mathbf{u}|$.

Maximum/Minimum value of the Directional derivative

Let f represent a function of either two or three variables. Since the directional derivative is a dot product, we see from the definition of dot or scalar product that

$$D_{\mathbf{b}}(f)=\nabla f\cdot\hat{\mathbf{b}}=|\nabla f|\,|\hat{\mathbf{b}}|\cos\theta=|\nabla f|\cos\theta, \qquad \text{as } (|\hat{\mathbf{b}}|=1),$$

where θ is the angle between the vector ∇f and **b**. Because $0\le\theta\le\pi$, we have $-1\le\cos\theta\le 1$ and consequently $-|\nabla f|\le D_{\mathbf{b}}(f)\le|\nabla f|$.

Thus *the maximum value of the directional derivative is* $|\nabla f|$ *and it occurs when* $\theta = 0$*, that is,* $\hat{\mathbf{b}}$ *has the same direction as* ∇f*. This direction is the direction of the normal vector. The minimum value of the directional derivative is* $-|\nabla f|$ *and it occurs when* $\theta=\pi$ *that is* $\hat{\mathbf{b}}$ *and* ∇f *have opposite directions.* We may also say that the function f increases most rapidly in the direction of gradient vector ∇f, and decreases most rapidly in the direction of $-\nabla f$. In any direction orthogonal to ∇f, the directional derivative in zero.

EXAMPLE 1.29 Find the directional derivative of the function $f(x, y, z) = xy^2 + yz^3$ at the point $P(2, -1, 1)$ in the direction of the vector $\mathbf{i} + 2\mathbf{j} + 3\mathbf{k}$. (*AMIETE, Dec. 2009; Kurukshetra, 2006; AMIE, S-2005*)

Solution We have $\nabla f(x, y, z) = \nabla(xy^2 + yz^3) = y^2\mathbf{i} + (2xy + z^3)\mathbf{j} + 3yz^2\mathbf{k}$ and $\nabla f(2, -1, 1) = \mathbf{i} - 3\mathbf{j} - 3\mathbf{k}$.

The component of ∇f in the direction of a unit vector $\hat{\mathbf{b}}$ is given by $\nabla f\cdot\hat{\mathbf{b}}$ and is called the **directional** derivative of f in the direction of **b**.

The unit vector in the direction of $\mathbf{i} + 2\mathbf{j} + 3\mathbf{k}$ is given by $\hat{\mathbf{b}}=\dfrac{(\mathbf{i}+2\mathbf{j}+3\mathbf{k})}{\sqrt{(1)^2+(2)^2+(3)^2}}=\dfrac{\mathbf{i}+2\mathbf{j}+3\mathbf{k}}{\sqrt{14}}.$

Therefore, $D_{\mathbf{b}}f(2,-1,1)=\nabla f\cdot\hat{\mathbf{b}}=(\mathbf{i}-3\mathbf{j}-3\mathbf{k})\cdot\dfrac{(\mathbf{i}+2\mathbf{j}+3\mathbf{k})}{\sqrt{14}}=\dfrac{(1)(1)+(-3)(2)+(-3)(3)}{\sqrt{14}}=\dfrac{-14}{\sqrt{14}}=-\sqrt{14}.$

EXAMPLE 1.30 Find the directional derivative of $f(x, y) = 2x^2y^3 + 6xy$ at $(1, 1)$, in the direction of a unit vector which makes an angle of $\pi/6$ with x-axis.

Solution We have $\nabla f(x, y) = (4xy^3 + 6y)\,\mathbf{i} + (6x^2y^2 + 6x)\mathbf{j}$ and at the point $(1, 1)$, $\nabla f = 10\mathbf{i} + 12\mathbf{j}$.

The unit vector is given by $\hat{\mathbf{b}} = (\cos\theta)\,\mathbf{i} + (\sin\theta)\mathbf{j} = \dfrac{\sqrt{3}}{2}\mathbf{i}+\dfrac{1}{2}\mathbf{j}$, since $\theta = \pi/6$.

Therefore $$D_{\mathbf{b}}f(1,1)=\nabla f\cdot\hat{\mathbf{b}}=(10\mathbf{i}+12\mathbf{j})\cdot\left(\frac{\sqrt{3}}{2}\mathbf{i}+\frac{1}{2}\mathbf{j}\right)=(10)\left(\frac{\sqrt{3}}{2}\right)+(12)\left(\frac{1}{2}\right)=5\sqrt{3}+6.$$

EXAMPLE 1.31 (*i*) In what direction from the point $(3, 1, -2)$ is the directional derivative of $f(x, y, z) = x^2y^2z^4$ a maximum? (*ii*) What is the magnitude of this maximum? (*AMIE, W-2005; MDU 2004*)

Solution We have $\nabla f(x, y, z) = \nabla(x^2y^2z^4) = 2xy^2z^4\mathbf{i} + 2x^2yz^4\mathbf{j} + 4x^2y^2z^3\mathbf{k}$.

At the point $(3, 1, -2)$, we have $\nabla f = 96\mathbf{i} + 288\mathbf{j} - 288\mathbf{k}$.

(*i*) The maximum value of the directional derivative is $|\nabla f(3, 1, -2)|$ and it occurs when $\hat{\mathbf{b}}$ has the same direction as ∇f (when $\cos\theta = 1$).

(*ii*) The maximum value of the directional derivative at (3, 1, –2) is $|\nabla f(3, 1, -2)|$

$$= \sqrt{(96)^2 + (288)^2 + (-288)^2} = \sqrt{175104} = 96\sqrt{19}.$$

Remark 5

The minimum value of $D_{\mathbf{b}} f(3, 1, -2)$ is then $-96\sqrt{19}$.

EXAMPLE 1.32 (*i*) Find the directional derivative of the scalar function $f(x, y, z) = 2x^3y - 3y^2z$ at the point P (1, 2, –1) in the direction of the line PQ, where Q has coordinates (3, –1, 5).

(*ii*) In what direction from P is the directional derivative a maximum?

(*iii*) What is the magnitude of the maximum directional derivative?

Solution (*i*) We have $\nabla f(x, y, z) = 6x^2y\mathbf{i} + (2x^3 - 6yz)\mathbf{j} - 3y^2\mathbf{k}$, and at the point (1, 2, –1),

$\nabla f = 12\mathbf{i} + 14\mathbf{j} - 12\mathbf{k}$.

The vector $\mathbf{b}$ = position vector of Q – position vector of P = $(3 - 1)\mathbf{i} + (-1-2)\mathbf{j} + [5 - (-1)]\mathbf{k}$
$= 2\mathbf{i} - 3\mathbf{j} + 6\mathbf{k}$.

The unit vector is given by $$\hat{\mathbf{b}} = \frac{2\mathbf{i} - 3\mathbf{j} + 6\mathbf{k}}{\sqrt{(2)^2 + (-3)^2 + (6)^2}} = \frac{2\mathbf{i} - 3\mathbf{j} + 6\mathbf{k}}{7}.$$

Therefore, directional derivative at P, $D_{\mathbf{b}} f(1, 2, -1) = \nabla f \cdot \hat{\mathbf{b}}$

$$= (12\mathbf{i} + 14\mathbf{j} - 12\mathbf{k}) \cdot \left(\frac{2\mathbf{i} - 3\mathbf{j} + 6\mathbf{k}}{7}\right) = \frac{24 - 42 - 72}{7} = -\frac{90}{7}.$$

(*ii*) The directional derivative is a maximum when $\mathbf{b}$ has the direction of $\nabla f = 12\mathbf{i} + 14\mathbf{j} - 12\mathbf{k}$.

(*iii*) The magnitude of the maximum value of the directional derivative is

$$|\nabla f(1, 2, -1)| = |12\mathbf{i} + 14\mathbf{j} - 12\mathbf{k}| = \sqrt{144 + 196 + 144} = 22.$$

EXAMPLE 1.33 If the directional derivative of $f(x, y, z) = axy^2 + byz + cz^2x^3$ at (1, 2, –1) has a maximum magnitude of 64 in the direction parallel to z-axis, find the values of constants a, b, c. (*AMIETE, June 2001*)

Solution We have $\nabla f(x, y, z) = \frac{\partial f}{\partial x}\mathbf{i} + \frac{\partial f}{\partial y}\mathbf{j} + \frac{\partial f}{\partial z}\mathbf{k} = (ay^2 + 3cz^2x^2)\mathbf{i} + (2axy + bz)\mathbf{j} + (by + 2czx^3)\mathbf{k}$ and

$\nabla f(1, 2, -1) = (4a + 3c)\mathbf{i} + (4a - b)\mathbf{j} + (2b - 2c)\mathbf{k}$.

The direction of unit vector $\hat{\mathbf{u}}$ parallel to the z-axis = $o\mathbf{i} + o\mathbf{j} + \mathbf{k}$. Therefore, $\nabla f \cdot \hat{\mathbf{u}} = 2b - 2c = 64$ (given). Directional derivative of f is maximum along $\hat{\mathbf{u}}$, this means that $\hat{\mathbf{u}}$ is parallel to ∇f at (1, 2, –1).

Thus $4a + 3c = 0$ and $4a - b = 0$. Solving these equations, we get $a = 6$, $b = 24$, $c = -8$.

EXAMPLE 1.34 Find the directional derivative of div grad ϕ (that is, $\nabla \cdot \nabla\phi$) at the point (1, –2, 1) in the direction of the normal to the surface $xy^2z = 3x + z^2$, where $\phi = 2x^3y^2z^4$. (*AMIE, June 2011*)

Solution Let $f(x, y, z) = xy^2z - 3x - z^2 = 0$ be the surface. Then, the normal vector is given by

$$\nabla f = \left(\mathbf{i}\frac{\partial}{\partial x} + \mathbf{j}\frac{\partial}{\partial y} + \mathbf{k}\frac{\partial}{\partial z}\right)(xy^2z - 3x - z^2) = (y^2z - 3)\mathbf{i} + 2xyz\,\mathbf{j} + (xy^2 - 2z)\mathbf{k} = 0.$$

The normal vector of f at (1, –2, 1) is given by $\nabla f(1, -2, 1) = \mathbf{i} - 4\mathbf{j} + 2\mathbf{k}$. The unit vector in the direction of $\mathbf{i} - 4\mathbf{j} + 2\mathbf{k}$ is given by

$$\hat{\mathbf{b}} = \frac{\mathbf{i} - 4\mathbf{j} + 2\mathbf{k}}{\sqrt{(1)^2 + (-4)^2 + (2)^2}} = \frac{\mathbf{i} - 4\mathbf{j} + 2\mathbf{k}}{\sqrt{21}}$$

We have $$\nabla\phi = \left(\mathbf{i}\frac{\partial}{\partial x} + \mathbf{j}\frac{\partial}{\partial y} + \mathbf{k}\frac{\partial}{\partial z}\right)2x^3y^2z^4 = 6x^2y^2z^4\mathbf{i} + 4x^3yz^4\mathbf{j} + 8x^3y^2z^3\mathbf{k}.$$

Therefore, $\nabla \cdot \nabla\phi = \left(\frac{\partial}{\partial x}\mathbf{i} + \frac{\partial}{\partial y}\mathbf{j} + \frac{\partial}{\partial z}\mathbf{k}\right) \cdot (6x^2y^2z^4\mathbf{i} + 4x^3yz^4\mathbf{j} + 8x^3y^2z^3\mathbf{k})$

$$= 12xy^2z^4 + 4x^3z^4 + 24x^3y^2z^2 = F(x, y, z) \text{ say.}$$

$$\nabla F(x, y, z) = \nabla(12xy^2z^4 + 4x^3z^4 + 24x^3y^2z^2)$$

$$= \mathbf{i}\frac{\partial}{\partial x}(12xy^2z^4 + 4x^3z^4 + 24x^3y^2z^2) + \mathbf{j}\frac{\partial}{\partial y}(12xy^2z^4 + 4x^3z^4 + 24x^3y^2z^2)$$

$$+ \mathbf{k}\frac{\partial}{\partial z}(12xy^2z^4 + 4x^3z^4 + 24x^3y^2z^2)$$

$$= (12y^2z^4 + 12x^2z^4 + 72x^2y^2z^2)\mathbf{i} + (24xyz^4 + 48x^3yz^2)\mathbf{j} + (48xy^2z^3 + 16x^3z^3 + 48x^3y^2z)\mathbf{k}$$

and $\nabla F(1, -2, 1) = (48 + 12 + 288)\mathbf{i} + (-48 - 96)\mathbf{j} + (192 + 16 + 192)\mathbf{k} = 348\,\mathbf{i} - 144\,\mathbf{j} + 400\,\mathbf{k}$.

Therefore, $D_{\mathbf{b}}F(1, -2, 1) = \nabla F \cdot \hat{\mathbf{b}} = (348\mathbf{i} - 144\mathbf{j} + 400\mathbf{k}) \cdot \frac{(\mathbf{i} - 4\mathbf{j} + 2\mathbf{k})}{\sqrt{21}}$

$$= \frac{(348)(1) + (-144)(-4) + (400)(2)}{\sqrt{21}} = \frac{1724}{\sqrt{21}}.$$

EXAMPLE 1.35 Find the directional derivative of $f(x, y, z) = x^2yz^3$ along the curve $x = e^{-u}$, $y = 2\sin u + 1$, $z = u - \cos u$ at the point P where $u = 0$.

Solution The point P corresponding to $u = 0$ is $(1, 1, -1)$.

$$\nabla f(x, y, z) = 2xyz^3\mathbf{i} + x^2z^3\mathbf{j} + 3x^2yz^2\mathbf{k}, \quad \text{and} \quad \nabla f(1, 1, -1) = -2\mathbf{i} - \mathbf{j} + 3\mathbf{k}.$$

The position vector $\mathbf{r}$ of any point on the given curve is given by

$$\mathbf{r} = e^{-u}\,\mathbf{i} + (2\sin u + 1)\mathbf{j} + (u - \cos u)\mathbf{k}.$$

A tangent vector to the curve is

$$\mathbf{r}'(u) = \frac{d\mathbf{r}}{du} = \frac{d}{du}\{e^{-u}\mathbf{i} + (2\sin u + 1)\mathbf{j} + (u - \cos u)\mathbf{k}\}$$

$= -e^{-u}\,\mathbf{i} + 2\cos u\,\mathbf{j} + (1 + \sin u)\mathbf{k}$. Therefore, $(d\mathbf{r}/du)_{u=0}$, that is, $\mathbf{r}'(0) = -\mathbf{i} + 2\mathbf{j} + \mathbf{k}$.

The unit tangent vector is given by $\hat{\mathbf{b}} = \frac{-\mathbf{i} + 2\mathbf{j} + \mathbf{k}}{\sqrt{1 + 4 + 1}} = \frac{-\mathbf{i} + 2\mathbf{j} + \mathbf{k}}{\sqrt{6}}$.

Therefore, $D_{\mathbf{b}}f(1, 1, -1) = \nabla f \bullet \hat{\mathbf{b}} = (-2\mathbf{i} - \mathbf{j} + 3\mathbf{k}) \bullet \left(\frac{-\mathbf{i} + 2\mathbf{j} + \mathbf{k}}{\sqrt{6}}\right) = \frac{1}{\sqrt{6}}(2 - 2 + 3) = \frac{3}{\sqrt{6}} = \frac{\sqrt{6}}{2}$.

1.13 PROPERTIES OF GRADIENT

If f and g are differentiable scalar functions of position (x, y, z), then

(*i*) $\nabla(kf) = k\,\nabla f$ (any number k),

(*ii*) $\nabla(f \pm g) = \nabla f \pm \nabla g$,

(*iii*) $\nabla(c_1 f + c_2 g) = c_1\,\nabla f + c_2\,\nabla g$, c_1, c_2 arbitrary constants,

(*iv*) $\nabla(fg) = f\,\nabla g + g\,\nabla f$,

(*v*) $\nabla\left(\frac{f}{g}\right) = \frac{g\nabla f - f\nabla g}{g^2}$, $g \neq 0$.

The proof of properties (*ii*), (*iv*) and (*v*) is given below.

(*ii*) $\nabla(f \pm g) = \left(\mathbf{i}\frac{\partial}{\partial x} + \mathbf{j}\frac{\partial}{\partial y} + \mathbf{k}\frac{\partial}{\partial z}\right)(f \pm g) = \mathbf{i}\frac{\partial}{\partial x}(f \pm g) + \mathbf{j}\frac{\partial}{\partial y}(f \pm g) + \mathbf{k}\frac{\partial}{\partial z}(f \pm g)$

$$= \left(\mathbf{i}\frac{\partial f}{\partial x} + \mathbf{j}\frac{\partial f}{\partial y} + \mathbf{k}\frac{\partial f}{\partial z}\right) \pm \left(\mathbf{i}\frac{\partial g}{\partial x} + \mathbf{j}\frac{\partial g}{\partial y} + \mathbf{k}\frac{\partial g}{\partial z}\right) = \nabla f \pm \nabla g.$$

(iv) $$\nabla(fg) = \left(\mathbf{i}\frac{\partial}{\partial x}+\mathbf{j}\frac{\partial}{\partial y}+\mathbf{k}\frac{\partial}{\partial z}\right)fg = \mathbf{i}\left(f\frac{\partial g}{\partial x}+g\frac{\partial f}{\partial x}\right)+\mathbf{j}\left(f\frac{\partial g}{\partial y}+g\frac{\partial f}{\partial y}\right)+\mathbf{k}\left(f\frac{\partial g}{\partial z}+g\frac{\partial f}{\partial z}\right)$$

$$= f\left(\mathbf{i}\frac{\partial g}{\partial x}+\mathbf{j}\frac{\partial g}{\partial y}+\mathbf{k}\frac{\partial g}{\partial z}\right)+g\left(\mathbf{i}\frac{\partial f}{\partial x}+\mathbf{j}\frac{\partial f}{\partial y}+\mathbf{k}\frac{\partial f}{\partial z}\right) = f\nabla g + g\nabla f.$$

(v) $$\nabla\left(\frac{f}{g}\right) = \left(\mathbf{i}\frac{\partial}{\partial x}+\mathbf{j}\frac{\partial}{\partial y}+\mathbf{k}\frac{\partial}{\partial z}\right)\left(\frac{f}{g}\right)$$

$$= \mathbf{i}\frac{\left(g\frac{\partial f}{\partial x}-f\frac{\partial g}{\partial x}\right)}{g^2}+\mathbf{j}\frac{\left(g\frac{\partial f}{\partial y}-f\frac{\partial g}{\partial y}\right)}{g^2}+\mathbf{k}\frac{\left(g\frac{\partial f}{\partial z}-f\frac{\partial g}{\partial z}\right)}{g^2},\ \text{if } g \neq 0$$

$$= \frac{1}{g^2}\left[g\left(\mathbf{i}\frac{\partial f}{\partial x}+\mathbf{j}\frac{\partial f}{\partial y}+\mathbf{k}\frac{\partial f}{\partial z}\right)-f\left(\mathbf{i}\frac{\partial g}{\partial x}+\mathbf{j}\frac{\partial g}{\partial y}+\mathbf{k}\frac{\partial g}{\partial z}\right)\right] = \frac{g\nabla f - f\nabla g}{g^2},\ g \neq 0.$$

EXERCISE 1.2

1. Compute the gradient of the following scalar fields.

(a) $f(x, y) = y^2 - 4xy$, (b) $f(x, y) = 5y - x^3y^2$,
(c) $f(x, y, z) = xy^2 + 3x^2 - z^3$, (d) $f(x, y, z) = xy^2/z^3$, (e) $f(x, y, z) = x^2 + yz$,
(f) $f(x, y, z) = x^3 + y^3 + z^3 - 3xyz$, (g) $f(x, y, z) = \log(x^2 + y^2 + z^2)$.

Ans. (a) $-4y\mathbf{i} + (2y - 4x)\mathbf{j}$, (b) $-3x^2y^2\mathbf{i} + (5 - 2x^3y)\mathbf{j}$, (c) $(y^2 + 6x)\mathbf{i} + 2xy\mathbf{j} - 3z^2\mathbf{k}$,
(d) $(y^2/z^3)\mathbf{i} + (2xy/z^3)\mathbf{j} - (3xy^2/z^4)\mathbf{k}$, (e) $2x\mathbf{i} + z\mathbf{j} + y\mathbf{k}$, (f) $3(x^2 - yz)\mathbf{i} + 3(y^2 - zx)\mathbf{j} + 3(z^2 - xy)\mathbf{k}$.

[Hint: (g) $\nabla \log(x^2 + y^2 + z^2) = \nabla \log r^2 = \frac{2}{r}\cdot\frac{\mathbf{r}}{r} = \frac{2\mathbf{r}}{r^2} = \frac{2(x\mathbf{i} + y\mathbf{j} + 3\mathbf{k})}{x^2 + y^2 + z^2}$.**]**

2. Find the gradient of the given function at the indicated point.

(a) $f(x, y) = x^2 - 4y^2$, (2, 4), (b) $f(x, y, z) = 3x^2y - y^3z^2$, (1, –2, –1),
(c) $f(x, y, z) = 2xz^4 - x^2y$, (2, –2, –1), (d) $f(x, y, z) = x^2y + y^2x + z^2$, (1, 1, 1),
(e) $f(x, y, z) = x^3 - y^3 + xz^2$, (1, 1, 2). *(AMIETE, Dec. 2010)*

Ans. (a) $4\mathbf{i} - 32\mathbf{j}$, (b) $-12\mathbf{i} - 9\mathbf{j} - 16\mathbf{k}$, (c) $10\mathbf{i} - 4\mathbf{j} - 16\mathbf{k}$, (d) $3\mathbf{i} + 3\mathbf{j} + 2\mathbf{k}$, (e) $7\mathbf{i} - 3\mathbf{j} + 4\mathbf{k}$.

3. (a) Evaluate $\nabla\left(3r^2 - 4\sqrt{r} + \frac{6}{\sqrt[3]{r}}\right)$. (b) Show that grad $\left(\frac{1}{r^2}\right) = \frac{-2\mathbf{r}}{r^4}$. *(MDU 2002)*

Ans. (a) $(6 - 2r^{-3/2} - 2r^{-7/3})\mathbf{r}$.

4. (a) Find the normal vector and the unit normal vector to the given curve/surface at the indicated point.

(i) $x^2 - y^2 = 12$, (4, 2), (ii) $z = x^2 + y^2$, (1, 2, 5),
(iii) $xy^3z^2 = 4$, (–1, –1, 2), (iv) $x^4 - 3xyz + z^2 + 1 = 0$, (1, 1, 1),
(v) $x^3 + y^3 + 3xyz = 3$, (1, 2, –1), *(AMIETE, Dec. 2012, June 2002)*
(vi) $f(x, y, z) = xy^2 + 2yz - 8 = 0$ at (3, –2. 1), *(AMIE, W-2007)*
(vii) $(x - 1)^2 + y^2 + (z + 2)^2 = 9$, (3, 1, –4). (viii) $x^3 - xyz = 1$, (1, 1, 1), *(AMIETE, Dec. 2004)*

(b) Find the magnitude of the vector drawn perpendicular to the surface $x^2 + 2y^2 + z^2 = 7$ at the point (1, –1, 2). *(AMIE, S 2009)*

Ans. (a)(i) $4(2\mathbf{i} - \mathbf{j})$; $(2\mathbf{i} - \mathbf{j})/\sqrt{5}$, (ii) $(2\mathbf{i} + 4\mathbf{j} - \mathbf{k})$; $(2\mathbf{i} + 4\mathbf{j} - \mathbf{k})/\sqrt{21}$,
(iii) $4(\mathbf{i} + 3\mathbf{j} - \mathbf{k})$; $-(\mathbf{i} + 3\mathbf{j} - \mathbf{k})/\sqrt{11}$, (iv) $\mathbf{i} - 3\mathbf{j} - \mathbf{k}$; $(\mathbf{i} - 3\mathbf{j} - \mathbf{k})/\sqrt{11}$,
(v) $(-\mathbf{i} + 3\mathbf{j} + 2\mathbf{k})$; $(-\mathbf{i} + 3\mathbf{j} + 2\mathbf{k})/\sqrt{14}$, (vi) $4\mathbf{i} - 10\mathbf{j} - 4\mathbf{k}$; $(2\mathbf{i} - 5\mathbf{j} - 2\mathbf{k})/\sqrt{33}$,
(vii) $4\mathbf{i} + 2\mathbf{j} - 4\mathbf{k}$; $(2\mathbf{i} + \mathbf{j} - 2\mathbf{k})/3$, (viii) $2\mathbf{i} - \mathbf{j} - \mathbf{k}$; $(2\mathbf{i} - \mathbf{j} - \mathbf{k})/\sqrt{6}$. (b) 6.

5. If $u = x + y + z$, $v = x^2 + y^2 + z^2$, $w = xy + yz + zx$, show that $\nabla u \bullet [\nabla v \times \nabla w] = 0$. *(UPTU, 2002)*
[**Alternatively:** If $u = x + y + z$, $v = x^2 + y^2 + z^2$, $w = yz + zx + xy$, prove that grad u grad v and grad w are coplanar.]

6. What is the angle between the normals to the surface $xy = z^2$ at the points (1, 4, 2) and (–3, –3, 3)?
Ans. $\cos^{-1}(1/\sqrt{22})$.

7. Find the angle between the two surfaces at the indicated point of intersection.
(*a*) $xy^2z = 3x + z^2$, $3x^2 - y^2 + 2z = 1$; (1, –2, 1).
(*b*) $x \log z = y^2 - 1$, $x^2y = 2 - z$, (1, 1, 1). *(AMIETE, Dec. 2006, JNTU 2003)*
Ans. (*a*) $\cos^{-1}(\sqrt{6}/14)$, (*b*) $\cos^{-1}(1/\sqrt{30})$.

8. Find the equation of the tangent plane to the ellipsoid $x^2 + 4y^2 + z^2 = 18$ at the point (1, 2, 1), and determine the acute angle that this plane makes with the *xy*-plane. **Ans.** $x + 8y + z = 18$; $\approx 83°$.

9. Find the values of constants λ and μ so that the surfaces $\lambda x^2 - \mu yz = (\lambda + 2)x$ and $4x^2y + z^3 = 4$ intersect orthogonally at the point (1, –1, 2). *(AMIETE, June 2011-2009, W-2008, MDU 2005)*
Ans. $\lambda = 5/2$, $\mu = 1$.

10. Find the directional derivative of
(*a*) $f(x, y) = xe^y + \cos(xy)$ at (2, 0) in the direction of 3**i** – 4**j**.
(*b*) $f(x, y, z) = 4xz^3 - 3x^2y^2z^2$ at (2, –1, 2) along *z*-axis.
(*c*) $f(x, y, z) = x^2yz + 4xz^2$ at (1, 2, –1) in the direction 2**i** – **j** – 2**k**.
(AMIE, W-2008; VTU 2007; JNTU 2006; MDU 2006)
(*d*) $f(x, y, z) = xy^2 - 4x^2y + z^2$ at (1, –1, 2) in the direction of 6**i** + 2**j** + 3**k**.
(*e*) $f(x, y, z) = 4xz^3 - 3x^2y^2z$ at (2, –1, 2) in the direction 2**i** – 3**j** + 6**k**.
(*f*) $\phi(x, y, z) = 4e^{2x-y+z}$ at (1, 1, –1) in the direction of vector **j**.
(*g*) $\phi = 3e^{2x-y+z}$ at $A(1, 1, -1)$ in the direction $\overrightarrow{AB}$ where B is the point (–3, 5, 6). *(AMIETE, Dec.2001)*
(*h*) $\phi(x, y) = x/(x^2 + y^2)$ in the direction of a line making an angle of 30° with the positive x-axis at the point (0, 1). *(AMIETE, Dec. 2002)*

Ans. (*a*) –1, (*b*) 48, (*c*) 37/3, (*d*) 54/7, (*e*) 376/7, (*f*) –4, (*g*) –5/3, (*h*) $\sqrt{3}/2$.

11. Find the directional derivative of the scalar function $\phi(x, y, z) = xy^2 + yz^3$ at the point (2, –1, 1) in the direction of normal to the surface $x \log z - y^2 + 4 = 0$ at (–1, 2, 1).
(MDU 2007; JNTU-2005; AMIETE, W-2004) **Ans.** $15/\sqrt{17}$.

12. Find the directional derivative of $\vec{V}^2$ where $\vec{V} = xy^2\mathbf{i} + zy^2\mathbf{j} + xz^2\mathbf{k}$ at the point (2, 0, 3) in the direction of the outward normal to the sphere $x^2 + y^2 + z^2 = 14$ at the point (3, 2, 1). *(AMIETE Dec. 2007)*
Ans. $702\sqrt{14}/7$.

[**Hint:** Let $f(x, y, z) = \vec{V}^2 = x^2y^4 + z^2y^4 + x^2z^4$. At the point (2, 0, 3), $\nabla f = 324\mathbf{i} + 432\mathbf{k}$.
The unit vector is given by $\hat{\mathbf{b}} = (3\mathbf{i} + 2\mathbf{j} + \mathbf{k})/\sqrt{14}$.
Therefore, $D_{\mathbf{b}}f(2, 0, 3) = \nabla f \cdot \hat{\mathbf{b}} = (324\mathbf{i} + 432\mathbf{k}) \cdot \dfrac{(3\mathbf{i} + 2\mathbf{j} + \mathbf{k})}{\sqrt{14}} = \dfrac{702\sqrt{14}}{7}$].

13. Find the directional derivative of div **u** at the point (2, 2, 1) in the direction of the outer normal to the sphere $x^2 + y^2 + z^2 = 9$, where $\mathbf{u} = x^2z\,\mathbf{i} + y^2x\,\mathbf{j} + z^2y\,\mathbf{k}$. **Ans.** 32/3

14. (*a*) Find the directional derivative of the function $f(x, y, z) = 4e^{2x-y+z}$ at the point $A(1, 1, -1)$ in the direction towards the point $B(-3, 5, 6)$.
(*b*) In what direction from the point (1, 1, –1) is the directional derivative of $f(x, y, z) = x^2 - 2y^2 + 4z^2$ a maximum? Also find the value of this maximum directional derivative.
Ans. (*a*) –20/9, (*b*) $2\mathbf{i} - 4\mathbf{j} - 8\mathbf{k}$; $2\sqrt{21}$.

15. What is the greatest rate of increase of
(*a*) $\phi(x, y, z) = x^2 + yz^2$ at (1, –1, 3)? (*AMIE, S-2004*) (*b*) $f(x, y, z) = xyz^2$ at (1, 0, 3)?
Ans. (*a*) 11; (*b*) 9.

16. Find the directions in which $\phi(x, y) = (x^2 + y^2)/2$ increases most rapidly and decreases most rapidly at the point (1, 1). **Ans.** $\frac{1}{\sqrt{2}}\mathbf{i} + \frac{1}{\sqrt{2}}\mathbf{j};\ -\frac{1}{\sqrt{2}}\mathbf{i} - \frac{1}{\sqrt{2}}\mathbf{j}$.

17. (*a*) Find the directional derivative of $f(x, y, z) = 2xy - z^2$ at (2, –1, 1) in a direction toward (3, 1, –1).
(*b*) In what direction is the directional derivative a maximum?
(*c*) What is the value of this maximum?
Ans. (*a*) 10/3, (*b*) $-2\mathbf{i} + 4\mathbf{j} - 2\mathbf{k}$, (*c*) $2\sqrt{6}$.

18. (*a*) Given that the directional derivative of $f(x, y, z)$ at the point (3, –2, 1) in the direction of $\mathbf{a} = 2\mathbf{i} - \mathbf{j} - 2\mathbf{k}$ is –5 and that $|\nabla f(3, -2, 1)| = 5$, find $\nabla f(3, -2, 1)$.
(*b*) Find the directional derivative of $\phi(x, y, z) = 5x^2y - 5y^2z + 2.5z^2x$ at the point *P*(1, 1, 1) in the direction of the line $(x-1)/2 = (y-3)/-2 = z$. (*UPTU 2004*)
Ans. (*a*) $-5(2\mathbf{i} - \mathbf{j} - 2\mathbf{k})/3$, (*b*) 35/3.

19. The temperature at a point (*x*, *y*, *z*) in a space is given by $T(x, y, z) = x^2 + y^2 - z$. A mosquito located at (1, 1, 2) desires to fly in such a direction that it will get warm as soon as possible. In what direction should it fly? (*AMIE, S-2005*) **Ans.** $(2\mathbf{i} + 2\mathbf{j} - \mathbf{k})/3$.

20. The temperature at a point (*x*, *y*, *z*) in a space is given by $T(x, y, z) = x^2 + y^2 - z$. A fly located at (4, 4, 2) desires to fly in a direction that gets cooler faster. Find the direction in which it should fly. Also find the rate decrease of temperature in the direction of flight. (*AMIETE, June 2005*)
Ans. In the direction of maximum rate of decrease, $-(8\mathbf{i} + 8\mathbf{j} - \mathbf{k})$.

21. Find the values of a, b, c so that the directional derivative of $f(x, y, z) = ax^2 + by^2 + cz^2$ at the point (1, 1, 2) has a maximum magnitude 4 in the direction of *y*-axis. [**Hint:** ∇f at (1, 1, 2) = 4**j**]
Ans. $a = 0$, $b = 2$, $c = 0$.

22. If the directional derivative of $f(x, y, z) = ax^2y + by^2z + cz^2x$ at (1, 1, 1) has maximum magnitude 15 in the direction parallel to line $\frac{x-1}{2} = \frac{y-3}{-2} = z$, find the value of *a*, *b*, *c*. (*AMIETE, June 2007, 2003*)
[**Hint:** The maximum directional derivative of $f(x, y, z)$ at any point is the value of ∇f at the point.

$$\nabla f = (2axy + cz^2)\mathbf{i} + (2byz + ax^2)\mathbf{j} + (by^2 + 2czx)\mathbf{k}.$$

$$\nabla f\,(1, 1, 1) = (2a + c)\mathbf{i} + (2b + a)\mathbf{j} + (b + 2c)\mathbf{k} = 15\left\{\frac{(2\mathbf{i} - 2\mathbf{j} + \mathbf{k})}{\sqrt{4+4+1}}\right\} = 10\mathbf{i} - 10\mathbf{j} + 5\mathbf{k}.$$

$\Rightarrow$ $2a + c = 10$, $2b + a = -10$, $b + 2c = 5$. Solving, we get $a = 20/9$; $b = -55/9$, $c = 50/9$].

23. Find the directional derivative of the function $f(x, y, z) = x^2 - y^2 + 2z^2$ at the point *P*(1, 2, 3) in the direction of the line *PQ*, where *Q* has coordinates (5, 0, 4).
(*AMIE, Dec 2011; MDU 2004; UPTU 2001; DU 2001*) **Ans.** $4\sqrt{21}/3$.

24. Find the directional derivative of $f(x, y) = x^2y^3 + xy$ at (2, 1) in the direction of a unit vector which makes an angle of π/3 with *x*-axis. (*AMIETE, June 2009*)
[**Hint:** $\nabla f = (2xy^3 + y)\mathbf{i} + (3x^2y^2 + x)\mathbf{j}$. $\nabla f(2, 1) = 5\mathbf{i} + 14\mathbf{j}$. $\hat{\mathbf{b}} = \cos\theta\,\mathbf{i} + \sin\theta\,\mathbf{j} = \mathbf{i}/2 + (\sqrt{3}/2)\mathbf{j}$.
Therefore direction derivative $= (5\mathbf{i} + 14\mathbf{j}) \cdot (\mathbf{i}/2 + (\sqrt{3}/2)\mathbf{j}) = (5 + 14\sqrt{3})/2$.]

25. Find the directional derivative of 1/*r* in the direction **r** where $\mathbf{r} = x\mathbf{i} + y\mathbf{j} + z\mathbf{k}$.
(*UPTU 2003, 2002*) **Ans.** $D_{\mathbf{b}}f = -1/(x^2 + y^2 + z^2)$.

26. Find the directional derivative of the scalar point function $\phi = x^2y + y^2z + z^2x$ at the point (2, 2, 2) in the direction of the normal to the surface $4x^2y + 2z^2 = 2$ at the point (2, –1, 3). (*AMIETE, Dec 2006*)

27. Show that the field of force given by $\mathbf{F}(x, y, z) = (y^2 \cos x + z^3)\mathbf{i} + (2y \sin x - 4)\mathbf{j} + (3xz^2 + 2)\mathbf{k}$ is conservative and find its scalar potential. Also compute the work done in moving the particle from (1, 1, –1) to (π/2, –1, 2). **Ans.** $f(x, y, z) = y^2 \sin x + xz^3 - 4y + 2z + k_1$.

28. If $\nabla f = 2xy\, z^3\mathbf{i} + x^2z^3\, \mathbf{j} + 3x^2yz^2\, \mathbf{k}$, find $f(x, y, z)$ if $f(1, -2, 2) = 4$. **Ans.** $f(x, y, z) = x^2yz^3 + 20$.

29. Find $f(r)$ such that $\nabla f = \mathbf{r}/r^5$ and $f(1) = 0$. **Ans.** $\frac{1}{3}\left(1 - \frac{1}{r^3}\right)$.

30. If $f(x, y, z) = x^2z + e^{y/x}$ and $g(x, y, z) = 2z^2y - xy^2$, find (*a*) $\nabla(f + g)$ and (*b*) $\nabla(fg)$ at the point (1, 0, –2). **Ans.** (*a*) $-4\mathbf{i} + 9\mathbf{j} + \mathbf{k}$, (*b*) $-8\mathbf{j}$.

31. If $f(x, y) = \log_e \sqrt{x + y^2}$, show that grad $f = \dfrac{\mathbf{r} - (\mathbf{k} \cdot \mathbf{r})\mathbf{k}}{\{\mathbf{r} - (\mathbf{k} \cdot \mathbf{r})\mathbf{k}\} \cdot \{\mathbf{r} - (\mathbf{k} \cdot \mathbf{r})\mathbf{k}\}}$, where $\mathbf{r} = x\mathbf{i} + y\mathbf{j} + z\mathbf{k}$.

[**Hint:** We have $\mathbf{k} \bullet \mathbf{r} = z$. Therefore, $\mathbf{r} - (\mathbf{k} \cdot \mathbf{r})\, \mathbf{k} = (x\mathbf{i} + y\mathbf{j} + z\mathbf{k}) - z\mathbf{k} = x\mathbf{i} + y\mathbf{j} = \mathbf{r}_1$ say).

We have to show that $\nabla f = \dfrac{\mathbf{r}_1}{|\mathbf{r}_1^2|} = \dfrac{x\mathbf{i} + y\mathbf{j}}{x^2 + y^2}$. Now $f(x, y) = \frac{1}{2} \log (x^2 + y^2)$.

$$\text{Therefore, } \nabla f = \left(\mathbf{i}\frac{\partial}{\partial x} + \mathbf{j}\frac{\partial}{\partial y}\right)\frac{1}{2}\log (x^2 + y^2) = \frac{1}{2}\left[\frac{2x\mathbf{i}}{x^2 + y^2} + \frac{2y\mathbf{j}}{x^2 + y^2}\right] = \left.\frac{x\mathbf{i} + y\mathbf{j}}{x^2 + y^2}\right]$$

32. If $f(x, y) = 3x^2y, \phi(x, y, z) = xz^2 - 2y$; evaluate grad [(grad f) • (grad ϕ)]. **Ans.** $(6yz^2 - 12x)\mathbf{i} + 6xz^2\mathbf{j} + 12xyz\mathbf{k}$.

33. If $f(x, y, z) = x^2z + e^{y/x}$ and $g(x, y, z) = 2z^2y - xy^2$, find (*a*) $\nabla(f + g)$ and (*b*) $\nabla(fg)$ at the point (1, 0, – 2). **Ans.** (*a*) $-4\mathbf{i} + 9\mathbf{j} + \mathbf{k}$, (*b*) $-8\mathbf{j}$.

1.14 DIVERGENCE AND CURL OF A VECTOR FIELD

Let $\mathbf{v}(x, y, z)$ be a differentiable vector field in 3 space and let v_1, v_2, v_3 be the components of $\mathbf{v}$.

1.14.1 Divergence of Vector Field v

The divergence of **v** or the divergence of the vector field **v**, denoted be div **v**, is defined as the scalar function.

$$\text{div } \mathbf{v} = \frac{\partial v_1}{\partial x} + \frac{\partial v_2}{\partial y} + \frac{\partial v_3}{\partial z}. \qquad \text{...(1.18)}$$

Another common notation for the divergence of **v** is $\nabla \bullet \mathbf{v}$, that is, del dot **v**.

$$\text{div } \mathbf{v} = \nabla \cdot \mathbf{v} = \left(\mathbf{i}\frac{\partial}{\partial x} + \mathbf{j}\frac{\partial}{\partial y} + \mathbf{k}\frac{\partial}{\partial z}\right) \cdot (v_1\mathbf{i} + v_2\mathbf{j} + v_3\mathbf{k}) = \frac{\partial v_1}{\partial x} + \frac{\partial v_2}{\partial y} + \frac{\partial v_3}{\partial z}.$$

Note that $\nabla \cdot \mathbf{v} \neq \mathbf{v} \cdot \nabla$.

EXAMPLE 1.36 Find the divergence of the following vector field:

(*i*) $\mathbf{v} = 3xyz^2\mathbf{i} + 2xy^3\mathbf{j} - x^2yz\mathbf{k}$ at the point (1, –1, 1),

(*ii*) $\mathbf{v} = (x^2y^2 - z^3)\mathbf{i} + 2xyz\mathbf{j} + e^{xyz}\mathbf{k}$. (*AMIE-S 2012, W-2010*)

Solution (*i*) $\text{div } \mathbf{v} = \nabla \bullet \mathbf{v} = \left(\mathbf{i}\frac{\partial}{\partial x} + \mathbf{j}\frac{\partial}{\partial y} + \mathbf{k}\frac{\partial}{\partial z}\right) \bullet (3xyz^2\mathbf{i} + 2xy^3\mathbf{j} - x^2yz\mathbf{k})$

$$= \frac{\partial}{\partial x}(3xyz^2) + \frac{\partial}{\partial y}(2xy^3) + \frac{\partial}{\partial z}(-x^2yz) = 3yz^2 + 6xy^2 - x^2y.$$

div **v** at the point (1, –1, 1) = 3(–1) (1)^2 + 6 (1) (–1)^2 – 1^2(–1) = –3 + 6 + 1 = 4.

(*ii*) We have $\text{div } \mathbf{v} = \frac{\partial}{\partial x}(x^2y^2 - z^3) + \frac{\partial}{\partial y}(2xyz) + \frac{\partial}{\partial z}(e^{xyz}) = 2xy^2 + 2xz + xye^{xyz}$.

1.14.2 Curl of Vector Field v

The **curl** of a vector field **v** denoted by curl **v**, is defined as the vector field.

$$\text{curl } \mathbf{v} = \left(\frac{\partial v_3}{\partial y} - \frac{\partial u_2}{\partial z}\right)\mathbf{i} + \left(\frac{\partial v_1}{\partial z} - \frac{\partial v_3}{\partial x}\right)\mathbf{j} + \left(\frac{\partial v_2}{\partial x} - \frac{\partial v_1}{\partial y}\right)\mathbf{k}. \quad \text{...(1.19)}$$

Another common notation for the curl of **v** is $\nabla \times \mathbf{v}$ that is, del cross **v**.

$$\text{curl } \mathbf{v} = \nabla \times \mathbf{v} = \left(\mathbf{i}\frac{\partial}{\partial x} + \mathbf{j}\frac{\partial}{\partial y} + \mathbf{k}\frac{\partial}{\partial z}\right) \times (v_1\mathbf{i} + v_2\mathbf{j} + v_3\mathbf{k})$$

$$= \begin{vmatrix} \mathbf{i} & \mathbf{j} & \mathbf{k} \\ \partial/\partial x & \partial/\partial y & \partial/\partial z \\ v_1 & v_2 & v_3 \end{vmatrix} = \begin{vmatrix} \partial/\partial y & \partial/\partial z \\ v_2 & v_3 \end{vmatrix}\mathbf{i} - \begin{vmatrix} \partial/\partial x & \partial/\partial z \\ v_1 & v_3 \end{vmatrix}\mathbf{j} + \begin{vmatrix} \partial/\partial x & \partial/\partial y \\ v_1 & v_2 \end{vmatrix}\mathbf{k}$$

$$= \left(\frac{\partial v_3}{\partial y} - \frac{\partial v_2}{\partial z}\right)\mathbf{i} + \left(\frac{\partial v_1}{\partial z} - \frac{\partial v_3}{\partial x}\right)\mathbf{j} + \left(\frac{\partial v_2}{\partial x} - \frac{\partial v_1}{\partial y}\right)\mathbf{k}.$$

Note that $\nabla \times \mathbf{v}$ is not equal to $\mathbf{v} \times \nabla$. Sometimes curl **v** is also written as

$$\text{curl } \mathbf{v} = \Sigma\left(\frac{\partial v_3}{\partial y} - \frac{\partial v_2}{\partial z}\right)\mathbf{i},$$

where Σ denotes summation obtained by the cyclic rotation of the unit vectors **i, j, k,** the components v_1, v_2, v_3 and the independent variables x, y, z respectively.

Remark 6

(a) If $\mathbf{r} = x\mathbf{i} + y\mathbf{j} + z\mathbf{k}$, then div $\mathbf{r} = 3$ and curl $\mathbf{r} = \mathbf{0}$.

(b) If **a** is a constant vector, then div $\mathbf{a} = 0$ and curl $\mathbf{a} = \mathbf{0}$.

(c) A vector **A** is called *solenoidal* if its divergence is zero, that is, $\nabla \cdot \mathbf{A} = 0$.

(d) A vector **A** is called irrotational if its curl is zero, that is, curl $\mathbf{A} = \nabla \times \mathbf{A} = \mathbf{0}$.

EXAMPLE 1.37 Find the curl of the vector field $\mathbf{v} = xyz\,\mathbf{i} + 2x^2y\,\mathbf{j} + (xz^2 - y^2z)\mathbf{k}$ and verify that div (curl **v**) = 0.

Solution We have $\text{curl } \mathbf{v} = \nabla \times \mathbf{v} = \begin{vmatrix} \mathbf{i} & \mathbf{j} & \mathbf{k} \\ \partial/\partial x & \partial/\partial y & \partial/\partial z \\ xyz & 2x^2y & xz^2 - y^2z \end{vmatrix}$

$$= \left[\frac{\partial}{\partial y}(xz^2 - y^2z) - \frac{\partial}{\partial z}(2x^2y)\right]\mathbf{i} + \left[\frac{\partial}{\partial z}(xyz) - \frac{\partial}{\partial x}(xz^2 - y^2z)\right]\mathbf{j} + \left[\frac{\partial}{\partial x}(2x^2y) - \frac{\partial}{\partial y}(xyz)\right]\mathbf{k}$$

$$= -2yz\mathbf{i} + (xy - z^2)\mathbf{j} + (4xy - xz)\mathbf{k} = -\{2yz\mathbf{i} + (z^2 - xy)\mathbf{j} + (xz - 4xy)\mathbf{k}\}.$$

$$\text{div (curl } \mathbf{v}) = \nabla \cdot (\nabla \times \mathbf{v}) = \nabla \cdot [-\{2yz\,\mathbf{i} + (-xy + z^2)\mathbf{j} + (-4xy + xz)\mathbf{k}\}]$$

$$= \frac{\partial}{\partial x}(-2yz) + \frac{\partial}{\partial y}(xy - z^2) + \frac{\partial}{dz}(4xy - xz) = 0 + x - x = 0.$$

EXAMPLE 1.38 If $f(x, y, z) = x^2yz^3$ and $\mathbf{A} = xz\mathbf{i} - y^2\mathbf{j} + 2x^2y\mathbf{k}$, find (*i*) div ($f$**A**), (*ii*) curl ($f$**A**).

Solution (*i*) div $(f\mathbf{A}) = \nabla \cdot (f\mathbf{A}) = \nabla \cdot (x^3yz^4\mathbf{i} - x^2y^3z^3\mathbf{j} + 2x^4y^2z^3\mathbf{k})$

$$= \nabla \cdot (x^3yz^4\mathbf{i} - x^2y^3z^3\mathbf{j} + 2x^4y^2z^3\mathbf{k}) = \left(\mathbf{i}\frac{\partial}{\partial x} + \mathbf{j}\frac{\partial}{\partial y} + \mathbf{k}\frac{\partial}{\partial z}\right) \cdot (x^3yz^4\mathbf{i} - x^2y^3z^3\mathbf{j} + 2x^4y^2z^3\mathbf{k})$$

$$= \frac{\partial}{\partial x}(x^3yz^4) + \frac{\partial}{\partial y}(-x^2y^3z^3) + \frac{\partial}{\partial z}(2x^4y^2z^3) = 3x^2yz^4 - 3x^2y^2z^3 + 6x^4y^2z^2.$$

(*ii*) Curl (f**A**) $=\nabla\times(f\mathbf{A})=\nabla\times(x^3yz^4\mathbf{i}-x^2y^3z^3\mathbf{j}+2x^4y^2z^3\mathbf{k})$

$$=\begin{vmatrix}\mathbf{i} & \mathbf{j} & \mathbf{k}\\ \partial/\partial x & \partial/\partial y & \partial/\partial z\\ x^3yz^4 & -x^2y^3z^3 & 2x^4y^2z^3\end{vmatrix}=(4x^4yz^3+3x^2y^3z^2)\mathbf{i}+(4x^3yz^3-8x^3y^2z^3)\mathbf{j}-(2xy^3z^3+x^3z^4)\mathbf{k}.$$

EXAMPLE 1.39 If **a** is constant vector and $\mathbf{r}=x\mathbf{i}+y\mathbf{j}+z\mathbf{k}$, show that

(*i*) grad($\mathbf{a}\cdot\mathbf{r}$) = **a**, (*AMIETE, Dec. 2012, June 2002*) (*ii*) $\nabla\cdot(\mathbf{a}\times\mathbf{r})=0$, (*AMIETE, June 2013*)

(*ii*) curl ($\mathbf{a}\times\mathbf{r}$) = 2**a**. (*AMIETE Dec. 2009, 2005; UPTU 2002*)

Solution (*i*) Let $\mathbf{a}=a_1\mathbf{i}+a_2\mathbf{j}+a_3\mathbf{k}$, where a_1, a_2, a_3 are constants.

We have $\mathbf{a}\cdot\mathbf{r}=(a_1\mathbf{i}+a_2\mathbf{j}+a_3\mathbf{k})\cdot(x\mathbf{i}+y\mathbf{j}+z\mathbf{k})=a_1x+a_2y+a_3z.$

$$\text{Now } \nabla(\mathbf{a}\cdot\mathbf{r})=\left(\mathbf{i}\frac{\partial}{\partial x}+\mathbf{j}\frac{\partial}{\partial y}+\mathbf{k}\frac{\partial}{\partial z}\right)(a_1x+a_2y+a_3z)=a_1\mathbf{i}+a_2\mathbf{j}+a_3\mathbf{k}=\mathbf{a}.$$

$$(ii)\quad \mathbf{a}\times\mathbf{r}=\begin{vmatrix}\mathbf{i} & \mathbf{j} & \mathbf{k}\\ a_1 & a_2 & a_3\\ x & y & z\end{vmatrix}=\mathbf{i}(a_2z-a_3y)+\mathbf{j}(a_3x-a_1z)+\mathbf{k}(a_1y-a_2x).$$

$$\text{Therefore, } \nabla\cdot(\mathbf{a}\times\mathbf{r})=\left(\mathbf{i}\frac{\partial}{\partial x}+\mathbf{j}\frac{\partial}{\partial y}+\mathbf{k}\frac{\partial}{\partial z}\right)\cdot[\mathbf{i}(a_2z-a_3y)+\mathbf{j}(a_3x-a_1z)+\mathbf{k}(a_1y-a_2x)]$$

$$=\frac{\partial}{\partial x}(a_2z-a_3y)+\frac{\partial}{\partial y}(a_3x-a_1z)+\frac{\partial}{\partial z}(a_1y-a_2x)=0.$$

$$(iii)\quad \text{We have } \nabla\times(\mathbf{a}\times\mathbf{r})=\left(\mathbf{i}\frac{\partial}{\partial x}+\mathbf{j}\frac{\partial}{\partial y}+\mathbf{k}\frac{\partial}{\partial z}\right)\times(\mathbf{a}\times\mathbf{r})=\mathbf{i}\times\frac{\partial}{\partial x}(\mathbf{a}\times\mathbf{r})+\mathbf{j}\times\frac{\partial}{\partial y}(\mathbf{a}\times\mathbf{r})+\mathbf{k}\times\frac{\partial}{\partial z}(\mathbf{a}\times\mathbf{r})$$

$$=\mathbf{i}\times\left(\mathbf{a}\times\frac{\partial\mathbf{r}}{\partial x}\right)+\mathbf{j}\times\left(\mathbf{a}\times\frac{\partial\mathbf{r}}{\partial y}\right)+\mathbf{k}\times\left(\mathbf{a}\times\frac{\partial\mathbf{r}}{\partial z}\right)\quad\text{(Since }\mathbf{a}\text{ is a constant vector)}$$

$$=\mathbf{i}\times(\mathbf{a}\times\mathbf{i})+\mathbf{j}\times(\mathbf{a}\times\mathbf{j})+\mathbf{k}\times(\mathbf{a}\times\mathbf{k})=(\mathbf{i}\cdot\mathbf{i})\mathbf{a}-(\mathbf{i}\cdot\mathbf{a})\mathbf{i}+(\mathbf{j}\cdot\mathbf{j})\mathbf{a}-(\mathbf{j}\cdot\mathbf{a})\mathbf{j}+(\mathbf{k}\cdot\mathbf{k})\mathbf{a}-(\mathbf{k}\cdot\mathbf{a})\mathbf{k}$$

$$=(\mathbf{i}\cdot\mathbf{i})\mathbf{a}+(\mathbf{j}\cdot\mathbf{j})\mathbf{a}+(\mathbf{k}\cdot\mathbf{k})\mathbf{a}-(\mathbf{i}\cdot\mathbf{a})\mathbf{i}-(\mathbf{j}\cdot\mathbf{a})\mathbf{j}-(\mathbf{k}\cdot\mathbf{a})\mathbf{k}=3\mathbf{a}-(a_1\mathbf{i}+a_2\mathbf{j}+a_3\mathbf{k})=2\mathbf{a}.$$

EXAMPLE 1.40 If $\mathbf{A}=2xz^2\mathbf{i}-yz\mathbf{j}+3xz^3\mathbf{k}$ and $f(x, y, z)=x^2yz$, find curl curl **A**, curl grad f at the point (1, 1, 1), and div (grad f) or $\nabla\cdot\nabla f$.

Solution We have $\nabla f=\left(\mathbf{i}\frac{\partial}{\partial x}+\mathbf{j}\frac{\partial}{\partial y}+\mathbf{k}\frac{\partial}{\partial z}\right)x^2yz=2xyz\,\mathbf{i}+x^2z\,\mathbf{j}+x^2y\,\mathbf{k}.$ Curl curl $\mathbf{A}=\nabla\times(\nabla\times\mathbf{A})$

$$=\nabla\times\begin{vmatrix}\mathbf{i} & \mathbf{j} & \mathbf{k}\\ \partial/\partial x & \partial/\partial y & \partial/\partial z\\ 2xz^2 & -yz & 3xz^3\end{vmatrix}=\nabla\times[y\mathbf{i}-(3z^3-4xz)\mathbf{j}]=\begin{vmatrix}\mathbf{i} & \mathbf{j} & \mathbf{k}\\ \partial/\partial x & \partial/\partial y & \partial/\partial z\\ y & -3z^3+4xz & 0\end{vmatrix}$$

$=(9z^2-4x)\mathbf{i}+(4z-1)\mathbf{k}.$ Therefore, Curl curl **A** at the point (1, 1, 1) = 5**i** + 3**k**.

$$\text{Curl grad } f=\nabla\times\nabla f=\begin{vmatrix}\mathbf{i} & \mathbf{j} & \mathbf{k}\\ \partial/\partial x & \partial/\partial y & \partial/\partial z\\ 2xyz & x^2z & x^2y\end{vmatrix}=(x^2-x^2)\mathbf{i}-(2xy-2xy)\mathbf{j}+(2xz-2xz)\mathbf{k}=0.$$

$$\text{div (grad } f)=\nabla\cdot\nabla f=\left(\mathbf{i}\frac{\partial}{\partial x}+\mathbf{j}\frac{\partial}{\partial y}+\mathbf{k}\frac{\partial}{\partial z}\right)\cdot\left(\mathbf{i}\frac{\partial f}{\partial x}+\mathbf{j}\frac{\partial f}{\partial y}+\mathbf{k}\frac{\partial f}{\partial z}\right)$$

$$= \frac{\partial}{\partial x}\left(\frac{\partial f}{\partial x}\right) + \frac{\partial}{\partial y}\left(\frac{\partial f}{\partial y}\right) + \frac{\partial}{\partial z}\left(\frac{\partial f}{\partial z}\right) = \frac{\partial^2 f}{\partial x^2} + \frac{\partial^2 f}{\partial y^2} + \frac{\partial^2 f}{\partial z^2} = \left(\frac{\partial^2}{\partial x^2} + \frac{\partial^2}{\partial y^2} + \frac{\partial^2}{\partial z^2}\right) f$$

$= \nabla^2 f$. The operator ∇^2 denote the *Laplacian operator*.

Therefore, $\nabla \cdot \nabla f = \left(\mathbf{i}\frac{\partial}{\partial x} + \mathbf{j}\frac{\partial}{\partial y} + \mathbf{k}\frac{\partial}{\partial z}\right) \cdot (2xyz\,\mathbf{i} + x^2 z\,\mathbf{j} + x^2 y\,\mathbf{k}) = \frac{\partial}{\partial x}(2xyz) + \frac{\partial}{\partial y}(x^2 z) + \frac{\partial}{\partial z}(x^2 y) = 2yz.$

Alternative Solution:

$$\text{div (grad } f) = \nabla \cdot \nabla f = \nabla^2 f = \left(\frac{\partial^2}{\partial x^2} + \frac{\partial^2}{\partial y^2} + \frac{\partial^2}{\partial z^2}\right) f = \left(\frac{\partial^2}{\partial x^2} + \frac{\partial^2}{\partial y^2} + \frac{\partial^2}{\partial z^2}\right) x^2 yz = 2yz.$$

EXAMPLE 1.41 If $\mathbf{r} = x\mathbf{i} + y\mathbf{j} + z\mathbf{k}$ and $r = |\mathbf{r}|$, show that div $(\mathbf{r}/r^3) = 0$.

Solution We have $\nabla \cdot \left(\frac{\mathbf{r}}{r^3}\right) = \left(\mathbf{i}\frac{\partial}{\partial x} + \mathbf{j}\frac{\partial}{\partial y} + \mathbf{k}\frac{\partial}{\partial z}\right) \cdot \left(\frac{x}{r^3}\mathbf{i} + \frac{y}{r^3}\mathbf{j} + \frac{z}{r^3}\mathbf{k}\right)$

$$= \frac{\partial}{\partial x}\left(\frac{x}{r^3}\right) + \frac{\partial}{\partial y}\left(\frac{y}{r^3}\right) + \frac{\partial}{\partial z}\left(\frac{z}{r^3}\right) = \frac{3}{r^3} - \frac{3}{r^4}\left(x\frac{\partial r}{\partial x} + y\frac{\partial r}{\partial y} + z\frac{\partial r}{\partial z}\right).$$

Since $r^2 = x^2 + y^2 + z^2$, we obtain $\frac{\partial r}{\partial x} = \frac{x}{r}$, $\frac{\partial r}{\partial y} = \frac{y}{r}$ and $\frac{\partial r}{\partial z} = \frac{z}{r}$.

Therefore, $\nabla \cdot \left(\frac{\mathbf{r}}{r^3}\right) = \frac{3}{r^3} - \frac{3}{r^5}(x^2 + y^2 + z^2) = \frac{3}{r^3} - \frac{3}{r^3} = 0.$

EXAMPLE 1.42 If $\mathbf{A} = x^2y\mathbf{i} + y^2x^3\mathbf{j} - 3\,x^2y\,\mathbf{k}$, $\mathbf{B} = 2xz^2\mathbf{i} - yz\mathbf{j} + x^2y^3\mathbf{k}$ and $f(x, y, z) = x^2yz$, find the values of the following at (1, 2, 1) (*i*) $(\mathbf{A} \cdot \nabla) f$, (*ii*) $(\mathbf{A} \cdot \nabla)\,\mathbf{B}$.

Solution (*i*) At (1, 2, 1) $\mathbf{A} = 2\mathbf{i} + 4\mathbf{j} - 6\,\mathbf{k}$.

$$\mathbf{A} \cdot \nabla = (2\mathbf{i} + 4\mathbf{j} - 6\mathbf{k}) \cdot \left(\mathbf{i}\frac{\partial}{\partial x} + \mathbf{j}\frac{\partial}{\partial y} + \mathbf{k}\frac{\partial}{\partial z}\right) = 2\frac{\partial}{\partial x} + 4\frac{\partial}{\partial y} - 6\frac{\partial}{\partial z}.$$

Hence $(\mathbf{A} \cdot \nabla) f = \left(2\frac{\partial}{\partial x} + 4\frac{\partial}{\partial y} - 6\frac{\partial}{\partial z}\right) x^2 yz = 4xyz + 4x^2 z - 6x^2 y$ and

$(\mathbf{A} \cdot \nabla) f$ at $(1, 2, 1) = 8 + 4 - 12 = 0$.

(*ii*) $(\mathbf{A} \cdot \nabla)\mathbf{B} = \left(2\frac{\partial}{\partial x} + 4\frac{\partial}{\partial y} - 6\frac{\partial}{\partial z}\right)(2xz^2\,\mathbf{i} - yz\,\mathbf{j} + x^2y^3\,\mathbf{k})$

$= 2(2z^2\,\mathbf{i} + 2xy^3\,\mathbf{k}) + 4(-z\,\mathbf{j} + 3x^2y^2\,\mathbf{k}) - 6(4xz\,\mathbf{i} - y\,\mathbf{j})$

$= (4z^2 - 24xz)\mathbf{i} + (6y - 4z)\mathbf{j} + (4xy^3 + 12x^2y^2)\mathbf{k}$ and $(\mathbf{A} \cdot \nabla)\mathbf{B}$ at $(1, 2, 1) = -20\mathbf{i} + 8\mathbf{j} + 80\mathbf{k}$.

EXAMPLE 1.43 If $\mathbf{A} = x^2z\mathbf{i} + yz^3\mathbf{j} - 3\,xy\,\mathbf{k}$, $\mathbf{B} = y^2\mathbf{i} - yz\mathbf{j} + 2x\mathbf{k}$ and $f(x, y, z) = 2x^2 + yz$, find

(*i*) $\mathbf{A} \cdot (\nabla f)$, (*ii*) $\mathbf{A} \times \nabla f$, (*iii*) $(\mathbf{A} \times \nabla) f$.

Solution (*i*) We have $\nabla f = \mathbf{i}\frac{\partial f}{\partial x} + \mathbf{j}\frac{\partial f}{\partial y} + \mathbf{k}\frac{\partial f}{\partial z} = \mathbf{i}\frac{\partial}{\partial x}(2x^2 + yz) + \mathbf{j}\frac{\partial}{\partial y}(2x^2 + yz) + \mathbf{k}\frac{\partial}{\partial z}(2x^2 + yz)$

$$= \mathbf{i}\,(4x) + \mathbf{j}\,(z) + \mathbf{k}\,(y) = 4x\,\mathbf{i} + z\,\mathbf{j} + y\,\mathbf{k}.$$

Therefore, $\mathbf{A} \cdot (\nabla f) = (x^2z\,\mathbf{i} + yz^3\,\mathbf{j} - 3xy\mathbf{k}) \cdot (4x\,\mathbf{i} + z\,\mathbf{j} + y\,\mathbf{k})$

$$= (x^2 z)(4x) + (yz^3)(z) + (-3xy)(y) = 4x^3 z + yz^4 - 3xy^2.$$

(*ii*) $\nabla f = 4x\mathbf{i} + z\mathbf{j} + y\mathbf{k}$.

Therefore, $\mathbf{A} \times \nabla f = (x^2 z\mathbf{i} + yz^3\mathbf{j} - 3xy\mathbf{k}) \times (4x\mathbf{i} + z\mathbf{j} + y\mathbf{k})$

$$= \begin{vmatrix} \mathbf{i} & \mathbf{j} & \mathbf{k} \\ x^2 z & yz^3 & -3xy \\ 4x & z & y \end{vmatrix} = (y^2z^3 + 3xyz)\mathbf{i} - (x^2yz + 12x^2y)\mathbf{j} + (x^2z^2 - 4xyz^3)\mathbf{k}.$$

(*iii*) $$\mathbf{A} \times \nabla = \begin{vmatrix} \mathbf{i} & \mathbf{j} & \mathbf{k} \\ x^2 z & yz^3 & -3xy \\ \partial/\partial x & \partial/\partial y & \partial/\partial z \end{vmatrix} = \mathbf{i}\left(yz^3\frac{\partial}{\partial z} + 3xy\frac{\partial}{\partial y}\right) - \mathbf{j}\left(x^2z\frac{\partial}{\partial z} + 3xy\frac{\partial}{\partial x}\right) + \mathbf{k}\left(x^2z\frac{\partial}{\partial y} - yz^3\frac{\partial}{\partial x}\right).$$

Therefore, $$(\mathbf{A} \times \nabla)f = \left[\mathbf{i}\left(yz^3\frac{\partial}{\partial z} + 3xy\frac{\partial}{\partial y}\right) - \mathbf{j}\left(x^2z\frac{\partial}{\partial z} + 3xy\frac{\partial}{\partial x}\right) + \mathbf{k}\left(x^2z\frac{\partial}{\partial y} - yz^3\frac{\partial}{\partial x}\right)\right](2x^2 + yz)$$

$$= \{(yz^3)(y) + (3xy)(z)\}\mathbf{i} - \{(x^2z)(y) + (3xy)(4x)\}\mathbf{j} + \{(x^2z)(z) - (yz^3)(4x)\}\mathbf{k}$$
$$= (y^2z^3 + 3xyz)\mathbf{i} - (x^2yz + 12x^2y)\mathbf{j} + (x^2z^2 - 4xyz^3)\mathbf{k}.$$

Remark 7

Note that $(\mathbf{A} \bullet \nabla)f = \mathbf{A} \bullet \nabla f$, and $(\mathbf{A} \times \nabla)f = \mathbf{A} \times \nabla f$.

1.15 PHYSICAL INTERPRETATION OF DIVERGENCE

We shall present an interpretation in fluid mechanics. Consider the flow of a compressible fluid of density $\rho(x, y, z, t)$ and velocity

$$\mathbf{v}(x, y, z, t) = v_1(x, y, z, t)\,\mathbf{i} + v_2(x, y, z, t)\,\mathbf{j} + v_3(x, y, z, t)\,\mathbf{k}.$$

Since the fluid is not assumed to be incompressible the density as well as the velocity may vary from point to point.

Consider an infinitesimal volume element ΔV (parallelopipied of sides Δx, Δy, Δz with edges parallel to the coordinate axes, the volume $\Delta V = \Delta x\, \Delta y\, \Delta z$) placed in a mass of moving fluid. There will be flow through each of its faces, and as a result the amount of fluid within the element may vary. To measure this variation, let us compute the loss of fluid from the element in the time Δt.

The volume of fluid which passes through one face of the element ΔV in time Δt is approximately equal to the component of the fluid velocity normal to the face times the area of the face times Δt, and the corresponding mass flow is the product of this volume and the density of fluid ρ.

Consider the flow through the left face whose area is $\Delta x \Delta z$.

Mass of fluid entering through this face during a short time interval Δt is given approximately by

$$-\rho v_2\, \Delta x\, \Delta z\, \Delta t.$$

The mass of fluid leaving through the right face during the same time interval is approximately

$$\left[\rho v_2 + \frac{\partial(\rho v_2)}{\partial y}\Delta y\right]\Delta x\, \Delta z\, \Delta t.$$

Two similar expressions are obtained by considering the other two pairs of parallel faces. We have

$$\begin{cases} \text{Front face:} & \left[\rho v_1 + \dfrac{\partial(\rho v_1)}{\partial x}\Delta x\right]\Delta y\, \Delta z\, \Delta t \\ \text{Rear face:} & -\rho v_1\, \Delta y\, \Delta z\, \Delta t \end{cases}$$

$$\begin{cases} \text{Top face:} & \left[\rho v_3 + \dfrac{\partial(\rho v_3)}{\partial z}\Delta z\right]\Delta x\, \Delta y\, \Delta t \\ \text{Bottom face:} & -\rho v_3\, \Delta x\, \Delta y\, \Delta t. \end{cases}$$

Adding these, we obtain the total loss of mass of the fluid during the time Δt as

$$\left[\frac{\partial(\rho v_1)}{\partial x} + \frac{\partial(\rho v_2)}{\partial y} + \frac{\partial(\rho v_3)}{\partial z}\right]\Delta V\, \Delta t \quad ...(1.20) \qquad [\because \Delta V = \Delta x\, \Delta y\, \Delta z.]$$

This loss of mass is due to the rate of change of density with respect to time and hence is equal to

$$-\frac{\partial \rho}{\partial t} \Delta V \Delta t. \qquad ...(1.21)$$

Equating (1.20) and (1.21) expressions, divide the resulting equation by $\Delta V \Delta t$ and let Δx, Δy, Δz and Δt approach zero, then we obtain

$$\frac{\partial}{\partial x}(\rho v_1) + \frac{\partial}{\partial y}(\rho v_2) + \frac{\partial}{\partial z}(\rho v_3) = -\frac{\partial \rho}{\partial t} \quad \text{or} \quad \frac{\partial \rho}{\partial t} + \text{div}(\rho \mathbf{v}) = 0. \qquad ...(1.22)$$

This equation is called the *continuity equation* of a compressible fluid. If the flow is steady, that is, independent of time, then $\partial\rho/\partial t = 0$ and the continuity equation is

$$\text{div}(\rho v) = 0. \qquad ...(1.23)$$

If the density ρ is constant, so that the fluid is incompressible, then Eq. (1.23) becomes

$$\text{div } \mathbf{v} = \frac{\partial v_1}{\partial x} + \frac{\partial v_2}{\partial y} + \frac{\partial v_3}{\partial z} = 0. \qquad ...(1.24)$$

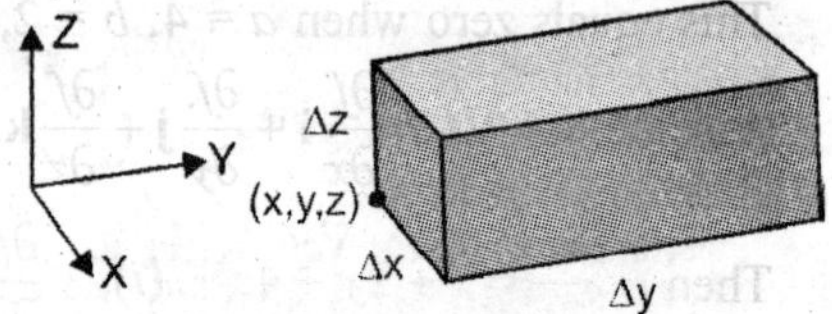

Fig. 1.11: Physical interpretation of the divergence

Therefore, when the fluid is incompressible, the divergence of the velocity vector vanishes. It states that the amount of fluid that enters and leaves a given volume is same, that is, there is no loss in the mass of the fluid.

Remark 8

Let **v** denote the velocity of a fluid in a medium. If div(**v**) = 0, then the fluid is said to be *incompressible.* In electromagnetic theory, if div(**v**) = 0, then the vector field **v** is said to be *solenoidal*.

1.16 PHYSICAL INTERPRETATION OF CURL

Consider a rigid body rotate with the uniform angular velocity $\boldsymbol{\omega} = \omega_1\mathbf{i} + \omega_2\mathbf{j} + \omega_3\mathbf{k}$ of magnitude ω about a fixed axis l through O, the origin of co-ordinates. Let the position vector of any point $P(x, y, z)$ on the rotating body be $\mathbf{r} = x\mathbf{i} + y\mathbf{j} + z\mathbf{k}$. The tangential (linear) velocity **v** of the point $P(x, y, z)$ is given by $\mathbf{v} = \boldsymbol{\omega} \times \mathbf{r} = (\omega_1\mathbf{i} + \omega_2\mathbf{j} + \omega_3\mathbf{k}) \times (x\mathbf{i} + y\mathbf{j} + z\mathbf{k})$

$$= \begin{vmatrix} \mathbf{i} & \mathbf{j} & \mathbf{k} \\ \omega_1 & \omega_2 & \omega_3 \\ x & y & z \end{vmatrix} = (\omega_2 z - \omega_3 y)\mathbf{i} + (\omega_3 x - \omega_1 z)\mathbf{j} + (\omega_1 y - \omega_2 x)\mathbf{k}.$$

Now $$\text{curl } \mathbf{v} = \nabla \times \mathbf{v} = \begin{vmatrix} \mathbf{i} & \mathbf{j} & \mathbf{k} \\ \partial/\partial x & \partial/\partial y & \partial/\partial z \\ \omega_2 z - \omega_3 y & \omega_3 x - \omega_1 z & \omega_1 y - \omega_2 x \end{vmatrix}$$

$$= 2(\omega_1\mathbf{i} + \omega_2\mathbf{j} + \omega_3\mathbf{k}) = 2\boldsymbol{\omega} \text{ or } \boldsymbol{\omega} = \left(\frac{1}{2}\right) \text{curl } \mathbf{v}. \qquad ...(1.25)$$

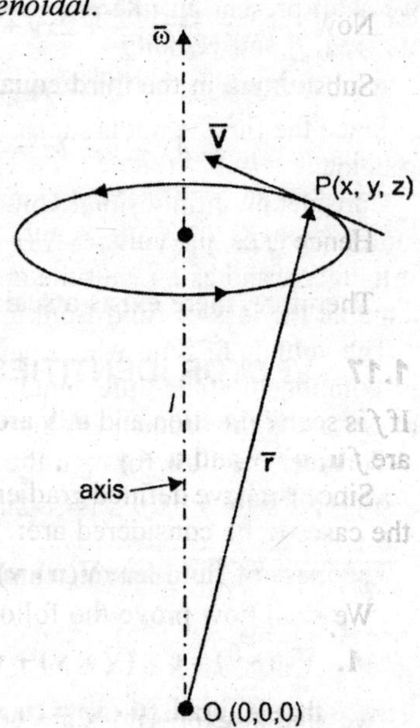

Fig. 1.12: Angular velocity

Hence the angular velocity of a uniformly rotating body is equal to one-half of the curl of the linear velocity.

Remark 9

(a) A force field **F** in 2-space or 3-space is said to be *conservative* in a region if it is obtained from a scalar potential function f in that region, that is **F** = grad *f*. Then, curl (**F**) = curl (grad f) = **0**.
Thus, if **F** is conservative then curl (**F**) = **0** and there exists a scalar potential function *f* such that **F** = grad *f*.

(b) Let **v** denote the velocity of a fluid in a medium. If curl **v** = **0**, then the flow of the fluid is said to be *irotational*, which mean that it is free of vortices. Some books use the word rotation or rot instead of curl. The symbol curl **v** is then replaced by rot **v**.

EXAMPLE 1.44 Find constants a, b and c so that $\mathbf{v} = (x + 2y + az)\,\mathbf{i} + (bx - 3y - z)\mathbf{j} + (4x + cy + 2z)\mathbf{k}$ is irrotational and hence find its scalar potential $f(x, y, z)$. *(AMIE, W-2008)*

Solution A vector **v** is called irrotational if curl **v** = **0**.

$$\text{curl } \mathbf{v} = \nabla \times \mathbf{v} = \begin{vmatrix} \mathbf{i} & \mathbf{j} & \mathbf{k} \\ \partial/\partial x & \partial/\partial y & \partial/\partial z \\ x+2y+az & bx-3y-z & 4x+cy+2z \end{vmatrix} = (c+1)\,\mathbf{i} + (a-4)\,\mathbf{j} + (b-2)\,\mathbf{k}.$$

This equals zero when $a = 4$, $b = 2$, $c = -1$ and $v = (x + 2y + 4z)i + (2x - 3y - z)j + (4x - y + 2z)k$.

Assume $\mathbf{v} = \nabla f = \dfrac{\partial f}{\partial x}\mathbf{i} + \dfrac{\partial f}{\partial y}\mathbf{j} + \dfrac{\partial f}{\partial z}\mathbf{k}$.

Then $\dfrac{\partial f}{\partial x} = x + 2y + 4z$ ...(*i*), $\dfrac{\partial f}{\partial y} = 2x - 3y - z$ ...(*ii*), $\dfrac{\partial f}{\partial z} = 4x - y + 2z$ ...(*iii*).

Integrating (*i*) partially with respect to x, keeping y and z constant. $f(x, y, z) = \dfrac{x^2}{2} + 2xy + 4xz + f(y,z)$.

Substituting in the second equation, we get

$$2x - 3y - z = 2x + \frac{\partial f}{\partial y}, \quad \text{or} \quad \frac{\partial f}{\partial y} = -3y - z, \quad \text{or} \quad f(y, z) = \frac{-3y^2}{2} - yz + f(z).$$

Now $f(x, y, z) = \dfrac{x^2}{2} + 2xy + 4xz - \dfrac{3y^2}{2} - yz + f(z)$.

Substituting in the third equation, we get

$$4x - y + 2z = 4x - y + \frac{\partial f}{\partial z}, \quad \text{or} \quad \frac{\partial f}{\partial z} = 2z. \text{ or } f = z^2 + c, \text{ (} c \text{ being constant).}$$

Hence $f(x, y, z) = \dfrac{x^2}{2} + 2xy + 4xz - \dfrac{3y^2}{2} - yz + z^2 + c$.

Therefore, there exists a scalar function $f(x, y, z)$ such that $\nabla f = v$.

1.17 VECTOR IDENTITIES

If f is scalar function and **u**, **v** are vector functions, then the possible forms of products between these functions are f**u**, **u** • **v** and **u** × **v**.

Since we have defined gradient of scalar point functions and divergence, curl of vector point functions. Now the cases to be considered are:

$$\nabla(\mathbf{u} \cdot \mathbf{v}), \quad \nabla \cdot (f\mathbf{v}), \quad \nabla \cdot (\mathbf{u} \times \mathbf{v}), \quad \nabla \times (f\,\mathbf{v}), \quad \nabla \times (\mathbf{u} \times \mathbf{v}).$$

We shall now prove the following expansion formulae:

1. $\nabla(\mathbf{u} \cdot \mathbf{v}) = \mathbf{u} \times (\nabla \times \mathbf{v}) + \mathbf{v} \times (\nabla \times \mathbf{u}) + (\mathbf{u} \cdot \nabla)\mathbf{v} + (\mathbf{v} \cdot \nabla)\mathbf{u}$

that is, grad $(\mathbf{u} \cdot \mathbf{v}) = (\mathbf{u} \times \text{curl } \mathbf{v}) + (\mathbf{v} \times \text{curl } \mathbf{u}) + (\mathbf{u} \cdot \nabla)\mathbf{v} + (\mathbf{v} \cdot \nabla)\mathbf{u}$.

2. $\nabla \cdot (f\,\mathbf{v}) = f(\nabla \cdot \mathbf{v}) + (\nabla f) \cdot \mathbf{v}$ *(UPTU 2004)*

that is, $\text{div}(f\,\mathbf{v}) = f \text{ div } \mathbf{v} + (\text{grad} f) \cdot \mathbf{v}$.

3. $\nabla \cdot (\mathbf{u} \times \mathbf{v}) = \mathbf{v} \cdot (\nabla \times \mathbf{u}) - \mathbf{u} \cdot (\nabla \times \mathbf{v})$ *(AMIETE, June 2013)*

that is, div $(\mathbf{u} \times \mathbf{v}) = \mathbf{v} \cdot (\text{curl } \mathbf{u}) - \mathbf{u} \cdot (\text{curl } \mathbf{v})$. *(AMIETE, Dec. 2012, 2011, 2009; UPTU 2003)*

4. $\nabla \times (f\,\mathbf{v}) = f(\nabla \times \mathbf{v}) + (\nabla f) \times \mathbf{v}$,

that is, $\text{curl}(f\,\mathbf{v}) = f \text{ curl } \mathbf{v} + (\text{grad} f) \times \mathbf{v}$. *(UPTU 2003)*

5. $\nabla \times (\mathbf{u} \times \mathbf{v}) = \mathbf{u}(\nabla \cdot \mathbf{v}) - \mathbf{v}(\nabla \cdot \mathbf{u}) + (\mathbf{v} \cdot \nabla)\mathbf{u} - (\mathbf{u} \cdot \nabla)\mathbf{v}$,

that is, curl $(\mathbf{u} \times \mathbf{v}) = \mathbf{u}(\text{div } \mathbf{v}) - \mathbf{v}(\text{div } \mathbf{u}) + (\mathbf{v} \cdot \nabla)\mathbf{u} - (\mathbf{u} \cdot \nabla)\mathbf{v}$.

Proof:

1. $\nabla(\mathbf{u}\cdot\mathbf{v})=\left(\mathbf{i}\frac{\partial}{\partial x}+\mathbf{j}\frac{\partial}{\partial y}+\mathbf{k}\frac{\partial}{\partial z}\right)(\mathbf{u}\cdot\mathbf{v})=\sum\mathbf{i}\frac{\partial}{\partial x}(\mathbf{u}\cdot\mathbf{v})=\sum\mathbf{i}\left(\mathbf{u}\cdot\frac{\partial\mathbf{v}}{\partial x}+\frac{\partial\mathbf{u}}{\partial x}\cdot\mathbf{v}\right)$

$$=\sum\left(\mathbf{u}\cdot\frac{\partial\mathbf{v}}{\partial x}\right)\mathbf{i}+\sum\left(\mathbf{v}\cdot\frac{\partial\mathbf{u}}{\partial x}\right)\mathbf{i}. \qquad ...(i)$$

We know that $\mathbf{u}\times(\mathbf{v}\times\mathbf{w})=(\mathbf{u}\cdot\mathbf{w})\mathbf{v}-(\mathbf{u}\cdot\mathbf{v})\mathbf{w}$

or $(\mathbf{u}\cdot\mathbf{v})\mathbf{w}=(\mathbf{u}\cdot\mathbf{w})\mathbf{v}-\mathbf{u}\times(\mathbf{v}\times\mathbf{w})$.

Therefore, $\left(\mathbf{u}\cdot\frac{\partial\mathbf{v}}{\partial x}\right)\mathbf{i}=(\mathbf{u}\cdot\mathbf{i})\frac{\partial\mathbf{v}}{\partial x}-\mathbf{u}\times\left(\frac{\partial\mathbf{v}}{\partial x}\times\mathbf{i}\right)=(\mathbf{u}\cdot\mathbf{i})\frac{\partial\mathbf{v}}{\partial x}+\mathbf{u}\times\left(\mathbf{i}\times\frac{\partial\mathbf{v}}{\partial x}\right).$ $[\because \mathbf{a}\times\mathbf{b}=-\mathbf{b}\times\mathbf{a}]$

Hence, $$\sum\mathbf{i}\left(\mathbf{u}\cdot\frac{\partial\mathbf{v}}{\partial x}\right)=\sum(\mathbf{u}\cdot\mathbf{i})\frac{\partial\mathbf{v}}{\partial x}+\mathbf{u}\times\left(\sum\mathbf{i}\times\frac{\partial\mathbf{v}}{\partial x}\right)=(\mathbf{u}\cdot\nabla)\,\mathbf{v}+\mathbf{u}\times(\nabla\times\mathbf{v}). \qquad ...(ii)$$

Interchanging **u** and **v**, we have

$$\sum\mathbf{i}\left(\mathbf{v}\cdot\frac{\partial\mathbf{u}}{\partial x}\right)=(\mathbf{v}\cdot\nabla)\,\mathbf{u}+\mathbf{v}\times(\nabla\times\mathbf{u}). \qquad ...(iii)$$

Using (*ii*) and (*iii*), we can write (*i*) in the form

$$\nabla(\mathbf{u}\cdot\mathbf{v})=\mathbf{u}\times(\nabla\times\mathbf{v})+\mathbf{v}\times(\nabla\times\mathbf{u})+(\mathbf{u}\cdot\nabla)\,\mathbf{v}+(\mathbf{v}\cdot\nabla)\,\mathbf{u}$$

or $\text{grad}\,(\mathbf{u}\cdot\mathbf{v})=(\mathbf{u}\times\text{curl}\,\mathbf{v})+(\mathbf{v}\times\text{curl}\,\mathbf{u})+(\mathbf{u}\cdot\nabla)\,\mathbf{v}+(\mathbf{v}\cdot\nabla)\,\mathbf{u}$.

2. $\nabla\cdot(f\mathbf{V})=\nabla\cdot(f\,v_1\mathbf{i}+f\,v_2\mathbf{j}+f\,v_3\mathbf{k})=\frac{\partial}{\partial x}(f\,v_1)+\frac{\partial}{\partial y}(f\,v_2)+\frac{\partial}{\partial z}(f\,v_3)$

$$=\frac{\partial f}{\partial x}v_1+f\frac{\partial v_1}{\partial x}+\frac{\partial f}{\partial y}v_2+f\frac{\partial v_2}{\partial y}+\frac{\partial f}{\partial z}v_3+f\frac{\partial v_3}{\partial z}$$

$$=\frac{\partial f}{\partial x}v_1+\frac{\partial f}{\partial y}v_2+\frac{\partial f}{\partial z}v_3+f\left(\frac{\partial v_1}{\partial x}+\frac{\partial v_2}{\partial y}+\frac{\partial v_3}{\partial z}\right)$$

$$=\left(\frac{\partial f}{\partial x}\mathbf{i}+\frac{\partial f}{\partial y}\mathbf{j}+\frac{\partial f}{\partial z}\mathbf{k}\right)\cdot(v_1\mathbf{i}+v_2\mathbf{j}+v_3\mathbf{k})+f\left(\frac{\partial}{\partial x}\mathbf{i}+\frac{\partial}{\partial y}\mathbf{j}+\frac{\partial}{\partial z}\mathbf{k}\right)\cdot(v_1\mathbf{i}+v_2\mathbf{j}+v_3\mathbf{k})$$

$$=(\nabla f)\cdot\mathbf{v}+f(\nabla\cdot\mathbf{v}) \text{ or } \text{div}\,(f\,\mathbf{v})=f\,\text{div}\,\mathbf{v}+(\text{grad}\,f)\cdot\mathbf{v}.$$

3. $\nabla\cdot(\mathbf{u}\times\mathbf{v})=\left(\mathbf{i}\frac{\partial}{\partial x}+\mathbf{j}\frac{\partial}{\partial y}+\mathbf{k}\frac{\partial}{\partial z}\right)\cdot(\mathbf{u}\times\mathbf{v})=\mathbf{i}\cdot\frac{\partial}{\partial x}(\mathbf{u}\times\mathbf{v})+\mathbf{j}\cdot\frac{\partial}{\partial y}(\mathbf{u}\times\mathbf{v})+\mathbf{k}\cdot\frac{\partial}{\partial z}(\mathbf{u}\times\mathbf{v})$

$$=\mathbf{i}\cdot\left(\mathbf{u}\times\frac{\partial\mathbf{v}}{\partial x}+\frac{\partial\mathbf{u}}{\partial x}\times\mathbf{v}\right)+\mathbf{j}\cdot\left(\mathbf{u}\times\frac{\partial\mathbf{v}}{\partial y}+\frac{\partial\mathbf{u}}{\partial y}\times\mathbf{v}\right)+\mathbf{k}\cdot\left(\mathbf{u}\times\frac{\partial\mathbf{v}}{\partial z}+\frac{\partial\mathbf{u}}{\partial z}\times\mathbf{v}\right)$$

$$=\left\{\mathbf{i}\cdot\left(\mathbf{u}\times\frac{\partial\mathbf{v}}{\partial x}\right)+\mathbf{j}\cdot\left(\mathbf{u}\times\frac{\partial\mathbf{v}}{\partial y}\right)+\mathbf{k}\cdot\left(\mathbf{u}\times\frac{\partial\mathbf{v}}{\partial z}\right)\right\}+\left\{\mathbf{i}\cdot\left(\frac{\partial\mathbf{u}}{\partial x}\times\mathbf{v}\right)+\mathbf{j}\cdot\left(\frac{\partial\mathbf{u}}{\partial y}\times\mathbf{v}\right)+\mathbf{k}\cdot\left(\frac{\partial\mathbf{u}}{\partial z}\times\mathbf{v}\right)\right\}$$

$$=-\left\{\mathbf{i}\cdot\left(\frac{\partial\mathbf{v}}{\partial x}\times\mathbf{u}\right)+\mathbf{j}\cdot\left(\frac{\partial\mathbf{v}}{\partial y}\times\mathbf{u}\right)+\mathbf{k}\cdot\left(\frac{\partial\mathbf{v}}{\partial z}\times\mathbf{u}\right)\right\}+\left\{\mathbf{i}\cdot\left(\frac{\partial\mathbf{u}}{\partial x}\times\mathbf{v}\right)+\mathbf{j}\cdot\left(\frac{\partial\mathbf{u}}{\partial y}\times\mathbf{v}\right)+\mathbf{k}\cdot\left(\frac{\partial\mathbf{u}}{\partial z}\times\mathbf{v}\right)\right\}$$

$[\because \mathbf{a}\times\mathbf{b}=-\mathbf{b}\times\mathbf{a}]$

$$=-\left\{\left(\mathbf{i}\times\frac{\partial\mathbf{v}}{\partial x}\right)\cdot\mathbf{u}+\left(\mathbf{j}\times\frac{\partial\mathbf{v}}{\partial y}\right)\cdot\mathbf{u}+\left(\mathbf{k}\times\frac{\partial\mathbf{v}}{\partial z}\right)\cdot\mathbf{u}\right\}+\left\{\left(\mathbf{i}\times\frac{\partial\mathbf{u}}{\partial x}\right)\cdot\mathbf{v}+\left(\mathbf{j}\times\frac{\partial\mathbf{u}}{\partial y}\right)\cdot\mathbf{v}+\left(\mathbf{k}\times\frac{\partial\mathbf{u}}{\partial z}\right)\cdot\mathbf{v}\right\}$$

$$=-\left\{\mathbf{i}\times\frac{\partial\mathbf{v}}{\partial x}+\mathbf{j}\times\frac{\partial\mathbf{v}}{\partial y}+\mathbf{k}\times\frac{\partial\mathbf{v}}{\partial z}\right\}\cdot\mathbf{u}+\left\{\mathbf{i}\times\frac{\partial\mathbf{u}}{\partial x}+\mathbf{j}\times\frac{\partial\mathbf{u}}{\partial y}+\mathbf{k}\times\frac{\partial\mathbf{u}}{\partial z}\right\}\cdot\mathbf{v}$$

$$=-(\nabla\times\mathbf{v})\cdot\mathbf{u}+(\nabla\times\mathbf{u})\cdot\mathbf{v}=\mathbf{v}\cdot(\nabla\times\mathbf{u})-\mathbf{u}\cdot(\nabla\times\mathbf{v}) \text{ or } \text{div}\,(\mathbf{u}\times\mathbf{v})=\mathbf{v}\cdot(\text{curl}\,\mathbf{u})-\mathbf{u}\cdot(\text{curl}\,\mathbf{v}).$$

4. $\nabla\times(f\nabla)=\nabla\times(f v_1\mathbf{i}+f v_2\mathbf{j}+f v_3\mathbf{k})=\begin{vmatrix}\mathbf{i} & \mathbf{j} & \mathbf{k}\\ \partial/\partial x & \partial/\partial y & \partial/\partial z\\ f v_1 & f v_2 & f v_3\end{vmatrix}$

$$=\left[\frac{\partial}{\partial y}(f v_3)-\frac{\partial}{\partial z}(f v_2)\right]\mathbf{i}+\left[\frac{\partial}{\partial z}(f v_1)-\frac{\partial}{\partial x}(f v_3)\right]\mathbf{j}+\left[\frac{\partial}{\partial x}(f v_2)-\frac{\partial}{\partial y}(f v_1)\right]\mathbf{k}$$

$$=\left[f\frac{\partial v_3}{\partial y}+\frac{\partial f}{\partial y}v_3-f\frac{\partial v_2}{\partial z}-\frac{\partial f}{\partial z}v_2\right]\mathbf{i}+\left[f\frac{\partial v_1}{\partial z}+\frac{\partial f}{\partial z}v_1-f\frac{\partial v_3}{\partial x}-\frac{\partial f}{\partial x}v_3\right]\mathbf{j}$$

$$+\left[f\frac{\partial v_2}{\partial x}+\frac{\partial f}{\partial x}v_2-f\frac{\partial v_1}{\partial y}-\frac{\partial f}{\partial y}v_1\right]\mathbf{k}$$

$$=f\left[\left(\frac{\partial v_3}{\partial y}-\frac{\partial v_2}{\partial z}\right)\mathbf{i}+\left(\frac{\partial v_1}{\partial z}-\frac{\partial v_3}{\partial x}\right)\mathbf{j}+\left(\frac{\partial v_2}{\partial x}-\frac{\partial v_1}{\partial y}\right)\mathbf{k}\right]$$

$$+\left[\left(\frac{\partial f}{\partial y}v_3-\frac{\partial f}{\partial z}v_2\right)\mathbf{i}+\left(\frac{\partial f}{\partial z}v_1-\frac{\partial f}{\partial x}v_3\right)\mathbf{j}+\left(\frac{\partial f}{\partial x}v_2-\frac{\partial f}{\partial y}v_1\right)\mathbf{k}\right]$$

$$=f(\nabla\times\mathbf{v})+\begin{vmatrix}\mathbf{i} & \mathbf{j} & \mathbf{k}\\ \partial f/\partial x & \partial f/\partial y & \partial f/\partial z\\ v_1 & v_2 & v_3\end{vmatrix}=f(\nabla\times\mathbf{v})+(\nabla f)\times\mathbf{v}$$

or curl $(f\mathbf{v})=f$ curl $\mathbf{v}$ + (grad f) × $\mathbf{v}$.

5. We have $\nabla\times(\mathbf{u}\times\mathbf{v})=\left(\mathbf{i}\frac{\partial}{\partial x}+\mathbf{j}\frac{\partial}{\partial y}+\mathbf{k}\frac{\partial}{\partial z}\right)\times(\mathbf{u}\times\mathbf{v})$

$$=\mathbf{i}\times\frac{\partial}{\partial x}(\mathbf{u}\times\mathbf{v})+\mathbf{j}\times\frac{\partial}{\partial y}(\mathbf{u}\times\mathbf{v})+\mathbf{k}\times\frac{\partial}{\partial z}(\mathbf{u}\times\mathbf{v}).\quad\text{...(i)}$$

Now $\mathbf{i}\times\frac{\partial}{\partial x}(\mathbf{u}\times\mathbf{v})=\mathbf{i}\times\left(\mathbf{u}\times\frac{\partial\mathbf{v}}{\partial x}+\frac{\partial\mathbf{u}}{\partial x}\times\mathbf{v}\right)=\mathbf{i}\times\left(\mathbf{u}\times\frac{\partial\mathbf{v}}{\partial x}\right)+\mathbf{i}\times\left(\frac{\partial\mathbf{u}}{\partial x}\times\mathbf{v}\right)$

$$=\left[\left(\mathbf{i}\cdot\frac{\partial\mathbf{v}}{\partial x}\right)\mathbf{u}-(\mathbf{i}\cdot\mathbf{u})\frac{\partial\mathbf{v}}{\partial x}\right]+\left[(\mathbf{i}\cdot\mathbf{v})\frac{\partial\mathbf{u}}{\partial x}-\left(\mathbf{i}\cdot\frac{\partial\mathbf{u}}{\partial x}\right)\mathbf{v}\right].\quad\text{...(ii)}$$

$[\because \mathbf{a}\times(\mathbf{b}\times\mathbf{c})=(\mathbf{a}\cdot\mathbf{c})\mathbf{b}-(\mathbf{a}\cdot\mathbf{b})\mathbf{c}]$

Similarly, we have

$$\mathbf{j}\times\frac{\partial}{\partial y}(\mathbf{u}\times\mathbf{v})=\left[\left(\mathbf{j}\cdot\frac{\partial\mathbf{v}}{\partial y}\right)\mathbf{u}-(\mathbf{j}\cdot\mathbf{u})\frac{\partial\mathbf{v}}{\partial y}\right]+\left[(\mathbf{j}\cdot\mathbf{v})\frac{\partial\mathbf{u}}{\partial y}-\left(\mathbf{j}\cdot\frac{\partial\mathbf{u}}{\partial y}\right)\mathbf{v}\right].\quad\text{...(iii)}$$

$$\mathbf{k}\times\frac{\partial}{\partial z}(\mathbf{u}\times\mathbf{v})=\left[\left(\mathbf{k}\cdot\frac{\partial\mathbf{v}}{\partial z}\right)\mathbf{u}-(\mathbf{k}\cdot\mathbf{u})\frac{\partial\mathbf{v}}{\partial z}\right]+\left[(\mathbf{k}\cdot\mathbf{v})\frac{\partial\mathbf{u}}{\partial z}-\left(\mathbf{k}\cdot\frac{\partial\mathbf{u}}{\partial z}\right)\mathbf{v}\right].\quad\text{...(iv)}$$

Adding (*ii*), (*iii*) and (*iv*), we obtain

$$\nabla\times(\mathbf{u}\times\mathbf{v})=\left[\left\{\mathbf{i}\cdot\frac{\partial\mathbf{v}}{\partial x}+\mathbf{j}\cdot\frac{\partial\mathbf{v}}{\partial y}+\mathbf{k}\cdot\frac{\partial\mathbf{v}}{\partial z}\right\}\mathbf{u}-\left\{(\mathbf{i}\cdot\mathbf{u})\frac{\partial\mathbf{v}}{\partial x}+(\mathbf{j}\cdot\mathbf{u})\frac{\partial\mathbf{v}}{\partial y}+(\mathbf{k}\cdot\mathbf{u})\frac{\partial\mathbf{v}}{\partial z}\right\}\right]$$

$$+\left[\left\{(\mathbf{i}\cdot\mathbf{v})\frac{\partial\mathbf{u}}{\partial x}+(\mathbf{j}\cdot\mathbf{v})\frac{\partial\mathbf{u}}{\partial y}+(\mathbf{k}\cdot\mathbf{v})\frac{\partial\mathbf{u}}{\partial z}\right\}-\left\{\mathbf{i}\cdot\frac{\partial\mathbf{u}}{\partial x}+\mathbf{j}\cdot\frac{\partial\mathbf{u}}{\partial y}+\mathbf{k}\cdot\frac{\partial\mathbf{u}}{\partial z}\right\}\mathbf{v}\right]$$

$$=\mathbf{u}\left[\mathbf{i}\cdot\frac{\partial\mathbf{v}}{\partial x}+\mathbf{j}\cdot\frac{\partial\mathbf{v}}{\partial y}+\mathbf{k}\cdot\frac{\partial\mathbf{v}}{\partial z}\right]-\left[u_1\frac{\partial\mathbf{v}}{\partial x}+u_2\frac{\partial\mathbf{v}}{\partial y}+u_3\frac{\partial\mathbf{v}}{\partial z}\right]$$

$$+\left[v_1\frac{\partial\mathbf{u}}{\partial x}+v_2\frac{\partial\mathbf{u}}{\partial y}+v_3\frac{\partial\mathbf{u}}{\partial z}\right]-\mathbf{v}\left[\mathbf{i}\cdot\frac{\partial\mathbf{u}}{\partial x}+\mathbf{j}\cdot\frac{\partial\mathbf{u}}{\partial y}+\mathbf{k}\cdot\frac{\partial\mathbf{u}}{\partial z}\right]$$

$$= \mathbf{u}\,(\nabla \cdot \mathbf{v}) - (u_1\mathbf{i} + u_2\mathbf{j} + u_3\mathbf{k}) \cdot \left(\mathbf{i}\frac{\partial}{\partial x} + \mathbf{j}\frac{\partial}{\partial y} + \mathbf{k}\frac{\partial}{\partial z}\right)\mathbf{v}$$

$$+ (v_1\mathbf{i} + v_2\mathbf{j} + v_3\mathbf{k}) \cdot \left(\mathbf{i}\frac{\partial}{\partial x} + \mathbf{j}\frac{\partial}{\partial y} + \mathbf{k}\frac{\partial}{\partial z}\right)\mathbf{u} - \mathbf{v}\,(\nabla \cdot \mathbf{u})$$

$$= \mathbf{u}\,(\nabla \cdot \mathbf{v}) - (\mathbf{u} \cdot \nabla)\,\mathbf{v} + (\mathbf{v} \cdot \nabla)\,\mathbf{u} - \mathbf{v}\,(\nabla \cdot \mathbf{u}).$$

Rearranging the terms on R.H.S., we obtain

$$\nabla \times (\mathbf{u} \times \mathbf{v}) = \mathbf{u}\,(\nabla \cdot \mathbf{v}) - \mathbf{v}\,(\nabla \cdot \mathbf{u}) + (\mathbf{v} \cdot \nabla)\,\mathbf{u} - (\mathbf{u} \cdot \nabla)\,\mathbf{v}$$

or $$\text{Curl}\,(\mathbf{u} \times \mathbf{v}) = \mathbf{u}\,(\text{div}\,\mathbf{v}) - \mathbf{v}\,(\text{div}\,\mathbf{u}) + (\mathbf{v} \cdot \nabla)\,\mathbf{u} - (\mathbf{u} \cdot \nabla)\,\mathbf{v}.$$

EXAMPLE 1.45 Prove the following identities:

(*i*) curl (grad f) = **0** or $\nabla \times (\nabla f) = 0$, (*PTU 2000*)

(*ii*) div (curl **v**) = 0 or $\nabla \cdot (\nabla \times \mathbf{v}) = 0$, (*MDU 2003*)

(*iii*) curl (curl **v**) *i.e.*, $\nabla \times (\nabla \times \mathbf{v}) = \nabla(\nabla \cdot \mathbf{v}) - \nabla^2\mathbf{v}$,

or grad (div **v**) = $\nabla \times (\nabla \times \mathbf{v}) + \nabla^2\mathbf{v}$ where **v**, a differentiable vector field.

(*PTU 2006; AMIETE Dec 2004; UPTU 2003; MDU 2002*)

Solution (*i*) From the definition, we have

$$\nabla \times (\nabla f) = \nabla \times \left(\frac{\partial f}{\partial x}\mathbf{i} + \frac{\partial f}{\partial y}\mathbf{j} + \frac{\partial \mathbf{i}}{\partial z}\mathbf{k}\right) = \begin{vmatrix} \mathbf{i} & \mathbf{j} & \mathbf{k} \\ \partial/\partial x & \partial/\partial y & \partial/\partial z \\ \partial f/\partial x & \partial f/\partial y & \partial f/\partial z \end{vmatrix}$$

$$= \left[\frac{\partial}{\partial y}\left(\frac{\partial f}{\partial z}\right) - \frac{\partial}{\partial z}\left(\frac{\partial f}{\partial y}\right)\right]\mathbf{i} + \left[\frac{\partial}{\partial z}\left(\frac{\partial f}{\partial x}\right) - \frac{\partial}{\partial x}\left(\frac{\partial f}{\partial z}\right)\right]\mathbf{j} + \left[\frac{\partial}{\partial x}\left(\frac{\partial f}{\partial y}\right) - \frac{\partial}{\partial y}\left(\frac{\partial f}{\partial x}\right)\right]\mathbf{k}$$

$$= \left(\frac{\partial^2 f}{\partial y \partial z} - \frac{\partial^2 f}{\partial z \partial y}\right)\mathbf{i} + \left(\frac{\partial^2 f}{\partial z \partial x} - \frac{\partial^2 f}{\partial x \partial z}\right)\mathbf{j} + \left(\frac{\partial^2 f}{\partial x \partial y} - \frac{\partial^2 f}{\partial y \partial x}\right)\mathbf{k} = \mathbf{0}.$$

(*ii*) We have **v**, a differentiable vector field = $v_1\mathbf{i} + v_2\,\mathbf{j} + v_3\mathbf{k}$.

From the definition $$\text{div}\,(\text{curl}\,\mathbf{v}) = \nabla \cdot (\nabla \times \mathbf{v}) = \nabla \cdot \begin{vmatrix} \mathbf{i} & \mathbf{j} & \mathbf{k} \\ \partial/\partial x & \partial/\partial y & \partial/\partial z \\ v_1 & v_2 & v_3 \end{vmatrix}$$

$$= \nabla \cdot \left[\left(\frac{\partial v_3}{\partial y} - \frac{\partial v_2}{\partial z}\right)\mathbf{i} + \left(\frac{\partial v_1}{\partial z} - \frac{\partial v_3}{\partial x}\right)\mathbf{j} + \left(\frac{\partial v_2}{\partial x} - \frac{\partial v_1}{\partial y}\right)\mathbf{k}\right]$$

$$= \frac{\partial}{\partial x}\left(\frac{\partial v_3}{\partial y} - \frac{\partial v_2}{\partial z}\right) + \frac{\partial}{\partial y}\left(\frac{\partial v_1}{\partial z} - \frac{\partial v_3}{\partial x}\right) + \frac{\partial}{\partial z}\left(\frac{\partial v_2}{\partial x} - \frac{\partial v_1}{\partial y}\right)$$

$$= \frac{\partial^2 v_3}{\partial x \partial y} - \frac{\partial^2 v_2}{\partial x \partial z} + \frac{\partial^2 v_1}{\partial y \partial z} - \frac{\partial^2 v_3}{\partial y \partial x} + \frac{\partial^2 v_2}{\partial z \partial x} - \frac{\partial^2 v_1}{\partial z \partial y} = 0.$$

(*iii*) $$\text{curl}\,(\text{curl}\,\mathbf{v}) = \nabla \times (\nabla \times \mathbf{v}) = \left(\sum \mathbf{i}\frac{\partial}{\partial x}\right) \times \left[\sum \mathbf{i}\left(\frac{\partial v_3}{\partial y} - \frac{\partial v_2}{\partial z}\right)\right]$$

$$= \sum \mathbf{i}\left[\frac{\partial}{\partial y}\left(\frac{\partial v_2}{\partial x} - \frac{\partial v_1}{\partial y}\right) - \frac{\partial}{\partial z}\left(\frac{\partial v_1}{\partial z} - \frac{\partial v_3}{\partial x}\right)\right] = \sum \mathbf{i}\left[\frac{\partial^2 v_2}{\partial y \partial x} + \frac{\partial^2 v_3}{\partial z \partial x} - \left(\frac{\partial^2 v_1}{\partial y^2} + \frac{\partial^2 v_1}{\partial z^2}\right)\right]$$

$$= \sum \mathbf{i}\left[\frac{\partial}{\partial x}\left(\frac{\partial v_2}{\partial y} + \frac{\partial v_3}{\partial z}\right) - \left(\frac{\partial^2 v_1}{\partial y^2} + \frac{\partial^2 v_1}{\partial z^2}\right)\right] = \sum \mathbf{i}\left[\frac{\partial}{\partial x}\left(\frac{\partial v_1}{\partial x} + \frac{\partial v_2}{\partial y} + \frac{\partial v_3}{\partial z}\right) - \left(\frac{\partial^2 v_1}{\partial x^2} + \frac{\partial^2 v_1}{\partial y^2} + \frac{\partial^2 v_1}{\partial z^2}\right)\right]$$

$$=\left(\sum \mathbf{i}\frac{\partial}{\partial x}\right)(\nabla\cdot\mathbf{v})-\left(\frac{\partial^2}{\partial x^2}+\frac{\partial^2}{\partial y^2}+\frac{\partial^2}{\partial z^2}\right)\left(\sum \mathbf{i}v_1\right)=\nabla(\nabla\cdot\mathbf{v})-\nabla^2\mathbf{v}.$$

Hence, $\text{grad}(\text{div}\,\mathbf{v}) = \text{curl}\,(\text{curl}\,\mathbf{v}) + \nabla^2\mathbf{v}$.

EXAMPLE 1.46 Show that div grad r^n or $\nabla^2(r^n) = n(n+1)r^{n-2}$, where $r^2 = x^2 + y^2 + z^2$. Hence show that div grad $\left(\frac{1}{r}\right)$ or $\nabla^2\left(\frac{1}{r}\right) = 0$. *(AMIETE, June 2010; SVTU, 2006; JNTU 2006; UPTU 2005)*

Solution $\nabla^2(r^n) = \nabla\cdot(\nabla r^n) = \nabla\cdot\left(nr^{n-1}\frac{\mathbf{r}}{r}\right) = n\nabla\cdot(r^{n-2}\mathbf{r}) = n[(\nabla r^{n-2})\cdot\mathbf{r} + r^{n-2}(\nabla\cdot\mathbf{r})]$

[by Art. 1.17, formula (2)]

$$= n\left[(n-2)r^{n-3}\cdot\frac{\mathbf{r}}{r}\cdot\mathbf{r} + r^{n-2}(3)\right] = n[(n-2)r^{n-2} + 3r^{n-2}] = n(n+1)r^{n-2}.$$

Setting $n = -1$ in the above result, we obtain $\nabla^2(r^{-1}) = 0$ or $\nabla^2\left(\frac{1}{r}\right) = 0$.

EXAMPLE 1.47 If $\nabla\cdot\mathbf{E} = 0$, $\nabla\cdot\mathbf{H} = 0$, $\nabla\times\mathbf{E} = -\frac{\partial\mathbf{H}}{\partial t}$, $\nabla\times\mathbf{H} = \frac{\partial\mathbf{E}}{\partial t}$. Show that the vectors **E** and **H** satisfy the wave equation $\nabla^2\mathbf{u} = \frac{\partial^2\mathbf{u}}{\partial t^2}$. *(AMIETE, June 2011, 2008)*

Solution We have $\nabla\times(\nabla\times\mathbf{E}) = \nabla\times\left(-\frac{\partial\mathbf{H}}{\partial t}\right) = -\frac{\partial}{\partial t}(\nabla\times\mathbf{H}) = -\frac{\partial}{\partial t}\left(\frac{\partial\mathbf{E}}{\partial t}\right) = -\frac{\partial^2\mathbf{E}}{\partial t^2}$.

By example 1.45 (*iii*) $\nabla\times(\nabla\times\mathbf{E}) = -\nabla^2\mathbf{E} + \nabla(\nabla\cdot\mathbf{E}) = -\nabla^2\mathbf{E}$. Then $\nabla^2\mathbf{E} = \frac{\partial^2\mathbf{E}}{\partial t^2}$. ...(*i*)

Similarly, $\nabla\times(\nabla\times\mathbf{H}) = \nabla\times\left(\frac{\partial\mathbf{E}}{\partial t}\right) = \frac{\partial}{\partial t}(\nabla\times\mathbf{E}) = \frac{\partial}{\partial t}\left(-\frac{\partial\mathbf{H}}{\partial t}\right) = -\frac{\partial^2\mathbf{H}}{\partial t^2}$.

But $\nabla\times(\nabla\times\mathbf{H}) = -\nabla^2\mathbf{H} + \nabla(\nabla\cdot\mathbf{H}) = -\nabla^2\mathbf{H}$. Then $\nabla^2\mathbf{H} = \frac{\partial^2\mathbf{H}}{\partial t^2}$. ...(*ii*)

From (*i*) and (*ii*) it is obvious that **E** and **H** satisfy the wave equation $\nabla^2\mathbf{u} = \frac{\partial^2\mathbf{u}}{\partial t^2}$.

EXAMPLE 1.48 If r is the distance of a point (x, y, z) from the origin, prove that

$$\text{curl}\left(\mathbf{k}\times\text{gard}\ \frac{1}{r}\right) + \text{grad}\left(\mathbf{k}\cdot\text{grad}\ \frac{1}{r}\right) = 0,$$

where **k** is a unit vector in the direction of OZ. *(AMIETE, June 2012, Dec. 2009; UPTU 2001)*

Solution We have $\text{curl}\left(\mathbf{k}\times\text{gard}\ \frac{1}{r}\right) = \nabla\times\left(\mathbf{k}\times\nabla\frac{1}{r}\right)$

$$= \mathbf{k}\left(\nabla\cdot\nabla\frac{1}{r}\right) - \nabla\frac{1}{r}(\nabla\cdot\mathbf{k}) + \left(\nabla\frac{1}{r}\cdot\nabla\right)\mathbf{k} - (\mathbf{k}\cdot\nabla)\nabla\frac{1}{r}$$ [By art 1.17 Formula (5)]

$$= \mathbf{k}\left(\nabla\cdot\nabla\frac{1}{r}\right) - \nabla\frac{1}{r}(\nabla\cdot\mathbf{k}) + 0 - 0 = \mathbf{k}\left(\nabla\cdot\nabla\frac{1}{r}\right) - \nabla\frac{1}{r}(\nabla\cdot\mathbf{k})$$ ...(*i*)

Now $\text{grad}\left(\mathbf{k}\cdot\text{grad}\ \frac{1}{r}\right) = \nabla\left(\mathbf{k}\cdot\nabla\frac{1}{r}\right) = \mathbf{k}\times\left(\nabla\times\nabla\frac{1}{r}\right) + \nabla\frac{1}{r}\times(\nabla\times\mathbf{k}) + (\mathbf{k}\cdot\nabla)\nabla\frac{1}{r} + \left(\nabla\frac{1}{r}\cdot\nabla\right)\mathbf{k}$

[By art 1.17 Formula (1)]

$$= 0 + 0 + (\mathbf{k}\cdot\nabla)\nabla\frac{1}{r} + 0 = (\mathbf{k}\cdot\nabla)\nabla\frac{1}{r}.$$ ...(*ii*)

Adding (*i*) and (*ii*) we obtain $\text{curl}\left(\mathbf{k}\times\text{gard}\,\dfrac{1}{r}\right)+\text{grad}\left(\mathbf{k}\cdot\text{grad}\,\dfrac{1}{\mathrm{r}}\right)$

$$=\mathbf{k}\left(\nabla\cdot\nabla\frac{1}{r}\right)=\mathbf{k}\left[\nabla\cdot\left(-\frac{\mathbf{r}}{r^3}\right)\right]=-\mathbf{k}\left[\nabla\cdot\left(\frac{\mathbf{r}}{r^3}\right)\right]=-\mathbf{k}\left[\frac{1}{r^3}\nabla\cdot\mathbf{r}+\nabla\frac{1}{r^3}\cdot\mathbf{r}\right]$$

$$=-\mathbf{k}\left[\frac{3}{r^3}-\frac{3}{r^4}\frac{\mathbf{r}}{r}\cdot\mathbf{r}\right]=-\mathbf{k}\left[\frac{3}{r^3}-\frac{3}{r^3}\right]=0.\ \text{Hence proved.}$$

EXAMPLE 1.49 If r and $\mathbf{R}$ have their usual meanings and $\mathbf{A}$ is a constant vector, prove that

$$\text{curl}\left(\frac{\mathbf{A}\times\mathbf{R}}{r^n}\right)=\frac{2-n}{r^n}\mathbf{A}+n\frac{\mathbf{A}\cdot\mathbf{R}}{r^{n+2}}\mathbf{R}.$$

(*AMIETE, June 2008; JNTU, 2005; Andhra 2000*)

Solution $\text{curl}\left(\dfrac{\mathbf{A}\times\mathbf{R}}{r^n}\right)=\nabla\times[r^{-n}(\mathbf{A}\times\mathbf{R})]=r^{-n}[\nabla\times(\mathbf{A}\times\mathbf{R})]+\nabla r^{-n}\times(\mathbf{A}\times\mathbf{R})$ [Art. 1.17 formula 4]

$$=r^{-n}[(\nabla\cdot\mathbf{R})\mathbf{A}-(\mathbf{A}\cdot\nabla)\mathbf{R}]+\left[-nr^{-n-1}\frac{\mathbf{R}}{r}\right]\times(\mathbf{A}\times\mathbf{R})=r^{-n}(3\mathbf{A}-\mathbf{A})-nr^{-(n+2)}\mathbf{R}\times(\mathbf{A}\times\mathbf{R})$$

[$\because \nabla\cdot\mathbf{R}=3$ and $(\mathbf{A}\cdot\nabla)\mathbf{R}=\mathbf{A}$]

$$=2\mathbf{A}r^{-n}-nr^{-(n+2)}\{(\mathbf{R}\cdot\mathbf{R})\mathbf{A}-(\mathbf{A}\cdot\mathbf{R})\mathbf{R}\}$$ (expanding the triple vector product)

$$=\frac{2\mathbf{A}}{r^n}-nr^{-(n+2)}\left[\mathbf{A}r^2-(\mathbf{A}\cdot\mathbf{R})\mathbf{R}\right]=\frac{2\mathbf{A}}{r^n}-\frac{n\mathbf{A}}{r^n}+nr^{-(n+2)}(\mathbf{A}\cdot\mathbf{R})\mathbf{R}=\frac{2-n}{r^n}\mathbf{A}+n\frac{(\mathbf{A}\cdot\mathbf{R})}{r^{n+2}}\mathbf{R}.$$

EXAMPLE 1.50 Show that div grad $r^n=n(n+1)\,r^{n-2}$, where $r^2=x^2+y^2+z^2$.

Solution We have $\partial r/\partial x=x/r,\ \ \partial r/\partial y=y/r,\ \ \partial r/\partial z=z/r.$

$$\text{grad}\,r^n=\nabla r^n=n\frac{r^{n-1}}{r}\mathbf{r}=nr^{n-2}(x\mathbf{i}+y\mathbf{j}+z\mathbf{k}).$$

Now, $\text{div}\,(\text{grad}\,r^n)=\nabla\cdot nr^{n-2}(x\mathbf{i}+y\mathbf{j}+z\mathbf{k})=n\left[\dfrac{\partial}{\partial x}(xr^{n-2})+\dfrac{\partial}{\partial y}(yr^{n-2})+\dfrac{\partial}{\partial z}(zr^{n-2})\right]$

$$=n\left[3r^{n-2}+x(n-2)\,r^{n-3}\frac{x}{r}+y\,(n-2)\,r^{n-3}\frac{y}{r}+z\,(n-2)\,r^{n-3}\frac{z}{r}\right]$$

$$=n\left[3r^{n-2}+(n-2)\,r^{n-2}\right]=n\,(n+1)\,r^{n-2}.$$

Note: As a special case, $\text{div grad}\left(\dfrac{1}{r}\right)=0$ (Put $n=-1$).

EXAMPLE 1.51 A central field $\mathbf{A}$ in space is defined by $\mathbf{A}=\mathbf{r}f(r)$, where $\mathbf{r}=x\mathbf{i}+y\mathbf{j}+z\mathbf{k}$ and $r=(x^2+y^2+z^2)^{1/2}$. Determine $f(r)$ so that the field may be irrotational and solenoidal. (*AMIETE June 2005*)

Solution $\nabla\times[\mathbf{r}f(r)]=f(r)\,(\nabla\times\mathbf{r})+\nabla f(r)\times\mathbf{r}$ [by Art. 1.17, formula 4]

$$=f(r)\,(\nabla\times\mathbf{r})+f'(r)\cdot\frac{\mathbf{r}}{r}\times\mathbf{r}=\mathbf{0}.$$ ($\because \nabla\times\mathbf{r}=0$ and $\mathbf{r}\times\mathbf{r}=0$)

Hence $\mathbf{r}f(r)$ is irrotational for all $f(r)$. Again since $\mathbf{r}f(r)$ is solenoidal, therefore, $\nabla\cdot[\mathbf{r}f(r)]=0$

or $f(r)(\nabla\cdot\mathbf{r})+\mathbf{r}\cdot\nabla f(r)=0$ [by Art. 1.17, formula 2]

or $3f(r)+\mathbf{r}\cdot f'(r)\dfrac{\mathbf{r}}{r}=0$ ($\because\ \nabla\cdot\mathbf{r}=3$) or $3f(r)+f'(r)\,r=0$ ($\because\ \mathbf{r}\cdot\mathbf{r}=r^2$)

or $\dfrac{f'(r)}{f(r)}+\dfrac{3}{r}=0$ or $\log f(r)+3\log r=\log c$

or $\quad f(r)(r^3) = c$. Therefore $f(r) = c/r^3$, where c is an arbitrary constant.

Hence, if $f(r) = c/r^3$, then $\mathbf{r}\,f(r)$ is both irrotational and solenoidal.

EXAMPLE 1.52 (*i*) If $u = f(r)$ and $r^2 = x^2 + y^2 + z^2$, show that $\nabla^2 u = f''(r) + \frac{2}{r} f'(r)$.

(*ii*) Also show that $u = A + \frac{B}{r}$ is a possible solution of $\nabla^2 u = 0$ that is Laplace's equation, where A and B are arbitrary constants.

(*AMIETE Dec 2003*)

Solution (*i*) We have $\nabla^2 u = \nabla^2 f(r) = \nabla \cdot \nabla\,[f(r)] = \nabla \cdot \left(\frac{f'(r)}{r}\mathbf{r}\right) = \frac{f'(r)}{r}(\nabla \cdot \mathbf{r}) + \nabla\left(\frac{f'(r)}{r}\right) \cdot \mathbf{r}$

$$= \frac{3f'(r)}{r} + \mathbf{r} \cdot \left[\frac{rf''(r) - f'(r)}{r^2}\right]\frac{\mathbf{r}}{r} = \frac{3f'(r)}{r} + \frac{rf''(r) - f'(r)}{r} = \frac{2}{r} f'(r) + f''(r).$$

(*ii*) From the result of the part (*i*), we have

$$\nabla^2\left(A + \frac{B}{r}\right) = \frac{d^2}{dr^2}\left(A + \frac{B}{r}\right) + \frac{2}{r}\frac{d}{dr}\left(A + \frac{B}{r}\right) = \frac{2B}{r^3} + \frac{2}{r}\left(-\frac{B}{r^2}\right) = 0,$$

showing that $A + (B/r)$ is a solution of Laplace's equation.

EXAMPLE 1.53 Prove that $\operatorname{div}(r^n\,\mathbf{R}) = (n+3)r^n$, where $\mathbf{R} = x\mathbf{i} + y\mathbf{j} + z\mathbf{k}$ and $r = |\mathbf{R}|$. Hence show that $\mathbf{R}/r^3$ is solenoidal when $n = -3$. (*VTU, 2006; UPTU, 2006; PTU, 2005*)

Solution We have $\mathbf{R} = x\mathbf{i} + y\mathbf{j} + z\mathbf{k}$ and $r = |\mathbf{R}| = \sqrt{x^2 + y^2 + z^2}$.

$\therefore \operatorname{div}(r^n\mathbf{R}) = \nabla \cdot (x^2 + y^2 + z^2)^{n/2}(x\mathbf{i} + y\mathbf{j} + z\mathbf{k})$

$$= \left(\frac{\partial}{\partial x}\mathbf{i} + \frac{\partial}{\partial y}\mathbf{j} + \frac{\partial}{\partial z}\mathbf{k}\right) \cdot (x\mathbf{i} + y\mathbf{j} + z\mathbf{k})\,(x^2 + y^2 + z^2)^{n/2}$$

$$= \frac{\partial}{\partial x}[x(x^2 + y^2 + z^2)^{n/2}] + \frac{\partial}{\partial y}[y(x^2 + y^2 + z^2)^{n/2}] + \frac{\partial}{\partial z}[z(x^2 + y^2 + z^2)^{n/2}]$$

$$= \Sigma\left[1.(x^2 + y^2 + z^2)^{n/2} + x\frac{n}{2}(x^2 + y^2 + z^2)^{\frac{n}{2}-1} 2x\right]$$

$$= \Sigma r^n + n\Sigma x^2(x^2 + y^2 + z^2)^{\frac{n}{2}-1} = 3r^n + n(r^2) \cdot r^{2\left(\frac{n}{2}-1\right)} = r^n(n+3).$$

Thus $\operatorname{div}(r^n\,\mathbf{R}) = (n+3)r^n$. Substitute $n = -3$ in above equation, we get $\operatorname{div}(\mathbf{R}/r^3) = 0$, that is, $\mathbf{R}/r^3$ is solenoidal.

EXAMPLE 1.54 Show that $\mathbf{v}(x, y, z) = (z - y)\,\mathbf{i} + (x - z)\,\mathbf{j} + (y - x)\,\mathbf{k}$ is solenoidal and find $\mathbf{w}(x, y, z)$ such that $\mathbf{v} = \operatorname{curl} \mathbf{w}$.

Solution Since $\nabla \cdot \mathbf{v} = \frac{\partial}{\partial x}(z - y) + \frac{\partial}{\partial y}(x - z) + \frac{\partial}{\partial z}(y - x) = 0 + 0 + 0 = 0.$

Therefore, $\mathbf{v}$ is solenoidal. Let $\mathbf{w} = w_1\mathbf{i} + w_2\mathbf{j} + w_3\mathbf{k}$.

Therefore, $\nabla \times \mathbf{w} = \begin{vmatrix} \mathbf{i} & \mathbf{j} & \mathbf{k} \\ \partial/\partial x & \partial/\partial y & \partial/\partial z \\ w_1 & w_2 & w_3 \end{vmatrix} = \left(\frac{\partial}{\partial y}w_3 - \frac{\partial}{\partial z}w_2\right)\mathbf{i} + \left(\frac{\partial}{\partial z}w_1 - \frac{\partial}{\partial x}w_3\right)\mathbf{j} + \left(\frac{\partial}{\partial x}w_2 - \frac{\partial}{\partial y}w_1\right)\mathbf{k}.$

If **v** = curl **w**, we get

$$\frac{\partial}{\partial y} w_3 - \frac{\partial}{\partial z} w_2 = z - y \quad ...(i), \qquad \frac{\partial}{\partial z} w_1 - \frac{\partial}{\partial x} w_3 = x - z \quad ...(ii), \qquad \frac{\partial}{\partial x} w_2 - \frac{\partial}{\partial y} w_1 = y - x. \quad ...(iii)$$

From (*i*), we have $\frac{\partial w_3}{\partial y} = z$ and $\frac{\partial w_2}{\partial z} = y$

or $\quad w_3 = yz + f(x, z) \quad ...(iv) \quad$ and $\quad w_2 = yz + f(x, y). \quad ...(v)$

Similarly from (*ii*) $\quad w_1 = xz + f(x, y) \quad ...(vi) \quad$ and $\quad w_3 = xz + f(y, z). \quad ...(vii)$

From (*iv*) and (*vii*), we conclude that $f(x, z) = xz$ and $f(y, z) = yz$.

Therefore, $w_3 = xz + yz = (x + y)z$. Similarly, $w_1 = (y + z)x$ and $w_2 = (z + x)y$.

Hence, $\quad \mathbf{w} = x(y + z)\mathbf{i} + y(z + x)\mathbf{j} + z(x + y)\mathbf{k}$.

Note: If we take $\frac{\partial w_3}{\partial y} = -y, \frac{\partial w_2}{\partial z} = -z$ in (*i*) and so on then, we get

$$\mathbf{w} = -\frac{1}{2}(y^2 + z^2)\mathbf{i} - \frac{1}{2}(z^2 + x^2)\mathbf{j} - \frac{1}{2}(x^2 + y^2)\mathbf{k}.$$

EXERCISE 1.3

1. If $\mathbf{r} = x\mathbf{i} + y\mathbf{j} + z\mathbf{k}$ and $r = |\mathbf{r}|$, show that (*a*) $\nabla \cdot \mathbf{r} = 3$, (*b*) $\nabla \times \mathbf{r} = \mathbf{0}$.
   (*AMIETE, June 2009; UPTU, 2006; PTU, 2006*)
2. Compute div **v** where $\mathbf{v} = 2x^2z\mathbf{i} - xy^2z\mathbf{j} + 3yz^2\mathbf{k}$. **Ans.** $4xz - 2xyz + 6yz$.
3. Find div **F** and curl **F** where $\mathbf{F} = \nabla(x^3 + y^3 + z^3 - 3xyz)$. (*AMIETE, June 2009; Kurukshetra, 2006*)
   Ans. $6(x + y + z)$; **0**
4. If $\mathbf{v} = xyz\,\mathbf{i} + 3x^2y\,\mathbf{j} + (xz^2 - y^2z)\mathbf{k}$, find (*a*) div **v**, (*b*) curl **v**.
   Ans. (*a*) $yz + 3x^2 + 2xz - y^2$, (*b*) $-2yz\,\mathbf{i} - (z^2 - xy)\,\mathbf{j} + (6xy - xz)\mathbf{k}$.
5. If $\mathbf{A} = yz^2\mathbf{i} + zx^2\mathbf{j} + xy^2\mathbf{k}$, evaluate curl **A** and show that $\mathbf{A} \cdot \text{curl}\,\mathbf{A} = xyz\,(xy + yz + zx)$.
6. Find the divergence and curl of the vector field given by
   (*a*) $xy^2\,\mathbf{i} + 2x^2yz\,\mathbf{j} - 3yz^2\mathbf{k}$, (*b*) $(x^2 - y^2)\mathbf{i} + 2xy\,\mathbf{j} + (y^2 - xy)\mathbf{k}$.
   Ans. (*a*) $y^2 + 2x^2z - 6yz; -(3z^2 + 2x^2y)\mathbf{i} + (4xyz - 2xy)\mathbf{k}$, (*b*) $4x; (2y - x)\mathbf{i} + y\mathbf{j} + 4y\mathbf{k}$.
7. If $\mathbf{A} = x^2z\,\mathbf{i} + yz^3\mathbf{j} - 3xy\,\mathbf{k}$, $\mathbf{B} = y^2\mathbf{i} - yz\,\mathbf{j} + 2x\,\mathbf{k}$ and $f(x, y, z) = 2x^2 + yz$, find
   (*a*) $(\mathbf{A} \cdot \nabla) f$, (*b*) $\mathbf{A} \cdot \nabla f$, (*c*) $(\mathbf{A} \cdot \nabla)\mathbf{B}$.
   Ans. (*a*) $4x^3z + 4z^4 - 3xy^2$, (*b*) same as (*a*), (*c*) $2y^2z^3\mathbf{i} + (3xy^2 - yz^4\mathbf{j}) + 2x^2z\mathbf{k}$.
8. If $\mathbf{v} = (x + z)\mathbf{i} - 3x^2y\,\mathbf{j} + 2yz\,\mathbf{k}$, find div **v**, curl **v**, grad (div **v**) and div (curl **v**).
   Ans. $1 - 3x^2 + 2y$; $2z\mathbf{i} + \mathbf{j} - 6xy\mathbf{k}$; $-6x\mathbf{i} + 2\mathbf{j}$; 0.
9. Evaluate curl grad $(x^2 + y^2 + z^2)$. **Ans. 0.**
10. If $\mathbf{A} = 3xyz^2\mathbf{i} + 2xy^3\mathbf{j} - x^2yz\,\mathbf{k}$ and $f(x, y\, z) = 3x^2 - yz$, find
    (*a*) $\nabla \times \mathbf{A}$, (*b*) $\nabla \cdot (\nabla \times \mathbf{A})$, (*c*) $\nabla \times (\nabla f)$. **Ans.** (*a*) $-x^2z\,\mathbf{i} + 8xyz\,\mathbf{j} + (2y^3 - 3xz^2)\mathbf{k}$, (*b*) 0, (*c*) **0**.
11. If $f(x, y, z) = xy + yz + zx$ and $\mathbf{A} = x^2y\,\mathbf{i} + y^2z\,\mathbf{j} + z^2x\,\mathbf{k}$, find
    (*a*) $\mathbf{A} \cdot \nabla f$, (*b*) $f\nabla \cdot \mathbf{A}$, and (*c*) $(\nabla f) \times \mathbf{A}$ at the point (3, –1, 2).
    Ans. (*a*) 25, (*b*) 2, (*c*) $56\mathbf{i} - 30\mathbf{j} + 47\mathbf{k}$.
12. If $\vec{A}$ and $\vec{B}$ be the vector joining the fixed points (x_1, y_1, z_1) and (x_2, y_2, z_2) respectively to a variable point (x, y, z). Prove that
    (*a*) grad $(\vec{A} \cdot \vec{B}) = \vec{A} + \vec{B}$, (*b*) div $(\vec{A} \times \vec{B}) = 0$, (*c*) curl $(\vec{A} \times \vec{B}) = 2\,(\vec{A} - \vec{B})$. (*AMIETE, Dec. 2011*)

[**Hint:** (a) We have $\vec{A} = (x - x_1)\mathbf{i} + (y - y_1)\mathbf{j} + (z - z_1)\mathbf{k}$

and $\vec{B} = (x - x_2)\mathbf{i} + (y - y_2)\mathbf{j} + (z - z_2)\mathbf{k}$.

Therefore, $\vec{A} \bullet \vec{B} = (x - x_1)(x - x_2) + (y - y_1)(y - y_2) + (z - z_1)(z - z_2)$.

$$\text{grad}\,(\vec{A} \bullet \vec{B}) = \left(\mathbf{i}\frac{\partial}{\partial x} + \mathbf{j}\frac{\partial}{\partial y} + \mathbf{k}\frac{\partial}{\partial z}\right)(\vec{A} \bullet \vec{B}) = \sum \mathbf{i}\frac{\partial}{\partial x}(x^2 - xx_2 - xx_1 + x_1x_2)$$

$$= \sum \mathbf{i}\,(2x - x_2 - x_1) = \sum \mathbf{i}\,\{(x - x_1) + (x - x_2)\} = \vec{A} + \vec{B}].$$

13. If **a** is a constant vector, prove that $\nabla \times (\mathbf{a} \times \mathbf{u}) = \mathbf{a}\,(\nabla \bullet \mathbf{u}) - (\mathbf{a} \bullet \nabla)\,\mathbf{u}$. *(AMIE, W-2011)*

[**Hint:** We have $\nabla \times (\mathbf{a} \times \mathbf{u}) = \sum \mathbf{i} \times \frac{\partial}{\partial x}(\mathbf{a} \times \mathbf{u}) = \sum \mathbf{i} \times \left(\mathbf{a} \times \frac{\partial \mathbf{u}}{\partial x}\right) = \left(\sum \mathbf{i} \bullet \frac{\partial \mathbf{u}}{\partial x}\right) a - \left(\sum \mathbf{i} \bullet \mathbf{a}\right)\frac{\partial \mathbf{u}}{\partial x}$

$$= (\nabla \bullet \mathbf{u})\,\mathbf{a} - \left(\mathbf{a} \bullet \sum \mathbf{i}\frac{\partial}{\partial x}\right)\mathbf{u} = \mathbf{a}\,(\nabla \bullet \mathbf{u}) - (\mathbf{a} \bullet \nabla)\,\mathbf{u}\Bigg].$$

14. If $\mathbf{F}(x, y, z) = 3xz^2\mathbf{i} - yz\mathbf{j} + (x + 2z)\mathbf{k}$, find curl (curl **F**). **Ans.** $-6x\,\mathbf{i} + (6z - 1)\mathbf{k}$.

15. Given $f(x, y, z) = 2x^3y^2z^4$. Find (*a*) div (grad f) or $\nabla \bullet \nabla f$ (*b*) $\nabla^2 f$.

Ans. (*a*) and (*b*) $12xy^2z^4 + 4x^3z^4 + 24x^3y^2z^2$.

16. Show that:

(*a*) $\nabla^2(1/r) = 0$, where $r = |\mathbf{R}| = \sqrt{x^2 + y^2 + z^2}$. (*b*) $\nabla^2(\log r) = 1/r^2$,

(*c*) $\nabla^2 r^n = n\,(n + 1)\,r^{n-2}$, where n is a constant.

[**Hint:** (*a*) $$\nabla^2(1/r) = \left(\frac{\partial^2}{\partial x^2} + \frac{\partial^2}{\partial y^2} + \frac{\partial^2}{\partial z^2}\right)\left(\frac{1}{\sqrt{x^2 + y^2 + z^2}}\right)$$

$$\frac{\partial}{\partial x}\left(\frac{1}{\sqrt{x^2 + y^2 + z^2}}\right) = -x(x^2 + y^2 + z^2)^{-3/2}.$$

Therefore, $$\frac{\partial^2}{\partial x^2} = \frac{\partial}{\partial x}\left(\frac{\partial}{\partial x}\right) = \frac{\partial}{\partial x}[-x(x^2 + y^2 + z^2)^{-3/2}] = \frac{2x^2 - y^2 - z^2}{(x^2 + y^2 + z^2)^{-5/2}}.$$

Thus $$\sum \frac{\partial^2}{\partial x^2}\left(\frac{1}{\sqrt{x^2 + y^2 + z^2}}\right) = 0.]$$

17. Show that the vector $\mathbf{v} = (x + 3y)\mathbf{i} + (y - 3z)\mathbf{j} + (x - 2z)\mathbf{k}$ is solenoidal.

18. Show that the vector $\mathbf{F} = (x^2 - yz)\mathbf{i} + (4y + z^2x)\mathbf{j} - (2xz + 4z)\mathbf{k}$ is solenoidal.

19. Find the value of the constant p so that the vector $\mathbf{A} = (x + 3y)\mathbf{i} + (y - 2z)\mathbf{j} + (x + pz)\mathbf{k}$ is solenoidal. **Ans.** –2.

20. Show that the vector field **A** where $\mathbf{A} = (x^2 - y^2 + x)\mathbf{i} - (2xy + y)\mathbf{j}$ is irrotational; and find a scalar function $f(x, y)$ such that **A** = grad f. **Ans.** $(2x^3 - 6xy^2 + 3x^2 - 3y^2)/6$.

21. (*a*) Prove that $\mathbf{v} = 2xyz\,\mathbf{i} + x^2z\,\mathbf{j} + x^2y\,\mathbf{k}$ is irrotational and find $f(x, y, z)$ so that **v** = grad f.

Ans. $f(x, y, z) = x^2yz$.

(*b*) Show that the vector field defined by $\mathbf{F} = 2xyz^3\mathbf{i} + x^2z^3\mathbf{j} + 3x^2yz^2\mathbf{k}$ is irrotational. Find the scalar potential f such that **F** = grad f. **Ans.** $f(x, y, z) = x^2yz^3$.

22. A vector field **F** is given by $\mathbf{F} = (x^2 + xy^2)\mathbf{i} + (y^2 + x^2y)\mathbf{j}$. Show that the field is irrotational (or conservative) and find the scalar potential ϕ. **Ans.** $\phi = \frac{x^3}{3} + \frac{x^2y^2}{2} + \frac{y^3}{3} + c$.

23. A fluid motion is given by $\bar{v} = (y\sin z - \sin x)\hat{\mathbf{i}} + (x\sin z + 2yz)\hat{\mathbf{j}} + (xy\cos z + y^2)\hat{\mathbf{k}}$. Is the motion irrotational? If so, find the velocity potential. *(AMIETE, Dec. 2007; PTU 2006)*

Ans. Yes, $xy\sin z + \cos x + zy^2 + c$.

24. Show that $\mathbf{F} = (y^2 + 2xz^2)\mathbf{i} + (2xy - z)\mathbf{j} + (2x^2z - y + 2z)\mathbf{k}$ is irrotational and hence find its scalar potential. *(AMIETE Dec 2004)*

25. Show that vector field represented by $\mathbf{F} = (z^2 + 2x + 3y)\mathbf{i} + (3x + 2y + z)\mathbf{j} + (y + 2zx)\mathbf{k}$ is irrotational but not solenoidal. Also obtain a scalar f function such that grad $f = \mathbf{F}$. *(AMIETE Dec. 2003)*

26. A vector field is given by $\mathbf{F} = (x^2 - y^2 + x)\mathbf{i} - (2xy + y)\mathbf{j}$. Show that the field is irrotational and find its scalar potential. *(AMIETE June 2003)* **Ans.** $\frac{1}{3}x^3 - xy^2 + \frac{1}{2}(x^2 - y^2) + c$.

27. Show that the vector field given by $\mathbf{A} = 3x^2y\mathbf{i} + (x^3 - 2yz^2)\mathbf{j} + (3z^2 - 2y^2z)\mathbf{k}$ is irrotational but not solenoidal. Also find a scalar function f such that grad $f = \mathbf{A}$. *(AMIETE Dec. 2002)*

28. If $\nabla f = (y^2 - 2xyz^3)\mathbf{i} + (3 + 2xy - x^2z^3)\mathbf{j} + (6z^3 - 3x^2yz^2)\mathbf{k}$, find $f(x, y, z)$.

Ans. $f(x, y, z) = xy^2 - x^2yz^3 + 3y + \frac{3}{2}z^4 + c$.

29. If ϕ and ψ are scalar functions, prove that $\nabla\phi \times \nabla\psi$ is solenoidal.

30. If $u\mathbf{F} = \nabla v$, where u, v are scalar fields and $\mathbf{F}$ is vector field, show that

$\mathbf{F} \cdot \nabla\mathbf{F} = 0$ and $\mathbf{F} \cdot (\nabla \times \mathbf{F}) = 0$. *(AMIETE, June 2011, Dec. 2009)*

31. If $r = |\mathbf{r}| = |x\mathbf{i} + y\mathbf{j} + z\mathbf{k}|$, prove that $\nabla \cdot (r^3\mathbf{r}) = 6r^3$.

32. If **a** and **b** are constant vectors and **r** is the position vector with modulus r, prove that

(a) $\nabla \times (\mathbf{a} \cdot \mathbf{r})\,\mathbf{a} = 0$, *(b)* $\nabla \times [\mathbf{a} \times (\mathbf{b} \times \mathbf{r})] = \mathbf{a} \times \mathbf{b}$,

(c) $\nabla \times \left(\frac{\mathbf{a} \times \mathbf{r}}{r}\right) = \frac{\mathbf{a}}{r} + \frac{(\mathbf{a} \cdot \mathbf{r})\mathbf{r}}{r^3}$, *(d)* curl $\left(\frac{\mathbf{r} \times \mathbf{a}}{r^3}\right) = \frac{\mathbf{a}}{r^3} - \frac{3\mathbf{r}}{r^5}(\mathbf{r} \cdot \mathbf{a})$, *(MDU 2007)*

(e) $\nabla \cdot \left(\frac{\mathbf{a} \times \mathbf{r}}{r^n}\right) = 0$, *(f)* $\nabla\left(\frac{\mathbf{a} \cdot \mathbf{r}}{r^n}\right) = \frac{\mathbf{a}}{r^n} - \frac{n(\mathbf{a} \cdot \mathbf{r})\mathbf{r}}{r^{n+2}}$,

(g) $r^5\left[\mathbf{b} \cdot \nabla\left\{\mathbf{a} \cdot \nabla\left(\frac{1}{r}\right)\right\}\right] = 3\,(\mathbf{a} \cdot \mathbf{r})\,(\mathbf{b} \cdot \mathbf{r}) - (\mathbf{a} \cdot \mathbf{b})\,r^2$. *(AMIETE Dec. 2006)*

[**Hint:** *(g)* Let $\mathbf{a} = a_1\mathbf{i} + a_2\mathbf{j} + a_3\mathbf{k}$, $\mathbf{r} = x\mathbf{i} + y\mathbf{j} + z\mathbf{k}$.

$$\mathbf{a} \cdot \nabla\left(\frac{1}{r}\right) = \mathbf{a} \cdot \left(-\frac{\mathbf{r}}{r^3}\right) = -\frac{\mathbf{a} \cdot \mathbf{r}}{r^3} = -\left(\frac{a_1x + a_2y + a_3z}{r^3}\right).$$

Now $$\nabla\left\{\mathbf{a} \cdot \nabla\left(\frac{1}{r}\right)\right\} = \nabla\left\{-\left(\frac{a_1x + a_2y + a_3z}{r^3}\right)\right\} = \sum \mathbf{i}\frac{\partial}{\partial x}\left(-\frac{a_1x + a_2y + a_3z}{r^3}\right)$$

$$= \sum \mathbf{i}\left[-\frac{a_1}{r^3} + \frac{3(a_1x + a_2y + a_3z)}{r^4}\left(\frac{x}{r}\right)\right] = -\frac{1}{r^3}\sum \mathbf{i}\,a_1 + \frac{3(a_1x + a_2y + a_3z)}{r^5}\sum \mathbf{i}x = -\frac{\mathbf{a}}{r^3} + \frac{3(\mathbf{a} \cdot \mathbf{r})}{r^5}\mathbf{r}.$$

Therefore, $\mathbf{b} \cdot \nabla\left\{\mathbf{a} \cdot \nabla\left(\frac{1}{r}\right)\right\} = \frac{3(\mathbf{a} \cdot \mathbf{r})(\mathbf{b} \cdot \mathbf{r})}{r^5} - \frac{(\mathbf{a} \cdot \mathbf{b})}{r^3}$, on taking dot product of **b** and $\nabla\left\{\mathbf{a} \cdot \nabla\left(\frac{1}{r}\right)\right\}$.]

33. If $u = x^2 + y^2 + z^2$, $\vec{v} = x\mathbf{i} + y\mathbf{j} + z\mathbf{k}$ then find the value of $\nabla \cdot (u\vec{v})$.

(AMIETE June 2008, 2007, 2004; PTU 2005) **Ans.** $5u$.

[**Hint:** div $u\vec{v} = u$ div $\vec{v}$ + (grad u) $\cdot\, \vec{v}$].

34. If **E** and **H** are irrotational fields, prove that **E** × **H** is solenoidal.

[**Hint:** As **E** and **H** are irrotational. Therefore, $\nabla \times \mathbf{E} = \mathbf{0}$ and $\nabla \times \mathbf{H} = \mathbf{0}$.

Now $\nabla \cdot (\mathbf{E} \times \mathbf{H}) = (\nabla \times \mathbf{E}) \cdot \mathbf{H} - (\nabla \times \mathbf{H}) \cdot \mathbf{E} = 0$.]

35. If $\nabla \cdot \left\{ \frac{f(r)\,\mathbf{r}}{r} \right\} = 0$, show that $f(r) = A/r^2$, where A is an arbitrary constant.

36. Show that $r^{\alpha}\mathbf{r}$ is any irrotational vector for any value of α but is solenoidal if α + 3 = 0, where $\mathbf{r} = x\mathbf{i} + y\mathbf{j} + z\mathbf{k}$ and r is the magnitude of **r**. *(VTU, 2006; Kottayam, 2005)*

[**Hint:** Let $\mathbf{A} = r^{\alpha}\mathbf{r} = (x^2+y^2+z^2)^{\alpha/2}(x\mathbf{i}+y\mathbf{j}+z\mathbf{k}) = \sum x(x^2+y^2+z^2)^{\alpha/2}\mathbf{i}$

Therefore, $$\text{curl } \mathbf{A} = \begin{vmatrix} \mathbf{i} & \mathbf{j} & \mathbf{k} \\ \partial/\partial x & \partial/\partial y & \partial/\partial z \\ x(x^2+y^2+z^2)^{\alpha/2} & y(x^2+y^2+z^2)^{\alpha/2} & z(x^2+y^2+z^2)^{\alpha/2} \end{vmatrix}$$

$$= \sum \mathbf{i} \left\{ \frac{\alpha z}{2}(x^2+y^2+z^2)^{\frac{\alpha}{2}-1}(2y) - \frac{\alpha y}{2}(x^2+y^2+z^2)^{\frac{\alpha}{2}-1}(2z) \right\} = 0.$$ Hence **A** is irrotational for any value of α.

Now $\text{div } \mathbf{A} = \nabla \cdot (r^{\alpha}\mathbf{r}) = (\alpha+3)r^{\alpha}$ which is zero when α + 3 = 0, that is, **A** is solenoidal if α + 3 = 0.]

37. Prove that $\nabla \cdot \left(\frac{\mathbf{r}}{r^3} \right) = 0$.

[**Hint:** Use the formula div (f**A**) = f(div **A**) + (grad f) · **A**]

38. Evaluate $\nabla^2 \left[\nabla \cdot \left(\frac{\mathbf{r}}{r^2} \right) \right]$. **Ans.** $2r^{-4}$.

39. Show that curl (**u** + **v**) = curl **u** + curl **v**.

40. If $\mathbf{r} = x\mathbf{i} + y\mathbf{j} + z\mathbf{k}$ show that div ($\mathbf{r}f$) = $3f + \mathbf{r} \cdot \text{grad } f$.

41. If r is the distance of a point (x, y, z) from the origin, find the value of the expression $\hat{j} \times \text{grad } \frac{1}{\mathbf{r}}$.

(AMIETE, Dec. 2010) **Ans.** $(x^2+y^2+z^2)^{-3/2}(\hat{k}x - \hat{i}z)$.

[**Hint:** $\hat{j} \times \nabla\left(\frac{1}{\mathbf{r}}\right) = \hat{j} \times \left(\frac{-\mathbf{r}}{r^3}\right) = -\frac{1}{r^3}(0\hat{i} + 1\hat{j} + 0\hat{k}) \times (x\hat{i} + y\hat{j} + z\hat{k})$

$$= -\frac{1}{r^3}\begin{vmatrix} \hat{i} & \hat{j} & \hat{k} \\ 0 & 1 & 0 \\ x & y & z \end{vmatrix} = \frac{1}{r^3}(\hat{k}x - \hat{i}z) = (x^2+y^2+z^2)^{-3/2}(\hat{k}x - \hat{i}z)$$].

42. Find $\vec{f} \times (\nabla \times \vec{g})$ at the point (1, –1, 2) if $\vec{f} = xz^2\hat{i} + 2y\hat{j} - 3xz\hat{k}$, $\vec{g} = 3xz\hat{i} + 2yz\hat{j} - z^2\hat{k}$.

(AMIETE, Dec. 2010)

[**Hint:** We have $$\nabla \times \vec{g} = \begin{vmatrix} \hat{i} & \hat{j} & \hat{k} \\ \partial/\partial x & \partial/\partial y & \partial/\partial z \\ 3xz & 2yz & -z^2 \end{vmatrix} = 3x\hat{j} - 2y\hat{i}.$$

Therefore $$\vec{f} \times (\nabla \times \vec{g}) = \begin{vmatrix} \hat{i} & \hat{j} & \hat{k} \\ xz^2 & 2y & -3xz \\ -2y & 3x & 0 \end{vmatrix} = (9x^2z)\,\hat{i} - (-6xyz)\,\hat{j} + (3x^2z^2 + 4y^2)\,\hat{k}.$$

At the point (1, –1, 2), $\vec{f} \times (\nabla \times \vec{g}) = 18\hat{i} - 12\hat{j} + 16\hat{k}$].

43. If $\mathbf{A} = \nabla \times (\phi \mathbf{i})$, where $\nabla^2 \phi = 0$, show that $\mathbf{A} \bullet \nabla \times \mathbf{A} = \dfrac{\partial \phi}{\partial z}\dfrac{\partial^2 \phi}{\partial y \partial x} - \dfrac{\partial \phi}{\partial y}\dfrac{\partial^2 \phi}{\partial z \partial x}$. *(AMIETE, June 2007)*

44. Prove the following (f and g are scalar functions).

(*a*) $\nabla^2(fg) = g\nabla^2 f + 2\nabla f \bullet \nabla g + f\nabla^2 g$, (*b*) $\text{div}\,(f\,\nabla g) = f\nabla^2 g + \nabla f \bullet \nabla g$.

1.18 LINE INTEGRALS

Consider a curve C. Let A and B denote the initial and terminal point of a portion of curve C. See Fig. 1.13. Let any point on C be represented by $\mathbf{r}(s) = x(s)\,\mathbf{i} + y(s)\,\mathbf{j} + z(s)\,\mathbf{k}$, where s in the arc length measured from the initial point A on the curve.

Let $\mathbf{r}(s)$ be a continuous and differential vector function of s. We shall regard that orientation of the curve *positive* in which a point P on the curve moves as s increases. Orientation opposite to the above will be regarded *negative*. Let $f(x, y, z)$ be a given function, defined for all points on C and let it be a continuous function of s. The curve between A and B is divided into n parts in an arbitrary manner with the points $A = P_0, P_1, P_2, \ldots P_{k-1}, P_k, \ldots, P_{n-1}, P_n = B$ the points of division of C. Let the lengths of the arcs $P_0P_1, P_1P_2, \ldots, P_{k-1}\,P_k, \ldots\, P_{n-1}\,P_n$ be $s_1, s_2, \ldots, s_k, \ldots, s_n$ measured from a fixed point on C. On each arc joining P_{k-1} to P_k (where k goes from 1 to n) choose an arbitrary point $Q_k\,(x_k, y_k, z_k)$. Let $f(x_k, y_k, z_k)$ be the value of the function of f at an arbitrary point in the arc $P_{k-1}P_k$. We form the sum

Fig. 1.13

$$S_n = \sum_{k=1}^{n} f(x_k, y_k, z_k)\,\Delta s_k, \qquad \text{where } \Delta s_k = s_k - s_{k-i}.$$

Let this be done from $n = 2, 3, 4, \ldots$ in an independent manner such that the greatest of $\Delta s_k \to 0$ as $n \to \infty$. This gives a sequence $S_2, S_3, \ldots S_n, \ldots$. The limit of this sequence is called the line integral of f along C from A to B and is denoted by $\int_C f(x, y, z)\,ds$. ...(1.26). The curve C is called the path of integration.

Evaluation of Line Integrals

A line integral can be evaluated by converting it into a definite integral. Let C be a simple curve. Let the parametric representation of C be written as

$$x = x(t), \quad y = y(t), \quad z = z(t), \qquad t_0 \le t \le t_1.$$

The position vector of a point on the curve C can be written as

$$\mathbf{r}(t) = x(t)\,\mathbf{i} + y(t)\,\mathbf{j} + z(t)\,\mathbf{k}, \qquad t_0 \le t \le t_1.$$

The line integral of f over C with respect to the arc length s is given by

$$\int_C f(x, y, z)\,ds = \int_{t_0}^{t_1} f(x(t), y(t), z(t))\frac{ds}{dt}\,dt = \int_{t_0}^{t_1} f(x(t), y(t), z(t))\sqrt{\left(\frac{dx}{dt}\right)^2 + \left(\frac{dy}{dt}\right)^2 + \left(\frac{dz}{dt}\right)^2}\,dt.$$

EXAMPLE 1.55 Evaluate $\int_C y\,ds$ along the curve C given by $y = 2\sqrt{x}$ from $x = 3$ to $x = 24$.

Solution We know that $\dfrac{ds}{dx} = \sqrt{1 + \left(\dfrac{dy}{dx}\right)^2} = \sqrt{1 + \dfrac{1}{x}} = \dfrac{\sqrt{x+1}}{\sqrt{x}}$.

Therefore, $\displaystyle\int_C y\,ds = \int_C y\frac{ds}{dx}\,dx = \int_3^{24} 2\sqrt{x} \bullet \frac{\sqrt{x+1}}{\sqrt{x}}\,dx = 2\int_3^{24} \sqrt{x+1}\,dx = \frac{4}{3}[(x+1)^{3/2}]_3^{24} = \frac{4}{3}(125 - 8) = 156.$

EXAMPLE 1.56 Evaluate $\int_C (x + y^2 + yz)\, ds$, where C is the curve defined by $y = 2x$, $z = 2$ from (1, 2, 2) to (3, 6, 2).

Solution Let $x = t$. Then $y = 2t$, $z = 2$. Therefore, the curve C is represented by $x = t, y = 2t, z = 2, 1 \le t \le 3$.

We have $\frac{ds}{dt} = \sqrt{5}$. Therefore, $\int_C (x + y^2 + yz) \frac{ds}{dt} dt = \sqrt{5} \int_1^3 (t + 4t^2 + 4t)\, dt = \sqrt{5} \left[5\frac{t^2}{2} + \frac{4t^3}{3} \right]_1^3$

$$= \sqrt{5} \left(\frac{45}{2} + 36 - \frac{5}{2} - \frac{4}{3} \right) = \frac{164}{3} \sqrt{5}.$$

Line Integral of Vector Fields

Let C be a smooth curve. Let $\mathbf{F}(x, y, z) = F_1(x, y, z)\mathbf{i} + F_2(x, y, z)\mathbf{j} + F_3(x, y, z)\mathbf{k}$ be a vector field, that is, continuous on C. We define $d\mathbf{r} = \mathbf{i}\, dx + \mathbf{j}\, dy + \mathbf{k}\, dz$.

Then, the line integral of $\mathbf{F}$ over C is defined by

$$\int_C \mathbf{F} \cdot d\mathbf{r} = \int_C F_1\, dx + F_2\, dy + F_3\, dz. \qquad \text{...(1.27)}$$

Let the parametric representation of simple curve C be

$$x = x(t), \quad y = y(t), \quad z = z(t), \quad t_0 \le t \le t_1.$$

Then

$$\int_C \mathbf{F} \cdot d\mathbf{r} = \int_{t_0}^{t_1} F[x(t), y(t), z(t)] \cdot \frac{dr}{dt}\, dt. \qquad \text{...(1.28)}$$

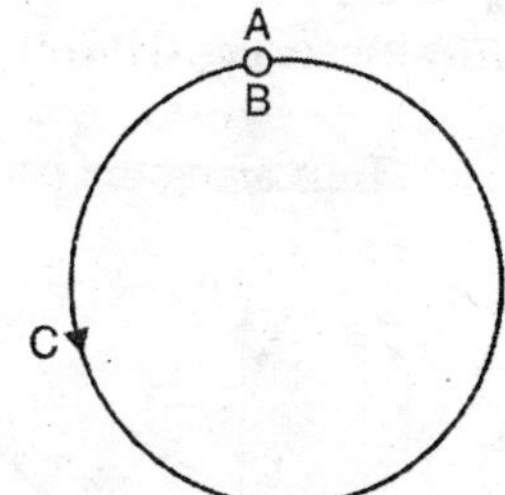

Fig. 1.14

If the path of integration C in eq. (1.28) is a closed curve. (Ref. Fig. 1.14), then instead of $\int_C$ we write $\oint_C$.

If the curve C is piecewise smooth Fig. (1.15) containing the arcs $C_1, C_2, \ldots C_n$ joined end to end, we define a line integral along a piecewise smooth curve C to be the sum of the integrals along the sections.

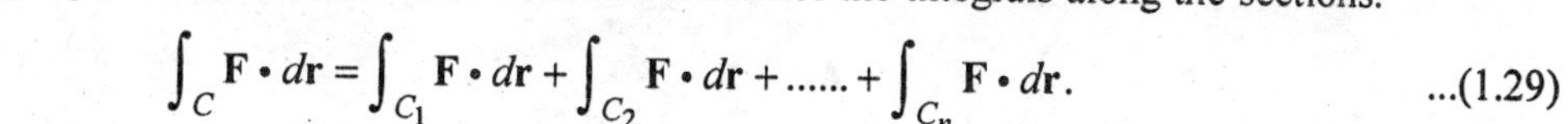

$$\int_C \mathbf{F} \cdot d\mathbf{r} = \int_{C_1} \mathbf{F} \cdot d\mathbf{r} + \int_{C_2} \mathbf{F} \cdot d\mathbf{r} + \ldots\ldots + \int_{C_n} \mathbf{F} \cdot d\mathbf{r}. \qquad \text{...(1.29)}$$

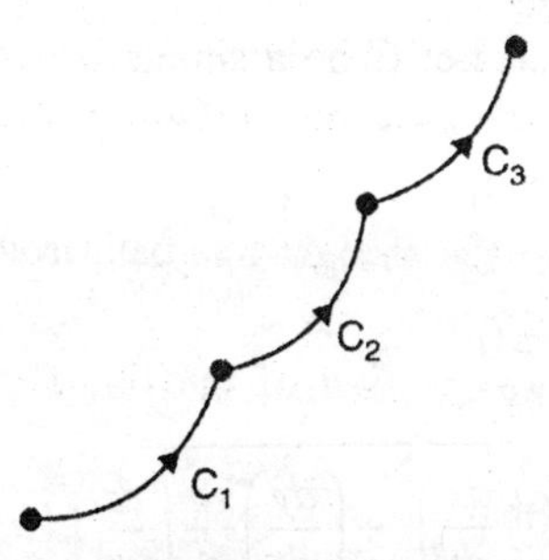

Fig. 1.15

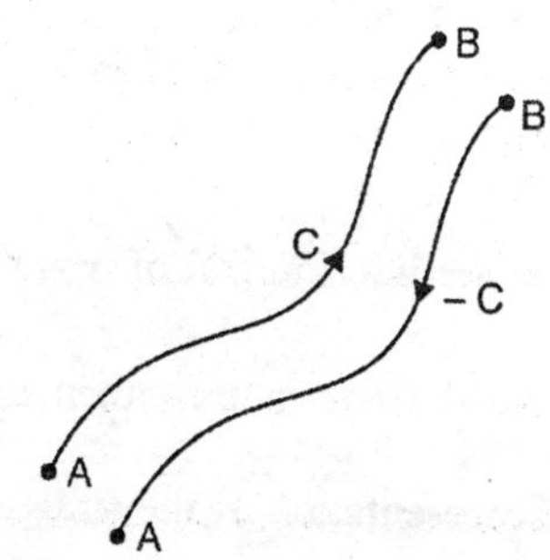

Fig. 1.16

Remark 10

Reversal of the path of integration changes the sign of the line integral. (Fig. 1.16)

EXAMPLE 1.57 If $\mathbf{F} = x^2\mathbf{i} + xy\mathbf{j}$, evaluate $\int_C \mathbf{F} \cdot d\mathbf{r}$ along the curve C in the xy plane consisting of:

(*i*) the straight line joining (0, 0) to (1, 1),
(*ii*) the parabola $y^2 = x$ joining the same points,
(*iii*) the straight lines from (0, 0) to (1, 0), then to (1, 1).

Solution Since $\mathbf{r} = x\,\mathbf{i} + y\,\mathbf{j}$, $d\mathbf{r} = dx\,\mathbf{i} + dy\,\mathbf{j}$.

Therefore, $\int_C \mathbf{F} \cdot d\mathbf{r} = \int_C (x^2\mathbf{i} + xy\,\mathbf{j}) \cdot (dx\,\mathbf{i} + dy\,\mathbf{j}) = \int_C x^2 dx + xy\,dy.$

(*i*) Along the straight line joining (0, 0) to (1, 1). $y = x \Rightarrow dy = dx$, and x varies from 0 to 1. [Fig. 1.17 (*a*)]

Therefore, $$\int_C \mathbf{F} \cdot d\mathbf{r} = \int_0^1 x^2 dx + x(x)dx = \int_0^1 (x^2 + x^2)\,dx = 2\int_0^1 x^2 dx = \frac{2}{3}.$$

(*ii*) Along the parabola $y^2 = x$, Fig. 1.17 (*b*) we have $dx = 2y\,dy$ and y goes from 0 to 1. Therefore,

$$\int_C \mathbf{F} \cdot d\mathbf{r} = \int_C x^2 dx + xy\,dy = \int_0^1 y^4(2y\,dy) + y^2 \cdot y\,dy = \int_0^1 (2y^5 + y^3)\,dy = \left[\frac{y^6}{3} + \frac{y^4}{4}\right]_0^1 = \frac{7}{12}.$$

Alternative Solution: Let $y = t$ in $y^2 = x$. Then the parametric equations of C are $y = t$, $x = t^2$. Points (0, 0), (1, 1) correspond to $t = 0$ and $t = 1$ respectively.

Then $$\int_C \mathbf{F} \cdot d\mathbf{r} = \int_0^1 (t^2)^2 2t\,dt + t^2(t)\,dt = \int_0^1 (2t^5 + t^3)\,dt = \frac{7}{12}.$$

(*iii*) Refer Fig. 1.17(*c*). Along C_1, $y = 0$, $dy = 0$. Along C_2, $x = 1$, $dx = 0$.

Thus along the path OMP, $$\int_C \mathbf{F} \cdot d\mathbf{r} = \int_{C_1} \mathbf{F} \cdot d\mathbf{r} + \int_{C_2} \mathbf{F} \cdot d\mathbf{r} = \int_0^1 x^2 dx + \int_0^1 y\,dy = \frac{1}{3} + \frac{1}{2} = \frac{5}{6}.$$

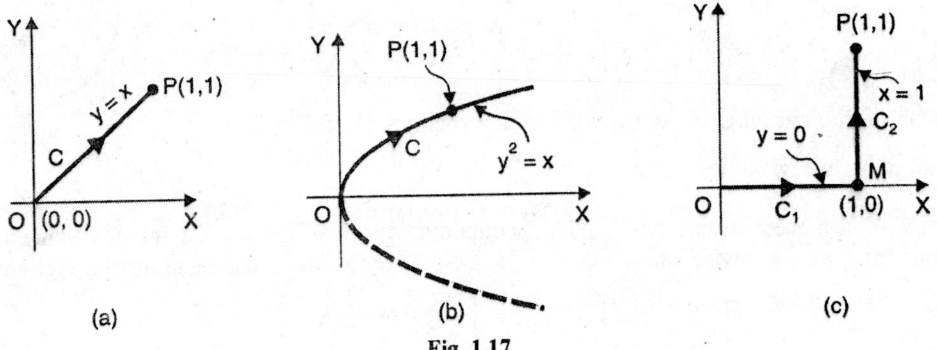

Fig. 1.17

Note: The three results are different, although the end points are the same. This shows that *the value of a line integral will in general depend not only on* **F** *and on the end points O, P of the path but also on the path along which we integrate from O to P.*

Example 1.58 Evaluate the line integral of $\mathbf{v} = x^2\mathbf{i} + yz\,\mathbf{j} + z\,\mathbf{k}$ over the straight line path from (1, 2, 2) to (3, 4, 2).

Solution If **a** and **b** are the two points, then the parametric representation of the line joining them is $\mathbf{r} = \mathbf{a} + t\,(\mathbf{b} - \mathbf{a})$.

Thus the parametric representation of the straight line is given by

$$\mathbf{r}(t) = (\mathbf{i} + 2\mathbf{j} + 2\mathbf{k}) + t\,(2\mathbf{i} + 2\mathbf{j}) = (1 + 2t)\mathbf{i} + (2 + 2t)\mathbf{j} + 2\mathbf{k}. \quad 0 \le t \le 1.$$

Therefore, $$d\mathbf{r}/dt = 2\mathbf{i} + 2\mathbf{j} \text{ and } \int_C \mathbf{v} \cdot d\mathbf{r} = \int_C \mathbf{v} \cdot \frac{d\mathbf{r}}{dt}\,dt = \int_0^1 [2(1 + 2t)^2 + 2(2 + 2t)\,2]\,dt$$

$$= \int_0^1 (8t^2 + 16t + 10)\,dt = \left[\frac{8t^3}{3} + 8t^2 + 10t\right]_0^1 = \frac{62}{3}.$$

EXAMPLE 1.59 If $\mathbf{F} = (3x^2 + 6y)\mathbf{i} - 14yz\,\mathbf{j} + 20xz^2\,\mathbf{k}$, evaluate $\int_C \mathbf{F} \cdot d\mathbf{r}$ from (0, 0, 0) to (1, 1, 1) along the following paths C:

(*i*) $x = t,\ y = t^2,\ z = t^3$, ***(VTU, 2001)***
(*ii*) **the straight lines from (0, 0, 0) to (1, 0, 0),** then to (1, 1, 0), and then to (1, 1, 1),
(*iii*) **the straight line joining (0, 0, 0) to (1, 1, 1).** (*AMIE, S-2006*)

Solution We have

$$\int_C \mathbf{F} \cdot d\mathbf{r} = \int_C [(3x^2 + 6y)\mathbf{i} - 14yz\,\mathbf{j} + 20xz^2\mathbf{k}] \cdot (dx\,\mathbf{i} + dy\,\mathbf{j} + dz\,\mathbf{k})$$

$$= \int_C (3x^2 + 6y)\,dx - 14yzdy + 20xz^2 dz.$$

(*i*) If $x = t,\ y = t^2,\ z = t^3$, points (0, 0, 0) and (1, 1, 1) correspond to $t = 0$ and t = 1, respectively. Then

$$\int_C \mathbf{F} \cdot d\mathbf{r} = \int_{t=0}^{1} (3t^2 + 6t^2)\,dt - 14t^5(2t\,dt) + 20t^7(3t^2 dt)$$

$$= \int_{t=0}^{1} (60t^9 - 28t^6 + 9t^2)\,dt = [6t^{10} - 4t^7 + 3t^3]_0^1 = 5.$$

(*ii*) Along the straight line from (0, 0, 0) to (1, 0, 0), $y = 0$, $z = 0$, $dy = 0$, $dz = 0$ while x varies from 0 to 1. Then the integral over this part of the path is

$$\int_{x=0}^{1} \{3x^2 + 6(0)\}\,dx = \int_0^1 3x^2 dx = [x^3]_0^1 = 1.$$

Along the straight line from (1, 0, 0) to (1, 1, 0), $x = 1$, $z = 0$, $dx = 0$, $dz = 0$ while y varies from 0 to 1. Then the integral over this part of the path is

$$\int_{y=0}^{1} -14y(0)\,dy = 0.$$

Along the straight line from (1, 1, 0) to (1, 1, 1), $x = 1$, $y = 1$, $dx = 0$, $dy = 0$ while z varies from 0 to 1. Then the integral over this part of the path is

$$\int_{z=0}^{1} 20(1)z^2 dz = \frac{20}{3}[z^3]_0^1 = \frac{20}{3}.\ \text{Therefore, } \int_C \mathbf{F} \cdot d\mathbf{r} = 1 + 0 + (20/3) = 23/3.$$

(*iii*) The straight line joining (0, 0, 0) and (1, 1, 1) is given in parametric form by $x = t$, $y = t$, $z = t$. Then

$$\int_C \mathbf{F} \cdot d\mathbf{r} = \int_{t=0}^{1} (3t^2 + 6t - 14t^2 + 20t^3)\,dt$$

$$= \int_{t=0}^{1} (20t^3 - 11t^2 + 6t)\,dt = \left[5t^4 - \frac{11}{3}t^3 + 3t^2\right]_0^1 = \frac{13}{3}.$$

EXAMPLE 1.60 Evaluate $\int_C (x + y)\,dx - x^2 dy + (y + z)dz$, where C: $x^2 = 4y$, $z = x$, $0 \le x \le 2$.

(*AMIE; June 2012, W-2010; AMIETE, June 2008*)

Solution We parametrise C as $x = t$, $y = t^2/4$, $z = t$, $0 \le t \le 2$.

$$\text{Therefore, } \int_C (x + y)\,dx - x^2 dy + (y + z)\,dz = \int_0^2 \left(t + \frac{t^2}{4}\right)dt - t^2 \cdot d\left(\frac{t^2}{4}\right) + \left(\frac{t^2}{4} + t\right)dt$$

$$= \int_0^2 \left(t + \frac{t^2}{4} - \frac{t^3}{2} + \frac{t^2}{4} + t\right)dt = \int_0^2 \left(2t + \frac{t^2}{2} - \frac{t^3}{2}\right)dt = \left[t^2 + \frac{t^3}{6} - \frac{t^4}{8}\right]_0^2 = 4 + \frac{4}{3} - 2 = \frac{10}{3}.$$

EXAMPLE 1.61 Evaluate $\oint_C (3x^2 - 8y^2)\,dx + (4y - 6xy)\,dy$ on the closed curve C defined by $x^2 = y$ and $y^2 = x$.

Solution Since C is piecewise smooth, we express the integral as a sum of integrals. We write

$$\oint_C Pdx + Qdy = \int_{C_1} Pdx + Qdy + \int_{C_2} Pdx + Qdy,$$

where C_1, C_2 are the curves shown in Fig. 1.18.

On C_1, we use x as a parameter. Since $y = x^2$, $dy = 2xdx$, we have

$$\int_{C_1} (3x^2 - 8y^2)\,dx + (4y - 6xy)\,dy$$

$$= \int_0^1 (3x^2 - 8x^4)\,dx + (4x^2 - 6x^3)2x dx$$

$$= \int_0^1 (-20x^4 + 8x^3 + 3x^2)\,dx = [-4x^5 + 2x^4 + x^3]_0^1 = -1.$$

Fig. 1.18

On C_2, we use y as a parameter. From $x = y^2$, we get $dx = 2y\,dy$ and so

$$\int_{C_2} (3x^2 - 8y^2)\,dx + (4y - 6xy)\,dy = \int_1^0 (3y^4 - 8y^2)2y dy + (4y - 6y^3)\,dy$$

$$= \int_1^0 (6y^5 - 22y^3 + 4y)\,dy = \left[y^6 - \frac{11}{2}y^4 + 2y^2\right]_1^0 = \frac{5}{2}.$$

Hence $\oint_C (3x^2 - 8y^2)\,dx + (4y - 6xy)\,dy = -1 + \frac{5}{2} = \frac{3}{2}.$

1.19 APPLICATIONS OF LINE INTEGRALS

Work done by a force (*a*) If **F** is a variable force acting on a particle which moves along the curve C. Then, the work done by the force **F** in displacing the particle from the point P to the point Q along the curve C is given by

$$W = \int_P^Q \mathbf{F} \cdot d\mathbf{r} = \int_{C^*} \mathbf{F} \cdot d\mathbf{r}, \qquad \text{...(1.30)}$$

where C^* is the part of C, whose initial and terminal points are A and B.

(*b*) Suppose that **F** is a *conservative* vector field. Then, **F** can be written as $\mathbf{F} = \text{grad} f = \nabla f$, where f is a scalar potential (field). Then, the work done

$$W = \int_{C^*} \mathbf{F} \cdot d\mathbf{r} = \int_{C^*} (\nabla f) \cdot d\mathbf{r} = \int_{C^*} \left(\frac{\partial f}{\partial x}dx + \frac{\partial f}{\partial y}dy + \frac{\partial f}{\partial z}dz\right)$$

$$= \int_A^B df = [f(x, y, z)]_A^B. \qquad \text{...(1.31)}$$

Therefore, work done depends only on the initial and terminal points of the curve C^*, that is, the work done is independent of the path of integration. It is same for all paths.

Circulation

A line integral of a vector field **v** around a simple closed curve C is said to be the *circulation* of **v** around C

$$\text{circulation} = \oint_C \mathbf{v} \cdot d\mathbf{r} = \oint_C \mathbf{v} \cdot \frac{d\mathbf{r}}{ds}\,ds = \oint_C \mathbf{v} \cdot \mathbf{T}\,ds, \qquad \text{...(1.32)}$$

where **T** is the tangent vector to C. If **v** is the velocity field of a fluid and C be a closed curve in its domain, then the, circulation is a measure of the amount by which the fluid tends to turn the curve C by rotating or circulating around it. If $\oint_C \mathbf{v} \cdot \mathbf{T}\,ds > 0$, then the fluid tends to rotate C in the counter clockwise direction, while if $\oint_C \mathbf{v} \cdot \mathbf{T}\,ds < 0$, then the fluid tends to rotate C in the clockwise direction. If **v** acts in the direction perpendicular to **T** at every point on C, then $\oint_C \mathbf{v} \cdot \mathbf{T}\,ds = 0$, and the curve does not move at all.

EXAMPLE 1.62 Find the work done in moving a particle once round the circle $x^2 + y^2 = 9, z = 0$ if the vector field is given by $\mathbf{F} = (2x - y - z)\mathbf{i} + (x + y - z^2)\mathbf{j} + (3x - 2y + 4z)\mathbf{k}$.

Solution Since the particle moves in the plane $z = 0$ (*i.e.*, *xy*-plane),

$$\mathbf{F} = (2x - y)\mathbf{i} + (x + y)\mathbf{j} + (3x - 2y)\mathbf{k} \text{ and } d\mathbf{r} = dx\,\mathbf{i} + dy\,\mathbf{j}.$$

$$\text{Work done} = \oint_C \mathbf{F}\cdot d\mathbf{r} = \oint_C [(2x - y)\mathbf{i} + (x + y)\mathbf{j} + (3x - 2y)\mathbf{k}]\cdot(dx\,\mathbf{i} + dy\,\mathbf{j})$$

$$= \oint_C (2x - y)\,dx + (x + y)\,dy.$$

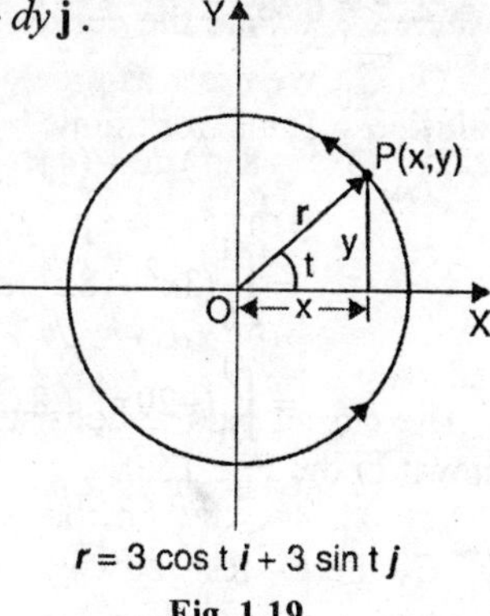

$r = 3\cos t\, i + 3\sin t\, j$

Fig. 1.19

Now C is the circle $x^2 + y^2 = 9$.

Choose the parametric equations of the circle as $x = 3\cos t$, $y = 3\sin t$, where t varies from 0 to 2π (See Fig. 1.19). Then

$$\oint_C \mathbf{F}\cdot d\mathbf{r} = \int_0^{2\pi} [2(3\cos t) - 3\sin t](-3\sin t)\,dt + [3\cos t + 3\sin t](3\cos t)\,dt$$

$$= \int_0^{2\pi} (9 - 9\sin t\cos t)\,dt = \left[9t - 9\frac{\sin^2 t}{2}\right]_0^{2\pi} = 18\pi.$$

Note: In traversing C we have chosen the counterclockwise direction indicted in the Fig. 1.19. We call this the *positive direction*.

EXAMPLE 1.63 Find the work done is moving the particle in the force field

$\mathbf{F}(x, y, z) = 3x^2\,\mathbf{i} + (2xz - y)\,\mathbf{j} + z\mathbf{k}$ along (*SVTU, 2007; JNTU, 2002*)

(*i*) the straight line from $A(0, 0, 0)$ to $B(2, 1, 3)$,

(*ii*) the space curve $x = 2t^2, y = t, z = 4t^2 - t$ from $t = 0$ to $t = 1$,

(*iii*) the curve defined by $x^2 = 4y, 3x^3 = 8z$ from $x = 0$ to $x = 2$.

(*AMIETE, June 2011, Dec 2009, June 2006; MDU 2002*)

Solution $\int_C \mathbf{F}\cdot d\mathbf{r} = \int_C [3x^2\mathbf{i} + (2xz - y)\mathbf{j} + z\mathbf{k}]\cdot(dx\,\mathbf{i} + dy\,\mathbf{j} + dz\,\mathbf{k}) = \int_C 3x^2dx + (2xz - y)\,dy + z\,dz.$

(*i*) The equations of line AB are $x/2 = y = z/3 = t$ (say).

Therefore, $x = 2t$, $y = t$ and $z = 3t$ are the parametric equations of the line AB.

The points (0, 0, 0) to (2, 1, 3) correspond to $t = 0$ and $t = 1$ respectively.

$$\text{Work done} = \int_C \mathbf{F}\cdot d\mathbf{r} = \int_0^1 3(2t)^2 d(2t) + [2(2t)(3t) - t]\,dt + 3t\,d(3t)$$

$$= \int_0^1 24t^2dt + (12t^2 - t)\,dt + 9t\,dt = \int_0^1 (36t^2 + 8t)\,dt = 16.$$

(*ii*) $$\text{Work done} = \int_C \mathbf{F}\cdot d\mathbf{r} = \int_0^1 3(2t^2)^2 d(2t^2) + [2(2t^2)(4t^2 - t) - t]\,dt + (4t^2 - t)\,d\,(4t^2 - t)$$

$$= \int_0^1 48t^5dt + (16t^4 - 4t^3 - t)\,dt + (32t^3 - 12t^2 + t)\,dt$$

$$= \int_0^1 (48t^5 + 16t^4 + 28t^3 - 12t^2)\,dt = \left[8t^6 + \frac{16}{5}t^5 + 7t^4 - 4t^3\right]_0^1 = 14.2.$$

(*iii*) Let $x = t$ in $x^2 = 4y$, $3x^3 = 8z$. Then the parametric equations of C are $x = t$, $y = t^2/4$, $z = 3t^3/8$.

Obviously t varies from 0 to 2. Hence the work done $= \int_C \mathbf{F}\cdot d\mathbf{r}$

$$= \int_0^2 3t^2dt + \left[2(t)\left(\frac{3t^3}{8}\right) - \frac{t^2}{4}\right]d\left(\frac{t^2}{4}\right) + \frac{3t^3}{8}\,d\left(\frac{3t^3}{8}\right)$$

$$= \int_0^2 \left(3t^2 - \frac{t^3}{8} + \frac{51}{64}t^5\right)dt = \left[t^3 - \frac{t^4}{32} + \frac{17t^6}{128}\right]_0^2 = 16.$$

EXAMPLE 1.64 Find the work done by the force $\mathbf{F} = x\,\mathbf{i} - z\,\mathbf{j} + 2y\,\mathbf{k}$ in the displacement along the closed path C consisting of the segments C_1, C_2 and C_3 where on C_1, $0 \le x \le 1$, $y = x$, $z = 0$, on C_2, $0 \le z \le 1$, $x = 1$, $y = 1$ on C_3, $1 \ge x \ge 0$, $y = z = x$.

Solution Total work done $= \oint_C \mathbf{F} \cdot d\mathbf{r}$

$$= \oint_C (x\mathbf{i} - z\mathbf{j} + 2y\mathbf{k}) \cdot (dx\,\mathbf{i} + dy\,\mathbf{j} + dz\,\mathbf{k})$$

$$= \oint_C x\,dx - z\,dy + 2y\,dz.$$

The closed path C consisting of segments C_1, C_2 and C_3 is shown in the Fig. 1.20.

Fig. 1.20

Let W_1, W_2, W_3 be the work done in displacement along C_1, C_2 and C_3 respectively.

Since on C_1, $0 \le x \le 1$, $y = x$, $z = 0$, $dy = dx$, $dz = 0$.

Therefore, $\quad W_1 = \int_{C_1} x\,dx = \int_0^1 x\,dx = 1/2.$

On C_2, $0 \le z \le 1$, $x = 1$, $y = 1$, $dx = 0$, $dy = 0$. Therefore, $W_2 = \int_{C_2} 2dz = 2\int_0^1 dz = 2.$

On C_3, $1 \ge x \ge 0$, $y = z = x$, $dy = dz = dx$. Therefore, $W_3 = \int_{C_3} x\,dx - x\,dx + 2x\,dx = 2\int_1^0 x\,dx = -1.$

Thus, total work done $= W_1 + W_2 + W_3 = (1/2) + 2 - 1 = 3/2.$

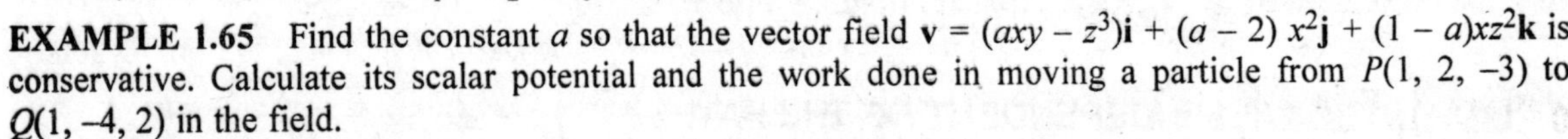

EXAMPLE 1.65 Find the constant a so that the vector field $\mathbf{v} = (axy - z^3)\mathbf{i} + (a-2)x^2\mathbf{j} + (1-a)xz^2\mathbf{k}$ is conservative. Calculate its scalar potential and the work done in moving a particle from $P(1, 2, -3)$ to $Q(1, -4, 2)$ in the field.

Solution Since $\mathbf{v}$ is a conservative field so curl $(\mathbf{v}) = \nabla \times \mathbf{v} = \mathbf{0}$.

$$\text{curl}(\mathbf{v}) = \begin{vmatrix} \mathbf{i} & \mathbf{j} & \mathbf{k} \\ \partial/\partial x & \partial/\partial y & \partial/\partial z \\ axy - z^3 & (a-2)x^2 & (1-a)xz^2 \end{vmatrix} = (0)\mathbf{i} - [z^2(1-a) + 3z^2]\mathbf{j} + [2x(a-2) - ax]\mathbf{k} = \mathbf{0},$$

or $z^2(4 - a) = 0$ and $x(a - 4) = 0$, whence we have $a = 4$.

Now $\quad \mathbf{v} = (4xy - z^3)\mathbf{i} + 2x^2\mathbf{j} - 3xz^2\mathbf{k}. \quad$ We have $\quad \mathbf{v} = \text{grad}\,\phi = \dfrac{\partial \phi}{\partial x}\mathbf{i} + \dfrac{\partial \phi}{\partial y}\mathbf{j} + \dfrac{\partial \phi}{\partial z}\mathbf{k}.$

Hence $\quad \dfrac{\partial \phi}{\partial x} = 4xy - z^3 \quad ...(i), \qquad \dfrac{\partial \phi}{\partial y} = 2x^2 \quad ...(ii), \qquad \dfrac{\partial \phi}{\partial z} = -3xz^2 \quad ...(iii).$

Integrating, we get from (*i*), (*ii*) and (*iii*) respectively,

$$\phi(x, y, z) = 2x^2y - xz^3 + f(y, z), \quad ...(iv) \qquad \phi(x, y, z) = 2x^2y + g(x, z), \quad ...(v)$$

and $\quad \phi(x, y, z) = -xz^3 + h(x, y). \quad ...(vi)$

The results (*iv*), (*v*) and (*vi*) agree if we choose

$$f(y, z) = c, \qquad g(x, z) = -xz^3 + c, \qquad h(x, y) = 2x^2y + c.$$

Therefore, the scalar potential is given by $\phi(x, y, z) = 2x^2y - xz^3 + c.$

Hence, the work done by $\mathbf{v}$ in moving a particle from $P(1, 2, -3)$ to $Q(1, -4, 2)$ is given by

$$W = \int_P^Q \mathbf{v} \cdot d\mathbf{r} = [\phi(x, y, z)]_P^Q = [2x^2y - xz^3 + c]_{(1,2,-3)}^{(1,-4,2)}$$

$$= (-8 - 8 + c) - (4 + 27 + c) = -16 - 31 = -47.$$

EXAMPLE 1.66 Show that the vector field $\mathbf{F} = 2x(y^2 + z^3)\,\mathbf{i} + 2x^2y\mathbf{j} + 3\,x^2z^2\mathbf{k}$ is conservative. Find its scalar potential and the work done by it in moving a particle from $P(-1, 2, 1)$ to $Q(2, 3, 4)$.

(*AMIETE., June 2010, 2009, Dec 2008, 2005*)

Solution A necessary and sufficient condition that a vector field **F** will be conservative is that curl $\mathbf{F} = \nabla \times \mathbf{F} = \mathbf{0}$.

$$\text{Now, curl}(\mathbf{F}) = \nabla \times \mathbf{F} = \begin{vmatrix} \mathbf{i} & \mathbf{j} & \mathbf{k} \\ \partial/\partial x & \partial/\partial y & \partial/\partial z \\ 2x(y^2+z^3) & 2x^2y & 3x^2z^2 \end{vmatrix} = \mathbf{i}(0) - \mathbf{j}(6xz^2 - 6xz^2) + \mathbf{k}(4xy - 4xy) = \mathbf{0}.$$

Thus, the vector field **F** is conservative. We have $\mathbf{F} = \text{grad}\,\phi = \nabla\phi$. Hence

$$\frac{\partial \phi}{\partial x} = 2x(y^2 + z^3) \quad ...(i), \qquad \frac{\partial \phi}{\partial y} = 2x^2y \quad ...(ii), \qquad \frac{\partial \phi}{\partial z} = 3x^2z^2 \quad ...(iii).$$

Integrating, we obtain from (*i*), (*ii*) and (*iii*) respectively,

$$\phi(x, y, z) = x^2(y^2 + z^3) + f(y, z), \quad ...(iv) \qquad f(x, y, z) = x^2y^2 + g(x, z), \quad ...(v)$$

$$\phi(x, y, z) = x^2z^3 + h(x, y). \quad ...(vi)$$

The results (*iv*), (*v*) and (*vi*) agree if we choose

$$\phi(y, z) = c, \quad g(x, z) = x^2z^3 + c, \quad h(x, y) = x^2y^2 + c.$$

Therefore, the scalar potential is given by $\phi(x, y, z) = x^2(y^2 + z^3) + c$.

Hence, the work done by **F** in moving a particle from $P(-1, 2, 1)$ to $Q(2, 3, 4)$ is

$$W = \int_P^Q \mathbf{F} \cdot d\mathbf{r} = [\phi(x, y, z)]_P^Q = [x^2y^2 + x^2z^3]_{(-1,2,1)}^{(2,3,4)} = (36 + 256 - 4 - 1) = 287.$$

1.20 LINE INTEGRALS INDEPENDENT OF THE PATH

We have seen that the value of $\int_C \mathbf{F} \cdot d\mathbf{r}$ or $\int_C f\,dx + g\,dy + h\,dz$ depends not only on the end point A and B of the curve C but also on the path of C. We shall now derive the conditions under which the line integral is independent of the path of integration, that is, it depends only on the end points A and B of the curve C.

Let $\phi(x, y, z)$ be a differential scalar function. The differential of $f(x, y, z)$ is defined by

$$d\phi = \frac{\partial \phi}{\partial x} dx + \frac{\partial \phi}{\partial y} dy + \frac{\partial \phi}{\partial z} dz = (\nabla\phi) \cdot d\mathbf{r}.$$

Therefore, a differential expression $f(x, y, z)dx + g(x, y, z)dy + h(x, y, z)dz$ is an exact differential, if there exists a scalar function $\phi(x, y, z)$ such that

$$d\phi = f(x, y, z)\,dx + g\,(x, y, z)\,dy + h\,(x, y, z)\,dz.$$

Let $f\,dx + g\,dy + h\,dz$ is an exact differential in D. Then, there exists a scalar function ϕ such that

$$f = \frac{\partial \phi}{\partial x}, \quad g = \frac{\partial \phi}{\partial y}, \quad \text{and} \quad h = \frac{\partial \phi}{\partial z}.$$

Let the parametric form of C be $\mathbf{r}(t) = x(t)\mathbf{i} + y(t)\mathbf{j} + z(t)\mathbf{k}$, $a \le t \le b$. The initial and terminal points A and B of C corresponds to $t = a$ and $t = b$ respectively. Then

$$\int_C f dx + g dy + h dz = \int_A^B \left(\frac{\partial \phi}{\partial x} dx + \frac{\partial \phi}{\partial y} dy + \frac{\partial \phi}{\partial z} dz \right) = \int_a^b \left(\frac{\partial \phi}{\partial x}\frac{dx}{dt} + \frac{\partial \phi}{\partial y}\frac{dy}{dt} + \frac{\partial \phi}{\partial z}\frac{dz}{dt} \right) dt$$

$$= \int_a^b \frac{d\phi}{dt} dt = [\phi(x(t), y(t), z(t))]_a^b = \phi(B) - \phi(A).$$

This shows that the value of the integral is simply the difference of the values of ϕ at the two end points of curve C and is, therefore, independent of the path C.

Test for Path Independence in Space

Let C be a curve in a simply connected domain D in space. Let f, g and h be continuous and have continuous first order partial derivatives in a domain D in space. Then $\int_C f\,dx + g\,dy + h\,dz$ is independent of path C if and only if $\dfrac{\partial f}{\partial y} = \dfrac{\partial g}{\partial x}$, $\dfrac{\partial f}{\partial z} = \dfrac{\partial h}{\partial x}$, and $\dfrac{\partial g}{\partial z} = \dfrac{\partial h}{\partial y}$.

Remark 11

(a) If the line integral path is independent, then $\mathbf{F} = \text{grad}\,(\phi)$.
Hence curl $(\mathbf{F}) = \text{curl}(\text{grad}\,\phi) = \mathbf{0}$. ...(1.33)
The given vector field $\mathbf{F}$ is called a *gradient field* and the function ϕ is called the potential function for $\mathbf{F}$. Therefore, in a gradient force field, the work done by the force $\mathbf{F}$ in moving a particle from a position A to a position B is independent of the path of integration. Such a force field is also called a *conservative field*, that is, total energy = Kinetic energy + potential energy = constant.

(b) If $\mathbf{F}$ is a conservative force field, the work done along any simple closed path is zero. Refer Fig. 1.21.

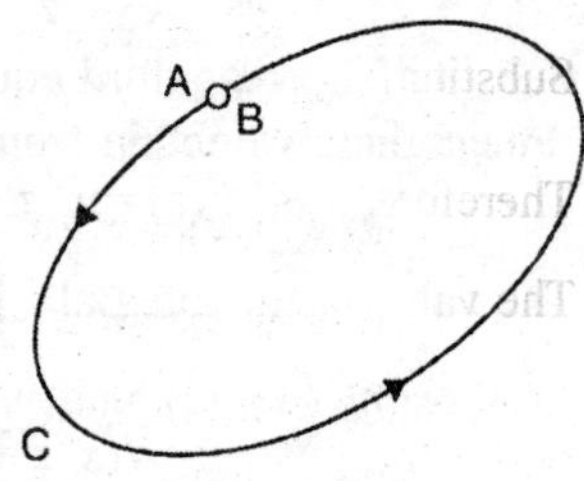

Fig. 1.21

EXAMPLE 1.67 Show that the integral $\int_{(1,2)}^{(3,4)} (xy^2 + y^3)\,dx + (x^2y + 3xy^2)\,dy$ is independent of the path joining the points (1, 2) and (3, 4). Hence evaluate the integral. *(AMIETE, June 2007)*

Solution We have $f(x, y) = xy^2 + y^3$ and $g(x, y) = x^2y + 3xy^2$. Now, $\dfrac{\partial f}{\partial y} = 2xy + 3y^2$ and $\dfrac{\partial g}{\partial x} = 2xy + 3y^2$.

Since $\partial f/\partial y = \partial g/\partial x$, the integral is independent of the path. Also the integrand is an exact differential. Therefore, there exists a function $\phi\,(x, y)$ such that

$$\partial\phi/\partial x = f(x, y) = xy^2 + y^3 \text{ and } \partial\phi/\partial y = g(x, y) = x^2y + 3xy^2.$$

Integrating, the first equation with respect to x, we get $\phi(x, y) = (x^2y^2/2) + xy^3 + g(y)$.

Taking the partial derivative of this last equation with respect to y and equating the result equal to $g\,(x, y)$, we have, $\dfrac{\partial\phi}{\partial y} = x^2y + 3xy^2 + g'(y) = x^2y + 3xy^2$, which implies $g'(y) = 0$ and so $g(y) = c$, constant.

$$\text{Hence, } \phi(x, y) = \frac{x^2y^2}{2} + xy^3 + c.$$

$$\text{Therefore, } \int_C (xy^2 + y^3)\,dx + (x^2y + 3xy^2)\,dy = \int_{(1,2)}^{(3,4)} d\left(\frac{x^2y^2}{2} + xy^3\right)$$

$$= \left[\frac{x^2y^2}{2} + xy^3\right]_{(1,2)}^{(3,4)} = (72 + 192) - (2 + 8) = 264 - 10 = 254.$$

EXAMPLE 1.68 Show that the following line integral is independent of path C from points $P(-1, 2, 3)$ to $Q(2, 2, 4)$ and hence evaluate the integral

$$\int_C (2xz + y)\,dx + (x + z)\,dy + (x^2 + y)\,dz.$$ *(AMIETE, June 2005)*

Solution We have $f(x, y, z) = 2xz + y$, $g(x, y, z) = x + z$, and $h(x, y, z) = x^2 + y$.

We find $\dfrac{\partial f}{\partial y} = 1 = \dfrac{\partial g}{\partial x}$, $\dfrac{\partial f}{\partial z} = 2x = \dfrac{\partial h}{\partial x}$, and $\dfrac{\partial g}{\partial z} = 1 = \dfrac{\partial h}{\partial y}$.

The integral is independent of path of integration. Also, the integrand is an exact differential. Therefore, there exists a function $\phi\ (x, y, z)$ such that

$$\frac{\partial \phi}{\partial x} = 2xz + y, \quad \frac{\partial \phi}{\partial y} = x + z, \quad \text{and} \quad \frac{\partial \phi}{\partial z} = x^2 + y.$$

Integrating the first equation with respect to x, we obtain $\phi(x, y, z) = x^2 z + xy + h(y, z)$.

Substituting in the second equation, we get $\dfrac{\partial \phi}{\partial y} = x + z = x + \dfrac{\partial h}{\partial y}(y, z)$, or $\dfrac{\partial h}{\partial y} = z$.

Integrating, we get $h(y, z) = yz + s(z)$ and $\phi(x, y, z) = x^2 z + xy + yz + s(z)$.

Substituting in the third equation, we get $\dfrac{\partial \phi}{\partial z} = x^2 + y = x^2 + y + \dfrac{ds}{dz}$, or $\dfrac{ds}{dz} = 0$ or $s = c$, constant.

Therefore, $\phi(x, y, z) = x^2 z + xy + yz + c$.

The value of the integral $\displaystyle\int_C (2xz + y)\,dx + (x + z)\,dy + (x^2 + y)\,dz$

$$= \int_{(-1,2,3)}^{(2,2,4)} d(x^2 z + xy + yz + c) = [x^2 z + xy + yz + c]_{(-1,2,3)}^{(2,2,4)} = 16 + 4 + 8 - 3 + 2 - 6 = 30 - 9 = 21.$$

EXAMPLE 1.69 Show that $\displaystyle\int_C (yz - 1)\,dx + (z + xz + z^2)\,dy + (y + xy + 2yz)\,dz$ is independent of the path of integration from (1, 2, 2) to (2, 3, 4) and hence evaluate the integral. *(AMIETE, June 2009; PTU 2006)*

Solution We have $f(x, y, z) = yz - 1$, $g(x, y, z) = z + xz + z^2$, and $h(x, y, z) = y + xy + 2yz$.

We find $\dfrac{\partial f}{\partial y} = z = \dfrac{\partial g}{\partial x}$, $\dfrac{\partial f}{\partial z} = y = \dfrac{\partial h}{\partial x}$, and $\dfrac{\partial g}{\partial z} = 1 + x + 2z = \dfrac{\partial h}{\partial y}$.

The integral is independent of the path of integration. Also, the integrand is an exact differential and so there exists a function $\phi(x, y, z)$ such that

$$\frac{\partial \phi}{\partial x} = yz - 1 \quad ...(i), \qquad \frac{\partial \phi}{\partial y} = z + xz + z^2 \quad ...(ii), \quad \text{and} \quad \frac{\partial \phi}{\partial z} = y + xy + 2yz \quad ...(iii).$$

Integrating the first equation with respect to x, we obtain $\phi(x, y, z) = xyz - x + h(y, z)$.

Substituting in the second equation, we get $\dfrac{\partial \phi}{\partial y} = z + xz + z^2 = xz + \dfrac{\partial h}{\partial y}(y\ z)$, or $\dfrac{\partial h}{\partial y} = z + z^2$.

Integrating, we get $h(y, z) = yz + yz^2 + s(z)$ and $\phi(x, y, z) = xyz - x + yz + yz^2 + s(z)$.
Substituting in the third equation, we obtain

$$\frac{\partial \phi}{\partial z} = y + xy + 2yz = xy + y + 2yz + \frac{ds}{dz} \quad \text{or} \quad \frac{ds}{dz} = 0 \quad \text{or} \quad s = c, \text{constant}.$$

Therefore, $\phi(x, y, z) = xyz - x + yz + yz^2 + c$.

The value of the integral $= \displaystyle\int_C (yz - 1)\,dx + (z + xz + z^2)\,dy + (y + xy + 2yz)\,dz$

$$= \int_{(1,2,2)}^{(2,3,4)} d(xyz - x + yz + yz^2 + c) = [xyz - x + yz + yz^2 + c]_{(1,2,2)}^{(2,3,4)}$$

$$= (24 - 2 + 12 + 48 + c) - (4 - 1 + 4 + 8 + c) = 82 - 15 = 67.$$

EXAMPLE 1.70 Show that the vector field $\mathbf{F} = (yze^{xyz} - 4x)\mathbf{i} + (xze^{xyz} + z)\mathbf{j} + (xye^{xyz} + y)\mathbf{k}$ is conservative. Hence evaluate the line integral

$$\int (yze^{xyz} - 4x)\,dx + (xze^{xyz} + z)\,dy + (xye^{xyz} + y)\,dz$$

along a path joining the points (0, 0, 0) to (1, 1, 1). Give a physical interpretation of the results.

(AMIETE, Dec 2004)

Solution A necessary and sufficient condition that **F** be conservative is that curl (**F**) = $\nabla \times \mathbf{F} = \mathbf{0}$.

We have curl (F) $= \nabla \times F = \begin{vmatrix} \mathbf{i} & \mathbf{j} & \mathbf{k} \\ \partial/\partial x & \partial/\partial y & \partial/\partial z \\ yze^{xyz} - 4x & xze^{xyz} + z & xye^{xyz} + y \end{vmatrix}$

$$= \mathbf{i}[(xe^{xyz} + x^2zye^{xyz} + 1) - xe^{xyz} - x^2zye^{xyz} - 1] - \mathbf{j}[ye^{xyz} + xy^2ze^{xyz} - ye^{xyz} - xy^2ze^{xyz}]$$
$$+ \mathbf{k}[ze^{xyz} + x^2zye^{xyz} - ze^{xyz} - xyz^2e^{xyz}] = \mathbf{0}.$$

Therefore, the vector field **F** is conservative.

$\mathbf{F}\cdot d\mathbf{r} = (yze^{xyz} - 4x)\,dx + (xze^{xyz} + z)\,dy + (xye^{xyz} + y)\,dz$ is an exact differential $d\phi$, where ϕ is such that

(*i*) $\dfrac{d\phi}{dx} = yze^{xyz} - 4x$, (*ii*) $\dfrac{d\phi}{dy} = xze^{xyz} + z$, (*iii*) $\dfrac{d\phi}{dz} = xye^{xyz} + y$.

From these we obtain, respectively,

$$\phi = e^{xyz} - 2x^2 + f(y, z), \quad \phi = e^{xyz} + yz + g(x, z), \quad \phi = e^{xyz} + yz + h(x, y).$$

These are consistent if $f(y, z) = yz + c$, $g(x, z) = -2x^2 + c$, $h(x, y) = -2x^2 + c$.

Therefore, the scalar potential is given by $\phi(x, y, z) = e^{xyz} - 2x^2 + yz + c$

Thus, $$\int_{(0,0,0)}^{(1,1,1)} \mathbf{F}\cdot d\mathbf{r} = [e^{xyz} - 2x^2 + yz + c]_{(0,0,0)}^{(1,1,1)} = e - 2.$$

Physically $\int_C \mathbf{F}\cdot d\mathbf{r}$ represents the work done in moving an object from (0, 0, 0) to (1, 1, 1) along C. In a conservative force field the work done is independent of the path C joining these points.

EXERCISE 1.4

1. Calculate the work done when a force $\mathbf{F} = 3xy\mathbf{i} - y^2\mathbf{j}$ moves a particle in the xy-plane from (0, 0) to (1, 2) along the parabola $y = 2x^2$. **Ans.** $-7/6$.

   Alternative: If $\mathbf{F} = 3xy\mathbf{i} - y^2\mathbf{j}$ find $\int_C \mathbf{F}\cdot d\mathbf{r}$ where C is the curve in xy plane, $y = 2x^2$ from (0, 0) to (1, 2). *(AMIETE, Dec 2010)* **Ans.** $-7/6$.

2. Evaluate the line integral of $\mathbf{v} = x^2\mathbf{i} - 2y\mathbf{j} + z^2\mathbf{k}$ over the straight line path from (–1, 2, 3) to (2, 3, 5).

   [**Hint:** $\mathbf{r} = (-1 + 3t)\mathbf{i} + (2 + t)\mathbf{j} + (3 + 2t)\mathbf{k}$, $0 \le t \le 1$. $d\mathbf{r}/dt = 3\mathbf{i} + \mathbf{j} + 2\mathbf{k}$. Therefore

   $$\int_C \mathbf{v}\cdot\frac{d\mathbf{r}}{dt}\,dt = \int_0^1 [3(-1+3t)^2 - 2(2+t) + 2(3+2t)^2]\,dt = \int_0^1 (17 + 4t + 35t^2)\,dt = \frac{92}{3}.]$$ *(AMIE, S-2012)*

3. A vector field is given by $\mathbf{F} = (\sin y)\mathbf{i} + x(1 + \cos y)\mathbf{j}$. Evaluate the line integral $\int_C \mathbf{F}\cdot d\mathbf{r}$, where C is a circular path given by $x^2 + y^2 = a^2$, $z = 0$. *(AMIETE Dec. 2007; MDU 2006; PTU 2003)* **Ans.** πa^2.

4. Show that $\int_C \mathbf{F} \cdot d\mathbf{r} = 3\pi$, given that $\mathbf{F} = z\mathbf{i} + x\mathbf{j} + y\mathbf{k}$ and C being the arc of the curve

$\mathbf{r} = \cos t\,\mathbf{i} + \sin t\,\mathbf{j} + t\,\mathbf{k}$ from $t = 0$ to $t = 2\pi$.

5. Evaluate the line integral $\oint_C (x^2 + xy)\,dx + (x^2 + y^2)\,dy$, where C is the square formed by the lines $y = \pm 1$ and $x = \pm 1$. (*Kerala 2001*)

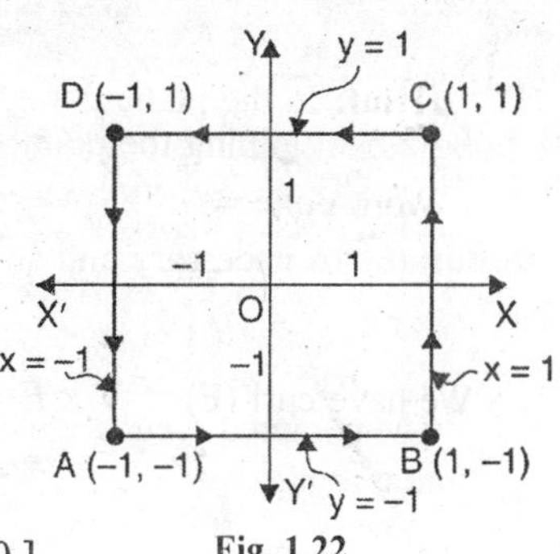

Fig. 1.22

[Hint: $\oint_C (x^2 + xy)\,dx + (x^2 + y^2)\,dy$

$$= \int_{AB} (x^2 + xy)\,dx + (x^2 + y^2)\,dy + \int_{BC} (x^2 + xy)\,dx + (x^2 + y^2)\,dy$$

$$+ \int_{CD} (x^2 + xy)\,dx + (x^2 + y^2)\,dy + \int_{DA} (x^2 + xy)\,dx + (x^2 + y^2)\,dy$$

$$= \int_{-1}^{1} (x^2 - x)\,dx + \int_{-1}^{1} (1 + y^2)\,dy + \int_{1}^{-1} (x^2 + x)\,dx + \int_{1}^{-1} (1 + y^2)\,dy = 0.]$$

6. Evaluate $\oint_C (y^2 dx - x^2 dy)$, on the given closed curve C: triangle with vertices P(1, 0), Q(0, 1) and R(–1, 0) traversed counter clockwise. **Ans.** –2/3.

[Hint: Refer Fig. 1.23. The equation of curve C_1: Line PQ is $x + y = 1$ $\Rightarrow dx + dy = 0$.

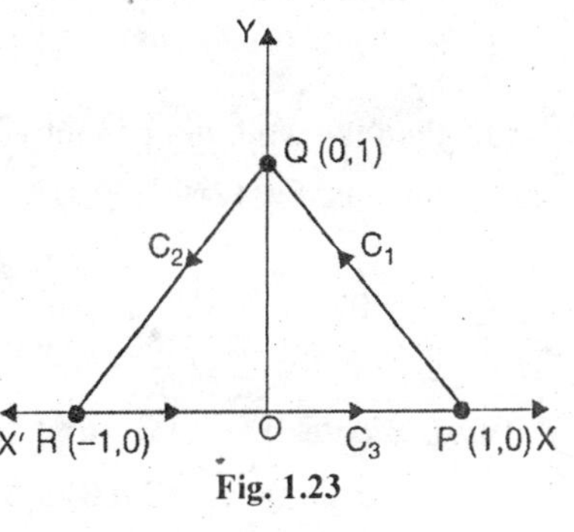

Fig. 1.23

$$\therefore \int_{C_1} (y^2 dx - x^2 dy) = \int_{C_1} (1 - x)^2 dx - x^2(-dx)$$

$$= \int_{C_1} (2x^2 - 2x + 1)\,dx = \int_{1}^{0} (2x^2 - 2x + 1)\,dx = \frac{-2}{3}.$$

The equation of curve C_2 : line QR is $-x + y = 1 \Rightarrow dx = dy$.

$$\int_{C_2} (y^2 dx - x^2 dy) = \int_0^{-1} (1 + x)^2 dx - x^2 dx = \int_0^{-1} (2x + 1)\,dx = 0.$$

The equation of curve C_3: line RP is $y = 0 \Rightarrow dy = 0$.

$$\int_{C_3} (y^2 dx - x^2 dy) = 0.$$

Hence $\oint_C (y^2 dx - x^2 dy) = \int_{C_1} (y^2 dx - x^2 dy) + \int_{C_2} (y^2 dx - x^2 dy)$

$$+ \int_{C_3} (y^2 dx - x^2 dy) = (-2/3) + 0 + 0 = -2/3.]$$

7. Evaluate $\oint_C \mathbf{F} \cdot d\mathbf{r}$ where $\mathbf{F} = xy\mathbf{i} + (x^2 + y^2)\mathbf{j}$ and C is the rectangle in the xy-plane bounded by the lines $y = 2$, $x = 4$, $y = 10$ and $x = 1$. **Ans.** 60.

8. Evaluate the line integral of $\mathbf{F} = (2x^2 y + y + z^2)\mathbf{i} + (2 + 2yz)\mathbf{j} + (2z + 3y^2 z^2)\mathbf{k}$ along the curve $C : a^2 - y^2 - z^2 = 0,\ x = 0$. **Ans.** zero.

9. Find the total work done in moving a particle in a force field given by $\mathbf{F} = 3xy\mathbf{i} - 5z\,\mathbf{j} + 10x\mathbf{k}$ along the curve $x = t^2 + 1$, $y = 2t^2$, $z = t^3$ from $t = 1$ to $t = 2$.

[Hint: Total work done $= \int_C \mathbf{F} \cdot d\mathbf{r} = \int_C 3xy\,dx - 5z\,dy + 10x\,dz$

$$= \int_{t=1}^{2} 3(t^2 + 1)(2t^2)\,d(t^2 + 1) - 5(t^3)\,d(2t^2) + 10(t^2 + 1)\,d(t^3)$$

$$= \int_1^2 (12t^5 + 10t^4 + 12t^3 + 30t^2)\,dt = 303.]$$

10. Find the work done in moving a particle once round an ellipse $x^2/16 + y^2/9 = 1$ in the xy plane (traced in the anticlockwise direction) if the force field is given by

$$\mathbf{F} = (3x - 4y + 2z)\mathbf{i} + (4x + 2y - 3z^2)\mathbf{j} + (2xz - 4y^2 + z^3)\mathbf{k}.$$ **Ans:** 96π.

[**Hint:** In the plane $z = 0$, $\mathbf{F} = (3x - 4y)\mathbf{i} + (4x + 2y)\mathbf{j} - 4y^2\,\mathbf{k}$ and $d\mathbf{r} = dx\,\mathbf{i} + dy\,\mathbf{j}$.

Work done $= \oint_C \mathbf{F} \cdot d\mathbf{r} = \int_C \{(3x - 4y)\mathbf{i} + (4x + 2y)\mathbf{j} - 4y^2\mathbf{k}\} \cdot (dx\mathbf{i} + dy\mathbf{j})$

$$= \int_C (3x - 4y)\,dx + (4x + 2y)\,dy.$$

The parametric equations of the ellipse are $x = 4\cos t$, $y = 3\sin t$, where t varies from 0 to 2π. Then the line integral

$$= \int_{t=0}^{2\pi} (12\cos t - 12\sin t)(-4\sin t)\,dt + (16\cos t + 6\sin t)(3\cos t)\,dt$$

$$= \int_{t=0}^{2\pi} (48 - 30\sin t\cos t)\,dt = 96\pi.\ ]$$

11. Find the work done when a force $\mathbf{F} = (x^2 - y^2 + x)\mathbf{i} - (2xy + y)\mathbf{j}$ moves a particle in the xy-plane from (0, 0) to (1, 1) along the parabola $y^2 = x$. Is the work done different when the path is the straight line $y = x$? **Ans.** –2/3, work done is same.

12. Find the work done in moving a particle once round the circle $x^2 + y^2 = 9$ in the xy plane if the field of force is $\mathbf{F} = (2x - y - z)\mathbf{i} + (x + y - z^2)\mathbf{j} + (3x - 2y + 4z)\mathbf{k}$. Also find the scalar potential of **F**. **Ans.** 18π.

13. Compute the work done by the force $\mathbf{F} = (2y + 3)\mathbf{i} + xz\,\mathbf{j} + (yz - x)\mathbf{k}$. when it moves a particle from the point (0, 0, 0) to the point (2, 1, 1) along the curve $x = 2t^2$, $y = t$, $z = t^3$. (*Madras 2000*) **Ans.** 288/35.

14. Evaluate $\int_C (y + 3z)\,dx + (2z + x)\,dy + (3x + 2y)\,dz$, where C is the arc of the helix $x = a\cos\theta$, $y = a\sin\theta$, $z = (2a\theta)/\pi$ between the points $(a, 0, 0)$ and $(0, a, a)$. **Ans.** $2a^2$.

15. Find the circulation of **F** around the curve C, where $\mathbf{F} = e^x[(\sin y)\mathbf{i} + (\cos y)\mathbf{j}]$ and C is the rectangle whose vertices are (0, 0), (1, 0), (1, $\pi/2$) and (0, $\pi/2$). **Ans.** zero.

16. Find the circulation of **F** round the curve C, where $\mathbf{F} = (2x + y^2)\mathbf{i} + (3y - 4x)\mathbf{j}$ and C is the curve $y = x^2$ from (0, 0) to (1, 1) and the curve $y^2 = x$ from (1, 1) to (0, 0). **Ans.** –49/30.

17. Show that vector field defined by the vector function $\mathbf{v} = xyz(yz\,\mathbf{i} + xz\,\mathbf{j} + xy\,\mathbf{k})$ is conservative. (*AMIE, W-2010; AMIETE June 2008*)

18. Show that $\mathbf{F} = (2xy + z^3)\mathbf{i} + x^2\mathbf{j} + 3xz^2\mathbf{k}$ is a conservative force field. Find its scalar potential. Find also the work done in moving an object in this field from (1, –2, 1) to (3, 1, 4). (*AMIETE, June 2006*) **Ans.** $\phi = x^2y + xz^3 + c$; 202.

19. Prove that $\mathbf{F} = (2xy + z^3)\mathbf{i} + x^2\mathbf{j} + 3xz^2\mathbf{k}$ is a conservative force field. Also, find the scalar potential. (*AMIE, W-2007*) **Ans.** $x^2y + xz^3 + c$.

20. Show that $\mathbf{F} = (e^x\cos y + yz)\mathbf{i} + (xz - e^x\sin y)\mathbf{j} + (xy + z)\mathbf{k}$ is conservative and find a potential function for it. **Ans.** $e^x\cos y + xyz + (z^2/2) + c$.

21. Show that the integral $\int_{(1,2)}^{(3,4)} (6xy^2 - y^3)\,dx + (6x^2y - 3xy^2)\,dy$ is independent of the path joining (1, 2) and (3, 4). Hence evaluate the integral. **Ans.** 236.

22. Show that the line integral $\int_C (3x^2 dx + 2yz\,dy + y^2 dz)$ is independent of the path in any domain in space and find its value, if C has the initial point $A(0, 1, 2)$ and terminal point $B(1, -1, 7)$.

(*AMIE S-2009*) **Ans.** 6.

[Hint: $\int_C d(x^3 + y^2 z) = x^3 + y^2 z$. The value of the integral is $[x^3 + y^2 z]_{(0,1,2)}^{(1,-1,7)} = (1+7) - (2) = 6.$]

23. Show that the integral $\int_C (x^2 - 2y^3)dx + (x + 5y)dy$ is not independent of the path C.

[Hint: From, $P(x, y) = x^2 - 3y^3$ and $Q(x, y) = 3x + 6y$, we find $\partial P/\partial y = -9y^2$ and $\partial Q/\partial x = 3$. Since $\partial P/\partial y \neq \partial Q/\partial x$, therefore the integral is not independent of the path.]

24. Determine whether the line integral $\int 2xyz^2 dx + (x^2 z^2 + z\cos yz)dy + (2x^2 yz + y\cos yz)dz$ is independent of the path of integration? If so, then evaluate it from $(1, 0, 1)$ to $(0, \pi/2, 1)$.

Ans. Independent of path; 1.

$$\left[\textbf{Hint: } \int_C (2xyz^2 dx + x^2 z^2 dy + 2x^2 yz\,dz) + (z\cos yz\,dy + y\cos yz\,dz)\right.$$

$$= \int_C d(x^2 yz^2) + \{(\cos yz)\cdot d(yz)\} = \int_C d(x^2 yz^2 + \sin yz) = x^2 yz^2 + \sin yz.$$

$$\left.\text{The value of the integral is } [x^2 yz^2 + \sin yz]_{(1,0,1)}^{(0,\pi/2,1)} = 1\right].$$

25. Show that $ydx + xdy + 4dz$ is exact, and evaluate the integral $\int_{(1,1,1)}^{(2,3,-1)} ydx + xdy + 4dz$ over the line segment from $(1, 1, 1)$ to $(2, 3, -1)$.

Ans. –3.

26. Calculate the integral $\oint_C [(y^2 + 2xz^2 - 1)dx + 2xydy + 2x^2 zdz]$, if C is the ellipse $x^2/a^2 + y^2/b^2 = 1;\ z = 0$.

Ans. zero.

27. If $\mathbf{F} = 2y\mathbf{i} - z\mathbf{j} + x\mathbf{k}$, evaluate $\int_C \mathbf{F} \times d\mathbf{r}$ along the curve $x = \cos t$, $y = \sin t$, $z = 2\cos t$, from $t = 0$ to $t = \pi/2$.

$$\left[\textbf{Hint: } \mathbf{F} \times d\mathbf{r} = \begin{vmatrix} \mathbf{i} & \mathbf{j} & \mathbf{k} \\ 2y & -z & x \\ dx & dy & dz \end{vmatrix} = -(z\,dz + x\,dy)\mathbf{i} - (2y\,dz - x\,dx)\mathbf{j} + (2y\,dy + z\,dx)\mathbf{k}\right.$$

$$= -(-4\cos t\sin t + \cos^2 t)dt\,\mathbf{i} - (\cos t\sin t - 4\sin^2 t)dt\,\mathbf{j} + (2\sin t\cos t - 2\cos t\sin t)dt\,\mathbf{k}.$$

$$\left.\therefore \int_C \mathbf{F} \times d\mathbf{r} = \int_0^{\pi/2} \{(4\cos t\sin t - \cos^2 t)\mathbf{i} + (4\sin^2 t - \cos t\sin t)\mathbf{j}\}dt = \left(2 - \frac{\pi}{4}\right)\mathbf{i} + \left(\pi - \frac{1}{2}\right)\mathbf{j}\right].$$

1.21 GREEN'S THEOREM IN THE PLANE (TRANSFORMATION BETWEEN DOUBLE INTEGRALS AND LINE INTEGRALS)

Let C be piecewise smooth simple closed curves bounding a region R. If $f(x, y)$ and $g(x, y)$ be functions which are continuous and have continuous partial derivatives $\partial f/\partial y$ and $\partial g/\partial x$ everywhere in some domain containing R. Then

$$\iint_R \left(\frac{\partial g}{\partial x} - \frac{\partial f}{\partial y}\right) dx\,dy = \oint_C (f(x, y)dx + g(x, y))dy \qquad \text{...(1.34)}$$

the integration being carried in the positive direction (counter clockwise direction) of C.

Remark 12

(a) A curve "traversed in counterclockwise direction" is also called traversed in positive direction. Unless otherwise stated, integral along a closed curve C will imply integral along C described in positive direction.

(b) $\oint_C (f(x, y)dx + g(x, y))dy$ is equal to zero around every simple closed curve in a simply connected region if and only if $(\partial f/\partial y) = (\partial g/\partial x)$.

EXAMPLE 1.71 Verify the Green's theorem in the plane for $\oint_C (xy + y^2)dx + x^2 dy$, where C is the closed curve of the region bounded by $y = x$ and $y = x^2$.

(*AMIETE, Dec 2010; SVTU, 2006S; PTU 2005, VTU 2005, MDU 2003*)

Solution The curves $y = x$ and $y = x^2$ intersect at (0, 0) and (1, 1). The positive direction in traversing C is shown in the Fig. 1.24. Now $\oint_C = \int_{C_1} + \int_{C_2}$.

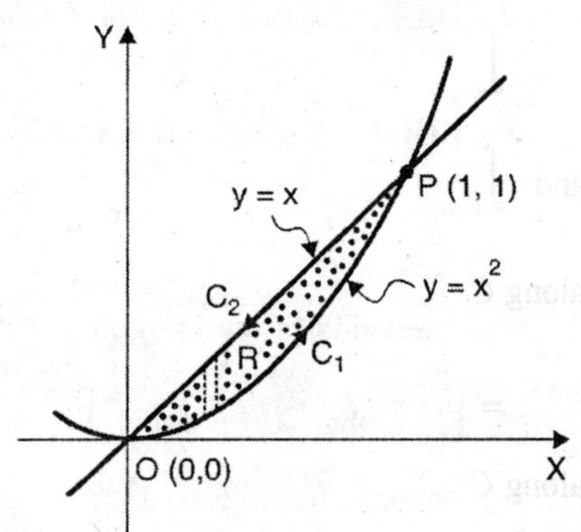

Fig. 1.24

Along C_1: $y = x^2$ from (0, 0) to (1, 1), the line integral

$$\int_{C_1} (xy + y^2)dx + x^2 dy$$

$$= \int_0^1 [x(x^2) + x^4]dx + x^2(2x)dx = \int_0^1 (3x^3 + x^4)dx = \frac{19}{20}.$$

Along C_2: $y = x$ from (1, 1) to (0, 0) line integral

$$\int_{C_2} (xy + y^2)dx + x^2 dy = \int_1^0 [x(x) + x^2]dx + x^2 dx = \int_1^0 3x^2 dx = -1.$$

Therefore, $\oint_C (xy + y^2)dx + x^2 dy = (19/20) - 1 = -(1/20)$. ...(*i*)

Also $$\iint_R \left(\frac{\partial g}{\partial x} - \frac{\partial f}{\partial y}\right) dydx = \iint_R \left[\frac{\partial}{\partial x}(x^2) - \frac{\partial}{\partial y}(xy + y^2)\right] dydx = \iint_R (x - 2y)\, dydx$$

$$= \int_{x=0}^1 \int_{y=x^2}^x (x - 2y)\, dydx = \int_0^1 [xy - y^2]_{x^2}^x \, dx = \int_0^1 (x^4 - x^3)dx = \left[\frac{x^5}{5} - \frac{x^4}{4}\right]_0^1 = -\frac{1}{20}.$$

This agrees with the result obtained in (*i*). Hence, the Green's theorem is verified.

EXAMPLE 1.72 Use Green's theorem to evaluate $\oint_C (y - \sin x)dx + \cos x\, dy$, where C is a plane triangle enclosed by the lines $y = 0$, $x = \pi/2$ and $y = 2x/\pi$.

(*AMIETE, June 2013, Dec. 2009; MDU 2007; PTU 2006; AMIE-2005; JNTU 2005*)

Solution We have $f(x, y) = y - \sin x$, $g(x, y) = \cos x$. Using the Green's theorem, we obtain

$$\oint_C (y - \sin x)dx + \cos x\, dy = \iint_R \left[\frac{\partial}{\partial x}(\cos x) - \frac{\partial}{\partial y}(y - \sin x)\right] dydx$$

$$= \iint_R (-\sin x - 1)\, dydx = -\int_{x=0}^{\pi/2}\left[\int_{y=0}^{2x/\pi} (1 + \sin x)\, dy\right] dx$$

$$= -\int_0^{\pi/2} [(1 + \sin x)\, y]_0^{2x/\pi} dx = -\int_0^{\pi/2} (1 + \sin x)\frac{2x}{\pi} dx$$

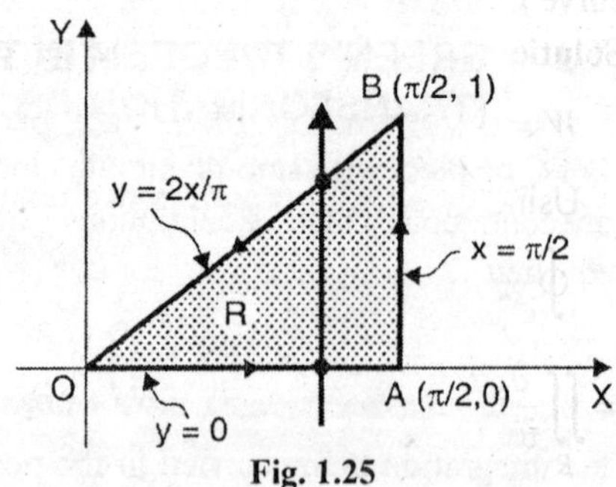

Fig. 1.25

$$= \frac{-2}{\pi}\int_0^{\pi/2}(x + x\sin x)\,dx = \frac{-2}{\pi}\left[\frac{x^2}{2} + \{x(-\cos x) + \sin x\}\right]_0^{\pi/2}$$

$$= -\left(\frac{\pi}{4} + \frac{2}{\pi}\right), \text{ the required value.}$$

EXAMPLE 1.73 Verify the Green's theorem for $f(x, y) = e^{-x}\sin y$, $g(x, y) = e^{-x}\cos y$ and C is the square with vertices at $(0, 0)$, $(\pi/2, 0)$, $(\pi/2, \pi/2)$, $(0, \pi/2)$. *(AMIETE-June 2010, Dec. 2005; AMIE S-2008)*

Solution We can write the line integral as

$$\oint_C f(x, y)\,dx + g(x, y)\,dy = \left[\int_{C_1} + \int_{C_2} + \int_{C_3} + \int_{C_4}\right](f(x, y)\,dx + g(x, y)\,dy)$$

where C_1, C_2, C_3 and C_4 are the boundary lines shown in Fig. 1.26. We have along C_1: $y = 0$, $0 \le x \le \pi/2$ and

$$\int_{C_1} e^{-x}(\sin y\,dx + \cos y\,dy) = 0, \text{ along } C_2\text{: } x = \pi/2,\ 0 \le y \le \pi/2$$

and $$\int_{C_2} e^{-x}(\sin y\,dx + \cos y\,dy) = \int_0^{\pi/2} e^{-\pi/2}\cos y\,dy = e^{-\pi/2},$$

along C_3: $y = \pi/2$, $\pi/2 \le x \le 0$ and $\int_{C_3} e^{-x}(\sin y\,dx + \cos y\,dy)$

$$= \int_{\pi/2}^{0} e^{-x}dx = [-e^{-x}]_{\pi/2}^{0} = [e^{-x}]_0^{\pi/2} = e^{-\pi/2} - 1,$$

along C_4: $x = 0$, $\pi/2 \le y \le 0$ and

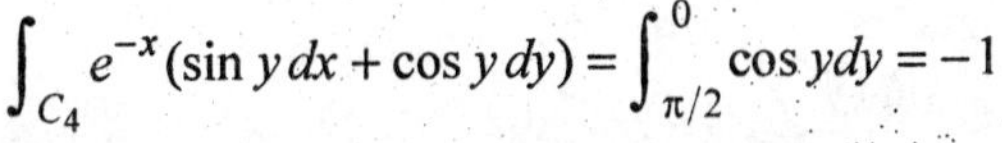

$$\int_{C_4} e^{-x}(\sin y\,dx + \cos y\,dy) = \int_{\pi/2}^{0}\cos y\,dy = -1.$$

Fig. 1.26

Therefore, $\oint_C f(x, y)\,dx + g(x, y)\,dy = e^{-\pi/2} + e^{-\pi/2} - 1 - 1 = 2(e^{-\pi/2} - 1).$

Using Green's theorem, we obtain

$$\oint_C f(x, y)\,dx + g(x, y)\,dy = \iint_R (-e^{-x}\cos y - e^{-x}\cos y)\,dydx = \int_{x=0}^{\pi/2}\int_{y=0}^{\pi/2}(-2e^{-x}\cos y)\,dy\,dx$$

$$= -2e^{-x}\int_{x=0}^{\pi/2}[\sin y]_0^{\pi/2}dx = -2\int_0^{\pi/2} e^{-x}dx = 2[e^{-x}]_0^{\pi/2} = 2(e^{-\pi/2} - 1).$$

Hence, the Green's theorem is verified.

EXAMPLE 1.74 Find the work done by the force $\mathbf{F} = (-16y + \sin x^2)\mathbf{i} + (4e^y + 3x^2)\mathbf{j}$ acting along the simple closed curve C shown in Fig. 1.27.

Solution The work done by the force **F** is given by

$$W = \oint_C \mathbf{F}\cdot d\mathbf{r} = \oint(-16y + \sin x^2)\,dx + (4e^y + 3x^2)\,dy.$$

Using the Green's theorem, we obtain

$$\oint_C (-16y + \sin x^2)\,dx + (4e^y + 3x^2)\,dy$$

$$= \iint_R \frac{\partial}{\partial x}(4e^y + 3x^2) - \frac{\partial}{\partial y}(-16y + \sin x^2) = \iint_R (6x + 16)\,dxdy.$$

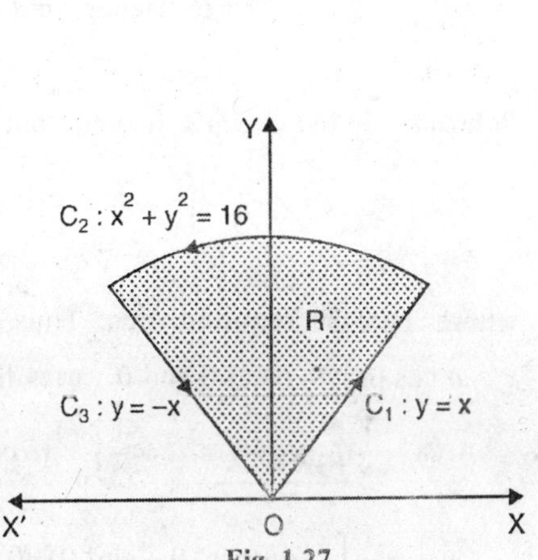

Fig. 1.27

It is convenient to use polar coordinates to evaluate the integral. The region R is given by

$$R: x = r\cos\theta,\ y = r\sin\theta,\ 0 \le r \le 4,\ \pi/4 \le \theta \le 3\pi/4.$$

$$\text{Therefore, } W = \iint_R (6x+16)\,dx dy = \int_{\pi/4}^{3\pi/4}\int_0^4 (6r\cos\theta + 16)\, r\, dr\, d\theta$$

$$= \int_{\pi/4}^{3\pi/4}\left[\int_0^4 (6r^2\cos\theta + 16r)\,dr\right]d\theta = \int_{\pi/4}^{3\pi/4} [2r^3\cos\theta + 8r^2]_0^4\, d\theta$$

$$= \int_{\pi/4}^{3\pi/4} (128\cos\theta + 128)\,d\theta = [128\sin\theta + 128\theta]_{\pi/4}^{3\pi/4} = 64\pi.$$

EXAMPLE 1.75 Find the work done by moving a particle once around the circle $x^2 + y^2 = a^2$, $z = b$ in the force field

$$\mathbf{F} = (xy + \sin xy)\mathbf{i} + \left(\frac{x^2}{2} + \frac{x}{y}\sin xy + \frac{1}{y^2}\cos xy\right)\mathbf{j} + (e^x \cos xy)\mathbf{k}.$$

Solution Work done $= \oint_C \mathbf{F} \cdot d\mathbf{r}$

$$= \oint_C (xy + \sin xy)\mathbf{i} + \left(\frac{x^2}{2} + \frac{x}{y}\sin xy + \frac{1}{y^2}\cos xy\right)\mathbf{j} + (e^x \cos xy)\mathbf{k} \cdot (dx\,\mathbf{i} + dy\,\mathbf{j})$$

$$= \oint_C (xy + \sin xy)\,dx + \left(\frac{x^2}{2} + \frac{x}{y}\sin xy + \frac{1}{y^2}\cos xy\right)dy.$$

Here $f(x, y) = xy + \sin xy$ and $g(x, y) = \dfrac{x^2}{2} + \dfrac{x}{y}\sin xy + \dfrac{\cos xy}{y^2}$.

Now $\dfrac{\partial f}{\partial y} = x + x\cos xy$, and $\dfrac{\partial g}{\partial x} = x + \dfrac{1}{y}\sin xy + x\cos xy - \dfrac{\sin xy}{y} = x + x\cos xy.$

Since $\partial f/\partial y = \partial g/\partial x$, then by the Green's theorem

$$\oint_C f(x, y)\,dx + g(x, y)\,dy = \iint_R \left(\frac{\partial g}{\partial x} - \frac{\partial f}{\partial y}\right) dx\,dy = 0,$$

where R is the region bounded by C. (See Remark 2. Article 1.21.)

EXAMPLE 1.76 Apply the Green's theorem to show that the area bounded by a simple closed curve C is given by $\dfrac{1}{2}\oint_C x dy - y dx$. Hence, find the area of an ellipse whose semi-major and minor axes are of lengths a and b. *(PTU 2006; Kerala 2005; Madras 2003; MDU 2002; VTU 2000S).*

Solution In the Green's theorem, put $f = -y$, $g = x$. Then

$$\oint_C x\,dy - y\,dx = \iint_R \left(\frac{\partial}{\partial x}(x) - \frac{\partial}{\partial y}(-y)\right) dy\,dx = 2\iint_R dx\,dy = 2A,$$

where A is the required area. Thus, $A = \dfrac{1}{2}\oint_C x\,dy - y\,dx.$ The parametric equations of an ellipse are $x = a\cos\theta$, $y = b\sin\theta$ and θ varies from 0 to 2π.

$$\text{Area} = \frac{1}{2}\oint_C x\,dy - y\,dx = \frac{1}{2}\int_0^{2\pi} (a\cos\theta)(b\cos\theta)\,d\theta - (b\sin\theta)(-a\sin\theta)\,d\theta$$

$$= \frac{1}{2}\int_0^{2\pi} ab(\cos^2\theta + \sin^2\theta)\,d\theta = \frac{1}{2}\int_0^{2\pi} ab\,d\theta = \pi ab.$$

Simple Closed Curve, Simply and Multiply Connected Domain

A curve is called a *simple closed* curve if it does not intersect itself between its end points. For example, the curve in Fig. 1.28(a) is simple closed while the curve in Fig. 1.28(b) is closed but not simple and is known as *multiple curve*.

A domain D is called *simply connected* if every simple closed contour C lying entirely in D can be shrunk, or contracted to a point without leaving D. In other words, in a simply connected domain every simple closed contour C lying entirely within it encloses only points of the domain D. A domain which is not simply connected is called *multiply connected* domain. For example, the domain bounded by a circle or a square is simply connected domain. Fig. 1.28(c).

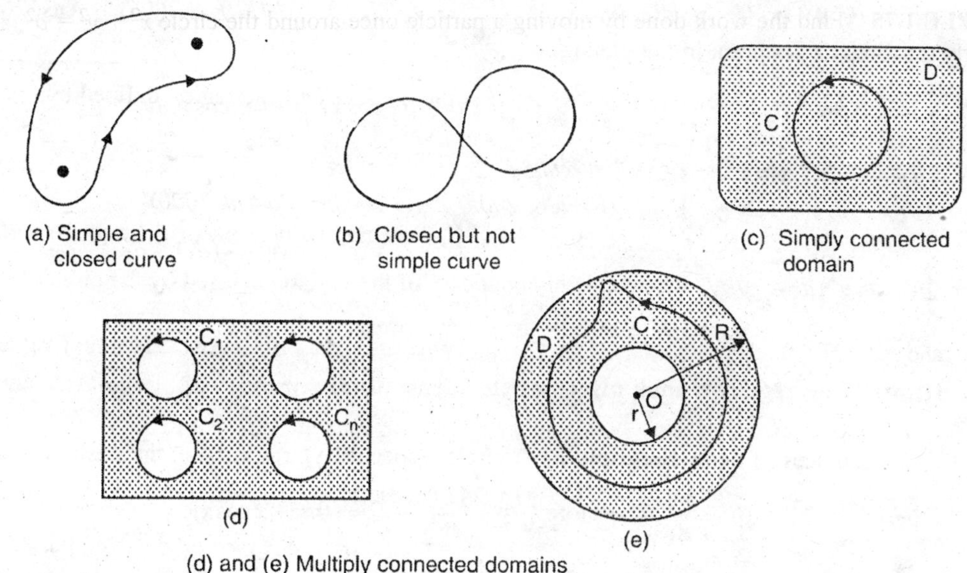

(a) Simple and closed curve
(b) Closed but not simple curve
(c) Simply connected domain
(d)
(e)
(d) and (e) Multiply connected domains

Fig. 1.28 (a to e)

A plain sheet of paper from which some interior parts are removed is a multiply connected domain Fig. 1.28(d). Another simple example of a multiply connected domain is the annulus $r < |z| < R$, that is, the domain between two concentric circles $|z| = R$ and $|z| = r$ cannot be shrunk to a point. Fig. 1.28(e). In plain terms a *simply connected* domain is one which has no "holes" in it. A domain with one "hole" is called *doubly connected* and a domain with two "holes" is called *triply connected* and so on.

EXERCISE 1.5

1. Apply Green's theorem to evaluate line integral $\oint_C [\sin y\, dx + x(1 + \cos y)\, dy]$ over a circular path C, $x^2 + y^2 = a^2$, $z = 0$. *(AMIETE, Dec. 2011)*

[Hint: In this problem $f(x, y) = \sin y$, $g(x, y) = x(1 + \cos y)$. $\frac{\partial g}{\partial x} - \frac{\partial f}{\partial y} = 1 + \cos y - \cos y = 1$.

By Green's theorem, the given line integral is

$$\iint_R 1\, dy\, dx = \iint_{x^2+y^2=a^2} dx dy = \text{area of circle of radius } a = \pi a^2 \Bigg].$$

2. Using Green's theorem to evaluate $\oint_C [(2x^2 - y^2)\,dx + (x^2 + y^2)\,dy]$, where C is the boundary in the xy-plane of the area enclosed by the x-axis and the semi-circle $x^2 + y^2 = 1$ in the upper half xy-plane. *(AMIETE, Dec. 2012; AMIE S-2010)* **Ans.** 4/3.

3. Evaluate $\oint_C (x^2 - 2xy)\,dx + (x^2 y + 3)\,dy$ around the boundary of the region defined by $y^2 = 8x$ and $x = 2$, using the Green's theorem. **Ans.** 128/5.

4. Evaluate by the Green's theorem $\oint_C e^{-x}(\sin y\,dx + \cos y\,dy)$, where C is the boundary of the rectangle with vertices $(0, 0)$, $(\pi, 0)$, $(\pi, \pi/2)$, $(0, \pi/2)$. *(AMIE S-2006)* **Ans.** $2(e^{-\pi} - 1)$.

5. Verify the Green's Theorem in the plane for

   (a) $\oint_C (3x^2 - 8y^2)\,dx + (4y - 6xy)\,dy$, where C is the boundary of the region defined by:

   (i) $y = \sqrt{x}$, $y = x^2$, *(AMIE W-2005)*

   (ii) $x = 0$, $y = 0$, $x + y = 1$. *(Kerala, 2005; Anna 2003S; Madras 2000)*

   Ans. *(i)* Common value = 3/2, *(ii)* Common value = 5/3.

   (b) $\oint_C (2x - y^3)\,dx - xy\,dy$, where C is the boundary of the region enclosed by the circles $x^2 + y^2 = 1$ and $x^2 + y^2 = 9$. **Ans.** Common value = 60 π.

   [**Hint:** The region is multiply connected. Therefore, $\oint_C = \oint_{C_2} - \oint_{C_1}$, where C_2 and C_1 are circumferences of outer circle and inner circle respectively, traversed in the positive sense.

   $\oint_{C_2} = \dfrac{243\pi}{4}$ and $\oint_{C_1} = \dfrac{3\pi}{4}$. Therefore, $\oint_C \dfrac{243\pi}{4} - \dfrac{3\pi}{4} = 60\pi$].

6. Verify Green's theorem in the plane for $\oint_C (2xy - x^2)\,dx + (x + y^2)\,dy$, where C is the closed curve of the region bounded by $y = x^2$ and $y^2 = x$. **Ans.** Common value = 1/30.

7. Verify the Green's theorem in the xy-plane for $\oint_C [(xy^2 - 2xy)\,dx + (x^2 y + 3)\,dy]$ around the boundary C of the region enclosed by $y^2 = 8x$ and $x = 2$.

8. Using the Green's theorem to evaluate $\int_C (x^2 y\,dx + x^2 dy)$, where C is the boundary described counter clockwise of the triangle with vertices $(0, 0)$, $(1, 0)$, $(1, 1)$. *(UPTU 2004)* **Ans.** 5/12.

9. Apply the Green's theorem to evaluate $\oint_C [(2x^2 - y^2)\,dx + (x^2 + y^2)\,dy]$, where C is the boundary of the area enclosed by the x-axis and the upper half of circle $x^2 + y^2 = a^2$. *(AMIE-W-2010; UPTU 2005)* **Ans.** $4a^3/3$.

   [**Hint:** Here $P(x, y) = 2x^2 - y^2$, $Q(x, y) = x^2 + y^2$. By Green's theorem

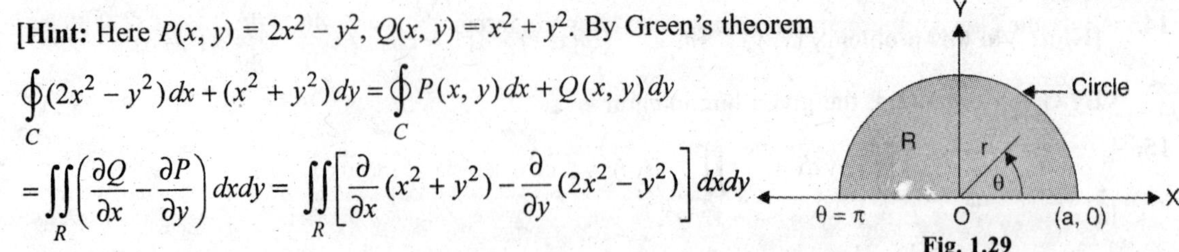

$$\oint_C (2x^2 - y^2)\,dx + (x^2 + y^2)\,dy = \oint_C P(x, y)\,dx + Q(x, y)\,dy$$

$$= \iint_R \left(\frac{\partial Q}{\partial x} - \frac{\partial P}{\partial y}\right) dxdy = \iint_R \left[\frac{\partial}{\partial x}(x^2 + y^2) - \frac{\partial}{\partial y}(2x^2 - y^2)\right] dxdy$$

Fig. 1.29

$$= 2\iint_R (x+y)\,dxdy,$$ where R is the region shown in Fig. 1.29.

$$= 2\int_{r=0}^{a}\int_{\theta=0}^{\pi} r(\cos\theta + \sin\theta)\cdot r\,d\theta\,dr$$

[Changing to polar coordinates (r, θ), r varies from 0 to a and θ varies from 0 to π]

$$= 2\int_{r=0}^{a}\int_{\theta=0}^{\pi} r^2(\cos\theta + \sin\theta)\,d\theta\,dr = 2\int_{r=0}^{a} r^2[\sin\theta - \cos\theta]_0^{\pi}\,dr = 4\left[\frac{r^3}{3}\right]_0^a = \frac{4}{3}a^3.$$

10. Evaluate: $\int_{(0,0)}^{(2,1)} (10x^4 - 2xy^3)\,dx - 3x^2y^2dy$ along the path $x^4 - 6xy^3 = 4y^2$. **Ans.** 60.

[**Hint:** It is difficult to evaluate directly. We note that $f(x, y) = 10x^4 - 2xy^3$ and $g(x, y) = -3x^2y^2$ and $\partial f/\partial y = -6xy^2 = \partial g/\partial x$. Hence the integral is independent of path *i.e.* we can use any path. One simple path consists of the straight line segments from (0, 0) to (2, 0) and then from (2, 0) to (2, 1).

Along the straight line path from (0, 0) to (2, 0), the integral is equal to $\int_0^2 10x^4dx = 64$.

Along the straight line path from (2, 0) to (2, 1) the integral $= \int_0^1 -12y^2dy = -4$.

$\therefore$ Required value of line integral = 64 – 4 = 60.

Alternative Solution: $\frac{\partial f}{\partial y} = \frac{\partial}{\partial y}(10x^4 - 2xy^3) = -6xy^2$ and $\frac{\partial g}{\partial x} = \frac{\partial}{\partial x}(-3x^2y^2) = -6xy^2$.

Since $\frac{\partial f}{\partial y} = \frac{\partial g}{\partial x}$. $(10x^4 - 2xy^3)\,dx - 3x^2y^2\,dy$ is an exact differential of $(2x^5 - x^2y^3)$.

Then $\int_{(0,0)}^{(2,1)} (10x^4 - 2xy^3)\,dx - 3x^2y^2\,dy = \int_{(0,0)}^{(2,1)} d\,(2x^5 - x^2y^3) = |\,2x^5 - x^2y^3\,|_{(0,0)}^{(2,1)} = 60.$]

11. Prove that the line integral $\int_C \frac{xdy - ydx}{x^2+y^2}$, taken in the positive direction over any closed contour with the origin inside it is equal to 2π. *(AMIE, W-2008)*

12. Use Green's theorem to evaluate the line integral $\oint_C (e^{-x^2} + y^2)\,dx + (\log y - x^2)\,dy$, where C is the square with vertices (0, 1), (1, 1) (1, 2) and (0, 2). *(AMIE, W-2006)* **Ans.** –4.

13. Find the work done by the force field $\mathbf{F}(x, y) = (e^x - y^3)\mathbf{i} + (\cos y + x^3)\mathbf{j}$ on a particle that travels once around the unit circle $x^2 + y^2 = 1$ in the counterclockwise direction. **Ans.** $3\pi/2$.

14. Apply the Green's theorem to evaluate the following integral: $\oint_C (y^2dx + x^2dy)$ C: the triangle bounded by $x = 0$, $x + y - 1$, $y - 0$. **Ans.** Zero.

15. Use Green's theorem to evaluate $\oint_C (x^2 + xy)\,dx + (x^2 + y^2)\,dy$, where C is the square formed by the lines $y = \pm 1$, $x = \pm 1$. *(AMIETE, June 2012)* **Ans.** Zero.

16. Evaluate $\oint_C (2xy^3 - y^2 \cos x)\,dx + (1 - 2y\sin x + 3x^2y^2)\,dy$ around a parallelogram with vertices at (0, 0), (3, 0), (5, 2), (2, 2). **Ans.** zero.

[**Hint:** Here $f(x, y) = 2xy^3 - y^2 \cos x$, and $g(x, y) = 1 - 2y \sin x + 3x^2y^2$.
Since $\partial f/\partial y = 6xy^2 - 2y\cos x = \partial g/\partial x$, so the line integral around any closed path, in particular the parallelogram is zero.]

1.22 SURFACE INTEGRALS

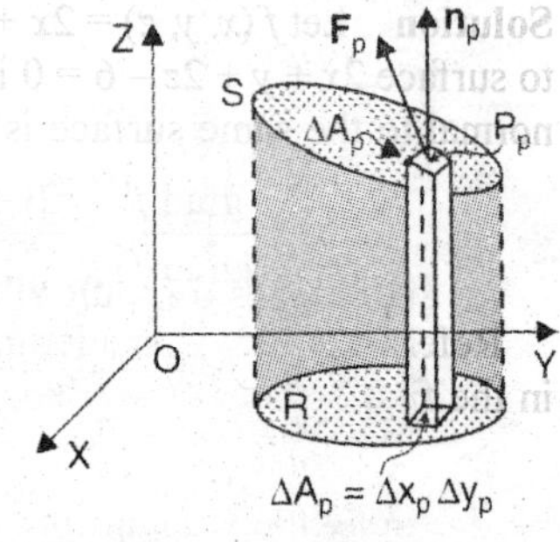

Fig. 1.30

Let **S** be a two-sided* surface having projection R on the xy-plane. (See fig. 1.30). We consider any one side of S as positive. This is the outer side if S is a closed surface. Let **F** be a vector function. Denote by **n** the unit vector normal to any point of the positive side surface S. Divide the surface S into m parts $S_1, S_2, S_3, \ldots S_p, \ldots S_m$ of areas $\Delta A_1, \Delta A_2, \Delta A_3, \ldots, \Delta A_p, \ldots \Delta A_m$. Let P_p be any point within ΔA_p whose coordinates are (x_p, y_p, z_p). Define $\mathbf{F}(x_p, y_p, z_p) = \mathbf{F}_p$. Let $\mathbf{n}_p$ be the positive unit normal to ΔA_p at P_p. Form the sum

$$\sum_{p=1}^{m} \mathbf{F}_p \cdot \mathbf{n}_p\, \Delta A_p, \qquad \ldots (1.35),$$

where $\mathbf{F}_p \cdot \mathbf{n}_p$ is the normal component of $\mathbf{F}_p$ at P_p.

Now take the limit of this sum as $m \to \infty$ in such a way that the largest element of the surface area shrinks to a point. This implies that as $m \to \infty$, the length of the largest dimension of each ΔA_p approaches zero. This limit, if it exists, is called the surface integral of the normal component of **F** over S and is denoted by $\iint_S \mathbf{F} \cdot \mathbf{n}\, dA$.

Let the surface S has projection R on the xy-plane. The projection of ΔA_p on the xy-plane is $(\mathbf{n}_p\, \Delta A_p) \cdot \mathbf{k}$ or $\mathbf{n}_p \cdot \mathbf{k}\, \Delta A_p$ which is equal to $\Delta x_p \Delta y_p$ so that $\Delta A_p = \dfrac{\Delta x_p \Delta y_p}{\mathbf{n}_p \cdot \mathbf{k}}$. Thus the sum (1.34) become

$$\sum_{p=1}^{m} \mathbf{F}_p \cdot \mathbf{n}_p \frac{\Delta x_p \Delta y_p}{\mathbf{n}_p \cdot \mathbf{k}}.$$

By the fundamental theorem of integral calculus the limit of this sum as $m \to \infty$ in such a manner that the largest $\Delta x_p \Delta y_p$ approach zero is

$$\iint_R \mathbf{F} \cdot \mathbf{n} \frac{dx\,dy}{\mathbf{n} \cdot \mathbf{k}}. \quad \text{Thus} \quad \iint_S \mathbf{F} \cdot \mathbf{n}\, dA = \iint_R \mathbf{F} \cdot \mathbf{n} \frac{dx\,dy}{\mathbf{n} \cdot \mathbf{k}}. \qquad \ldots (1.36)$$

If R is the projection of S on the yz-plane, then

$$\iint_S \mathbf{F} \cdot \mathbf{n}\, dA = \iint_R \mathbf{F} \cdot \mathbf{n} \frac{dy\,dz}{\mathbf{n} \cdot \mathbf{i}}. \qquad \ldots (1.37)$$

* An example of one sided surface is the Möbius strip. A model of this surface can be made by taking a long strip of rectangular paper, give one end a half twist, and then attach the ends with tape.

If we tried to colour only one side of the surface, we would find the whole thing coloured. This is sometimes called a non-orientale surface. A two-sided surface is orientable.

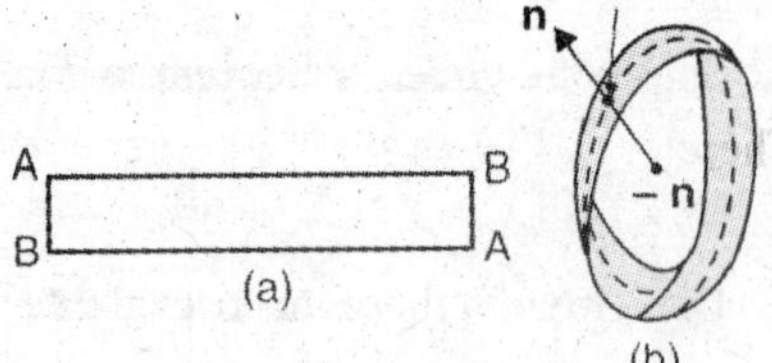

Möbius strip
Non-orientable surface–
one-sided surface

If R is the projection of S on the zx-plane, then

$$\iint_S \mathbf{F}\cdot\mathbf{n}\, dA = \iint_R \mathbf{F}\cdot\mathbf{n}\, \frac{dz\, dx}{\mathbf{n}\cdot\mathbf{j}}. \quad \text{....(1.38)}$$

While evaluating surface integrals, we should choose the most convenient of the results (1.36), (1.37), (1.38).

EXAMPLE 1.77 Evaluate the surface integral $\iint_S \mathbf{F}\cdot\mathbf{n}\, dA$ where $\mathbf{F} = (x + y^2)\mathbf{i} - 2x\mathbf{j} + 2yz\mathbf{k}$ and S is the surface of the plane $2x + y + 2z = 6$ in the first octant.

Solution Let $f(x, y, z) = 2x + y + 2z - 6 = 0$ be the surface. Vector normal to surface $2x + y + 2z - 6 = 0$ is given by grad $f = 2\mathbf{i} + \mathbf{j} + 2\mathbf{k}$. A unit vector normal to the same surface is

$$\mathbf{n} = \frac{\text{grad } f}{|\text{grad } f|} = \frac{2\mathbf{i}+\mathbf{j}+2\mathbf{k}}{\sqrt{4+1+4}} = \frac{1}{3}(2\mathbf{i}+\mathbf{j}+2\mathbf{k}).$$

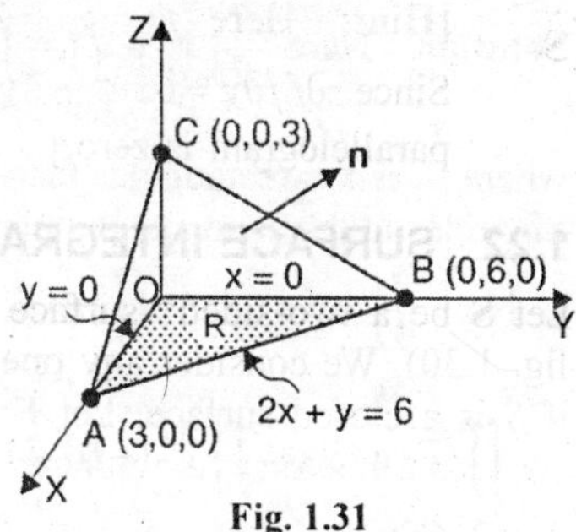

Fig. 1.31

Refer Fig. 1.31. Consider the projection of S on the xy-plane. The projection of the portion of the plane ABC in the first octant is the triangle bounded by $x = 0$, $y = 0$ and $2x + y = 6$. We have

$$dA = \frac{dx\, dy}{\mathbf{n}\cdot\mathbf{k}} = \frac{dx\, dy}{(1/3)(2\mathbf{i}+\mathbf{j}+2\mathbf{k})\cdot\mathbf{k}} = \frac{dx\, dy}{2/3}.$$

Therefore, $$\iint_S \mathbf{F}\cdot\mathbf{n}\, dA = \iint_S \frac{2(x+y^2) - 2x + 4yz}{3}\, dA = \frac{2}{3}\iint_S (y^2 + 2yz)\, dA.$$

From the equation of the plane, we get $2z = 6 - 2x - y$. Hence,

$$\iint_S \mathbf{F}\cdot\mathbf{n}\, dA = \frac{2}{3}\iint_S (y^2 + 6y - 2xy - y^2)\, dA = 2\iint_R (3y - xy)\, dx\, dy$$

$$= 2\int_{x=0}^{3}\left[\int_{y=0}^{6-2x} (3y - xy)\, dy\right] dx = 2\int_0^3 \left[(3-x)\frac{y^2}{2}\right]_0^{6-2x} dx = 4\int_0^3 (3-x)^3\, dx = 81.$$

EXAMPLE 1.78 Evaluate the surface integral $\iint_S \mathbf{F}\cdot\mathbf{n}\, dA$, where $\mathbf{F} = z\mathbf{i} + x\mathbf{j} - 3y^2z\mathbf{k}$ and S is the portion of the surface of the cylinder $x^2 + y^2 = 16$ included in the first octant between $z = 0$ and $z = 5$.

(AMIETE Dec 2003)

Solution Let $f(x, y, z) = x^2 + y^2 - 16 = 0$ be the surface. Then

$$\text{grad } f = 2x\mathbf{i} + 2y\mathbf{j}, \quad \mathbf{n} = \frac{\text{grad } f}{|\text{grad } f|} = \frac{2(x\mathbf{i} + y\mathbf{j})}{\sqrt{4(x^2+y^2)}} = \frac{1}{4}(x\mathbf{i} + y\mathbf{j}).$$

We cannot take the projection of S on xy-plane. Project S on the yz-plane. The projection is a rectangle with sides of lengths 4 and 5. We have

$$dA = \frac{dy\, dz}{\mathbf{n}\cdot\mathbf{i}} = \frac{dy\, dz}{x/4}.$$

Therefore, $$\iint_S \mathbf{F}\cdot\mathbf{n}\, dA = \iint_S \frac{1}{4}(zx + xy)\, dA = \int_{z=0}^{5}\int_{y=0}^{4} \frac{1}{4}x(y+z)\frac{dy\, dz}{(x/4)}$$

$$= \int_0^5\left[\int_0^4 (y+z)\, dy\right] dz = \int_0^5 \left[\frac{y^2}{2} + yz\right]_0^4 dz = \int_0^5 (4z + 8)\, dz$$

$$= [8z + 2z^2]_0^5 = 40 + 50 = 90.$$

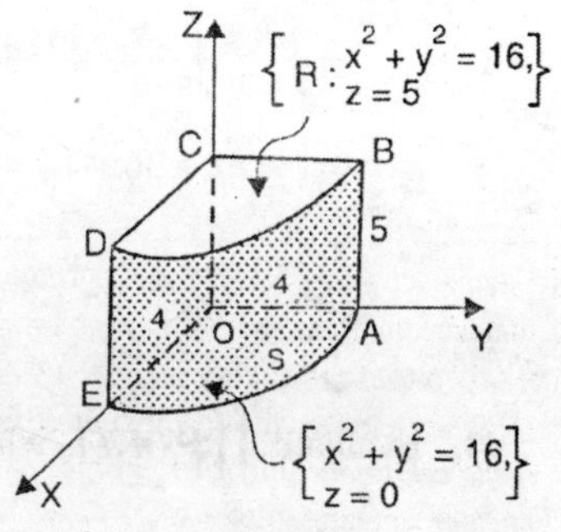

Fig. 1.32

EXAMPLE 1.79 Evaluate $\iint_S \mathbf{F}\cdot\mathbf{n}\,dA$, where $\mathbf{F} = 2x^2y\mathbf{i} - y^2\mathbf{j} + 4xz^2\mathbf{k}$ and S is the closed surface of the region in the first octant bounded by the cylinder $y^2 + z^2 = 9$ and the planes $x = 0, x = 2, y = 0$ and $z = 0$.

(*AMIETE, Dec 2006*)

Solution Here $\iint_S \mathbf{F}\cdot\mathbf{n}\,dA = \iint_{S_1} (\mathbf{F}\cdot\mathbf{n})\,dA + \iint_{S_2} (\mathbf{F}\cdot\mathbf{n})\,dA + \iint_{S_3} (\mathbf{F}\cdot\mathbf{n})\,dA = \iint_{S_4} (\mathbf{F}\cdot\mathbf{n})\,dA + \iint_{S_5} (\mathbf{F}\cdot\mathbf{n})\,dA,$

where S_1 is the rectangular face in *xy*-plane; S_2 the rectangular face in *xz*-plane; faces S_3 and S_4 are quadrants of circles in *yz*-plane and the plane $x = 2$ respectively; S_5 is the curved surface to the cylinder in the first octant.

Now $\iint_{S_1} (\mathbf{F}\cdot\mathbf{n})\,dA = \iint_{S_1} \{\mathbf{F}\cdot(-\mathbf{k})\}\,dA = \iint_{S_1} -4xz^2\,dA = 0$, because $z = 0$ in the *xy*-plane.

$\iint_{S_2} (\mathbf{F}\cdot\mathbf{n})\,dA = \iint_{S_2} \mathbf{F}\cdot(-\mathbf{j})\}\,dA = \iint_{S_2} y^2\,dA = 0$, because $y = 0$ in the *xz*-plane.

$\iint_{S_3} (\mathbf{F}\cdot\mathbf{n})\,dS = \iint_{S_3} \mathbf{F}\cdot(-\mathbf{i})\}\,dA = \iint_{S_3} -2x^2y\,dA = 0$, because $x = 0$ in yz-plane.

$\iint_{S_4} \mathbf{F}\cdot\mathbf{n}\,dS = \iint_{S_4} \mathbf{F}\cdot\mathbf{i}\,dA = \iint_{S_4} 2x^2y\,dA$

$= \int_{z=0}^{3}\int_{y=0}^{\sqrt{9-z^2}} 8y\,dy\,dz = 4\int_0^3 (9 - z^2)\,dz = 72.$

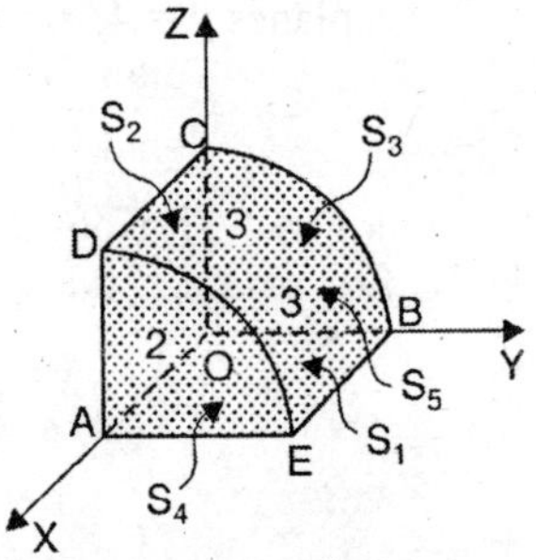

Fig. 1.33

To get $\mathbf{n}$ in S_5 we note that $\nabla(y^2 + z^2) = 2y\mathbf{i} + 2z\mathbf{k}$. Thus $\mathbf{n} = \dfrac{2y\mathbf{j} + 2z\mathbf{k}}{\sqrt{4y^2 + 4z^2}} = \dfrac{y\mathbf{j} + z\mathbf{k}}{3}$. $\quad (\because y^2 + z^2 = 9)$

$$\therefore \qquad \mathbf{F}\cdot\mathbf{n} = (2x^2y\mathbf{i} - y^2\mathbf{j} + 4xz^2\mathbf{k})\cdot\frac{y\mathbf{j} + z\mathbf{k}}{3} = \frac{-y^3 + 4xz^3}{3}, \text{ and } \mathbf{n}\cdot\mathbf{k} = z/3.$$

Hence $\iint_{S_5} (\mathbf{F}\cdot\mathbf{n})\,dA = \frac{1}{3}\int_0^2\int_0^3 (-y^3 + 4xz^3)\cdot\frac{dy\,dx}{z/3} = \int_{x=0}^{2}\int_{y=0}^{3}\left(\frac{-y^3}{z} + 4xz^2\right)dy\,dx,$

$$= \int_{x=0}^{2}\int_{y=0}^{3}\left[\frac{-y^3}{\sqrt{9-y^2}} + 4x(9 - y^2)\right]dy\,dx \qquad (\because z^2 = 9 - y^2)$$

$$= \int_{x=0}^{2}\int_{\theta=0}^{\pi/2}\left[\frac{-27\sin^3\theta}{3\cos\theta} + 4x(9\cos^2\theta)\right]3\cos\theta\,d\theta\,dx \qquad (\text{Putting } y = 3\sin\theta)$$

$$= \int_{x=0}^{2}\int_{\theta=0}^{\pi/2} (-27\sin^3\theta + 108x\cos^3\theta)\,d\theta\,dx = \int_0^2\left(-27\times\frac{2}{3} + 108x\times\frac{2}{3}\right)dx = \int_0^2 (-18 + 72x)\,dx$$

$= [-18x + 36x^2]_0^2 = -36 + 144 = 108.$ Therefore, $\iint_S \mathbf{F}\cdot\mathbf{n}\,dA = 0 + 0 + 0 + 72 + 108 = 180.$

EXERCISE 1.6

1. Evaluate $\iint_S \mathbf{F}\cdot\mathbf{n}\,dA$, where $\mathbf{F} = 18z\mathbf{i} - 12\mathbf{j} + 3y\mathbf{k}$ and S is that part of the plane $2x + 3y + 6z = 12$ which is located in the first octant (*MDU, 2001*) **Ans.** 24.

[**Hint:** $\mathbf{n}=\frac{1}{7}(2\mathbf{i}+3\mathbf{j}+6\mathbf{k})$. $\mathbf{F}\cdot\mathbf{n}=\frac{36z-36+18y}{7}=\frac{36-12x}{7}$, using $z=(12-2x-3y)/6$, from the equation of the surface. We have $\mathbf{n}\cdot\mathbf{k}=\left(\frac{2}{7}\mathbf{i}+\frac{3}{7}\mathbf{j}+\frac{6}{7}\mathbf{k}\right)\cdot\mathbf{k}=\frac{6}{7}$.

Therefore, $$\iint_S \mathbf{F}\cdot\mathbf{n}\,dA=\iint_S \frac{36-12x}{7}\,dA=\iint_R\left(\frac{36-12x}{7}\right)\frac{dy\,dx}{\mathbf{n}\cdot\mathbf{k}}=\iint_R(6-2x)\,dydx$$

$$=\int_{x=0}^{6}\left[\int_{y=0}^{(12-2x)/3}(6-2x)\,dy\right]dx=24.\Big]$$

2. If $\mathbf{F}=2y\mathbf{i}-z\mathbf{j}+x^2\mathbf{k}$ and S is surface of the parabolic cylinder $y^2=8x$ in the first octant bounded by the planes $y=4$, $z=6$, evaluate the surface integral $\iint_S \mathbf{F}\cdot\mathbf{n}\,dA$.

[**Hint:** $\mathbf{n}=\frac{8\mathbf{i}-2y\mathbf{j}}{\sqrt{64+4y^2}}$, $\mathbf{F}\cdot\mathbf{n}=\frac{16y+2yz}{\sqrt{64+4y^2}}$, $\mathbf{n}\cdot\mathbf{i}=\frac{8}{\sqrt{64+4y^2}}$. Projecting the surface S on yz plane, we obtain

$\iint_S \mathbf{F}\cdot\mathbf{n}\,dA=\iint_R \mathbf{F}\cdot\mathbf{n}\frac{dydz}{\mathbf{n}\cdot\mathbf{i}}$, (where R is the rectangular region in the yz plane bounded by the lines $y=0$, $z=0$, $y=4$ and $z=6$) $=\frac{1}{4}\int_{z=0}^{6}\int_{y=0}^{4}(8y+yz)\,dydz=132.\Big]$

3. Evaluate $\iint_S \mathbf{F}\cdot\mathbf{n}\,dA$, where $\mathbf{F}=xy\mathbf{i}-x^2\mathbf{j}+(x+z)\,\mathbf{k}$, S is that portion of the plane $2x+2y+z=6$ included in the first octant and $\mathbf{n}$ is a unit normal to S. (*AMIETE June 2004, Dec 2001*) **Ans.** 27/4.

4. If $\mathbf{F}=4xz\mathbf{i}-y^2\mathbf{j}+yz\mathbf{k}$ evaluate $\iint_S \mathbf{F}\cdot\mathbf{n}\,dA$, where S is the surface of the cube bounded by the planes, $x=0$, $x=1$, $y=0$, $y=1$, $z=0$, $z=1$. **Ans.** 3/2.

5. Evaluate $\iint_S \mathbf{F}\cdot\mathbf{n}\,dA$, where $\mathbf{F}(x,y,z)=xy\mathbf{i}+z^2\mathbf{j}+2yz\mathbf{k}$, S is the surface of the plane $x+y+z=1$ in the first octant and $\mathbf{n}$ is unit outward normal to the surface S. **Ans.** 1/8.

6. Show that $\iint_S (yz\,\mathbf{i}+zx\,\mathbf{j}+xy\,\mathbf{k})\cdot\mathbf{n}\,dA=3/8$, where S is the surface of the sphere $x^2+y^2+z^2=1$ in the first octant.

1.23 DIVERGENCE THEOREM OF GAUSS (TRANSFORMATION BETWEEN VOLUME INTEGRALS AND SURFACE INTEGRALS)

Let D be a closed and bounded region in the three dimensional space whose boundary is a piecewise smooth surface S, that is oriented outward. Let $\mathbf{F}(x,y,z)=F_1(x,y,z)\mathbf{i}+F_2(x,y,z)\mathbf{j}+F_3(x,y,z)\mathbf{k}$ be a vector field for which F_1, F_2, F_3 are continuous and have continuous first order partial derivatives in some domain containing D. Then

$$\iiint_D \text{div}\,(\mathbf{F})\,dV=\iiint_D(\nabla\cdot\mathbf{F})\,dV=\iint_S(\mathbf{F}\cdot\mathbf{n})\,dA, \quad ...(1.39)$$

where $\mathbf{n}$ is the outer unit normal vector to S. [Refer fig. 1.34.]

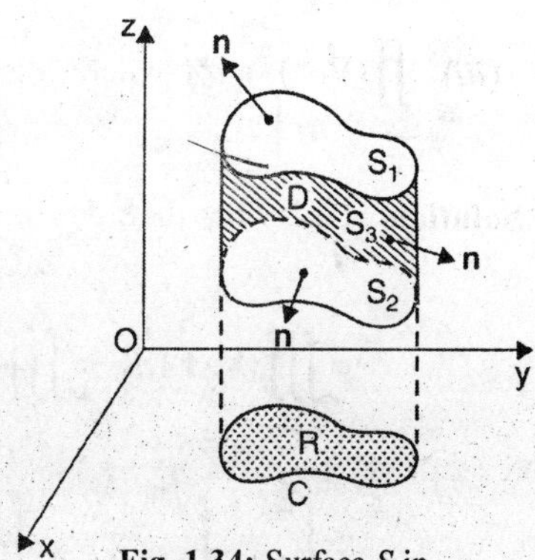

Fig. 1.34: Surface S in divergence theorem.

The region D shown in Fig. 1.34, whose surface consists of three pieces, S_1: Top surface, S_2: Bottom surface, S_3: Side (vertical). R is the projection of D onto the xy-plane and C surface is the boundary of R.

Expressed in words the divergence theorem states that "the surface integral of the normal component of a vector point function **F**, taken over a closed surface is equal to the integral of the divergence of **F**, taken over the volume enclosed by the surface".

1.23.1 Cartesian Form of Divergence Theorem

Let vector function **F**(x, y, z) has components (F_1, F_2, F_3).

Therefore, $\mathbf{F} = F_1\mathbf{i} + F_2\mathbf{j} + F_3\mathbf{k}$. Now

$$\text{div } \mathbf{F} = \nabla \cdot \mathbf{F} = \frac{\partial F_1}{\partial x} + \frac{\partial F_2}{\partial y} + \frac{\partial F_3}{\partial z} \text{ and } (\mathbf{F}\cdot\mathbf{n})\,dA = [F_i(\mathbf{n}\cdot\mathbf{i}) + F_2(\mathbf{n}\cdot\mathbf{j}) + F_3(\mathbf{n}\cdot\mathbf{k})]\,dA.$$

If α, β, γ are the angles which **n** the unit normal makes with the positive directions of x, y, z axes or **i, j, k** directions respectively. Then $\mathbf{n} \cdot \mathbf{i} = \cos\alpha$, $\mathbf{n} \cdot \mathbf{j} = \cos\beta$, and $\mathbf{n} \cdot \mathbf{k} = \cos\gamma$.

$$\text{Therefore } (\mathbf{F}\cdot\mathbf{n})\,dA = F_1 \cos\alpha\, dA + F_2 \cos\beta\, dA + F_3 \cos\gamma\, dA \quad \text{...(1.40)}$$

$$= F_1 dy\, dz + F_2 dz\, dx + F_3 dx\, dy.$$

Therefore, $\iiint_D \nabla\cdot\mathbf{F}\, dV = \iint_S (\mathbf{F}\cdot\mathbf{n})\, dA$ may be written as

$$\iiint_D \left(\frac{\partial F_1}{\partial x} + \frac{\partial F_2}{\partial y} + \frac{\partial F_3}{\partial z}\right) dx\, dy\, dz = \iint_S F_1 dy\, dz + F_2 dz\, dx + F_3 dx\, dy, \quad \text{...(1.41)}$$

which is the cartesian form of the divergence theorem.

Note: If (ℓ, m, n) are direction cosines of normal **n** then (1.39) can be written as $(F_1\ell + F_2 m + F_3 n)\,dA$. Then the divergence theorem may be written in the form

$$\iiint_D \left(\frac{\partial F_1}{\partial x} + \frac{\partial F_2}{\partial y} + \frac{\partial F_3}{\partial z}\right) dx\, dy\, dz = \iint_S (\ell F_1 + mF_2 + nF_3)\, dA. \quad \text{...(1.42)}$$

EXAMPLE 1.80 Use divergence theorem to evaluate the surface integral:

(*i*) $\iint_S (\mathbf{F}\cdot\mathbf{n})\, dA$ where $\mathbf{F} = 4x\mathbf{i} - 2y^2\mathbf{j} + z^2\mathbf{k}$ and S is the surface bounding the region $x^2 + y^2 = 4$, $z = 0$, $z = 3$. (*SVTU 2007S; JNTU '06; Kerala '05*)

(*ii*) $\iint_S (\mathbf{v}\cdot\mathbf{n})\, dA$ where $\mathbf{v} = x\mathbf{i} + y\mathbf{j} + z\mathbf{k}$ and S is the surface of the sphere $x^2 + y^2 + z^2 = 9$.

(*iii*) $\iint_S (\nabla r^2)\cdot\mathbf{n}\, dA$ where S is a closed surface enclosing a volume V and $r^2 = x^2 + y^2 + z^2$. (*AMIE S-2006*)

Solution (*i*) By the divergence theorem $\iint_S (\mathbf{F}\cdot\mathbf{n})\, dA = \iiint_D (\nabla\cdot\mathbf{F})\, dV$

$$= \iiint_D (\nabla\cdot\mathbf{F})\, dV = \iiint_D \left[\frac{\partial}{\partial x}(4x) + \frac{\partial}{\partial y}(-2y^2) + \frac{\partial}{\partial z}(z^2)\right] dV = \iiint_D (4 - 4y + 2z)\, dV$$

$$= \int_{x=-2}^{2} \int_{y=-\sqrt{4-x^2}}^{\sqrt{4-x^2}} \int_{z=0}^{3} (4 - 4y + 2z)\, dz\, dy\, dx$$

$$= \int_{x=-2}^{2} \int_{y=-\sqrt{4-x^2}}^{\sqrt{4-x^2}} [4z - 4yz + z^2]_0^3 \, dy\, dx = \int_{x=-2}^{2} \int_{y=-\sqrt{4-x^2}}^{\sqrt{4-x^2}} (21 - 12y)\, dy\, dx$$

$$= 42\int_{x=-2}^{2} \sqrt{4-x^2}\, dx = 84\int_{0}^{2} \sqrt{4-x^2}\, dx = 84\left[\frac{x}{2}\sqrt{4-x^2} + \frac{4}{2}\sin^{-1}\frac{x}{2}\right]_0^2 = 84\pi.$$

(*ii*) $\iint_S (\mathbf{v}\cdot\mathbf{n})\, dA = \iiint_D (\nabla\cdot\mathbf{v})\, dV$ (divergence theorem)

$$= \iiint_D \left(\frac{\partial}{\partial x}\mathbf{i} + \frac{\partial}{\partial y}\mathbf{j} + \frac{\partial}{\partial z}\mathbf{k}\right)\cdot(x\mathbf{i} + y\mathbf{j} + z\mathbf{k})\, dV = \iiint_V \left(\frac{\partial x}{\partial x} + \frac{\partial y}{\partial y} + \frac{\partial z}{\partial z}\right) dV$$

$$= 3\iiint_D dV = 3V = 3(4/3)\pi 3^3 = 108\pi.$$

As we know that volume of a sphere is equal to $(4/3)\,\pi\,(\text{radius})^3$.

(*iii*) Using the divergence theorem

$$\iint_S (\nabla r^2)\cdot\mathbf{n}\, dA = \iiint_D (\nabla\cdot\nabla r^2)\, dV = \iiint_D \nabla\cdot\left(2r\frac{\mathbf{r}}{r}\right) dV \qquad \left[\because\ \nabla f(r) = f'(r)\frac{\mathbf{r}}{r}\right]$$

$$= 2\iiint_D \nabla\cdot\mathbf{r}\, dV = 2\iiint_D \left[\frac{\partial}{\partial x}(x) + \frac{\partial}{\partial y}(y) + \frac{\partial}{\partial z}(z)\right] dV,$$ since $\mathbf{r} = x\mathbf{i} + y\mathbf{j} + z\mathbf{k}$

$$= 6\iiint_D dV = 6V.$$

EXAMPLE 1.81 Find $\iint_S (\mathbf{F}\cdot\mathbf{n})\, dA$, where $\mathbf{F} = (2x + 3z)\mathbf{i} - (3x + y)\mathbf{j} + (y + 2z)\mathbf{k}$ and S is the surface of the sphere having center at (3, –1, 2) and radius 3 units. (*AMIETE, Dec. 2010*).

Solution Let V be the volume enclosed by the surface S. Thus by divergence theorem, we have

$$\iint_S (\mathbf{F}\cdot\mathbf{n})\, dA = \iiint_D (\nabla\cdot\mathbf{F})\, dV = \iiint_D \left[\frac{\partial}{\partial x}(2x + 3z) + \frac{\partial}{\partial y}(-3x - y) + \frac{\partial}{\partial z}(y + 2z)\right] dV$$

$$= \iiint_D (2 - 1 + 2)\, dV = 3\iiint_D dV = 3V.$$

The volume of a sphere of radius 3 units $= (4/3)\pi\,(3)^3 = 36\pi$ cubic units.

Hence $\iint_S (\mathbf{F}\cdot\mathbf{n})\, dA = 3(36\pi) = 108\pi.$

EXAMPLE 1.82 Verify divergence theorem for $\mathbf{F} = 4xz\mathbf{i} - y^2\mathbf{j} + yz\mathbf{k}$ taken over the cube bounded by the planes $x = 0$, $x = 2$, $y = 0$, $y = 2$, $z = 0$, $z = 2$. (*MDU 2002*)

Solution We have $\nabla\cdot\mathbf{F} = \frac{\partial}{\partial x}(4xz) + \frac{\partial}{\partial y}(-y^2) + \frac{\partial}{\partial z}(yz) = 4z - 2y + y = 4z - y.$

Now $\iiint_D (\nabla \cdot \mathbf{F})\, dV = \iiint_D (4z - y)\, dV = \int_{x=0}^{2} \int_{y=0}^{2} \int_{z=0}^{2} (4z - y)\, dz\, dy\, dx$

$$= \int_{x=0}^{2} \int_{y=0}^{2} [2z^2 - yz]_0^2\, dy\, dx = 2\int_{x=0}^{2} \int_{y=0}^{2} (4 - y)\, dy\, dx$$

$$= 2\int_{x=0}^{2} \left[4y - \frac{y^2}{2}\right]_0^2 dx = 12\int_0^2 dx = 24. \qquad \text{...}(i)$$

To evaluate surface integral $\iint_S (\mathbf{F} \cdot \mathbf{n})\, dA$, we note that S consists of six faces.

Over the surface $BOCL$: $\mathbf{n} = -\mathbf{i}$, $x = 0$. Then

$$\iint_{BOCL} (\mathbf{F} \cdot \mathbf{n})\, dA = \int_0^2 \int_0^2 (-y^2\mathbf{j} + yz\mathbf{k}) \cdot (-\mathbf{i})\, dy\, dz = 0.$$

Over the face $OAMC$: $\mathbf{n} = -\mathbf{j}$, $y = 0$. Then

$$\iint_{OAMC} (\mathbf{F} \cdot \mathbf{n})\, dA = \int_0^2 \int_0^2 (4xz\mathbf{i}) \cdot (-\mathbf{j})\, dx\, dz = 0.$$

Over the face $BQAO$: $\mathbf{n} = -\mathbf{k}$, $z = 0$. Then

$$\iint_{BQAO} (\mathbf{F} \cdot \mathbf{n})\, dA = \int_0^2 (-y^2\mathbf{j}) \cdot (-\mathbf{k})\, dx\, dy = 0.$$

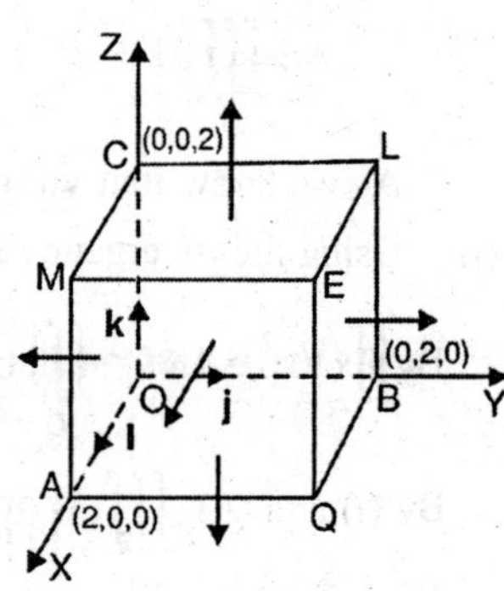

Fig. 1.35

Over the face $MEQA$: $\mathbf{n} = \mathbf{i}$, $x = 2$. Then

$$\iint_{MEQA} (\mathbf{F} \cdot \mathbf{n})\, dA = \int_0^2 \int_0^2 (8z\mathbf{i} - y^2\mathbf{j} + yz\mathbf{k}) \cdot \mathbf{i}\, dy\, dz = 8\int_0^2 \int_0^2 z\, dy\, dz = 32.$$

Over the face $ELBQ$: $\mathbf{n} = \mathbf{j}$, $y = 2$. Then

$$\iint_{ELBQ} (\mathbf{F} \cdot \mathbf{n})\, dA = \int_0^2 \int_0^2 (4xz\mathbf{i} - 4\mathbf{j} + 2z\mathbf{k}) \cdot \mathbf{j}\, dx\, dz = -4\int_0^2 \int_0^2 dx\, dz = -16.$$

Over the face $LEMC$: $\mathbf{n} = \mathbf{k}$, $z = 2$. Then

$$\iint_{LEMC} (\mathbf{F} \cdot \mathbf{n})\, dA = \int_0^2 \int_0^2 (8x\mathbf{i} - y^2\mathbf{j} + 2y\mathbf{k}) \cdot \mathbf{k}\, dx\, dy = 2\int_0^2 \int_0^2 y\, dx\, dy = 8.$$

Adding, we see that over the whole surface

$$\iint_S (\mathbf{F} \cdot \mathbf{n})\, dA = 0 + 0 + 0 + 32 - 16 + 8 = 24. \qquad \text{...}(ii)$$

From (*i*) and (*ii*), we have $\iint_S (\mathbf{F} \cdot \mathbf{n})\, dA = \iiint_D (\nabla \cdot \mathbf{F})\, dV$ and thus the theorem is verified,

EXAMPLE 1.83 Verify the divergence theorem for the function $\mathbf{F} = 2x^2y\mathbf{i} - y^2\mathbf{j} + 4xz^2\mathbf{k}$ taken over the region in the first octant bounded by $y^2 + z^2 = 9$ and $x = 2$.

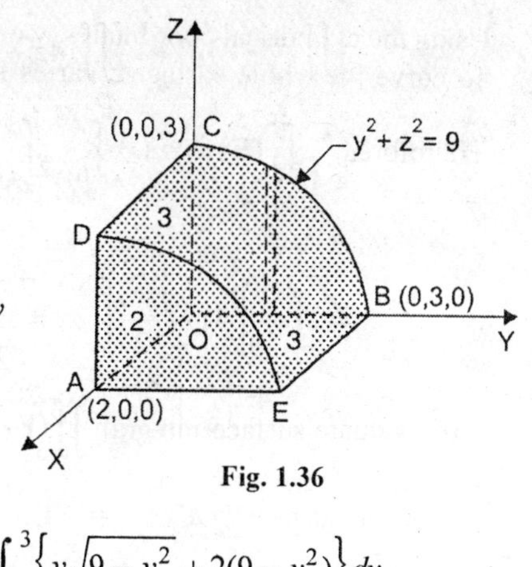

Fig. 1.36

Solution Volume integral $= \iiint_D (\nabla \cdot \mathbf{F})\,dV$

$$= \iiint_D \left[\frac{\partial}{\partial x}(2x^2 y) + \frac{\partial}{\partial y}(-y^2) + \frac{\partial}{\partial z}(4xz^2)\right] dV$$

$$= \iiint_D (4xy - 2y + 8xz)\,dV = \int_0^3 \int_0^{\sqrt{9-y^2}} \int_0^2 (4xy - 2y + 8xz)\,dx\,dz\,dy$$

$$= \int_0^3 \int_0^{\sqrt{9-y^2}} [2x^2 y - 2yx + 4x^2 z]_0^2\,dz\,dy$$

$$= 4\int_0^3 \int_0^{\sqrt{9-y^2}} (y + 4z)\,dz\,dy = 4\int_0^3 \left[yz + 2z^2\right]_0^{\sqrt{9-y^2}} dy = 4\int_0^3 \left\{y\sqrt{9-y^2} + 2(9-y^2)\right\} dy$$

$$= 4\left[-\frac{1}{3}(9-y^2)^{3/2} + 18y - \frac{2}{3}y^3\right]_0^3 = 180. \qquad ...(i)$$

By Example 1.79, surface integral $\iint_S (\mathbf{F} \cdot \mathbf{n})\,dA = 180.$ ...(*ii*)

By (*i*) and (*ii*) $\iiint_D (\nabla \cdot \mathbf{F})\,dV = \iint_S (\mathbf{F} \cdot \mathbf{n})\,dA$ and thus, the divergence theorem is verified.

EXAMPLE 1.84 Verify the divergence theorem for $\mathbf{F} = 4x\mathbf{i} - 2y^2\mathbf{j} + z^2\mathbf{k}$ taken over the region bounded by the cylinder $x^2 + y^2 = 4$, $z = 0$ and $z = 3$. (*MDU 2004; AMIETE-June 2003, June 2001*)

Solution By Example 1.80(*i*), volume integral $= \iiint_D (\nabla \cdot \mathbf{F})\,dV = 84\pi.$

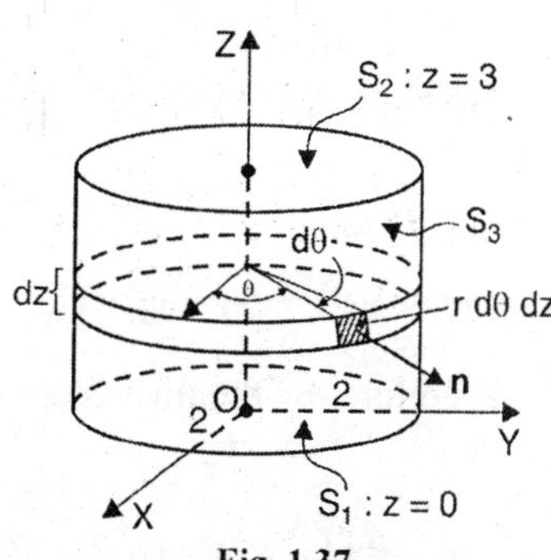

Fig. 1.37

The surface S of the cylinder consists of a base $S_1 (z = 0)$, the top $S_2 (z = 3)$ and the convex portion $S_3 (x^2 + y^2 = 4)$. Refer Fig. 1.37.

Then, surface integral

$$= \iint_S (\mathbf{F} \cdot \mathbf{n})\,dA = \iint_{S_1} (\mathbf{F} \cdot \mathbf{n})\,dA + \iint_{S_2} (\mathbf{F} \cdot \mathbf{n})\,dA + \iint_{S_3} (\mathbf{F} \cdot \mathbf{n})\,dA.$$

Over the surface S_1: $z = 0$, $\mathbf{n} = -\mathbf{k}$, $\mathbf{F} = 4x\mathbf{i} - 2y^2\mathbf{j}$ and $\mathbf{F} \cdot \mathbf{n} = 0$.

Therefore, $\iint_{S_1} (\mathbf{F} \cdot \mathbf{n})\,dA = 0.$

Over the surface S_2: $z = 3$, $\mathbf{n} = \mathbf{k}$, $\mathbf{F} = 4x\mathbf{i} - 2y^2\mathbf{j} + 9\mathbf{k}$ so that $\mathbf{F} \cdot \mathbf{n} = 9$.

Therefore, $\iint_{S_2} (\mathbf{F} \cdot \mathbf{n})\,dA = 9 \iint_{S_2} dA = 9$ (area o]f circular region with radius 2) $= 36\pi$.

Over the surface S_3: $x^2 + y^2 = 4$. A vector normal to $x^2 + y^2 - 4 = 0$ is given by $\nabla(x^2 + y^2 - 4) = 2x\mathbf{i} + 2y\mathbf{j}$.

A unit vector normal to S_3 is $\mathbf{n} = \dfrac{2x\mathbf{i} + 2y\mathbf{j}}{\sqrt{4x^2 + 4y^2}} = \dfrac{x\mathbf{i} + y\mathbf{j}}{2}$, $\quad (\because x^2 + y^2 = 4.)$

Therefore, $\mathbf{F} \cdot \mathbf{n} = (4x\mathbf{i} - 2y^2\mathbf{j} + z^2\mathbf{k}) \cdot \left(\dfrac{x\mathbf{i} + y\mathbf{j}}{2}\right) = 2x^2 - y^3.$

Using the cylindrical coordinates, we write $x = r\cos\theta = 2\cos\theta$, $y = r\sin\theta = 2\sin\theta$, $dA = r\,d\theta\,dz = 2\,d\theta\,dz$. To curve the whole surface z varies from 0 to 3 and θ varies from 0 to 2π.

Therefore, $$\iint_{S_3} (\mathbf{F}\cdot\mathbf{n})\,dA = \int_{\theta=0}^{2\pi}\int_{z=0}^{3} [2(2\cos\theta)^2 - (2\sin\theta)^3]2dz\,d\theta$$

$$= 16\int_{\theta=0}^{2\pi}\int_{z=0}^{3} (\cos^2\theta - \sin^3\theta)\,dz\,d\theta = 48\int_0^{2\pi} (\cos^2\theta - \sin^3\theta)\,d\theta$$

$$= 48\int_0^{2\pi}\left[\frac{1}{2}(1+\cos 2\theta) - \frac{1}{4}(3\sin\theta - \sin 3\theta)\right]d\theta = 48\pi.$$

$$\left[\because \int_0^{2\pi} \sin nx\,dx = \int_0^{2\pi} \cos nx\,dx = 0, n \text{ being an integer.}\right]$$

Adding we find that $\iint_S (\mathbf{F}\cdot\mathbf{n})\,dA = 0 + 36\pi + 48\pi = 84\pi$.

Therefore, $\iiint_D (\nabla\cdot\mathbf{F})\,dV = \iint_S (\mathbf{F}\cdot\mathbf{n})\,dA$ and the divergence theorem is verified.

EXAMPLE 1.85 Evaluate $\iint_S (a^2x^2 + b^2y^2 + c^2z^2)^{-1/2}\,dA$, where S in the surface of ellipsoid $ax^2 + by^2 + cz^2 = 1$.

Solution Taking $f(x,y,z) = ax^2 + by^2 + cz^2 - 1 = 0$.

$$\nabla f = \left(\mathbf{i}\frac{\partial}{\partial x} + \mathbf{j}\frac{\partial}{\partial y} + \mathbf{k}\frac{\partial}{\partial z}\right)(ax^2 + by^2 + cz^2 - 1) = 2ax\mathbf{i} + 2by\mathbf{j} + 2cz\mathbf{k}.$$

Therefore, unit vector normal to the ellipsoid $= \mathbf{n} = \dfrac{\nabla f}{|\nabla f|}\dfrac{ax\mathbf{i} + by\mathbf{j} + cz\mathbf{k}}{\sqrt{a^2x^2 + b^2y^2 + c^2z^2}}$.

Since $\mathbf{F}\cdot\mathbf{n} = (a^2x^2 + b^2y^2 + c^2z^2)^{-1/2}$, therefore, $\mathbf{F}\cdot(ax\mathbf{i} + by\mathbf{j} + cz\mathbf{k}) = 1$.

Obviously $\mathbf{F} = x\mathbf{i} + y\mathbf{j} + z\mathbf{k}$. $[\because ax^2 + by^2 + cz^2 = 1]$.

Therefore, by the divergence theorem $\iint_S (\mathbf{F}\cdot\mathbf{n})\,dA = \iiint_D (\nabla\cdot\mathbf{F})\,dV$

or $$\iint_S (a^2x^2 + b^2y^2 + c^2z^2)^{-1/2} dA = \iiint_D \left(\frac{\partial}{\partial x}(x) + \frac{\partial}{\partial y}(y) + \frac{\partial}{\partial z}(z)\right)dV$$

$$= 3\iiint_D dV = 3V = 3\cdot\frac{4\pi}{3}\frac{1}{\sqrt{abc}} = \frac{4\pi}{\sqrt{abc}}. \qquad \left[\because \text{Volume of ellipsoid} = \frac{4\pi}{3}\frac{1}{\sqrt{abc}}.\right]$$

EXAMPLE 1.86 Use the divergence theorem to evaluate $\iint_S (\mathbf{v}\cdot\mathbf{n})dA$; where $\mathbf{v} = x^2z\,\mathbf{i} + y\,\mathbf{j} - xz^2\mathbf{k}$ and S is the boundary of the region bounded by the paraboloid $z = x^2 + y^2$ and the plane $z = 4y$.

(AMIETE June, 2012, 2011, 2009, 2008)

Solution We have,

$$\iint_S (\mathbf{v}\cdot\mathbf{n})dA = \iiint_D (\nabla\cdot\mathbf{v})dV = \iiint_D \left[\frac{\partial}{\partial x}(x^2 z) + \frac{\partial}{\partial y}(y) + \frac{\partial}{\partial z}(-xz^2)\right] dV$$

$$= \iiint_D (2xz + 1 - 2xz)\,dV = \iiint_D dV = \int_{y=0}^{4}\int_{x=-\sqrt{4y-y^2}}^{\sqrt{4y-y^2}}\int_{z=x^2+y^2}^{4y} dz\,dx\,dy,$$

since the projection of S on the xy-plane is $x^2 + y^2 = 4y$. Therefore,

$$\iint_S (\mathbf{v}\cdot\mathbf{n})dA = \int_{y=0}^{4}\int_{x=-\sqrt{4y-y^2}}^{\sqrt{4y-y^2}} (4y - x^2 - y^2)\,dxdy = \int_{y=0}^{4}\left[2(4y-y^2)(4y-y^2)^{\frac{1}{2}} - \frac{2}{3}(4y-y^2)^{3/2}\right]dy$$

$$= \int_{y=0}^{4}\frac{4}{3}(4y-y^2)^{3/2}\,dy = \frac{4}{3}\int_{y=0}^{4}[4-(y-2)^2]^{3/2}\,dy.$$ Put $y - 2 = 2\sin\theta$, we obtain

$$\iint_S (\mathbf{v}\cdot\mathbf{n})dA = \frac{4}{3}\int_{-\pi/2}^{\pi/2} 16\cos^4\theta\,d\theta = \frac{4}{3}(32)\frac{3}{4}\cdot\frac{1}{2}\cdot\frac{\pi}{2} = 8\pi.$$

EXAMPLE 1.87 By transforming to triple integral, evaluate $I = \iint_S x^3 dy\,dz + x^2 y\,dz\,dx + x^2 z\,dx\,dy$ where S is the closed surface consisting of the cylinder $x^2 + y^2 = a^2$ $(0 \le z \le b)$ and the circular discs $z = 0$ and $z = b$.

(Burdwan, 2003)

Solution Since $dy\,dz = \cos\alpha\,dA$, $dz\,dx = \cos\beta\,dA$, $dx\,dy = \cos\gamma\,dA$. The given integral can be written as

$$I = \iint_S (x^3\cos\alpha + x^2 y\cos\beta + x^2 z\cos\gamma)\,dA = \iint_S (\mathbf{F}\cdot\mathbf{n})\,dA,$$

where $\mathbf{F} = x^3\mathbf{i} + x^2y\mathbf{j} + x^2z\mathbf{k}$ and $\mathbf{n} = (\cos\alpha)\,\mathbf{i} + (\cos\beta)\,\mathbf{j} + (\cos\gamma)\,\mathbf{k}$, the outward drawn unit normal.

$$\text{Volume integral} = \iiint_D (\nabla\cdot\mathbf{F})dV = \iiint_D \left\{\frac{\partial}{\partial x}(x^3) + \frac{\partial}{\partial y}(x^2 y) + \frac{\partial}{\partial z}(x^2 z)\right\}dV$$

$$= \iiint_D (3x^2 + x^2 + x^2)\,dV = \iiint_D 5x^2\,dx\,dy\,dz,$$

where D is the region bounded by the cylinder and the circular discs.

Making use of the symmetry of the region D

$$I = 4\int_{z=0}^{b}\int_{y=0}^{a}\int_{x=0}^{\sqrt{a^2-y^2}} 5x^2\,dx\,dy\,dz = 20\int_0^b\int_0^a\left[\frac{x^3}{3}\right]_0^{\sqrt{a^2-y^2}} dy\,dz = \frac{20}{3}\int_0^b\boxed{\int_0^a (a^2-y^2)^{3/2}\,dy}\,dz.$$

Set $y = a\sin\theta$, we obtain

$$\int_0^a (a^2-y^2)^{3/2}\,dy = \int_0^{\pi/2}(a^3\cos^3\theta)\,a\cos\theta\,d\theta = a^4\int_0^{\pi/2}\cos^4\theta\,d\theta = \frac{3\pi a^4}{16}.$$

Therefore, $I = \dfrac{20}{3}\displaystyle\int_0^b \frac{3\pi a^4}{16}dz = \frac{5}{4}\pi a^4 b.$

EXAMPLE 1.88 Use the divergence theorem to evaluate $\iint_S (\ell x^2 + my^2 + nz^2)\,dA$ taken over the sphere $(x-a)^2 + (y-b)^2 + (z-c)^2 = \rho^2$ and ℓ, m, n are the direction cosines of the external normal to the sphere.

(AMIETE June, 2010, 2009, W 2008)

Solution *Employing Coordinates Conversion Formulas*: Spherical to rectangular of the given sphere having centre at (a, b, c) and radius ρ.

We have, $x = a + \rho\sin\phi\cos\theta$, $y = b + \rho\sin\phi\sin\theta$, $z = c + \rho\cos\phi$.

Corresponding volume elements $dV = \rho^2 \sin\phi\, d\rho\, d\phi\, d\theta$, radius varies from 0 to ρ, θ varies from 0 to 2π and ϕ varies from 0 to π.

Refer Art 1.23.1. $\iint_S (\ell x^2 + my^2 + nz^2)\,dA$ may be written in the form

$$\iiint_D \left\{\frac{\partial}{\partial x}(x^2) + \frac{\partial}{\partial y}(y^2) + \frac{\partial}{\partial z}(z^2)\right\} dx\,dy\,dz = 2\iiint_D (x+y+z)\,dV$$

$$= 2\int_{\theta=0}^{2\pi}\int_{\phi=0}^{\pi}\int_{\rho=0}^{\rho} [(a+b+c) + \rho(\sin\phi\cos\theta + \sin\phi\sin\theta + \cos\phi)]\rho^2 \sin\phi\, d\rho\, d\phi\, d\theta$$

$$= 2\int_{\theta=0}^{2\pi}\int_{\phi=0}^{\pi}\int_{\rho=0}^{\rho} (a+b+c)\rho^2 \sin\phi\, d\rho\, d\phi\, d\theta$$

$$+ 2\int_{\theta=0}^{2\pi}\int_{\phi=0}^{\pi}\int_{\rho=0}^{\rho} \{\sin^2\phi\,(\cos\theta + \sin\theta) + \sin\phi\cos\phi\}\rho^3\, d\rho\, d\phi\, d\theta$$

$$= \frac{2}{3}\int_{\theta=0}^{2\pi}\int_{\phi=0}^{\pi} \rho^3 (a+b+c)\sin\phi\, d\phi\, d\theta + \frac{\rho^4}{2}\int_{\theta=0}^{2\pi}\int_{\phi=0}^{\pi} (\cos\theta + \sin\theta)\sin^2\phi\, d\phi\, d\theta$$

$$+ \frac{\rho^4}{4}\int_{\theta=0}^{2\pi}\int_{\phi=0}^{\pi} \sin 2\phi\, d\phi\, d\theta$$

$$= \frac{2}{3}\rho^3(a+b+c)\int_{\theta=0}^{2\pi} [-\cos\phi]_0^{\pi} d\theta + \frac{\rho^4}{4}(\cos\theta + \sin\theta)\int_{\theta=0}^{2\pi}\left[\phi - \frac{\sin 2\phi}{2}\right]_0^{\pi} d\theta - \frac{\rho^4}{8}\int_{\theta=0}^{2\pi} [\cos 2\phi]_0^{\pi} d\theta$$

$$= \frac{4}{3}\rho^3(a+b+c)[\theta]_0^{2\pi} + \frac{\pi\rho^4}{4}\int_{\theta=0}^{2\pi}(\cos\theta + \sin\theta)\,d\theta + 0 \qquad [\because\ [\cos 2\theta]_0^{\pi} = 0.]$$

$$= \frac{8}{3}\pi\rho^3(a+b+c) + 0 = \frac{8}{3}\pi\rho^3(a+b+c).$$

1.24 GREEN'S IDENTITIES (FORMULAS)

Two important identities derived from divergence theorems, called Green's identities which are of great importance in the fields of electro dynamics and hydrodynamics. Let S be a piecewise smooth surface bounding a domain D in some region. Let $f(x, y, z)$ and $g(x, y, z)$ are scalar functions which are continuous and have continuous first and second order partial derivatives in this region of the three dimensional space. Then

(a) $\iiint_D (f\nabla^2 g + \nabla f \cdot \nabla g)\,dV = \iint_S f(\nabla g \cdot \mathbf{n})\,dA.$

This result is known as Green's first identity or Theorem.

(b) $\iiint_D (f\nabla^2 g - g\nabla^2 f)\,dV = \iint_S (f\nabla g - g\nabla f)\cdot \mathbf{n}\,dA.$

This result is called Green's second identity.

Proof:

(*a*) Let the functions f and g be such that $\mathbf{F} = f \text{ grad } g = f\nabla g$. If we put $\mathbf{F} = f\nabla g$ in the divergence theorem

$$\iiint_D (\nabla \cdot \mathbf{F})\, dV = \iint_S (\mathbf{F} \cdot \mathbf{n})\, dA, \text{ we obtain } \iiint_D \nabla \cdot (f\nabla g)\, dV = \iint_S f(\nabla g \cdot \mathbf{n})\, dA. \quad ...(i)$$

Using formula (2) Art 1.17, $\nabla \cdot (f\nabla g) = f(\nabla \cdot \nabla g) + (\nabla g \cdot \nabla f) = f\nabla^2 g + \nabla f \cdot \nabla g$.

Thus (*i*) can be written as

$$\iiint_D (f\nabla^2 g + \nabla f \cdot \nabla g)\, dV = \iint_S f(\nabla g \cdot \mathbf{n})\, dA, \quad ...(ii)$$

which proves *Green's first identity.*

(*b*) Interchanging f and g in (*ii*), we obtain

$$\iiint_D (g\nabla^2 f + \nabla g \cdot \nabla f)\, dV = \iint_S g(\nabla f \cdot \mathbf{n})\, dA. \quad ...(iii)$$

Subtracting (*iii*) from (*ii*), we have

$$\iiint_D (f\nabla^2 g - g\nabla^2 f)\, dV = \iint_S (f\nabla g - g\nabla f) \cdot \mathbf{n}\, dA,$$

which proves *Green's second identity.*

EXERCISE 1.7

1. Show that for the vector $\mathbf{F} = y\cos(2x + y^2)\mathbf{i} - \cos(2x + y^2)\mathbf{j}$, $\iint_S (\mathbf{F} \cdot \mathbf{n})\, dA = 0$, where S is a closed surface bounding a volume V.

   [**Hint:** $\iint_S (\mathbf{F} \cdot \mathbf{n})\, dA = \iiint_D (\nabla \cdot \mathbf{F})\, dV = 0$. Therefore, $\nabla \cdot \mathbf{F} = -2y\sin(2x + y^2) + 2y\sin(2x + y^2) = 0$].

2. Verify the divergence theorem for $\mathbf{F} = (x^2 - yz)\mathbf{i} + (y^2 - zx)\mathbf{j} + (z^2 - xy)\mathbf{k}$ taken over the rectangular parallelepiped $0 \le x \le a$, $0 \le y \le b$, $0 \le z \le c$. **Ans.** Common value = $abc(a + b + c)$

   (*AMIETE, Dec. 2012; MDU 2006; PTU 2003; Madras 2003; UPTU 2002; VTU 2000s*)

3. Use divergence theorem to evaluate $\iint_S (\mathbf{F} \cdot \mathbf{n})\, dA$, where $\mathbf{F} = x^3\mathbf{i} + y^3\mathbf{j} + z^3\mathbf{k}$ and S is the surface of the sphere $x^2 + y^2 + z^2 = a^2$. (*AMIETE, Dec. 2009; PTU 2004*)

   [**Hint:** $\iint_S (\mathbf{F} \cdot \mathbf{n})\, dA = \iiint_D (\nabla \cdot \mathbf{F})\, dV$, where V is the volume of the sphere

$$= \iiint_D (3x^2 + 3y^2 + 3z^2)\, dx\, dy\, dz = 3\iiint_D (x^2 + y^2 + z^2)\, dx\, dy\, dz$$

$$= 3\iiint_D r^2 \cdot r^2 \sin\theta\, dr\, d\theta\, d\phi \text{ (by transforming to spherical polar coordinates)}$$

$$= 3\int_0^{2\pi}\int_0^{\pi}\int_0^{a} r^4 \sin\theta\, dr\, d\theta\, d\phi = \frac{12}{5}\pi a^5.\Bigg]$$

4. Show that if **R** is the position vector of any point of a closed surface S bounding a volume V, then $\iint_S (\mathbf{R}\cdot\mathbf{n})\,dA = 3V$. *(AMIETE, June 2013, Dec. 2007)*

   [**Hint:** $\iint_S (\mathbf{R}\cdot\mathbf{n})\,dA = \iiint_D (\nabla\cdot\mathbf{R})\,dV$. By divergence theorem $\iint_S (\mathbf{R}\cdot\mathbf{n})\,dA = 3\iiint_D dV = 3V$.]

5. Find the value of $\iint_S (\mathbf{F}\cdot\mathbf{n})\,dA$ where $\mathbf{F} = r^2(\mathbf{i}x + \mathbf{j}y + \mathbf{k}z)$ and $r^2 = x^2 + y^2 + z^2$ and S is the surface of the sphere $x^2 + y^2 + z^2 = a^2$. Compute the integral with the aid of divergence theorem.

   [**Hint:** $\nabla\cdot\mathbf{F} = \nabla\cdot(\mathbf{r}r^2) = r^2\nabla\cdot\mathbf{r} + \mathbf{r}\cdot\nabla r^2 = 3r^2 + \mathbf{r}\cdot 2r\frac{\mathbf{r}}{r} = 3r^2 + 2r^2 = 5r^2$.

   Therefore, $\iint_S (\mathbf{F}\cdot\mathbf{n})\,dA = \iiint_D (\nabla\cdot\mathbf{F})\,dV = \iiint_D 5r^2\,dV = 5\left(\frac{4}{5}\pi a^5\right)$, by problem (3) above $= 4\pi a^5$.]

6. The vector field $\mathbf{F} = x^2\mathbf{i} + z\,\mathbf{j} + yz\,\mathbf{k}$ is defined over the volume of the cuboid given by $0 \le x \le a$, $0 \le y \le b$, $0 \le z \le c$. Evaluate the surface integral $\iint_S (\mathbf{F}\cdot\mathbf{n})\,dA$, where S is the surface of the cuboid.

   (AMIETE June 2007, Dec. 2006; UPTU 2002) **Ans.** $abc\,(2a + b)/2$.

7. Verify Gauss Divergence theorem for $\mathbf{F} = x^2\mathbf{i} + y^2\mathbf{j} + z^2\mathbf{k}$ on the surface S of the cuboid formed by the planes $x = 0$, $x = a$, $y = 0$, $y = b$, $z = 0$ and $z = c$. *(AMIETE Dec. 2004)*

8. Verify the divergence theorem for $\mathbf{F} = (2x - z)\mathbf{i} + x^2y\,\mathbf{j} - xz^2\mathbf{k}$ taken over the region bounded by $x = 0$, $x = 1$, $y = 0$, $y = 1$, $z = 0$, $z = 1$. **Ans.** Common value = 11/6.

9. Find the value of the surface integral, by divergence theorem, $\iint_S (2x^2y\,dy\,dz - y^2\,dz\,dx + 4xz^2\,dx\,dy)$ where S is the curved surface of the cylinder $y^2 + z^2 = 9$ bounded by the planes $x = 0$, $x = 2$.

   Ans. 180.

10. Verify the divergence theorem for $\mathbf{F}(x, y, z) = x\mathbf{i} + y\mathbf{j} + z\mathbf{k}$ on the surface S of the sphere $x^2 + y^2 + z^2 = a^2$. *(AMIETE June 2006)*

11. Verify the divergence theorem for the function $\mathbf{F} = y\mathbf{i} + x\,\mathbf{j} + z^2\mathbf{k}$ taken over the cylindrical region bounded by $x^2 + y^2 = a^2$, $z = 0$, $z = h$. *(MDU 2005)* **Ans.** Common value = πa^2h^2.

12. Verify the divergence theorem for $\mathbf{F} = \frac{x}{r}\mathbf{i} + \frac{y}{r}\mathbf{j} + \frac{z}{r}\mathbf{k}$, where $r = \sqrt{x^2 + y^2 + z^2}$, over the region $x^2 + y^2 + z^2 \le a^2$.

    [**Hint:** $\mathbf{F} = \mathbf{r}/r$. $\nabla\cdot\mathbf{F} = \frac{1}{r}\nabla\cdot\mathbf{r} + \nabla\frac{1}{r}\cdot\mathbf{r} = \frac{2}{r}$.

    Therefore, $\iiint_D (\nabla\cdot\mathbf{F})\,dV = \iiint_D \frac{2}{r}\,dV = \int_0^{2\pi}\int_0^{\pi}\int_0^{a} \frac{2}{r}\cdot r^2\sin\theta\;dr\,d\theta\,d\phi = 4\pi a^2$.

    ($\because$ in spherical coordinates $dV = r^2\sin\theta\,dr\,d\theta\,d\phi$)

    Also $\mathbf{N} = \nabla f = \nabla(x^2 + y^2 + z^2 - a^2) = 2x\mathbf{i} + 2y\mathbf{j} + 2z\mathbf{k}$.

    $\mathbf{n} = \frac{\nabla f}{|\nabla f|} = \frac{x\mathbf{i} + y\mathbf{j} + z\mathbf{k}}{a}$ and $\mathbf{F}\cdot\mathbf{n} = 1$ [$\because$ on the surface $r = a$]

$\therefore \iint_S (\mathbf{F}\cdot\mathbf{n})\, dA = \iint_S 1 dA = 4\pi a^2$. Hence divergence theorem is verified.]

13. Evaluate:

(*a*) $\iint_S x^2\, dy\, dz + y^2\, dz\, dx + z^2\, dx\, dy$, where S is the surface of the cube $0 \le x \le 1, 0 \le y \le 1, 0 \le z \le 1$.

(*b*) $\iint_S yz\, dy\, dz + zx\, dz\, dx + xy\, dx\, dy$, S being the surface of the sphere $x^2 + y^2 + z^2 = 4$.

Ans. (*a*) 3, (*b*) zero.

14. Evaluate $\iint_S xz^2\, dy\, dz + (x^2 y - z^3)\, dz\, dx + (2xy + y^2 z)\, dx\, dy$, where S is the entire surface of the hemispherical region bounded by $z = \sqrt{a^2 - x^2 - y^2}$ and $z = 0$ by using the divergence theorem.

Ans. $2\pi a^5/5$.

[**Hint:** $\iint_S xz^2\, dy\, dz + (x^2 y - z^3)\, dz\, dx + (2xy + y^2 z)\, dx\, dy$

$$= \iint_S [\{xz^2\mathbf{i} + (x^2 y - z^3)\mathbf{j} + (2xy + y^2 z)\mathbf{k}\}\cdot\{(\cos\alpha)\mathbf{i} + (\cos\beta)\mathbf{j} + (\cos\gamma)\mathbf{k}\}]\, dA$$

$= \iint_S (\mathbf{F}\cdot\mathbf{n})\, dA = \iiint_D (\nabla\cdot\mathbf{F})\, dV$, by divergence theorem, where $\mathbf{n}$ is outward drawn unit normal.

$$= \iiint_D \left\{\frac{\partial}{\partial x}(xz^2) + \frac{\partial}{\partial y}(x^2 y - z^3) + \frac{\partial}{\partial z}(2xy + y^2 z)\right\} dV = \iiint_D (x^2 + y^2 + z^2)\, dV,$$

where D is the region bounded by the hemisphere and the xy-plane.

By use of spherical coordinates, above integral is equal to $4\int_{\theta=0}^{\pi/2}\int_{\phi=0}^{\pi/2}\int_{\rho=0}^{a} \rho^2\cdot\rho^2 \sin\phi\, d\rho\, d\phi\, d\theta = 2\pi a^5/5$.]

15. If $\mathbf{F} = \nabla\phi$ and $\nabla^2\phi = -4\pi\rho$, prove that $\iint_S (\mathbf{F}\cdot\mathbf{n})\, dA = -4\pi\rho \iiint_D dV$, where the symbols have their usual meanings.

16. Prove that $\iiint_D \frac{dV}{r^2} = \iint_S \left(\frac{\mathbf{R}\cdot\mathbf{n}}{r^2}\right) dA$, where $r = |\mathbf{R}| = |x\mathbf{i} + y\mathbf{j} + z\mathbf{k}|$, other symbols having their usual meanings.

[**Hint:** R.H.S. $= \iint_S \left(\frac{\mathbf{R}\cdot\mathbf{n}}{r^2}\right) dA = \iiint_D \left(\nabla\cdot\frac{\mathbf{R}}{r^2}\right) dV$.

Now $\nabla\cdot\frac{\mathbf{R}}{r^2} = \frac{1}{r^2}\nabla\cdot\mathbf{R} + \mathbf{R}\cdot\nabla\frac{1}{r^2} = \frac{3}{r^2} + \mathbf{R}\cdot\left(-2r^{-3}\frac{\mathbf{R}}{r}\right) = \frac{1}{r^2}$ etc.]

17. Evaluate the surface integral $\iint_S (yz\mathbf{i} + zx\mathbf{j} + xy\mathbf{k}) \cdot dA$, where S is the surface of the sphere $x^2 + y^2 + z^2 = 1$ in the first octant.

[**Hint:** Here $\mathbf{F} = yz\mathbf{i} + zx\mathbf{j} + xy\mathbf{k}$. $\iiint_D (\nabla \cdot \mathbf{F})\, dV = \iint_{S_1} (\mathbf{F} \cdot \mathbf{n})\, dA$,

where S_1 is the closed surface encloing the volume of the sphere in first octant. (Refer Fig. 1.38). Since $\nabla \cdot \mathbf{F} = 0$.

Therefore, $\iiint_D (\nabla \cdot \mathbf{F})\, dV = 0$. By divergence theorem,

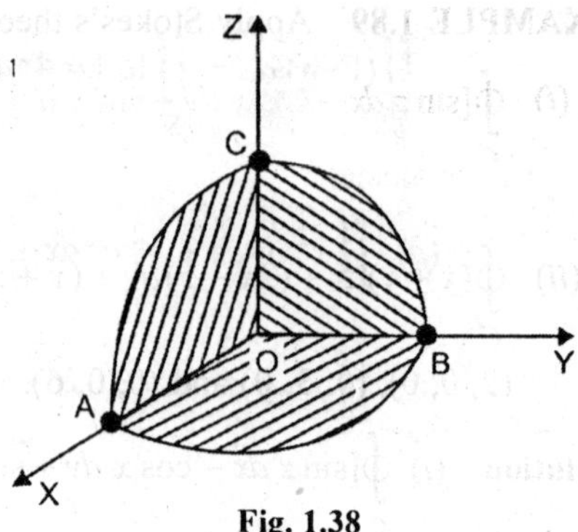

Fig. 1.38

$$\iiint_D \text{div } \mathbf{F}\, dV = \iint_{S_1} (\mathbf{F} \cdot \mathbf{n})\, dA = \iint_{OBC} (\mathbf{F} \cdot \mathbf{n})\, dA + \iint_{OAC} (\mathbf{F} \cdot \mathbf{n})\, dA + \iint_{OAB} (\mathbf{F} \cdot \mathbf{n})\, dA + \iint_S (\mathbf{F} \cdot \mathbf{n})\, dA,$$

where S is the curved surface of the sphere in first octant.

$$\Rightarrow 0 = -\frac{1}{8} - \frac{1}{8} - \frac{1}{8} + \iint_S (\mathbf{F} \cdot \mathbf{n})\, dA. \text{ Therefore, } \iint_S (\mathbf{F} \cdot \mathbf{n})\, dA = \frac{3}{8}.$$]

1.25 STOKES'S THEOREM (TRANSFORMATION BETWEEN SURFACE INTEGRALS AND LINE INTEGRALS)

Let S be a piecewise smooth orientable surface bounded by a piecewise smooth simple closed curve C. Let $\mathbf{F}(x, y, z) = \mathbf{F}_1(x, y, z)\, \mathbf{i} + \mathbf{F}_2(x, y, z)\, \mathbf{j} + \mathbf{F}_3(x, y, z)\, \mathbf{k}$ be a vector field for which F_1, F_2 and F_3 are continuous and have continuous first order partial derivatives in a domain which contains S. If C is traversed in the positive direction, then

$$\iint_S (\text{curl } \mathbf{F}) \cdot \mathbf{n})\, dA = \iint_S (\nabla \times \mathbf{F}) \cdot \mathbf{n}\, dA = \oint_C \mathbf{F} \cdot d\mathbf{r} \qquad \text{...(1.43)}$$

where $\mathbf{n}$ is the unit normal vector to S in the direction of orientation of C. (Refer Fig. 1.39)

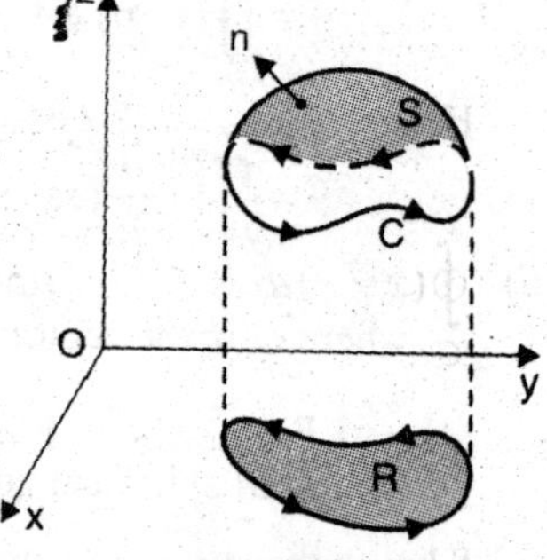

Fig. 1.39

Remark 13

(a) Expressed in words, the Stokes's theorem states that "The line integral of the tangential component of a vector **F** taken around a simple closed curve C is equal to the surface integral of the normal component of curl **F** taken over any surface S having C as its boundary".

(b) The cartesian form of Stokes's theorem is

$$\iint_S \left(\frac{\partial F_3}{\partial y} - \frac{\partial F_2}{\partial z}\right) dydz + \left(\frac{\partial F_1}{\partial z} - \frac{\partial F_3}{\partial x}\right) dzdx + \left(\frac{\partial F_2}{\partial x} - \frac{\partial F_1}{\partial y}\right) dxdy$$

$$= \oint_C (F_1 dx + F_2 dy + F_3 dz). \qquad \text{...(1.43)}$$

Also by the definition of surface integral, relation (1.43) is equivalent to

$$\iint_S \left[\left(\frac{\partial F_3}{\partial y} - \frac{\partial F_2}{\partial z}\right)\cos\alpha + \left(\frac{\partial F_1}{\partial z} - \frac{\partial F_3}{\partial x}\right)\cos\beta + \left(\frac{\partial F_2}{\partial x} - \frac{\partial F_1}{\partial y}\right)\cos\gamma\right] dA = \oint_C (F_1 dx + F_2 dy + F_3 dz),$$

where $\cos\alpha$, $\cos\beta$, $\cos\gamma$ are the direction cosines of the normal at any point to the surface.

EXAMPLE 1.89 Apply Stokes's theorem to evaluate:

(*i*) $\oint_C [\sin z\, dx - \cos x\, dy + \sin y\, dz]$, where C is the boundary of the rectangle $0 \le x \le \pi$, $0 \le y \le 1$, $z = 3$.

(*AMIETE, Dec 2011, 2010 June 2009; AMIE S-2009*)

(*ii*) $\oint_C (x+y)dx + (2x-z)dy + (y+z)dz$, where C is the boundary of the triangle with vertices (2, 0, 0), (0, 3, 0) and (0, 0, 6). (*AMIE, W-2011*)

Solution (*i*) $\oint_C [\sin z\, dx - \cos x\, dy + \sin y\, dz] = \oint_C [(\sin z)\mathbf{i} - (\cos x)\mathbf{j} + (\sin y)\mathbf{k}] \cdot [(dx)\mathbf{i} + (dy)\mathbf{j} + (dz)\mathbf{k}]$

$= \oint_C \mathbf{F} \cdot d\mathbf{r}$, where $\mathbf{F} = (\sin z)\mathbf{i} - (\cos x)\mathbf{j} + (\sin y)\mathbf{k} = \iint_S (\nabla \times \mathbf{F}) \cdot \mathbf{n}\, dA$, [by Stokes's theorem]

where S is the surface of the rectangle $0 \le x \le \pi$, $0 \le y \le 1$, $z = 3$.

$$\text{Now } \nabla \times \mathbf{F} = \begin{vmatrix} \mathbf{i} & \mathbf{j} & \mathbf{k} \\ \partial/\partial x & \partial/\partial y & \partial/\partial z \\ \sin z & -\cos x & \sin y \end{vmatrix} = (\cos y)\mathbf{i} + (\cos z)\mathbf{j} + (\sin x)\mathbf{k}.$$

Also for the surface of the rectangle, $\boldsymbol{n} = \boldsymbol{k}$. $\therefore$ $(\nabla \times \mathbf{F}) \cdot \mathbf{n} = [(\cos y)\mathbf{i} + (\cos z)\mathbf{j} + (\sin x)\mathbf{k}] \cdot \mathbf{k} = \sin x$.

$$\text{Thus, } \iint_S (\nabla \times \mathbf{F}) \cdot \mathbf{n}\, dA = \iint_S \sin x\, dA = \int_0^{\pi} \int_0^1 \sin x\, dy\, dx = \int_0^{\pi} \sin x\, dx = 2.$$

$$\text{Hence } \oint_C [\sin z\, dx - \cos x\, dy + \sin y\, dz] = 2.$$

(*ii*) $\oint_C (x+y)dx + (2x-z)dy + (y+z)dz = \oint_C [(x+y)\mathbf{i} + (2x-z)\mathbf{j} + (y+z)\mathbf{k}] \cdot d\mathbf{r}$

$= \oint_C \mathbf{F} \cdot d\mathbf{r}$ where $\mathbf{F} = (x+y)\mathbf{i} + (2x-z)\mathbf{j} + (y+z)\mathbf{k} = \iint_S (\nabla \times \mathbf{F}) \cdot \mathbf{n}\, dA$. [By Stokes's Theorem],

S being the surface of the triangle ABC (see Fig. 1.40.).

$$\text{Now } \nabla \times \mathbf{F} = \begin{vmatrix} \mathbf{i} & \mathbf{j} & \mathbf{k} \\ \partial/\partial x & \partial/\partial y & \partial/\partial z \\ x+y & 2x-z & y+z \end{vmatrix} = 2\mathbf{i} + \mathbf{k}.$$

To obtain **n** we note that a vector normal to the plane of the triangle ABC, that is, $x/2 + y/3 + z/6 = 1$, that is, $3x + 2y + z - 6 = 0$ is $\nabla(3x + 2y + z - 6) = 3\mathbf{i} + 2\mathbf{j} + \mathbf{k}$.

A unit normal at any point of the triangle ABC is

$$\mathbf{n} = \frac{3\mathbf{i} + 2\mathbf{j} + \mathbf{k}}{\sqrt{9+4+1}} = \frac{3\mathbf{i} + 2\mathbf{j} + \mathbf{k}}{\sqrt{14}}.$$

$$\text{Now } (\nabla \times \mathbf{F}) \cdot \mathbf{n} = (2\mathbf{i} + \mathbf{k}) \cdot \frac{3\mathbf{i} + 2\mathbf{j} + \mathbf{k}}{\sqrt{14}} = \frac{7}{\sqrt{14}}.$$

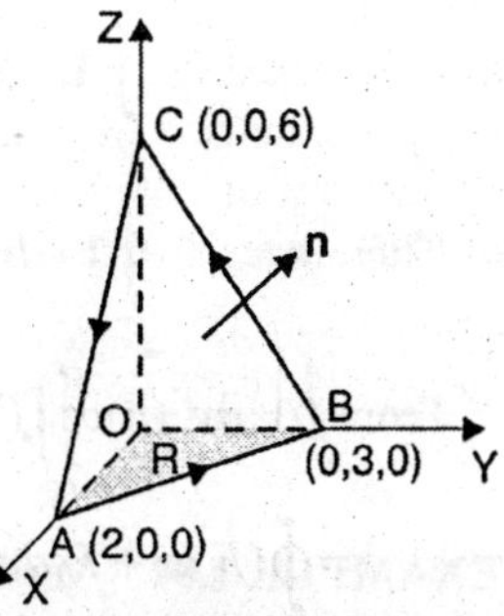

Fig. 1.40

Therefore, $$\iint_S (\nabla \times \mathbf{F}) \cdot \mathbf{n} dA = \iint_R \frac{7}{\sqrt{14}} \frac{dx\,dy}{\mathbf{n}\cdot\mathbf{k}} = \iint_R 7 dx\,dy \quad \left(\because \mathbf{n}\cdot\mathbf{k} = \frac{1}{\sqrt{14}}\right),$$

where R is the projection of S on the xy-plane, that is, triangle OAB = 7 (area of triangle OAB)
$= 7(1/2)(2)(3) = 21$ square units.

Hence $$\oint_C (x + y)\,dx + (2x - z)\,dy + (y + z)\,dz = 21.$$

EXAMPLE 1.90 Verify Stokes's theorem for the function $\mathbf{F} = x^2\mathbf{i} + xy\mathbf{j}$ integrated round the square in the plane $z = 0$ and bounded by the lines $x = 0$, $y = 0$, $x = a$ and $y = a$. (*PTU 2002*)

Solution $$\text{curl}\,\mathbf{F} = \nabla \times \mathbf{F} = \begin{vmatrix} \mathbf{i} & \mathbf{j} & \mathbf{k} \\ \partial/\partial x & \partial/\partial y & \partial/\partial z \\ x^2 & xy & 0 \end{vmatrix} = \mathbf{k}\,y.$$

Therefore, $$\iint_S (\nabla \times \mathbf{F}) \cdot \mathbf{n}\; dA = \int_0^a \int_0^a \mathbf{k}y \cdot \mathbf{k}\; dydx = \int_{x=0}^{a} \int_{y=0}^{a} y\,dy\,dx = \frac{a^3}{2}. \quad \ldots(i)$$

Now $$\oint_C \mathbf{F} \cdot d\mathbf{r} = \int_{OA} \mathbf{F} \cdot d\mathbf{r} + \int_{AB} \mathbf{F} \cdot d\mathbf{r} + \int_{BC} \mathbf{F} \cdot d\mathbf{r} + \int_{CO} \mathbf{F} \cdot d\mathbf{r}$$

Along OA, $y = 0$, $dy = 0$. $\mathbf{F} = x^2\mathbf{i}$, $d\mathbf{r} = dx\mathbf{i}$ and x varies from 0 to a.

Therefore, $$\int_{OA} \mathbf{F} \cdot d\mathbf{r} = \int_0^a x^2\mathbf{i} \cdot dx\,\mathbf{i} = \int_0^a x^2 dx = \frac{a^3}{3}.$$

Along AB, $x = a$, $dx = 0$, $\mathbf{F} = a^2\mathbf{i} + ay\mathbf{j}$, $d\mathbf{r} = dy\mathbf{j}$ and y changes from 0 to a.

Therefore, $$\int_{AB} \mathbf{F} \cdot d\mathbf{r} = \int_0^a (a^2\mathbf{i} + ay\mathbf{j}) \cdot dy\,\mathbf{j} = a\int_0^a y\,dy = \frac{a^3}{2}.$$

Similarly $$\int_{BC} \mathbf{F} \cdot d\mathbf{r} = \int_a^0 (x^2\mathbf{i} + xa\mathbf{j}) \cdot dx\,\mathbf{i} = \int_a^0 x^2\,dx = -\frac{a^3}{3}.$$

$$\int_{CO} \mathbf{F} \cdot d\mathbf{r} = 0 \quad (\because \text{along } CO,\ x = 0. \text{ Therefore, } \mathbf{F} = 0)$$

Therefore, $$\oint_C \mathbf{F} \cdot d\mathbf{r} = \frac{a^3}{3} + \frac{a^3}{2} - \frac{a^3}{3} + 0 = \frac{a^3}{2}. \quad \ldots(ii)$$

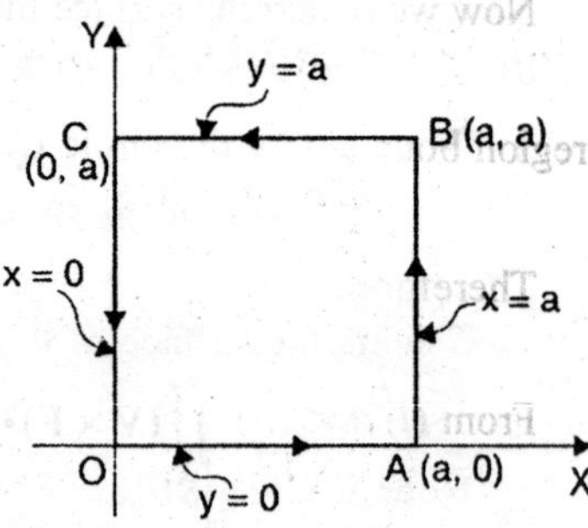

Fig. 1.41

From (*i*) and (*ii*) $\iint_S (\nabla \times \mathbf{F}) \cdot \mathbf{n}\; dA = \oint_C \mathbf{F} \cdot d\mathbf{r}$. Hence Stokes's theorem is verified.

EXAMPLE 1.91 Verify Stokes's theorem for the vector field $\mathbf{F} = (2x - y)\mathbf{i} - yz^2\mathbf{j} - y^2z\,\mathbf{k}$ over the upper half of the surface of $x^2 + y^2 + z^2 = 1$ bounded by its projection on the xy-plane.
(*Madras 2006; SVTU 2006; PTU 2001*)

Solution Here the curve C bounding the surface of the sphere is the circle in the xy-plane with radius one and centre at (0, 0). The parametric equations of the circle are: $x = \cos\theta$, $y = \sin\theta$, $z = 0$ $\quad 0 \le \theta \le 2\pi$.

Therefore, $$\oint_C \mathbf{F} \cdot d\mathbf{r} = \oint_C (2x - y)\,dx - yz^2\,dy - y^2z\,dz = \int_0^{2\pi} (2\cos\theta - \sin\theta)(-\sin\theta\, d\theta) = \pi. \quad \ldots(i)$$

Now, $$\nabla\times\mathbf{F}=\begin{vmatrix}\mathbf{i} & \mathbf{j} & \mathbf{k}\\ \partial/\partial x & \partial/\partial y & \partial/\partial z\\ 2x-y & -yz^2 & -y^2z\end{vmatrix}=\mathbf{k}.$$

Therefore, $$\iint_S(\nabla\times\mathbf{F})\cdot\mathbf{n}\,dA=\iint_S\mathbf{k}\cdot\mathbf{n}\,dA=\iint_R dy\,dx=\pi(1)^2=\pi. \qquad ...(ii)$$

(where **R** is the projection of S on the xy-plane, that is., a circle with radius unity)

From (*i*) and (*ii*) $\iint_S(\nabla\times\mathbf{F})\cdot\mathbf{n}\,dA=\oint_C\mathbf{F}\cdot d\mathbf{r}$ and the Stokes's theorem is verified.

EXAMPLE 1.92 Evaluate $\iint_S(\nabla\times\mathbf{F})\cdot\mathbf{n}\,dA$, where $\mathbf{F}=y\mathbf{i}+z\mathbf{j}+x\mathbf{k}$ and S the portion of the surface $x^2+y^2-2ax+az=0$ above the xy-plane. Also Verify Stokes's theorem

Solution We have $\iint_S(\nabla\times\mathbf{F})\cdot\mathbf{n}\,dA=\oint_C\mathbf{F}\cdot d\mathbf{r}$ [By Stokes's theorem]

where C is the circle given by $x^2+y^2-2ax+az=0$, $z=0$, that is, $x^2+y^2-2ax=0$ in the xy-plane.

The equation of the circle can be written in the form $(x-a)^2+y^2=a^2$ or $x=a+a\cos\theta$, $y=a\sin\theta$, the parametric equations of the circle.

In the xy-plane, that is $z=0$, $\mathbf{F}\cdot d\mathbf{r}=(y\mathbf{i}+x\mathbf{k})\cdot(dx\,\mathbf{i}+dy\,\mathbf{j})=y\,dx.$

Therefore, $$\oint_C\mathbf{F}\cdot d\mathbf{r}=\int_0^{2\pi}(a\sin\theta)(-a\sin\theta)\,d\theta=-a^2\int_0^{2\pi}\sin^2\theta\,d\theta=-\pi a^2. \qquad ...(i)$$

Now we obtain the surface integral, we note that $\iint_S(\nabla\times\mathbf{F})\cdot\mathbf{n}\,dA=\iint_{S_1}(\nabla\times\mathbf{F})\cdot\mathbf{n}\,dA$, where S_1 is the plane region bounded by the circle C. Here $\mathbf{n}=\mathbf{k}$. Now $(\nabla\times\mathbf{F})\cdot\mathbf{n}=-(\mathbf{i}+\mathbf{j}+\mathbf{k})\cdot\mathbf{k}=-1.$

Therefore, $$\iint_{S_1}(\nabla\times\mathbf{F})\cdot\mathbf{n}\,dA=-\iint_{S_1}1\,dA=-\pi a^2. \qquad ...(ii)$$

From (*i*) and (*ii*) $\iint_S(\nabla\times\mathbf{F})\cdot\mathbf{n}\,dA=\oint_C\mathbf{F}\cdot d\mathbf{r}$ and the Stokes's theorem is verified.

EXAMPLE 1.93 Apply Stokes's theorem to prove that $\oint_C y\,dx+z\,dy+x\,dz=-2\sqrt{2}\,\pi a^2$, where C is the curve given by $x^2+y^2+z^2-2ax-2ay=0$, $x+y=2a$ and begins at the point $(2a, 0, 0)$ and goes at first below the z-plane.

Solution Obviously the curve C is a great circle of the sphere $x^2+y^2+z^2-2ax-2ay=0$ (the circle is shown dotted in Fig. 1.42) in the plane $x+y=2a$ which passes through A and B and is perpendicular to the z-plane.

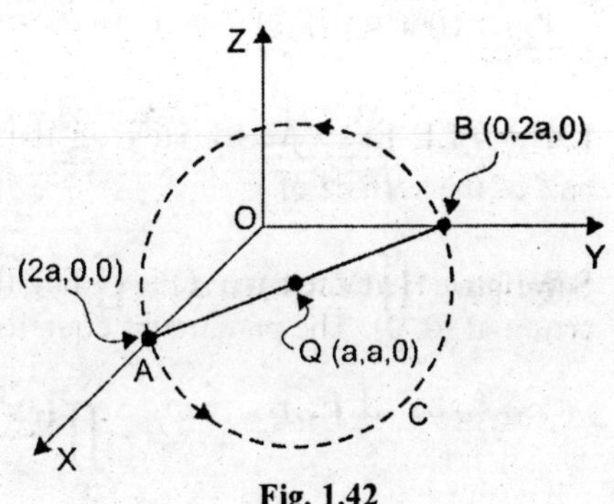

Fig. 1.42

Now $$\oint_C y\,dx+z\,dy+x\,dz=\int_C(y\mathbf{i}+z\mathbf{j}+x\mathbf{k})\cdot d\mathbf{r}$$

$$=\iint_S\{\nabla\times(y\mathbf{i}+z\mathbf{j}+x\mathbf{k})\}\cdot\mathbf{n}\,dA,$$

where S is the surface of the circle referred to above.

$$\nabla \times (y\mathbf{i} + z\mathbf{j} + x\mathbf{k}) = -(\mathbf{i} + \mathbf{j} + \mathbf{k}), \quad f(x, y) = x + y - 2a = 0.$$

Therefore, $\mathbf{N} = \nabla f = \mathbf{i} + \mathbf{j}$. Hence $\mathbf{n} = \dfrac{\mathbf{N}}{|\mathbf{N}|} = \dfrac{\mathbf{i}+\mathbf{j}}{\sqrt{2}}$.

$$\text{Thus the given integral} = \iint_S \left[-(\mathbf{i}+\mathbf{j}+\mathbf{k}) \cdot \frac{\mathbf{i}+\mathbf{j}}{\sqrt{2}}\right] dA$$

$$= \frac{-2}{\sqrt{2}} \iint_S dA = \frac{-2}{\sqrt{2}} \text{ (area of the circle) } = \frac{-2}{\sqrt{2}} \pi (\sqrt{2}a)^2 = -2\sqrt{2}\pi a^2.$$

$$\text{Hence } \oint_C y\,dx + z\,dy + x\,dz = -2\sqrt{2}\,\pi\,a^2.$$

EXAMPLE 1.94 Verify Stokes's theorem for the function $\mathbf{F} = 2y^3\mathbf{i} + x^3\mathbf{j} + z\mathbf{k}$, where C is the curve of intersection of cone $z = \sqrt{x^2 + y^2}$ by the plane $z = 4$ and S is surface of cone below $z = 4$.

(PTU 2006; AMIETE Dec 2005)

Solution Refer Figure 1.43. The bounding curve C of S is a circle with equations $x^2 + y^2 = 16$, $z = 4$. Setting $x = 4\cos\theta$, $y = 4\sin\theta$, we obtain

$$\oint_C \mathbf{F} \cdot d\mathbf{r} = \oint_C 2y^3 dx + x^3 dy + z\,dz = \oint_C 2y^3 dx + x^3 dy \quad [\because dz = 0]$$

$$= \int_{2\pi}^{0} 64[2\sin^3\theta(-4\sin\theta) + \cos^3\theta(4\cos\theta)]\,d\theta$$

$$= -256\int_0^{2\pi} (\cos^4\theta - 2\sin^4\theta)\,d\theta = -1024\int_0^{\pi/2} (\cos^4\theta - 2\sin^4\theta)\,d\theta$$

$$= -1024\left[\frac{3}{4}\cdot\frac{1}{2}\cdot\frac{\pi}{2} - 2\left(\frac{3}{4}\cdot\frac{1}{2}\cdot\frac{\pi}{2}\right)\right] = 192\pi. \quad ...(i)$$

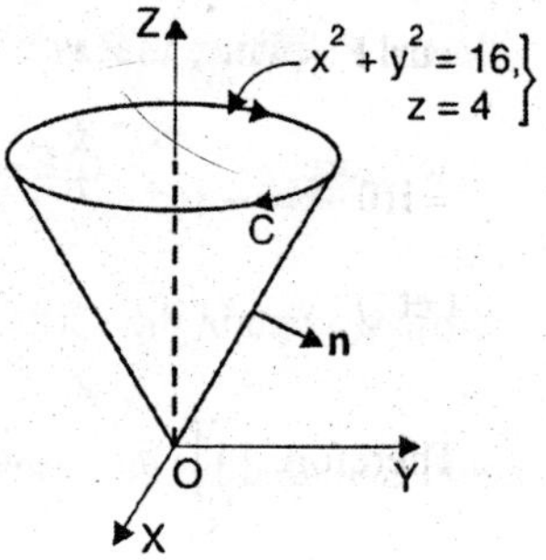

Fig. 1.43

$$\text{We have curl } \mathbf{F} = \nabla \times \mathbf{F} = \begin{vmatrix} \mathbf{i} & \mathbf{j} & \mathbf{k} \\ \partial/\partial x & \partial/\partial y & \partial/\partial z \\ 2y^3 & x^3 & z \end{vmatrix} = \mathbf{i}(0) - \mathbf{j}(0) + \mathbf{k}(3x^2 - 6y^2) = \mathbf{k}(3x^2 - 6y^2).$$

If the outward normal to S is taken, then it points downwards. The orientation of C is shown in Fig. 1.43. Let $f(x, y, z) = \sqrt{x^2 + y^2} - z = 0$ be taken as the equation of the surface. Then unit normal is given by $\mathbf{n} = \mathbf{N}/|\mathbf{N}|$.

$$\mathbf{N} = \nabla f = \frac{x\mathbf{i} + y\mathbf{j}}{\sqrt{x^2 + y^2}} - \mathbf{k} = \frac{x\mathbf{i} + y\mathbf{j} - z\mathbf{k}}{z}, \text{ and } \mathbf{n} = \frac{(x\mathbf{i} + y\mathbf{j} - z\mathbf{k})/z}{\sqrt{(x^2 + y^2 + z^2)/z^2}} = \frac{x\mathbf{i} + y\mathbf{j} - z\mathbf{k}}{\sqrt{2}\,z}, \text{ except at the origin.}$$

$$\text{We have } \iint_S (\nabla \times \mathbf{F}) \cdot \mathbf{n}\,dA = \iint_S -\frac{(3x^2 - 6y^2)}{\sqrt{2}}\,dA = -\iint_R \frac{(3x^2 - 6y^2)}{\sqrt{2}} \frac{dx\,dy}{\mathbf{n}\cdot\mathbf{k}}$$

$$= -\iint_R \frac{(3x^2 - 6y^2)}{\sqrt{2}} \frac{dx\,dy}{(-1/\sqrt{2})} = \iint_R (3x^2 - 6y^2)\,dx\,dy.$$

Substituting $x = r\cos\theta$, $y = r\sin\theta$, we obtain

$$\iint_S (\nabla\times\mathbf{F})\cdot\mathbf{n}\,dA = \int_{\theta=2\pi}^{0}\int_{r=0}^{4}(3\cos^2\theta - 6\sin^2\theta)\,r^3 dr\,d\theta = 64\int_{2\pi}^{0}(3\cos^2\theta - 6\sin^2\theta)\,d\theta$$

$$= 96\int_{2\pi}^{0}[(1+\cos 2\theta) - 2(1-\cos 2\theta)]d\theta$$

$$= 96\int_{2\pi}^{0}(3\cos 2\theta - 1)\,d\theta = 96\left[\frac{3}{2}\sin 2\theta - \theta\right]_{2\pi}^{0} = 96(2\pi) = 192\pi. \qquad ...(ii)$$

From (*i*) and (*ii*) $\iint_S (\nabla\times\mathbf{F})\cdot\mathbf{n}\,dA = \oint_C \mathbf{F}\cdot d\mathbf{r}$. Hence, Stokes's theorem is verified.

EXAMPLE 1.95 Verify Stokes's theorem for the vector field $\mathbf{F}(x, y, z) = xy\mathbf{i} + yz\mathbf{j} + zx\mathbf{k}$, taking S to be part of the cylinder $z = 1 - x^2$ for which $0 \le x \le 1$, $-2 \le y \le 2$.

Solution The surface S, the curve C (which is composed of the union of C_1, C_2, C_3 and C_4), and the region R are shown in Fig. 1.44.

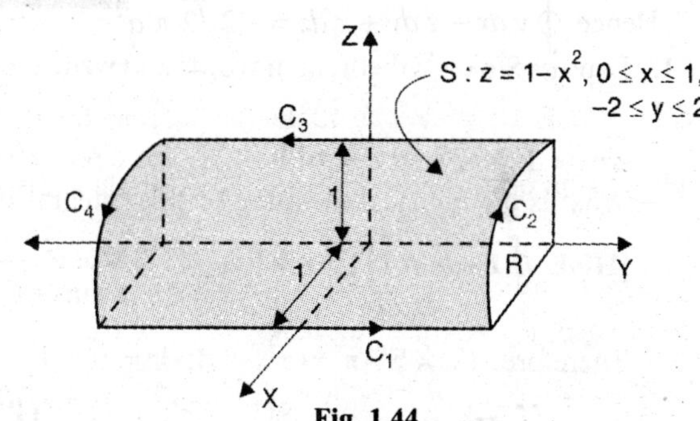

Fig. 1.44

Surface Integral:

$$\text{curl }\mathbf{F} = \nabla\times\mathbf{F} = \begin{vmatrix} \mathbf{i} & \mathbf{j} & \mathbf{k} \\ \partial/\partial x & \partial/\partial y & \partial/\partial z \\ xy & yz & zx \end{vmatrix}$$

$$= \mathbf{i}(0 - y) - \mathbf{j}(z - 0) + \mathbf{k}(0 - x) = -y\mathbf{i} - z\mathbf{j} - x\mathbf{k}.$$

Let $f(x,y,z) = x^2 + z - 1 = 0$ define the cylinder, then unit normal is given by $\mathbf{n} = \dfrac{\nabla f}{|\nabla f|} = \dfrac{2x\mathbf{i} + \mathbf{k}}{\sqrt{4x^2+1}}$.

Therefore, $\iint_S (\nabla\times\mathbf{F})\cdot\mathbf{n}\,dA = \iint_S \dfrac{-2xy - x}{\sqrt{4x^2+1}}\,dA = \iint_R \dfrac{-2xy - x}{\sqrt{4x^2+1}}\cdot\dfrac{dx\,dy}{\mathbf{n}\cdot\mathbf{k}}$

$$= \int_{x=0}^{1}\int_{y=-2}^{2}(-2xy - x)\,dy\,dx = \int_0^1[-xy^2 - xy]_{-2}^{2}\,dx = \int_0^1(-4x)\,dx = -2, \qquad ...(i)$$

Line Integral:

We write $\oint_C = \int_{C_1} + \int_{C_2} + \int_{C_3} + \int_{C_4}$.

We have $\oint_C \mathbf{F}\cdot d\mathbf{r} = \oint_C xy\,dx + yz\,dy + zx\,dz$. On C_1: $x = 1$, $z = 0$, $dx = 0$, $dz = 0$, so $\oint_{C_1} y(0) + y(0)\,dy + 0 = 0$.

On C_2: $y = 2$, $z = 1 - x^2$, $dy = 0$, $dz = -2x\,dx$, so

$$\oint_{C_2} 2x\,dx + 2(1-x^2)0 + x(1-x^2)(-2x\,dx) = \int_1^0 (2x - 2x^2 + 2x^4)\,dx$$

$$= \int_0^1 (2x^2 - 2x - 2x^4)\,dx = \left[\frac{2}{3}x^3 - x^2 - 2\frac{x^5}{5}\right]_0^1 = \frac{2}{3} - 1 - \frac{2}{5} = -\frac{11}{15}.$$

On C_3: $x = 0$, $z = 1$, $dx = 0$, $dz = 0$, so $\oint_{C_3} 0 + y\,dy + 0 = \int_2^{-2} y\,dy = 0.$

On C_4: $y = -2$, $z = 1 - x^2$, $dy = 0$, $dz = -2x\, dx$, so

$$\oint_{C_4} -2x\,dx - 2(1-x^2)0 + x(1-x^2)(-2x\,dx) = \int_0^1 (-2x - 2x^2 + 2x^4)\,dx = \left[-x^2 - \frac{2}{3}x^3 + \frac{2}{5}x^5\right]_0^1 = -\frac{19}{15}.$$

Hence, $\oint_C xy\,dx + yz\,dy + zx\,dz = 0 - \frac{11}{15} + 0 - \frac{19}{15} = -2.$ ...(*ii*)

From (*i*) and (*ii*) $\iint_S (\text{curl } \mathbf{F} \cdot \mathbf{n})\, dA = \oint_C \mathbf{F} \cdot d\mathbf{r}$.

Hence, Stroke's theorem is verified.

EXERCISE 1.8

1. Express Stokes's theorem in words and write it in Cartesian form. Evaluate $\oint_C \mathbf{F} \cdot d\mathbf{r}$ by Stokes's theorem, where $\mathbf{F} = y\mathbf{i} + xz^3\mathbf{j} - zy^3\mathbf{k}$, C is the circle $x^2 + y^2 = 4$, $z = 1.5$. **Ans.** $19\pi/2$.
What is the physical meaning of this integral if **F** defines a field of force.

[Hint: $\oint_C \mathbf{F} \cdot d\mathbf{r} = \iint_S (\nabla \times \mathbf{F}) \cdot \mathbf{n}\, dA$. $\nabla \times \mathbf{F} = -(3y^2z + 3xz^2)\mathbf{i} + (z^3 - 1)\mathbf{k}$ and $\mathbf{n} = \mathbf{k}$.

Therefore, $(\nabla \times \mathbf{F}) \cdot \mathbf{n} = (z^3 - 1)\mathbf{k} \cdot \mathbf{k} = z^3 - 1$.

Thus $\iint_S (\nabla \times \mathbf{F}) \cdot \mathbf{n}\, dA = \iint_S (z^3 - 1)\, dA = \frac{19}{8} \iint_S dA = \frac{19}{8} \times 4\pi = \frac{19\pi}{2}$.

($\because z = 1.5$ on the surface of the given circle)]

2. Verify Stokes's theorem for a vector field defined by $\mathbf{F} = (x^2 - y^2)\mathbf{i} + 2xy\mathbf{j}$ in the rectangular region in the xy-plane bounded by the lines $x = 0$, $x = a$, $y = 0$ and $y = b$. **Ans.** Common value $= 2\, ab^2$.

3. Verify Stokes's theorem for the vector field $\mathbf{F} = 4xz\mathbf{i} - y^2\mathbf{j} + yz\mathbf{k}$ over the area in the plane $z = 0$ bounded by $x = 0$, $y = 0$ and $x^2 + y^2 = 1$.

4. Apply Stokes's theorem to evaluate $\iint_S (\nabla \times \mathbf{F}) \cdot \mathbf{n}\, dA$, where $\mathbf{F} = z\mathbf{i} + x\mathbf{j} + y\mathbf{k}$ and S is the sphere $x^2 + y^2 + z^2 = 1$ above the xy-plane. Also verify Stokes's theorem. **Ans.** π.

5. Verify Stokes's theorem for the vector function $\mathbf{F}(x, y, z) = z\mathbf{i} + x\mathbf{j} + y\mathbf{k}$. Assume that the surface S is oriented upward. S is the portion of the plane $2x + y + 2z = 6$ in the first octant.

6. Verify Stokes's theorem for $\mathbf{F} = (x^2 + y^2)\,\mathbf{i} - 2xy\mathbf{j}$ taken round the rectangle bounded by the lines $x = \pm a$, $y = 0$, $y = b$. (*VTU 2007; Raipur 2005; UPTU 2003; PTU 2003; JNTU 2003*)
Ans. Common value $= -4ab^2$.

7. Using Strokes's theorem, evaluate $\oint_C \mathbf{F} \cdot d\mathbf{r}$, where vector field $\mathbf{F} = (2x + y - 2z)\mathbf{i} + (2x - 4y + z^2)\mathbf{j} + (x - 2yz^2)\mathbf{k}$ and C is the circle with centre (0, 0, 3) and radius 5 in the plane $z = 3$. **Ans.** 25π.

8. Evaluate $\oint_C \mathbf{F} \cdot d\mathbf{r}$ by Stokes's theorem, where $\mathbf{F} = y^2\mathbf{i} + x^2\mathbf{j} - (x + z)\mathbf{k}$ and C is the boundary of the triangle with vertices at (0, 0, 0), (1, 0, 0) and (1, 1, 0).
(*AMIETE, Dec. 2007; AMIE, S-2005, UPTU 2001*) **Ans.** 1/3.

[Hint: $\oint_C \mathbf{F} \cdot d\mathbf{r} = \iint_C \text{curl } \mathbf{F} \cdot \mathbf{n}\, dA = 2\int_0^1 \int_{y=0}^{x} (x - y)\,dy\,dx = \frac{1}{3}$.]

9. State Stokes's theorem. Evaluate the surface integral $\iint_S (\text{curl}\,\mathbf{F})\cdot\mathbf{n}\,dA$ by transforming it into line integral, S being that part of the surface of the paraboloid $z = 1 - x^2 - y^2$ for which $z \geq 0$ and $\mathbf{F} = y\mathbf{i} + z\mathbf{j} + x\mathbf{k}$.
(AMIETE, Dec. 2010) **Ans.** $-\pi$.

10. Verify Stokes's theorem for $\mathbf{F} = -y\mathbf{i} + 2yz\mathbf{j} + y^2\mathbf{k}$, where S is the surface of upper half of the sphere $x^2 + y^2 + z^2 = a^2$, and C is the circular boundary on XOY-plane. *(AMIETE, June 2005)*

11. Use Stokes's theorem to evaluate $\iint_S (\nabla\times\mathbf{F})\cdot\mathbf{n}\,dA$, where $\mathbf{F} = y\mathbf{i} + (x - 2xz)\mathbf{j} - xy\mathbf{k}$ and S is the surface of the sphere $x^2 + y^2 + z^2 = a^2$, above the xy-plane. *(AMIE, S-2011)* **Ans.** 0

12. If $\mathbf{F} = 3y\mathbf{i} - xz\mathbf{j} + yz^2\mathbf{k}$, where S is the surface of the paraboloid $2z = x^2 + y^2$ bounded by $z = 2$ and C is its boundary. Evaluate $\iint_S (\nabla\times\mathbf{F})\cdot\mathbf{n}\,dA$ using Stokes's theorem. *(AMIETE, June 2011, Dec. 2009)* **Ans.** 20π.

13. Find the value of $\iint_S (\nabla\times\mathbf{F})\cdot\mathbf{n}\,dA$ taken over the upper portion of the surface $x^2 + y^2 - 2ax + az = 0$ and the bounding curve lies in the plane $z = 0$, where $\mathbf{F} = (y^2 + z^2 - x^2)\mathbf{i} + (z^2 + x^2 - y^2)\mathbf{j} + (x^2 + y^2 - z^2)\mathbf{k}$.
(AMIETE Dec 2002) **Ans.** $2\pi a^3$.

[**Hint:** By Stokes's Theorem $\iint_S (\nabla\times\mathbf{F})\cdot\mathbf{n}\,dA, = \oint_C \mathbf{F}\cdot d\mathbf{r}$ where C is the circle $x^2 + y^2 - 2ax = 0$, that is, $(x - a)^2 + y^2 = a^2$ with parametric equations $x = a + a\cos\theta$, $y = a\sin\theta$.
Also $\mathbf{F}\cdot d\mathbf{r} = \{(y^2 - x^2)\mathbf{i} + (x^2 - y^2)\mathbf{j}\}\cdot(\mathbf{i}\,dx + \mathbf{j}\,dy)$]

14. Use Stokes's theorem or otherwise, evaluate $\oint_C (2x - y)\,dx - yz^2 dy - y^2 z\,dz$, where C is the circle $x^2 + y^2 = 1$, corresponding to the surface of sphere of unit radius.
(MDU 2007; PTU 2003; UPTU 2002) **Ans.** π.

15. Verify the Stokes's theorem for the function $\mathbf{F} = (x - y - z)\mathbf{i} + (y - z - x)\mathbf{j} + (z - x - y)\mathbf{k}$ over the enclosed surface of the cylinder $x^2/a^2 + y^2/b^2 = 1$, bounded by the plane $z = h$ and open at the end $z = 0$. **Ans.** Common value = 0

16. Apply Stokes's theorem to evaluate $\oint_C y\,dx + z\,dy + x\,dz$, where C is the curve of intersection of $x^2 + y^2 + z^2 = a^2$ and $x + z = a$. **Ans.** $-\pi a^2/\sqrt{2}$.

17. Compute the circulation of vector field $\mathbf{v} = x^2y^3\mathbf{i} + \mathbf{j} + z\mathbf{k}$ along the circle $x^2 + y^2 = a^2$, z = 0, with the help of Stokes's theorem.

[**Hint:** Circulation of the field $= \oint_C \mathbf{v}\cdot d\mathbf{r} = \iint_S (\nabla\times\mathbf{v})\cdot\mathbf{n}\,dA$. [by Stokes's theorem]

Now $\nabla\times\mathbf{v} = \begin{vmatrix} \mathbf{i} & \mathbf{j} & \mathbf{k} \\ \partial/\partial x & \partial/\partial y & \partial/\partial z \\ x^2y^3 & 1 & z \end{vmatrix} = -3x^2y^2\mathbf{k}$, and $(\nabla\times\mathbf{v})\cdot\mathbf{n} = -3x^2y^2\mathbf{k}\cdot\mathbf{k} = -3x^2y^2$.

$$\therefore \quad \iint_S (\nabla\times\mathbf{v})\cdot\mathbf{n}\,dA = -4\int_0^a\int_0^{\sqrt{a^2-x^2}} 3x^2y^2\,dy\,dx = -\frac{\pi a^6}{8}.]$$

18. If $\mathbf{F} = 2yz\mathbf{i} - (x + 3y - 2)\mathbf{j} + (x^2 + z)\mathbf{k}$, evaluate $\iint_S (\nabla \times \mathbf{F}) \cdot \mathbf{n}\, dA$ over the surface of intersection of the cylinders $x^2 + y^2 = a^2$, $x^2 + z^2 = a^2$ which is included in the first octant.

[Hint: Refer Fig. 1.45. $\iint_S (\nabla \times \mathbf{F}) \cdot \mathbf{n}\, dA = \oint_C \mathbf{F} \cdot d\mathbf{r}$

$$= \oint_{C_1} \mathbf{F} \cdot d\mathbf{r} + \oint_{C_2} \mathbf{F} \cdot d\mathbf{r} + \oint_{C_3} \mathbf{F} \cdot d\mathbf{r} + \oint_{C_4} \mathbf{F} \cdot d\mathbf{r}$$

$$= \frac{a^2}{2} + \left(\frac{3a^2}{2} - 2a\right) + \left(\frac{2a^3}{3} - \frac{a^2}{2}\right) + \left(-\frac{\pi a^2}{4} - \frac{3}{2}a^2 + 2a\right)$$

$$= -\frac{a^2}{12}(3\pi + 8a)].$$

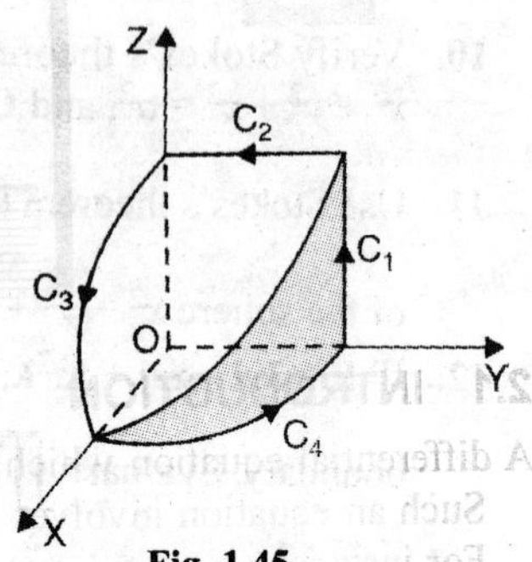

Fig. 1.45

19. Verify Stokes's theorem for the vector $\mathbf{v} = (3x - y)\,\mathbf{i} - 2yz^2\,\mathbf{j} - 2y^2z\,\mathbf{k}$, where S is the surface of the sphere $x^2 + y^2 + z^2 = 16$, $z > 0$. *(AMIE-S, 2012)*

CHAPTER 2

Partial Differential Equations

2.1 INTRODUCTION

A differential equation which involves partial derivatives is called a partial differential equation (PDE). Such an equation involves one dependent variable and two or more than two independent variables. For instance

$$x\frac{\partial z}{\partial x}+y\frac{\partial z}{\partial y}=z, \;...(i) \qquad \frac{\partial^2 z}{\partial x^2}+\frac{\partial^2 z}{\partial y^2}=0, \;...(ii) \qquad \frac{\partial u}{\partial t}=c^2\left(\frac{\partial^2 u}{\partial x^2}+\frac{\partial^2 u}{\partial y^2}\right), \;...(iii) \qquad \frac{\partial^2 u}{\partial x\partial y}=\left(\frac{\partial u}{\partial z}\right)^3 \;...(iv)$$

are examples of partial differential equations.

The *order* of a partial differential equation is the order of the highest partial derivative in the equation. The *degree* of an equation is the degree of the highest order derivative occurring in the equation.

Thus equation (*i*) is a partial differential equation of the first order while equations (*ii*), (*iii*) and (*iv*) are partial differential equations of second order. The degree of all the above equations is one.

A *solution* of a partial differential equation is a function z of x and y which when substituted in the differential equation identically satisfies the equation. Sometimes, the solution can be of the form $f(x, y, z) = 0$ indicating that z is implicitly a function of x and y.

The *general solution* of a partial differential equation contains, arbitrary constants or arbitrary functions and sometimes both.

In the ensuing discussion, we shall restrict ourselves to partial differential equations involving one dependent variable z and only two independent variables x and y so that $z = f(x, y)$ and we shall use the following notations.

The first order partial derivatives $p=\frac{\partial z}{\partial x}=z_x$ and $q=\frac{\partial z}{\partial y}=z_y$, and the second order partial derivatives

$$r=\frac{\partial^2 z}{\partial x^2}=z_{xx},\; s=\frac{\partial^2 z}{\partial x\partial y}=z_{xy},\; t=\frac{\partial^2 z}{\partial y^2}=z_{yy} \text{ etc.}$$

An equation containing x, y, z, p, q defines a first order partial differential equation. A first order partial differential equation can be written as

$$f(x, y, z, p, q) = 0. \qquad ...(2.1)$$

An equation containing x, y, z, p, q, r, s, t defines a second order partial differential equation, and is written as

$$g(x, y, z, p, q, r, s, t) = 0. \qquad ...(2.2)$$

2.2 FORMATION OF PARTIAL DIFFERENTIAL EQUATIONS

Just like ordinary differential equations, partial differential equations also arise as a result of elimination of arbitrary constants from a given relation between the variables. Besides, partial differential equations are also formed by elimination of arbitrary functions from given functional relations between variables. We shall note from the solved examples of next section that if the number of constants to be eliminated is equal to the number of independent variables, the partial differential equations obtained are of first order. In case number of constants is more than the number of independent variables, the differential equations that arise are of second or higher order. When partial differential equations are obtained by eliminating arbitrary functions the order of differential equations is, in general, equal to the number of arbitrary functions eliminated.

EXAMPLE 2.1 Derive partial differential equations by eliminating the arbitrary constants a and b, from the following equations

(*i*) $z = ax + by + ab$, (*ii*) $z = ax + a^2y^2 + b$.

Solution (*i*) Differentiating z partially with respect to x and y, we have $p = \partial z/\partial x = a$, $q = \partial z/\partial y = b$.

Substituting for a and b in $z = ax + by + ab$, we obtain $z = px + qy + pq$, a partial differential equation of first order.

(*ii*) Differentiating z partially w.r.t x and y, we get $p = \partial z/\partial x = a$, $q = \partial z/\partial y = 2a^2 y$.

Eliminating 'a' between these results, we obtain

$q = 2p^2y$, the required partial differential equation of first order.

EXAMPLE 2.2 Form a partial differential equation by eliminating the arbitrary constant from $z = (x^2 + a)(y^2 + b)$.

Solution Differentiating z partially w.r.t. x and y respectively, we get

$$p = \partial z/\partial x = 2x(y^2 + b), \quad ...(i) \quad \text{and} \quad q = \partial z/\partial y = 2y(x^2 + a). \quad ...(ii)$$

Multiplying (*i*) and (*ii*), we obtain $pq = 4xy(x^2 + a)(y^2 + b)$ or $pq = 4xyz$, which is a partial differential equation of the first order.

EXAMPLE 2.3 Form the partial differential equation by eliminating the arbitrary functions f and g from $z = f(2x + 3y) + g(2x + y)$.

Solution Differentiating z partially with respect to x and y, we obtain

$p = \partial z/\partial x = 2f'(2x+3y) + 2g'(2x+y)$, ...(*i*) and $q = \partial z/\partial y = 3f'(2x+3y) + g'(2x+y)$. ...(*ii*)

Now differentiating (*i*) partially w.r.t. x $\quad r = \partial^2 z/\partial x^2 = 4f''(2x+3y) + 4g''(2x+y)$. ...(*iii*)

Differentiating (*ii*) partially w.r.t. x $\quad s = \partial^2 z/\partial x\partial y = 6f''(2x+3y) + 2g''(2x+y)$. ...(*iv*)

Differentiating (*ii*) partially w.r.t. y $\quad t = \partial^2 z/\partial y^2 = 9f''(2x+3y) + g''(2x+y)$. ...(*v*)

Solving (*iii*) and (*iv*), we get $\quad f''(2x+3y) = (2s - r)/8,\ g''(2x+y) = (3r - 2s)/8$.

Substituting these values in (*v*), we have

$$t = \frac{9(2s-r)}{8} + \frac{(3r-2s)}{8} = 2s - \frac{3r}{4} \text{ or } 3r - 8s + 4t = 0, \text{ that is, } 3\frac{\partial^2 z}{\partial x^2} - 8\frac{\partial^2 z}{\partial x \partial y} + 4\frac{\partial^2 z}{\partial y^2} = 0, \text{ the required equation.}$$

EXAMPLE 2.4 Form a partial differential equation by eliminating arbitrary function from the following equations:

(*i*) $z = (x+y)\,\phi(x^2 - y^2)$, (*ii*) $z = f(xy/z)$. (*Anna, 2001*)

Solution (*i*) We have $z = (x + y)\,\phi\,(x^2 - y^2)$.

Differenciating partialy with respect to x and y, we obtain

$$\frac{\partial z}{\partial x} = 1.\phi + (x+y)\,2x.\phi'. \quad ...(1) \qquad \frac{\partial z}{\partial y} = 1.\phi + (x+y)(-2y)\phi'. \quad ...(2)$$

From (2), $\phi' = \dfrac{\phi - (\partial z/\partial y)}{2y(x+y)}$. ...(3). Substituting (3) in (1), we obtain, $\dfrac{\partial z}{\partial x} = \phi + 2x(x+y)\left[\dfrac{\phi - (\partial z/\partial y)}{2y(x+y)}\right]$

$$\text{or} \quad p = \phi + \frac{x}{y}(\phi - q) = \left(\frac{x+y}{y}\right)\phi - \frac{x}{y}q. \quad ...(4)$$

From the given equation, we have $\phi = z/(x+y)$. Substituting this value in (4), we obtain

$$p = \left(\frac{x+y}{y}\right)\left(\frac{z}{x+y}\right) - \frac{x}{y}q \quad \text{or} \quad yp + xq = z.$$

(*ii*) Differentiating partially w.r.t x and y, we get

$$\frac{\partial z}{\partial x} = p = f'\left(\frac{xy}{z}\right) y\left(\frac{1}{z} - \frac{x}{z^2}\cdot p\right), \quad ...(5) \quad \text{and} \quad \frac{\partial z}{\partial y} = q = f'\left(\frac{xy}{z}\right) x\left(\frac{1}{z} - \frac{y}{z^2}\cdot q\right). \quad ...(6)$$

On dividing (5) by (6), we obtain

$$\frac{p}{q} = \frac{y(z - xp)}{x(z - yq)} \Rightarrow px - qy = 0.$$

Remark 1

(a) In example 2.1 (*i*), the given relation $z = ax + by + ab$ is called the primitive or general solution of $z = px + qy + pq$, a differential equation of first order.
Similarly in example 2.1(*ii*) $z = ax + a^2y^2 + b$ is the general solution of $q = 2p^2y$ a differential equation of first order. *Thus the general solution of a partial differential equation of first order contains, in general, two arbitrary constants.*

(b) In example 2.3, $z = f(2x + 3y) + g(2x + y)$ is the general solution of $3\frac{\partial^2 z}{\partial x^2} - 8\frac{\partial^2 z}{\partial x\partial y} + 4\frac{\partial^2 z}{\partial y^2} = 0$, (a partial differential equation of second order). This solution contains two arbitrary functions f and g. *In general, the general solution of a differential equation of nth order contains n arbitrary functions.*

EXAMPLE 2.5 Find the partial differential equation by eliminating ϕ from $\phi(u, v) = 0$ where u, v are known functions of x, y and z and ϕ is some arbitrary function of u and v.

Solution Differentiating $\phi(u, v) = 0$ partially w.r.t. x and y respectively, we get

$$\frac{\partial\phi}{\partial u}\left(\frac{\partial u}{\partial x} + p\frac{\partial u}{\partial z}\right) + \frac{\partial\phi}{\partial v}\left(\frac{\partial v}{\partial x} + p\frac{\partial v}{\partial z}\right) = 0, \quad ...(i) \quad \text{and} \quad \frac{\partial\phi}{\partial u}\left(\frac{\partial u}{\partial y} + q\frac{\partial u}{\partial z}\right) + \frac{\partial\phi}{\partial v}\left(\frac{\partial v}{\partial y} + q\frac{\partial v}{\partial z}\right) = 0. \quad ...(ii)$$

Eliminating $\frac{\partial\phi}{\partial u}, \frac{\partial\phi}{\partial v}$ from (*i*) and (*ii*), we have $\begin{vmatrix} \frac{\partial u}{\partial x} + p\frac{\partial u}{\partial z} & \frac{\partial v}{\partial x} + p\frac{\partial v}{\partial z} \\ \frac{\partial u}{\partial y} + q\frac{\partial u}{\partial z} & \frac{\partial v}{\partial y} + q\frac{\partial v}{\partial z} \end{vmatrix} = 0$

or $$\left(\frac{\partial u}{\partial x} + p\frac{\partial u}{\partial z}\right)\left(\frac{\partial v}{\partial y} + q\frac{\partial v}{\partial z}\right) - \left(\frac{\partial v}{\partial x} + p\frac{\partial v}{\partial z}\right)\left(\frac{\partial u}{\partial y} + q\frac{\partial u}{\partial z}\right) = 0$$

or $$p\left(\frac{\partial u}{\partial z}\frac{\partial v}{\partial y} - \frac{\partial u}{\partial y}\frac{\partial v}{\partial z}\right) + q\left(\frac{\partial u}{\partial x}\frac{\partial v}{\partial z} - \frac{\partial u}{\partial z}\frac{\partial v}{\partial x}\right) = \left(\frac{\partial u}{\partial y}\frac{\partial v}{\partial x} - \frac{\partial u}{\partial x}\frac{\partial v}{\partial y}\right). \quad ...(iii)$$

The Equation (iii) can be written as a first order partial differential equation given by

$$p\frac{\partial(u, v)}{\partial(y, z)} + q\frac{\partial(u, v)}{\partial(z, x)} = \frac{\partial(u, v)}{\partial(x, y)} \quad \text{or} \quad P(x, y, z)\,p + Q(x, y, z)\,q = R(x, y, z)$$

where $P(x, y, z) = \frac{\partial(u, v)}{\partial(y, z)}$, $Q(x, y, z) = \frac{\partial(u, v)}{\partial(z, x)}$, $R(x, y, z) = \frac{\partial(u, v)}{\partial(x, y)}$ are the Jacobians.

EXAMPLE 2.6 Obtain the partial differential equation governing the equations $\phi(u, v) = 0$, $u = xyz$, $v = x + y + z$.

Solution We can use $Pp + Qq = R$ directly to obtain the partial differential equation.

We have $$P(x, y, z) = \frac{\partial(u, v)}{\partial(y, z)} = \begin{vmatrix} \partial u/\partial y & \partial u/\partial z \\ \partial v/\partial y & \partial v/\partial z \end{vmatrix} = \begin{vmatrix} xz & xy \\ 1 & 1 \end{vmatrix} = xz - xy = x(z - y),$$

$$Q(x, y, z) = \frac{\partial(u, v)}{\partial(z, x)} = \begin{vmatrix} \partial u/\partial z & \partial u/\partial x \\ \partial v/\partial z & \partial v/\partial x \end{vmatrix} = \begin{vmatrix} xy & yz \\ 1 & 1 \end{vmatrix} = xy - yz = y(x - z),$$

$$R(x, y, z) = \frac{\partial(u, v)}{\partial(x, y)} = \begin{vmatrix} \partial u/\partial x & \partial u/\partial y \\ \partial v/\partial x & \partial v/\partial y \end{vmatrix} = \begin{vmatrix} yz & xz \\ 1 & 1 \end{vmatrix} = yz - xz = z(y - x).$$

Hence, we obtain the linear partial differential equation

$$x(y - z)\,p + y(z - x)\,q = z(x - y).$$

EXAMPLE 2.7 Find the differential equation of all planes which are at a constant distance '*a*' from the origin. *(AMIETE, Dec. 2012; Kurukshetra, 2008)*

Solution Let the equation of the plane be $Ax + By + Cz = D$, ... (1)
where A, B and C are the direction ratios of the normal to the plane.

Since the perpendicular distance from the origin to the plane (1) is equal to '*a*'.

Therefore, $$\left|\frac{A(0)+B(0)+C(0)-D}{\sqrt{A^2+B^2+C^2}}\right| = a$$

$$\Rightarrow \quad D = a\sqrt{A^2+B^2+C^2} \qquad ...(2)$$

Substituting the value of D in Eq. (1), we obtain

$$Ax + By + Cz - a\sqrt{A^2+B^2+C^2} = 0. \qquad ...(3)$$

Differentiating partially (3) with respect to x and y, we obtain

$$A + C\frac{\partial z}{\partial x} = 0, \quad \text{and} \quad B + C\frac{\partial z}{\partial y} = 0,$$

or

$$A + Cp = 0, \quad \text{and} \quad B + Cq = 0.$$

On substituting values of A and B in terms of C in Eq. (3), we obtain

$$-Cpx - Cqy + Cz - a\sqrt{C^2p^2 + C^2q^2 + C^2} = 0$$

$$\Rightarrow \quad z = px + qy + a\sqrt{1+p^2+q^2},$$ the required partial differential equation.

EXERCISE 2.1

Form the partial differential equations (by eliminating the arbitrary constants) from the following equations, 1 to 7.

1. $z = ax + by + a^2 + b^2$. **Ans.** $z = px + qy + p^2 + q^2$.
2. $2z = \dfrac{x^2}{a^2} + \dfrac{y^2}{b^2}$. *(AMIETE, June, 2011; Dec, 2009)* **Ans.** $xp + yq = 2z$.
3. $z = (x-a)^2 + (y-b)^2$. *(Osmania, 2003)* **Ans.** $p^2 + q^2 = 4z$.
4. $(x-a)^2 + (y-b)^2 + z^2 = c^2$. *(AMIETE, June, 2012, 2008; Kottayam, 2005)* **Ans.** $z^2(p^2 + q^2 + 1) = c^2$.
5. $z = xy + y\sqrt{x^2 - a^2} + b$. **Ans.** $px + qy = pq$.
6. $z = \dfrac{x^2}{a^2} + \dfrac{y^2}{b^2} + \dfrac{z^2}{c^2} - 1$. **Ans.** $xz\left(\partial^2 z/\partial x^2\right) + x(\partial z/\partial x)^2 - z(\partial z/\partial x) = 0$.
7. $z = a(x+y) + b(x-y) + abt + c$. **Ans.** $\dfrac{\partial z}{\partial t} = \dfrac{1}{4}\left[\left(\dfrac{\partial z}{\partial x}\right)^2 - \left(\dfrac{\partial z}{\partial y}\right)^2\right]$.
8. Find the differential equation of all spheres of radius 3 units having their centres in the *xy*-plane. *(Madras, 2000S)* **Ans.** $z^2(p^2 + q^2 + 1) = 9$.
9. Find the differential equation of all spheres whose centres lie on the z–axis. *(AMIETE, Dec. 2009)* **Ans.** $py - qx = 0$.
   [**Hint:** The equation of such a sphere is $x^2 + y^2 + (z-c)^2 = a^2$, where a and c are arbitrary constants.]
10. Find the PDE satisfied by all right circular cones whose axes coincide with the z-axis. **Ans.** $py - qx = 0$.
    [**Hint:** The equation of such a cone with a semi-vertical angle α is given by $x^2 + y^2 = (z-c)^2 \tan^2\alpha$, where c and α are arbitrary coustants.]

In the following problems, eliminate the arbitrary functions to obtain a partial differential equation.

11. (*a*) $z = f(x^2 - y^2)$. (*AMIETE; Dec, 2011, June 2008; Madras, 2003; Osmania, 2003*) **Ans.** $py + qx = 0$.

(*b*) $\phi(x + y + z, x^2 + y^2 - z^2) = 0$. **Ans.** $p(y + z) - q(z + x) = x - y$.

(*c*) $f(xy + z^2, x + y + z) = 0$. (*VTU, 2006*) **Ans.** $p(x - 2z) + q(2z - y) = y - x$.

(*d*) $z = f(xy/z)$. **Ans.** $px - qy = 0$.

(*e*) $z = yf(x) + xg(y)$. (*AMIE-S-2012; VTU, 2004*) **Ans.** $xy\dfrac{\partial^2 z}{\partial x\partial y} = x\dfrac{\partial z}{\partial x} + y\dfrac{\partial z}{\partial y} - z$.

(*f*) $u = f(x + 4t) + g(x - 4t)$. **Ans.** $16\dfrac{\partial^2 u}{\partial x^2} - \dfrac{\partial^2 u}{\partial t^2} = 0$.

(*g*) $xyz = \phi(x + y + z)$. **Ans.** $px(y - z) + qy(z - x) = z(x - y)$.

(*h*) $z = y^2 + 2f(x^{-1} + \log y)$. (*KTU, 2007; JNTU, 2005; Madras 2000*)

[**Hint:** $p = \partial z/\partial x = 2f'(x^{-1} + \log y)(-x^{-2}) \Rightarrow -px^2 = 2f'(x^{-1} + \log y)$ ...(*i*)

$q = \partial z/\partial y = 2y + 2f'(x^{-1} + \log y)\, y^{-1} \Rightarrow qy - 2y^2 = 2f'(x^{-1} + \log y)$ ...(*ii*)

From (*i*) and (*ii*), $px^2 + qy = 2y^2$].

(*i*) $z = f_1(y + 2x) + f_2(y - 3x)$. (*Kuruksherta, 2005*) **Ans.** $\dfrac{\partial^2 z}{\partial x^2} + \dfrac{\partial^2 z}{\partial x\partial y} - 6\dfrac{\partial^2 z}{\partial y^2} = 0$.

(*j*) $z = xf_1(x + y) + f_2(x + y)$. **Ans.** $r - 2s + t = 0$.

(*k*) $y = f(x - at) + F(x + at)$. **Ans.** $\dfrac{\partial^2 y}{\partial t^2} = a^2\dfrac{\partial^2 y}{\partial x^2}$.

(*l*) $\phi(x^2 + y^2 + z^2, z^2 - 2xy) = 0$. (*AMIE-S-2012*)

2.3 EQUATIONS SOLVABLE BY DIRECT INTEGRATION

Some partial differential equations which contains only one partial derivative can be solved with the help of direct integration. In place of the constants of integration, we use arbitrary functions of the variable kept fixed. The method is best explained with the help of following examples.

EXAMPLE 2.8 Solve $\dfrac{\partial^3 z}{\partial x^2\partial y} = \cos(2x + 3y)$.

Solution We have $\dfrac{\partial^3 z}{\partial x^2\partial y} = \dfrac{\partial}{\partial x}\left(\dfrac{\partial^2 z}{\partial x\partial y}\right) = \cos(2x + 3y)$.

Integrating with respect to x, we obtain $\dfrac{\partial^2 z}{\partial x\partial y} = \dfrac{1}{2}\sin(2x + 3y) + f(y)$.

Again integrating with respect to x, $\dfrac{\partial z}{\partial y} = -\dfrac{1}{4}\cos(2x + 3y) + xf(y) + g(y)$.

Now, integrating with respect to y, we get $z = -\dfrac{1}{12}\sin(2x + 3y) + x\cdot\int f(y)dy + \int g(y)dy + \phi(x)$

$$= -\frac{1}{12}\sin(2x + 3y) + x\cdot h(y) + \Psi(y) + \phi(x),$$

where h, Ψ, ϕ are arbitrary functions.

EXAMPLE 2.9 Solve $\dfrac{\partial^3 z}{\partial x^2\partial y} + 18xy^2 + \sin(2x - y) = 0$.

Solution Integrating twice with respect to x, keeping y fixed.

$$\frac{\partial^2 z}{\partial x \partial y} + 9x^2y^2 - \frac{1}{2}\cos(2x - y) = f(y).$$

$$\frac{\partial z}{\partial y} + 3x^3y^2 - \frac{1}{4}\sin(2x - y) = xf(y) + g(y).$$

Now integrating with respect to y (keeping x fixed)

$$z + x^3y^3 - \frac{1}{4}\cos(2x - y) = x\int f(y)dy + \int g(y)dy + \phi(x) = xh(y) + \Psi(y) + \phi(x).$$

Thus $z = \frac{1}{4}\cos(2x - y) - x^3y^3 + xh(y) + \Psi(y) + \phi(x)$, where h, Ψ, ϕ are arbitrary functions.

EXAMPLE 2.10 Solve $\frac{\partial^2 z}{\partial x \partial y} = \sin x \sin y$, for which $\frac{\partial z}{\partial y} = -2\sin y$ when $x = 0$ and $z = 0$ when y is an odd multiple of $\pi/2$. *(VTU, 2004)*

Solution We have $\frac{\partial^2 z}{\partial x \partial y} = \frac{\partial}{\partial x}\left(\frac{\partial z}{\partial y}\right) = \sin x \sin y.$

Integrating with respect to x, keeping y as constant, we obtain

$$\frac{\partial z}{\partial y} = -\cos x \sin y + f(y). \quad ...(i)$$

When $x = 0$, $\frac{\partial z}{\partial y} = -2\sin y = -\sin y \cos 0 + f(y) \Rightarrow f(y) = -\sin y.$

Therefore, $\frac{\partial z}{\partial y} = -\cos x \sin y - \sin y.$

Integrating with respect to y, keeping x as constant, we obtain

$$z = \cos x \cos y + \cos y + g(x). \quad ...(ii)$$

When y is an odd multiple of $\pi/2$, $z = 0$.

Therefore, $0 = 0 + 0 + g(x) \Rightarrow g(x) = 0.$ $\left[\because \cos(2n+1)\frac{\pi}{2} = 0\right]$

Setting $g(x) = 0$ in *(ii)*, we obtain $z = \cos x \cos y + \cos y = \cos y(1 + \cos x)$, the required particular solution.

EXAMPLE 2.11 Solve the equation $\frac{\partial^2 z}{\partial x \partial y} = x^2 y$, for which $z(x, 0) = x^2$, $z(1, y) = \cos y$.

Solution We have $\frac{\partial}{\partial x}\left(\frac{\partial z}{\partial y}\right) = x^2 y$. Integrating will respect to x, we find

$$\frac{\partial z}{\partial y} = \frac{1}{3}x^3 y + f(y). \quad ...(i)$$

Now integrating with respect to y,

$$z = \frac{1}{6}x^3y^2 + \int f(y)\,dy + g(x)$$

$$= \frac{1}{6}x^3y^2 + h(y) + g(x) \quad ...(ii)$$

where $h(y)$ and $g(x)$ are arbitrary functions and is therefore a general solution.

Since $z(x, 0) = x^2$, we have from *(ii)*

$$x^2 = 0 + h(0) + g(x) \quad \text{or} \quad g(x) = x^2 - h(0).$$

Thus $z = \frac{1}{6}x^3y^2 + h(y) + x^2 - h(0). \quad ...(iii)$

Since $z(1, y) = \cos y$, we have from (*iii*)

$$\cos y = \frac{1}{6}y^2 + h(y) + 1 - h(0)$$

$$\text{or } h(y) = \cos y - \frac{1}{6}y^2 - 1 + h(0). \quad ...(iv)$$

Setting (*iv*) in (*iii*), the required solution is

$$z = \frac{1}{6}x^3y^2 + \cos y - \frac{1}{6}y^2 - 1 + h(0) + x^2 - h(0)$$

$$\text{or } z = \frac{1}{6}x^3y^2 + \cos y - \frac{1}{6}y^2 + x^2 - 1.$$

EXAMPLE 2.12 Solve $\dfrac{\partial^2 z}{\partial x^2} + z = 0$, given that when $x = 0$, $z = e^y$ and $\partial z/\partial x = 1$.

Solution If z were a functions of x alone, the solution would have been $z = c_1 \cos x + c_2 \sin x$, where c_1 and c_2 are arbitrary constants.

But here z is a functions of x and y, therefore, c_1 and c_2 are arbitrary functions of y. Hence, the solution of the given equation is $z = f(y) \cos x + \phi(y) \sin x$. ...(1)

Therefore, $\dfrac{\partial z}{\partial x} = -f(y) \sin x + \phi(y) \cos x.$ ...(2)

When $x = 0$, $z = e^y$, we have from (1) $\quad z = e^y = f(y) \cdot 1 \Rightarrow f(y) = e^y.$

When $x = 0$, $\partial z/\partial x = 1$, we have from (2) $\quad \partial z/\partial x = 1 = -e^y(0) + \phi(y) \cdot 1 \Rightarrow \phi(y) = 1.$

Hence, the required solution is $z = e^y \cos x + \sin x$.

EXAMPLE 2.13 Solve $\partial^2 z/\partial x^2 = a^2 z$ given that when $x = 0$, $\partial z/\partial x = a \sin y$ and $\partial z/\partial y = 0$.

Solution If z is a function of x alone, then solution is given by $z = c_1 \sinh(ax) + c_2 \cosh(ax)$, where c_1 and c_2 are constant.

Here z is a function of x and y both, so c_1 and c_2 will be functions of y alone.

Let $z = f(y) \sinh(ax) + g(y) \cosh(ax)$ be the solution of the given equation.

Therefore, $\quad \partial z/\partial x = a f(y) \cosh(ax) + a\, g(y) \sinh(ax)$ ...(1) and

$\partial z/\partial y = f'(y) \sinh(ax) + g'(y) \cosh(ax).$ ...(2)

When $x = 0$, we have from (1)

$\partial z/\partial x = a \sin y = af(y) \cosh(0) + ag(y) \sinh(0) = af(y)$. Therefore $f(y) = \sin y$.

When $x = 0$, we have from (2)

$\partial z/\partial y = 0 = f'(y) \sinh(0) + g'(y) \cosh(0) \Rightarrow g'(y) = 0 \Rightarrow g(y) = \text{constant} = k$ (say).

Therefore, $\quad z = \sin y \sinh(ax) + k \cosh(ax)$ is the required solution.

EXERCISE 2.2

Solve the following differential equations.

1. $\dfrac{\partial^2 z}{\partial x^2} = xy.$ **Ans.** $z = \dfrac{1}{6}x^3 y + xf(y) + g(y).$

2. $\dfrac{\partial^2 z}{\partial x \partial y} = \dfrac{1}{xy}.$ **Ans.** $z = \log x \log y + f(x) + g(y)$. [**Note:** The problem may be written as $xy\,s = 1$]

3. $\dfrac{\partial^2 z}{\partial y^2} = \sin xy.$ (*AMIETE, June, 2011; 2010; Dec., 2009*) **Ans.** $z = -(1/x^2)\sin(xy) + yf(x) + g(x).$

4. $\dfrac{\partial^2 z}{\partial y^2} = x^4 \sin(xy)$. (*AMIETE, Dec., 2008*) **Ans.** $z = -x^2 \sin(xy) + yf(x) + g(x)$.

5. $\dfrac{\partial^2 u}{\partial x \partial t} = e^{-t} \cos x$. **Ans.** $u = -e^{-t} \sin x + \phi(t) + g(x)$, where $\phi(t) = \int f(t)\, dt$ and $g(x)$ is an arbitrary function.

6. $\dfrac{\partial^2 u}{\partial y \partial x} = 4x \sin(3xy)$. **Ans.** $u = -\dfrac{4}{9y} \sin(3xy) + h(x) + \Psi(y)$.

7. $\dfrac{\partial^3 z}{\partial x^2 \partial y} = \sin(3x + 2y)$. **Ans.** $z = f(x) + x\,\phi_1(y) + \phi_2(y) + \dfrac{1}{18} \cos(3x + 2y)$.

8. $xy^2 s = 1 - 4x^2 y$. **Ans.** $z = -\dfrac{1}{y} \log x - 2x^2 \log y + f(y) + \phi(x)$.

9. $r = x^2 / e^y$. **Ans.** $z = \dfrac{x^4}{12e^y} + x\phi_1(y) + \phi_2(y)$.

10. $r = x^2 + y^2$. **Ans.** $z = \dfrac{1}{3}(x^3 y + xy^3) + \phi_1(x) + \phi_2(y)$.

2.4 LINEAR PARTIAL DIFFERENTIAL EQUATIONS OF THE FIRST ORDER

Partial differential equations in which p and q occur in the first degree and they are also not multiplied together are defined as the *linecr partial differential equations.*

Differential equations which are not linear are called *non-linear differential equations.*

A linear partial differential equation of the first order, commonly known *as Lagrange's linear equation* is of the form

$$Pp + Qq = R,$$

where P, Q and R are functions of x, y, z or constants, $p = \partial z/\partial x$ and $q = \partial z/\partial y$. Basically speaking this equation is quasi-linear equation by virtue of the presence of z in P, Q and R. When P, Q and R do not involve z, the equation is known as *linear equation.*

In what follows, we shall discuss methods for solving some important types of linear and non-linear differential equations of the first and second order.

The procedure for solving linear partial differential equation of the first order will be clear from the following theorem.

Theorem 2.1 *If $u(x, y, z) = a$ and $v(x, y, z) = b$ are two independent solutions of* $\dfrac{dx}{P} = \dfrac{dy}{Q} = \dfrac{dz}{R}$*, *where P, Q, R are functions of x, y, z then $\phi(u, v) = 0$ with ϕ arbitrary, is the general solution of $Pp + Qq = R$.*

Such an equation is obtained by the elimination of an arbitrary function ϕ from the equation $\phi(u, v) = 0$ where u and v are functions of x, y, z.

Proof: Taking the differential of $u(x, y, z) = a$ and $v(x, y, z) = b$, we get

$$\frac{\partial u}{\partial x}dx + \frac{\partial u}{\partial y}dy + \frac{\partial u}{\partial z}dz = 0, \qquad \frac{\partial v}{\partial x}dx + \frac{\partial v}{\partial y}dy + \frac{\partial v}{\partial z}dz = 0.$$

As u and v are independent functions, we solve for the ratios and get

$$\frac{dx}{\dfrac{\partial u}{\partial y}\dfrac{\partial v}{\partial z} - \dfrac{\partial u}{\partial z}\dfrac{\partial v}{\partial y}} = \frac{dy}{\dfrac{\partial u}{\partial z}\dfrac{\partial v}{\partial x} - \dfrac{\partial u}{\partial x}\dfrac{\partial v}{\partial z}} = \frac{dz}{\dfrac{\partial u}{\partial x}\dfrac{\partial v}{\partial y} - \dfrac{\partial u}{\partial y}\dfrac{\partial v}{\partial x}} \quad \text{or} \quad \frac{dx}{P} = \frac{dy}{Q} = \frac{dz}{R}.$$

Since the denominators are the relations defining P, Q, R in the equation $Pp + Qq = R$ whose general solution is $\phi(u, v) = 0$ (See solved Example 2.5), the result follows.

---

* These equations are called auxiliary or subsidiary equations

Rule to solve Pp + Qq = R Lagrange's Equation

(*i*) Form the *auxiliary or subsidiary equations*

$$\frac{dx}{P} = \frac{dy}{Q} = \frac{dz}{R}. \qquad ...(2.3)$$

(*ii*) Find any two independent solutions of the subsidiary equations say $u = a$, $v = b$, (a, b being arbitrary constants).

(*iii*) Then $\phi(u, v) = 0$ or $u = \phi(v)$ is a solution of the equation $Pp + Qq = R$.
Here $\phi(v)$ is an arbitrary function of v.

The method of solving the Lagrange's equation will be clear from the solved examples of next section. In this process we have to get two independent solutions ($u = a$, $v = b$) of the simultaneous equations (2.3) and this may pose some difficulty. The readers are advised to note carefully the methods used to get these solutions of the system (2.3), given below.

(a) Method of grouping: We take two terms from the simultaneous Eqs. (2.3), say $\frac{dx}{P} = \frac{dy}{Q}$ when z is absent or can be cancelled. Thus we have a differential equation in x and y only which can be solved and we get one solution.

Similarly it may be convenient to solve $\frac{dx}{P} = \frac{dz}{R}$ or $\frac{dy}{Q} = \frac{dz}{R}$ and get the second solution.

(b) Method of multipliers: We use multipliers l, m, n (which need not always be constants) and get

$$\frac{dx}{P} = \frac{dy}{Q} = \frac{dz}{R} = \frac{l dx + m dy + n dz}{lP + mQ + nR} \qquad ...(2.4)$$

These multipliers are so chosen that $lP + mQ + nR = 0$. Then $l dx + m dy + n dz = 0$.

[$\because$ If $\frac{l dx + m dy + n dz}{lP + mQ + nR} = k$, then $l dx + m dy + n dz = k(lP + mQ + nR)$.

Now if $lP + mQ + nR = 0$, then $l dx + m dy + n dz = 0$]. Integrating this we get one solution.
It may be possible to get another set of multipliers l, m, n and get another solution.

(c) Combination of methods (*a*) and (*b*).

EXAMPLE 2.14 Find the general solution of following partial differential equations

(*i*) $2p + 3q = 1$, (*ii*) $yq - xp = z$.

Solution (*i*) Comparing the given equation with $Pp + Qq = R$, we obtain $P = 2$, $Q = 3$, $R = 1$. The subsidiary simultaneous equations are $dx/2 = dy/3 = dz/1$.

From $dx/2 = dy/3$ we get $3dx = 2dy$ whose solution is $3x - 2y = c_1$, a constant.

Similarly, solving $dy/3 = dz/1$, we get $y - 3z = c_2$, a constant.

The general solution is given by $\phi(c_1, c_2) = 0$ or $\phi(3x - 2y, y - 3z) = 0$.

Remark The general solution can also be written in the alternative forms $3x - 2y = \psi(y - 3z)$ or $y - 3z = f(3x - 2y)$.

(*ii*) Here $P = -x$, $Q = y$, $R = z$. The subsidiary equations are $dx/-x = dy/y = dz/z$.

From the first pair of terms, we get $dx/-x = dy/y$, whose solution is

$$\log x + \log y = \log c_1, \text{ that is, } xy = c_1. \qquad ...(1)$$

From the second and the third terms, we get $dy/y = dz/z$ whose solution is

$\log\left(\frac{y}{z}\right) = \log c_2$, that is, $\frac{y}{z} = c_2$. ...(2) From eqs. (1) and (2), the general solution is $\phi\left(xy, \frac{y}{z}\right) = 0$.

EXAMPLE 2.15 Solve the following partial differential equations

(*i*) $y^2p - xyq = x(z - 2y)$, (*Kuk., 2008; AMIE, S-2001*)

(*ii*) $x(y - z)p + y(z - x)q = z(x - y)$. (*Osmania, 2003; AMIE, W-2001*)

Solution (*i*) The subsidiary equations are $\dfrac{dx}{y^2} = \dfrac{dy}{-xy} = \dfrac{dz}{x(z-2y)}$.

From the first pair of terms, we get $dx/y^2 = dy/-xy$ or $dx/y = -dy/x$ or $x\,dx + y\,dy = 0$ whose solution is $x^2/2 + y^2/2 = c$ or $x^2 + y^2 = c_1$, where $c_1 = 2c$.

From the second and the third terms, we get

$$\frac{dy}{-xy} = \frac{dz}{x(z-2y)} \quad\text{or}\quad \frac{-dy}{y} = \frac{dz}{z-2y} \quad\text{or}\quad (z-2y)\,dy + y\,dz = 0, \quad\text{that is,}\quad (z\,dy + y\,dz) - 2y\,dy = 0.$$

This can be written in the form $d(yz) - 2y\,dy = 0$.

Solving this differential equation, we have $yz - y^2 = c_2$.

The general solution is given by $\phi(x^2 + y^2, yz - y^2) = 0$.

(*ii*) The subsidiary equations are given by

$$\frac{dx}{x(y-z)} = \frac{dy}{y(z-x)} = \frac{dz}{z(x-y)}. \qquad \text{...(1)}$$

Taking the multipliers 1, 1, 1, we note that

$$\text{each term} = \frac{dx + dy + dz}{x(y-z) + y(z-x) + z(x-y)}.$$

Since $x(y - z) + y(z - x) + z(x - y) = 0$, therefore, $dx + dy + dz = 0$.

Integration gives, $x + y + z = c_1$.

The equations (1) can be written in the form $\dfrac{dx/x}{y-z} = \dfrac{dy/y}{z-x} = \dfrac{dz/z}{x-y}$. Again using the multipliers 1, 1, 1

we have
$$\text{each term} = \frac{dx/x + dy/y + dz/z}{y - z + z - x + x - y}.$$

Since $y - z + z - x + x - y = 0$. Therefore, $\dfrac{dx}{x} + \dfrac{dy}{y} + \dfrac{dz}{z} = 0$. Integration gives,

$$\log x + \log y + \log z = \log b \quad\text{or}\quad xyz = c_2.$$

The general solution is given by $\phi(c_1, c_2) = 0$, or $\phi(x + y + z, xyz) = 0$.

EXAMPLE 2.16 Find the solution of the partial differential equation

$$(mz - ny)\frac{\partial z}{\partial x} + (nx - lz)\frac{\partial z}{\partial y} = ly - mx \qquad (\textit{VTU, 2004; AMIE S-2002; W-2001})$$

Solution: The subsidiary equations are $\dfrac{dx}{mz - ny} = \dfrac{dy}{nx - lz} = \dfrac{dz}{ly - mx}$.

Using the multipliers *l*, *m*, *n*, each term $= \dfrac{l\,dx + m\,dy + n\,dz}{l(mz - ny) + m(nx - lz) + n(ly - mx)}$.

Since $l(mz - ny) + m(nx - lz) + n(ly - mx) = 0$. Therefore,

$l\,dx + m\,dy + n\,dz = 0$. Intergration gives $lx + my + nz = c_1$. ...(*i*)

Similarly, using multipliers *x*, *y*, *z*, we have each term $= \dfrac{x\,dx + y\,dy + z\,dz}{0}$

$\therefore$ $xdx + ydy + zdz = 0$, which on integration gives $\dfrac{x^2}{2} + \dfrac{y^2}{2} + \dfrac{z^2}{2} = k$ or $x^2 + y^2 + z^2 = c_2$, where $c_2 = 2k$.

The general solution is given by $\phi(c_1, c_2) = 0$, or $\phi(lx + my + nz, x^2 + y^2 + z^2) = 0$.

EXAMPLE 2.17 Solve the following partial differential equations

(*i*) $(x^2 - y^2 - z^2)p + 2xyq = 2xz$, *(Rajsthan 2006; JNTU 2005; AMIE, S-2004; Madras, 2000 S)*

(*ii*) $x^2p + y^2q = (x + y)z$. *(AMIE W-2006, S-2002)*

Solution: (*i*) The subsidiary equations are

$$\frac{dx}{x^2 - y^2 - z^2} = \frac{dy}{2xy} = \frac{dz}{2xz}. \qquad ...(1)$$

From the last pair of terms, we get $\dfrac{dy}{2xy} = \dfrac{dz}{2xz}$, that is, $\dfrac{dy}{y} = \dfrac{dz}{z}$. The solution is $y/z = c_2$.

Using multipliers x, y, z, we obtain

$$\text{each term} = \frac{xdx + ydy + zdz}{x(x^2 - y^2 - z^2) + y(2xy) + z(2xz)} = \frac{xdx + ydy + zdz}{x(x^2 + y^2 + z^2)}. \qquad ...(2)$$

From the second term and the above term, we get $\dfrac{dy}{2xy} = \dfrac{xdx + ydy + zdz}{x(x^2 + y^2 + z^2)}$ or $\dfrac{dy}{y} = \dfrac{2xdx + 2ydy + 2zdz}{x^2 + y^2 + z^2}$.

Integration gives, $\log y = \log(x^2 + y^2 + z^2) + \log c$

or $\log \dfrac{x^2 + y^2 + z^2}{y} = -\log c = \log \dfrac{1}{c}$ or $\dfrac{x^2 + y^2 + z^2}{y} = c_1$, where $c_1 = c^{-1}$.

The general solution is given by $\phi(c_1, c_2) = 0$, or $\phi\left(\dfrac{x^2 + y^2 + z^2}{y}, \dfrac{y}{z}\right) = 0$.

(*ii*) The subsidiary equations are $\dfrac{dx}{x^2} = \dfrac{dy}{y^2} = \dfrac{dz}{(x + y)z}$. ...(1)

From the first pair of terms in (1),

we have $\dfrac{dx}{x^2} = \dfrac{dy}{y^2}$, which on integration gives $-\dfrac{1}{x} = -\dfrac{1}{y} + a$ or $\dfrac{1}{y} - \dfrac{1}{x} = a$.

Equations (1) can be written in the form $\dfrac{dx/x}{x} = \dfrac{dy/y}{y} = \dfrac{dz/z}{x + y} = \dfrac{dx/x + dy/y - dz/z}{x + y - x - y}$.

(using 1, 1, –1 as multipliers)

Therefore, $\dfrac{dx}{x} + \dfrac{dy}{y} - \dfrac{dz}{z} = 0$ or $\log x + \log y - \log z = \log b$ or $\dfrac{xy}{z} = b$.

The general solution is given by $\phi(a, b) = 0$, or $\phi\left[\dfrac{1}{y} - \dfrac{1}{x}, \dfrac{xy}{z}\right] = 0$.

EXAMPLE 2.18 Find the general solution of the following partial differential equations

(*i*) $(y + z)p + (z + x)q = x + y$, (*ii*) $x\dfrac{\partial z}{\partial x} + y\dfrac{\partial z}{\partial y} + t\dfrac{\partial z}{\partial t} = xyt$, z being the dependent variable.

Solution (*i*) The subsidiary equations are $\dfrac{dx}{y + z} = \dfrac{dy}{z + x} = \dfrac{dz}{x + y}$.

$$\text{Now, each term} = \frac{dx - dy}{y - x} = \frac{dy - dz}{z - y} = \frac{dx + dy + dz}{2(x + y + z)}. \qquad ...(1)$$

From the first pair of terms, we get $\dfrac{dx-dy}{(x-y)} = \dfrac{dy-dz}{(y-z)}$.

Integration gives, $\log(x-y) = \log(y-z) + \log c$ or $\dfrac{(x-y)}{(y-z)} = c$. ...(2)

From the first term and last term in (1), we have $-\dfrac{dx-dy}{x-y} = \dfrac{dx+dy+dz}{2(x+y+z)}$.

Integration gives, $-\log(x-y) = \dfrac{1}{2}\log(x+y+z) - \log b_1$ ...(3)

or $(x-y)^2(x+y+z) = b$. The general solution is given by

$$\phi(b,c) = 0, \text{ or } \phi\left((x-y)^2(x+y+z), \frac{x-y}{y-z}\right) = 0.$$

(*ii*) In the differential equation $Pp + Qq = R$, there are two independent variables x and y. To solve this we have to find two independent solutions satisfying the equations $\dfrac{dx}{P} = \dfrac{dy}{Q} = \dfrac{dz}{R}$.

This procedure can be extended to first order first degree differential equations involving more than two independent variables. In the present case the number of independent variables is three, that is, x, y, t. Therefore, the subsidiary equations are

$$\frac{dx}{x} = \frac{dy}{y} = \frac{dt}{t} = \frac{dz}{xyt}. \qquad ...(1)$$

From the first pair of terms, $dx/x = dy/y$, we get $\log x = \log y + \log a$ or $x/y = a$.

Similarly, $\dfrac{dy}{y} = \dfrac{dt}{t}$ gives $\dfrac{y}{t} = b$.

Using the multipliers yt, xt, xy, -3, we get each term $= \dfrac{ytdx + xtdy + xydt - 3dz}{ytx + xty + xyt - 3xyt}$.

Since $ytx + xty + xyt - 3xyt = 0$. Therefore, $ytdx + xtdy + xydt - 3dz = 0$.
Integration gives $xyt - 3z = c$.

Hence, the general solution is $\phi\left(\dfrac{x}{y}, \dfrac{y}{t}, xyt - 3z\right) = 0$.

EXAMPLE 2.19 Find the general solution of the following partial differential equations

(*i*) $x\dfrac{\partial z}{\partial y} = y\dfrac{\partial z}{\partial x} + xe^{(x^2+y^2)}$, (*ii*) $2x(y+z^2)p + y(2y+z^2)q = z^3$. (*AMIE, W-2003*)

Solution (*i*) The given equation is $-yp + xq = xe^{x^2+y^2}$

The subsidiary equations are $\dfrac{dx}{-y} = \dfrac{dy}{x} = \dfrac{dz}{xe^{x^2+y^2}}$. From $\dfrac{dx}{-y} = \dfrac{dy}{x}$, we get $xdx + ydy = 0$.

Integration gives $\dfrac{x^2}{2} + \dfrac{y^2}{2} = \dfrac{a}{2}$ (say). Therefore, $x^2 + y^2 = a$. ...(1)

From $\dfrac{dy}{x} = \dfrac{dz}{xe^{x^2+y^2}}$, we have $e^a dy = dz$. (putting $x^2 + y^2 = a$)

Solving $e^a y = z + b$ or $e^a y - z = b$ or $ye^{x^2+y^2} - z = b$.

Therefore, solution required is $\phi(x^2 + y^2, ye^{x^2+y^2} - z) = 0$.

(*ii*) The subsidiary equations are $\dfrac{dx}{2x(y+z^2)} = \dfrac{dy}{y(2y+z^2)} = \dfrac{dz}{z^3}$.

Using the multipliers, $1/x, -1/y, -1/z$, we have each ratio $= \dfrac{x^{-1}dx - y^{-1}dy - z^{-1}dz}{2(y+z^2)-(2y+z^2)-z^2}$.

Since $2(y + z^2) - (2y + z^2) - z^2 = 0$, therefore, $x^{-1}dx - y^{-1}dy - z^{-1}dz = 0$.

Integration gives $\log x - \log y - \log z = \log a \Rightarrow x/yz = a$.

Using the last pair of terms, we get $2y^2dz + z^2(ydz - zdy) = 0$. Dividing by y^2z^2, we obtain

$\dfrac{2dz}{z^2} + \dfrac{ydz - zdy}{y^2} = 0$ or $\dfrac{2dz}{z^2} + \dfrac{d}{dz}\left(\dfrac{z}{y}\right) = 0$. Integration gives, $-\dfrac{2}{z} + \dfrac{z}{y} = b$.

Hence, the general solution is $\phi\left(\dfrac{x}{yz}, \dfrac{z}{y} - \dfrac{2}{z}\right) = 0$.

EXAMPLE 2.20 Find the general solution of the partial differential equation

$(x^2 - yz)\, p + (y^2 - zx)\, q = z^2 - xy$. *(AMIETE, June 2012, VTU, 2006; Bhopal, 2002)*

Solution The subsidiary equations are

$$\frac{dx}{x^2 - yz} = \frac{dy}{y^2 - zx} = \frac{dz}{z^2 - xy}. \qquad ...(i)$$

Now, each term $= \dfrac{dx - dy}{x^2 - y^2 - (y - x)z} = \dfrac{dy - dz}{y^2 - z^2 - x(z - y)}$,

that is, $\dfrac{d(x-y)}{(x-y)(x+y+z)} = \dfrac{d(y-z)}{(y-z)(x+y+z)}$ or $\dfrac{d(x-y)}{x-y} = \dfrac{d(y-z)}{y-z}$.

Integration gives, $\log(x - y) = \log(y - z) + \log a$ or $\dfrac{x-y}{y-z} = a$. ...(*ii*)

Now use multipliers x, y, z. Each term of subsidiary equations (*i*)

$$= \frac{xdx + ydy + zdz}{x^3 + y^3 + z^3 - 3xyz} = \frac{xdx + ydy + zdz}{(x+y+z)(x^2+y^2+z^2-xy-yz-zx)}. \qquad ...(iii)$$

Also each term of the subsidiary equations (*i*) $= \dfrac{dx + dy + dz}{x^2 + y^2 + z^2 - xy - yz - zx}$. ...(*iv*)

Equating (*iii*) and (*iv*) and cancelling the common factor, we get $\dfrac{xdx + ydy + zdz}{x+y+z} = dx + dy + dz$

or $\displaystyle\int (xdx + ydy + zdz) = \int (x+y+z)\,d(x+y+z) + c_1$

or $x^2 + y^2 + z^2 = (x + y + z)^2 + 2c_1$ or $xy + yz + zx = b$. ...(*v*)

Combining (*ii*) and (*v*), the general solution is $\dfrac{x-y}{y-z} = f(xy + yz + zx)$.

EXERCISE 2.3

Find the general solution of the following partial differential equations.

1. $3p + 4q = 2$. **Ans.** $\phi(3y - 4x, 3z - 2x) = 0$.

2. $xp + yq = 3z$. (*AMIETE, June 2012, Dec, 2009*)

Ans. $\phi(z/x^3, y/x) = 0$. This may be written as $z = x^3 f(x/y)$.

3. $p\sqrt{x} + q\sqrt{y} = \sqrt{z}$. (*AMIETE, June, 2011, 2008*) **Ans.** $\sqrt{x} - \sqrt{y} = f(\sqrt{x} - \sqrt{z})$.

4. $px^2 + qy^2 = z^2$. **Ans.** $x - y = xy\phi(1/x - 1/z)$.

5. $p\cos(x + y) + q\sin(x + y) = z$. **Ans.** $[\cos(x + y) + \sin(x + y)]e^{y-x} = \phi\left[z^{\sqrt{2}}\tan\left(\frac{x+y}{2} + \frac{\pi}{8}\right)\right]$.

6. $p - q = \log(x + y)$. **Ans.** $x\log(x + y) - z = f(x + y)$.

7. $px - qy = y^2 - x^2$. (*JNTU, 2002 S*) **Ans.** $x^2 + y^2 + 2z = f[\log(xy)]$.

[**Hint:** The auxiliary equations are $\frac{dx}{x} = \frac{dy}{-y} = \frac{dz}{(y^2 - x^2)}$. Use multipliers x, y, 1].

8. $p\tan x + q\tan y = \tan z$. (*Kuk., 2000*) **Ans.** $\phi\left(\frac{\sin x}{\sin y}, \frac{\sin y}{\sin z}\right) = 0$.

9. $\frac{y-z}{yz}p + \frac{z-x}{zx}q = \frac{x-y}{xy}$. (*AMIE S-2003, W-2002, W-2001*) **Ans.** $\phi(x + y + z, xyz) = 0$.

[**Hint:** Multiplying throughout by xyz, we get $x(y - z)p + y(z - x)q = z(x - y)$.
Now proceed as in solved Example 2.14(*ii*)]

10. $2yzp + zxq = 3xy$. **Ans.** $\phi(x^2 - 2y^2, 3y^2 - z^2) = 0$.

11. $x(z^2 - y^2)\frac{\partial z}{\partial x} + y(x^2 - z^2)\frac{\partial z}{\partial y} = z(y^2 - x^2)$.

(*AMIETE, Dec, 2009; AMIE S-2008, 2007; Madras 2000; VTU, 2000S*)

[**Hint:** The subsidiary equations are $\frac{dx}{x(z^2 - y^2)} = \frac{dy}{y(x^2 - z^2)} = \frac{dz}{z(y^2 - x^2)}$.

Using multipliers x, y, z, each term $= \frac{xdx + ydy + zdz}{x^2(z^2 - y^2) + y^2(x^2 - z^2) + z^2(y^2 - x^2)}$.

Since, $x^2(z^2 - y^2) + y^2(x^2 - z^2) + z^2(y^2 - x^2) = 0$. Therefore, $xdx + ydy + zdz = 0$. Integration gives $x^2 + y^2 + z^2 = c_1$ where $c_1 = 2a$.

Now, Using multipliers $1/x, 1/y, 1/z$ we get $dx/x + dy/y + dz/z = 0$ which on integration gives $xyz = c_2$. The general solution is $\phi(x^2 + y^2 + z^2, xyz) = 0$].

12. $x^2(y - z)p + y^2(z - x)q = z^2(x - y)$. (*AMIETE, June 2011; 2010; SVTU, 2007*)

Ans. $\phi\left(\frac{1}{x} + \frac{1}{y} + \frac{1}{z}, xyz\right) = 0$.

13. $x(y^2 + z)p - y(x^2 + z)q = z(x^2 - y^2)$. **Ans.** $\phi(x^2 + y^2 - 2z, xyz) = 0$.

[**Hint:** Use multipliers x, y, -1 and then $1/x, 1/y, 1/z$].

14. $xy^2p + y^3q = (zxy^2 - 4x^3)$. **Ans.** $\phi\left[\frac{y}{x}, x - \log\left(z - \frac{4x^2}{y^2}\right)\right] = 0$.

15. $(y - z)p + (z - x)q = x - y$. (*AMIETE, June 2013, 2010*) **Ans.** $f(x + y + z, x^2 + y^2 + z^2) = 0$.

[**Hint:** Use multipliers 1, 1, 1 and then x, y, z]

16. $px(z-2y^2)=(z-qy)(z-y^2-2x^3)$. *(AMIE, S-2000)*

[Hint: The given equation is $px(z-2y^2)+qy(z-y^2-2x^3)=(z-y^2-2x^3)z$.

The subsidiary equations are $\dfrac{dx}{x(z-2y^2)}=\dfrac{dy}{y(z-y^2-2x^3)}=\dfrac{dz}{z(z-y^2-2x^3)}$.

From the last pair of terms, we get $dy/y=dz/z$. Integration gives $y/z=a$. ...(1)

Now using the first and third terms $\dfrac{dx}{x(z-2y^2)}=\dfrac{dz}{z(z-y^2-2x^3)}$

or $\dfrac{dx}{x(z-2a^2z^2)}=\dfrac{dz}{z(z-a^2z^2-2x^3)}$ [Putting $y=az$ from (1)]

or $\dfrac{dx}{x(1-2a^2z)}=\dfrac{dz}{z-a^2z^2-2x^3}$ or $(z-a^2z^2-2x^3)dx-x(1-2a^2z)dz=0$

or $(zdx-xdz)-a^2(z^2dx-2xzdz)-2x^3dx=0$

or $\dfrac{xdz-zdx}{x^2}-a^2\left(\dfrac{2xzdz-z^2dx}{x^2}\right)+2xdx=0$. Integration gives

$\dfrac{z}{x}-a^2\left(\dfrac{z^2}{x}\right)+x^2=b$ or $\dfrac{z}{x}-\dfrac{y^2}{x}+x^2=b$. Therefore, $f\left(\dfrac{y}{z},\dfrac{z}{x}-\dfrac{y^2}{x}+x^2\right)=0$**]**.

17. $zx\dfrac{\partial z}{\partial x}-zy\dfrac{\partial z}{\partial y}=y^2-x^2$. *(AMIE, S-2005)* **Ans.** $\phi(xy, x^2+y^2+z^2)=0$.

18. $(y-z)p+(x-y)q=z-x$. **Ans.** $\phi(x+y+z, x^2+2yz)=0$.

19. $(y-z)p-(x+z)q=x-y$. **Ans.** $\phi[x+y+z, (x+z)(y+z)]=0$.

20. $(z^2-2yz-y^2)p+(xy+zx)q=xy-zx$. *(Kerala, 2005; AMIE, W-2000)*

[Hint: The subsidiary equations are $\dfrac{dx}{z^2-2yz-y^2}=\dfrac{dy}{xy+zx}=\dfrac{dz}{xy-zx}$.

Using x, y, z as multipliers, we have $xdx+ydy+zdz=0$, which on integration gives $x^2+y^2+z^2=a$, where $a=2c_1$. ...(1)

From the last two terms, we have $\dfrac{dy}{y+z}=\dfrac{dz}{y-z}\Rightarrow(y-z)dy=(y+z)dz\Rightarrow ydy-(zdy+ydz)-zdz=0$

or, $ydy-d(yz)-zdz=0$, which on integration gives $y^2-2yz-z^2=b$. ...(2)

From (1) and (2), the general solution is $\phi(x^2+y^2+z^2, y^2-2yz-z^2)=0$**]**.

21. $2xzp+2yzq=z^2-x^2-y^2$.

[Hint: The subsidiary equations are $\dfrac{dx}{2xz}=\dfrac{dy}{2yz}=\dfrac{dz}{z^2-x^2-y^2}$.

From the first pair of terms, we get $\log y=\log ax$ or $y=ax$. Using multipliers x, y, z. Now each term is equal to $\dfrac{xdx+ydy+zdz}{2x^2z+2y^2z+z^3-x^2z-y^2z}=\dfrac{xdx+ydy+zdz}{z(x^2+y^2+z^2)}$.

From the first term and above term, we get $\dfrac{dx}{x}=\dfrac{2(xdx+ydy+zdz)}{x^2+y^2+z^2}$.

Integration gives $\log(x^2+y^2+z^2)=\log x+\log b$ or $x^2+y^2+z^2=bx$.

The general solution is given by $\phi\left(\dfrac{y}{x},\dfrac{x^2+y^2+z^2}{x}\right)=0$**]**.

22. $z = px + qy + a\sqrt{x^2 + y^2 + z^2}$. **Ans.** $\dfrac{z + \sqrt{x^2 + y^2 + z^2}}{x^{1-a}} = \phi\left(\dfrac{y}{x}\right)$.

23. $z(x + y)\dfrac{\partial z}{\partial x} + z(x - y)\dfrac{\partial z}{\partial y} = x^2 + y^2$. **Ans.** $f(x^2 - y^2 - z^2, 4xy - z^2)$.

24. $\dfrac{dx}{x^2 + y^2} = \dfrac{dy}{2xy} = \dfrac{dz}{(x + y)z}$ **Ans.** $\phi\left(\dfrac{x + y}{z}, \dfrac{2y}{y^2 - x^2}\right) = 0$.

25. $z(p - q) = z^2 + (x + y)^2$. **Ans.** $\phi[x + y, \log_e(x^2 + y^2 + z^2 + 2xy) - 2x] = 0$.

2.5 NON-LINEAR EQUATIONS OF THE FIRST ORDER

In this section we shall discuss special methods for solving four important types (or forms) of non-linear differential equations of first order. These will be followed by a general method for solving this type of equations. The complete solution of such equations contains only two arbitrary constants, that is, equal to the number of independent variables and the particular solution is obtained by assigning particular values to the constants.

Type 1 $f(p, q) = 0$ ...(*i*) *that is, equations involving p and q only.*

This type of equation does not contain x, y, z, explicitly. For example $p^2 + q^2 = 9$ is an equation of this type.

Method of Solution We assume $z = ax + by + c$ (*ii*)

as a solution, where a and b are connected by the relation $f(a, b) = 0$. (*iii*)

Since $p = \partial z/\partial x = a$, $q = \partial z/\partial y = b$ which when substituted in (*iii*) gives (*i*).

Therefore, $z = ax + by + c$ is a solution where $f(a, b) = 0$ or $b = \phi(a)$ or $z = ax + \phi(a)y + c$ is the complete solution, where a and c are arbitrary constants.

EXAMPLE 2.21 Solve $p^2 + q^2 = 9$.

Solution $z = ax + by + c$ is a solution where $a^2 + b^2 = 9$. [By putting $p = a$ and $q = b$ in $p^2 + q^2 = 9$]

Now $b = \sqrt{9 - a^2}$. Therfore, $z = ax + \sqrt{9 - a^2}\, y + c$ is the complete solution required.

Type 2 $f(z, p, q) = 0$, *that is, equations containing p, q, z only.*

Method of Solution We assume $z = \phi(y + ax) = \phi(u)$ as a trial solution.

Then $p = \partial z/\partial x = \phi'(y + ax)\cdot a = a\phi'(u) = a\,dz/du$ and $q = \partial z/\partial y = \phi'(y + ax) = \phi'(u) = dz/du$.

This type of equation does not contain x, y explicitly.

Substituting these values of p and q in the given equation, we get $f\left(z, a\dfrac{dz}{du}, \dfrac{dz}{du}\right) = 0$, which is an ordinary differential equation in u and z.

Let $z = F(u, b)$ be its solution. Then $z = F(y + ax, b)$ is the complete solution required.

EXAMPLE 2.22 Solve $p(1 + q^2) = q(z - a)$.

Solution Let $z = \phi(y + ax) = \phi(u)$.

Then $p = \partial z/\partial x = a\phi'(y + ax) = a\phi'(u) = a\,dz/du$, $q = \partial z/\partial y = \phi'(y + ax) = \phi'(u) = dz/du$.

Substituting the values of p and q in the given equation, we get $a\dfrac{dz}{du}\left[1 + \left(\dfrac{dz}{du}\right)^2\right] = \dfrac{dz}{du}(z - a)$.

Now either $dz/du = 0$ which gives $z = c$ (*a* constant)

or $a\left[1 + (dz/du)^2\right] = z - a$, that is, $(dz/du)^2 = (z - 2a)/a$.

Therefore, $dz/du = \sqrt{z-2a}/\sqrt{a}$. Integration gives $\sqrt{a}\int \frac{dz}{\sqrt{z-2a}} = u + b$

or $2\sqrt{a}\sqrt{z-2a} = y + ax + b$, or $4a(z-2a) = (y+ax+b)^2$, is the required solution.

Type 3 *$f_1(x, p) = f_2(y, q)$, that is, variables separable form.*

This type of equation does not contains z explicitly and the terms involving p and x can be separated from those involving q and y and the equation can be written in the form $f_1(x, p) = f_2(y, q)$.

Method of Solution Let $f_1(x, p) = f_2(y, q) = a$ (an arbitrary constant).

Let $p = F_1(x, a)$ be the solution of $f_1(x, p) = a$ and $q = F_2(y, a)$ be that of $f_2(y, q) = a$.

Since z is a function of x and y, we have $dz = \frac{\partial z}{\partial x}dx + \frac{\partial z}{\partial y}dy$ or $dz = pdx + qdy$.

Putting the values of p and q, we get $dz = F_1(x, a)dx + F_2(y, a)dy$.

Integration, gives $z = \int F_1(x, a)dx + \int F_2(y, a)dy + b$, which is the required solution.

EXAMPLE 2.23 Solve $p - x^2 = q + y^2$.

Solution Let $p - x^2 = a$ and $q + y^2 = a$. Therefore, $p = a + x^2$ and $q = a - y^2$.

Putting values of p and q in $dz = pdx + qdy$, we obtain $dz = (a + x^2)dx + (a - y^2)dy$.

Solving, we get $z = ax + (x^3/3) + ay - (y^3/3) + b$.

Type 4 $z = px + qy + f(p, q)$ *(Clairaut's form)*

Method of Solution It can be easily verified that $z = ax + by + f(a, b)$ is a solution, because $\partial z/\partial x = p = a$ and $\partial z/\partial y = q = b$.

EXAMPLE 2.24 Solve $z = px + qy + 4p^2q^2$.

Solution The given equation is of Clairaut's form. Hence the complete solution is $z = ax + by + 4a^2b^2$.

EXAMPLE 2.25 Solve $p^2q^2(px + qy - z) = 2$.

Solution We write the given equation as $z = px + qy - (2/p^2q^2)$, which is a Clairaut's equation. Hence the solution is $z = ax + by - (2/a^2b^2)$.

EXAMPLE 2.26 Solve (*i*) $\sqrt{p} + \sqrt{q} = 1$, (*ii*) $pq + p + q = 0$.

Solution (*i*) It is of the type $f(p, q) = 0$. Hence the complete solution is $z = ax + by + c$,

where $\sqrt{a} + \sqrt{b} = 1$ or $b = (1 - \sqrt{a})^2$. Thus $z = ax + (1 - \sqrt{a})^2 y + c$, is the required solution.

(*ii*) The complete solution is $z = ax + by + c$, where $ab + a + b = 0$ or $b = -a/(a+1)$.

Therefore, $z = ax - \frac{a}{a+1}y + c$ is the required solution.

EXAMPLE 2.27 Solve $x^2p^2 + y^2q^2 = z^2$. *(Kerala, 2005; Kurukshetra, 2005; Madras, 2003)*

Solution The given equation can be written as

$$\left(\frac{x}{z}\cdot\frac{\partial z}{\partial x}\right)^2 + \left(\frac{y}{z}\cdot\frac{\partial z}{\partial y}\right)^2 = 1. \quad ...(i)$$

Setting, $\frac{\partial x}{x} = \partial X, \quad \frac{\partial y}{y} = \partial Y, \quad \frac{\partial z}{z} = \partial Z \Rightarrow X = \log x, \quad Y = \log y, \quad Z = \log z.$

Hence $\frac{x}{z}\cdot\frac{\partial z}{\partial x} = \frac{x}{z}\cdot\frac{z\partial Z}{x\partial X} = \frac{\partial Z}{\partial X}$ and $\frac{y}{z}\cdot\frac{\partial z}{\partial y} = \frac{\partial Z}{\partial Y}$.

Then (*i*) becomes $\left(\frac{\partial Z}{\partial X}\right)^2 + \left(\frac{\partial Z}{\partial Y}\right)^2 = 1$ or $P^2 + Q^2 = 1$, ...(*ii*) where $P = \partial Z/\partial X$ and $Q = \partial Z/\partial Y$.

The complete solution of (*ii*) is given by $Z = aX + bY + c$, ...(*iii*) where $a^2 + b^2 = 1$ or $b = \sqrt{1-a^2}$.

Thus (*iii*) reduces to $Z = aX + \sqrt{1-a^2}\, Y + c$ or $\log z = a\log x + \sqrt{1-a^2}\log y + c$

$$= \log x^a + \log y^{\sqrt{1-a^2}} + \log k, \text{ where } c = \log k$$

$$= \log kx^a y^{\sqrt{1-a^2}}, \quad \text{or} \quad z = kx^a y^{\sqrt{1-a^2}}, \text{ the required solution.}$$

EXAMPLE 2.28 Solve:

(*i*) $z = p^2 + q^2$, (*JNTU, 2002S; Madras, 2000*) (*ii*) $p(1+q) = qz$. (*Madras, 2000S*)

Solution (*i*) Let $z = \phi(y + ax) = \phi(u)$ be a solution.

Then $p = \partial z/\partial x = a\phi'(u) = a\,dz/du$ and $q = \partial z/\partial y = \phi'(u) = dz/du$.

Substituting in the given equation, we have

$$z = a^2\left(\frac{dz}{du}\right)^2 + \left(\frac{dz}{du}\right)^2 = (a^2+1)\left(\frac{dz}{du}\right)^2 \text{ or } \frac{dz}{du} = \frac{1}{\sqrt{a^2+1}} z^{1/2}.$$

Solving, we get $2\sqrt{a^2+1}\, z^{1/2} = u + b$ or $4(a^2+1)z = (u+b)^2 = (y + ax + b)^2$, as the required solution.

(*ii*) Proceeding as in part (*i*), we get

$$a \bullet \frac{dz}{du}\left(1 + \frac{dz}{du}\right) = \frac{dz}{du} \bullet z$$

Therefore, either $dz/du = 0$ which gives $z = c$ (a constant) or $a[1 + (dz/du)] = z$ or $dz/du = (z-a)/a$.

Solving, $a\log(z-a) = u + b$ or $a \log (z - a) = y + ax + b$ is the solution.

EXAMPLE 2.29 Solve $z^2(p^2x^2 + q^2) = 1$. (*Madras 2000S*)

Solution This question can be reduced to the type II by putting $X = \log x$, then

$$p = \frac{\partial z}{\partial x} = \frac{\partial z}{\partial X} \bullet \frac{\partial X}{\partial x} = \frac{1}{x} \bullet \frac{\partial z}{\partial X} \text{ so that } px = \frac{\partial z}{\partial X}.$$

Now the given equation reduces to $z^2\left\{\left(\frac{\partial z}{\partial X}\right)^2 + \left(\frac{\partial z}{\partial Y}\right)^2\right\} = 1.$ ...(1)

Put $z = \phi(y + aX) = \phi(u)$. Then $\frac{\partial z}{\partial X} = a\phi'(u) = a\frac{dz}{du}$ and $\frac{\partial z}{\partial y} = \phi'(u) = \frac{dz}{du}$.

Now (1) becomes $z^2\left\{a^2\left(\frac{dz}{du}\right)^2 + \left(\frac{dz}{du}\right)^2\right\} = 1$ or $z^2(a^2+1)\left(\frac{dz}{du}\right)^2 = 1$.

Therefore, $(a^2+1)^{1/2}\, z\frac{dz}{du} = 1$ that is, $\sqrt{a^2+1}\cdot z\,dz = du$.

Solving, $\sqrt{a^2+1}\,\frac{z^2}{2} = u + b$ or $\sqrt{a^2+1}\, z^2 = 2(y + aX + b)$

or $\sqrt{a^2+1}\, z^2 = 2(y + a\log x + b)$ is the required solution.

EXAMPLE 2.30 Solve $yp + xq + pq = 0$.

Solution The given equation can be written in the form $\frac{y}{q} + \frac{x}{p} + 1 = 0$. (variable separable form)

Now let $\frac{y}{q} = -\left(\frac{x}{p} + 1\right) = \frac{1}{a}$, so that $q = ay$ and $p = -ax/(a+1)$.

Substituting in $dz = pdx + qdy$ we get $dz = \frac{-ax}{a+1}dx + ay\,dy$.

Solving $z = -\frac{ax^2}{2(a+1)} + \frac{a}{2}y^2 + b$.

EXAMPLE 2.31 Solve : $z = px + qy + \sqrt{(l + p^2 + q^2)}$. (*Anna, 2005 S*)

Solution This is Clairaut's form. Hence the solution is $z = ax + by + \sqrt{1 + a^2 + b^2}$.

EXERCISE 2.4

Obtain the complete solution of the following equations

1. $p^3 - q^3 = 0$. **Ans.** $z = a(x + y) + c$.
2. $p^2 + q^2 = 2$. **Ans.** $z = ax + \sqrt{2 - a^2}\, y + c$.
3. $p^2 + q^2 = npq$. **Ans.** $z = ax + \frac{a}{2}(n \pm \sqrt{n^2 - 4})\, y + c$.
4. $z^2 = 1 + p^2 + q^2$. **Ans.** $z = \cosh\left[(y + ax + b)/\sqrt{a^2 + 1}\right]$.
5. $z^2(p^2 + q^2 + 1) = 1$. **Ans.** $(a^2 + 1)(1 - z^2) = (y + ax + b)^2$.
6. $p + q = \sin x + \sin y$. **Ans.** $z = a(x - y) - (\cos x + \cos y) + b$.
7. $\sqrt{p} + \sqrt{q} = x - y$. **Ans.** $3z = (x + a)^3 + (y + a)^3 + b$.
8. $z = px + qy + \sqrt{\frac{pq}{(p+q)}}$. **Ans.** $z = ax + by + \sqrt{\frac{ab}{(a+b)}}$.
9. $p^2 - x = q^2 - y$. **Ans.** $z = \frac{2}{3}\left[(x + a)^{3/2} + (y + a)^{3/2}\right] + b$.
10. $p^2 y(1 + x^2) = qx^2$. **Ans.** $z = a\sqrt{1 + x^2} + \frac{a}{2}y^2 + b$.
11. $p(1 - q^2) = q(1 - z)$. (*AMIE W-2004*) **Ans.** $4(1 - a + az) = (x + ay + b)^2$.
12. $z^2 = pqxy$. (*AMIETE, Dec 2009; VTU, 2004*) **Ans.** $z = ax^b y^{1/b}$.

    [Hint: The given equation is $1 = \left(\frac{x}{z}\frac{\partial z}{\partial x}\right)\left(\frac{y}{z}\frac{\partial z}{\partial y}\right)$. Put $Z = \log z$, $X = \log x$, $Y = \log y$. Then equation reduces to $1 = PQ$. This is of the standard type I, Putting therefore, $P = b$ and $Q = 1/b$, we have $Z = bX + \frac{1}{b}Y + a \Rightarrow \log z = b\log x + \frac{1}{b}\log y + \log a \Rightarrow z = ax^b y^{1/b}$.]

13. $\sqrt{p} - \sqrt{q} = 2x$. **Ans.** $z = [(a + 2x)^3/6] + a^2 y + b$.
14. $z = px + qy - 2\sqrt{pq}$. **Ans.** $ax + by - 2\sqrt{ab}$.
15. $(p - q)(z - px - qy) = 1$.

    [Hint: Rewriting this in Clairant's form: $z = px + qy + 1/(p - q)$. The complete solution is $z = ax + by + 1/(a - b)$].

2.6 CHARPIT'S METHOD

Charpit's method is a general method for solving first order non-linear partial differential equations in two independent variables.

Let
$$f(x, y, z, p, q) = 0 \quad ...(i)$$
be the given differential equation.

If we can find another relation between p, q, x, y and z say,
$$F(x, y, z, p, q) = 0 \quad ...(ii)$$
then we can solve (*i*) and (*ii*) for p and q and substitute them in
$$dz = p\,dx + q\,dy. \quad ...(2.5)$$

Solution of Eq. (2.5), if it exists, is the complete solution of (*i*).

We shall now discuss a method to determine the relation (*ii*).

Let us assume the relation (*ii*) between x, y, z, p and q.

Differentiating (*i*) and (*ii*) partially w.r.t. x and y we get
$$\frac{\partial f}{\partial x}+\frac{\partial f}{\partial z}p+\frac{\partial f}{\partial p}\frac{\partial p}{\partial x}+\frac{\partial f}{\partial q}\frac{\partial q}{\partial x}=0, \quad ...(2.6)$$
$$\frac{\partial F}{\partial x}+\frac{\partial F}{\partial z}p+\frac{\partial F}{\partial p}\frac{\partial p}{\partial x}+\frac{\partial F}{\partial q}\frac{\partial q}{\partial x}=0, \quad ...(2.7)$$
$$\frac{\partial f}{\partial y}+\frac{\partial f}{\partial z}q+\frac{\partial f}{\partial p}\frac{\partial p}{\partial y}+\frac{\partial f}{\partial q}\frac{\partial q}{\partial y}=0, \quad ...(2.8)$$
$$\frac{\partial F}{\partial y}+\frac{\partial F}{\partial z}q+\frac{\partial F}{\partial p}\frac{\partial p}{\partial y}+\frac{\partial F}{\partial q}\frac{\partial q}{\partial y}=0. \quad ...(2.9)$$

Eliminating $\partial p/\partial x$ between the equations (2.6) and (2.7) and $\partial q/\partial y$ between the equations (2.8) and (2.9), we get
$$\left(\frac{\partial f}{\partial x}\frac{\partial F}{\partial p}-\frac{\partial F}{\partial x}\frac{\partial f}{\partial p}\right)+\left(\frac{\partial f}{\partial z}\frac{\partial F}{\partial p}-\frac{\partial F}{\partial z}\frac{\partial f}{\partial p}\right)p+\left(\frac{\partial f}{\partial q}\frac{\partial F}{\partial p}-\frac{\partial F}{\partial q}\frac{\partial f}{\partial p}\right)\frac{\partial q}{\partial x}=0, \quad ...(2.10)$$
and
$$\left(\frac{\partial f}{\partial y}\frac{\partial F}{\partial q}-\frac{\partial F}{\partial y}\frac{\partial f}{\partial q}\right)+\left(\frac{\partial f}{\partial z}\frac{\partial F}{\partial q}-\frac{\partial F}{\partial z}\frac{\partial f}{\partial q}\right)q+\left(\frac{\partial f}{\partial p}\frac{\partial F}{\partial q}-\frac{\partial F}{\partial p}\frac{\partial f}{\partial q}\right)\frac{\partial p}{\partial y}=0. \quad ...(2.11)$$

Adding the above equations and noting that
$$\frac{\partial q}{\partial x}=\frac{\partial^2 z}{\partial x\partial y}=\frac{\partial p}{\partial y}, \quad \text{we get}$$
$$\left(\frac{\partial f}{\partial x}+p\frac{\partial f}{\partial z}\right)\frac{\partial F}{\partial p}+\left(\frac{\partial f}{\partial y}+q\frac{\partial f}{\partial z}\right)\frac{\partial F}{\partial q}+\left(-\frac{\partial f}{\partial p}\right)\frac{\partial F}{\partial x}+\left(-\frac{\partial f}{\partial q}\right)\frac{\partial F}{\partial y}+\left(-p\frac{\partial f}{\partial p}-q\frac{\partial f}{\partial q}\right)\frac{\partial F}{\partial z}=0 \quad ...(2.12)$$
that is,
$$\left(-\frac{\partial f}{\partial p}\right)\frac{\partial F}{\partial x}+\left(-\frac{\partial f}{\partial q}\right)\frac{\partial F}{\partial y}+\left(-p\frac{\partial f}{\partial p}-q\frac{\partial f}{\partial q}\right)\frac{\partial F}{\partial z}+\left(\frac{\partial f}{\partial x}+p\frac{\partial f}{\partial z}\right)\frac{\partial F}{\partial p}+\left(\frac{\partial f}{\partial y}+q\frac{\partial f}{\partial z}\right)\frac{\partial F}{\partial q}=0.$$

This is a Lagrange's form of linear differential equation of first order in which F is the dependent variable and x, y, z, p, q are independent variables.

The auxiliary equations are
$$\frac{dx}{-\frac{\partial f}{\partial p}}=\frac{dy}{-\frac{\partial f}{\partial q}}=\frac{dz}{-p\frac{\partial f}{\partial p}-q\frac{\partial f}{\partial q}}=\frac{dp}{\frac{\partial f}{\partial x}+p\frac{\partial f}{\partial z}}=\frac{dq}{\frac{\partial f}{\partial y}+q\frac{\partial f}{\partial z}}=\frac{dF}{0}. \quad ...(2.13)$$

These equations are known as *Charpit's equations*. Any integral of (2.13) will satisfy (2.12) and if that integral involves p or q or both it can be taken as the relation (*ii*). In practice we choose the simplest of the **integrals of (2.13).** The relation (2.13) should be committed to memory to solve problems by Charpit's method.

EXAMPLE 2.32 Use Charpit's method to solve the equation $pxy + pq + qy = yz$.

(AMIETE, June 2013, Dec 2011; JNTU, 2006; Kurushetra, 2006)

Solution Let $f(x, y, z, p, q) = pxy + pq + qy - yz = 0$. ...(*i*)

We have $\dfrac{\partial f}{\partial x} = py, \dfrac{\partial f}{\partial y} = px + q - z, \dfrac{\partial f}{\partial z} = -y, \dfrac{\partial f}{\partial p} = xy + q, \dfrac{\partial f}{\partial q} = p + y.$

The Charpit's equations become

$$\frac{dx}{-(xy+q)} = \frac{dy}{-(p+y)} = \frac{dz}{-p(xy+q)-q(p+y)} = \frac{dp}{py+p(-y)} = \frac{dq}{px+q-z-qy} \Rightarrow dp = 0 \text{ or } p = a \quad ...(ii)$$

From (*i*) and (*ii*) $axy + aq + qy - yz = 0$ or $q = \dfrac{y(z-ax)}{y+a}$.

Substituting these values of p and q in $dz = p\,dx + q\,dy$, we have

$$dz = a\,dx + y\frac{(z-ax)}{y+a}dy \quad \text{or} \quad \frac{dz - a\,dx}{z-ax} = \frac{y}{y+a}dy = \left(1 - \frac{a}{y+a}\right)dy.$$

Integrating, $\log(z - ax) = y - a\log(y + a) + \log b$

or $$\frac{(z-ax)(y+a)^a}{b} = e^y \quad \text{or} \quad z = ax + be^y(y+a)^{-a}.$$

EXAMPLE 2.33 Use Charpit's method to solve the equation $z = p^2x + q^2y$.

(AMIETE, Dec 2012; NIT Kurukshetra, 2008)

Solution Let $f(x, y, z, p, q) = p^2x + q^2y - z = 0$. ...(*i*)

We have $\dfrac{\partial f}{\partial x} = p^2, \quad \dfrac{\partial f}{\partial y} = q^2, \quad \dfrac{\partial f}{\partial z} = -1, \quad \dfrac{\partial f}{\partial p} = 2px, \quad \dfrac{\partial f}{\partial q} = 2qy.$

The Charpit's equations become

$$\frac{dx}{-2px} = \frac{dy}{-2qy} = \frac{dz}{-2(p^2x+q^2y)} = \frac{dp}{p^2-p} = \frac{dq}{q^2-q}.$$

From above, we have

$$\frac{p^2dx + 2px\,dp}{p^2x} = \frac{q^2dy + 2qy\,dq}{q^2y} \quad \text{or} \quad \frac{d(p^2x)}{p^2x} = \frac{d(q^2y)}{q^2y}.$$

On integration, $\log(p^2x) = \log(q^2y) + \log a \Rightarrow p^2x = aq^2y$, ...(*ii*)

where a is an arbitrary constant.

Substituting value of p^2x from Eq. (*ii*) in Eq. (*i*), we obtain

$$aq^2y + q^2y = z \quad \text{or} \quad q = \sqrt{\frac{z}{(1+a)y}} \quad ...(iii)$$

Therefore, from (*i*) $$p = \sqrt{\frac{az}{(1+a)x}}. \quad ...(iv)$$

Thus, $$dz = p\,dx + q\,dy = \sqrt{\frac{az}{(1+a)x}}\,dx + \sqrt{\frac{z}{(1+a)y}}\,dy$$

or $$\sqrt{1+a}\,\frac{dz}{\sqrt{z}} = \sqrt{a}\,\frac{dx}{\sqrt{x}} + \frac{dy}{\sqrt{y}}$$

or $$\sqrt{(1+a)z} = \sqrt{ax} + \sqrt{y} + b, \text{ where } b \text{ is an arbitrary constant.}$$

EXAMPLE 2.34 Find the complete integral of the partial differential equation $q = -xp + p^2$.

Solution Let $f(x, y, z, p, q) = p^2 - xp - q = 0$. ...(*i*)

We have, $$\frac{\partial f}{\partial x} = -p, \frac{\partial f}{\partial y} = 0, \frac{\partial f}{\partial z} = 0, \frac{\partial f}{\partial p} = 2p - x, \frac{\partial f}{\partial q} = -1.$$

The Charpit's equations become
$$\frac{dx}{-2p + x} = \frac{dy}{1} = \frac{dz}{-2p^2 + xp + q} = \frac{dp}{-p} = \frac{dq}{0}. \quad ...(ii)$$

Using the last two terms, we get $\log p = -y + \log a$ or $p = ae^{-y}$.

Substituting in the given differential equation, we obtain $q = -axe^{-y} + a^2e^{-2y}$.

Then, $dz = pdx + qdy$ becomes $dz = ae^{-y}dx + (-axe^{-y} + a^2e^{-2y})\,dy$.

Integrating, we obtain $$z = 2axe^{-y} - (a^2e^{-2y})/2 + b.$$

EXAMPLE 2.35 Solve $(p^2 + q^2)\,y = qz$. (*VTU, 2007; Hissar, 2005; Osmania, 2000S*)

Solution Let $f(x, y, z, p, q) = (p^2 + q^2)\,y - qz = 0$. ...(*i*)

The Charpit's equations become $$\frac{dx}{-2py} = \frac{dy}{z - 2qy} = \frac{dz}{-qz} = \frac{dp}{-pq} = \frac{dq}{p^2}. \quad ...(ii)$$

The last two ratios give $pdp + qdq = 0$. Integration gives, $p^2 + q^2 = c^2$, where c is an arbitrary constant.

Now to solve (*i*) and (*ii*), put $p^2 + q^2 = c^2$ in (*i*), so that $q = c^2y/z$.

Substituting this value of q in (*ii*), we get $p = c\sqrt{(z^2 - c^2y^2)}/z$.

Hence, $$dz = pdx + qdy = \frac{c}{z}\sqrt{(z^2 - c^2y^2)}\,dx + \frac{c^2y}{z}dy$$

or $$zdz - c^2y\,dy = c\sqrt{(z^2 - c^2y^2)}\,dx \text{ or } \frac{\frac{1}{2}d(z^2 - c^2y^2)}{\sqrt{(z^2 - c^2y^2)}} = c\,dx.$$

Integration gives $$\sqrt{(z^2 - c^2y^2)} = cx + a$$

or $$z^2 = (a + cx)^2 + c^2y^2,$$ where a is an arbitrary constant.

EXAMPLE 2.36 Solve $2zx - px^2 - 2qxy + pq = 0$. (*Rajasthan, 2006*) ...(*i*)

Solution Let $$f(x, y, z, p, q) = 2zx - px^2 - 2qxy + pq = 0.$$

Therefore, $$\partial f/\partial x = 2z - 2px - 2qy, \quad \partial f/\partial y = -2qx, \quad \partial f/\partial z = 2x,$$
$$\partial f/\partial p = -x^2 + q, \quad \partial f/\partial q = -2xy + p.$$

The Charpit's equations become
$$\frac{dx}{-\frac{\partial f}{\partial p}} = \frac{dy}{-\frac{\partial f}{\partial q}} = \frac{dz}{-p\frac{\partial f}{\partial p} - q\frac{\partial f}{\partial q}} = \frac{dp}{\frac{\partial f}{\partial x} + p\frac{\partial f}{\partial z}} = \frac{dq}{\frac{\partial f}{\partial y} + q\frac{\partial f}{\partial z}}$$

or $$\frac{dx}{x^2 - q} = \frac{dy}{2xy - p} = \frac{dz}{px^2 - 2pq + 2xyq} = \frac{dp}{2z - 2qy} = \frac{dq}{0}.$$

Using the last two terms, we get, $dq = 0$ or $q = a$, where a is an arbitrary constant.

Putting $q = a$ in (*i*), we get $p = \dfrac{2x(z - ay)}{x^2 - a}$. Substituting for p and q in the relation $dz = pdx + qdy$, we get
$$dz = \frac{2x(z - ay)\,dx}{x^2 - a} + a\,dy \text{ or } \frac{dz - a\,dy}{z - ay} = \frac{2x\,dx}{x^2 - a}.$$

Integration gives, $\log(z-ay)=\log(x^2-a)+\log b=\log b(x^2-b)$

or $z=ay+b(x^2-a)$ where b is an arbitrary constant.

EXAMPLE 2.37 Solve $2z+p^2+qy+2y^2=0$. *(JNTU, 2005; Kurukshetra, 2005)*

Solution Let $f(x,y,z,p,q)=2z+p^2+qy+2y^2=0$. ...(*i*)

The Charpit's equations become

$$\frac{dx}{-2p}=\frac{dy}{-y}=\frac{dz}{-(2p^2+qy)}=\frac{dp}{2p}=\frac{dq}{4y+3q}.$$

Using the first and fourth terms, we obtain $dp=-dx$ or $p=-x+a$, where a is an arbitrary constant. Putting $p=a-x$ in (*i*), we get

$$q=\frac{1}{y}[-2z-2y^2-(a-x)^2].$$

Therefore, $$dz=pdx+qdy=(a-x)dx-\frac{1}{y}[2z+2y^2+(a-x)^2]dy$$

or $$ydz=y(a-x)dx-2zdy-2y^2dy-(a-x)^2dy.$$

Multiplying both sides by $2y$, we obtain

$$2y^2dz+4yzdy=2y^2(a-x)\,dx-2y\,(a-x)^2\,dy-4y^3dy.$$

Integration gives, $2zy^2=-[2y^2(a-x)^2+y^4]+b$, where b is an arbitrary constant

or $y^2\left[2(x-a)^2+2z+y^2\right]=b$, the required solution.

EXAMPLE 2.38 Fing the complete integrals of the partial differential equation $2(z+xp+yq)=yp^2$.

Solution We have $f(x,y,z,p,q)=2(z+xp+yq)-yp^2=0$. ...(*i*)

Therefore, $$\frac{\partial f}{\partial p}=2x-2yp,\quad \frac{\partial f}{\partial q}=2y,\quad \frac{\partial f}{\partial x}=2p,\quad \frac{\partial f}{\partial y}=2q-p^2,\quad \frac{\partial f}{\partial z}=2.$$

The Charpit's equations become

$$\frac{dx}{-2x+2yp}=\frac{dy}{-2y}=\frac{dz}{-p(2x-2yp)-2qy}=\frac{dp}{2p+2p}=\frac{dq}{2q-p^2+2q}.$$

Using the second and the fourth terms, we have

$$\frac{dp}{p}+2\frac{dy}{y}=0\Rightarrow py^2=a,$$ where a is an arbitrary constant.

Substituting $p=a/y^2$ in (*i*), we obtain

$$q=-\frac{z}{y}-\frac{ax}{y^3}+\frac{a^2}{2y^4}.$$ Hence $dz=pdx+qdy$

$$=\left(\frac{a}{y^2}\right)dx+\left(-\frac{z}{y}-\frac{ax}{y^3}+\frac{a^2}{2y^4}\right)dy$$

or $$ydz+zdy=a\frac{(ydx-xdy)}{y^2}+\frac{a^2}{2y^3}dy.$$

Integrating, we get $$yz=a\frac{x}{y}-\frac{a^2}{4y^2}+b$$

or $$z=\frac{ax}{y^2}-\frac{a^2}{4y^3}+\frac{b}{y},$$ where b is an arbitrary constant.

EXERCISE 2.5

Find the complete integrals of the following partial differential equations.

1. $z = pq$. **Ans.** $4az = (ax + y + b)^2$.
2. $px + qy + z = xq^2$. **Ans.** $xz = -\dfrac{a^2}{2x^2} + a\left(\dfrac{y}{x}\right) + b$.
3. $p^2 + qy - z = 0$. **Ans.** $z = ay + \dfrac{1}{4}(x+b)^2$.
4. $p(q^2 + 1) + (b - z)q = 0$. **Ans.** $4[a(z-b) - 1] = (x + ay + b)^2$.
5. $1 + p^2 = qz$. **Ans.** $\dfrac{z^2}{2} \pm \left[\dfrac{z}{2}\sqrt{z^2 - 4a^2} - 2a^2 \log(z + \sqrt{z^2 - 4a^2})\right] = 2ax + 2y + b$.
6. $2xz - px^2 - 2xyq + pq = 0$. **Ans.** $z = ay + b(x^2 - a)$.
7. $(p^2 + q^2)x = pz$. **Ans.** $z^2 = a^2x^2 + (ay + b)^2$.
8. $z^2 = pq\,xy$. *(VTU, 2004)* **Ans.** $z = ax^b y^{1/b}$.
9. $p^2 - q^2 = x - y$. *(AMIETE, Dec 2010)* **Ans.** $z = \dfrac{2}{3}(x+a)^{3/2} + \dfrac{2}{3}(y+a)^{3/2} + b$.
10. $q = (z + px)^2$.

2.7 LINEAR HOMOGENEOUS EQUATIONS WITH CONSTANT COEFFICIENTS

An equation of the form

$$\frac{\partial^n z}{\partial x^n} + a_1 \frac{\partial^n z}{\partial x^{n-1}\partial y} + a_2 \frac{\partial^n z}{\partial x^{n-2}\partial y^2} + \ldots + a_n \frac{\partial^n z}{\partial y^n} = f(x, y) \qquad \ldots(2.14)$$

where a's are constants and $f(x, y)$ is a function of x and y is called a *linear homogeneous partial differential equation of nth order with constant coefficients.* The equation is called homogeneous because all derivatives involved are of the same order.

If we write $D^r = \dfrac{\partial^r}{\partial x^r}$ and $D'^r = \dfrac{\partial^r}{\partial y^r}$ then (2.14) can be written in the form

$$D^n z + a_1 D^{n-1} D' z + a_2 D^{n-2} D'^2 z \ldots + a_n D'^n z = f(x, y)$$

or

$$(D^n + a_1 D^{n-1} D' + a_2 D^{n-2} D'^2 \ldots + a_n D'^n) z = f(x, y)$$

or briefly

$$F(D, D')z = f(x, y).$$

The readers already know how to solve ordinary linear differential equations with constant coefficients. As in the case of ordinary differential equations, the complete solution of (2.14) consists of two parts, *complementary function* (C.F.) and the *particular integral* (P.I.).

(*a*) The complementary function is the solution of $F(D, D')z = 0$. It must contain n arbitrary functions, where n is the order of the differential equation.

(*b*) The particular integral is the particular solution of

$$F(D, D')z = f(x, y).$$

The complete solution of (2.14) is z = C.F. + P.I.

2.8 RULES FOR FINDING THE COMPLEMENTARY FUNCTION (C.F.)

The procedure for finding the complementary function will be explained by considering a differential equations of second order. The rule can be easily extended to equations of higher order.

Let us consider the equation

$$\frac{\partial^2 z}{\partial x^2} + a_1 \frac{\partial^2 z}{\partial x \partial y} + a_2 \frac{\partial^2 z}{\partial y^2} = 0. \qquad \ldots(1)$$

This equation can be written in the symbolic notation (or operator form) as

$$(D^2 + a_1 DD' + a_2 D'^2)z = 0. \quad ...(2)$$

The equation obtained by equating the operator on z equal to zero that is,

$$D^2 + a_1 DD' + a_2 D'^2 = 0 \quad ...(3)$$

is called the *auxiliary equation* or (A.E.).

Case I Roots of A.E. real and different

Let m_1, m_2 be roots of (3) considered as a quadratic in D/D'.

Then (2) can be put in the form $(D - m_2D')(D - m_1D')z = 0$. ...(4)

Now the solution of $(D - m_1D')z = 0$ is also a solution of (4).

But $(D - m_1 D')z = 0$ is equivalent to $\dfrac{\partial z}{\partial x} - m_1 \dfrac{\partial z}{\partial y} = 0$, that is, $p - m_1 q = 0$ (Lagrange's form).

The subsidiary equations for this are

$$\frac{dx}{1} = \frac{dy}{-m_1} = \frac{dz}{0}. \quad \text{(Art. 2.4)}$$

From these equations, we get $y + m_1x = a$ and $z = b$.

Therefore, $z = \phi_1(y + m_1x)$, a solution of $(D - m_1D')z = 0$ is also a solution of (4).

Similarly $z = \phi_2(y + m_2x)$ is a solution of (4).

Hence $z = \phi_1(y + m_1x) + \phi_2(y + m_2x)$ is the complete solution of (4).

Case II Roots of A.E. equal

Let the two roots of A.E. be each equal to m.

Then (2) becomes $(D - mD')(D - mD')z = 0$ or $(D - mD')^2 z = 0$. ...(5)

Let $(D - mD')z = u$, then (5) becomes $(D - mD')u = 0$ which has $u = \phi(y + mx)$ as its solution.

Hence, $(D - mD')z = \phi(y + mx)$ or $p - mq = \phi(y + mx)$. [Lagrange's form]

The auxiliary system for this is

$$\frac{dx}{1} = \frac{dy}{-m} = \frac{dz}{\phi(y + mx)}. \text{ Now } \frac{dx}{1} = \frac{dy}{-m} \text{ gives } y + mx = a. \quad ...(6)$$

Then $\dfrac{dx}{1} = \dfrac{dz}{\phi(a)}$ or $z = x\phi(a) + b$ or $z = x\phi(y + mx) + b$. ...(7)

From (6) and (7), $z - x\phi(y + mx) = \psi(y + mx)$

or $z = x\phi(y + mx) + \psi(y + mx)$ is the required solution.

Remark 2

If $\dfrac{\partial^2 z}{\partial x^2} + a_1 \dfrac{\partial^2 z}{\partial x \partial y} + a_2 \dfrac{\partial^2 z}{\partial y^2} = 0$, that is, $(D^2 + a_1 DD' + a_2 D'^2)z = 0$

be the differential equation, then $D^2 + a_1DD' + a_2D'^2 = 0$ is the auxiliary equation. If we put $D = m$, $D' = 1$ in this, we get $m^2 + a_1m + a_2 = 0$.

Now roots of $D^2 + a_1DD' + a_2D'^2 = 0$, considered as quadratic in D/D', are the same as those of $m^2 + a_1 m + a_2 = 0$. Hence this quadratic in m may also be called the A.E. Henceforth we shall write the A.E. in this form.

2.9 RULE TO WRITE THE COMPLEMENTARY FUNCTION

Let the given differential equation be

$$\frac{\partial^2 z}{\partial x^2} + a_1 \frac{\partial^2 z}{\partial x \partial y} + a_2 \frac{\partial^2 z}{\partial y^2} = 0.$$

In the symbolic form, this can be written as $(D^2 + a_1 DD' + a_2 D'^2) z = 0$. ...(1)

1. *Write the A.E., that is,* $m^2 + a_1 m + a_2 = 0$, that is, replace D by m and D' by 1.
2. *If the roots of the A.E. are real and distinct, say* m_1 *and* m_2 *then* $z = \phi_1(y + m_1 x) + \phi_2(y + m_2 x)$ is the C.F.
3. *If the roots of the A.E. are equal, say m each, then*

$$z = \phi_1(y + mx) + x\phi_2(y + mx) \text{ is the C.F.}$$

Remarks 3

(a) The rule for finding the C.F. can be easily extended to homogeneous differential equations of higher order.

For example if the differential equation is of 3rd order and m_1, m_2, m_3 are roots of A.E., then

$$z = \phi_1(y + m_1 x) + \phi_2(y + m_2 x) + \phi_3(y + m_3 x) \text{ is the C.F.}$$

In case all the three roots are equal and each equal to m (say), then

$$z = \phi_1(y + mx) + x\phi_2(y + mx) + x^2\phi_3(y + mx) \text{ is the C.F. etc.}$$

(b) If the given differential equation is $f(D, D')z = 0$, then C.F. is the complete solution.

EXAMPLE 2.39 Find the complementary function of following partial differential equations

(*i*) $\frac{\partial^2 z}{\partial x^2} - a^2 \frac{\partial^2 z}{\partial y^2} = 0$, (*ii*) $\frac{\partial^2 z}{\partial x^2} - 2\frac{\partial^2 z}{\partial x \partial y} + \frac{\partial^2 z}{\partial y^2} = 0$.

Solution (*i*) The given equation in the operator form is $(D^2 - a^2 D'^2) z = 0$.

The A.E. is $m^2 - a^2 = 0$. Its roots are $m = \pm a$.

Therefore, C.F. is $z = \phi_1(y + ax) + \phi_2(y - ax)$.

(*ii*) The A.E. is $m^2 - 2m + 1 = 0$ or $(m - 1)^2 = 0$. Therefore, $m = 1, 1$.

Since each root is 1, the C.F. is $z = \phi_1(y + x) + x\phi_2(y + x)$.

2.10 PARTICULAR INTEGRAL (P.I.) (*SPECIAL METHODS*)

If the given differential equation is $F(D, D')z = f(x, y)$ ...(1)

then its particular integral is denoted by $z = \frac{1}{F(D, D')} f(x, y)$.

[$\because$ If z is replaced by $\frac{1}{F(D, D')} f(x, y)$ in (1), then

$$\text{L.H.S.} = F(D, D')\frac{1}{F(D, D')} f(x, y) = f(x, y) = \text{R.H.S.},$$

bearing in mind that $\frac{1}{F(D, D')}$ and $F(D, D')$ are inverse operators]

We shall give below, without proof, methods for obtaining P.I. of (1) for different forms of $f(x, y)$. These are analogous to the short methods for finding P.I. in ordinary differential equations.

(*a*) If $F(D, D')z = e^{ax+by}$

then $\text{P.I.} = \frac{1}{F(D, D')} e^{ax+by} = \frac{1}{F(a, b)} e^{ax+by}$,

provided $F(a, b) \neq 0$. $F(a, b) = 0$ is called a case of failure.

(b) If $F(D^2, DD', D'^2)z = \sin(ax + by)$

then $\text{P.I.} = \dfrac{1}{F(-a^2, -ab, -b^2)}\sin(ax + by)$ provided $F(-a^2, -ab, -b^2) \neq 0$.

The same is the rule when $f(x, y) = \cos(ax + by)$. For the case of failure, that is, $F(-a^2, -ab, -b^2) = 0$. See example 2.47 (*ii*).

(c) If $F(D, D')z = x^p y^q$, (p, q being positive integers), then

$$\text{P.I.} = \frac{1}{F(D, D')}x^p y^q = [F(D, D')]^{-1} x^p y^q.$$

We find the particular integral by expanding $[F(D, D')]^{-1}$ in ascending powers of D or D' and then operate on $x^p y^q$.

EXAMPLE 2.40 Find the P.I. of following partial differential equations

(*i*) $\dfrac{\partial^2 z}{\partial x^2} + 3\dfrac{\partial^2 z}{\partial x \partial y} + \dfrac{\partial^2 z}{\partial y^2} = e^{2x+3y}$, (*ii*) $(D^2 - 3DD' + 2D'^2)z = \sin(x - 2y)$,

(*iii*) $(D^2 - DD' - 6D'^2)z = x + y$.

Solution (*i*) The operator form of the differential equation is

$$(D^2 + 3DD' + D'^2)z = e^{2x+3y}.$$

Therefore, $$\text{P.I.} = \frac{1}{D^2 + 3DD' + D'^2}e^{2x+3y}.$$

Here $a = 2$, $b = 3$. Putting $D = 2$, $D' = 3$, we get

$$\text{P.I.} = \frac{e^{2x+3y}}{2^2 + 3\cdot 2\cdot 3 + 3^2} = \frac{e^{2x+3y}}{31}.$$

(*ii*) $$\text{P.I.} = \frac{1}{D^2 - 3DD' + 2D'^2}\sin(x - 2y).$$

Now a = 1, b = –2. Putting $D^2 = -1^2$, $DD' = -(1)(-2)$, $D'^2 = -(-2)^2$, we get

$$\text{P.I.} = \frac{1}{-1 - 3(2) + 2(-4)}\sin(x - 2y) = -\frac{1}{15}\sin(x - 2y).$$

(*iii*) $$\text{P.I.} = \frac{1}{D^2 - DD' - 6D'^2}(x + y) = \frac{1}{D^2}\left(1 - \frac{D'}{D} - 6\frac{D'^2}{D^2}\right)^{-1}(x + y)$$

$$= \frac{1}{D^2}\left[1 - \left(\frac{D'}{D} + 6\frac{D'^2}{D^2}\right)\right]^{-1}(x + y) = \frac{1}{D^2}\left(1 + \frac{D'}{D} + \ldots\right)(x + y)$$

(we expand by Binomial Theorem upto first degree term in D'/D, because $x + y$ contains first degree term in y)

$$= \frac{1}{D^2}(x + y + x)\left[\because \frac{D'}{D}(x + y) = \frac{1}{D}(0 + 1) = \frac{1}{D}1 = \int 1\,dx = x\right]$$

$$= \frac{1}{D^2}(2x + y) = 2\cdot\frac{x^3}{6} + y\cdot\frac{x^2}{2} = \frac{x^3}{3} + \frac{x^2 y}{2}.$$

$$\left[\because \frac{1}{D^2}x = \frac{1}{D}\int x\,dx = \frac{1}{D}\cdot\frac{x^2}{2} = \int\frac{x^2}{2}dx = \frac{x^3}{6}\right].$$

Notes: 1. In part (*iii*) above we could take D'^2 common in place of D^2. In that case the result will be different but the difference can be included in C.F.

2. To evaluate $\frac{1}{f(D, D')}x^p y^q$ we should expand $[f(D, D')]^{-1}$ in powers of D'/D, if power of y, that is, q is less than power of x (that is, p.). We expand in powers of D/D' if $p < q$.

EXAMPLE 2.41 Find the solutions of the following partial differential equations

(*i*) $\frac{\partial^2 z}{\partial x^2} + \frac{\partial^2 z}{\partial x \partial y} - 2\frac{\partial^2 z}{\partial y^2} = 0$, (*ii*) $(2D^2 + 5DD' + 2D'^2) z = 0$.

Solution (*i*) The given differential equation is $(D^2 + DD' - 2D'^2) z = 0$.
The auxiliary equation is $m^2 + m - 2 = 0$ whose roots are $m = 1, -2$.
The solution required is $z = \phi_1(y + x) + \phi_2(y - 2x)$.

(*ii*) The A.E. is $2m^2 + 5m + 2 = 0$. Its roots are $m = -2, -1/2$.
Hence the solution is $z = \phi_1(y - 2x) + \phi_2\left(y - \frac{x}{2}\right)$ or $z = \phi_1(y - 2x) + \phi_2(2y - x)$.

EXAMPLE 2.42 Solve the following partial differential equations

(*i*) $(D - 2D')(D + 3D')^2 z = 0$, (*ii*) $\frac{\partial^3 z}{\partial x^3} - 3\frac{\partial^3 y}{\partial x^2 \partial y} + 3\frac{\partial^3 z}{\partial x \partial y^2} - \frac{\partial^3 z}{\partial y^3} = 0$.

Solution (*i*) The A.E. is $(m - 2)(m + 3)^2 = 0$. Therefore, $m = 2, -3, -3$.
Here two roots of A.E. are equal. Therefore, the solution is

$$z = \phi_1(y + 2x) + \phi_2(y - 3x) + x\phi_3(y - 3x).$$

(*ii*) The given equation can be written as $(D^3 - 3D^2D' + 3DD'^2 - D'^3) z = 0$.
The A.E. is $m^3 - 3m^2 + 3m - 1 = 0$ or $(m - 1)^3 = 0$. Therefore, $m = 1, 1, 1$.
Since all the three roots are equal, the solution is

$$z = \phi_1(y + x) + x\phi_2(y + x) + x^2\phi_3(y + x).$$

EXAMPLE 2.43 Solve the following partial differential equations

(*i*) $\frac{\partial^2 z}{\partial x^2} - 5\frac{\partial^2 z}{\partial x \partial y} + 6\frac{\partial^2 z}{\partial y^2} = e^{x+y}$, (*AMIE W-2002*) (*ii*) $\frac{\partial^3 z}{\partial x^3} - 3\frac{\partial^3 z}{\partial x^2 \partial y} + 4\frac{\partial^3 z}{\partial y^3} = e^{x+2y}$.

Solution (*i*) The given differential equation is $(D^2 - 5DD' + 6D'^2) z = e^{x+y}$.
The auxiliary equation in $m^2 - 5m + 6 = 0$. Therefore, $m = 2, 3$.
Hence the C.F. $= \phi_1(y + 2x) + \phi_2(y + 3x)$.

$$\text{P.I.} = \frac{1}{D^2 - 5DD' + 6D'^2}e^{x+y} = \frac{1}{1^2 - 5(1)(1) + 6(1)^2}e^{x+y} = \frac{1}{2}e^{x+y} \quad (\text{Putting } D = 1, D' = 1)$$

Therefore, the complete solution is $z = \text{C.F.} + \text{P.I.} = \phi_1(y + 2x) + \phi_2(y + 3x) + (e^{x+y})/2$.

(*ii*) The given equation can be written in the symbolic form

$$(D^3 - 3D^2D' + 4D'^3) z = e^{x+2y}.$$

A.E. is $m^3 - 3m^2 + 4 = 0$ or $(m - 2)^2(m + 1) = 0$. Therefore, $m = 2, 2, -1$.

Hence, the C.F. $= \phi_1(y + 2x) + x\phi_2(y + 2x) + \phi_3(y - x)$.

$$\text{P.I.} = \frac{1}{D^3 - 3D^2D' + 4D'^3}e^{x+2y} = \frac{1}{1^3 - 3 \cdot 1^2 \cdot 2 + 4 \cdot 2^3}e^{x+2y} \quad (\text{Putting } D = 1, D' = 2)$$

$= \frac{e^{x+2y}}{27}$. The complete solution is $z = \phi_1(y + 2x) + x\phi_2(y + 2x) + \phi_3(y - x) + e^{(x+2y)}/27$.

EXAMPLE 2.44 Solve the following partial differential equations

(*i*) $(D^3 - 4D^2D' + 4DD'^2)\,z = 6\sin(3x + 2y)$, (*AMIE, W-2008*)

(*ii*) $\dfrac{\partial^2 z}{\partial x^2} + \dfrac{\partial^2 z}{\partial x\,\partial y} - 6\dfrac{\partial^2 z}{\partial y^2} = \cos(3x + y)$, (*iii*) $\dfrac{\partial^2 z}{\partial x^2} - \dfrac{\partial^2 z}{\partial x\,\partial y} = \sin x\cos 2y$.

Solution (*i*) The A.E. is $m^3 - 4m^2 + 4m = 0$ or $m(m^2 - 4m + 4) = 0$,

that is, $m(m-2)^2 = 0$. Therefore, $m = 0, 2, 2$.

$$\text{C.F.} = \phi_1(y) + \phi_2(y + 2x) + x\phi_3(y + 2x).$$

$$\text{P.I.} = \frac{1}{D^3 - 4D^2D' + 4DD'^2}6\sin(3x + 2y)$$

$$= \frac{1}{D(D^2 - 4DD' + 4D'^2)}6\sin(3x + 2y).$$

Putting $D^2 = -3^2 = -9,\ DD' = -(3)(2) = -6,\ D'^2 = -2^2 = -4$, we obtain

$$\text{P.I.} = \frac{1}{D[-9 - 4(-6) + 4(-4)]}6\sin(3x + 2y) = -\frac{1}{D}6\sin(3x + 2y)$$

$$= -6\left[\frac{-\cos(3x + 2y)}{3}\right] = 2\cos(3x + 2y).$$

The complete solution is $z = \phi_1(y) + \phi_2(y + 2x) + x\phi_3(y + 2x) + 2\cos(3x + 2y)$.

(*ii*) The symbolic form of the equation is $(D^2 + DD' - 6D'^2)z = \cos(3x + y)$.

The A.E. is $m^2 + m - 6 = 0$. Therefore, $m = -3, 2$.

$$\text{C.F.} = \phi_1(y - 3x) + \phi_2(y + 2x).$$

$$\text{P.I.} = \frac{1}{D^2 + DD' - 6D'^2}\cos(3x + y)$$

$$= \frac{1}{-3^2 - (3)(1) - 6(-1^2)}\cos(3x + y) = -\frac{1}{6}\cos(3x + y).$$

Therefore, required solution is $z = \phi_1(y - 3x) + \phi_2(y + 2x) - \{\cos(3x + y)\}/6$.

(*iii*) The given equation is $(D^2 - DD')\,z = \sin x\cos 2y$. A.E. is $m^2 - m = 0$ or $m = 0, 1$.

$$\text{C.F.} = \phi_1(y) + \phi_2(y + x).$$

$$\text{P.I.} = \frac{1}{D^2 - DD'}\sin x\cos 2y = \frac{1}{2}\frac{1}{D^2 - DD'}[\sin(x + 2y) + \sin(x - 2y)]$$

$$= \frac{1}{2}\left[\frac{1}{D^2 - DD'}\sin(x + 2y) + \frac{1}{D^2 - DD'}\sin(x - 2y)\right]$$

$$= \frac{1}{2}\left[\frac{1}{-1^2 + 1.2}\sin(x + 2y) + \frac{1}{-1^2 - (-1)(-2)}\sin(x - 2y)\right]$$

$$= \frac{1}{2}\left[\sin(x + 2y) - \frac{1}{3}\sin(x - 2y)\right] = \frac{1}{2}\sin(x + 2y) - \frac{1}{6}\sin(x - 2y).$$

Therefore, the complete solution is

$$z = \phi_1(y) + \phi_2(y + x) + \frac{1}{2}\sin(x + 2y) - \frac{1}{6}\sin(x - 2y).$$

EXAMPLE 2.45 Solve the following equations

(*i*) $\dfrac{\partial^2 z}{\partial x^2} - 3\dfrac{\partial^2 z}{\partial x \partial y} + 2\dfrac{\partial^2 z}{\partial y^2} = e^{2x+3y} + \sin(x-2y)$,

(*ii*) $(D^3 - 7DD'^2 - 6D'^3)z = \sin(x+2y) + e^{2x+y}$. (*AMIE W-2008, S-2001*)

Solution (*i*) The given equation in operator form is $(D^2 - 3DD' + 2D'^2)z = e^{2x+3y} + \sin(x-2y)$.

A.E. is $m^2 - 3m + 2 = 0$. Therefore, $m = 1, 2$. C.F. $= \phi_1(y+x) + \phi_2(y+2x)$.

$$\text{P.I.} = \frac{1}{D^2 - 3DD' + 2D'^2}\left[e^{2x+3y} + \sin(x-2y)\right]$$

$$= \frac{1}{D^2 - 3DD' + 2D'^2} e^{2x+3y} + \frac{1}{D^2 - 3DD' + 2D'^2}\sin(x-2y)$$

$$= \frac{1}{2^2 - 3(2)(3) + 2(3)^2} e^{2x+3y} + \frac{\sin(x-2y)}{-1^2 - 3(-1)(-2) + 2\left[-(-2)^2\right]}$$

$$= \frac{1}{4}e^{2x+3y} - \frac{1}{15}\sin(x-2y).$$

The complete solution is

$$z = \phi_1(y+x) + \phi_2(y+2x) + \frac{1}{4}e^{2x+3y} - \frac{1}{15}\sin(x-2y).$$

(*ii*) A.E. is $m^3 - 7m - 6 = 0$. Solving, we get $m = -1, -2, 3$.

Therefore, C.F. $= \phi_1(y-x) + \phi_2(y-2x) + \phi_3(y+3x)$.

$$\text{P.I.} = \frac{1}{D^3 - 7DD'^2 - 6D'^3}\left[\sin(x+2y) + e^{2x+y}\right]$$

$$= \frac{1}{D^3 - 7DD'^2 - 6D'^3}\sin(x+2y) + \frac{1}{D^3 - 7DD'^2 - 6D'^3}e^{2x+y}.$$

$$\text{P.I. of first term} = \frac{1}{D^3 - 7DD'^2 - 6D'^3}\sin(x+2y)$$

$$= \frac{1}{D\cdot D^2 - 7(D\cdot D'^2) - 6D'\cdot D'^2}\sin(x+2y)$$

(writing $D^2 = -1^2 = -1, D'^2 = -2^2 = -4$)

$$= \frac{1}{D(-1) - 7D(-4) - 6D'(-4)}\sin(x+2y) = \frac{1}{3(9D + 8D')}\sin(x+2y) \quad ...(A)$$

$$= \frac{9D - 8D'}{3(81D^2 - 64D'^2)}\sin(x+2y) = \frac{9D - 8D'}{3\left[81(-1) - 64(-4)\right]}\sin(x+2y)$$

$$= \frac{9D - 8D'}{3(175)}\sin(x+2y) = \frac{1}{525}\left[9\cos(x+2y) - 16\cos(x+2y)\right]$$

$$= \frac{-7}{525}\cos(x+2y) = -\frac{1}{75}\cos(x+2y).$$

$$\text{P.I. of second term} = \frac{1}{D^3 - 7DD'^2 - 6D'^3} e^{2x+y} = \frac{1}{2^3 - 7(2)(1^2) - 6(1)^3} e^{2x+y} = -\frac{1}{12} e^{2x+y}.$$

Therefore, the complete solution is $z = \phi_1(y-x) + \phi_2(y-2x) + \phi_3(y+3x) - \frac{1}{75}\cos(x+2y) - \frac{1}{12}e^{2x+y}$.

Important Note An alternate and simpler method for evaluation of P.I. after step (A) in the above solution is as follows:

$$\text{P.I.} = \frac{1}{3(9D+8D')}\sin(x+2y) = \frac{D}{3(9D^2+8DD')}\sin(x+2y) \qquad \text{(note this step)}$$

$$= \frac{D}{3[9(-1)+8(-2)]}\sin(x+2y) = -\frac{1}{75}\cos(x+2y).$$

EXAMPLE 2.46 Solve the following equations

(*i*) $\frac{\partial^2 z}{\partial x^2} + \frac{\partial^2 z}{\partial x \partial y} - 6\frac{\partial^2 z}{\partial y^2} = x + y$, (*AMIE S-2005, S-2003*)

(*ii*) $\frac{\partial^2 z}{\partial x^2} + 3\frac{\partial^2 z}{\partial x \partial y} + 2\frac{\partial^2 z}{\partial y^2} = x + y$. (*AMIE W-2004*)

Solution (*i*) The given equation is $(D^2 + DD' - 6D'^2)z = x + y$.

The A.E. is $m^2 + m - 6 = 0$ so that $m = 2, -3$.

$$\text{C.F.} = \phi_1(y+2x) + \phi_2(y-3x).$$

$$\text{P.I.} = \frac{1}{D^2 + DD' - 6D'^2}(x+y) = \frac{1}{D^2}\left[1 + \left(\frac{D'}{D} - 6\frac{D'^2}{D^2}\right)\right]^{-1}(x+y)$$

$$= \frac{1}{D^2}\left(1 - \frac{D'}{D}\right)(x+y) = \frac{1}{D^2}\left[x + y - \frac{1}{D}(0+1)\right]$$

$$= \frac{1}{D^2}(x + y - x) = \frac{1}{D^2}y = \frac{x^2 y}{2}.$$

$$\left[\because \frac{D'}{D}(x+y) = \frac{1}{D}\frac{\partial}{\partial y}(x+y) = \frac{1}{D}(1) = \int 1\,dx = x\right]$$

Therefore, the complete solution is $z = \text{C.F.} + \text{P.I.} = \phi_1(y+2x) + \phi_2(y-3x) + \frac{x^2 y}{2}$.

(*ii*) The operator form of the differential equation is $(D^2 + 3DD' + 2D'^2)z = x + y$.

The A.E. is $m^2 + 3m + 2 = 0$ so that $m = -1, -2$.

$$\text{C.F.} = \phi_1(y-x) + \phi_2(y-2x).$$

$$\text{P.I.} = \frac{1}{D^2 + 3DD' + 2D'^2}(x+y) = \frac{1}{D^2}\left[1 + \frac{3D'}{D} + 2\frac{D'^2}{D^2}\right]^{-1}(x+y)$$

$$= \frac{1}{D^2}\left(1 - \frac{3D'}{D}\right)(x+y) = \frac{1}{D^2}(x + y - 3x) = \frac{1}{D^2}(y - 2x) = \frac{x^2}{2}y - \frac{x^3}{3}.$$

Therefore, the complete solution is $z = \text{C.F.} + \text{P.I} = \phi_1(y-x) + \phi_2(y-2x) + \frac{x^2 y}{2} - \frac{x^3}{3}$.

EXAMPLE 2.47 Solve the following partial differential equations

(*i*) $\dfrac{\partial^2 z}{\partial x^2} - a^2 \dfrac{\partial^2 z}{\partial y^2} = x^2$, (*ii*) $\dfrac{\partial^3 z}{\partial x^3} - 2\dfrac{\partial^3 z}{\partial x^2 \partial y} = 2e^{2x} + 3x^2 y$. (*SVTU, 2007; AMIE S-2004, S-2001*)

Solution (*i*) The given equation is $(D^2 - a^2 D'^2)z = x^2$.

A.E. is $m^2 - a^2 = 0$ has $m = \pm a$ as its roots. C.F. $= \phi_1(y + ax) + \phi_2(y - ax)$.

$$\text{P.I.} = \frac{1}{D^2 - a^2 D'^2} x^2 = \frac{1}{D^2}\left(1 - a^2 \frac{D'^2}{D^2}\right)^{-1} x^2$$

$$= \frac{1}{D^2}\left(1 + a^2 \frac{D'^2}{D^2}\right) x^2 = \frac{1}{D^2}(x^2) = \frac{x^4}{12}.$$

Therefore, the complete solution is $z = \phi_1(y + ax) + \phi_2(y - ax) + \left(\dfrac{x^4}{12}\right)$.

(*ii*) A.E. is $m^3 - 2m^2 = 0$ has $m = 0, 0, 2$ as its roots.

$$\text{C.F.} = \phi_1(y) + x\phi_2(y) + \phi_3(y + 2x).$$

$$\text{P.I.} = \frac{1}{D^3 - 2D^2 D'}(2e^{2x} + 3x^2 y) = \frac{1}{D^3 - 2D^2 D'} 2e^{2x} + \frac{1}{D^3 - 2D^2 D'} 3x^2 y$$

$$= \frac{1}{2^3 - 2 \cdot 2^2 \cdot 0} 2e^{2x} + \frac{3}{D^3}\left(1 - 2\frac{D'}{D}\right)^{-1} x^2 y$$

$$= \frac{e^{2x}}{4} + \frac{3}{D^3}\left(1 + 2\frac{D'}{D}\right) x^2 y = \frac{e^{2x}}{4} + \frac{3}{D^3}\left(x^2 y + 2\frac{1}{D} x^2\right)$$

$$= \frac{e^{2x}}{4} + \frac{3}{D^3}\left(x^2 y + \frac{2x^3}{3}\right) = \frac{e^{2x}}{4} + 3y\frac{1}{D^3}x^2 + 2 \cdot \frac{1}{D^3}x^3$$

$$= \frac{e^{2x}}{4} + 3y \cdot \frac{x^5}{3 \cdot 4 \cdot 5} + 2 \cdot \frac{x^6}{4 \cdot 5 \cdot 6} = \frac{e^{2x}}{4} + \frac{x^5 y}{20} + \frac{x^6}{60}.$$

Therefore, the complete solution is z = C.F. + P.I.

$$= \phi_1(y) + x\phi_2(y) + \phi_3(y + 2x) + \frac{e^{2x}}{4} + \frac{x^5 y}{20} + \frac{x^6}{60}.$$

2.11 GENERAL METHOD FOR FINDING P.I.

So far we know how to find the P.I. of $F(D, D')z = f(x, y)$ when $f(x, y)$ is of the form $e^{ax + by}$, sin or cos $(ax + b)$, $x^p y^q$.

When $f(x, y)$ is of a different form, the P.I. can be found by using the general method, *viz.*

$$\frac{1}{D - mD'} f(x, y) = \int f(x, c - mx)\,dx$$

in which c is to be replaced by y + mx after integration.

In fact this method can be used in all cases discussed in Art. 2.10 including cases of failure, that is, in the case

$$\frac{1}{F(D, D')} e^{ax+by} \text{ when } F(a, b) = 0$$

or
$$\frac{1}{F(D^2, DD', D'^2)} \sin \text{ or } \cos (ax + by) \text{ when } F(-a^2, -ab, -b^2) = 0 .$$

The following examples will make the procedure clear.

EXAMPLE 2.48 Solve the following equations

(*i*) $(D^2 - DD' - 2D'^2)z = (y-1)e^x$,

(*ii*) $\frac{\partial^2 z}{\partial x^2} + \frac{\partial^2 z}{\partial x \partial y} - 6\frac{\partial^2 z}{\partial y^2} = y \cos x$ or $y(\sec x)^{-1}$. *(Anna, 2005S; UPTU, 2003; AMIE, S-2003; S-2000)*

Alternatively: $r + s - 6t = y \cos x$ or $y (\sec x)^{-1}$.

Solution (*i*) The A.E. is $m^2 - m - 2 = 0$ has roots $m = -1, 2$.

Therefore, $\text{C.F.} = \phi_1(y - x) + \phi_2(y + 2x)$.

$$\text{P.I.} = \frac{1}{D^2 - DD' - 2D'^2}(y-1)e^x = \frac{1}{(D-2D')(D+D')}(y-1)e^x$$

$$= \frac{1}{D-2D'} \cdot \frac{1}{D+D'}(y-1)e^x = \frac{1}{D-2D'}\int (c+x-1)e^x \, dx$$

$$\left[\because \frac{1}{D-mD'} f(x, y) = \int f(x, c-mx)dx,\right.$$

where c is replaced by $y + mx$ after integration].

$$= \frac{1}{D-2D'}[(c+x-1)e^x - e^x] = \frac{1}{D-2D'}[(y-x+x-1)e^x - e^x]$$

(replacing c by $y - x$)

$$= \frac{1}{D-2D'}[(y-2)e^x] = \int (c-2x-2)e^x dx$$ (replacing y by $c - 2x$)

$$= (c-2x-2)e^x - (-2)e^x = (c-2x)e^x$$ (replacing c by $y + 2x$)

$$= (y + 2x - 2x)\, e^x = ye^x.$$

Hence, complete solution is $z = \phi_1(y-x) + \phi_2(y+2x) + ye^x$.

(*ii*) The given equation in *symbolic form* is $(D^2 + DD' - 6 D'^2)\, z = y \cos x$ and the A.E. is $m^2 + m - 6 = 0$. Solving, the roots are $m = 2, -3$.

Therefore, $\text{C.F.} = \phi_1(y + 2x) + \phi_2(y - 3x)$.

$$\text{P.I.} = \frac{1}{D^2 + DD' - 6D'^2} y \cos x = \frac{1}{(D+3D')(D-2D')} y \cos x$$

$$= \frac{1}{D+3D'} \cdot \frac{1}{D-2D'} y \cos x$$ (replacing y by $(c - 2x)$)

$$= \frac{1}{D+3D'}\int (c-2x)\cos x \, dx = \frac{1}{D+3D'}[(c-2x)(\sin x) - (-2)(-\cos x)]$$

$$= \frac{1}{D+3D'}[(c-2x)\sin x - 2\cos x]$$ (replacing c by $y + 2x$)

$$= \frac{1}{D+3D'}[y\sin x - 2\cos x] \qquad \text{(replacing } y \text{ by } c+3x)$$

$$= \int [(c+3x)\sin x - 2\cos x]\,dx$$

$$= (c+3x)(-\cos x) - (3)(-\sin x) - 2\sin x = (c+3x)(-\cos x) + \sin x$$

$$= -y\cos x + \sin x. \qquad (\because \text{ on replacing } c \text{ by } y-3x)$$

Hence, the complete solution is $z = \phi_1(y+2x) + \phi_2(y-3x) - y\cos x + \sin x$.

EXAMPLE 2.49 Obtain the solutions of following equations

(*i*) $(D-2D')(D-D')^2 z = e^{x+y}$, (*ii*) $r + s - 6t = \cos(2x+y)$,

(*iii*) $4r - 4s + t = 16\log(x+2y)$.

Solution (*i*) The auxiliary equation $(m-2)(m-1)^2 = 0$ has $m = 2, 1, 1$ as its roots.

Therefore, C.F. $= \phi_1(y+2x) + \phi_2(y+x) + x\phi_3(y+x)$.

Here $F(D, D') = (D-2D')(D-D')^2$.

Since $a = 1, b = 1, F(a, b) = 0$, and, therefore, it is a case of failure and we cannot find P.I. by the method of Art. 2.10. Applying the general method

$$\text{P.I.} = \frac{1}{(D-2D')(D-D')^2}(e^{x+y}) = \frac{1}{(1-2)(D-D')^2}e^{x+y}$$

$$= -\frac{1}{D-D'}\cdot\frac{1}{D-D'}e^{x+y} = -\frac{1}{D-D'}\int e^{x+c-x}\,dx$$

$$= -\frac{1}{D-D'}\int e^c\,dx = -\frac{1}{D-D'}\cdot xe^c = -\frac{1}{D-D'}xe^{x+y}$$

$$= -\int xe^{x+c-x}dx = -e^c\int x\,dx = -\frac{x^2}{2}e^c = -\frac{x^2}{2}e^{x+y}.$$

Therefore, the complete solution is $z = \phi_1(y+2x) + \phi_2(y+x) + x\phi_3(y+x) - \frac{x^2}{2}e^{x+y}$.

(*ii*) We know that $p = \frac{\partial z}{\partial x}, q = \frac{\partial z}{\partial y}; r = \frac{\partial^2 z}{\partial x^2}, s = \frac{\partial^2 z}{\partial x \partial y}, t = \frac{\partial^2 z}{\partial y^2}$.

Therefore, the given equation is $\frac{\partial^2 z}{\partial x^2} + \frac{\partial^2 z}{\partial y \partial x} - 6\frac{\partial^2 z}{\partial y^2} = \cos(2x+y)$.

A.E. is $m^2 + m - 6 = 0$ and has $m = -3, 2$ as its roots. Therefore, C.F. $= \phi_1(y-3x) + \phi_2(y+2x)$.

Here $a = 2, b = 1$ and $f(D^2, DD', D'^2) = D^2 + DD' - 6D'^2$.

Now $f(-a^2, -ab, -b^2) = -2^2 - (2)(1) - 6(-1^2) = 0$.

Therefore, it is a case of failure and we apply the general method.

$$\text{Now } \frac{1}{D^2 + DD' - 6D'^2}\cos(2x+y) = \frac{1}{(D+3D')(D-2D')}\cos(2x+y)$$

$$= \frac{1}{D+3D'}\int \cos(2x+c-2x)\,dx = \frac{1}{D+3D'}\int \cos c\,dx = \frac{1}{D+3D'}x\cos c$$

$$= \frac{1}{D+3D'} x\cos(y+2x) = \int x\cos(c+3x+2x)dx$$

$$= \int x\cos(5x+c)dx = \frac{x\sin(5x+c)}{5} + \frac{\cos(5x+c)}{25}$$

$$= \frac{x}{5}\sin(5x+y-3x) + \frac{\cos(5x+y-3x)}{25} = \frac{x}{5}\sin(2x+y) + \frac{\cos(2x+y)}{25}.$$

Hence, complete solution is $z = \phi_1(y-3x) + \phi_2(y+2x) + \frac{x}{5}\sin(2x+y) + \frac{1}{25}\cos(2x+y)$.

(*iii*) We have $(4D^2 - 4DD' + D'^2)z = 16\log(x+2y)$.

The auxiliary equation is $4m^2 - 4m + 1 = 0$ and has $m = ½, ½$ as its roots.

Therefore, C.F. $= \phi_1(2y+x) + x\,\phi_2(2y+x)$.

$$\text{P.I.} = \frac{1}{(2D-D')^2} 16\log(x+2y) = 4\frac{1}{\left(D-\frac{D'}{2}\right)}\left\{\frac{1}{\left(D-\frac{D'}{2}\right)}\log(x+2y)\right\}$$

$$= 4\cdot\frac{1}{\left(D-\frac{D'}{2}\right)}\left[\int \log\left\{x+2\left(c-\frac{x}{2}\right)\right\}dx\right] = 4\frac{1}{\left(D-\frac{D'}{2}\right)}\int \log(2c)dx$$

$$= 4\frac{1}{\left(D-\frac{D'}{2}\right)} x\log(2c) = 4\frac{1}{\left(D-\frac{D'}{2}\right)} x\log(x+2y) \qquad \left(\because \text{on replacing } c \text{ by } y+\frac{x}{2}\right)$$

$$= 4\left[\int x\log\left\{x+2\left(c-\frac{1}{2}x\right)\right\}dx\right] = 4\int x\log 2c\,dx = 2x^2\log 2c = 2x^2\log(x+2y).$$

$$\left(\text{on replacing } c \text{ by } y+\frac{1}{2}x\right)$$

Hence, the complete solution is $z = \phi_1(2y+x) + x\phi_2(2y+x) + 2x^2\log(x+2y)$.

EXERCISE 2.6

Solve the following partial differential equations:

1. $\frac{\partial^2 z}{\partial x^2} + 7\frac{\partial^2 z}{\partial x\partial y} + 12\frac{\partial^2 z}{\partial y^2} = 0$. **Ans.** $z = \phi_1(y-3x) + \phi_2(y-4x)$.
2. $(2D^2 + 5DD' + 2D'^2)z = 0$. **Ans.** $z = \phi_1(y-2x) + \phi_2(2y-x)$.
3. $\frac{\partial^3 z}{\partial x^3} - 3\frac{\partial^3 z}{\partial x^2\partial y} + 4\frac{\partial^3 z}{\partial y^3} = 0$. **Ans.** $y = \phi_1(y-x) + \phi_2(y+2x) + x\phi_3(y+2x)$.
4. $(D^2 - 2DD' + D'^2)z = e^{x+2y}$. **Ans.** $z = \phi_1(y+x) + x\phi_2(y+x) + e^{x+2y}$.
5. $(D^2 - 7DD' + 12D'^2)z = e^{x-y}$. **Ans.** $z = \phi_1(y+3x) + \phi_2(y+4x) + \frac{e^{x-y}}{20}$.
6. $(D-2D')^2(D+3D')z = e^{2x+y}$. **Ans.** $z = \phi_1(y+2x) + x\phi_2(y+2x) + \phi_3(y-3x) + \frac{1}{10}x^2e^{2x+y}$.
7. $r - 4s + 4t = e^{2x+y}$. **Ans.** $z = f_1(y+2x) + xf_2(y+2x) + \frac{1}{2}x^2e^{y+2x}$.

8. $\dfrac{\partial^2 z}{\partial x^2}-2\dfrac{\partial^2 z}{\partial x\partial y}+\dfrac{\partial^2 z}{\partial y^2}=\sin x$. **Ans.** $z=\phi_1(y+x)+x\phi_2(y+x)-\sin x$.

9. $(2D^2-5DD'+2D'^2)z=5\sin(2x-y)$. **Ans.** $z=\phi_1(y+2x)+\phi_2(2y+x)-\dfrac{1}{4}\sin(2x-y)$.

10. $(2D^2-5DD'+2D'^2)z=5\sin(2x+y)$. **Ans.** $z=\phi_1(y+2x)+\phi_2(2y+x)-(5/3)x\cos(2x+y)$.

11. $\dfrac{\partial^2 z}{\partial x^2}-\dfrac{\partial^2 z}{\partial x\partial y}=\cos(x+2y)$. **Ans.** $z=\phi_1(y)+\phi_2(y+x)+\cos(x+2y)$.

12. $(D^2-6DD'+9D'^2)z=6x+2y$. *(AMIE W-2003)* **Ans.** $z=\phi_1(y+3x)+x\phi_2(y+3x)+x^2(y+3x)$.

13. $\dfrac{\partial^2 z}{\partial x^2}+4\dfrac{\partial^2 z}{\partial x\partial y}-5\dfrac{\partial^2 z}{\partial y^2}=y^2+x$. *(AMIE, S-2000)*

Ans. $z=\phi_1(y+x)+\phi_2(y-5x)+\dfrac{y^2x^2}{2}+\dfrac{x^3}{6}-\dfrac{4x^3y}{3}+\dfrac{7x^4}{4}$.

14. $\dfrac{\partial^2 z}{\partial x^2}+\dfrac{\partial^2 z}{\partial y^2}=12(x+y)$. **Ans.** $z=\phi_1(y+ix)+\phi_2(y-ix)+2(x^3+3x^2y)$.

15. (*a*) $\dfrac{\partial^2 z}{\partial x^2}-\dfrac{\partial^2 z}{\partial y^2}=x^2+y^2$. **Ans.** $z=\phi_1(y-x)+\phi_2(y+x)+\left(\dfrac{x^4}{6}+\dfrac{x^2y^2}{2}\right)$.

(*b*) $\dfrac{\partial^2 z}{\partial x^2}+3\dfrac{\partial^2 z}{\partial x\partial y}+2\dfrac{\partial^2 z}{\partial x^2}=12xy$. **Ans.** $z=\phi_1(y-x)+\phi_2(y-2x)+2x^3\left(y-\dfrac{3}{4}x\right)$.

(*c*) $3\dfrac{\partial^2 z}{\partial x^2}-4\dfrac{\partial^2 z}{\partial x\partial y}+\dfrac{\partial^2 z}{\partial y^2}=3x^2y$. **Ans.** $z=\phi_1(x+3y)+\phi_2(x+y)+\dfrac{1}{45}x^5+\dfrac{1}{12}x^4y$.

16. $(D^2-6DD'+9D'^2)z=12x^2+36xy$. *(AMIE W-2001)* **Ans.** $z=\phi_1(y+3x)+x\phi_2(y+3x)+10x^4+6x^3y$.

17. $r+2s+t=2(y-x)+\sin(x-y)$. **Ans.** $z=\phi_1(y-x)+x\phi_2(y-x)+x^2y-x^3+(x^2/2)\sin(x-y)$.

18. $(D^2+2DD'+D'^2)z=2\sin y-x\cos y$. **Ans.** $z=\phi_1(y-x)+x\phi_2(y-x)+x\cos y-4\sin y$.

19. (*a*) $\dfrac{\partial^2 z}{\partial x^2}-\dfrac{\partial^2 z}{\partial x\partial y}=\cos x\cos 2y$. **Ans.** $z=\phi_1(y)+\phi_2(y+x)+\dfrac{1}{2}\cos(x+2y)-\dfrac{1}{6}\cos(x-2y)$.

(*b*) $\dfrac{\partial^2 z}{\partial x^2}-\dfrac{\partial^2 z}{\partial y^2}=x-y$. **Ans.** $z=\phi_1(y+x)+\phi_2(y-x)+\dfrac{x^3}{3}-\dfrac{x^2y}{2}$.

20. (*a*) $(D^3+D^2D'-DD'^2-D'^3)z=e^x\cos 2y$. *(Kuk., 2000)*

Ans. $z=\phi_1(y+x)+\phi_2(y-x)+x\phi_3(y-x)+\dfrac{e^x}{25}(\cos 2y+2\sin 2y)$.

(*b*) $\dfrac{\partial^2 z}{\partial x^2}-\dfrac{\partial^2 z}{\partial y^2}=e^{x+2y}$. **Ans.** $z=\phi_1(y+x)+\phi_2(y-x)-\dfrac{1}{3}e^{x+2y}$.

$\left[\textbf{Hint: } (a)\ \dfrac{1}{f(D,D')}e^{ax}X=e^{ax}\dfrac{1}{f(D+a,D')}X, \text{etc.}\right]$

2.12 NON-HOMOGENEOUS LINEAR EQUATIONS WITH CONSTANT COEFFICIENTS

In the equation
$$F(D,D')z=f(x,y) \qquad ...(2.15)$$
if the polynomial expression $F(D, D')$ is not homogeneous then (2.15) is defined as non-homogeneous linear partial differential equation. As in the case of homogeneous linear partial differential equations, the complete solution (C.S.) of (2.15) is given by

C.S. = C.F. + P.I.

The procedure of getting the P.I. is the same as in the case of homogeneous linear equations. However, to evaluate the C.F., we factorise the polynomial expression $f(D, D')$ into factors of the type $(D - mD' - c)$. This gives us an equation $(D - mD' - c)\, z = 0$. This equation may be written as

$$p - mq = cz, \qquad \text{...(2.16)}$$

which is the Lagrange's first order partial differential equations. The subsidiary equations are:

$$\frac{dx}{1} = \frac{dy}{-m} = \frac{dz}{cz}.$$

The integrals are given by $y + mx = a$ and $z = be^{cx}$.

Taking $b = \phi(a)$, the solution of eq. (2.16) is given by $z = \phi(a)e^{cx} = \phi(y + mx)e^{cx}$.
Such solutions obtained from various factors when added, give the C.F. of (2.15).

EXAMPLE 2.50 Solve $(D^2 + DD' + D' - 1)z = e^{-x}$.

Solution Here $F(D, D')z = (D^2 + DD' + D' - 1)z = (D+1)(D + D' - 1)z$.

For C.F., we write $(D+1)(D + D' - 1)z = 0$. ...(1)

As given above, the solution corresponding to the factor $(D - mD' - c)$ is known to be $z = e^{cx}\phi(y + mx)$.

Therefore, $\qquad \text{C.F.} = e^{x}\phi(y - x) + e^{-x}\psi(y)$.

$$\text{P.I.} = \frac{1}{(D+1)(D+D'-1)}e^{-x} = -\frac{1}{2}\frac{1}{D+1}e^{-x} = -\frac{x}{2}e^{-x}.$$

Hence, the complete solution is $z = e^{x}\phi(y - x) + e^{-x}\psi(y) - \dfrac{x}{2}e^{-x}$.

EXAMPLE 2.51 Solve $(D^2 - DD' + D' - 1)\, z = \cos(x + 2y)$.

Solution Here $F(D, D')\, z = (D^2 - DD' + D' - 1)\, z = (D - 1)(D - D' + 1)\, z$.

Since the solution corresponding to the factor $(D - mD' - c)$ is known to be $z = \phi(y + mx)e^{cx}$, the C.F. of the given equation is $z = e^{x}\phi(y) + e^{-x}\psi(y + x)$.

$$\text{P.I.} = \frac{1}{D^2 - DD' + D' - 1}\cos(x + 2y) = \frac{1}{-1 + 2 + D' - 1}\cos(x + 2y)$$

$$= \frac{1}{D'}\cos(x + 2y) = \frac{1}{2}\sin(x + 2y).$$

Therefore, the complete solution is given by $z = e^{x} \cdot \phi(y) + e^{-x}\psi(y + x) + \dfrac{1}{2}\sin(x + 2y)$.

EXAMPLE 2.52 Solve $(2DD' + D'^2 - 3D')\, z = 3\cos(3x - 2y)$.

Solution Here $f(D, D')\, z = D'(D' + 2D - 3)$.

Since the solution corresponding to the factor $D' - mD - c$ is known to be $z = e^{cy}\phi(x + my)$.

Therefore, $\qquad \text{C.F.} = \phi(x) + e^{3y}\psi(x - 2y)$.

$$\text{The particular integral} = \frac{1}{2DD' + D'^2 - 3D'}3\cos(3x - 2y)$$

$$= \frac{1}{-2(3)(-2) - (-2)^2 - 3D'}3\cos(3x - 2y)$$

$$= 3 \cdot \frac{1}{8 - 3D'}\cos(3x - 2y) = 3 \cdot \frac{8 + 3D'}{64 - 9D'^2}\cos(3x - 2y)$$

$$= 3 \cdot \frac{8+3D'}{64-9\{-(-2)^2\}} \cos(3x-2y) = \frac{3}{100}\left[8\cos(3x-2y)+6\sin(3x-2y)\right]$$

$$= \frac{3}{50}[4\cos(3x-2y)+3\sin(3x-2y)].$$

Hence, the complete solution is

$$z = \phi(x) + e^{3y}\psi(x-2y) + \frac{3}{50}\left[4\cos(3x-2y)+3\sin(3x-2y)\right].$$

EXERCISE 2.7

Solve the following partial differential equations

1. $(D-D'-1)(D-D'-2)z = 0.$ **Ans.** $z = e^x\phi(y+x) + e^{2x}\psi(y+x).$
2. $(2D^2 - DD' - D'^2 + 6D + 3D')z = 0.$ **Ans.** $z = \phi(x-2y) + e^{-3x}\psi(y+x).$
3. $r + 2s + t + 2p + 2q + z = 0.$ **Ans.** $z = e^{-x}\phi_1(y-x) + xe^{-x}\phi_2(y-x).$
4. $DD'(D-2D'-3)z = 0.$ **Ans.** $z = f_1(y) + f_2(x) + e^{3x}f_3(y+2x).$
5. $(D^2 + 2DD' + D'^2 - 2D - 2D')z = \sin(x+2y).$ *(UPTU, 2004)*

   Ans. $z = \phi(y-x) + e^{2x}\psi(y-x) + \frac{1}{39}\{2\cos(x+2y) - 3\sin(x+2y)\}.$
6. $(D^2 - DD' + D' - 1)z = \cos(x+2y) + e^y.$ **Ans.** $z = e^x\phi(y) + e^{-x}\psi(y+x) + \frac{1}{2}\sin(x+2y) - xe^y.$
7. $\frac{\partial^2 z}{\partial x^2} - \frac{\partial^2 z}{\partial x \partial y} + \frac{\partial z}{\partial x} = x^2 + y^2.$ *(Madras 2000S)*

   Ans. $z = \phi(y) + e^{-x}\psi(y+x) + \frac{1}{3}x^3 - x^2 + xy^2 + 2xy + 4x.$
8. $(D-D'-1)(D-D'-2)z = e^{2x-y} + x.$ **Ans.** $z = \phi(y+x)e^x + \psi(y+x)e^{2x} + \frac{1}{2}e^{2x-y} + \frac{x}{2} + \frac{3}{4}.$
9. $(D+D'-1)(D+2D'-3)z = 4 + 3x + 6y.$ **Ans.** $z = e^x\phi(y-x) + e^{3x}\psi(y-2x) + x + 2y + 6.$
10. $\frac{\partial^2 z}{\partial x^2} - \frac{\partial^2 z}{\partial y^2} + \frac{\partial z}{\partial x} + 3\frac{\partial z}{\partial y} - 2z = e^{x-y} - x^2 y.$

    Ans. $z = e^{-2x}\phi(y+x) + e^x\psi(y-x) - \frac{1}{4}e^{x-y} + \frac{1}{2}\left(x^2y + xy + \frac{3}{2}x^2 + \frac{3}{2}y + 3x + \frac{21}{4}\right).$
11. $(D^2 - D'^2 - 3D + 3D')z = xy + e^{x+2y}.$

    Ans. $z = e^{3x}\phi(y-x) + \psi(y+x) - ye^{x+2y} - \frac{1}{162}(27x^2y + 9x^3 + 18x^2 + 18xy + 18x + 6y + 8).$

[Hint: $\quad \text{P.I}_1 = \frac{1}{(D-D')(D+D'-3)}xy = \frac{1}{-3D\left(1-\frac{D'}{D}\right)\left[1-\left(\frac{D}{3}+\frac{D'}{3}\right)\right]}xy$

$$= -\frac{1}{3D}\left(1-\frac{D'}{D}\right)^{-1}\left[1-\left(\frac{D}{3}+\frac{D'}{3}\right)\right]^{-1}xy$$

$$= -\frac{1}{3}\left[\frac{1}{D}+\frac{1}{3}+\frac{2}{3}\frac{D'}{D}+\frac{1}{9}D+\frac{1}{3}D'+\frac{4}{27}DD'+\frac{D'}{D^2}+\ldots\right]xy$$

$$= -\frac{1}{3}\left(\frac{x^2y}{2}+\frac{1}{3}xy+\frac{x^2}{3}+\frac{y}{9}+\frac{x}{3}+\frac{4}{27}+\frac{x^3}{6}\right)$$

$$= -\frac{1}{162}(9x^3 + 27x^2y + 18x^2 + 18xy + 18x + 6y + 8).$$

$$\text{PI}_2 = \frac{1}{(D - D')(D + D' - 3)} e^{x+2y} = -ye^{x+2y}.]$$

12. $(D^2 - D'^2 + D + 3D' - 2)\,z = x^2y$.

Ans. $z = e^{-2x}\phi(y + x) + e^x\psi(y - x) + \frac{1}{4}[3 - x^2(2y + 3) - x(2y + 6) - 3y]$.

2.13 NON-LINEAR DIFFERENTIAL EQUATIONS OF THE SECOND ORDER

In the preceding sections, we discuss the solutions of linear as well as non-linear equations of the first order as also the solution of linear homogeneous equations of higher order. We now discuss the method due to *Monge* for integrating the equation.

$$Rr + Ss + Tt = V \qquad ...(2.17)$$

where R, S, T and V are functions of x, y, z, p and q.

Since

$$dp = \frac{\partial p}{\partial x}dx + \frac{\partial p}{\partial y}dy = rdx + sdy, \qquad ...(2.18)$$

and

$$dq = \frac{\partial q}{\partial x}dx + \frac{\partial q}{\partial y}dy = sdx + tdy. \qquad ...(2.19)$$

We have from (2.18) and (2.19)

$$r = \frac{dp - sdy}{dx} \quad \text{and} \quad t = \frac{dq - sdx}{dy}.$$

Substituting for r and t from (2.18) and (2.19) in (2.17) and rearranging the terms, we obtain

$$R\,dp\,dy + T\,dq\,dx - Vdx\,dy - s\,[R\,dy^2 - S\,dx\,dy + Tdx^2] = 0 \qquad ...(2.20)$$

If any relation between x, y, z, p and q will make each of the brackted expression vanish, then this will satisfy (2.17) and therefore, we have

$$R\,dp\,dy + T\,dq\,dx - Vdx\,dy = 0 \qquad ...(2.21)$$

and

$$R\,(dy)^2 - S\,dx\,dy + T\,(dx)^2 = 0. \qquad ...(2.22)$$

Equations (2.21) and (2.22) are called *Monge's Subsidiary Equations*. Equation (2.22) can be resolved into two equations of the form

$$dy - m_1dx = 0 \quad \text{and} \quad dy - m_2dx = 0. \qquad ...(2.23)$$

Combining one of these two relations with (2.17), we obtain two integrals of the type $u_1 = a$ and $v_1 = b$. This gives $u_1 = f_1(v_1)$ where f_1 is an arbitrary function and is an intermediary integral.

From the second of the equations (2.23) in the same way we obtain another pair of integrals, $u_2 = a$ and $v_2 = b$; so that $u_2 = f_2(v_2)$ is another intermediary integrals, f_2 being arbitrary.

To get the final integral, either of the two intermediary integrals may be integrated in a situation when $m_1 = m_2$. However, when m_1 and m_2 are unequal, the two intermediate integrals are solved for p and q and their values substituted in

$$dz = pdx + qdy.$$

This relation on integration gives the complete integral.

EXAMPLE 2.53 Solve $q\,(q^2 + s) = pt$.

Solution Comparing with the standard equation, we have

$$R = 0,\ S = q,\ T = -p,\ V = -q^3.$$

The Monge's equations are $q\,dx\,dy + p(dx)^2 = 0 \Rightarrow q\,dy + p\,dx = 0$ ...(*i*)

and $p\,dq\,dx - q^3dx\,dy = 0 \Rightarrow p\,dq - q^3\,dy = 0.$ ...(*ii*)

From the first equation, $q\,dy + p\,dx = 0$; then $dz = p\,dx + q\,dy = 0 \Rightarrow z = a$.

Substituting $q\,dy = -p\,dx$ in the second equation, yields

$$p\,dq - q^2(-p\,dx) = 0 \text{ or } dq + q^2 dx = 0 \text{ or } \frac{dq}{q^2} + dx = 0.$$

Thus $-\frac{1}{q} + x = b$ and $-\frac{1}{q} + x = f(z)$ or $\frac{\partial y}{\partial z} - x = -f(z)$.

Integrating, $$y = xz - \int f(z)dz + \phi_2(x)$$

or $y = xz + \phi_1(z) + \phi_2(x)$, which is the required equation of the problem.

EXAMPLE 2.54 Solve $x(r + 2xs + x^2 t) = p + 2x^3$.

Solution Here $R = x$, $S = 2x^2$, $T = x^3$, $V = p + 2x^3$.

The Monge's equations are $(dy)^2 - 2x\,dx\,dy + x^2(dx)^2 = 0$ or $(dy - x\,dx)^2 = 0$

and $x\,dp\,dy + x^3\,dq\,dx - (p + 2x^3)\,dx\,dy = 0$. We seek a solution of the system $dy - x\,dx = 0$,

$$x\,dp\,dy + x^3\,dq\,dx - (p + 2x^3)\,dx\,dy = 0.$$

From the first equation, $x^2 - 2y = a$. Substituting $dy = x\,dx$ in the second, we obtain

$$x\,dp + x^2 dq - (p + 2x^3)\,dx = 0.$$

Using the integrating factor $1/x^2$, we obtain the intermediate integral $p + xq = x^3 + xf(x^2 - 2y)$.

The Lagrange system is $\frac{dx}{1} = \frac{dy}{x} = \frac{dz}{x^3 + xf(x^2 - 2y)}$.

The first two members yield $x^2 - 2y = c$ and the first and third become $\frac{dx}{1} = \frac{dz}{x^3 + xf(c)}$.

Solving, we obtain $z = \frac{x^4}{4} + \frac{x^2}{2} f(c) + \phi(c)$ or $z = \frac{1}{4}x^4 + \frac{1}{2}x^2 f(x^2 - 2y) + \phi(x^2 - 2y)$.

EXAMPLE 2.55 Solve $r - t\cos^2 x + p\tan x = 0$.

Solution Comparing the given equation with $Rr + Ss + Tt = V$, we have $R = 1$, $S = 0$, $T = -\cos^2 x$, $V = -p\tan x$.

Substituting in Monge's equations, we get

$$dp\,dy - \cos^2 x\,dq\,dx + p\tan x\,dx\,dy = 0 \quad \text{...}(i)$$

and $$dy^2 - \cos^2 x\,dx^2 = 0 \quad \text{or} \quad (dy + \cos x\,dx)(dy - \cos x\,dx) = 0 \quad \text{...}(ii)$$

Equation (*ii*) gives $$dy - \cos x\,dx = 0, \quad \text{...}(iii)$$

$$dy + \cos x\,dx = 0. \quad \text{...}(iv)$$

From (*iii*), we have $$y - \sin x = c_1. \quad \text{...}(v)$$

From (*iii*) and (*i*), we have

$$dp(\cos x\,dx) - \cos^2 x\,dq\,dx + p\tan x\,dx(\cos x\,dx) = 0$$

or $$dp - \cos x\,dq + p\tan x\,dx = 0 \text{ or } \sec x\,dp - dq + p\tan x\sec x\,dx = 0$$

or $$d(p\sec x) - dq = 0. \text{ Integrating, } p\sec x - q = c_2. \quad \text{...}(vi)$$

From (*v*) and (*vi*), we get the intermediate solution as

$$p\sec x - q = f_1(y - \sin x). \quad \text{...}(vii)$$

Similarly, from (*iv*) and (*i*), we get the second intermediate solution as

$$p\sec x + q = f_2(y + \sin x). \quad \text{...}(viii)$$

From (*vii*) and (*viii*), we obtain

$$p = \frac{1}{2}\cos x\{f_1(y - \sin x) + f_2(y + \sin x)\}$$

and $$q = \frac{1}{2}\{f_2(y + \sin x) - f_1(y - \sin x)\}.$$

Substituting these values of p and q in $dz = p\,dx + q\,dy$, we obtain

$$dz = \frac{1}{2}\cos x\,\{f_1(y-\sin x)+f_2(y+\sin x)\}\,dx+\frac{1}{2}\{f_2(y+\sin x)-f_1(y-\sin x)\}\,dy$$

$$= \frac{1}{2}f_2(y+\sin x)\,(dy+\cos x\,dx)-\frac{1}{2}f_1(y-\sin x)\,(dy-\cos x\,dx). \quad ...(ix)$$

Integrating (ix), $z = \phi_1(y - \sin x) + \phi_2(y + \sin x)$ is the solution of the problem.

EXAMPLE 2.56 Solve $r - a^2 t = 0$.

Solution Here $R = 1$, $S = 0$, and $T = -a^2$. With these values, the Monge's equations are given by

$$dp\,dy - a^2\,dq\,dx = 0, \quad ...(1)$$

$$dy^2 - a^2\,dx^2 = 0. \quad ...(2)$$

Eq. (2) on resolution gives

$$dy - a\,dx = 0 \quad \text{and} \quad dy + a\,dx = 0.$$

Hence, $$y - ax = c_1 \quad \text{and} \quad y + ax = c_2.$$

Combining the first of these two equations with (1), we get $a\,dx\,(dp - a\,dq) = 0$, whence

$$p - a\,q = c_1' = \phi\,(y - ax). \quad ...(3)$$

Similarly the second of the above relation when combined with (1) and integrated, gives

$$p + aq = c_2' = \psi\,(y + ax). \quad ...(4)$$

Eqs. (3) and (4) when solved for p and q, give

$$p = \frac{1}{2}[\phi(y-ax)+\psi(y+ax)].$$

$$q = \frac{1}{2a}[\psi(y+ax)-\phi(y-ax)].$$

Substituting the values of p and q in the relation $dz = pdx + qdy$, we get

$$dz = \frac{1}{2a}[\psi(y+ax)\,(dy+a\,dx)-\phi(y-ax)\,(dy-a\,dx)].$$

Integration of this equation gives

$z = f_1(y + ax) + f_2(y - ax)$, as the solution of the problem.

EXAMPLE 2.57 Solve $(x - y)(xr - xs - ys + yt) = (x + y)(p - q)$. *(SVTU, 2007)*

Solution Here $R = x(x - y)$, $S = -(x - y)(x + y)$, $T = y(x - y)$, $V = (x + y)(p - q)$.

With these values, the Monge's equations are

$$x\,dy^2 + (x + y)\,dxdy + y\,dx^2 = 0 \quad ...(1)$$

$$x\,dp\,dy + y\,dq\,dx - \frac{x+y}{x-y}(p-q)\,dx\,dy = 0. \quad ...(2)$$

Eq. (1) on resolution gives

$$(xdy + ydx)\,(dx + dy) = 0. \quad ...(3)$$

The integrals of (3) are $xy = c_1$, $x + y = c_2$.

Taking $xy = c_1$ and dividing (2) by xdy or its equivalent $-y\,dx$, we get

$$dp - dq - \frac{dx-dy}{x-y}(p-q) = 0$$

or $\dfrac{dp-dq}{p-q} = \dfrac{dx-dy}{x-y}$. This relation on integration gives $\dfrac{p-q}{x-y} = c$. Hence the first integral is

$$p - q = (x - y)\,\phi\,(xy). \quad ...(4)$$

Equation (4) is Lagrange's equation. Its subsidiary equations are

$$\frac{dx}{1} = \frac{dy}{-1} = \frac{dz}{(x-y)\,\phi(xy)}. \quad ...(5)$$

The first two relations in (5) give $y + x = a$.
Using this relation, we get

$$dz = -\phi(ax - x^2)(a - 2x)\,dx.$$

Integrating this, we get $z = \phi_1(ax - x^2) + b$.

Writing $b = \phi_2(a)$ and $a = x + y$, we get

$z = \phi_1(xy) + \phi_2(x + y)$ as the solution of the problem.

EXERCISE 2.8

Solve the following equations by Monge's method.

1. $r + (b + c)s + bct = xy$. **Ans.** $z = \phi_1(y - bx) + \phi_2(y - cx) + \frac{1}{6}x^3y - (b + c)\frac{x^4}{24}$.
2. $y^2r - 2ys + t = p + 6y$. (*Osmania. 2002*) **Ans.** $z = y^3 - y\phi_1(y^2 + 2x) + \phi_2(y^2 + 2x)$.
3. $r - t\cos^2 x + p\tan x = 0$. **Ans.** $z = \phi_1(y - \sin x) + \phi_2(y + \sin x)$.
4. $(r - s)y + (s - t)x + q - p = 0$. **Ans.** $z = \phi_1(x + y) + \phi_2(x^2 - y^2)$.
5. $(q + 1)s = (p + 1)t$. **Ans.** $z = \phi_1(x) + \phi_2(x + y + z)$.
6. $q^2r - 2pqs + p^2t = pq^2$. **Ans.** $y = \phi_1(z) + e^x\phi_2(z)$, where $\phi_2(z) = -1/f(z)$.
7. $(1 + q)^2r - 2(1 + p + q + pq)s + (1 + p)^2t = 0$. **Ans.** $y = \phi_1(x + y + z) + x\phi_2(x + y + z)$.
8. $xy(t - r) + (x^2 - y^2)(s - 2) = py - qx$. **Ans.** $z = \phi_1\left(\frac{y}{x}\right) + \phi_2(x^2 + y^2) + xy$.
9. $x^2r + 2xys + y^2t = 0$. **Ans.** $z = \phi_1(y/x) + x\phi_2(y/x)$.

CHAPTER 3

Applications of Partial Differential Equations

3.1 INTRODUCTION

The mathematical formulation of many problems in science and engineering gives rise to partial differential equations. Some of the important partial differential equations thus obtained are:

1. $\dfrac{\partial^2 y}{\partial t^2}=c^2\dfrac{\partial^2 y}{\partial x^2}$.

This is known asn one-dimensional *wave equation* or *vibrating string equation.* This is the simplest example of an hyperbolic equation.

2. $\dfrac{\partial u}{\partial t}=c^2\dfrac{\partial^2 u}{\partial x^2}$.

This equation occurs in problems of *one-dimensional heat flow.* This is a parabolic equation and is also known as *diffusion equation.*

3. $\dfrac{\partial^2 u}{\partial x^2}+\dfrac{\partial^2 u}{\partial y^2}=0$.

This is known as *Laplace equation* in two dimensions. It occurs in problems of two-dimensional heat flow in steady state, that is, when temperature does not change with time. This is an example of two dimensional elliptic equation.

4. *Transmission Line Equations*

(a) *Telephone Equations:*

(i) $\dfrac{\partial V}{\partial x}=-L\dfrac{\partial i}{\partial t}-Ri;\quad \dfrac{\partial i}{\partial x}=-C\dfrac{\partial V}{\partial t}-GV.$ *(ii)* $\dfrac{\partial^2 V}{\partial x^2}=LC\dfrac{\partial^2 V}{\partial t^2}+(RC+LG)\dfrac{\partial V}{\partial t}+RGV.$

(iii) $\dfrac{\partial^2 i}{\partial x^2}=LC\dfrac{\partial^2 i}{\partial t^2}+(RC+LG)\dfrac{\partial i}{\partial t}+RGi$

where i and V denote the current and voltage respectively and L, C, R and G are the inductance, capacitance, Resistance and conductance respectively.

(b) *Telegraph Equations:* If L and G are small ($L = G = 0$) which is precisely the case in submarine cables, the telephone equations reduces to telegraph equations and the resulting equalions takes the form:

(i) $\dfrac{\partial V}{\partial x}=-Ri;\quad \dfrac{\partial i}{\partial x}=-C\dfrac{\partial V}{\partial t}.$ *(ii)* $\dfrac{\partial^2 V}{\partial x^2}=RC\dfrac{\partial V}{\partial t}.$ *(iii)* $\dfrac{\partial^2 i}{\partial x^2}=RC\dfrac{\partial i}{\partial t}.$

Equations *(ii)* and *(iii)* are similar to one-dimensional heat flow equations.

(c) *Radio Equations:* In the case of high frequency lines where the effects of conductance and resistance are negligible ($G = R = 0$), the telephone equations assume the form:

(i) $\dfrac{\partial V}{\partial x}=-L\dfrac{\partial i}{\partial t};\quad \dfrac{\partial i}{\partial x}=-C\dfrac{\partial V}{\partial t}.$ *(ii)* $\dfrac{\partial^2 V}{\partial x^2}=LC\dfrac{\partial^2 V}{\partial t^2}$ *(iii)* $\dfrac{\partial^2 i}{\partial x^2}=LC\dfrac{\partial^2 i}{\partial t^2}.$

The last two of the above equations are similar to the *wave equation.*

3.2 INITIAL AND BOUNDARY CONDITIONS

In the case of physical problems giving rise to partial differential equations, we have to find solutions which satisfy certain initial* and boundary** conditions. Such problems are called *boundary value problems.* In the previous chapter we found solutions of differential equations which involved arbitrary functions. These arbitrary functions are difficult to adjust so as to satisfy given boundary conditions. A method suitable for such problems is the *method of separation of variables* also known as *product method.*

3.3 METHOD OF SEPARATION OF VARIABLES (PRODUCT METHOD)

This method describes the solution which expresses the dependent variable as a product of functions each of which depends only on one variable. For example if z is the dependent variable and depends on two independent variables x and y, that is, $z(x, y)$, then we assume the solution in the form

$$z = X(x)\ Y(y). \quad ...(3.1)$$

where X is a function of x alone and Y is a function of y alone. Substitution for z from (3.1) in the given differential equation reduces it to ordinary differential equations which can be conveniently solved for the solution. The procedure will be clear from the following solved examples.

EXAMPLE 3.1 Using method of separation of variables, solve $3\frac{\partial u}{\partial x} + 2\frac{\partial u}{\partial y} = 0$ given that $u(x, 0) = 4e^{-x}$.

(AMIETE, June 2013, Dec 2011)

Alternative form: $3u_x + 2u_y = 0$ with $u(x, o) = 4e^{-x}$.

Solution Let $u(x, y) = X(x) \cdot Y(y)$ be the solution.

Substituting in the given equation, we have $3X'Y + 2XY' = 0$ or $\frac{X'}{X} = -\frac{2Y'}{3Y} = k\ (say)$. ...(*i*)

From (i) $\frac{X'}{X} = k$. Solving $\log X = kx + c'$ or $X(x) = c_1 e^{kx}$. (where $c_1 = ec'$)

Similarly, from $Y' + \frac{3}{2}kY = 0$, we get $Y(y) = c_2 e^{-\frac{3}{2}ky}$.

So $u(x, y) = X(x)\ Y(y) = c_1 e^{kx} \cdot c_2 e^{-\frac{3}{2}ky} = ce^{\frac{k}{2}(2x-3y)}$, where $c = c_1 c_2$.

But $u(x, o) = 4e^{-x}$ (given). Therefore, $ce^{kx} = 4e^{-x}$.

Comparing the two sides, we get $c = 4$ and $k = -1$.

Hence, the required solution is $u(x, y) = 4e^{-\frac{1}{2}(2x-3y)}$.

EXAMPLE 3.2 Solve by the method of separation of variables

$$\frac{\partial^2 z}{\partial x^2} - 2\frac{\partial z}{\partial x} + \frac{\partial z}{\partial y} = 0.$$

(AMIETE, June 2011, 2010, 2004)

Solution Let $z = X(x) \bullet Y(y)$ ...(*i*)

be the solution of the given differential equation, where X is a function of x and Y a function of y only.

Let $\frac{dX}{dx} = X'$, $\frac{d^2X}{dx^2} = X''$, $\frac{dY}{dy} = Y'$ and $\frac{d^2Y}{dy^2} = Y''$.

Substituting (*i*) in the differential equation we get

$$X''Y - 2X'Y + XY' = 0, \quad \text{or} \quad (X'' - 2X')Y + XY' = 0. \quad ...(ii)$$

Dividing (*ii*) by XY and separating the variables, this equation becomes $\frac{X'' - 2X'}{X} = -\frac{Y'}{Y}$. ...(*iii*)

---

* The conditions which are given for time $t = 0$, are called *initial conditions.*

** The conditions given at the boundary of the region or interval are called *boundary conditions.*

Now x and y are independent variables.

If we vary y, the right side of (*iii*) will change while the left hand side will remain unaffected and vice-versa. In order that (*iii*) holds, each side must be equal to a constant because X and Y are respectively functions of x and y only.

Let $\dfrac{X''-2X'}{X}=-\dfrac{Y'}{Y}=k$, a constant.

This gives $X''-2X'-kX=0$ ...(*iv*) $Y'+kY=0$. ...(*v*) These are two ordinary differential equations.

To solve (*iv*), we know that $m^2-2m-k=0$ is the auxiliary equation.

Its roots are $m=\dfrac{2\pm\sqrt{4+4k}}{2}=1\pm\sqrt{1+k}$. Let $1+\sqrt{1+k}=\alpha$ and $1-\sqrt{1+k}=\beta$.

Therefore, $X=c_1e^{\alpha x}+c_2e^{\beta x}$. From (v), we have $Y'/Y=-k$ or $\log Y=-ky+c$. ...(*vi*)

Therefore, $Y=e^{-ky+c}=c_3e^{-ky}$ (where $c_3=e^c$). ...(*vii*)

From (*vi*) and (*vii*) the required solution is

$$z(x,y)=XY=(c_1e^{\alpha x}+c_2e^{\beta x})\,c_3e^{-ky}=(Ae^{\alpha x}+Be^{\beta x})\,e^{-ky},$$

where c_1c_3 and c_2c_3 are replaced by A and B respectively.

EXAMPLE 3.3 Using the method of separation of variables, solve the partial differential equation $\dfrac{\partial u}{\partial x}=2\dfrac{\partial u}{\partial t}+u$, under the condition $u(x, 0) = 6\,e^{-3x}$. (*AMIETE, Dec. 2010, June 2009, Kurukshetra 2006*)

[Before we solve the problem we should note that u is a function of x and t and therefore may be written as $u(x, t)$. The value of u when $t = 0$ is written as $u(x, 0)$ and its value when $x = 0$ is denoted by $u(0, t)$].

Solution Let $u(x, t) = X(x) \bullet T(t)$ be the solution. Substituting in the given equation, we have

$$X'T=2XT'+XT \quad ...(3.2) \quad \text{or} \quad (X'-X)T=2XT'$$

or $\dfrac{X'-X}{2X}=\dfrac{T'}{T}=k\,(\text{say})$ (after dividing by XT).

Therefore $X'-X-2kX=0$ or $X'-(1+2k)X=0$ ...(*i*) and $T'-kT=0$. ...(*ii*)

From (*i*) $X'/X=1+2k$. Solving $\log X=(1+2k)x+c'$ or $X=c_1e^{(1+2k)x}$. (where $c_1=e^{c'}$) ...(*iii*)

From (*ii*) $T'/T=k$ or $\log T=kt+c''$. Therefore, $T=c_2\,e^{kt}$ where $(c_2=e^{c''})$. ...(*iv*)

From (*iii*) and (*iv*) we get $u=XT=c_1c_2e^{(1+2k)x}\bullet e^{kt}=A\,e^{(1+2k)x}\bullet e^{kt}$. ...(v)

Now $u(x,0)=A\,e^{(1+2k)x}$. $(\because e^{ok}=e^0=1)$

But $u(x,0)=6e^{-3x}$ (given). Therefore, $A\,e^{(1+2k)x}=6e^{-3x}$.

Comparing the two sides, we get $A = 6$ and $1 + 2k = -3$ or $k = -2$.

Putting these values in (v). Thus $u(x,t)=6\,e^{-3x}\,e^{-2t}=6\,e^{-(3x+2t)}$ is the required solution of the given equation.

Remark 1

Eq. (3.2) can also be written in the form $X'T=(2T'+T)X$, that is, $\dfrac{X'}{X}=\dfrac{2T'+T}{T}=k$.

Solving $X'-kX=0$ and $2T'+(1-k)T=0$, we get the same result.

EXAMPLE 3.4 Using the method of separation of variables, solve the parabolic partial differential equation

$$\frac{\partial^2u}{\partial x^2}=16\frac{\partial u}{\partial y}.$$

(*AMIETE, Dec 2005*)

Solution Using $u(x, y) = X(x)\,Y(y)$, the equation becomes

$$X''\,Y = 16\,XY' \quad \text{or} \quad \frac{X''}{16X} = \frac{Y'}{Y}.$$

Since the left-hand side of the last equation is independent of y and is equal to the right-hand side, which is independent of x, we conclude that both sides of the equation are independent of x and y. In other words, each side of the equation must be equal to a constant, say k^2 or $-k^2$. We have the following three cases:

Case 1 If $k^2 = 0$. Then, we obtain $X'' = 0$ and $Y' = 0$.

Integrating, we obtain $X = c_1 x + c_2$ and $Y = c_3$.

Hence, $u(x, y) = (c_1 x + c_2)\,c_3 = A_1 x + B_1$,

where $A_1 = c_1 c_3$ and $B_1 = c_2 c_3$ are arbitrary constants.

Case 2 If $k^2 > 0$, the two equalities $\dfrac{X''}{16X} = \dfrac{Y'}{Y} = k^2$ lead to $X'' - 16k^2 X = 0$ and $Y' - k^2 Y = 0$.

The solution of these equations are

$$X(x) = c_4 \cosh(4\,kx) + c_5 \sinh(4\,kx) \text{ and } Y(y) = c_6\,e^{k^2 y}.$$

Hence,
$$u(x, y) = [c_4 \cosh(4\,kx) + c_5 \sinh(4\,kx)]\,c_6\,e^{k^2 y}$$
$$= A_2 e^{k^2 y} \cosh 4\,kx + B_2\,e^{k^2 y} \sinh 4\,kx,$$

where $A_2 = c_4 c_6$ and $B_2 = c_5 c_6$ are arbitrary constants.

Case 3 If $-k^2 < 0$, the two equalities $\dfrac{X''}{16X} = \dfrac{Y'}{Y} = -k^2$ give $X'' + 16\,k^2 X = 0$ and $Y' + k^2 Y = 0$.

The solutions of these equations are

$$X(x) = c_7 \cos(4\,kx) + c_8 \sin(4\,kx) \text{ and } Y(y) = c_9\,e^{-k^2 y}.$$

Hence,
$$u(x, y) = [c_7 \cos(4\,kx) + c_8 \sin(4\,kx)]\,c_9\,e^{-k^2 y}$$
$$= [A_3 \cos(4\,kx) + B_3 \sin(4\,kx)]\,e^{-k^2 y},$$

where $A_3 = c_7 c_9$ and $B_3 = c_8 c_9$ are arbitrary constants.

3.4 CLASSIFICATION OF LINEAR SECOND ORDER PARTIAL DIFFERENTIAL EQUATIONS

Consider the second order, linear homogeneous partial differential equation (pde)

$$A\frac{\partial^2 u}{\partial x^2} + B\frac{\partial^2 u}{\partial x \partial y} + C\frac{\partial^2 u}{\partial y^2} + D\frac{\partial u}{\partial x} + E\frac{\partial u}{\partial y} + Fu = 0,$$

where the coefficients A, B, C, D, E, F are functions of x, y, or are real constants. The *pde* is said to be of a *parabolic type* if $B^2 - 4AC = 0$ (example: *heat equation*)

an *hyperbolic type* if $B^2 - 4AC > 0$ (example: *wave equation*)

and an *elliptic type* if $B^2 - 4AC < 0$ (example: *Laplace equation*)

The importance of this classification is that it governs the number and type of conditions to be prescribed in order that the problem is well posed and has a unique solution. Note that the lower order (first order) terms do not play any role in the classification of the equation.

Elliptic Equation

The Poisson's equation: $\dfrac{\partial^2 u}{\partial x^2} + \dfrac{\partial^2 u}{\partial y^2} = f(x, y)$

and the Laplace's equation: $\dfrac{\partial^2 u}{\partial x^2} + \dfrac{\partial^2 u}{\partial y^2} = 0$

Fig. 3.1

are examples of two dimensional elliptic equations. The solution of these equations is a function $u(x, y)$ which is satisfied at every point of the region R enclosed by a closed curve C as also by the boundary conditions specified on the boundary of the closed curve C. (See Fig. 3.1.)

Parabolic Equation

The one-dimensional diffusion equation (heat conduction equation):

$\frac{\partial u}{\partial t} = c^2 \frac{\partial^2 u}{\partial x^2}$ is a well known example of the parabolic equation.

The solution of the equation is a temperature function $u(x, t)$ defind for the values of space co-ordinate x between $x = 0$ and $x = l$ and for the values of the time coordinate t between 0 and ∞. The solution is not defind in a closed domain but propagates in an open-ended region with prescribed conditions on an open boundary which is a semi-infinite area bounded by the x-axis and the two parallel lines $x = 0$ and $x = l$. (See Fig. 3.2)

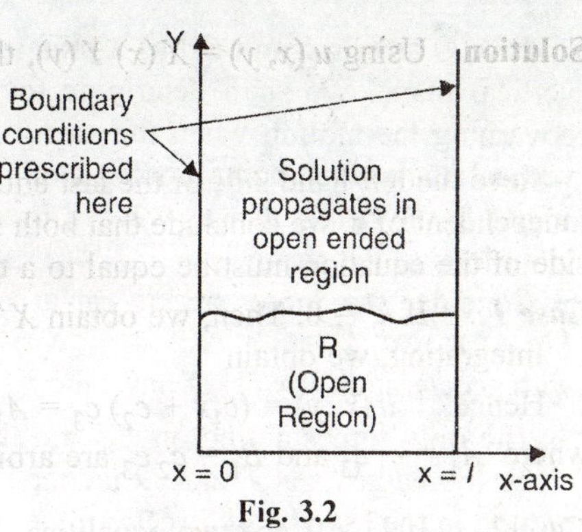

Fig. 3.2

Hyperbolic Equation

The one-dimensional wave equation: $\frac{\partial^2 u}{\partial t^2} = c^2 \frac{\partial^2 u}{\partial x^2}$ is the simplest example of the hyperbolic partial differential equation. The solution of the equation is a displacement function $u(x, t)$ defined for values of x between 0 and l and for t between 0 and ∞, which satisfies the initial and boundary conditions of the problem. The solution propagates in an open-ended region.

EXAMPLE 3.5 Classify the following partial differential equations (*pde*)

(*i*) $f_{xx} + 2f_{xy} + 4f_{yy} = 0$, (*AMIETE, Dec 2007*) (*ii*) $y^2 u_{xx} - 2xyu_{xy} + x^2 u_{yy} = \frac{y^2}{x} u_x + \frac{x^2}{y} u_y$,

(*iii*) $2\frac{\partial^2 u}{\partial x^2} + 4\frac{\partial^2 u}{\partial x \partial y} + \frac{\partial^2 u}{\partial y^2} = 0$, (*iv*) $(x^2 - 1)\frac{\partial^2 z}{\partial x^2} + 2xy\frac{\partial^2 z}{\partial x \partial y} + (y^2 - 1)\frac{\partial^2 z}{\partial y^2} - x\frac{\partial z}{\partial x} - y\frac{\partial z}{\partial y} = 0$.

Solution (*i*) We may write the given equation as $\frac{\partial^2 f}{\partial x^2} + 2\frac{\partial^2 f}{\partial x \partial y} + 4\frac{\partial^2 f}{\partial y^2} = 0$.

We have $A = 1$, $B = 2$, $C = 4$ and $B^2 - 4AC = 4 - 16 = -12 < 0$.
Hence, the given *pde* is an elliptic equation.

(*ii*) We may write the given equation as $y^2 \frac{\partial^2 u}{\partial x^2} - 2xy\frac{\partial^2 u}{\partial x \partial y} + x^2 \frac{\partial^2 u}{\partial y^2} = \frac{y^2}{x}\frac{\partial u}{\partial x} + \frac{x^2}{y}\frac{\partial u}{\partial y}$.

We have $A = y^2$, $B = -2xy$, $C = x^2$, $D = \frac{-y^2}{x}$, $E = \frac{-x^2}{y}$, and $B^2 - 4AC = 4x^2y^2 - 4yx^2y^2 = 0$.

Hence, the given *pde* is a parabolic equation.

(*iii*) We have $A = 2$, $B = 4$, $C = 1$ and $B^2 - 4AC = 16 - 8 = 8 > 0$.
Hence, the given *pde* is an hyperbolic equation.

(*iv*) We have $A = x^2 - 1$, $B = 2xy$, $C = y^2 - 1$, $D = -x$, $E = -y$,
and $B^2 - 4AC = 4x^2y^2 - 4(x^2y^2 - x^2 - y^2 + 1) = 4(x^2 + y^2 - 1)$.

If $x^2 + y^2 - 1 > 0$, that is, outside the unit circle, the given pde is an hyperbolic equation. If $x^2 + y^2 - 1 = 0$, that is, on the unit circle, the given *pde* is a parabolic equation. If $x^2 + y^2 - 1 < 0$, that is, inside the unit circle, the given *pde* is an elliptic equation.

3.5 EQUATION OF VIBRATING STRING

$$\frac{\partial^2 \mathbf{y}}{\partial \mathbf{t}^2} = \mathbf{c}^2 \frac{\partial^2 \mathbf{y}}{\partial \mathbf{x}^2}.$$

Consider a perfectly flexible homogeneous string of length l tightly stretched between two points 0 and A. We assume the tension in the string to be so large that gravity may be neglected in comparison with it.

Take O as origin and OA as X-axis and let XY-plane be the vertical plane. We shall determine the differential equation governing the motion when the string is set vibrating in the vertical plane, each point moving along a straight line at right angles to the string, that is, we will investigate the transverse vibrations of the string.

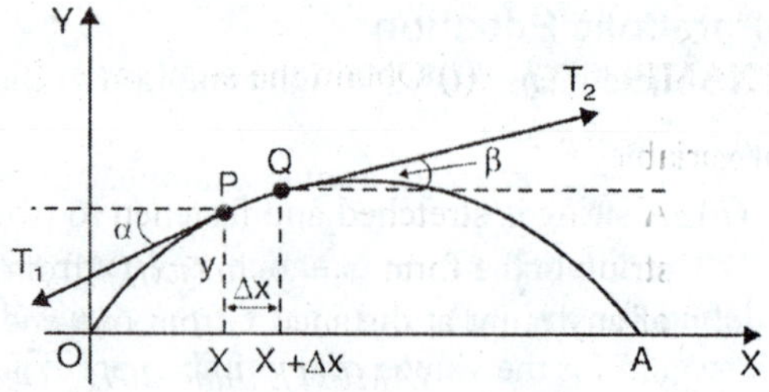

Fig. 3.3: Vibrating string.

Consider the forces acting on a small portion between $P(x, y)$ and $Q(x + \Delta x, y + \Delta y)$ moving vertically, with acceleration $\partial^2 y/\partial t^2$ at any time t.

The only forces acting on PQ are the tensions T_1 and T_2 along the tangents at P and Q inclined at angles α and β respectively with the horizontal. Since there is no motion in the horizontal direction, the horizontal components of tension must be constant.

Therefore, $T_1 \cos\alpha = T_2 \cos\beta = T$, a constant. ...(3.3)

Let Δs be the length of the string PQ. Hence its mass is ρΔs, where ρ is the mass per unit length of the string. By Newton's second law of motion,

$$\rho\Delta s \frac{\partial^2 y}{\partial t^2} = T_2 \sin\beta - T_1 \sin\alpha.$$

Then, $$\frac{\rho\Delta s}{T}\frac{\partial^2 y}{\partial t^2} = \frac{T_2 \sin\beta}{T_2\cos\beta} - \frac{T_1\sin\alpha}{T_1\cos\alpha}$$ ...[By (3.3)]

$= \tan\beta - \tan\alpha$, ...(3.4) that is, difference of slopes at P and Q.

tan α and tan β are slopes of the tangents at $P(x, y)$ and $Q(x + \Delta x, y + \Delta y)$.

Therefore, $$\tan\alpha = \left(\frac{\partial y}{\partial x}\right)_P \text{ and } \tan\beta = \left(\frac{\partial y}{\partial x}\right)_Q.$$

Thus (3.4), can be written in the form

$$\frac{\rho}{T}\Delta s\frac{\partial^2 y}{\partial t^2} = \left(\frac{\partial y}{\partial x}\right)_Q - \left(\frac{\partial y}{\partial x}\right)_P \quad \text{or} \quad \frac{\rho}{T}\frac{\partial^2 y}{\partial t^2} = \frac{1}{\Delta s}\left[\left(\frac{\partial y}{\partial x}\right)_Q - \left(\frac{\partial y}{\partial x}\right)_P\right].$$

Slope of the curve being small $\Delta s \simeq \Delta x$. If now we let Δx approach zero, the right hand reduces to $\partial^2 y/\partial x^2$ and thus we obtain the linear homogeneous partial differential equation

$$\frac{\partial^2 \mathbf{y}}{\partial \mathbf{t}^2} = \mathbf{c}^2\frac{\partial^2 \mathbf{y}}{\partial \mathbf{x}^2}, \quad ...(3.5)$$ where $c^2 = T/\rho$ is the governing wave equation.

3.6 BOUNDARY CONDITIONS

Eq. (3.5) is obliged to satisfy the following initial and boundary conditions.

(1) At the ends O and A, there is no motion, that is,

(*i*) $y = 0$ at $x = 0$; and (*ii*) $y = 0$ at $x = l$. These conditions must hold for all values of time t i.e. $t \geq 0$. Alternately $y(0, t) = y(l, t) = 0$ for $t \geq 0$.

(2) If the string is initially held in the form of the curve $y = f(x)$ and then released from this position, the initial condition of vibration will be

(*i*) $y = f(x)$ at $t = 0$ and (*ii*) $\partial y/\partial t = 0$ at $t = 0$.

(3) If every element of the string is moving with a definite velocity and the motion is reckoned when the string is at the equilibrium position, then the initial conditions are represented by

(*i*) $\partial y/\partial t = \phi(x)$ at $t = 0$ and (*ii*) $y = 0$ at $t = 0$.

(4) If the string is moving with a specified velocity and is also given a displacement, the conditions can be specified accordingly.

In the following examples we shall solve this equation by the method of separation of variables.

EXAMPLE 3.6 (*i*) Obtain the solution of the wave equation $\frac{\partial^2 y}{\partial t^2} = c^2 \frac{\partial^2 y}{\partial x^2}$, using the method of separation of variables.

(*ii*) A string is stretched and fastened to two points at a distance l apart. Motion is started by displacing the string in the form $y = a\sin(\pi x/l)$ from which it is released at a time $t = 0$. Show that the displacement of any point at distance x from one end at time t is given by $y(x, t) = a\sin(\pi x/l)\cos(\pi ct/l)$.

(*AMIETE June 2011; Dec 03; SVTU 2007; Kerala 2005; UPTU 2004; AMIE, S-2002*)

Solution (*i*) Let $Y(x, t) = X(x) \cdot T(t)$ ...(1)

be the solution of the given equation. Substituting (1) in the given equation, we obtain

$$XT'' = c^2 X''T$$

or

$$\frac{X''}{X} = \frac{1}{c^2}\frac{T''}{T} = h(\text{say}) \quad \text{(after dividing both sides of above by } XT).$$

Therefore, $X'' - hX = 0$ ...(2) and $T'' - c^2 hT = 0$. ...(3)

Now three cases arise.

Case I ($h > 0$): Let $h = k^2$. Then (2) and (3) become respectively

$$X'' - k^2 X = 0 \text{ and } T'' - c^2 k^2 T = 0.$$

Solving these ordinary equations, we get

$$X = c_1 e^{kx} + c_2 e^{-kx} \text{ and } T = c_3 e^{ckt} + c_4 e^{-ckt}. \quad ...(4)$$

Case II $h = 0$. Then (2) and (3) become $X'' = 0$ and $T'' = 0$.

Solving, we get $X = c_5 x + c_6$ and $T = c_7 t + c_8$. ...(5)

Case III ($h < 0$): Let $h = -k^2$. From (2) and (3), we get

$$X'' + k^2 X = 0 \text{ and } T'' + c^2 k^2 T = 0.$$

Solving, we get $X = c_9 \cos kx + c_{10} \sin kx$,

and $T = c_{11} \cos ckt + c_{12} \sin ckt$. ...(6)

Thus depending upon the value of h we have from (4), (5) and (6) the solution as

$$y = (c_1 e^{kx} + c_2 e^{-kx})(c_3 e^{ckt} + c_4 e^{-ckt}) \quad ...(7)$$

$$y = (c_5 x + c_6)(c_7 t + c_8) \quad ...(8)$$

$$y = (c_9 \cos kx + c_{10} \sin kx)(c_{11} \cos ckt + c_{12} \sin ckt). \quad ...(9)$$

Since it is a problem on vibrating string every position of which is repeated as time passes, the solution should be a periodic function of t. Hence only (9) can be taken as the solution of the given equation.

(*ii*) The equation of the vibrating string is

$$\frac{\partial^2 y}{\partial t^2} = c^2 \frac{\partial^2 y}{\partial x^2}. \quad ...(10)$$

Let O and A be the points to which the ends are fastened so that $OA = l$ (Refer Fig. 3.3).

Since the ends of the string are fixed, $y = 0$ when $x = 0$ and $x = l$, for all values of t.

Boundary conditions are $y(0, t) = 0$ and $y(l, t) = 0$. ...(11)

The string is in the shape $y = a\sin(\pi x/l)$ and at rest when $t = 0$.

The *initial conditions* are

$$y(x, 0) = a\sin\frac{\pi x}{l}, \quad \left(\frac{\partial y}{\partial t}\right)_{t=0} = 0. \quad ...(12)$$

Now we have to solve (10) subject to the conditions (11) and (12).

As shown in part (*i*), the solution is of the form

$$y(x, t) = (a_1 \cos kx + a_2 \sin kx)(a_3 \cos ckt + a_4 \sin ckt). \quad ...(13)$$

Since $y(0, t) = 0$ for all t (first boundary condition), therefore,

$$0 = a_1(a_3 \cos ckt + a_4 \sin ckt). \text{ Hence } a_1 = 0.$$

Therefore, (13) reduces to

$$y(x, t) = a_2 \sin kx\,(a_3 \cos ckt + a_4 \sin ckt)$$

$$= \sin kx(b_1 \cos ckt + b_2 \sin ckt). \qquad \text{(where } b_1 = a_2a_3,\ b_2 = a_2a_4).$$

From the second boundary condition $y(l, t) = 0$ for all t, we get

$$0 = \sin kl(b_1 \cos ckt + b_2 \sin ckt).$$

Therefore, $\sin kl = 0$ or $kl = n\pi$ (n being an integer) or $k = n\pi/l$.

$$y(x,t) = \sin\frac{n\pi x}{l}\left(b_1 \cos\frac{cn\pi}{l}t + b_2 \sin\frac{cn\pi}{l}t\right).$$

Now
$$\frac{\partial y}{\partial t} = \sin\frac{n\pi x}{l}\left(-b_1\frac{cn\pi}{l}\sin\frac{cn\pi t}{l} + b_2\frac{cn\pi}{l}\cos\frac{cn\pi t}{l}\right).$$

By second initial condition $\left(\dfrac{\partial y}{\partial t}\right)_{t=0} = 0.$

$$0 = \sin\frac{n\pi x}{l}\left(b_2\frac{cn\pi}{l}\right) \text{ which gives } b_2 = 0.$$

Hence, $y(x,t) = b_1 \sin\dfrac{n\pi x}{l}\cos\dfrac{cn\pi}{l}t.$ ...(14)

From this equation $y(x,0) = b_1 \sin\dfrac{n\pi x}{l}$ at $t = 0$.

But $y(x,0) = a\sin\dfrac{\pi x}{l}$ (first initial condition).

$a\sin\dfrac{\pi x}{l} = b_1 \sin\dfrac{n\pi x}{l}$. This holds when $b_1 = a$ and $n = 1$.

Hence, $y(x,t) = a\sin\dfrac{\pi x}{l}\cos\dfrac{\pi ct}{l}$ is the required solution.

Important note We know that if $y_1, y_2, ..., y_n$ are solutions of a differential equation, then $A_1y_1 + A_2y_2 + ..., A_ny_n$ where A's are arbitrary constants, is also a solution.
This is known as **superposition principle.**

From (14) above $y = b_1 \sin\dfrac{n\pi x}{l}\cos\dfrac{cn\pi t}{l}$ or $y = \sin\dfrac{n\pi x}{l}\cos\dfrac{cn\pi t}{l}$ is a solution of the wave equation, n being an integer.

Putting $n = 1, 2, 3, ...$

$y_1 = \sin\dfrac{\pi x}{l}\cos\dfrac{c\pi t}{l},\ y_2 = \sin\dfrac{2\pi x}{l}\cos\dfrac{2c\pi t}{l}, ...$ are also solutions. Hence by superposition principle

$$y = A_1 \sin\frac{\pi x}{l}\cos\frac{c\pi t}{l} + A_2 \sin\frac{2\pi x}{l}\cos\frac{2c\pi t}{l} + ... + A_n \sin\frac{n\pi x}{l}\cos\frac{nc\pi t}{l} + ...$$

or
$$y(x,t) = \sum_{n=1}^{\infty} A_n \sin\frac{n\pi x}{l}\cos\frac{cn\pi t}{l} \text{ is also a solution.}$$

We could have taken this to be the solution in the above example. Then using the first initial condition, that is, $y(x,0) = a\sin\dfrac{\pi x}{l}$, we get

$$a\sin\frac{\pi x}{l} = \sum A_n \sin\frac{n\pi x}{l} = A_1 \sin\frac{\pi x}{l} + A_2 \sin\frac{2\pi x}{l} + ...$$

which gives $A_1 = a,\ A_2 = A_3 = ... = 0$. Therefore, $y(x,t) = a\sin\dfrac{\pi x}{l}\cos\dfrac{c\pi t}{l}$ is a solution.

EXAMPLE 3.7 A tightly stretched string with fixed ends at $x = 0$ and $x = l$ is initially in equilibrium position. It is set vibrating by giving to each of its points an initial velocity $\left(\frac{\partial y}{\partial t}\right)_{t=0} = b\sin^3\left(\frac{\pi x}{l}\right)$. Find the displacement $y(x, t)$. *(AMIETE, June 2009; UPTU, 2003; AMIE, S-2001)*

Solution The equation of the vibrating string is $\frac{\partial^2 y}{\partial t^2} = c^2 \frac{\partial^2 y}{\partial x^2}$.

We have to solve this equation subject to the *boundary conditions:*

$$y(0, t) = 0 \text{ and } y(l, t) = 0$$

and subject to the *initial conditions:*

$$y(x, 0) = 0 \text{ and } \left(\frac{\partial y}{\partial t}\right)_{t=0} = b\sin^3\frac{\pi x}{l}.$$

Assuming the solution to be of the form $y(x, t) = X(x)T(t)$, we get

$$XT'' = c^2X''T$$

or

$$\frac{X''}{X} = \frac{1}{c_2}\cdot\frac{T''}{T} = -k^2 \text{ (a constant)}.$$

(As is clear from example 3.6 we have taken the constant as $-k^2$ so that solution is a periodic function of t)

Therefore, $X'' + k^2X = 0$ and $T'' + c^2k^2T = 0$.

Solving the differential equations we get

$$X = c_1 \cos kx + c_2 \sin kx \text{ and } T = c_3 \cos ckt + c_4 \sin ckt.$$

Hence, the solution is of the form

$$y = (c_1 \cos kx + c_2 \sin kx)(c_3 \cos ckt + c_4 \sin ckt).$$

Now $y(0, t) = 0$ (first boundary condition).

$$0 = c_1(c_3 \cos ckt + c_4 \sin ckt).$$

This is true for all t and this gives $c_1 = 0$.

Therefore, $y(x, t) = c_2 \sin kx(c_3 \cos ckt + c_4 \sin ckt)$

$= \sin kx(a_1 \cos ckt + a_2 \sin ckt)$. (where $a_1 = c_2c_3$ and $a_2 = c_2c_4$).

Also $y(l, t) = 0$ (second boundary condition).

$$0 = \sin kl(a_1 \cos ckt + a_2 \sin ckt) \text{ for all } t.$$

This gives $kl = n\pi$ or $k = n\pi/l$. (n being an integer).

Hence, $y(x, t) = \sin\frac{n\pi x}{l}\left(a_1 \cos\frac{cn\pi}{l}t + a_2 \sin\frac{cn\pi}{l}t\right)$.

By first initial condition $y(x, 0) = 0$.

$$0 = a_1 \sin\frac{n\pi x}{l} \quad \text{for all } x \quad \text{or} \quad a_1 = 0.$$

Thus $y(x, t) = a_2 \sin\frac{n\pi x}{l} \sin\frac{cn\pi}{l}t.$

Since this is a solution for $n = 1, 2, 3, \ldots$, therefore the sum of these solutions is also a solution.

Hence, $y(x, t) = \sum A_n \sin\frac{n\pi x}{l} \sin\frac{n\pi ct}{l}$ ...(*i*)

is also a solution (superposition principle).

Now, $\frac{\partial y}{\partial t} = \sum A_n \sin\frac{n\pi x}{l}\cdot\frac{cn\pi}{l}\cos\frac{cn\pi t}{l}.$

$$\left(\frac{\partial y}{\partial t}\right)_{t=0} = \sum A_n \frac{cn\pi}{l} \sin\frac{n\pi x}{l}. \quad \text{But} \quad \left(\frac{\partial y}{\partial t}\right)_{t=0} = b\sin^3\frac{\pi x}{l} \quad \text{(given).}$$

Therefore, $$b\sin^3\frac{\pi x}{l} = \sum A_n \frac{cn\pi}{l}\sin\frac{n\pi x}{l}. \quad \text{Now } \sin^3\frac{\pi x}{l} = \frac{1}{4}\left(3\sin\frac{\pi x}{l} - \sin\frac{3\pi x}{l}\right).$$

$(\because \sin 3\theta = 3\sin\theta - 4\sin^3\theta)$.

Therefore, $$\frac{3b}{4}\sin\frac{\pi x}{l} - \frac{b}{4}\sin\frac{3\pi x}{l} = \sum A_n \frac{cn\pi}{l}\sin\frac{n\pi x}{l}$$

$$= A_1\frac{c\pi}{l}\sin\frac{\pi x}{l} + A_2\frac{2c\pi}{l}\sin\frac{2\pi x}{l} + A_3\frac{3c\pi}{l}\sin\frac{3\pi x}{l} + \ldots$$

Equating cofficients, $A_1\frac{c\pi}{l} = \frac{3b}{4}$, $A_2 = 0$, $A_3\frac{3c\pi}{l} = -\frac{b}{4}$, $\quad A_4 = A_5 = \ldots 0$,

we obtain, $A_1 = \frac{3bl}{4c\pi}$, $A_3 = -\frac{bl}{12c\pi}$.

Substituting in (*i*), the required solution is

$$y(x, t) = \frac{3bl}{4c\pi}\sin\frac{\pi x}{l}\sin\frac{c\pi}{l}t - \frac{bl}{12c\pi}\sin\frac{3\pi x}{l}\sin\frac{3c\pi}{l}t$$

$$= \frac{bl}{12c\pi}\left[9\sin\frac{\pi x}{l}\sin\frac{\pi ct}{l} - \sin\frac{3\pi x}{l}\sin\frac{3\pi ct}{l}\right].$$

Important Note: In the following solved examples and others that we shall encounter in this chapter, we shall be required to determine constants $b_n (n = 1, 2, 3, 4, \ldots)$ subject to the condition

$$f(x) = \sum_{n=1}^{\infty} b_n \sin\frac{n\pi x}{l} \quad (0 < x < l).$$

We know that the series on the right is a sine series in which $f(x)$ can be expressed in the interval $(0, l)$ and is known *half-range Fourier sine series.*

Therefore b_n can be determined by the formula

$$b_n = \frac{2}{l}\int_0^l f(x)\sin\frac{n\pi x}{l}\,dx.$$

The value of the integral on the right is a function of n.

By putting $n = 1, 2, 3, \ldots$, we can get b_1, b_2, b_3 etc. Similarly, if we have

$$\phi(x) = \sum_{n=0}^{\infty} b_n \cos\frac{n\pi x}{l} \quad (0 < x < l)$$

Then, $$b_0 = \frac{2}{l}\int_0^l \phi(x)\,dx \quad \text{and} \quad b_n = \frac{2}{l}\int_0^l \phi(x)\cos\frac{n\pi x}{l}\,dx.$$

EXAMPLE 3.8 A string is stretched and fastened to two points l apart. Motion is started by displacing the string into the form $y = k(lx - x^2)$ from which it is released at time $t = 0$. Find the displacement of any point on the string at a distance x from one end at time t.

Solution The equation of the vibrating string is

$$\frac{\partial^2 y}{\partial t^2} = c^2\frac{\partial^2 y}{\partial x^2}. \qquad \ldots(i)$$

Since the end points are fixed, $y = 0$ when $x = 0$ and $x = l$ for all t.

Therefore, $y(0, t) = 0$ and $y(l, t) = 0$ are the *boundary conditions.* ...(*ii*)

Also at $t = 0$, the string is at rest in the shape $y = k(lx - x^2)$.

Therefore, *initial conditions* are $(\partial y/\partial t)_{t=0} = 0$ and $y(x, 0) = k(lx - x^2)$. ...(*iii*)

Now we have to solve (*i*) subject to the conditions (*ii*) and (*iii*). Proceeding as in example 3.7 and using the conditions (*ii*) and also $(\partial y/\partial t)_{t=0} = 0$, we get the solution in the form

$$y(x, t) = \sum_{n=1}^{\infty} A_n \sin\frac{n\pi x}{l}\cos\frac{cn\pi t}{l}. \quad ...(iv)$$

Since $y(x, 0) = k(lx - x^2)$, we have

$$k(lx - x^2) = \sum_{n=1}^{\infty} A_n \sin\frac{n\pi x}{l}. \quad ...(v)$$

Now (*v*) is the half-range sine series representing $k(lx - x^2)$ in the interval (o, l).

Therefore, $A_n = \frac{2}{l}\int_0^l k(lx - x^2)\sin\frac{n\pi x}{l}\,dx$

$$= \frac{2k}{l}\left[(lx - x^2)\left(-\frac{l}{n\pi}\cos\frac{n\pi x}{l}\right) - (l - 2x)\left(-\frac{l^2}{n^2\pi^2}\sin\frac{n\pi x}{l}\right) + (-2)\left(\frac{l^3}{n^3\pi^3}\cos\frac{n\pi x}{l}\right)\right]_0^l$$

$$= \begin{cases} \dfrac{4kl^2}{n^3\pi^3}(1 - \cos n\pi) = 0, \text{ when } n \text{ is even}, \\ \dfrac{8kl^2}{n^3\pi^3}, \text{ when } n \text{ is odd}. \end{cases}$$

Putting $n = 2m - 1$, the solution (*iv*) can be written in the form

$$y = \frac{8kl^2}{\pi^3}\sum_{m=1}^{\infty}\left[\frac{1}{(2m-1)^3}\sin(2m-1)\frac{\pi x}{l}\cos(2m-1)\frac{c\pi t}{l}\right].$$

EXAMPLE 3.9 A tightly stretched string with end points fixed at $x = 0$ and $x = L$ is initially at rest in equilibrium state. If it is set vibrating by giving to each of its points a velocity $\mu x(L - x)$, where μ is a constant. Find the displacement of the string at any distance x from one end at any of time t.

(*AMIETE, Dec 2006; UPTU 2002*)

Solution We have to solve the equation $\dfrac{\partial^2 y}{\partial t^2} = c^2\dfrac{\partial^2 y}{\partial x^2}$ subject to the *boundary conditions* $y(0, t) = 0$, $y(L, t) = 0$ and *initial conditions* $y = 0$ and $\partial y/\partial t = \mu x(L - x)$ when $t = 0$.

Proceeding as in example 3.7, the solution satisfying the boundary conditions is

$$y = \sin\frac{n\pi x}{L}\left(b_1\cos\frac{cn\pi}{L}t + b_2\sin\frac{cn\pi}{L}t\right).$$

Since $y = 0$ when $t = 0$, we have $b_1 = 0$.

Hence, $$y = b_n\sin\frac{n\pi x}{L}\sin cn\pi\frac{t}{L} \text{ (replacing } b_2 \text{ by } b_n).$$

Adding up the solutions for different values of n.

$$y = \sum_{n=1}^{\infty} b_n\sin\frac{n\pi x}{L}\sin\frac{cn\pi t}{L}. \quad ...(i)$$

Therefore, $$\frac{\partial y}{\partial t} = \sum_{n=1}^{\infty}\frac{c\pi}{L}nb_n\sin\frac{n\pi x}{L}\cos\frac{cn\pi t}{L}.$$

$\frac{\partial y}{\partial t} = \mu x(L - x)$ when $t = 0$. Therefore, $\mu x(L - x) = \frac{c\pi}{L}\sum nb_n \sin\frac{n\pi x}{L}$. ...(ii)

Since (ii) is half-range sine-series, we have

$$\frac{c\pi}{L}nb_n = \frac{2}{L}\int_0^l \lambda x(L - x)\sin\frac{n\pi x}{L}\,dx = \frac{4\lambda L^2}{n^3\pi^3}(1 - \cos n\pi) = \begin{cases} 0, & \text{when n is even,} \\ \frac{8\mu L^2}{n^3\pi^3}, & \text{when } n \text{ is odd.} \end{cases}$$

or
$$b_n = \frac{8\lambda L^3}{c\pi^4 n^4} \text{ when } n \text{ is odd.}$$

Putting the value of b_n in (i) we get the solution

$$y(x, t) = \frac{8\lambda L^3}{c\pi^4}\sum_{n=1,3,5,\ldots}^{\infty}\left[\frac{1}{n^4}\sin\frac{n\pi x}{L}\sin\frac{n\pi ct}{L}\right].$$

EXERCISE 3.1

1. Classify the following partial differential eqs. (i) to (vi).

(i) $x^2\frac{\partial^2 u}{\partial y^2} + 2xy\frac{\partial^2 u}{\partial x\partial y} + y^2\frac{\partial^2 u}{\partial x^2} = 0,$ (*AMIETE, Dec. 2006*)

(ii) $\frac{\partial^2\phi}{\partial x^2} + 2\frac{\partial^2\phi}{\partial y^2} - \frac{\partial\phi}{\partial x} = 0,$

(iii) $\frac{\partial^2 f}{\partial x^2} + \frac{\partial^2 f}{\partial y^2} + 2\frac{\partial^2 f}{\partial x\partial y} = 0,$

(iv) $\frac{\partial^2 u}{\partial x^2} - 2\frac{\partial^2 u}{\partial x\partial y} + 2\frac{\partial^2 u}{\partial y^2} = x + 3y,$

(v) $\frac{\partial u}{\partial x} + \frac{\partial^2 u}{\partial x\partial y} = 4,$

[Hint: $A = 0$, $B = 1$, $C = 0$, $B^2 - 4AC = 1 > 0$.**]**

(vi) $\frac{\partial z}{\partial t} = c^2\frac{\partial^2 z}{\partial x^2} + \frac{\partial z}{\partial x},$

(vii) $(1 - y)\frac{\partial^2 f}{\partial x^2} + 2x\frac{\partial^2 f}{\partial x\partial y} + (1 + y)\frac{\partial^2 f}{\partial y^2} = 0.$

(viii) Show that the *pde* $y\frac{\partial^2 u}{\partial x^2} + 2x\frac{\partial^2 u}{\partial x\partial y} + y\frac{\partial^2 u}{\partial y^2} = 0$ is elliptic if $x^2 < y^2$. (*AMIETE, Dec. 2004*)

Ans. (i) Parabolic (ii) Elliptic (iii) Parabolic (iv) Elliptic (v) hyperbolic
(vi) Parabolic (vii) $x^2 + y^2 > 1$: hyperbolic; $x^2 + y^2 = 1$: parabolic; $x^2 + y^2 < 1$: elliptic.

Using the method of separation of variables **find** the solution of the equations 2 to 4

2. $u_x + u_y = 2(x + y)u$. (*AMIETE, Dec. 2010*) **Ans.** $u = ce^{x^2+y^2+kx-ky}$.

3. $2x\frac{\partial z}{\partial x} - 3y\frac{\partial z}{\partial y} = 0$. **Ans.** $z = Ax^{h/2}\,y^{h/3}$.

4. $x^2u_{xy} + 3y^2u = 0$. **Ans.** $u = Ae^{-\left(\frac{k}{x} + \frac{y^3}{k}\right)}$

Solve each of the following problems 5 to 11 by the method of separation of variables.

5. $\frac{\partial u}{\partial x} = 4\frac{\partial u}{\partial y}$, $u(0, y) = 8e^{-3y}$. **Ans.** $u(x, y) = 8e^{-12x - 3y}$.

6. $\frac{\partial u}{\partial x} = 4\frac{\partial u}{\partial y}$, $u(0, y) = 8e^{-3y} + 4e^{-5y}$. **Ans.** $u(x, y) = 8e^{-12x-3y} + 4e^{-20x-5y}$.

7. $4\frac{\partial u}{\partial x} + \frac{\partial u}{\partial y} = 3u$, $u(0, y) = 4e^{-y} - e^{-5y}$. **Ans.** $u = 4e^{x-y} - e^{2x-5y}$.

8. $\dfrac{\partial u}{\partial x} - 2\dfrac{\partial u}{\partial y} = u$, given that $u(x, 0) = 3e^{-5x} + 2e^{-3x}$. *(AMIETE, June 2001)*

Ans. $u(x, y) = 3e^{-5x-3y} + 2e^{-3x-2y}$.

9. $4\dfrac{\partial u}{\partial x} + \dfrac{\partial u}{\partial y} = 3u$, given that $u = 3e^{-y} - e^{-5y}$ at $x = 0$. *(AMIETE, Dec. 2002)* **Ans.** $u(x, y) = 3e^{x-y} - e^{2x-5y}$.

10. $2\dfrac{\partial u}{\partial x} = 3\dfrac{\partial u}{\partial y}$ under the condition $u(0, y) = e^{-5y}$. **Ans.** $u(x, y) = e^{-5(3x+2y)/2}$.

11. $\dfrac{\partial^2 u}{\partial x^2} = \dfrac{\partial u}{\partial y} + 2u$, given that $u = 0$ and $\dfrac{\partial u}{\partial x} = 1 + e^{-3y}$, when $x = 0$ for all values of y.

Ans. $u = \dfrac{1}{\sqrt{2}}\sinh\sqrt{2}x + e^{-3y}\sin x$.

12. A tightly stretched string with fixed end points $x = 0$ and $x = l$ is initially in a position given by $y = y_0 \sin^3(\pi x/l)$. If it is released from rest from this position, find the displacement $y(x, t)$.

(AMIETE, June 2012, Dec. 2010) **Ans.** $y(x,t) = \dfrac{y_0}{4}\left[3\sin\dfrac{\pi x}{l}\cos\dfrac{\pi ct}{l} - \sin\dfrac{3\pi x}{l}\cos\dfrac{3\pi ct}{l}\right]$.

13. Solve the boundary value problem $\dfrac{\partial^2 y}{\partial t^2} = 4\dfrac{\partial^2 y}{\partial x^2}$, $y(0, t) = y(5, t) = 0$, $y(x, 0) = 0$, $\left(\dfrac{\partial y}{\partial t}\right)_{t=0} = f(x)$.

If (*i*) $f(x) = 5\sin \pi x$ (*ii*) $f(x) = 3\sin 2\pi x - 2\sin 5\pi x$. *(AMIE., S-2004)*

Ans. (*i*) $y(x, t) = (5\sin \pi x \sin 2\pi t)/2\pi$. (*ii*) $y(x, t) = (15\sin 2\pi x \sin 4\pi t - 4\sin 5\pi x \sin 10\pi t)/20\pi$.

14. Show how the wave equation $\dfrac{\partial^2 y}{\partial t^2} = c^2\dfrac{\partial^2 y}{\partial x^2}$ can be solved by the method of separation of variables. If the initial displacement and velocity of a string stretched between the points $x = 0$ and $x = l$ are given by $y = f(x)$ and $\dfrac{\partial y}{\partial t} = g(x)$, determine the displacement of any point x of the string at any time $t > 0$.

Ans. $y = \displaystyle\sum_{n=1}^{\infty} \sin\dfrac{n\pi x}{l}\left(A_n \cos\dfrac{cn\pi t}{l} + B_n \sin\dfrac{cn\pi t}{l}\right)$,

where $A_n = \dfrac{2}{l}\displaystyle\int_0^l f(x)\sin\dfrac{n\pi x}{l}\,dx$ and $B_n = \dfrac{2}{n\pi c}\displaystyle\int_0^l g(x)\sin\dfrac{n\pi x}{l}\,dx$.

15. A tightly stretched string with fixed end points $x = 0$ and $x = L$ is initially at rest in its equilibrium position. If it is set vibrating by giving to each of its points a velocity

$$\left(\frac{\partial y}{\partial t}\right)_{t=0} = 3(Lx - x^2),\ \text{find } y(x, t).$$

Ans. $y(x,t) = \dfrac{24L^3}{c\pi^4}\displaystyle\sum_{m=1}^{\infty}\left[\dfrac{1}{(2m-1)^4}\sin(2m-1)\dfrac{\pi x}{L} \times \sin(2m-1)\dfrac{c\pi t}{L}\right]$.

16. A tightly stretched string of length l is fastened at both ends. The mid-point of the string is taken to a height b and then released from rest in that position. Show that the displacement at any time t is given by

$$y = \frac{8b}{\pi^2}\left[\frac{1}{1^2}\sin\frac{\pi x}{l}\cos\frac{\pi ct}{l} - \frac{1}{3^2}\sin\frac{3\pi x}{l}\cos\frac{3\pi ct}{l} + \ldots\right].$$

17. A string of length l is fastened at both ends. One end is taken at the origin and at a distance b from this end, the string is displaced a distance d transversely and is released from rest from this position. Find the equation of subsequent motion.

$$\textbf{Ans. } y(x,t)=\frac{2dl^2}{b(l-b)\pi^2}\sum_{n=1}^{\infty}\frac{1}{n^2}\sin\frac{n\pi b}{l}\sin\frac{n\pi x}{l}\cos\frac{n\pi ct}{l}.$$

18. A string is stretched between the fixed points (0, 0) and (l, 0) and released at rest from the initial deflection given by

$$f(x)=\begin{cases}\dfrac{2k}{l}x & 0<x<\dfrac{l}{2}\\[2mm] \dfrac{2k}{l}(l-x) & \dfrac{l}{2}<x<l.\end{cases}$$

Find the deflection of the string at any point x and at any time t. *(AMIETE, June 2010, Dec. 2008)*

$$\textbf{Ans. } y(x,t)=\frac{8k}{\pi^2}\left[\frac{1}{1^2}\sin\frac{\pi x}{l}\cos\frac{\pi ct}{l}-\frac{1}{3^2}\sin\frac{3\pi x}{l}\cos\frac{3\pi ct}{l}+\frac{1}{5^2}\sin\frac{5\pi x}{l}\cos\frac{5\pi ct}{l}-\ldots\right].$$

3.7 ONE-DIMENSIONAL HEAT FLOW: $\dfrac{\partial u}{\partial t}=c^2\dfrac{\partial^2 u}{\partial x^2}$.

Consider a homogeneous bar of uniform cross section with its sides insulated so that heat flows in the direction perpendicular to the cross-section of area A. We take the origin at one end O of the bar and X-axis along the direction of heat flow.

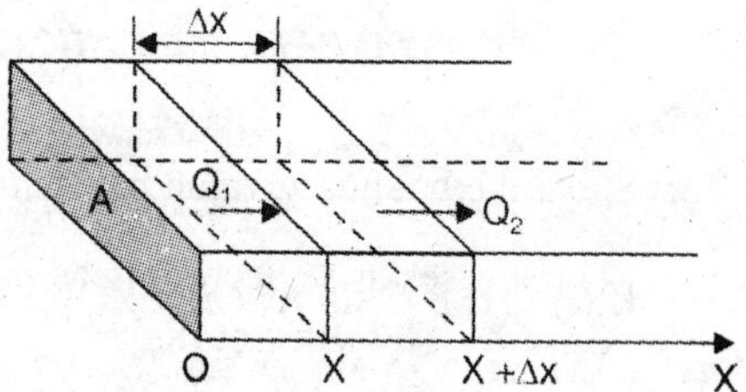

Fig. 3.4: Flow of heat in a bar.

The temperature u at any point of the bar depends on the distance x from one end and time t. Moreover the temperature of all points of any cross-section is the same.

Experimentally we know that the rate of flow of heat across any area A is proportional to the area and the temperature gradient *i.e.*, rate of change of temperature wth respect to distance normal to the area.

Hence Q_1, the quantity of heat flowing across the section at a distance x is given by

$$Q_1=-kA\left(\frac{\partial u}{\partial x}\right)_x \text{ cal per sec,}$$

where k is *coefficient of conductivity*, (also known as *thermal conductivity*), A the area of cross section; the negative sign on the right is attached because as x increases, u decreases.

Similarly Q_2, the quantity of heat flowing across the section at a distance $x + \Delta x$ is

$$Q_2=-kA\left(\frac{\partial u}{\partial x}\right)_{x+\Delta x} \text{ cal per sec.}$$

Hence, the amount of heat gained per second by the slab with thickness Δx is

$$Q_1-Q_2=kA\left[\left(\frac{\partial u}{\partial x}\right)_{x+\Delta x}-\left(\frac{\partial u}{\partial x}\right)_x\right]. \qquad \text{...(3.6)}$$

But the rate of increase of heat in the slab $= s\rho A\,\Delta x\dfrac{\Delta u}{\Delta t}$, ...(3.7)

where s is the specific heat and ρ, the density of the material of the bar.

Therefore, from Eqs. (3.6) and (3.7)

$$s\rho A\Delta x\frac{\Delta u}{\Delta t}=kA\left[\left(\frac{\partial u}{\partial x}\right)_{x+\Delta x}-\left(\frac{\partial u}{\partial x}\right)_x\right]$$

or

$$s\rho\frac{\Delta u}{\Delta t}=k\bullet\frac{1}{\Delta x}\left[\left(\frac{\partial u}{\partial x}\right)_{x+\Delta x}-\left(\frac{\partial u}{\partial x}\right)_x\right].$$

Now taking the limit $\Delta x \to 0$, $\Delta t \to 0$, $s\rho \dfrac{\partial u}{\partial t} = k \cdot \dfrac{\partial^2 u}{\partial x^2}$,

that is, $$\frac{\partial u}{\partial t} = c^2 \frac{\partial^2 u}{\partial x^2}, \quad \text{where } c^2 = k/s\rho. \qquad ...(3.8)$$

c^2 is known as *diffusivity* of the material of the bar. When the steady state in reached, that is, as $t \to \infty$, $\partial u/\partial t = 0$ and the heat equation is $d^2u/dx^2 = 0$.

We shall solve this equation by the method of separation of variables.

EXAMPLE 3.10 (*i*) Obtain the solution of heat flow equation $\dfrac{\partial u}{\partial t} = c^2 \dfrac{\partial^2 u}{\partial x^2}$ by the method of separation of variables.

(*ii*) A rod of length l with insulated sides is initially at a uniform temperature u_0. Its ends are suddenly cooled to 0°C and are kept at that temperature. Find the temperature function $u(x, t)$.

Solution (*i*) Let $u(x, t) = X(x)\, T(t)$ be the solution of the equation

$$\frac{\partial u}{\partial t} = c^2 \frac{\partial^2 u}{\partial x^2}. \qquad ...(1)$$

Substituting $u = XT$ in this we get $XT' = c^2X''T$

or $\dfrac{X''}{X} = \dfrac{1}{c^2}\dfrac{T'}{T} = h(\text{say})$, where h is a constant. This gives

$$X'' - hX = 0 \quad ...(2) \quad \text{and} \quad T' - c^2hT = 0. \qquad ...(3)$$

Three cases arise according as h is positive, zero or negative.

Case I ($h > 0$): Let $h = k^2$.

Then (2) and (3) become

$$X'' - k^2X = 0 \text{ and } T' - c^2k^2T = 0.$$

Solving these differential equations, we get

$$X = c_1e^{kx} + c_2e^{-kx} \text{ and } T = c_3e^{c^2k^2t}. \qquad ...(4)$$

Case II $h = 0$.

Then (2) and (3) become $X'' = 0$ and $T' = 0$.

Solving these differential equations, we get

$$X = c_4 x + c_5 \text{ and } T = c_6. \qquad ...(5)$$

Case III ($h < 0$): Let $h = -k^2$.

Then (2) and (3) reduce to $X'' + k^2X = 0$ and $T' + c^2k^2T = 0$.

Solving these differential equations, we get

$$X = (c_7 \cos kx + c_8 \sin kx) \text{ and } T = c_9e^{-c^2k^2t}. \qquad ...(6)$$

Thus depending upon the values of h, we have from (4), (5) and (6), the solution of (1) in the form

$$u(x, t) = (b_1e^{kx} + b_2e^{-kx})\, e^{c^2k^2t} \qquad ...(7)$$

$$u(x, t) = b_3x + b_4 \qquad ...(8)$$

and $$u(x,t) = (b_5 \cos kx + b_6 \sin kx)\, e^{-c^2k^2t}. \qquad ...(9)$$

Solution (7) will have to be rejected as $u(x, t)$ will become infinitely large for large values of t and this will violate the physical requirement of the problem.

Solution (8) is a trivial solution as it gives a non-zero temperature for all times. Hence solution (9) can justificably be taken as a correct viable solution of the heat conduction Eq. (1). However, in certain situations in heat flow problems, the combination of (8) and (9) may give a desired solution, as will be seen later in solved example.

(*ii*) Let OA be the rod with length l. We take O as the origin and OA as X-axis. The temperature $u(x, t)$ at any point satisfies the differential equation

$$\frac{\partial u}{\partial t} = c^2 \frac{\partial^2 u}{\partial x^2}.$$

By part (*i*) the solution of this differential equation is of the form

$$u(x, t) = (A \cos kx + B \sin kx)\, e^{-c^2 k^2 t}. \quad ...(1)$$

Since the ends are cooled at 0° C and kept at that temperature throughout, the *boundary conditions* are

$$u(0, t) = 0, \qquad u(l, t) = 0. \quad ...(2)$$

The rod is initially at temperature u_o, therefore the *initial condition* is $u(x, 0) = u_0$. ...(3)

Now we have to find the constants in (1) subject to the boundary conditions (2) and initial condition (3).

Since $u(0, t) = 0$ (first boundary condition), we have from (1)

$$0 = Ae^{-c^2 k^2 t}, \text{ or } A = 0.$$

Thus Eq. (1) becomes $\quad u(x, t) = B \sin kx e^{-c^2 k^2 t}$. ...(4)

Also $u(l, t) = 0$ (second boundary condition), therefore, from (4)

$0 = B \sin kl e^{-c^2 k^2 t}$. Since $B \neq 0$, therefore, $kl = n\pi$ or $k = \frac{n\pi}{l}$ (n being an integer).

The solution (4) now reduces to $u(x, t) = B \sin \frac{n\pi x}{l} e^{-c^2 n^2 \pi^2 t / l^2}$.

Since this is a solution holding for all values of n, *i.e.*, $n = 1, 2, 3, ...$, the sum of these is also a solution (superposition principle). Hence

$$u(x, t) = \sum_{n=1}^{\infty} b_n \sin \frac{n\pi x}{l}\, e^{-c^2 n^2 \pi^2 t / l^2}, \text{ is the general solution.}$$

Now by Eq. (3) $\quad u(x, 0) = u_o$. ...(5)

$$\therefore \quad u_0 = \sum_{n=1}^{\infty} b_n \sin \frac{n\pi x}{l}. \quad ...(6)$$

Since this is half-range sine series for u_o,

$$b_n = \frac{2}{l} \int_0^l u_0 \sin \frac{n\pi x}{l}\, dx = \begin{cases} 0, & \text{when } n \text{ is even}, \\ 4u_0 / n\pi, & \text{when } n \text{ is odd}. \end{cases}$$

Hence,
$$u(x, t) = \frac{4u_0}{\pi} \sum_{n=1,3,5}^{\infty} \frac{1}{n} \sin \frac{n\pi x}{l} \exp\left\{\frac{-c^2 n^2 \pi^2 t}{l^2}\right\}.$$

EXAMPLE 3.11 Solve $\frac{\partial^2 u}{\partial x^2} = h^2 \frac{\partial u}{\partial t}$ when $u(0, t) = u(l, t) = 0$ and $u(x, 0) = \sin\left(\frac{\pi x}{l}\right)$, by the method of separation of variables.

Solution Proceeding as in example 3.10, the solution satisfying the condition $u(0, t) = 0$ and $u(l, t) = 0$ is

$$u(x, t) = \sum_{n=1}^{\infty} b_n \sin \frac{n\pi x}{l} e^{-n^2 \pi^2 t / h^2 l^2} \left(\because \text{here } c^2 = \frac{1}{h^2}\right).$$

Since $u(x, 0) = \sin \frac{\pi x}{l}$, we get $\sin \frac{\pi x}{l} = \sum_{n=1}^{\infty} b_n \sin \frac{n\pi x}{l}$.

Comparing coefficients of different terms on both sides, $b_1 = 1$, $b_2 = b_3 = ... = 0$.

Hence $u(x, t) = \sin \frac{\pi x}{l} e^{-\pi^2 t / h^2 l^2}$ is the required solution.

EXAMPLE 3.12 A rod of length l has its ends A and B kept at 0°C and 100°C respectively, until, steady state conditions prevail. If the temperature at B is reduced suddenly to 0°C and kept so, while that of A is maintained, find the temperature $u(x, t)$ at a distance x from A and at time t. *(AMIETE, June 2002)*

Solution Before the temperature at B is reduced to 0°C, the temperature at any point of the rod did not change with time (steady state conditions prevailed). When it depends on x only, the equation of heat conduction, tha is,

$$\frac{\partial u}{\partial t} = c^2 \frac{\partial^2 u}{\partial x^2} \quad \text{...(1)}$$

reduces to $\dfrac{d^2u}{dx^2} = 0$ whose general solution is $u = c_1 x + c_2$. ...(2)

At that time we had $u = 0$ when $x = 0$ and $u = 100$ when $x = l$.

Putting these values in (2) we get $c_2 = 0$ and $c_1 = 100/l$.

Therefore, $u = 100x/l$ when $t = 0$, that is, $u(x, 0) = 100x/l$.

Thus we have to solve (1) subject to the *boundary conditions* $u(0, t) = 0$, $u(l, t) = 0$ and the *initial condition* $u(x, 0) = 100x/l$.

Proceeding as in example 3.10 the solution satisfying the boundary conditions is

$$u(x,t) = \sum b_n \sin\frac{n\pi x}{l}\, e^{-\frac{n^2\pi^2c^2t}{l^2}}$$

Since $u(x, 0) = \dfrac{100x}{l}$ we get $\dfrac{100x}{l} = \sum_{n=1}^{\infty} b_n \sin\dfrac{n\pi x}{l}$,

where $b_n = \dfrac{2}{l}\displaystyle\int_0^l \frac{100x}{l}\sin\frac{n\pi x}{l}\,dx = -\frac{200}{n\pi}\cos n\pi = \frac{200}{n\pi}(-1)^{n+1}$.

Hence, $u(x,t) = \dfrac{200}{\pi}\displaystyle\sum_{n=1}^{\infty}\frac{(-1)^{n+1}}{n}\sin\frac{n\pi x}{l}\, e^{-n^2\pi^2c^2t/l^2}$.

Rod with non-zero end temperatures

In the following example we shall consider the case when the ends of the bar are maintained at non-zero temperatures, that is, $u(0, t)$ and $u(l, t) \neq 0$.

EXAMPLE 3.13 The ends A and B of a rod 20 cm long have their temperatures at 30°C and 80°C until steady state conditions prevails. The temperatures of the ends are changed to 40°C and 60°C respectively. Find the temperature distribution in the rod at any time t. *(AMIETE, Dec. 2003)*

Solution When the steady-state conditions prevail, the temperature at any point P of the rod depends only on its distance from the point A and is independent of the time. The heat equation becomes

$$\frac{d^2u}{dx^2} = 0. \quad \text{...(1)}$$

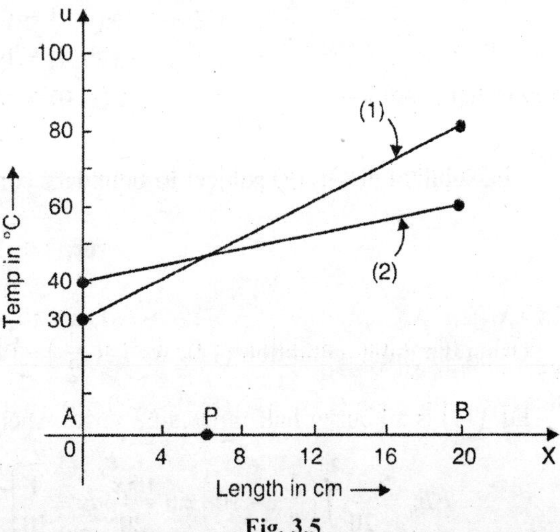

Fig. 3.5

The solution of (1) is given by $u = ax + b$. Using the conditions at the end points that is, $u = 30$°C at A and $u = 80$°C at B, the values of a and b are given by $b = 30$ and $a = 5/2$. Hence when the steady-state conditions prevail, the temperature distribution is given by

$$u = 30 + (5x/2). \quad \text{...(2)}$$

The temperature distribution given in (2) acts as the initial temperature for the next stage of the problem. When the temperatures at A and B are suddenly changed; let the temperature at any point of the rod at a distance x from the end A be $u(x, t)$. Then $u(x, t)$ satisfies the differential equation

$$\frac{\partial u}{\partial t} = c^2 \frac{\partial^2 u}{\partial x^2} \quad ...(3)$$

subject to the *boundary* and *initial conditions* as

$$u(0, t) = 40 \quad ...(4)$$

$$u(20, t) = 60 \quad ...(5)$$

$$u(x, 0) = 30 + (5x/2). \quad ...(6)$$

Now since the temperatures at the ends of the rod are not zero as is borne out from (4) and (5), we express $u(x, t)$ as

$$u(x, t) = v(x, t) + g(x). \quad ...(7)$$

where $v(x, t)$ denotes the transient part of the solution that is, a function of both x and t and $g(x)$ denotes the steady state part of the solution which is a function of x alone and is independent of t.

Substituting from above in (3), we get $\frac{\partial v}{\partial t} = c^2 \frac{\partial^2 v}{\partial x^2} + c^2 g''(x)$.

Now choose $g(x)$ such that: (*i*) $g''(x) = 0$ (*ii*) $v(0, t) = 0$ (*iii*) $v(20, t) = 0$

From (*i*) on integration, we get $g(x) = ax + b$

and $u(0, t) = v(0, t) + g(0)$ (From 7) or $40 = 0 + b$, therefore, $b = 40$.

$u(20, t) = v(20, t) + g(20)$ or $60 = 0 + g(20) = 20a + 40$. Therefore, $a = 1$

This gives us $g(x) = 40 + x$.

Hence, $u(x, t) = v(x, t) + 40 + x \quad ...(8)$

and $v(x, t)$ satisfies the differential equation $\frac{\partial v}{\partial t} = c^2 \frac{\partial^2 v}{\partial x^2}$. ...(9)

With the *boundary conditions*

$$v(0, t) = 0 \quad ...(10)$$

$$v(20, t) = 0 \quad ...(11)$$

and initial condition

$$v(x, 0) = u(x, 0) - 40 - x$$

$$= 30 + (5x/2) - 40 - x = (3x/2) - 10. \quad ...(12)$$

The solution of Eq. (9) subject to boundary conditions (10) and (11) is given by

$$v(x,t) = \sum_{n=1}^{\infty} B_n \sin \frac{n\pi x}{20} e^{-(c^2\pi^2 n^2 t/400)}.$$ [refer example 3.10 (*ii*)]

Using the initial condition (12), we get $\frac{3}{2}x - 10 = \sum_{n=1}^{\infty} B_n \sin \frac{n\pi x}{20}$. ...(13)

Eq. (13) is a Fourier half range sine series where

$$B_n = \frac{2}{20}\int_0^{20}\left(\frac{3}{2}x - 10\right)\sin\frac{n\pi x}{20}\,dx = \frac{1}{10}\left[\frac{-20}{n\pi}(10 + 20\cos n\pi)\right] = \frac{-20}{\pi}\left(\frac{1 + 2\cos n\pi}{n}\right).$$

Therefore, the complete solution $u(x, t) = v(x, t) + 40 + x$ is given by

$$u(x,t) = 40 + x - \frac{20}{\pi}\sum_{n=1}^{\infty}\left(\frac{1 + 2\cos n\pi}{n}\right)\sin\frac{n\pi x}{20}\, e^{-(c^2\pi^2 n^2 t/400)}.$$

EXAMPLE 3.14 An insulated metal rod of length l has one end A kept at 0°C and the other end B at 100°C until steady state is attained. At $t = 0$, the temperature at A is then suddenly raised to 50°C and thereafter maintained while at the same time $t = 0$, the end B is insulated. Find the temperature at any point of the rod at any time t.

Solution When the steady state temperature is reached, the temperature at any point of the rod depends upon its distance from the point A. The heat equation reduces to $\dfrac{d^2u}{dx^2} = 0$. Its solution is $u = ax + b$. With the *boundary conditions* $u = 0$ at A and $u = 100$ at B, the values of a and b are $100/l$ and 0 respectively. The initial temperature distribution defining the *initial condition* is given by

$$u(x, 0) = \frac{100x}{l}. \quad \text{...(1)}$$

Since one of the boundary conditions subsequent to temperature raise is not zero, we consider the solution $u(x, t)$ as the sum of the steady state solution and the transient solution, that is,

$$u(x, t) = v(x, t) + u_s(x). \quad \text{...(2)}$$

The equation of heat conduction together with the boundary and initial conditions is given by

$$\frac{\partial u}{\partial t} = c^2\frac{\partial^2 u}{\partial x^2} \quad \text{...(3)} \qquad u(0, t) = 50 \quad \text{...(4)},$$

$$\frac{\partial u(l,t)}{\partial x} = 0 \quad \text{...(5)} \qquad u(x, 0) = \frac{100x}{l} \quad \text{...(6)}.$$

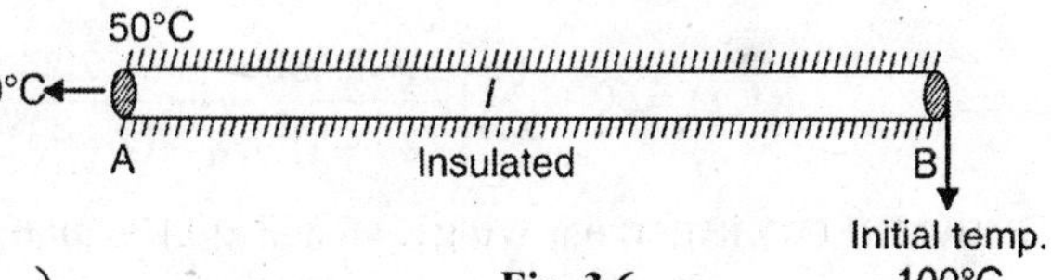

50°C
0°C
l
A
Insulated
B
Initial temp. 100°C

Fig. 3.6

Eqs. (2) and (3) together give $\dfrac{\partial v}{\partial t} = c^2\left(\dfrac{\partial^2 v}{\partial x^2} + \dfrac{d^2u_0(x)}{dx^2}\right)$. ...(7)

Now choosing $u_s(x)$ such that (*i*) $\dfrac{d^2u_s(x)}{dx^2} = 0$, (*ii*) $v(0, t) = 0$, (*iii*) $\dfrac{\partial v}{\partial x}(l, t) = 0$.

Eq. (*i*) on integration gives $u_s(x) = ax + b$. The constants a and b are evaluated from the boundary conditions that is, $u_s(0) = 50$; $\dfrac{du_s(x)}{dx} = 0$ at $x = l$. This gives $a = 0$, $b = 50$. Thus the steady-state part $u_s(x) = 50$.

The transient part $v(x, t)$ satisfies the differential equation $\dfrac{\partial v}{\partial t} = c^2\dfrac{\partial^2 v}{\partial x^2}$ ...(8)

and the *initial* and *boundary conditions* given by

(*i*) $v(x, 0) = u(x, 0) - u_s(x) = \dfrac{100x}{l} - 50$ (*ii*) $v(0, t) = 0$ and (*iii*) $\dfrac{\partial v(l, t)}{\partial x} = 0$.

The solution of (8) is given by $v(x, t) = (A\cos kx + B\sin kx)\, e^{-c^2k^2t}$.

Using condition (*ii*) that is, $x = 0$, we get $A = 0$. Hence $v(x, t) = B\sin kx\, e^{-c^2k^2t}$.

Using condition (*iii*) that is, $\left.\dfrac{\partial v}{\partial x}\right|_{x=l} = 0 = Bk\cos kl\, e^{-c^2k^2t}$, we get $\cos kl = 0$

(B cannot be zero as A is already zero)

This gives $kl = \dfrac{(2n-1)\pi}{2}$ or $k = \dfrac{(2n-1)\pi}{2l}$, $n = 1, 2, 3, \ldots$

Therefore, $$v(x, t) = \sum_{n=1}^{\infty} B_n \sin\frac{(2n-1)\pi x}{2l}\, e^{-c^2\frac{(2n-1)^2\pi^2 t}{4l^2}}. \quad \text{...(9)}$$

To evaluate B_n, we put $t = 0$, This gives $\dfrac{100x}{l} - 50 = \displaystyle\sum_{n=1}^{\infty} B_n \sin\frac{(2n-1)\pi x}{2l}$. ...(10)

This equation is a Fourier half range sine series in $(0, l)$. B_n is given by

$$B_n = \frac{2}{l}\int_0^l \left(\frac{100x}{l} - 50\right)\sin\frac{(2n-1)\pi x}{2l}\,dx$$

$$= \frac{2}{l}\left[\left(\frac{100x}{l} - 50\right)\left(-\cos\frac{(2n-1)\pi x}{2l} \times \frac{2l}{(2n-1)\pi}\right) - \frac{100}{l}\left(-\sin\frac{(2n-1)\pi x}{2l} \times \frac{4l^2}{(2n-1)^2\pi^2}\right)\right]_0^l$$

$$= \frac{2}{l}\left[\frac{400l}{(2n-1)^2\pi^2}\sin\frac{(2n-1)\pi}{2} - \frac{100l}{(2n-1)\pi}\right].$$

Substituting for B_n in Eq. (9), we get

$$v(x, t) = \sum_{n=1}^{\infty}\left(\frac{800(-1)^n}{(2n-1)^2\pi^2} - \frac{200}{(2n-1)\pi}\right)\sin\frac{(2n-1)\pi x}{2l}\, e^{\frac{-c^2(2n-1)^2\pi^2 t}{4l^2}}. \qquad ...(11)$$

Combining Eqs. (2) and (11), the complete-solution of the problem is given by

$$u(x, t) = 50 + \sum_{n=1}^{\infty}\left(\frac{(-1)^n 800}{(2n-1)^2\pi^2} - \frac{200}{(2n-1)\pi}\right)\sin\frac{(2n-1)\pi x}{2l}\, e^{\frac{-c^2(2n-1)^2\pi^2 t}{4l^2}}.$$

EXAMPLE 3.15 A bar with insulated sides is initially at temperature 0°C throughout. The end x = 0 is kept at 0°C for all time and the heat is suddenly applied so that $\partial u/\partial x = 10$ at $x = l$ for all time. Find the temperature function $u(x, t)$.

Solution The differential equation governing the flow of heat in the bar is given by

$$\frac{\partial u}{\partial t} = c^2\frac{\partial^2 u}{\partial x^2} \qquad ...(1)$$

subject to the *boundary* and *initial conditions* presscribed as

$$u(0, t) = 0 \qquad ...(2)$$

$$\frac{\partial u}{\partial x}(l, t) = 10 \qquad ...(3)$$

$$u(x, 0) = 0. \qquad ...(4)$$

Since one of the conditions prescribed on boundary is other than zero, we express the solution as the combination of the steady-state solution and the transient solution, the later ultimately vanishing with large time. Therefore, the solution of Eq. (1) is given by

$$u(x, t) = v(x, t) + u_s(x). \qquad ...(5)$$

As in the previous problem, the steady-state solution $u_s(x)$ is given as

$$u_s(x) = ax + b.$$

Using boundary conditions Eqs. (2) and (3), $b = 0$, $a = 10$. Therefore, $u_s(x) = 10x$. ...(6)

The transient solution $v(x, t)$ is obtained by solving the differential equation

$$\frac{\partial v}{\partial t} = c^2\frac{\partial^2 v}{\partial x^2}, \qquad ...(7)$$

with the *boundary conditions*

$$v(0, t) = 0, \qquad ...(8)$$

$$\frac{\partial v}{\partial t}(0, t) = 0, \qquad ...(9)$$

$$v(x, 0) = 0. \qquad ...(10)$$

The solution of Eq. (7) is given by

$$v = (A\cos kx + B\sin kx)\,e^{-c^2k^2t}. \qquad \text{(refer example 3.10 (ii)).}$$

Using conditions Eqs. (8) and (9), A = 0, $k = \dfrac{(2n-1)\pi}{2l}$. (refer previous example)

Therefore, $$v = \sum_{n=1}^{\infty} B_n \sin\frac{(2n-1)\pi x}{2l}\, e^{\frac{-c^2(2n-1)^2\pi^2 t}{4l^2}} \qquad \text{...(11)}$$

The complete solution is

$$u = 10x + \sum_{n=1}^{\infty} B_n \sin\frac{(2n-1)\pi x}{2l}\, e^{\frac{-c^2(2n-1)^2\pi^2 t}{4l^2}}$$

Now $u(x, 0) = v(x, 0) = 0$. Therefore, B_n, using half range sine series, is given by

$$B_n = \frac{2}{l}\int_0^l -10x\sin\frac{(2n-1)\pi x}{2l}\,dx = \frac{-80l}{(2n-1)^2\pi^2}\sin\frac{(2n-1)\pi}{2}.$$

Therefore, the complete solution $u(x, t)$ is given by

$$u(x,t) = 10x - \frac{80l}{\pi^2}\sum_{n=1}^{\infty}\frac{(-1)^n}{(2n-1)^2}\sin\frac{(2n-1)\pi x}{2l}\bullet e^{-\frac{c^2(2n-1)^2\pi^2 t}{4l^2}}$$

EXERCISE 3.2

1. Using the method of separation of variables show that one set of solution of $\dfrac{\partial u}{\partial t} = c^2\dfrac{\partial^2 u}{\partial x^2}$ is $u = (A\cos px + B\sin px)e^{-c^2p^2t}$, where *A, B, p* are constants.

2. Solve the equation $\dfrac{\partial u}{\partial t} = 4\dfrac{\partial^2 u}{\partial x^2}$, $u(0, t) = 0$, $u(\pi, t) = 0$, $u(x, 0) = 2\sin 3x - 4\sin 5x$ by the method of separation of variables. **Ans.** $u = 2e^{-36t}\sin 3x - 4e^{-100t}\sin 5x$.

3. Find by the method of separation of variables the solution $u(x, t)$ of the boundary value problem

$$\frac{\partial u}{\partial t} = 3\frac{\partial^2 u}{\partial x^2},\quad t > 0,\ 0 < x < 2.$$

$u(0, t) = 0$, $u(2, t) = 0$, $t > 0$ $\quad u(x, 0) = x$, $0 < x < 2$. **Ans.** $u(x,t) = -\dfrac{4}{\pi}\displaystyle\sum_{n=1}^{\infty}\frac{(-1)^n}{n}\sin\frac{n\pi x}{2}\,e^{-3k^2t}$.

4. Solve the equation $\dfrac{\partial u}{\partial t} = \dfrac{\partial^2 u}{\partial x^2}$ with boundary conditions:
$u(x, 0) = 3\sin n\pi x$, $u(0, t) = 0$ and $u(1, t) = 0$, where $0 < x < 1$, $t > 0$. (*AMIETE, Dec. 2010, Dec. 2009*)

Ans. $u(x,t) = 3\displaystyle\sum_{n=1}^{\infty} e^{-n^2\pi^2 t}\sin n\pi x$.

5. Solve the boundary value problem

$$\frac{\partial u}{\partial t} = \frac{\partial^2 u}{\partial x^2},\quad \left(\frac{\partial u}{\partial x}\right)_{x=0} = 0,\ u(2,t) = 0,\ u(x,0) = 8\cos\frac{3\pi x}{4} - 6\cos\frac{9\pi x}{4}$$

by the method of separation of variables.

Ans. $u = 8\exp(-9\pi^2 t/16)\cos\dfrac{3\pi x}{4} - 6\exp\left(-\dfrac{81\pi^2 t}{16}\right)\cos\dfrac{9\pi x}{4}$.

6. Solve the differential equation $\frac{\partial u}{\partial t} = \alpha^2 \frac{\partial^2 u}{\partial x^2}$ for the conduction of heat along a rod without radiation, subject to the following conditions:

(a) u is not infinite for $t \to \infty$, (b) $\partial u/\partial x = 0$ for $x = 0$ and $x = l$,

(c) $u = lx - x^2$ for $t = 0$, between $x = 0$ and $x = l$. *(AMIETE, June-2004)*

Ans. $u = \frac{l^2}{6} - \frac{l^2}{\pi^2} \sum \frac{1}{n^2} e^{-4n^2\pi^2\alpha^2 t/l^2} \cos \frac{2\pi n x}{l}$.

7. The ends A and B of a rod 30 cm long have their temperatures kept at 20°C and 80°C until steady state conditions prevail. The temperature of the end B is suddenly reduced to 60°C and kept so while the end A is raised to 40°C. Find the temperature distribution in the rod after time t. *(AMIETE, W-2002)*

Ans. $u(x, t) = 40 + \frac{2x}{3} - \frac{40}{\pi} \sum_{n=1}^{\infty} \frac{1}{n} \sin \frac{n\pi x}{15} \exp(-n^2\pi^2 c^2 t/225)$.

8. The ends A and B of a rod l cm long have the temperature 0°C and 100°C respectively until steady state prevails. The temperature of A is then suddenly raised to 25°C while that of B is reduced to 75°C. Find the temperature distribution in the rod at time t.

Ans. $u = 25 + \frac{50x}{l} - \frac{100}{\pi}\left[\frac{1}{2} e^{-4bt} \sin \frac{2\pi x}{l} + \frac{1}{2} e^{-16bt} \sin \frac{4\pi x}{l} + \ldots\right]$, where $b = (\pi c/l)^2$.

9. The equation for the conduction of heat along a bar of length l is $\frac{\partial \theta}{\partial t} = k \frac{\partial^2 \theta}{\partial x^2}$, neglecting radiation. Find an expression for θ if the ends of the bar are maintained at zero temperature initially and if initially the temperature is T at the centre of the bar and falls uniformally to zero at the ends.

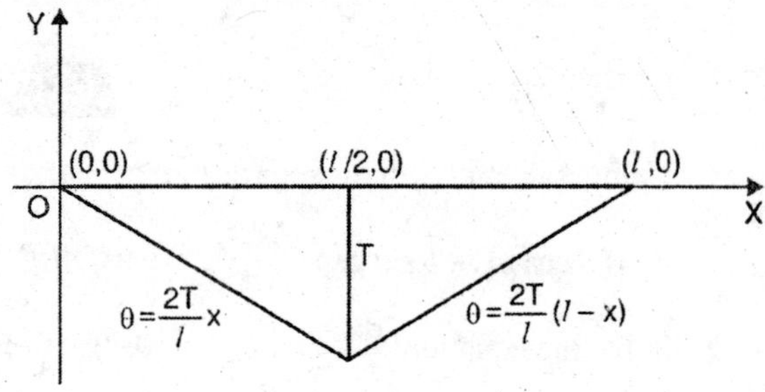

Fig. 3.7

Ans. $\theta(x, t) = \frac{8T}{\pi^2} \sum_{n=1}^{\infty} \frac{1}{n^2} \sin \frac{n\pi}{2} \sin \frac{n\pi x}{l} \; e^{\frac{-kn^2\pi^2 t}{l^2}}$

[**Hint:** The initial temperature distribution is given by

$$\theta(x, 0) = \frac{2T}{l} x \text{ for } 0 \le x \le \frac{l}{2} = \frac{2T}{l}(l - x) \text{ for } \frac{l}{2} \le x \le l].$$

10. A homogeneous rod of conducting material of length 100 cm has its ends kept at zero temperature and the temperature initially is

$$u(x, 0) = \begin{cases} x, & 0 \le x \le 50, \\ 100 - x, & 50 \le x \le 100. \end{cases}$$

Find the temperature $u(x, t)$ at any time. *(AMIE, W-2003)*

Ans. $u(x, t) = \frac{400}{\pi^2} \sum_{n=0}^{\infty} \frac{(-1)^n}{(2n+1)^2} e^{\frac{-c^2(2n+1)^2\pi^2 t}{100^2}} \sin \frac{(2n+1)\pi x}{100}$.

11. A uniform rod of length l, whose surface is thermally insulated, is initially at temperature q_0. At time $t = 0$, one end is suddenly cooled to temperature 0°C and subsequentlly maintained to this temperature and at the same time the other end is thermally insulated. Find the temperature at end $x = l$ at any time t.

Ans. $u(x, t) = \frac{4\theta_0}{\pi} \sum_{n=1}^{\infty} \frac{1}{2n-1} \sin \frac{(2n-1)\pi x}{2l} e^{\frac{-c^2(2n-1)^2\pi^2 t}{4l^2}}$, $u(l, t) = \frac{4\theta_0}{\pi} \sum_{n=1}^{\infty} \frac{(-1)^{n-1}}{2n-1} e^{\frac{-c^2(2n-1)^2\pi^2 t}{4l^2}}$

12. A rod of length l has its ends A and B which are kept at 0°C and 75°C, until steady-state conditions prevail. If the temperature of A is suddenly raised to 75°C and that of B to 175°C and maintained thereafter, find the subsequent temperature distribution of the rod.

$$\textbf{Ans. } u(x,t)=75+\frac{100}{l}x-\frac{300}{\pi}\sum_{n=1}^{\infty}\frac{1}{2n-1}\sin\frac{(2n-1)\pi x}{l}\,e^{\frac{-c^2(2n-1)^2\pi^2 t}{l^2}}.$$

13. The temperatures at the ends $x = 0$ and $x = 50$ cm in length of a rod are held at 0°C and 50°C respectively until steady-state conditions prevail. The two ends of the rod are suddenly insulated. Find the temperature distribution of the rod, assuming that the surface of the rod is impervious to heat flow.

$$\textbf{Ans. } u(x,t)=25-\frac{200}{\pi^2}\sum_{n=1}^{\infty}\frac{1}{(2n-1)^2}\cos\frac{(2n-1)\pi x}{50}\,e^{\frac{-c^2(2n-1)^2\pi^2 t}{2500}}.$$

14. A rod of length l has its ends A and B are maintained at 20°C and 40°C respectively until steady-state conditions prevail. The temperature at A is suddenly raised to 50°C while that of B is lowered to 10°C and maintained thereafter. Find the subsequent temperature distribution of the rod.

$$\textbf{Ans. } u(x,t)=50-\frac{40x}{l}-\frac{60}{\pi}\sum_{n=1}^{\infty}\frac{1}{n}\sin\frac{2n\pi x}{l}\,e^{\frac{-4c^2n^2\pi^2 t}{l^2}}.$$

15. A rod of length l has its lateral surface insulated and is so thin that heat flow in the rod can be regarded as one dimensional. Initially the rod is at the temperature 100°C throughout. At $t = 0$ the temperature at the left end of the rod is suddenly reduced to 50°C and maintained thereafter at this value, while the right end is maintained at 100°C. Let $u(x, t)$ be the temperature at point x in the rod at any subsequent time t.

(*i*) Write down the appropriate partial differential equation for $u(x, t)$ with initial and boundary conditions.

(*ii*) Solve the differential equation in (*i*) above using method of separation of variables and show that

$$u(x,t)=50+\frac{50x}{l}+\frac{100}{\pi}\sum_{n=1}^{\infty}\frac{1}{n}\sin\frac{n\pi x}{l}\exp\frac{-n^2\pi^2 t}{a^2l^2},$$

where a^2 is the constant involved in the partial differential equation. *(AMIETE, Dec. 2004)*

[**Hint:** Here we solve the one-dimensional heat equation $\frac{\partial^2 u}{\partial x^2}=a^2\frac{\partial u}{\partial t}$ subject to the boundary that is, end conditions $u(0, t) = 50$, $u(l, t) = 100$ and the initial condition $u(x, 0) = 100$].

3.8 TWO DIMENSIONAL HEAT FLOW

- $$\frac{\partial^2 u}{\partial x^2}+\frac{\partial^2 u}{\partial y^2}=a^2\frac{\partial u}{\partial t}.$$

If the temperature at any point of a metal plate in the XY-plane depends upon the x and y co-ordinates and time t so that there is no heat-flow normal to the XY-plane, then heat flow is called two dimensional. Proceeding as in Art. 3.7 we can prove that the temperature u at any point of the plate at any time t is governed by the differential equation

$$\frac{\partial^2 u}{\partial x^2}+\frac{\partial^2 u}{\partial y^2}=a^2\frac{\partial u}{\partial t}. \qquad \text{...(3.9)}$$

In the steady state, that is, when temperature u does not change with time, $\partial u/\partial t = 0$ and Eq. (3.9) reduces to the form

Y
D(x, y+δy)
C(x+δx, y+δy)
A(x, y)
B(x+δx, y)
O
X

Fig. 3.8

$$\frac{\partial^2 u}{\partial x^2}+\frac{\partial^2 u}{\partial y^2}=0. \qquad \text{...(3.10)}$$

The above equation is known as **Laplace equation** *in two dimensions.*

This equation is called the Laplace equation in *cartesian co-ordinates* or two dimensional heat flow in steady-state conditions.

Another form of this equation is **Polar form.** This equation is obtained by substituting $x = r\cos\theta$, $y = r\sin\theta$ in the above equation and in the (r, θ) coordinates the Laplace equation is written as

$$r^2\frac{\partial^2 u}{\partial r^2}+r\frac{\partial u}{\partial r}+\frac{\partial^2 u}{\partial \theta^2}=0. \qquad ...(3.11)$$

The polar form of the Laplace equation is useful in solving problems where the configuration is circular or semi-circular or a sector of a circle.

Method of Solving Laplace Equation (Cartesian Form)

The equation is given by

$$\frac{\partial^2 u}{\partial x^2}+\frac{\partial^2 u}{\partial y^2}=0. \qquad ...(1)$$

Let $u = XY$ be the solution of Eq. (1). Substituting for u in Eq. (1) and using the method of separation of variables, we get

$$Y\frac{d^2X}{dx^2}+X\frac{d^2Y}{dy^2}=0 \text{ or } YX''+XY''=0.$$

This gives $$\frac{X''}{X}+\frac{Y''}{Y}=0 \text{ or } \frac{X''}{X}=-\frac{Y''}{Y}. \qquad ...(2)$$

Since X''/X is a function of x alone and Y''/Y is function of y alone, and because x and y are independent of each other, relation (2) will hold if each side is constant (otherwise there will be a functional relation between x and y).

The constant can be taken either positive or negative depending upon the *non-zero* boundary conditions.

(*i*) If the non-zero condition is in terms of x, we assume the constant as negative and write this as $-k^2$.

(*ii*) If the non-zero condition is in terms of y, we assume the constant as positive and write this as $+k^2$.

Note: If the plate is infinite that is, it is too long in comparison to its width, then the temperature at infinity is always taken as zero. For example

(*i*) If the plate is infinite along Y-direction, we write $u(x, \infty) = 0$ for all x and

(*ii*) If the plate is infinite in X-direction, we write $u(\infty, y) = 0$ for all y.

The procedure will be clear from the following problem.

EXAMPLE 3.16 A rectangular plate with insulated surfaces is 8 cm wide and so long compared to its width that it may be considered infinite in length. If the temperature along one short edge $y = 0$ is given by $u(x, 0) = 100\sin(\pi x/8)$, $0 < x < 8$, while the two long edges, $x = 0$ and $x = 8$, as well as the other short edge are kept at 0°C, find the steady state temperature $u(x, y)$.

Solution In the steady state, the temperature u satisfies the equation $\frac{\partial^2 u}{\partial x^2}+\frac{\partial^2 u}{\partial y^2}=0.$

Here the *boundary conditions* are

$$u(0, y) = 0, \qquad ...(1)$$
$$u(8, y) = 0, \qquad ...(2)$$
$$u(x, 0) = 100\sin\frac{\pi x}{8}, \qquad ...(3)$$
$$\lim_{y\to\infty} u(x, y) = 0, \text{ that is, } u(x,\infty)=0. \qquad ...(4)$$

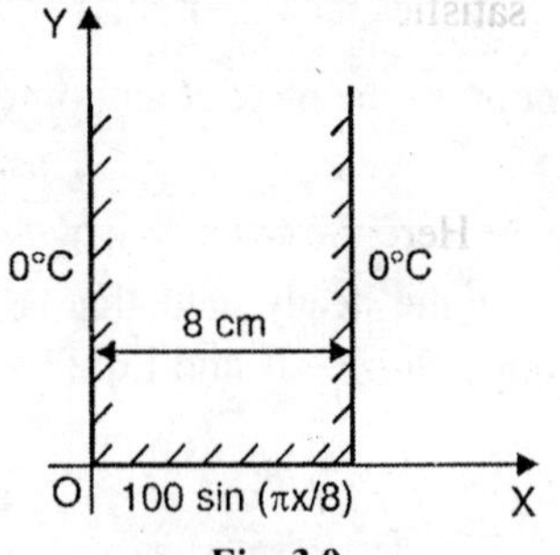

Fig. 3.9

Let $u = XY$ be the solution of the given equation. Substituting this in the given equation, using method of separation of variables, we get

$$X''Y+Y''X=0 \text{ or } \frac{X''}{X}=-\frac{Y''}{Y}=-k^2 \quad \text{(because of non-zero condition).}$$

This gives $X'' = -k^2X$ and $Y'' = K^2Y$.

The solutions of these equations are

$$X = c_1 \cos kx + c_2 \sin kx,$$
$$Y = c_3 e^{ky} + c_4 e^{-ky}.$$

Therefore, $u = XY = (c_1 \cos kx + c_2 \sin kx)(c_3 e^{ky} + c_4 e^{-ky})$. ...(5)

Now applying the condition (1) in (5) that is, $u(0, y) = 0$, we get

$$0 = c_1(c_3e^{ky} + c_4e^{-ky}). \text{ This gives } c_1 = 0.$$

Relation Eq. (5) with $c_1 = 0$, gives

$$u = \sin kx(c_5e^{ky} + c_6e^{-ky}); (c_5 = c_2c_3;\ c_6 = c_2c_4). \quad ...(6)$$

Again using condition (4) in (6) that is, $u(x, \infty) = 0$, we get $c_5 = 0$.

$\therefore$ Eq. (6) with $c_5 = 0$ can be written as

$$u(x, y) = c_6 \sin kx\, e^{-ky}. \quad ...(7)$$

Using condition (2) in (7) that is, $u(8, y) = 0$, we obtain

$$0 = c_6 \sin 8k\, e^{-ky}. \quad ...(8)$$

Now either c_6 is zero or $\sin 8k$ is zero. ($\because$ condition is on $x = 8$).

But c_6 which is equal to $c_2 c_4$ from Eq. (6) above cannot be zero.

$\therefore$ $\sin 8k = 0$ (If c_2 is zero, the whole X solution vanishes).

Now $\sin 8k = 0$ gives $8k = n\pi$. Therefore, $k = n\pi/8$.

Substituting for $k = n\pi/8$. in Eq. (7), we get

$$u = \sum_{n=1}^{\infty} B_n \sin\frac{n\pi x}{8} e^{-\frac{n\pi y}{8}}. \quad ...(9)$$

Using the third boundary condition, we get from Eq. (9)

$$100\sin\frac{\pi x}{8} = \sum_{n=1}^{\infty} B_n \sin\frac{n\pi x}{8}.$$

Equating the coefficients, $B_1 = 100$, $B_2 = B_3 = \ldots\ldots = 0$; $n = 1$.

Therefore, from (9) the solution is given by

$$u(x, y) = 100\sin\frac{\pi x}{8} e^{\frac{-\pi y}{8}}.$$

EXAMPLE 3.17 An infinitely long plane uniform plate is bounded by two parallel edges and an end at right angles to them. The breadth is π; this end is maintained at a temperature u_0 at all points and other edges are at zero temperature. Determine the temperature at any point of the plate in steady state.

(AMIETE, June 2012, Dec. 2009, 2008, Dec. 2005; JNTU 2002 S)

Solution In the steady state, the temperature $u(x, y)$ at any point $P(x, y)$ satisfies the differential equation

$$\frac{\partial^2 u}{\partial x^2} + \frac{\partial^2 u}{\partial y^2} = 0. \quad ...(A)$$

Here the *boundary conditions* are

$$u(0, y) = 0, \quad ...(1)$$
$$u(\pi, y) = 0, \quad ...(2)$$
$$u(x, 0) = u_0, \quad ...(3)$$
$$\lim_{y\to\infty} u(x, y) = 0. \quad ...(4)$$

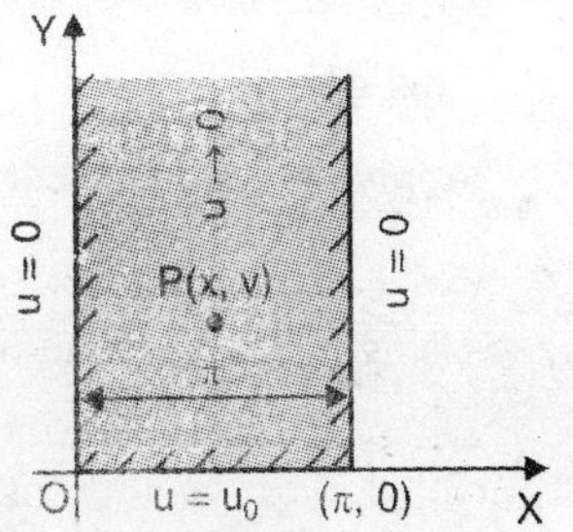

Fig. 3.10

By example 3.16, the solution of (*A*) which satisfies the boundary conditions (1), (2) and (4) is

$$u(x,y)=\sum_{n=1}^{\infty} B_n (\sin nx)e^{-ny}. \quad ...(5) \qquad \text{(because here width is } \pi \text{ in place of 8)}$$

Therefore, $$u(x,0)=\sum_{n=1}^{\infty} B_n \sin nx. \quad \text{But } u(x,0)=u_0. \qquad \text{[boundary condition (3)].}$$

Therefore, $$u_0=\sum_{n=1}^{\infty} B_n \sin nx. \qquad ...(6)$$

This is half-range sine series in $(0, \pi)$.

Therefore, $$B_n=\frac{2}{\pi}\int_0^{\pi} u_0 \sin nx\,dx=\begin{cases}0, & (n \text{ even}),\\ \dfrac{4u_0}{n\pi}, & (n \text{ odd}).\end{cases}$$

Therefore, $$u(x,y)=\frac{4u_0}{\pi}\left[e^{-y}\sin x+\frac{1}{3}e^{-3y}\sin 3x+\frac{1}{5}e^{-5y}\sin 5x+...\right].$$

EXAMPLE 3.18 An infinitely long metal plate in the form of area is enclosed between $y = 0$ and $y = l$ for positive values of x. The temperature is zero along the edges $y = 0$, $y = l$ and at ∞. If edge $x = 0$ is kept at a temperature u_0, find the steady state temperature distribution in the plate.

Solution In the steady-state, the temperature u satisfies the equation

$$\frac{\partial^2 u}{\partial x^2}+\frac{\partial^2 u}{\partial y^2}=0. \qquad ...(1)$$

Here the *boundary conditions* are

$$u(x,0)=0, \qquad ...(2)$$
$$u(x,l)=0, \qquad ...(3)$$
$$u(\infty,y)=0, \qquad ...(4)$$
$$u(0,y)=u_0. \qquad ...(5)$$

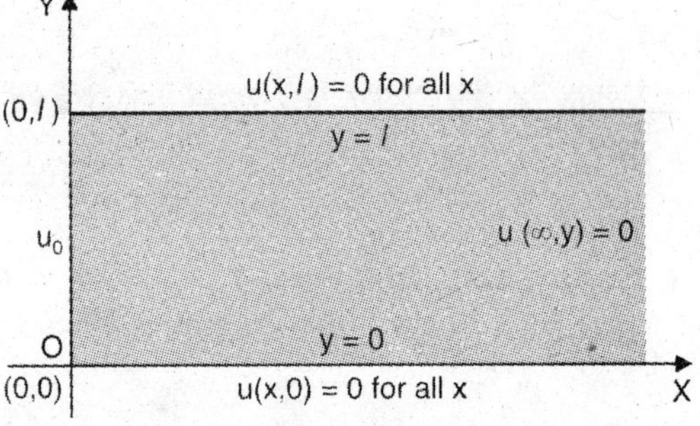

Fig. 3.11

Let $u = XY$ be the solution of the given equation. Substituting this in the given equation and separating the variables, we get

$$\frac{X''}{X}=-\frac{Y''}{Y}=k^2 \quad \text{(Non-zero condition in } y).$$

Solving the two ordinary equations viz $X'' = k^2X$ and $Y'' = -k^2Y$, we get the solution

$$u=XY=(c_1e^{kx}+c_2e^{-kx})(c_3\cos ky+c_4\sin ky). \qquad ...(6)$$

Applying boundary condition Eq. (2) in Eq. (6), we get $c_3 = 0$.

This give $$u=(c_1'e^{kx}+c_2'e^{-kx})\sin ky;\ (c_1'=c_1c_4;\ c_2'=c_2c_4). \qquad ...(7)$$

Applying boundary condition, Eq. (4) in Eq. (7), we get $c_1'=0$. Hence,

$$u=c_2'e^{-kx}\sin ky. \qquad ...(8)$$

Using boundary condition Eq. (3) in Eq. (8), we obtain

$$0=c_2'e^{-kx}\sin kl.$$

Since $c_2'\neq 0$, $\sin kl=0$, giving $k=\dfrac{n\pi}{l}$.

Therefore, Eq. (8) for all values of n can be written as

$$u = \sum c_n \sin \frac{n\pi y}{l} e^{\frac{-n\pi x}{l}}. \qquad \text{...(9)}$$

To evalute c_n, we use the boundary condition (5). This gives $c_n = \frac{2}{l}\int_0^l u_0 \sin \frac{n\pi y}{l} dy = \frac{2u_0}{n\pi}(1-(-1)^n)$.

Therefore, $u(x, y) = \frac{4u_0}{\pi} \sum_{n=1}^{\infty} \frac{1}{2n-1} \sin \frac{(2n-1)\pi y}{l} e^{\frac{-(2n-1)\pi x}{l}}$ by replacing n by $(2n - 1)$ in (9).

EXAMPLE 3.19 Evaluate the steady temperature in a rectangular plate of length a and width b, the sides of which are kept at temperature zero, the lower end is kept at temperature $f(x)$ and the upper edge is kept insulated.

Solution In steady state temperature is governed by

$$\frac{\partial^2 u}{\partial x^2} + \frac{\partial^2 u}{\partial y^2} = 0. \qquad \text{...(A)}$$

We have to solve the above equation subject to the *conditions*:

$$u(0, y) = 0, \text{ and } u(a, y) = 0 \qquad \text{...(1)}$$

$$(\partial u/\partial y)_{y=b} = 0 \quad \text{and } u(x, 0) = f(x). \qquad \text{...(2)}$$

[We should note the condition $(\partial u/\partial y)_{y=b} = 0$ on the upper edge $y = b$ which is kept insulated]

The solution of (A) satisfying first two conditions is

$$u(x, y) = \sin \frac{n\pi x}{a}(b_1 e^{n\pi y/a} + b_2 e^{-n\pi y/a})$$

$$= \sin \frac{n\pi x}{a}\left(c_1 \cosh \frac{n\pi y}{a} + c_2 \sinh \frac{n\pi y}{a}\right)^*.$$

$$\frac{\partial u}{\partial y} = \sin \frac{n\pi x}{a} \cdot \frac{n\pi}{a}\left[c_1 \sinh \frac{n\pi y}{a} + c_2 \cosh \frac{n\pi y}{a}\right].$$

$$\left(\frac{\partial u}{\partial y}\right)_{y=b} = 0 \text{ gives } c_2 = \frac{-c_1 \sinh \frac{n\pi b}{a}}{\cosh \frac{n\pi b}{a}}.$$

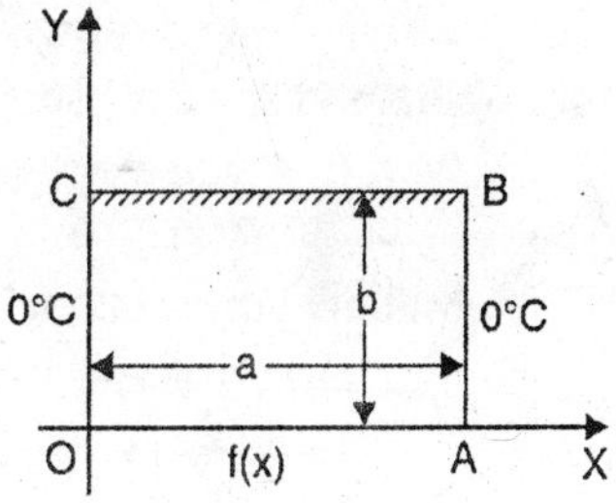

Fig. 3.12

Therefore, $$u(x, y) = \sum c_n \sin \frac{n\pi x}{a}\left[\cosh \frac{n\pi y}{a} \cosh \frac{n\pi b}{a} - \sinh \frac{n\pi y}{a} \sinh \frac{n\pi b}{a}\right] \operatorname{sech} \frac{n\pi b}{a}.$$

Thus $$u(x, y) = \sum c_n \sin \frac{n\pi x}{a} \cosh \frac{n\pi}{a}(y-b) \operatorname{sech} \frac{n\pi b}{a}. \qquad \text{...(3)}$$

[$\because \cosh(\theta - \phi) = \cosh\theta \cosh\phi - \sinh\theta \sinh\phi$]

Now, $$u(x, 0) = \sum c_n \sin \frac{n\pi x}{a} \qquad \left(\because \cosh \frac{n\pi b}{a} \operatorname{sech} \frac{n\pi b}{a} = 1\right)$$

* **Note:** While solving problems on plates which are bounded, we prefer to use hyperbolic function instead of exponential functions. This is mainly due to the reason that hyperbolic functions are convenient to handle while applying boundary conditions in such problems. $e^{ky} = \cosh ky + \sinh ky$; $e^{-ky} = \cosh ky - \sinh ky$.

But $u(x, 0) = f(x)$.

Therefore, $$f(x) = \sum c_n \sin\frac{n\pi x}{a}, \text{ where } c_n = \frac{2}{a}\int_0^a f(x)\sin\frac{n\pi x}{a}\,dx. \quad \text{...(4)}$$

Hence $u(x, y)$ is given by (3) when c_n is given by (4).

Note *While studying steady state temperature distribution in rectangular sheets it is very convenient and natural that we use Laplace equation in rectangular co-ordinates, that is,* $\frac{\partial^2 u}{\partial x^2} + \frac{\partial^2 u}{\partial y^2} = 0$. *However, in some problems polar co-ordinates are more useful. In such problems we require Laplace equation in polar form with r, θ as independent variables in place of x, y. The polar form of Laplace equation* $\frac{\partial^2 u}{\partial x^2} + \frac{\partial^2 u}{\partial y^2} = 0$ *is*

$$\frac{\partial^2 u}{\partial x^2} + \frac{1}{r}\frac{\partial u}{\partial r} + \frac{1}{r^2}\frac{\partial^2 u}{\partial \theta^2} = 0 \quad \text{or} \quad r^2\frac{\partial^2 u}{\partial x^2} + r\frac{\partial u}{\partial r} + \frac{\partial^2 u}{\partial \theta^2} = 0. \quad \text{...(3.12)}$$

EXAMPLE 3.20 The bounding diameter of a semi-circular plate of radius 10 cm is kept at 0°C and the temperature along the semi-circular boundary at u_0 until steady state conditions prevail. Find the temperature distribution in the plate.

Solution Taking the centre O of the circle as pole and diameter as initial line, let the temperature at $P(r, \theta)$ be $u(r, \theta)$.

Now $u(r, \theta)$ satisfies the Laplace equation

$$r^2\frac{\partial^2 u}{\partial r^2} + r\frac{\partial u}{\partial r} + \frac{\partial^2 u}{\partial \theta^2} = 0. \quad \text{...(1)}$$

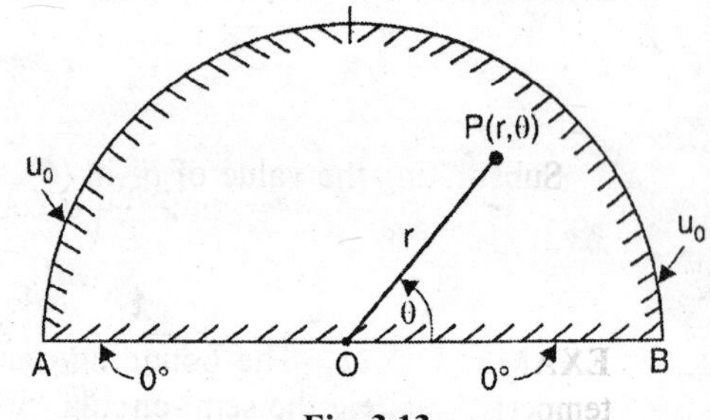

Fig. 3.13

We have to solve (1) subject to the *boundary conditions*

$$u(r, 0) = 0,\ u(r, \pi) = 0,\ u(0, \theta) = 0,\ u(10, \theta) = u_0.$$

Let $u = R(r)\,T(\theta)$.

Substituting this in Eq. (1) we get

$$r^2R''\,T + rR'\,T + RT'' = 0$$

or $$(r^2R'' + rR')T + RT'' = 0$$

or $$\frac{r^2R'' + rR'}{R} = -\frac{T''}{T} = h \text{ (say)}.$$

Therefore, $r^2R'' + rR' - hR = 0$ ...(2) [Cauchy's Homogeneous equation]

and $T'' + hT = 0.$ ...(3)

Putting $r = e^z$ in (2) we get $$\frac{d^2R}{dz^2} - hR = 0. \quad \text{....(4)}$$

Case I ($h > 0$). Let $h = k^2$.

Then, the solutions of Eqs. (3) and (4) are

$$T = c_1 \cos k\theta + c_2 \sin k\theta$$

and $$R = c_3 e^{kz} + c_4 e^{-kz} = c_3 r^k + c_4 r^{-k}.$$

Case II $h = 0$.

Then, $$T = c_5\theta + c_6.$$
$$R = c_7 z + c_8 = c_7 \log r + c_8.$$

Case III ($h < 0$). Let $h = -k^2$.

Then the solutions are

$$T = c_9 e^{k\theta} + c_{10} e^{-k\theta}.$$
$$R = c_{11} \cos kz + c_{12} \sin kz = c_{11} \cos (k \log r) + c_{12} \sin (k \log r).$$

Thus the possible solutions are

$$u(r, \theta) = (c_1 \cos k\theta + c_2 \sin k\theta)(c_3 r^k + c_4 r^{-k}) \qquad \text{...(A)}$$

$$u(r, \theta) = (c_5\theta + c_6)(c_7 \log r + c_8) \qquad \text{...(B)}$$

and $$u(r, \theta) = (c_9 e^{k\theta} + c_{10} e^{-k\theta})[c_{11} \cos(k \log r) + c_{12} \sin(k \log r)]. \qquad \text{...(C)}$$

Since (*B*) and (*C*) don't satisfy the boundary condition $u(0, \theta) = 0$, these are rejected.

[In dealing with such problems the constant *h* is to be chosen positive as this will give trigonometric function in θ]

Hence, $u(r, \theta) = (c_1 \cos k\theta + c_2 \sin k\theta)(c_3 r^k + c_4 r^{-k})$. ...(5)

As, $u = 0$ when $r \to 0$, $c_4 = 0$.

Therefore, $u = c_3 r^k (c_1 \cos k\theta + c_2 \sin k\theta) = r^k (b_1 \cos k\theta + b_2 \sin k\theta)$.

Again $u(r, 0) = 0$, we get $b_1 = 0$.

Hence, $u = b_2 r^k \sin k\theta$.

Using $u(r, \pi) = 0$, we get $0 = b_2 r^k \sin k\pi$.

Therefore, $k = n$ (an integer).

Thus $u(r, \theta) = b_n r^n \sin n\theta$ (replacing b_2 by b_n).

By superposition principle $u(r, \theta) = \Sigma\, b_n r^n \sin n\theta$. ...(6)

Using the last boundary condition, that is, $u(10, \theta) = u_0$, we get

$$u_0 = \sum_{n=1}^{\infty} b_n 10^n \sin n\theta .$$

Since the right hand side is half-range sine series representing u_0 in $(0, \pi)$

$$b_n = \frac{2}{10^n \pi} \int_0^{\pi} u_0 \sin n\theta \, d\theta = \begin{cases} \dfrac{4u_0}{n\pi} \cdot \dfrac{1}{10^n}, & (n \text{ odd}). \\ 0. & (n \text{ even}). \end{cases}$$

Substituting the value of b_n in (6), the required temperature distribution is given by

$$u(r, \theta) = \frac{4u_0}{\pi} \left[\frac{r}{10} \sin\theta + \frac{1}{3}\left(\frac{r}{10}\right)^3 \sin 3\theta + \dots \right].$$

EXAMPLE 3.21 The bounding diameter of a semi-circular plate of radius *a* cm is kept at 0°C and the temperature along the semi-circular boundary is given by

$$u(a, \theta) = \begin{cases} 50\,\theta, & \text{when } 0 < \theta \le \pi/2, \\ 50\,(\pi - \theta), & \text{when } \pi/2 < \theta < \pi. \end{cases}$$

Find the steady state temperature function $u(r, \theta)$. (*Madras, 2003*)

Solution Here we have to solve

$$r^2 \frac{\partial^2 u}{\partial r^2} + r \frac{\partial u}{\partial r} + \frac{\partial^2 u}{\partial \theta^2} = 0 \qquad \text{...(1)}$$

subject to the *boundary conditions*

$$u(r, 0) = 0 \qquad \text{...(A)}$$

$$u(r, \pi) = 0 \qquad \text{...(B)}$$

$$u(a, \theta) = \begin{cases} 50\,\theta & 0 < \theta \le \pi/2 \\ 50\,(\pi - \theta) & \pi/2 < \theta < \pi. \end{cases} \qquad \text{...(C)}$$

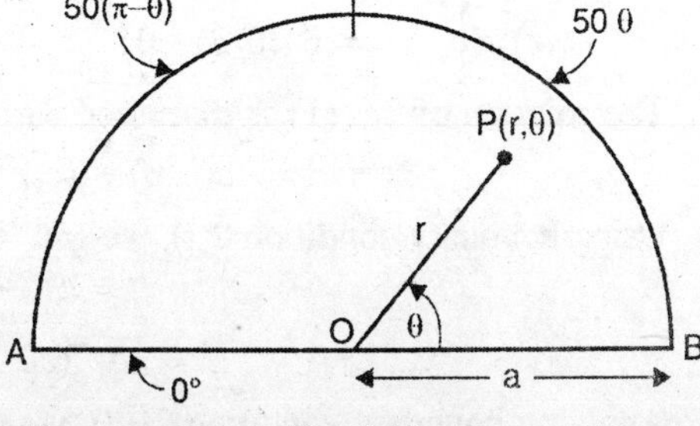

Fig. 3.14

The solution of Eq. (1) satisfying boundary conditions (A) and (B) by example 3.20, is

$$u(r, \theta) = \sum_{n=1}^{\infty} b_n r^n \sin n\theta. \qquad \text{...(2)}$$

Using the boundary condition (C), we get

$$u(a, \theta) = \sum_{n=1}^{\infty} b_n a^n \sin n\theta \qquad ...(3)$$

where $\qquad u(a, \theta) = 50\,\theta,\quad$ when $0 < \theta \le \pi/2$

$\qquad\qquad = 50(\pi - \theta),$ when $\pi/2 < \theta < \pi.$

But Eq. (3) is the half-range Fourier sine series for $u(a, \theta)$ in $(0. \pi)$.

Therefore, $$b_n a^n = \frac{2}{\pi}\int_0^{\pi} u(a, \theta)\sin n\theta\, d\theta$$

$$= \frac{2}{\pi}\left[\int_0^{\pi/2} 50\theta \sin n\theta\, d\theta + \int_{\pi/2}^{\pi} 50\,(\pi - \theta)\sin n\theta\, d\theta\right] \text{[by (C)]}$$

$$= \frac{200}{\pi n^2}\sin\frac{n\pi}{2} \quad \text{or} \quad b_n = \frac{200}{\pi a^n n^2}\sin\frac{n\pi}{2}.$$

Putting the value of b_n in (2), the required temperature distribution is

$$u(r, \theta) = \frac{200}{\pi}\sum_{n=1,3,5...}\left[\frac{1}{a^n n^2}\sin\frac{n\pi}{2} r^n \sin n\theta\right]$$

$$= \frac{200}{\pi}\left[\frac{1}{1^2}\left(\frac{r}{a}\right)\sin\theta - \frac{1}{3^2}\left(\frac{r}{a}\right)^3 \sin 3\theta + \frac{1}{5^2}\left(\frac{r}{a}\right)^5 \sin 5\theta - ...\right] = \frac{200}{\pi}\sum\frac{1}{a^n n^2}\sin\frac{n\pi}{2} r^n \sin n\theta.$$

EXAMPLE 3.22 A plate in the shape of a truncated quadrant of a circle is bounded by $r = a$, $r = b$, $\theta = 0$ and $\theta = \pi/2$. Its faces are insulated and maintained at temperature 0°C along three edges while the edge $r = a$ is kept at temperature $\theta((\pi/2) - \theta)$. Find the steady-state temperature distribution.

Solution In the polar form the steady state temperature distribution is given by

$$r^2\frac{\partial^2 u}{\partial r^2} + r\frac{\partial u}{\partial r} + \frac{\partial^2 u}{\partial \theta^2} = 0. \qquad ...(1)$$

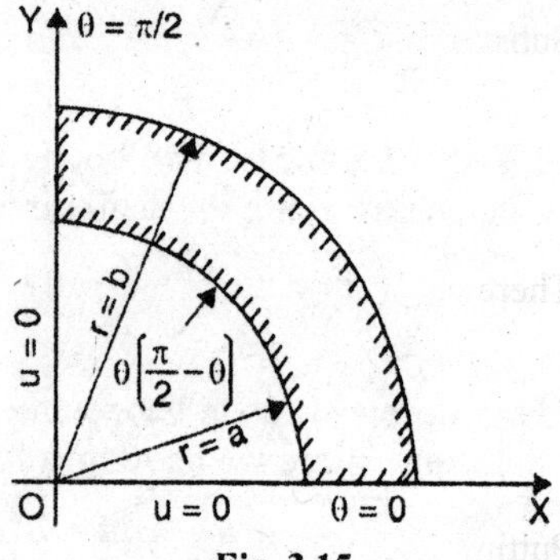

Y $\theta = \pi/2$; $u = 0$; $r = b$; $\theta\left[\frac{\pi}{2} - \theta\right]$; $r = a$; O ; $u = 0$; $\theta = 0$; X

Fig. 3.15

We have to solve (1) subject to the *boundary conditions*

$$\left.\begin{aligned} &(i)\ u(r, 0) = 0 && a \le r \le b \\ &(ii)\ u(r, \pi/2) = 0 && a \le r \le b \\ &(iii)\ u(b, \theta) = 0 && 0 \le \theta \le \pi/2 \\ &(iv)\ u(a, \theta) = \theta\,((\theta/2) - \theta) && 0 \le \theta \le \pi/2 \end{aligned}\right\} \qquad ...(2)$$

The solution of Eq. (1) as discussed earlier in example (3.20) is given by

$$u(r, \theta) = (c_1 r^k + c_2 r^{-k})\,(c_3 \cos k\theta + c_4 \sin k\theta). \qquad ...(3)$$

Using boundary condition 2(*i*), we get

$$0 = (c_1 r^k + c_2 r^{-k})\, c_3. \text{ Therefore, } c_3 = 0.$$

$$u = (c_1' r^k + c_2' r^{-k})\sin k\theta. \quad (c_1' = c_1 c_4; c_2' = c_2 c_4) \qquad ...(4)$$

Applying boundary condition 2(*iii*), we get

$$0 = (c_1' b^k + c_2' b^{-k})\sin k\theta.$$

This gives $$c_1' b^k + c_2' b^{-k} = 0 \quad \text{or} \quad c_2' = -c_1' b^k / b^{-k} = -c_1' b^{2k}.$$

With this value of c_2', Eq. (4) can be written as

$$u(r,\theta) = c_1'(r^k - b^{2k}r^{-k})\sin k\theta = c_1'b^k\left((r/b)^k - (r/b)^{-k}\right)\sin k\theta. \quad ...(5)$$

Again using boundary condition 2(*ii*) i.e., at $\theta = \pi/2$ in Eq. (5) above, we get

$$0 = c_1'b^k[(r/b)^k - (r/b)^{-k}]\sin\frac{k\pi}{2}.$$

This gives $\sin(k\pi/2) = 0 = \sin n\pi$ or $k = 2n$ for all n.

Therefore, $$u(r,\theta) = \sum_{n=1}^{\infty} B_n b^{2n}\left[(r/b)^{2n} - (r/b)^{-2n}\right]\sin 2n\theta. \quad ...(6)$$

To evaluate the constant B_n, we use the boundary condition 2 (*iv*) i.e., at $r = a$ in Eq. (6). This gives

$$u(a,\theta) = \theta\left(\frac{\pi}{2} - \theta\right) = \sum_{n=1}^{\infty} B_n b^{2n}\left[\left(\frac{a}{b}\right)^{2n} - \left(\frac{a}{b}\right)^{-2n}\right]\sin 2n\theta = \sum_{r=1}^{\infty} B_n' \sin 2n\theta \quad ...(7)$$

where $$B_n' = B_n b^{2n}[(a/b)^{2n} - (a/b)^{-2n}].$$

Relation Eq. (7) is a Fourier half range sine series in $(0, \pi/2)$.

Therefore, $$B_n' = \frac{2}{\pi/2}\int_0^{\pi/2} \theta\,(\pi/2 - \theta)\sin 2n\theta\, d\theta$$

$$= \frac{4}{\pi}\left[\left(\frac{\pi\theta}{2} - \theta^2\right)\left(-\frac{\cos 2n\theta}{2n}\right) - \left(\frac{\pi}{2} - 2\theta\right)\left(-\frac{\sin 2n\theta}{4n^2}\right) + (-2)\frac{\cos 2n\theta}{8n^3}\right]_0^{\pi/2}$$

$$= \frac{1}{\pi n^3}(1 - (-1)^n) = \begin{cases} 0, & \text{for } n \text{ even} \\ 2/\pi n^3, & \text{for } n \text{ odd.} \end{cases}$$

Substituting for B_n', B_n is given by

$$B_n b^{2n} = \frac{2}{\pi n^3}\frac{1}{(a/b)^{2n} - (a/b)^{-2n}}.$$

Therefore, $$u(r,\theta) = \sum \frac{2}{\pi n^3}\frac{\left[(r/b)^{2n} - (r/b)^{-2n}\right]}{\left[(a/b)^{2n} - (a/b)^{-2n}\right]}\sin 2n\theta, \qquad n = 1, 3, 5, ...$$

$$= \sum \frac{2}{\pi n^3}\left(\frac{a}{r}\right)^{2n}\left(\frac{r^{4n} - b^{4n}}{a^{4n} - b^{4n}}\right)\sin 2n\theta.$$

Putting $n = 2m - 1$, where $m = 1, 2, 3, ...$, we get

$$u(r,\theta) = \sum_{m=1}^{\infty} \frac{2}{\pi}\frac{1}{(2m-1)^3}\left(\frac{a}{r}\right)^{4m-2}\left(\frac{r^{8m-4} - b^{8m-4}}{a^{8m-4} - b^{8m-4}}\right)\sin(4m-2)\theta.$$

EXAMPLE 3.23 A plate in the form of a sector of a circle is bounded by lines $\theta = 0$, $\theta = \pi/3$ and $r = a$. Its surfaces are insulated and temperature along the boundaries are; $u(r, 0) = 0$; $u(r, \pi/3) = 0$ and $u(a, \theta) = \lambda\theta(\pi/3 - \theta)$. Find the steady state temperature distribution.

Solution Let $u(r, \theta)$ denote the temperature distribution in the plate. The equation governing the heat flow in polar coordinates is given by

$$r^2\frac{\partial^2 u}{\partial r^2} + r\frac{\partial u}{\partial r} + \frac{\partial^2 u}{\partial \theta^2} = 0. \quad ...(1)$$

We seek the solution of (1) subject to the *boundary conditions*.

(*i*) $u(r, 0) = 0$ for all r (*ii*) $u(0, \theta) = 0$, $0 < \theta < \pi/3$

(*iii*) $u(r, \pi/3) = 0$ for all r (*iv*) $u(a, \theta) = \lambda\theta(\pi/3 - \theta)$, for $0 < \theta < \pi/3$.

As discuss in the previous example, the solution of (1) is given by

$$u(r, \theta) = (c_1 r^k + c_2 r^{-k})(c_3 \cos k\theta + c_4 \sin k\theta). \quad ...(2)$$

Using boundary condition (*i*) *i.e.*, $u = 0$ at $\theta = 0$, we get

$$0 = (c_1 r^k + c_2 r^{-k})\, c_3 \text{ which gives } c_3 = 0.$$

Therefore, $u = (c_1' r^k + c_2' r^{-k}) \sin k\theta \cdot \quad (c_1' = c_1 c_4;\ c_2' = c_2 c_4) \quad ...(3)$

Using boundary conditions (*ii*) *i.e.*, $u = 0$ at $r = 0$, we get $c_2' = 0$.

Hence (3) is written as

$$u = c_1' r^k \sin k\theta. \quad ...(4)$$

Again using condition (*iii*) *i.e.*, $u = 0$ at $\theta = \pi/3$, we get $0 = c_1' r^k \sin(\pi k/3)$.

Since $c_1' \neq 0$, $\sin(\pi k/3) = 0 = \sin n\pi$. This gives $k = 3n$.

With $k = 3n$, the solution is given by

$$u = \sum_{n=1}^{\infty} c_n' r^{3n} \sin 3n\theta. \quad ...(5)$$

Using condition (*iv*), that is, $u = \lambda\theta(\pi/3 - \theta)$ when $r = a$, we get

$$\lambda\theta\left(\frac{\pi}{3} - \theta\right) = \sum_{n=1}^{\infty} c_n' a^{3n} \sin 3n\theta. \quad ...(6)$$

Relation (6) is a Fourier half range sine series in $(0, \pi/3)$.

The constant $c_n' a^{3n}$ is given by $c_n' a^{3n} = \dfrac{2}{\pi/3}\displaystyle\int_0^{\pi/3} \lambda\theta(\pi/3 - \theta)\sin 3n\theta\, d\theta$.

Integrating the above integral and substituting the boundary conditions, we get

$$c_n' a^{3n} = \frac{6\lambda}{\pi}\left[\frac{-2(-1)^n}{27n^3} + \frac{2}{27n^3}\right] = \begin{cases} \dfrac{12\lambda}{27\pi n^3}(1 - (-1)^n) = 0, & \text{when } n \text{ is even} \\[2ex] \dfrac{24\lambda}{27\pi n^3}, & \text{when } n \text{ is odd.} \end{cases}$$

Therefore, $c_n' = \dfrac{8\lambda}{9\pi a^{3n} n^3}$ when n is odd.

Putting $n = 2m - 1$ and substituting for c_n' in (5), the solution is given by

$$u(r, \theta) = \frac{8\lambda}{9\pi}\sum_{m=1}^{\infty}\left(\frac{r}{a}\right)^{6m-3}\frac{\sin(6m-3)\theta}{(2m-1)^3}.$$

Y, $\theta = \pi/2$, $\theta = \pi/3$, $u = 0$, $r = a$, $u = \lambda\theta(\pi/3 - \theta)$, $\theta = 0$, O, $u = 0$, X

Fig. 3.16

EXERCISE 3.3

1. A rectangular plate with insulated surfaces is 10 cm wide and so long compared to its width that it may be considered infinite in length without introducing any appreciable error. If the temperature of the short edge $y = 0$ is given by

$$u(x, 0) = \begin{cases} 20x, & 0 \le x \le 5 \\ 20(10 - x), & 5 < x \le 10 \end{cases}$$

while the two long edges $x = 0$ and $x = 10$ as well as the other short edge are kept at 0°C. Find the steady state temperature at any point (x, y) of the plate.

$$\textbf{Ans. } u(x,y) = \frac{800}{\pi^2}\left(e^{-\pi y/10}\sin\frac{\pi x}{10} - \frac{1}{9}e^{-3\pi y/10}\sin\frac{3\pi x}{10} + \ldots\right).$$

2. Solve $\dfrac{\partial^2 u}{\partial x^2} + \dfrac{\partial^2 u}{\partial y^2} = 0$, subject to the conditions $u(0, y) = u(l, y) = u(x, 0) = 0$, and $u(x, a) = \sin(n\pi x)/l$.

[Hint: By example 3.16, the solution is of the form $u = (a_1 \cos kx + a_2 \sin kx)(a_3 e^{ky} + a_4 e^{-ky})$. Now $u(0, y) = 0$ gives $a_1 = 0$, then $u = \sin kx(b_1 e^{ky} + b_2 e^{-ky})$. $u(l, y) = 0$ gives $k = p\pi/l$ (p being an integer).

Therefore, $\quad u = \sin\dfrac{p\pi x}{l}(b_1 e^{p\pi y/l} + b_2 e^{-p\pi y/l})$.

Again $u(x, 0) = 0$ gives $0 = \sin(p\pi x/l)[b_1 + b_2]$.

Therefore, $\quad b_1 + b_2 = 0$ or $b_2 = -b_1$.

Thus $\quad u = b_1 \sin\dfrac{p\pi x}{l}(e^{p\pi y/l} - e^{-p\pi y/l}) = c_p \sin\dfrac{p\pi x}{l}\sinh\dfrac{p\pi y}{l}$, where $2b_1 = c_p$.

Hence, the general solution is $u = \displaystyle\sum_{p=1}^{\infty} c_p \sin\frac{p\pi x}{l}\sinh\frac{p\pi y}{l}$.

Lastly $u(x, a) = \sin\dfrac{n\pi x}{l}$. $\quad$ Therefore, $\sin\dfrac{n\pi x}{l} = \displaystyle\sum_{p=1}^{\infty} c_p \sin\frac{p\pi x}{l}\sinh\frac{p\pi a}{l}$.

Equating coefficients, $c_n \sinh\dfrac{n\pi a}{l} = 1$ or $c_n = 1\Big/\sinh\dfrac{n\pi a}{l}$, while other c's are zero.

Therefore, $\quad u(x, y) = \sin\dfrac{n\pi x}{l}\left(\sinh\dfrac{n\pi y}{l}\Big/\sinh\dfrac{n\pi a}{l}\right)$.

3. A rectangular plate is bounded by the lines $x = 0$, $y = 0$, $x = a$, $y = b$. Its surfaces are insulated and the temperature along the upper horizontal edge is 100°C while the other three edges are kept at 0°C. Find the steady state temperature function $u(x, y)$.

$$\textbf{Ans. } u = \frac{400}{\pi}\sum_{n=1}^{\infty}\frac{\sin\{(2n-1)\pi x/a\}\sinh\{(2n-1)\pi y/a\}}{(2n-1)\sinh\{(2n-1)\pi b/a\}}.$$

4. Find the steady state temperature at any point of a square plate if two adjacent edges are kept at 0°C and the other two at 100°C.

$$\textbf{Ans. } u(x,y) = \frac{400}{\pi}\sum_{1}^{\infty}\frac{1}{(2n-1)\sinh(2n-1)\pi} \times$$

$$\left[\sin\frac{(2n-1)\pi x}{l}\sinh\frac{(2n-1)\pi y}{l} + \sin(2n-1)\frac{\pi y}{l}\sinh(2n-1)\frac{\pi x}{l}\right],$$

where l is the length of one side.

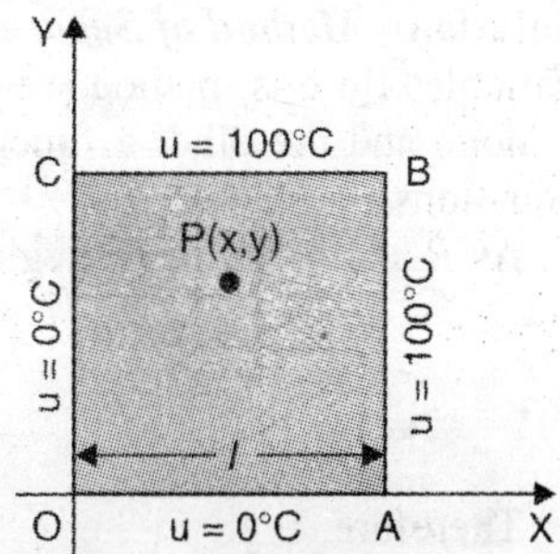

Fig. 3.17

[Hint: Assume $u(x, y) = u_1(x, y) + u_2(x, y)$ where $u_1(x, y)$ and $u_2(x, y)$ are solutions of $\dfrac{\partial^2 u}{\partial x^2} + \dfrac{\partial^2 u}{\partial y^2} = 0$ and further $u_1(x, y)$ is the temperature at $P(x, y)$ with edge BC kept at 100°C and the other three edges at 0°C, while $u_2(x, y)$ is the temperature at $P(x, y)$ with edge AB kept at 100°C and the other three edges at 0°C].

5. A rectangular metal plate has sides of lengths a and b. The sides $x = 0, x = a$, and $y = b$ are insulated while the edge $y = 0$ is kept at temperature $u_0 \cos(\pi x/a)$. Find the temperature $u(x, y)$ in steady-state.

Ans. $u(x, y) = u_0 \cos\frac{\pi x}{a} \operatorname{sech}\frac{\pi b}{a} \cosh\left\{\frac{\pi(y-b)}{a}\right\}$. [**Note:** $\cosh a \cosh b \pm \sinh a \sinh b = \cosh(a \pm b)$].

6. A thin semi-circular plate of radius a has its bounding diameter kept at temperature zero and its circumference at k. Find the temperature distribution in the steady state.

Ans. $u(r, \theta) = \frac{4k}{\pi} \sum_{n=1,3,\ldots}^{\infty} \left[\left(\frac{r}{a}\right)^n \frac{1}{n} \sin n\theta\right]$.

7. Obtain the steady state temperature distribution in a semi-circular plate, with radius a whose bounding diameter is kept at 0°C while the circumference is kept at 50°C.

Ans. $u = \frac{200}{\pi} \sum_{n=1}^{\infty} \frac{(r/a)^{2n-1}}{2n-1} \sin(2n-1)\theta$.

8. A semi-circular plate of radius 10 cm has insulated faces and heat flows in plane curves. The bounding diameter is kept at 0°C and on the circumference the temperature distribution is maintained at $u(10, \theta) = 400(\pi\theta - \theta^2)/\pi, \quad 0 \le \theta \le \pi$. Determine the temperature distribution $u(r, \theta)$.

Ans. $u(r, \theta) = \frac{3200}{\pi^2} \sum_{n=1}^{\infty} \left[\frac{1}{(2n-1)^3} \left(\frac{r}{10}\right)^{2n-1} \sin(2n-1)\theta\right]$.

9. A semi-circular plate of radius a cm has insulated faces and heat flows in plane curves. The bounding diameter is kept at 0° C and the semi-circumference is maintained at a temperature given by

$$k\theta/\pi \text{ when } 0 \le \theta \le \pi/2 \quad \text{and} \quad k(\pi - \theta)/\pi \quad \text{when } \pi/2 \le \theta \le \pi.$$

Find the temperature distribution in the steady state.

Ans. $u(r, \theta) = \frac{4k}{\pi^2} \sum_{n=1}^{\infty} \frac{(-1)^{n-1}}{(2n-1)^2} \left(\frac{r}{a}\right)^{2n-1} \sin(2n-1)\theta$.

Solution of Transmission Line Equations Problems by the Method of Separation of Variables

EXAMPLE 3.24 Using the method of separation of variables find the current i and voltatge V in a transmission line of l length, t seconds after the ends are suddenly grounded, given that $I(x, 0) = I_0$, $V(x, 0) = \sin(\pi x/l)$ and that R and G are negligible.

Solution ***Method of Separation of Variables.*** Let z be the dependent variable and x and y two independent variables. In this method we assume the solution to be product of two functions, one of them a function of x alone and the other a function of y alone. In this way the solution is reduced to the solution of ordinary equations.

As R and G are to be neglected we use the equations (Radio equations)

$$\frac{\partial V}{\partial x} = -L\frac{\partial i}{\partial t} \text{ and } \frac{\partial i}{\partial x} = -C\frac{\partial V}{\partial t}.$$

Therefore,

$$\frac{\partial^2 V}{\partial x^2} = -L\frac{\partial^2 i}{\partial x \partial t} \text{ and } \frac{\partial^2 i}{\partial x \partial t} = -C\frac{\partial^2 V}{\partial t^2}.$$

Eliminating i between these equations, we get

$$\frac{\partial^2 V}{\partial x^2} = LC\frac{\partial^2 V}{\partial t^2}. \quad \ldots(1)$$

Since the ends are suddenly grounded, we have

$$V(0, t) = 0, V(l, t) = 0. \quad ...(2)$$

Also the *initial conditions* are $V(x,0) = V_0 \sin\frac{\pi x}{l}, \quad i(x,0) = i_0$. ...(3)

Therefore, $\frac{\partial i(x,0)}{\partial x} = 0$ gives $\frac{\partial V(x,0)}{\partial t} = 0 \quad \left(\because \quad \frac{\partial i}{\partial x} = -C\frac{\partial V}{\partial t}\right)$. ...(4)

Let $V = X(x)\,T(t)$ be the solution of (1).

Substituting in (1) we get $X''T = LCXT''$.

Therefore, $\frac{X''}{X} = LC\frac{T''}{T} = -k^2$ (say). $\Rightarrow X'' + k^2X = 0$ and $LCT'' + K^2T = 0$.

Solving these equations, we get

$$V = (c_1 \cos kx + c_2 \sin kx)\left(c_3 \cos\frac{k}{\sqrt{LC}}t + c_4 \sin\frac{k}{\sqrt{LC}}t\right).$$

Using the boundary conditions (2) we get $c_1 = 0$ and $k = n\pi/l$.

Therefore, $$V = \sin\frac{n\pi x}{l}\left(A_n \cos\frac{n\pi}{l\sqrt{LC}}t + B_n \sin\frac{n\pi}{l\sqrt{LC}}t\right).$$

Using the initial conditions (4) we get $B_n = 0$.

Therefore, $$V = A_n \sin\frac{n\pi x}{l}\cos\frac{n\pi t}{l\sqrt{LC}}.$$

By the initial condition (3) $V_0 \sin\frac{\pi x}{l} = A_n \sin\frac{n\pi x}{l}$. Therefore, $A_n = V_0$, $n = 1$.

Hence $V(x,t) = V_0 \sin\frac{\pi x}{l}\cos\frac{\pi t}{l\sqrt{LC}}$ is the required voltage.

To get i we use the Eq. (1).

From the first, we get $\frac{\partial i}{\partial t} = -\frac{1}{L}\frac{\partial V}{\partial x} = -\frac{V_0\pi}{Ll}\cos\frac{\pi x}{l}\cos\frac{\pi t}{l\sqrt{LC}}$. ...(5)

From the second $\frac{\partial i}{\partial x} = -C\frac{\partial V}{dt} = \frac{CV_0\pi}{l\sqrt{LC}}\sin\frac{\pi x}{l}\sin\frac{\pi t}{l\sqrt{LC}}$.. ...(6)

Integrating (6) w.r.t. 't' we get

$$i(x,t) = -V_0\sqrt{\frac{C}{L}}\cos\frac{\pi x}{l}\sin\frac{\pi t}{l\sqrt{LC}} + \phi(x),$$

where $\phi(x)$ is an arbitrary function of x.

This value satisfies (6) if $\phi'(x) = 0$, that is, $\phi(x)$ is constant. From (3) it is i_0.

Therefore, $$i(x,t) = i_0 - V_0\sqrt{\frac{C}{L}}\cos\frac{\pi x}{l}\sin\frac{\pi t}{l\sqrt{LC}}.$$

EXAMPLE 3.25 Neglecting R and G, find the e.m.f. $V(x, t)$ in a line of length l, t seconds after the ends were suddenly grounded, given that

$$i(x,0) = i_0 \text{ and } V(x,0) = E_1 \sin\frac{\pi x}{l} + E_5 \sin\frac{5\pi x}{l}. \qquad (SVTU, 2007)$$

Solution Since R and G are negligible, we use the *Radio equation* $\frac{\partial^2 V}{\partial x^2} = LC\frac{\partial^2 V}{\partial t^2}$. ...(*i*)

Since the ends are suddenly grounded, we have the boundary conditions *viz.* $V(0, t) = 0$, $V(l, t) = 0$. ...(*ii*)

Also the initial conditions are $i(x, 0) = i_0$ and $V(x, 0) = E_1 \sin(\pi x/l) + E_5 \sin(5\pi x/l)$. ...(*iii*)

Now $\frac{\partial i}{\partial x}(x, 0) = -C\frac{\partial V}{\partial t}$ gives $\frac{\partial V}{\partial t}(x, 0) = 0$. ...(*iv*)

Let $V = X(x)\,T(t)$ be the solution of (*i*). Therefore, (*i*) gives $X''T = LCXT''$ or $\frac{X''}{X} = LC\frac{T''}{T} = -k^2$ (say).

Therefore, $X'' + k^2X = 0$ and $T'' + (k^2/LC)\,T = 0$. Solving these equations, we get

$$V = (c_1 \cos kx + c_2 \sin kx)\left(c_3 \cos\frac{k}{\sqrt{LC}}\cdot t + c_4 \sin\frac{k}{\sqrt{LC}}t\right).$$

Using the boundary conditions (*ii*), we get $c_1 = 0$ and $k = n\pi/l$.

Therefore, $$V = \sin\frac{n\pi x}{l}\left(a_n \cos\frac{n\pi}{l\sqrt{LC}}t + b_n \sin\frac{n\pi}{l\sqrt{LC}}t\right).$$

Using the initial condition $\frac{\partial V}{\partial t}(x, 0) = 0$ gives $b_n = 0$.

Therefore, $V = a_n \sin\frac{n\pi x}{l}\cos\frac{n\pi t}{l\sqrt{LC}}$ is a solution.

Thus the most general solution of (*i*) is $V = \sum_{n=1}^{\infty} a_n \sin\frac{n\pi x}{l}\cos\frac{n\pi t}{l\sqrt{LC}}$.

Finally by the initial condition (*iii*), we have

$$E_1 \sin\frac{\pi x}{l} + E_5 \sin\frac{5\pi x}{l} = \sum a_n \sin\frac{n\pi x}{l}.$$

Therefore, $a_1 = E_1$, $a_5 = E_5$, while all other a's are zero.

Hence, $V(x, t) = E_1 \sin\frac{\pi x}{l}\cos\frac{\pi t}{l\sqrt{LC}} + E_5 \sin\frac{5\pi x}{l}\cos\frac{5\pi t}{l\sqrt{LC}}$, is the required solution.

EXAMPLE 3.26 Solve the telephone equation $\frac{\partial^2 E}{\partial x^2} = LC\frac{\partial^2 E}{\partial t^2} + RC\frac{\partial E}{\partial t}$ when $E(0, t) = E_0 \sin qt$, $E = 0$ as $x \to \infty$ assuming that qL/R is large compared with unity. (*AMIETE, June 2011, 2008*)

Solution Let $E(x, t) = Ae^{px}e^{iqt}$ be the solution of the given equation. Substituting in the given equation, we obtain

$$p^2 = -LCq^2\left(1 - \frac{iR}{qL}\right).$$

Hence, $$p = \pm iq\sqrt{LC}\left(1 - \frac{iR}{qL}\right)^{1/2} = \pm iq\sqrt{LC}\left(1 - \frac{1}{2}\frac{iR}{qL} + \textit{higher power of } \frac{R}{qL}\right)$$

$$\approx \pm\left\{iq\sqrt{LC} + \frac{1}{2}R\sqrt{C/L}, \text{ since } \frac{R}{qL} \text{ is small.}\right.$$

It p is positive, substitution in the assumed solution shows that $E(x, t) \to \infty$ as $x \to \infty$, which is contrary to the boundary condition $E = 0$ for $x = \infty$. Hence, we omit the positive sign. Since q can also be negative, we have the general solution

$$E(x, t) = e^{-\frac{xR\sqrt{C}}{2\sqrt{L}}}\left\{Ae^{iqt + iqx\sqrt{LC}} + Be^{-iqt - iqx\sqrt{LC}}\right\}.$$

At $x = 0$, $E = E_0 \sin qt$. Hence $E_0 \sin qt = Ae^{iqt} + Be^{-iqt}$ giving $A = -B = E_0/2i$.

Therefore, the required solution is

$$E(x,t) = E_0\, e^{-\frac{xR\sqrt{C}}{2\sqrt{L}}} \left\{ e^{i(qt + qx\sqrt{LC})} - e^{-i(qt + qx\sqrt{LC})} \right\}$$

or
$$E(x,t) = E_0\, e^{-x\frac{R\sqrt{C}}{2\sqrt{L}}} \sin\,(qt + qx\sqrt{LC}).$$

EXAMPLE 3.27 A transmission line 1000 km long is initially under steady-state conditions with potential 1300 volts at the sending end ($x = 0$) and 1200 volts at the receiving end ($x = 1000$). The terminal end of the line is suddenly grounded, but the potential at the source is kept at 1300 volts. Assuming the inductance and leakance to be negligible, find the potential $V(x, t)$. *(AMIETE, June 2012, Dec. 2009; Andhra, 2000)*

Solution We use the telegraph equation

$$\frac{\partial^2 V}{\partial x^2} = RC\frac{\partial V}{\partial t} \quad \text{or} \quad \frac{\partial V}{\partial t} = \frac{1}{RC}\cdot\frac{\partial^2 V}{\partial x^2}. \qquad ...(i)$$

V_s = initial steady voltage satisfying $\dfrac{\partial^2 V}{\partial x^2} = 0 = 1300 - \dfrac{1300 - 1200}{1000}x = 1300 - 0.1x = V(x, 0)$. ...(*ii*)

V'_s = Steady voltage (after grounding the terminal end) when steady conditions are ultimately reached $= 1300 - 1.3x$.

Therefore, $V(x,t) = V'_s(x) + V_t(x,t)$, where $V_t(x, t)$ is the transient part

$$= (1300 - 1.3x) + \sum_{n=1}^{\infty} b_n \sin\frac{n\pi x}{l} \exp\left\{\frac{-n^2\pi^2 t}{l^2 RC}\right\}. \qquad ...(iii)$$

Putting $t = 0$ in (*iii*), we have

or
$$1300 - 0.1x = V(x,0) = 1300 - 1.3x + \sum_{n=1}^{\infty} b_n \sin\frac{n\pi x}{l}$$

or
$$1.2x = \sum_{n=1}^{\infty} b_n \sin\frac{n\pi x}{l}, \quad \text{where } b_n = \frac{2}{l}\int_0^l 1.2 \sin\frac{n\pi x}{l}\,dx = \frac{2400}{\pi}\frac{(-1)^{n+1}}{n}.$$

Hence
$$V(x,t) = 1300 - 1.3x + \frac{2400}{\pi}\sum_{n=1}^{\infty}\frac{(-1)^{n+1}}{n}\exp\left\{\frac{-n^2\pi^2 t}{1000^2 RC}\right\}\sin\frac{n\pi x}{1000}.$$

EXERCISE 3.4

1. In a telegraph wire, the sending end of the line is at a potential e_o, the far end being earthed until steady state conditions prevail. The sending end is suddenly earthed. Show that the potential at a point distant x from the sending end is given by

$$e(x,t) = \frac{2e_0}{\pi}\sum_{n=1}^{\infty}\frac{1}{n}\exp\left\{-\frac{n^2\pi^2 t}{CRl^2}\right\}\sin\frac{n\pi x}{l},$$

where l is the length of the wire and C, R have their usual meanings.
[**Hint:** Writing e in place of V in the formula of Art. 3.1-4(b).]

we have to solve $\dfrac{\partial^2 e}{\partial x^2} = RC\dfrac{\partial e}{\partial t}$ or $\dfrac{\partial e}{\partial t} = \dfrac{1}{RC}\dfrac{\partial^2 e}{\partial x^2} = c^2\dfrac{\partial^2 e}{\partial x^2}$ (say).

In steady state $e = \frac{e_0}{l}(l - x)$. Hence the *boundary conditions* under which $\frac{\partial e}{\partial t} = c^2 \frac{\partial^2 e}{\partial x^2}$ is to be solved are $e(0\ t) = e(l, t) = 0\ (t \geq 0)$ and $e(x, 0) = e_0(l - x)/l$ for all x in $(0, l)$.

Just like the solution of one dimensional heat equation, we have $e(x, t) = (A \cos kx + B \sin kx)e^{-k^2c^2t}$.

Using the first two boundary conditions the general solution is $e(x, t) = \sum b_n \sin \frac{n\pi x}{l} \exp\left(-\frac{n^2\pi^2c^2t}{l^2}\right)$.

Last boundary condition gives $\frac{e_0}{l}(l - x) = \sum b_n \sin \frac{n\pi x}{l}$, where $b_n = 2e_0/(n\pi)$.

Therefore, $e(x, t) = \sum_{n=1}^{\infty} \frac{2e_0}{n\pi} \sin \frac{n\pi x}{l} \exp\left(-\frac{n^2\pi^2c^2t}{l^2}\right) = \frac{2e_0}{n\pi} \sum_{n=1}^{\infty} \frac{1}{n} \sin \frac{n\pi x}{l} \exp\left\{\frac{-n^2\pi^2t}{RCl^2}\right\}$.]

2. A telegraph cable is of length l. Initially the line is uncharged so that $V(x, 0) = 0$. If at $t = 0$ the end $x = l$ is connected to a constant e.m.f. E, find $V(x, t)$ and $i(x, t)$. In particular, show that the current at the receiving end $x = 0$ is given by

$$i(0, t) = \frac{-E}{lR} + \frac{2E}{lR} \sum_{n=1}^{\infty} \left\{(-1)^{n+1} \exp\left(\frac{-n^2\pi^2t}{l^2RC}\right)\right\}.$$

Ans. $V(x, t) = \frac{Ex}{l} + 2E \sum_{n=1}^{\infty} \frac{(-1)^n}{n\pi} \sin \frac{n\pi x}{l} \exp\left\{-\frac{n^2\pi^2t}{l^2RC}\right\}$;

$$i(x, t) = -\frac{E}{lR} + \frac{2E}{lR} \sum_{n=1}^{\infty} (-1)^{n+1} \cos \frac{n\pi x}{l} \exp\left\{-\frac{n^2\pi^2t}{l^2RC}\right\}.$$

[**Hint:** Here we have to solve the telegraph equation $\frac{\partial^2 V}{\partial x^2} = RC \frac{\partial V}{\partial t}$. ...(1)

Initially $V(x, 0) = 0$ and finally when the steady state conditions prevail, $V_s = Ex/l$.

To get V in the intermediate period, reckoning time from the instant when the end is connected to e.m.f. E, we assume

$$V(x, t) = V_t(x, t) + V_s(x)$$

where V_s is the steady state solution at the end and V_t is the transient part of the solution which satisfies (1) and tends to zero as $t \to \infty$ and which satisfies the conditions

$$V_t(0, t) = 0 \text{ and } V_t(l, t) = E].$$

3. A telephone wire of length l is maintained at a steady state voltage distribution of 20 volts at one end and 12 volts at the other end. The latter end is suddenly grounded at time $t = 0$. Neglect leakage and inductance find the voltage and the current in the wire t seconds later.

Ans. $V(x, t) = \frac{20(l - x)}{l} + \frac{24}{\pi} \sum_{n=1}^{\infty} \frac{(-1)^{n+1}}{n} \sin \frac{n\pi x}{l} \cdot e^{-n^2\pi^2t/Rcl^2}$,

$$i(x, t) = \frac{20}{lR} + \frac{24}{lR} \sum_{n=1}^{\infty} (-1)^n \cos \frac{n\pi x}{l} e^{-n^2\pi^2t/RCl^2}.$$

4. Obtain the solution of the *radio equation*, $\frac{\partial^2 V}{\partial x^2} = LC \frac{\partial^2 V}{\partial t^2}$ appropriate to the case when a periodic e.m.f. $V_0 \cos pt$ is applied at the end $x = 0$ of the line. **Ans.** $V(x, t) = V_0 \cos(pt - p\sqrt{LC}\, x)$.

CHAPTER

4

Complex Analysis and Complex Integration

4.1 INTRODUCTION

The theory of functions of a complex variable is fundamental to the subject of applied mathematics and forms a solid and an essential background of engineers, scientists and mathematicians, Its increased application is of tremendous value in the solution of problems concerning areas of fluid mechanics, heat flow, electric fields, aerodynamics, elasticity and many other branches of science and engineering.

4.2 FUNCTIONS OF A COMPLEX VARIABLE : ANALYTIC FUNCTIONS

We are familiar with some aspects of complex numbers and defined a complex number as an expression of the form $z = x + iy$, where x and y are real numbers and i is represented by $i = \sqrt{-1}$ and possesses the property that $i^2 = -1$. The number x is called the *real part* of z and is denoted by $x = Re\ (z)$; the number y is the *imaginary part* of z and is represented by $y = Im(z)$. If either x or y or both vary, then z is defined as a *complex variable.*

Consider two complex variables z and w defined by $z = x + iy$ and $w = u + iv$. If for every value of z in a certain region R of the z-plane there corresponds one or more values of w in the w-plane, then w is said to be a *function* of z and is represented by

$$w = f(z)$$

or

$$w = u + iv = f(z) = f(x + iy) \qquad ...(4.1)$$

where u and v are real functions of x and y. It may be noted the complex variables z and w are represented on separate complex planes, called respectively as the z-plane and the w-plane. The functional relationship $w = f(z)$, therefore, establishes a correspondence between the points (x, y) of the z-plane and the points (u, v) of the w-plane.

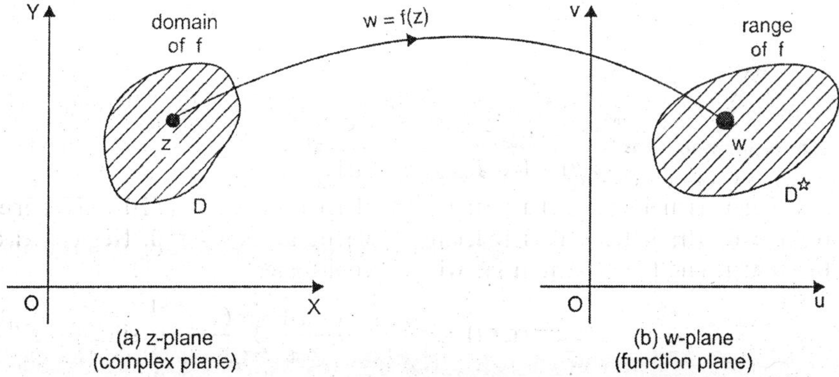

Fig. 4.1 and Fig. 4.2
z-plane and the w-plane. A function $w = f(z)$.

If one value of z corresponds precisely to a unique value of w, according to relation (4.1), then w is said to be a *single-valued function* of z. If however, one value of z yields more values of w, then $w = f(z)$ is called a *many-valued function* or *multi-valued function* of z. For example $w = z^2$, $w = \dfrac{1}{z^3}$, $w = \dfrac{z}{|z| - 4}$ are single-valued function of z while $w = z^{1/2}$, $w = \arg(z)$ are multi or many-valued function of z. In the relation $w = z^{1/2}$ one value of z results in two values of w, except at $z = 0$, whereas a single value of z in the relation $w = \arg(z)$ yields an infinite number of values of w, except at $z = 0$. It may be remarked that the function $w = f(z)$ may not be defined at all points of the z-plane. The function given by the relation $w = \dfrac{z}{|z| - 4}$ is defined at all points of the z-plane except for points on the circle $|z| = 4$ or $x^2 + y^2 = 16$ where it is not defined.

4.3 LIMITS OF A FUNCTION

Let $f(z)$ be a single-valued function of z defined in some neighbourhood of the complex point $z = z_0$ except possible at the point $z = z_0$ itself. We say that $f(z)$ tends to the limit ℓ as z tends to the point z_0 if for any positive number ε, however small, there exists a positive real number δ such that

$$|f(z) - \ell| < \varepsilon \quad \text{whenever} \quad 0 < |z - z_0| < \delta.$$

In other words, we can say that $f(z) \to \ell$ as $z \to z_0$, or $\lim_{z \to z_0} f(z) = \ell$.

Since ε is arbitrary and chosen to be small, the above definition implies that the difference in the absolute value of $f(z)$ and l can be made as small as we wish by choosing the point z sufficiently close to the point z_0, no matter in whichever manner z approaches the point z_0.

EXAMPLE 4.1 If $f(z) = \dfrac{\bar{z}}{z}$, show that when $z \to 0$, $f(z)$ does not exist.

Solution We have $f(z) = \dfrac{\bar{z}}{z} = \dfrac{x - iy}{x + iy}$. Now let $z \to 0$ along the real axis. Then $y = 0$ and $z = x + iy = x$. Hence the required limit is

$$\lim_{z \to 0} f(z) = \frac{x}{x} = 1.$$

Again let $z \to 0$ along the imaginary axis, so that $x = 0$ and $z = x + iy = iy$. Hence the limit is given by

$$\lim_{z \to 0} f(z) = \frac{-iy}{iy} = -1.$$

Thus we notice that these limits are different when z tends to zero along two different paths. But if the limit is to exist it must be unique and have the same value in whichever manner z approaches zero. Hence the limit does not exist.

4.4 CONTINUITY OF A FUNCTION

The concept of continuity is closely linked with the concept of the limit. Suppose $f(z)$ be a single-valued function of z defined in a certain neighbourhood of the point $z = z_0$ and also at $z = z_0$ (that is, in a δ neighbourhood of z_0). The function $f(z)$ is said to be continuous at the point $z = z_0$ when it satisfies each of the following conditions.

(*i*) $f(z_0)$ must exist.

(*ii*) $\lim_{z \to z_0} f(z)$ must exist, that is, $f(z)$ is defined at z_0.

(*iii*) $\lim_{z \to z_0} f(z) = f(z_0)$.

In other words, a function $f(z)$ is continuous at z_0 if, given any positive number ε, we can find a number δ, depending upon ε and z_0, such that

$$|f(z) - f(z_0)| < \varepsilon$$

for all points of the domain R satisfying the condition that $|z - z_0| < \delta$.

We state below a few theorems concerning continuous functions.

(*i*) A function $f(z)$ is said to be *continuous* in a certain *region* R of the complex z-plane if it is continuous at every point of the region.

(*ii*) By definition, $w = f(z) = u(x, y) + iv(x, y)$. If $f(z)$ is continuous at the point $z = z_0$, then $u(x, y)$ and

$v(x, y)$ which are the real and imaginary part of $f(z)$ will also be continuous at $z = z_0$, that is, at $x = x_0$ and $y = y_0$. Conversely if $u(x, y)$ and $v(x, y)$ are continuous functions at $z = z_0$, then $f(z)$ will be a continuous at $z = z_0$.

(*iii*) The sum, the difference and the product of continuous functions is also continuous. The quotient of two continuous functions is also a continuous function except for those values of z which render the denominator zero.

Note: A function $f(x, y)$ is continuous at a point (x_0, y_0), if

(*i*) $f(x_0, y_0)$ is finite, and

(*ii*) $\lim\limits_{\substack{x \to x_0 \\ y \to y_0}} f(x, y) = f(x_0, y_0)$, in whatever manner $x \to x_0$ and $y \to y_0$.

EXAMPLE 4.2 Show that $f(z) = \dfrac{xy^3}{x^2 + y^6}$ is discontinuous at (0, 0), given that $f(0) = 0$.

Solution Let $x \to 0$ first and then $y \to 0$, we have

$$\lim_{y \to 0} \lim_{x \to 0} \frac{xy^3}{x^2 + y^6} = \lim_{y \to 0} \frac{y^3 \times 0}{y^6} = 0.$$

Again let $y \to 0$ first and then $x \to 0$, we get

$$\lim_{x \to 0} \lim_{y \to 0} \frac{xy^3}{x^2 + y^6} = \lim_{x \to 0} \frac{x \times 0}{x^2} = 0.$$

Next let x and y both tend to zero simultaneously along the curve $x = y^3$. Then

$$\lim_{\substack{x \to y^3 \\ y \to 0}} \frac{xy^3}{x^2 + y^6} = \lim_{y \to 0} \frac{y^6}{2y^6} = \frac{1}{2} \neq 0.$$

Hence $\lim\limits_{x \to 0} \dfrac{xy^3}{x^2 + y^6}$ is not zero. When x and y simultaneously approach zero in any manner. The function $f(z)$ is therefore, not continuous at the origin.

EXAMPLE 4.3 Examine the function $f(z) = \begin{cases} \dfrac{\operatorname{Im} z}{|z|} & \text{when } z \neq 0 \\ 0 & \text{when } z = 0 \end{cases}$ for continuity at $z = 0$.

Solution Let $z = x + iy$. We have $f(z) = f(x + iy) = \dfrac{y}{\sqrt{x^2 + y^2}}$, when $z \neq 0$.

Let z approach 0 along the line $y = mx$ so that $z = x + imx$.

Then $f(z) = f(x + imx) = \dfrac{mx}{\sqrt{x^2 + m^2x^2}} = \dfrac{m}{\sqrt{1 + m^2}}$.

This takes different values for different values of m. Hence the $\lim\limits_{z \to 0} f(z)$ does not exist.

Thus, $f(z)$ is not continuous at $z = 0$.

EXAMPLE 4.4 If $f(z) = \dfrac{x^3 y (y - ix)}{x^6 + y^2}$, $z \neq 0$, and $f(0) = 0$. Prove that $\dfrac{f(z) - f(0)}{z} \to 0$ as $z \to 0$ along any radius vector, but not as $z \to 0$ in any manner.

Solution Since $f(0) = 0$, we have

$$\frac{f(z)-f(0)}{z}=\frac{x^3y\,(y-ix)}{x^6+y^2}\cdot\frac{1}{x+iy}$$ [since $z = x + iy$]

or $$\frac{f(z)-f(0)}{z}=\frac{-ix^3y\,(x+iy)}{x^6+y^2}\cdot\frac{1}{x+iy}=-\frac{ix^3y}{x^6+y^2}.$$

Let $z \to 0$ along any radius vector $y = mx$, then

$$\lim_{z\to 0}\frac{f(z)-f(0)}{z}=\lim_{\substack{y=mx\\x\to 0}}\left\{\frac{-ix^3y}{x^6+y^2}\right\}=\lim_{x\to 0}\frac{-imx^4}{x^6+m^2x^2}=0.$$

Again let $z \to 0$ along the path $y = x^3$, then

$$\lim_{z\to 0}\frac{f(z)-f(0)}{z}=\lim_{\substack{y=x^3\\x\to 0}}\left\{\frac{-ix^3y}{x^6+y^2}\right\}=\lim_{x\to 0}\frac{-ix^6}{2x^6}=-\frac{i}{2}.$$

Hence the result.

EXERCISE 4.1

1. Show that the following limits donot exist

(a) $\lim_{z\to 0}\dfrac{[\text{Re}\,z-\text{Im}\,z]^2}{|z|^2}$, (b) $\lim_{z\to 0}\dfrac{z}{|z|}$.

2. Prove that $\lim_{z\to i}\dfrac{3z^4-2z^3+8z^2-2z+5}{z-i}=4\,(1+i)$.

3. Find the limit of each of the following:

(a) $\lim_{z\to i}\dfrac{2z}{z+i}$, (b) $\lim_{z\to 1+i}(z^2-5z+10)$,

(c) $\lim_{z\to 2e^{i\pi/3}}\dfrac{z^3+8}{z^4+4z^2+16}$, (d) $\lim_{z\to e^{i\pi/3}}(z-e^{i\pi/3})\dfrac{z}{z^3+1}$,

(e) $\lim_{z\to\infty}\left(\dfrac{2z^4+1}{z^4+1}\right)$, (f) $\lim_{z\to i\pi/2}z^2\cosh(4z/3)$,

(g) $\lim_{z\to i/2}\dfrac{(2z-3)(4z+i)}{(iz-1)^2}$, (h) $\lim_{z\to 1+2i}\left(\dfrac{z^2-2z+5}{z^2+3-4i}\right)$.

$$\left[\textbf{Hint (h)}: f(z)=\frac{(z-1)^2-(2i)^2}{z^2-(1+2i)^2}=\frac{(z-1+2i)(z-1-2i)}{(z+1+2i)(z-1-2i)}\cdot\ \lim_{z\to 1+2i}f(z)=\frac{4i}{2+4i}=\frac{2}{5}(2+i).\right]$$

Ans. (a) $1-i$, (b) $5-3i$, (c) $\dfrac{3}{8}-\dfrac{\sqrt{3}}{8}i$, (d) $\dfrac{1}{6}-i\dfrac{\sqrt{3}}{6}$, (e) 2, (f) $\pi^2/8$, (g) $-\dfrac{4}{3}-4i$, (h) $\dfrac{2}{5}(2+i)$.

4. If $f(z)=\dfrac{2z-1}{3z+2}$, prove that $\lim_{h\to 0}\dfrac{f(z_0+h)-f(z_0)}{h}=\dfrac{7}{(3z_0+2)^2}$, provided $z_0\neq -2/3$.

5. Prove that if $f(z)=3z^2+2z$, then $\lim_{z\to z_0}\dfrac{f(z)-f(z_0)}{z-z_0}=6z_0+2$.

6. Show that the function $f(z)=\dfrac{x^2y\,(y-x)}{(x^6+y^2)(x+y)}$ is discontinuous at (0, 0), given that $f(0) = 0$.

7. Show that the following functions are discontinuous at the origin

(a) $f(z)=\dfrac{2xy}{x^2+y^2}$, $z \neq 0$; $f(0)=0$, (b) $f(z)=\dfrac{5xy^2}{x^2+y^4}$, $z \neq 0$; $f(0)=0$.

8. Discuss the continuity of the following functions at the origin

(a) $f(x, y)=\dfrac{3xy}{\sqrt{x^2+y^2}}$ for every x and y excepting (0, 0) and $f(0, 0) = 0$.

(b) $f(z)=\dfrac{2xy\,(x+y)}{x^2+y^2}$, $z \neq 0$ and $f(0) = 0$.

9. Prove that $f(z)=\dfrac{z}{z^4+1}$ is continuous at all points inside and on the unit circle $|z| = 1$ except at four points, and determine these points.

10. Show that $f(z)=\dfrac{z^2+1}{z^2-3z+2}$ is continuous for all z outside the circle $|z| = 2$.

4.5 COMPLEX DIFFERENTIATION

Let $w = f(z)$ be a single-valued function of z defined in a certain region R of the complex z-plane. Then, the *derivative* or the *differential coefficient* of w with respect to z is defined as

$$\frac{dw}{dz} = f'(z) = \lim_{\Delta z \to 0} \frac{f(z+\Delta z) - f(z)}{\Delta z} \qquad \text{...(4.2)}$$

provided the above limit exists uniquely and has the same value for all the ways $\Delta z \to 0$. If $[f(z+\Delta z) - f(z)]$ is denoted by Δw, then the derivative of $w = f(z)$ is also defined as

$$\frac{dw}{dz} = \lim_{\Delta z \to 0} \frac{\Delta w}{\Delta z}. \qquad \text{...(4.3)}$$

In the complex z-plane Δz may tend to zero in infinitely many ways. To show this let P be any point of the region R corresponding to $z = x + iy$ and Q be the neighbouring point given by $z + \Delta z$ where $\Delta z = \Delta x + i\Delta y$ and Δx, Δy are small increments in x and y respectively. As $\Delta z \to 0$, the point Q approaches P by moving either along a straight path or along any curved path. Fig. 4.3.

As a particular case, suppose Q approaches P along the path QAP in which case $\Delta y \to 0$ first and then $\Delta x \to 0$. In other words the point Q may be assumed to be lying on the horizontal line through P. Alternatively, Q may approach P along the path QBP where first Δx and then Δy approach zero, i.e. the point Q may be assumed to be lying on the vertical line through P.

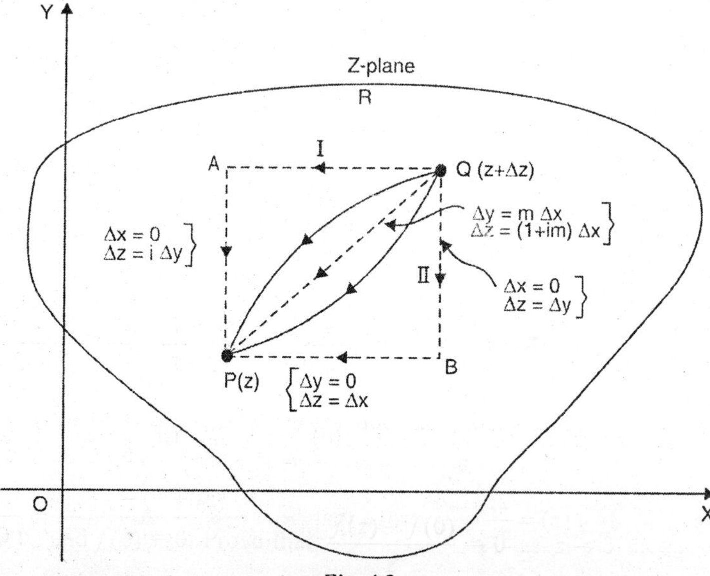

Fig. 4.3

There are, thus several ways in which $Q \to P$. For each one of the choices of position of Q and the approach of Q towards P, we will obtain as many distinct values of the derivative dw/dz as we wish. But for the existence of dw/dz or $f'(z)$, it is imperative that the value of dw/dz must be the same for all such ways in which $\Delta z \to 0$.

In other words the function $w = f(z)$ will be differentialble if the limit represented in Eq. (4.2) is unique, no matter how $\Delta z \to 0$.

It is clear from the foregoing that dw/dz may not only depend on z but also on the manner in which $\Delta z \to 0$. Also when $\Delta z \to 0$ in any manner dw/dz may not exist. It, therefore, becomes a settled fact to determine the necessary and sufficient conditions in order that dw/dz exist. This question is answered by the following:

Theorem 4.1 Given $w = f(z) = u(x, y) + iv(x, y)$. The *necessary and sufficient* conditions for the existence of dw/dz for every value of z in a region R of the z-plane are

(*i*) the four partial derivatives, namely $\dfrac{\partial u}{dx}, \dfrac{\partial u}{\partial y}, \dfrac{\partial v}{\partial x}, \dfrac{\partial v}{\partial y}$ are continuous functions of x and y in R, and

(*ii*) $\dfrac{\partial u}{dx} = \dfrac{\partial v}{\partial y}, \quad \dfrac{\partial u}{\partial y} = -\dfrac{\partial v}{\partial x}$.

The relation at (*ii*) above are known as **Cauchy-Riemann differential equations**, after the two mathematicians and are sometime abbreviated as **C-R equations.**

(A) *Conditions are necessary* Let Δu and Δv be the increments in the values of u and v respectively corresponding to small increments Δx and Δy in x and y, so that

$$\Delta w = f(z + \Delta z) - f(z) = u + \Delta u + i(v + \Delta v) - (u + iv) = \Delta u + i\Delta v.$$

$$\Delta z = x + \Delta x + i(y + \Delta y) - (x + iy) = \Delta x + i\Delta y.$$

$$\text{By definition } f'(z) = \lim_{\Delta z \to 0} \frac{\Delta w}{\Delta z} = \lim_{\Delta z \to 0} \frac{\Delta u + i\Delta v}{\Delta z}$$

$$= \lim_{\substack{\Delta x \to 0 \\ \Delta y \to 0}} \frac{\Delta u + i\Delta v}{\Delta x + i\Delta y}. \qquad \text{...(4.4)}$$

Now since u and v are functions of the real variables x and y and if they have continuous partial derivatives of the first order, then to a first approximation we have by Taylor's expansion

$$\Delta u = \left(\frac{\partial u}{\partial x} + \alpha_1\right)\Delta x + \left(\frac{\partial u}{\partial y} + \alpha_2\right)\Delta y \qquad \text{...(4.5)}$$

$$\Delta v = \left(\frac{\partial v}{\partial x} + \alpha_3\right)\Delta x + \left(\frac{\partial v}{\partial y} + \alpha_4\right)\Delta y \qquad \text{...(4.6)}$$

where $\lim\limits_{\substack{\Delta x \to 0 \\ \Delta y \to 0}} \alpha_t = 0, \; t = 1, 2, 3, 4.$

Substituting from Eqs. (4.5) and (4.6) in Eq. (4.4), we have

$$f'(z) = \lim_{\substack{\Delta x \to 0 \\ \Delta y \to 0}} \frac{\left(\dfrac{\partial u}{\partial x} + i\dfrac{\partial v}{\partial x} + \alpha_1 + i\alpha_3\right)\Delta x + \left(\dfrac{\partial u}{\partial y} + i\dfrac{\partial v}{\partial y} + \alpha_2 + i\alpha_4\right)\Delta y}{\Delta x + i\Delta y}$$

$$= \lim_{\substack{\Delta x \to 0 \\ \Delta y \to 0}} \frac{\dfrac{\partial u}{\partial x} + i\dfrac{\partial v}{\partial x} + \alpha_1 + i\alpha_3 + \left(\dfrac{\partial u}{\partial y} + i\dfrac{\partial v}{\partial y} + \alpha_2 + i\alpha_4\right)\dfrac{\Delta y}{\Delta x}}{1 + i\dfrac{\Delta y}{\Delta x}} \qquad \text{...(4.7)}$$

$$= \frac{S + mT}{1 + i\,m},$$

where $S = \dfrac{\partial u}{\partial x} + i\dfrac{\partial v}{\partial x}, \quad T = \dfrac{\partial u}{\partial y} + i\dfrac{\partial v}{\partial y}$ and $m = \dfrac{dy}{dx}$.

If $f'(z)$ is to be unique, it must have the same value for all such ways in which $\Delta z \to 0$ and hence it must be independent of $\frac{dy}{dx}$ that is, m. In other words $\frac{d}{dm} f'(z) = 0$. Applying this criterion to Eq. (4.7) we get

$$\frac{d}{dm} f'(z) = 0 = \frac{d}{dm}\left(\frac{S+mt}{1+im}\right) = \frac{(1+im)T-(S+mT)i}{(1+im)^2}. \quad ...(4.8)$$

Eq. (4.8) gives $T = iS$ or $\frac{\partial u}{\partial y} + i\frac{\partial v}{\partial y} = i\frac{\partial u}{\partial x} - \frac{\partial v}{\partial x}$, ...(4.9)

Equating the real and imaginary terms in Eq. (4.9), we get

$$\left.\begin{aligned} \frac{\partial u}{\partial x} &= \frac{\partial v}{\partial y} \\ \frac{\partial u}{\partial y} &= -\frac{\partial v}{\partial x} \end{aligned}\right\}. \quad ...(4.10)$$

Eq. (4.10) are the conditions which the real and imaginary parts of $w = u + iv$ must satisfy in order that $w = f(z) = u + iv$ must have a unique derivative dw/dz. The differential equations given in Eqs. (4.10) above are called *Cauchy-Riemann equations.*

Thus for a function w(z) to have a unique derivative dw/dz, it is necessary that the four partial derivatives $\frac{\partial u}{\partial x}, \frac{\partial u}{\partial y}, \frac{\partial v}{\partial x}, \frac{\partial v}{\partial y}$ must exist and they must satisfy C-R equations.

(B) *Conditions are sufficient* We will now establish that if $w(z) = f(z) = u(x, y) + iv(x, y)$ is a single-valued function of z and if the real and imaginary parts $u(x, y)$ and $v(x, y)$ have continuous partial derivatives $\frac{\partial u}{\partial x}, \frac{\partial u}{\partial y}, \frac{\partial v}{\partial x}, \frac{\partial v}{\partial y}$ satisfying the *C-R* equations, then the function $w(z)$ will have the unique derivative dw/dz.

Relations Eqs. (4.5) and (4.6) can be rewritten as

$$\Delta u = \frac{\partial u}{\partial x}\Delta x + \frac{\partial u}{\partial y}\Delta y + \alpha_1 \Delta x + \alpha_2 \Delta y \quad ...(4.11)$$

and $$\Delta v = \frac{\partial v}{\partial x}\Delta x + \frac{\partial v}{\partial y}\Delta y + \alpha_3 \Delta x + \alpha_4 \Delta y \quad ...(4.12)$$

where α_t $(t = 1, 2, 3, 4)$ are infinitesimals which tend to zero when Δx and Δy approach zero.

Making use of the assumptions that both u and v satisfy *C-R* equations that is, $\frac{\partial u}{\partial x} = \frac{\partial v}{\partial y}, \frac{\partial u}{\partial y} = -\frac{\partial v}{\partial x}$, Eqs. (4.11) and (4.12) can be written as

$$\Delta u = \frac{\partial u}{\partial x}\Delta x - \frac{\partial v}{\partial x}\Delta y + \alpha_1 \Delta x + \alpha_2 \Delta y$$

and $$\Delta v = \frac{\partial v}{\partial x}\Delta x + \frac{\partial u}{\partial x}\Delta y + \alpha_3 \Delta x + \alpha_4 \Delta y.$$

Consequently

$$\begin{aligned} \Delta u + i\Delta v &= \left(\frac{\partial u}{\partial x} + i\frac{\partial v}{\partial x}\right)\Delta x + \left(i\frac{\partial u}{\partial x} - \frac{\partial v}{\partial x}\right)\Delta y + (\alpha_1 + i\alpha_3)\Delta x + (\alpha_2 + i\alpha_4)\Delta y \\ &= \left(\frac{\partial u}{\partial x} + i\frac{\partial v}{\partial x}\right)\Delta x + i\left(\frac{\partial u}{\partial x} + i\frac{\partial v}{\partial x}\right)\Delta y + \varepsilon \\ &= \left(\frac{\partial u}{\partial x} + i\frac{\partial v}{\partial x}\right)(\Delta x + i\Delta y) + \varepsilon, \quad ...(4.13) \end{aligned}$$

where $\varepsilon = (\alpha_1 + i\alpha_3)\Delta x + (\alpha_2 + i\alpha_4)\Delta y$.

Eq. (4.13) on division by $\Delta x + i\Delta y$ can be written as

$$\frac{\Delta u + i\Delta v}{\Delta x + i\Delta y} = \frac{\partial u}{\partial x} + i\frac{\partial v}{\partial x} + \frac{\varepsilon}{\Delta x + i\Delta y}. \qquad ...(4.14)$$

Now in the limit $\dfrac{\varepsilon}{\Delta x + i\Delta y} \to 0$ when Δx and Δy approach zero, ($\because$ the denominator contains infinitesimals of the first order while the numerator ε contains quantities of infinitesimals of the second and higher orders.) Then,

$$\lim_{\substack{\Delta x \to 0 \\ \Delta y \to 0}} \frac{\Delta u + i\Delta v}{\Delta x + i\Delta y} = f'(z). \qquad \text{[From (4.4) above]}$$

Hence with the foregoing assumptions Eq. (4.14) can be written as

$$f'(z) = \frac{\partial u}{\partial x} + i\frac{\partial v}{\partial x}. \qquad ...(4.15)$$

Eq. (4.15) shows that $f'(z)$ exists and this holds for any manner of approach of $\Delta z \to 0$, assuming first that $\Delta y = 0$ and then $\Delta x = 0$, so that Δz is wholly real.

Important: Using C-R equations the Eq. (4.15) can be written in any one of the following forms:

$$f'(z) = \left\{\begin{array}{l} \dfrac{\partial u}{\partial x} + i\dfrac{\partial v}{\partial x} \\ \dfrac{\partial v}{\partial y} - i\dfrac{\partial u}{\partial y} \\ \dfrac{\partial u}{\partial x} - i\dfrac{\partial u}{\partial y} \\ \dfrac{\partial v}{\partial y} + i\dfrac{\partial v}{\partial x} \end{array}\right\}. \qquad ...(4.16)$$

4.6 ANALYTIC FUNCTIONS

A function $f(z)$ which is single-valued and is differentiable at every point in some neighbourhood of the point $z = z_0$ is said to be **analytic** at the point $z = z_0$. If the function is analytic at every point z of a certain region R of the complex z-plane, it is said to be analytic in that region.

A function analytic in R is also sometimes termed as a **regular function** or a **holomorphic function** in R.

A point at which an analytic function $f(z)$ ceases to possess a derivative is defined as a **singular point** or the **singularity** of the function. For example, the point $z = a$ is the singularity of $\dfrac{1}{z-a}$.

From the definition of analytic functions we observe that if a function $f(z) = u(x, y) + iv(x, y)$ is analytic in a certain region R, it must satisfy the following conditions:

(*i*) To every point z in R, there corresponds a definite value of $f(z)$ that is $f(z)$ is single-valued.

(*ii*) $f(z)$ is continuous function of z in R.

(*iii*) $f(z)$ possesses a unique derivative at every point z in R.

In other words if $u(x, y)$ and $v(x, y)$ are the real and imaginary parts of $f(z)$, they must satisfy *C-R* equations, that is,

$$\frac{\partial u}{\partial x} = \frac{\partial v}{\partial y}, \qquad \frac{\partial u}{\partial y} = -\frac{\partial v}{\partial x}. \qquad ...(4.17)$$

The real and imaginary parts of an analytic function $f(z)$, namely u and v are called the ***conjugate functions***. The relation between the conjugate functions u and v is given by the *C-R* equations represented at (4.17) above.

EXAMPLE 4.5 Determine whether the following functions are analytic or not.

(*i*) $w = x^2 - y^2 - 2xy + i(x^2 - y^2 + 2xy)$, (*ii*) $w = \dfrac{x - iy}{x^2 y^2}$,

(*iii*) $w = x^2y^2 + 2ix^2y^2$.

Solution (*i*) We have $u(x, y) = x^2 - y^2 - 2xy$ and $v(x, y) = x^2 - y^2 + 2xy$.

Therefore,
$$\frac{\partial u}{\partial x} = 2x - 2y, \quad \frac{\partial v}{\partial y} = -2y + 2x,$$

$$\frac{\partial u}{\partial y} = -2y - 2x, \quad \frac{\partial v}{\partial x} = 2x + 2y.$$

Since
$$\frac{\partial u}{\partial x} = \frac{\partial v}{\partial y} \text{ and } \frac{\partial u}{\partial y} = -\frac{\partial v}{\partial x}.$$

The Cauchy-Riemann equations are satisfied for all (x, y) and the first order partial derivatives of $u(x, y)$, $v(x, y)$ are continuous everywhere. Hence, the given function is analytic.

(*ii*) We have
$$u(x, y) = \frac{x}{x^2 y^2} = \frac{1}{xy^2} \text{ and } v(x, y) = \frac{-y}{x^2 y^2} = \frac{-1}{x^2 y}.$$

Therefore,
$$\frac{\partial u}{\partial x} = -\frac{1}{x^2 y^2} \text{ and } \frac{\partial v}{\partial y} = \frac{1}{x^2 y^2}.$$

As $\dfrac{\partial u}{\partial x} \neq \dfrac{\partial v}{\partial y}$. The Cauchy-Riemann equations are not satisfied.

Therefore, the given function is not analytic.

(*iii*) We have $u(x, y) = x^2y^2$ and $v(x, y) = 2x^2y^2$.

Therefore,
$$\frac{\partial u}{\partial x} = 2xy^2, \quad \frac{\partial v}{\partial y} = 4x^2 y, \quad \frac{\partial u}{\partial y} = 2x^2 y \text{ and } \frac{\partial v}{\partial x} = 4xy^2.$$

The Cauchy-Riemann equations gives $2xy^2 = 4x^2y$ and $2x^2y = -4xy^2$.

These equations are satisfied only along the lines $x = 0$, $y = 0$.

Since there is no region (neighbourhood) in which w is analytic, the given function is not analytic.

***Remark* 1**

As stated in the definition of analyticity at a point, a function $f(z)$ is analytic at a point if there is some neighbourhood containing the point in which $f(z)$ is analytic. Similarly a function is analytic along a curve if the function is analytic in some neighbourhood of each point on the curve, that is, if there is some region containing the curve in which the function is analytic.

EXAMPLE 4.6 Show that the function $e^x(\cos y + i \sin y)$ is analytic. Find also its derivative.

Solution Let $e^x(\cos y + i \sin y) = f(z) = u + iv$, so that $e^x \cos y = u$ and $e^x \sin y = v$.

Now
$$\frac{\partial u}{\partial x} = e^x \cos y, \quad \frac{\partial u}{\partial y} = -e^x \sin y, \quad \frac{\partial v}{\partial x} = e^x \sin y, \quad \frac{\partial v}{\partial y} = e^x \cos y.$$

Clearly from the above relations. *C-R* equations are satisfied. Hence $f(z)$ is analytic. Also

$$f'(z) = \frac{\partial u}{\partial x} + i\frac{\partial v}{\partial x} = e^x \cos y + i\, e^x \sin y = e^x(\cos y + i \sin y) = e^x e^{iy} = e^{x+iy} = e^z.$$

EXAMPLE 4.7 Show that the function $f(z) = \bar{z}$ is continuous at the point $z = 0$ but not differentiable at $z = 0$. (*AMIETE, June 2011, 2010, 2008*)

Solution Let $z = x + iy$. Then $\bar{z} = x - iy$ and $f(z) = \bar{z} = x - iy$, $\Delta z = \Delta x + i\Delta y$, $\overline{\Delta z} = \Delta x - i\Delta y$.

Now $\lim\limits_{z \to 0} f(z) = \lim\limits_{\substack{x \to 0 \\ y \to 0}} (x - iy) = 0 = f(0)$. Therefore, $f(z) = \bar{z}$ is continuous at $z = 0$.

Now, at $z = 0$ $\quad \lim_{\Delta z \to 0} \dfrac{f(\Delta z) - f(0)}{\Delta z} = \lim_{\Delta z \to 0} \dfrac{\overline{\Delta z}}{\Delta z} = \lim_{\substack{\Delta x \to 0 \\ \Delta y \to 0}} \left[\dfrac{\Delta x - i\Delta y}{\Delta x + i\Delta y}\right].$

Consider the path $y = mx$. We have $\Delta y = m\Delta x$. Hence, $\Delta x \to 0$ implies $\Delta y \to 0$.

Therefore, $\lim_{\Delta z \to 0} \dfrac{\overline{\Delta z}}{\Delta z} = \lim_{\Delta x \to 0} \dfrac{\Delta x - im\Delta x}{\Delta x + im\Delta x} = \dfrac{1 - im}{1 + im}$, which depends on m. Thus, the limit does not exist.

Therefore, the function $f(z) = \bar{z}$ is not differentiable at $z = 0$.

EXAMPLE 4.8 Show that the function $f(z) = |z|^2$ is differentiable only at $z = 0$ and no where else.

Solution We have $f(z) = |z|^2 = z\bar{z}$. Therefore,

$$\lim_{\Delta z \to 0} \frac{f(z + \Delta z) - f(z)}{\Delta z} = \lim_{\Delta z \to 0} \frac{(z + \Delta z)(\bar{z} + \overline{\Delta z}) - z\bar{z}}{\Delta z} = \lim_{\Delta z \to 0} \left[z\frac{\overline{\Delta z}}{\Delta z} + \bar{z} + \overline{\Delta z}\right] = 0 \text{ for } z = 0.$$

Hence, the function $f(z) = |z|^2$ is differentiable at $z = 0$ and $f'(0) = 0$.

For $z \neq 0$, $\lim_{\Delta z \to 0} (\bar{z} + \overline{\Delta z}) = \bar{z}$, but $\lim_{\Delta z \to 0} \dfrac{\overline{\Delta z}}{\Delta z}$ does not exist (refer Example 4.7). Hence, the function $f(z) = |z|^2$ is not differentiable at $z \neq 0$.

4.7 POLAR FORM OF THE CAUCHY-RIEMANN EQUATIONS

Show that the Cauchy-Riemann equations in polar form are given by

$$\frac{\partial u}{\partial r} = \frac{1}{r}\frac{\partial v}{\partial \theta}, \quad \frac{\partial v}{\partial r} = -\frac{1}{r}\frac{\partial u}{\partial \theta}$$ *(AMIETE Dec. 2012; VTU 2006, Kurukshetra 2005)*

Deduce that

$$\frac{\partial^2 u}{\partial r^2} + \frac{1}{r}\frac{\partial u}{\partial r} + \frac{1}{r^2}\frac{\partial^2 u}{\partial \theta^2} = 0.$$ *(Kurukshetra 2005; MDU 2002)*

Let (r, θ) be the polar co-ordinates of a point whose cartesian co-ordinates are (x, y). Then

$$x = r\cos\theta, \qquad y = r\sin\theta.$$

Now $\quad z = x + iy = r\cos\theta + ir\sin\theta = r(\cos\theta + i\sin\theta) = re^{i\theta}$.

Consider

$$w = f(z) = u + iv$$

or

$$u + iv = f(re^{i\theta}), \qquad \text{...(4.18)}$$

where u and v are functions of r and θ. Differentiating Eq. (4.18) partially with respect to r and θ, successively we obtain

$$\frac{\partial u}{\partial r} + i\frac{\partial v}{\partial r} = f'(re^{i\theta})\, e^{i\theta} \qquad \text{...(4.19)}$$

$$\frac{\partial u}{\partial \theta} + i\frac{\partial v}{\partial \theta} = f'(re^{i\theta}) \times ire^{i\theta}. \qquad \text{...(4.20)}$$

From Eqs. (4.19) and (4.20), we get

$$\frac{\partial u}{\partial \theta} + i\frac{\partial v}{\partial \theta} = ir\left(\frac{\partial u}{\partial r} + i\frac{\partial v}{\partial r}\right) = -r\frac{\partial v}{\partial r} + ir\frac{\partial u}{\partial r}. \qquad \text{...(4.21)}$$

Equating the coefficients of the real and imaginary parts in Eq. (4.21), we have

$$\frac{\partial u}{\partial r} = \frac{1}{r}\frac{\partial v}{\partial \theta} \quad \text{...(4.22)} \qquad \text{and} \qquad \frac{\partial v}{\partial r} = -\frac{1}{r}\frac{\partial u}{\partial \theta}. \qquad \text{...(4.23)}$$

which are the required C–R equations in polar form.

Differentiating Eq. (4.22) partially with respect to r, we get

$$\frac{\partial^2 u}{\partial r^2} = -\frac{1}{r^2}\frac{\partial v}{\partial \theta} + \frac{1}{r}\frac{\partial^2 v}{\partial r \partial \theta}. \qquad \text{...(4.24)}$$

Differentiating Eq. (4.23) partially with respect to θ, we get

$$\frac{\partial^2 v}{\partial\theta\partial r} = -\frac{1}{r}\frac{\partial^2 u}{\partial\theta^2}. \qquad ...(4.25)$$

Eq. (4.24) with the help of Eqs. (4.22) and (4.25), can be written as

$$\frac{\partial^2 u}{\partial r^2} = -\frac{1}{r}\left(\frac{\partial u}{\partial r}\right) + \frac{1}{r}\left(-\frac{1}{r}\frac{\partial^2 u}{\partial\theta^2}\right)$$

or

$$\frac{\partial^2 u}{\partial r^2} + \frac{1}{r}\frac{\partial u}{\partial r} + \frac{1}{r^2}\frac{\partial^2 u}{\partial\theta^2} = 0. \qquad ...(4.26)$$

4.8 DERIVATIVE OF $w = f(z) = u + iv$ IN POLAR FORM

We know that $f'(z) = \frac{dw}{dz} = \frac{\partial u}{\partial x} + i\frac{\partial v}{\partial x} = \frac{\partial w}{\partial x}$ (Art 4.5 Eq. 4.16).

Also the relation between polar co-ordinates and cartesian co-ordinates are given by

$$r^2 = x^2 + y^2, \qquad ...(4.27) \qquad \theta = \tan^{-1}(y/x). \qquad ...(4.28)$$

Now

$$\frac{\partial w}{\partial x} = \frac{\partial w}{\partial r}\frac{\partial r}{\partial x} + \frac{\partial w}{\partial\theta}\frac{\partial\theta}{\partial x} = \frac{\partial w}{\partial r}\cdot\frac{x}{r} - \frac{\partial w}{\partial\theta}\left(\frac{1}{1+(y^2/x^2)}\right)\frac{y}{x^2}$$

$$= \frac{\partial w}{\partial r}\cos\theta - \frac{\partial w}{\partial\theta}\frac{1}{r}\sin\theta. \qquad ...(4.29)$$

But

$$w = u + iv = f(re^{i\theta}).$$

Therefore,

$$\frac{\partial w}{\partial r} = f'(re^{i\theta})\, e^{i\theta}, \qquad ...(4.30)$$

and

$$\frac{\partial w}{\partial\theta} = f'(re^{i\theta})\, ire^{i\theta} = ir\frac{\partial w}{\partial r}. \qquad \text{[From Eq. (4.30)]} \qquad ...(4.31)$$

Substituting from Eq. (4.31) in Eq. (4.29), we obtain

$$\frac{dw}{dz} = f'(z) = \frac{\partial w}{\partial x} = \frac{\partial w}{\partial r}\cos\theta - \frac{1}{r}\sin\theta\left(ir\frac{\partial w}{\partial r}\right)$$

$$= (\cos\theta - i\sin\theta)\frac{\partial w}{\partial r}. \qquad ...(4.32)$$

Alternatively

$$\frac{dw}{dz} = f'(z) = \frac{1}{ir}\frac{\partial w}{\partial\theta}\cos\theta - \frac{\partial w}{\partial r}\frac{1}{r}\sin\theta = \left(-\frac{i}{r}\cos\theta + \frac{i^2}{r}\sin\theta\right)\frac{\partial w}{\partial\theta}$$

$$= -\frac{i}{r}(\cos\theta - i\sin\theta)\frac{\partial w}{\partial\theta}. \qquad ...(4.33)$$

Thus we obtain from relations (4.32) and (4.33)

$$\frac{dw}{dz} = f'(z) = (\cos\theta - i\sin\theta)\frac{\partial w}{\partial r} = -\frac{i}{r}(\cos\theta - i\sin\theta)\frac{\partial w}{\partial\theta}.$$

EXAMPLE 4.9 Show that the function

$$f(z) = \begin{cases} \dfrac{x^3(1+i) - y^3(1-i)}{x^2+y^2}, & z \neq 0 \\ 0, & z = 0 \end{cases}.$$

satisfies Cauchy-Riemann equations at $z = 0$ but $f'(0)$ does not exist.

(AMIETE, June 2009; UPTU, 2003; VTU 2001; JNTU, 2000)

Solution Let $f(z)=\dfrac{x^3(1+i)-y^3(1-i)}{x^2+y^2}=u+iv.$

Separating real and imaginary parts, we get

$$u=\frac{x^3-y^3}{x^2+y^2},\qquad v=\frac{x^3+y^3}{x^2+y^2}.$$

Since u and v, both are rational functions of x and y and non-zero for all values of z except for $z=0$, we find that u and v are continuous for all those points at which $z\neq 0$.

Again $f(0)=0$ implies that $u(0, 0)=0$ and $v(0, 0)=0$. Hence u and v are both continuous at the origin, giving thereby that $f(z)$ is continuous at the origin. Consequently $f(z)$ is continuous everywhere.

Again, at the origin

$$\frac{\partial u}{\partial x}=\lim_{\substack{y=0\\x\to 0}}\frac{u(x,0)-u(0,0)}{x-0}=\lim_{x\to 0}\left(\frac{x}{x}\right)=1,$$

$$\frac{\partial u}{\partial y}=\lim_{\substack{x=0\\y\to 0}}\frac{u(0,y)-u(0,0)}{y-0}=\lim_{y\to 0}\left(\frac{-y}{y}\right)=-1.$$

$$\frac{\partial v}{\partial x}=\lim_{\substack{y=0\\x\to 0}}\frac{v(x,0)-v(0,0)}{x-0}=\lim_{x\to 0}\left(\frac{x}{x}\right)=1,$$

$$\frac{\partial v}{\partial y}=\lim_{\substack{x=0\\y\to 0}}\frac{v(0,y)-v(0,0)}{y-0}=\lim_{y\to 0}\left(\frac{y}{y}\right)=1.$$

Thus we find that at the origin

$$\frac{\partial u}{\partial x}=\frac{\partial v}{\partial y}\text{ and }\frac{\partial u}{\partial y}=-\frac{\partial v}{\partial x}.$$

Hence we see that Cauchy-Riemann equations are satisfied at z = 0.

Again $$f'(0)=\lim_{\Delta z\to 0}\frac{f(0+\Delta z)-f(0)}{\Delta z}=\lim_{\Delta z\to 0}\frac{f(\Delta z)-0}{\Delta z}$$

$$=\lim_{z\to 0}\frac{f(z)}{z}=\lim_{z\to 0}\frac{x^3(1+i)-y^3(1-i)}{x^2+y^2}\frac{1}{z}=\lim_{z\to 0}\frac{x^3(1+i)-y^3(1-i)}{(x^2+y^2)(x+iy)}.$$

Now let $z\to 0$ along $x=0$, so that

$$f'(0)=\lim_{y\to 0}\frac{-y^3(1-i)}{iy^3}=1+i. \qquad \text{...(A)}$$

Suppose again that $z\to 0$ along the radius vector $y=x$,

then, $$f'(0)=\lim_{x\to 0}\frac{x^3(1+i)-x^3(1-i)}{x^2+x^2}\cdot\frac{1}{x+ix}$$

$$=\lim_{x\to 0}\frac{2ix^3}{2x^3(1+i)}=\frac{(1+i)^2}{2(1+i)}=\frac{1+i}{2}. \qquad \text{...(B)} \qquad [\because 2i=(1+i)^2]$$

Thus we find from relations (A) and (B) that $f'(0)$ is not unique. Hence $f'(0)$ does not exist at the origin and therefore, $f(z)$ is not analytic at the origin although C–R equations are satisfied there.

***Remark* 2**

It must be noted that for a function to be analytic, the C–R equations are only necessary conditions. They alone are not sufficient for analyticity but the four partial derivatives $\dfrac{\partial u}{\partial x},\dfrac{\partial u}{\partial y},\dfrac{\partial v}{\partial x},\dfrac{\partial v}{\partial y}$ must also be continuous.

Thus for the function $f(z) = u(x, y) + iv(x, y)$ to be analytic in a region R of the z-plane, the C–R equations namely

$$\frac{\partial u}{\partial x} = \frac{\partial v}{\partial y}, \quad \frac{\partial u}{\partial y} = -\frac{\partial v}{\partial x}$$

will be both necessary and sufficient conditions only when the four partial derivatives of u and v are continuous in R. The derivative of $f(z)$ will exist and given by either of the expressions

$$f'(z) = \frac{\partial u}{\partial x} + i\frac{\partial v}{\partial x} \quad \text{or} \quad f'(z) = \frac{\partial v}{\partial y} - i\frac{\partial u}{\partial y}.$$

EXAMPLE 4.10 Show that the function $w = f(z) = z\bar{z}$ is differentiable but not analytic at $z = 0$.

Solution $w = f(z) = u + iv = z\bar{z} = (x + iy)(x - iy) = x^2 + y^2$.

Separating real and imaginary parts, we get $u = x^2 + y^2$, $v = 0$.

Therefore, $\frac{\partial u}{\partial x} = 2x, \quad \frac{\partial u}{\partial y} = 2y, \quad \frac{\partial v}{\partial x} = 0 \quad \text{and} \quad \frac{\partial v}{\partial y} = 0.$

Thus, we find that

$$\frac{\partial u}{\partial x} = \frac{\partial v}{\partial y} = 0 \quad \text{when} \quad x = 0,$$

and

$$\frac{\partial u}{\partial y} = -\frac{\partial v}{\partial x} = 0 \quad \text{when} \quad y = 0.$$

Thus $C - R$ equations are satisfied only when both x and y are zero. Hence $f(z)$ has a derivative at the origin.

Again for points $z \neq 0$, the $C - R$ equations are not satisfied. As such there is no neighbourhood of the point $z = 0$ where the function is differentiable. Hence $f(z)$ is not analytic at $z = 0$.

4.9 PROPERTIES OF ANALYTIC FUNCTIONS

1. *If $f(z) = u + iv$ is an analytic function, then both u and v satisfy Laplace's equation.*

$$\frac{\partial^2 \varphi}{\partial x^2} + \frac{\partial^2 \varphi}{\partial y^2} = 0. \quad \text{...(4.34)}$$

If $f(z) = u + iv$ is analytic in some region R of the z-plane, then u and v satisfy Cauchy-Riemann equations

$$\frac{\partial u}{\partial x} = \frac{\partial v}{\partial y} \quad \text{...(4.35)} \qquad \text{and} \qquad \frac{\partial u}{\partial y} = -\frac{\partial v}{\partial x}. \quad \text{...(4.36)}$$

Differentiating Eq. (4.35) with respect to x and Eq. (4.36) with respect to y partially, we get on addition

$$\frac{\partial^2 u}{\partial x^2} + \frac{\partial^2 u}{\partial y^2} = \frac{\partial^2 v}{\partial x \partial y} - \frac{\partial^2 v}{\partial y \partial x} = 0. \quad \text{...(4.37)} \qquad \left[\because \frac{\partial^2 v}{\partial x \partial y} = \frac{\partial^2 v}{\partial y \partial x}\right]$$

Similarly differentiating Eq. (4.35) with respect to y and Eq. (4.36) with respect to x partially and taking negative addition, we obtain

$$\frac{\partial^2 v}{\partial y^2} + \frac{\partial^2 v}{\partial x^2} = \frac{\partial^2 u}{\partial y \partial x} - \frac{\partial^2 u}{\partial x \partial y} = 0. \quad \text{...(4.38)}$$

Eqs. (4.37) and (4.38) show that the real and imaginary parts, u and v, of $f(z)$ satisfy Laplace's Eq. (4.34) in two variables, x and y. If a function of two variables is a solution of Laplace's equation, it is called a **Harmonic Function.** Thus both $u(x, y)$ and $v(x, y)$ are harmonic functions. If two harmonic functions $u(x, y)$ and $v(x, y)$ satisfy $C - R$ equations in a domain R of the z-plane, then $u(x, y)$ is said to be a **conjugate harmonic function** of $v(x, y)$ and both $u(x, y)$ and $v(x, y)$ are called a **pair of conjugate harmonic functions** in R.

Eq. (4.34) can be written as $\nabla^2\varphi = 0$, where $\nabla^2 = \frac{\partial^2}{\partial x^2} + \frac{\partial^2}{\partial y^2}$ is called the Laplace operator in two variables.

2. *If $f(z) = u + iv$ is an analytic function, the curves of the family $u(x, y)$ = constant = a cut the curves of the family $v(x, y)$ = constant = b orthogonally at their points of intersection.* (*AMIETE, June 2013*)
Two curves are said to cut each other orthogonally when at each of their points of intersection they are at right angle.
Now $u = a$ represents a family of curves for different values of a. Similarly $v = b$ defines another family of curves for various values of b. Taking total differentials, we obtain

$$du = 0 = \frac{\partial u}{\partial x}dx + \frac{\partial u}{\partial y}dy. \quad [\because u = a]$$

or
$$\frac{du}{dx} = \frac{\partial u}{\partial x} + \frac{\partial u}{\partial y}\frac{dy}{dx} = 0, \quad ...(4.39)$$

and similarly
$$\frac{dv}{dx} = \frac{\partial v}{\partial x} + \frac{\partial v}{\partial y}\frac{dy}{dx} = 0. \quad ...(4.40)$$

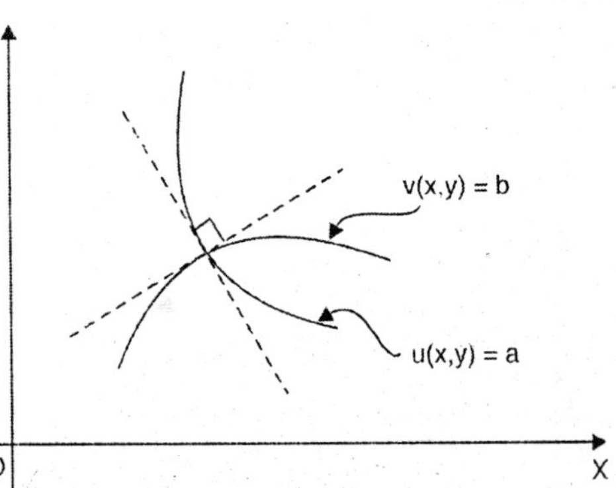

Fig. 4.4

Eqs. (4.39) and (4.40) give respectively the slope of the curves $u = a$ and $v = b$ as

$$\frac{dy}{dx} = -\frac{\partial u/\partial x}{\partial u/\partial y} = m_1 \text{ (say)}, \quad ...(4.41)$$

$$\frac{dy}{dx} = -\frac{\partial v/\partial x}{\partial v/\partial y} = m_2 \text{ (say)}. \quad ...(4.42)$$

Since $f(z)$ is analytic, u and v satisfy C–R equations, that is,

$$\frac{\partial u}{\partial x} = \frac{\partial v}{\partial y}, \quad \frac{\partial u}{\partial y} = -\frac{\partial v}{\partial x}.$$

Now the product of the slopes of the two families is

$$m_1 m_2 = \frac{\partial u/\partial x}{\partial u/\partial y} \times \frac{\partial v/\partial x}{\partial v/\partial y} = \frac{\partial u/\partial x}{\partial u/\partial y} \times \left(-\frac{\partial u/\partial y}{\partial u/\partial x}\right) = -1. \quad \text{[using } C\text{–}R \text{ equations]}$$

Since the product of the slopes is –1, the curves cut each other orthogonally. The two families $u = a$ and $v = b$ are said to be *orthogonal trajectories* of one another.

4.10 METHOD OF CONSTRUCTION OF THE CONJUGATE FUNCTION

Let $f(z) = u + iv$ be an analytic function where both $u(x, y)$ and $v(x, y)$ are conjugate functions. Being given one of these say $u(x, y)$, it is required to determine the other that is, $v(x, y)$. The method is as follows:

1. *First Method* Since v depends on x, y we have

$$dv = \frac{\partial v}{\partial x}dx + \frac{\partial v}{\partial y}dy = -\frac{\partial u}{\partial y}dx + \frac{\partial u}{\partial x}dy. \quad \text{[by Cauchy-Riemann equations]} \quad ...(4.43)$$

The above equation is of the form

$$dv = M\,dx + N\,dy, \quad \text{where } M = -\frac{\partial u}{\partial y} \text{ and } N = \frac{\partial u}{\partial x}.$$

Now
$$\frac{\partial M}{\partial y} = -\frac{\partial^2 u}{\partial y^2} \quad \text{and} \quad \frac{\partial N}{\partial x} = \frac{\partial^2 u}{\partial x^2}.$$

Therefore,
$$\frac{\partial N}{\partial x} - \frac{\partial M}{\partial y} = \frac{\partial^2 u}{\partial x^2} + \frac{\partial^2 u}{\partial y^2} = 0. \quad \text{[as } u \text{ satisfies Laplace's equation]}$$

This gives $\partial M/\partial y = \partial N/\partial x$ and this is a condition for Eq. (4.43) to be an exact equation. Hence integrating Eq. (4.43) under the condition of exactness, we get

$$v = \int_{y \text{ constant}} -\frac{\partial u}{\partial y}\,dx + \int_{y \text{ terms only}} \frac{\partial u}{\partial x}\,dy + c, \qquad ...(4.44)$$

where c is and arbitrary constant. Having known v, we can find $f(z) = u + iv$ by substituting for u and v.

2. *Second Method* (Milne Thomson's Method)

Let $\quad f(z) = u(x, y) + iv(x, y)$. ...(4.45)

Since $\quad z = x + iy$ and $\bar{z} = x - iy$, we have

$2x = z + \bar{z}$ and $2iy = z - \bar{z}$. ...(4.46)

Substituting for x and y from Eq. (4.46) in Eq. (4.45) we get

$$f(z) = u\left(\frac{z+\bar{z}}{2}, \frac{z-\bar{z}}{2i}\right) + iv\left(\frac{z+\bar{z}}{2}, \frac{z-\bar{z}}{2i}\right). \qquad ...(4.47)$$

Eq. (4.47) is a formal identity in the two independent variables z and $\bar{z}$. Replacing $\bar{z}$ by z, we get $x = z$, $y = 0$ and then $\quad f(z) = u(z, 0) + iv(z, 0)$. ...(4.48)

Eq. (4.48) is the same as Eq. (4.45), if we replace x by z and y by 0.

Now $\quad f'(z) = \dfrac{dw}{dz} = \dfrac{\partial u}{\partial x} + i\dfrac{\partial v}{\partial x} = \dfrac{\partial u}{\partial x} - i\dfrac{\partial u}{\partial y}$. [C – R equations]

Let $\quad \dfrac{\partial u}{\partial x} = \phi_1(x, y), \quad \dfrac{\partial u}{\partial y} = \phi_2(x, y)$.

Therefore, $\quad f'(z) = \phi_1(x, y) - i\phi_2(x, y) = \phi_1(z, 0) - i\phi_2(z, 0)$. ...(4.49)

Integrating Eq. (4.49) we get

$$f(z) = \int \{\phi_1(z, 0) - i\phi_2(z, 0)\}\,dz + c_1 \qquad ...(4.50)$$

Thus $f(z)$ is constructed if $u(x, y)$ is known. Similarly if $v(x, y)$ is given, we have

$$f(z) = \int \{\psi_1(z, 0) + i\,\psi_2(z, 0)\}\,dz + c_2, \qquad ...(4.51)$$

where $\psi_1(x, y) = \dfrac{\partial v}{\partial y}$ and $\psi_2(x, y) = \dfrac{\partial v}{\partial x}$.

EXAMPLE 4.11 Determine the analytic function $f(z) = u + iv$ if

(*i*) $u(x, y) = e^x (x \cos y - y \sin y)$, (*ii*) $v(x, y) = \log(x^2 + y^2) + x - 2y$,

(*iii*) $u(x, y) = x^3 - 3xy^2 + 3x^2 - 3y^2 + 2x + 1$.

Solution (*i*) We have $u(x, y) = e^x (x \cos y - y \sin y)$.

Therefore, $\quad \dfrac{\partial u}{\partial x} = e^x(x\cos y - y\sin y) + e^x \cos y = e^x(x\cos y - y\sin y + \cos y)$ ...(1)

and $\quad \dfrac{\partial u}{\partial y} = e^x(-x\sin y - \sin y - y\cos y) = -e^x(x\sin y + \sin y + y\cos y)$ ...(2)

From the Cauchy-Riemann equation $\dfrac{\partial v}{\partial x} = -\dfrac{\partial u}{\partial y}$, we get

$$\frac{\partial v}{\partial x} = -\frac{\partial u}{\partial y} = e^x(x\sin y + \sin y + y\cos y).$$

Integrating partially with respect to x, we obtain

$$v = e^x(x\sin y + \sin y + y\cos y) - e^x(\sin y) + \phi(y)$$

$$= e^x(x\sin y + y\cos y) + \phi(y), \text{ where } \phi(y) \text{ is an arbitrary function of } y.$$

Using the Cauchy-Riemann equation $\frac{\partial v}{\partial y} = \frac{\partial u}{\partial x}$, we obtain

$$e^x(x\cos y + \cos y - y\sin y) + \phi'(y) = e^x(x\cos y - y\sin y + \cos y) \text{ or } \phi'(y) = 0.$$

Integrating with respect to y, we obtain $f(y) = c$, a constant.

Hence, we obtain $v(x, y) = e^x(x \sin y + y \cos y) + c$ and

$$f(z) = u(x, y) = iv(x, y) = e^x(x\cos y - y\sin y) + ie^x(x\sin y + y\cos y) + ic$$

$$= e^x[(x + iy)\cos y + i(x + iy)\sin y] + ic$$

$$= e^x[(z\cos y + iz\sin y) + ic = ze^x(\cos y + i\sin y) + ic$$

$$= ze^x \cdot e^{iy} + ic = ze^{x+iy} + ic = ze^z + ic.$$

Alternative Solution: Using Milne-Thomson's Method

We have $f(z) = u + iv$.

Therefore, $f'(z) = \frac{\partial u}{\partial x} + i\frac{\partial v}{\partial x} = \frac{\partial u}{\partial x} - i\frac{\partial u}{\partial y}$

$$= e^x(x\cos y - y\sin y + \cos y) + ie^x(x\sin y + \sin y + y\cos y).$$

Putting $x = z$ and $y = 0$ in the right side, we obtain $f'(z) = e^z(z + 1)$. Integrating, we obtain

$$f(z) = (z + 1)e^z - e^z + c = ze^z + c, \text{ where } c \text{ is a complex constant.}$$

(*ii*) We have $v(x, y) = \log(x^2 + y^2) + x - 2y$.

Therefore, $\frac{\partial v}{\partial x} = \frac{2x}{x^2 + y^2} + 1, \quad \frac{\partial v}{\partial y} = \frac{2y}{x^2 + y^2} - 2.$

Since $f(z) = u + iv$. Therefore $f'(z) = \frac{\partial u}{\partial x} + i\frac{\partial v}{\partial x} = \frac{\partial v}{\partial y} + i\frac{\partial v}{\partial x}$

$$= \left(\frac{2y}{x^2 + y^2} - 2\right) + i\left(\frac{2x}{x^2 + y^2} + 1\right) \quad ...(3)$$

Employing Milne Thomson's method, on replacing x by z and y by 0 on right hand side of Eq. (3), we obtain

$$f'(z) = -2 + i\left(\frac{2}{z} + 1\right) = (i - 2) + \frac{2i}{z}.$$

Integrating with respect to z, we get $f(z) = (i - 2)z + 2i\log z + c$, where c is a complex constant.

(*iii*) We have $u(x, y) = x^3 - 3xy^2 + 3x^2 - 3y^2 + 2x + 1$.

Therefore, $\frac{\partial u}{\partial x} = 3x^2 - 3y^2 + 6x + 2$ and $\frac{\partial u}{\partial y} = -6xy - 6y$.

From the Cauchy-Riemann equation $\frac{\partial v}{\partial x} = -\frac{\partial u}{\partial y}$, we get $\frac{\partial v}{\partial x} = -\frac{\partial u}{\partial y} = 6xy + 6y$.

Integrating partially with respect to x, we obtain

$$v(x, y) = 3x^2y + 6xy + \phi(y), \text{ where } \phi(y) \text{ is an arbitrary function of } y.$$

Using the Cauchy-Riemann equation $\frac{\partial v}{\partial y} = \frac{\partial u}{\partial x}$, we obtain

$$3x^2 + 6x + \phi'(y) = 3x^2 - 3y^2 + 6x + 2 \text{ or } \phi'(y) = -3y^2 + 2.$$

Integrating with respect to y, we obtain $\phi(y) = -y^3 + 2y + c$.

Hence, we obtain $v(x, y) = 3x^2y + 6xy - y^3 + 2y + c$, where c is an arbitrary constant.

Now, $f(z) = u(x, y) + iv(x, y) = (x^3 - 3xy^2 + 3x^2 - 3y^2 + 2x + 1) + i(3x^2y + 6xy - y^3 + 2y + c)$

$$= (x^3 - 3xy^2 + 3ix^2y - iy^3) + 3(x^2 + 2ixy - y^2) + 2(x + iy) + (1 + ic)$$

$$= (x + iy)^3 + 3(x + iy)^2 + 2(x + iy) + (1 + ic)$$

$$= z^3 + 3z^2 + 2z + (1 + ic),$$ where $1 + ic$ is a complex constant.

EXAMPLE 4.12 If $f(z) = u + iv$ is an analytic function of $z = x + iy$, find u, v and the analytic function $f(z)$, if

(*i*) $u - v = (x - y)(x^2 + 4xy + y^2)$,

(*ii*) $u - v = \dfrac{\cos x + \sin x - e^{-y}}{2(\cos x - \cosh y)}$ and $f(\pi/2) = 0$,

(*AMIETE, Dec. 2012, June 2005, Osmania 2003; Madras 2000S*)

(*iii*) $u - v = e^{-x}[(x - y)\sin y - (x + y)\cos y]$,

(*iv*) $3u + 2v = y^2 - x^2 + 16xy$. (*AMIETE, Dec 2004*)

Solution (*i*) We have $u - v = (x - y)(x^2 + 4xy + y^2)$.

Therefore, $$\frac{\partial u}{\partial x} - \frac{\partial v}{\partial x} = 1(x^2 + 4xy + y^2) + (x - y)(2x + 4y) = 3x^2 - 3y^2 + 6xy \quad ...(1)$$

and $$\frac{\partial u}{\partial y} - \frac{\partial v}{\partial y} = -1(x^2 + 4xy + y^2) + (x - y)(4x + 2y) = 3x^2 - 3y^2 - 6xy. \quad ...(2)$$

From Cauchy-Riemann equations since given function $f(z)$ is analytic, we have

$$\frac{\partial u}{\partial y} = -\frac{\partial v}{\partial x}, \quad \frac{\partial v}{\partial y} = \frac{\partial u}{\partial x}. \quad ...(3)$$

From Eqs. (2) and (3), we obtain

$$-\frac{\partial u}{\partial x} - \frac{\partial v}{\partial x} = 3x^2 - 3y^2 - 6xy. \quad ...(4)$$

Adding Eqs. (1) and (4), we obtain $\dfrac{\partial v}{\partial x} = 3y^2 - 3x^2$. Thus $\dfrac{\partial u}{\partial y} = 3x^2 - 3y^2$.

Substituting Eq. (4) from Eq. (1), we obtain $\dfrac{\partial u}{\partial x} = 6xy$. Thus, $\dfrac{\partial v}{\partial y} = 6xy$.

Now, $$du = \frac{\partial u}{\partial x}dx + \frac{\partial u}{\partial y}dy = 6xy\,dx + (3x^2 - 3y^2)\,dy = d(3x^2y - y^3)$$

$\Rightarrow$ $u = 3x^2y - y^3 + c_1$.

$$dv = \frac{\partial v}{\partial x}dx + \frac{\partial v}{\partial y}dy = 3y^2dx - 3x^2dx + 6xydy = d(3xy^2 - x^3)$$

$\Rightarrow$ $v = 3xy^2 - x^3 + c_2$.

Therefore, $$f(z) = u(x, y) + iv(x, y)$$
$$= (3x^2y - y^3 + c_1) + i(3xy^2 - x^3 + c_2)$$
$$= (-y + ix)^3 + (c_1 + ic_2) = -i(x + iy)^3 + (c_1 + ic_2)$$
$$= -iz^3 + c,$$ where c is a complex constant.

(*ii*) Let $f(z) = u + iv$. Therefore, $if(z) = iu - v$. Adding these, we get $(1 + i)f(z) = (u - v) + i(u + v)$.

Setting $(1 + i)f(z) = F(z)$, $u - v = U$, $u + v = V$, then $F(z) = U + iV$.

From $u - v = U = \dfrac{\cos x + \sin x - e^{-y}}{2(\cos x - \cosh y)}$, we get

$$\frac{\partial U}{\partial x} = \frac{\partial u}{\partial x} - \frac{\partial v}{\partial x} = \frac{1}{2}\frac{[(-\sin x + \cos x)(\cos x - \cosh y) - (\cos x + \sin x - e^{-y})(-\sin x)]}{(\cos x - \cosh y)^2}$$

and
$$\frac{\partial U}{\partial y} = \frac{\partial u}{\partial y} - \frac{\partial v}{\partial y} = \frac{1}{2}\frac{[e^{-y}(\cos x - \cosh y) - (\cos x + \sin x - e^{-y})\sinh y]}{(\cos x - \cosh y)^2}.$$

Thus,
$$F'(z) = \frac{\partial U}{\partial x} + i\frac{\partial V}{\partial x} = \frac{\partial U}{\partial x} - i\frac{\partial U}{\partial y} \qquad \text{[using Cauchy-Riemann equations]}$$

$$= \frac{1}{2(\cos x - \cosh y)^2}[(\cos x - \cosh y)(-\sin x + \cos x) + \sin x(\cos x + \sin x - e^{-y})$$

$$-i\{(\cos x - \cosh y)e^{-y} + \sinh y(\cos x + \sin x - e^{-y})\}].$$

By Milne-Thomson's method we express $F'(z)$ in terms of z, on replacing x by z and y by zero, we obtain

$$F'(z) = \frac{1}{2(\cos z - 1)^2}[(\cos z - 1)(-\sin z + \cos z) + \sin z(\cos z + \sin z - 1) - i(\cos z - 1)]$$

$$= \frac{1}{2(\cos z - 1)^2}[-\sin z\cos z + \cos^2 z + \sin z - \cos z + \sin z\cos z + \sin^2 z - \sin z - i\cos z + i]$$

$$= \frac{1}{2(\cos z - 1)^2} - (1 + i)(\cos z - 1) = -\left(\frac{1+i}{2}\right)\cdot\frac{1}{\cos z - 1}$$

$$= \left(\frac{1+i}{2}\right)\frac{1}{1 - \cos z} = \left(\frac{1+i}{2}\right)\cdot\frac{1}{2\sin^2 z/2} = \left(\frac{1+i}{4}\right)\operatorname{cosec}^2(z/2).$$

Integrating, we get $F(z) = -\left(\dfrac{1+i}{4}\right)\left(2\cot\dfrac{z}{2}\right) + c$ or $(1+i)f(z) = c - \left(\dfrac{1+i}{2}\right)\cot\left(\dfrac{z}{2}\right)$

or
$$f(z) = \frac{c}{1+i} - \frac{1}{2}\cot\frac{z}{2} = c_1 - \frac{1}{2}\cot\left(\frac{z}{2}\right), \text{ where } c_1 = \frac{c}{1+i}.$$

Now when $z = \pi/2$, $f(z) = 0$. Therefore, $f\left(\dfrac{\pi}{2}\right) = 0 = c_1 - \dfrac{1}{2}\cot\left(\dfrac{\pi}{4}\right) \Rightarrow c_1 = \dfrac{1}{2}.$

Hence, $f(z) = \dfrac{1}{2}\left(1 - \cot\dfrac{z}{2}\right).$

(*iii*) From $u - v = e^{-x}[(x - y)\sin y - (x + y)\cos y]$, we get

$$\frac{\partial u}{\partial x} - \frac{\partial v}{\partial x} = -e^{-x}[(x - y)\sin y - (x + y)\cos y] + e^{-x}(\sin y - \cos y)$$

$$= e^{-x}[(1 - x + y)\sin y + (x + y - 1)\cos y]. \qquad ...(5)$$

$$\frac{\partial u}{\partial y} - \frac{\partial v}{\partial y} = e^{-x}[x\cos y - (\sin y + y\cos y) + x\sin y - (\cos y - y\sin y)]$$

$$= e^{-x}[(x + y - 1)\sin y + (x - y - 1)\cos y]. \qquad ...(6)$$

Using the Cauchy-Riemann equations, $\dfrac{\partial u}{\partial x} = \dfrac{\partial v}{\partial y}$ and $\dfrac{\partial u}{\partial y} = -\dfrac{\partial v}{\partial x}$, we obtain from Eq. (6)

$$\frac{\partial u}{\partial x}+\frac{\partial v}{\partial x}=e^{-x}[(1-x-y)\sin y+(1-x+y)\cos y]. \qquad ...(7)$$

From Eqs. (5) and (7), we obtain

$$\frac{\partial u}{\partial x}=e^{-x}[(1-x)\sin y+y\cos y], \qquad ...(8)$$

and $$\frac{\partial v}{\partial x}=e^{-x}[(1-x)\cos y-y\sin y]. \qquad ...(9)$$

Integrating $\frac{\partial u}{\partial x}$ and $\frac{\partial v}{\partial x}$ partially with respect to x, we get respectively

$$u(x, y) = e^{-x}[x \sin y - y \cos y] + \phi_1(y),$$

and $v(x, y) = e^{-x}[y \sin y + x \cos y] + \phi_2(y)$, where $\phi_1(y)$ and $\phi_2(y)$ are arbitrary functions of y. Substituting in Eqs. (8) and (9), we obtain respectively

$$e^{-x}[(1-x)\sin y+y\cos y]=e^{-x}[\sin y+y\cos y-x\sin y]+\phi_1'(y),$$

and $$e^{-x}[(1-x)\cos y-y\sin y]=-e^{-x}[x\cos y-\cos y+y\sin y]+\phi_2'(y).$$

Simplifying, we obtain $\phi_1'(y)=0$ and $\phi_2'(y)=0$ which gives $\phi_1(y)=c_1$ and $\phi_2(y)=c_2$, where c_1 and c_2 are arbitrary constants.

Hence, we have $u(x, y) = e^{-x}[x \sin y - y \cos y] + c_1, \quad v(x, y) = e^{-x}[y \sin y + x \cos y] + c_2$

and
$$\begin{aligned} f(z) &= u(x, y) + iv(x, y) = e^{-x}(x \sin y - y \cos y) + c_1 + i[e^{-x}(y \sin y + x \cos y) + c_2] \\ &= e^{-x}[(x + iy) \sin y + i(x + iy) \cos y] + c_1 + ic_2 \\ &= ie^{-x}(x + iy)(\cos y - i \sin y) + c_1 + ic_2 \\ &= i(x + iy) e^{-x} e^{-iy} + c_1 + ic_2 = i(x + iy) e^{-(x + iy)} + c_1 + ic_2 \\ &= i z e^{-z} + c, \text{ where } c = c_1 + ic_2 \text{ is a complex constant.} \end{aligned}$$

(*iv*) From $3u + 2v = y^2 - x^2 + 16xy$, we get

$$3\frac{\partial u}{\partial x}+2\frac{\partial v}{\partial x}=-2x+16y \qquad ...(10)$$

and $$3\frac{\partial u}{\partial y}+2\frac{\partial v}{\partial y}=2y+16x. \qquad ...(11)$$

Using the Cauchy-Riemann equations, $\frac{\partial u}{\partial x}=\frac{\partial v}{\partial y}$ and $\frac{\partial u}{\partial y}=-\frac{\partial v}{\partial x}$, we obtain from Eq. (11)

$$2\frac{\partial u}{\partial x}-3\frac{\partial v}{\partial x}=2y+16x. \qquad ...(12)$$

From Eqs. (10) and (12), we obtain

$$\frac{\partial u}{\partial x}=2x+4y \quad ...(13) \quad \text{and} \quad \frac{\partial v}{\partial x}=-4x+2y. \quad ...(14)$$

Integrating $\frac{\partial u}{\partial x}$ and $\frac{\partial v}{\partial x}$ partially with respect to x, we obtain respectively

$u(x, y) = x^2 + 4xy + \phi_1(y)$ and $v(x, y) = -2x^2 + 2xy + \phi_2(y)$, where $\phi_1(y)$ and $\phi_2(y)$ are arbitrary functions of y. Substituting in Eqs. (13) and (14), we obtain respectively

$$2x+4y+\phi_1'(y)=2x+4y \quad \text{and} \quad -4x+2y+\phi_2'(y)=-4x+2y.$$

Simplifying, we obtain $\phi_1'(y)=0$ and $\phi_2'(y)=0$ which gives $\phi_1(y) = c_1$ and $\phi_2(y) = c_2$, where c_1 and c_2 are arbitrary constants.

Hence, we have $u(x, y) = x^2 + 4xy + c_1$ and $v(x, y) = -2x^2 + 2xy + c_2$.

$$f'(z) = \frac{\partial u}{\partial x} + i\frac{\partial v}{\partial x} = \frac{\partial u}{\partial x} - i\frac{\partial u}{\partial y}.$$

By Milne-Thomson's method, we express $f'(z)$ in terms of z by putting $x = z$ and $y = 0$. Thus

$$f'(z) = \frac{\partial u}{\partial x} + i\frac{\partial v}{\partial x} = 2z - 4iz.$$

Integrating, we get $f(z) = z^2 - 2iz^2 + c = z^2(1 - 2i) + c$, where c is a complex constant.

EXAMPLE 4.13 Find analytic function $f(z) = u(r, \theta) + iv(r, \theta)$ where $v(r, \theta) = r^2 \cos 2\theta - r\cos\theta + 2$.

(AMIETE, June 2007)

Solution We have $z = re^{i\theta}$ and $v(r, \theta) = r^2 \cos 2\theta - r\cos\theta + 2$.

Therefore, $\dfrac{\partial v}{\partial r} = 2r\cos 2\theta - \cos\theta$, ...(*i*) $\quad \dfrac{\partial v}{\partial \theta} = -2r^2 \sin 2\theta + r\sin\theta$, ...(*ii*)

Using Cauchy-Riemann equation $\dfrac{\partial u}{\partial r} = \dfrac{1}{r}\dfrac{\partial v}{\partial \theta}$, we get $\dfrac{\partial u}{\partial r} = -2r\sin 2\theta + \sin\theta$.

Integrating partially with respect to r, we obtain $u(r, \theta) = -r^2 \sin 2\theta + r\sin\theta + \phi(\theta)$, ...(*iii*)

where $\phi(\theta)$ is an arbitrary function of θ.

Using the Cauchy-Riemann equation $\dfrac{\partial v}{\partial r} = -\dfrac{1}{r}\dfrac{\partial u}{\partial \theta}$, that is, $\dfrac{\partial u}{\partial \theta} = -r\dfrac{\partial v}{\partial r}$, we get

$$-2r^2\cos 2\theta + r\cos\theta + \phi'(\theta) = -2r^2\cos 2\theta + r\cos\theta \text{ [from (i) and (iii)] or } \phi'(\theta) = 0.$$

Integrating with respect to θ, we get $\phi(\theta) = c$, where c is an arbitrary constant.

Hence, we obtain $u(r, \theta) = -r^2 \sin 2\theta + r\sin\theta + c$.

Therefore,

$$\begin{aligned} f(z) = u(r,\theta) + iv(r,\theta) &= -r^2\sin 2\theta + r\sin\theta + c + i[r^2\cos 2\theta - r\cos\theta + 2] \\ &= r^2(-\sin 2\theta + i\cos 2\theta) + r(\sin\theta - i\cos\theta) + c + 2i \\ &= r^2(i^2\sin 2\theta + i\cos 2\theta) - r(i\cos\theta - \sin\theta) + c + 2i \\ &= r^2 i(\cos 2\theta + i\sin 2\theta) - ri(\cos\theta + i\sin\theta) + c + 2i \\ &= r^2 i \cdot e^{i2\theta} - rie^{i\theta} + c + 2i = i[(re^{i\theta})^2 - (re^{i\theta})] + c + 2i \\ &= i(z^2 - z) + c + 2i. \end{aligned}$$

EXAMPLE 4.14 Find an analytic function $f(z)$ such that Re $[f'(z)] = 3x^2 - 4y - 3y^2$ and $f(1 + i) = 0$.

Solution We have $f'(z) = \dfrac{\partial u}{\partial x} + i\dfrac{\partial v}{\partial x}$. Therefore, $\text{Re}[f'(z)] = \dfrac{\partial u}{\partial x} = 3x^2 - 4y - 3y^2$.

Integrating with respect to x, we get $u(x, y) = x^3 - 4xy - 3xy^2 + \phi_1(y)$, where $\phi_1(y)$ is an arbitrary function.

Using Cauchy-Riemann equation $\dfrac{\partial u}{\partial x} = \dfrac{\partial v}{\partial y}$, we get $\dfrac{\partial v}{\partial y} = 3x^2 - 4y - 3y^2$.

Integrating with respect to y, we obtain $v(x, y) = 3x^2 y - 2y^2 - y^3 + \phi_2(x)$, where $\phi_2(x)$ is an arbitrary function.

Now $\dfrac{\partial u}{\partial y} = -\dfrac{\partial v}{\partial x}$ gives $-4x - 6xy + \phi_1'(y) = -[6xy + \phi_2'(x)] \Rightarrow \phi_1'(y) = 4x - \phi_2'(x)$.

Since $\phi_1(y)$ is a function of y alone, we get $\phi_2'(x) - 4x = 0$ and $\phi_1'(y) = 0$.

Integration gives $\phi_1(y) = c_1$ and $\phi_2(x) = 2x^2 + c_2$, where c_1 and c_2 are constants.

Now $f(z) = u(x, y) + iv(x, y) = x^3 - 4xy - 3xy^2 + c_1 + i(3x^2 y - 2y^2 - y^3 + 2x^2 + c_2)$

$$= (x^3 + 3ix^2 y - 3xy^2 - iy^3) + 2i(x^2 + 2ixy - y^2) + c_1 + ic_2 = (x + iy)^3 + 2i(x + iy)^2 + (c_1 + ic_2)$$

$$= z^3 + 2iz^2 + (c_1 + ic_2).$$

Now $f(1 + i) = 0 = (1 + i)^3 + 2i(1 + i)^2 + (c_1 + ic_2) = (c_1 - 6) + i(2 + c_2)$ gives $c_1 = 6$, $c_2 = -2$.

Hence, $f(z) = z^3 + 2iz^2 + (6 - 2i)$.

EXAMPLE 4.15 Show that the given function $u(x, y)$ is harmonic. Find the corresponding conjugate harmonic function $v(x, y)$.

(*i*) $\sinh x \sin y$, (*ii*) $\frac{1}{2}\log(x^2 + y^2)$, (*AMIETE, Dec 2010; UPTU 2003; MDU 2000*)

(*iii*) $2x - 3x^3 + 9xy^2$.

Solution (*i*) We have $u(x, y) = \sinh x \sin y$.

Therefore,
$$\frac{\partial u}{\partial x} = \cosh x \sin y, \quad \frac{\partial^2 u}{\partial x^2} = \sinh x \sin y,$$
$$\frac{\partial u}{\partial y} = \sinh x \cos y, \quad \frac{\partial^2 u}{\partial y^2} = -\sinh x \sin y.$$

Since $\frac{\partial^2 u}{\partial x^2} + \frac{\partial^2 u}{\partial y^2} = 0$, the function $u(x, y)$ is harmonic.

From the Cauchy-Riemann equation $\frac{\partial v}{\partial y} = \frac{\partial u}{\partial x}$, we get $\frac{\partial v}{\partial y} = \frac{\partial u}{\partial x} = \cosh x \sin y$.

Integrating partially with respect to y, we get

$v = -\cosh x \cos y + \phi(x)$, where $\phi(x)$ is an arbitrary function of x.

Using the Cauchy-Riemann equation $\frac{\partial v}{\partial x} = -\frac{\partial u}{\partial y}$, we get

$$-\sinh x \cos y + \phi'(x) = -\sinh x \cos y \quad \text{or} \quad \phi'(x) = 0.$$

Integrating with respect to x, we obtain $\phi(x) = c$, where c is a constant.

Hence, we obtain $v(x, y) = -\cosh x \cos y + c$.

(*ii*) We have $u(x, y) = \frac{1}{2}\log(x^2 + y^2)$.

Therefore,
$$\frac{\partial u}{\partial x} = \frac{x}{x^2 + y^2}, \quad \frac{\partial^2 u}{\partial x^2} = \frac{1(x^2 + y^2) - x(2x)}{(x^2 + y^2)^2} = \frac{y^2 - x^2}{(x^2 + y^2)^2},$$

and
$$\frac{\partial u}{\partial y} = \frac{y}{x^2 + y^2} \quad \text{and} \quad \frac{\partial^2 u}{\partial y^2} = \frac{x^2 - y^2}{(x^2 + y^2)^2}.$$

Since $\frac{\partial^2 u}{\partial x^2} + \frac{\partial^2 u}{\partial y^2} = 0$, the function $u(x, y)$ is harmonic.

Let $v(x, y)$ be the harmonic conjugate of $u(x, y)$.

We have
$$dv = \frac{\partial v}{\partial x} dx + \frac{\partial v}{\partial y} dy = -\frac{\partial u}{\partial y} dx + \frac{\partial u}{\partial x} dy$$
$$= \frac{-y}{x^2 + y^2} dx + \frac{x}{x^2 + y^2} dy \quad \text{or} \quad dv = \frac{x\,dy - y\,dx}{x^2 + y^2}.$$

Integrating, we obtain $v = \tan^{-1}(y/x) + c$, a constant.

(iii) We have $u(x, y) = 2x - 3x^3 + 9xy^2$.

Therefore, $\frac{\partial u}{\partial x} = 2 - 9x^2 + 9y^2, \quad \frac{\partial^2 u}{\partial x^2} = -18x, \quad \frac{\partial u}{\partial y} = 18xy, \quad \frac{\partial^2 u}{\partial y^2} = 18x.$

Since $\frac{\partial^2 u}{\partial x^2} + \frac{\partial^2 u}{\partial y^2} = -18x + 18x = 0$, the function $u(x, y)$ is harmonic.

Let $v(x, y)$ be the conjugate function of u, we have

$$dv = \frac{\partial v}{\partial x}dx + \frac{\partial v}{\partial y}dy = -\frac{\partial u}{\partial y}dx + \frac{\partial u}{\partial x}dy \qquad \text{[By C.R. equations]}$$

$$= -18\,xy\,dx + (2 - 9x^2 + 9y^2)\,dy = -9d\,(x^2 y) + d(2y) + d(3y^3).$$

Integrating, we obtain

$$v = -9x^2 y + 2y + 3y^3 + c, \text{ a constant.}$$

EXAMPLE 4.16 Show that the given function is harmonic. Find the corresponding conjugate harmonic function and construct the analytic function $f(z) = u + iv$.

(i) $u(x, y) = y^3 - 3x^2y + 2x$, (ii) $v(x, y) = e^x \sin y$, (*AMIETE, June 2009, June 2008*)

(iii) $u(x, y) = x^2 - y^2 - 2xy - 2x + 3y$.

Solution (i) We have $u(x, y) = y^3 - 3x^2y + 2x$.

Therefore, $\frac{\partial u}{\partial x} = -6xy + 2, \quad \frac{\partial^2 u}{\partial x^2} = -6y, \quad \frac{\partial u}{\partial y} = 3y^2 - 3x^2, \quad \frac{\partial^2 u}{\partial y^2} = 6y.$

Since $\frac{\partial^2 u}{\partial x^2} + \frac{\partial^2 u}{\partial y^2} = -6y + 6y = 0$, the function $u(x, y)$ is harmonic.

From the Cauchy-Riemann equation $\frac{\partial v}{\partial y} = \frac{\partial u}{\partial x}$, we get $\frac{\partial v}{\partial y} = -6xy + 2$.

Integrating partially with respect to y, we obtain

$$v = -3xy^2 + 2y + \phi(x), \text{ where } \phi(x) \text{ is an arbitrary function of } x.$$

Using the Cauchy-Riemann equation $\frac{\partial v}{\partial x} = -\frac{\partial u}{\partial y}$, we obtain

$$-3y^2 + \phi'(x) = -(3y^2 - 3x^2) \quad \text{or} \quad \phi'(x) = 3x^2.$$

Integrating with respect to x, we obtain $\phi(x) = x^3 + c$, where c is a constant.

Hence, we obtain $v(x, y) = -3xy^2 + 2y + x^3 + c$, and

$$\begin{aligned} f(z) &= u(x, y) + iv(x, y) \\ &= y^3 - 3x^2y + 2x + i(-3xy^2 + 2y + x^3 + c) \\ &= 2(x + iy) + y^3 - 3x^2y + i(x^3 - 3xy^2) + ic \\ &= 2(x + iy) + i[x^3 - 3xy^2 - i(y^3 - 3x^2y)] + ic \\ &= 2(x + iy) + i(x + iy)^3 + ic = 2z + iz^3 + ic. \end{aligned}$$

(ii) We have $v(x, y) = e^x \sin y$.

Therefore, $\frac{\partial v}{\partial x} = e^x \sin y, \quad \frac{\partial^2 v}{\partial x^2} = e^x \sin y, \quad \frac{\partial v}{\partial y} = e^x \cos y, \quad \frac{\partial^2 v}{\partial y^2} = -e^x \sin y.$

Since $\frac{\partial^2 v}{\partial x^2} + \frac{\partial^2 v}{\partial y^2} = e^x \sin y - e^x \sin y = 0$, the given function $v(x, y)$ is harmonic.

From the Cauchy-Riemann equation $\frac{\partial u}{\partial x} = \frac{\partial v}{\partial y}$, we get $\frac{\partial u}{\partial x} = e^x \cos y$.

Integrating partially with respect to x, we get

$$u(x, y) = e^x \cos y + \phi(y), \text{ where } \phi(y) \text{ is an arbitrary function of } y.$$

Using the Cauchy-Riemann equation $\frac{\partial u}{\partial y} = -\frac{\partial v}{\partial x}$, we get

$$-e^x \sin y + \phi'(y) = -e^x \sin y \quad \text{or} \quad \phi'(y) = 0.$$

Integrating with respect to y, we obtain $\phi(y) = c$, where c being any arbitrary constant.

Hence, we obtain $u(x, y) = e^x \cos y + c$, and

$$\begin{aligned} f(z) &= u(x, y) + iv(x, y) = e^x \cos y + ie^x \sin y + c \\ &= e^x (\cos y + i \sin y) + c = e^x e^{iy} + c \\ &= e^{x+iy} + c = e^z + c. \end{aligned}$$

(*iii*) We have $u(x, y) = x^2 - y^2 - 2xy - 2x + 3y$.

Therefore, $\frac{\partial u}{\partial x} = 2x - 2y - 2, \quad \frac{\partial^2 u}{\partial x^2} = 2, \quad \frac{\partial u}{\partial y} = -2y - 2x + 3$, and $\frac{\partial^2 u}{\partial y^2} = -2.$

Since $\frac{\partial^2 u}{\partial x^2} + \frac{\partial^2 u}{\partial y^2} = 0$. Hence the function $u(x, y)$ is harmonic.

From the Cauchy-Riemann equation $\frac{\partial v}{\partial y} = \frac{\partial u}{\partial x}$, we get $\frac{\partial v}{\partial y} = \frac{\partial u}{\partial x} = 2x - 2y - 2.$

Integrating partially with respect to y, we obtain

$$v = 2xy - y^2 - 2y + \phi(x), \text{ where } \phi(x) \text{ is an arbitrary function of } x.$$

Using the Cauchy-Riemann equation $\frac{\partial v}{\partial x} = -\frac{\partial u}{\partial y}$, we get

$$2y + \phi'(x) = -(-2y - 2x + 3) \quad \text{or} \quad \phi'(x) = 2x - 3.$$

Integrating with respect to x, we obtain $\phi(x) = x^2 - 3x + c$, where c is a constant.

Hence, we obtain $v(x, y) = x^2 - y^2 + 2xy - 3x - 2y + c$,

and

$$\begin{aligned} f(z) &= u(x, y) + iv(x, y) \\ &= x^2 - y^2 - 2xy - 2x + 3y + i(x^2 - y^2 + 2xy - 3x - 2y + c) \\ &= (1 + i)(x^2 + 2ixy - y^2) - (2 + 3i)(x + iy) + c \\ &= (1 + i)(x + iy)^2 - (2 + 3i)(x + iy) + c \\ &= (1 + i)z^2 - (2 + 3i)z + c, \text{ where } c \text{ is complex constant.} \end{aligned}$$

EXAMPLE 4.17 Show that the function $u(r, \theta) = r^2 \cos 2\theta$ is harmonic. Find its conjugate harmonic function and the corresponding analytic function $f(z)$.

Solution We have $z = re^{i\theta}$ and $u(r, \theta) = r^2 \cos 2\theta$.

Therefore, $\frac{\partial u}{\partial r} = 2r \cos 2\theta, \quad \frac{\partial^2 u}{\partial r^2} = 2\cos 2\theta, \quad \frac{\partial u}{\partial \theta} = -2r^2 \sin 2\theta, \quad \frac{\partial^2 u}{\partial \theta^2} = -4r^2 \cos 2\theta.$

Since $\frac{\partial^2 u}{\partial r^2} + \frac{1}{r}\frac{\partial u}{\partial r} + \frac{1}{r^2}\frac{\partial^2 u}{\partial \theta^2} = 2\cos 2\theta + 2\cos 2\theta - 4\cos 2\theta = 0$, the function $u(r, \theta)$ satisfies the Laplace equation in polar form and therefore is harmonic.

From the Cauchy-Riemann equation $\frac{\partial v}{\partial \theta} = r\frac{\partial u}{\partial r}$, we obtain $\frac{\partial v}{\partial \theta} = r\frac{\partial u}{\partial r} = 2r^2 \cos 2\theta.$

Integrating partially with respect to θ, we obtain

$$v(r, \theta) = r^2 \sin 2\theta + \phi(r), \text{ where } \phi(r) \text{ is an arbitrary function of } r.$$

Using the Cauchy-Riemann equation $\frac{\partial v}{\partial r} = -\frac{1}{r}\frac{\partial u}{\partial \theta}$, we get

$$2r \sin 2\theta + \phi'(r) = 2r \sin 2\theta \quad \text{or} \quad \phi'(r) = 0.$$

Integrating with respect to r, $\phi(r) = c$, where c is an arbitrary constant.

Hence, we obtain $\quad v(r, \theta) = r^2 \sin 2\theta + c.$

Therefore,
$$\begin{aligned} f(z) &= u(r, \theta) + iv(r, \theta) = r^2 \cos 2\theta + i(r^2 \sin 2\theta + c) \\ &= r^2(\cos 2\theta + i \sin 2\theta) + ic = r^2 e^{2i\theta} + ic \\ &= (r e^{i\theta})^2 + ic = z^2 + ic. \end{aligned}$$

EXAMPLE 4.18 If $f(z)$ is a regular function of z, then prove that

$$\left(\frac{\partial^2}{\partial x^2} + \frac{\partial^2}{\partial y^2}\right)|f(z)|^2 = 4|f'(z)|^2.$$ *(UPTU 2007; MDU 2007; AMIETE, June 2005)*

Solution Let $f(z) = u + iv$ then $|f(z)|^2 = u^2 + v^2$.

Therefore,
$$\left(\frac{\partial^2}{\partial x^2} + \frac{\partial^2}{\partial y^2}\right)|f(z)|^2 = \left(\frac{\partial^2}{\partial x^2} + \frac{\partial^2}{\partial y^2}\right)(u^2 + v^2) = \frac{\partial^2}{\partial x^2}(u^2 + v^2) + \frac{\partial^2}{\partial y^2}(u^2 + v^2).$$

Now
$$\frac{\partial}{\partial x}(u^2 + v^2) = 2u\frac{\partial u}{\partial x} + 2v\frac{\partial v}{\partial x}.$$

Therefore,
$$\frac{\partial^2}{\partial x^2}(u^2 + v^2) = \frac{\partial}{\partial x}\left(2u\frac{\partial u}{\partial x} + 2v\frac{\partial v}{\partial x}\right)$$

$$= 2\left[u\frac{\partial^2 u}{\partial x^2} + \left(\frac{\partial u}{\partial x}\right)^2 + v\frac{\partial^2 v}{\partial x^2} + \left(\frac{\partial v}{\partial x}\right)^2\right]. \quad ...(i)$$

Similarly
$$\frac{\partial^2}{\partial y^2}(u^2 + v^2) = 2\left[u\frac{\partial^2 u}{\partial y^2} + \left(\frac{\partial u}{\partial y}\right)^2 + v\frac{\partial^2 v}{\partial y^2} + \left(\frac{\partial v}{\partial y}\right)^2\right]. \quad ...(ii)$$

Adding (*i*) and (*ii*), we obtain

$$\left(\frac{\partial^2}{\partial x^2} + \frac{\partial^2}{\partial y^2}\right)(u^2 + v^2) = 2\left[u\left(\frac{\partial^2 u}{\partial x^2} + \frac{\partial^2 u}{\partial y^2}\right) + v\left(\frac{\partial^2 v}{\partial x^2} + \frac{\partial^2 v}{\partial y^2}\right) + \left(\frac{\partial u}{\partial x}\right)^2 + \left(\frac{\partial u}{\partial y}\right)^2 + \left(\frac{\partial v}{\partial x}\right)^2 + \left(\frac{\partial v}{\partial y}\right)^2\right]. \quad ...(iii)$$

Since $f(z)$ is a regular function, u and v satisfy Laplace's equation.

Therefore,
$$\left(\frac{\partial^2 u}{\partial x^2} + \frac{\partial^2 u}{\partial y^2}\right) = 0, \quad \text{and} \quad \left(\frac{\partial^2 v}{\partial x^2} + \frac{\partial^2 v}{\partial y^2}\right) = 0.$$

Also by Cauchy's Riemann equations $\frac{\partial u}{\partial x} = \frac{\partial v}{\partial y}$ and $\frac{\partial u}{\partial y} = -\frac{\partial v}{\partial x}$.

Thus Eq. (*iii*) can be written as

$$\left(\frac{\partial^2}{\partial x^2} + \frac{\partial^2}{\partial y^2}\right)(u^2 + v^2) = 2\left[\left(\frac{\partial u}{\partial x}\right)^2 + \left(\frac{\partial v}{\partial x}\right)^2 + \left(\frac{\partial v}{\partial x}\right)^2 + \left(\frac{\partial u}{\partial x}\right)^2\right]$$

$$= 4\left[\left(\frac{\partial u}{\partial x}\right)^2 + \left(\frac{\partial v}{\partial x}\right)^2\right] = 4|f'(z)|^2. \qquad \left[\because f'(z) = \frac{\partial u}{\partial x} + i\frac{\partial v}{\partial x}\right]$$

4.11 COMPLEX POTENTIAL FUNCTION

The theory of conjugate functions is of great utility in the solution of two-dimensional steady state flow problems in hydrodynamics, thermodynamics and electronics. In such applications $\Omega(z) = \phi(x, y) + i\psi(x, y)$ where $\Omega(z)$ is analytic, is called **complex potential function** or simply **complex potential.** The real part of $\Omega(z)$, that is, $\phi(x, y)$ is called **velocity potential function** or **velocity potential** and the imaginary part $\psi(x, y)$ is called the **stream function** or lines of force.

The equation $\phi(x, y) = c_1$ and $\psi(x, y) = c_2$ where c_1 and c_2 are constants, represent equipotential lines and stream lines. Also $\phi(x, y)$ and $\psi(x, y)$ are harmonic and we have seen in the last section how to get one when the other is known.

EXAMPLE 4.19 An electrostatic field in the xy-plane is given by the potential function $\phi = 3x^2y - y^3$, find the stream function.

Solution Let $\psi(x, y)$ be a stream function. Using Cauchy-Riemann equations

$$\frac{\partial \psi}{\partial x} = -\frac{\partial \phi}{\partial y} \text{ and } \frac{\partial \psi}{\partial y} = \frac{\partial \phi}{\partial x}.$$

As $$\frac{\partial \phi}{\partial y} = 3x^2 - 3y^2 \text{ and } \frac{\partial \phi}{\partial x} = 6xy.$$

Therefore, $$\frac{\partial \psi}{\partial x} = 3y^2 - 3x^2 \quad ...(i) \quad \text{and} \quad \frac{\partial \psi}{\partial y} = 6xy. \quad ...(ii)$$

Integrating (i) partially with respect to x, we get $\psi = 3xy^2 - x^3 + F(y)$.

Substituting this value of ψ in (ii), we obtain $6xy + F'(y) = 6xy$.

Therefore, $F'(y) = 0$ or $F(y) = c$, a constant. Hence, $\psi = 3xy^2 - x^3 + c$.

EXAMPLE 4.20 In a two-dimensional flow of a field, the stream function is given by $\psi = \tan^{-1}(y/x)$, find the velocity potential ϕ.

Solution By Cauchy-Riemann equations;

$$\frac{\partial \phi}{\partial x} = \frac{\partial \psi}{\partial y} \text{ and } \frac{\partial \phi}{\partial y} = -\frac{\partial \psi}{\partial x}.$$

We have $$\frac{\partial \psi}{\partial y} = \frac{1}{1+(y^2/x^2)}\left(\frac{1}{x}\right) = \frac{x}{x^2+y^2} \text{ and } \frac{\partial \psi}{\partial x} = \frac{1}{1+(y^2/x^2)}\left(\frac{-y}{x^2}\right) = \frac{-y}{x^2+y^2}.$$

Therefore, $$\frac{\partial \phi}{\partial x} = \frac{x}{x^2+y^2} \quad ...(1) \quad \text{and} \quad \frac{\partial \phi}{\partial y} = \frac{y}{x^2+y^2}. \quad ...(2)$$

Integrating (1) partially with respect to x, we get $\phi = \frac{1}{2}\log(x^2+y^2) + F(y)$, ...(3),

where $F(y)$ is a function of y.

Substituting this value of ϕ in (2), we obtain

$$\frac{y}{x^2+y^2} + F'(y) = \frac{y}{x^2+y^2}.$$ Therefore $F'(y) = 0$ or $F(y) = c$, a constant.

Hence $\phi = \frac{1}{2}\log(x^2+y^2) + c$.

EXAMPLE 4.21 If $w = \phi + i\psi$ represents the complex potential for an electric field and $\psi = x^2 - y^2 + \frac{x}{x^2+y^2}$, determine the function ϕ. *(Bhopal 2002S; JNTU, 2000)*

Solution Since w represents the complex potential, it is analytic and satisfies Cauchy-Riemann equations.

$$\frac{\partial \phi}{\partial x} = \frac{\partial \psi}{\partial y} \text{ and } \frac{\partial \phi}{\partial y} = -\frac{\partial \psi}{\partial x}.$$ We have $$\psi = x^2 - y^2 + \frac{x}{x^2+y^2}. \quad ...(i)$$

Using $\frac{\partial\phi}{\partial x}=\frac{\partial\psi}{\partial y}$, we get $\frac{\partial\phi}{\partial x}=\frac{\partial\psi}{\partial y}=-2y-\frac{2xy}{(x^2+y^2)^2}$. [from (*i*)]

Integrating this partially with respect to *x*, we obtain

$$\phi=-2xy+\frac{y}{x^2+y^2}+F(y), \quad ...(ii), \text{ where } F(y) \text{ is an arbitrary function.}$$

Using the Riemann equation $\frac{\partial\phi}{\partial y}=-\frac{\partial\psi}{\partial x}$ ...(*iii*), we have

$$\frac{\partial\phi}{\partial y}=-2x+\frac{x^2-y^2}{(x^2+y^2)^2}+F'(y). \quad [\text{from } (ii)] \quad \text{And} \quad \frac{\partial\psi}{\partial x}=2x+\frac{y^2-x^2}{(x^2+y^2)^2}. \quad [\text{from } (i)]$$

Therefore, from eq. (*iii*), we obtain

$$-2x+\frac{x^2-y^2}{(x^2+y^2)^2}+F'(y)=-2x+\frac{x^2-y^2}{(x^2+y^2)^2} \quad \text{or} \quad F'(y)=0 \Rightarrow F(y)=c, \text{ an arbitrary constant.}$$

Thus, $\phi=-2xy+\frac{y}{x^2+y^2}+c.$

EXERCISE 4.2

1. Show that the following functions are analytic functions of *z*.

(*a*) $w=\frac{x-iy}{x^2+y^2}$, (*b*) $w=\cosh z$,

(*c*) $w=e^z$, (*d*) $w=2x-3y+i(3x+2y)$.

2. Determine α such that the function $f(z)=\frac{1}{2}\log_e(x^2+y^2)+i\tan^{-1}\left(\frac{\alpha x}{y}\right)$ be an analytic function.

(*AMIETE, Dec 2009; JNTU 2003*) **Ans.** $\alpha=-1$.

3. Find the values of the constants *a*, *b*, *c*, *d* such that the function $f(z)=x^2+axy+by^2+i(cx^2+dxy+y^2)$ is analytic.

[**Hint:** Let $f(z)=u(x,y)+iv(x,y)$. $u(x,y)=x^2+axy+by^2$ and $v(x,y)=cx^2+dxy+y^2$.

Therefore, $\frac{\partial u}{\partial x}=2x+ay,\ \frac{\partial u}{\partial y}=ax+2by,\ \frac{\partial v}{\partial x}=2cx+dy,\ \frac{\partial v}{\partial y}=dx+2y.$

Using the Cauchy-Riemann equations $\frac{\partial u}{\partial x}=\frac{\partial v}{\partial y}$ and $\frac{\partial u}{\partial y}=-\frac{\partial v}{\partial x}$, we obtain

$$2x+ay=dx+2y \quad \text{and} \quad ax+2by=-2cx-dy.$$

Solving these equations, we get $a=2$, $b=-1$, $c=-1$ and $d=2$.]

4. Determine the analytic function $f(z)=u+iv$, such that

(*a*) $\text{Re}\,[f(z)]=x^3-3xy^2+3x^2-3y^2+1$, (*b*) $u(x,y)=e^x\cos y$,

(*c*) $v(x,y)=\frac{x-y}{x^2+y^2}$, (*d*) $u(x,y)=e^{2x}(x\cos 2y-y\sin 2y)$.

Ans. (*a*) $f(z)=z^3+3z^2+c'$, where c' is complex constant.

(*b*) $f(z)=e^z+ic$, (*c*) $f(z)=\frac{1+i}{z}+c$, (*d*) $f(z)=ze^{2z}+ic$.

5. If $f(z)$ is an analytic function with constant modulus, show that $f(z)$ is constant.

(*AMIETE, Dec. 2009; UPTU, 2006; Madras 2003*)

6. Show that the function $w = \frac{x}{x^2+y^2} - \frac{iy}{x^2+y^2}$, $(x, y) \neq (0, 0)$ is an analytic function.

(*AMIETE, Dec. 2001*)

[**Hint:** We have $u(x, y) = x/(x^2+y^2)$, $v(x, y) = -y/(x^2+y^2)$.

Therefore, $\frac{\partial u}{\partial x} = \frac{y^2-x^2}{(x^2+y^2)^2}$, $\frac{\partial u}{\partial y} = -\frac{2xy}{(x^2+y^2)^2}$, $\frac{\partial v}{\partial x} = \frac{2xy}{(x^2+y^2)^2}$ and $\frac{\partial v}{\partial y} = \frac{y^2-x^2}{(x^2+y^2)^2}$.

Since, $\frac{\partial u}{\partial x} = \frac{\partial v}{\partial y}$ and $\frac{\partial u}{\partial y} = -\frac{\partial v}{\partial x}$. Therefore, the function w is analytic. $f(z) = u + iv$.

Therefore, $f'(z) = \frac{\partial u}{\partial x} + i\frac{\partial v}{\partial x} = \frac{\partial u}{\partial x} - i\frac{\partial u}{\partial y} = \frac{y^2-x^2}{(x^2+y^2)^2} + 2\frac{ixy}{(x^2+y^2)^2} = \frac{(y^2-x^2)+i(2xy)}{(x^2+y^2)^2}$.

On replacing x by z and y by 0, on right hand side, we get

$$f'(z) = \frac{-z^2}{z^4} = -\frac{1}{z^2}.\Big]$$

7. Find analytic function whose real part is $\frac{\sin 2x}{(\cosh 2y - \cos 2x)}$.

(*JNTU 2005; AMIETE, June 2004, June 2003; Anna 2003*) **Ans.** $f(z) = \cot z + ic$.

Alternative form: $u(x, y) = \frac{2\sin 2x}{e^{2y} + e^{-2y} - 2\cos 2x}$.

8. If $f(z) = \begin{cases} \frac{x^2 y(y - ix)}{x^4 + y^2}, & \text{when } z \neq 0, \\ \text{zero} & \text{when } z = 0. \end{cases}$ Prove that $\frac{f(z) - f(0)}{z} \to 0$, as $z \to 0$, along any radius vector but not as $z \to 0$ in any manner. (*AMIETE, Dec. 2010*)

9. Show that the function $f(z) = \sqrt{|xy|}$ is not analytic at the origin even though CR equations are satisfied thereof. (*AMIETE, June 2011, Dec. 2009, PTU, 2006; UPTU, 2005*)

[**Hint:** Let $f(z) = u(x, y) + iv(x, y) = \sqrt{|xy|}$, then $u(x, y) = \sqrt{|xy|}$, $v(x, y) = 0$.

At the origin, we have

$$\frac{\partial u}{\partial x} = \lim_{x\to 0} \frac{u(x, 0) - u(0, 0)}{x} = \lim_{x\to 0} \frac{0-0}{x} = 0,$$

$$\frac{\partial u}{\partial y} = \lim_{y\to 0} \frac{u(0, y) - u(0, 0)}{y} = \lim_{y\to 0} \frac{0-0}{y} = 0,$$

$$\frac{\partial v}{\partial x} = \lim_{x\to 0} \frac{v(x, 0) - v(0, 0)}{x} = \lim_{x\to 0} \frac{0-0}{x} = 0,$$

$$\frac{\partial v}{\partial y} = \lim_{y\to 0} \frac{v(0, y) - v(0, 0)}{y} = \lim_{y\to 0} \frac{0-0}{y} = 0.$$

From the above results, it is evident that $\frac{\partial u}{\partial x} = \frac{\partial v}{\partial y}$ and $\frac{\partial u}{\partial y} = -\frac{\partial v}{\partial x}$.

Hence Cauchy-Riemann equations are satisfied at the origin.

Now $$f'(0) = \lim_{z\to 0} \frac{f(z) - f(0)}{z} = \lim_{z\to 0} \frac{\sqrt{|xy|} - 0}{x + iy}.$$

Let $z \to 0$ along the line $y = mx$, then

$$f'(0) = \lim_{x \to 0} \frac{\sqrt{|mx^2| - 0}}{x(1+im)} = \frac{\sqrt{|m|}}{1+im} \text{ which is not unique.}$$

Therefore, $f'(0)$ does not exist. Hence $f(z)$ is not analytic at the origin.]

10. Determine the analytic function $f(z) = u + iv$, where $u + v = e^x (\cos y + \sin y)$.

Ans. $e^z + c$, c is a complex constant.

11. Show that the given functions is harmonic. Find the corresponding conjugate harmonic function.

(a) $u(x, y) = 2x - 3x^3 + 9xy^2$, (b) $v(x, y) = \dfrac{-y}{(x^2+y^2)}$, (c) $v(x, y) = \log(x^2 + y^2) + x + y$.

Ans. (a) $v(x, y) = -9x^2y + 3y^3 + 2y + c$; (b) $u(x, y) = \dfrac{x}{x^2+y^2} + c$;

(c) $u(x, y) = x - y - 2\tan^{-1}(y/x) + c$.

12. If u is a harmonic function, then show that $w = u^2$ is not a harmonic function, unless u is a constant.

(AMIETE, June 2010, Dec. 2008, Dec 2006)

[**Hint:** Since u is a harmonic function, $\dfrac{\partial^2 u}{\partial x^2} + \dfrac{\partial^2 u}{\partial y^2} = 0.$ ...(*i*)

Now $w = u^2$ will be a harmonic function if $\dfrac{\partial^2 w}{\partial x^2} + \dfrac{\partial^2 w}{\partial y^2} = 0.$

We have $\dfrac{\partial w}{\partial x} = 2u\dfrac{\partial u}{\partial x}$, $\dfrac{\partial w}{\partial y} = 2u\dfrac{\partial u}{\partial y}$ and

$$\frac{\partial^2 w}{\partial x^2} = 2\left(\frac{\partial u}{\partial x}\right)^2 + 2u\frac{\partial^2 u}{\partial x^2}, \quad \frac{\partial^2 w}{\partial y^2} = 2\left(\frac{\partial u}{\partial y}\right)^2 + 2u\frac{\partial^2 u}{\partial y^2}.$$

Therefore,
$$\frac{\partial^2 w}{\partial x^2} + \frac{\partial^2 w}{\partial y^2} = 2\left[\left(\frac{\partial u}{\partial x}\right)^2 + \left(\frac{\partial u}{\partial y}\right)^2\right] + 2u\left(\frac{\partial^2 u}{\partial x^2} + \frac{\partial^2 u}{\partial y^2}\right) = 2\left[\left(\frac{\partial u}{\partial x}\right)^2 + \left(\frac{\partial u}{\partial y}\right)^2\right] \quad \text{[From (i)]}$$

Now $\dfrac{\partial^2 w}{\partial x^2} + \dfrac{\partial^2 w}{\partial y^2} = 0$, only when $\dfrac{\partial u}{\partial x} = 0$ and $\dfrac{\partial u}{\partial y} = 0$, that is u is a constant function.

Hence $w = u^2$ is not a harmonic function unless u is a constant.]

13. Show that the given function $u(x, y)$ is harmonic. Find the corresponding conjugate harmonic function $v(x, y)$ and construct the analytic function $f(z) = u + iv$

$u(x, y) = e^{-x}(x \cos y + y \sin y)$. **Ans.** $v(x, y) = e^{-x}(y \cos y - x \sin y) + c$;

$f(z) = ze^{-z} + ic$, c a real constant.

14. Show that the function $u(x, y) = 4xy - 3x + 2$ is harmonic. Construct the corresponding analytic function $w = f(z)$ in terms of complex variable z. *(AMIETE, Dec. 2007)*

15. Show that the function $v(r, \theta) = 3r^2 \sin 2\theta - 2r \sin \theta$ in polar coordinates (r, θ) is harmonic. Find the corresponding harmonic conjugate function and construct the analytic function $f(z) = u + iv$ such that $f(1) = 1$. *(AMIETE, June 2006)*

16. Find the regular function whose $Im[f(z)]$ are

(a) $e^{-x}(x \sin y - y \cos y)$, (b) $\dfrac{x-y}{x^2+y^2}$. *(UPTU, 2006; MDU, 2001)*

Ans. (a) $f(z) = -ze^{-z} + c$; (b) $\dfrac{1+i}{z} + c$, where c is complex constant.

17. Given that $u(x, y) = x^2 - y^2$ and $v(x, y) = \dfrac{-y}{x^2 + y^2}$. Prove that both u and v are harmonic functions but $u + iv$ is not analytic function of z. *(UPTU 2005; MDU 2002; JNTU 2000)*

18. Show that the polar form of Cauchy-Riemann equation is

$$\frac{\partial u}{\partial r} = \frac{1}{r}\frac{\partial v}{\partial \theta}, \quad \frac{\partial v}{\partial r} = -\frac{1}{r}\frac{\partial u}{\partial \theta}.$$

Use the result to show that log z is analytic.

[Hint: $f(z) = \log z = \log (re^{i\theta}) = \log r + i\theta = u + iv$.

Here $u = \log r$, $v = \theta$. Therefore, $\dfrac{\partial u}{\partial r} = \dfrac{1}{r}$, $\dfrac{\partial u}{\partial \theta} = 0$, $\dfrac{\partial v}{\partial r} = 0$, $\dfrac{\partial v}{\partial \theta} = 1$.

Thus, $\dfrac{\partial u}{\partial r} = \dfrac{1}{r} = \dfrac{1}{r} \cdot 1 = \dfrac{1}{r}\dfrac{\partial v}{\partial \theta}$ and $-\dfrac{1}{r}\dfrac{\partial u}{\partial \theta} = -\dfrac{1}{r}0 = 0 = \dfrac{\partial v}{\partial r}$. Hence log z is analytic.]

19. In a two dimensional fluid flow, the stream function is $\psi = -y/(x^2 + y^2)$, find the velocity potential ϕ.

Ans. $\dfrac{x}{x^2 + y^2} + c.$

INTEGRATION OF COMPLEX FUNCTIONS

4.12 CONTOUR INTEGRALS (LINE INTEGRALS IN THE COMPLEX PLANE)

Let $f(z)$ be continuous function of z defined in a region R and let C be a smooth curve in the same region with end points a and b.

Subdivide C into n segments by point z_i $(i = 1, 2, ..., n-1)$ and let $a = z_0$ and $b = z_n$.

On each segment we choose an arbitrary point, say, ξ_1 between z_0 and z_1 and ξ_2 between z_1 and z_2 and so on and form the sum

$$S_n = \sum_{k=1}^{n} f(\xi_k)\, \Delta z_k \qquad ...(4.52)$$

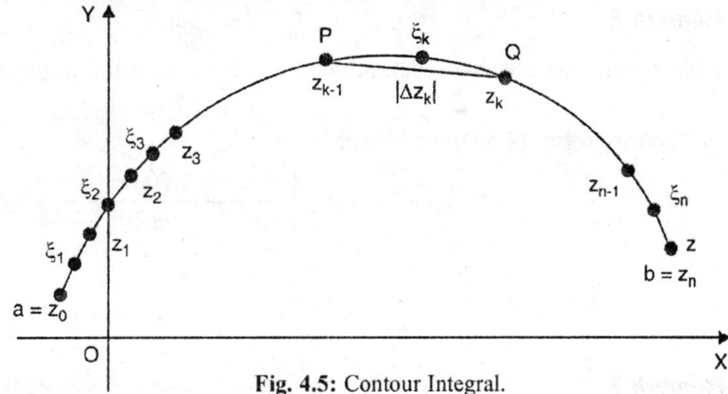

Fig. 4.5: Contour Integral.

where $\Delta z_k = z_k - z_{k-1}$ and $|\Delta z_k| = |z_k - z_{k-1}|$ is the length of the chord PQ.

Then, if it exists, the limit of S_n as $n \to \infty$ such that the length of each chord $\Delta z_k \to 0$ is denoted by

$$\int_C f(z)\, dz \text{ or } \int_a^b f(z)\, dz \qquad ...(4.53)$$

which is called the *complex line integral* or simply the *line integral* of $f(z)$ along the curve C.

In case the point z_0 and z_n coincide so that C is a closed curve, the integral (4.53) is called **contour integral** and is symbolically written as $\oint_C f(z)\, dz$.

Remark 3

This result also holds when C is a piecewise smooth curve and $f(z)$ is a piecewise continuous function defined at each point of C.

Remark 4

If $f(z) = u(x, y) + iv(x, y)$ then, since $dz = dx + idy$, the complex integral (4.53) can be expressed in terms of real line integrals as

$$\int_C f(z)\,dz = \int_C (u+iv)(dx+idy) \qquad \text{...(4.54)}$$

$$= \int_C (udx - vdy) + i\int_C vdx + udy. \qquad \text{...(4.55)}$$

From Eq. (4.55), we observe that a complex line integral can be expressed in terms of two real line integrals, which can be evaluated by using the rules of integration for real functions.

Remark 5

It can be seen from the definition of complex line integral that when the same path of integration is used in each integral then

$$\int_a^b f(z)\,dz = -\int_b^a f(z)\,dz.$$

If c is a point on the arc joining a and b then

$$\int_a^b f(z)\,dz = \int_a^c f(z)\,dz + \int_c^b f(z)\,dz.$$

Remark 6

Suppose that the piecewise smooth curve C is represented in the parametric form as

$$C : z(t) = x(t) + i\,y(t), \qquad a \le t \le b.$$

Then from Eq. (4.54), we obtain

$$\int_C f(z)\,dz = \int_C [u(t)+iv(t)]\left[\frac{dx(t)}{dt} + i\frac{dy(t)}{dt}\right]dt = \int_C [u(t)+iv(t)]\frac{dz}{dt}\,dt$$

or

$$\int_C f(z)\,dz = \int_a^b f[z(t)]\,z'(t)\,dt. \qquad \text{...(4.56)}$$

Remark 7

For a smooth curve $C : z = z(t) = x(t) + iy(t)$, $a \le t \le b$ the length of the curve C is given by

$$L = \int_a^b |z'(t)|\,dt = \int_a^b \sqrt{[x'(t)]^2 + [y'(t)]^2}\,dt$$

$$= \int_a^b \sqrt{\left(\frac{dx}{dt}\right)^2 + \left(\frac{dy}{dt}\right)^2}\,dt. \qquad \text{...(4.57)}$$

Theorem 4.2

If f is continuous on a smooth curve C and if $|f(z)| \le M$ for all z on C, (M a constant) then

$$\left|\int_C f(z)\,dz\right| \le ML,$$

where L is the length of the curve C.

Proof

From the definition of contour integral [see Eq. (4.52)] taking the absolute value and applying the triangle inequality, we obtain

$$|S_n| = \left|\sum_{k=1}^{n} f(\xi_k)\,\Delta z_k\right| \le \sum_{k=1}^{n} |f(\xi_k)||\Delta z_k|$$

$$\le M \sum_{k=1}^{n} |z_k - z_{k-1}|, \qquad \text{...(4.58)}$$

where $|f(z)| \le M$ on C. Now $\sum_{k=1}^{n} |z_k - z_{k-1}| = \overset{*}{L}$, is the sum of the lengths of the chords whose end points are $z_0, z_1, \ldots, z_{n-1}, z_n$ (Fig. 4.5). Since a straight line path is the shortest distance between any two points, $|z_k - z_{k-1}|$ does not exceed the length of the arc joining the points z_{k-1} and z_k. Therefore, as $n \to \infty$ such that $\max |\Delta z_k| \to 0$, we have $\overset{*}{L} \to L$; the length of the curve C. Hence from Eq. (4.58), we obtain

$$\left|\int_C f(z)\,dz\right| = \lim_{n\to\infty} |S_n| \le ML.$$

This inequality is called the ML-inequality.

EXAMPLE 4.22 Evaluate the integral $I = \int_{1+i}^{2+4i} z^2\,dz$

(*i*) along the parabola $x = t$, $y = t^2$ where $1 \le t \le 2$,
(*ii*) along the straight line joining $1 + \mathbf{i}$ and $2 + 4i$,
(*iii*) along straight line from $1 + i$ to $2 + i$ and then to $2 + 4i$.

Solution We have $I = \int_{1+i}^{2+4i} z^2\,dz = \int_{(1,1)}^{(2,4)} (x+iy)^2\,(dx + idy) = \int_{(1,1)}^{(2,4)} (x^2 - y^2 + 2ixy)\,(dx + i\,dy)$

$$= \int_{(1,1)}^{(2,4)} (x^2 - y^2)\,dx - 2\,xy\,dy + i\int_{(1,1)}^{(2,4)} 2\,xy\,dx + (x^2 - y^2)\,dy.$$

(*i*) The point (1, 1) and (2, 4) correspond to $t = 1$ and $t = 2$ respectively. Then the above line integral become

$$\int_{t=1}^{2} [(t^2 - t^4)\,dt - 2\,(t)\,(t^2)\,2t\,dt] + i\int_{t=1}^{2} [2(t)\,(t^2)\,dt + (t^2 - t^4)\,2t\,dt]$$

$$= \int_{t=1}^{2} (t^2 - 5t^4)\,dt + i\int_{t=1}^{2} (4t^3 - 2t^5)\,dt = \left[\frac{t^3}{3} - t^5\right]_1^2 + i\left[t^4 - \frac{t^6}{3}\right]_1^2$$

$$= \left(\frac{7}{3} - 31\right) + i\,(15 - 21) = -\frac{86}{3} - 6i.$$

(*ii*) The line joining (1, 1) and (2, 4) has the equation $y-1=\frac{4-1}{2-1}(x-1)$ or $y=3x-2$. We find,

$$I=\int_{x=1}^{2}[\{x^2-(3x-2)^2\}\,dx-2x\,(3x-2)\,3dx]+i\int_{x=1}^{2}[\{2x\,(3x-2)\}\,dx+\{x^2-(3x-2)^2\}\,3dx]$$

$$=\int_{x=1}^{2}(-26x^2+24x-4)\,dx+i\int_{x=1}^{2}(-18x^2+32x-12)\,dx$$

$$=\left[-\frac{26x^3}{3}+12x^2-4x\right]_1^2+i\left[-6x^3+16x^2-12x\right]_1^2=\frac{-86}{3}-6i.$$

(*iii*) From $1+i$ to $2+i$ [or (1, 1) to (2, 1)], $y=1$, $dy=0$, we have

$$\int_{x=1}^{2}(x^2-1^2)\,dx+i\int_{x=1}^{2}(2x)(1)\,dx=\left[\frac{x^3}{3}-x\right]_1^2+i\left[x^2\right]_1^2=\frac{4}{3}+3i. \quad \text{...(1)}$$

From $2+i$ to $2+4i$ [or (2, 1) to (2, 4)], $x=2$, $dx=0$ and we have

$$\int_{y=1}^{4}-4y\,dy+i\int_{y=1}^{4}(4-y^2)\,dy=\left[-2y^2\right]_1^4+i\left[4y-\frac{y^3}{3}\right]_1^4=-30-9i. \quad \text{...(2)}$$

Adding (1) and (2), we obtain $I=\left(\frac{4}{3}-30\right)+i\,(3-9)=\frac{-86}{3}-6i.$

Note: In the above example, the value of the integral is the same whether the path of integration is along the curve *AB* or along straight line *AB* or first along *AC* and then *CB*. In fact the value will be same even if some other path is chosen. It is so because the integrand, that is, z^2 is analytic in the z-plane. If the integrand is not analytic, the value depends on the path of integration as we shall see in the following example.

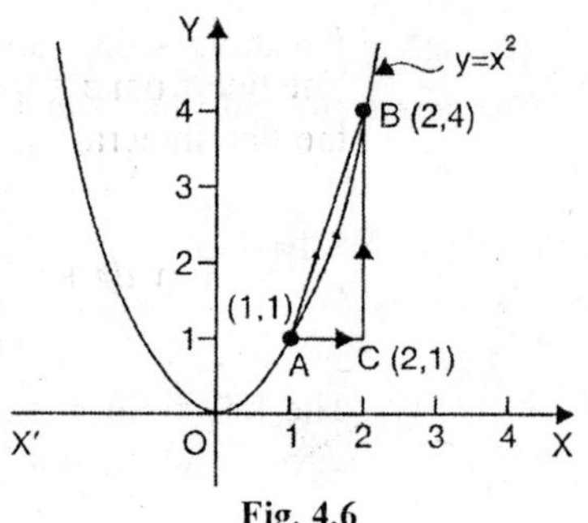

Fig. 4.6

EXAMPLE 4.23 Evaluate the integral

$$I=\int_{0}^{1+i}(x^2-iy)\,dz,$$

(*i*) along the straight line $y=x$,

(*ii*) along the curve $y=x^2$.

Solution We have $I=\int_0^{1+i}(x^2-iy)\,dz=\int_{(0,0)}^{(1,1)}(x^2-iy)\,(dx+idy)=\int_{(0,0)}^{(1,1)}x^2dx+y\,dy+i\int_{(0,0)}^{(1,1)}x^2dy-y\,dx.$

(*i*) Along the line C_1: $y=x$, $dy=dx$ and x varies from 0 to 1.

Therefore, $\int_{C_1}(x^2-iy)\,dz=\int_{x=0}^{1}(x^2dx+x\,dx)+i\int_0^1(x^2dx-x\,dx)=\left[\frac{x^3}{3}+\frac{x^2}{2}\right]_0^1+i\left[\frac{x^3}{3}-\frac{x^2}{2}\right]_0^1=\frac{5}{6}-\frac{i}{6}.$

(*ii*) Along the curve $C: y=x^2$, $dy=2x\,dx$ and x varies from 0 to 1.

Therefore, $\int_C (x^2 - iy)\,dz = \int_{x=0}^{1} x^2 dx + 2x^3 dx + i\int_{x=0}^{1} 2x^3 dx - x^2 dx$

$$= \left[\frac{x^3}{3} + \frac{x^4}{2}\right]_0^1 + i\left[\frac{x^4}{2} - \frac{x^3}{3}\right]_0^1 = \frac{5}{6} + \frac{i}{6}.$$

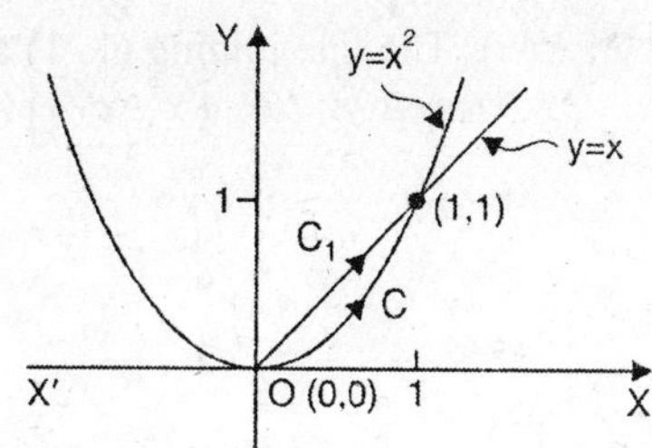

Fig. 4.7

EXAMPLE 4.24 Evaluate the following integrals:

(*i*) $I = \int_C \bar{z}\,dz$ from $z = 0$ to $z = 4 + 2i$ along the curve C given by

(*a*) $z = t^2 + it$, (*b*) the line from $z = 0$ to $z = 2i$ and then the line from $z = 2i$ to $z = 4 + 2i$.

(*ii*) $I = \int_C^{2+i} (\bar{z})^2\,dz$, along (*a*) the line $y = x/2$, (*b*) the real axis to 2 and then vertically to $2 + i$.

(*UPTU, 2002; VTU 2001*)

Solution (*i*) (*a*) The point $z = 0$ and $z = 4 + 2i$ on C correspond to $t = 0$ and $t = 2$ respectively. Then the line integral

$$= \int_{t=0}^{2} \overline{(t^2 + it)}\,d\,(t^2 + it) = \int_0^2 (t^2 - it)\,(2t + i)\,dt = \int_0^2 (2t^3 - it^2 + t)\,dt = \left[\frac{t^4}{2} - i\frac{t^3}{3} + \frac{t^2}{2}\right]_0^2 = 10 - \frac{8i}{3}.$$

(*b*) The given line integral $= \int_C (x - iy)\,(dx + i\,dy) = \int_C x\,dx + y\,dy + i\int_C x\,dy - y\,dx.$

The line from $z = 0$ to $z = 2i$ is the same as the line from (0, 0) to (0, 2) for which $x = 0$, $dx = 0$ and the line integral

$$= \int_{y=0}^{2} y\,dy + i\int_{y=0}^{2} (0)\,dy - y\,(0) = \int_{y=0}^{2} y\,dy = 2.$$

The line from $z = 2i$ to $z = 4 + 2i$ is the same as the line from (0, 2) to (4, 2) for which $y = 2$, $dy = 0$ and the line integral

$$= \int_{x=0}^{4} x\,dx + (2)\,(0) + i\int_0^4 x\,(0) - 2dx = 8 - 8i.$$ The required value $= 2 + (8 - 8i) = 10 - 8i$.

(*ii*) Refer Fig. 4.8. Along the line OA, $x = 2y$, $dx = 2dy$, $z = x + iy = 2y + iy = y\,(2 + i)$, $\bar{z} = (2 - i)y$ and $dz = (2 + i)\,dy$.

$$\therefore\ I = \int_0^{2+i} (\bar{z})^2\,dz = \int_0^1 (2 - i)^2 y^2 \cdot (2 + i)\,dy$$

$$= (10 - 5i)\left[\frac{y^3}{3}\right]_0^1 = \frac{10 - 5i}{3}.$$

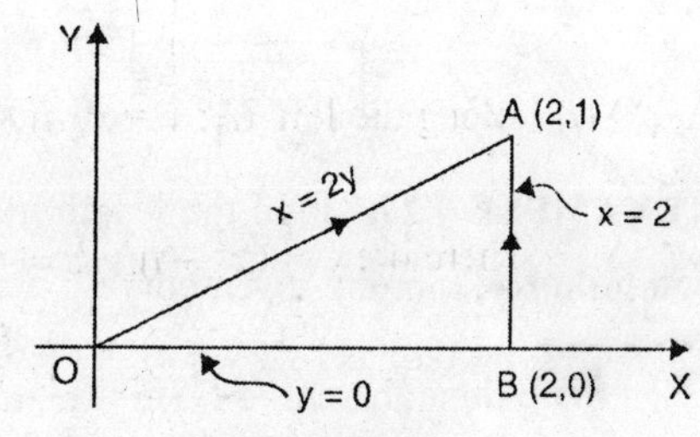

Fig. 4.8

Now along OB, $z = x$, $\bar{z} = x$, $dz = dx$ and along BA, $z = 2 + iy$, $\bar{z} = 2 - iy$, $dz = i\,dy$.

$$\text{Therefore, } I = \int_0^2 x^2 dx + \int_0^1 (2-iy)^2\, i dy = \left[\frac{x^3}{3}\right]_0^2 + \int_0^1 [4y + (4-y^2)\, i]\, dy$$

$$= \frac{8}{3} + \left[2y^2 + \left(4y - \frac{y^3}{3}\right) i\right]_0^1 = \frac{8}{3} + 2 + \left(4 - \frac{1}{3}\right) i = \frac{1}{3}(14 + 11i).$$

EXAMPLE 4.25 Evaluate $\oint \frac{1}{z}\, dz$ where C is the circle $|z| = r$.

Solution We have $\oint_C \frac{1}{z}\, dz = \oint_C \frac{1}{x+iy}(dx + i\, dy) = \oint_C \frac{x - iy}{x^2 + y^2}(dx + i\, dy).$ ...(*i*)

The cartesian form of $C : |z| = r$ is $x^2 + y^2 = r^2$. Any point on the circle is $x = r \cos \theta$ $y = r \sin \theta$.

Therefore, Eq. (*i*) can be written in the form $\int_0^{2\pi} \frac{r\cos\theta - i\, r \sin\theta}{r^2}(-r\sin\theta + ir\cos\theta)\, d\theta = \int_0^{2\pi} i\, d\theta = 2\pi i.$

($\because$ as z goes round the circle, θ varies from 0 to 2π)

Alternative Solution : The equation of the circle $C : |z| = r$ can be written in the form $z = re^{i\theta}$

Therefore, putting $z = re^{i\theta}$, $dz = rie^{i\theta} d\theta$, we have $\oint \frac{1}{z}\, dz = \int_0^{2\pi} \frac{rie^{i\theta}}{re^{i\theta}}\, d\theta = i\int_0^{2\pi} d\theta = 2\pi i.$

EXAMPLE 4.26 Evaluate $\oint_C \frac{dz}{(z-a)^n}$, where n is an integer and C is the circle $|z - a| = r$.

Solution The equation $|z - a| = r$ can be written as $z - a = re^{i\theta}$. Therefore, $dz = r\, ie^{i\theta} d\theta$.

$$\text{Thus} \quad \oint \frac{dz}{(z-a)^n} = \int_0^{2\pi} \frac{rie^{i\theta}\, d\theta}{r^n e^{ni\theta}} = \frac{i}{r^{n-1}} \int_0^{2\pi} e^{i\theta(1-n)}\, d\theta. \qquad \text{...}(i)$$

Case 1: When $n = 1$, Eq. (*i*) becomes $i\int_0^{2\pi} d\theta = 2\pi i.$

Case 2: When $n \neq 1$, Eq. (*i*) can be written in the form

$$\frac{1}{r^{n-1}} \int_0^{2\pi} [\cos(1-n)\theta + i \sin(1-n)\theta]\, d\theta = \frac{1}{r^{n-1}}\left[\frac{\sin(1-n)\theta}{1-n} - \frac{i\cos(1-n)\theta}{1-n}\right]_0^{2\pi} = 0.$$

Therefore, when C is the circle $|z - a| = r$.

$$\oint_C \frac{dz}{(z-a)^n} = \begin{cases} 2\pi i & \text{if } n = 1, \\ 0 & \text{if } n = -1, \pm 2, \pm 3, \ldots \text{ that is, an integer} \neq 1. \end{cases}$$

EXAMPLE 4.27 Find the length of the curve $C : z(t) = (t + \sin t) + i(1 + \cos t)$, $0 \leq t \leq 2\pi$.

Solution From the given curve $C : z(t) = (t + \sin t) + i(1 + \cos t)$, we obtain

$z'(t) = (1 + \cos t) + i(-\sin t)$ and $|z'(t)| = |(1 + \cos t) + i(-\sin t)|$

$$= \sqrt{(1+\cos t)^2 + \sin^2 t} = 2|\cos(t/2)|.$$

Hence, the length of the curve C is given by

$$L = \int_C |z'(t)|\,dt = 2\int_0^{2\pi} |\cos(t/2)|\,dt = 2\int_0^{\pi}\cos(t/2)\,dt - 2\int_{\pi}^{2\pi}\cos(t/2)\,dt$$

$$= 4[\sin(t/2)]_0^{\pi} - 4[\sin(t/2)]_{\pi}^{2\pi} = 4\left(\sin\frac{\pi}{2}\right) - 4(\sin\pi - \sin\pi/2) = 8\sin\frac{\pi}{2} = 8.$$

EXAMPLE 4.28 Evaluate the integral $I = \int_C (x + y^2 - ixy)\,dz$ where $C : z = z(t) = \begin{cases} t - 2i, & 1 \le t \le 2, \\ 2 - i(4 - t), & 2 < t \le 3. \end{cases}$

(*AMIETE, June 2011, Dec. 2006*)

Solution The curve C is continuous but not differentiable at $z = 2$, as

$\dfrac{dz}{dt} = \begin{cases} 1, & 1 < t \le 2 \\ i, & 2 < t < 3. \end{cases}$ Also $dz/dt \ne 0$ for any t. Therefore, the curve C is piecewise smooth.

On the interval $[1, 2)$, we have $z = t - 2i$, that is, $x = t$, $y = -2$, $z'(t) = 1$ and $f(z) = x + y^2 - ixy = t + 4 + 2it$.

On the interval $[2, 3)$, we have $z = 2 - i(4 - t)$, that is, $x = 2$, $y = t - 4$, $z'(t) = i$ and $f(z) = x + y^2 - ixy = 2 + (t - 4)^2 - 2i(t - 4)$.

Hence, $$I = \int_C (x + y^2 - ixy)\,dz = \int_1^2 f(z)\,z'(t)\,dt + \int_2^3 f(z)\,z'(t)\,dt$$

$$= \int_1^2 [4 + t + 2it]\,dt + \int_2^3 [2 + (t-4)^2 - 2i(t-4)]\,i\,dt$$

$$= \int_1^2 [4 + t + 2it]\,dt + \int_2^3 [t^2 - (8 + 2i)t + 18 + 8i]\,i\,dt$$

$$= \left[4t + \frac{1}{2}(1 + 2i)t^2\right]_1^2 + i\left[\frac{1}{3}t^3 - (4 + i)t^2 + (18 + 8i)t\right]_2^3$$

$$= 4 + \frac{3}{2}(1 + 2i) + i\left[\frac{19}{3} - 5(4 + i) + (18 + 8i)\right] = \left(4 + \frac{3}{2} - 3\right) + i\left(3 + \frac{19}{3} - 2\right) = \frac{5}{2} + \frac{22}{3}i.$$

EXAMPLE 4.29 Evaluate the integral $I = \int_C \left(\dfrac{z}{\bar{z}}\right) dz$, where C is the boundary of the half annulus as shown in Fig. 4.9.

Solution Let $z = x + iy$ and $f(z) = \dfrac{z}{\bar{z}}$. We have

along AB: $y = 0$, $z = x$, $f(z) = 1$, $-3 \le x \le -2$, $dz = dx$

along C_1: $z = 2e^{i\theta}$, $f(z) = \dfrac{2e^{i\theta}}{2e^{-i\theta}} = e^{2i\theta}$, $\pi \le \theta \le 0$, $dz = 2ie^{i\theta}d\theta$

along CD: $y = 0$, $z = x$, $f(z) = 1$, $2 \le x \le 3$, $dz = dx$

along C_2: $z = 3e^{i\theta}$, $f(z) = \dfrac{3e^{i\theta}}{3e^{-i\theta}} = e^{2i\theta}$, $0 \le \theta \le \pi$, $dz = 3ie^{i\theta}d\theta$.

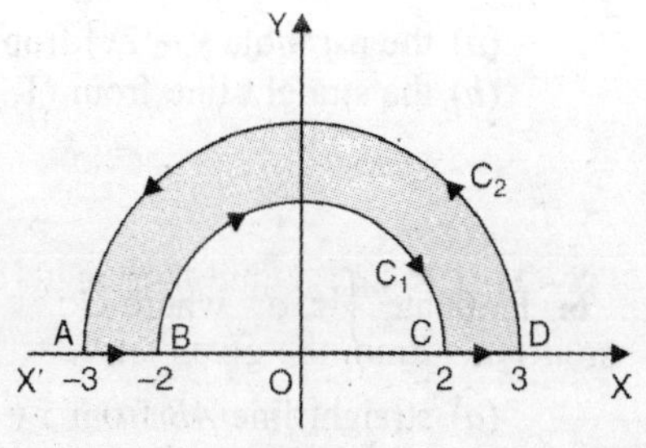

Fig. 4.9

Hence, we obtain $I = \int_{-3}^{-2} dx + 2i\int_{\pi}^{0} e^{3i\theta}\, d\theta + \int_{2}^{3} dx + 3i\int_{0}^{\pi} e^{3i\theta}\, d\theta = (-2+3) + \frac{2i}{3i}[e^{3i\theta}]_{\pi}^{0} + 1 + [e^{3i\theta}]_{0}^{\pi}$

$$= 1 + \frac{2}{3}[\cos 3\theta + i\sin 3\theta]_{\pi}^{0} + 1 + [\cos 3\theta + i\sin 3\theta]_{0}^{\pi}$$

$$= 1 + \frac{2}{3}[1-(-1)] + 1 + [-1-1] = 1 + \frac{4}{3} + 1 - 2 = \frac{4}{3}.$$

EXERCISE 4.3

1. Evaluate the integral $I = \int_{0}^{3+i} z^2 dz$ along

   (*a*) the line $y = x/3$ (*b*) the real axis to 3 and then vertically to $3 + i$. **Ans.** Each equal to $6 + (26/3)i$.

2. Evaluate the integral $I = \int z^2 dz$ from A (1, 1) to B (2, 4) along :

   (*a*) the line segment AD parallel to x-axis and DB parallel to y-axis.

   (*b*) the straight line AB joining the two points.

   (*c*) the curve $C : x = t, y = t^2$.

   Ans. $I = \frac{-86}{3} - 6i$, the three integrals are equal.

Fig. 4.10

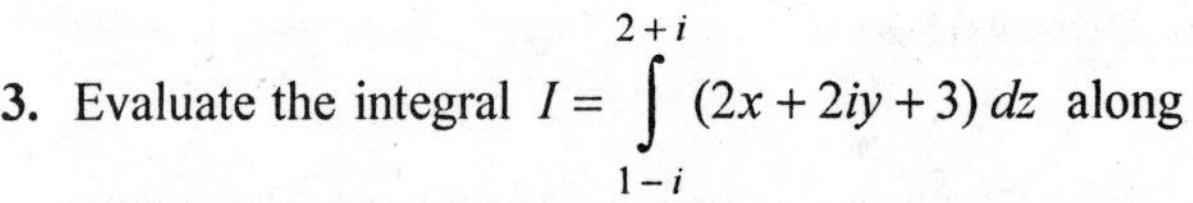

3. Evaluate the integral $I = \int_{1-i}^{2+i} (2x + 2iy + 3)\, dz$ along

   (*a*) the path $x = t + 1, y = 2t^2 - 1$,

   (*b*) the straight line joining $1 - i$ and $2 + i$. **Ans.** (*a*) $6 + 12i$, (*b*) $6 + 12i$.

4. Find the value of the integral $\int_{0}^{1+i} (x - y + ix^2)\, dz$

   (*a*) along the straight line from $z = 0$ to $z = 1 + i$,

   (*b*) along the real axis from $z = 0$ to $z = 1$ and then along a line parallel to imaginary axis from $z = 1$ to $z = 1 + i$. (*AMIETE, Dec 2010; UPTU 2003*) **Ans.** (*a*) $\frac{1}{3}(-1+i)$, (*b*) $-\frac{1}{2} + \frac{5i}{6}$.

5. Evaluate the integral $I = \int_{C} (x^2 - iy^2)\, dz$ along

   (*a*) the parabola $y = 2x^2$ from (1, 2) to (2, 8),

   (*b*) the straight line from (1, 1) to (1, 8) and then from (1, 8) to (2, 8).

   Ans. (*a*) $\frac{511}{3} - \frac{49}{5}i$, (*b*) $\frac{518}{3} - 57i$.

6. Evaluate $\int_{C} |z|\, dz$ where C :

   (*a*) straight line AB from $z = -i$ to $z = i$,

   (*b*) the left half of the unit circle $|z| = 1$ from $z = -i$ to $z = i$. **Ans.** (*a*) i, (*b*) $2i$.

7. Evaluate the line integral $\int_C (y - x - 3ix^2)\,dz$ where the contour C consists of two straight lines, one from $z = 0$ to $z = i$ and the other from $z = i$ to $z = 1 + i$. **Ans.** $(1 - i)/2$.

8. Evaluate the integral $\oint_C (z^2 - 2z + 1)\,dz$ where C is the circle $x^2 + y^2 = 2$. **Ans.** Zero.

9. Evaluate the integral $\oint_C (5z^4 - z^3 + 2)\,dz$ around

   (*a*) circle : $|z| = 2$, (*b*) square with vertices (0, 0), (3, 0) (3, 3) and (0, 3). **Ans.** Zero in both case.

10. Evaluate the integral $I = \int_0^{1+i} z^2 dz$. **Ans.** $-\frac{2}{3} + \frac{2}{3}i$.

11. Find the length of the curve $C : z = (1 - i)\,t^2,\ -1 \le t \le 1$. **Ans.** $2\sqrt{2}$ unit.

4.13 CAUCHY INTEGRAL THEOREM

Simple Closed Curve, Simply and multiply connected Domain

A curve is called a *simple closed* curve if it does not intersect itself between its end points. For example, the curve in Fig 4.11 (a) is simple closed while the curve in Fig 4.11 (b) is closed but not simple and is known as *multiple curve.*

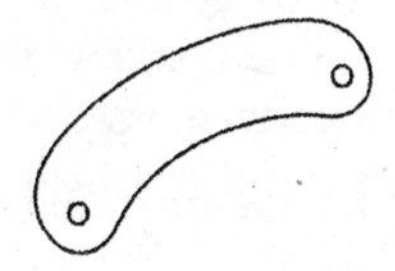

(a) Simple and closed curve

(b) Closed but not simple curve

Fig. 4.11

A domain D is called *simply connected* if every simple closed contour C lying entirely in D can be shrunk, or contracted to a point without leaving D. In other words, in a simply connected domain every simple closed contour C lying entirely within it encloses only points of the domain D. A domain which is not simply connected is called *multiply connected* domain. For example, the domain bounded by a circle or a square is simply connected domain Fig 4.12 (a). A plain sheet of paper from which some interior parts are removed is a multiply connected domain Fig. 4.12 (b). Another simple example of a multiply connected domain is the annulus $r < |z| < R$ that is the domain between two concentric circles $|z| = R$ and $|z| = r$ cannot be shrunk to a point. Fig. 4.12 (c).

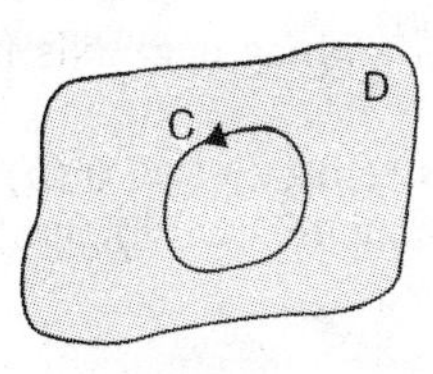

(a) Simply connected domain

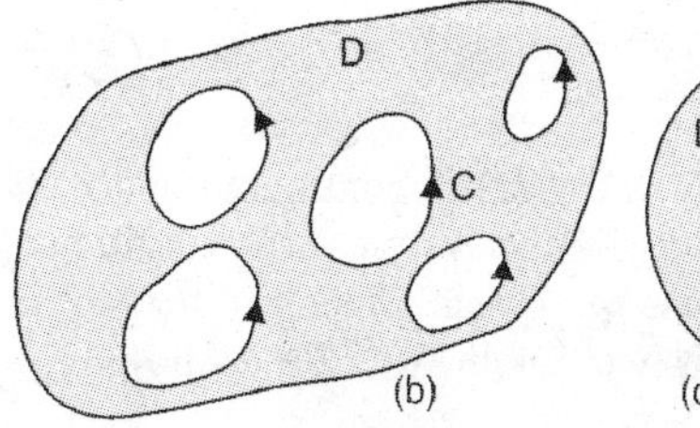

(b) and (c) multiply connected domains

Fig. 4.12

In plain terms a *simply connected* domain is one which has no "holes" in it. A domain with one "hole" is called *doubly connected* and a domain with two "holes" is called *triply connected* and so on.

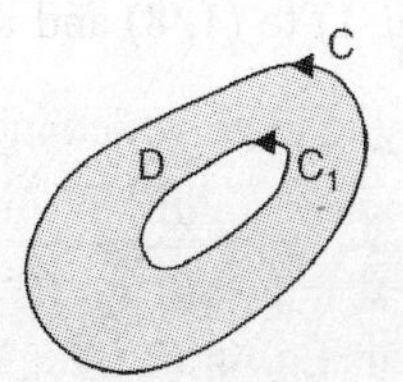

(a) Doubly connected doamin

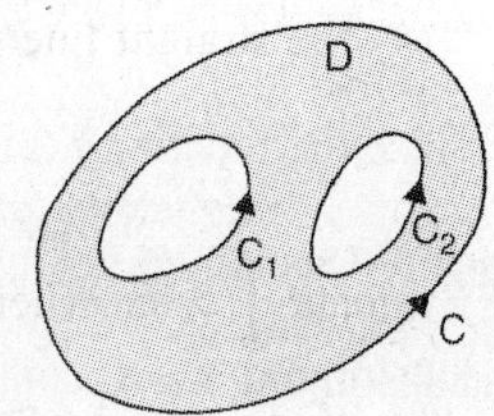

(b) Triply connected domain

Fig. 4.13

Any multiply connected domain can be converted into a simply connected domain by introducing sufficient number of cuts in the domain. Consider a doubly connected domain as shown in Fig. 4.14 (a). Introduce a cut in the domain along PQ Fig. 4.14 (b).

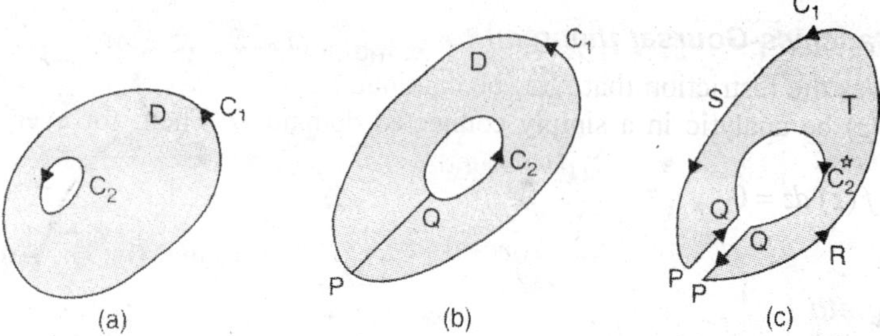

Fig. 4.14: Doubly connected domain and its conversion to a simply connected domain.

The new domain in Fig. 4.14 (c) is simply connected with the bounding curve as $PRSPQTQP$. If the domain is triply connected, then we introduce two cuts to convert it into a simply connected domain. (Fig. 4.15)

In general, by introducing the required number of cuts, a muliply connected domain can be converted into a simply connected domain. This concept is very useful in evaluating integrals along a curve C which is bounding a muliply connected domain.

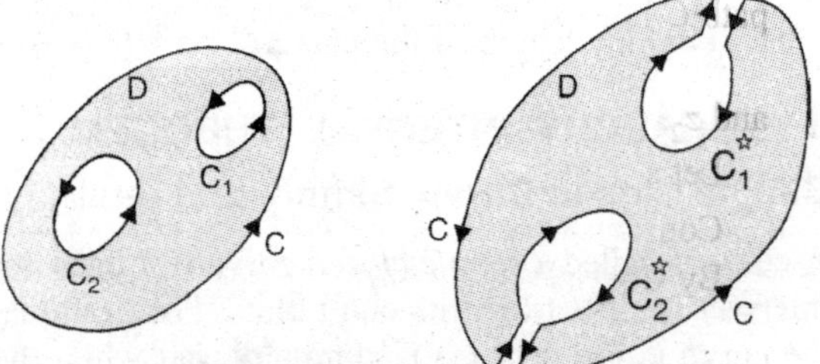

Fig. 4.15: Triply connected domain and its conversion to a simply connected domain.

Theorem 4.3 (Cauchy's integral theorem)

Statement : If a function $f(z)$ is analytic in a simply connected domain D and that $f'(z)$ is continuous in D. Then for every simple closed coutour C in D

$$\oint_C f(z)\,dz = 0. \quad \text{...(4.59)}$$

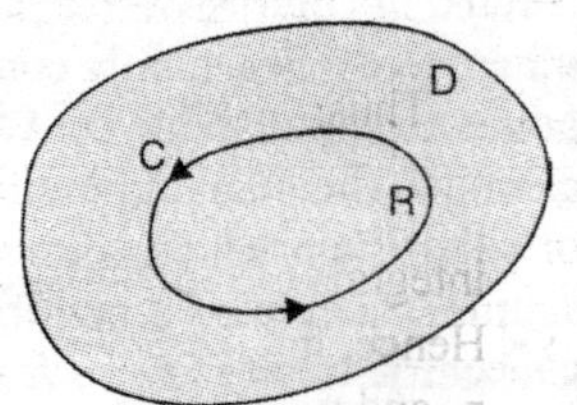

Fig. 4.16: Cauchy integral theorem.

Proof: Let D be a simply connected domain containing a simple, smooth closed curve C inside it (Fig. 4.16). Let $f(z) = u(x, y) + iv(x, y)$. Then

$$\oint_C f(z)\,dz = \oint_C (u + iv)\,(dx + idy) = \oint_C (u\,dx - v\,dy) + i\oint_C v dx + u dy. \quad \text{...(4.60)}$$

Since $f(z)$ is analytic in D and $f'(z)$ is continuous in D, the real and imaginary parts $u(x, y)$ and $v(x, y)$ together with their first order partial derivatives are continuous in D. To evaluate the real integrals on the right hand side of Eq. (4.60), we use the Green's theorem for real integrals.

Hence by Green's theorem, we obtain from Eq. (4.60)

$$\oint_C f(z)\,dz = \iint_R \left(-\frac{\partial v}{\partial x} - \frac{\partial u}{\partial y}\right) dxdy + i\iint_R \left(\frac{\partial u}{\partial x} - \frac{\partial v}{\partial y}\right) dxdy. \quad \text{...(4.61)}$$

Since the function $f(z)$ is analytic in D, which include R where R is the region bounded by C, it satisfies the Cauchy-Rieman equations $\dfrac{\partial u}{\partial x} = \dfrac{\partial v}{\partial y}$ and $\dfrac{\partial u}{\partial y} = -\dfrac{\partial v}{\partial x}$.

Substituting these conditions in Eq. (4.61), the integrands of each of the double integrals vanish. Hence,

$$\oint_C f(z)\,dz = 0.$$

Remark 8

The above theorem is also known as Cauchy's theorem. The Cauchy integral theorem uses two conditions: (*i*) the domain D is simply connected and (*ii*) $f(z)$ is analytic and $f'(z)$ is continuous in D. The condition that D is simply connected is a necessary condition.

Theorem 4.4 (Cauchy's-Goursat theorem)

This theorem removes the restriction that $f'(z)$ be continuous.

Statement: Let $f(z)$ be analytic in a simply connected domain D. Then, for every simple smooth, closed contour C in D, $\oint_C f(z)\,dz = 0.$

Independence of path

Let $f(z)$ be analytic in a simply connected domain D and C be any path joining two points z_1 and z_2 in D. The path C lies entirely in D. Then $\int_C f(z)\,dz$ is independent of the path C and depends only on the end points z_1 and z_2.

Let z_1 and z_2 be any two points in D. (Fig. 4.17).

Consider two paths C_1 and C_2 in D from z_1 to z_2.

By Cauchy's theorem,

$$\int_{ABCEA} f(z)\,dz = 0 \quad \text{or} \quad \int_{ABC} f(z)\,dz + \int_{CEA} f(z)\,dz = 0.$$

Hence $$\int_{ABC} f(z)\,dz = -\int_{CEA} f(z)\,dz = \int_{AEC} f(z)\,dz.$$

Thus, $$\int_{C_1} f(z)\,dz = \int_{C_2} f(z)\,dz = \int_{z_1}^{z_2} f(z)\,dz,$$ which shows that the integrals of $f(z)$ along C_1 and C_2 are equal. But, the paths C_1 and C_2 joining the points z_1 and z_2 are arbitrary. Hence, the integral of $f(z)$ is independent of the path of integration in D and depends only on the end points z_1 and z_2.

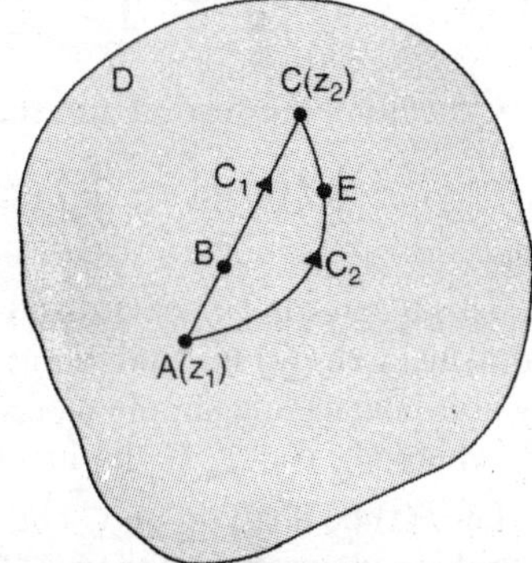

Fig. 4.17: Independence of path.

Deformation of Path

Let z_1, z_2 be any two points in D. Consider any two paths C_1 and C_2 in D joining the points z_1 and z_2. When $\int_C f(z)\,dz$ is independent of the path of integration, we can imagine that the path C_2 was obtained from C_1 by continuously moving C_1 (with ends fixed) until it coincides with C_2. Fig. 4.18 shows three of the infinitely many intermediate paths for which the integral always retains its value. This deformation of the path of integration, keeping the end points fixed, can be continued so long as the deformed path always contains only points at which $f(z)$ is analytic. The value of the integral remains the same under the deformation. This is called the *principle of deformation of path or contour*.

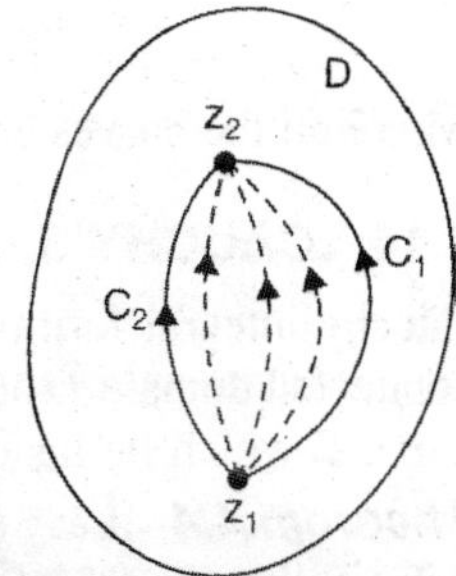

Fig. 4.18: Deformation of path.

4.13.1 Extension of Cauchy's Integral Theorem for Multiply Connected Domains

Consider a doubly connected domain D, whose outer and inner boundary curves are C_1 and C_2 respectively. (Fig. 4.19).

Let $f(z)$ be analytic in the doubly connected domain D and on C_1 and C_2. Then, we can prove

that $$\oint_{C_1} f(z)\,dz = \oint_{C_2} f(z)\,dz \quad ...(4.62)$$

where both the curve C_1 and C_2 are traversed in the same direction. Refer Fig. 4.19 (iii), doubly connected domain has been converted into a simply connected domain by introducing a cut in the domain.

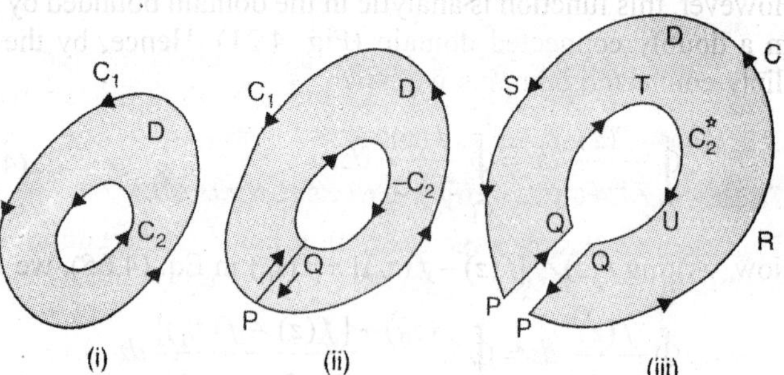

Fig. 4.19: Doubly connected domain.

The simply connected domain (Fig. 4.19 (iii)) is bounded by a simple closed curve $PRSPQTUQP$.

Since $f(z)$ is analytic in D, we can apply the Cauchy integral theorem to obtain

$$\int_{PRSPQTUQP} f(z)\,dz = 0 \text{ or } \left[\int_{PRSP} + \int_{PQ} + \int_{QTUQ} + \int_{QP}\right] f(z)\,dz = 0$$

or $$\oint_{C_1} f(z)\,dz + \oint_{PQ} f(z)\,dz + \oint_{C_2^*} f(z)\,dz - \oint_{PQ} f(z)\,dz = 0. \quad (\because PQ \text{ and } QP \text{ have opposite directions})$$

Cancelling the second and the fourth integrals on the left hand side, we obtain

$$\oint_{C_1} f(z)\,dz = -\oint_{C_2^*} f(z)\,dz = \oint_{C_2} f(z)\,dz, \text{ where } C_1 \text{ and } C_2 \text{ are traversed in the same direction}$$

This result can be generalised to multiply connected domains. Let $f(z)$ be analytic in a domain D bounded by non-intersecting simple closed curves $C, C_1, C_2, ..., C_n$ where $C_1, C_2, ..., C_n$ lie inside C.

Let $f(z)$ be analytic on C and C_i, $i = 1, 2, ..., n$. We introduce n cuts and convert D to a simply connected domain (Fig. 4.20). Applying the Cauchy integral theorem, we obtain

$$\oint_C f(z)\,dz = \oint_{C_1} f(z)\,dz + \oint_{C_2} f(z)\,dz + \dots + \oint_{C_n} f(z)\,dz \quad ...(4.63)$$

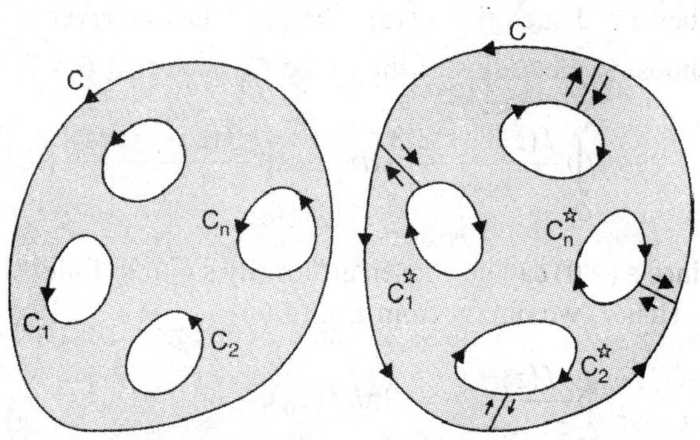

Fig. 4.20: Multiply connected domain.

where all the curves are traversed in the anticlockwise direction.

4.14 CAUCHY'S INTEGRAL FORMULA

(AMIETE, June 2013)

Cauchy integral formula is a representation of an analytic function $f(z)$ at any interior point z_0 of a simply connected domain D as a contour integral, that is, an integral evaluated along the boundary of a simple closed curve C which lie inside D and encloses the point z_0.

Theorem 4.4 Let $f(z)$ be analytic in a simply connected domain D, and let C be a simple closed contour lying entirely within D. If z_0 is any point within C, then

$$f(z_0) = \frac{1}{2\pi i}\oint \frac{f(z)}{z - z_0}\,dz \quad ...(4.64)$$ where C is traversed in the anticlockwise direction.

Proof: The given function $f(z)$ is analytic in a simply connected domain D and C is any simple closed contour in D enclosing the point $z = z_0$. Let $C_1 : |z - z_0| = \rho$ be a circle with infinitesimal radius ρ, such that it lies entirely inside C. Consider the function $\dfrac{f(z)}{z - z_0}$. This function is analytic everywhere in D except at the point $z = z_0$.

However, this function is analytic in the domain bounded by the simple closed curve C and the circle C_1, that is, in a doubly connected domain (Fig. 4.21). Hence, by the extension of the Cauchy integral theorem for mutliply connected domains we have

$$\oint_C \frac{f(z)}{z-z_0}\,dz = \oint_{C_1} \frac{f(z)}{z-z_0}\,dz. \qquad \text{...(4.65)}$$

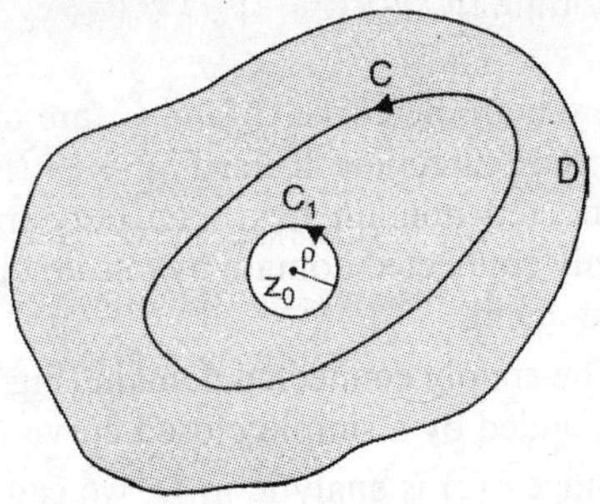

Fig. 4.21: Cauchy integral formula.

Now, writing $f(z) = [f(z) - f(z_0)] + f(z_0)$ in Eq. (4.65), we obtain

$$\oint_C \frac{f(z)}{z-z_0}\,dz = \oint_{C_1} \frac{f(z_0) + [f(z) - f(z_0)]}{z-z_0}\,dz$$

$$= f(z_0) \oint_{C_1} \frac{dz}{z-z_0} + \oint_{C_1} \frac{f(z)-f(z_0)}{z-z_0}\,dz$$

$$= 2\pi i\, f(z_0) + \oint_{C_1} \frac{f(z)-f(z_0)}{z-z_0}\,dz. \qquad \text{...(4.66)}$$

We have $\left|\oint_{C_1} \frac{f(z)-f(z_0)}{z-z_0}\,dz\right| \le \oint_{C_1} \left|\frac{f(z)-f(z_0)}{z-z_0}\right| |dz| = \oint_{C_1} \frac{|f(z)-f(z_0)|}{|z-z_0|}|dz|.$ ...(4.67)

Since $f(z)$ is analytic, it is continuous in D. Therefore, for any given real number $\varepsilon > 0$, there exists a real number $\delta > 0$ such that $|f(z) - f(z_0)| < \varepsilon$, whenever $|z - z_0| < \delta$.

Choosing the radius of the circle C_1 such that $\rho < \delta$, we obtain from Eq. (4.67)

$$\left|\oint_{C_1} \frac{f(z)-f(z_0)}{z-z_0}\,dz\right| \le \oint_{C_1} \frac{|f(z)-f(z_0)|}{|z-z_0|}|dz| < \oint_{C_1} \frac{\varepsilon}{\rho}|dz| = \frac{\varepsilon}{\rho}(2\pi\rho) = 2\pi\varepsilon.$$

Since ε (> 0) can be chosen arbitrarily small, it follows that the last integral in Eq. (4.66) must have the value zero. Hence, we obtain from Eq. (4.66)

$$\oint_C \frac{f(z)}{z-z_0}\,dz = 2\pi i\, f(z_0), \quad \text{or} \quad f(z_0) = \frac{1}{2\pi i}\oint_C \frac{f(z)}{z-z_0}\,dz.$$

EXAMPLE 4.30 Evaluate the following integrals, using Cauchy's integral formula:

(*i*) $\dfrac{1}{2\pi i}\oint_C \dfrac{z^2+5}{z-3}\,dz$, (*a*) $C : |z| = 4$, (*b*) $C : |z| = 1$.

(*ii*) $\oint_C \dfrac{z^2-4z+4}{z+i}\,dz$, where C is the circle $|z| = 2$.

(*iii*) $\oint_C \dfrac{\sin 3z}{z+\pi/2}\,dz$, where C is the circle $|z| = 5$.

(*iv*) $\oint_C \dfrac{e^{2z}}{(z-1)(z-2)}\,dz$, where C is the circle $|z| = 3$. (*Madras, 2003*)

Solution (*i*) (*a*) The integrand $(z^2+5)/(z-3)$ is not analytic at the point $z=3$. This point lies inside the circle $C:|z|=4$. Using the Cauchy's integral formula with $f(z)=z^2+5$ and $z_0=3$, we obtain

$$\frac{1}{2\pi i}\oint_C \frac{z^2+5}{z-3}\,dz = f(3) = [z^2+5]_{z=3} = 3^2+5 = 14.$$

(*b*) In this case, the point $z_0=3$ does not lie within the circle $|z|=1$. Hence, the function $(z^2+5)/(z-3)$ is analytic within and on C. Therefore, by Cauchy's integral formual, we obtain

$$\frac{1}{2\pi i}\oint_C \frac{z^2+5}{z-3}\,dz = 0.$$

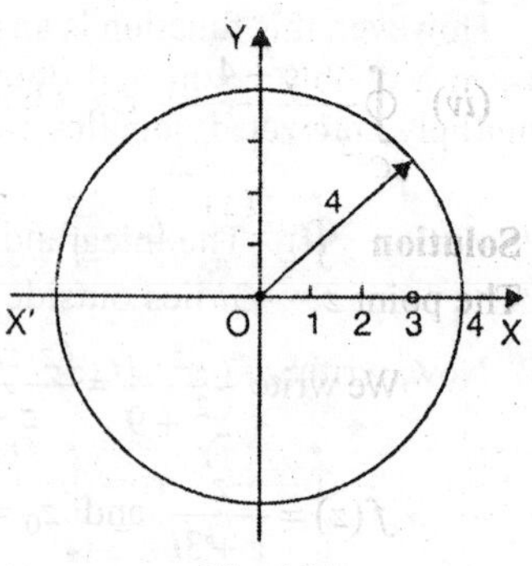

Fig. 4.22

(*ii*) The integrand $(z^2-4z+4)/(z+i)$ is not analytic at the point $z=-i$. This point lies inside the circle $C:|z|=2$. Using the Cauchy's integral formula with $f(z)=z^2-4z+4$ and $z_0=-i$, we obtain

$$\oint_C \frac{z^2-4z+4}{z+i}\,dz = 2\pi i f(-i) = 2\pi i\,\{[z^2-4z+4]_{z=-i}\} = 2\pi i\,(3+4i) = 2\pi\,(-4+3i).$$

(*iii*) The integrand $\dfrac{\sin 3z}{z+\pi/2}$ is not analytic at the point $z=-\pi/2$. This point lies inside the circle $C:|z|=5$. Using the Cauchy's integral formula with $f(z)=\sin 3z$ and $z_0=-\pi/2$, we obtain

$$\oint_C \frac{\sin 3z}{z+\pi/2}\,dz = 2\pi i\, f(-\pi/2) = \{2\pi i\,[\sin 3z]_{z=-\pi/2}\} = 2\pi i\left[\sin 3\left(\frac{-\pi}{2}\right)\right] = 2\pi i.$$

(*iv*) Denote the given integral by I. The integrand of I is not analytic at the points $z=1$ and $z=2$ both of which lie inside C. We write $\dfrac{e^{2z}}{(z-1)(z-2)} = e^{2z}\left[\dfrac{1}{z-2}-\dfrac{1}{z-1}\right]$.

Therefore, $$\oint_C \frac{e^{2z}}{(z-1)(z-2)}\,dz = \oint_C \frac{e^{2z}}{z-2}\,dz - \oint_C \frac{e^{2z}}{z-1}\,dz = I_1 - I_2.$$

Now, the integrands, of I_1 and I_2 are not analytic at a point which lies inside C. Using the Cauchy integral formula to I_1 and I_2, we get

$$I_1 = \oint_C \frac{e^{2z}}{z-2}\,dz = \{2\pi i\,[e^{2z}]_{z=2}\} = 2\pi i e^4, \qquad I_2 = \oint_C \frac{e^{2z}}{z-1}\,dz = \{2\pi i\,[e^{2z}]_{z=1}\} = 2\pi i e^2.$$

Hence, $I = I_1 - I_2 = 2\pi i e^4 - 2\pi i e^2 = 2\pi i\,(e^4 - e^2)$.

EXAMPLE 4.31 Evaluate the following integrals, using Cauchy's integral formula :

(*i*) $\oint_C \dfrac{z}{z^2+9}\,dz$, where C is the circle $|z-2i|=4$.

(*ii*) $\oint_C \dfrac{dz}{z(z^2+4)}$, $C:|z|=1$. (*AMIETE, June 2011, Dec 2009*)

(*iii*) $\dfrac{1}{2\pi i}\oint_C \dfrac{\cos \pi z}{z^2-1}\,dz$ around a rectangle with vertices at: (*a*) $2\pm i$, $-2\pm i$, (*b*) $-i$, $2-i$, $2+i$, i.

(iv) $\oint_C \frac{z+4}{z^2+2z+5}\,dz$, where C is the circle $|z+1-i| = 2$.

Solution (i) The integrand $z/(z^2+9)$ is not analytic at the point $z = \pm 3i$. The point $z = -3i$ lies outside C.

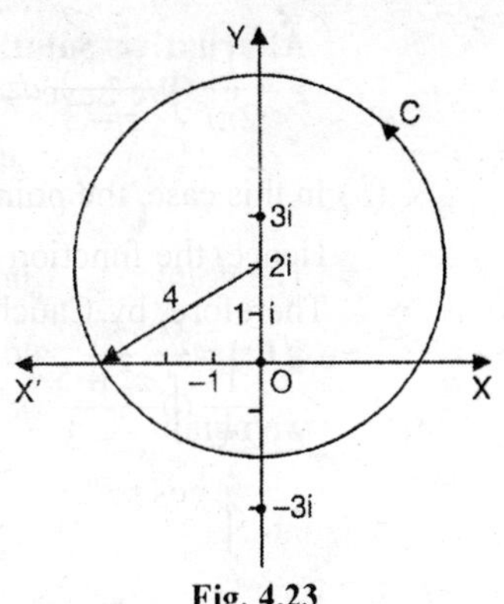

Fig. 4.23

We write $\frac{z}{z^2+9} = \frac{(z/z+3i)}{z-3i}$. Using the Cauchy's integral formula with

$f(z) = \frac{z}{z+3i}$ and $z_0 = 3i$, we obtain

$$\oint_C \frac{z}{z^2+9}\,dz = \oint_C \frac{z/(z+3i)}{z-3i}\,dz = 2\pi i\, f(3i)$$

$$= 2\pi i\left[\frac{z}{z+3i}\right]_{z=3i} = 2\pi i\,\frac{3i}{6i} = \pi i.$$

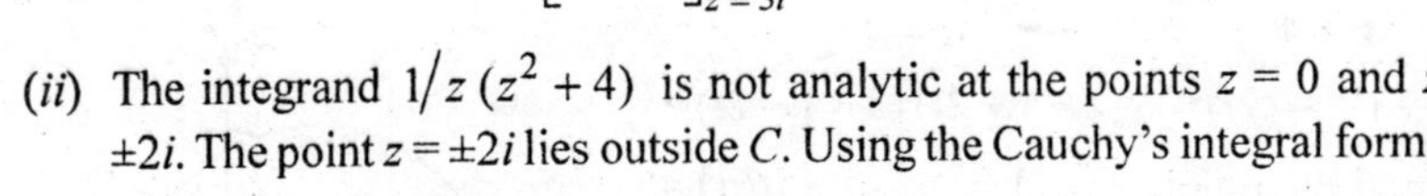

(ii) The integrand $1/z\,(z^2+4)$ is not analytic at the points $z = 0$ and $z = \pm 2i$. The point $z = \pm 2i$ lies outside C. Using the Cauchy's integral formula with $f(z) = \frac{1}{z^2+4}$ and $z_0 = 0$, we obtain

$$\oint_C \frac{dz}{z\,(z^2+4)} = \oint_C \frac{1/(z^2+4)}{z}\,dz = 2\pi i\, f(0)$$

$$= \left\{2\pi i\left[\frac{1}{z^2+4}\right]_{z=0}\right\} = \frac{2\pi i}{4} = \frac{\pi i}{2}.$$

Fig. 4.24

(iii) Let the given integral be denoted by I.

(a) The integrand is not analytic at the points $z = 1$ and $z = -1$ both of which lie inside C. We write

$\frac{1}{z^2-1} = \frac{1}{2}\left(\frac{1}{z-1} - \frac{1}{z+1}\right)$. Therefore, $\oint_C \frac{\cos \pi z}{z^2-1}\,dz = \frac{1}{2}\oint_C \frac{\cos \pi z}{z-1}\,dz - \frac{1}{2}\oint_C \frac{\cos \pi z}{z+1}\,dz = I_1 - I_2.$

Using Cauchy's integral formula to I_1 and I_2, we get

$$I_1 = \frac{1}{2}\oint_C \frac{\cos \pi z}{z-1}\,dz = \frac{1}{2}\cdot 2\pi i\, f(1) = \frac{1}{2}\{2\pi i\,[\cos \pi z]_{z=1}\} = \pi i \cos \pi.$$

$$I_2 = \frac{1}{2}\oint_C \frac{\cos \pi z}{z+1}\,dz = \frac{1}{2}\cdot 2\pi i\, f(-1) = \frac{1}{2}\{2\pi i\,[\cos \pi z]_{z=-1}\} = \pi i \cos \pi.$$

Hence $I = I_1 - I_2 = \pi i \cos \pi - \pi\, i \cos \pi = 0.$

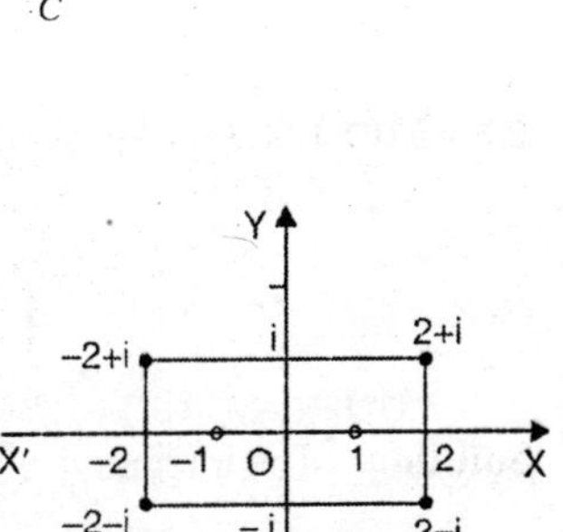

Fig. 4.25

(b) Refer Fig. 4.26, $z_0 = 1$ lies within the square but $z_0 = -1$ does not lie in that region.

As in part (a) $\oint_C \frac{\cos \pi z}{z^2-1}\,dz = \frac{1}{2}\oint_C \frac{\cos \pi z}{z-1}\,dz - \frac{1}{2}\oint_C \frac{\cos \pi z}{z+1}\,dz.$

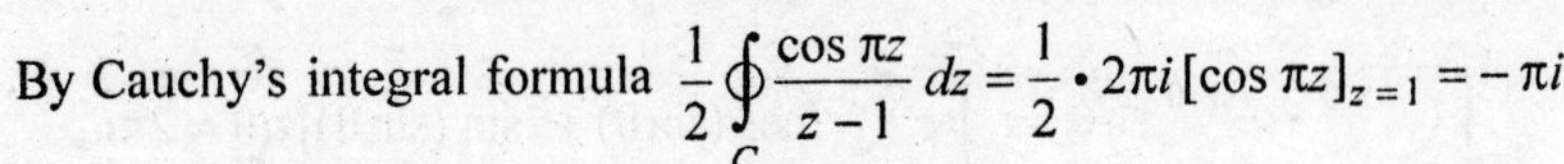

By Cauchy's integral formula $\frac{1}{2}\oint_C \frac{\cos \pi z}{z-1}\,dz = \frac{1}{2}\cdot 2\pi i\,[\cos \pi z]_{z=1} = -\pi i,$

and $\frac{1}{2}\oint_C \frac{\cos \pi z}{z+1}\, dz = 0$. Hence $\oint_C \frac{\cos \pi z}{z^2-1}\, dz = -\pi i + 0 = -\pi i$.

Alternative Solution: The integrand is not analytic at the points $z = \pm 1$. We have

$$\oint_C \frac{\cos \pi z}{z^2-1}\, dz = \oint_C \frac{\cos \pi z/(z+1)}{z-1}\, dz.$$

The point $z = -1$ lie outside C. Using Cauchy's integral formula with

$f(z) = \frac{\cos \pi z}{z+1}$ and $z_0 = 1$,

we obtain

$$\oint_C \frac{\cos \pi z}{z^2-1}\, dz = \oint_C \frac{\cos \pi z/(z+1)}{z-1}\, dz = 2\pi i\, f(1) = 2\pi i \left[\frac{\cos \pi z}{z+1}\right]_{z=1} = 2\pi i \left(\frac{\cos \pi}{2}\right) = -\pi i.$$

Fig. 4.26

(*iv*) We write $\frac{z+4}{z^2+2z+5} = \frac{z+4}{(z+1)^2-(2i)^2} = \frac{z+4}{(z+1+2i)(z+1-2i)}$.

The integrand is not analytic at the points $z = -1 + 2i, -1 - 2i$. The point $z = -1 - 2i$ lies outside C. (Refer Fig. 4.27).

Taking $f(z) = \frac{z+4}{z+1+2i}$, $z_0 = -1 + 2i$ and using the Cauchy's integral formula, we obtain

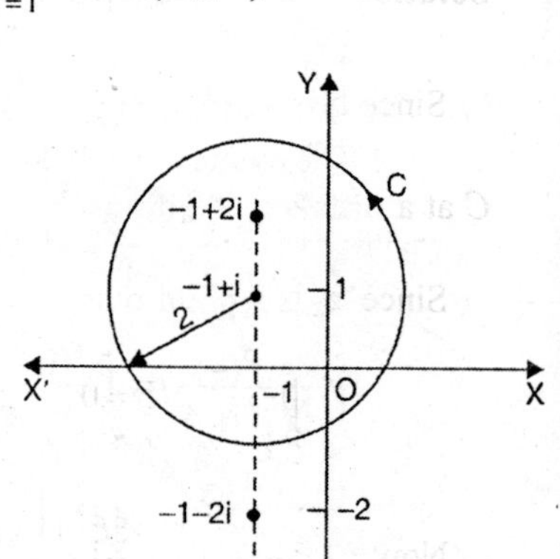

Fig. 4.27

$$\oint_C \frac{(z+4)/(z+1+2i)}{z+1-2i)}\, dz = 2\pi i\, f(-1+2i)$$

$$= \left\{2\pi i \left[\frac{z+4}{z+1+2i}\right]_{z=-1+2i}\right\}$$

$$= 2\pi i \left(\frac{-1+2i+4}{-1+2i+1+2i}\right) = 2\pi i \left(\frac{3+2i}{4i}\right) = \frac{\pi}{2}(3+2i).$$

EXAMPLE 4.32 Show that $\oint_C \frac{e^z}{z}\, dz = 2\pi i$, $C: |z| = 1$. Hence, show that

$$\int_0^{2\pi} e^{\cos\theta} \cos(\sin\theta)\, d\theta = 2\pi, \text{ and } \int_0^{2\pi} e^{\cos\theta} \sin(\sin\theta)\, d\theta = 0.$$

Solution The integrand e^z/z is not analytic at the point $z = 0$ which lies inside C. Using the Cauchy's integral formula with $f(z) = e^z$ and $z_0 = 0$, we obtain $\oint_C \frac{e^z}{z}\, dz = 2\pi i\, f(0) = 2\pi i\, \{[e^z]_{z=0}\} = 2\pi i$.

Let $z = e^{i\theta}$, we have $\oint_C \frac{e^z}{z}\, dz = \int_0^{2\pi} \frac{e^{e^{i\theta}}(ie^{i\theta})}{e^{i\theta}}\, d\theta = i\int_0^{2\pi} e^{(\cos\theta + i\sin\theta)}\, d\theta$

or $i\int_0^{2\pi} (e^{\cos\theta} \cdot e^{i\sin\theta})\, d\theta = i\int_0^{2\pi} e^{\cos\theta}\, [\cos(\sin\theta) + i\sin(\sin\theta)]\, d\theta = 2\pi i.$

Equating the real and imaginary parts, we get

$$\int_0^{2\pi} e^{\cos\theta} \sin(\sin\theta)\, d\theta = 0 \quad \text{and} \quad \int_0^{2\pi} e^{\cos\theta} \cos(\sin\theta)\, d\theta = 2\pi.$$

Poisson's Integral Formula for a circle

EXAMPLE 4.33 Let $f(z)$ be analytic inside and on a circle C defined by $|z| = R$, and let $z = re^{i\theta}$ be any point inside C. Choosing a point $\overset{*}{z}$ outside the circle C at a distance of R^2/r from the origin on the same ray as z, prove the *Poisson integral formula* for the circle :

$$u(r, \theta) = \frac{1}{2\pi} \int_0^{2\pi} \frac{(R^2 - r^2)\, u(R, \phi)\, d\phi}{R^2 - 2rR\cos(\phi - \theta) + r^2}, \quad r < R.$$

Solution $z = re^{i\theta}$ is a point inside C. By Cauchy's integral formula, we get $f(z) = \dfrac{1}{2\pi i} \oint_C \dfrac{f(\xi)}{\xi - z}\, d\xi.$ ...(1)

Since ξ is a point on C, let $\xi = Re^{i\phi}$. Therefore, $|z| = r$ and $|\xi| = R$. Now, $\overset{*}{z}$ is a point outside the circle C at a distance R^2/r from the origin on the same ray as z. We have $\overset{*}{z} = \dfrac{R^2}{r} e^{i\theta} = \dfrac{R^2}{re^{-i\theta}} = \dfrac{\xi\bar{\xi}}{\bar{z}}.$

Since $\overset{*}{z}$ is a point outside C, we have by Cauchy integral theorem

$$\oint_C \frac{f(\xi)}{\xi - \overset{*}{z}}\, d\xi = 0 \quad \text{or} \quad \frac{1}{2\pi i} \oint_C \frac{f(\xi)}{\xi - \overset{*}{z}}\, d\xi = 0. \qquad \text{...(2)}$$

Now $\quad \xi - \overset{*}{z} = \xi - \dfrac{\xi\bar{\xi}}{\bar{z}} = \dfrac{\xi(\bar{z} - \bar{\xi})}{\bar{z}}.$

Fig. 4.28

Subtracting Eq. (2) from Eq. (1), we obtain

$$f(z) = \frac{1}{2\pi i} \oint_C \left[\frac{1}{\xi - z} - \frac{1}{\xi - \overset{*}{z}} \right] f(\xi)\, d\xi$$

$$= \frac{1}{2\pi i} \oint_C \left[\frac{1}{\xi - z} - \frac{\bar{z}}{\xi(\bar{z} - \bar{\xi})} \right] f(\xi)\, d\xi = \frac{1}{2\pi i} \oint_C \left[\frac{z\bar{z} - \xi\bar{\xi}}{\xi(\xi - z)(\bar{z} - \bar{\xi})} \right] f(\xi)\, d\xi.$$

Substituting $z = re^{i\theta}$, $\xi = Re^{i\phi}$, we obtain

$$f(re^{i\theta}) = \frac{1}{2\pi i} \int_0^{2\pi} \frac{(r^2 - R^2)\, f(Re^{i\phi})\, Rie^{i\phi}}{(Re^{i\phi} - re^{i\theta})(re^{-i\theta} - Re^{-i\phi})\, Re^{i\phi}}\, d\phi = \frac{1}{2\pi} \int_0^{2\pi} \frac{(R^2 - r^2)\, f(Re^{i\phi})}{(Re^{i\phi} - re^{i\theta})(Re^{-i\phi} - re^{-i\theta})}\, d\phi$$

$$= \frac{1}{2\pi} \int_0^{2\pi} \frac{(R^2 - r^2)\, f(Re^{i\phi})}{R^2 - 2rR\cos(\phi - \theta) + r^2}\, d\phi. \qquad \text{...(3)}$$

Since $\quad f(re^{i\theta}) = u(r, \theta) + iv(r, \theta) \quad$ and $\quad f(Re^{i\phi}) = u(R, \phi) + iv(R, \phi).$

Comparing the real parts in Eq. (3), we obtain

$$u(r, \theta) = \frac{1}{2\pi} \int_0^{2\pi} \frac{(R^2 - r^2)\, u(R, \phi)}{R^2 - 2rR\cos(\phi - \theta) + r^2}\, d\phi.$$

4.15 CAUCHY'S INTEGRAL FORMULA FOR DERIVATIVES

Using the Cauchy's integral formula, we now establish that an analytic function possesses derivatives of all orders, that is, if f is analytic at a point z_0, then, f', f'', f''', and so on are also analytic at z_0. This is a very fundamental result in the case of functions of a complex variable. Such a result does not exist for functions of real variables.

Theorem 4.5 Let $f(z)$ be analytic in a simply connected domain D. Let z_0 be any point in D and C be any simple closed curve in D enclosing the point $z = z_0$. Then, $f(z)$ has derivatives of all orders in D which are also analytic in D. Further

$$f^{(n)}(z_0) = \frac{ni}{2\pi i}\oint_C \frac{f(z)}{(z-z_0)^{n+1}}\,dz, \quad n = 1, 2, \ldots\ldots \qquad \ldots(4.68)$$

where C is traversed in the anti-clockwise direction.

Proof: The result can be proved by the Principle of Mathematical Induction. Consider the case $n = 1$. We are to prove that

$$f'(z_0) = \frac{1}{2\pi i}\oint_C \frac{f(z)}{(z-z_0)^2}\,dz. \qquad \ldots(4.69)$$

By the definition of a derivative, we have

$$f'(z_0) = \lim_{\Delta z_0 \to 0} \frac{f(z_0+\Delta z_0) - f(z_0)}{\Delta z_0}.$$

Now if z_0 is inside C, then $z_0 + \Delta z_0$ is also inside C for $|\Delta z_0|$ sufficiently small. Using the Cauchy's integral formula, we obtian

$$f(z_0) = \frac{1}{2\pi i}\oint_C \frac{f(z)}{z-z_0}\,dz \quad \text{and} \quad f(z_0+\Delta z_0) = \frac{1}{2\pi i}\oint_C \frac{f(z)}{z-(z_0+\Delta z_0)}\,dz$$

Hence, $$\frac{f(z_0+\Delta z_0) - f(z_0)}{\Delta z_0} = \frac{1}{2\pi i \Delta z_0}\oint_C \left[\frac{1}{z-(z_0+\Delta z_0)} - \frac{1}{z-z_0}\right] f(z)\,dz$$

$$= \frac{1}{2\pi i}\oint_C \frac{f(z)\,dz}{(z-z_0)(z-z_0-\Delta z_0)}.$$

We now write $$\frac{1}{(z-z_0)(z-z_0-\Delta z_0)} = \frac{1}{(z-z_0)^2}\left[\frac{(z-z_0-\Delta z_0)+\Delta z_0}{z-z_0-\Delta z_0}\right] = \frac{1}{(z-z_0)^2}\left[1+\frac{\Delta z_0}{z-z_0-\Delta z_0}\right]$$

$$= \frac{1}{(z-z_0)^2} + \frac{\Delta z_0}{(z-z_0)^2(z-z_0-\Delta z_0)}.$$

Therefore, $$f'(z_0) = \frac{1}{2\pi i}\lim_{\Delta z_0 \to 0}\left[\oint_C \frac{f(z)\,dz}{(z-z_0)^2} + \oint_C \frac{\Delta z_0 f(z)}{(z-z_0)^2(z-z_0-\Delta z_0)}\,dz\right]$$

$$= \frac{1}{2\pi i}\oint_C \frac{f(z)}{(z-z_0)^2}\,dz + \frac{1}{2\pi i}\lim_{\Delta z_0 \to 0}\oint_C \frac{\Delta z_0 f(z)}{(z-z_0)^2(z-z_0-\Delta z_0)}\,dz. \qquad \ldots(4.70)$$

Since $f(z)$ is analytic and it is continuous in C, it is bounded on C, that is $|f(z)| \le M$ for all z on C. Let d be the smallest distance from z_0 to any point z on C. (Fig. 4.29). Therefore, for any z on C, we have the following:

(*i*) $|z - z_0| \geq d$ or $|z - z_0|^2 \geq d^2$ or $\dfrac{1}{|z - z_0|^2} \leq \dfrac{1}{d^2}$.

(*ii*) $|z - z_0 - \Delta z_0| \geq \big||z - z_0| - |\Delta z_0|\big| \geq d - |\Delta z_0|$

or $$\left|\frac{1}{z - z_0 - \Delta z_0}\right| \leq \frac{1}{d - |\Delta z_0|}.$$

Let L be the length of the curve C. Using the ML-inequality, we obtain

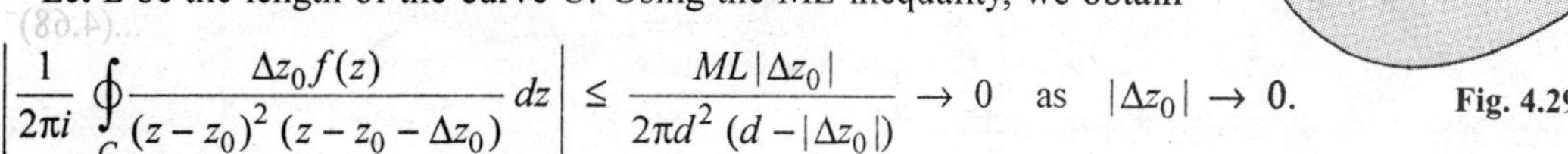

$$\left|\frac{1}{2\pi i}\oint_C \frac{\Delta z_0 f(z)}{(z - z_0)^2 (z - z_0 - \Delta z_0)} dz\right| \leq \frac{ML|\Delta z_0|}{2\pi d^2 (d - |\Delta z_0|)} \to 0 \text{ as } |\Delta z_0| \to 0.$$

Fig. 4.29

Hence, $f'(z_0)$ exists and from Eq. (4.70), we get $f'(z_0) = \dfrac{1}{2\pi i}\displaystyle\oint_C \frac{f(z)}{(z - z_0)^2} dz$.

Since z_0 is any point in D, $f'(z)$ is analytic in D. Hence, we have the result that if $f(z)$ is analytic, then $f'(z)$ is also analytic. Repeating the above procedure with $f(z)$ replaced $f'(z)$, we establish the existence and analyticity of $f''(z)$. Continuing this procedure, we prove that derivatives of all orders of an analytic function $f(z)$ exist and these derivatives are all analytic. We obtain

$$f''(z_0) = \frac{2!}{2\pi i}\oint_C \frac{f(z)}{(z - z_0)^3} dz. \qquad f'''(z_0) = \frac{3!}{2\pi i}\oint_C \frac{f(z)}{(z - z_0)^4} dz.$$

$$f^{(n)}(z_0) = \frac{n!}{2\pi i}\oint_C \frac{f(z)}{(z - z_0)^{n+1}} dz, \quad n = 1, 2, \ldots\ldots$$

Remark 9

These result can be obtained by repeated differentiation with respect to z_0 of the Cauchy's integral formula.

Theorem 4.6 Morera's Theorem

If $f(z)$ be continuous in a simple connected domain D and if $\displaystyle\oint_C f(z) = 0$ around every simple closed curve C in D. Then $f(z)$ is analytic in D. It can be extended to multiply connected domains.

Proof: Let z be an arbitrary point in D and consider a fixed point z_0 in D. Since $\displaystyle\int_C f(z)\, dz = 0$, therefore the value of $\displaystyle\int_{z_0}^{z} f(\xi)\, d\xi$ is independent of the path joining z_0 to z and depends on z only. Let this value be denoted by $F(z)$, then

$$F(z) = \int_{z_0}^{z} f(\xi)\, d\xi. \qquad \ldots(4.71)$$

Let $f(z) = u + iv$ and $F(z) = U + iV$. Then, from Eq. (4.71), we obtain

$$U + iV = \int_{z_0}^{z} (u + iv)(dx + idy) = \int_{(x_0, y_0)}^{(x, y)} (udx - vdy) + i\int_{(x_0, y_0)}^{(x, y)} (udy + vdx).$$

Comparing the real and imaginary parts, we obtain

$$U = \int_{(x_0, y_0)}^{(x, y)} (u\, dx - v\, dy) \text{ and } V = \int_{(x_0, y_0)}^{(x, y)} (u\, dy + v\, dx).$$

Differentiating partially with respect to x and y, we get

$$\frac{\partial U}{\partial x} = u, \quad \frac{\partial U}{\partial y} = -v \quad \text{and} \quad \frac{\partial V}{\partial x} = v, \quad \frac{\partial V}{\partial y} = u.$$

We have $\frac{\partial U}{\partial x} = \frac{\partial V}{\partial y} = u$ and $\frac{\partial U}{\partial y} = -\frac{\partial V}{\partial x} = -v$.

Hence, U and V satisfy the Cauchy-Riemann equations and $\frac{\partial U}{\partial x}, \frac{\partial U}{\partial y}, \frac{\partial V}{\partial x}, \frac{\partial V}{\partial y}$ are continuous so $F(z)$ is analytic in D.

Now as $F(z)$ is analytic in D and $F'(z) = \frac{\partial U}{\partial x} + i\frac{\partial V}{\partial x}$.

Also $F'(z) = \frac{\partial U}{\partial x} + i\frac{\partial V}{\partial x} = u + iv = f(z)$. Since $F(z)$ is analytic, $F'(z) = f(z)$ is also analytic in D. Finally $f(z)$ is analytic in D.

Morera's theorem is often used to establish the analyticity of functions given in integral form.

Cauchy's Inequality

Let $f(z)$ be analytic inside and on a circle $C : |z - z_0| = r$. Then $|f^{(n)}(z_0)| \le \frac{M(n!)}{r^n}$, $n = 0, 1, 2, 3, \ldots\ldots$ where M is a constant such that $|f(z)| \le M$ on C.

Proof: Using the Cauchy's integral formula for derivatives, we have

$$f^{(n)}(z_0) = \frac{n!}{2\pi i} \oint_C \frac{f(z)}{(z - z_0)^{n+1}}\, dz, \quad n = 0, 1, 2, 3, \ldots\ldots$$

Now using ML-inequaltiy, we obtain

$$|f^{(n)}(z_0)| = \left| \frac{n!}{2\pi i} \oint_C \frac{f(z)}{(z - z_0)^{n+1}}\, dz \right| \le \frac{n!}{2\pi} \bullet \frac{M}{r^{n+1}} \bullet (2\pi r) = \frac{n!M}{r^n} \quad \text{(as the length of } C \text{ is } 2\pi r.)$$

For $n = 0$, Cauchy's inequality gives $|f(z_0)| \le M$. Note that z_0 is the centre of the circle C and M the bound of $f(z)$ on C.

EXAMPLE 4.34 Evaluate the following integrals:

(*i*) $\oint_C \frac{\sin^2 z}{\{z - (\pi/6)\}^3}\, dz$, if C is the circle $|z| = 1$. *(Rohtak, 2005)*

(*ii*) If $F(\xi) = \oint_C \frac{3z^2 + 7z + 1}{z - \xi}\, dz$, where C is the circle $x^2 + y^2 = 4$, find the values $F(3)$, $F(1)$, $F'(1 - i)$ and $F''(1 - i)$. *(AMIETE, June 2005)*

(*iii*) $I = \int_C \frac{e^{2z}}{(z + 1)^4}\, dz$, where C is the circle $|z| = 3$.

(*iv*) $\oint_C \frac{ze^z}{(z - 3)^3}\, dz$, where the point $z_0 = 3$ lies within the closed curve C.

(v) $\oint_C \frac{e^z}{(z-1)^2(z^2+4)}\,dz$, for any contour for which 1 lies inside and $\pm 2i$ lies outside (counter clockwise).

(vi) $\oint_C \frac{e^z}{z^2(z+1)^3}\,dz$, $C: |z| = 2$. *(AMIETE, June, 2010, 2009, Dec. 2008)*

Solution (*i*) The integrand is not analytic at the point $z = \pi/6$ which lies inside C. Using the Cauchy integral formula for derivatives ($n = 2$) with $f(z) = \sin^2 z$, we obtain

$$\oint_C \frac{\sin^2 z}{\{z-(\pi/6)\}^3}\,dz = \frac{2\pi i}{2!}\left[\frac{d^2}{dz^2}(\sin^2 z)\right]_{z=\pi/6} = \{\pi i\,[2\cos 2z]_{z=\pi/6}\} = \pi i.$$

(*ii*) We have $F(\xi) = \oint_C \frac{3z^2+7z+1}{z-\xi}\,dz$. Taking $\xi = 3$, we obtain $F(3) = \oint_C \frac{3z^2+7z+1}{z-3}\,dz$.

The integrand is not analytic at the point $z = 3$ which lies outside the circle $C: |z| = 2$. Therefore, by Cauchy's integral theorem $\oint_C \frac{3z^2+7z+1}{z-3}\,dz = 0$. Hence, $F(3) = 0$.

Now $F(1) = \oint_C \frac{3z^2+7z+1}{z-1}\,dz$. The integrand is not analytic at the point $z = 1$ which lies inside C. Using the Cauchy's integral formula with $f(z) = 3z^2 + 7z + 1$ and $z_0 = 1$, we obtain

$$\oint_C \frac{3z^2+7z+1}{z-1}\,dz = 2\pi i\, f(1) = \{2\pi i\,[3z^2+7z+1]_{z=1}\} = 2\pi i\,(3+7+1) = 22\pi i.$$

Therefore, $F(1) = 22\pi i$.

The point $z = \xi = 1 - i$ lies inside C, therefore by Cauchy's integral formula

$$\oint_C \frac{3z^2+7z+1}{z-\xi}\,dz = \{2\pi i\,[f(z)]_{z=\xi}\} = 2\pi i\,(3\xi^2+7\xi+1).$$

That is, $F(\xi) = 2\pi i\,(3\xi^2 + 7\xi + 1)$.

Therefore, $F'(\xi) = 2\pi i\,(6\xi + 7)$ and $F''(\xi) = 2\pi i\,(6) = 12\pi i$.

Thus $F'(1-i) = 2\pi i\,(6 - 6i + 7) = 2\pi i\,(13 - 6i) = 2\pi\,(6 + 13i)$ and $F''(1-i) = 12\pi i$.

(*iii*) The integrand is not analytic at the point $z = -1$ which lies inside C. Using the Cauchy integral formula for derivatives ($n = 3$) with $f(z) = e^{2z}$, we obtain

$$I = \frac{2\pi i}{3!}\left[\frac{d^3}{dz^3}(e^{2z})\right]_{z=-1} = \frac{\pi i}{3}[8e^{2z}]_{z=-1} = \frac{8\pi i}{3}e^{-2}.$$

(*iv*) The integrand is not anlaytic at the point $z = 3$ which lies inside C. Using the Cauchy integral formula for derivatives ($n = 2$) with $f(z) = ze^z$, we obtain

$$I = \frac{2\pi i}{2!}\left[\frac{d^2}{dz^2}(ze^z)\right]_{z=3} = \pi i\,[ze^z + 2e^z]_{z=3} = \pi i\,[3e^3 + 2e^3] = 5\pi i\, e^3.$$

(v) $\oint_C \frac{e^z}{(z-1)^2(z^2+4)}dz = \oint_C \frac{e^z/(z^2+4)}{(z-1)^2}dz = \frac{2\pi i}{1!}\left[\frac{d}{dz}\left(\frac{e^z}{z^2+4}\right)\right]_{z=1} = 2\pi i\left[\frac{e^z(z^2+4)-e^z 2z}{(z^2+4)^2}\right]_{z=1}$

$$= 2\pi i\left\{\frac{e(5)-e(2)}{25}\right\} = \frac{6e\pi}{25}i.$$

(vi) The integrand is not analytic at the point $z = 0$ and $z = -1$ both of which lie inside $C : |z| = 2$. Enclose point $z = 0$ and $z = -1$ by circles $C_1 : |z| = \delta_1$ and $C_2 : |z+1| = \delta_2$ respectively, such that they do not intersect and lie inside C. Using the extension of the Cauchy integral theorem for multiply connected domains, we obtain

$$\oint_C \frac{e^z}{z^2(z+1)^3}dz = \oint_{C_1}\frac{[e^z/(z+1)^3]}{z^2}dz + \oint_{C_2}\frac{[e^z/z^2]}{(z+1)^3}dz.$$

Using the Cauchy's integral formula for derivatives, we obtain

$$\oint_C \frac{e^z}{z^2(z+1)^3}dz = \frac{2\pi i}{1!}\left[\frac{d}{dz}\left\{\frac{e^z}{(z+1)^3}\right\}\right]_{z=0} + \frac{2\pi i}{2!}\left[\frac{d^2}{dz^2}\left(\frac{e^z}{z^2}\right)\right]_{z=-1}$$

$$= 2\pi i\left[\frac{e^z\{(z+1)^3-3(z+1)^2\}}{(z+1)^6}\right]_{z=0} + \pi i\left[\frac{d}{dz}\left\{\frac{(z-2)e^z}{z^3}\right\}\right]_{z=-1}$$

$$= 2\pi i(-2) + \pi i\left[\frac{\{z^3(z-1)-3z^2(z-2)\}e^z}{z^6}\right]_{z=-1} = -4\pi i + 11\pi i e^{-1} = (11e^{-1}-4)\pi i.$$

EXERCISE 4.4

Evaluate the following integrals :

1. (a) $\oint_C \frac{dz}{z-a}$, $C : |z-a| = r$, (*AMIETE, Dec. 2012*) (b) $\oint_C \frac{e^z}{z+1}dz$, $C : \left|z+\frac{1}{2}\right| = 1$,

(c) $\oint_C \frac{z^2-z+1}{z-1}dz$, (i) $C : |z| = 1$, (ii) $C : |z| = 1/2$. (*AMIETE, Dec. 2012*)

Ans. (a) $2\pi i$, (b) $2\pi i/e$, (c) (i) $2\pi i$, (ii) 0.

2. $\oint_C \frac{\sin\pi z^2+\cos\pi z^2}{(z-1)(z-2)}dz$, where C is the circle $|z| = 3.2$.

[**Hint:** Let the given integral be denoted by I. The integrand of I is not analytic at the points $z = 1$ and $z = 2$ both of which lie inside C.

Since $\frac{1}{(z-1)(z-2)} = \frac{1}{z-2} - \frac{1}{z-1}$, we have

$$\oint_C \frac{\sin\pi z^2+\cos\pi z^2}{(z-1)(z-2)}dz = \oint_C \frac{\sin\pi z^2+\cos\pi z^2}{z-2}dz - \oint_C \frac{\sin\pi z^2+\cos\pi z^2}{z-1}dz = I_1 - I_2.$$

Using the Cauchy's integral formula, we obtain

$$I_1 = \{2\pi i[\sin\pi z^2+\cos\pi z^2]_{z=2}\} = 2\pi i(\sin 4\pi+\cos 4\pi) = 2\pi i(0+1) = 2\pi i.$$

$$I_2 = \{2\pi i[\sin\pi z^2+\cos\pi z^2]_{z=1}\} = 2\pi i(\sin\pi+\cos\pi) = 2\pi i(0-1) = -2\pi i.$$

The required integral = $2\pi i - (-2\pi i) = 4\pi i$].

3. $\oint_C \frac{\cos(\pi z^2)}{(z-1)(z-2)} dz$, where C is the circle $|z| = 3$. **Ans.** $4\pi i$.

4. $\oint_C \frac{dz}{z(z+\pi i)}$, $C : |z+3i| = 1$. *(AMIETE, Dec. 2010)* **Ans.** 0.

5. $\oint_C \frac{dz}{z(z^2+4)}$, $C : |z| = 1$.

[**Hint:** The integrand $1/z(z^2+4)$ is not analytic at the points $z = 0$, $z = \pm 2i$. The points $z = \pm 2i$ lie outside C. Using the Cauchy's integral formula with $f(z) = 1/(z^2+4)$ and $z_0 = 0$, we obtain

$$\oint_C \frac{dz}{z(z^2+4)} dz = \oint_C \frac{[1/(z^2+4)]}{z} dz = 2\pi i\, f(0) = 2\pi i \left[\frac{1}{z^2+4}\right]_{z=0} = \frac{\pi i}{2}\Bigg].$$

6. $\oint_C \frac{e^{-z}(3z^2+z)}{z^2-1} dz$, $C : |z| = 2$.

[**Hint:** We write $\frac{3z^2+z}{z^2-1} = 3 + \frac{2}{z-1} - \frac{1}{z+1}$.

Therefore, $\oint_C \frac{e^{-z}(3z^2+z)}{z^2-1} dz = 3\oint_C e^{-z} dz + 2\oint_C \frac{e^{-z}}{z-1} dz - \oint_C \frac{e^{-z}}{z+1} dz = I_1 + I_2 - I_3$.

$I_1 = 0$ (by Cauchy integral theorem).

Since $z = \pm 1$ are inside C, using Cauchy's integral formula with $f(z) = e^{-z}$ and $z_0 = 1, -1$. We obtain

$$I_2 = \{2(2\pi i)[e^{-z}]_{z=1}\} = 4\pi i e^{-1}, \qquad I_3 = \{2\pi i [e^{-z}]_{z=-1}\} = 2\pi i e.$$

The required integral $= 0 + 4\pi i e^{-1} - 2\pi i e = 2\pi i\left(\frac{2}{e} - e\right)\Bigg]$.

7. $\oint_C \frac{z^2+z+1}{z^2-7z+2} dz$, where C is the ellipse $4x^2 + 9y^2 = 1$. **Ans.** 0

8. $\oint_C \frac{e^z}{z^2+1} dz$, (*a*) $C : |z| = 1/2$, (*b*) $C : |z-i| = 1$. **Ans.** (*a*) 0, (*b*) $\pi(\cos 1 + i \sin 1)$

9. $\oint_C \frac{3z^2+z}{z^2-1} dz$, where $C : |z-1| = 1$.

[**Hint:** $\oint_C \frac{3z^2+z}{z^2-1} dz = \frac{1}{2}\oint_C \frac{3z^2+z}{z-1} dz - \frac{1}{2}\oint_C \frac{3z^2+z}{z+1} dz$

$$= \frac{1}{2} \cdot 2\pi i [3z^2+z]_{z=1} - 0 = \pi i (4) = 4\pi i.$$

($\because$ $z = 1$ lies inside C, $z = -1$ lies outside C)]

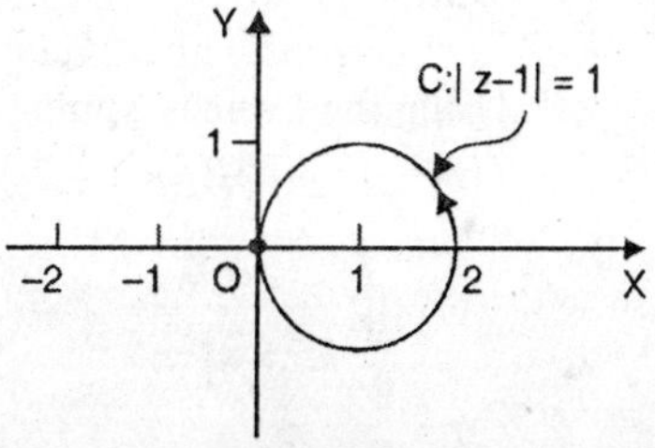

Fig. 4.30

10. $\oint_C \frac{dz}{z^2+4}$ where (*a*) $C : |z-2i| = 1$, (*b*) $C : |z+2i| = 1$, (*c*) $|z| = 4$.

Ans. (*a*) $= \pi/2$, (*b*) $= -\pi/2$, (*c*) $= 0$.

11. $\oint_C \frac{z^3-6}{2z-i}\,dz, \quad C:|z|=1.$

[**Hint:** $\oint_C \frac{z^3-6}{2z-i}\,dz = \oint_C \frac{\frac{1}{2}z^3-3}{z-\frac{i}{2}}\,dz$. The integrand $\frac{\left(\frac{1}{2}z^3-3\right)}{z-\frac{i}{2}}$ is not analytic at the point $z = i/2$ which lies within C. Using the Cauchy's integral formula wiht $f(z) = \frac{1}{2}z^3 - 3$ and $z_0 = i/2$, we obtain

$$\oint_C \frac{z^3-6}{2z-i}\,dz = \oint_C \frac{\frac{1}{2}z^3-3}{z-\frac{i}{2}}\,dz = 2\pi i\, f(i/2) = 2\pi i\left[\frac{1}{2}z^3-3\right]_{z=i/2} = 2\pi i\left(\frac{i^3}{16}-3\right) = \frac{\pi}{8} - 6\pi i.\Bigg]$$

12. $\oint_C \frac{dz}{2-\bar{z}}, \quad C=|z|=1.$

[**Hint:** We write $2-\bar{z} = 2-\frac{z\bar{z}}{z} = 2-\frac{|z|^2}{z} = 2-\frac{1}{z} = \frac{2z-1}{z}$, since $|z|^2 = 1$ on C.

Therefore, $\oint_C \frac{dz}{2-\bar{z}} = \oint_C \frac{z\,dz}{2z-1} = \frac{1}{2}\oint_C \frac{z}{z-(1/2)}\,dz.$

The integrand $\frac{z}{z-(1/2)}$ is not analytic at the point $z = 1/2$ which lies within C. Using the Cauchy's integral formula wiht $f(z) = z$ and $z_0 = 1/2$, we obtain

$$\frac{1}{2}\oint_C \frac{z}{z-(1/2)}\,dz = \frac{1}{2}[2\pi i\, f(1/2)] = \pi i\,[z]_{z=1/2} = \frac{\pi i}{2}.\Bigg]$$

13. $\oint_C \frac{z+4}{z^2+2z+5}\,dz, \quad C:|z+1-i|=2.$

[**Hint:** The integrand $\frac{z+4}{z^2+2z+5}$ is not anlaytic at the points $z = -1 \pm 2i$. The point $z = -1 - 2i$ lie outside C.

$$\oint_C \frac{z+4}{z^2+2z+5}\,dz = \oint_C \frac{z+4}{(z+1+2i)(z+1-2i)}\,dz.$$

Using the Cauchy's integral formula with $f(z) = (z+4)/(z+1+2i)$ and $z_0 = -1+2i$, we obtain

$$\oint_C \frac{z+4}{z^2+2z+5}\,dz = \oint_C \frac{(z+4)/(z+1+2i)}{z+1-2i}\,dz$$

$$= 2\pi i\, f(-1+2i) = 2\pi i\left[\frac{z+4}{z+1+2i}\right]_{z=-1+2i}$$

$$= 2\pi i\left[\frac{3+2i}{4i}\right] = \frac{\pi}{2}(3+2i).\Bigg]$$

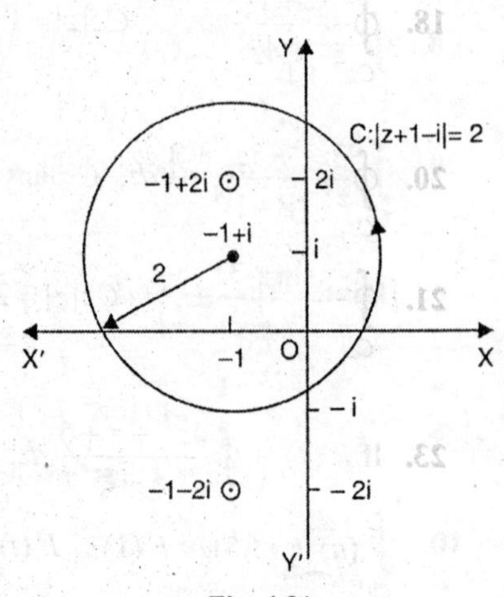

Fig. 4.31

14. $\oint_C \frac{e^z}{(z+1)^2}\,dz, \quad C:|z-1|=3.$ **Ans.** $2\pi i/e$

15. $\oint_C \frac{e^{-z}}{z^2}\,dz, \quad C:|z|=1.$ **Ans.** $-2\pi i$

16. $\oint_C \frac{dz}{(z^2+4)^2}, \quad C:|z-i|=2.$

[**Hint:** Refer Fig. 4.32, the point $z=-2i$ lie outside C. We have

$$\oint_C \frac{dz}{(z^2+4)^2} = \oint_C \frac{dz}{\{z^2-(2i)^2\}^2} = \oint_C \frac{dz}{(z+2i)^2\,(z-2i)^2} = \oint_C \frac{1/(z+2i)^2}{(z-2i)^2}\,dz$$

$$= \{2\pi i\,[f'(z)]_{z=2i}\} = 2\pi i\left[\frac{-2}{(z+2i)^3}\right]_{z=2i} = 2\pi i\left[\frac{-2}{(4i)^3}\right] = \frac{-4\pi i}{64\,i^2 i} = \frac{\pi}{16}.\Big]$$

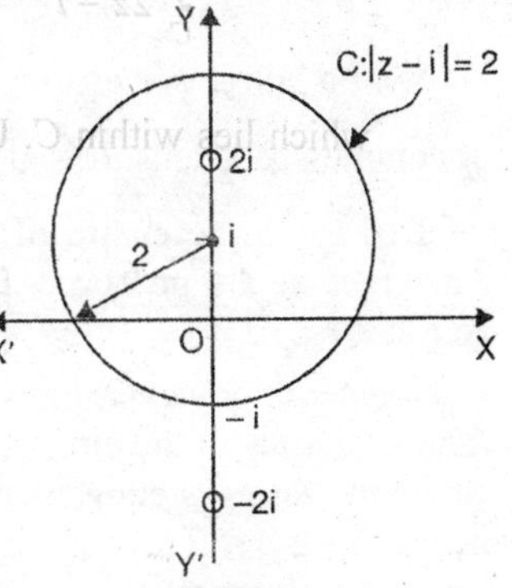

Fig. 4.32

17. $\oint_C \frac{3z+1}{z\,(z-2)^2}\,dz$; C is given in Fig. 4.33

[**Hint:** The points $z=0$ and $z=2$ lies inside the curves C_1 and C_2.

We write $\oint_C \frac{3z+1}{z\,(z-2)^2}\,dz = -\oint_C \frac{(3z+1)/(z-2)^2}{z}\,dz + \oint_C \frac{(3z+1)/z}{(z-2)^2}$

$$= -2\pi i\, f(0) + \frac{2\pi i}{1!}\left[\frac{d}{dz}\left(\frac{3z+1}{z}\right)_{z=2}\right]$$

$$= -2\pi i\left[\frac{3z+1}{(z-2)^2}\right]_{z=0} + 2\pi i\left[\frac{3z-(3z+1)1}{z^2}\right]_{z=2}$$

$$= -\frac{\pi i}{2} - \frac{\pi i}{2} = -\pi i\,]$$

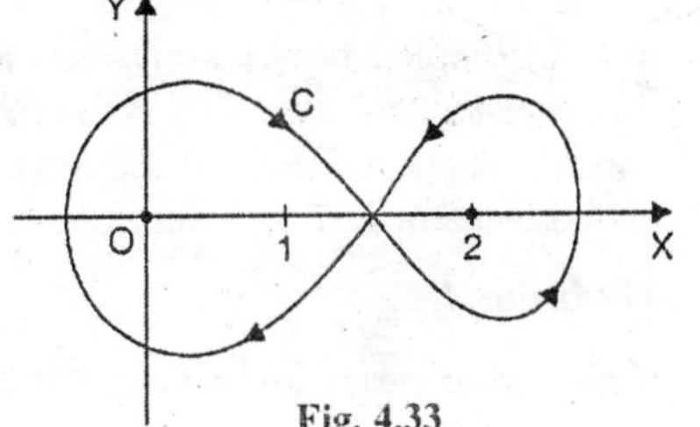

Fig. 4.33

18. $\oint_C \frac{z+1}{z^4+4z^3}\,dz, \quad C:|z|=1.$ **Ans.** $\frac{-3\pi i}{32}.$

19. $\oint_C \frac{\sin^6 z}{\left(z-\frac{\pi}{6}\right)^3}\,dz, \quad C:|z|=1.$ **Ans.** $21\pi i/16.$

20. $\oint_C \frac{3z^4+5z^2+2}{(z+1)^4}\,dz$, C : any simple closed contour coutaining the point $z=-1$ inside it. **Ans.** $-24\pi i.$

21. $\oint_C \frac{\sin z}{\left(z-\frac{\pi}{2}\right)^5}\,dz, \quad C:|z|=2.$ **Ans.** $\pi i/12.$

22. $\oint_C \frac{e^{-2z}}{(z+1)^6}\,dz, \quad C:|z|=2.$ **Ans.** $-8\pi e^2 i/15.$

23. If $F(\xi) = \oint_C \frac{4z^2+z+5}{z-\xi}\,dz$, where C is the ellipse $\frac{x^2}{4}+\frac{y^2}{9}=1$, find the values of :

(*a*) $F(3.5)$, $F(1)$, $F(i)$; (*b*) $F'(-1)$, $F''(-i)$. (*AMIETE, June 2004*)

Ans. (*a*) 0, $20\,\pi i$, $2\pi\,(i-1)$; (*b*) $-14\,\pi i$, $16\,\pi i$.

POWER SERIES, TAYLOR AND LAURENT SERIES

4.16 POWER SERIES

An infinite series of the form

$$\sum_{n=0}^{\infty} a_n (z - z_0)^n = a_0 + a_1 (z - z_0) + a_2 (z - z_0)^2 + \ldots + a_n (z - z_0)^n + \ldots \qquad \ldots(4.72)$$

is called a *power series* in $z - z_0$ or about $z = z_0$ where $a_0, a_1, \ldots$ are real or complex constants. It is also written as $\sum a_n (z - z_0)^n$. The point z_0 is referred to as the centre of the power series. The power series always converges at the point $z = z_0$, since for $z = z_0$, the series reduces to a constant a_0.

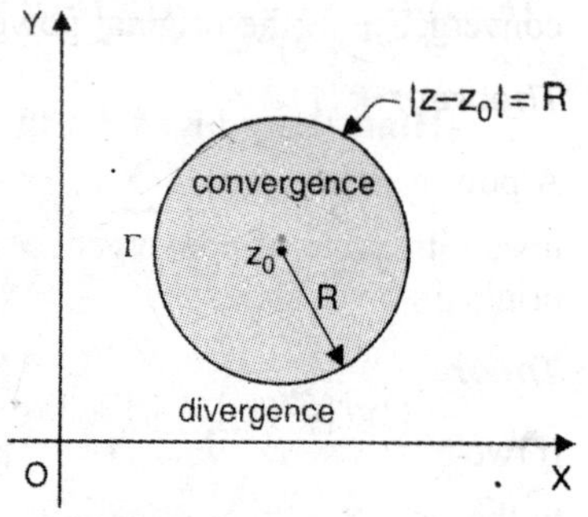

Fig. 4.34: Circle of Convergence

In general, however, the series converges for other points as well. Let R be the radius of the circle with centre at z_0 that contains all the points at which the series is convergent. Then, the series is convergent for all z for which $|z - z_0| < R$ and divergent for all z for which $|z - z_0| > R$, while for $|z - z_0| = R$ it may or may not converge. (see Fig. 4.34)

Geometrically if Γ is a circle of radius R with centre at $z = z_0$, then the series (4.72) converges at all points inside Γ and diverges at all points outside Γ, while it may or may not converge on the circle Γ.

Circle of Convergence

Every Complex power series has *radius of convergence R*. The circle $|z - z_0| = R$ is called the *circle of convergence* and its radius R the radius of convergence of the power series. If $R = 0$, the series is convergent only at the point $z = z_0$. If $R = \infty$, the series is convergent for all z. No general statements can be made about the convergence of a power series (4.72) on the circle of convergence itself.

Theorem 4.7

If the power series given in Eq. (4.72) converges at a point $z = \overset{*}{z} \neq z_0$, then it converges absolutly for all z in the disk $|z - z_0| < |\overset{*}{z} - z_0|$.

***Remark* 10**

If the power series given in Eq. (4.72) diverges at point $z = \overset{*}{z}$, then it diverges for all z, where $|z - z_0| > |\overset{*}{z} - z_0|$.

Theorem 4.8

Let $\sum a_n (z - z_0)^n$ be a power series and let $\overset{*}{L} = \lim\limits_{n \to \infty} \left| \dfrac{a_{n+1}}{a_n} \right|$. Then, the radius of convergence of the power series is $R = 1/\overset{*}{L}$.

If $\lim\limits_{n \to \infty} \left| \dfrac{a_{n+1}}{a_n} \right| = 0$, the radius of convergence is ∞, that is, the power series Eq. (4.72) converges for all z.

If $\lim\limits_{n \to \infty} \left| \dfrac{a_{n+1}}{a_n} \right| = \infty$, the radius of convergence is $R = 0$ at the centre z_0.

We can also use the Cauchy's root test, we write $\lim\limits_{n \to \infty} [|a_n|]^{1/n} = \overset{*}{L}$ and similar remarks can be made for the root test.

Theorem 4.9

A power series $\sum a_n (z-z_0)^n$ represents an analytic function within its circle of convergence.

Theorem 4.10

A power series $f(z) = \sum a_n (z-z_0)^n$ can be integrated term by term along any curve C which lies entirely inside its circle of convergence. The radius of convergence of the integrated series is same as the radius of convergence of the original power series.

Theorem 4.11

A power series $f(z) = \sum a_n (z-z_0)^n$ can be differentiated term by term in any region which lies entirely inside its circle of convergence. The radius of convergence of the differentiated series is same as that of the original power series.

Theorem 4.12

If two power series $\sum a_n (z-z_0)^n$ and $\sum b_n (z-z_0)^n$ have the same radius of convergence R and converge to the same sum within the circle of convergence $|z-z_0| = R$, then these two series are identical.

Theorem 4.13

If $f(z) = \sum a_n (z-z_0)^n$ and $g(z) = \sum b_n (z-z_0)^n$ are two power series with radius of convergence R_1 and R_2 respectively, then the radius of convergence of the series

$$f(z) \pm g(z) = \sum a_n (z-z_0)^n \pm \sum b_n (z-z_0)^n \text{ is } R = \min(R_1, R_2).$$

Theorem 4.14

If $f(z) = \sum a_n (z-z_0)^n$ and $g(z) = \sum b_n (z-z_0)^n$ be two power series with radii of convergence R_1 and R_2 respectively. The *Cauchy product* of these series is given by

$$f(z)\, g(z) = \sum_{n=0}^{\infty} \left(\sum_{r=0}^{\infty} a_r\, b_{n-r} \right) (z-z_0)^n = \sum_{n=0}^{\infty} C_n (z-z_0)^n,$$

where $C_n = \sum_{r=0}^{\infty} a_r\, b_{n-r}$. The radius of convergence of the product series is $R = \min(R_1, R_2)$.

EXAMPLE 4.35 Find the radius of convergence and the circle of convergence of the following power series:

(*i*) $\sum_{n=0}^{\infty} \frac{(2n)!}{(n!)^2} (z-3i)^n$, (*ii*) $\sum \frac{n!}{2^n} (z+1-i)^n$, (*iii*) $\sum \left(1 - \frac{\pi}{n}\right)^{n^2} z^n$,

(*iv*) $\sum \left(\frac{6n+1}{2n+5} \right)^n (z-2i)^n$, (v) $\sum n^{\log n} z^n$.

Solution (*i*) $R = \lim_{n\to\infty} \left| \frac{a_n}{a_{n+1}} \right| = \lim_{n\to\infty} \left[\frac{(2n)!}{(n!)^2} \bullet \frac{\{(n+1)!\}^2}{(2n+2)!} \right] = \lim_{n\to\infty} \frac{(n+1)^2}{(2n+2)(2n+1)} = \frac{1}{4}.$

Therefore, the radius of convergence is $R = 1/4$. The series converges in the open disk $|z-3i| < 1/4$ of radius 1/4 and centrė $3i$. The circle of convergence is $|z-3i| = 1/4$.

(*ii*) $R = \lim_{n\to\infty} \left| \frac{a_n}{a_{n+1}} \right| = \lim_{n\to\infty} \left[\frac{n!}{2^n} \bullet \frac{2^{n+1}}{(n+1)!} \right] = \lim_{n\to\infty} \frac{2}{n+1} = 0.$

The power series is convergent only at its centre, that is, at $z_0 = -1 + i$.

(iii) $\dfrac{1}{R} = \lim_{n\to\infty} [|a_n|^{1/n}] = \lim_{n\to\infty}\left[\left(1-\dfrac{\pi}{n}\right)^{n^2}\right]^{1/n} = \lim_{n\to\infty}\left(1-\dfrac{\pi}{n}\right)^{n} = \lim_{n\to\infty}\left[\left(1-\dfrac{\pi}{n}\right)^{\frac{n}{\pi}}\right]^{\pi} = \dfrac{1}{e^{\pi}}.$

Therefore, the radius of convergence is $R = e^{\pi}$. Since the centre of the power series is $z_0 = 0$, the circle of convergence is $|z| = e^{\pi}$.

(iv) $a_n = \left(\dfrac{6n+1}{2n+5}\right)^n$, the root test in the form $\lim_{n\to\infty} [|a_n|^{1/n}] = \lim_{n\to\infty}\left(\dfrac{6n+1}{2n+5}\right) = 3.$

The radius of convergence of the series is $R = 1/3$. The circle of convergence is $|z - 2i| = 1/3$; the series converges absolutely for $|z - 2i| < 1/3$.

(v) $\overset{*}{L} = \lim_{n\to\infty} |a_n|^{1/n} = \lim_{n\to\infty} [n^{\log n}]^{1/n}.$

Taking logarithms on both sides, we obtain $\log \overset{*}{L} = \lim_{n\to\infty} \dfrac{1}{n}\log[n^{\log n}] = \lim_{n\to\infty} \dfrac{[\log n]^2}{n}.$

Using L' Hospital rule, we obtain

$$\log \overset{*}{L} = \lim_{n\to\infty} \frac{\frac{d}{dn}[\log n]^2}{\frac{d}{dn}(n)} = \lim_{n\to\infty}\left(\frac{2\log n}{n}\right) = \lim_{n\to\infty}\frac{2}{n} = 0.$$

Therefore, $\overset{*}{L} = e^0 = 1$ and radius of convergence is $R = \dfrac{1}{\overset{*}{L}} = 1.$

Since the centre of the power series is $z_0 = 0$, the circle of convergence is $|z| = 1$.

EXAMPLE 4.36 Find the radius of convergence of the following power series:

(i) $\sum (n+1+2^n)\, z^n$, (ii) $\sum \dfrac{(-1)^n}{\sqrt{n}}(2z+1)^n.$

Solution (i) We write the given power series as the sum of the two power series $\sum (n+1)\, z^n$ and $\sum 2^n z^n$.

The radius of convergence of the first series is $R_1 = \lim_{n\to\infty}\left(\dfrac{a_n}{a_{n+1}}\right) = \lim_{n\to\infty}\dfrac{(n+1)}{(n+2)} = 1,$

and of second series is $R_2 = \lim_{n\to\infty}\left(\dfrac{a_n}{a_{n+1}}\right) = \lim_{n\to\infty}\left(\dfrac{2^n}{2^{n+1}}\right) = \dfrac{1}{2}.$

Therefore, the radius of convergence of the given series is $R = \min(R_1, R_2) = \dfrac{1}{2}.$

(ii) $R = \lim_{n\to\infty}\left|\dfrac{a_n}{a_{n+1}}\right| = \lim_{n\to\infty}\left|\dfrac{(-1)^n 2^n}{\sqrt{n}}\dfrac{\sqrt{n+1}}{(-1)^{n+1}2^{n+1}}\right| = \dfrac{1}{2}.$

The radius of convergence is $R = 1/2$. The centre of the power series is $z_0 = -1/2$, the circle of convergence is $\left|z+\dfrac{1}{2}\right| = \dfrac{1}{2}$. The series converges when $\left|z+\dfrac{1}{2}\right| < \dfrac{1}{2}$.

EXERCISE 4.5

Find the center and the radius of convergence of the following power series:

1. $\sum_{n=0}^{\infty} \dfrac{n+5i}{(2n)!}(z-i)^n.$
2. $\sum_{n=0}^{\infty} (n+2i)^n z^n.$
3. $\sum_{n=0}^{\infty} \dfrac{1}{(1+i)^n}(z+2-i)^n.$

4. $\sum_{n=0}^{\infty}\left(\frac{4-2i}{1+5i}\right)^n z^n.$ 5. $\sum_{n=0}^{\infty}\frac{n!}{n^n}(z+\pi)^n.$ 6. $\sum\frac{(1+i)^n}{n+2}(z+3i)^n.$

7. $\sum\frac{i^n}{3^{n+1}}(z-i)^n.$ 8. $\sum_{n=0}^{\infty}\frac{1}{n!}\left(\frac{iz-1}{2+i}\right)^n.$

$$\left[\textbf{Hint of problem 8: } \sum\frac{1}{n!}\left(\frac{iz-1}{2+i}\right)^n=\sum\frac{1}{n!}\frac{i^n}{(2+i)^n}\left(z-\frac{1}{i}\right)^n=\sum\frac{1}{n!}\left(\frac{i}{2+i}\right)^n(z+i)^n\right]$$

Ans. 1. i, ∞ 2. 0, 0 3. $-2+i, \sqrt{2}.$ 4. $0, \sqrt{1.3}.$ 5. $-\pi, e$ 6. $-3i, \frac{\sqrt{2}}{2}.$ 7. $i, 3$ 8. $-i, \infty$

4.17 TAYLOR SERIES

A power series with a nonzero radius of convergence represents an analytic function. The converse if this statement is also true, that is, every function which is analytic inside a certain circle $|z-z_0|<R$ can be expressed as a power series which converges to this function inside this circle. This series is called the *Taylor's series.*

Theorem 4.15 (Taylor's Theorem)

A function $f(z)$ that is, analytic inside a circle $|z-z_0|=R$ may be represented inside this circle as a convergent power series

$$f(z)=\sum a_n(z-z_0)^n \text{ where } a_n=\frac{1}{n!}f^{(n)}(z_0).$$

Proof: Let z be a fixed point within the circle $C: |z-z_0|=R$. Construct a circle C_1 of radius $r<R$ centred at the point z_0 and containing the point z (Fig. 4.35). Since z is an interior point of the domain $|z-z_0|=r$ in which the function $f(z)$ is analytic. We use the Cauchy integral formula to obtain the value of f at z.

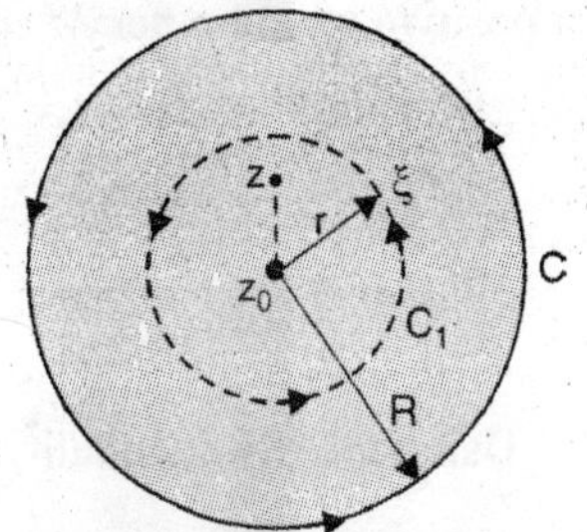

Fig. 4.35: Taylor's theorem

$$f(z)=\frac{1}{2\pi i}\oint_{C_1}\frac{f(\xi)}{\xi-z}d\xi \qquad ...(4.73)$$

where ξ is a variable point on C_1. Since $\xi\neq z$, we can write

$$\frac{1}{\xi-z}=\frac{1}{(\xi-z_0)-(z-z_0)}=\frac{1}{\xi-z_0}\left[1-\frac{z-z_0}{\xi-z_0}\right]^{-1}$$

$$=\frac{1}{\xi-z_0}\left[1+\frac{z-z_0}{\xi-z_0}+\left(\frac{z-z_0}{\xi-z_0}\right)^2+\cdots+\left(\frac{z-z_0}{\xi-z_0}\right)^{n-1}+\left(\frac{z-z_0}{\xi-z_0}\right)^n\right]+T_n(z)$$

where $$T_n(z)=\frac{1}{(\xi-z_0)}\left[\left(\frac{z-z_0}{\xi-z_0}\right)^{n+1}+\left(\frac{z-z_0}{\xi-z_0}\right)^{n+2}+\cdots\right]=\frac{1}{(\xi-z_0)}\left[\frac{\{(z-z_0)/(\xi-z_0)\}^{n+1}}{1-\{(z-z_0)/(\xi-z_0)\}}\right]$$

$$=\left(\frac{z-z_0}{\xi-z_0}\right)^{n+1}\cdot\frac{1}{(\xi-z_0)}\left[\frac{\xi-z_0}{\xi-z}\right]=\frac{1}{\xi-z}\left[\frac{z-z_0}{\xi-z_0}\right]^{n+1}.$$

Now Eq. (4.73) becomes

$$f(z)=\frac{1}{2\pi i}\oint_{C_1}\left[\frac{1}{\xi-z_0}\left\{1+\left(\frac{z-z_0}{\xi-z_0}\right)+\left(\frac{z-z_0}{\xi-z_0}\right)^2+\cdots+\left(\frac{z-z_0}{\xi-z_0}\right)^n\right\}+T_n(z)\right]f(\xi)\,d\xi$$

$$= \frac{1}{2\pi i} \oint_{C_1} \frac{f(\xi)}{\xi - z_0} d\xi + \frac{z - z_0}{2\pi i} \oint_{C_1} \frac{f(\xi)}{(\xi - z_0)^2} d\xi + \cdots + \frac{(z - z_0)^n}{2\pi i} \oint_{C_1} \frac{f(\xi)}{(\xi - z_0)^{n+1}} d\xi + R_n(z) \quad ...(4.74)$$

where $$R_n(z) = \frac{1}{2\pi i} \oint_{C_1} T_n(z) f(\xi) d\xi = \frac{(z - z_0)^{n+1}}{2\pi i} \oint_{C_1} \frac{f(\xi)}{(\xi - z_0)^{n+1} (\xi - z)} d\xi.$$

Using the Cauchy's integral formula for derivatives, we obtain from Eq. (4.74)

$$f(z) = f(z_0) + \frac{f'(z_0)}{1!}(z - z_0) + \frac{f''(z_0)}{2!}(z - z_0)^2 + \cdots + \frac{f^{(n)}(z_0)}{n!}(z - z_0)^n + \cdots$$

or $$f(z) = \sum_{n=0}^{\infty} a_n (z - z_0)^n, \text{ where } a_n = \frac{1}{n!} f^{(n)}(z_0) \quad ...(4.75) \quad \text{provided } \lim_{n \to \infty} R_n(z) = 0.$$

The series given by Eq. (4.75) is called the Taylor's series expansion of $f(z)$ about the point z_0. We now wish to show that $R_n(z) \to 0$ as $n \to \infty$.

Since $\xi - z \neq 0$ and $f(z)$ is analytic within and on C_1, it follows that $\dfrac{f(\xi)}{\xi - z}$ is bounded. Therefore, there exists a positive real number M such that $\left|\dfrac{f(\xi)}{\xi - z}\right| \le M$ for all ξ on C_1.

We also have $|\xi - z_0| = r$ and the length of C_1 is $2\pi r$. We have

$$|R_n(z)| = \frac{|z - z_0|^{n+1}}{|2\pi i|} \left| \oint_{C_1} \frac{f(\xi)}{(\xi - z_0)^{n+1}(\xi - z)} d\xi \right| \le \frac{|z - z_0|^{n+1}}{2\pi} \oint_{C_1} \frac{|f(\xi)|}{|\xi - z_0|^{n+1} |\xi - z|} |d\xi|.$$

Using the ML-inequality, we obtain

$$|R_n(z)| \le \frac{|z - z_0|^{n+1}}{2\pi} \cdot \frac{M}{r^{n+1}} (2\pi r) = Mr \left|\frac{z - z_0}{r}\right|^{n+1}.$$

Now, z is any point inside C_1. Therefore $|(z - z_0)/r| < 1$. Hence, $|R_n(z)| \to 0$ as $n \to \infty$.

Remark 11

(a) If a function $f(z)$ is not analytic at a point $z = z_0$, it cannot be expanded in Taylor's series about the point $z = z_0$.

(b) If $z_0 = 0$ in eq. (4.75), then $f(z) = \sum_{n=0}^{\infty} a_n z^n$, where $a_n = \dfrac{f^{(n)}(0)}{n!}$.

This series is called the *Maclaurin's series.*

EXAMPLE 4.37 Expand $f(z) = \sin z$ in a Taylor series about $z = \pi/4$.

Solution We have

$f(z) = \sin z,$ $\quad f(\pi/4) = \sqrt{2}/2.$

$f'(z) = \cos z,$ $\quad f'(\pi/4) = \sqrt{2}/2.$

$f''(z) = -\sin z,$ $\quad f''(\pi/4) = -\sqrt{2}/2.$

$f'''(z) = -\cos z,$ $\quad f'''(\pi/4) = -\sqrt{2}/2.$

$f^{iv}(z) = \sin z,$ $\quad f^{iv}(\pi/4) = \sqrt{2}/2.$

....

....

Since $z_0 = \pi/4$, $$f(z) = f(z_0) + (z - z_0)\, f'(z_0) + \frac{(z - z_0)^2}{2!} f''(z_0) + \cdots + \frac{1}{n!}(z - z_0)^n f^{(n)}(z_0) + \cdots$$

$$= \frac{\sqrt{2}}{2} + \frac{\sqrt{2}}{2}\left(z - \frac{\pi}{4}\right) - \frac{\sqrt{2}}{2 \cdot 2!}\left(z - \frac{\pi}{4}\right)^2 - \frac{\sqrt{2}}{2 \cdot 3!}\left(z - \frac{\pi}{4}\right)^3 + \cdots$$

$$= \frac{\sqrt{2}}{2}\left\{1 + \left(z - \frac{\pi}{4}\right) - \frac{[z - (\pi/4)]^2}{2!} - \frac{[z - (\pi/4)]^3}{3!} + \cdots\right\}$$

EXAMPLE 4.38 Expand $f(z) = \dfrac{1}{1 - z}$ in a Taylor's series with center $z_0 = 2i$.

Solution We have

$$f(z) = \frac{1}{1 - z}, \qquad f(2i) = \frac{1}{1 - 2i}.$$

$$f'(z) = \frac{1}{(1 - z)^2}, \qquad f'(2i) = \frac{1}{(1 - 2i)^2}.$$

$$f''(z) = \frac{2}{(1 - z)^3}, \qquad f''(2i) = \frac{2}{(1 - 2i)^3}.$$

$$f'''(z) = \frac{3.2}{(1 - z)^4}, \qquad f'''(2i) = \frac{6}{(1 - 2i)^4}.$$

....

....

$$f^{(n)}(z) = \frac{n!}{(1 - z)^{n+1}}, \qquad f^{(n)}(2i) = \frac{n!}{(1 - 2i)^{n+1}}.$$

We obtain the Taylor series

$$\frac{1}{1 - z} = \sum_{n=0}^{\infty} \frac{1}{(1 - 2i)^{n+1}} (z - 2i)^n. \qquad \text{...}(i)$$

Alternative Solution: We write

$$\frac{1}{1 - z} = \frac{1}{1 - z + 2i - 2i} = \frac{1}{1 - 2i - (z - 2i)} = \frac{1}{(1 - 2i)\left[1 - \left(\dfrac{z - 2i}{1 - 2i}\right)\right]} = \frac{1}{(1 - 2i)\left(1 - \dfrac{z - 2i}{1 - 2i}\right)}.$$

We now write $\dfrac{1}{1 - \dfrac{z - 2i}{1 - 2i}}$ as a power series.

Therefore, $$\frac{1}{1 - z} = \frac{1}{1 - 2i}\left(1 - \frac{z - 2i}{1 - 2i}\right)^{-1} = \frac{1}{1 - 2i}\left[1 + \frac{z - 2i}{1 - 2i} + \left(\frac{z - 2i}{1 - 2i}\right)^2 + \left(\frac{z - 2i}{1 - 2i}\right)^3 + \cdots\right]$$

$$= \frac{1}{1 - 2i} + \frac{1}{(1 - 2i)^2}(z - 2i) + \frac{1}{(1 - 2i)^3}(z - 2i)^2 + \frac{1}{(1 - 2i)^4}(z - 2i)^3 + \cdots \qquad \text{...}(ii)$$

The series (*ii*) is exactly the same as that (*i*). This series has centre $2i$ and radius of convergence $\sqrt{5}$.

EXAMPLE 4.39 Find the Taylor's expansion for the function $f(z)=\dfrac{1}{(z+1)^2}$ about the point $z=-i$.

(AMIETE, June 2011)

Solution We have

$$f(z)=\frac{1}{(z+1)^2},\qquad f(-i)=\frac{1}{(-i+1)^2}=\frac{i}{2}.$$

$$f'(z)=\frac{-2}{(z+1)^3},\qquad f'(-i)=\frac{-2}{(-i+1)^3}=\frac{1-i}{2}.$$

$$f''(z)=\frac{6}{(z+1)^4},\qquad f''(-i)=-\frac{3}{2}.$$

$$f'''(z)=\frac{-24}{(z+1)^5},\qquad f'''(-i)=3\,(1+i).$$

....

....

Substituting these values in the Taylor's series, we obtain the series

$$\frac{1}{(z+1)^2}=\frac{i}{2}+(z+i)\frac{(1-i)}{2}-\frac{(z+i)^2}{2!}\bullet\frac{3}{2}+\frac{(z+i)^3}{3!}\bullet 3\,(1+i)+\cdots$$

$$=\frac{i}{2}\left[1-(1+i)\,(z+i)+\frac{3}{4}(1+i)^2\,(z+i)^2-\frac{1}{2}(1+i)^3\,(z+i)^3+\cdots\right]$$

EXAMPLE 4.40 Find Taylor's series expansion of $f(z)=\dfrac{2z^3+1}{z^2+z}$ about the point $z=i$.

(AMIETE, June 2003)

Solution We have $f(z)=\dfrac{2z^3+1}{z^2+z}=2\,(z-1)+\dfrac{1}{z}+\dfrac{1}{z+1},$ or $f(z)=f_1(z)+f_2(z)+f_3(z),$

where $\qquad f_1(z)=2(z-1),\quad f_2(z)=\dfrac{1}{z},\quad f_3(z)=\dfrac{1}{z+1}.$

Now Taylor's expansion for $f_1(z)=2z-2$ about $z=i$ is given by

$$f_1(z)=2z-2,\qquad f_1(i)=2i-2.$$

$$f_1'(z)=2,\qquad f_1'(i)=2.$$

$$f_1''(z)=0,\qquad f_1''(i)=0.$$

Hence, $f_1(z)=2z-2=2\,(i-1)+2\,(z-i)$.

Again Taylor's expansion for $f_2(z)=(1/z)$ about $z=i$ is given by

$$f_2(z)=z^{-1},\qquad f_2(i)=\frac{1}{i}$$

$$f_2'(z)=-z^{-2},\qquad f_2'(i)=-\frac{1}{i^2}=1.$$

$$f_2''(z)=2z^{-3},\qquad f_2''(i)=\frac{2}{i^3}=2i.$$

....

....

$$f_2^{(n)}(z) = \frac{(-1)^n n!}{z^{n+1}}, \qquad f_2^{(n)}(i) = \frac{(-1)^n n!}{i^{n+1}}.$$

Hence, $f_2(z) = \frac{1}{z} = \sum_{n=0}^{\infty} \frac{(-1)^n}{i^{n+1}} (z-i)^n$. Similarly, it can be easily shown that

$$f_3(z) = \frac{1}{z+1} = \sum_{n=0}^{\infty} \frac{(-1)^n}{(1+i)^{n+1}} (z-i)^n.$$

Thus the Taylor's expansion for $f(z)$ about the point $z = i$ is given by

$$f(z) = \frac{2z^3+1}{z^2+z} = 2(i-1) + 2(z-i) + \sum_{n=0}^{\infty} \left[\frac{(-1)^n}{i^{n+1}} + \frac{(-1)^n}{(1+i)^{n+1}} \right] (z-i)^n.$$

EXAMPLE 4.41 Find the Taylor's series expansion of $f(z) = \frac{1}{z^2 + (1+2i)\,z + 2i}$ about $z = 0$. Also find its radius of convergence. *(AMIETE, June 2009)*

Solution We have $f(z) = \frac{1}{z^2 + (1+2i)\,z + 2i} = \frac{1}{(z+2i)\,(z+1)} = \frac{1}{(1-2i)} \left[\frac{1}{z+2i} - \frac{1}{z+1} \right].$

The function is analytic at $z = 0$, we have

$$\frac{1}{z+2i} = \frac{1}{2i}\left[1 + \frac{z}{2i}\right]^{-1} = \frac{1}{2i} \sum_{n=0}^{\infty} (-1)^n \left(\frac{z}{2i}\right)^n, \quad \left|\frac{z}{2i}\right| < 1, \quad \text{or } |z| < 2, \ R_1 = 2$$

and

$$\frac{1}{z+1} = (1+z)^{-1} = \sum_{n=0}^{\infty} (-1)^n z^n, \quad |z| < 1, \ R_2 = 1.$$

Hence,

$$f(z) = \frac{1}{(1-2i)} \left[\frac{1}{2i} \sum_{n=0}^{\infty} (-1)^n \left(\frac{z}{2i}\right)^n - \sum_{n=0}^{\infty} (-1)^n z^n \right]$$

$$= \frac{1}{(1-2i)} \left[\sum_{n=0}^{\infty} \left\{ \left(\frac{1}{2i}\right)^{n+1} - 1 \right\} (-1)^n z^n \right].$$

The radius of convergence $R = \min(R_1, R_2) = 1$, which is the distance between the point $z = 0$ and the nearest point $z = -1$ at which the function $f(z)$ is not analytic.

4.18 LAURENT SERIES

In many cases we are required to expand a function $f(z)$ about a point where $f(z)$ is not analytic. In such cases Taylor's series does not exist but we can expand $f(z)$ in a series which contains both the positive and negative powers of $z - z_0$. This series is called the *Laurent's series.*

Let a function $f(z)$ be analytic inside the anulus domain D defined by $r < |z - z_0| < R$, $r < R$ and on the bounding circles $C_1 : |z - z_0| = r$ and $C_2 : |z - z_0| = R$. Then, $f(z)$ can be represented by the Laurent series as

$$f(z) = c_0 + c_1 (z - z_0) + c_2 (z - z_0)^2 + \ldots + \frac{c_{-1}}{z - z_0} + \frac{c_{-2}}{(z - z_0)^2} + \cdots$$

$$= \sum_{n=0}^{\infty} c_n (z - z_0)^n + \sum_{n=1}^{\infty} \frac{c_{-n}}{(z - z_0)^n} = \sum_{n=-\infty}^{\infty} c_n (z - z_0)^n.$$

The coefficients c_n are given by

$$c_n = \frac{1}{2\pi i} \oint_C \frac{f(\xi)}{(\xi - z_0)^{n+1}} d\xi, \qquad n = 0, \pm 1, \pm 2, \ldots \qquad \ldots(4.76)$$

where C is a simple closed curve that lies entirely within D and has z_0 in the interior. The circles C_1, C_2 and $C : |z - z_0| = r_1,\ r < r_1 < R$ are traversed in the anticlockwise direction.

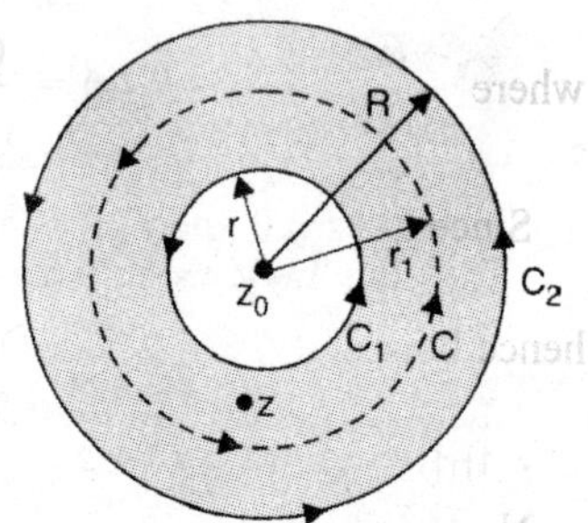

Fig. 4.36: The Laurent series.

Proof: Let C be the circle $|z - z_0| = r_1,\ r < r_1 < R$ inside the annulus. Let z be an arbitrary fixed point inside the annulus $r < |z - z_0| < R$ and inside C. (Fig. 4.36)

Using the extension of the Cauchy integral formula for multiply connected domains, we get

$$f(z) = \frac{1}{2\pi i} \oint_{C_2} \frac{f(\xi)}{\xi - z} d\xi - \frac{1}{2\pi i} \oint_{C_1} \frac{f(\xi)}{\xi - z} d\xi = I_1 + I_2 \qquad \ldots(4.77)$$

where both C_1 and C_2 are traversed in the anti-clockwise direction. We also have

$$\left| \frac{z - z_0}{\xi - z_0} \right| < 1 \text{ for } \xi \text{ on } C_2 \quad \text{and} \quad \left| \frac{\xi - z_0}{z - z_0} \right| < 1 \text{ for } \xi \text{ on } C_1.$$

Consider the integral along the path C_2, we have

$$\frac{1}{\xi - z} = \frac{1}{(\xi - z_0) - (z - z_0)} = \frac{1}{(\xi - z_0)} \left[1 - \left(\frac{z - z_0}{\xi - z_0} \right) \right]^{-1}$$

$$= \frac{1}{(\xi - z_0)} \left[1 + \frac{z - z_0}{\xi - z_0} + \left(\frac{z - z_0}{\xi - z_0} \right)^2 + \cdots \left(\frac{z - z_0}{\xi - z_0} \right)^n + \sum_{k=n+1}^{\infty} \left(\frac{z - z_0}{\xi - z_0} \right)^k \right]$$

$$= \frac{1}{(\xi - z_0)} \left[\sum_{k=0}^{\infty} \left(\frac{z - z_0}{\xi - z_0} \right)^k \right] + T_n(z) \qquad \ldots(4.78)$$

$$\text{where } T_n(z) = \frac{1}{(\xi - z_0)} \left[\left(\frac{z - z_0}{\xi - z_0} \right)^{n+1} + \left(\frac{z - z_0}{\xi - z_0} \right)^{n+2} + \left(\frac{z - z_0}{\xi - z_0} \right)^{n+3} + \cdots \right]$$

$$= \left(\frac{z - z_0}{\xi - z_0} \right)^{n+1} \frac{1}{(\xi - z_0)} \left[1 + \left(\frac{z - z_0}{\xi - z_0} \right) + \left(\frac{z - z_0}{\xi - z_0} \right)^2 + \cdots \right] = \left(\frac{z - z_0}{\xi - z_0} \right)^{n+1} \left(\frac{1}{\xi - z_0} \right) \left[\frac{1}{1 - \left(\frac{z - z_0}{\xi - z_0} \right)} \right]$$

$$= \left(\frac{z - z_0}{\xi - z_0} \right)^{n+1} \cdot \frac{1}{\xi - z_0} \cdot \frac{\xi - z_0}{\xi - z} = \left(\frac{z - z_0}{\xi - z_0} \right)^{n+1} \cdot \frac{1}{\xi - z}.$$

Integrating Eq. (4.78), term by term, we get

$$I_1 = \frac{1}{2\pi i}\oint_{C_2}\frac{f(\xi)}{\xi - z}\,d\xi = \frac{1}{2\pi i}\oint_{C_2}\frac{f(\xi)}{\xi - z_0}\,d\xi + \frac{(z-z_0)}{2\pi i}\oint_{C_2}\frac{f(\xi)}{(\xi - z_0)^2}\,d\xi + \cdots$$

$$\cdots + \frac{(z-z_0)^n}{2\pi i}\oint_{C_2}\frac{f(\xi)}{(\xi - z_0)^{n+1}}\,d\xi + R_n(z) \qquad ...(4.79)$$

where $$R_n(z) = \frac{(z-z_0)^{n+1}}{2\pi i}\oint_{C_2}\frac{f(\xi)}{(\xi - z_0)^{n+1}(\xi - z)}\,d\xi. \qquad ...(4.80)$$

Since $\xi - z \neq 0$ and $f(z)$ is analytic in $r < |z - z_0| < R$, it follows that $f(\xi)/(\xi - z)$ is continuous on C_2 and hence bounded. That is, there exists a real positive number $\overset{*}{M}$ such that $\left|\frac{f(\xi)}{\xi - z}\right| \leq \overset{*}{M}$ for all ξ on C_2.

Now $$|R_n(z)| = \left|\frac{(z-z_0)^{n+1}}{2\pi i}\oint_{C_2}\frac{f(\xi)}{(\xi - z_0)^{n+1}(\xi - z)}\,d\xi\right|$$

$$\leq \frac{|(z-z_0)^{n+1}|}{2\pi}\frac{\overset{*}{M}}{R^{n+1}}(2\pi R) = \left|\frac{z-z_0}{R}\right|^{n+1}(\overset{*}{M}R) \to 0 \text{ as } n \to \infty. \qquad [\because |(z-z_0)/R| < 1]$$

Since the integrand in Eq. (4.79) are analytic in the annulus $r < |z - z_0| < R$, the value of the integrand does not change under an arbitrary deformation of path in the domain of analyticity of the integrand function. Therefore, we can integrate over C instead of C_2.

Hence, Eq. (4.79) may be written as

$$I_1 = \frac{1}{2\pi i}\oint_C \frac{f(\xi)}{\xi - z_0}\,d\xi + \frac{z-z_0}{2\pi i}\oint_C\frac{f(\xi)}{(\xi - z_0)^2}\,d\xi + \cdots + \frac{(z-z_0)^n}{2\pi i}\oint_C\frac{f(\xi)}{(\xi - z_0)^{n+1}}\,d\xi + \cdots = \sum_{n=0}^{\infty} c_n (z-z_0)^n \qquad ...(4.81)$$

where $$c_n = \frac{1}{2\pi i}\oint_C\frac{f(\xi)}{(\xi - z_0)^{n+1}}\,d\xi.$$

Consider, now, integration along the path C_1. Since ξ is on C_1 and z is inside the annulus, we have $\left|\frac{\xi - z_0}{z - z_0}\right| < 1$. Therefore, we can write

$$\frac{1}{\xi - z} = \frac{1}{(\xi - z_0) - (z - z_0)} = -\frac{1}{(z-z_0)}\left[1 - \left(\frac{\xi - z_0}{z - z_0}\right)\right]^{-1}$$

$$= -\frac{1}{(z-z_0)}\left[1 + \frac{\xi - z_0}{z - z_0} + \left(\frac{\xi - z_0}{z - z_0}\right)^2 + \left(\frac{\xi - z_0}{z - z_0}\right)^3 + \cdots + \left(\frac{\xi - z_0}{z - z_0}\right)^n\right] + T_n^*(z) \qquad ...(4.82)$$

where $$T_n^*(z) = -\frac{1}{(z-z_0)}\left[\left(\frac{\xi - z_0}{z - z_0}\right)^{n+1} + \left(\frac{\xi - z_0}{z - z_0}\right)^{n+2} + \cdots\right]$$

$$= -\left(\frac{\xi - z_0}{z - z_0}\right)^{n+1}\frac{1}{(z-z_0)}\left[1 + \frac{\xi - z_0}{z - z_0} + \left(\frac{\xi - z_0}{z - z_0}\right)^2 + \cdots\right]$$

$$= -\left(\frac{\xi - z_0}{z - z_0}\right)^{n+1} \frac{1}{(z - z_0)} \left[\frac{1}{1 - \{(\xi - z_0)(z - z_0)\}}\right] = -\left(\frac{\xi - z_0}{z - z_0}\right)^{n+1} \frac{1}{(z - \xi)}.$$

Integratomg Eq. (4.82) term by term, we obtain

$$I_2 = -\frac{1}{2\pi i} \oint_{C_1} \frac{f(\xi)}{\xi - z} d\xi = \frac{1}{2\pi i} \oint_{C_1} \frac{f(\xi)}{z - z_0} d\xi + \frac{1}{2\pi i} \oint_{C_1} \frac{(\xi - z_0) f(\xi)}{(z - z_0)^2} d\xi + \cdots$$

$$\cdots + \frac{1}{2\pi i} \oint_{C_1} \frac{(\xi - z_0)^n}{(z - z_0)^{n+1}} f(\xi)\, d\xi + R_n^*(z). \quad \text{...(4.83)}$$

where $$R_n^*(z) = \frac{1}{2\pi i} \oint_{C_1} \left(\frac{\xi - z_0}{z - z_0}\right)^{n+1} \frac{f(\xi)}{z - \xi} d\xi.$$

Since $f(z)$ is analytic in the annulus $r < |z - z_0| < R$ and $z - \xi \neq 0$, it follows that $f(\xi)/(z - \xi)$ is continuous and hence bounded on C_1. Therefore, there exists a real positive number N, such that $\left|\frac{f(\xi)}{z - \xi}\right| \le N$ for all ξ on C_1.

Hence, $$|R_n^*(z)| \le \frac{n}{2\pi} \left|\left(\frac{r}{z - z_0}\right)^{n+1}\right| (2\pi r) \to 0 \text{ as } n \to \infty, \text{ since } |r/(z - z_0)| < 1.$$

Since the integrands in Eq. (4.83) are analytic in the annulus $r < |z - z_0| < R$, by the principle of deformation of path, we can integrate along any contour C in the domain of analyticity of the integrand function instead of C_1. Therefore, we obtain from Eq. (4.83)

$$I_2 = \frac{1}{2\pi i} \oint_C \frac{f(\xi)}{z - z_0} d\xi + \frac{1}{2\pi i} \oint_C \frac{(\xi - z_0)}{(z - z_0)^2} f(\xi)\, d\xi + \cdots + \frac{1}{2\pi i} \oint_C \frac{(\xi - z_0)^{n-1}}{(z - z_0)^n} f(\xi)\, d\xi + \cdots = \sum_{n=1}^{\infty} \frac{c_{-n}}{(z - z_0)^n} \quad \text{...(4.84)}$$

where $$c_{-n} = \frac{1}{2\pi i} \oint_C (\xi - z_0)^{n-1} f(\xi)\, d\xi.$$

Combining Eqs. (4.81) and (4.84), we obtain

$$f(z) = \sum_{n=0}^{\infty} c_n (z - z_0)^n + \sum_{n=1}^{\infty} \frac{c_{-n}}{(z - z_0)^n} = \sum_{n=-\infty}^{\infty} c_n (z - z_0)^n. \quad \text{...(4.85)}$$

Remark 12

It is difficult to obtain the Laurent series expansions using Eq. (4.76). Since the Laurent series is unique, it can be obtained by various other methods like binomial expansions etc.

EXAMPLE 4.42 Expand the following function in Laurent's series :

(*i*) $f(z) = \dfrac{1}{z - 2}$ valid for (*a*) $|z| < 2$, (*b*) $|z| > 2$.

(*ii*) $f(z) = \dfrac{1}{z^2 - 3z + 2}$ in the region (*a*) $|z| < 1$, (*b*) $1 < |z| < 2$, (*c*) $|z| > 2$.

(UPTU 2006; MDU 2001; Madras 2000S; VTU 2000)

Solution (*i*) (*a*) When $|z| < 2$, we have $|z/2| < 1$.

Therefore, $$\frac{1}{z - 2} = -\frac{1}{2\left(1 - \frac{z}{2}\right)} = -\frac{1}{2}\left(1 - \frac{z}{2}\right)^{-1} = -\frac{1}{2}\left(1 + \frac{z}{2} + \frac{z^2}{4} + \frac{z^3}{8} + \cdots\right)$$

[by binomial expansion]

(*b*) When $|z| > 2$, that is, $|2/z| < 1$, we have

$$\frac{1}{z-2} = \frac{1}{z\left(1-\frac{2}{z}\right)} = \frac{1}{z}\left(1-\frac{2}{z}\right)^{-1} = \frac{1}{z}\left(1+\frac{2}{z}+\frac{4}{z^2}+\frac{8}{z^3}+\cdots\right)$$

(*ii*) Resolving into partial fractions

$$\frac{1}{z^2-3z+2} = \frac{1}{(z-1)(z-2)} = \frac{1}{z-2} - \frac{1}{z-1}.$$

(*a*) When $|z| < 1$, then $|z/2| < 1$.

Therefore, $$\frac{1}{(z-1)(z-2)} = \frac{1}{z-2} - \frac{1}{z-1} = -\frac{1}{2\left(1-\frac{z}{2}\right)} + \frac{1}{(1-z)} = -\frac{1}{2}\left(1-\frac{z}{2}\right)^{-1} + (1-z)^{-1}$$

$$= -\frac{1}{2}\left(1+\frac{z}{2}+\frac{z^2}{4}+\frac{z^3}{8}+\cdots\right) + (1+z+z^2+z^3+\cdots)$$

$$= \frac{1}{2} + \frac{3}{4}z + \frac{7}{8}z^2 + \frac{15}{16}z^3 + \cdots\cdots$$

(*b*) When $1 < |z| < 2$, that is, $|1/z|$ and $|z/2|$ are both less than 1.

Therefore, $$\frac{1}{(z-1)(z-2)} = \frac{1}{z-2} - \frac{1}{z-1} = -\frac{1}{2\left(1-\frac{z}{2}\right)} - \frac{1}{z\left(1-\frac{1}{z}\right)} = -\frac{1}{2}\left(1-\frac{z}{2}\right)^{-1} - \frac{1}{z}\left(1-\frac{1}{z}\right)^{-1}$$

$$= -\frac{1}{2}\left(1+\frac{z}{2}+\frac{z^2}{4}+\frac{z^3}{8}+\cdots\right) - \frac{1}{z}\left(1+\frac{1}{z}+\frac{1}{z^2}+\cdots\right)$$

$$= -\frac{1}{2}\left(1+\frac{z}{2}+\frac{z^2}{4}+\cdots\right) - \left(\frac{1}{z}+\frac{1}{z^2}+\frac{1}{z^3}+\cdots\right).$$

(*c*) When $|z| > 2$, then, $|2/z| < 1$ and $|1/z| < 1$.

Therefore, $$\frac{1}{(z-1)(z-2)} = \frac{1}{z-2} - \frac{1}{z-1} = \frac{1}{z\left(1-\frac{2}{z}\right)} - \frac{1}{z\left(1-\frac{1}{z}\right)} = \frac{1}{z}\left(1-\frac{2}{z}\right)^{-1} - \frac{1}{z}\left(1-\frac{1}{z}\right)^{-1}$$

$$= \frac{1}{z}\left(1+\frac{2}{z}+\frac{4}{z^2}+\frac{8}{z^3}+\cdots\right) - \frac{1}{z}\left(1+\frac{1}{z}+\frac{1}{z^2}+\frac{1}{z^3}+\cdots\right)$$

$$= \frac{1}{z^2} + \frac{3}{z^3} + \frac{7}{z^4} + \cdots.$$

EXAMPLE 4.43 Expand $f(z) = \dfrac{1}{(z+1)(z+3)}$ in a Laurent's series valid for

(*i*) $1 < |z| < 3$, (*ii*) $|z| > 3$, (*iii*) $0 < |z+1| < 2$, (*AMIETE, June 2007*) (*iv*) $|z| < 1$.

Solution We write $\dfrac{1}{(z+1)(z+3)} = \dfrac{1}{2}\left[\dfrac{1}{z+1} - \dfrac{1}{z+3}\right]$. [resolving into partial fractions] (*i*) If $z > 1$, we have $\dfrac{1}{2(z+1)} = \dfrac{1}{2z\left(1+\dfrac{1}{z}\right)} = \dfrac{1}{2z}\left(1+\dfrac{1}{z}\right)^{-1} = \dfrac{1}{2z}\left(1-\dfrac{1}{z}+\dfrac{1}{z^2}-\dfrac{1}{z^3}+\cdots\right)$

$$= \frac{1}{2z} - \frac{1}{2z^2} + \frac{1}{2z^3} - \frac{1}{2z^4} + \cdots$$

If $|z| < 3$, $\dfrac{1}{2(z+3)} = \dfrac{1}{6\left(1+\dfrac{z}{3}\right)} = \dfrac{1}{6}\left(1+\dfrac{z}{3}\right)^{-1} = \dfrac{1}{6}\left(1-\dfrac{z}{3}+\dfrac{z^2}{9}-\dfrac{z^3}{27}+\cdots\right) = \dfrac{1}{6}-\dfrac{z}{18}+\dfrac{z^2}{54}-\dfrac{z^3}{162}+\cdots$

Thus, the required Laurent expansion valid for both $|z| > 1$ and $|z| < 3$, that is, $1 < |z| < 3$ is

$$\cdots - \frac{1}{2z^4} + \frac{1}{2z^3} - \frac{1}{2z^2} + \frac{1}{2z} - \frac{1}{6} + \frac{z}{18} - \frac{z^2}{54} + \frac{z^3}{162} - \cdots$$

(*ii*) If $|z| > 1$, we have by part (*i*) $\dfrac{1}{2(z+1)} = \dfrac{1}{2z} - \dfrac{1}{2z^2} + \dfrac{1}{2z^3} - \dfrac{1}{2z^4} + \cdots$

If $|z| > 3$, $\dfrac{1}{2(z+3)} = \dfrac{1}{2z\left(1+\dfrac{3}{z}\right)} = \dfrac{1}{2z}\left(1+\dfrac{3}{z}\right)^{-1} = \dfrac{1}{2z}\left(1-\dfrac{3}{z}+\dfrac{9}{z^2}-\dfrac{27}{z^3}+\cdots\right)$

$$= \frac{1}{2z} - \frac{3}{2z^2} + \frac{9}{2z^3} - \frac{27}{2z^4} + \cdots$$

The required Laurent expansion valid for both $|z| > 1$ and $|z| > 3$, that is, $|z| > 3$ is by subtraction

$$\frac{1}{z^2} - \frac{4}{z^3} + \frac{13}{z^4} - \cdots$$

(*iii*) Let $z + 1 = t$. Then $\dfrac{1}{(z+1)(z+3)} = \dfrac{1}{t(t+2)} = \dfrac{1}{2t\left(1+\dfrac{t}{2}\right)} = \dfrac{1}{2t}\left(1+\dfrac{t}{2}\right)^{-1} = \dfrac{1}{2t}\left[1-\dfrac{t}{2}+\dfrac{t^2}{4}-\dfrac{t^3}{8}+\cdots\right]$

$$= \frac{1}{2(z+1)} - \frac{1}{4} + \frac{1}{8}(z+1) - \frac{1}{16}(z+1)^2 + \cdots$$

valid for $|t| < 2$, $t \neq 0$ or $0 < |z+1| < 2$.

(*iv*) If $|z| < 1$, $\dfrac{1}{2(z+1)} = \dfrac{1}{2}(1 - z + z^2 - z^3 + z^4 - \cdots) = \dfrac{1}{2} - \dfrac{1}{2}z + \dfrac{1}{2}z^2 - \dfrac{1}{2}z^3 - \dfrac{1}{2}z^4 - \cdots$

If $|z| < 3$, we have by part (*i*), $\dfrac{1}{2(z+3)} = \dfrac{1}{6} - \dfrac{z}{18} + \dfrac{z^2}{54} - \dfrac{z^3}{162} + \cdots$

Then the required Laurent expansion, valid for both $|z| < 1$ and $|z| < 3$, that is, $|z| < 1$, is by subtraction

$\dfrac{1}{3} - \dfrac{4}{9}z + \dfrac{13}{27}z^2 - \dfrac{40}{81}z^3 + \cdots$. This is a Taylor series.

EXAMPLE 4.44 Expand $f(z) = \dfrac{1}{(z-1)^2 (z-3)}$ in a Laurent's series valid for

(*i*) $0 < |z - 1| < 2$ and (*ii*) $0 < |z - 3| < 2$.

Solution (*i*) We want only powers of $z - 1$, and so we need to express $z - 3$ in terms of $z - 1$. This can be done by writing

$$f(z) = \frac{1}{(z-1)^2(z-3)} = \frac{1}{(z-1)^2[(z-1)-2]} = -\frac{1}{2}\left[\frac{1}{(z-1)^2\left[1-\dfrac{(z-1)}{2}\right]}\right] = -\frac{1}{2(z-1)^2}\left[1-\frac{(z-1)}{2}\right]^{-1}$$

$$= -\frac{1}{2(z-1)^2}\left[1+\frac{z-1}{2}+\frac{(z-1)^2}{4}+\frac{(z-1)^3}{8}+\cdots\right] = -\frac{1}{2(z-1)^2} - \frac{1}{4(z-1)} - \frac{1}{8} - \frac{(z-1)}{16} - \cdots$$

(*ii*) To obtain powers of $z - 3$ we write $z - 1 = (z - 3) + 2$ and

$$f(z) = \frac{1}{(z-1)^2(z-3)} = \frac{1}{(z-3)}[2+(z-3)]^{-2} = \frac{1}{4(z-3)}\left[1+\frac{z-3}{2}\right]^{-2}$$

$$= \frac{1}{4(z-3)}\left[1+(-2)\left(\frac{z-3}{2}\right)+\frac{(-2)(-3)}{2!}\left(\frac{z-3}{2}\right)^2+\frac{(-2)(-3)(-4)}{3!}\left(\frac{z-3}{2}\right)^3+\cdots\right]$$

$$= \frac{1}{4(z-3)}\left[1-(z-3)+\frac{3}{4}(z-3)^2-\frac{1}{2}(z-3)^3+\cdots\right]$$

$$= \frac{1}{4(z-3)} - \frac{1}{4} + \frac{3}{16}(z-3) - \frac{1}{8}(z-3)^2 + \cdots$$

EXAMPLE 4.45 Find the Laurent's expansion of the function $f(z) = \dfrac{7z-2}{(z+1)\,z\,(z-2)}$ in the annulus

(*i*) $0 < |z + 1| < 1$, (*ii*) $|z + 1| > 3$.

Solution (*i*) We write $f(z) = \dfrac{7z-2}{(z+1)\,z\,(z-2)} = \dfrac{-3}{z+1} + \dfrac{1}{z} + \dfrac{2}{z-2}$ [resolving into partial fractions]

We want only powers of $z + 1$, and so we need to express z and $z - 2$ in terms of $z + 1$. This can be done by writing

$$f(z) = \frac{-3}{(z+1)} + \frac{1}{(z+1)-1} + \frac{2}{(z+1)-3} = \frac{-3}{z+1} - [1-(z+1)]^{-1} - \frac{2}{3}\left[1-\left(\frac{z+1}{3}\right)\right]^{-1} \quad \text{for } 0<|z+1|<1$$

$$= \frac{-3}{z+1} - [1+(z+1)+(z+1)^2+(z+1)^3+\cdots] - \frac{2}{3}\left[1+\frac{z+1}{3}+\frac{(z+1)^2}{9}+\frac{(z+1)^3}{27}+\cdots\right]$$

$$= \frac{-3}{z+1} - \frac{5}{3} - \frac{11}{9}(z+1) - \frac{29}{27}(z+1)^2 - \frac{83}{81}(z+1)^3 - \cdots \quad \text{for } 0 < |z+1| < 1.$$

(*ii*)
$$f(z) = \frac{-3}{z+1} + \frac{1}{(z+1)\left[1-\dfrac{1}{z+1}\right]} + \frac{2}{(z+1)\left[1-\dfrac{3}{z+1}\right]}, \quad \text{for } |z+1|>3$$

$$= -3(z+1)^{-1} + \frac{1}{z+1}\left[1-\frac{1}{z+1}\right]^{-1} + \frac{2}{z+1}\left[1-\frac{3}{z+1}\right]^{-1}$$

$$= -3(z+1)^{-1} + \frac{1}{z+1}\left[1+\frac{1}{(z+1)}+\frac{1}{(z+1)^2}+\frac{1}{(z+1)^3}+\cdots\right]$$

$$+\frac{2}{z+1}\left[1+\frac{3}{z+1}+\frac{9}{(z+1)^2}+\frac{27}{(z+1)^3}+\cdots\right]$$

$$= -3(z+1)^{-1} + (z+1)^{-1} + (z+1)^{-2} + (z+1)^{-3} + (z+1)^{-4} + \ldots$$

$$+2[(z+1)^{-1} + 3(z+1)^{-2} + 9(z+1)^{-3} + 27(z+1)^{-4} + \ldots]$$

$$= 7(z+1)^{-2} + 19(z+1)^{-3} + 55(z+1)^{-4} + \cdots \quad \text{for } |z+1| > 3.$$

EXAMPLE 4.46 Expand the following functions in Laurent's series :

(*i*) $f(z) = \dfrac{z}{(z+1)(z-2)}$ about $z=-2$ (*ii*) $f(z) = \dfrac{e^z}{(z+1)^2}$ about $z=1$.

Solution (*i*) As $z = -1$ and $z = -2$ are the two singular points of the function, we can expand the function in Laurent's series about these points.

To expand $f(z) = \dfrac{z}{(z+1)(z+2)}$ about z = –2, that is, in powers of z + 2, we put $z + 2 = u$.

Then
$$\frac{z}{(z+1)(z+2)} = \frac{u-2}{(u-1)u} = \frac{2-u}{u}\cdot\frac{1}{1-u} = \frac{2-u}{u}(1-u)^{-1}$$

$$= \frac{2-u}{u}(1+u+u^2+u^3+\ldots) \quad \text{for } 0<|u|<1$$

$$= \frac{1}{u}(2+u+u^2+u^3+\ldots) = \frac{2}{u}+1+u+u^2+\ldots$$

$$= \frac{2}{z+2}+1+(z+2)+(z+2)^2+\ldots \quad \text{for } 0<|z+2|<1.$$

(*ii*) Setting $z - 1 = u$, we obtain

$$f(z) = \frac{e^z}{(z-1)^2} = \frac{e^{1+u}}{u^2} = \frac{e}{u^2}\bullet e^u = \frac{e}{u^2}\left(1+u+\frac{u^2}{2!}+\frac{u^3}{3!}+\frac{u^4}{4!}+\cdots\right)$$

$$= e\left(\frac{1}{u^2}+\frac{1}{u}+\frac{1}{2!}+\frac{u}{3!}+\frac{u^2}{4!}+\cdots\right) = e\left(\frac{1}{(z-1)^2}+\frac{1}{z-1}+\frac{1}{2!}+\frac{1}{3!}(z-1)+\frac{1}{4!}(z-1)^2+\cdots\right).$$

EXAMPLE 4.47 Obtain the first three terms of the Laurent series expansion of the function $f(z) = \dfrac{1}{e^z-1}$ about the point $z = 0$ valid in the region $0<|z|<2\pi$. (*AMIETE, June 2010, 2009, Dec. 2008, 2005*)

Solution The given function is not analytic when $e^z = 1$, that is, at $z = 0$ and $z = 2n\pi i$, $n = \pm1, \pm2, \ldots$. The required Laurent series expansion is about the point $z = 0$.

Its region of convergence is $0<|z|<2\pi$.

We have
$$f(z) = \frac{1}{e^z - 1} = \frac{1}{z + \frac{z^2}{2!} + \frac{z^3}{3!} + \cdots + \frac{z^n}{n!} + \cdots} = \frac{1}{z}\left[1 + \left(\frac{z}{2} + \frac{z^2}{6} + \frac{z^3}{24} + \cdots\right)\right]^{-1}$$

$$= \frac{1}{z}\left[1 - \left(\frac{z}{2} + \frac{z^2}{6} + \frac{z^3}{24} + \cdots\right) + \left(\frac{z}{2} + \frac{z^2}{6} + \frac{z^3}{24} + \cdots\right)^2 - \left(\frac{z}{2} + \frac{z^2}{6} + \frac{z^3}{24} + \cdots\right)^3 + \cdots\right]$$

$$= \frac{1}{z}\left[1 - \left(\frac{z}{2} + \frac{z^2}{6} + \frac{z^3}{24} + \cdots\right) + \left(\frac{z^2}{4} + \frac{z^3}{6} + \cdots\right) - \left(\frac{z^3}{8} + \cdots\right)\right]$$

$$= \frac{1}{z}\left[1 - \frac{z}{2} + \left(-\frac{1}{6} + \frac{1}{4}\right)z^2 + \left(-\frac{1}{24} + \frac{1}{6} - \frac{1}{8}\right)z^3 + \cdots\right] \approx \frac{1}{z} - \frac{1}{2} + \frac{z}{12}.$$

EXAMPLE 4.48 Find all the possible Taylor's and Laurent series expansions of the function $f(z) = \frac{1}{(z+1)(z+2)^2}$ about the point z = 1. *(AMIETE, Dec. 2006)*

Solution The given function is not analytic at the points $z = -1$ and $z = -2$. The distances between the point $z = 1$ and the points $z = -1$, $z = -2$ are 2 and 3 respectively. Therefore, we consider the regions (*i*) $|z - 1| < 2$, (*ii*) $2 < |z - 1| < 3$ and (*iii*) $|z - 1| > 3$.

In the region, $|z - 1| < 2$, the function $f(z)$ is analytic. Therefore, we obtain a Taylor series expansion in this region. In the other regions, we obtain Laurent series expansions.

We write
$$f(z) = \frac{1}{(z+1)(z+2)^2} = \frac{1}{z+1} - \frac{1}{z+2} - \frac{1}{(z+2)^2} \quad \text{[resolving into partial fractions]}$$

$$= \frac{1}{(z-1)+2} - \frac{1}{(z-1)+3} - \frac{1}{[(z-1)+3]^2}.$$

(*i*) In the region $|z - 1| < 2$, we write

$$f(z) = \frac{1}{2\left[1 + \frac{z-1}{2}\right]} - \frac{1}{3\left[1 + \frac{z-1}{3}\right]} - \frac{1}{9\left[1 + \frac{z-1}{3}\right]^2}$$

$$= \frac{1}{2}\left[1 + \frac{z-1}{2}\right]^{-1} - \frac{1}{3}\left[1 + \frac{z-1}{3}\right]^{-1} - \frac{1}{9}\left[1 + \frac{z-1}{3}\right]^{-2}$$

$$= \frac{1}{2}\sum_{n=0}^{\infty}(-1)^n\left(\frac{z-1}{2}\right)^n - \frac{1}{3}\sum_{n=0}^{\infty}(-1)^n\left(\frac{z-1}{3}\right)^n - \frac{1}{9}\sum_{n=0}^{\infty}(-1)^n(n+1)\left(\frac{z-1}{3}\right)^n$$

$$= \sum_{n=0}^{\infty}(-1)^n(z-1)^n[2^{-n-1} - 3^{-n-1} - 3^{-n-2}(n+1)]$$

$$= \sum_{n=0}^{\infty}(-1)^n[2^{-n-1} - 3^{-n-2}(n+4)](z-1)^n.$$

The first series is valid in $|z - 1| < 2$ and the second and third series are valid in $|z - 1| < 3$. Hence, the sum is valid in $|z - 1| < 2$.

(*ii*) In the region $2 < |z-1| < 3$, we write

$$f(z) = \frac{1}{(z-1)}\left[1+\frac{2}{z-1}\right]^{-1} - \frac{1}{3}\left[1+\frac{z-1}{3}\right]^{-1} - \frac{1}{9}\left[1+\frac{z-1}{3}\right]^{-2}$$

$$= \frac{1}{(z-1)}\left[\sum_{n=0}^{\infty}(-1)^n\left(\frac{2}{z-1}\right)^n\right] - \frac{1}{3}\sum_{n=0}^{\infty}(-1)^n\left(\frac{z-1}{3}\right)^n - \frac{1}{9}\sum_{n=0}^{\infty}(-1)^n(n+1)\left(\frac{z-1}{3}\right)^n$$

$$= \sum_{n=0}^{\infty}(-1)^n\frac{2^n}{(z-1)^{n+1}} - \frac{1}{9}\sum_{n=0}^{\infty}(-1)^n(n+4)\left(\frac{z-1}{3}\right)^n.$$

The first series is valid in $|z-1| > 2$ and the second and third series are valid in $|z-1| < 3$. Hence, the sum is valid in the annulus $2 < |z-1| < 3$.

(*iii*) In the region $|z-1| > 3$, we write

$$f(z) = \frac{1}{(z-1)}\left[1+\frac{2}{z-1}\right]^{-1} - \frac{1}{(z-1)}\left[1+\frac{3}{z-1}\right]^{-1} - \frac{1}{(z-1)^2}\left[1+\frac{3}{z-1}\right]^{-2}$$

$$= \frac{1}{(z-1)}\sum_{n=0}^{\infty}(-1)^n\left(\frac{2}{z-1}\right)^n - \frac{1}{(z-1)}\sum_{n=0}^{\infty}(-1)^n\left(\frac{3}{z-1}\right)^n - \frac{1}{(z-1)^2}\sum_{n=0}^{\infty}(-1)^n(n+1)\left(\frac{3}{z-1}\right)^n$$

$$= \sum_{n=0}^{\infty}(-1)^n\left[\frac{(2^n-3^n)}{(z-1)^{n+1}} - \frac{3^n(n+1)}{(z-1)^{n+2}}\right].$$

EXERCISE 4.6

1. Expand $f(z) = \cos z$ in a Taylor's series about $z = \pi/4$.

Ans. $\frac{\sqrt{2}}{2}\left[1-\left(z-\frac{\pi}{4}\right)-\frac{1}{2!}\left(z-\frac{\pi}{4}\right)^2+\frac{1}{3!}\left(z-\frac{\pi}{4}\right)^3+\cdots\right]$

2. (*a*) Expand $\log(1+z)$ in a Taylor's series about $z = 0$ and determine the region of convergence for the resulting series. (*AMIETE, Dec. 2010*)

(*b*) Expand the function $f(z) = 1/z$ about $z = 2$ in Taylor's series. Obtain its radius of convergence.

Ans. (*a*) $z - \frac{z^2}{2} + \frac{z^3}{3} - \frac{z^4}{4} + \cdots + (-1)^{n-1}\frac{z^n}{n} + \cdots$; series converges for all values of z within the circle $|z| = 1$.

(*b*) [**Hint:** Taylor's series of function $f(z) = 1/z$ about $z = 2$ is obtain as

$$\frac{1}{z} = \frac{1}{2+z-2} = \frac{1}{2\left(1+\frac{z-2}{2}\right)} = \frac{1}{2}\left(1+\frac{z-2}{2}\right)^{-1} = \frac{1}{2}\left[1-\frac{z-2}{2}+\frac{z-2}{2^2}-\frac{z-2}{2^3}+\cdots\right]$$

$= \frac{1}{2} - \frac{z-2}{2^2} + \frac{z-2}{2^3} - \frac{z-2}{2^4} + \cdots$ which converges when $\left|\frac{z-2}{2}\right| < 1$ or $|z-2| < 2$.

Radius of convergence is $R = 2$.]

3. Find the Taylor's series which represent the function $f(z)=\dfrac{z^2-1}{(z+2)(z+3)}$ in the region $|z|<2$.

(*AMIETE, Dec. 2010*) **Ans.** $1+\sum_{n=0}^{\infty}\left[(-1)^n \cdot \left\{\dfrac{3}{2^{n+1}}-\dfrac{8}{3^{n+1}}\right\} z^n\right]$.

4. Expand $f(z)=\dfrac{z}{(z+1)(z+2)}$ in a Taylor's series about (*a*) $z=0$, (*b*) $z=2$.

[**Hint:** (*a*) We write $\dfrac{z}{(z+1)(z+2)}=\dfrac{2}{z+2}-\dfrac{1}{z+1}$ [resolving into partial fractions]

$$=\left(1+\frac{z}{2}\right)^{-1}-(1+z)^{-1}=\frac{z}{2}-\frac{3}{4}z^2+\frac{7}{8}z^3-\cdots \text{ where } |z|<1.$$

(*b*) To expand $f(z)$ about $z=2$, that is, in powers of $(z-2)$, we put $z-2=t$ and have

$$\frac{z}{(z+1)(z+2)}=\frac{2+h}{(h+3)(h+4)}=\frac{2}{h+4}-\frac{1}{h+3}=\frac{1}{2}\left(1+\frac{h}{4}\right)^{-1}-\frac{1}{3}\left(1+\frac{h}{3}\right)^{-1}$$

$$=\frac{1}{2}\left(1-\frac{h}{4}+\frac{h^2}{4^2}-\cdots\right)-\frac{1}{3}\left(1-\frac{h}{3}+\frac{h^2}{3^2}-\cdots\right)$$

$$=\left(\frac{1}{2}-\frac{1}{3}\right)-\left(\frac{1}{2^3}-\frac{1}{3^2}\right)h+\left(\frac{1}{2^5}-\frac{1}{3^3}\right)h^2-\cdots$$

$$=\left(\frac{1}{2}-\frac{1}{3}\right)-\left(\frac{1}{2^3}-\frac{1}{3^2}\right)(z-2)+\left(\frac{1}{2^5}-\frac{1}{3^3}\right)(z-2)^2-\cdots \quad \text{for } |z-2|<3.$$

5. Show that when $|z+1|<1$, $\dfrac{1}{z^2}=1+\sum_{n=1}^{\infty}(n+1)(z+1)^n$.

[**Hint:** $\dfrac{1}{z^2}=\dfrac{1}{[(z+1)-1]^2}=\dfrac{1}{[1-(z+1)]^2}=[1-(z+1)]^{-2}$. As $|z+1|<1$, we have by binomial expansion

$$[1-(z+1)]^{-2}=1+2(z+1)+3(z+1)^2+\cdots$$

Therefore, $\dfrac{1}{z^2}=1+2(z+1)+3(z+1)^2+\cdots=1+\sum_{n=1}^{\infty}(n+1)(z+1)^n$.

6. Expand the following functions in Laurent's series :

(*a*) $f(z)=\dfrac{z-1}{z^2}$ for $|z-1|>1$.

(*b*) $f(z)=\dfrac{1}{(z-1)(z-2)}$ for (*i*) $|z-1|>1$, (*ii*) $0<|z-2|<1$.

(*c*) $f(z)=\dfrac{7z-2}{z(z-2)(z+1)}$ in the region $1<|z+1|<3$. (*AMIETE, June 2010*)

(*d*) $f(z)=\dfrac{\sin z}{z-\pi}$ about $z=\pi$, (*e*) $\dfrac{1}{z^2-4z+3}$ for $1<|z|<3$. (*AMIETE, Dec. 2012*)

Ans. (*a*) $\dfrac{1}{z-1}-\dfrac{2}{(z-1)^2}+\dfrac{3}{(z-1)^3}-\cdots$;

(*b*) (*i*) $\dfrac{1}{(z-1)^2}+\dfrac{1}{(z-1)^3}+\dfrac{1}{(z-1)^4}+\cdots$; (*ii*) $\dfrac{1}{z-2}-1+(z-2)-(z-2)^2+(z-2)^3-\cdots$;

(*c*) $\cdots+(z+1)^{-2}-2(z+1)^{-1}-\dfrac{2}{3}-\dfrac{2}{9}(z+1)-\dfrac{2}{27}(z+1)^2-\cdots$;

(*d*) $-1+\dfrac{(z-\pi)^2}{3!}-\dfrac{(z-\pi)^4}{5!}+\cdots$; (*e*) $-\dfrac{1}{2}\sum_{n=1}^{\infty}\dfrac{1}{z^n}-\dfrac{1}{2}\sum_{n=0}^{\infty}\dfrac{z^n}{3^{n+1}}$.

7. Find Laurent's series expansion of $f(z)=\dfrac{1}{(z+1)(z+3)}$ in the powers of $(z+1)$ for the range of $0<|z+1|<2$. *(AMIETE, June 2007)*

Ans. $\dfrac{1}{(z+1)(z+3)}=\dfrac{1}{2(z+1)}-\dfrac{1}{4}+\dfrac{1}{8}(z+1)-\dfrac{1}{16}(z+1)^2+\cdots$

8. Expand $f(z)=\dfrac{1}{z(z-1)}$ in a Laurent series valid for $1<|z-2|<2$.

Ans. $f(z)=-\dfrac{1}{(z-2)^4}+\dfrac{1}{(z-2)^3}-\dfrac{1}{(z-2)^2}+\dfrac{1}{z-2}-\dfrac{1}{2}+\dfrac{z-2}{4}-\dfrac{(z-2)^2}{8}+\dfrac{(z-2)^3}{16}-\cdots$

9. Obtain the first four terms of the Laurent's series expansion of function $f(z)=\dfrac{e^z}{z(z^2+1)}$.

Ans. $\dfrac{1}{z}+1+\dfrac{1}{2}z-\dfrac{5}{6}z^2+\cdots \quad (0<|z|<1)$.

10. Prove the each of the following Laurent's expansions is valid in the region indicated

(*a*) $f(z)=\dfrac{1}{(z-1)(2-z)}=\sum_{n=0}^{\infty}\dfrac{z^n}{2^n+1}+\sum_{n=1}^{\infty}\dfrac{1}{z^n}$ if $1<|z|<2$,

(*b*) $f(z)=\dfrac{1}{(z-1)(2-z)}=\sum_{n=2}^{\infty}\dfrac{1-2^{n-1}}{z^n}$, if $|z|>2$.

11. Find all Taylor and Laurent series of $f(z)=\dfrac{-2z+3}{z^2-3z+2}$ with center 0.

[**Hint:** We write $f(z)=\dfrac{-2z+3}{z^2-3z+2}=-\dfrac{1}{z-1}-\dfrac{1}{z-2}$.

Consider the regions (*a*) $|z|<1$, (*b*) $1<|z|<2$ and (*c*) $|z|>2$.

(*a*) If $|z|<1$. We have $-\dfrac{1}{z-1}=(1-z)^{-1}=1+z+z^2+z^3+\cdots$ ...(1)

If $|z|<2$, $-\dfrac{1}{z-2}=\dfrac{1}{2\left(1-\dfrac{z}{2}\right)}=\dfrac{1}{2}\left(1-\dfrac{z}{2}\right)^{-1}=\dfrac{1}{2}\left(1+\dfrac{z}{2}+\dfrac{z^2}{4}+\dfrac{z^3}{8}+\cdots\right)$ ...(2)

Adding (1) and (2) $f(z)=\frac{3}{2}+\frac{5}{4}z+\frac{9}{8}z^2+\cdots.$

(*b*) If $|z|>1$, $-\frac{1}{z-1}=-\frac{1}{z\left(1-\frac{1}{z}\right)}=-\frac{1}{z}\left(1-\frac{1}{z}\right)^{-1}=-\frac{1}{z}\left(1+\frac{1}{z}+\frac{1}{z^2}+\frac{1}{z^3}+\cdots\right)$

$$=-\frac{1}{z}-\frac{1}{z^2}-\frac{1}{z^3}-\frac{1}{z^4}-\cdots$$

If $|z|<2$, $-\frac{1}{z-2}=\frac{1}{2\left(1-\frac{z}{2}\right)}=\frac{1}{2}+\frac{z}{4}+\frac{z^2}{8}+\frac{z^3}{16}+\cdots$

Hence, $f(z)=\frac{1}{2}+\frac{z}{4}+\frac{z^2}{8}+\cdots-\frac{1}{z}-\frac{1}{z^2}-\cdots.$

(*c*) If $|z|>1$. We have $-\frac{1}{z-1}=-\frac{1}{z}-\frac{1}{z^2}-\frac{1}{z^3}-\frac{1}{z^4}-\cdots$

If $|z|>2$. We have $-\frac{1}{z-2}=-\frac{1}{z\left(1-\frac{2}{z}\right)}=-\frac{1}{z}\left(1-\frac{2}{z}\right)^{-1}=-\frac{1}{z}\left(1+\frac{2}{z}+\frac{4}{z^2}+\frac{8}{z^3}+\cdots\right)$

$$=-\frac{1}{z}-\frac{2}{z^2}-\frac{4}{z^3}-\frac{8}{z^4}-\cdots$$

Hence, $f(z)=-\frac{2}{z}-\frac{3}{z^2}-\frac{5}{z^3}-\frac{9}{z^4}-\cdots.$]

ZEROS, SINGULARITIES AND RESIDUES

4.19 INTRODUCTION

In this section, we shall define zeroes and singularities of complex functions. There are different types of singularities which can be classified using the Laurent series. Residue of a function f(z) at a singularity is also defined using the Laurent series. The concept of residues has been utilised in the evaluation of complex integrals and certain real integrals.

4.20 ZEROS AND SINGULARITIES OF COMPLEX FUNCTIONS

Let $f(z)$ be a complex function. A point $z=z_0$ is called a zero of $f(z)$ if $f(z)$ is analytic at z_0 and $f(z_0)=0$.

If $f(z_0)=0$ but $f'(z_0)\neq 0$, then z_0 is called a *simple zero* or a *zero of the first order*.

If $f(z_0)=0, f'(z_0)=0, f''(z_0)=0, \ldots\ldots, f^{(n-1)}(z_0)=0$, but $f^n(z_0)\neq 0$, then z_0 is called a zero of order *n*. For example, the function

(*i*) $f(z)=1+z^2$ has simple zeros at $z=\pm i$.

$f(z)=\sin z$ has simple zeros at $z=n\pi$, $n=0,\pm1,\pm2,\ldots,$ since at each of these points $\sin z=0$ but its derivative $\cos z\neq 0$.

(*ii*) $f(z)=(1-e^z)^2$ has zeros of order 2 (double zeros) at $z=2n\pi i$, n = $0,\pm1,\pm2,\ldots,$ since at each of these points $f(z)=(1-e^z)^2=0$ and $f'(z)=-2e^z(1-e^z)=0$ but $f'''(z)=-2[e^z-2e^{2z})$ $=-2e^z(1-2e^z)\neq 0$.

(*iii*) $f(z)=(z-4)^3$ has zeros of order 3 at $z=4$. We see that $f(4)=0, f'(4)=0, f''(4)=0$, but $f'''(4)=6\neq 0$.

Zeros may be discussed by means of Taylor Series

If the function $f(z)$ is analytic at $z = z_0$, we can write the Taylor's series expansion of $f(z)$ about the point z_0 as

$$f(z) = f(z_0) + (z - z_0) f'(z_0) + \ldots + \frac{(z - z_0)^m}{m!} f^{(m)}(z_0) + \frac{(z - z_0)^{m+1}}{(m+1)!} f^{(m+1)}(z_0) + \ldots$$

when $z = z_0$ is a zero of order m, we obtain

$$f(z) = \frac{(z - z_0)^m}{m!} f^{(m)}(z_0) + \frac{(z - z_0)^{m+1}}{(m+1)!} f^{(m+1)}(z_0) + \ldots$$

$$= (z - z_0)^m \left[\frac{f^{(m)}(z_0)}{m!} + \frac{(z - z_0)}{(m+1)!} f^{(m+1)}(z_0) + \ldots \right] = (z - z_0)^m \, h(z) \qquad \ldots(4.86)$$

where $h(z)$ is analytic at z_0 and $h(z_0) \neq 0$. When a function $f(z)$ has a zero at $z = z_0$ but has no other zero in some neighbourhood of the point z_0, then the zero $z = z_0$ is called an *isolated zero.*

Singular Points of a Function

A point $z = z_0$, at which the function $f(z)$ is not defined or the function is not analytic but every neighbourhood of $z = z_0$ contains points at which $f(z)$ is analytic is called a *singular point* or singularity $f(z)$. A rational function $p(z)/q(z)$ has a singular point at $z = z_0$, whenever $q(z_0) = 0$ and $p(z_0) \neq 0$.

For example, the function $f(z) = \dfrac{z^2 - z - 6}{z^2 - z - 12} = \dfrac{(z+2)(z-3)}{(z-4)(z+3)}$

(*i*) has simple zeros at $z = -2$, and $z = 3$,
(*ii*) has singular point at $z = -3$ and $z = 4$.

If the singular point z_0 is such that there is no other singular point in its neighbourhood, that is, if $f(z)$ is analytic in some deleted neighbourhood of z_0, then the point z_0 is called an *isolated singular point or isolated singularity of $f(z)$.*

For example, the function $f(z) = \dfrac{1}{z - a}$ is analytic everywhere except at $z = a$. Thus $f(z)$ has an isolated singularity at $z = a$.

The function $f(z) = \dfrac{z^2 + 4}{(z-1)^2 (z^2 + 9)}$ has three isolated singular points, namely $z = 1$, $z = 3i$ and $z = -3i$. tan z has isolated singularities at $\pm \pi/2$, $\pm 3\pi/2$. tan $(1/z)$ has a non-isolated singularity at 0.

Laurent series expansion of the function $f(z)$ can be used to classify the isolated singular points.

Let $z = z_0$ is an isolated singularity of a function $f(z)$ and that

$$f(z) = \sum_{n=-\infty}^{\infty} a_n (z - z_0)^n = \sum_{n=0}^{\infty} a_n (z - z_0)^n + \sum_{n=1}^{\infty} \frac{b_n}{(z - z_0)^n} \qquad \ldots(4.87)$$

is the Laurent series representation $f(z)$ valid for $0 < |z - z_0| < R$. The sum of first series in equation (4.87) is called the **analytic part** of the Laurent series, whereas the sum of the second series involving negative powers of $(z - z_0)$ is called the **principal part** of the Laurent series.

Removable Singularity

If the function $f(z)$ is not defined at $z = z_0$, but $\lim\limits_{z \to z_0} f(z)$ exists, then the point $z = z_0$ is called a removable singular point. Note that in this case, the principal part of the Laurent series is zero, that is, all the coefficients b_n are zero.

For example, the function $f(z) = \dfrac{\sin z}{z}$ has a removable singularity at $z = 0$, since $\lim\limits_{z \to z_0} \left(\dfrac{\sin z}{z} \right) = 1$ exists.

Pole

If the principal part of the Laurent series expansion of the function $f(z)$ has only a finite number of terms, that is, it is of the form

$$\frac{b_1}{z-z_0}+\frac{b_2}{(z-z_0)^2}+\cdots+\frac{b_m}{(z-z_0)^m}$$

where m is a finite integer and $b_{m+1}, b_{m+2}, \ldots$ are all zero, then $z = z_0$ is called a *pole* of order m.

If $m = 1$, then $z = z_0$ is called a *simple pole.*

For example, the function $f(z)=\dfrac{2}{(z-3)^4}+\dfrac{3}{(z+3)^3}$ has a pole of order 4 at $z = 3$ and a pole of order 3 at $z = -3$.

Essential Singular Point

If the principal part of the Laurent series expansion of the function $f(z)$ has infinite number of terms, then the point $z = z_0$ is called an *essential singular point.* Examples of functions having an essential singularity at $z = 0$ are

$$e^{1/z}=1+\frac{1}{z}+\frac{1}{2!z^2}+\ldots,\qquad \sin(1/z)=\frac{1}{z}-\frac{1}{3!z^3}+\frac{1}{5!z^5}-\ldots$$

Singular at Infinity

To determine whether the point $z = \infty$ is a singular point of $f(z)$, we substitute $z = 1/w$ in the given function and discuss the singularity of the resulting function $f(1/w)$ at $w = 0$.

EXAMPLE 4.49 Find the nature and location of singularities of the following functions:

(*i*) $(z+1)\sin\left(\dfrac{1}{z-2}\right)$, (*ii*) $\dfrac{z-\sin z}{z^2}$, (*iii*) $\dfrac{e^{2z}}{(z-1)^4}$, (*iv*) ze^{1/z^2}.

Solution (*i*) We have $f(z)=(z+1)\sin\left(\dfrac{1}{z-2}\right)=(t+2+1)\sin\left(\dfrac{1}{t}\right)$, by writing t = z – 2

$$=(t+3)\left\{\frac{1}{t}-\frac{1}{3!t^3}+\frac{1}{5!t^5}-\cdots\right\}$$

$$=\left\{1-\frac{1}{3!t^2}+\frac{1}{5!t^4}-\cdots\right\}+\left\{\frac{3}{t}-\frac{1}{2t^3}+\frac{3}{5!t^5}-\cdots\right\}$$

$$=1+\frac{3}{t}-\frac{1}{6t^2}-\frac{1}{2t^3}+\frac{1}{120t^4}-\cdots\cdots$$

$$=1+\frac{3}{z-2}-\frac{1}{6(z-2)^2}-\frac{1}{2(z-2)^3}+\cdots$$

Since there are infinite numbers of terms in the negative powers of $(z-2)$ which shows that $z = 2$ is an essential singularity of $f(z)$.

(*ii*) We have $f(z)=\dfrac{z-\sin z}{z^2}=\dfrac{1}{z^2}\left[z-\left(z-\dfrac{z^3}{3!}+\dfrac{z^5}{5!}-\dfrac{z^7}{7!}+\cdots\right)\right]$

$$=\frac{1}{z^2}\left(\frac{z^3}{3!}-\frac{z^5}{5!}+\frac{z^7}{7!}-\cdots\right)=\frac{z}{3!}-\frac{z^3}{5!}+\frac{z^5}{7!}-\cdots\cdots$$

Since, there are no negative powers of z in the expansion of the function $f(z)$, therefore, $z = 0$ is a removable singularity.

(*iii*) We have $f(z) = \dfrac{e^{2z}}{(z-1)^4} = \dfrac{e^{2(t+1)}}{t^4}$ [by writing $t = z - 1$]

$$= \frac{e^2}{t^4} \cdot e^{2t} = \frac{e^2}{t^4}\left[1 + \frac{2t}{1!} + \frac{(2t)^2}{2!} + \frac{(2t)^3}{3!} + \frac{(2t)^4}{4!} + \frac{(2t)^5}{5!} + \cdots\right]$$

$$= e^2\left(\frac{1}{t^4} + \frac{2}{t^3} + \frac{2}{t^2} + \frac{4}{3t} + \frac{2}{3} + \frac{4}{15}t + \cdots\right)$$

$$= e^2\left(\frac{1}{(z-1)^4} + \frac{2}{(z-1)^3} + \frac{2}{(z-1)^2} + \frac{4}{3(z-1)} + \frac{2}{3} + \frac{4}{15}(z-1) + \cdots\right)$$

Since, there are four number of terms containing negative power of $(z - 1)$, therefore, $z = 1$ is a pole of fourth order.

(*iv*) We have $f(z) = ze^{1/z^2} = z\left[1 + \dfrac{1}{1!z^2} + \dfrac{1}{2!z^4} + \dfrac{1}{3!z^6} + \dfrac{1}{4!z^8} + \cdots\right]$

$$= z + \frac{1}{z} + \frac{1}{2z^3} + \frac{1}{6z^5} + \frac{1}{24z^7} + \cdots \infty = z + z^{-1} + \frac{z^{-3}}{2} + \frac{z^{-5}}{6} + \frac{z^{-7}}{24} + \cdots \infty$$

Since, there are infinite number of terms in the negative powers of z therefore, $z = 0$ is an essential singularity of $f(z)$.

Residues

If a function $f(z)$ is analytic inside and on a simple closed contour C, then by Cauchy integral theorem, we have $\oint_C f(z)\,dz = 0$. However, when the integrand $f(z)$ has one or more isolated singular points inside C, then the Cauchy integral theorem cannot be used and the value of the complex integral $\oint_C f(z)\,dz$ may not be zero. Each of these isolated singular points inside C contributes to the value of the complex integral. These contributions are called *residues*. The theory of residues enables us a simpler way of evaluating complex integrals.

Residue of an analytic function at an isolated singular point

Let $f(z)$ be analytic for all z except at the point $z = z_0$. Let z_0 be an isolated singular point of $f(z)$. Then $f(z)$ has a Laurent series

$$f(z) = a_0 + a_1(z - z_0) + a_2(z - z_0)^2 + \cdots + \frac{b_1}{z - z_0} + \frac{b_2}{(z - z_0)^2} + \cdots$$

that converges in some region R: $0 < |z - z_0| < R$. The coefficients b_n are defined by

$$b_n = \frac{1}{2\pi i}\oint_C \frac{f(z)}{(z - z_0)^{-n+1}}\,dz, \quad n = 1, 2, \ldots$$

In particular, when $n = 1$, we get

$$b_1 = \frac{1}{2\pi i}\oint_C f(z)\,dz, \quad \text{or} \quad \oint_C f(z)\,dz = 2\pi i\, b_1. \qquad \ldots(4.88)$$

The coefficient of $(z - z_0)^{-1}$, that is, b_1 in the Laurent series expansion of $f(z)$, given in Eq. (4.88) is called the *residue* of $f(z)$ at $z = z_0$ and we denote it by

$$b_1 = \operatorname*{Res}_{z = z_0} f(z).$$

Thus, if the residue b_1 at $z = z_0$ can be determined by some method, than the value of the contour integral $\oint_C f(z)\,dz$ is given by $2\pi i b_1$.

Now, we discuss methods to find the residue of a function $f(z)$ at an isolated singular point.

Residue at a removable singular point

If $z = z_0$ is a removable singular point of $f(z)$, then there is no term in the principal part of the Laurent series expansion of $f(z)$ about z_0. Therefore, residue of $f(z)$ at z_0 is zero and $\oint_C f(z)\,dz = 0$.

Residue at a simple pole

Let $z = z_0$ be a simple pole of $f(z)$. Then, in the neighbourhood of the point z_0, we can write $f(z)$ as

$$f(z) = \sum_{n=0}^{\infty} a_n (z - z_0)^n + \frac{b}{z - z_0}. \qquad ...(4.89)$$

By multiplying both sides by $(z - z_0)$ and then taking the limits as $z \to z_0$, we obtain

$$b_1 = \text{Res}\,[f(z), z_0] = \lim_{z \to z_0} (z - z_0)\, f(z). \qquad ...(4.90)$$

An alternative method for computing a residue at a simple pole

Since $z = z_0$ is a simple pole, $f(z)$ is of the form

$$f(z) = \frac{p(z)}{q(z)} \qquad [p,\, q \text{ analytic at } z = z_0]$$

where $p(z_0) \neq 0$ and $q(z)$ has a simple zero at z_0. Since $q(z)$ is analytic, we can write its Taylor series expansion as

$$\begin{aligned} q(z) &= q(z_0) + (z - z_0)\, q'(z_0) + \frac{1}{2!}(z - z_0)^2 q''(z_0) + \ldots \\ &= (z - z_0)\, q'(z_0) + \text{higher power of } (z - z_0). \end{aligned}$$

Therefore, $$b_1 = \underset{z \to z_0}{\text{Res}}\; f(z) = \lim_{z \to z_0} [(z - z_0)\, f(z)]$$

$$= \lim_{z \to z_0} \left[(z - z_0) \frac{p(z)}{q(z)} \right] = \lim_{z \to z_0} \frac{(z - z_0)\, p(z)}{(z - z_0)\,[q'(z) + \text{higher power of } (z - z_0)]}, \; (z - z_0) \text{ cancels.}$$

Hence our second formula for the residue at a simple pole is

$$b_1 = \underset{z = z_0}{\text{Res}}\; f(z) = \frac{p(z_0)}{q'(z_0)},\; q'(z_0) \neq 0. \qquad ...(4.91)$$

Residue at a pole of order m

Let $z = z_0$ be a pole of order m of $f(z)$. Then, in the neighbourhood of the point $z = z_0$, we can write

$$f(z) = \sum_{n=0}^{\infty} a_n (z - z_0)^n + \frac{b_1}{z - z_0} + \frac{b_2}{(z - z_0)^2} + \cdots + \frac{b_m}{(z - z_0)^m}.$$

Multiplying both sides by $(z - z_0)^m$, we get

$$(z - z_0)^m f(z) = \sum_{n=0}^{\infty} a_n (z - z_0)^{m+n} + b_1 (z - z_0)^{m-1} + b_2 (z - z_0)^{m-2} + \ldots\ldots$$

Differentiating both sides $(m - 1)$ times and then taking limits as $z \to z_0$, we obtain

$$b_1 = \text{Res}\,[f(z), z_0] = \frac{1}{(m-1)!} \lim_{z \to z_0} \frac{d^{m-1}}{dz^{m-1}} [(z - z_0)^m f(z)]. \qquad ...(4.92)$$

Residue at an isolated essential singular point

When $z = z_0$ is an isolated essential singular point, we need to expand $f(z)$ in Laurent series about the point $z = z_0$ and obtain the residue at z_0 directly from the coefficient of the term $(z - z_0)^{-1}$.

Residue at z = ∞

Let $z = \infty$ be an isolated singular point of $f(z)$. Let C be the circle $|z| = r$ traversed in the clockwise direction such that all the finite singularities of $f(z)$ lie inside C. The isolated singularity at $z = \infty$ is the only singularity in the domain $|z| > r$.

By definition the residue at $z = \infty$ is equal to $\dfrac{1}{2\pi i} \oint_C f(z)\, dz$.

Let $z = 1/w$. Thus $dz = -(1/w^2)\, dw$ and $\dfrac{1}{2\pi i} \oint_C f(z)\, dz = \dfrac{1}{2\pi i} \oint_{\overset{*}{C}} f\left(\dfrac{1}{w}\right)\left(-\dfrac{1}{w^2}\right) dw = -\dfrac{1}{2\pi i} \oint_{\overset{*}{C}} g(w)\, dw$

where $\overset{*}{C}$ is a circle describe about the origin in the anticlockwise direction (note that z describes the circle in clockwise direction and w describe the circle in anticlockwise direction) and

$$g(w) = \frac{1}{w^2} f\left(\frac{1}{w}\right). \qquad ...(4.93)$$

Now, $z = \infty$ gives $w = 0$ and $w = 0$ is the only isolated singularity of $g(w)$ inside $\overset{*}{C}$. Hence we obtain

$$\underset{z=\infty}{\text{Res}}\, f(z) = -\underset{w=0}{\text{Res}}\, g(w)$$

$= -$ coefficient of term $\left(\dfrac{1}{w}\right)$ in the Laurent series expansion of $g(w)$ about $w = 0$. ...(4.94)

EXAMPLE 4.50 Obtain the residues at the singular points for the function $f(z)$, where $f(z)$ is given by

(*i*) $\dfrac{z+1}{z^2 - 2z}$, (*ii*) $\dfrac{\exp(z)}{z^2 + \pi^2}$, (*iii*) $\dfrac{z}{z^2 - 3z + 2}$, (*iv*) $\dfrac{ze^{iz}}{z^2 + a^2}$, (*v*) $\dfrac{1}{z^3 + z^5}$.

Solution (*i*) The poles of the function $f(z) = \dfrac{z+1}{z^2 - 2z}$ are given by $z^2 - 2z = 0$

$\Rightarrow z(z - 2) = 0$ giving $z = 0, 2$. The function $f(z)$ has simple poles at $z = 0$ and $z = 2$.

Hence, $$\underset{z=0}{\text{Res}}\, f(z) = \lim_{z \to 0} [z \cdot f(z)] = \lim_{z \to 0} \left[\frac{z+1}{z-2}\right] = -\frac{1}{2}.$$

$$\underset{z=2}{\text{Res}}\, f(z) = \lim_{z \to 2} [(z-2) \cdot f(z)] = \lim_{z \to 2} \left[\frac{z+1}{z}\right] = \frac{3}{2}.$$

(*ii*) The function $f(z) = \dfrac{e^z}{z^2 + \pi^2}$ has simple poles at $z = \pm \pi i$.

Hence, $$\underset{z=\pi i}{\text{Res}}\, f(z) = \lim_{z \to \pi i} [(z - \pi i)\, f(z)] = \lim_{z \to \pi i} \left[\frac{e^z}{z + \pi i}\right] = \frac{e^{\pi i}}{2\pi i} = \frac{i}{2\pi}. \qquad [\because e^{i\pi} = -1 = i^2]$$

$$\underset{z=-\pi i}{\text{Res}}\, f(z) = \lim_{z \to -\pi i} [(z + \pi i)\, f(z)] = \lim_{z \to -\pi i} \frac{e^z}{z - \pi i} = \frac{e^{-\pi i}}{-2\pi i} = \frac{-i}{2\pi}.$$

(*iii*) The singular point of $f(z)$ are the roots of the equation $z^2 - 3z + 2 = 0$. The function $f(z)$ has simple poles at $z = 1$ and $z = 2$.

Hence, $$\operatorname*{Res}_{z=1} f(z) = \lim_{z\to 1} [(z-1)\, f(z)] = \lim_{z\to 1}\left[\frac{z}{z-2}\right] = \frac{1}{-1} = -1.$$

$$\operatorname*{Res}_{z=2} f(z) = \lim_{z\to 2} [(z-2)\, f(z)] = \lim_{z\to 2}\left[\frac{z}{z-1}\right] = \frac{2}{1} = 2.$$

(*iv*) $f(z)$ has simple poles at $z = \pm\, ia$.

Hence, $$\operatorname*{Res}_{z=ia} f(z) = \lim_{z\to ia} [(z-ia)\, f(z)] = \lim_{z\to ia}\left[\frac{ze^{iz}}{z+ia}\right] = \frac{iae^{-a}}{2ia} = \frac{e^{-a}}{2}.$$

$$\operatorname*{Res}_{z=-ia} f(z) = \lim_{z\to -ia} [(z+ia)\, f(z)] = \lim_{z\to -ia}\left[\frac{ze^{iz}}{z-ia}\right] = \frac{(-ia)e^{a}}{-2ia} = \frac{e^{a}}{2}.$$

(*v*) We may write $f(z) = \dfrac{1}{z^3+z^5} = \dfrac{1}{z^3(1+z^2)} = \dfrac{1}{z^3(z+i)(z-i)}$.

The function has simple poles at $z = \pm\, i$ and a pole of order 3 at $z = 0$.

Hence, $$\operatorname*{Res}_{z=i} f(z) = \lim_{z\to i} [(z-i)\, f(z)] = \lim_{z\to i}\left[\frac{1}{z^3(z+i)}\right] = \frac{1}{i^3(2i)} = \frac{1}{2}.$$

$$\operatorname*{Res}_{z=-i} f(z) = \lim_{z\to -i} [(z+i)\, f(z)] = \lim_{z\to -i}\left[\frac{1}{z^3(z-i)}\right] = \lim_{z\to -i}\left[\frac{1}{-i^3(-2i)}\right] = \frac{1}{2}.$$

$$\operatorname*{Res}_{z=0} f(z) = \frac{1}{2!}\lim_{z\to 0}\frac{d^2}{dz^2}[z^3 f(z)] = \frac{1}{2}\lim_{z\to 0}\frac{d^2}{dz^2}\left[\frac{1}{1+z^2}\right]$$

$$= \frac{1}{2}\lim_{z\to 0}\frac{d^2}{dz^2}[1 - z^2 + z^4 - z^6 + \ldots]$$

$$= \frac{1}{2}\lim_{z\to 0}[-2 + 12z^2 - 30z^4 + \ldots] = -1.$$

EXAMPLE 4.51 Find the residues at all the singular points of

(*i*) $f(z) = \dfrac{z+2}{(z+1)^2(z-2)}$, (*ii*) $f(z) = \dfrac{z^2}{(z-1)^2(z+2)}$, (*UPTU, 2003*) (*iii*) $f(z) = \dfrac{ze^z}{(z-a)^3}$,

(*iv*) $f(z) = \dfrac{z^2-2z}{(z+1)^2(z^2+4)}$, (*v*) $f(z) = \dfrac{z^2}{(z^2+1)^2}$, (*vi*) $f(z) = \dfrac{\log(1+z)}{(1+z^2)^2}$, (*vii*) $f(z) = \dfrac{z}{\cos z}$.

Solution (*i*) The function $f(z)$ has a double pole at $z = -1$ and simple pole at $z = 2$.

Hence, $$\operatorname*{Res}_{z=2} f(z) = \lim_{z\to 2} [(z-2)\, f(z)] = \lim_{z\to 2}\left[\frac{z+2}{(z+1)^2}\right] = \frac{4}{9}.$$

$$\operatorname*{Res}_{z=-1} f(z) = \lim_{z\to -1}\frac{1}{1!}\frac{d}{dz}[(z+1)^2 f(z)] = \lim_{z\to -1}\frac{d}{dz}\left(\frac{z+2}{z-2}\right)$$

$$= \lim_{z\to -1}\left[\frac{(z-2)-(z+2)}{(z-2)^2}\right] = \lim_{z\to -1}\frac{-4}{(z-2)^2} = -\frac{4}{9}.$$

(*ii*) The function $f(z)$ has a double pole at $z = 1$ and simple pole at $z = -2$.

Hence, $$\operatorname*{Res}_{z=1} f(z) = \lim_{z\to 1} \frac{1}{1!}\frac{d}{dz}[(z-1)^2 f(z)] = \lim_{z\to 1} \frac{d}{dz}\left(\frac{z^2}{z+2}\right)$$

$$= \lim_{z\to 1}\left[\frac{2z(z+2)-z^2}{(z+2)^2}\right] = \lim_{z\to 1}\left[\frac{z^2+4z}{(z+2)^2}\right] = \frac{5}{9}.$$

$$\operatorname*{Res}_{z=-2} f(z) = \lim_{z\to -2}[(z+2)f(z)] = \lim_{z\to -2}\left[\frac{z^2}{(z-1)^2}\right] = \frac{4}{9}.$$

(*iii*) The function $f(z)$ has a pole of the third order at $z = a$.

Hence, $$\operatorname*{Res}_{z=a} f(z) = \lim_{z\to a}\frac{1}{2!}\frac{d^2}{dz^2}[(z-a)^3 f(z)] = \frac{1}{2}\lim_{z\to a}\frac{d^2}{dz^2}(ze^z) = \frac{1}{2}\lim_{z\to a}\frac{d}{dz}(e^z + ze^z)$$

$$= \frac{1}{2}\lim_{z\to a}(2e^z + ze^z) = \frac{1}{2}(2e^a + ae^a) = \frac{1}{2}e^a(a+2).$$

(*iv*) The function $f(z)$ has a double pole at $z = -1$ and simple poles at $z = \pm 2i$.

Hence, $$\operatorname*{Res}_{z=-1} f(z) = \lim_{z\to -1}\frac{1}{1!}\frac{d}{dz}[(z+1)^2 f(z)] = \lim_{z\to -1}\frac{d}{dz}\left[\frac{z^2-2z}{z^2+4}\right]$$

$$= \lim_{z\to -1}\left[\frac{(z^2+4)(2z-2)-(z^2-2z)\,2z}{(z^2+4)^2}\right] = -\frac{14}{25}.$$

$$\operatorname*{Res}_{z=2i} f(z) = \lim_{z\to 2i}[(z-2i)f(z)] = \lim_{z\to 2i}\left[\frac{z^2-2z}{(z+1)^2(z+2i)}\right]$$

$$= \frac{-4-4i}{(2i+1)^2 4i} = \frac{1+i}{4+3i} = \frac{(1+i)(4-3i)}{(4+3i)(4-3i)} = \frac{7+i}{25}.$$

$$\operatorname*{Res}_{z=-2i} f(z) = \lim_{z\to -2i}[(z+2i)f(z)] = \lim_{z\to -2i}\left[\frac{z^2-2z}{(z+i)^2(z-2i)}\right] = \frac{-4+4i}{(-2i+1)^2(-4i)} = \frac{7-i}{25}.$$

(*v*) The function $f(z) = \dfrac{z^2}{(z^2+1)^2} = \dfrac{z^2}{(z+i)^2(z-i)^2}$ has poles of order 2 at $z = \pm i$.

Hence, $$\operatorname*{Res}_{z=i} f(z) = \lim_{z\to i}\frac{d}{dz}[(z-i)^2 f(z)] = \lim_{z\to i}\frac{d}{dz}\left[\frac{z^2}{(z+i)^2}\right]$$

$$= \lim_{z\to i}\left[\frac{(z+i)^2(2z)-z^2[2(z+i)]}{(z+i)^4}\right] = -\frac{4i}{16} = -\frac{i}{4}.$$

$$\operatorname*{Res}_{z=-i} f(z) = \lim_{z\to -i}\frac{d}{dz}[(z+i)^2 f(z)] = \lim_{z\to -i}\frac{d}{dz}\left[\frac{z^2}{(z-i)^2}\right]$$

$$= \lim_{z\to -i}\left[\frac{(z-i)^2(2z)-z^2[2(z-i)]}{(z-i)^4}\right] = \frac{i}{4}.$$

(*vi*) The function $f(z) = \dfrac{\text{Log}\,(1+z)}{(1+z^2)^2} = \dfrac{\text{Log}\,(1+z)}{(z+i)^2\,(z-i)^2}$ has poles of order 2 at $z = \pm\, i$.

Hence, $$\underset{z=i}{\text{Res}}\, f(z) = \frac{1}{1!} \lim_{z\to i} \frac{d}{dz}\, [(z-i)^2 f(z)] = \lim_{z\to i} \frac{d}{dz}\left[\frac{\text{Log}\,(1+z)}{(z+i)^2}\right]$$

$$= \lim_{z\to i}\left[\frac{1}{(z+i)^2}\cdot\frac{1}{z+1} + \text{Log}\,(1+z)\cdot\left(\frac{-2}{(z+i)^3}\right)\right]$$

$$= \left(-\frac{1}{4}\right)\left(\frac{1-i}{2}\right) + \text{Log}\,(1+i)\cdot\left(\frac{-2}{-8i}\right) = \left(\frac{i-1}{8}\right) + \left(\text{Log}\,\sqrt{2} + i\frac{\pi}{4}\right)\left(\frac{1}{4i}\right)$$

$$= \left(-\frac{1}{8} + \frac{\pi}{16}\right) + i\left(\frac{1}{8} - \frac{\text{Log}\,\sqrt{2}}{4}\right).$$

Similarly $$\underset{z=-i}{\text{Res}}\, f(z) = \left(-\frac{1}{8} + \frac{\pi}{16}\right) + i\left(-\frac{1}{8} + \frac{\text{Log}\,\sqrt{2}}{4}\right).$$

(*vii*) The denominator of $f(z) = \dfrac{z}{\cos z}$ vanishes at $z = (2n+1)\dfrac{\pi}{2}$, $n = 0, \pm 1, \pm 2, \ldots$

Denoting this value by z_0, we have

$$\lim_{z\to z_0} (z - z_0) f(z) = \lim_{z\to z_0} \frac{z}{\cos z}\left(z - n\pi - \frac{\pi}{2}\right) = \lim_{z\to z_0} \frac{2z - n\pi - \frac{\pi}{2}}{-\sin z}, \quad \text{(by L'Hospital's rule)}$$

$$= \frac{2n\pi + \pi - n\pi - \frac{\pi}{2}}{-\sin\left(n\pi + \frac{\pi}{2}\right)} = \frac{n\pi + \frac{\pi}{2}}{-\sin\left(n\pi + \frac{\pi}{2}\right)} = (-1)^{n+1}\left(n + \frac{1}{2}\right)\pi.$$

Thus $z = (2n+1)\dfrac{\pi}{2}$ gives the poles of $\dfrac{z}{\cos z}$. They are simple poles with residues $(-1)^{n+1}\left(n + \dfrac{1}{2}\right)\pi$.

EXAMPLE 4.52 Determine the residues at all the singular points of $f(z)$ where $f(z)$ is given by

(*i*) $\cot z$ (*ii*) $\sec z$ (*iii*) $z \sin\left(\dfrac{1}{z}\right)$.

Solution (*i*) The function $f(z) = \cot z = \dfrac{\cos z}{\sin z} = \dfrac{\phi(z)}{\psi(z)}$ has singular points where $\sin z = 0$, that is, at

$z = z_k = k\pi$, $k = 0, \pm 1, \pm 2, \ldots$

Since $\psi'(z) = \dfrac{d}{dz}(\sin z) = \cos z \neq 0$ at $z = z_k$, $f(z)$ has simple poles at z_k.

Employing the formula $\underset{z=z_k}{\text{Res}}\, f(z) = \dfrac{\phi(z_k)}{\psi'(z_k)} = \dfrac{\cos z_k}{\cos z_k} = 1$ for all k.

Note: The residue of $f(z) = \tan z$ at $z = \pi/2$ is -1.

(*ii*) We have $f(z) = \sec z = \dfrac{1}{\cos z} = \dfrac{\phi(z)}{\psi(z)}$ has singular points where

$\cos z = 0$ or at $z = z_k = (2k+1)\dfrac{\pi}{2}$, $k = 0, \pm 1, \pm 2, \ldots$

Since $\psi'(z) = \frac{d}{dz}(\cos z) = -\sin z \neq 0$ at $z = z_k$, $f(z)$ has simple poles at z_k.

Using the formula $\operatorname*{Res}_{z=z_k} f(z) = \frac{\phi(z_k)}{\psi'(z_k)} = \frac{1}{-\sin z_k} = -\frac{1}{\sin\left[(2k+1)\frac{\pi}{2}\right]} = (-1)^{k+1}$.

(*iii*) The point $z = 0$ is an isolated essential singular point of $f(z)$. We write the Laurent series expansion of $f(z)$ about $z = 0$ to compute the residue

$$f(z) = z \sin\left(\frac{1}{z}\right) = z\left[\frac{1}{z} - \frac{1}{3!z^3} + \cdots\right] = 1 - \frac{1}{3!z^2} + \cdots$$

Therefore, $\operatorname*{Res}_{z=0} f(z) =$ coefficient of $\left(\frac{1}{z}\right)$ in the Laurent series = 0.

EXAMPLE 4.53 (*i*) Find the sum of the residues of $f(z) = \frac{\sin z}{z \cos z}$ at its pole inside $|z| = 2$.

(*ii*) Find the residue at $z = 0$ of the function $f(z) = \frac{1 + e^z}{\sin z + z \cos z}$.

Solution (*i*) The poles of $f(z) = \frac{\sin z}{z \cos z}$ are $z = 0, \pm\frac{\pi}{2} \pm \frac{3\pi}{2}, \cdots$. Of these $z = 0, \pm \pi/2$ lie inside $|z| = 2$.

Hence, $\operatorname*{Res}_{z=0} f(z) = \lim_{z \to 0} [z \cdot f(z)] = \lim_{z \to 0}\left(\frac{\sin z}{\cos z}\right) = 0.$

$$\operatorname*{Res}_{z=\pi/2} f(z) = \lim_{z \to \pi/2} [(z - \pi/2) f(z)] = \lim_{z \to \pi/2} (z - \pi/2)\frac{\sin z}{z \cos z} \quad \left(\frac{0}{0}\right).$$

$$= \lim_{z \to \pi/2} \frac{(z - \pi/2)\cos z + \sin z}{\cos z - z \sin z} \quad \text{(By L' Hospital's rule)}$$

$$= \frac{1}{-(\pi/2)} = -\frac{2}{\pi}.$$

Similarly, $\operatorname*{Res}_{z=-\pi/2} f(z) = \frac{2}{\pi}$. Therefore, sum of residues $= 0 - \frac{2}{\pi} + \frac{2}{\pi} = 0$.

(*ii*) $\operatorname*{Res}_{z=0} f(z) = \lim_{z \to 0} [z\, f(z)] = \lim_{z \to 0}\left[\frac{z(1 + e^z)}{\sin z + z \cos z}\right] = \lim_{z \to 0}\left[\frac{1 + e^z}{\frac{\sin z}{z} + \cos z}\right] = \frac{1+1}{1+1} = 1.$

EXAMPLE 4.54 Find the residues at $z = \infty$ for the function $f(z)$, where $f(z)$ is given by

(*i*) $\frac{z^2}{z^3 + 2}$, (*ii*) $z^3 \cos\left(\frac{1}{z}\right)$, (*iii*) $\frac{\sin 2z}{(z+1)^3}$.

Solution Let $z = \frac{1}{w}$. Then $g(w) = \left(\frac{1}{w^2}\right) f\left(\frac{1}{w}\right)$. We have $\operatorname*{Res}_{z=\infty} f(z) = -$ (residue of $g(w)$ at $w = 0$).

(*i*) $g(w) = \left(\frac{1}{w^2}\right) f\left(\frac{1}{w}\right) = \frac{(1/w^2)}{w^2\left(\frac{1}{w^3} + 2\right)} = \frac{1}{w^4\left(\frac{1}{w^3} + 2\right)} = \frac{1}{w + 2w^4} = \frac{1}{w(1 + 2w^3)}.$

$$= \frac{1}{w}[1+2w^3]^{-1} = \frac{1}{w}[1-2w^3+4w^6-\ldots] = \frac{1}{w} - 2w^2 + \cdots$$

Coefficient of $(1/w)$ in $g(w) = 1$. Therefore, $\underset{w=0}{\text{Res}}\, g(w) = 1$ and $\underset{z=\infty}{\text{Res}}\, f(z) = -1$.

(ii) $$g(w) = \frac{1}{w^2} f\left(\frac{1}{w}\right) = \frac{1}{w^5}\cos w = \frac{1}{w^5}\left[1 - \frac{w^2}{2!} + \frac{w^4}{4!} - \frac{w^6}{6!} + \cdots\right] = \frac{1}{w^5} - \frac{1}{2w^3} + \frac{1}{24w} - \frac{w}{6!} + \cdots$$

Coefficient of $(1/w)$ in $g(w) = 1/24$.

Therefore, $\underset{w=0}{\text{Res}}\, g(w) = \dfrac{1}{24}$ and $\underset{z=\infty}{\text{Res}}\, f(z) = -\dfrac{1}{24}$.

(iii) $$g(w) = \left(\frac{1}{w^2}\right) f\left(\frac{1}{w}\right) = \frac{\sin\left(\frac{2}{w}\right)}{w^2\left(1+\frac{1}{w}\right)^3} = \frac{w\sin\left(\frac{2}{w}\right)}{(1+w)^3} = w\sin\left(\frac{2}{w}\right)(1+w)^{-3}$$

$$= w\left[\frac{2}{w} - \frac{1}{3!}\left(\frac{2}{w}\right)^3 + \frac{1}{5!}\left(\frac{2}{w}\right)^5 - \frac{1}{7!}\left(\frac{2}{w}\right)^7 + \cdots\right][1 - 3w + 6w^2 - 10w^3 + 15w^4 - 21w^5 + \ldots]$$

$$= \left[2 - \frac{1}{3!}\frac{2^3}{w^2} + \frac{1}{5!}\frac{2^5}{w^4} - \frac{1}{7!}\frac{2^7}{w^6} + \cdots\cdots\right][1 - 3w + 6w^2 - 10w^3 + 15w^4 - 21w^5 + \ldots]$$

Coefficient of $(1/w)$ in $g(w) = (3)\dfrac{2^3}{3!} - (10)\dfrac{2^5}{5!} + (21)\dfrac{2^7}{7!} - \cdots$

$$= \frac{2^3}{2!} - \frac{2^6}{4!} + \frac{3.2^7}{6!} - \cdots = 2\left[2 - \frac{2^3}{3!} + \frac{2^5}{5!} - \cdots\right] = 2\sin 2.$$

Therefore, $\underset{w=0}{\text{Res}}\, g(w) = 2\sin 2$ and $\underset{z=\infty}{\text{Res}}\, f(z) = -2\sin 2$.

4.21 EVALUATION OF CONTOUR INTEGRALS USING RESIDUES

In the section, we shall discuss the application of residues in evaluating contour integrals of the form $I = \oint_C f(z)\,dz$ where C is a simple, piecewise smooth closed curve. We use the result given in the following theorem for evaluating the contour integrals.

Theorem 4.16 (Residue Theorem)

Let $f(z)$ be analytic inside and on a simple closed curve C except for a finite number of singular points $z_1, z_2, z_3, \ldots, z_k$ lying inside C. Then the integral of $f(z)$ taken counter clockwise around C equal to $2\pi i$ times the sum of the residues of $f(z)$ at $z_1, z_2, z_3, \ldots, z_k$.

That is, $$\oint_C f(z)\,dz = 2\pi i \sum_{j=1}^{k} \underset{z=z_j}{\text{Res}}\, f(z) \qquad \ldots(4.95)$$

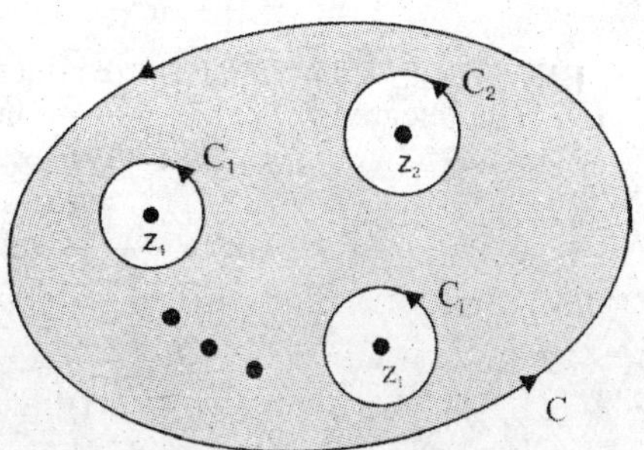

Fig. 4.37: Residue Theorem

Proof

Around each of the singular points z_j, $j = 1, 2, \ldots, k$ draw non-intersecting circles c_j with radius small enough that these k circles and C are all separated (Fig 4.37). Then $f(z)$ is analytic in the multiply connected region bounded by C and $C_1, C_2, ..., C_k$. By the extension of the Cauchy's integral theorem for multiply connected domains, we can write

$$\oint_C f(z)\,dz = \oint_{C_1} f(z)\,dz + \oint_{C_2} f(z)\,dz + \oint_{C_3} f(z)\,dz + \cdots + \oint_{C_k} f(z)\,dz \qquad ...(4.96)$$

where all the curves $C, C_1, C_2, ..., C_k$ are traversed in the counter clockwise direction.

But
$$\oint_{C_j} f(z)\,dz = 2\pi i\left[\operatorname*{Res}_{z=z_j} f(z)\right] \qquad j = 1, 2, \ldots, k.$$

Hence, from Eq. (4.96)
$$\oint_C f(z)\,dz = 2\pi i \sum_{j=1}^{k} \operatorname*{Res}_{z=z_j} f(z).$$

This theorem is very useful in evaluating complex and real integrals.

EXAMPLE 4.55 Evaluate the following integrals by the residue theorem.

(*i*) $I = \oint_C \frac{2z-1}{z(z+1)(z-3)}\,dz$, where c is $|z| = 2$. (*ii*) $I = \oint_C \frac{e^z}{(z+1)^n}\,dz$, $c: |z-1| = 3$.

(*iii*) $I = \oint_C \frac{z^2-2z}{(z+1)^2(z^2+4)}\,dz$, $c: |z| = 3$.

(*iv*) $I = \oint_C \frac{e^z-1}{z(z-1)(z-i)^2}\,dz$, where (*a*) $c: |z| = 1/2$, (*b*) $c: |z| = 2$.

(*v*) $I = \oint_C \frac{e^z}{\cos \pi z}\,dz$, where $c: |z| = 1$. (*vi*) $I = \oint_C (z+1)\cot\left(\frac{z}{2}\right)dz$, $c: |z| = 1.5$.

(*AMIETE, June 2011, Dec. 2009; MDU, May 2007*)

Solution (*i*) The integrand $f(z) = \frac{2z-1}{z(z+1)(z-3)}$ has three simple poles at $z = 0, -1, 3$. Of these only $z = 0, -1$ lies inside C.

We get
$$\operatorname*{Res}_{z=0} f(z) = \lim_{z\to 0}\,[z\,f(z)] = \lim_{z\to 0}\left[\frac{2z-1}{(z+1)(z-3)}\right] = \frac{1}{3}.$$

$$\operatorname*{Res}_{z=-1} f(z) = \lim_{z\to -1}\,[(z+1)\,f(z)] = \lim_{z\to -1}\left[\frac{2z-1}{z(z-3)}\right] = \frac{-3}{4}.$$

Hence,
$$I = 2\pi i\left[\operatorname*{Res}_{z=0} f(z) + \operatorname*{Res}_{z=-1} f(z)\right] = 2\pi i\left(\frac{1}{3} - \frac{3}{4}\right) = \frac{-5\pi i}{6}.$$

(*ii*) The integrand $f(z) = \frac{e^z}{(z+1)^n}$ has a pole of order n at $z = -1$, which lies inside C, we get

$$\operatorname*{Res}_{z=-1} f(z) = \frac{1}{(n-1)!}\lim_{z\to -1}\frac{d^{n-1}}{dz^{n-1}}[(z+1)^n f(z)] = \frac{1}{(n-1)!}\lim_{z\to -1}\frac{d^{n-1}}{dz^{n-1}}[e^z]$$

$$= \frac{1}{(n-1)!}\lim_{z\to -1}\,[e^z] = \frac{e^{-1}}{(n-1)!}.$$

Hence,
$$I = 2\pi i\left[\operatorname*{Res}_{z=-1} f(z)\right] = \frac{2\pi i}{e(n-1)!}.$$

(*iii*) The integrand $f(z) = \dfrac{z^2 - 2z}{(z+1)^2(z^2+4)}$ has simple poles at $z = \pm 2i$ and a pole of order 2 at $z = -1$. All the singular points lies inside $|z| = 3$. We get

$$\operatorname*{Res}_{z=-1} f(z) = \frac{1}{1!} \lim_{z \to -1} \frac{d}{dz}[(z+1)^2 f(z)] = \lim_{z \to -1} \frac{d}{dz}\left[\frac{z^2 - 2z}{z^2 + 4}\right]$$

$$= \lim_{z \to -1}\left[\frac{(z^2+4)(2z-2) - (z^2 - 2z)\,2z}{(z^2+4)^2}\right] = \lim_{z \to -1}\left[\frac{2z^2 + 8z - 8}{(z^2+4)^2}\right] = \frac{-14}{25}.$$

$$\operatorname*{Res}_{z=2i} f(z) = \lim_{z \to 2i} [(z - 2i) f(z)] = \lim_{z \to 2i}\left[\frac{z^2 - 2z}{(z+1)^2(z+2i)}\right]$$

$$= \frac{-4 - 4i}{(2i+1)^2\,(4i)} = \frac{i+1}{3i+4} = \frac{7+i}{25}.$$

$$\operatorname*{Res}_{z=-2i} f(z) = \lim_{z \to -2i} [(z + 2i) f(z)] = \lim_{z \to -2i}\left[\frac{z^2 - 2z}{(z+1)^2(z-2i)}\right]$$

$$= \frac{4i^2 + 4i}{(-2i+1)^2\,(-4i)} = \frac{1-i}{-3i+4} = \frac{7-i}{25}.$$

Hence,
$$I = 2\pi i\left[\operatorname*{Res}_{z=-1} f(z) + \operatorname*{Res}_{z=2i} f(z) + \operatorname*{Res}_{z=-2i} f(z)\right]$$

$$= 2\pi i\left[-\frac{14}{25} + \frac{7}{25} + \frac{i}{25} + \frac{7}{25} - \frac{i}{25}\right] = 2\pi i(0) = 0.$$

(*iv*) The integrand $f(z) = \dfrac{e^z - 1}{z(z-1)(z-i)^2}$ has a removable singularity at $z = 0$, simple pole at $z = 1$ and a pole of order 2 at $z = i$.

(*a*) Since $z = 0$ is the only singular points which lies inside C and $z = 0$ is a removable singular point, we get

$$I = 2\pi i\left[\operatorname*{Res}_{z \to 0} f(z)\right] = 2\pi i(0) = 0.$$

(*b*) All the singular points $z = 0$, $z = 1$ and $z = i$ lie inside C. We get

$$\operatorname*{Res}_{z=0} f(z) = 0,\ \operatorname*{Res}_{z=1} f(z) = \lim_{z \to 1} [(z-1) f(z)] = \lim_{z \to 1}\left[\frac{e^{z-1}}{z\,(z-i)^2}\right] = \frac{e-1}{(1-i)^2} = \frac{i(e-1)}{2}.$$

$$\operatorname*{Res}_{z=i} f(z) = \frac{1}{1!} \lim_{z \to i} \frac{d}{dz}[(z-i)^2 f(z)] = \lim_{z \to i} \frac{d}{dz}\left[\frac{e^z - 1}{z\,(z-1)}\right]$$

$$= \lim_{z \to i}\left[\frac{z(z-1)e^z - (e^z - 1)(2z-1)}{z^2(z-1)^2}\right] = \frac{i(i-1)e^i - (e^i - 1)(2i-1)}{i^2(i-1)^2} = -\frac{1}{2}(3e^i - 2 - i).$$

Hence,
$$I = 2\pi i\left[\operatorname*{Res}_{z=0} f(z) + \operatorname*{Res}_{z=1} f(z) + \operatorname*{Res}_{z=i} f(z)\right]$$

$$= 2\pi i\left[0 + \frac{i(e-1)}{2} - \frac{1}{2}(3e^i - 2 - i)\right] = \pi i\,(ie - 3e^i + 2).$$

(*v*) The integrand $f(z) = \dfrac{e^z}{\cos \pi z}$ has simple poles at $z = \pm\frac{1}{2}, \pm\frac{3}{2}, \pm\frac{5}{2}, \cdots$. Out of these only the poles at $z = 1/2$ and $z = -1/2$ lie inside the given circle $|z| = 1$.

$$\operatorname*{Res}_{z=1/2} f(z) = \lim_{z\to 1/2}\left[\left(z - \frac{1}{2}\right)f(z)\right] = \lim_{z\to 1/2}\left[\left(z - \frac{1}{2}\right)\frac{e^z}{\cos \pi z}\right] \qquad \left(\frac{0}{0}\text{ form}\right)$$

$$= \lim_{z\to 1/2}\frac{\dfrac{d}{dz}\left(z - \dfrac{1}{2}\right)e^z}{\dfrac{d}{dz}\cos \pi z} = \lim_{z\to 1/2}\frac{\left(z - \dfrac{1}{2}\right)e^z + e^z}{-\pi \sin \pi z} = \frac{e^{1/2}}{-\pi}.$$

Similarly $$\operatorname*{Res}_{z=-1/2} f(z) = \lim_{z\to -1/2}\left[\left(z + \frac{1}{2}\right)f(z)\right] = \lim_{z\to -1/2}\left[\left(z + \frac{1}{2}\right)\frac{e^z}{\cos \pi z}\right] = \frac{e^{-1/2}}{\pi}.$$

Hence, $$I = 2\pi i\left[\operatorname*{Res}_{z=1/2} f(z) + \operatorname*{Res}_{z=-1/2} f(z)\right] = 2\pi i\left[\frac{-e^{1/2}}{\pi} + \frac{e^{-1/2}}{\pi}\right]$$

$$= -4i\left[\frac{e^{1/2} - e^{-1/2}}{2}\right] = -4i \sinh\frac{1}{2}.$$

(*vi*) The integrand $f(z) = (z+1)\cot\left(\dfrac{z}{2}\right)$ has simple pole at $z = 0$, which lies inside C. We obtain

$$\operatorname*{Res}_{z=0} f(z) = \lim_{z\to 0}\left[\frac{z(z+1)\cos(z/2)}{\sin(z/2)}\right] = 2\lim_{z\to 0}\left[\frac{(z/2)}{\sin(z/2)}\right]\lim_{z\to 0}\left[(z+1)\cos\frac{z}{2}\right] = 2\cdot 1.1 = 2.$$

Hence, $$I = 2\pi i\left[\operatorname*{Res}_{z=0} f(z)\right] = 2\pi i\,(2) = 4\pi i.$$

EXERCISE 4.7

1. Find the poles and the residues at the poles for each of the following functions:

(*a*) $f(z) = \dfrac{2z+1}{z^2 - z - 2}$, (*b*) $f(z) = \left(\dfrac{z+1}{z-1}\right)^2$,

(*c*) $f(z) = \dfrac{z+1}{z^2(z-2)}$, (*d*) $f(z) = \dfrac{z^2+4}{z^3 + 2z^2 + 2z}$.

Ans. (*a*) Simple poles at $z = -1, 2$, Res $(z = -1) = 1/3$, Res $(z = 2) = 5/3$.
(*b*) Pole of order 2 at $z = 1$, Res $(z = 1) = 4$.
(*c*) Poles $z = 0$ and $z = 2$; Residue $(z = 0) = -3/4$, Residue $(z = 2) = 3/4$.
(*d*) Poles $z = 0, -1 \pm i$, Residue $2, -1/2\,(1 \mp 3i)$.

2. Find the residue of $\dfrac{1}{(z^2 + a^2)^3}$ at $z = ia$. **Ans.** Res $(z = ia) = \dfrac{3}{16a^5 i}$.

3. Find the residues at $z = 0$ of the functions:

(*a*) $f(z) = \operatorname{cosec}^2 z$, (*b*) $f(z) = \dfrac{1}{z^3}\operatorname{cosec}(z^2)$, (*c*) $f(z) = z\cos\left(\dfrac{1}{z}\right)$.

[Hint: (*a*) $f(z) = \operatorname{cosec}^2 z = (\sin z)^{-2} = \left(z - \dfrac{z^3}{6} + \dfrac{z^5}{120} - \cdots\right)^{-2}$

$$= z^{-2}\left[1 - \left(\frac{z^2}{6} - \frac{z^4}{120} + \cdots\right)\right]^{-2} = \frac{1}{z^2}\left[1 + \frac{z^2}{3} + \text{terms involving } z^4 \text{ and higher powers}\right].$$

Therefore, $\operatorname*{Res}_{z=0} f(z) = \text{coefficient of } \left(\dfrac{1}{z}\right)$ in the Laurent series = 0.

(*b*) $f(z) = \dfrac{1}{z^3 \sin z^2} = \dfrac{1}{z^3}\left(z^2 - \dfrac{z^6}{3!} + \dfrac{z^{10}}{5!} - \cdots\right)^{-1} = \dfrac{1}{z^5}\left[1 - \left(\dfrac{z^4}{3!} - \dfrac{z^8}{5!} + \cdots\right)\right]^{-1}$

$$= \frac{1}{z^5}\left(1 + \frac{z^4}{3!} + \cdots\right) = \frac{1}{z^5} + \frac{1}{6z} + \cdots$$

Therefore, $\operatorname*{Res}_{z=0} f(z) = \text{coefficient of } \left(\dfrac{1}{z}\right)$ in the Laurent series $= \dfrac{1}{6}$.

(*c*) $f(z) = z \cos \dfrac{1}{z} = z\left(1 - \dfrac{1}{2!z^2} + \dfrac{1}{4!z^4} - \cdots\right) = z - \dfrac{1}{2!z} + \dfrac{1}{4!z^3} - \cdots$

Therefore, $\operatorname*{Res}_{z=0} f(z) = \text{coefficient of } \left(\dfrac{1}{z}\right)$ in the Laurent series $= -\dfrac{1}{2}$.**]**

4. Find the zeros and poles of $f(z) = \dfrac{z^2+4}{z^3+2z^2+2z}$ and determine the residues at the poles.

Ans. zeros: $z = \pm 2i$, $\operatorname*{Res}_{z=0} f(z) = 2$, $\operatorname*{Res}_{z=-1+i} f(z) = -\dfrac{1}{2}(1-3i)$, $\operatorname*{Res}_{z=-1-i} f(z) = -\dfrac{1}{2}(1+3i)$.

5. Evaluate the following integrals using the residue theorem

(*a*) $I = \oint_C \dfrac{e^z}{(z+1)^2}$ around the circle $|z - 1| = 3$. **Ans.** $(2\pi i)/e$.

(*b*) $I = \oint_C \dfrac{z-3}{z^2+2z+5}\, dz$, where (*i*) $C : |z + 1 - i| = 2$, (*ii*) $C : |z + 1 + i| = 2$. (*JNTU, 2003*)

Ans. $\pi(-2 + i)$, $\pi(2 + i)$.

[Hint: The poles of $f(z)$ are given by $z^2 + 2z + 5 = 0$, that is, $z = \dfrac{-2 \pm \sqrt{4-20}}{2} = -1 \pm 2i$.

(*i*) Out of the two poles only $-1 + 2i$ lie inside C.

$$\operatorname*{Res}_{z=-1+2i} f(z) = \lim_{z\to(-1+2i)} [(z+1-2i)\, f(z)] = \lim_{z\to(-1+2i)} \left(\frac{z-3}{z+1+2i}\right) = \frac{-4+2i}{4i} = \frac{2i+1}{2}.$$

Hence, $I = 2\pi i\left[\operatorname*{Res}_{z=-1+2i} f(z)\right] = \dfrac{2\pi i\,(2i+1)}{2} = \pi(-2+i)$.

(*ii*) The only pole that lie inside $|z + 1 + i| = 2$ is $z = -1 - 2i$.

$$\operatorname*{Res}_{z=-1-2i} f(z) = \lim_{z\to-1-2i} [(z+1+2i)\, f(z)] = \lim_{z\to-1-2i} \left(\frac{z-3}{z+1-2i}\right) = \frac{-4-2i}{-4i} = \frac{2+i}{2i}.$$

Hence, $I = 2\pi i\left[\operatorname*{Res}_{z=-1-2i} f(z)\right] = \pi(2+i)$.**]**

(c) $I = \oint_C \frac{\sin \pi z^2 + \cos \pi z^2}{(z-1)^2 (z-2)} dz$, where C is the circle $|z| = 3$. (*VTU, 2006, Anna 2003, UPTU 2002*)

Ans. $4\pi (\pi + 1)i$

(d) $I = \oint \tan z \, dz$, where the contour C is the circle $|z| = 2$. **Ans.** $-4\pi i$

(e) $I = \oint_C \frac{1}{(z-1)^2 (z-3)} dz$, where the contour C is the rectangle defined by $x = 0, x = 4, y = -1, y = 1$.
Ans. 0

[**Hint:** Refer Fig. 4.38. Both pole $z = 1$ and $z = 3$ lie within the rectangle. We found that

$$\oint_C \frac{1}{(z-1)^2 (z-3)} dz = 2\pi i \left(-\frac{1}{4} + \frac{1}{4}\right) = 0.]$$

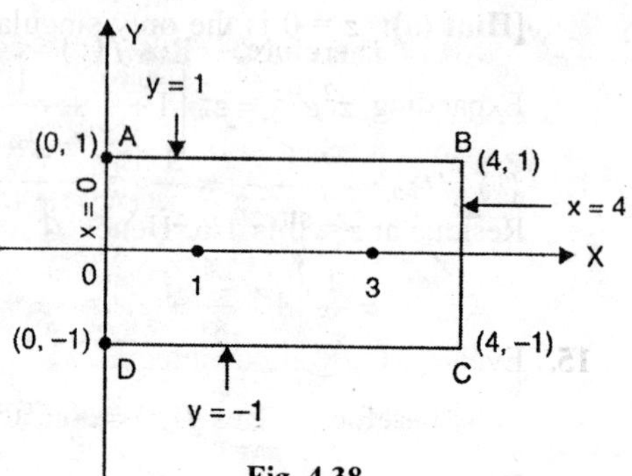

Fig. 4.38

(*f*) $\oint_C \frac{e^z}{(z-1)(z+3)^2} dz, \; C : |z| = 10.$

[**Hint:** $\operatorname*{Res}_{z=1} f(z) = \lim_{z \to 1} \left[\frac{e^z}{(z+3)^2}\right] = \frac{e}{16}.$

$$\operatorname*{Res}_{z=-3} f(z) = \lim_{z \to -3} \frac{d}{dz}\left[\frac{e^z}{z-1}\right] = \lim_{z \to -3} \left[\frac{(z-1)e^z - e^z}{(z-1)^2}\right] = \frac{-5e^{-3}}{16}.$$

The required integral $= 2\pi i \left[\frac{e}{16} - \frac{5e^{-3}}{16}\right] = \frac{\pi i}{8}(e - 5e^{-3}).]$

6. If C be the circle $|z| = 3$, show that $\oint_C \frac{z+2}{(z+1)^2 (z-2)} dz = 0.$

7. Find the value of $\oint_C \frac{12z - 7}{(z-1)^2 (2z+3)} dz$, where C is the circle $|z| = 2$. **Ans.** $-4\pi i$.

Evaluate the following integrals:

8. $\oint_C \frac{z^2 + 2z - 2}{z - 4} dz$, where C containing the point $z = 4$ in its interior. **Ans.** $44\pi i$.

9. $\oint_C \frac{1 - 2z}{z(z-1)(z-2)} dz, \quad C : |z| = 1.5.$ **Ans.** $3\pi i$.

10. $\oint_C \frac{z - 1}{(z+1)^2 (z-2)} dz, \quad C : |z - i| = 2.$ **Ans.** $-2\pi i/9$.

11. $\oint_C \frac{dz}{1 + z^4}, \quad C : |z - 1| = 1.$ **Ans.** $-\pi i/\sqrt{2}$.

12. $\oint_C \frac{e^z}{(z^2 + a^2)} dz$ where C encloses both the poles. **Ans.** $\frac{\pi}{a}(e^{ai} - e^{-ai})$.

13. (*a*) $\oint\limits_C \frac{dz}{\sinh z}$, $C:|z|=4$, (*b*) $\oint\limits_C \frac{dz}{\cosh z}$, $C:|z|=2$. **Ans.** (*a*) $-2\pi i$, (*b*) 0.

14. (*a*) $\oint\limits_C z^2 e^{1/z} dz$, $C:|z|=1$, (*b*) $\oint\limits_C \sin\left(\frac{1}{z}\right) dz$, $C:|z|=1$. **Ans.** $2\pi i$.

[**Hint (*a*):** $z = 0$ is the only singular point.

Expanding $z^2 e^{1/z} = z^2\left[1+\frac{1}{z}+\frac{1}{2!z^2}+\frac{1}{3!z^3}+\cdots\right] = z^2+z+\frac{1}{2}+\frac{1}{6z}+\cdots$

Residue at $z = 0$ is 1/6. Hence $\oint\limits_C z^2 e^{1/z} dz = 2\pi i\left(\frac{1}{6}\right) = \pi i/3.$]

15. Evaluate the contour integral $I = \oint\limits_C \frac{dz}{e^z - 1}$, $C:|z|=1$.

[**Hint:** The integrand $f(z) = \frac{1}{e^z-1}$ has singularities when $e^z - 1 = 0$ or $e^z = 1 = e^{2n\pi i}$ or $z = 2n\pi i$, $n = 0, \pm 1, \pm 2, \ldots$, which are simple poles. Only the simple pole $z = 0$ lies inside C. Hence by residue theorem

$$I = 2\pi i\left[\operatorname*{Res}_{z=0} f(z)\right] = 2\pi i \lim_{z\to 0}\left(\frac{z}{e^z-1}\right) = 2\pi i \lim_{z\to 0}\frac{1}{e^z} = 2\pi i.$$]

16. Evaluate $\oint\limits_C \frac{e^z}{z^4+5z^3} dz$, where the contour C is the circle $|z| = 2$. **Ans.** $\frac{17\pi}{125} i$.

4.22 EVALUATION OF REAL INTEGRALS USING RESIDUES

Many types of real integrals can be evaluated with the help of residue theorem. These integrals are first transformed to associated contour integrals. The contour integrals are then evaluated by using the residue theorem. The values of the real integrals are then obtained from these results. Here we consider three types.

Type 1: Real Definite Integrals Involving Trigonometric Functions

Consider an integral of the form

$$I = \int_0^{2\pi} F(\cos\theta, \sin\theta)\, d\theta \qquad \text{...(4.97)}$$

where F is a real rational function of $\cos\theta$ and $\sin\theta$. Such integrals can be converted into complex integral where the contour C is the unit circle centered at the origin. Setting $z = e^{i\theta} = \cos\theta + i\sin\theta$, we get

$$\cos\theta = \frac{e^{i\theta}+e^{-i\theta}}{2} = \frac{z+z^{-1}}{2} = \frac{z^2+1}{2z}, \qquad \sin\theta = \frac{e^{i\theta}-e^{-i\theta}}{2i} = \frac{(z-z^{-1})}{2i} = \frac{z^2-1}{2zi},$$

and $dz = ie^{i\theta}\, d\theta$, or $d\theta = \frac{dz}{ie^{i\theta}} = \frac{dz}{iz}$.

As θ varies from 0 to 2π, z traverses the unit circle in the anticlockwise direction. Then, the integral given in eq. (4.97) can be written as

$$I = \oint\limits_C F\left(\frac{z+z^{-1}}{2}, \frac{z-z^{-1}}{2i}\right)\frac{dz}{iz}, \quad C \text{ is } |z|=1. \qquad \text{...(4.98)}$$

The contour integral given in Eq. (4.98) can be evaluated by using the residue theorem.

EXAMPLE 4.56 Evaluate the following real definite integrals using contour integration.

(*i*) $\int_0^{2\pi} \frac{d\theta}{2+\cos\theta}$, (*AMIETE, June 2002*) (*ii*) $\int_0^{2\pi} \frac{d\theta}{a+b\sin\theta}$, $a>|b|$,

(*iii*) $\int_0^{\pi} \frac{d\theta}{a+b\cos\theta}$, $a>|b|$, (*iv*) $\int_0^{\pi} \frac{\cos 2\theta}{1-2a\cos\theta+a^2}d\theta$, $(|a|<1)$, (*v*) $\int_0^{2\pi} \frac{d\theta}{3-2\cos\theta+\sin\theta}$.

Solution (*i*) Let $z = e^{i\theta}$. Then $\cos\theta = \frac{e^{i\theta}+e^{-i\theta}}{2} = \frac{z+z^{-1}}{2}$, $dz = ie^{i\theta}d\theta = iz\,d\theta$

so that $\int_0^{2\pi} \frac{d\theta}{2+\cos\theta} = \oint_C \frac{dz/iz}{2+\frac{(z+z^{-1})}{2}} = \frac{2}{i}\oint_C \frac{dz}{z^2+4z+1} = \oint_C f(z)\,dz$,

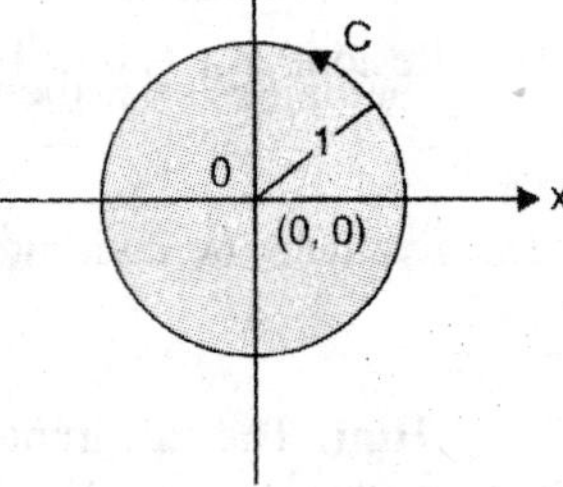

Fig. 4.39

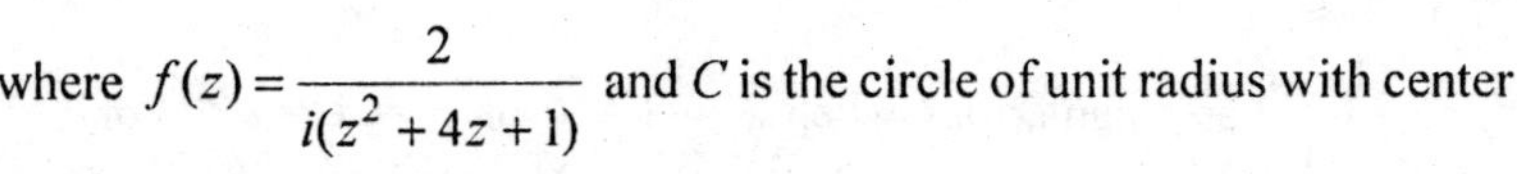

where $f(z) = \frac{2}{i(z^2+4z+1)}$ and C is the circle of unit radius with center at the origin as shown in Fig. 4.39.

Setting $z^2 + 4z + 1 = 0$, we find that the integrand has simple poles at $z_1 = -2+\sqrt{3}$ and $z_2 = -2-\sqrt{3}$. Of these only the pole $z_1 = -2+\sqrt{3}$ lies inside the $C : |z| = 1$.

Hence, $\underset{z=z_1}{\text{Res}}\, f(z) = \frac{2}{i}\lim_{z\to z_1}\left[\frac{z-z_1}{(z-z_1)(z-z_2)}\right] = \frac{2}{i}\frac{1}{(z_1-z_2)} = \frac{2}{i2\sqrt{3}} = \frac{1}{i\sqrt{3}}$

and $\int_0^{2\pi} \frac{d\theta}{2+\cos\theta} = 2\pi i\left[\underset{z=z_1}{\text{Res}}\, f(z)\right] = \frac{2}{3}\pi\sqrt{3}.$

(*ii*) Setting $z = e^{i\theta}$. Then $\sin\theta = \frac{e^{i\theta}-e^{-i\theta}}{2i} = \frac{z-z^{-1}}{2i}$, $dz = ie^{i\theta} = iz\,d\theta$ so that

$\int_0^{2\pi} \frac{d\theta}{a+b\sin\theta} = \oint_C \frac{1}{a+\frac{b}{2i}(z-z^{-1})}\cdot\frac{dz}{iz} = 2\oint_C \frac{dz}{bz^2+2iaz-b} = \oint_C f(z)\,dz$, where $f(z) = \frac{2}{bz^2+2iaz-b}$

and C is the circle of unit radius with centre at the origin. Setting $bz^2 + 2iaz - b = 0$ we find that the integrand has simple poles at $z_1 = \frac{i}{b}(-a+\sqrt{a^2+b^2})$ and $z_2 = \frac{i}{b}(-a-\sqrt{a^2-b^2})$. Of these only the pole $z_1 = \frac{i}{b}(-a+\sqrt{a^2-b^2})$ lies inside $|z| = 1$.

Hence, $\underset{z=z_1}{\text{Res}}\, f(z) = 2\lim_{z\to z_1}\left[\frac{(z-z_1)}{bz^2+2iaz-b}\right] = 2\lim_{z\to z_1}\frac{1}{2bz+2ia}$, by L' Hospital's rule

$$= \frac{1}{bz_1+ia} = \frac{1}{b\frac{i}{b}(-a+\sqrt{a^2-b^2})+ia} = \frac{1}{i\sqrt{a^2-b^2}}.$$

$$\therefore \quad \int_0^{2\pi} \frac{d\theta}{a+b\sin\theta} = 2\pi i\left[\underset{z=z_1}{\text{Res}}\, f(z)\right] = 2\pi i\cdot\frac{1}{i\sqrt{a^2-b^2}} = \frac{2\pi}{\sqrt{a^2-b^2}}.$$

(*iii*) Since $\int_0^{2a} f(x)\,dx = 2\int_0^{a} f(x)\,dx$, if $f(2a-x) = f(x)$, we have

$$\int_0^{2\pi} \frac{d\theta}{a + b\cos\theta} = 2\int_0^{\pi} \frac{d\theta}{a + b\cos\theta} \qquad [\because \cos(2\pi - \theta) = \cos\theta]$$

or
$$\int_0^{\pi} \frac{d\theta}{a + b\cos\theta} = \frac{1}{2}\int_0^{2\pi} \frac{d\theta}{a + b\cos\theta}. \qquad ...(i)$$

Setting $e^{i\theta} = z$, then $\cos\theta = \dfrac{e^{i\theta} + e^{-i\theta}}{2} = \dfrac{z + z^{-1}}{2}$, $dz = ie^{i\theta}$, $d\theta = iz\,d\theta$,

so that
$$\int_0^{2\pi} \frac{d\theta}{a + b\cos\theta} = \oint_C \frac{dz/iz}{\left[a + b\left(\dfrac{z + z^{-1}}{2}\right)\right]} = \frac{2}{i}\oint_C \frac{dz}{bz^2 + 2az + b} = \oint_C f(z)\,dz, \text{ where}$$

$$f(z) = \frac{2}{i}\left(\frac{1}{bz^2 + 2az + b}\right)$$ and C is the circle of unit radius with centre at the origin.

Setting $bz^2 + 2az + b = 0$, we find that the integrand has simple poles at

$$z_1 = \frac{-a + \sqrt{a^2 - b^2}}{b} \quad \text{and} \quad z_2 = \frac{-a - \sqrt{a^2 - b^2}}{b}.$$

Since $a > |b|$, only the pole $z_1 = \{-a + \sqrt{a^2 - b^2}\}/b$ lie inside the $|z| = 1$.

Hence,
$$\operatorname*{Res}_{z = z_1} f(z) = \frac{2}{i}\lim_{z \to z_1}\left[\frac{(z - z_1)}{bz^2 + 2az + b}\right]\left(\frac{0}{0}\right) = \frac{2}{i}\lim_{z \to z_1}\frac{1}{2bz + 2a}, \text{ by L' Hospital's rule}$$

$$= \frac{2}{i}\frac{1}{2bz_1 + 2a} = \frac{1}{i\sqrt{a^2 - b^2}}.$$

Thus,
$$\int_0^{\pi} \frac{d\theta}{a + b\cos\theta} = \frac{1}{2}\cdot 2\pi i\left[\operatorname*{Res}_{z = z_1} f(z)\right] = \frac{\pi i}{i\sqrt{a^2 - b^2}} = \frac{\pi}{\sqrt{a^2 - b^2}}.$$

(*iv*) Substitute $z = e^{i\theta}$. We have $d\theta = dz/iz$, $\cos\theta = \dfrac{z^2 + 1}{2z}$, $\cos 2\theta = \dfrac{e^{2i\theta} + e^{-2i\theta}}{2} = \dfrac{z^4 + 1}{2z^2}$ and

$$1 - 2a\cos\theta + a^2 = 1 + a^2 - \frac{2a(z^2 + 1)}{2z} = (z - az^2 - a + a^2 z)/z.$$

Therefore, the given integral,

$$I = \oint_C \frac{(z^4 + 1)\cdot z}{2z^2\,(z - az^2 - a + a^2 z)}\frac{dz}{iz} = \frac{1}{2i}\oint_C \frac{(z^4 + 1)\,dz}{z^2\,(z - az^2 - a + a^2 z)} = \frac{1}{2i}\oint_C \frac{(z^4 + 1)\,dz}{z^2 (z - a)(1 - az)} = \oint_C f(z)\,dz,$$

where
$$f(z) = \frac{1}{2i}\cdot\frac{z^4 + 1}{z^2 (z - a)(1 - az)}$$ and C is $|z| = 1$.

Setting $z^2(z - a)(1 - az) = 0$, we find simple poles at $z = a$, $1/a$ and pole of order 2 at $z = 0$. Of these the poles $z = 0$ and $z = a$ lie inside $|z| = 1$.

$$\underset{z=0}{\text{Res}}\, f(z) = \lim_{z \to 0} \frac{d}{dz}[z^2 f(z)] = \frac{1}{2i} \lim_{z \to 0} \frac{d}{dz}\left[\frac{z^4+1}{z - az^2 - a + a^2 z}\right]$$

$$= \frac{1}{2i} \lim_{z \to 0}\left[\frac{(z - az^2 - a + a^2 z)\, 4z^3 - (z^4+1)(1 - 2az + a^2)}{(z - az^2 - a + a^2 z)^2}\right] = -\frac{(1+a^2)}{2ia^2}.$$

$$\underset{z=a}{\text{Res}}\, f(z) = \lim_{z \to a} [(z-a) f(z)] = \frac{1}{2i} \lim_{z \to a}\left[\frac{z^4+1}{z^2(1-az)}\right] = \frac{a^4+1}{2ia^2(1-a^2)}.$$

Hence, $$\int_0^{2\pi} \frac{\cos 2\theta}{1 - 2a\cos\theta + a^2}\, d\theta = 2\int_0^{\pi} \frac{\cos 2\theta}{1 - 2a\cos\theta + a^2}\, d\theta = 2\pi i \left[\underset{z=0}{\text{Res}}\, f(z) + \underset{z=a}{\text{Res}}\, f(z)\right]$$

$$= 2\pi i\left[-\frac{1+a^2}{2ia^2} + \frac{a^4+1}{2ia^2(1-a^2)}\right] = \frac{2\pi a^2}{1-a^2},$$

or $$\int_0^{\pi} \frac{\cos 2\theta}{1 - 2a\cos\theta + a^2}\, d\theta = \frac{1}{2} \bullet \frac{2\pi a^2}{1-a^2} = \frac{\pi a^2}{1-a^2}.$$

(*v*) Let $z = e^{i\theta}$. Then $\sin\theta = \dfrac{z - z^{-1}}{2i}$, $\cos\theta = \dfrac{z + z^{-1}}{2}$, $dz = iz d\theta$ so that

$$\int_0^{2\pi} \frac{d\theta}{3 - 2\cos\theta + \sin\theta} = \oint_C \frac{dz/iz}{3 - 2\dfrac{(z+z^{-1})}{2} + \dfrac{(z - z^{-1})}{2i}} = \oint_C \frac{2dz}{(1-2i)z^2 + 6iz - 1 - 2i} = \oint_C f(z)\, dz,$$

where $f(z) = \dfrac{2}{(1-2i)z^2 + 6iz - 1 - 2i}$ and C is the circle of unit radius with centre at the origin.

The poles of $\dfrac{2}{(1-2i)z^2 + 6iz - (1+2i)}$ are the simple poles at

$$z = -6i \pm \sqrt{36i^2 + 4(1-2i)(1+2i)}\Big/2(1-2i) = \frac{-6i \pm 4i}{2(1-2i)}, \text{ that is } (2-i)/5 \text{ and } 2 - i.$$

Only $(2-i)/5$ lies inside C.

$$\underset{z=(2-i)/5}{\text{Res}}\, f(z) = 2 \lim_{z \to (2-i)/5}\left[\frac{z - \{(2-i)/5\}}{(1-2i)z^2 + 6iz - 1 - 2i}\right] \left(\frac{0}{0}\right)$$

$$= 2 \lim_{z \to (2-i)/5}\left(\frac{1}{2(1-2i)z + 6i}\right) = \frac{1}{2i}. \qquad \text{[by L' Hospital's rule]}$$

Hence, $$\oint_C \frac{2dz}{(1-2i)z^2 + 6iz - 1 - 2i} = 2\pi i\left(\frac{1}{2i}\right) = \pi.$$

EXAMPLE 4.57 Evaluate the following real definite integrals using contour integration.

(*i*) $\displaystyle\int_0^{2\pi} \frac{\cos 3\theta}{5 - 4\cos\theta}\, d\theta$, (*Madras, 2003; Delhi 2002*) (*ii*) $\displaystyle\int_0^{2\pi} \frac{d\theta}{(a + b\cos\theta)^2}$, (*iii*) $\int_0^{2\pi} e^{\cos\theta} \cos(\sin\theta)\, d\theta$, (*iv*) $\int_0^{\pi} \frac{1+2\cos\theta}{5+4\cos\theta}\, d\theta$.

Solution (*i*) Let $z = e^{i\theta}$. Then $\cos\theta = \frac{z+z^{-1}}{2}$, $\cos 3\theta = \frac{e^{3i\theta}+e^{-3i\theta}}{2} = \frac{z^3+z^{-3}}{2}$, $dz = iz\, d\theta$.

Therefore, $$\int_0^{2\pi} \frac{\cos 3\theta}{5-4\cos\theta}\, d\theta = \oint_C \frac{(z^3+z^{-3})/2}{5-4\left(\frac{z+z^{-1}}{2}\right)} \frac{dz}{iz} = -\frac{1}{2i}\oint_C \frac{z^6+1}{z^3(2z-1)(z-2)}\, dz = \oint_C f(z)\, dz,$$

where $f(z) = -\frac{1}{2i} \cdot \frac{z^6+1}{z^3(2z-1)(z-2)}$ and C is the circle of unit radius with center at the origin.

The integrand has a pole of order 3 at $z = 0$ and a simple pole $z = 1/2$ and $z = 2$. Of these only the poles $z = 0$ and $z = 1/2$ lies inside the $C : |z| = 1$.

$$\operatorname*{Res}_{z=0} f(z) = -\frac{1}{2i} \lim_{z\to 0} \frac{1}{2!} \frac{d^2}{dz^2}[z^3 \cdot f(z)] = -\frac{1}{4i} \lim_{z\to 0} \frac{d^2}{dz^2}\left[\frac{z^6+1}{2z^2-5z+2}\right]$$

$$= -\frac{1}{4i} \lim_{z\to 0} \frac{d}{dz}\left[\frac{6z^5(2z^2-5z+2)-(z^6+1)(4z-5)}{(2z^2-5z+2)^2}\right]$$

$$= -\frac{1}{4i} \lim_{z\to 0} \frac{d}{dz}\left[\frac{8z^7-25z^6+12z^5-4z+5}{(2z^2-5z+2)^2}\right]$$

$$= -\frac{1}{4i} \lim_{z\to 0}\left[\frac{\{(2z^2-5z+2)^2(56z^6-150z^5+60z^4-4) - 2(8z^7-25z^6+12z^5-4z+5)(2z^2-5z+2)(4z-5)\}}{(2z^2-5z+2)^4}\right]$$

$$= -\frac{1}{4i}\left[\frac{4(-4)-2(5)(-10)}{16}\right] = -\frac{21}{16i}.$$

$$\operatorname*{Res}_{z=1/2} f(z) = -\frac{1}{2i} \lim_{z\to 1/2}\left[\left(z-\frac{1}{2}\right) f(z)\right] = -\frac{1}{4i} \lim_{z\to 1/2}\left[\frac{z^6+1}{z^3(z-2)}\right] = \frac{65}{48i}.$$

Hence, $$\int_0^{2\pi} \frac{\cos 3\theta}{5-4\cos\theta}\, d\theta = 2\pi i\left[\operatorname*{Res}_{z=0} f(z) + \operatorname*{Res}_{z=1/2} f(z)\right] = 2\pi i\left[-\frac{2i}{16i} + \frac{65}{48i}\right] = \frac{\pi}{12},$$ the required result.

(*ii*) In example 4.56 (*iii*), we have proved that

$$\int_0^{\pi} \frac{1}{a+b\cos\theta}\, d\theta = \frac{\pi}{\sqrt{a^2-b^2}}. \text{ Therefore, } \int_0^{2\pi} \frac{d\theta}{a+b\cos\theta} = 2\int_0^{\pi} \frac{d\theta}{a+b\cos\theta} = \frac{2\pi}{\sqrt{a^2-b^2}}.$$

Differentiating both sides with respect to a.

$$\frac{d}{da}\int_0^{2\pi} \frac{1}{a+b\cos\theta}\, d\theta = \frac{d}{da}\frac{2\pi}{\sqrt{a^2-b^2}}$$

or $$\int_0^{2\pi}\left[\frac{\partial}{\partial a}\left(\frac{1}{a+b\cos\theta}\right)\right] d\theta = 2\pi\left[-\frac{1}{2}(a^2-b^2)^{-3/2} \cdot 2a\right]$$ [Leibnitz Rule] or $\displaystyle\int_0^{2\pi} -\frac{1}{(a+b\cos\theta)^2}\,d\theta = \frac{-2\pi a}{(a^2-b^2)^{3/2}} \Rightarrow \int_0^{2\pi} \frac{d\theta}{(a+b\cos\theta)^2} = \frac{2\pi a}{(a^2-b^2)^{3/2}}.$

(*iii*) The integrand can be written as

$$e^{\cos\theta}\cos(\sin\theta) = \frac{1}{2}e^{\cos\theta}\,[e^{i\sin\theta} + e^{-i\sin\theta}] = \frac{1}{2}[e^{\cos\theta + i\sin\theta} + e^{\cos\theta - i\sin\theta}]$$

$$= \frac{1}{2}[e^{e^{i\theta}} + e^{e^{-i\theta}}] = \frac{1}{2}[e^z + e^{1/z}],$$ where $z = e^{i\theta}$. We have $d\theta = dz/iz$ and

$$I = \int_0^{2\pi} e^{\cos\theta}\cos(\sin\theta)\,d\theta = \frac{1}{2}\oint_C (e^z + e^{1/z})\frac{dz}{iz} = \oint_C f(z)\,dz,$$

where $f(z) = (e^z + e^{1/z})/2iz$ and C is $|z| = 1$.
Now, $z = 0$ is an essential singularity of the integrand. The Laurent series expansion of $f(z)$ is given by

$$f(z) = \frac{1}{2iz}\left[\left(1 + z + \frac{z^2}{2!} + \cdots\right) + \left(1 + \frac{1}{z} + \frac{1}{2!z^2} + \cdots\right)\right] = \frac{1}{2i}\left[\left(1 + \frac{z}{2!} + \cdots\right) + \left(\frac{2}{z} + \frac{1}{z^2} + \cdots\right)\right].$$

Hence, $\displaystyle\operatorname*{Res}_{z=0} f(z) = \frac{1}{i}$. Therefore, $\displaystyle I = 2\pi i\left[\operatorname*{Res}_{z=0} f(z)\right] = 2\pi i \cdot \frac{1}{i} = 2\pi.$

(*iv*) We have $\displaystyle\int_0^{\pi} \frac{1+2\cos\theta}{5+4\cos\theta}\,d\theta = \frac{1}{2}\int_0^{2\pi}\frac{1+2\cos\theta}{5+4\cos\theta}\,d\theta$. Setting $z = e^{i\theta}$. Then $\cos\theta = \dfrac{z+z^{-1}}{2}$, $dz = iz\,d\theta$, so that

$$\int_0^{\pi}\frac{1+2\cos\theta}{5+4\cos\theta}\,d\theta = \frac{1}{2}\oint_C \frac{1+(z+z^{-1})}{5+2(z+z^{-1})}\frac{dz}{iz} = \frac{1}{2i}\oint_C \frac{z^2+z+1}{z(2z^2+5z+2)}\,dz$$

$$= \frac{1}{2i}\oint_C \frac{(z^2+z+1)}{z(z+2)(2z+1)}\,dz = \oint_C f(z)\,dz,$$ where $f(z) = \dfrac{z^2+z+1}{2iz(z+2)(2z+1)}$ and $C : |z| = 1$.

The integrand has simple poles at $z = 0, -1/2$ and -2. Of these the first two poles lies inside the $C : |z| = 1$.

Hence, $\displaystyle\operatorname*{Res}_{z=0} f(z) = \lim_{z\to 0}\,[z\,f(z)] = \frac{1}{2i}\lim_{z\to 0}\left[\frac{z^2+z+1}{(z+2)(2z+1)}\right] = \frac{1}{4i}.$

$$\operatorname*{Res}_{z=-1/2} f(z) = \frac{1}{4i}\lim_{z\to -1/2}\left[\frac{z^2+z+1}{z(z+2)}\right] = \frac{1}{4i}\left[\frac{\frac{1}{4}-\frac{1}{2}+1}{-\frac{1}{2}\left(-\frac{1}{2}+2\right)}\right] = -\frac{1}{4i}.$$

Hence, $\displaystyle\int_0^{\pi}\frac{1+2\cos\theta}{5+4\cos\theta}\,d\theta = 2\pi i\left[\operatorname*{Res}_{z=0} f(z) + \operatorname*{Res}_{z=-1/2} f(z)\right] = 2\pi i\left[\frac{1}{4i} - \frac{1}{4i}\right] = 0.$

Type II Improper Real Integrals of the Form $\int_{-\infty}^{\infty} f(x)dx$

We now consider real integrals of the type $\displaystyle\int_{-\infty}^{\infty} f(x)dx.$ ...(4.99)

Such an integral, for which the interval of integration is not finite, is called an **improper integral** and is defined in terms of two distinct limits:

$$\int_{-\infty}^{\infty} f(x)\,dx = \lim_{a\to-\infty}\int_{a}^{0} f(x) + \lim_{b\to\infty}\int_{0}^{b} f(x)\,dx \qquad ...(4.100)$$

If both limits exist, we write

$$\int_{-\infty}^{\infty} f(x)\,dx = \lim_{R\to\infty}\int_{-R}^{R} f(x)\,dx \qquad ...(4.101)$$

We assume that the function $f(x)$ in Eq. (4.99) is a real rational function whose denominator is different from zero for all real x and is of degree at least two units higher than the degree of the numerator. Then the limits in Eq. (4.100) exist, and we may start from Eq. (4.101). The expression on the right hand side of Eq. (4.101) is called the ***Cauchy principal value*** of the integral and is written C.P.V

$$\int_{-\infty}^{\infty} f(x)\,dx = \lim_{R\to\infty}\int_{-R}^{R} f(x)\,dx.$$

It may exist even if the limits in Eq. (4.100) do not exist. For example

$$\lim_{R\to\infty}\int_{-R}^{R} x\,dx = \lim_{R\to\infty}\left[\frac{R^2}{2} - \frac{(-R)^2}{2}\right] = 0$$

but

$$\lim_{R\to\infty}\int_{0}^{R} x\,dx = \lim_{R\to\infty}\frac{R^2}{2} = \infty.$$

To evaluate an integral $\int_{-\infty}^{\infty} f(x)\,dx$, where $f(x) = \dfrac{p(x)}{q(x)}$ is continuous on $(-\infty, \infty)$, by residue theory we replace x by the complex variable z and integrate the complex function f over a closed contour C consisting of a semi-circular C_R with radius R, large enough to enclose the poles of $f(z)$ in the upper half plane, and the part of the real axis from $x = -R$ to $x = R$. The path traversed in the anti clockwise direction. See Figure 4.40.

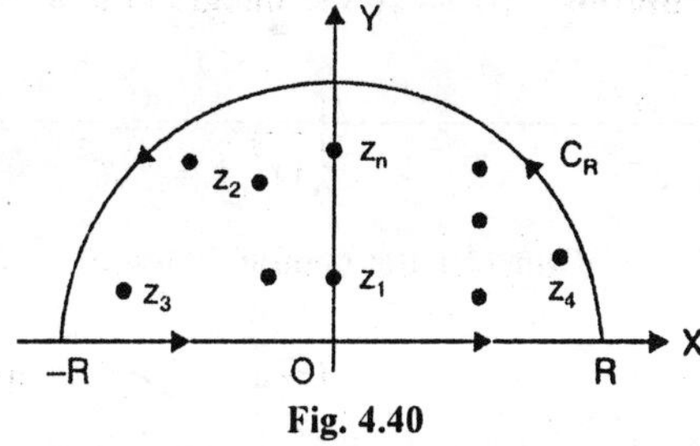

Fig. 4.40

By Cauchy's Residue theorem, we have

$$\oint_C f(z)\,dz = \oint_{C_R} f(z)\,dz + \int_{-R}^{R} f(x)\,dx = 2\pi i \sum_{k=1}^{n} \text{Res}\,(f(z), z_k) \qquad ...(4.102)$$

where z_k, $k = 1, 2, ..., n$ denotes poles in the upper half plane.

We now determine the integral $\oint_{C_R} f(z)\,dz$.

If we put $z = Re^{i\theta}$ in the integral $\int f(z)\,dz = \int \dfrac{p(z)}{q(z)}\,dz$, then R = constant on C_R and as z moves along C_R, θ varies from 0 to π. Since by assumption the degree of $q(z)$ is greater than that of $p(z)$ by at least two, we have $|f(z)| < \dfrac{k}{|z|^2}$, $(|z| = R > R_0)$ for sufficiently large constants k and R_0.

By the ML-in equality*

$$\left|\int_{C_R} f(z)\,dz\right| < \frac{k}{R^2}\cdot \pi R = \frac{k\pi}{R} \quad (R > R_0) \text{ which tends to zero as } R \to \infty.$$

Therefore, as $R \to \infty$, the value of the integral $\int_{C_R} f(z)\,dz$ approachs zero and from Eq. (4.102), we have

$$\int_{-\infty}^{\infty} f(x)\,dx = 2\pi i \sum_{k=1}^{n} \text{Res}\,(f(z), z_k) \qquad ...(4.103)$$

EXAMPLE 4.58 Evaluate the following by contour integration:

(*i*) $\displaystyle\int_0^{\infty} \frac{x^2}{(x^2+a^2)(x^2+b^2)}\,dx \ (a, b > 0)$ (*Anna 2003; Delhi 2002, MDU, 2002*)

(*ii*) $\displaystyle\int_{-\infty}^{\infty} \frac{dx}{x^4+1}$, (*AMIETE, Dec 2008; UPTU, 2007; JNTU, 2006*)

(*iii*) $\displaystyle\int_{-\infty}^{\infty} \frac{x^2-x+2}{x^4+10x^2+9}\,dx$, (*AMIETE, June 2007, Dec. 2003, Delhi 2002*)

(*iv*) $\displaystyle\int_0^{\infty} \frac{x^2}{(x^2+a)^2}\,dx,\quad a > 0,$ (*AMIETE, June 2009*) (*v*) $\displaystyle\int_{-\infty}^{\infty} \frac{x^2\,dx}{(x^2+1)^2(x^2+2x+2)}$.

Solution (*i*) Since the integrand is an even function, we have

$$\int_0^{\infty} \frac{x^2}{(x^2+a^2)(x^2+b^2)}\,dx = \frac{1}{2}\int_{-\infty}^{\infty} \frac{x^2}{(x^2+a^2)(x^2+b^2)}\,dx.$$

Consider the contour integral

$$I = \int_C f(z)\,dz, \text{ where } f(z) = \frac{z^2}{2(z^2+a^2)(z^2+b^2)} = \frac{z^2}{2(z+ia)(z-ia)(z+ib)(z-ib)}$$

and C be the closed contour consisting of the line from $-R$ to R on the x-axis and the semicircle $C_R : |z| = R$ of sufficiently large radius R in upper half plane, traversed in the positive (counter clockwise) sense shown in Fig. 4.41.

The integrand has simple poles at $z = \pm\, ai$ and $z = \pm\, bi$. Therefore, $f(z)$ is analytic in the upper half plane except for the simple poles at $z_1 = ai$ and $z_2 = bi$. We find that

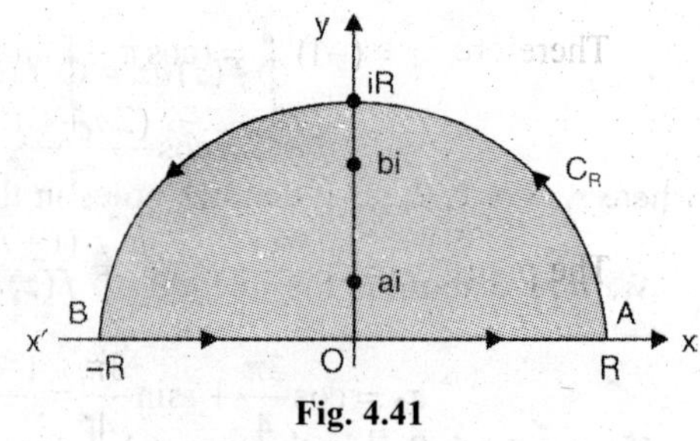

Fig. 4.41

* If f is continuous on a smooth curve C and if $|f(z)| \le M$ for all z on C, then $\left|\int_C f(z)\,dz\right| \le ML$, where L is the length of C and M a constant.

$$\operatorname*{Res}_{z=ai} f(z) = \lim_{z \to ai} \left[\frac{z^2}{2(z^2+b^2)(z+ai)} \right] = \frac{a}{4i(a^2-b^2)}$$

and $$\operatorname*{Res}_{z=bi} f(z) = \lim_{z \to bi} \left[\frac{z^2}{2(z^2+a^2)(z+bi)} \right] = \frac{-b}{4i(a^2-b^2)}.$$

We now write $\int_C f(z)\,dz = \int_{-R}^{R} f(x)\,dx + \int_{C_R} f(z)\,dz.$ ...(1)

Along C_R : $|z| = R$ in the upper half plane encircling the poles $z = ai, bi$, we have

$$|f(z)| = \frac{|z^2|}{2|z^2+a^2||(z^2+b^2)|} \le \frac{R^2}{2(R^2-a^2)(R^2-b^2)}.$$

Since $|z^2 + a^2| \ge |z|^2 - a^2 = R^2 - a^2$ and $|z^2 + b^2| \ge |z|^2 - b^2 = R^2 - b^2$.

Hence, $$\left| \int_{C_R} f(z)\,dz \right| \le \frac{R^2\ (\pi R)}{2(R^2-a^2)(R^2-b^2)} \to 0 \text{ as } R \to \infty.$$

Setting $R \to \infty$ in equation (1), we obtain

$$I = \frac{1}{2}\int_{-\infty}^{\infty} \frac{x^2 dx}{(x^2+a^2)(x^2+b^2)} = \int_{0}^{\infty} \frac{x^2 dx}{(x^2+a^2)(x^2+b^2)}$$

$$= 2\pi i \left[\operatorname*{Res}_{z=ai} f(z) + \operatorname*{Res}_{z=bi} f(z) \right] = 2\pi i \left[\frac{a}{4i(a^2-b^2)} - \frac{b}{4i(a^2-b^2)} \right] = \frac{\pi}{2(a+b)}.$$

Therefore, $$\int_{0}^{\infty} \frac{x^2 dx}{(x^2+a^2)(x^2+b^2)} = \frac{\pi}{2(a+b)}.$$

(*ii*) Consider the contour integral $I = \int_C f(z)\,dz$, where $f(z) = \dfrac{1}{z^4+1}$ and C be the closed contour consisting of the line from $-R$ to R on the x-axis and the semi-circle C_R of radius > 1.

The poles of $f(z) = \dfrac{1}{z^4+1}$ are obtained by solving $z^4 + 1 = 0$.

Therefore, $z = (-1)^{1/4} = (\cos\pi + i\sin\pi)^{1/4} = [\cos(2n+1)\pi + i\sin(2n+1)\pi]^{1/4}$

$= \cos\dfrac{(2n+1)\pi}{4} + i\sin\dfrac{(2n+1)\pi}{4}$ (by DeMoivre's theorem), where $n = 0, 1, 2, 3$.

The poles $z_1 = \cos\dfrac{\pi}{4} + i\sin\dfrac{\pi}{4} = \dfrac{(1+i)}{\sqrt{2}} = e^{\pi i/4}$ and

$z_2 = \cos\dfrac{3\pi}{4} + i\sin\dfrac{3\pi}{4} = \dfrac{(-1+i)}{\sqrt{2}} = e^{3\pi i/4}$ lie inside the upper half plane.

We obtain $$\operatorname*{Res}_{z=z_1} f(z) = \lim_{z \to z_1}\left[\frac{z-z_1}{z^4+1}\right] \left(\frac{0}{0}\right) = \lim_{z\to z_1}\left[\frac{1}{4z^3}\right] = \frac{1}{4z_1^3} = \frac{z_1}{4z_1^4} = -\frac{(1+i)}{4\sqrt{2}}$$

and $$\operatorname*{Res}_{z=z_2} f(z) = \lim_{z \to z_2}\left[\frac{z-z_2}{z^4+1}\right] \left(\frac{0}{0}\right) = \lim_{z\to z_2}\left[\frac{1}{4z^3}\right] = \frac{1}{4z_2^3} = \frac{z_2}{4z_2^4} = -\frac{(-1+i)}{4\sqrt{2}}.$$

We now write $\quad \int_C f(z)\,dz = \int_{-R}^{R} f(x)\,dx + \int_{C_R} f(z)\,dz.$ ...(2)

Let $R > 1$, so that the singular points z_1 and z_2 lie inside $C : |z| = R$. Then $|f(z)| = \dfrac{1}{|1+z^4|} \le \dfrac{1}{R^4 - 1}$.

Using the ML–inequality, we obtain

$$\left|\int_{C_R} f(z)\,dz\right| \le \frac{1}{R^4-1}(\pi R) \to 0 \text{ as } R \to \infty.$$

Setting $R \to \infty$ in (2), we get

$$\int_{-\infty}^{\infty} f(x)\,dx = \int_{-\infty}^{\infty} \frac{dx}{1+x^4} = 2\pi i\left[\underset{z=z_1}{\text{Res}}\ f(z) + \underset{z=z_2}{\text{Res}}\ f(z)\right] = \frac{-2\pi i}{4\sqrt{2}}[(1+i)+(-1+i)] = \frac{\pi}{\sqrt{2}} = \frac{\pi\sqrt{2}}{2}.$$

Note: $\displaystyle\int_0^{\infty} \frac{dx}{1+x^4} = \frac{1}{2}\int_{-\infty}^{\infty} \frac{dx}{1+x^4} = \frac{\pi\sqrt{2}}{4}.$

(*iii*) Consider the contour integral $I = \int_C f(z)\,dz$, where $f(z) = \dfrac{z^2 - z + 2}{z^4 + 10z^2 + 9}$ and C be the closed contour consisting of real axis from $-R$ to R and semi-circle $C_R : |z| = R$ of sufficiently large radius R in upper half plane. See Fig. 4.41.

Now $f(z)$ has simple poles at points for which $z^4 + 10z^2 + 9 = 0 \Rightarrow (z^2 + 9)(z^2 + 1) = 0 \Rightarrow z = \pm i, \pm 3i$. Of these poles, $z_1 = i$ and $z_2 = 3i$ lie in upper half-plane.

We obtain $\displaystyle \underset{z=i}{\text{Res}}\ f(z) = \lim_{z\to i}\left[\frac{z^2 - z + 2}{(z^2+9)(z+i)}\right] = \frac{i^2 - i + 2}{(i^2+9)(2i)} = \frac{1-i}{16i}$

and $\displaystyle \underset{z=3i}{\text{Res}}\ f(z) = \lim_{z\to 3i}\left[\frac{z^2 - z + 2}{(z^2+1)(z+3i)}\right] = \frac{9i^2 - 3i + 2}{(9i^2+1)\,6i} = \frac{7+3i}{48i}.$

We now write $\quad \int_C f(z)\,dz = \int_{-R}^{R} f(x)\,dx + \int_{C_R} f(z)\,dz$ ...(3)

Along $C_R : |z| = R$ in the upper half plane, encircling the poles $z = i, 3i$, we have

$$|f(z)| = \frac{|z^2| + |-z| + 2}{|z^4| - 10|z^2| - 9} \le \frac{R^2 + R + 2}{R^4 - 10R^2 - 9}.$$

[$\because$ $|a + b| \le |a| + |b|$ and $|a - b| \ge |a| - |b|$]

Hence, $\displaystyle \left|\int_{C_R} f(z)\,dz\right| \le \frac{(R^2 + R + 2)\,\pi R}{R^4 - 10R^2 - 9} \to 0 \text{ as } R \to \infty.$

Setting $R \to \infty$ in (3), we get

$$\int_{-\infty}^{\infty} \frac{x^2 - x + 2}{x^4 + 10x^2 + 9}\,dx = 2\pi i\left[\underset{z=i}{\text{Res}}\ f(z) + \underset{z=3i}{\text{Res}}\ f(z)\right] = 2\pi i\left[\frac{1-i}{16i} + \frac{7+3i}{48i}\right] = 2\pi i \cdot \frac{5}{24i} = \frac{5\pi}{12}.$$

(*iv*) Since the integral is an even function, we can write $\displaystyle \int_0^{\infty} \frac{x^2 dx}{(x^2 + a^2)^2} = \frac{1}{2}\int_{-\infty}^{\infty} \frac{x^2 dx}{(x^2 + a^2)^2}.$

Consider the contour integral $I = \int_C f(z)\, dz$, where $f(z) = \dfrac{z^2}{2(z^2 + a^2)^2}$ and C be the closed contour consisting of real axis from $-R$ to R and semicircle $C_R : |z| = R$ of sufficiently large radius R in upper half plane shown in Fig. 4.41. The integrand $f(z)$ is analytic in the upper half plane except for the pole of order 2 at $z = ai$. We find that

$$\operatorname*{Res}_{z=ai} f(z) = \frac{1}{1!} \lim_{z \to ai} \frac{d}{dz}[(z - ai)^2 f(z)] = \lim_{z \to ai} \frac{d}{dz}\left[\frac{z^2}{2(z + ai)^2}\right]$$

$$= \lim_{z \to ai}\left[\frac{4z(z + ai)^2 - 4z^2(z + ai)}{4(z + ai)^4}\right] = -\frac{i}{8a}.$$

We now write $$\int_C f(z)\, dz = \int_{-R}^{R} f(x)\, dx + \int_{C_R} f(z)\, dz \qquad ...(4)$$

Let $R > a$, Then $|f(z)| = \dfrac{|z|^2}{2|z^2 + a^2|^2} \le \dfrac{R^2}{2(R^2 - a^2)^2}$, since $|z^2 + a^2| \ge R^2 - a^2$.

Hence, $$\left|\int_{C_R} f(z)\, dz\right| \le \frac{R^2}{2(R^2 - a^2)^2}(\pi R) \to 0 \text{ as } R \to \infty.$$

Setting $R \to \infty$ in Eq. (4), we obtain

$$\frac{1}{2}\int_{-\infty}^{\infty} \frac{x^2 dx}{(x^2 + a^2)^2} = \int_0^{\infty} \frac{x^2 dx}{(x^2 + a^2)^2} = 2\pi i\left[\operatorname*{Res}_{z=ai} f(z)\right] = -2\pi i\left(\frac{-i}{8a}\right) = \frac{\pi}{4a}.$$

(v) Consider the contour integral $I = \int_C f(z)\, dz$, where $f(z) = \dfrac{z^2}{(z^2 + 1)^2(z^2 + 2z + 2)}$ and C be the closed contour consisting of real axis from $-R$ to R and semicircle $C_R : |z| = R$ of sufficiently large radius R in upper half plane shown in Fig. 4.41. The integrand $f(z)$ is analytic in the upper half plane except for the poles of order 2 at $z = i$ and simple pole at $z = -1 + i$.

$$\operatorname*{Res}_{z=i} f(z) = \lim_{z \to i} \frac{d}{dz}\left[\frac{z^2}{(z + i)^2(z^2 + 2z + 2)}\right]$$

$$= \lim_{z \to i}\left[\frac{2z\,(z + i)^2(z^2 + 2z + 2) - z^2[2(z + i)\,(z^2 + 2z + 2) + (z + i)^2(2z + 2)]}{(z + i)^4\,(z^2 + 2z + 2)^2}\right]$$

$$= \frac{-3i}{4(4i - 3)} = \frac{-3i(4i + 3)}{4(4i - 3)\,(4i + 3)} = \frac{9i - 12}{100}.$$

$$\operatorname*{Res}_{z=-1+i} f(z) = \lim_{z \to -1+i}\left[\frac{z^2}{(z^2 + 1)^2(z + 1 + i)}\right] = \frac{(-1 + i)^2}{(1 - 2i)^2(2i)} = \frac{1}{3 + 4i} = \frac{3 - 4i}{25}.$$

We now write $$\int_C f(z)\, dz = \int_{-R}^{R} f(x)\, dx + \int_{C_R} f(z)\, dz$$

But $$\int_{C_R} f(z)\, dz \to 0 \text{ as } R \to \infty.$$

We get $$\int_{-\infty}^{\infty} f(x)\,dx = 2\pi i\left[\operatorname*{Res}_{z=i} f(z) + \operatorname*{Res}_{z=-1+i} f(z)\right]$$

or $$\int_{-\infty}^{\infty} \frac{x^2\,dx}{(x^2+1)^2(x^2+2x+2)} = 2\pi i\left[\frac{9i-12}{100} + \frac{3-4i}{25}\right] = \frac{7\pi}{50}.$$

Remark 13

It is often tedious to show that the contour integral along C_R approaches zero as $R \to \infty$. Sufficient conditions under which this is always true are as follows. Suppose $f(z) = P(z)/Q(z)$, where the degree of $P(z)$ is n and the degree of $Q(z)$ is $m \geq n + 2$. If C_R is a semi-circular contour $z = Re^{i\theta}$, $0 \leq \theta \leq \pi$, then

$$\int_{C_R} f(z)\,dz \to 0 \text{ as } R \to \infty.$$

Type III Improper Real Integrals of the Form $\int_{-\infty}^{\infty} f(x)\cos(mx)\,dx$ and $\int_{-\infty}^{\infty} f(x)\sin(mx)\,dx$.

We now consider the evaluation of improper real integrals of the forms

$$\int_{-\infty}^{\infty} f(x)\cos(mx)\,dx \text{ and } \int_{-\infty}^{\infty} f(x)\sin(mx)\,dx, \quad m > 0. \qquad \text{...(4.104)}$$

These integrals are called *Fourier integrals.* Since

$$\int_{-\infty}^{\infty} f(x)\cos(mx)\,dx = \int_{-\infty}^{\infty} f(x)\,\mathrm{Re}(e^{imx})\,dx = \mathrm{Re}\int_{-\infty}^{\infty} f(x)\,e^{imx}\,dx,$$

and $$\int_{-\infty}^{\infty} f(x)\sin(mx)\,dx = \int_{-\infty}^{\infty} f(x)\,\mathrm{Im}(e^{imx})\,dx = \mathrm{Im}\int_{-\infty}^{\infty} f(x)\,e^{imx}\,dx.$$

We now consider the evaluation of the integral $I = \int_{-\infty}^{\infty} f(x)\,e^{imx}\,dx.$ ...(4.105)

The real and imaginary parts of I give the values of the integrals in Eq. (4.104).

To evaluate the improper real integral given in Eq. (4.105), we consider the associated contour integral

$$I = \int_C f(z)\,e^{imz}\,dz. \qquad \text{...(4.106)}$$

where $f(z) = p(z)/q(z)$ and C is the contour consisting semicircular C_R with radius R, large enough to enclose the poles of $f(z)$ in the upper half plane and the part of the real axis from $x = -R$ to $x = R$. The path traversed in the anticlockwise direction. We assume that $f(z)$ satisfies the following conditions:

(*i*) $f(z)$ is analytic in the upper half of the z-plane except at a finite number of poles $z_1, z_2, \ldots, z_n$ in this half plane, none of which lies on the real axis.

(*ii*) The degree of the denominator $q(z)$ in $f(z)$ is at least two units higher than the degree of the numerator $p(z)$.

Now write I as

$$I = \int_C f(z)\,e^{imz}\,dz = \int_{C_R} f(z)\,e^{imz}\,dz + \int_{-R}^{R} f(x)e^{imx}\,dx. \qquad \text{...(4.107)}$$

Using the residue theorem, we obtain

$$\int_C f(z)\, e^{imz}\, dz = \int_{C_R} f(z)\, e^{imz}\, dz + \int_{-R}^{R} f(x) e^{imx} dx = 2\pi i \left\{ \sum_{k=1}^{n} \text{Res}\,[f(z)\, e^{imz}, z_k] \right\}$$

where z_k, $k = 1, 2, ..., n$ denotes poles in the upper half plane. Now consider the integral $\int_{C_R} f(z)\, e^{imz}\, dz$.

We have $\left| \int_{C_R} f(z)\, e^{imz}\, dz \right| \le \int_{C_R} |f(z)\, e^{imz}|\, |dz|.$

Now, $|e^{imz}| = |e^{im(x+iy)}| = |e^{imx}||e^{-my}| = e^{-my} \le 1. \quad (m > 0,\ y \ge 0)$ ...(4.108)

Hence, $\left| \int_{C_R} f(z)\, e^{imz}\, dz \right| \le \int_{C_R} |f(z)\, e^{-my}|\, |dz| \le \int_{C_R} |f(z)|\, |dz|.$ ...(4.109)

As $R \to \infty$ the integral $\int_{C_R} f(z)\, e^{imz}\, dz$ is zero. Therefore, as $R \to \infty$, we obtain from Eq. (4.107)

$$\lim_{R\to\infty} \int_C f(z)\, e^{imz}\, dz = \lim_{R\to\infty} \int_{-R}^{R} f(x)\, e^{imx} dx.$$

Hence, $\int_{-\infty}^{\infty} f(x) e^{imx} dx = 2\pi i \left\{ \sum_{k=1}^{n} \text{Res}\,[f(z)\, e^{imz}, z_k] \right\}.$

Comparing the real and imaginary parts on both sides, we get

$$\int_{-\infty}^{\infty} f(x) \cos(mx)\, dx = \text{Re} \left\{ 2\pi i \sum_{k=1}^{n} \text{Res}\,[f(z)\, e^{imz}, z_k] \right\}$$

and $$\int_{-\infty}^{\infty} f(x) \sin(mx)\, dx = \text{Im} \left\{ 2\pi i \sum_{k=1}^{n} \text{Res}\,[f(z)\, e^{imz}, z_k] \right\}$$

where z_k, $k = 1, 2, ..., n$ denotes poles in the upper half plane.

EXAMPLE 4.59 Obtain the Cauchy principal value of the following improper real integrals using contour integration:

(*i*) $\int_0^{\infty} \frac{\cos mx}{a^2 + x^2} dx, \quad m \ge 0,$ (*ii*) $\int_0^{\infty} \frac{x \sin x}{x^2 + 9} dx,$

(*iii*) $\int_{-\infty}^{\infty} \frac{\sin^2 2x}{1 + x^2} dx,$ (*AMIETE, Dec. 2006*) (*iv*) $\int_0^{\infty} \frac{x \sin x}{(x^2 + a^2)^2} dx.$ (*AMIETE, Dec. 2005, 2004*)

Solution (*i*) Consider the contour integral

$$I = \int_C \frac{e^{imz}}{a^2 + z^2} dz = \int_C e^{imz}\, f(z)\, dz, \text{ where } f(z) = \frac{1}{a^2 + z^2}$$

and C be the closed contour consisting of real axis from $-R$ to R and semicircle $C_R : |z| = R$ of sufficiently large radius R in upper half plane shown in Fig. 4.41. The function $f(z)$ is analytic in the upper half plane except for the simple pole at $z_1 = ai$. We now write

$$\int_C \frac{e^{imz}}{a^2+z^2}\,dz = \int_{C_R} \frac{e^{imz}}{a^2+z^2}\,dz + \int_{-R}^{R} \frac{e^{imx}}{a^2+x^2}\,dx = 2\pi i\Big[\operatorname*{Res}_{z=ai} e^{imz} f(z)\Big].$$

We find that $\operatorname*{Res}_{z=ai}\,[e^{miz} f(z)] = \lim_{z\to ai}\left[\dfrac{(z-ai)\,e^{miz}}{(z+ai)(z-ai)}\right] = \dfrac{e^{-am}}{2ai}.$

Now, $$\left|\int_{C_R} e^{imz} f(z)\,dz\right| \le \int_{C_R} |e^{imz}|\;|f(z)|\;|dz| \le \int_{C_R} |f(z)|\;|dz|,$$

since $|e^{imz}| = |e^{im(x+iy)}| = |e^{imx}\cdot e^{-my}| = e^{-my} \le 1.$ Also $|z^2+a^2| \ge |z|^2 - a^2 = R^2 - a^2.$

Therefore, $$\int_{C_R} |f(z)|\;|dz| = \int_{C_R} \frac{|dz|}{|z^2+a^2|} \le \int_{C_R} \frac{|dz|}{R^2-a^2} = \frac{\pi R}{R^2-a^2} \to 0 \text{ as } R\to\infty.$$

Hence, as $R\to\infty$, we obtain

$$\int_{-\infty}^{\infty} \frac{e^{imx}}{a^2+x^2}\,dx = 2\pi i\Big[\operatorname*{Res}_{z=ai} e^{imz} f(z)\Big] = 2\pi i\left(\frac{e^{-am}}{2ai}\right) = \frac{\pi}{a}e^{-am}.$$

Therefore, $$\int_0^{\infty} \frac{\cos mx}{a^2+x^2}\,dx,\quad m>0 = \frac{1}{2}\int_{-\infty}^{\infty} \frac{\cos mx}{a^2+x^2}\,dx = \frac{\pi}{2a}e^{-am}.$$

(*ii*) Since the integrand is an even function of x, we can write

$$\int_0^{\infty} \frac{x\sin x}{x^2+9}\,dx = \frac{1}{2}\int_{-\infty}^{\infty} \frac{x\sin x}{x^2+9}\,dx = \frac{1}{2}\,\mathrm{Im}\int_{-\infty}^{\infty} \frac{xe^{ix}}{x^2+9}\,dx.$$

Now, consider the contour integral

$$I = \int_C f(z)\,e^{iz}\,dz, \text{ where } f(z) = \frac{z}{z^2+9}$$

and C be the closed contour consisting of real axis from $-R$ to R and semicircle $C_R : |z| = R$ of sufficiently large radius R in upper half plane shown in Fig. 4.41. The function $f(z)$ is analytic in the upper half plane except for the simple pole at $z = 3i$. We now write

$$\int_C \frac{z}{z^2+9}e^{iz}\,dz = \int_{C_R} \frac{ze^{iz}}{z^2+9}\,dz + \int_{-R}^{R} \frac{x}{x^2+9}e^{ix}\,dx = 2\pi i\Big[\operatorname*{Res}_{z=3i} f(z)\,e^{iz}\Big]$$

$$= 2\pi i \lim_{z=3i}\left[\frac{(z-3i)\,z\cdot e^{-3}}{(z+3i)(z-3i)}\right] = 2\pi i\cdot\frac{3i}{6i}e^{-3} = \frac{\pi i}{e^3}.$$

Now, $$\left|\int_{C_R} \frac{ze^{iz}}{z^2+9}\,dz\right| \le \frac{R(\pi R)}{R^2-9} \to 0 \text{ as } R\to\infty,$$ since $|e^{iz}| = e^{-y} \le 1$ and $|z^2+9| \ge R^2-9.$

Hence, as $R\to\infty$, we obtain $I = \displaystyle\int_{-\infty}^{\infty} f(x)\,e^{ix}\,dx = \int_{-\infty}^{\infty} \frac{xe^{ix}}{x^2+9}\,dx = \frac{\pi}{e^3}i.$

Therefore, $\displaystyle\int_0^{\infty} \frac{x \sin x}{x^2+9}\, dx = \frac{1}{2} \text{Im} \int_{-\infty}^{\infty} \frac{x e^{ix}}{x^2+9}\, dx = \frac{\pi}{2e^3}.$

(*iii*) We have $\quad \sin^2 2x = \dfrac{1-\cos 4x}{2} = -\dfrac{1}{2} \text{Re}\,[e^{4ix} - 1].$

Therefore, $\displaystyle\int_{-\infty}^{\infty} \frac{\sin^2 2x}{1+x^2}\, dx = -\frac{1}{2} \text{Re} \int_{-\infty}^{\infty} \frac{(e^{4ix}-1)}{1+x^2}\, dx.$

Consider the contour integral

$$I = \int_C \frac{e^{4iz}-1}{1+z^2}\, dz = \int_C (e^{4iz}-1)\, f(z)\, dz, \text{ where } f(z) = \frac{1}{1+z^2}$$

and C be the closed path consisting of real axis from $-R$ to R and semicircle $C_R : |z| = R$ of sufficiently large radius R in upper half plane as shown in Fig. 4.41. The function $f(z)$ is analytic in the upper half plane except for the simple pole at $z = i$. We find that

$$\underset{z=i}{\text{Res}}\,(e^{4iz}-1)\, f(z) = \lim_{z \to i} \left[\frac{(z-i)(e^{4iz}-1)}{(z+i)(z-i)}\right] = \lim_{z \to i} \left[\frac{e^{4iz}-1}{z+i}\right] = \frac{e^{-4}-1}{2i}.$$

We now write $\displaystyle I = \int_{C_R} \frac{e^{4iz}-1}{1+z^2}\, dz + \int_{-R}^{R} \frac{e^{4ix}-1}{1+x^2}\, dx = 2\pi i \left[\underset{z=i}{\text{Res}} \left(\frac{e^{4iz}-1}{1+z^2}\right)\right] = \pi\,(e^{-4}-1).$

Now, $\displaystyle\left|\int_{C_R} \frac{e^{4iz}-1}{1+z^2}\, dz\right| \le \int_{C_R} \frac{|e^{4iz}-1|}{|z^2+1|} |dz| \le \frac{2(\pi R)}{R^2-1} \to 0 \text{ as } R \to \infty,$

since $\quad |e^{4iz}-1| \le |e^{4iz}| + 1 = e^{-4y} + 1 \le 2$ and $|z^2+1| \ge |z|^2 - 1 = R^2 - 1.$

Hence, as $R \to \infty$, we obtain $\displaystyle\int_{-\infty}^{\infty} \frac{e^{4ix}-1}{1+x^2}\, dx = \pi\,(e^{-4}-1).$

Therefore, $\displaystyle\int_{-\infty}^{\infty} \frac{\sin^2 2x}{1+x^2}\, dx = -\frac{1}{2} \text{Re} \int_{-\infty}^{\infty} \frac{(e^{4ix}-1)}{1+x^2}\, dx = \frac{\pi}{2}(1-e^{-4}).$

(*iv*) Since the integrand is an even function, we write

$$\int_0^{\infty} \frac{x \sin x}{(x^2+a^2)^2}\, dx = \frac{1}{2} \int_{-\infty}^{\infty} \frac{x \sin x}{(x^2+a^2)^2}\, dx = \frac{1}{2} \text{Im} \int_{-\infty}^{\infty} \frac{x\, e^{ix}}{(x^2+a^2)^2}\, dx.$$

Now, consider the contour integral

$$I = \int_C f(z)\, e^{iz}\, dz, \text{ where } f(z) = \frac{z}{(z^2+a^2)^2},$$

and C is the closed path consisting of real axis from $-R$ to R and semicircle $C_R : |z| = R$ of sufficiently large radius R in upper half plane as shown in Fig. 4.41. The function $f(z)$ is analytic in the upper half plane except for the pole of order 2 at $z = ai$. We find that

$$\underset{z=ai}{\text{Res}}\,[f(z)\, e^{iz}] = \frac{1}{1!} \lim_{z \to ai} \frac{d}{dz}\left[\frac{(z-ai)^2 \cdot z e^{iz}}{(z+ai)^2 (z-ai)^2}\right] = \lim_{z \to ai} \frac{d}{dz}\left[\frac{z e^{iz}}{(z+ai)^2}\right]$$

$$= \lim_{z \to ai} \left[\frac{(e^{iz} + ize^{iz})(z+ai)^2 - 2ze^{iz}(z+ai)}{(z+ai)^4} \right] = \frac{e^{-a}}{4a}.$$

We now write $I = \int_{C_R} f(z)\, e^{iz} dz + \int_{-R}^{R} f(x)\, e^{ix} dx = 2\pi i \left[\operatorname*{Res}_{z=ai} (f(z)\, e^{iz}) \right] = \dfrac{\pi i e^{-a}}{2a}.$

Now, $$\left| \int_{C_R} \frac{ze^{iz}}{(z^2+a^2)^2} dz \right| \le \frac{R(\pi R)}{(R^2-a^2)^2} \to 0 \text{ as } R \to \infty.$$

since $|e^{iz}| = e^{-y} \le 1$ and $|z^2 + a^2| \ge R^2 - a^2$.

Hence, as $R \to \infty$, we obtain $I = \int_{-\infty}^{\infty} f(x)\, e^{ix} dx = \int_{-\infty}^{\infty} \dfrac{x\, e^{ix}}{(x^2+a^2)^2} dx = \dfrac{\pi i e^{-a}}{2a}.$

Therefore, $$\int_0^{\infty} \frac{x \sin x}{(x^2+a^2)^2} dx = \frac{1}{2} \operatorname{Im} \int_{-\infty}^{\infty} \frac{xe^{ix}}{(x^2+a^2)^2} dx = \frac{\pi e^{-a}}{4a}.$$

EXERCISE 4.8

Use contour integration to evaluate each of the following definite integrals.

1. $\int_0^{2\pi} \dfrac{d\theta}{5 + 3\sin\theta}$,
2. $\int_0^{2\pi} \dfrac{d\theta}{5 - 3\cos\theta}$,
3. $\int_0^{2\pi} \dfrac{d\theta}{(2 + \cos\theta)^2}$,
4. $\int_0^{2\pi} \dfrac{d\theta}{2 + \sin\theta}$,
5. $\int_0^{2\pi} \dfrac{d\theta}{1 - 2a\cos\theta + a^2}$ $(0 < a < 1)$, *(Kerala 2005)*
6. $\int_0^{2\pi} \dfrac{d\theta}{(5 - 3\sin\theta)^2}$,
7. $\int_0^{2\pi} \dfrac{\sin^2\theta}{5 + 4\cos\theta} d\theta$,
8. $\int_0^{2\pi} \dfrac{\cos 2\theta}{5 + 4\cos\theta} d\theta$,
9. $\int_0^{2\pi} \dfrac{d\theta}{(1 + \sin^2\theta)^2}$,
10. $\int_0^{\pi} \dfrac{\sin^2\theta}{5 - 4\cos\theta} d\theta$,
11. $\int_0^{\pi} \dfrac{a}{a^2 + \sin^2\theta} d\theta$, $a > 0$, *(UPTU, 2002)*
12. $\int_0^{\pi} \dfrac{d\theta}{2 - \cos\theta}$,
13. $\int_0^{2\pi} \dfrac{d\theta}{(5 - 3\cos\theta)^2}$,
14. $\int_0^{2\pi} e^{\cos\theta} \cos(n\theta - \sin\theta)\, d\theta$,
15. $\int_0^{2\pi} e^{\cos\theta} \sin(n\theta - \sin\theta)\, d\theta$.

Ans. 1. $\pi/2$, 2. $\pi/2$, 3. $4\pi\sqrt{3}/9$, 4. $2\pi/\sqrt{3}$, 5. $2\pi/(1-a)^2$, 6. $3\pi/32$, 7. $\pi/4$, 8. $\pi/6$, 9. $3\pi/2\sqrt{2}$, 10. $\pi/8$, 11. $\pi/\sqrt{a^2+1}$, 12. $2\pi/\sqrt{3}$, 13. $5\pi/32$, 14. $2\pi/n!$, 15. 0.

[**Hint:** Problems 14 and 15.

Let $I = \int_0^{2\pi} e^{\cos\theta} \bullet e^{-(n\theta - \sin\theta)i} d\theta = \int_0^{2\pi} e^{\cos\theta + i\sin\theta} \bullet e^{-in\theta} d\theta = \int_0^{2\pi} e^{i\theta} \bullet e^{-in\theta} d\theta = \frac{1}{i}\oint_C \frac{e^z}{z^{n+1}} dz = \oint_C f(z)\, dz.$

Setting $z = e^{i\theta}$, $dz = ie^{i\theta}\, d\theta = izd\theta$, where $f(z) = e^z/iz^{n+1}$ and C is the unit circle $|z| = 1$. $f(z)$ has a pole of order $(n + 1)$ at the origin.

$$\operatorname*{Res}_{z=0} f(z) = \frac{1}{in!}\left[\frac{d^n}{dz^n}(e^z)\right]_{z=0} = \frac{1}{in!}. \text{ Hence } I = 2\pi i \bullet \frac{1}{in!} = \frac{2\pi}{n!}.$$

Equating real and imaginary parts, we obtain

$$\int_0^{2\pi} e^{\cos\theta} \cos(n\theta - \sin\theta)\, d\theta = \frac{2\pi}{n!} \text{ and } \int_0^{2\pi} e^{\cos\theta} \sin(n\theta - \sin\theta)\, d\theta = 0.]$$

Use contour integration to verify each of the following integrals.

16. $\int_0^{\infty} \frac{dx}{1+x^2} = \frac{\pi}{2},$

17. $\int_0^{\infty} \frac{dx}{1+x^4} = \frac{\pi}{2\sqrt{2}}.$

18. $\int_{-\infty}^{\infty} \frac{dx}{(x^2+1)(x^2+9)} = \frac{\pi}{12},$

19. $\int_0^{\infty} \frac{dx}{1+x^6} = \frac{\pi}{3},$

20. $\int_0^{\infty} \frac{dx}{x^4+x^2+1} = \frac{\pi\sqrt{3}}{6},$

21. $\int_{-\infty}^{\infty} \frac{x^2 dx}{(x^2+1)(x^2+4)} = \frac{\pi}{3},$

22. $\int_{-\infty}^{\infty} \frac{dx}{(1+x^2)^2} = \frac{3\pi}{8},$

23. $\int_0^{\infty} \frac{x^2 dx}{(x^2+9)(x^2+4)} = \frac{\pi}{200},$

24. $\int_{-\infty}^{\infty} \frac{x^2 dx}{(x^2+a^2)^3},\ a > 0 = \frac{\pi}{8a^3}.$

(AMIETE, June 2005)

Use the method of contour integration to prove that

25. $\int_{-\infty}^{\infty} \frac{\sin x}{x^2+4x+5} dx = \frac{-\pi}{e} \sin 2.$

26. $\int_0^{\infty} \frac{\cos mx}{1+x^2} dx \quad m > 0 = \frac{\pi}{2} e^{-m}.$

27. $\int_{-\infty}^{\infty} \frac{\cos x\, dx}{(x^2+a^2)(x^2+b^2)} = \frac{\pi}{a^2-b^2}\left(\frac{e^{-b}}{b} - \frac{e^{-a}}{a}\right) \quad (a > b > 0).$

28. $\int_0^{\infty} \frac{x^2 \sin x}{(x^2+a^2)^2} dx \quad (a > 0) = -\frac{\pi}{4}(a-2)e^{-a}.$ *(AMIETE, Dec. 2002)*

29. $\int_0^{\infty} \frac{x^3 \sin x}{(x^2+a^2)(x^2+b^2)} dx \quad (a > 0,\ b > 0) = \frac{\pi}{2(a^2-b^2)}(a^2 e^{-a} - b^2 e^{-b}).$ *(AMIETE, June 2001)*

BILINEAR TRANSFORMATIONS AND CONFORMAL MAPPING

4.23 INTRODUCTION

We know that the real function $y = \phi(x)$ can be represented graphically by a curve in the cartesian xy-plane. Similarly the real function $z = F(x, y)$ can be exhibited by a surface in space of three dimensions. This mode of graphical representation fails in the case of complex functions because a complex function $w = f(z)$, i.e.,

$u + iv = f(x + iy)$ involves four real variables, two independent variables x and y and two dependent variables u and v, and we need a space of four dimensions to represent it graphically in the cartesion fashion.

Although we cannot draw a graph of a complex function $w = f(z)$, there is, however, another way in which the relationship $w = f(z)$ can be interpreted as a mapping or transformation from the z-plane to the w-plane. Choose two complex planes and call them z-plane and w-plane. In the z-plane plot the point $z = x + iy$ and in the w-plane plot the corresponding point $w = u + iv$. Thus the function $w = f(z)$ defines a correspondence between points of these two planes. For any point P in the z-plane, $w = f(z)$ determines a point Q in the w-plane. We say that the point P *maps* into Q or Q is the *image* of P.

If P describes a curve C in the z-plane, Q moves along some curve C_1, called the image of C, in the w-plane. Similalry different regions in the z-plane will be mapped into regions in the w-plane. *The function $w = f(z)$ thus defines a mappling or transformation of the z-plane into the w-plane or is represented geometrically by this transformation.*

If we separate $w = f(z)$ into real and imaginary parts we get

$$u = u(x, y)$$

$$v = v(x, y)$$

which are the **equations of transformation** that map a given set of points in the z-plane onto another set of points in the w-plane.

We define a few important transformations.

Linear Transformation

A *mapping* or a transformation of the form

$$w = az + b, \qquad ...(4.110)$$

where a, b are real or complex constants, is called a *linear mapping* or *linear transformation.* We write the above linear mapping in the form $w = |a| e^{i\phi} z + b$ and consider it as the composition of the following mappings:

(*i*) $w_1 = |a| z$, (*ii*) $w_2 = e^{i\phi} w_1$ and (*iii*) $w = w_2 + b$.

we discuss each of these mapping separately.

Translation Mapping w = z + c

Let $z = x + iy$, $c = c_1 + ic_2$ and $w = u + iv$. Then, we get

$$u + iv = (x + iy) + (c_1 + ic_2) = (x + c_1) + i(y + c_2)$$

or $u = x + c_1$ and $v = y + c_2$.

Therefore, the point (x, y) in the z-plane is mapped as the point $(u, v) = (x + c_1, y + c_2)$ in the w-plane. As a result of the transformation figure in z-plane is translated through the vector $c_1 + ic_2$ with their shape and sizes unchanged.

Magnification/Contraction Mapping w = kz, k being real and positive constant

We write $z = re^{i\theta}$. Now $w = kre^{i\theta}$, $k > 0$.

Therefore $|w| = kr$ and $\arg(w) = \theta$.

Under this mapping, a point in the z-plane is moved radially to a new position at a distance kr from the origin. Thus, a vector in the z-plane is magnified (stretched) by a factor k, if $k > 1$ and contracted by a factor k, if $k < 1$.

Rotation Mapping $w = e^{i\phi}z$

We write $z = re^{i\theta}$, and obtain $w = re^{i(\theta + \phi)}$, $|w| = r$ and $\arg(w) = \theta + \phi$.

Under this mapping, a point $z(r, \theta)$ in the z-plane is mapped as a point $w(r, \theta + \phi)$ in the w-plane. Therfore, this mapping rotates a region in the z-plane through an angle ϕ in the w-plane in the anticlockwise direction if $\phi > 0$ and in the clockwise direction if $\phi < 0$.

EXAMPLE 4.60 Let a rectangular region *ABCD* with vertices at *A* (1, 1), *B* (3, 1), *C* (3, 3), *D* (1, 3) be defined in the z-plane. Find the image of this region in the *w*-plane, under the mapping $w = z + (1 + 2i)$.

Solution Writing $z = x + iy$ and $w = u + iv$, we get $u + iv = (x + iy) + (1 + 2i) = (x + 1) + i(y + 2)$.

Comparing real and imaginary parts, we obtain $u = x + 1$ and $v = y + 2$.

The point (x, y) in the *z*-plane is mapped as the point $(u, v) = (x + 1, y + 2)$ in the *w*-plane. Therefore, the point

A (1, 1) is mapped as *A′* (2, 3), *B*(3, 1) is mapped as *B′* (4, 3),

C (3, 3) is mapped as *C′* (4, 5), *D*(1, 3) is mapped as *D′* (2, 5).

Thus, the image of the rectangular region *ABCD* in the *z*-plane is the rectangular region *A′ B′ C′ D′* in the w-plane. (Fig. 4.42).

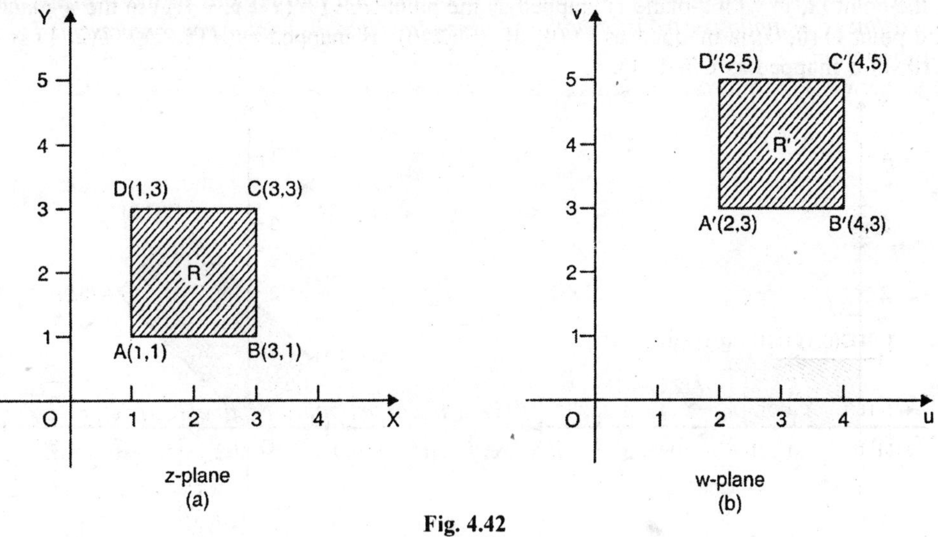

Fig. 4.42

EXAMPLE 4.61 Let a triangular region *OAB* with vertices at *O* (0, 0), *A* (1, 0), *B* (0, 1), be defined in the z-plane. Find the image of this region in the *w*-plane under the mapping. $w = 3z$.

Solution Writing $z = x + iy$ and $w = u + iv$, we get $u = 3x$ amd $v = 3y$. The point (x, y) in the z-plane is mapped as the point $(u, v) = (3x, 3y)$ in the *w*-plane. Therefore, the point *O* (0, 0) is mapped as *O* (0, 0) *A* (1, 0) is mapped as *A′*(3, 0) *B* (0, 1) is mapped as *B′*(0, 3).

Thus, the image of the triangular region *OAB*, in the *z*-plane is another triangular region *OA′B′* in the *w*-plane Fig. 4.43.

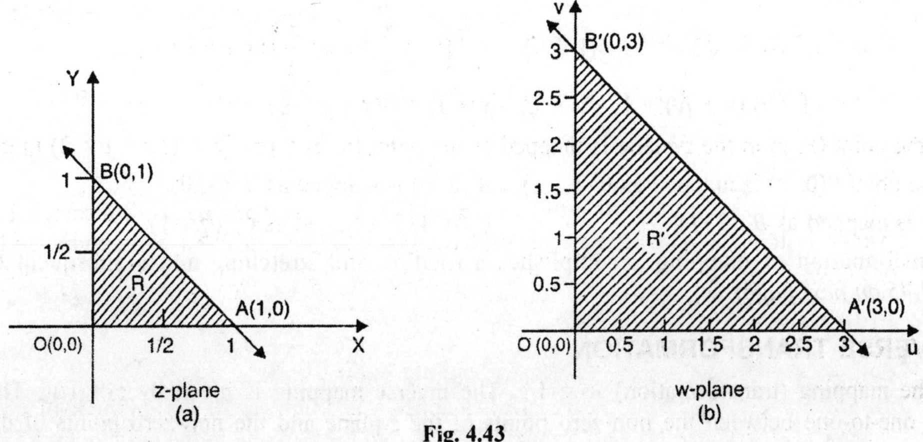

Fig. 4.43

Remark: The region is magnified in both *u* and *v* directions.

EXAMPLE 4.62 Let a rectangular region $OABC$ with vertices at $O(0, 0)$, $A(2, 0)$, $B(2, 1)$, $C(0, 1)$ be defined in the z-plane. Find the image of this region in the w-plane under the mapping $w = \sqrt{2}\, e^{\pi i/4} z$.

Solution Writing $z = re^{i\theta}$, we obtain $w = \sqrt{2}\, re^{i(\theta + \frac{\pi}{4})}$.

Therefore, any point in the z-plane is rotated through an angle $\pi/4$ in the w-plane. Writing $z = x + iy$, $w = u + iv$, we get

$$w = u + iv = \sqrt{2}\left(\cos\frac{\pi}{4} + i\sin\frac{\pi}{4}\right)(x + iy) = (1 + i)(x + iy) = x - y + i(x + y).$$

Hence, the point (x, y) with z-plane is mapped as the point $(u, v) = (x - y, x + y)$ in the w-plane.

Thus the point O (0, 0) is mapped as O (0, 0), $A(2, 0)$ is mapped as $A'(2, 2)$, $B(2, 1)$ is mapped as $B'(1, 3)$ $C(0, 1)$ is mapped as $C'(-1, 1)$.

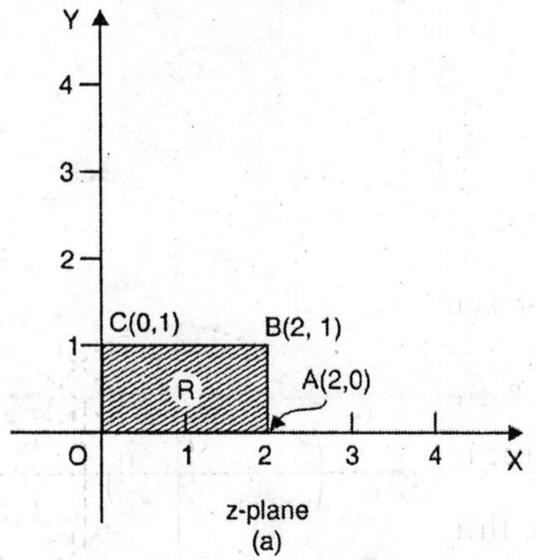

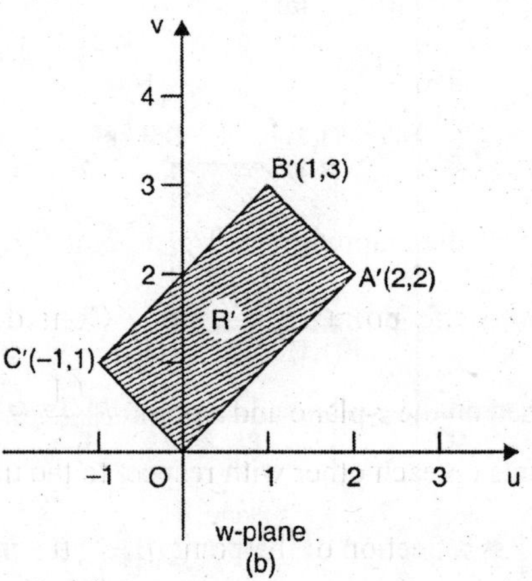

Fig. 4.44

The rectangular region $OABC$ in the z-plane is rotated by an angle $\pi/4$ and is mapped onto another rectangular region $O\,A'B'C'$ in the w-plane. The lengths of the sides of rectangular $O\,A'B'C'$ are stretched of magnitude $\sqrt{2}$.

EXAMPLE 4.63 Let a rectangular region $OABC$ with vertices at $O(0, 0)$, $A(2, 0)$, $B(2, 1)$, $C(0, 1)$ be defined in the z-plane. Find the image of the region in the w-plane under the mapping $w = \sqrt{2}\, e^{\pi i/4} z + (1 - 2i)$.

Solution Writing $z = x + iy$ and $w = u + iv$, we get

$$w = u + iv = \sqrt{2}\, e^{\pi i/4} z + (1 - 2i) = \sqrt{2}\left(\cos\frac{\pi}{4} + i\sin\frac{\pi}{4}\right)(x + iy) + 1 - 2i$$

$$= (1 + i)(x + iy) + 1 - 2i = (x - y + 1) + i(x + y - 2).$$

Hence, the point (x, y) in the z-plane is mapped as the point $(u, v) = (x - y + 1, x + y - 2)$ in the w-plane.

Thus, the point 0(0, 0) is mapped as $0'(1, -2)$, $A(2, 0)$ is mapped as $A'(3, 0)$,

$B(2, 1)$ is mapped as $B'(2, 1)$, $C(0, 1)$ is mapped as $C'(0, -1)$.

The transformation or mapping accomplishes a rotation and stretching and a subsquent translation. (*See Fig. 4.45 on next page*).

4.24 INVERSE TRANSFORMATION

Consider the mapping (transformation) $w = 1/z$. The inverse mapping is given by $z = 1/w$. The mapping $w = 1/z$ is one-to-one between the non-zero points of the z-plane and the non-zero points of the w-plane. Substituting $z = re^{i\theta}$ and $w = \rho e^{i\phi}$ in the mapping $w = 1/z$, we obtain $\rho e^{i\phi} = (1/r)e^{-i\theta}$. Comparing the real and imaginary parts, we obtain $\rho = 1/r$ and $\phi = -\theta$.

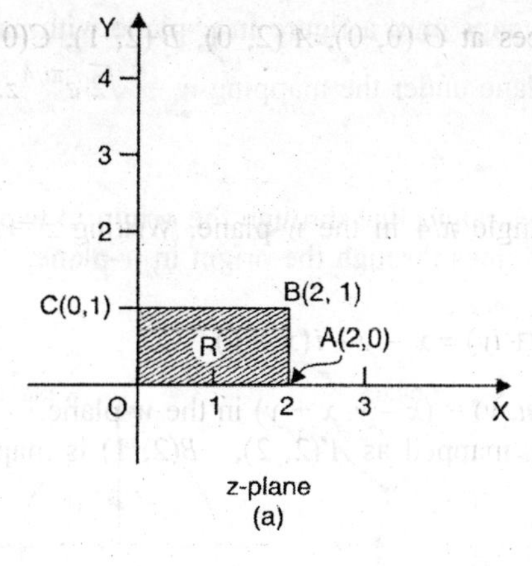

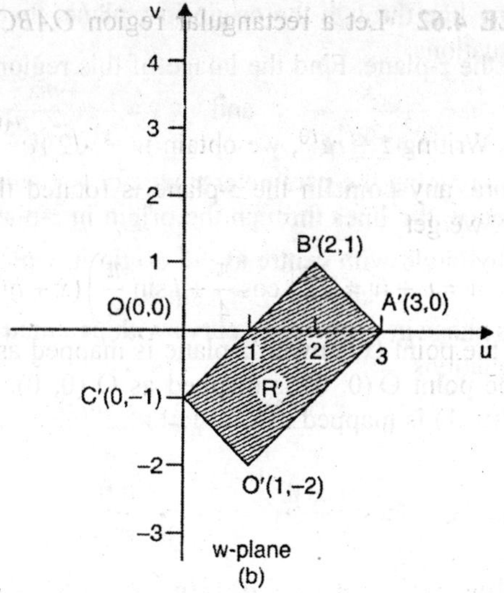

Fig. 4.45

Thus, under the mapping $w = 1/z$, a point $P(r, \theta)$ in the z-plane is mapped onto the point $P'\left(\dfrac{1}{r}, -\theta\right)$. Considering the w-plane superimposed on the z-plane and a point $P_1\left(\dfrac{1}{r}, \theta\right)$, then P and P_1 are inverse points on each other with respect to the unit circle. The point $P'\left(\dfrac{1}{r}, -\theta\right)$ is reflection of the point $P_1\left(\dfrac{1}{r}, \theta\right)$ in the real axis. Thus the transformation $w = 1/z$, maps a point $P(z) = P(r, \theta)$ into its inverse point with respect to the circle $|z| = 1$ followed by the reflection of this inverse point in the real axis.

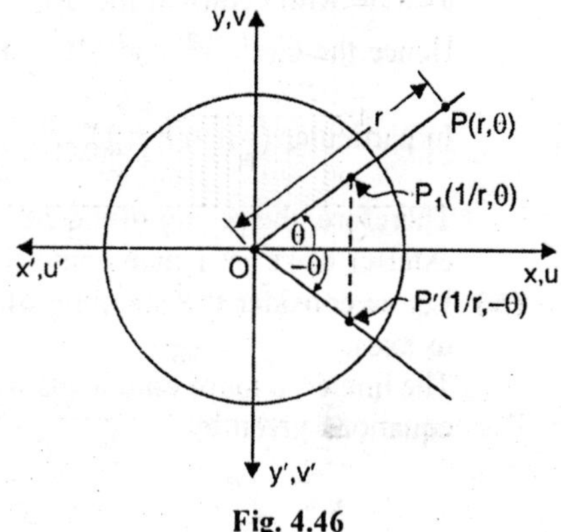

Fig. 4.46

Obviously points exterior to the unit circle $|z| = 1$ are mapped onto the non-zero points interior to the unit circle $|w| = 1$ under the transformation $w = 1/z$.

Hence, the transformation $w = 1/z$ maps interior of the unit circle $|z| = 1$ onto the exterior of unit circle $|w| = 1$ and vice versa. However, the point O, that is, $z = 0$ is mapped to the point $w = \infty$, called the point at infinity and the point $z = \infty$ in the z-plane is mapped onto $w = 0$.

EXAMPLE 4.64 Discuss the way in which the z-plane is mapped onto the w-plane by the function $w = 1/z$.

Solution We have $w = 1/z$ or $u + iv = \dfrac{1}{x+iy} = \dfrac{1}{x+iy} \cdot \dfrac{x-iy}{x-iy} = \dfrac{x-iy}{x^2+y^2}$.

Comparing real and imaginary parts, we obtain

$$u = \frac{x}{x^2+y^2} \quad \text{and} \quad v = \frac{-y}{x^2+y^2}. \qquad \text{...(1)}$$

With the help of these equations many features of transformation can be inferred.

(*i*) When y is +ve, v is –ve and when y is –ve, v is +ve. Hence, the z-plane above x-axis is mapped onto the half w-plane below u-axis and the lower half of z-plane maps into upper half of w-plane.

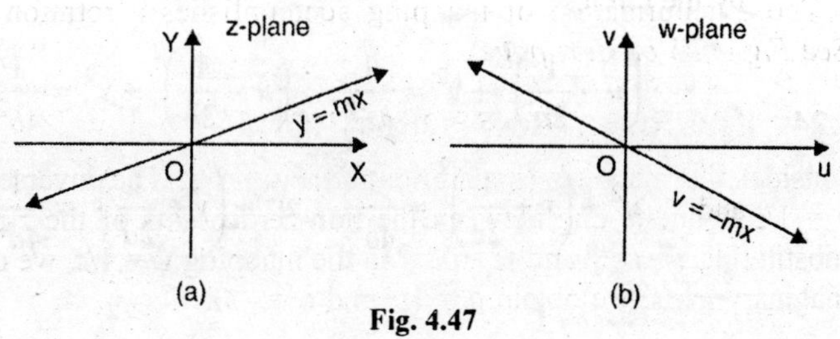

Fig. 4.47

(*ii*) Any line through the origin in z-plane, that is, $y = mx$ maps onto a figure in w-plane with parametric equations

$$u = \frac{x}{x^2 + m^2x^2} \quad \text{and} \quad v = \frac{-mx}{x^2 + m^2x^2}.$$

Eliminating the parameter x we get $v = -mu$ which is a straight line through the origin in w-plane. Hence, the lines through the origin in z-plane map onto lines through the origin in w-plane.

(*iii*) Any circle with centre at the origin in z-plane is $x^2 + y^2 = r^2$,

or $\quad x = r \cos \theta, \quad y = r \sin \theta.$

Its image in w-plane is a curve with parametric equations

$$u = \frac{r \cos \theta}{r^2}, \quad v = -\frac{r \sin \theta}{r^2}$$

that is, $\quad u = \frac{\cos \theta}{r}, \quad v = -\frac{\sin \theta}{r}.$

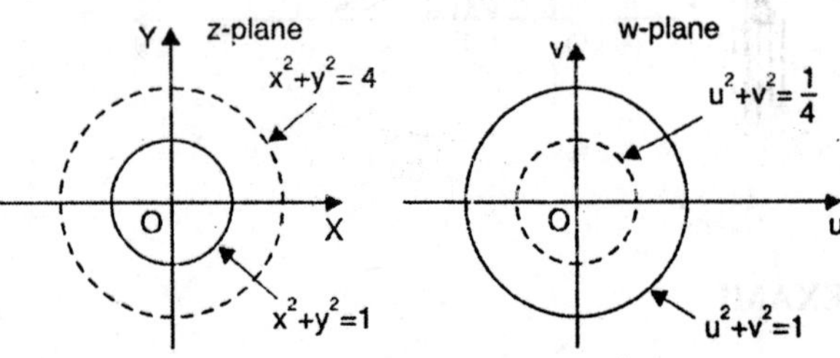

Fig. 4.48

Eliminating θ, we obtain $u^2 + v^2 = \frac{1}{r^2}$,

a circle with centre at the origin.

Hence the circle $x^2 + y^2 = r^2$ on z = plane maps onto the circle $u^2 + v^2 = (1/r^2)$ on the w-plane.

In particular (when $r = 1$), $x^2 + y^2 = 1$ maps onto the circle $u^2 + v^2 = 1$. When $r > 1$, $\frac{1}{r} < 1$.

Therefore, the region of the z-plane defined by $x^2 + y^2 > 1$ maps onto the region $u^2 + v^2 < 1$, that is, the exterior of $|z| = 1$ maps onto the interior of $|w| = 1$.

(*iv*) Let us consider the mapping of lines parallel to axes.

The line $x = a$ maps onto locus with parametric equations given by

$$u = \frac{a}{a^2 + y^2}, \quad v = \frac{-y}{a^2 + y^2}.$$

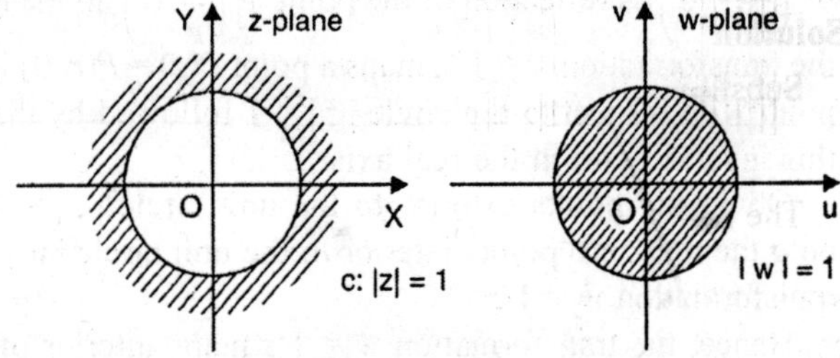

Fig. 4.49

Eliminating y we get

$$u^2 + v^2 = \frac{1}{a^2 + y^2} = \frac{u}{a}$$

or $\quad u^2 + v^2 - \frac{u}{a} = 0 \quad$ or $\quad \left(u - \frac{1}{2a}\right)^2 + v^2 = \frac{1}{4a^2}$, a circle which passes through the origin and has centre on u-axis.

Similarly we can show that lines parallel to real axis (x-axis) in z-plane map onto circles in w-plane which pass through the origin and have centres on v-axis.

The rectangle bounded by $x = a$, $x = b$ and $y = c$, $y = d$ is mapped onto the region in w-plane bounded by the circles

$$\left(u - \frac{1}{2a}\right)^2 + v^2 = \frac{1}{4a^2}, \quad \left(u - \frac{1}{2b}\right)^2 + v^2 = \frac{1}{4b^2},$$

and $\quad u^2 + \left(v + \frac{1}{2c}\right)^2 = \frac{1}{4c^2}, \quad u^2 + \left(v + \frac{1}{2d}\right)^2 = \frac{1}{4d^2}.$ (See Fig. 4.50)

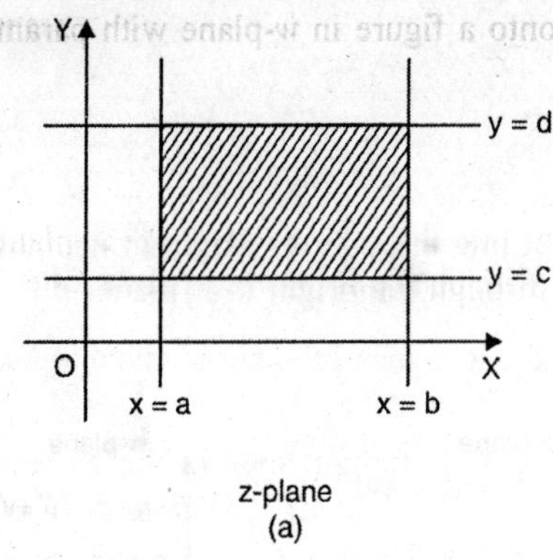

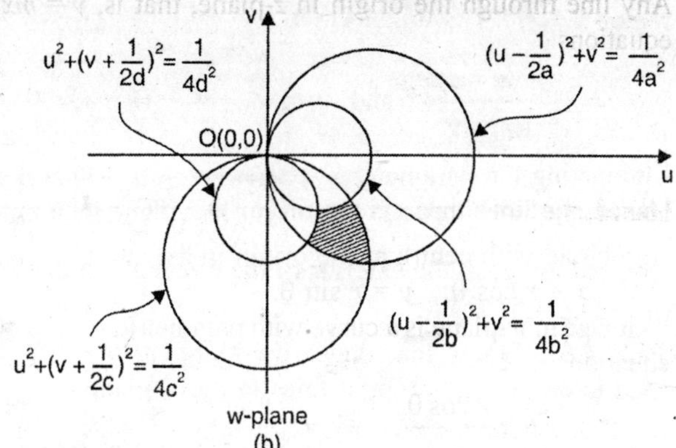

Fig. 4.50

EXAMPLE 4.65 Find the image in the w-plane of the disk $|z - 1| \le 1$, under the mapping $w = 1/z$.
(*AMIETE, June 2012, Dec. 2009*)

Solution We have $w = 1/z$ or $z = 1/w$. The given region $|z - 1| \le 1$ is mapped as $\left|\dfrac{1}{w} - 1\right| \le 1$ or $|1 - w| \le |w|$.

Let $w = u + iv$. Then, we obtain $|(1 - u) - iv| \le |u + iv|$

or $\qquad (1 - u)^2 + v^2 \le u^2 + v^2 \qquad$ or $\qquad 1 - 2u \le 0 \quad$ or $\quad u \ge (1/2)$.

Thus, the image of the region $|z - 1| \le 1$ under the mapping $w = 1/z$ is the half plane $u \ge 1/2$ in the w-plane. The bounding circle $|z - 1| = 1$ is mapped as the line $u = 1/2$.

EXAMPLE 4.66 Find the image of the region bounded by the lines $x - y < 1$ and $x + y > 1$ under the mapping $w = 1/z$.

Solution The given region is shown in Fig. 4.51 (a).

Substituting $z = x + iy$ and $w = u + iv$ in $w = 1/z$ and comparing the real and imaginary parts, we obtain

$$x = u/(u^2 + v^2) \quad \text{and} \quad y = -v/(u^2 + v^2).$$

The region $x - y < 1$ is transformed as

$$\frac{u+v}{u^2+v^2} < 1, \quad \text{or} \quad u^2 + v^2 - (u+v) > 0 \quad \text{or} \quad \left(u - \frac{1}{2}\right)^2 + \left(v - \frac{1}{2}\right)^2 > \frac{1}{2}.$$

The boundary of this region is the circle $\left|w - \dfrac{1}{2}(1+i)\right| = \dfrac{1}{\sqrt{2}}$ ...(*i*) with centre $\left(\dfrac{1}{2}, \dfrac{1}{2}\right)$ and radius $\dfrac{1}{\sqrt{2}}$.

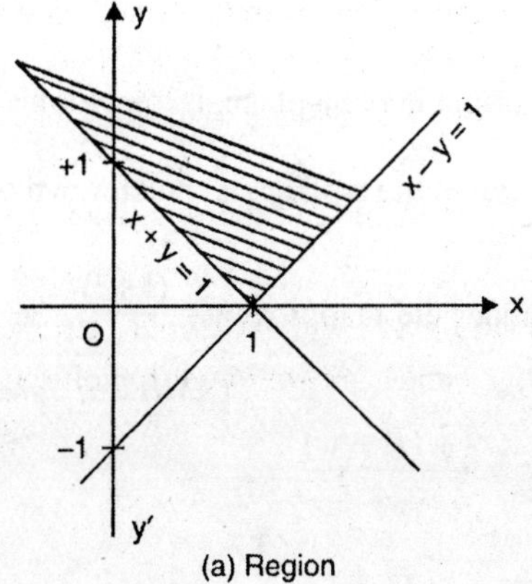

(a) Region

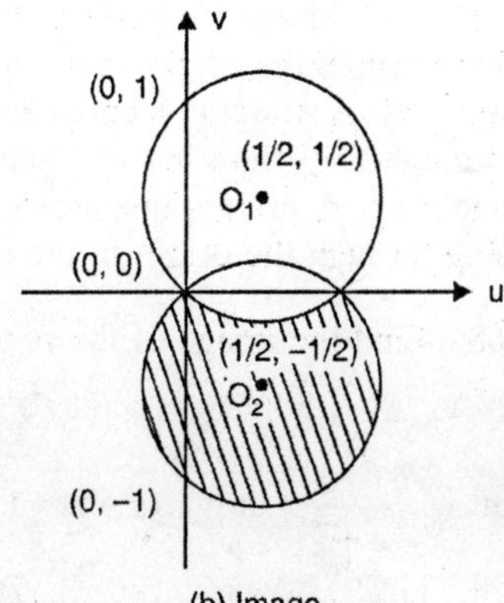

(b) Image

Fig. 4.51

The region $x + y > 1$ is transformed to

$$\frac{u-v}{u^2+v^2} > 1, \quad \text{or} \quad u^2+v^2-(u-v)<0 \quad \text{or} \quad \left(u-\frac{1}{2}\right)^2+\left(v+\frac{1}{2}\right)^2 < \frac{1}{2}.$$

The boundary of this region is the circle $\left|w-\frac{1}{2}(1-i)\right| = \frac{1}{\sqrt{2}}$ ...(*ii*) with centre at $\left(\frac{1}{2}, -\frac{1}{2}\right)$ and radius $\frac{1}{\sqrt{2}}$.

The required image in the exterior of the circle given in eq. (*i*) and the interior of the circle given in Eq. (*ii*) as shown in Fig. 4.51 (b).

EXAMPLE 4.67 Show that under the mapping $w = 1/z$, all circles and straight lines in the z-plane are transformed to circles and straight lines in the w-plane. (*AMIETE, June 2011, 2008*)

Solution The equation $a(x^2 + y^2) + gx + fy + c = 0$ ...(*i*) represents a circle if $a \neq 0$ and a straight line if $a = 0$, in the z-plane. Substituting $z = x + iy$ and $w = u + iv$ in $w = 1/z$, we get

$$x+iy = \frac{1}{u+iv} = \frac{1}{u+iv} \times \frac{u-iv}{u-iv} = \frac{u-iv}{u^2+v^2}.$$

Comparing the real and imaginary parts, we obtain $x = \frac{u}{u^2+v^2}$ and $y = \frac{-v}{u^2+v^2}$.

Substituting in Eq. (*i*), we obtain

$$\frac{a}{u^2+v^2} + \frac{gu}{u^2+v^2} - \frac{fv}{u^2+v^2} + c = 0 \quad \text{or} \quad a + gu - fv + c\,(u^2+v^2) = 0. \qquad ...(ii)$$

If $c \neq 0$, eq. (*ii*) can be written as

$$u^2 + \frac{g}{c}u + v^2 - \frac{f}{c}v + \frac{a}{c} = 0 \quad \text{or} \quad \left(u+\frac{g}{2c}\right)^2 + \left(v-\frac{f}{2c}\right)^2 = \frac{g^2+f^2-4ac}{4c^2}$$

which is the equation of a circle with centre at $\left(\frac{-g}{2c}, \frac{f}{2c}\right)$ and radius $= \sqrt{(g^2+f^2-4ac)}/2c$.

If $c = 0$, we obtain from eq. (*ii*) $gu - fv + a = 0$, which is the equation of a straight line. We observe the following:

(1) If $a \neq 0$ and $c \neq 0$. A circle with centre at $\left(\frac{-g}{2a}, \frac{-f}{2a}\right)$ and radius $= \sqrt{(g^2+f^2-4ac)}/2a$ in the z-plane is transformed onto a circle with centre at $\left(\frac{-g}{2c}, \frac{f}{2c}\right)$ and radius $= \sqrt{(g^2+f^2-4ac)}/2c$ in the w-plane but the centre of the original circle is not mapped onto the centre of the image circle and their radii are unequal.

(2) If $a \neq 0$ and $c = 0$. The circle passing through the origin in the z-plane is transformed onto a straight line not passing through the origin in the w-plane.

(3) If $a = 0$ and $c \neq 0$. A straight line not passing through the origin in the z-plane, is transformed onto a circle passing through the origin in the w-plane.

(4) If $a = 0$ and $c = 0$. A straight line passing through the origin in the z-plane, is transformed onto a straight line passing through the origin in the w-plane.

EXAMPLE 4.68 Find the image of the region $|z - i| < 2$, under the mapping $w = \frac{(1+i)}{z+i}$.

(*AMIETE, June 2011, 2009*)

Solution From $w = \frac{1+i}{z+i}$ or $w(z+i) = 1+i$, we have $z = \frac{1+i(1-w)}{w}$.

Hence, $|z-i| = \left|\frac{1+i(1-w)}{w} - i\right| = \left|\frac{1+i(1-2w)}{w}\right| < 2$ gives

$$|1 + i(1 - 2w)| < 2|w| \quad \text{or} \quad |1 + i(1 - 2u - 2iv)| < 2|u + iv|$$

or $$|(1 + 2v) + i(1 - 2u)| < 2|u + iv| \quad \text{or} \quad (1 + 2v)^2 + (1 - 2u)^2 < 4(u^2 + v^2)$$

or $$2v - 2u + 1 < 0, \quad \text{where } w = u + iv.$$

Therefore, interior of the disk $|z - i| < 2$ is mapped as the half plane $2v - 2u + 1 < 0$. (Fig. 4.52)

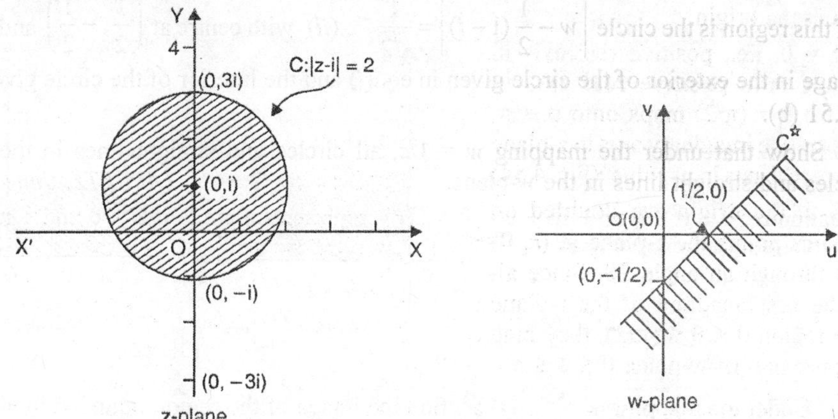

Fig. 4.52

The bounding circle C is mapped as the line $\overset{*}{C}$.

EXAMPLE 4.69 Discuss the way in which the z-plane is mapped onto the w-plane by the function $w = z^2$.

Solution We have $w = z^2$ or $u + iv = x^2 - y^2 + 2ixy$

Therefore, the equations of transformation are $u = x^2 - y^2$, $v = 2xy$.

(*i*) Any line parallel to x-axis, say $y = c$, maps onto a curve in w-plane with parametric equations

$$u = x^2 - c^2, \quad v = 2xc.$$

Eliminating x we get $v^2 = 4c^2(u + c^2)$ which is a parabola, whose vertex is at $(-c^2, 0)$ and focus at origin. Any line parallel to y-axis, say $x = a$, maps onto a curve with parametric equations $u = a^2 - y^2$, $v = 2ay$. Eliminating y we get $v^2 = -4a^2(u - a^2)$, a parabola whose vertex is at $(a^2, 0)$ and focus at origin. The rectangular region bounded by the lines $x = (1/2)$, $x = 1$ and $y = (1/2)$, $y = 1$ maps onto the region bounded by the parabolas

$$v^2 = -u + \frac{1}{4}, \; v^2 = -4(u - 1) \quad \text{and} \quad v^2 = u + \frac{1}{4}, \; v^2 = 4(u + 1) \quad \text{(See Fig. 4.53)}$$

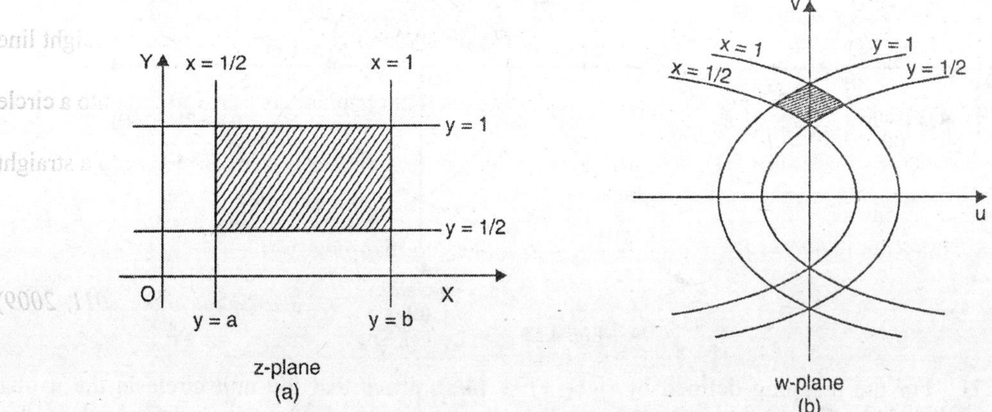

Fig. 4.53: Images of x-constant y-constant under $w = z^2$.

(*ii*) Expressing z and w in their polar forms, the mapping function $w = z^2$ becames

$$\rho e^{i\phi} = (re^{i\theta})^2 = r^2 e^{2i\theta}.$$

Comparing moduli and arguments, we obtain $\rho = r^2$ and $\phi = 2\theta$.

The circle $r = a$ maps onto $\rho = a^2$ in w-plane. Thus, the circles with centre at the origin map onto circles with centre at the origin.

(*iii*) If $\theta = 0$, $\phi = 0$, i.e., positive real axis in z-plane maps onto positive real axis in w-plane. Also $\theta = (\pi/2)$ maps onto $\phi = \pi$, that is, the positive imaginary axis in z-plane onto negative real axis in w-palne. (Fig. 4.54)

(*iv*) The angles at the origin are doubled on mapping. Thus points the z-plane at (r, θ) are rotated through an angle 2θ. Since all points in the first quadrant of the z-plane occupy the region $0 \le \theta \le (\pi/2)$, they map onto the upper half of w-plane $0 \le \phi \le \pi$.

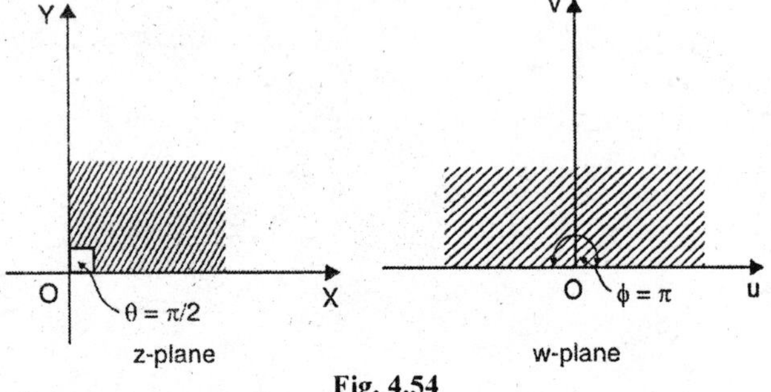

Fig. 4.54

EXAMPLE 4.70 Under the mapping $w = f(z) = z^2$, find the image of the region bounded by the lines $x = 1$, $y = 1$ and $x + y = 1$. Is the mapping conformal?

Solution $w = z^2$ is equivalent to $u + iv = (x + iy)^2 = x^2 - y^2 + 2ixy$.

Comparing the real and imaginary parts, we obtain $u = x^2 - y^2$, $v = 2xy$. The line x = 1 maps onto $u = 1 - y^2$, $v = 2y$ or $u = 1 - (v^2/4)$ or $v^2 = -4(u - 1)$, a parabola with vertex at (1, 0) and opening to left. The line $y = 1$ maps into $u = x^2 - 1$, $v = 2x$ or $v^2 = 4(u + 1)$, a parabola with vertex at (–1, 0) and opening to right. The line $x + y = 1$ maps into

$$u = x^2 - (1 - x)^2 = 2x - 1,\ v = 2x(1 - x) = 2x - 2x^2$$

or $v = (u + 1) - 2\left(\dfrac{u+1}{2}\right)^2$ or $u^2 = -2[v - 1/2]$, a parabola with vertex at (0, 1/2) and opening downward.

The regions are shown shaded in Fig. 4.55 (a) and (b), where points A, B, C map into A', B', C'. Since the point $z = 0$ at which $f'(z) = 0$ is not inside the region considered, the mapping is conformal.

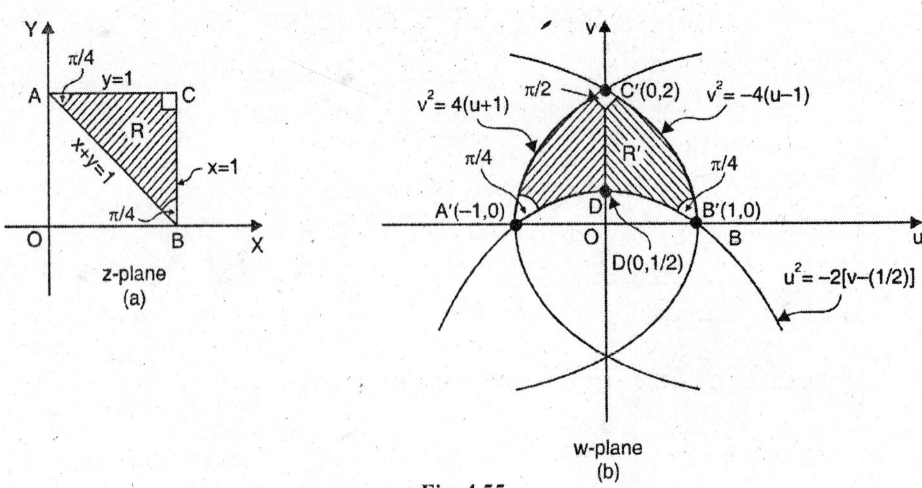

Fig. 4.55

EXAMPLE 4.71 For the mapping defined by $(w + 1)^2 = (4/z)$, prove that the unit circle in the w-plane corresponds to a parabola in the z-plane, and corresponding to any point inside the unit circle in w-plane, there is a point outside the parabola in the z-plane.

Solution (*i*) We have $w + 1 = 2/\sqrt{z}$. ...(*i*) Consider the unit circle in *w*-plane, $C : |w| = 1$.

Any point on this circle is $w = \rho e^{i\phi} = e^{i\phi} = \cos\phi + i\sin\phi$.
Substituting the value of *w* in (*i*), we obtain

$$1 + \cos\phi + i\sin\phi = \frac{2}{\sqrt{z}} = \frac{2}{\sqrt{re^{i\theta}}} = \frac{2}{\sqrt{r}}\left(\cos\frac{\theta}{2} - i\sin\frac{\theta}{2}\right).$$

Comparing the real and imaginary parts, we obtain

$$1 + \cos\phi = \frac{2}{\sqrt{r}}\cos\frac{\theta}{2} \quad \text{and} \quad \sin\phi = \frac{-2}{\sqrt{r}}\sin\frac{\theta}{2}, \text{ that is}$$

$$\cos\phi = \frac{2}{\sqrt{r}}\cos\frac{\theta}{2} - 1 \quad \text{and} \quad \sin\phi = \frac{-2}{\sqrt{r}}\sin\frac{\theta}{2}.$$

Squaring and adding, we obtain $\frac{4}{r}\cos^2\frac{\theta}{2} - \frac{4}{\sqrt{r}}\cos\frac{\theta}{2} + 1 + \frac{4}{r}\sin^2\frac{\theta}{2} = 1$

or $$\frac{4}{r} - \frac{4}{\sqrt{r}}\cos\frac{\theta}{2} = 0 \Rightarrow r\cos^2\frac{\theta}{2} = 1 \Rightarrow \frac{2}{r} = 1 + \cos\theta,$$

which is the equation of a parabola in the z-plane.

(*ii*) Any point inside the circle, $|w| = 1$ is $w = \rho e^{i\phi}$, where $\rho < 1$.
Substituting this value of *w* in (*i*), we have

$$(1 + \rho e^{i\phi}) = \frac{2}{\sqrt{re^{i\theta}}} \quad \text{or} \quad 1 + \rho\cos\phi + i\rho\sin\phi = \frac{2}{\sqrt{r}}\left(\cos\frac{\theta}{2} - i\sin\frac{\theta}{2}\right).$$

Comparing the real and imaginary parts, we obtain

$$1 + \rho\cos\phi = \frac{2}{\sqrt{r}}\cos\frac{\theta}{2}, \quad \text{that is,} \quad \rho\cos\phi = \frac{2}{\sqrt{r}}\cos\frac{\theta}{2} - 1 \quad \text{and} \quad \rho\sin\phi = \frac{-2}{\sqrt{r}}\sin\frac{\theta}{2}.$$

Squaring and adding, we obtain $\frac{4}{r}\cos^2\frac{\theta}{2} - \frac{4}{\sqrt{r}}\cos\frac{\theta}{2} + 1 + \frac{4}{r}\sin^2\frac{\theta}{2} = \rho^2$

or $$\frac{4}{r} - \frac{4}{\sqrt{r}}\cos\frac{\theta}{2} = \rho^2 - 1 < 0, \quad \text{since } \rho < 1$$

or $$\frac{1}{\sqrt{r}} - \cos\frac{\theta}{2} < 0, \quad \text{or} \quad \frac{1}{r} < \cos^2\frac{\theta}{2}, \quad \text{or} \quad r > \frac{2}{1 + \cos\theta}$$

which represents a point outside the parabola.
Hence, corresponding to any point inside the unit circle in *w*-plane, there is a point outside the parabola in the *z*-plane.

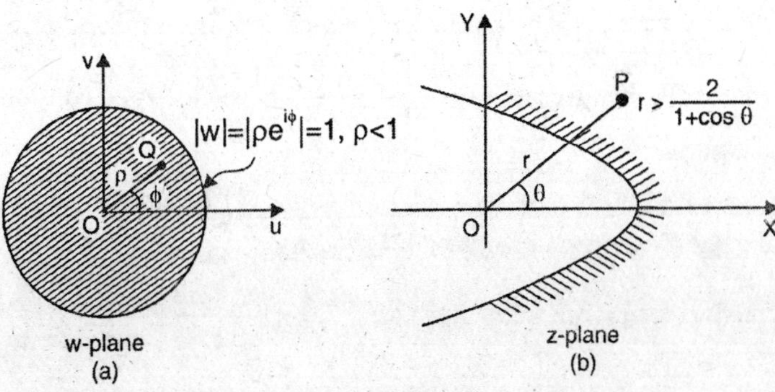

Fig. 4.56

EXAMPLE 4.72 Find the image of infinite strip $\frac{1}{4} \le y \le \frac{1}{2}$, under the mapping $w = 1/z$.

(AMIETE, Dec. 2004)

Solution From the mapping $w = 1/z$, we get $z = 1/w$. Writing $w = u + iv$ and $z = x + iy$, we have

$$z = x + iy = \frac{1}{u+iv} = \frac{u-iv}{(u+iv)(u-iv)} = \frac{u-iv}{u^2+v^2}.$$

Comparing the real and imaginary parts,

we obtain $x = \dfrac{u}{u^2+v^2}$ and $y = \dfrac{-v}{u^2+v^2}$.

If $y > \frac{1}{4}$, then $\dfrac{-v}{u^2+v^2} > \dfrac{1}{4}$,

that is, $u^2 + v^2 + 4v < 0$ or $u^2 + (v+2)^2 < 4$.

If $y < \frac{1}{2}$, then $\dfrac{-v}{u^2+v^2} < \dfrac{1}{2}$,

that is, $u^2 + v^2 + 2v > 0$ or $u^2 + (v+1)^2 > 1$.

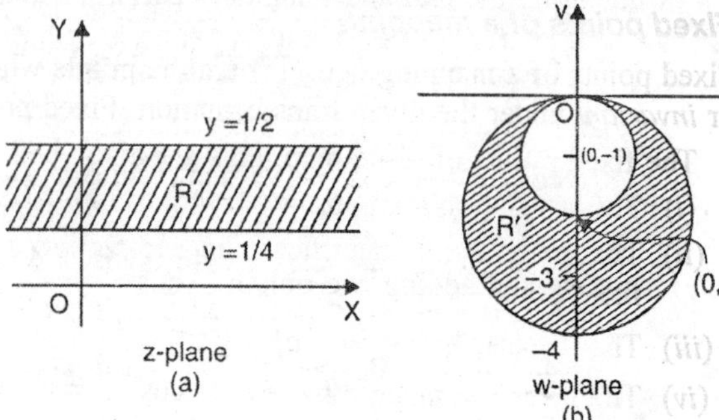

Fig. 4.57

Finally, $\frac{1}{4} \le y \le \frac{1}{2}$ imply $u^2 + (v+2)^2 \le 4$ and $u^2 + (v+1)^2 \ge 1$.

Hence, the infinite strip $\frac{1}{4} \le y \le \frac{1}{2}$ is transformed into the region R' in w-plane which is bounded by the two circles $u^2 + (v+2)^2 = 4$ and $u^2 + (v+1)^2 = 1$ such that the region is exterior to the circle $u^2 + (v+1)^2 = 1$ and interior to the circle $u^2 + (v+2)^2 = 4$ as shown in the Fig. 4.57 (b).

4.25 BILINEAR TRANSFORMATION

The transformation
$$w = f(z) = \frac{az+b}{cz+d}, \quad ad - bc \ne 0, \qquad \text{...(4.111)}$$
where a, b, c, d are real or complex constants, is called a *bilinear* or a *linear fractional* or *Möbius transformation.* We note the following:

(*i*) The condition $ad - bc \ne 0$ is essential, because if $ad - bc = 0$, then $\frac{a}{b} = \frac{c}{d}$ and the transformation becomes $w = \dfrac{az+b}{cz+d} = \dfrac{b}{d} \bullet \dfrac{(a/b)z+1}{(c/d)z+1} = \dfrac{b}{d}$ which is a constant.

(*ii*) If both $c = 0$ and $d = 0$, then w is not defined for any z.

(*iii*) If $c = 0$ and $d \ne 0$, then we get $w = (az+b)/d$, which is a linear mapping.

(*iv*) If $a = 0$, $b \ne 0$ and $c \ne 0$, then we get $w = \dfrac{b}{c}\left[\dfrac{1}{z+(d/c)}\right]$ which can be considered as a combination of the following mappings.

(*a*) translation : $w_1 = z + (d/c)$ (*b*) inversion : $w_2 = 1/w_1$

(*c*) rotation and magnification/contraction : $w = (b/c)\, w_2$.

(*v*) If $a \ne 0$, $c \ne 0$, then we can write Eq. (4.111) as $w = \dfrac{a}{c} + \dfrac{bc-ad}{c}\left[\dfrac{1}{cz+d}\right] = \dfrac{a}{c} + q\left[\dfrac{1}{z+(d/c)}\right]$, where $q = \dfrac{bc-ad}{c^2}$. This mapping can be considered as the composition of the following mappings.

(*a*) translation : $w_1 = z + (d/c)$ (*b*) inversion : $w_2 = 1/w_1$

(*c*) rotation and magnification/contraction : $w_3 = q\, w_2$ (*d*) translation : $w = w_3 + (a/c)$.

Therefore, the general bilinear transformation can be considered as a composition of linear and inversion mappings.

The inverse mapping of the bilinear transformation given in Eq. (4.111) is $z = \dfrac{b - wd}{cw - a}$.

Fixed points of a mapping

Fixed points of a mapping $w = f(z)$ are the points which are mapped onto themselves. These points are *fixed* or *invariant* under the given transformation. Fixed points are obtained by solving the equation $w = f(z) = z$.

The fixed points of some mappings are given below:

(*i*) The translation mapping $w = z + c$ has no fixed points in the finite *z*-plane.

(*ii*) The rotation and magnification / contraction mapping $w = az$, a any complex constant has $z = 0$ as the fixed point.

(*iii*) The linear mapping $w = az + c$ has one fixed point in the finite plane at $z = c/(1 - a)$, $a \neq 1$.

(*iv*) The inversion mapping $w = 1/z$ has two fixed points at $z = \pm 1$.

The fixed points of bilinear transformation are given by $z = \dfrac{az + b}{cz + d}$ or $cz^2 - (a - d)\, z - b = 0$. ...(4.112)

This is a quadratic equation in z and has at the most two roots. Hence, the bilinear transformation has at most two fixed points. If the roots of Eq. (4.112) are equal, then $(a - d)^2 + 4bc = 0$. In this case, the mapping has a single fixed point at $z = (a - d)/2c$.

If $c = 0$, the bilinear transformation Eq. (4.111) reduces to a linear transformation $w = \dfrac{az + b}{d}$ and has one fixed point at $z = b/(d - a)$, $d \neq a$ in the finite plane.

EXAMPLE 4.73 Show that the transformation $w = f(z)$ which maps three distinct points z_1, z_2, z_3, in the *z*-plane onto three given distinct points w_1, w_2, w_3 in the *w*-plane is given implicity by

$$\frac{w - w_1}{w - w_3} \bullet \frac{w_2 - w_3}{w_2 - w_1} = \frac{z - z_1}{z - z_3} \bullet \frac{z_2 - z_3}{z_2 - z_1}.$$

If one of these points is the point ∞, then the ratio of the two differences contaning this point is replaced by 1.

Solution Let the required transformation be $w = f(z) = \dfrac{az + b}{cz + d}$, $ad - bc \neq 0$.

If w_j corresponds to z_j, $j = 1, 2, 3$ we have

$$w - w_j = \frac{az + b}{cz + d} - \frac{az_j + b}{cz_j + d} = \frac{(ad - bc)(z - z_j)}{(cz + d)(cz_j + d)}$$

Then, $$w - w_1 = \frac{(ad - bc)(z - z_1)}{(cz + d)(cz_1 + d)}, \qquad w - w_3 = \frac{(ad - bc)(z - z_3)}{(cz + d)(cz_3 + d)}.$$

Replacing w by w_2 and z by z_2, we obtain

$$w_2 - w_1 = \frac{(ad - bc)(z_2 - z_1)}{(cz_2 + d)(cz_1 + d)}, \qquad w_2 - w_3 = \frac{(ad - bc)(z_2 - z_3)}{(cz_2 + d)(cz_3 + d)}.$$

Hence, we have $$\frac{w - w_1}{w - w_3} = \frac{(z - z_1)(cz_3 + d)}{(z - z_3)(cz_1 + d)} \quad \text{and} \quad \frac{w_2 - w_1}{w_2 - w_3} = \frac{(z_2 - z_1)(cz_3 + d)}{(z_2 - z_3)(cz_1 + d)}.$$

Dividing these two equations, we obtain

$$\frac{w-w_1}{w-w_3}\bullet\frac{w_2-w_3}{w_2-w_1}=\frac{(z-z_1)\bullet(z_2-z_3)}{(z-z_3)(z_2-z_1)}. \qquad ...(i)$$

Solving for w in terms of z gives the required transformation. The right-hand side of (*i*) is called the *cross ratio* of z_1, z_2, z_3 and z.

If one of the points is ∞, say $z_1=\infty$, then $\dfrac{z-z_1}{z_2-z_1}=\lim\limits_{z_1\to\infty}\left[\dfrac{(z/z_1)-1}{(z_2/z_1)-1}\right]=1.$

If $w_1=\infty$, we obtain $\dfrac{w-w_1}{w_2-w_1}=\lim\limits_{w_1\to\infty}\left[\dfrac{(w/w_1)-1}{(w_2/w_1)-1}\right]=1.$

EXAMPLE 4.74 Find the fixed or invariant points of the transformation $w=\dfrac{3z-5}{z+1}$.

Solution The fixed points are solutions to $z=\dfrac{3z-5}{z+1}$ or $z^2-2z+5=0$, that is, $z=\dfrac{2\pm\sqrt{4-20}}{2}=1\pm 2i.$

EXAMPLE 4.75 Find the bilinear transformation which maps point $z=-1, 0, 1$ into $w=0, i, 3i$ respectively.

Solution The bilinear transformation which maps the point z_1, z_2, z_3 in the z-plane onto the points w_1, w_2, w_3 in the w-plane is given by

$$\frac{w-w_1}{w-w_3}\bullet\frac{w_2-w_3}{w_2-w_1}=\frac{z-z_1}{z-z_3}\bullet\frac{z_2-z_3}{z_2-z_1}.$$

We have the transformation as $\dfrac{w-0}{w-3i}\bullet\dfrac{i-3i}{i-0}=\dfrac{z-(-1)}{z-1}\bullet\dfrac{0-1}{0-(-1)}$

or $$\frac{-2iw}{(w-3i)i}=-\left(\frac{z+1}{z-1}\right) \text{ or } \frac{2w}{w-3i}=\frac{z+1}{z-1}.$$

Solving for w in terms of z, we obtain $w=-3i\dfrac{z+1}{z-3}$.

EXAMPLE 4.76 Find the bilinear transformation which maps the points $z=1, i, 2+i$ in the z-plane onto the points $w=i, 1, \infty$ in the w-plane.

Solution The bilinear transformation which maps the points z_1, z_2, z_3 in the z-plane onto the points w_1, w_2, w_3 in the w-plane is given by

$$\frac{w-w_1}{w-w_3}\bullet\frac{w_2-w_3}{w_2-w_1}=\frac{z-z_1}{z-z_3}\bullet\frac{z_2-z_3}{z_2-z_1}. \qquad ...(i)$$

We may write (*i*) as

$$\frac{w-w_1}{(w/w_3)-1}\bullet\frac{(w_2/w_3)-1}{w_2-w_1}=\frac{z-z_1}{z-z_3}\bullet\frac{z_2-z_3}{z_2-z_1}.$$ Substituting for the points, we obtain

$$\frac{w-i}{0-1}\bullet\frac{0-1}{1-i}=\frac{z-1}{z-(2+i)}\bullet\frac{i-(2+i)}{i-1}$$

or $$\frac{w-i}{1-i}=\frac{z-1}{z-(2+i)}\bullet\frac{-2}{i-1} \text{ or } w-i=\frac{2(z-1)}{z-(2+i)}$$

or $$w=\frac{2(z-1)}{z-(2+i)}+i=\frac{2z-1+iz-2i}{z-(2+i)}=\frac{(2+i)z-(2i+1)}{z-(2+i)}.$$

EXAMPLE 4.77 Find the bilinear transformation which maps the points $z_1 = \infty$, $z_2 = i$, $z_3 = 0$ in the z-plane onto the points $w_1 = 0$, $w_2 = i$ and $w_3 = \infty$ in the w-plane.

Solution The bilinear transformation which maps the points z_1, z_2, z_3 in the z-plane onto the points w_1, w_2, w_3 in the w-plane is given by

$$\frac{w-w_1}{w-w_3} \bullet \frac{w_2-w_3}{w_2-w_1} = \frac{z-z_1}{z-z_3} \bullet \frac{z_2-z_3}{z_2-z_1}. \qquad ...(i)$$

We may write Eq. (*i*) as

$$\frac{w-w_1}{(w/w_3)-1} \bullet \frac{(w_2/w_3)-1}{w_2-w_1} = \frac{(z/z_1)-1}{z-z_3} \bullet \frac{z_2-z_3}{(z_2/z_1)-1}.$$ Substituting for the points, we have

$$\frac{w-0}{0-1} \bullet \frac{0-1}{i-0} = \frac{0-1}{z-0} \bullet \frac{i-0}{0-1} \quad \text{or} \quad \frac{w}{i} = \frac{i}{z} \quad \text{or} \quad w = -\frac{1}{z}.$$

EXAMPLE 4.78 Show that the bilinear transformation $w = (iz+2)/(4z+i)$ transforms the real axis in the z-plane onto a circle in the w-plane. Find the centre and the radius of the circle in the w-plane. Find the point in the z-plane which is mapped onto the centre of the circle in the w-plane.

(*AMIETE, June 2010, 2009, Dec. 2003*)

Solution From $w = \dfrac{iz+2}{4z+i}$, we obtain $z = \dfrac{2-iw}{4w-i}$.

Substituting $z = x + iy$ and $w = u + iv$, we get

$$x + iy = \frac{2-i(u+iv)}{4(u+iv)-i} = \frac{[(2+v)-iu][4u-i(4v-1)]}{[4u+i(4v-1)][4u-i(4v-1)]}$$

$$= \frac{[4u(2+v)-u(4v-1)]-i[(2+v)(4v-1)+4u^2]}{16u^2+(4v-1)^2} = \frac{9u-i(4u^2+4v^2+7v-2)}{16u^2+(4v-1)^2}.$$

Comparing the real and imaginary parts, we get

$$x = \frac{9u}{16u^2+(4v-1)^2}, \quad \text{and} \quad y = -\frac{4u^2+4v^2+7v-2}{16u^2+(4v-1)^2}.$$

The image of the real axis, that is, $y = 0$ is obtained as

$$4u^2+4v^2+7v-2=0 \quad \text{or} \quad u^2+v^2+\frac{7}{4}v-\frac{1}{2}=0 \quad \text{or} \quad (u-0)^2+\left(v+\frac{7}{8}\right)^2=\left(\frac{9}{8}\right)^2$$

which is a circle with centre at $\left(0, -\dfrac{7}{8}\right)$ and radius $\dfrac{9}{8}$.

For $u = 0$ and $v = -7/8$, we obtain $x = 0$ and $y = -\dfrac{(49/16)-(49/8)-2}{81/4} = \dfrac{1}{4}$.

Thus, the centre of the circle in the w-plane in the image of the point (0, 1/4) in the z-plane.

EXAMPLE 4.79 Find the image of the closed half disk $|z| \le 1$, $I_m(z) \ge 0$ under the bilinear transformation $w = z/(z+1)$.

(*AMIETE, Dec. 2010*)

Solution The given half disk is the intersection of the unit disk $|z| \le 1$ and the upper half plane $I_m(z) \ge 0$. (*See Fig. 4.58 on next page.*)

From $w = \dfrac{z}{z+1}$, we obtain $z = \dfrac{w}{1-w}$.

Hence, $|z| = 1$ gives

$$|z| = \frac{|w|}{|1-w|} = 1 \quad \text{or} \quad |w| = |1-w|$$

or $u^2 + v^2 = (1-u)^2 + v^2$ or $u = 1/2$.

Since $z = 0$ is mapped as $w = 0$, the closed unit disk $|z| \leq 1$, is mapped as $u \leq 1/2$.

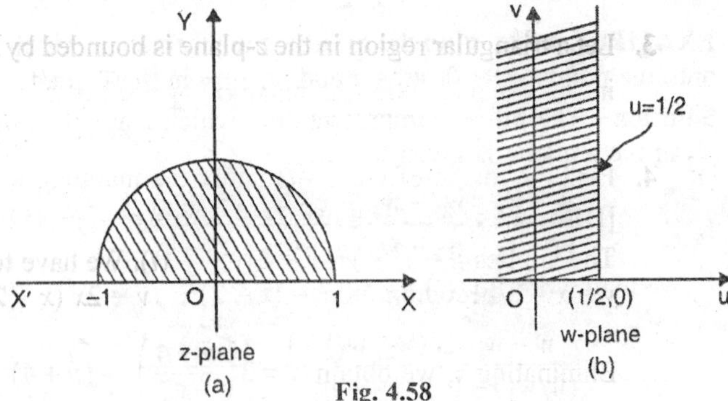

Fig. 4.58

Substituting $z = x + iy$ and $w = u + iv$ in the mapping $z = w/(1-w)$, we get

$$x + iy = \frac{u+iv}{1-(u+iv)} = \frac{(u+iv)[(1-u)+iv]}{[(1-u)-iv][(1-u)+iv)]} = \frac{(u-u^2-v^2)+iv}{(1-u)^2+v^2}.$$

Comparing the real and imaginary parts, we obtain $x = \dfrac{u-u^2-v^2}{(1-u)^2+v^2}$ and $y = \dfrac{v}{(1-u)^2+v^2}$.

Therefore, the image of the region $I_m(z) \geq 0$, that is, $y \geq 0$ is obtained as $v \geq 0$. Hence, the required image is the region $u \leq 1/2$ and $v \geq 0$.

EXAMPLE 4.80 Find the image of the half plane $x + y > 0$ under the bilinear transformation $w = (z-1)/(z+i)$.

Solution From the mapping $w = \dfrac{z-1}{z+i}$, we get $z = \dfrac{1+iw}{1-w}$. ...(*i*)

Substituting $z = x + iy$ and $w = u + iv$ in Eq. (*i*), we obtain

$$x + iy = \frac{1+i(u+iv)}{1-(u+iv)}$$

$$= \frac{(1-v)+iu}{(1-u)-iv} \times \frac{(1-u)+iv}{(1-u)+iv}$$

$$= \frac{(1-u-v)+i(v+u-u^2-v^2)}{(1-u)^2+v^2}.$$

Fig. 4.59

Comparing the real and imaginary parts, we obtain

$$x = \frac{1-u-v}{(1-u)^2+v^2}, \quad \text{and} \quad y = \frac{v+u-u^2-v^2}{(1-u)^2+v^2}.$$

Therefore, $x + y > 0$ gives $1 - u^2 - v^2 > 0$ or $u^2 + v^2 < 1$ or $|w| < 1$. The line $x + y = 0$ is mapped as $|w| = 1$. Now consider any point in the given half plane, say $z = 1$. Its image is $w = 0$, which lies inside $|w| = 1$. Hence, the image of the region $x + y > 0$ is the interior of the unit disk. (Fig. 4.59).

EXERCISE 4.9

1. Let a rectangular region with vertices at $A(1, 1)$, $B(3, 1)$, $C(3, 3)$, $D(1, 3)$ be defined in the z-plane. Find the image of the region in the w-plane under the mapping $w = z + (1 + 3i)$.

   Ans. Maps as a rectangle with vertices at (2, 4), (4, 4), (4, 6), (2, 6).

2. Let a triangular region with vertices at $A(0, 1)$, $B(1, -1)$, $C(1, 1)$ be defined in the z-plane. Find the image of this region in the w-plane under the mapping $w = iz + 4$.

   Ans. Maps as a triangle with vertices at (3, 0), (5, 1), (3, 1).

3. Let a triangular region in the z-plane is bounded by $x = 0, y = 0, x = 2, y = 3$. Find the image of this region in the w-plane under the mapping $w = \sqrt{2}\, e^{i\pi/4} z$.

Ans. Maps as a rectangle with vertices at (0, 0), (2, 2), (–1, 5), (–3, 3).

4. Find the image of $y = x - 2$ under the mapping $w = z^2$.
[**Hint:** $w = z^2 \Rightarrow u + iv = (x + iy)^2 = x^2 - y^2 + 2ixy$.
The implies $u = x^2 - y^2$, $v = 2xy$, ...(*i*). We have to find the image of $y = x - 2$. Eq. (*i*) with the help of $y = x - 2$ become $u = x^2 - (x - 2)^2$, $v = 2x(x - 2)$ or $u = 4x - 4$, $v = 2x^2 - 4x$.
Eliminating x, we obtain $v = 2\left(\frac{u+4}{4}\right)^2 - (u+4)$ or $u^2 = 8(v+2)$.
This is parabola in w-plane having vertex at (0, –2) and latus-rectum = 8.
Thus, the image of $y = x - 2$ in z-plane under the mapping $w = z^2$ is $u^2 = 8(v + 2)$ a parabola in w-plane.]

5. Find the image of $|z - 2i| = 2$ under the mapping $w = 1/z$. (*AMIETE, June 2001*)

Ans. The line $v + (1/4) = 0$.

6. Find the image of the closed unit disk $|z| \le 1$ under the mapping $w = (1 - i)z - 2i$.
[**Hint:** We have $w = (1 - i)z - 2i$
or $z = (w + 2i) / (1 - i)$.
Hence,
$$|z| = \left|\frac{w+2i}{1-i}\right| = \frac{|w+2i|}{|1-i|} = \frac{1}{\sqrt{2}}|w+2i|.$$
Therefore, the unit disk $|z| \le 1$ is mapped as $|w + 2i| \le \sqrt{2}$ or $u^2 + (v+2)^2 \le 2$.]

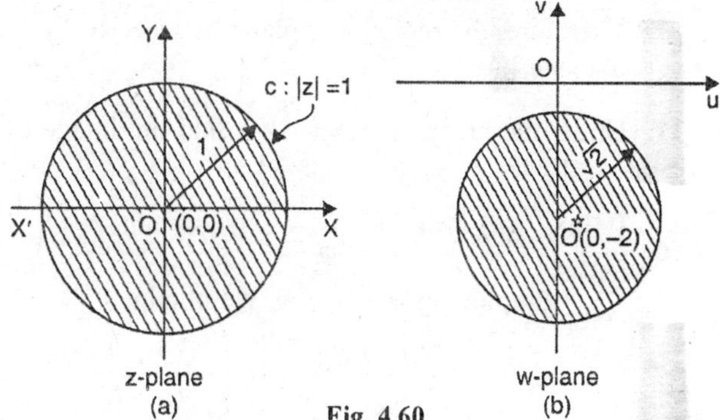

Fig. 4.60

7. Show that under the transformation $w = \frac{z-i}{z+i}$, real axis in the z-plane is mapped onto the circle $|w| = 1$. What portion of the z-plane corresponds to the interior of the circle.
(*AMIETE, June 2004, JNTU, 2003*)

[**Hint:** We have $|w| = \left|\frac{z-i}{z+i}\right| = \left|\frac{x+i(y-1)}{x+i(y+1)}\right| = \sqrt{\frac{x^2+(y-1)^2}{x^2+(y+1)^2}}$.
The real axis in z-plane, that is, $y = 0$ transforms onto $|w| = 1$.
Hence, the real axis in the z-plane is mapped onto the circle $|w| = 1$.
The interior of the circle, that is, $|w| < 1$ gives $\frac{x^2+(y-1)^2}{x^2+(y+1)^2} < 1$, that is, $-4y < 0$ or $y > 1$.
Thus the upper half of the z-plane corresponds to the interior of the circle $|w| = 1$.]

8. Find the mapping of the circle $|z - 3 - 4i| = 2$ by the transformation $w = az$, where $a = \sqrt{2}\,(1 + i)$.

Ans. The given circle in z-plane maps onto a circle with radius 4 and centre at $\left(10, \frac{\pi}{4} + \theta\right)$ on w-plane, where $\theta = \tan^{-1}(4/3)$.

9. Show that the transformation $w = \sin z$ maps the families of lines x = const and y = const onto two families of confocal central conics. (*AMIETE, Dec. 2011, Dec. 2009*)
[**Hint:** We have $w = \text{siz}\, z \Rightarrow u + iv = \sin(x + iy) = \sin x \cosh y + i \cos x \sinh y$.
Comparing the real and imaginary parts, we obtain $u = \sin x \cosh y$, $v = \cos x \sinh y$.

When $x = a$, $u = \sin a \cosh y$, $v = \cos a \sinh y$.

Therefore, $$\frac{u^2}{\sin^2 a} - \frac{v^2}{\cos^2 a} = \cosh^2 y - \sinh^2 y = 1. \quad ...(i) \text{ (Hyperbols)}$$

Similarly if $y = b$, $$\frac{u^2}{\cosh^2 b} + \frac{v^2}{\sinh^2 b} = \sin^2 x + \cos^2 x = 1. \quad ...(ii) \text{ (Ellipses)}$$

Eqs. (*i*) and (*ii*) represent two families of confocal conics.]

10. Discuss the transformation $w = e^z$, and show that it transforms the region between the real axis and a line parallel to real axis at $y = \pi$ into the upper half of the w-plane. (*JNTU, 2000S*)

[Hint: We have $w = e^z$ or $\rho e^{i\phi} = e^{x+iy} = e^x \cdot e^{iy}$.

Therefore, $\rho = e^x$ and $\phi = y$. Consider straight line parallel to y-axis at a distance 'c' and another line parallel to x-axis at a distance 'a'.

Hence, $x = c$ maps onto the circle $\rho = e^c$. Further, $y = a$ maps onto the radial line $\phi = a$.

As y varies from 0 to π, ϕ varies from 0 to π.

Therefore, the region in z-plane between $0 < y < \pi$ maps onto the region $v > 0$ that is, upper half of the w-plane.]

11. Under the mapping $w = \dfrac{i}{z-1}$, find the image of the following regions (*a*) $I_m z < 0$, (*b*) $|z| > 1$.

[Hint: (*a*) From $w = \dfrac{i}{z-1}$, we obtain $z = i\left(\dfrac{1+w}{w}\right)$.

Employing $w = u + iv$, we obtain

$$z = i\left[\frac{1+u+iv}{u+iv}\right] = \left[\frac{i(1+u+iv)(u-iv)}{(u+iv)(u-iv)}\right]$$

$$= \frac{i}{u^2+v^2}[u+u^2+v^2-iv] = \frac{v}{u^2+v^2} + \frac{i(u+u^2+v^2)}{u^2+v^2}.$$

The mapping of the region $I_m z < 0$ is given by

$$u + u^2 + v^2 < 0 \quad \text{or} \quad \left(u+\frac{1}{2}\right)^2 + (v-0)^2 < \frac{1}{4} \quad \text{or} \quad \left|w+\frac{1}{2}\right| < \frac{1}{2}$$

which is a circle with centre at $\left(-\frac{1}{2}, 0\right)$ and radius (1/2).

(*b*) We have $|z| = \left|\dfrac{i(1+w)}{w}\right| = \left|\dfrac{1+w}{w}\right|$.

Hence, $|z| > 1$ gives $|1 + w| > w$. Substituting $w = u + iv$, we obtain

$(1+u)^2 + v^2 > u^2 + v^2$ or $2u + 1 > 0$ or $u > -(1/2)$.]

12. Find the image of $|z - 2i| = 1$ under the mapping $w = 1/z$.

Ans. The required image is the circle $|3w + 2i| = 1$.

13. Show that the transformation $w = z + (1/z)$ maps the circles $|z| = r = \text{const} \neq 1$ onto an ellipse.

$u = (r + r^{-1})\cos\theta$, $v = (r - r^{-1})\sin\theta$.

In particular discuss the case when $r = 1$.

[Hint: Let $z = r(\cos\theta + i\sin\theta)$. Therefore,

$$w = z + z^{-1} = r(\cos\theta + i\sin\theta) + r^{-1}(\cos\theta - i\sin\theta)$$
$$= (r + r^{-1})\cos\theta + i(r - r^{-1})\sin\theta.$$

Hence, $u = (r + r^{-1}) \cos \theta$ and $v = (r - r^{-1}) \sin \theta$ are the parametrical equations of an ellipse. Hence circles $|z| = r = \text{const} \neq 1$ are mapped onto ellipses. The circle $r = 1$ is mapped onto the segment $-2 \le u \le 2$ of the u-axis.]

14. Show that the transformation $w = z + \dfrac{a^2 - b^2}{4z}$ transforms the circle $|z| = \dfrac{a+b}{2}$ in the z-plane onto an ellipse of semi-axes a, b in the w-plane.

15. Show that under the mapping $w = z^2$, the circle $C : |z - 1| = 1$ transforms onto the cardiod $\rho = 2(1 + \cos \phi)$ where $w = \rho e^{i\phi}$ in the w-plane.

16. Show that under the transformation $w = (1/z)$, the circle $C : |z - 3| = 5$ is mapped onto a circle $C : \left| w + \dfrac{3}{16} \right| = \dfrac{5}{16}$ is the w-plane.

17. For the mapping defined by $(w + 1)^2 = (4/z)$, prove that the unit circle in the w-plane corresponds to a parabola in the z-plane. *(AMIETE, Dec. 2002)*

18. Find the fixed or invariant points of the transformation $w = \dfrac{2z - 5}{z + 4}$. **Ans.** $-1 \pm 2i$.

19. The invariant points of the transformation $w = \dfrac{1+z}{1-z}$ are given by ....

(AMIETE, June 2010) **Ans.** $\pm$i.

20. Find the bilinear transform whose fixed points are 1 and 2.

[**Hint:** Let the transformation by $w = \dfrac{az + b}{cz + d}$. The fixed points of the transformation are given by

$z = \dfrac{az + b}{cz + d} \Rightarrow cz^2 - (a - d)z - b = 0$. We have $z_1 + z_2 = \dfrac{a - d}{c}$ and $z_1 z_2 = -\dfrac{b}{c}$.

Since fixed points are 1 and 2. Therefore, $3 = \dfrac{a - d}{c}$ and $2 = -\dfrac{b}{c}$.

This implies $\dfrac{d}{c} = \dfrac{a}{c} - 3$ and $\dfrac{b}{c} = -2$. Let $\dfrac{a}{c} = \lambda$, then required transformation is the family of curves given by $w = \left(\dfrac{a}{c} z + \dfrac{b}{c}\right) \Big/ \left(z + \dfrac{d}{c}\right) = (\lambda z - 2)/(z + \lambda - 3)$, λ arbitrary.

Thus, we obtain an infinite number of such transformations.]

Find a bilinear transformation which maps respectively the given points.

21. 0, $-i$, -1 in the z-plane onto the points i, 1, 0 in the w-plane. **Ans.** $w = -i\left(\dfrac{z+1}{z-1}\right)$.

22. 0, 1, ∞ in the z-plane onto -1, $-i$, 1 in the w-plane. **Ans.** $w = (z - i)/(z + i)$.

23. 0, 1, ∞ in the z-plane onto the points i, -1, $-i$ in the w-plane. **Ans.** $w = -i\left(\dfrac{z-1}{z+i}\right)$.

24. i, $-i$, 1 in the z-plane onto the points 0, 1, ∞ in the w-plane. **Ans.** $w = \dfrac{(1-i)(z-i)}{2(z-1)}$.

25. $1 + i$, $-i$, $2 - i$ in the z-plane onto the points 0, 1, i in the w-plane. **Ans.** $w = \dfrac{2z - 2 - 2i}{(i-1)z - 3 - 5i}$.

26. Find the bilinear transformation which maps the points $z = 1, i, -1$ onto the points $w = i, 0, -i$. *(AMIETE, June 2012, Dec. 2009; Bombay 2006; VTU, 2000)*

Hence find: (*a*) the image of $|z| < 1$, (*b*) the invariant points of this transformation.

[Hint: The bilinear transformation which maps the points z_1, z_2, z_3 in the *z*-plane onto the points w_1, w_2, w_3 in the *w*-plane is given by

$$\frac{w - w_1}{w - w_3} \cdot \frac{w_2 - w_3}{w_2 - w_1} = \frac{z - z_1}{z - z_3} \cdot \frac{z_2 - z_3}{z_2 - z_1}.$$

We have the transformation as

$$\frac{w - i}{w + i} \cdot \frac{0 + i}{0 - i} = \frac{z - 1}{z + 1} \cdot \frac{i + 1}{i - 1} \quad \text{or} \quad \frac{w - i}{w + i} = \frac{z - 1}{z + 1} \cdot \frac{1 + i}{1 - i}.$$

By componendo and dividendo, we get

$$\frac{2w}{-2i} = \frac{(z-1)(1+i) + (z+1)(1-i)}{(z-1)(1+i) - (z+1)(1-i)} = \frac{2z - 2i}{2zi - 2} \quad \text{or} \quad -\frac{w}{i} = \frac{z - i}{zi - 1} \quad \text{or} \quad w = \frac{1 + iz}{1 - iz}. \quad ...(i)$$

(*a*) We may write (*i*) as $z = i\dfrac{1 - w}{1 + w}$.

Therefore, $\left|\dfrac{i(1 - w)}{1 + w}\right| = |z| < 1$ or $|i|\,|1 - w| < |1 + w|$ $\quad [\because |i| = 1]$

or $|1 - u - iv| < |1 + u + iv|$ or $(1 - u)^2 + v^2 < (1 + u)^2 + v^2$ which reduces to $u > 0$.

Hence, the interior of the circle $x^2 + y^2 = 1$ in the *z*-plane is mapped onto the entire half of the *w*-plane to the right of the imaginary axis.

(*b*) Invariant point are solutions to $z = \dfrac{1 + iz}{1 - iz}$ or $iz^2 + (i - 1)z + 1 = 0$.

or $$z = \frac{1 - i \pm \sqrt{(i-1)^2 - 4i}}{2i} = -\frac{1}{2}[1 + i \pm \sqrt{6i}].]$$

27. Show that the bilinear transformation $w = \dfrac{2z + 3}{z - 4}$ maps the circle $x^2 + y^2 - 4x = 0$ onto the line $u + (3/4) = 0$.

28. Obtain the bilinear transformation which maps $z_1 = i, z_2 = 1, z_3 = -1$ onto $w_1 = 1, w_2 = 0, w_3 = \infty$, respectively. Also, obtain its inverse transformation. *(AMIETE, Dec. 2010)*

29. Show that the transformation $w = \dfrac{5 - 4z}{4z - 2}$ transforms the circle $C : |z| = 1$ onto a circle of radius unity in the *w*-plane and find the centre of the circle. **Ans.** $\left(-\dfrac{1}{2}, 0\right)$

30. Show that the transformation $w = \dfrac{3 - z}{z - 2}$ transforms the circle with centre $[(5/2), 0]$ and radius $1/2$ in the *z*-plane into the imaginary axis in the *w*-plane and the interior of the circle into the right half of the plane. *(AMIETE, June 2012)*

31. Find the image of the annulus $1 < |z| < 2$ under the mapping $w = z/(z - 1)$.

[Hint: From $w = \dfrac{z}{z - 1}$, we obtain $z = \dfrac{w}{w - 1} = \dfrac{u + iv}{(u - 1) + iv}$.

The region $|z| > 1$ is mapped onto the region

$$\left|\frac{u+iv}{u-1+iv}\right| > 1 \text{ or } u^2+v^2 > (u-1)^2+v^2 \text{ or } u > 1/2.$$

The region $|z| < 2$ is mapped onto the region

$$\left|\frac{u+iv}{u-1+iv}\right| < 2 \text{ or } u^2+v^2 < 4[(u-1)^2+v^2]$$

$$\text{or} \quad \left(u-\frac{4}{3}\right)^2 + v^2 > \left(\frac{2}{3}\right)^2.$$

Thus, the required region is on the right of the plane $u = (1/2)$ and outside the circle $\left(u-\frac{4}{3}\right)^2 + v^2 = \frac{4}{9}$.

The points on the line $u = (1/2)$ and on the boundary of the circle are not included.]

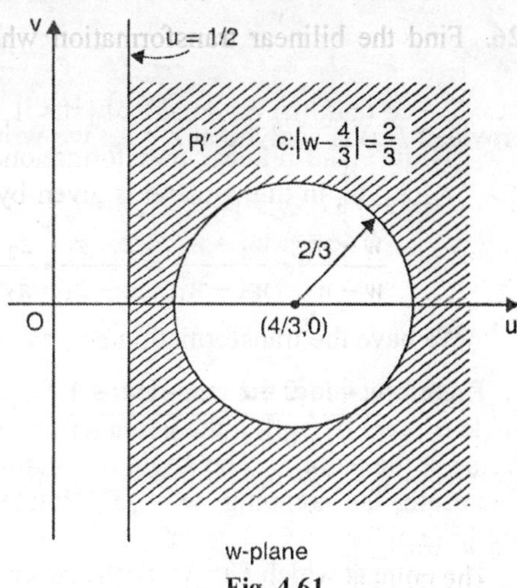

Fig. 4.61

4.26 CONFORMAL MAPPING

Let $w = f(z)$ be an analytic function and define a mapping from a region D in the z-plane onto a region $\overset{*}{D}$ in the w-plane. A *mapping* $w = f(z)$ *is said to be conformal if it preserves the angle between two curves both in magnitude and direction.*

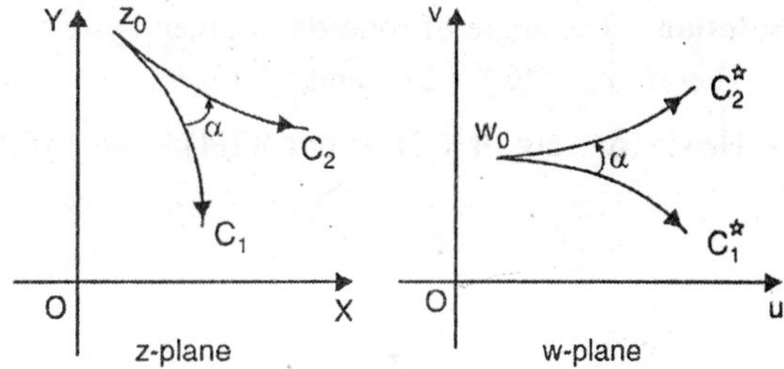

Fig. 4.62: Conformal mapping.

Suppose C_1 and C_2 be two smooth curves passing through the point z_0 and θ_1 and θ_2 be the angles of inclination of directed line tangents to C_1 and C_2 respectively at the point z_0. Under the mapping $w = f(z)$, let C_1^* and C_2^* be the images of C_1 and C_2 respectively. The image curve pass through the point $w_0 = f(z_0)$. Let ϕ_1 and ϕ_2 be the angles of inclination of the directed line tangent to C_1^* and C_2^* respectively, at w_0, which is the image of the point z_0. We obtain that under the mapping $w = f(z)$, the curves C_1 and C_2 are rotated in the same direction by the same angle, that is, $\alpha = \phi_2 - \phi_1 = \theta_2 - \theta_1$.

A mapping that preserves the magnitude of angle but not necessarily the direction is called an *isogonal mapping.*

Consider the transformation $w = f(z)$ *where* $f(z)$ *is analytic at* z_0 *and* $f'(z_0) \neq 0$. *Prove that under this transformation the tangent at* z_0 *to any curve C in the z-plane passing through* z_0 *(Fig. 4.63) is rotated through the angle* $\arg f'(z_0)$.

As a point moves from z_0 to $z_0 + \Delta z$ along C, the image point moves along $\overset{*}{C}$ in the w-plane from w_0 to $w_0 + \Delta w$. If the parameter used to describe the curve is t, then corresponding to the path $z = z(t)$ [or $x = x(t)$, $y = y(t)$] in the z-plane, we have the path $w = w(t)$ [or $u = u(t)$, $v = v(t)$] in the w-plane.

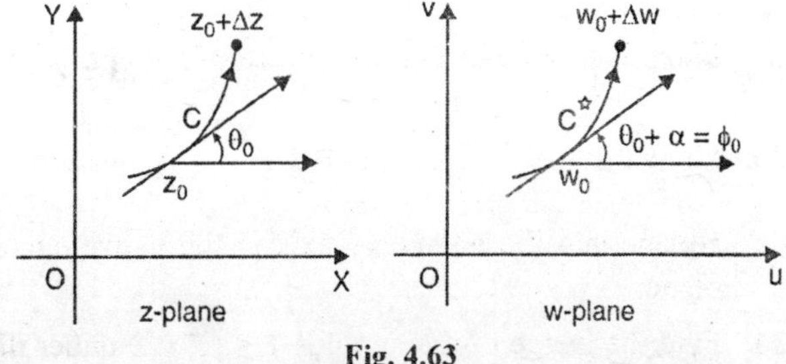

Fig. 4.63

The derivatives dz/dt and dw/dt represent tangent vectors to corresponding points on C and $\overset{*}{C}$.

Now, $\dfrac{dw}{dt} = \dfrac{dw}{dz} \cdot \dfrac{dz}{dt} = f'(z)\dfrac{dz}{dt}$ and, in particular at z_0 and w_0,

$$\left.\frac{dw}{dt}\right|_{w=w_0} = f'(z_0)\left.\frac{dz}{dt}\right|_{z=z_0} \qquad \text{...(4.113)}$$

provided $f(z)$ is analytic at $z = z_0$. we write

$$\left.\frac{dw}{dt}\right|_{w=w_0} = \rho_0\, e^{i\phi_0}, \quad f'(z_0) = \mathrm{Re}^{i\alpha}, \quad \left.\frac{dz}{dt}\right|_{z=z_0} = r_0\, e^{i\theta_0}, \text{ we have from Eq. (4.113).}$$

$$\rho_0\, e^{i\phi_0} = Re^{i\alpha} \bullet r_0\, e^{i\theta_0}$$

$$= Rr_0\, e^{i(\theta_0+\alpha)}, \quad \text{so that } \phi_0 = \theta_0 + \alpha = \theta_0 + \arg f'(z_0).$$

Therefore, under the mapping $w = f(z)$, the tangent line to a smooth curve C in the z-plane is rotated by an angle $\alpha = \arg\,[f'(z_0)]$ in the w-plane. The angle α is called the *angle of rotation* and $|f'(z_0)|$ is called the *scale factor* at the point z_0. The angle of rotation α is independent of C as $\arg\,[f'(z_0)]$ depends only on the part z_0. Therefore, the mapping, $w = f(z)$ rotates all the curves passing through the point z_0 by the same angle $\arg\,[f'(z_0)]$.

The point at which $f'(z_0) = 0$, that is, the points at which the mapping is not conformal are called the *critical points* of the mapping.

EXAMPLE 4.81 Determine the angle of rotation at the point $z = 2 + i$ under the mapping $w = z^2$. Find its scale factor also.

Solution The angle of rotation is given by $\alpha = \arg\,[f'(z_0)]$. We have $f(z) = z^2$ and $z_0 = 2 + i$.

Therefore, $f'(z) = 2z$ and $f'(2 + i) = 4 + 2i$.

Hence, $\alpha = \arg\,(4 + 2i) = \tan^{-1}(2/4) = \tan^{-1}\,(0.5)$ and scale factor $= |f'(z_0)| = |4 + 2i| = \sqrt{4^2 + 2^2} = 2\sqrt{5}$.

CHAPTER

5

Numerical Methods

5.1 INTRODUCTION

Let $y = f(x)$ be a discrete function of the independent variable x. When x takes up the values $x_0, x_0 + h, x_0 + 2h,...$ that is, $x_i = x_0 + ih$, $i = 0, 1, 2,..., n$ let the corresponding functional values of y be $y_0, y_1, y_2,..., y_n$. The values of x are called the *arguments* and those of y are defined as *entries*, the constant difference between any two consecutive values of the independent variable x, namely h, is called the *interval of differencing* or step length. The process of estimating the value of y for a value of x lying in the interval $x_0 \le x \le x_n$ is defined as ***interpolation***. If however, a value of y is calculated for a value of x lying outside the range (x_0, x_n), then the process of obtaining y is called *extrapolation*. The term interpolation is however, considered to include extrapolation.

5.2 DIFFERENCES

Let the function $y_i = f(x_i)$, $i = 0, 1, 2,..., n$ take the values $y_0, y_1, y_2,...,$ corresponding to the values $x_0, x_1, x_2,$ ... of x. The nodal points $x_0, x_1(=x_0 + h), x_2(= x_0 + 2h), ..., x_n(= x_0 + nh)$ be equi-spaced with step length h. For equally spaced nodes, we shall discuss three types of differences, namely, forward, backward and central.

5.2.1 Forward Differences

If $y_0, y_1, ..., y_n$ denote a set of values of y corresponding to values of the argument x as $x_0, x_1, ..., x_n$ respectively, then $y_1 - y_0, \ y_2 - y_1, \ y_3 - y_2, ..., y_n - y_{n-1}$ are defined as the differences of y. Denoting these differences by $\Delta y_0, \Delta y_1, \Delta y_2, ..., \Delta y_{n-1}$ respectively, we have

$$\Delta y_0 = y_1 - y_0 = f(x_0 + h) - f(x_0)$$

$$\Delta y_1 = y_2 - y_1 = f(x_1 + h) - f(x_1)$$

....

....

$$\Delta y_{n-1} = y_n - y_{n-1} = f(x_{n-1} + h) - f(x_{n-1}).$$

The symbol Δ is called the *forward difference operator* and $\Delta y_0, \Delta y_1, \Delta y_2, ...,$ are called the first forward differences of $y_0, y_1, y_2, ...,$ respectively. Thus we can think of Δ as an operator which when offered y_r as an input produces $y_{r+1} - y_r$ as an output for all r values under consideration.

$$\boxed{y_r} \rightarrow \boxed{\Delta} \rightarrow \boxed{y_{r+1} - y_r}$$

The differences of the first forward differences are called the second forward differences. Similarly we can define third forward differences, fourth forward differences etc. Thus

$$\Delta^2 y_0 = \Delta(\Delta y_0) = \Delta(y_1 - y_0) = \Delta y_1 - \Delta y_0 = y_2 - y_1 - (y_1 - y_0) = y_2 - 2y_1 + y_0.$$

$$\Delta^3 y_0 = \Delta(\Delta^2 y_0) = \Delta(y_2 - 2y_1 + y_0) = \Delta y_2 - 2\Delta y_1 + \Delta y_0$$
$$= y_3 - y_2 - 2(y_2 - y_1) + (y_1 - y_0) = y_3 - 3y_2 + 3y_1 - y_0.$$

$$\Delta^4 y_0 = \Delta(\Delta^3 y_0) = \Delta(y_3 - 3y_2 + 3y_1 - y_0) = \Delta y_3 - 3\Delta y_2 + 3\Delta y_1 - \Delta y_0$$
$$= y_4 - y_3 - 3(y_3 - y_2) + 3(y_2 - y_1) - (y_1 - y_0) = y_4 - 4y_3 + 6y_2 - 4y_1 + y_0 \text{ and so on.}$$

In general the nth forward difference of y_0 is the difference of $(n - 1)^{\text{th}}$ forward difference of y_0 and is denoted by

$$\Delta^n y_0 = \Delta(\Delta^{n-1} y_0).$$

The forward difference Table 5.1 shows the manner in which the various forward differences can be formed. The first entry is called the leading term and Δy_0, $\Delta^2 y_0$, $\Delta^3 y_0$ etc., are called the leading differences. The forward differences with respect to any point, say x_0, slope in the downward direction (y_0, Δy_0, $\Delta^2 y_0$, ...).

Table 5.1: Forward differences table

Argument x	*Entry* y	*First difference* Δy	*Second difference* $\Delta^2 y$	*Third difference* $\Delta^3 y$
x_0	y_0			
		$\Delta y_0 = y_1 - y_0$		
$x_0 + h$	y_1		$\Delta^2 y_0 = \Delta y_1 - \Delta y_0$	
		$\Delta y_1 = y_2 - y_1$		$\Delta^3 y_0 = \Delta^2 y_1 - \Delta^2 y_0$
$x_0 + 2h$	y_2		$\Delta^2 y_1 = \Delta y_2 - \Delta y_1$	
		$\Delta y_2 = y_3 - y_2$		
$x_0 + 3h$	y_3			

Remark 1

If $y = f(x)$ is a polynomial of degree n in x represented by $f(x) = \sum_{r=0}^{n} a_r x^r$ then the nth order finite differences of y are constant and their value is given by $n!\ a_n\ h^n$, h being the interval of differencing.

EXAMPLE 5.1 Build upto the constant third difference a difference table of the polynomial $f(x) = 2x^3 - 3x^2 + 4x + 5$ for $x = 0, 1, 2, 3, 4$.

Solution We have $y = f(x) = 2x^3 - 3x^2 + 4x + 5$. Putting $x = 0, 1, 2, 3, 4$ in the polynomial $f(x)$, we obtain $f(0) = 5, f(1) = 8, f(2) = 17, f(3) = 44, f(4) = 101$.

Table 5.2: Difference table for the function

x	$y = f(x)$	Δy	$\Delta^2 y$	$\Delta^3 y$	$\Delta^4 y$
0	5				
		3			
1	8		6		
		9		12	
2	17		18		0
		27		12	
3	44		30		
		57			
4	101				

5.2.2 Backward Differences

The differences $y_1 - y_0, y_2 - y_1, ..., y_n - y_{n-1}$ when represented by $\nabla y_1, \nabla y_2, ..., \nabla y_n$ respectively, are defined as first backward differences so that

$$\nabla y_1 = y_1 - y_0 = f(x_0 + h) - f(x_0)$$
$$\nabla y_2 = y_2 - y_1 = f(x_1 + h) - f(x_1)$$
$$\cdots\cdots\cdots\cdots\cdots\cdots$$
$$\cdots\cdots\cdots\cdots\cdots\cdots$$
$$\nabla y_n = y_n - y_{n-1} = f(x_{n-1} + h) - f(x_{n-1}).$$

The symbol ∇ is called the backward difference operator.

Similarly, the second and higher order backward differences are defined as

$$\nabla^2 y_i = \nabla(\nabla y_i) = \nabla(y_i - y_{i-1}) = \nabla y_i - \nabla y_{i-1} = y_i - 2y_{i-1} + y_{i-2}$$
$$\nabla^3 y_i = \nabla(\nabla^2 y_i) = \nabla y_i - 2\nabla y_{i-1} + \nabla y_{i-2} = (y_i - y_{i-1}) - 2(y_{i-1} - y_{i-2}) + (y_{i-2} - y_{i-3})$$
$$= y_i - 3y_{i-1} + 3y_{i-2} - y_{i-3},$$

and in general, the nth backward difference is the difference of $(n-1)^{th}$ difference, that is

$$\nabla^n y_i = \nabla^{n-1}(\nabla y_i) = \nabla^{n-1}[y_i - y_{i-1}] = \nabla^{n-1} y_i - \nabla^{n-1} y_{i-1} = \nabla[\nabla^{n-1} y_i].$$

The successive backward differences are shown in a Table 5.3.

Table 5.3: Backward differences table

Argument x	*Entry* y	*First difference* ∇y	*Second difference* $\nabla^2 y$	*Third difference* $\nabla^3 y$
x_0	y_0			
		$\nabla y_1 = y_1 - y_0 = \Delta y_0$		
x_1	y_1		$\nabla^2 y_2 = \nabla y_2 - \nabla y_1 = \Delta^2 y_0$	
		$\nabla y_2 = y_2 - y_1 = \Delta y_1$		$\nabla^3 y_3 = \nabla^2 y_3 - \nabla^2 y_2 = \Delta^3 y_0$
x_2	y_2		$\nabla^2 y_3 = \nabla y_3 - \nabla y_2 = \Delta^2 y_1$	
		$\nabla y_3 = y_3 - y_2 = \Delta y_2$		
x_3	y_3			

From the Table 5.3, we note that $\Delta^n y_i = \nabla^n y_{i+n}$. The backward differences with respect to any point, say x_n, slope in the upward direction $(y_n, \nabla y_n, \nabla^2 y_n, \ldots)$.

5.2.3 Central Differences

While the forward and backward differences are gainfully used for interpolating the values near the beginning and the end points of the give data, the central differences are best suited for interpolation near the middle point of the tabular values. The central difference operator δ is defined by the following relations:

$$\delta y_n = y_{n+1/2} - y_{n-1/2} = f\left(x_0 + \left(n + \frac{1}{2}\right)h\right) - f\left(x_0 + \left(n - \frac{1}{2}\right)h\right)$$

and

$$\delta y_{n+1/2} = y_{n+1} - y_n = f(x_0 + (n+1)h) - f(x_0 + nh).$$

Setting $n = 0, 1, 2, \ldots$ we obtain

$$\delta y_{1/2} = y_1 - y_0 = f(x_0 + h) - f(x_0)$$

$$\delta y_{3/2} = y_2 - y_1 = f(x_0 + 2h) - f(x_0 + h) \text{ and so on.}$$

In a similar manner higher order central differences can be defined as

$$\begin{aligned} \delta^2 y_n &= \delta(\delta y_n) = \delta(y_{n+1/2} - y_{n-1/2}) = \delta y_{n+1/2} - \delta y_{n-1/2} \\ &= y_{n+1} - y_n - y_n + y_{n-1} = y_{n+1} - 2y_n + y_{n-1} \\ \delta^3 y_n &= \delta(\delta^2 y_n) = \delta(y_{n+1} - 2y_n + y_{n-1}) \\ &= \delta y_{n+1} - 2\delta y_n + \delta y_{n-1} = y_{n+3/2} - 3y_{n+1/2} + 3y_{n-1/2} - y_{n-3/2} \end{aligned}$$

The central differences are exhibited in Table 5.4.

Table 5.4: Central differences table

x	y	δy	$\delta^2 y$	$\delta^3 y$
x_0	y_0			
		$\delta y_{1/2} = y_1 - y_0$		
x_1	y_1		$\delta^2 y_1 = \delta y_{3/2} - \delta y_{1/2}$	
		$\delta y_{3/2} = y_2 - y_1$		$\delta^3 y_{3/2} = \delta^2 y_2 - \delta^2 y_1$
x_2	y_2		$\delta^2 y_2 = \delta y_{5/2} - \delta y_{3/2}$	
		$\delta y_{5/2} = y_3 - y_2$		
x_3	y_3			

We note from the above table that the central differences on the same horizontal line have the same suffix. Furthermore the differences of odd order are manifested by half value of the suffix and those of even order by integral value of suffixes.

Alternatively, we can write the central differences centered about x_0 as given in Table 5.5.

Table 5.5: Central difference table

Argument x	*Entry* y	*Ist difference* δy	*2nd difference* $\delta^2 y$	*3rd difference* $\delta^3 y$	*4th difference* $\delta^4 y$
x_{-2}	y_{-2}				
		$\delta y_{-3/2}$			
x_{-1}	y_{-1}		$\delta^2 y_{-1}$		
		$\delta y_{-1/2}$		$\delta^3 y_{-1/2}$	
x_0	y_0		$\delta^2 y_0$		$\delta^4 y_0$
		$\delta y_{1/2}$		$\delta^3 y_{1/2}$	
x_1	y_1		$\delta^2 y_1$		
		$\delta y_{3/2}$			
x_2	y_2				

5.2.4 Other Difference Operators

We have already discussed the operators Δ, ∇ and δ. We now study another two operators namely the shift operator E and the averaging operator μ.

(a) Shift (Displacement) operator E

The shift operator E is defined as the operator that increases the argument of a function by one tabular interval. Thus, by definition,

$$Ey_n = y_{n+1} \quad \text{or} \quad Ef(x_n) = f(x_n + h).$$

Here the input to the operator is $f(x_n)$. The output is $f(x_n + h)$ or if the input to the operator is y_n. The output is y_{n+1}.

$$\boxed{f(x_n)} \longrightarrow \boxed{E} \longrightarrow \boxed{f(x_n + h)}$$

The effect of the operation of E on y_n or $f(x_n)$ is to shift or increase the functional value y_n or $f(x_n)$ by one interval to the next higher value y_{n+1}. The higher order operations of E can be similarly defined by

$$E^2 f(x_n) = E(Ef(x_n)) = Ef(x_n + h) = f(x_n + 2h) = y_{n+2}$$

$$E^3 f(x_n) = E^3 y_n = f(x_n + 3h) = y_{n+3}$$

....

....

In general the rth operation with E is represented by

$$E^r f(x_n) = E^r(y_n) = f(x_n + rh) = y_{n+r} = E^n(y_r)$$

In particular $E^{1/2}\, y_n = y_{n+(1/2)}$, $E^{-1/2} y_n = y_{n-(1/2)}$ and $E^r y_0 = y_r$.

(b) The Averaging (or Mean) Operator μ

The averaging operator μ is defined by the relation.

$$\mu y_n = \frac{1}{2}(y_{n+1/2} + y_{n-1/2}) = \frac{1}{2}(E^{1/2} y_n + E^{-1/2} y_n) = \frac{1}{2}(E^{1/2} + E^{-1/2}) y_n,$$

and

$$\mu\delta y_n = \mu(y_{n+1/2} - y_{n-1/2}) = \mu y_{n+1/2} - \mu y_{n-1/2}$$

$$= \frac{1}{2}(y_{n+1} + y_n) - \frac{1}{2}(y_n + y_{n-1})$$

$$= \frac{1}{2}(y_{n+1} + y_n - y_n - y_{n-1}) = \frac{1}{2}(y_{n+1} - y_{n-1}).$$

The averaging operator μ gives the mean value at n of the two adjacent values at $n + \frac{1}{2}$ and $n - \frac{1}{2}$.

5.3 RELATIONS BETWEEN THE OPERATORS

We know from the definition of finite differences that

$$y_1 - y_0 = \Delta y_0 = \nabla y_1 = \delta y_{1/2}. \quad \text{...(5.1)}$$

It is obvious that from the above relations that the value $y_1 - y_0$ is the forward difference of y_0, the backward difference of y_1 and the central difference of $y_{1/2}$. This means that is is not the difference $y_1 - y_0$ that changes but the change is affected only in the notations.

(a) Relation between *E* and Δ

Consider the relation $y_1 - y_0 = \Delta y_0$, we have $y_1 = y_0 + \Delta y_0 = (1 + \Delta)y_0$. But $y_1 = Ey_0$.
Therefore, $Ey_0 = (1 + \Delta)y_0$ and this gives $\boldsymbol{E = 1 + \Delta}$. ...(5.2)

(b) Relation between *E* and ∇

Again $y_1 - y_0 = \nabla y_1$ or $y_0 = y_1 - \nabla y_1 = (1 - \nabla)y_1$.

But $y_0 = E^{-1}y_1$. Therefore, $E^{-1}y_1 = (1 - \nabla)y_1$ or $1 - \nabla = E^{-1}$ or $\nabla = 1 - E^{-1}$ or $\boldsymbol{E = (1 - \nabla)^{-1}}$. ...(5.3)

Remark 2 From Eq. (5.2) and Eq. (5.3), we obtain $\nabla - \Delta = -\Delta\nabla$. ...(5.4)

(c) Relation between *E*, δ and Δ

From the relation $y_1 - y_0 = \delta y_{1/2}$, we have $E^{1/2}y_{1/2} - E^{-1/2}y_{1/2} = \delta y_{1/2}$

or $(E^{1/2} - E^{-1/2})y_{1/2} = \delta y_{1/2}$. This gives $\boldsymbol{\delta = E^{1/2} - E^{-1/2}}$. ...(5.5) *(AMIE, W-2012)*

Again $\delta = E^{1/2} - E^{-1/2} = E^{-1/2}(E - 1) = E^{-1/2}\Delta$ or $\boldsymbol{\Delta = \delta E^{1/2}}$. ...(5.6)

Remark 3

From Eq. (5.5), we have
$$\begin{cases} \delta = E^{1/2}(1 - E^{-1}) = \nabla E^{1/2} = \nabla(1 - \nabla)^{-1/2}. & \text{...(5.7)} \\ \delta = E^{-1/2}(E - 1) = \Delta E^{-1/2} = \Delta(1 + \Delta)^{-1/2}. & \text{...(5.8)} \end{cases}$$

(d) Relation between *E*, Δ and *D*

Defining the differential operator D by $\frac{d}{dx}$, we have

$$Df(x) = \frac{d}{dx}f(x) = f'(x).$$

Again $Ef(x) = f(x + h) = f(x) + hf'(x) + \frac{h^2}{2!}f''(x) + \frac{h^3}{3!}f'''(x) + \cdots$ [by Taylor's theorem]

$$= f(x) + hDf(x) + \frac{h^2D^2}{2!}f(x) + \frac{h^3D^3}{3!}f(x) + \ldots\ldots$$

$$= \left(1 + hD + \frac{h^2D^2}{2!} + \frac{h^3D^3}{3!} + \cdots\right)f(x) = e^{hD}f(x).$$

Therefore, $\boldsymbol{E = e^{hD} = 1 + D}$. ...(5.9) *(AMIE-W-2012)*

Taking logarithm on both sides, we obtain

$$hD = \log_e E = \log_e(1 + \Delta) = \Delta - \frac{1}{2}\Delta^2 + \frac{1}{3}\Delta^3 - \frac{\Delta^4}{4} + \cdots \quad \textit{(AMIE-W-2012)}$$

or
$$\boldsymbol{D = \frac{1}{h}\left[\Delta - \frac{\Delta^2}{2} + \frac{\Delta^3}{3} - \frac{\Delta^4}{4} + \cdots\right]}. \quad \text{...(5.10)}$$

Remark 4

From Eq. (5.3) $E = (1 - \nabla)^{-1} = e^{hD}$. ...(5.11)

Taking logarithms in the above equation, we obtain

$$hD = \log_e E = \log_e(1 - \nabla)^{-1} = -\log_e(1 - \nabla)$$

or
$$D = \frac{1}{h}\left[\nabla + \frac{1}{2}\nabla^2 + \frac{1}{3}\nabla^3 + \cdots\right]. \quad ...(5.12)$$

(e) Relation between μ, δ

We have $\delta = E^{1/2} - E^{-1/2} = e^{hD/2} - e^{-hD/2} = 2\sinh\left(\frac{hD}{2}\right)$. ...(5.13)

Again $\mu = \frac{1}{2}(E^{1/2} + E^{-1/2}) = \frac{1}{2}(e^{hD/2} + e^{-hD/2}) = \cosh\left(\frac{hD}{2}\right)$. ...(5.14)

Squaring Eqs. (5.13) and (5.14) and subtracting, we obtain

$$\mu^2 - \frac{\delta^2}{4} = \cosh^2\left(\frac{hD}{2}\right) - \sinh^2\left(\frac{hD}{2}\right) = 1 \quad \text{or} \quad \mu^2 = 1 + \frac{\delta^2}{4}. \quad ...(5.15)$$

(AMIETE, Dec. 2012)

Remark 5

We know that
$$\mu\delta y_n = \frac{1}{2}(\delta y_{n+1/2} + \delta y_{n-1/2}) = \frac{1}{2}(y_{n+1} - y_n + y_n - y_{n-1})$$
$$= \frac{1}{2}(y_{n+1} - y_{n-1}) = \frac{1}{2}(E - E^{-1})y_n.$$

Hence,
$$\mu\delta = \frac{1}{2}(E - E^{-1}) = \frac{1}{2}(e^{hD} - e^{-hD}) = \sinh(hD)$$

or
$$hD = \sinh^{-1}(\mu\delta). \quad ...(5.16)$$

EXAMPLE 5.2 Evaluate the following:

(*i*) $\Delta\left[\frac{1}{f_i}\right]$ (*AMIETE, June 2004*), (*ii*) $\Delta^2\left[\frac{5x+12}{x^2+5x+6}\right]$, (*iii*) $\Delta^n(x^n)$,

(*iv*) $\Delta^n(1/x)$, (*v*) $\Delta^n(abc^{x})$, (*vi*) $\Delta^n(e^{2x+5})$. (*AMIE, W-2000*)

the interval of differencing being unity.

Solution (*i*) $\Delta\left(\frac{1}{f_i}\right) = \frac{1}{f_{i+1}} - \frac{1}{f_i} = -\frac{f_{i+1} - f_i}{f_i f_{i+1}} = -\frac{\Delta f_i}{f_i f_{i+1}}$.

(*ii*)
$$\Delta^2\left(\frac{5x+12}{x^2+5x+6}\right) = \Delta^2\left\{\frac{5x+12}{(x+2)(x+3)}\right\} = \Delta^2\left\{\frac{2}{x+2} + \frac{3}{x+3}\right\}$$
$$= \Delta\left\{\Delta\left(\frac{2}{x+2}\right) + \Delta\left(\frac{3}{x+3}\right)\right\} = \Delta\left\{2\left(\frac{1}{x+3} - \frac{1}{x+2}\right) + 3\left(\frac{1}{x+4} - \frac{1}{x+3}\right)\right\}$$
$$= -2\Delta\left\{\frac{1}{(x+2)(x+3)}\right\} - 3\Delta\left\{\frac{1}{(x+3)(x+4)}\right\}$$
$$= -2\left\{\frac{1}{(x+3)(x+4)} - \frac{1}{(x+2)(x+3)}\right\} - 3\left\{\frac{1}{(x+4)(x+5)} - \frac{1}{(x+3)(x+4)}\right\}$$
$$= \frac{4}{(x+2)(x+3)(x+4)} + \frac{6}{(x+3)(x+4)(x+5)} = \frac{2(5x+16)}{(x+2)(x+3)(x+4)(x+5)}.$$

(*iii*) $\Delta^n(x^n) = \Delta^{n-1}\{\Delta x^n\} = \Delta^{n-1}\{(x+1)^n - x^n\} = \Delta^{n-1}\{{}^nc_1 x^{n-1}\}$

[(On binomial expansion), other terms vanish, having powers < (n – 1)]

$$= n\Delta^{n-2}\{\Delta x^{n-1}\} = n\Delta^{n-2}\{(x+1)^{n-1} - x^{n-1}\} = n \cdot {}^{n-1}c_1 \Delta^{n-2}(x^{n-2})$$

$$= n(n-1)\Delta^{n-2}(x^{n-2}) = n(n-1)(n-2) \ldots\ldots\ldots 3 \cdot 2\,\Delta x$$

$$= n(n-1)(n-2) \ldots\ldots\ldots 3 \cdot 2 \cdot 1 = n!. \qquad [\text{As } \Delta x = (x+1) - x = 1]$$

(*iv*) $\Delta^n\left(\frac{1}{x}\right) = \Delta^{n-1}\left\{\Delta\left(\frac{1}{x}\right)\right\} = \Delta^{n-1}\left\{\frac{1}{x+1} - \frac{1}{x}\right\} = \Delta^{n-1}\left\{\frac{-1}{x(x+1)}\right\}$

$$= \Delta^{n-2}\left[\Delta\left\{\frac{-1}{x(x+1)}\right\}\right] = \Delta^{n-2}(-1)\left\{\frac{1}{(x+1)(x+2)} - \frac{1}{x(x+1)}\right\}$$

$$= \Delta^{n-2}(-1)\left\{\frac{-2}{x(x+1)(x+2)}\right\} = \Delta^{n-2}\frac{(-1)^2 2}{x(x+1)(x+2)} = \frac{(-1)^n n!}{x(x+1)(x+2)\ldots\ldots(x+n)}.$$

(*v*) $\Delta^n(ab^{cx}) = \Delta^{n-1}[\Delta(ab^{cx})] = \Delta^{n-1}[ab^{c(x+1)} - ab^{cx}] = (b^c - 1)\,\Delta^{n-1}(ab^{cx})$

$$= (b^c - 1)\,\Delta^{n-2}[\Delta(ab^{cx})] = (b^c - 1)\,\Delta^{n-2}\,[ab^{c(x+1)} - ab^{cx})$$

$$= (b^c - 1)^2\,\Delta^{n-2}(ab^{cx})$$

....

....

$$= (b^c - 1)^{n-1}\,\Delta(ab^{cx}) = (b^c - 1)^{n-1}\,(ab^{c(x+1)} - ab^{cx}) = (b^c - 1)^n\,(ab^{cx}).$$

(*vi*) $\Delta^n(e^{2x+5}) = \Delta^{n-1}[\Delta e^{2x+5}] = \Delta^{n-1}[e^{2(x+1)+5} - e^{2x+5}]$

$$= (e^2 - 1)\,\Delta^{n-1}\,(e^{2x+5}) = (e^2 - 1)\,\Delta^{n-2}\,[\Delta e^{2x+5}]$$

$$= (e^2 - 1)^2\,\Delta^{n-2}(e^{2x+5})$$

....

....

$$= (e^2 - 1)^{n-1}\,\Delta e^{2x+5} = (e^2 - 1)^n\,e^{2x+5}.$$

Remark 6

If the interval of differencing being h, then $\Delta^n\,e^{2x+5} \quad = (e^{2h} - 1)^n\,e^{2x+5}$.

EXAMPLE 5.3 Evaluate the following:

(*i*) $\Delta^3[(1+x)(1+2x)(1+3x)]$, if the interval of differencing is 1. (*AMIETE, Dec. 2011*)

(*ii*) $\Delta^{10}[(1-ax)(1-bx^2)(1-cx^3)(1-dx^4)]$.

Solution (*i*) $\Delta^3[(1+x)(1+2x)(1+3x)] = \Delta^3(6x^3)$, other terms vanishing as they have powers < 3.

$= 6\Delta^3x^3 = 6\Delta^2(\Delta x^3) = 6\Delta^2\{(x+1)^3 - x^3\}$

$= 6\Delta^2(3x^2 + 3x + 1) = 18\Delta^2x^2$, other terms vanishing, having powers < 2

$= 18\Delta(\Delta x^2) = 18\,\Delta\{(x+1)^2 - x^2\} = 18\,\Delta(2x+1)$

$= 36\Delta x = 36\{(x+1) - x\} = 36.$ **[Note:** $\Delta^3x^3 = 3!$**]**

(*ii*) $\Delta^{10}[(1-ax)(1-bx^2)(1-cx^3)(1-dx^4)] = \Delta^{10}(abcd\,x^{10})$, other terms vanish having powers < 10

$= abcd\,\Delta^{10}(x^{10}) = abcd\,(10!)$. **[Note:** $\Delta^{10}(x^n) = 0$ for $n < 10$**]**

EXAMPLE 5.4 Prove the following relations

(*i*) $\Delta(f_i \cdot g_i) = f_i\,\Delta\,g_i + g_{i+1}\Delta\,f_i$, (*ii*) $\Delta(f_i/g_i) = \dfrac{g_i\Delta f_i - f_i\Delta g_i}{g_i g_{i+1}}$, (*AMIETE; June 2004*)

(*iii*) $\Delta(f_i^2) = (f_i + f_{i+1})\Delta f_i$. (*AMIETE; June 2005*)

Solution (*i*) $\Delta(f_i \cdot g_i) = f_{i+1}\ g_{i+1} - f_i g_i = f_{i+1}\ g_{i+1} - f_i\ g_i + f_i\ g_{i+1} - f_i\ g_{i+1}$

$= f_i(g_{i+1} - g_i) + g_{i+1}(f_{i+1} - f_i) = f_i\ \Delta g_i + g_{i+1}\ \Delta f_i$.

(*ii*) $\Delta\left(\dfrac{f_i}{g_i}\right) = \dfrac{f_{i+1}}{g_{i+1}} - \dfrac{f_i}{g_i} = \dfrac{g_i f_{i+1} - f_i g_{i+1}}{g_i\ g_{i+1}} = \dfrac{g_i\ (f_{i+1} - f_i) - f_i(g_{i+1} - g_i)}{g_i g_{i+1}} = \dfrac{g_i \Delta f_i - f_i \Delta g_i}{g_i g_{i+1}}$.

(*iii*) $\Delta(f_i^2) = f_{i+1}^2 - f_i^2 = (f_{i+1} + f_i)(f_{i+1} - f_i) = (f_{i+1} + f_i)\,\Delta f_i$.

EXAMPLE 5.5 Evaluate the following, the interval of differencing being h

(*i*) $\Delta \cot 2^x$, (*ii*) $\Delta^2 \cos 2x$, (*iii*) $\Delta e^{2x} \log 3x$, (*iv*) $\Delta\left(\dfrac{x^2}{\cos 2x}\right)$.

Solution (*i*) $\Delta \cot 2^x = \cot 2^{(x+h)} - \cot 2^x = \dfrac{\cos 2^{x+h}}{\sin 2^{x+h}} - \dfrac{\cos 2^x}{\sin 2^x}$

$$= \frac{\sin 2^x \cdot \cos 2^{x+h} - \cos 2^x \cdot \sin 2^{x+h}}{\sin 2^x \cdot \sin 2^{x+h}} = \frac{\sin(2^x - 2^{x+h})}{\sin 2^x \sin 2^{x+h}} = \frac{\sin\{2^x(1-2^h)\}}{\sin 2^x \sin 2^{x+h}}.$$

(*ii*) $\Delta^2 \cos 2x = \Delta(\Delta \cos 2x) = \Delta[\cos 2(x+h) - \cos 2x]$

$= \Delta \cos 2(x+h) - \Delta \cos 2x$

$= [\cos 2(x+2h) - \cos 2(x+h)] - [\cos 2(x+h) - \cos 2x]$

$= -2 \sin(2x+3h) \sin h + 2 \sin(2x+h) \sin h$

$= -2 \sin h[\sin(2x+3h) - \sin(2x+h)]$

$= -2 \sin h\,[(2 \cos(2x+2h) \sin h] = -4 \sin^2 h \cos(2x+2h)$.

(*iii*) Let $f(x) = e^{2x}$ and $g(x) = \log 3x$ so that

$$\Delta f(x) = e^{2(x+h)} - e^{2x} = e^{2x}(e^{2h} - 1) \text{ and}$$

$$\Delta g(x) = \log 3(x+h) - \log 3x = \log\frac{3(x+h)}{3x} = \log\left(1+\frac{h}{x}\right).$$

Now $\Delta[f(x)\,g(x)] = f(x+h)\,\Delta g(x) + g(x)\,\Delta f(x)$.

Therefore,
$$\Delta e^{2x} \log 3x = e^{2(x+h)} \log\left(1+\frac{h}{x}\right) + (\log 3x)\,e^{2x}(e^{2h} - 1)$$

$$= e^{2x}\left[e^{2h} \log\left(1+\frac{h}{x}\right) + (e^{2h} - 1)\log 3x\right].$$

(iv) Let $f(x) = x^2$ and $g(x) = \cos 2x$, so that

$$\Delta f(x) = (x+h)^2 - x^2 = h(2x+h) \text{ and}$$

$$\Delta g(x) = \cos 2(x+h) - \cos 2x = -2 \sin(2x+h) \sin h.$$

Now
$$\Delta\left[\frac{f(x)}{g(x)}\right] = \frac{g(x)\,\Delta f(x) - f(x)\,\Delta g(x)}{g(x+h)\,g(x)}.$$

Therefore,
$$\Delta\left[\frac{x^2}{\cos 2x}\right] = \frac{\cos 2x \cdot h(2x+h) - x^2[-2\sin(2x+h)\sin h]}{\cos(2x+2h)\cos 2x}$$

$$= \frac{h(2x+h)\cos 2x + 2x^2 \sin h \cdot \sin(2x+h)}{\cos(2x+2h)\cos 2x}.$$

EXAMPLE 5.6 Evaluate the following

(*i*) $\left(\dfrac{\Delta^2}{E}\right)x^3$, (*AMIE., W-2004*) (*ii*) $\dfrac{\Delta^2 x^3}{Ex^3}$,

(iii) $\Delta \log f(x)$, (AMIETE, June 2012; AMIE., S-2003)

(iv) $\left(\frac{\Delta^2}{E}\right)e^x \cdot \frac{Ee^x}{\Delta^2 e^x}$, the interval of differencing being h. (AMIE., W-2004; Bhopal 2002)

(v) $E\Delta f(x)$ and $E\nabla f(x)$ for $x = 1$ with $h = 0.2$. Given $f(x) = xe^x$, the interval of differencing being h. (AMIE, W-2004)

Solution (i) $\left(\frac{\Delta^2}{E}\right)x^3 = \left\{\frac{(E-1)^2}{E}\right\}x^3 = (E - 2 + E^{-1})x^3 = Ex^3 - 2x^3 + E^{-1}x^3$

$$= (x+h)^3 - 2x^3 + (x-h)^3 = 6h^2x.$$

(ii) $$\frac{\Delta^2 x^3}{Ex^3} = \frac{(E-1)^2 x^3}{Ex^3} = \frac{(E^2 - 2E + 1)x^3}{Ex^3} = \frac{(x+2h)^3 - 2(x+h)^3 + x^3}{(x+h)^3} = \frac{6xh^2 + 6h^3}{(x+h)^3} = \frac{6h^2}{(x+h)^2}.$$

(iii) $$\Delta \log f(x) = \log f(x+h) - \log f(x) = \log\left[\frac{f(x+h)}{f(x)}\right] = \log\left[\frac{E\,f(x)}{f(x)}\right] = \log\left[\frac{(1+\Delta)\,f(x)}{f(x)}\right]$$

$$= \log\left[\frac{f(x) + \Delta f(x)}{f(x)}\right] = \log\left[1 + \frac{\Delta f(x)}{f(x)}\right].$$

(iv) We have, $$\left(\frac{\Delta^2}{E}\right)e^x = \Delta^2 E^{-1} e^x = \Delta^2 e^{x-h} = \Delta(\Delta e^{x-h}) = \Delta(e^x - e^{x-h})$$

$$= \Delta e^x - \Delta e^{x-h} = e^{x+h} - e^x - (e^x - e^{x-h}) = e^{x+h} - 2e^x + e^{x-h}$$

$$= e^x(e^h - 2 + e^{-h}) = e^x(e^{2h} - 2e^h + 1)\ \frac{1}{e^h} = \frac{e^x(e^h - 1)^2}{e^h}. \qquad ...(1)$$

Again $$\frac{Ee^x}{\Delta^2 e^x} = \frac{e^{x+h}}{\Delta(\Delta e^x)} = \frac{e^{x+h}}{\Delta(e^{x+h} - e^x)} = \frac{e^{x+h}}{\Delta e^{x+h} - \Delta e^x}$$

$$= \frac{e^{x+h}}{e^{x+2h} - e^{x+h} - (e^{x+h} - e^x)} = \frac{e^{x+h}}{e^{x+2h} - 2e^{x+h} + e^x}$$

$$= \frac{e^{x+h}}{e^x(e^{2h} - 2e^h + 1)} = \frac{e^h}{(e^h - 1)^2}. \qquad ...(2)$$

Multiplying Eqs. (1) and (2), we get

$$\left(\frac{\Delta^2}{E}\right)e^x \cdot \frac{Ee^x}{\Delta^2 e^x} = \frac{e^x(e^h - 1)^2}{e^h} \cdot \frac{e^h}{(e^h - 1)^2} = e^x.$$

(v) $E\Delta f(x) = E\{f(x+h) - f(x)\} = Ef(x+h) - Ef(x) = f(x+2h) - f(x+h)$.

Therefore, $E\Delta(xe^x) = (x + 2h)\, e^{x+2h} - (x + h)e^{x+h}$.

$E\Delta(xe^x)$ at $(x = 1, h = 0.2) = (1.4)\, e^{1.4} - (1.2)\, e^{1.2} = (1.4)(4.0552) - (1.2)(3.32012)$

$= 5.67728 - 3.98414 = 1.69314$.

And $\quad E\,\nabla f(x) = E\{f(x) - f(x-h)\} = f(x+h) - f(x)$.

Therefore, $\quad E\,\nabla(xe^x) = (x + h)e^{x+h} - xe^x$.

Thus $E\nabla(xe^x)$ at $(x = 1, h = 0.2) = (1.2)e^{1.2} - e = (1.2)(3.32012) - 2.71828 = 1.26586$.

EXAMPLE 5.7 If $y_0 = 2$, $y_1 = -6$, $y_2 = 8$, $y_3 = 9$ and $y_4 = 17$, then, without using difference table, calculate $\Delta^4 y_0$. (AMIE. S-2011)

Solution We know that $\Delta = E - 1$. Therefore, $\Delta^4 y_0 = (E - 1)^4 y_0 = (E^4 - 4E^3 + 6E^2 - 4E + 1)y_0$

$= y_4 - 4y_3 + 6y_2 - 4y_1 + y_0 = 17 - 4(9) + 6(8) - 4(-6) + 2$

$= 17 - 36 + 48 + 24 + 2 = 55$.

EXAMPLE 5.8 Show that $y_{-3} = y_0 - 3\nabla y_0 + 3\nabla^2 y_0 - \nabla^3 y_0$.

Solution Using $\nabla = 1 - E^{-1}$ or $E = (1-\nabla)^{-1}$, we obtain

$$E^{-3} y_0 = \{(1-\nabla)^{-1}\}^{-3} y_0 = (1-\nabla)^3 y_0 = (1 - 3\nabla + 3\nabla^2 - \nabla^3) y_0$$

or $\quad y_{-3} = y_0 - 3\nabla y_0 + 3\nabla^2 y_0 - \nabla^3 y_0$.

EXAMPLE 5.9 Prove that $\delta^2 y_5 = y_6 - 2y_5 + y_4$.

Solution Employing $\quad \delta = (E^{1/2} - E^{-1/2})$ or $\delta^2 = (E^{1/2} - E^{-1/2})^2 = (E - 2 + E^{-1})$.

Therefore, $\quad \delta^2 y_5 = (E - 2 + E^{-1}) y_5 = y_6 - 2y_5 + y_4$.

EXAMPLE 5.10 Prove the following relations:

(*a*) $\Delta + \nabla = \dfrac{\Delta}{\nabla} - \dfrac{\nabla}{\Delta}$, (*AMIE, S-2012; AMIETE, June 2003*) (*ii*) $\Delta^3 y_2 = \nabla^3 y_5$, (*AMIETE, June 2006*)

(*iii*) $\Delta = \dfrac{1}{2}\delta^2 + \delta\sqrt{1 + \dfrac{1}{4}\delta^2}$, (*AMIETE, Dec. 2012, June 2004; Madras, 2003*)

(*iv*) $\mu\delta = \dfrac{1}{2}\Delta + \dfrac{1}{2}\Delta E^{-1}$, (*v*) $\sqrt{1+\mu^2\delta^2} = 1 + \dfrac{1}{2}\delta^2$. (*AMIETE, June 2005*)

Solution (*i*) Using operational relations: $\Delta = E - 1$ and $\nabla = 1 - E^{-1}$.

L.H.S. $= \Delta + \nabla = E - 1 + 1 - E^{-1} = E - E^{-1}$.

$$\text{R.H.S.} = \frac{\Delta}{\nabla} - \frac{\nabla}{\Delta} = \frac{E-1}{1-E^{-1}} - \frac{1-E^{-1}}{E-1} = \frac{E(1-E^{-1})}{1-E^{-1}} - \frac{E^{-1}(E-1)}{E-1} = E - E^{-1} = \text{L.H.S. Hence the result.}$$

(*ii*) $\Delta^3 y_2 = (E-1)^3 y_2 = (E^3 - 3E^2 + 3E - 1) y_2 = y_5 - 3y_4 + 3y_3 - y_2$.

$\nabla^3 y_5 = (1 - E^{-1})^3 y_5 = (1 - 3E^{-1} + 3E^{-2} - E^{-3}) y_5 = y_5 - 3y_4 + 3y_3 - y_2$.

From above it is obvious that $\Delta^3 y_2 = \nabla^3 y_5$.

(*iii*)
$$\begin{aligned}\text{R.H.S.} = \frac{1}{2}\delta^2 + \delta\sqrt{1+\frac{1}{4}\delta^2} &= \frac{1}{2}(E^{1/2} - E^{-1/2})^2 + (E^{1/2} - E^{-1/2}) \times \sqrt{1 + \frac{1}{4}(E^{1/2} - E^{-1/2})^2}\\
&= \frac{1}{2}(E - 2 + E^{-1}) + (E^{1/2} - E^{-1/2})\sqrt{\frac{E + 2 + E^{-1}}{4}}\\
&= \frac{1}{2}(E - 2 + E^{-1}) + \frac{1}{2}(E^{1/2} - E^{-1/2})(E^{1/2} + E^{-1/2})\\
&= \frac{1}{2}(E - 2 + E^{-1}) + \frac{1}{2}(E - E^{-1}) = \frac{1}{2}(2E - 2)\\
&= E - 1 = \Delta = \text{L.H.S. Hence the result.}\end{aligned}$$

(*iv*)
$$\begin{aligned}\text{L.H.S.} = \mu\delta &= \frac{1}{2}(E^{1/2} + E^{-1/2})(E^{1/2} - E^{-1/2}) = \frac{1}{2}(E - E^{-1}) = \frac{1}{2}\left(1 + \Delta - \frac{1}{E}\right)\\
&= \frac{1}{2}\Delta + \frac{1}{2}\left(\frac{E-1}{E}\right) = \frac{1}{2}\Delta + \frac{1}{2}\Delta E^{-1} = \text{R.H.S. Hence the result}\end{aligned}$$

(*v*) We know that $\delta = E^{1/2} - E^{-1/2}$ and $\mu = \dfrac{1}{2}(E^{1/2} + E^{-1/2})$.

Therefore, $\quad \mu\delta = \dfrac{1}{2}(E^{1/2} + E^{-1/2})(E^{1/2} - E^{-1/2}) = \dfrac{1}{2}(E - E^{-1})$,

and $\quad \sqrt{1+\mu^2\delta^2} = \sqrt{1 + \dfrac{1}{4}(E^2 - 2 + E^{-2})} = \dfrac{1}{2}(E + E^{-1})$. ...(1)

Now $\quad 1 + \dfrac{1}{2}\delta^2 = 1 + \dfrac{1}{2}(E^{1/2} - E^{-1/2})^2 = 1 + \dfrac{1}{2}(E - 2 + E^{-1}) = \dfrac{1}{2}(E + E^{-1})$. ...(2)

From Eqs. (1) and (2), we find that $\sqrt{1+\mu^2\delta^2} = 1 + \dfrac{1}{2}\delta^2$.

EXAMPLE 5.11 Find the missing values in the following table:

x :	0	1	2	3	4	5
$y = f(x)$:	10	—	8	10	—	50

Solution Let the missing values be y_1 and y_4. Since the table contains a set of four known values of x and $f(x)$, we can represent $f(x)$ as a polynomial of degree three in x. In this case, the third differences of $f(x)$ namely $\Delta^3 f(x)$ or $\Delta^3 y$ will be constant and all differences of order four and more will be zero. Therefore, $\Delta^4 y_x = 0$ and $\Delta^5 y_x = 0$ for all values of x. Considering the leading differences, we get $\Delta^4 y_0 = 0$ or $(E-1)^4 y_0 = 0$. Expanding this, we obtain $(E^4 - 4E^3 + 6E^2 - 4E + 1)y_0 = 0$

or
$$y_4 - 4y_3 + 6y_2 - 4y_1 + y_0 = 0.$$

Substituting the known values of y_0, y_2 and y_3 and simplifying, we obtain

$$y_4 - 4y_1 = -18. \qquad \text{...(i)}$$

Again taking $\Delta^5 y = 0$ or $(E-1)^5 y_0 = 0$, that is,

$$(E^5 - 5E^4 + 10E^3 - 10E^2 + 5E - 1)y_0 = 0 \quad \text{or} \quad y_5 - 5y_4 + 10y_3 - 10y_2 + 5y_1 - y_0 = 0.$$

Substituting the given values of y_x and simplifying, we get $y_4 - y_1 = 12$. ...(ii)

Solving (i) and (ii), we obtain $y_1 = 10$ and $y_4 = 22$.

5.4 FACTORIAL POLYNOMIAL

The continued product of n factors of the type $x(x-h)(x-2h), ..., (x - \overline{n-1}h)$ in which the first factor is x and the successive factors decrease by a constant is called the **factorial polynomial** of order n and is denoted by $x^{(n)}$. Thus

$$x^{(n)} = x(x-h)(x-2h), ..., [x-(n-1)h].$$

In particular

$$x^{(4)*} = x(x-h)(x-2h)(x-3h); \qquad x^{(3)} = x(x-h)(x-2h);$$
$$x^{(2)} = x(x-h); \qquad x^{(1)} = x \text{ and } x^{(0)} = 1.$$

The importance of these polynomials lies in the fact that they play the same role in calculus of finite differences as the function x^n play in ordinary calculus. It may be noted that various differences of factorial polynomials are again factorial polynomials. This is evident from the following results.

(A) $\Delta x^{(n)} = nhx^{(n-1)}, \quad n = 1, 2, 3, ...$

Proof:

$$\Delta x^{(n)} = (x+h)^{(n)} - x^{(n)}$$
$$= [(x+h)x(x-h), ..., (x-(n-2)h] - [x(x-h), ... (x-(n-1)h]$$
$$= x(x-h), ..., [x-(n-2)h][(x+h)-(x-(n-1)h)]$$
$$= x(x-h), ... [x-(n-2)h][nh] = nhx^{(n-1)}. \qquad \text{...(5.17)}$$

Similarly, $\Delta^2 x^{(n)} = \Delta(\Delta x^{(n)}) = \Delta[nhx^{(n-1)}] = nh\Delta x^{(n-1)} = n(n-1)h^2 x^{(n-2)}$

and $\Delta^n x^{(n)} = n(n-1)(n-2), ..., 1 \cdot h^n = n!h^n$.

Remarks 7

(a) If $h = 1$, then $x^{(n)} = x(x-1)(x-2), ..., (x-n+1)$ and $\Delta x^{(n)} = nx^{(n-1)}$; $\quad \Delta^n x^{(n)} = n!$

(b) The result proved at Eq. (5.17) above bears close analogy with the result $Dx^n = nx^{n-1}$ where D is the differential operator defined as $D = d/dx$.

(B) In ordinary calculus, $(1/D)\,x^n$ or $D^{-1}x^n$ defines a function whose derivative is x^n. Therefore, $\frac{1}{D}x^n = \frac{x^{n+1}}{n+1}$.

For this reason $\frac{1}{D}x^n$ or $D^{-1}x^n$ is called the anti-derivatives of x^n.

*Obviously $x^{(4)} = x(x-h)(x-2h)^{(2)}$.

By analogy with the above terminology of calculus, we obtain

$$\frac{1}{\Delta}x^{(n)} = \Delta^{-1}x^{(n)} = \frac{x^{(n+1)}}{n+1}.$$

$\frac{1}{\Delta}x^{(n)}$ is called the antidifference of $x^{(n)}$. It stands for a function whose first difference is $x^{(n)}$.

5.4.1 Methods of Expressing an Algebraic Polynomial in Factorial Notation

To transform a given polynomial in terms of factorial notation, we give below two methods and their procedures are explained by taking some examples.

EXAMPLE 5.12 Express $2x^3 - 3x^2 + 3x + 15$ in terms of factorial polynomials.

Solution First Method (*Direct Method*)

Let
$$y = 2x^3 - 3x^2 + 3x + 15 = A\,x^{(3)} + B\,x^{(2)} + C\,x^{(1)} + D$$
$$= Ax(x-1)(x-2) + Bx(x-1) + Cx + D, \qquad \text{...(1)}$$
where A, B, C, D are constants to be determined.

Putting $x = 0$, 1 and 2 successively on both sides in the above relation, we get

$$D = 15,\ C + D = 17,\ 2B + 2C + D = 25 \text{ giving } D = 15,\ C = 2,\ B = 3.$$

Equating the coefficient of x^3 on both sides, we get $A = 2$.

Therefore, $2x^3 - 3x^2 + 3x + 15 = 2x^{(3)} + 3x^{(2)} + 2x^{(1)} + 15$.

Second Method (Method of Synthetic Division)

In this method we write the given polynomial in terms of factorial notation, that is,

$$2x^3 - 3x^2 + 3x + 15 = Ax(x-1)(x-2) + Bx(x-1) + Cx + D.$$

It is clear from above that when we divide both sides by x, the quotient on the left side $2x^2 - 3x + 3$ and on the right is $A(x-1)(x-2) + B(x-1) + C$ while the remainders are 15 and D respectively.

Again when these quotients are divided by $(x-1)$, the resulting quotients of the two sides will be $2x - 1$ and $A(x-2) + B$ respectively. The corresponding remainders will be given by 2 and C.

Lastly when the quotients $2x - 1$ and $A(x-2) + B$ are divided by $x - 2$, the quotients will be 2 and A and remainders 3 and B. From the above discussions, we get $A = 2$, $B = 3$, $C = 2$ and $D = 15$.

Substituting the values of A, B, C and D, we get the desired polynomial in factorial notation.

The above procedure can be put in a very simple and compact form known as *synthetic division* as follows:

0	2	−3	3	15
		0	0	0
1	2	−3	3	$15 = D$
		2	−1	
2	2	−1	$2 = C$	
		4		
	$2 = A$	$3 = B$		

Therefore, $2x^3 - 3x^2 + 15 = 2x^{(3)} + 3x^{(2)} + 2x^{(1)} + 15$.

Remark 8

While using the method of synthetic division arrange the given polynomial in descending powers of x and write zero for the coefficient of the missing term.

EXAMPLE 5.13 Express $f(x) = 3x^3 + x^2 + x + 1$ in terms of factorial polynomial and hence show that $\Delta^3 f(x) = 18$.

Solution Using the method of synthetic division, we have

0	3	1	1	1
		0	0	0
1	3	1	1	1
		3	4	
2	3	4	5	
		6		
	3	10		

Thus in factorial notation, the given polynomial can be written as

$$f(x) = 3x^{(3)} + 10x^{(2)} + 5x^{(1)} + 1.$$

Therefore,
$$\Delta f(x) = 9x^{(2)} + 20x^{(1)} + 5, \qquad [\because \Delta x^{(n)} = nx^{(n-1)}]$$
$$\Delta^2 f(x) = 18x^{(1)} + 20, \qquad \Delta^3 f(x) = 18.$$

EXAMPLE 5.14 Represent the function $f(x) = x^4 - 5x^3 + 3x + 4$ and its successive differences in factorial notation. Hence show that $\Delta^5 f(x) = 0$.

Solution Using the method of synthetic division the factorial notation of given polynomial can be written as

0	1	–5	0	3	4
		0	0	0	0
1	1	–5	0	3	4
	0	1	–4	–4	
2	1	–4	–4	–1	
	0	2	–4		
3	1	–2	–8		
	0	3			
	1	1			

Therefore,
$$f(x) = x^{(4)} + x^{(3)} - 8x^{(2)} - x^{(1)} + 4,$$
$$\Delta f(x) = 4x^{(3)} + 3x^{(2)} - 16x^{(1)} - 1,$$
$$\Delta^2 f(x) = 12x^{(2)} + 6x^{(1)} - 16, \qquad \Delta^3 f(x) = 24x^{(1)} + 6$$
$$\Delta^4 f(x) = 24, \qquad \Delta^5 f(x) = 0$$

EXAMPLE 5.15 What is the anti-difference of the polynomial $f(x) = x^4 - 5x^3 + 3x + 4$?

Solution: Write the given polynomial in factorial notation. By example 5.14,

$f(x) = x^{(4)} + x^{(3)} - 8x^{(2)} - x^{(1)} + 4$. We have to determine $\frac{1}{\Delta} \cdot f(x)$ or $\Delta^{-1} f(x)$.

The anti-difference of $f(x)$ is $\frac{1}{\Delta} f(x) = \frac{1}{\Delta}[x^{(4)} + x^{(3)} - 8x^{(2)} - x^{(1)} + 4]$

$$= \frac{x^{(5)}}{5} + \frac{x^{(4)}}{4} - \frac{8x^{(3)}}{3} - \frac{x^{(2)}}{2} + 4x^{(1)} + c$$

where c is the constant of integration.

EXAMPLE 5.16 Obtain the function whose first difference is $2x^3 + 3x^2 - 5x + 4$.

Solution Let $f(x)$ be the required function so that

$$\Delta f(x) = 2x^3 + 3x^2 - 5x + 4.$$

Using the method of synthetic division, the factorial notation of $2x^3 + 3x^2 - 5x + 4$ is given by

$2x^{(3)} + 9x^{(2)} + 4$

0	2	3	–5	4
		0	0	0
1	2	3	–5	4
		2	5	
2	2	5	0	
		4		
	2	9		

Therefore, $\Delta f(x) = 2x^{(3)} + 9x^{(2)} + 4$

or
$$f(x) = \frac{1}{\Delta}(2x^{(3)} + 9x^{(2)} + 4) = \frac{x^{(4)}}{2} + 3x^{(3)} + 4x^{(1)} + c$$
$$= \frac{1}{2}x(x-1)(x-2)(x-3) + 3x(x-1)(x-2) + 4x + c$$
$$= \frac{1}{2}(x^2 - x)(x^2 - 5x + 6) + 3x(x^2 - 3x + 2) + 4x + c = \frac{1}{2}(x^4 - 7x^2 + 14x) + c.$$

EXERCISE 5.1

1. Evaluate the following:

(*a*) $\Delta\left(\frac{1}{x^2 + 5x + 6}\right)$, (*b*) $\Delta(x!)$, (*c*) $\Delta^n (e^{ax})$, (*AMIETE June 2009*)

(*d*) $\Delta^2 (ab^{cx})$ (*e*) $\Delta^4[x^3(x - 1)]$, (*AMIE, W-2002*), (*f*) $\Delta[2^x/x!]$.

Interval of differencing being unity.

Ans. (*a*) $-2/(x + 2)(x + 3)(x + 4)$, (*b*) $x(x!)$, (*c*) $(e^a - 1)^n e^{ax}$, (*d*) $(b^c - 1)^2 ab^{cx}$, (*e*) $4!$, (*f*) $\frac{(1-x)\,2^x}{(x+1)!}$

(*f*) [**Hint:** Let $f(x) = 2^x$ and $g(x) = x!$ so that $\Delta f(x) = 2^{x+1} - 2^x = 2^x(2 - 1) = 2^x$, and $\Delta g(x) = (x + 1)! - x! = (x + 1)\,x! - x! = x(x!)$.

Now, $$\Delta\left[\frac{f(x)}{g(x)}\right] = \frac{g(x)\,\Delta f(x) - f(x)\,\Delta g(x)}{g(x+h)\,g(x)}.$$

Therefore, $$\Delta\left[\frac{2^x}{x!}\right] = \frac{x!2^x - 2^x \cdot x(x!)}{(x+1)!\,x!} = \frac{(1-x)2^x}{(x+1)!}\Bigg].$$

2. Evaluate the following, interval of differencing being *h*:

(*a*) $\Delta \tan^{-1}x$, (*b*) $\Delta^2\left(\frac{1}{x^2 + 5x + 6}\right)$, (*c*) $\Delta(x^2 + \sin x)$, (*d*) $\Delta^2(ab^x)$.

Ans. (*a*) $\tan^{-1}\left\{\frac{h}{1 + hx + x^2}\right\}$; (*b*) $\frac{6}{(x+2)(x+3)(x+4)(x+5)}$;

(*c*) $h(h + 2x) + 2\sin\frac{h}{2}\cos\left(x + \frac{h}{2}\right)$; (*d*) $a(b^h - 1)^2 \cdot b^x$.

3. With the usual notation, prove the following operator relations:

(*a*) $E\Delta = \Delta E$, (*b*) $\Delta^2 = E^2 - 2E + 1$, and $\nabla^2 = E^{-2} - 2E^{-1} +$

(*c*) $\nabla E = E\nabla = \Delta = E - 1$, (*d*) $(1 + \Delta)(1 - \nabla) = 1$, (*AMIE, W-2001*)

(*e*) $\nabla = \Delta E^{-1}$,

(*f*) $(\nabla + \Delta)^2 (x^2 + x) = 8h^2$, the interval of differencing being *h*,

(*g*) $(E^{1/2} + E^{-1/2})(1 + \Delta)^{1/2} = 2 + \Delta$, (*h*) $\Delta\nabla = \Delta - \nabla = \nabla\Delta = \delta^2$, (*AMIETE, June 2006; Mumbai 2005*)

(*i*) $\delta^3 y_{1/2} = y_2 - 3y_1 + 3y_0 - y_{-1}$, (*j*) $\delta = \Delta(1+\Delta)^{-1/2} = \nabla(1-\nabla)^{-1/2}$,

(*k*) $\Delta^3 = \delta^3 E^{3/2}$, (*AMIE, S-2005*) (*l*) $E^{1/2} = \left(1+\dfrac{\delta^2}{4}\right)^{1/2} + \dfrac{\delta}{2}$,

(*m*) $E^{1/2} = \mu + \dfrac{1}{2}\delta$ and $E^{-1/2} = \mu - \dfrac{1}{2}\delta$, (*n*) $\nabla = 1 - E^{-1} = 1 - (1+\Delta)^{-1} = \delta E^{-1/2}$

(*o*) $\mu\delta = \dfrac{1}{2}(\Delta + \nabla)$, (*p*) $\nabla^2 = h^2D^2 - h^3D^3 + \dfrac{7}{12}h^4D^4 - \ldots\ldots$

(*NIT, Kurukshetra, 2010, 2009*)

4. Show that:

(*a*) $\dfrac{\Delta}{E}x^3$, $(h=1) = 3x^2 - 3x + 1$, (*b*) $\Delta^4\{x^3(x-1)\} = 24$, (*AMIE, W-2002*)

(*c*) $\nabla^2 y_2 = 2$, if $y_0 = 1$, $y_1 = 2$ and $y_2 = 5$, (*AMIE, S-2004*)

(*d*) $\delta^3 Y_{1/2} = Y_2 - 3Y_1 + 3Y_0 - Y_{-1}$. (*AMIETE, Dec. 2004*)

5. Given $u_0 = 3$, $u_1 = 12$, $u_2 = 81$, $u_3 = 200$, $u_4 = 100$, $u_5 = 8$. Find $\Delta^5 u_0$. **Ans.** 755.

6. If $f(x) = e^{ax}$, evaluate $\nabla^n f(x)$, where ∇ is the backward difference operator. (*AMIETE, June 2006*)
Ans. $(1 - e^{-a})^n e^{ax}$.

7. Given that $f(1) + f(2) + f(3) = 25$, $f(4) = 29$ and $f(5) + f(6) = 113$, find the value of $f(10)$. **Ans.** 191.

8. (*a*) Given $\log_{10} 100 = 2$, $\log_{10} 101 = 2.0043$, $\log_{10} 103 = 2.0128$, $\log_{10} 104 = 2.0170$, find $\log_{10} 102$. **Ans.** 2.0086 approx.

(*b*) Find the missing values in the following tables:

x	45	50	55	60	65
$f(x)$	3.0	—	2.0	–	–2.4

(*AMIE, W-2010; VTU,, 2001*) **Ans.** $f(50) = 2.925$, $f(60) = 0.225$.

(*c*)

x	0	1	2	3	4
$f(x)$	1	3	9	—	81

(*SVTU, 2007; Raipur 2005; WBTU, 2005*) **Ans.** 31.

9. Find $u_{1/2}$, given $u_{-1} = 202$, $u_0 = 175$, $u_1 = 82$ and $u_2 = 55$. **Ans.** 128.50.

10. Express the following polynomials in terms of factorial notation

(*a*) $f(x) = x^3 - x + 1$, (*b*) $f(x) = x^4 - 2x^3 - x$,

(*c*) $f(x) = 3x^3 - 7x^2 + 8x - 1$, (*d*) $f(x) = 2x^3 - 3x^2 + 3x - 10$ and hence show that $\Delta^3 f(x) = 12$,

(*e*) $f(x) = 3x^3 - 2x^2 + x + 1$ and hence show that $\Delta^3 f(x) = 18$. (*AMIE, W-2012*)

Ans. (*a*) $x^{(3)} + 3x^{(2)} + 1$, (*b*) $x^{(4)} + 4x^{(3)} + x^{(2)} - 2x^{(1)}$, (*c*) $3x^{(3)} + 2x^{(2)} + 4x^{(1)} - 1$,
(*d*) $2x^{(3)} + 3x^{(2)} + 2x^{(1)} - 10$, (*e*) $3x^{(3)} + 7x^{(2)} + 1$.

11. Represent the function $f(x) = x^4 - 12x^3 + 24x^2 - 30x + 9$ and its successive differences in factorial notation. Hence, show that $\Delta^5 f(x) = 0$.

Ans. $f(x) = x^{(4)} - 6x^{(3)} - 5x^{(2)} - 17x^{(1)} + 9$,
$\Delta f(x) = 4x^{(3)} - 18x^{(2)} - 10x^{(1)} - 17$, $\Delta^2 f(x) = 12x^{(2)} - 36x^{(1)} - 10$,
$\Delta^3 f(x) = 24x^{(1)} - 36$, $\Delta^4 f(x) = 24$, $\Delta^5 f(x) = 0$.

12. Express x^3 in terms of factorial function and hence evaluate $\Delta^2 x^3$. **Ans.** $x^{(3)} + 3x^{(2)} + x^{(1)}$; $6(x+1)$.

13. Obtain a function whose first difference is

(*a*) e^x, (*b*) $9x^2 + 11x + 5$, (*AMIE, S-2010*) (*c*) $x^3 + 3x^2 + 5x + 12$.

Ans. (*a*) $\left(\dfrac{e^x}{e-1}\right) + c$, (*b*) $3x^3 + x^2 + x + c$, (*c*) $\dfrac{1}{4}(x^4 + 2x^3 + 5x^2 + 40x) + c$.

14. Prove that $\left(\dfrac{\Delta^2}{E}\right)u_x \neq \dfrac{\Delta^2 u_x}{Eu_x}$. (*AMIE, S-2009*)

5.5 NEWTON'S FORMULAE FOR INTERPOLATION

We now derive two interpolation formulae which are very frequently used in many engineering and scientific problems. These formulae known as Newton's forward interpolation formula and Newton's backward interpolation formula are obtained by using the forward and backward differences of a function.

5.5.1 Newton's Forward Differences Formula

Let $y_x = f(x)$ be a function of x taking the values $y_0, y_1, y_2, y_3, ..., y_n$ corresponding to the values of $x = x_0$, $x_0 + h, x_0 + 2h, x_0 + 3h, ...$. Suppose it is required to evaluate $f(x)$ for a value of $x = x_0 + uh$, where u is any real number.

Writing
$$y(x) = y(x_0 + uh) = f(x_0 + uh)$$
$$= E^u y_{x_0} = y_u, \text{ denoting } E = 1 + \Delta.$$

Since $f(x)$ is approximated by polynomial $P_n(x)$.

We get $y_u \approx P_n(x) = E^u y_0 = (1 + \Delta)^u y_0$

$$= \left[1 + u\Delta + \frac{u(u-1)}{2!}\Delta^2 + \frac{u(u-1)(u-2)}{3!}\Delta^3 + \cdots + \frac{u(u-1)(u-2)\cdots(u-n+1)}{n!}\Delta^n\right] y_0$$

$$= y_0 + u\Delta y_0 + \frac{u(u-1)}{2!}\Delta^2 y_0 + \frac{u(u-1)(u-2)}{3!}\Delta^3 y_0 + \cdots + \frac{u(u-1)(u-2)\cdots(u-n+1)}{n!}\Delta^n y_0. \quad ...(5.18)$$

The formula (5.18) may also be written as

$$f(x) \approx P_n(x) = f_0 + u\Delta f_0 + \frac{u(u-1)}{2!}\Delta^2 f_0 + \frac{u(u-1)(u-2)}{3!}\Delta^3 f_0 + \cdots$$
$$\cdots + \frac{u(u-1)(u-2)\cdots(u-n+1)}{n!}\Delta^n f_0.$$

The result (5.18) is known as **Newton's forward interpolation formula** or **Newton's formula for equal interval/spacing.**

This formula is usually applied for interpolating the values of $f(x)$ near the top of the table (since all the forward differences are available) and for extrapolating values of y a little backward (that is, to the left) from y_0. The students should note that for speedy evaluation of the result, x_0 should be chosen such a value in the table that gives the value of u lying between 0 and 1.

5.5.2 Newton's Backward Differences Formula

Using backward difference notation, we have

$$y_u = f(x_n + uh) = E^u f(x_n) = (1 - \nabla)^{-u} y_n \qquad [\because (E^{-1})^{-u} = (1 - \nabla)^{-u}]$$

$$= 1 + u\nabla + \frac{u(u+1)}{2!}\nabla^2 + \frac{u(u+1)(u+2)}{3!}\nabla^3 + \cdots + \left[\frac{u(u+1)(u+2)\cdots(u+n-1)}{n!}\nabla^n\right] y_n$$

(using Binomial theorem for –ve index)

Therefore, we get
$$y_u = f(x_n + uh) = y_n + u\nabla y_n + \frac{u(u+1)}{2!}\nabla^2 y_n + \frac{u(u+1)(u+2)}{3!}\nabla^3 y_n + ... \quad ...(5.19)$$
$$\cdots + \frac{u(u+1)(u+2)\cdots(u+n-1)}{n!}\nabla^n y_n$$

where $u = (x - x_n)/h$.

The result (5.19) may also be written as

$$f(x) \approx P_n(x) = f_0 + u\nabla f_n + \frac{u(u+1)}{2!}\nabla^2 f_n + \cdots + \frac{u(u+1)(u+2)\cdots(u+n-1)}{n!}\nabla^n f_n.$$

The result (5.19) is known as **Newton's backward interpolation formula** and this formula is used for interpolating the values of y near the end of a set of tabular values (since all the backward differences are available) and also for extrapolating values of y a little distance ahead (that is, to the right) of y_n.

EXAMPLE 5.17 For the data

x:	−3	−2	−1	0	1	2
$f(x)$:	−2	−4	−4	−2	2	8

construct the forward and backward difference tables. Using the corresponding interpolation, show that the interpolating polynomial is same.

Solution The step length is $h = 1$ and $x_0 = -3$, $x_1 = -2$, $x_2 = -1$, $x_3 = 0$, $x_4 = 1$, $x_5 = 2$. Refer difference Table 5.6.

Table 5.6: Forward and backward difference table

x	$f(x)$	$\Delta f\,/\,\nabla f$	$\Delta^2 f\,/\,\nabla^2 f$	$\Delta^3 f\,/\,\nabla^3 f$
−3	**−2**			
		−2		
−2	−4		**2**	
		0		**0**
−1	−4		2	
		2		0
0	−2		2	
		4		**0**
1	2		**2**	
		6		
2	**8**			

(*i*) We have $u = \dfrac{x - x_0}{h} = \dfrac{x-(-3)}{1} = x + 3$. Using the Newton's forward difference interpolating polynomial, we obtain $P_2(x) = -2 + (x+3)(-2) + \dfrac{(x+3)(x+3-1)}{2!}(2) = -2 - 2x - 6 + x^2 + 5x + 6 = x^2 + 3x - 2$.

(*ii*) $u = \dfrac{x - x_n}{h} = \dfrac{x-2}{1} = x - 2$. Using the Newton's backward difference interpolating polynomial, we obtain $P_2(x) = 8 + 6(x-2) + \dfrac{(x-2)(x-2+1)}{2!}(2) = 8 + 6x - 12 + x^2 - 3x + 2 = x^2 + 3x - 2$.

EXAMPLE 5.18 For the data

x:	−4	−2	0	2	4	6
$f(x)$:	−67	−9	1	11	69	223

construct the forward difference table and determine the corresponding interpolating polynomial.

Solution The step length is $h = 2$, and $x_0 = -4$, $x_1 = -2$, $x_2 = 0$ $x_3 = 2$, $x_4 = 4$, $x_5 = 6$. The difference table is as under:

Table 5.7: Forward difference table

x	$f(x)$	Δf	$\Delta^2 f$	$\Delta^3 f$	$\Delta^4 f$
−4	**−67**				
		58			
−2	−9		**−48**		
		10		**48**	
0	1		0		**0**
		10		48	
2	11		48		0
		58		48	
4	69		96		
		154			
6	223				

We have $u = \dfrac{x - x_0}{h} = \dfrac{x+4}{2}$. The data represents a cubic polynomial.

Using the Newton's forward difference interpolating polynomial,

$$P_3(x) = f_0 + u\Delta f_0 + \frac{u(u-1)}{2!}\Delta^2 f_0 + \ldots\ldots + \frac{u(u-1)\ldots\ldots(u-n+1)}{n!}\Delta^n f_0,$$

we obtain $P_3(x) = -67 + \left(\frac{x+4}{2}\right)(58) + \frac{1}{2!}\left(\frac{x+4}{2}\right)\left(\frac{x+4}{2}-1\right)(-48) + \frac{1}{3!}\left(\frac{x+4}{2}\right)\left(\frac{x+4}{2}-1\right)\left(\frac{x+4}{2}-2\right)(48)$

$$= -67 + \left(\frac{x+4}{2}\right)(58) + \frac{1}{2}\frac{(x+4)(x+2)}{(4)}(-48) + \frac{1}{6}\frac{(x+4)(x+2)x}{(8)}(48)$$

$$= -67 + 29(x+4) - 6(x+2)(x+4) + x(x+2)(x+4) = x^3 + x + 1.$$

EXAMPLE 5.19 For the data

x:	–4	–2	0	2	4	6
$f(x)$:	261	19	1	15	253	1291

construct the backward difference table and determine the corresponding interpolating polynomial.

Solution The step length is $h = 2$ and $x_0 = -4$, $x_1 = -2$, $x_2 = 0$, $x_3 = 2$, $x_4 = 4$, $x_5 = 6$. We have the following difference table:

Table 5.8: Backward difference table

x	$f(x)$	∇f	$\nabla^2 f$	$\nabla^3 f$	$\nabla^4 f$	$\nabla^5 f$
–4	261					
		–242				
–2	19		224			
		–18		–192		
0	1		32		384	
		14		192		0
2	15		224		**384**	
		238		**576**		
4	253		**800**			
		1038				
6	**1291**					

We have $u = \frac{x - x_n}{2} = \frac{x-6}{2}$. The data represents a bi-quadratic polynomial. Using the Newton's backward difference interpolating polynomial,

$$P_4(x) = f_n + u\nabla f_n + \frac{u(u+1)}{2!}\nabla^2 f_n + \frac{1}{3!}u(u+1)(u+2)\nabla^3 f_n + \ldots + \frac{u(u+1)\ldots(u+n-1)}{n!}\nabla^n f_n,$$ we obtain

$$P_4(x) = 1291 + \left(\frac{x-6}{2}\right)(1038) + \left(\frac{x-6}{2}\right)\left(\frac{x-6}{2}+1\right)\frac{800}{2!} + \left(\frac{x-6}{2}\right)\left(\frac{x-6}{2}+1\right)\left(\frac{x-6}{2}+2\right)\frac{576}{3!}$$

$$+ \left(\frac{x-6}{2}\right)\left(\frac{x-6}{2}+1\right)\left(\frac{x-6}{2}+2\right)\left(\frac{x-6}{2}+3\right)\frac{384}{4!}$$

$= 1291 + (x-6)\,519 + (x-4)(x-6)\,100 + (x-2)(x-4)(x-6)\,12 + x(x-2)(x-4)(x-6)$

$= x^4 - x + 1.$

EXAMPLE 5.20 For the data

x:	0	0.5	1.0	1.5	2.0
$f(x)$:	–0.5	1.125	3.5	7.375	13.5

find an approximation to $f(0.25)$ and $f(0.75)$.

Solution The points 0.25 and 0.75 are near the top of the table of values, we shall use the Newton's forward difference interpolating polynomial.

The data represents a cubic polynomial. Here $h = 0.5$ and $u = (x - 0)/0.5 = 2x$.

Table 5.9: Forward difference table

j	x_j	f_j	Δf_j	$\Delta^2 f_j$	$\Delta^3 f_j$	$\Delta^4 f_j$
0	0	**–0.500**				
			1.625			
1	0.5	1.125		**0.75**		
			2.375		**0.75**	
2	1.0	3.500		1.50		0
			3.875		0.75	
3	1.5	7.375		2.25		
			6.125			
4	2.0	13.500				

$$f(x) \approx P_3(x) = -0.5 + 2x(1.625) + \frac{2x(2x-1)}{2!}(0.75) + \frac{2x(2x-1)(2x-2)(0.75)}{3!}$$

$$= -0.5 + 3.25x + x(2x-1)(0.75) + x(2x-1)(2x-2)(0.25).$$

Therefore, $f(0.25) \approx P_3(0.25) = -0.5 + 3.25(0.25) + 0.25(-0.5)(0.75) + 0.25(-0.5)(-1.5)(0.25)$
$= -0.5 + 0.8125 - 0.09375 + 0.046875 = 0.2656,$ and
$f(0.75) \approx P_3(0.75) = -0.5 + 1.5(1.625) + 0.75(0.5)(0.75) + 0.75(0.5)(-0.5)(0.25)$
$= -0.5 + 2.4375 + 0.28125 - 0.046875 = 2.1719.$

EXAMPLE 5.21 For the data

x:	0.0	0.2	0.4	0.6	0.8
$f(x)$:	0.500	1.108	1.764	2.516	3.412

find an approximation of $f(0.7)$ and $f(0.5)$.

Solution The points 0.5 and 0.7 are near the bottom of the table of values, with $h = 0.2$. We shall use the Newton's backward difference formula. We form the following backward difference table:

Table 5.10: Backward difference table

j	x_j	f_j	∇f_j	$\nabla^2 f_j$	$\nabla^3 f_j$	$\nabla^4 f_j$
0	0	0.500				
			0.608			
1	0.2	1.108		0.048		
			0.656		0.048	
2	0.4	1.764		0.096		0
			0.752		**0.048**	
3	0.6	2.516		**0.144**		
			0.896			
4	0.8	**3.412**				

The data represents a cubic polynomial. We shall use the following formula with $h = 0.2$.

$$f(x) \approx P_3(x) = f_n + u\nabla f_n + \frac{1}{2}u(u+1)\nabla^2 f_n + \frac{1}{6}u(u+1)(u+2)\nabla^3 f_n.$$

For $x = 0.7$, $u = \dfrac{x - x_n}{h} = \dfrac{0.7 - 0.8}{0.2} = -0.5$, we obtain

$$f(0.7) \approx P_3(0.7) = 3.412 - 0.5\,(0.896) + \frac{1}{2}(-0.5)\,(0.5)\,(0.144) + \frac{1}{6}(-0.5)\,(0.5)\,(1.5)\,(0.048)$$

$$= 3.412 - 0.448 - 0.018 - 0.003 = 2.943.$$

For $x = 0.5, u = \dfrac{x - x_n}{h} = \dfrac{0.5 - 0.8}{0.2} = -1.5,$ we obtain

$$f(0.5) \approx 3.412 - 1.5\,(0.896) + \frac{1}{2}(-1.5)\,(-0.5)\,(0.144) + \frac{1}{6}(-1.5)\,(-0.5)\,(0.5)\,(0.048)$$

$$= 3.412 - 1.344 + 0.054 + 0.003 = 2.125.$$

EXAMPLE 5.22 From the following table, estimate the number of students who have obtained marks between 40 and 45.

Marks:	30-40	40-50	50-60	60-70	70-80
No. of students:	31	42	51	35	31

(AMIETE, June 2010, SVTU, 2007; Madras 2006, VTU 2005; PTU 2003)

Solution First we prepare the cumulative frequency table,

Marks less than (x):	40	50	60	70	80
No of students f	31	73	124	159	190

The difference table is as under:

Table 5.11: Difference table, Example 5.22

x	y	Δy	$\Delta^2 y$	$\Delta^3 y$	$\Delta^4 y$
40	**31**				
		42			
50	73		**9**		
		51		**–25**	
60	124		–16		**37**
		35		12	
70	159		-4		
		31			
80	190				

We now calculate the number of students who got marks less than 45. Taking $x = 45$, $x_0 = 40$, $h = 10$ we have $u = \dfrac{x - x_0}{h} = \dfrac{45 - 40}{10} = 0.5.$ Using Newton's forward interpolating polynomial, we get

$$y_{0.5} = y_0 + u\Delta y_0 + \frac{u(u-1)}{2!}\Delta^2 y_0 + \frac{u(u-1)\,(u-2)}{3!}\Delta^3 y_0 + \frac{u(u-1)\,(u-2)\,(u-3)}{4!}\Delta^4 y_0$$

$$= 31 + 0.5\,(42) + \frac{0.5\,(-0.5)}{2}(9) + \frac{(0.5)\,(-0.5)\,(-1.5)}{6}(-25) + \frac{0.5\,(-0.5)\,(-1.5)\,(-2.5)}{24}(37)$$

$$= 31 + 21 - 1.125 - 1.5625 - 1.4453 \approx 47.87.$$

Therefore, the number of students with marks less than 45 is 48. As the number of students with marks less than 40 is 31 so the required number of students getting marks between 40 and 45 are 48 – 31 = 17.

EXAMPLE 5.23 Compute $f(21)$ from the given table:

x:	20	24	28	32
$f(x)$:	0.01427	0.01581	0.01772	0.01996

(AMIE, Dec 2011)

Solution The point 21 is near the beginning of the table of values, with $h = 4$, we shall use the Newton's forward interpolating polynomial. The forward difference table is as follows:

Table 5.12: Forward difference table

j	x_j	f_j	Δf_j	$\Delta^2 f_j$	$\Delta^3 f_j$
0	20	**0.01427**			
			0.00154		
1	24	0.01581		**0.00037**	
			0.00191		**–0.00004**
2	28	0.01772		0.00033	
			0.00224		
3	32	0.01996			

$$f(x) \approx P_3(x) = f_0 + u\Delta f_0 + \frac{1}{2}u(u-1)\Delta^2 f_0 + \frac{1}{6}u(u-1)(u-2)\Delta^3 f_0.$$

For $x = 21$, $h = 4$, $u = \dfrac{x - x_0}{h} = \dfrac{21-20}{4} = 0.25$, we have

$$f(21) \approx 0.01427 + 0.25\,(0.00154) + \frac{(0.25)(-0.75)}{2}(0.00037) + \frac{0.25\,(-0.75)(-1.75)}{6}(-0.00004)$$

$$= 0.01427 + 0.000385 - 0.0000346 - 0.0000021 = 0.0146183.$$

EXERCISE 5.2

1. For the data

x:	0	1	2	3	4
$f(x)$:	3	6	11	18	27

construct the forward and backward difference tables. Using the corresponding interpolation, show that the interpolating polynomial is same. *(AMIE W-2005)* **Ans.** $x^2 + 2x + 3$.

2. Estimate the values of $f(22)$ and $f(42)$ from the following available data:

x:	20	25	30	35	40	45
$f(x)$:	354	332	291	260	231	204

(AMIETE, Dec. 2009) **Ans.** 352; 219.

3. (*a*) The following data are taken from the steam table.

Temp.°C	140	150	160	170	180
pressure kgf/cm^2	3.685	4.854	6.302	8.076	10.225

Using Newton's formula, find the pressure of steam for temperature 142°C and 175°C. *[AMIETE 2011-2009, AMIE 2009)* **Ans.** 3.899; 9.10 kgf/cm^2.

(*b*) Construct the difference table from the following values of y. Compute $f(x)$ at (*i*) 1.05, and (*ii*) 1.45.

x:	1.0	1.1	1.2	1.3	1.4	1.5
y:	0.24197	0.21875	0.19419	0.17137	0.14973	0.12952

(AMIE., W-2005) **Ans.** (*i*) 0.23079, (*ii*) 0.13927.

4. Given sin 45° = 0.7071, sin 50° = 0.7660, sin 55° = 0.8192, sin 60° = 0.8660. Find sin 52°, using Newton's forward interpolation formula. **Ans.** 0.788.

5. Given that

x:	1	2	3	4	5
y:	2	5	10	17	26

find the value of $\nabla^2 y_5$. *(AMIE W-2005)* **Ans.** 2.

6. Using Newton's backward interpolation formula, find $f(3.8)$ from the following table.

x:	0	1	2	3	4
$f(x)$:	1.00	1.50	2.20	3.10	4.60

Ans. 4.36.

7. Apply Newton's backward formula to find a polynomial of degree three which includes the following data.

x:	3	4	5	6
$y = f(x)$:	6	24	60	120

Ans. $x^3 - 3x^2 + 2x$.

8. (*a*) The following data gives the melting point of an alloy of lead and zinc, where t is the temperature in °C and p is the percentage of lead in the alloy.

p(%):	60	70	80	90
t:	226	250	276	304

Using Newton's interpolation method, find the melting point of the alloy containing 84% of lead.

Ans. 286.96°C.

(*b*) The following table gives the values of tan x for interval $0.1 \le x \le 0.30$.

x:	0.10	0.15	0.20	0.25	0.30
y = tan$x$:	0.1003	0.1511	0.2027	0.2553	0.3093

Use Newton's formulae for interpolation to obtain the values of (*i*) tan 0.12, and (*ii*) tan 0.26.

Ans. (*i*) 0.125284, (*ii*) 0.266.

9. The following table gives the population of a town during the last six censuses. Estimate, using Newton's interpolation formula, the increase in the population during the period 1996 to 1998.

Year:	1961	1971	1981	1991	2001	2011
Population: (in thousands):	12	15	20	27	39	52

Ans. Populations in the year 1996 is 32.875 thousands and in the year 1998 is 35.295 thousands.
Increase in population during the period 1996 to 1998 = 2.42 thousands.

10. Values of $f(x) = \sqrt{x}$ are listed in following table. Use the data of this table to calculate $\sqrt{155}$.

x:	150	152	154	156
$f(x) = \sqrt{x}$:	12.247	12.329	12.410	12.490

(*AMIE, W-2006*) **Ans.** 12.453.

11. Use Newton's backward difference polynomial to interpolate at $x = 1.5$, from the following data.

x:	–2	–1	0	1	2
$f(x)$:	20	12	6	2	0

(*AMIETE, June 2006*) **Ans.** $x^2 - 5x + 6$, 0.75.

12. Find the cubic polynomial which takes on the following values.

x:	0	1	2	3
$f(x)$:	1	2	1	10

Hence or otherwise evaluate $f(4)$. (*AMIE, W-2010, WBTU, 2005, Rohtak, 2005*)

Ans. $2x^3 - 7x^2 + 6x + 1$; 41.

13. The table gives the distances in Kilometers of the visible horizon for the given heights in metres above the earth surface

x (= height):	100	150	200	250	300	350	400
y (= distance):	10.63	13.03	15.04	16.81	18.42	19.90	21.27

Find the values of y when x = 218 metres and 410 metres. **Ans.** 15.70 km; 21.53 km.

14. The data

x:	1	2	3	4	5
$f(x)$:	5	15	35	77	141

represents a third degree polynomial. It is known that $f(3)$ is in error. Find the correct value of $f(3)$.
(AMIETE, June-2006) **Ans.** 37.

15. Find the polynomial that fits the following data.

x:	0.1	0.2	0.3	0.4
$f(x)$:	1.11	1.24	1.39	1.56

(AMIETE, June 2006) **Ans.** $f(x) = x^2 + x + 1$.

16. The data

x:	0	1	2	3	4
$f(x)$:	1	2	10	28	65

represents a polynomial of degree ≤ 3. It is known that $f(2)$ is in error. Find the correct value of $f(2)$.
(AMIETE, Dec. 2002) **Ans.** 9.

17. Find the first term of the series whose second and subsequent terms are 8, 3, 0, –1, 0.
(AMIETE, Dec. 2011)

[**Hint:** The forward difference table is shown below:

Table 5.13

Term	Argument	Entry	Δf_2	$\Delta^2 f_2$	$\Delta^3 f_2$
f_2	$x_0 + h$	**8**			
			–5		
f_3	$x_0 + 2h$	3		**2**	
			–3		0
f_4	$x_0 + 3h$	0		2	
			–1		0
f_5	$x_0 + 4h$	–1		2	
			1		
f_6	$x_0 + 5h$	0			

Let $f_1 (= x_0)$ be the first term, f_2 the second term $= x_0 + h = 8$.

We know $\quad f_1 = E^{-1} f_2 = (1 + \Delta)^{-1} f_2 = (1 - \Delta + \Delta^2 - \ldots\ldots) f_2$

$= f_2 - \Delta f_2 + \Delta^2 f_2 = (8) - (-5) + 2 = 15)$, (from the above table)]

18. The area A of a circle of a diameter d is given for the following values.

d:	80	85	90	95	100
A:	5026	5674	6362	7088	7854

Calculate the area of a circle of diameter 105. **Ans.** 8666.

19. Using Newton's backward interpolation formula, find the value of $e^{-1.9}$ from the following table of values of e^{-x}.

x:	1.00	1.25	1.50	1.75	2.00
e^{-x}:	0.3679	0.2865	0.2231	0.1738	0.1353

Ans. 0.1496.

20. In the table below, the values of y are consecutive terms of a series of which 43 is the sixth term. Find the first and the tenth terms of the series.

x:	3	4	5	6	7	8	9
y:	4.8	8.4	14.5	23.6	36.2	52.8	73.9

Ans. $y(1) = 3.1$, $y(10) = 100$.

5.6 CENTRAL DIFFERENCE INTERPOLATION FORMULAE

In the preceding sections, we have discussed Newton forward and backward interpolation formulae which are suited for interpolation near the beginning and end of the tabulated values. The forward and backward formulae fail to give very accurate results when the interpolation is done to estimate the value of a function near the middle of the tabulated values. To achieve more accuracy near the middle of the table, we develope some central difference formulae which are appropriately employed for interpolation near the centre of the table of values. Let $y = y_0$ be the central ordinate corresponding to $x = x_0$. Let the other ordinates be y_r at $x = x_r$ $(r = \pm 1, \pm 2, \ldots)$ we can form a difference table.

Table 5.14: Paths of Gauss formulas across a diagonal difference table

x	y	Δy	$\Delta^2 y$	$\Delta^3 y$	$\Delta^4 y$	$\Delta^5 y$	$\Delta^6 y$
$x_{-3} = x_0 - 3h$	y_{-3}						
		Δy_{-3}					
$x_{-2} = x_0 - 2h$	y_{-2}		$\Delta^2 y_{-3}$				
		Δy_{-2}		$\Delta^3 y_{-3}$			
$x_{-1} = x_0 - h$	y_{-1}		$\Delta^2 y_{-2}$		$\Delta^4 y_{-3}$		
		Δy_{-1}		$\Delta^3 y_{-2}$		$\Delta^5 y_{-3}$	
x_0	y_0		$\Delta^2 y_{-1}$		$\Delta^4 y_{-2}$		$\Delta^6 y_{-3}$
		Δy_0		$\Delta^3 y_{-1}$		$\Delta^5 y_{-2}$	
$x_1 = x_0 + h$	y_1		$\Delta^2 y_0$		$\Delta^4 y_{-1}$		
		Δy_1		$\Delta^3 y_0$			
$x_2 = x_0 + 2h$	y_2		$\Delta^2 y_1$				
		Δy_2					
$x_3 = x_0 + 3h$	y_3						

– – – – – Central line

———→ Path of Gauss's forward formula

·········→ Path of Gauss's backward formula

The above table can also be written in term of central differences, using the operator δ. (See Table 5.15)

Table 5.15: Central differences

x	y	δy	$\delta^2 y$	$\delta^3 y$	$\delta^4 y$	$\delta^5 y$	$\delta^6 y$
$x_{-3} = x_0 - 3h$	y_{-3}						
		$\delta y_{-5/2}$					
$x_{-2} = x_0 - 2h$	y_{-2}		$\delta^2 y_{-2}$				
		$\delta y_{-3/2}$		$\delta^3 y_{-3/2}$			
$x_{-1} = x_0 - h$	y_{-1}		$\delta^2 y_{-1}$		$\delta^4 y_{-1}$		
		$\delta y_{-1/2}$		$\delta^3 y_{-1/2}$		$\delta^5 y_{-1/2}$	
x_0	y_0		$\delta^2 y_0$		$\delta^4 y_0$		$\delta^6 y_0$
		$\delta y_{1/2}$		$\delta^3 y_{1/2}$		$\delta^5 y_{1/2}$	
$x_1 = x_0 + h$	y_1		$\delta^2 y_1$		$\delta^4 y_1$		
		$\delta y_{3/2}$		$\delta^3 y_{3/2}$			
$x_2 = x_0 + 2h$	y_2		$\delta^2 y_2$				
		$\delta y_{5/2}$					
$x_3 = x_0 + 3h$	y_3						

– – – – – Central line

———→ Path of Gauss's forward formula

·········→ Path of Gauss's backward formula

By simply using the operator relation $\delta = \Delta E^{-1/2}$.

It may be seen that the differences given in both the tables are the same in corresponding positions.

For example $\delta^5 y_{-1/2} = (\Delta E^{-1/2})^5 y_{-1/2} = \Delta^5 E^{-5/2} y_{-1/2} = \Delta^5 y_{-3}$.

5.6.1 Gauss's Forward Formula for Central Differences

The Newtons forward interpolation formula with origin at $x = x_0$ is

$$y_u = y_0 + u\Delta y_0 + \frac{u(u-1)}{2!}\Delta^2 y_0 + \frac{u(u-1)(u-2)}{3!}\Delta^3 y_0 + \frac{u(u-1)(u-2)(u-3)}{4!}\Delta^4 y_0 + \ldots \quad \ldots(5.20)$$

where $u = (x - x_0)/h$.

We have $\Delta^2 y_0 = \Delta^2 y_{-1} + \Delta^3 y_{-1}$; $\Delta^3 y_0 = \Delta^3 y_{-1} + \Delta^4 y_{-1}$; $\Delta^4 y_0 = \Delta^4 y_{-1} + \Delta^5 y_{-1}$ etc.

Also $\Delta^3 y_{-1} = \Delta^3 y_{-2} + \Delta^4 y_{-2}$; $\Delta^4 y_{-1} = \Delta^4 y_{-2} + \Delta^5 y_{-2}$ etc.

Substituting for $\Delta^2 y_0$, $\Delta^3 y_0$, $\Delta^4 y_0$, ..., in (5.20), we obtain

$$y_u = y_0 + u\Delta y_0 + \frac{u(u-1)}{2!}(\Delta^2 y_{-1} + \Delta^3 y_{-1}) + \frac{u(u-1)(u-2)}{3!}(\Delta^3 y_{-1} + \Delta^4 y_{-1}) + \frac{u(u-1)(u-2)(u-3)}{4!}(\Delta^4 y_{-1} + \Delta^5 y_{-1}) + \ldots\ldots$$

Hence, $$y_u = y_0 + u\Delta y_0 + u(u-1)\frac{\Delta^2 y_{-1}}{2!} + (u+1)u(u-1)\frac{\Delta^3 y_{-1}}{3!} + (u+1)u(u-1)(u-2)\frac{\Delta^4 y_{-2}}{4!} + \frac{(u+2)(u+1)u(u-1)(u-2)}{5!}\Delta^5 y_{-2} + \cdots \quad \ldots(5.21)$$

This is known as ***Gauss's forward difference formula*** for equal intervals. It is used to interpolate the values of y_u for value of u lying between 0 and 1.

Introducing the central-difference operator δ, relation (5.20) may be written as

$$y_u = y_0 + u\delta y_{1/2} + u(u-1)\frac{\delta^2 y_0}{2!} + (u+1)u(u-1)\frac{\delta^3 y_{1/2}}{3!} + (u+1)u(u-1)(u-2)\frac{\delta^4 y_0}{4!} + \ldots\ldots + \frac{(u+2)(u+1)u(u-1)(u-2)}{5!}\delta^5 y_{1/2} + \cdots \quad \ldots(5.22)$$

Remark 9

The above formula involves *odd* differences just below the central line and *even* differences on the line.

y_0 — $\Delta^2 y_{-1}$ — $\Delta^4 y_{-2}$ — $\Delta^6 y_{-3}$ — **Central line**

Δy_0 $\quad \Delta^3 y_{-1}$ $\quad \Delta^5 y_{-2}$ $\quad \Delta^7 y_{-3}$

5.6.2 Gauss's Backward Formula for Central Differences

If we substitute

$\Delta y_0 = \Delta y_{-1} + \Delta^2 y_{-1}$; $\Delta^2 y_0 = \Delta^2 y_{-1} + \Delta^3 y_{-1}$; $\Delta^3 y_0 = \Delta^3 y_{-1} + \Delta^4 y_{-1}$ etc.

Also $\Delta^3 y_{-1} = \Delta^3 y_{-2} + \Delta^4 y_{-2}$; $\Delta^4 y_{-1} = \Delta^4 y_{-2} + \Delta^5 y_{-2}$ etc.

in Newton's forward interpolation formula, we get

$$y_u = y_0 + u(\Delta y_{-1} + \Delta^2 y_{-1}) + \frac{u(u-1)}{2!}(\Delta^2 y_{-1} + \Delta^3 y_{-1}) + \frac{u(u-1)(u-2)}{3!}(\Delta^3 y_{-1} + \Delta^4 y_{-1}) + \frac{u(u-1)(u-2)(u-3)}{4!}(\Delta^4 y_{-1} + \Delta^5 y_{-1}) + \ldots$$

$$= y_0 + u\Delta y_{-1} + \frac{u(u+1)}{2!}\Delta^2 y_{-1} + \frac{(u+1)\,u\,(u-1)}{3!}\Delta^3 y_{-1} + \frac{(u+1)\,u\,(u-1)\,(u-2)}{4!}\Delta^4 y_{-1} + \frac{u(u-1)\,(u-2)\,(u-3)}{5!}\Delta^5 y_{-1} + \ldots$$

$$= y_0 + u\Delta y_{-1} + \frac{(u+1)u}{2!}\Delta^2 y_{-1} + \frac{(u+1)\,u\,(u-1)}{3!}(\Delta^3 y_{-2} + \Delta^4 y_{-2}) + \frac{(u+1)\,u\,(u-1)\,(u-2)}{4!}(\Delta^4 y_{-2} + \Delta^5 y_{-2}) + \ldots$$

Hence, $$y_u = y_0 + u\Delta y_{-1} + u(u+1)\frac{\Delta^2 y_{-1}}{2!} + (u+1)\,u\,(u-1)\frac{\Delta^3 y_{-2}}{3!} + (u+2)\,(u+1)\,u\,(u-1)\frac{\Delta^4 y_{-2}}{4!} + \ldots$$

$$\cdots + \frac{(u+2)\,(u+1)\,u(u-1)\,(u-2)}{5!}\Delta^5 y_{-3} + \cdots \qquad \ldots(5.23)$$

This is called **Gauss's backward interpolation formula:** It is used to interpolate the values of y_u for a negative value of u lying between –1 and 0.

In central differences notation, this formula will be

$$y_u = y_0 + u\delta y_{-1/2} + (u+1)u\,\frac{\delta^2 y_0}{2!} + (u+1)\,u\,(u-1)\frac{\delta^3 y_{-1/2}}{3!} + (u+2)\,(u+1)\,u\,(u-1)\frac{\delta^4 y_0}{4!} + \ldots$$

$$\cdots + \frac{(u+2)\,(u+1)\,u(u-1)\,(u-2)}{5!}\delta^5 y_{-1/2} + \cdots \qquad \ldots(5.24)$$

Remark 10

The above formula involves *odd* differences above the central line and *even* differences on the central line as shown below:

Δy_{-1} $\Delta^3 y_{-2}$ $\Delta^5 y_{-3}$ $\Delta^7 y_{-4}$

y_0 — $\Delta^2 y_{-1}$ — $\Delta^4 y_{-2}$ — $\Delta^6 y_{-3}$ — — — Central line

EXAMPLE 5.24 Apply Gauss's formula to find $f(41)$.

Given that $f(30) = 3678.2, f(35) = 2995.1, f(40) = 2400.1, f(45) = 1876.2, f(50) = 1416.3$.

Solution Taking 40 as an origin, $h = 5$ and $u = \frac{x - x_0}{h} = \frac{41-40}{5} = 0.2$.

We are now required is find the value $y_{0.2}$. The central difference table is given below:

Table 5.16

x	$f(x)$	Δf	$\Delta^2 f$	$\Delta^3 f$	$\Delta^4 f$
30	3678.2				
		–683.1			
35	2995.1		88.1		
		–595.0		–17.0	
$x_0 = 40$	**2400.1**		**71.1**		**9.9**
		–523.9		**–7.1**	
45	1876.2		64.0		
		–459.9			
50	1416.3				

Central line

Using Gauss's forward central difference formula,

$$y_u = f_0 + u\Delta f_0 + \frac{u(u-1)}{2}\Delta^2 f_{-1} + \frac{(u+1)\,u\,(u-1)}{3!}\Delta^3 f_{-1} + \frac{(u+1)\,u\,(u-1)\,(u-2)}{4!}\Delta^4 f_{-2} + \cdots$$

Substituting $u = 0.2$, we obtain $y_{0.2} \approx f(41) \approx 2400.1 + (0.2)(-523.9) + \frac{(0.2)(0.2-1)}{2}(71.1)$

$$+\frac{(1.2)\,0.2\,(0.2-1)}{6}(-7.1) + \frac{(1.2)\,0.2\,(0.2-1)(0.2-2)}{24}(9.9)$$

$$= 2400.1 - 104.78 - 5.688 + 0.2272 + 0.14256 = 2290.$$

EXAMPLE 5.25 Apply Gauss's forward formula to find the value of $f(x)$ at $x = 3.75$ from the following table:

x:	2.5	3.0	3.5	4.0	4.5	5.0
$f(x)$:	24.145	22.043	20.225	18.644	17.262	16.047

(Bhopal 2002, Madras 2000)

Solution Taking $x_0 = 3.5$, $h = 0.5$ and $u = \frac{3.75 - 3.5}{0.5} = 0.5$, we have the following central difference table:

Table 5.17

x	$f(x)$	Δf	$\Delta^2 f$	$\Delta^3 f$	$\Delta^4 f$	$\Delta^5 f$
2.5	24.145					
		−2.102				
3.0	22.043		0.284			
		−1.818		−0.047		
$x_0 = 3.5$	**20.225**		**0.237**		**0.009**	Central line
		−1.581		**−0.038**		**−0.003**
4.0	18.644		0.199		0.006	
		−1.382		−0.032		
4.5	17.262		0.167			
		−1.215				
5.0	16.047					

Using Gauss's forward central difference formula

$$y_u = f_0 + u\Delta f_0 + \frac{u(u-1)}{2!}\Delta^2 f_{-1} + \frac{(u+1)\,u\,(u-1)}{3!}\Delta^3 f_{-1} + \frac{(u+1)\,u\,(u-1)(u-2)}{4!}\Delta^4 f_{-2} + \ldots \quad \text{...(A)}$$

Substituting the values in Gauss's forward formula, (A), we obtain

$$y_{0.5} = f(3.75) \approx 20.225 + (-1.581)\,0.5 + \frac{(0.5)(-0.5)}{2}(0.237) + \frac{(1.5)(0.5)(-0.5)}{6}(-0.038)$$

$$+\frac{(1.5)(0.5)(-0.5)(-1.5)}{24}(0.009) + \ldots$$

$$= 20.225 - 0.7905 - 0.029625 + 0.002375 + 0.0002109 + \ldots$$

$= 19.407$ correct to three decimal places.

EXAMPLE 5.26 Interpolate by means of Gauss's backward central difference formula, the population of a town for the year 1994, from the given data:

year x:	1959	1969	1979	1989	1999	2009
population $f(x)$: (in thousands)	12	15	20	27	39	52

Solution Taking $x_0 = 1999$, $h = 10$ and $u = \frac{1994 - 1999}{10} = -0.5$. The central difference table is shown below:

Table 5.18

x	$f(x)$	Δf	$\Delta^2 f$	$\Delta^3 f$	$\Delta^4 f$	$\Delta^5 f$
1959	12					
		3				
1969	15		2			
		5		0		
1979	20		2		3	
		7		3		−10
1989	27		5		−7	
		12		**−4**		
x_0 = 1999	**39**		**1**			
		13				
2009	52					

Central line

Employing Gauss's backward central difference formula

$$y_u = f(x_0) + u\Delta f_{-1} + \frac{u(u+1)}{2!}\Delta^2 f_{-1} + \frac{(u+1)\,u\,(u-1)}{3!}\Delta^3 f_{-2} + \frac{(u+2)\,(u+1)\,u(u-1)}{4!}\Delta^4 f_{-2} + \ldots$$

Substituting for u and the value of the central differences, we get

$$y_{0.5} = 39 + (-0.5)\,(12) + \frac{(-0.5)\,(0.5)}{2}(1) + \frac{(0.5)\,(-0.5)\,(-1.5)}{6}(-4)$$

$$= 39 - 6 - 0.125 - 0.25 = 32.625.$$

EXERCISE 5.3

1. Apply Gauss's forward formula to find the annuity value for 27 years from the following table:

No. of years:	15	20	25	30	35	40
Annuity value:	10.3797	12.4622	14.0939	15.3725	16.3742	17.1591

Ans. 14.6430.

2. Use appropriate Gauss's central differences formula to find the value of $f(32)$ given that: $f(25) = 0.2707$, $f(30) = 0.3027$, $f(35) = 0.3386$, $f(40) = 0.3794$. **Ans.** 0.3165.

3. Using Gauss's central differences formula, to estimate the value of cos 51°42′ from the following data:

x:	50°	51°	52°	53°	54°
cos x:	0.6428	0.6293	0.6157	0.6018	0.5878

Ans. 0.6198.

4. The following table gives the values of x and log x.

x:	310	320	330	340	350	360
log x:	2.4914	2.5052	2.5185	2.5315	2.5441	2.5563

Find the value of log 337.5, using an appropriate central differences formula. **Ans.** 2.5283.

5. Use appropriate Gauss's central differences formula to find $y(25)$ when the values of x and y are given by the following table:

x:	20	24	28	32
y:	14	32	35	40

Ans. 33.4921.

6. Find y_{30} from Gauss's forward formula, given that y_{21} = 18.4708, y_{25} = 17.8144, y_{29} = 17.1070, y_{33} = 16.3432, y_{37} = 15.5154. **Ans.** 16.9216.

7. Use Gauss's backward formula for central differences to fiind sin 45° for the following table:

θ°:	20	30	40	50	60	70	80
sin θ:	0.34202	0.50200	0.64279	0.76604	0.86603	0.93969	0.98481

Ans. 0.70711.

8. Use Gauss's backward central differences formula to find cos 45° for the following table:

θ°:	20	30	40	50	60	70	80
sin θ:	0.93969	0.86603	0.76604	0.64279	0.50000	0.34202	0.17365

Ans. 0.70711.

5.6.3 Stirling Formula

Gauss's forward formula is given by

$$y_u = y_0 + u\Delta y_0 + \frac{u(u-1)}{2}\Delta^2 y_{-1} + \frac{(u+1)\,u(u-1)}{6}\Delta^3 y_{-1} + \frac{(u+1)\,u(u-1)\,(u-2)}{24}\Delta^4 y_{-2}$$
$$+\frac{(u+2)\,(u+1)u\,(u-1)\,(u-2)}{120}\Delta^5 y_{-2} + \cdots \quad ...(5.25)$$

Gauss' backward formula is

$$y_u = y_0 + u\Delta y_{-1} + \frac{u(u+1)}{2}\Delta^2 y_{-1} + \frac{(u+1)\,u(u-1)}{6}\Delta^3 y_{-2} + \frac{(u+2)\,(u+1)\,u\,(u-1)}{24}\Delta^4 y_{-2}$$
$$+\frac{(u+2)\,(u+1)u\,(u-1)\,(u-2)}{120}\Delta^5 y_{-3} + \cdots \quad ...(5.26)$$

Taking the mean of relation (5.25) and relation (5.26), we obtain

$$y_u = y_0 + u\left(\frac{\Delta y_0 + \Delta y_{-1}}{2}\right) + \frac{u^2}{2}\Delta^2 y_{-1} + \frac{u(u^2-1^2)}{6}\left(\frac{\Delta^3 y_{-1} + \Delta^3 y_{-2}}{2}\right) +$$
$$\frac{u^2(u^2-1^2)}{24}\Delta^4 y_{-2} + \frac{u(u^2-1^2)\,(u^2-2^2)}{120}\left(\frac{\Delta^5 y_{-2} + \Delta^5 y_{-3}}{2}\right) + \frac{u^2(u^2-1^2)\,(u^2-2^2)}{720}\Delta^6 y_{-3} + \cdots \quad ...(5.27)$$

Relation (5.27) is known as stirling formula and it is the mean of two Gauss's formulae. It is obvious from relation (5.27) that this formula gives the mean of odd differences just above and just below the central line and even differences on the central line as shown below:

$$y_0 \cdots \begin{pmatrix}\Delta y_{-1}\\ \Delta y_0\end{pmatrix} \cdots \Delta^2 y_{-1} \cdots\cdots\cdots \begin{pmatrix}\Delta^3 y_{-2}\\ \Delta^3 y_{-1}\end{pmatrix} \cdots \Delta^4 y_{-2} \cdots\cdots\cdots \begin{pmatrix}\Delta^5 y_{-3}\\ \Delta^5 y_{-2}\end{pmatrix} \cdots \Delta^6 y_{-3} \cdots\cdots\cdots$$

In the central differences notation, this formula will be

$$y_u = y_0 + \frac{u}{2}(\delta y_{-1/2} + \delta y_{1/2}) + \frac{u^2}{2!}\delta^2 y_0 + \frac{u(u^2-1^2)}{2(3!)}(\delta^3 y_{-1/2} + \delta^3 y_{1/2}) + \frac{1}{4}!u^2(u^2-1^2)\delta^4 y_0 + \cdots \quad ...(5.28)$$

5.6.4 Bessel's Formula

Gauss's forward formula is

$$y_u = y_0 + u\Delta y_0 + \frac{u(u-1)}{2!}\Delta^2 y_{-1} + \frac{(u+1)u\,(u-1)}{3!}\Delta^3 y_{-1} + \frac{(u+1)u\,(u-1)\,(u-2)}{4!}\Delta^4 y_{-2} + \cdots$$

This formula can be alternately written as

$$y_u = \frac{y_0 + y_0}{2} + u\Delta y_0 + \frac{u(u-1)}{2!}\left(\frac{\Delta^2 y_{-1} + \Delta^2 y_{-1}}{2}\right) + \frac{(u+1)\,u(u-1)}{3!}\Delta^3 y_{-1}$$

$$+ \frac{(u+1)u\,(u-1)\,(u-2)}{4!}\left(\frac{\Delta^4 y_{-2} + \Delta^4 y_{-2}}{2}\right) + \cdots \qquad ...(5.29)$$

We know that

$$\Delta y_0 = y_1 - y_0 \quad \text{or} \quad y_0 = y_1 - \Delta y_0 \qquad ...(5.30)$$

$$\Delta^3 y_{-1} = \Delta^2 y_0 - \Delta^2 y_{-1} \quad \text{or} \quad \Delta^2 y_{-1} = \Delta^2 y_0 - \Delta^3 y_{-1} \qquad ...(5.31)$$

$$\Delta^5 y_{-2} = \Delta^4 y_{-1} - \Delta^4 y_{-2} \quad \text{or} \quad \Delta^4 y_{-2} = \Delta^4 y_{-1} - \Delta^5 y_{-2} \qquad ...(5.32)$$

Relation (5.29) with the help of (5.30), (5.31), (5.32) and after some adjustments can be written as

$$y_u = \frac{y_0}{2} + \frac{1}{2}(y_1 - \Delta y_0) + u\Delta y_0 + \frac{1}{2}\frac{u(u-1)}{2!}\Delta^2 y_{-1} + \frac{1}{2}\frac{u(u-1)}{2!}(\Delta^2 y_0 - \Delta^3 y_{-1}) + \frac{(u+1)u\,(u-1)}{3!}\Delta^3 y_{-1} +$$

$$\frac{(u+1)\,u(u-1)\,(u-2)}{(2)\,4!}\Delta^4 y_{-2} + \frac{(u+1)u\,(u-1)\,(u-2)}{(2)\,4!}(\Delta^4 y_{-1} - \Delta^5 y_{-2}) + \cdots$$

$$= \frac{y_0 + y_1}{2} + \left(u - \frac{1}{2}\right)\Delta y_0 + \frac{1}{2}\frac{u(u-1)}{2!}(\Delta^2 y_{-1} + \Delta^2 y_0) + \frac{u(u-1)}{2!}\left(-\frac{1}{2} + \frac{u+1}{3}\right)\Delta^3 y_{-1}$$

$$+ \frac{(u+1)u\,(u-1)\,(u-2)}{24}\left(\frac{\Delta^4 y_{-2} + \Delta^4 y_{-1}}{2}\right) + \cdots$$

$$= \frac{y_0 + y_1}{2} + \left(u - \frac{1}{2}\right)\Delta y_0 + \frac{u(u-1)}{2!}\left(\frac{\Delta^2 y_{-1} + \Delta^2 y_0}{2}\right) + \frac{u\left(u - \frac{1}{2}\right)(u-1)}{3!}\Delta^3 y_{-1}$$

$$+ \frac{u(u^2-1)\,(u-2)}{4!}\left(\frac{\Delta^4 y_{-2} + \Delta^4 y_{-1}}{2}\right) + \cdots \qquad ...(5.33)$$

This is one form of Bessel's formula. This formula involves odd differences below the central line and the mean of even differences just on and below the central line as exhibited below, where the brackets mean that the average of the values are to be taken.

$$\begin{array}{llllllll} x_0 & \begin{pmatrix} y_0 \\ \\ y_1 \end{pmatrix} & \Delta y_0 & \begin{pmatrix} \Delta^2 y_{-1} \\ \\ \Delta^2 y_0 \end{pmatrix} & \Delta^3 y_{-1} & \begin{pmatrix} \Delta^4 y_{-2} \\ \\ \Delta^4 y_{-1} \end{pmatrix} & \Delta^5 y_{-2} & \begin{pmatrix} \Delta^6 y_{-3} \\ \\ \Delta^6 y_{-2} \end{pmatrix} \text{— — Central line} \\ x_1 & & & & & & & \end{array}$$

The formula (5.33) is frequently written in a slightly different form. Since $\Delta y_0 = y_1 - y_0$, the first two term viz. $\frac{y_0 + y_1}{2} + \left(u - \frac{1}{2}\right)\Delta y_0$ can be transformed to $\frac{y_0 + y_0 + \Delta y_0}{2} + u\Delta y_0 - \frac{1}{2}\Delta y_0$ that is, $y_0 + u\Delta y_0$. Then (5.33) becomes

$$y_u = y_0 + u\Delta y_0 + \frac{u(u-1)}{2!}\cdot\frac{\Delta^2 y_{-1} + \Delta^2 y_0}{2} + \frac{u\left(u - \frac{1}{2}\right)(u-1)}{3!}\Delta^3 y_{-1} + \frac{u(u^2-1)\,(u-2)}{4!}\frac{\Delta^4 y_{-2} + \Delta^4 y_{-1}}{2} + \cdots \qquad ...(5.33A)$$

Remark 11

A more symmetrical and convenient form of Bessel's formula is obtained by putting $u - \frac{1}{2} = v$, or $u = v + \frac{1}{2}$. Making this substitution in formula (5.33), we obtain

$$y_u = \frac{y_0 + y_1}{2} + v\Delta y_0 + \left(\frac{v^2 - \frac{1}{4}}{2!}\right)\left(\frac{\Delta^2 y_{-1} + \Delta^2 y_0}{2}\right) + \frac{v\left(v^2 - \frac{1}{4}\right)}{3!}\Delta^3 y_{-1}$$

$$+ \frac{\left(v^2 - \frac{1}{4}\right)\left(v^2 - \frac{9}{4}\right)}{4!}\left(\frac{\Delta^4 y_{-2} + \Delta^4 y_{-1}}{2}\right) + \cdots \qquad \text{...(5.33B)}$$

The following examples illustrate the use of central difference formulae.

EXAMPLE 5.27 Employ Stirling's formula to compute $y_{12.2}$ from the following data ($y_x = 1 + \log_{10} \sin x$).

$x°$:	10	11	12	13	14
$10^5\, y_x$:	23967	28060	31788	35209	38368

Solution Taking the origin at $x_0 = 12$, $x = 12.2$ and $h = 1$, we have $u = \dfrac{x - x_0}{h} = \dfrac{12.2 - 12}{1} = 0.2$. We construct the central difference table for the given data:

Table 5.19

u	$x°$	$10^5 y_x$	$10^5\Delta(y_x)$	$10^5\Delta^2(y_x)$	$10^5\Delta^3(y_x)$	$10^5\Delta^4(y_x)$
–2	10	23967				
			4093			
–1	11	28060		–365		
			3728		**58**	
0	12	y_0 = **31788**		**–307**		**–13**
			3421		**45**	
1	13	35209		–262		
			3159			
2	14	38368				

Central line

Stirling's formula is

$$y_u = y_0 + u\left(\frac{\Delta y_0 + \Delta y_{-1}}{2}\right) + \frac{u^2}{2}\Delta^2 y_{-1} + \frac{u(u^2-1)}{6}\left(\frac{\Delta^3 y_{-1} + \Delta^3 y_{-2}}{2}\right) + \frac{u^2(u^2-1)}{24}\Delta^4 y_{-2} + \cdots$$

Therefore,

$$y_{0.2} = \frac{1}{10^5}\left[31788 + 0.2\left(\frac{3421+3728}{2}\right) + \frac{(0.2)^2}{2}(-307) + \frac{(0.2)(0.04-1)}{6}\left(\frac{45+58}{2}\right) + \frac{(0.04)(0.04-1)}{24}(-13)\right]$$

$$= 10^{-5}(31788 + 714.9 - 6.14 - 1.648 + 0.0208) = 10^{-5}(32495 \cdot 1328) = 0.32495.$$

EXAMPLE 5.28 Apply Bessel's formula to obtain $f(25)$, given $f(20) = 2854$, $f(24) = 3162$, $f(28) = 3544$, $f(32) = 3992$. *(SVTU, 2007; VTU, 2000S)*

Solution Taking the origin at $x_0 = 24$, $x = 25$, and $h = 4$. Therefore, $u = \dfrac{x - x_0}{h} = \dfrac{25-24}{4} = \dfrac{1}{4}$. We have the following central difference table:

Table 5.20: Central difference table

u	x	$f(x)$	$\Delta f(x)$	$\Delta^2 f(x)$	$\Delta^3 f(x)$
–1	20	2854			
			308		
0	24	**3162**		**74**	
			382		**–8**
1	28	**3544**		**66**	
			448		
2	32	3992			

Employing Bessel's formula (5.33)

$$f(x)=\frac{1}{2}(f_0+f_1)+\left(u-\frac{1}{2}\right)\Delta f_0+\frac{u(u-1)}{2!}\frac{(\Delta^2 f_{-1}+\Delta^2 f_0)}{2}$$
$$+\frac{u\left(u-\frac{1}{2}\right)(u-1)}{3!}\Delta^3 f_{-1}+\frac{(u+1)\,u\,(u-1)\,(u-2)}{4!}\frac{(\Delta^4 f_{-2}+\Delta^4 f_{-1})}{2}+\ldots$$

Substituting the values in the Bessel's formula, we get

$$f(25)=\frac{1}{2}(3162+3544)+\left(\frac{1}{4}-\frac{1}{2}\right)382+\left(\frac{1}{2}\right)\left(\frac{1}{4}\right)\left(\frac{1}{4}-1\right)\left(\frac{74+66}{2}\right)+\frac{1}{6}\left(\frac{1}{4}\right)\left(\frac{1}{4}-\frac{1}{2}\right)\left(\frac{1}{4}-1\right)(-8)$$
$$=3353-95.5-6.5625-0.0625=3250{\cdot}875.$$

5.6.5 Choice of Application of an Interpolation Formula

In the preceding sections we have studied a number of interpolation formulae. We are now required to know which one of these formulae will yield better and most accurate result when applied to a particular problem.

1. **The Newton's forward interpolation** formula is best suited if the interpolation is desired at the point which is near the beginning of a set of tabular values.
2. **The Newton's backward interpolation** formula will be the ideal choice, if the value to be interpolated is near the end of a set of tabular values.
3. For interpolation close to the middle of the table, **Stirling's** formula gives the best result if $u\left(=\frac{x-x_0}{h}\right)$ lying $-0.25\le u\le 0.25$ and **Bessel's** formula gives a very accurate result for u lying between 0.25 and 0.75.

EXERCISE 5.4

1. Use Stirling's formula to find y_{28}, given that
$y_{20}=49225,\quad y_{25}=48316,\quad y_{30}=47236,\quad y_{35}=45926,\quad y_{40}=44306.$ **Ans.** $y_{28}=47692.$
2. Use Stirling formula to find $y(32)$ when the values of x and $y(x)$ are given by the following table:

x:	20	25	30	35	40	45
$y(x)$:	14.035	13.674	13.257	12.734	12.089	11.309

Ans. 13.0622.

3. Given the table

x:	310	320	330	340	350	360
$y=\log x$	2.49136	2.50515	2.51851	2.53148	2.54407	2.55630

Find the value of log 337.5 by Stirling's formula. **Ans.** 2.52828.

4. Given:

θ:	0°	5°	10°	15°	20°	25°	30°
tan θ:	0	0.0875	0.1763	0.2679	0.3640	0.4663	0.5774

Using Stirling's formula, show that tan16° = 0.2867. *(Anna, 2005)*

5. The zero order Bessel function $J_0(x)$ is given by the following table:

x:	9.0	9.1	9.2	9.3	9.4	9.5
$J_0(x)$:	–0.0903	–0.1142	–0.1367	–0.1577	–0.1768	–0.1939

Find $J_0(9.274)$ by using Stirling's formula. **Ans.** $J_0(9.274)=-0.1524.$

6. Given:

x:	0	4	8	12
y:	143	158	177	199

Calculate y_5 by Bessel's formula. **Ans.** 162.41.

7. Use Bessel's formula to obtain y_{25}, given $y_{20} = 24$, $y_{24} = 32$, $y_{28} = 35$, $y_{32} = 40$. **Ans.** 32.945.

8. Employ Bessel's formula to find the value of y at $x = 1.95$ given that

x:	1.7	1.8	1.9	2.0	2.1	2.2	2.3
y:	2.979	3.144	3.283	3.391	3.463	3.997	4.491

Ans. 3.3471.

9. Use Bessel's formula to find $f(28)$ from the following data:

x:	15	20	25	30	35	40
$f(x)$:	10.3797	12.4622	14.0939	15.3725	16.3742	17.1591

Ans. 15.0818.

10. From the data:

x:	10°	20°	30°	40°	50°	60°	70°	80°
cos x:	0.9848	0.9397	0.8660	0.7660	0.6428	0.5000	0.3420	0.1736

Obtain cos 23°, using Bessel's formula. **Ans.** 0.9205.

5.7 INTERPOLATION WITH UNEVENLY SPACED POINTS

5.7.1 Introduction

The various interpolation formulae studied so far, are applicable only when the functions are given at equidistant intervals of the independent variable or arguments and the various differences of the function tend to become sinall. However, if the functional values are given at unequal intervals of the argument, and in situations where the differences of the dependent variable y do not diminish in magnitude, these formula cannot be used. In such situations it is desirable to have interpolation formulae which are applicable for data of unequally spaced nodal values of the arguments. Two such formulae are Lagrange's formula and Newton's formulae for unequal intervals of the argument. The former employs the function values while the latter utilises differences. The differences used in Newton's formula are called divided differences, which are differences obtained in the usual manner and then divided by certain differences of the argument.

5.7.2 Lagrange's Interpolation Formula

Let $y = f(x)$ be a function which assumes the values $y_0, y_1, y_2, ..., y_n$ for values of $x = x_0, x_1, x_2, ..., x_n$ respectively. Since there are $(n + 1)$ pairs of values of x and y, we can express the function $f(x)$ as a polynomial of degree n in x of the form:

$$y = f(x) = a_0 + a_1 x + a_2 x^2 + \cdots + a_n x^n.$$

This polynomial can alternatively be written as

$$y = f(x) = a_0(x - x_1)(x - x_2) \cdots (x - x_n) + a_1(x - x_0)(x - x_2) \cdots (x - x_n) + \cdots$$
$$\cdots + a_n(x - x_0)(x - x_1) \cdots (x - x_{n-1}). \qquad ...(5.34)$$

Eq. (5.34) holds for all the given values of x, that is, when $x = x_0$, $y = y_0$ and when $x = x_1$, $y = y_1$, and so on. Substituting these values in Eq. (5.34) we get

$$y_0 = a_0(x_0 - x_1)(x_0 - x_2) \cdots (x_0 - x_n).$$

From this we obtain

$$a_0 = \frac{y_0}{(x_0 - x_1)(x_0 - x_2) \cdots (x_0 - x_n)}.$$

Similarly, putting $y = y_1$ when $x = x_1$ in Eq. (5.34), we obtain

$$a_1 = \frac{y_1}{(x_1 - x_0)(x_1 - x_2) \cdots (x_1 - x_n)}.$$

And like wise

$$a_2 = \frac{y_2}{(x_2 - x_0)(x_2 - x_1) \cdots (x_2 - x_n)}.$$

....

....

$$a_n = \frac{y_n}{(x_n - x_0)(x_n - x_1)\cdots(x_n - x_{n-1})}.$$

Substituting the values of $a_0, a_1, a_2, ..., a_n$ in Eq. (5.34), we get

$$y = f(x) = \frac{(x - x_1)(x - x_2)...(x - x_n)}{(x_0 - x_1)(x_0 - x_2)...(x_0 - x_n)} y_0 + \frac{(x - x_0)(x - x_2)...(x - x_n)}{(x_1 - x_0)(x_1 - x_2)...(x_1 - x_n)} y_1$$

$$+ \cdots + \frac{(x - x_0)(x - x_1)...(x - x_{n-1})}{(x_n - x_0)(x_n - x_1)...(x_n - x_{n-1})} y_n. \quad ...(5.35)$$

Relation (5.35) is known as *Lagrange's interpolation formula.*

It can be used irrespective of the fact whether the values of x are equispaced or not and whether the differences of y become smaller or not.

The Lagrange's formula can be put in the form

$$y = f(x) = \sum_{r=0}^{n} \frac{\phi(x)}{(x - x_r)\phi'(x_r)} f(x_r), \ r = 0, 1, ...n, \quad ...(5.36)$$

where $\phi(x) = (x - x_0)(x - x_1)...(x - x_n)$ and $\phi'(x_r) = \frac{d}{dx}[\phi(x)]\Big|_{x = x_r}$

EXAMPLE 5.29 Find the Lagrange interpolating polynomial that fits the following data values.

(*i*)

x	2	2.5
$f(x)$	4	5.5

Interpolate at $x = 2.2$,

(*ii*)

x	–1	–2	2
$f(x)$	–1	–9	11

Compute $f(0)$. *(AMIE, W-2004)*

(*iii*)

x	–2	1	0	2
$f(x)$	3	–3	1	–1

Interpolate at $x = 1.5$.

Solution (*i*) We have two data values. Hence, we can fit a linear Lagrange interpolating polynomial. We have

$$P_1(x) = \frac{x - 2.5}{2 - 2.5}(4) + \frac{x - 2}{2.5 - 2}(5.5) = 3x - 2.$$

At $x = 2.2$, we get $f(2.2) \approx P_1(2.2) = 4.6$

(*ii*) We have three data values. We can fit an interpolating polynomial of degree ≤ 2. We have

$$P_2(x) = \frac{(x + 2)(x - 2)}{(-1 + 2)(-1 - 2)}(-1) + \frac{(x + 1)(x - 2)}{(-2 + 1)(-2 - 2)}(-9) + \frac{(x + 1)(x + 2)}{(2 + 1)(2 + 2)}(11)$$

$$= \frac{x^2 - 4}{3} - \frac{9}{4}(x^2 - x - 2) + \frac{11}{12}(x^2 + 3x + 2) = -x^2 + 5x + 5.$$

At $x = 0$, we get $f(0) \approx P_2(0) = 5$.

(*iii*) We have four nodal points. We can fit a polynomial of degree ≤ 3. We have

$$P_3(x) = \frac{(x - 1)(x - 0)(x - 2)}{(-2 - 1)(-2)(-2 - 2)}(3) + \frac{(x + 2)(x - 0)(x - 2)}{(1 + 2)(1 - 0)(1 - 2)}(-3)$$

$$+ \frac{(x + 2)(x - 1)(x - 2)}{(0 + 2)(0 - 1)(0 - 2)}(1) + \frac{(x + 2)(x - 1)(x - 0)}{(2 + 2)(2 - 1)(2 - 0)}(-1)$$

$$= -\frac{1}{8}(x^3 - 3x^2 + 2x) + (x^3 - 4x) + \frac{1}{4}(x^3 - x^2 - 4x + 4) - \frac{1}{8}(x^3 + x^2 - 2x) = x^3 - 5x + 1.$$

At $x = 1.5$, we get $f(1.5) \approx P_3(1.5) = (1.5)^3 - 5\,(1.5) + 1 = 3.375 - 7.5 + 1 = -3.125.$

EXAMPLE 5.30 The following table gives the viscosity of oil as a function of temperature. Use Lagrange's formula to find viscosity of oil at a temperature of 140°.

Temp. $x°$:	110°	130°	160°	190°
Viscosity $f(x)$:	10.8	8.1	5.5	4.8

(AMIETE, Dec. 2011)

Solution Here $x = 140°$, $x_0 = 110°$, $x_1 = 130°$, $x_2 = 160°$, $x_3 = 190°$. We can find viscosity by substituting the above values in the Lagrange formula, we obtain

$$\text{Viscosity at temperature } 140° = \frac{(140-130)(140-160)(140-190)}{(110-130)(110-160)(110-190)}(10.8)$$

$$+\frac{(140-110)(140-160)(140-190)}{(130-110)(130-160)(130-190)}(8.1)$$

$$+\frac{(140-110)(140-130)(140-190)}{(160-110)(160-130)(160-190)}(5.5)$$

$$+\frac{(140-110)(140-130)(140-160)}{(190-110)(190-130)(190-160)}(4.8)$$

$$= \frac{10(-20)(-50)}{(-20)(-50)(-80)}(10.8) + \frac{30\,(-20)(-50)}{20\,(-30)(-60)}(8.1)$$

$$+\frac{30(10)(-50)}{50\,(30)(-30)}(5.5) + \frac{30\,(10)(-20)}{80\,(60)(30)}(4.8)$$

$$= -1.35 + 6.75 + 1.83 - 0.2 = 7.03.$$

EXAMPLE 5.31 Given $\log_{10} 654 = 2.8156$, $\log_{10} 658 = 2.8182$, $\log_{10} 659 = 2.8189$, $\log_{10} 661 = 2.8202$, find $\log_{10} 656$.

Solution We have

x:	654	658	659	661
$\log_{10} x$:	2.8156	2.8182	2.8189	2.8202

Here $x = 656$, $x_0 = 654$, $x_1 = 658$, $x_2 = 659$, $x_3 = 661$.

Since the expression for the polynomial is not required, we can find $\log_{10} 656$, by substituting the above values in the Lagrange's interpolation formula, we obtain

$$\log_{10} 656 = \frac{(656-658)(656-659)(656-661)}{(654-658)(654-659)(654-661)}(2.8156) + \frac{(656-654)(656-659)(656-661)}{(658-654)(658-659)(658-661)}(2.8182)$$

$$+\frac{(656-654)(656-658)(656-661)}{(659-654)(659-658)(659-661)}(2.8189) + \frac{(656-654)(656-658)(656-659)}{(661-654)(661-658)(661-659)}(2.8202)$$

$$= \frac{(-2)(-3)(-5)}{(-4)(-5)(-7)}(2.8156) + \frac{(2)(-3)(-5)}{(4)(-1)(-3)}(2.8182) + \frac{(2)(-2)(-5)}{(5)(1)(-2)}(2.8189) + \frac{(2)(-2)(-3)}{(7)(3)(2)}(2.8202)$$

$$= 0.6033 + 7.0455 - 5.6378 + 0.8058 = 2.8168.$$

EXAMPLE 5.32 The following date represents e^{-x}.

x:	1	1.2	1.5
$f(x)$:	0.36788	0.30119	0.22313

Using Lagrange interpolation, obtain an approximate value of $e^{-1.3}$.

Solution We have three nodal points. We can fit an interpolating polynomial of degree ≤ 2. We write

$$P_2(x) = \frac{(x-1.2)(x-1.5)}{(1-1.2)(1-1.5)}(0.36788) + \frac{(x-1)(x-1.5)}{(1.2-1)(1.2-1.5)}(0.30119) + \frac{(x-1)(x-1.2)}{(1.5-1)(1.5-1.2)}(0.22313)$$

$$= (x^2 - 2.7x + 1.8)\, 3.6788 - (x^2 - 2.5x + 1.5)(5.0198) + (x^2 - 2.2x + 1.2)\, 1.4875$$

$$= 0.1465x^2 - 0.6557x + 0.8771.$$

Hence, $e^{-1.3} \approx P_2(1.3) = 0.1465\,(1.3)^2 - 0.6557\,(1.3) + 0.8771 = 0.27227.$

5.7.3 Divided Differences

Let the function $y = f(x)$ take the values $f(x_0), f(x_1), f(x_2), ..., f(x_n)$, corresponding to the values of $x = x_0, x_1, x_2, ..., x_n$ respectively where $x_1 - x_0, x_2 - x_1, ..., x_n - x_{n-1}$ are not necessarily equally spaced. The *first order divided difference* of $f(x)$ in the arguments x_0, x_1 is defined as follows:

$$f[x_0, x_1] = \frac{f(x_1) - f(x_0)}{x_1 - x_0}. \qquad ...(5.37)$$

Similarly the first order divided differences in the arguments $(x_1, x_2), (x_2, x_3), ... (x_{n-1}, x_n)$ are represented by

$$f[x_1, x_2] = \frac{f(x_2) - f(x_1)}{x_2 - x_1}$$

$$f[x_2, x_3] = \frac{f(x_3) - f(x_2)}{x_3 - x_2}$$

....

....

$$f[x_{n-1} - x_n] = \frac{f(x_n - x_{n-1})}{x_n - x_{n-1}}. \qquad ...(5.38)$$

With these first order differences of $f(x)$, we can form higher order differences. The *second order differences* in the arguments (x_0, x_1, x_2) (x_1, x_2, x_3), ..., can be defined as

$$f[x_0, x_1, x_2] = \frac{f[x_1, x_2] - f[x_0, x_1]}{x_2 - x_0}$$

$$f[x_1, x_2, x_3] = \frac{f[x_2, x_3] - f[x_1, x_2]}{x_3 - x_1}$$

....

....

In a similar fashion, the third order differences are defines as

$$f[x_0, x_1, x_2, x_3] = \frac{f[x_1, x_2, x_3] - f[x_0, x_1, x_2]}{x_3 - x_0}$$

$$f[x_1, x_2, x_3, x_4] = \frac{f[x_2, x_3, x_4] - f[x_1, x_2, x_3]}{x_4 - x_1}$$

The nth order divided difference is defined as

$$f[x_0, x_1, x_2, ..., x_n] = \frac{f[x_1, x_2, ..., x_n] - f[x_0, x_1, ..., x_{n-1}]}{x_n - x_0}. \qquad ...(5.39)$$

The divided differences are given in Table 5.21 for the data values (x_i, y_i), $i = 0, 1, 2, 3, 4$. *The divided differences are symmetric with respect to their arguments,* that is, $f[x_0, x_1] = f[x_1, x_0]$, $f[x_0, x_1, x_2) = f[x_2, x_1, x_0]$ etc.

Table 5.21: Divided differences (d.d) table

Argument x	Entry $f(x)$	*First d.d*	*Second d.d*	*Third d.d.*	*Fourth d.d*
x_0	$f(x_0)$				
		$f[x_0, x_1]$			
x_1	$f(x_1)$		$f[x_0, x_1, x_2]$		
		$f[x_1, x_2]$		$f[x_0, x_1, x_2, x_3]$	
x_2	$f(x_2)$		$f[x_1, x_2, x_3]$		$f[x_0, x_1, x_2, x_3, x_4]$
		$f[x_2, x_3]$		$f[x_1, x_2, x_3, x_4]$	
x_3	$f(x_3)$		$f[x_2, x_3, x_4]$		
		$f[x_3, x_4]$			
x_4	$f(x_4)$				

Note: The first order difference is in terms of two arguments x_0, x_1; the second order divided difference is in three arguments x_0, x_1, x_2 and a third order divided difference is in four arguments x_0, x_1, x_2, x_3 and so on.

5.7.4 Newton's Formula for Divided Differences

Consider the function $y = f(x)$ for arguments $x = x_0, x_1, x_2, ..., x_n$ which are not all equispaced. The first divided difference of (x, x_0) is given by

$$f[x, x_0] = \frac{f(x) - f(x_0)}{x - x_0}.$$

Solving this for $f(x)$, we have $f(x) = f(x_0) + (x - x_0) f[x, x_0]$. ...(5.40)

The second divided difference of x, x_0, x_1 is given by

$$f[x, x_0, x_1] = \frac{f[x, x_0] - f[x_0, x_1]}{x - x_1},$$

that is, $f[x, x_0] = f[x_0, x_1] + (x - x_1) f[x, x_0, x_1]$. ...(5.41)

Substituting for $f[x, x_0]$ from Eq. (5.41) in Eq. (5.40), we obtain

$$f(x) = f(x_0) + (x - x_0) f[x_0, x_1] + (x - x_0)(x - x_1) f[x, x_0, x_1]. \quad ...(5.42)$$

Again the third order divided difference x, x_0, x_1, x_2 is given by

$$f[x, x_0, x_1, x_2] = \frac{f[x, x_0, x_1] - f[x_0, x_1, x_2]}{x - x_2}$$ and from this we obtain

$$f[x, x_0, x_1] = f[x_0, x_1, x_2] + (x - x_2) f[x, x_0, x_1, x_2]. \quad ...(5.43)$$

Substituting the value of $f[x, x_0, x_1]$ from Eq. (5.43) in Eq. (5.42), we obtain

$$f(x) = f(x_0) + (x - x_0) f[x_0, x_1] + (x - x_0)(x - x_1) f[x_0, x_1, x_2] + (x - x_0)(x - x_1)(x - x_2) f[x, x_0, x_1, x_2]. \quad ...(5.44)$$

Continuing on similar lines, we obtain

$$f(x) = f(x_0) + (x - x_0) f[x_0, x_1] + (x - x_0)(x - x_1) f[x_0, x_1, x_2] + (x - x_0)(x - x_1)(x - x_2) f[x_0, x_1, x_2, x_3] + \cdots + \cdots + (x - x_0)(x - x_1) \cdots (x - x_{n-1}) f[x_0, x_1, ..., x_n]. \quad ...(5.45)$$

If $f(x)$ is a polynomial of degree n, the $(n + 1)$th divided difference will be zero and therefor, the last term in Eq. (5.45), that is, $f[x_0, x_1, x_2, ..., x_n]$ will be zero being the $(n + 1)$th divided difference with $(n + 2)$ arguments $x, x_0, x_1, x_2, ..., x_n$.

Hence Eq. (5.45) can be written as

$$f(x) \approx P_n(x) = f(x_0) + (x - x_0) f[x_0, x_1] + (x - x_0)(x - x_1) f[x_0, x_1, x_2] + (x - x_0)(x - x_1)(x - x_2) f[x_0, x_1, x_2, x_3] + \cdots + \cdots + (x - x_0)(x - x_1) \cdots (x - x_{n-1}) f[x_0, x_1, ..., x_n]. \quad ...(5.46)$$

Relation (5.46) is known as Newton's divided difference interpolating polynomial.

EXAMPLE 5.33 Find the third divided difference of $f(x) = 1/x$, using the points x_0, x_1, x_2, x_3.

Solution We have $f[x_0, x_1] = \dfrac{f_1 - f_0}{x_1 - x_0} = \dfrac{1}{(x_1 - x_0)}\left[\dfrac{1}{x_1} - \dfrac{1}{x_0}\right] = \dfrac{x_0 - x_1}{x_0 x_1 (x_1 - x_0)} = -\dfrac{1}{x_0 x_1}.$

$$f[x_1, x_2] = \frac{f_2 - f_1}{x_2 - x_1} = -\frac{1}{x_1 x_2}.$$

Now $f[x_0, x_1, x_2] = \dfrac{f[x_1, x_2] - f[x_0, x_1]}{x_2 - x_0} = \dfrac{1}{(x_2 - x_0)}\left[-\dfrac{1}{x_1 x_2} + \dfrac{1}{x_0 x_1}\right] = \dfrac{x_2 - x_0}{x_0 x_1 x_2 (x_2 - x_0)} = \dfrac{1}{x_0 x_1 x_2}.$

Finally

$$f[x_0, x_1, x_2, x_3] = \frac{f[x_1, x_2, x_3] - f[x_0, x_1, x_2]}{x_3 - x_0} = \frac{\dfrac{1}{x_1 x_2 x_3} - \dfrac{1}{x_0 x_1 x_2}}{x_3 - x_0} = \frac{x_0 - x_3}{x_0 x_1 x_2 x_3 (x_3 - x_0)} = \frac{-1}{x_0 x_1 x_2 x_3}.$$

EXAMPLE 5.34 Construct the divided difference table for the following data:

x:	–3	–1	0	2	3
$f(x)$:	–9	5	3	11	33

Hence, obtain the interpolating polynomial that fits the given data.

Solution The divided difference table is as follow:

Table 5.22: Divided differences (d.d) table

x	$f(x)$	*First d.d*	*Second d.d*	*Third d.d.*	*Fourth d.d*
–3	–9				
		$\frac{5+9}{-1+3} = 7$			
–1	5		$\frac{-2-7}{0+3} = -3$		
		$\frac{3-5}{0+1} = -2$		$\frac{2+3}{2+3} = 1$	
0	3		$\frac{4+2}{2+1} = 2$		0
		$\frac{11-3}{2-0} = 4$		$\frac{6-2}{3+1} = 1$	
2	11		$\frac{22-4}{3-0} = 6$		
		$\frac{33-11}{3-2} = 22$			
3	33				

Since the fourth order divided difference is zero, the data represents a third degree polynomial. Using Newton's divided difference interpolating polynomial, we obtain

$$P_3(x) = f_0 + (x - x_0) f[x_0, x_1] + (x - x_0)(x - x_1) f[x_0, x_1, x_2] + (x - x_0)(x - x_1)(x - x_2) f[x_0, x_1, x_2, x_3]$$
$$= -9 + (x + 3)\,7 + (x + 3)(x + 1)(-3) + (x + 3)(x + 1)(x)(1)$$
$$= -9 + 7x + 21 - 3x^2 - 12x - 9 + x^3 + 4x^2 + 3x = x^3 + x^2 - 2x + 3.$$

EXAMPLE 5.35 Using Newton's divided difference interpolation, obtain the polynomial which fits the data.

x:	−3	−2	−1	1	2	3
$f(x)$:	18	12	8	6	8	12

Also obtain the approximate value of $f(0)$ and $f(5)$. *(AMIETE, Dec. 2006)*

Solution We have $x_0 = -3, f_0 = 18; x_1 = -2, f_1 = 12; x_2 = -1, f_2 = 8; x_3 = 1, f_3 = 6; x_4 = 2, f_4 = 8; x_5 = 3,$ $f_5 = 12$. The divided differences are given by

$$f[x_0, x_1] = \frac{f_1 - f_0}{x_1 - x_0} = \frac{12 - 18}{-2 + 3} = -6, \quad f[x_1, x_2] = \frac{f_2 - f_1}{x_2 - x_1} = \frac{8 - 12}{-1 - (-2)} = -4,$$

$$f[x_2, x_3] = \frac{6 - 8}{1 - (-1)} = -1, \ f[x_3, x_4] = \frac{8 - 6}{2 - 1} = 2, \quad f[x_4, x_5] = \frac{12 - 8}{3 - 2} = 4,$$

$$f[x_0, x_1, x_2] = \frac{f[x_1, x_2] - f[x_0, x_1]}{x_2 - x_0} = \frac{-4 - (-6)}{-1 - (-3)} = 1,$$

$$f[x_1, x_2, x_3] = \frac{-1 - (-4)}{1 - (-2)} = 1, \ f[x_2, x_3, x_4] = \frac{2 + 1}{2 + 1} = 1, \quad f[x_3, x_4, x_5] = \frac{4 - 2}{3 - 1} = 1.$$

Since all the second order divided differences are same so third order divided differences are zero. Hence, we have the following divided difference table.

Table 5.23: Divided differences (d.d) table

x	$f(x)$	*First d.d*	*Second d.d*	*Third d.d.*
−3	18			
		−6		
−2	12		1	
		−4		0
−1	8		1	
		−1		0
1	6		1	
		2		0
2	8		1	
		4		
3	12			

The data represents a second degree polynomial. Using Newton's divided difference interpolating polynomial, we obtain

$$P_2(x) = f_0 + (x - x_0) f[x_0, x_1] + (x - x_0)(x - x_1) f[x_0, x_1, x_2]$$
$$= 18 + (x + 3)(-6) + (x + 3)(x + 2)(1) = x^2 - x + 6.$$

At $x = 0$, we obtain $f(0) \approx P_2(0) = 6$, and $f(5) \approx P_2(5) = 26$.

EXAMPLE 5.36 Calculate a fourth divided difference for the following data:

x:	0	1	2	4	5
$f(x)$:	0	16	48	88	0

Apply Newton's formula to find the polynomial that fits the given data. What value does this polynomial take at $x = 3$? Compare your results with the Lagrange method.

Solution We construct the divided difference table for the given data:

Table 5.24: Divided differences (d.d) table

x	$f(x)$	First d.d	Second d.d	Third d.d.	Fourth d.d
0	0				
		$\frac{16-0}{1-0}=16$			
1	16		$\frac{32-16}{2-0}=8$		
		$\frac{48-16}{2-1}=32$		$\frac{-4-8}{4-0}=-3$	
2	48		$\frac{20-32}{4-1}=-4$		$\frac{-8+3}{5-0}=-1$
		$\frac{88-48}{4-2}=20$		$\frac{-36+4}{5-1}=-8$	
4	88		$\frac{-88-20}{5-2}=-36$		
		$\frac{-88}{5-4}=-88$			
5	0				

Using Newton's divided difference interpolating polynomial, we obtain

$$P_4(x) = 0 + 16x + x(x-1)8 + x(x-1)(x-2)(-3) + x(x-1)(x-2)(x-4)(-1)$$
$$= -x^4 + 4x^3 + 3x^2 + 10x.$$

Hence, $f(3) \approx P_4(3) = -81 + 108 + 27 + 30 = 84.$

By Lagrange formula

$$P_4(3) = 0 + \frac{3(3-2)(3-4)(3-5)}{1(1-2)(1-4)(1-5)}(16) + \frac{3(3-1)(3-4)(3-5)}{2(2-1)(2-4)(2-5)}(48) + \frac{3(3-1)(3-2)(3-5)}{4(4-1)(4-2)(4-5)}(88) + 0$$

$$= -\frac{6}{12}(16) + \frac{12}{12}(48) + \frac{12}{24}(88) = 84.$$

EXAMPLE 5.37 Using Newton's divided difference formula, evaluate $f(8)$ and $f(15)$ given:

x:	4	5	7	10	11	13
$f(x)$:	48	100	294	900	1210	2028

(*SVTU, 2007; VTU 2004*)

Solution We construct the divided difference table for the given data:

Table 5.25: Divided differences (d.d) table

x	$f(x)$	First d.d	Second d.d	Third d.d.	Fourth d.d
4	48				
		$\frac{100-48}{5-4}=52$			
5	100		$\frac{97-52}{7-4}=15$		
		$\frac{294-100}{7-5}=97$		$\frac{21-15}{10-4}=1$	
7	294		$\frac{202-97}{10-5}=21$		0

		$\frac{900-294}{10-7}=202$		$\frac{27-21}{11-5}=1$	
10	900		$\frac{310-202}{11-7}=27$		0
		$\frac{1210-900}{11-10}=310$		$\frac{33-27}{13-7}=1$	
11	1210		$\frac{409-310}{13-10}=33$		
		$\frac{2028-1210}{13-11}=409$			
13	2028				

The data represents a third degree polynomial. Using Newton's divided difference interpolating polynomial, we obtain:

$$P_3(x) = 48 + (x - 4)\ 52 + (x - 4)\ (x - 5)\ 15 + (x - 4)\ (x - 5)\ (x - 7)1 = x^3 - x^2 = x^2\ (x - 1).$$

Hence, $f(8) \approx P_3(x) = 64(7) = 448.$

$f(15) \approx P_3(x) = 225(14) = 3150.$

EXAMPLE 5.38 Obtain the third degree polynomial passing through the four points given below. Hence, determine $f(2.5)$ and $\int_1^2 f(x)\,dx$.

x:	1.0	1.5	2.0	2.8
$f(x)$:	3.000	3.375	5.0	12.072

Solution We first construct the divided difference table. Hence, obtain the interpolating polynomial that fits the given data:

Table 5.26: Divided differences (d.d) table

x	$f(x)$	*First d.d*	*Second d.d*	*Third d.d.*
1.0	3.000			
		$\frac{3.375-3.000}{1.5-1.0}=0.75$		
1.5	3.375		$\frac{3.25-0.75}{2.0-1.0}=2.5$	
		$\frac{5.000-3.375}{2.0-1.5}=3.25$		$\frac{4.3-2.5}{2.8-1.0}=1.000$
2.0	5.000		$\frac{8.84-3.25}{2.8-1.5}=4.3$	
		$\frac{12.072-5.000}{2.8-2.0}=8.84$		
2.8	12.072			

$$f(x) \approx P_3(x) = 3.000 + (x - 1)\ (0.75) + (x - 1)\ (x - 1.5)\ (2.5) + (x - 1)\ (x - 1.5)\ (x - 2)\ (1.0)$$

$$= x^3 - 2x^2 + x + 3.$$

Now, $f_3(2.5) \approx P_3(2.5) = 3.000 + (2.5 - 1)\ (0.75)$

$$= (2.5 - 1)\ (2.5 - 1.5)\ (2.5) + (2.5 - 1)\ (2.5 - 1.5)\ (2.5 - 2)\ 1.0 = 8.625.$$

$$\int_1^2 f(x)\,dx = \int_1^2 (x^3 - 2x^2 + x + 3)\,dx = \left[\frac{x^4}{4} - \frac{2x^3}{3} + \frac{x^2}{2} + 3x\right]_1^2 = \frac{1}{4}(15) - \frac{2}{3}(7) + \frac{1}{2}(3) + 3(1) = \frac{43}{12} = 3.583.$$

EXAMPLE 5.39 Given the values:

x:	5	7	11	13	17
$f(x)$:	150	392	1452	2366	5202

Evaluate $f(9)$ (*i*) using Lagrange's formula. *(AMIETE, June 2013, W-2010; VTU, 2006; JNTU 2004S)*

(*ii*) Newton's divided difference formula. *(PTU, 2005)*

Solution Here $x_0 = 5, x_1 = 7, x_2 = 11, x_3 = 13, x_4 = 17$ and $y_0 = 150, y_1 = 392, y_2 = 1452, y_3 = 2366, y_4 = 5202$.

(*i*) Putting $x = 9$ and substituting the above values in Lagrange's formula, we obtain:

$$f(9) = \frac{(9-7)(9-11)(9-13)(9-17)}{(5-7)(5-11)(5-13)(5-17)}(150) + \frac{(9-5)(9-11)(9-13)(9-17)}{(7-5)(7-11)(7-13)(7-17)}(392)$$

$$+ \frac{(9-5)(9-7)(9-13)(9-17)}{(11-5)(11-7)(11-13)(11-17)}(1452) + \frac{(9-5)(9-7)(9-11)(9-17)}{(13-5)(13-7)(13-11)(13-17)}(2366)$$

$$+ \frac{(9-5)(9-7)(9-11)(9-13)}{(17-5)(17-7)(17-11)(17-13)}(5202)$$

$$= \frac{(2)(-2)(-4)(-8)}{(-2)(-6)(-8)(-12)}(150) + \frac{(4)(-2)(-4)(-8)}{2(-4)(-6)(-10)}(392) + \frac{4(2)(-4)(-8)}{6(4)(-2)(-6)}(1452)$$

$$+ \frac{(4)(2)(-2)(-8)}{(8)(6)(2)(-4)}(2366) + \frac{4(2)(-2)(-4)}{12(10)(6)(4)}(5202)$$

$$= -\frac{50}{3} + \frac{3136}{15} + \frac{3872}{3} - \frac{2366}{3} + \frac{578}{5} = 810.$$

(*ii*) We construct the divided difference table for the given data:

Table 5.27: Divided differences (d.d) table

x	$f(x)$	*First d.d*	*Second d.d*	*Third d.d.*
5	150			
		$\frac{392-150}{7-5} = 121$		
			$\frac{265-121}{11-5} = 24$	
7	392			
		$\frac{1452-392}{11-7} = 265$		$\frac{32-24}{13-5} = 1$
			$\frac{457-265}{13-7} = 32$	
11	1452			
		$\frac{2366-1452}{13-11} = 457$		$\frac{42-32}{17-7} = 1$
			$\frac{709-457}{17-11} = 42$	
13	2366			
		$\frac{5202-2366}{17-13} = 709$		
17	5202			

Using Newton's divided difference interpolating polynomial we obtain

$$P_3(x) = 150 + (x-5)\,121 + (x-5)(x-7)\,24 + (x-5)(x-7)(x-11).$$

At $x = 9$, we obtain

$$f(9) \approx P_3(x) = 150 + 4(121) + 4(2)24 + 4(2)\;-2)$$

$$= 150 + 484 + 192 - 16 = 810.$$

EXERCISE 5.5

1. Fit a polynomial of minimum degree to the following data:

x:	−1	1	2	4	5
$f(x)$:	13	15	13	33	67

Ans. $x^3 - 3x^2 + 17$.

2. (*a*) Use Lagrange interpolation to find the value of $f(x)$ at $x = 1.5$ from the following table:

x:	−2	−1	1	2
$f(x)$:	−5	1	1	7

(*b*) Using Lagrange's interpolation formula, find the missing term in the table:

x:	2	5	8	14
y:	94.8	87.9	—	68.7

(*AMIE, S-2009*) **Ans.** (*a*) $2\frac{7}{8}$, (*b*) 81.25.

3. Using Lagrange's interpolation formula, find the value of y when $x = 10$, from the following table:

x:	5	6	9	11
y:	12	13	14	16

(*AMIETE, Dec. 2012; Bombay 2004*) **Ans.** 13.47.

4. The following data are taken from the steam tables:

Temp°C (t):	140	160	170	190
Pressure kgf/cm^2 (P):	3.685	6.302	8.076	12.575

Use Lagrange's interpolation formula to compute the pressure at temperature 165°C.

(*AMIE, S-2007*) **Ans.** 7.14975.

5. Given $u_1 = 22$, $u_2 = 30$, $u_4 = 82$, $u_7 = 106$, $u_{12} = 206$, find u_8 using Lagrange interpolation formula.

Ans. 120.44.

6. Use Lagrange's formula to find the interpolating polynomial from the following table:

x:	0	2	3	6
$f(x)$:	648	704	729	792

Ans. $648 + 30x - x^2$.

7. Use Lagrange's interpolation formula to find x corresponding to $f(x) = 7$ from the following table:

x:	1	3	4
$f(x)$:	4	12	19

Ans. 1.86.

[**Hint:** Interchange the roles of x and y in the Lagrange interpolation formula].

8. Using the following table, find interpolating polynomial that fits the given data:

x:	−1	0	3	6	7
$f(x)$:	3	−6	39	822	1611

Ans. $x^4 - 3x^3 + 5x^2 - 6$.

9. The following table gives certain corresponding values of x and $\log_{10}x$:

x:	321.0	322.8	324.2	325.0
$\log_{10}x$:	2.50651	2.50893	2.51081	2.51188

Compute the value of $\log_{10}$ 323.5. **Ans.** 2.50987.

10. Find the polynomial of degree three which takes the values prescribed below:

x:	0	1	2	4
$y = f(x)$:	1	1	2	5

Ans. $\dfrac{(-x^3 + 9x^2 - 8x + 12)}{12}$.

11. (*a*) Find the first term of the series whose second and subsequent terms are 8, 3, 0, –1, 0. **Ans.** 15.

(*b*) In the table below the values of y are consecutive terms of a series of which the number 21.6 is the sixth terms. Find the first and the tenth terms of the series.

x:	3	4	5	6	7	8	9
y:	2.7	6.4	12.5	21.6	34.3	51.2	72.9

Ans. 0.1; 100.

12. Using Lagrange's interpolation formula, find a polynomial which passes through the points (0, –12), (1, 0), (3, 6), (4, 12). *(AMIE, S-2010, W-2007)* **Ans.** $P_3(x) = x^3 - 7x^2 + 18x - 12$.

13. From the following table, find the value of $f(x)$ at $x = 4$:

x:	1.5	3	6
$f(x)$:	–0.25	2	20

(AMIE, S-2008) **Ans.** $P_2(x) = x^2 - 3x + 2$; 6.

14. Fit a polynomial for the following data:

x:	–2	–1	1	2	4	5
$f(x)$:	–19	–9	–1	9	71	135

15. The followng table gives the values of x and $f(x)$. Use Lagrange's formula to find the form of the function $f(x)$.

x:	0	1	2	3	4
$f(x)$:	3	6	11	18	27

Ans. $f(x) = x^2 + 2x + 3$.

16. The following table gives the values of x and $f(x)$:

x:	0	1	2	5
$f(x)$:	2	3	12	147

Find the form of $f(x)$ using the above data. **Ans.** $f(x) = x^3 + x^2 - x + 2$.

17. Find the Lagrange's interpolation polynomial for $f(x)$ from the following data. Hence obtain $f(2)$.

x:	0	1	4	5
y:	8	11	68	123

Ans. $x^3 - x^2 + 3x + 8$; 18.

18. The observed values of a function are respectively 168, 120, 72 and 63 at the four positions 3, 7, 9 and 10 of the independent variable. What is the best estimate you can give for the value of the function at the position 6 of the independent variable? **Ans.** 147.

19. Find the value of y when $x = 5$, given that
$y(1) = 8$, $y(3) = 15$, $y(4) = 11$, $y(8) = 32$, $y(10) = 40$.

20. Find the second divided difference of $f(x) = 1/x$, using the points x_0, x_1, x_2. **Ans.** $1/x_0\, x_1\, x_2$.

21. Find the third divided difference of $f(x) = x^4 + x^2 + 1$, using the points 2, 4, 9, 10. **Ans.** 25.

22. (*a*) Find the third divided difference of the polynomial $x^3 + 3x^2 + 1$ based on the points x_0, x_1, x_2 and x_3. *(AMIETE, Dec. 2005)* **Ans.** 1.

(*b*) Given that:

x:	0	1	2	4
$f(x)$:	1	1	2	5

find the third order divided difference. *(AMIE, W-2006)* **Ans.** –0.083.

23. Construct the divided difference table for the following data:

x:	0	1	2	4	5	6
$f(x)$:	1	14	15	5	6	19

Hence, obtain the interpolating polynomial that fits the given data. **Ans.** $x^3 - 9x^2 + 21x + 1$.

24. Construct the divided difference table for the given data:

x:	0	1	2	5
$f(x)$:	2	3	12	147

Using divided differences interpolation, obtain the interpolating polynomial which fits this data.

Ans. $x^3 + x^2 - x + 2$.

25. Using Newton's divided difference interpolation formula, find from the following table the value of y for $x = 5$:

x:	4.0	4.5	5.5	6.0	7.0
y:	1.58740	1.65096	1.76517	1.81712	1.91293

26. Using Newton's divided difference interpolation, obtain the polynomial which fits the data:

x:	–3	–2	–1	1	2	3
$f(x)$:	18	12	8	6	8	12

27. Construct the divided difference table for the data:

x:	–3	–1	0	1	3	5
$f(x)$:	–23	1	1	1	25	121

Using divided differences interpolation, obtain the interpolating polynomial which fits this data.

28. Certain corresponding values of x and $\log_{10}x$ are given below:

x:	300	304	305	307
$\log_{10}x$:	2.4771	2.4829	2.4843	2.4871

Evaluate $\log_{10} 301$, using (*a*) Lagrange's formula, (*b*) Newton divided difference formula.

(*AMIE, S-2006*) **Ans.** 2.4786.

29. The number of members of a Civil Engineering Society are given below:

x:	1987	1988	1989	1991
y:	150	192	241	374

Make the best estimate you can of the number of member in 1990. **Ans.** 300.

5.8 INVERSE INTERPOLATION

Inverse interpolation is the process of finding the value of the argument x corresponding to a given value of the function y lying between two tabulated values of y. In fact inverse interpolation reverses the role of both x and y variables. The y variable becomes the argument and the x varaible as the value. The problem of inverse interpolation can be solved by several methods, we discuss the following three methods, namely:

(*i*) **Lagrange's** method when the values of x are not necessarily equally spaced.

(*ii*) The **method of successive approximation** or **the method of iteration** when the values of x are at equal length of the interval and

(*iii*) Reversion of series.

5.8.1 Lagrange's Method

This method is similar to Lagrange's interpolation formula the only difference being that in this method x is expressed as a function of y. Hence, on inter changing x and y in Eq. (5.35), the new formula known as inverse formula is given by

$$x = \Psi(y) = \frac{(y-y_1)(y-y_2)\ldots(y-y_n)}{(y_0-y_1)(y_0-y_2)\ldots(y_0-y_n)}x_0 + \frac{(y-y_0)(y-y_2)\ldots(y-y_n)}{(y_1-y_0)(y_1-y_2)\ldots(y_1-y_n)}x_1 + \ldots + \frac{(y-y_0)(y-y_1)\ldots(y-y_{n-1})}{(y_n-y_0)(y_n-y_1)\ldots(y_n-y_{n-1})}x_n. \quad \ldots(5.47)$$

EXAMPLE 5.40 Apply Lagrange's formula inversely to find the value of x corresponding to $y = 100$ from the following table:

x:	3	5	7	9	11
y:	6	24	58	108	174

Solution Using Lagrange's interpolation polynomial of x given in (5.47), we have

$$\psi(y) = \frac{(100-24)(100-58)(100-108)(100-174)}{(6-24)(6-58)(6-108)(6-174)}(3) + \frac{(100-6)(100-58)(100-108)(100-174)}{(24-6)(24-58)(24-108)(24-174)}(5)$$

$$+\frac{(100-6)(100-24)(100-108)(100-174)}{(58-6)(58-24)(58-108)(58-174)}(7) + \frac{(100-6)(100-24)(100-58)(100-174)}{(108-6)(108-24)(108-58)(108-174)}(9)$$

$$+\frac{(100-6)(100-24)(100-58)(100-108)}{(174-6)(174-24)(174-58)(174-108)}(11)$$

$$= \frac{(76)(42)(-8)(-74)}{(-18)(-52)(-102)(-168)}(3) + \frac{(94)(42)(-8)(-74)}{(18)(-34)(-84)(-150)}(5) + \frac{(94)(76)(-8)(-74)}{(52)(34)(-50)(-116)}(7)$$

$$+\frac{(94)(76)(42)(-74)}{(102)(84)(50)(-66)}(9) + \frac{(94)(76)(42)(-8)}{(168)(150)(116)(66)}(11)$$

$= 0.35344 - 1.51546 + 2.88703 + 7.06760 - 0.13686 = 8.65575.$

EXAMPLE 5.41 From the following table, find the Lagrange's interpolating polynomial which agrees with the values of x at the given values of y. Hence find the value of x when $y = 2$.

x:	1	19	49	101
y:	1	3	4	5

Solution Let $x = \psi(y)$ be the required polynomial. Then

$$\psi(y) = \frac{(y-3)(y-4)(y-5)}{(1-3)(1-4)(1-5)}(1) + \frac{(y-1)(y-4)(y-5)}{(3-1)(3-4)(3-5)}(19) + \frac{(y-1)(y-3)(y-5)}{(4-1)(4-3)(4-5)}(49) + \frac{(y-1)(y-3)(y-4)}{(5-1)(5-3)5-4)}(101)$$

$$= -\frac{1}{24}(y^3 - 12y^2 + 47y - 60) + \frac{19}{4}(y^3 - 10y^2 + 29y - 20) - \frac{49}{3}(y^3 - 9y^2 + 23y - 15)$$

$$+\frac{101}{8}(y^3 - 8y^2 + 19y - 12),$$

which on simplification reduces to $\quad \psi(y) = y^3 - y^2 + 1.$

Thus, the Lagrange's interpolating polynomial of x is given by $x = y^3 - y^2 + 1$.

Therefore, when $y = 2$, $x = 2^3 - 2^2 + 1 = 5$.

5.8.2 Method of Successive Approximations or the Method of Iteration

To see how this method is applied let us consider Newton's formula for forward interpolation, given by,

$$y = y_0 + u\Delta y_0 + \frac{u(u-1)}{2!}\Delta^2 y_0 + \frac{u(u-1)(u-2)}{3!}\Delta^3 y_0 + \frac{u(u-1)(u-2)(u-3)}{4!}\Delta^4 y_0 + \cdots\cdots$$

where $u = (x - x_0)/h$.

Transposing and dividing through by Δy_0, we have

$$u = \frac{y-y_0}{\Delta y_0} - \frac{u(u-1)}{2!}\frac{\Delta^2 y_0}{\Delta y_0} - \frac{u(u-1)(u-2)}{3!}\frac{\Delta^3 y_0}{\Delta y_0} - \frac{u(u-1)(u-2)(u-3)}{4!}\frac{\Delta^4 y_0}{\Delta y_0}\cdots \qquad \ldots(5.48)$$

To get the first approximation for u, we neglect all differences higher than the first and therefore have

$$u^{(1)} = \frac{y-y_0}{\Delta y_0}.$$

The second approximation is obtained by substituting $u^{(1)}$ in the right-hand side of Eq. (5.48). We then have

$$u^{(2)} = \frac{y - y_0}{\Delta y_0} - \frac{u^{(1)}(u^{(1)}-1)}{2!}\frac{\Delta^2 y_0}{\Delta y_0} - \frac{u^{(1)}(u^{(1)}-1)(u^{(1)}-2)}{3!}\frac{\Delta^3 y_0}{\Delta y_0}$$
$$- \frac{u^{(1)}(u^{(1)}-1)(u^{(1)}-2)(u^{(1)}-3)}{4!}\frac{\Delta^4 y_0}{\Delta y_0} \ldots \quad \ldots(5.49)$$

The third approximation is

$$u^{(3)} = \frac{y - y_0}{\Delta y_0} - \frac{u^{(2)}(u^{(2)}-1)}{2!}\frac{\Delta^2 y_0}{\Delta y_0} - \frac{u^{(2)}(u^{(2)}-1)(u^{(2)}-2)}{3!}\frac{\Delta^3 y_0}{\Delta y_0}$$
$$- \frac{u^{(2)}(u^{(2)}-1)(u^{(2)}-2)(u^{(2)}-3)}{4!}\frac{\Delta^4 y_0}{\Delta y_0} \ldots \quad \ldots(5.50)$$

and so on for higher approximation.

We shall now illustrate the method by working an example.

EXAMPLE 5.42 Given a table of values of the probability integral $\left(2/\sqrt{x}\right)\int_0^x e^{-x^2}\,dx$. By inverse interpolation, using the method of successive approximation or iteration, find the value of x for which this integral value equal to 0.5.

x:	0.45	0.46	0.47	0.48	0.49	0.50
$y = f(x)$:	0.4754818	0.4846555	0.4937452	0.5027498	0.5116683	0.5204999

Solution Here, we use a central difference Bessel's formula. We form the following central difference table.

Table 5.28

u	x	$f(x)$	$\Delta f(x) \times 10^{-7}$	$\Delta^2 f(x) \times 10^{-7}$	$\Delta^3 f(x) \times 10^{-7}$	$\Delta^4 f(x) \times 10^{-7}$	
–2	0.45	0.4754818					
			91737				
–1	0.46	0.4846555		–840			
			90897		–11		
0	0.47	**0.4937452**		**–851**		**1**	Central line
			90046		**–10**		
1	0.48	**0.5027498**		**–861**		**2**	
			89185		–8		
2	0.49	0.5116683		–869			
			88316				
3	0.50	0.5204999					

We take $x_0 = 0.47$, $h = 0.01$, $f(x) = 0.5$. Using Bessel's formula (5.33B), we obtain

$$0.5 = \frac{1}{2}(0.4937452 + 0.5027498) + 0.0090046\,v + \left(\frac{v^2 - 0.25}{2}\right)\left(\frac{-851 - 861}{2}\right)10^{-7}$$
$$+ \frac{v(v^2 - 0.25)}{6}(-0.0000010)$$

or

$$0.5 = 0.4982475 + 0.0090046\,v + \frac{(v^2 - 0.25)}{2}(-0.0000856) + \frac{v(v^2 - 0.25)}{6}(-0.0000010).$$

Transposing and dividing through by 0.0090046, we obtain

$$v = 0.194623 - (v^2 - 0.25)(-0.004753) - v(v^2 - 0.25)(-0.0000185) \qquad ...(1)$$

A first approximation for v is obtained by neglecting all terms beyond the first in the right-hand member of Eq. (1). Hence, $v^{(1)} = 0.194623$.

Substituting this for v in the right-hand member of (1), we find the second approximation to be

$$v^{(2)} = 0.194623 - [(0.194623)^2 - 0.25](-0.004753) - 0.194623\,[(0.194623)^2 - 0.25](-0.0000185)$$
$$= 0.194623 - 0.001008 - 0.000001 = 0.193614.$$

Now substituting this value for v in the right-hand member of (1), we find

$$v^{(3)} = 0.194623 - 0.0010101 - 0.000001 = 0.193612.$$

This value differs only slightly from the proceeding and we therefore, make no further approximations.

Since $u = v + \frac{1}{2} = 0.193612 + 0.5 = 0.693612$, thus $x = x_0 + hu = 0.47 + 0.01\,(0.693612) = 0.47693612$, correct to six decimal places.

5.8.3 By Reversion of Series

The most obvious method of solving the problem of inverse interpolation is by reversion of series, All the interpolation formulae developed are in the form of a power series, and any convegent power series can be reverted. Thus, the power series

$$y = f(x) = a_0 + a_1x + a_2x^2 + a_3x^3 + \ldots + a_nx^n + \ldots \qquad ...(5.51)$$

when reverted becomes

$$x = \left(\frac{y-a_0}{a_1}\right) + c_1\left(\frac{y-a_0}{a_1}\right)^2 + c_2\left(\frac{y-a_0}{a_1}\right)^3 + c_3\left(\frac{y-a_0}{a_1}\right)^4 + \ldots + c_{n-1}\left(\frac{y-a_0}{a_1}\right)^n + \ldots \qquad ...(5.52)$$

where

$$\left.\begin{aligned} c_1 &= -a_2/a_1,\\ c_2 &= -\frac{a_3}{a_1} + 2\left(\frac{a_2}{a_1}\right)^2,\\ c_3 &= -\frac{a_4}{a_1} + 5\left(\frac{a_2a_3}{a_1^2}\right) - 5\left(\frac{a_2}{a_1}\right)^3,\\ c_4 &= -\frac{a_5}{a_1} + 6\frac{a_2a_4}{a_1^2} + 3\left(\frac{a_3}{a_1}\right)^3 - 21\frac{a_2^2a_3}{a_1^3} + 14\left(\frac{a_2}{a_1}\right)^4,\\ c_5 &= -\frac{a_6}{a_1} + 7\left(\frac{a_2a_5 + a_3a_4}{a_1^2}\right) - 28\left(\frac{a_2^2a_4 + a_2a_3^2}{a_1^3}\right) + 84\frac{a_2^3a_3}{a_1^4} - 42\left(\frac{a_2}{a_1}\right)^5, \text{ etc.} \end{aligned}\right\} \qquad ...(5.53)$$

When reverting a series with numerical coefficients, it is better to compute the c's from equations (5.53) and then substitute their values in Eq. (5.52) to find x.

We shall now write Newton's, Stirling's and Bessel's formulas in the form of power series and then write down the values of $a_0, a_1, \ldots, a_4$ in each case. We stop with fourth differences, but it could be extended to higher differences if necessary.

(a) **Newton's Formula**

$$y = y_0 + u\Delta y_0 + \frac{u(u-1)}{2!}\Delta^2 y_0 + \frac{u(u-1)(u-2)}{3!}\Delta^3 y_0 + \frac{u(u-1)(u-2)(u-3)}{4!}\Delta^4 y_0 + \ldots \qquad ...(5.54)$$

$$= y_0 + \left(\Delta y_0 - \frac{\Delta^2 y_0}{2} + \frac{\Delta^3 y_0}{3} - \frac{\Delta^4 y_0}{4}\right)u + \left(\frac{\Delta^2 y_0}{2} - \frac{\Delta^3 y_0}{2} + \frac{11}{24}\Delta^4 y_0\right)u^2$$

$$+ \left(\frac{\Delta^3 y_0}{6} - \frac{\Delta^4 y_0}{4}\right)u^3 + \frac{\Delta^4 y_0}{24}u^4. \qquad ...(5.55)$$

On comparison of Eqs. (5.55) and (5.51), we obtain

$$\left.\begin{aligned} a_0 &= y_0, \qquad a_1 = \Delta y_0 - \frac{\Delta^2 y_0}{2} + \frac{\Delta^3 y_0}{3} - \frac{\Delta^4 y_0}{4}, \\ a_2 &= \frac{\Delta^2 y_0}{2} - \frac{\Delta^3 y_0}{2} + \frac{11}{24}\Delta^4 y_0, \qquad a_3 = \frac{\Delta^3 y_0}{6} - \frac{\Delta^4 y_0}{4}, \qquad a_4 = \frac{\Delta^4 y_0}{24}. \end{aligned}\right\} \quad ...(5.56)$$

(b) **Stirling's Formula**

$$y = y_0 + u\frac{\Delta y_{-1} + \Delta y_0}{2} + \frac{u^2}{2}\Delta^2 y_{-1} + \frac{u(u^2-1)}{6}\frac{\Delta^3 y_{-2} + \Delta^3 y_{-1}}{2} + \frac{u^2(u^2-1)}{24}\Delta^4 y_{-2} + \quad ...(5.57)$$

$$= y_0 + \left(\frac{\Delta y_{-1} + \Delta y_0}{2} - \frac{\Delta^3 y_{-2} + \Delta^3 y_{-1}}{12}\right)u + \left(\frac{\Delta^2 y_{-1}}{2} - \frac{\Delta^4 y_{-2}}{24}\right)u^2 + \left(\frac{\Delta^3 y_{-2} + \Delta^3 y_{-1}}{12}\right)u^3 + \frac{\Delta^4 y_{-2}}{24}u^4 \quad ...(5.58)$$

On comparing Eq. (5.58) with Eq. (5.51), we get

$$\left.\begin{aligned} a_0 &= y_0 \\ a_1 &= \frac{1}{2}(\Delta y_{-1} + \Delta y_0) - \frac{1}{12}(\Delta^3 y_{-2} + \Delta^3 y_{-1}), \\ a_2 &= \frac{1}{2}\Delta^2 y_{-1} - \frac{1}{24}\Delta^4 y_{-2}, \qquad a_3 = \frac{1}{12}(\Delta^3 y_{-2} + \Delta^3 y_{-1}), \qquad a_4 = \frac{1}{24}\Delta^4 y_{-2}. \end{aligned}\right\} \quad ...(5.59)$$

(c) **Bessel's Formula**

The general form of Bessel's formula is

$$y_u = f(x) = \frac{1}{2}(y_0 + y_1) + \left(u - \frac{1}{2}\right)\Delta y_0 + \frac{u(u-1)}{2}\frac{(\Delta^2 y_{-1} + \Delta^2 y_0)}{2}$$

$$+ \frac{u\left(u - \frac{1}{2}\right)(u-1)}{3!}\Delta^3 y_{-1} + \frac{u(u^2-1)(u-2)}{4!}\frac{(\Delta^4 y_{-2} + \Delta^4 y_{-1})}{2} + \ldots\ldots$$

A more symmetrical and convenient form of Bessel's formula is obtained by putting

$$u - \frac{1}{2} = v \quad \text{or} \quad u = v + \frac{1}{2} \text{ in the above formula.}$$

Making the substitution in the above formula, we get

$$y = f(x) = \frac{1}{2}(y_0 + y_1) + v\Delta y_0 + \left(\frac{v^2 - \frac{1}{4}}{2!}\right)\left(\frac{\Delta^2 y_{-1} + \Delta^2 y_0}{2}\right) + \frac{v\left(v^2 - \frac{1}{4}\right)}{3!}\Delta^3 y_{-1}$$

$$+ \frac{\left(v^2 - \frac{1}{4}\right)\left(v^2 - \frac{9}{4}\right)}{4!}\left(\frac{\Delta^4 y_{-2} + \Delta^4 y_{-1}}{2}\right) \text{ upto fourth differences.} \quad ...(5.60)$$

$$= \frac{1}{2}(y_0 + y_1) - \frac{1}{16}(\Delta^2 y_{-1} + \Delta^2 y_0) + \frac{3}{256}(\Delta^4 y_{-2} + \Delta^4 y_{-1}) + \left(\Delta y_0 - \frac{1}{24}\Delta^3 y_{-1}\right)v$$

$$+ \left[\frac{1}{4}(\Delta^2 y_{-1} + \Delta^2 y_0) - \frac{5}{96}(\Delta^4 y_{-2} + \Delta^4 y_{-1})\right]v^2 + \frac{1}{6}\Delta^3 y_{-1}v^3 + \frac{1}{48}(\Delta^4 y_{-2} + \Delta^4 y_{-1})v^4. \quad ...(5.61)$$

On comparing Eq. (5.61) with Eq. (5.51), we get

$$\left.\begin{aligned} a_0 &= \frac{1}{2}(y_0 + y_1) - \frac{1}{16}(\Delta^2 y_{-1} + \Delta^2 y_0) + \frac{3}{256}(\Delta^4 y_{-2} + \Delta^4 y_{-1}), \\ a_1 &= \Delta y_0 - \frac{1}{24}\Delta^3 y_{-1}, \quad a_2 = \frac{1}{4}(\Delta^2 y_{-1} + \Delta^2 y_0) - \frac{5}{96}(\Delta^4 y_{-2} + \Delta^4 y_{-1}), \\ a_3 &= \frac{1}{6}\Delta^3 y_{-1}, \quad a_4 = \frac{1}{48}(\Delta^4 y_{-2} + \Delta^4 y_{-1}). \end{aligned}\right\} \quad \text{...(5.62)}$$

EXAMPLE 5.43 The values of sinh x for certain equidistants values of x are given in the folloiwng table. If sinh x = 62, find x by the method of reversion of series. Use Stirling's formula.

x:	4.80	4.81	4.82	4.83	4.84
$y = f(x) = \sinh x$:	60.7511	61.3617	61.9785	62.6015	62.2307

Solution Forming a difference table as shown below, we find that all differences above the second are zero. We also notice from the given table that the required value of x is shightly greater thern 4.82. Hence, we take $x_0 = 4.82$.

Table 5.29

u	x	$y = \sinh x$	Δy	$\Delta^2 y$	$\Delta^3 y$
–2	4.80	60.7511			
			0.6106		
–1	4.81	61.3617		0.0062	
			0.6168		0
0	4.82	**61.9785**		**0.0062**	
			0.6230		0
1	4.83	62.6015		0.0062	
			0.6292		
2	4.84	63.2307			

Here $a_0 = y_0 = 61.9785$, $\quad a_1 = \frac{1}{2}(\Delta y_{-1} + \Delta y_0) - \frac{1}{12}(\Delta^3 y_{-2} + \Delta^3 y_{-1}) = \frac{1}{2}(0.6168 + 0.6230) = 0.6199,$

$$a_2 = \frac{1}{2}\Delta^2 y_{-1} - \frac{1}{24}\Delta^4 y_{-2} = \frac{1}{2}(0.0062) = 0.0031, \quad a_3 = a_4 = 0.$$

Since $y = f(x) = 62$, we have

$$\frac{y - a_0}{a_1} = \frac{62 - 61.9785}{0.6199} = \frac{0.0215}{0.6199} = 0.034683 = w(\text{say}).$$

$$c_1 = -\frac{a_2}{a_1} = -\frac{0.0031}{0.6199} = -0.005001,$$

$$c_2 = -\frac{a_3}{a_1} + 2\left(\frac{a_2}{a_1}\right)^2 = 0 + 2(0.005001)^2 = 0.00005002, \qquad c_3 = 0, \text{ practically.}$$

Substituting these values in $u = w + c_1 w^2 + c_2 w^3$, we have

$$u = 0.034683 - 0.005001\,(0.034683)^2 + 0.00005002\,(0.034683)^3 = 0.0347 \text{ approx.}$$

Therefore, $x\,(= x_0 + hu) = 4.82 + (0.01)(0.0347) = 4.8203.$

EXAMPLE 5.44 Given a table of values of the probability integral $f(x) = \left(\frac{2}{\sqrt{\pi}}\right)\int_0^x e^{-x^2}\,dx$. By the method of reversion of series, find the value of x for which this integral value equal to 0.5. Use Bessel's formula.

x:	0.45	0.46	0.47	0.48	0.49	0.50
$f(x)$:	0.4754818	0.4846555	0.4937452	0.5027498	0.5116683	0.5204999

Solution Forming a difference table as shown below. We take $x_0 = 0.47$, $h = 0.01$, $f(x) = 0.5$.

Table 5.30

u	x	$f(x)$	$\Delta f(x) \times 10^{-7}$	$\Delta^2 f(x) \times 10^{-7}$	$\Delta^3 f(x) \times 10^{-7}$	$\Delta^4 f(x) \times 10^{-7}$
–2	0.45	0.4754818				
			91737			
–1	0.46	0.4846555		–840		
			90897		–11	
0	0.47	**0.4937452**		**–851**		**1**
			90046		**–10**	
1	0.48	**0.5027498**		**–861**		**2**
			89185		–8	
2	0.49	0.5116683		–869		
			88316			
3	0.50	0.5204999				

Central line

From the table, we get

$$\frac{1}{2}(y_0 + y_1) = \frac{1}{2}(0.4937452 + 0.5027498) = 0.4982475,$$

$$\frac{1}{2}(\Delta^2 y_{-1} + \Delta^2 y_0) = \frac{1}{2}(-851 - 861) \times 10^{-7} = -0.0000856,$$

$$\frac{1}{2}(\Delta^4 y_{-2} + \Delta^4 y_{-1}) = \frac{1}{2}(1 + 2) \times 10^{-7} = 0.00000015.$$

Hence, $$a_0 = \frac{1}{2}(y_0 + y_1) - \frac{1}{8}\left(\frac{\Delta^2 y_{-1} + \Delta^2 y_0}{2}\right) = 0.4982475 + \frac{0.0000856}{8} = 0.4982582,$$

$$a_1 = \Delta y_0 - \frac{1}{24}\Delta^3 y_{-1} = 0.0090046 + \frac{0.0000010}{24} = 0.00900464,$$

$$a_2 = \frac{1}{4}(\Delta^2 y_{-1} + \Delta^2 y_0) = \frac{-0.0000856}{2} = -0.0000428,$$

$$a_3 = \frac{1}{6}\Delta^3 y_{-1} = -\frac{0.0000010}{6} = -0.00000017, \qquad a_4 = 0, \text{ practically.}$$

Also we have $$\frac{y - a_0}{a_1} = \frac{(0.5 - 0.4982582)}{0.00900464} = 0.1934336 = w \text{ (say)}.$$

$$c_1 = \frac{-a_2}{a_1} = \frac{0.0000428}{0.00900464} = 0.004753,$$

$$c_2 = \frac{-a_3}{a_1} + 2\left(\frac{a_2}{a_1}\right)^2 = \frac{0.00000017}{0.00900464} + 2(-0.004753)^2$$

$$= 0.00001888 + 2(0.000022591) = 0.00006406,$$

$$c_3 = \frac{-a_4}{a_1} + 5\left(\frac{a_2 a_3}{a_1^2}\right) - 5\left(\frac{a_2}{a_1}\right)^3 = 0 + 5\,\frac{(-0.0000428)(-0.00000017)}{(0.00900464)^2} - 5(-0.004753)^3$$

$$= 5\left(\frac{0.0000428}{0.00900464}\right)\left(\frac{(0.00000017)}{0.00900464}\right) + 5(0.0000001074)$$

$$= 5(0.004753)(0.00001888) + 5(0.0000001074) = 0.000000986.$$

Substituting these quantitites in $v = u - \frac{1}{2} = w + c_1 w^2 + c_2 w^3$, we obtain

$$u - \frac{1}{2} = 0.1934336 + 0.004753\,(0.1934336)^2 + (0.00006406)\,(0.1934336)^3$$

$= 0.1934336 + 0.0001778 + 0.00000046 = 0.193612$. We have therefore,

$$u = 0.5 + 0.193612 = 0.693612$$

$$x = x_0 + hu = 0.47 + 0.01\,(0.693612) = 0.47693612,$$

which is the same value as found by the method of successive approximations.

5.9 ERROR IN NEWTON'S INTERPOLATION FORMULAE

Since $f(x)$ is approximated by $P_n(x)$, the computed results contain errors. The difference $f(x) - P_n(x)$ is called the *error of interpolation* or *truncation error*. In Newton's forward difference interpolation formula, the error of interpolation

$$E_n(f, u) = f(x) - P_n(x) = \frac{u(u-1)(u-2)\ldots(u-n)}{(n+1)!}\,h^{n+1} f^{(n+1)}(\xi), \qquad \ldots(5.63)$$

where ξ is some value of x between 0 and n.

This form of the error is practically of no use in computations, as generally we donot know anything about $f^{(n+1)}(\xi)$. However, we can replace the derivative $f^{(n+1)}(\xi)$ as follows:

$$f^{(n+1)}(\xi) \approx \frac{1}{h^{n+1}}\Delta^{n+1} f_0. \qquad \ldots(5.64)$$

Using Eq. (5.64) in Eq. (5.63), we obtain

$$\text{Error} = E_n(f, u) = \frac{u(u-1)(u-2)\ldots(u-n)}{(n+1)!}\,h^{n+1}\cdot\frac{1}{h^{n+1}}\Delta^{n+1} f_0$$

or

$$E_n(f, u) = \frac{u(u-1)(u-2)\ldots\ldots(u-n)}{(n+1)!}\,\Delta^{n+1} f_0. \qquad \ldots(5.65)$$

In a similar manner we can express the error in Newton's backward difference interpolation formula in the form

$$E_n(f, u) = \frac{u(u+1)(u+2)\ldots.(u+n)}{(n+1)!}\,\nabla^{n+1} f_n. \qquad \ldots(5.66)$$

EXAMPLE 5.45 The following table gives the values of sin x from $x = 45°$ to $60°$ in steps of $5°$. Find sin $52°$ by using Newton's interpolation formula. Also find the estimate of the error.

x:	45°	50°	55°	60°
$f(x) = \sin x$:	0.70711	0.76604	0.81915	0.86603

Solution The point 52° is near the beginning of the table of values, with $h = 5°$. We shall use the Newton's forward interpolating polynomial. The forward difference table is given as under:

Table 5.31: Forward difference table

x	$f(x) = \sin x$	Δf_j	$\Delta^2 f_j$	$\Delta^3 f_j$
45°	**0.70711**			
		0.05893		
50°	0.76604		**– 0.00582**	
		0.05311		**– 0.00041**
55°	0.81915		– 0.00623	
		0.04688		
60°	0.86603			

The data represents a cubic polynomial.

$$f(x) \approx P_3(x) = f_0 + u\Delta f_0 + \frac{u(u-1)}{2!}\Delta^2 f_0 + \frac{u(u-1)(u-2)}{3!}\Delta^3 f_0 .$$

For $x = 52°$, $h = 5°$, $u = \dfrac{x - x_0}{h} = \dfrac{52-45}{5} = 1.4$, we have

$$f(52) = \sin 52° \approx P_3(x) = 0.70711 + 1.4(0.05893) + \frac{1.4(0.4)}{2}(-0.00582) + \frac{1.4(0.4)(-0.6)}{6}(-0.00041)$$

$$= 0.70711 + 0.082502 - 0.00163 + 0.000023 = 0.78800.$$

The error of interpolation $E_n(f, u) = f(x) - P_n(x) = \dfrac{u(u-1) \ldots\ldots (u-n)}{(n+1)!}\Delta^{n+1} f_0$ (taking $n = 2$)

$$= \frac{1.4(1.4-1)(1.4-2)}{3!}\Delta^3 f_0 = \frac{(1.4)(0.4)(-0.6)}{6}(-0.00041) = 0.0000229.$$

EXAMPLE 5.46 The following table gives the values of sin x from $x = 25°$ to $x = 30°$ in steps of 1°. Using Newton's interpolation formula to evaluate sin 28°24′ and estimate the error.

x:	25°	26°	27°	28°	29°	30′
$y = \sin x$:	0.42262	0.43837	0.45399	0.46947	0.48481	0.50000

Solution The point 28°24′ = 28.4° is near the end of the table of values. The backward difference table is as given below:

Table 5.32: Backward difference table

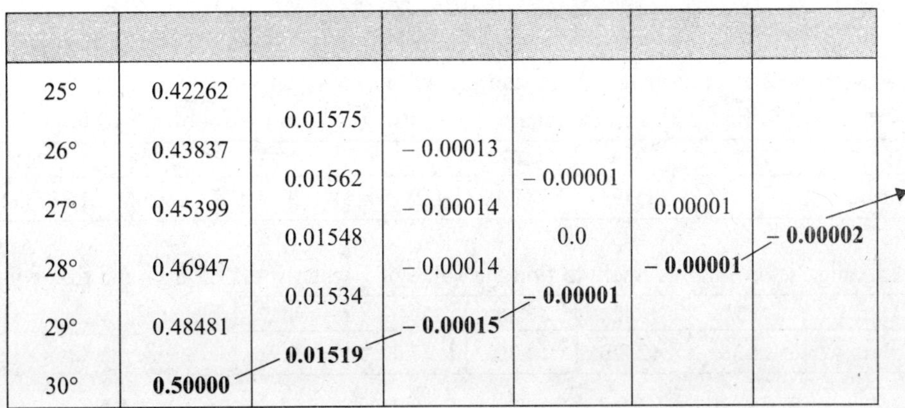

25°	0.42262					
		0.01575				
26°	0.43837		– 0.00013			
		0.01562		– 0.00001		
27°	0.45399		– 0.00014		0.00001	
		0.01548		0.0		**– 0.00002**
28°	0.46947		– 0.00014		**– 0.00001**	
		0.01534		**– 0.00001**		
29°	0.48481		**– 0.00015**			
		0.01519				
30°	**0.50000**					

Employing Newton's backward interpolation polynomial

$$f(x) \approx P_5(x) = y_n + u\nabla y_n + \frac{1}{2}u(u+1)\nabla^2 y_n + \frac{1}{6}u(u+1)(u+2)\nabla^3 y_n + \frac{1}{24}u(u+1)(u+2)(u+3)\nabla^4 y_n + \frac{1}{120}u(u+1)(u+2)(u+3)(u+4)\Delta^5 y_n + \cdots$$

We take $x_n = 30$, $y_n = 0.50000$, $x = 28°24' = 28.4°$. Then $u = \frac{x - x_n}{h} = \frac{28.4 - 30.0}{1} = -1.6.$

Therefore, $f(28.4°) \approx P_5(x) = 0.50000 - 1.6(0.01519) + \frac{1}{2}(-1.6)(-0.6)(-0.00015)$

$$+\frac{1}{6}(-1.6)(-0.6)(0.4)(-0.00001) + \frac{1}{24}(-1.6)(-0.6)(0.4)(1.4)(-0.00001)$$

$$+\frac{1}{120}(-1.6)(-0.6)(0.4)(1.4)(2.4)(-0.00002)$$

$$= 0.50000 - 0.024304 - 0.000072 - 0.0000006 - 0.0000002 - 0.0000002 = 0.475623.$$

The error of interpolation $E_n(f, u) = f(x) - P_n(x) = \frac{u(u+1)(u+2)+\cdots(u+n)}{(n+1)!}\nabla^{n+1} f_n$

$$= \frac{u(u+1)(u+2)(u+3)(u+4)}{5!}\nabla^5 f_n \quad (\text{taking } n = 4)$$

$$= \frac{(-1.6)(-0.6)(0.4)(1.4)(2.4)}{(5)(4)(3)(2)(1)}(-0.00002) = -0.0000002.$$

EXERCISE 5.6

1. Use Lagrange's formula inversely to obtain the value of t when $A = 85$ from the following table:

t:	2	5	8	14
A:	94.8	87.9	81.3	68.7

Ans. 6.5928

2. Apply Lagrange's formula inversely to obtain the root of the equation $f(x) = 0$, given that $f(30) = -30$, $f(34) = -13$, $f(38) = 3$ and $f(42) = 18$. **Ans.** 37.23

3. The following table gives the value of x and y:

x:	3	5	7	9	11
y:	6	24	58	108	174

Apply Lagrange's formula inversely to find the value of x when $y = 100$. **Ans.** 8.656

4. If $\cosh x = 1.285$, find x by inverse interpolation, using data in the following table:

x:	0.735	0.736	0.737	0.738	0.739	0.740	0.741	0.742
$\cosh x$:	1.2824937	1.2832974	1.2841023	1.2849085	1.2857159	1.2865247	1.2873348	1.2881461

Ans. 0.7381134

5. Use Lagrange's formula inversely to find the value of x when $y = 13.5$ from the following table:

x:	93	96.2	100.0	104.2	108.7
y:	11.38	12.80	14.70	17.07	19.91

Ans. 97.66

NUMERICAL DIFFERENTIATION

5.10 INTRODUCTION

Numerical differentiation is the process of calculating the derivatives of a function by means of a set of given values of that function. The problem may be solved by representing the function by an interpolation formula and then differentiating this formula as many times as desired.

If the function is given by a table of values for equal intervals of the independent variable, it should be represented by an interpolation formula employing differences, such as Newton's, Stirling's or Bessel's.

The choice of the interpolating formula for derivative employing differences, are the same as in the case of interpolation. If the values of the argument x are equispaced and we desire the derivative at a point near the *beginning* of a set of tabular values, we use Newton's forward interpolation formula. Whereas, if we desire the derivative at a point close to the *end* of the table, we use Newton's backward interpolation formula. For points which are nearer the middle values of the table, dy/dx is computed by using central differences formulae—Stirling's formula or Bessel's formula.

In case the values of x are not at equal length of the interval, we must express the function by Lagrange's formula or Newton's divided difference formula for obtaining the value of dy/dx. We shall now derive differentiation formulae for the derivatives using uniform nodal points. That is, when the difference between any two consecutive points is the same.

5.10.1 Derivative using Newton's Forward Difference Formula

Consider Newton's forward difference interpolating polynomial

$$y_u \approx P_n(x) = y_0 = u\Delta y_0 + \frac{u(u-1)}{2!}\Delta^2 y_0 + \frac{u(u-1)(u-2)}{3!}\Delta^3 y_0 + \frac{u(u-1)(u-2)(u-3)}{4!}\Delta^4 y_0 + \cdots \quad ...(5.67)$$

where $u = (x - x_0)/h$.

Differentiating (5.67) with respect to u, we get

$$\frac{dy}{du} = \Delta y_0 + \frac{2u-1}{2}\Delta^2 y_0 + \frac{3u^2-6u+2}{6}\Delta^3 y_0 + \frac{2u^3-9u^2+11u-3}{12}\Delta^4 y_0 + \cdots \quad ...(5.68)$$

Now $$\frac{dy}{dx} = \frac{dy}{du}\cdot\frac{du}{dx} = \frac{1}{h}\frac{dy}{du}$$

$$= \frac{1}{h}\left[\Delta y_0 + \frac{2u-1}{2}\Delta^2 y_0 + \frac{3u^2-6u+2}{6}\Delta^3 y_0 + \frac{2u^3-9u^2+11u-3}{12}\Delta^4 y_0 + \cdots\right]. \quad ...(5.69)$$

Formula (5.69) gives the value of dy/dx at any point x. Differentiating (5.69) further with respect to x

$$\frac{d^2y}{dx^2} = \frac{d}{dx}\left(\frac{dy}{dx}\right) = \frac{d}{du}\left(\frac{dy}{dx}\right)\frac{du}{dx} = \frac{1}{h}\frac{d}{du}\left(\frac{dy}{dx}\right)$$

$$= \frac{1}{h}\left[\frac{1}{h}\left(\Delta^2 y_0 + \frac{6u-6}{6}\Delta^3 y_0 + \frac{6u^2-18u+11}{12}\Delta^4 y_0 + \cdots\right)\right]$$

$$= \frac{1}{h^2}\left(\Delta^2 y_0 + (u-1)\Delta^3 y_0 + \frac{6u^2-18u+11}{12}\Delta^4 y_0 + \cdots\right). \quad ...(5.70)$$

$$\frac{d^3y}{dx^3} = \frac{d}{dx}\left(\frac{d^2y}{dx^2}\right) = \frac{d}{du}\left(\frac{d^2y}{dx^2}\right)\frac{du}{dx} = \frac{1}{h}\frac{d}{du}\left(\frac{d^2y}{dx^2}\right)$$

$$= \frac{1}{h^3}\left[\Delta^3 y_0 + \frac{12u-18}{12}\Delta^4 y_0 + \cdots\right]. \quad ...(5.71)$$

$$\frac{d^4y}{dx^4} = \frac{1}{h^4}[\Delta^4 y_0 + \cdots]. \qquad ...(5.72)$$

As a particular case, if $x = x_0$, then $u = 0$. Hence, on substituting this value of u in the above formulas, we find the sucessive derivatives at the tabular point x_0.

$$\left.\frac{dy}{dx}\right|_{x=x_0} = \frac{1}{h}\left[\Delta y_0 - \frac{1}{2}\Delta^2 y_0 + \frac{1}{3}\Delta^3 y_0 - \frac{1}{4}\Delta^4 y_0 + \cdots\right]. \qquad ...(5.73)$$

$$\left.\frac{d^2y}{dx^2}\right|_{x=x_0} = \frac{1}{h^2}\left[\Delta^2 y_0 - \Delta^3 y_0 + \frac{11}{12}\Delta^4 y_0 - \frac{5}{6}\Delta^5 y_0 + \frac{137}{180}\Delta^6 y_0 + \cdots\right]. \qquad ...(5.74)$$

$$\left.\frac{d^3y}{dx^3}\right|_{x=x_0} = \frac{1}{h^3}\left[\Delta^3 y_0 - \frac{3}{2}\Delta^4 y_0 + \cdots\right]. \qquad ...(5.75)$$

$$\left.\frac{d^4y}{dx^4}\right|_{x=x_0} = \frac{1}{h^4}\left[\Delta^4 y_0 - \cdots\right]. \qquad ...(5.76)$$

Proceeding on similar lines we can compute higher derivatives at $x = x_0$.

5.10.2 Derivative using Newton's Backward Difference Formula

Consider Newton's backward difference interpolating polynomial

$$y_u = y_n + u\nabla y_n + \frac{u(u+1)}{2!}\nabla^2 y_n + \frac{u(u+1)(u+2)}{3!}\nabla^3 y_n + \frac{u(u+1)(u+2)(u+3)}{4!}\nabla^4 y_n + \cdots \qquad ...(5.77)$$

where $u = (x - x_n) / h$.

Differentiating (5.77) with respect to u, we get

$$\frac{dy}{du} = \nabla y_n + \frac{2u+1}{2}\nabla^2 y_n + \frac{3u^2+6u+2}{6}\nabla^3 y_n + \frac{4u^3+18u^2+22u+6}{24}\nabla^4 y_n + \cdots$$

Now $$\frac{dy}{dx} = \frac{dy}{du}\cdot\frac{du}{dx} = \frac{1}{h}\left[\nabla y_n + \frac{2u+1}{2}\nabla^2 y_n + \frac{3u^2+6u+2}{6}\nabla^3 y_n + \frac{2u^3+9u^2+11u+3}{12}\nabla^4 y_n + \cdots\right]. \qquad ...(5.78)$$

Equation (5.78) gives the value of dy/dx at any point x in the interval which is a non-nodal.

Differentiating (5.78) successively, the higher order derivatives are given by

$$\frac{d^2y}{dx^2} = \frac{1}{h^2}\left[\nabla^2 y_n + (u+1)\nabla^3 y_n + \frac{6u^2+18u+11}{12}\nabla^4 y_n + \cdots\right]. \qquad ...(5.79)$$

$$\frac{d^3y}{dx^3} = \frac{1}{h^3}\left[\nabla^3 y_n + \frac{2u+3}{2}\nabla^4 y_n + \cdots\right]. \qquad ...(5.80)$$

$$\frac{d^4y}{dx^4} = \frac{1}{h^4}[\nabla^4 y_n + \cdots]. \qquad ...(5.81)$$

At the point $x = x_n$, we have $u = 0$.

Therefore, $$\left.\frac{dy}{dx}\right|_{x=x_0} = \frac{1}{h}\left[\nabla y_n + \frac{1}{2}\nabla^2 y_n + \frac{1}{3}\nabla^3 y_n + \frac{1}{4}\nabla^4 y_n + \cdots\right]. \qquad ...(5.82)$$

$$\left.\frac{d^2y}{dx^2}\right|_{x=x_0} = \frac{1}{h^2}\left[\nabla^2 y_n + \nabla^3 y_n + \frac{11}{12}\nabla^4 y_n + \frac{5}{6}\nabla^5 y_n + \cdots\right]. \quad ....(5.83)$$

$$\left.\frac{d^3y}{dx^3}\right|_{x=x_0} = \frac{1}{h^3}\left[\nabla^3 y_n + \frac{3}{2}\nabla^4 y_n + \cdots\right]. \quad ....(5.84)$$

$$\left.\frac{d^4y}{dx^4}\right|_{x=x_0} = \frac{1}{h^4}[\nabla^4 y_n + \cdots]. \quad ....(5.85)$$

Proceeding on similar lines, we can compute higher derivatives at $x = x_0$.

5.10.3 Derivative using Central Differences — Stirling's Formula

Consider the Stirling's formula

$$y_u = y_0 + u\left(\frac{\Delta y_0 + \Delta y_{-1}}{2}\right) + \frac{u^2}{2!}\Delta^2 y_{-1} + \frac{u(u^2-1)}{3!}\left(\frac{\Delta^3 y_{-1} + \Delta^3 y_{-2}}{2}\right) + \frac{u^2(u^2-1)}{4!}\Delta^4 y_{-2}$$
$$+ \frac{u(u^2-1)(u^2-2^2)}{5!}\left(\frac{\Delta^5 y_{-3} + \Delta^5 y_{-2}}{2}\right) + \frac{u^2(u^2-1)(u^2-2^2)}{6!}\Delta^6 y_{-3} + \cdots \quad ...(5.86)$$

where $u = \dfrac{x - x_0}{h}$ and $\dfrac{dy}{dx} = \dfrac{dy}{du}\cdot\dfrac{du}{dx} = \dfrac{1}{h}\dfrac{dy}{du}$.

Differentiating (5.86) with respect to x, we obtain

$$\frac{dy}{dx} = \frac{1}{h}\frac{dy}{du} = \frac{1}{h}\left[\frac{\Delta y_0 + \Delta y_{-1}}{2} + u\Delta^2 y_{-1} + \frac{3u^2-1}{6}\left(\frac{\Delta^3 y_{-1} + \Delta^3 y_{-2}}{2}\right) + \frac{4u^3-2u}{24}\Delta^4 y_{-2}\right.$$
$$\left. + \frac{5u^4 - 15u^2 + 4}{120}\left(\frac{\Delta^5 y_{-3} + \Delta^5 y_{-2}}{2}\right) + \frac{6u^5 - 20u^3 + 8u}{6!}\Delta^6 y_{-3} + \cdots\right. \quad ...(5.87)$$

Differentiating (5.87) successively, the higher order derivatives are given by

$$\frac{d^2y}{dx^2} = \frac{1}{h^2}\left[\Delta^2 y_{-1} + u\left(\frac{\Delta^3 y_{-1} + \Delta^3 y_{-2}}{2}\right) + \frac{12u^2-2}{24}\Delta^4 y_{-2} + \frac{20u^3-30u}{120}\left(\frac{\Delta^5 y_{-3} + \Delta^5 y_{-2}}{2}\right)\right.$$
$$\left. + \frac{30u^4 - 60u^2 + 8}{6!}\Delta^6 y_{-3} + \cdots\right]. \quad ....(5.88)$$

$$\frac{d^3y}{dx^3} = \frac{1}{h^3}\left[\frac{\Delta^3 y_{-1} + \Delta^3 y_{-2}}{2} + u\Delta^4 y_{-2} + \frac{2u^2-1}{4}\left(\frac{\Delta^5 y_{-3} + \Delta^5 y_{-2}}{2}\right) + \frac{u^3-u}{6}\Delta^6 y_{-3} + \cdots\right]. \quad ...(5.89)$$

$$\frac{d^4y}{dx^4} = \frac{1}{h^4}\left[\Delta^4 y_{-2} + u\left(\frac{\Delta^5 y_{-3} + \Delta^5 y_{-2}}{2}\right) + \frac{3u^2-1}{6}\Delta^6 y_{-3} + \cdots\right]. \quad ...(5.90)$$

$$\frac{d^5y}{dx^5} = \frac{1}{h^5}\left[\frac{\Delta^5 y_{-3} + \Delta^5 y_{-2}}{2} + u\Delta^6 y_{-3} + \cdots\right]. \quad ...(5.91)$$

$$\frac{d^6y}{dx^6} = \frac{1}{h^6}[\Delta^6 y_{-3} + \cdots]. \quad ...(5.92)$$

At the point $x = x_0$, $u = 0$ and therefore, (5.87) to (5.92) can be written as

$$\left.\frac{dy}{dx}\right|_{x=x_0} = \frac{1}{h}\left[\left(\frac{\Delta y_0 + \Delta y_{-1}}{2}\right) - \left(\frac{\Delta^3 y_{-1} + \Delta^3 y_{-2}}{12}\right) + \left(\frac{\Delta^5 y_{-3} + \Delta^5 y_{-2}}{60}\right) + \cdots\right]. \quad ...(5.93)$$

$$\left.\frac{d^2y}{dx^2}\right|_{x=x_0} = \frac{1}{h^2}\left[\Delta^2 y_{-1} - \frac{1}{12}\Delta^4 y_{-2} + \frac{1}{90}\Delta^6 y_{-3} + \cdots\right]. \quad ...(5.94)$$

$$\left.\frac{d^3y}{dx^3}\right|_{x=x_0} = \frac{1}{h^3}\left[\frac{\Delta^3 y_{-1} + \Delta^3 y_{-2}}{2} - \frac{1}{8}(\Delta^5 y_{-3} + \Delta^5 y_{-2}) + \cdots\right]. \quad ...(5.95)$$

$$\left.\frac{d^4y}{dx^4}\right|_{x=x_0} = \frac{1}{h^2}\left[\Delta^4 y_{-2} - \frac{1}{6}\Delta^6 y_{-3} + \cdots\right]. \quad ...(5.96)$$

$$\left.\frac{d^5y}{dx^5}\right|_{x=x_0} = \frac{1}{h^5}\left[\frac{\Delta^5 y_{-3} + \Delta^5 y_{-2}}{2} + \cdots\right]. \quad ...(5.97)$$

$$\left.\frac{d^6y}{dx^6}\right|_{x=x_0} = \frac{1}{h^6}[\Delta^6 y_{-3} + \cdots]. \quad ...(5.98)$$

EXAMPLE 5.47 Find the first and second derivatives of $f(x)$ at $x = 1.1$ from the following tabulated values

x:	1.0	1.2	1.4	1.6	1.8	2.0
$y = f(x)$:	0.0000	0.1280	0.5440	1.2960	2.4320	4.000

Solution Since we have to find the derivative at $x = 1.1$, we shall use the Newton's forward difference formula. The forward differences for the given data are given in Table 5.33.

Table 5.33

x	y	Δy	$\Delta^2 y$	$\Delta^3 y$	$\Delta^4 y$	$\Delta^5 y$
1.0	0.0					
		0.1280				
1.2	0.1280		**0.2880**			
		0.4160		**0.0480**		
1.4	0.5440		0.3360		0.000	
		0.7520		0.0480		0
1.6	1.2960		0.3840		0.000	
		1.1360		0.0480		
1.8	2.4320		0.4320			
		1.5680				
2.0	4.000					

We have $x_0 = 1.0$, $x = 1.1$ and $h = 0.2$.

Employing $u = \dfrac{x - x_0}{h}$, we get $u = \dfrac{1.1 - 1.0}{0.2} = 0.5$.

Substituting the value of u in (5.69), we obtain

$$\left.\frac{dy}{dx}\right|_{x=1.1} = \frac{1}{0.2}\left[0.1280 + \frac{2(0.5)-1}{2}(0.2880) + \frac{0.75-3+2}{6}(0.0480)\right]$$

$$= 5\left[0.1280 - \frac{0.25}{6}(0.048)\right] = 5[0.128 - 0.002] = 0.63.$$

Substituting the values of u, h and various differences in (5.70)

$$\left.\frac{d^2y}{dx^2}\right|_{x=1.1} = \frac{1}{(0.2)^2}[0.2880 + (0.5-1)\,0.048] = 25(0.2880 - 0.024) = 6.6.$$

EXAMPLE 5.48 Find the first, second and third derivatives of the function tabulated below, at the point $x = 1.5$.

x:	1.5	2.0	2.5	3.0	3.5	4.0
$y = f(x)$:	3.375	7.000	13.625	24.000	38.875	59.000

Solution The table of differences is as follows:

Table 5.34

x	y	Δy	$\Delta^2 y$	$\Delta^3 y$	$\Delta^4 y$
1.5	3.375				
		3.625			
2.0	7.000		**3.00**		
		6.625		**0.75**	
2.5	13.625		3.75		0
		10.375		0.75	
3.0	24.000		4.50		0
		14.875		0.75	
3.5	38.875		5.25		
		20.125			
4.0	59.000				

We have $h = 0.5$, $x = x_0 = 1.5$. Since $x = x_0 + uh$ therefore, $u = 0$.

Substituting the values of u and h in formulae (5.73), (5.74) and (5.75), we get

$$\left.\frac{dy}{dx}\right|_{x=1.5} = \frac{1}{0.5}\left[3.625 - \frac{1}{2}(3.0) + \frac{1}{3}(0.75)\right] = 4.75.$$

$$\left.\frac{d^2y}{dx^2}\right|_{x=1.5} = \frac{1}{(0.5)^2}(3.0 - 0.75) = 9.0, \quad \text{and} \quad \left.\frac{d^3y}{dx^3}\right|_{x=1.5} = \frac{1}{(0.5)^3}(0.75) = 6.0.$$

EXAMPLE 5.49 Find $f'(x)$, $f''(x)$ and $f'''(x)$ at $x = 1.30$ from the following table of values given below:

x:	1.00	1.05	1.10	1.15	1.20	1.25	1.30
$f(x)$:	1.00000	1.02470	1.04881	1.07238	1.09544	1.11803	1.14017

Solution Since it is desired to find the derivatives at the end point, we will use the backward difference formula. The backward difference table for the given data is given below.

Table 5.35

x	$f(x)$	$\Delta f(x)$	$\Delta^2 f(x)$	$\Delta^3 f(x)$	$\Delta^4 f(x)$	$\Delta^5 f(x)$
1.00	1.00000					
		0.02470				
1.05	1.02470		– 0.00059			
		0.02411		0.00005		
1.10	1.04881		– 0.00054		– 0.00001	
		0.02357		0.00004		– 0.00001
1.15	1.07238		– 0.00050		– 0.00002	
		0.02307		0.00002		**0.00003**
1.20	1.09545		– 0.00048		**0.00001**	
		0.02259		**0.00003**		
1.25	1.11804		**– 0.00045**			
		0.02214				
1.30	1.14018					

$x_n = 1.30$, $h = 0.05$, $x = 1.30$, we get $u = (x - x_n) / h = 0$.

Substituting the value of u in Eqs. (5.78, 5.79, 5.80), we get

$$f'(1.30) = \frac{1}{h}\left[\nabla y_n + \frac{1}{2}\nabla^2 y_n + \frac{1}{3}\nabla^3 y_n + \frac{1}{4}\nabla^4 y_n + \cdots\right]$$

$$= \frac{1}{0.05}\left[0.02214 + \frac{1}{2}(-0.00045) + \frac{1}{3}(0.00003) + \frac{1}{4}(0.00001)\right]$$

$$= 20(0.02214 - 0.000225 + 0.00001 + 0.0000025) = 20(0.0219275) = 0.43855.$$

$$f''(1.30) = \frac{1}{h^2}\left[\nabla^2 y_n + \nabla^3 y_n + \frac{11}{12}\nabla^4 y_n + \frac{5}{6}\nabla^5 y_n + \cdots\right]$$

$$= 400\left[-0.00045 + 0.00003 + \frac{11}{12}(0.00001) + \frac{5}{6}(0.00003)\right]$$

$$= 400[-0.000385] = -0.154.$$

$$f'''(1.30) = \frac{1}{h^3}\left[\nabla^3 y_n + \frac{3}{2}\nabla^4 y_n + \cdots\right] = 8000\left[0.00003 + \frac{3}{2}(0.00001)\right]$$

$$= 8000[0.000045] = 0.360.$$

EXAMPLE 5.50 Find the value of cos 1.747, using the value given in the table below:

x:	1.70	1.74	1.78	1.82	1.86
$f(x) = \sin x$:	0.9916	0.9857	0.9781	0.9691	0.9584

(Kuk. June 2009; Madras, 2000S)

Solution The table of differences is given below:

Table 5.36

x	$f(x)$	$\Delta f(x)$	$\Delta^2 f(x)$	$\Delta^3 f(x)$	$\Delta^4 f(x)$
1.70	0.9916				
		– 0.0059			
1.74	0.9857		**– 0.0017**		
		– 0.0076		**0.0003**	
1.78	0.9781		– 0.0014		**– 0.0006**
		– 0.0090		– 0.0003	
1.82	0.9691		– 0.0017		
		– 0.0107			
1.86	0.9584				

Taking $x_0 = 1.70$, $x = 1.747$ and the length of the interval $h = 0.04$.

We have $u = \dfrac{x - x_0}{h} = \dfrac{1.747 - 1.70}{0.04} = 1.75.$

Using Newton's forward difference interpolation formula

$$f(x_0 + uh) = f(x_0) + u\Delta f(x_0) + \frac{u(u-1)}{2}\Delta^2 f(x_0) + \frac{u(u-1)(u-2)}{6}\Delta^3 f(x) + \frac{u(u-1)(u-2)(u-3)}{24}\Delta^4 f(x_0) + \cdots \quad ...(1)$$

Using Eq. (1), the first derivative of $f(x_0 + uh)$ is given below.

$$f'(x_0 + hu) = \frac{1}{h}\left[\Delta f(x_0) + \frac{2u-1}{2}\Delta^2 f(x_0) + \frac{3u^2 - 6u + 2}{6}\Delta^3 f(x_0) + \frac{2u^3 - 9u^2 + 11u - 3}{12}\Delta^4 f(x_0) + \ldots\right] \quad ...(2)$$

Substituting the values in relation (2), we obtain

$$\frac{d}{dx}(\sin 1.747) = \cos 1.747 = \frac{1}{0.04}\left[-0.0059 + \left(\frac{1.35}{2}\right)(-0.0017) + \left(\frac{-0.908125}{6}\right)(0.0003) + \frac{0.743844}{12}(-0.0006)\right]$$

$$= 25[-0.0059 - 0.001148 - 0.000045 - 0.000037] = -0.17825.$$

EXAMPLE 5.51 The elevations above a datum line of seven points of roads 300 units apart, are 135, 149, 157, 183, 201, 205, 193 unit. Find the gradient of the road at the middle point.

Solution The difference table is given below:

Table 5.37

u	x	y	Δy	$\Delta^2 y$	$\Delta^3 y$	$\Delta^4 y$	$\Delta^5 y$
–3	0	135					
			14				
–2	300	149		–6			
			8		24		
–1	600	157		18		–50	
			26		**–26**		**70**
0	900	183		**–8**		**20**	
			18		**–6**		**–16**
1	1200	201		–14		4	
			4		–2		
2	1500	205		–16			
			–12				
3	1800	193					

It is required to determine the value of $\left.\dfrac{dy}{dx}\right|_{x=900}$. The length of the interval $h = 300$. Since $x = x_0 = 900$ therefore, $u = 0$.

Differentiating central difference Stirling formula, we get

$$\left.\frac{dy}{dx}\right|_{x=900} = \frac{1}{300}\left[\frac{1}{2}(26 + 18) - \frac{1}{12}(-26 - 6) + \frac{1}{60}(70 - 16)\right]$$

$$= \frac{1}{300}(22 + 2.66667 + 0.9) = \frac{25.56667}{300} = 0.085222.$$

EXAMPLE 5.52 Find the first and the second derivatives of the function tabulated below, at the point $x = 0.6$.

x:	0.4	0.5	0.6	0.7	0.8
$y = f(x)$:	1.5836494	1.7974426	2.0442376	2.3275054	2.6510818

Solution Taking the origin at 0.6 which is the central value of the data, the central difference table for the given data is given as follows:

Table 5.38

u	x	y	Δy	$\Delta^2 y$	$\Delta^3 y$	$\Delta^4 y$
–2	0.4	1.5836494				
			0.2137932			
–1	0.5	1.7974426		0.0330018		
			0.2467950		**0.0034710**	
0	0.6	**2.0442376**		**0.0364728**		**0.0003648**
			0.2832678		**0.0038358**	
1	0.7	2.3275054		0.0403086		
			0.3235764			
2	0.8	2.6510818				

We have $h = 0.1$, $x = x_0 = 0.6$. Therefore, $u = \dfrac{x - x_0}{h} = 0$. $\dfrac{du}{dx} = \dfrac{1}{h}$.

The central differences Stirling's formula is

$$y_u = y_0 + u\left(\frac{\Delta y_0 + \Delta y_{-1}}{2}\right) + \frac{u^2}{2!}\Delta^2 y_{-1} + \frac{u(u^2-1)}{3!}\left(\frac{\Delta^3 y_{-1} + \Delta^3 y_{-2}}{2}\right) + \cdots \quad \text{...(1)}$$

Differentiating (1) with respect to x, we get

$$\frac{dy}{dx} = \frac{dy}{du}\cdot\frac{du}{dx} = \frac{1}{h}\left[\frac{\Delta y_0 + \Delta y_{-1}}{2} + u\Delta^2 y_{-1} + \frac{3u^2-1}{6}\left(\frac{\Delta^3 y_{-1} + \Delta^3 y_{-2}}{2}\right) + \frac{4u^3 - 2u}{24}\Delta^4 y_{-2} + \frac{5u^4 - 15u^2 + 4}{120}\left(\frac{\Delta^5 y_{-3} + \Delta^5 y_{-2}}{2}\right) + \cdots\right].$$

Differentiating again, we obtain

$$\frac{d^2y}{dx^2} = \frac{1}{h^2}\frac{d^2y}{du^2} = \frac{1}{h^2}\left[\Delta^2 y_{-1} + u\left(\frac{\Delta^3 y_{-1} + \Delta^3 y_{-2}}{2}\right) + \frac{12u^2 - 2}{24}\Delta^4 y_{-2} + \frac{20u^3 - 30u}{120}\left(\frac{\Delta^5 y_{-3} + \Delta^5 y_{-2}}{2}\right) + \frac{30u^4 - 60u^2 + 8}{720}\Delta^6 y_{-3} + \cdots\right].$$

Now
$$\left.\frac{dy}{dx}\right|_{x=0.6} = \frac{1}{0.1}\left[\frac{1}{2}(0.2832678 + 0.2467950) - \frac{1}{12}(0.0038358 + 0.0034710)\right]$$
$$= 10\,(0.2650314 - 0.0006089) = 2.644225,$$

and
$$\left.\frac{d^2y}{dx^2}\right|_{x=0.6} = \frac{1}{0.01}\left[0.0364728 - \frac{1}{12}(0.0003648)\right] = 100\,(0.0364728 - 0.0000304)$$
$$= 100 \times 0.0364424 = 3.64424.$$

Note: The function from which the above table is obtained is $y = 2e^x - x - 1$.

Hence, $dy/dx = 2e^x - 1$, $d^2y/dx^2 = 2e^x$.

At $x = 0.6$, we get $dy/dx = 2.644238$ and $d^2y/dx^2 = 3.644238$.

The values found by numerical differentiation are therefore correct to 4 significant figures in both the cases.

EXAMPLE 5.53 A slider in a machine moves along a fixed straight rod. Its distance x units along the rod is given below for various values of the time t seconds.

t:	0.0	0.1	0.2	0.3	0.4	0.5	0.6
x:	30.13	31.62	32.87	33.64	33.95	33.81	33.24

Find the velocity of the slider and also its acceleration, at time $t = 0.3$ seconds. (*SVTU 2007; Madras 2006*)

Solution The above values and the central differences are tabulated below:

Table 5.39

u	t (sec)	x (unit)	*First diff.*	*Second diff.*	*Third diff.*	*Fourth diff.*	*Fifth diff.*	*Sixth diff.*
–3	0.0	30.13						
			1.49					
–2	0.1	31.62		– 0.24				
			1.25		– 0.24			
–1	0.2	32.87		– 0.48		0.26		
			0.77		**0.02**		**– 0.27**	
0	0.3	33.64		**– 0.46**		**– 0.01**		**0.29**
			0.31		**0.01**		**0.02**	
1	0.4	33.95		– 0.45		0.01		
			– 0.14		0.02			
2	0.5	33.81		– 0.43				
			– 0.57					
3	0.6	33.24						

We have $h = 0.1$, $x = x_0 = 33.64$. Therefore, $u = \dfrac{x - x_0}{h} = 0$. Differentiating central differences Stirling's formula and substituting the proper values, we get

$$\text{Velovity} = \left.\frac{dx}{dt}\right|_{at\ x = 0.3} = \frac{1}{0.1}\left[\frac{1}{2}(0.31 + 0.77) - \frac{1}{12}(0.01 + 0.02) + \frac{1}{60}(0.02 - 0.27)\right]$$

$= 10\,(0.54 - 0.0025 - 0.0042) = 5.33$. Thus the velocity = 5.33 units/sec.

$$\text{Acceleration} = \left.\frac{d^2x}{dt^2}\right|_{at\ x = 0.3} = 100\left[-0.46 - \frac{1}{12}(-0.01) + \frac{1}{90}(0.29)\right]$$

$$= 100\,(-0.46 + 0.0008 + 0.0032) = -45.6 \text{ unit/sec}^2,$$

–ve sign indicates that there is retardation.

EXAMPLE 5.54 Using Bessel's formula, find $f'(7.5)$ from the following table:

x:	7.47	7.48	7.49	7.50	7.51	7.52	7.53
$f(x)$:	0.193	0.195	0.198	0.201	0.203	0.206	0.208

Solution Bessel's formula for central differences is given by

$$y_u = y_0 + u\Delta y_0 + \frac{u(u-1)}{2!}\cdot\frac{\Delta^2 y_{-1} + \Delta^2 y_0}{2} + \frac{u\left(u - \frac{1}{2}\right)(u-1)}{3!}\Delta^3 y_{-1} + \frac{u(u^2-1)(u-2)}{4!}\cdot\frac{\Delta^4 y_{-2} + \Delta^4 y_{-1}}{2} + \cdots$$

Differentiating this with respect to u, we obtain

$$\frac{dy}{dx}=\frac{dy}{du}\cdot\frac{du}{dx}=\frac{1}{h}\left[\Delta y_0+\frac{2u-1}{2}\cdot\frac{\Delta^2 y_{-1}+\Delta^2 y_0}{2}+\frac{3u^2-3u+\frac{1}{2}}{6}\Delta^3 y_{-1}+\frac{4u^3-6u^2-2u+2}{24}\cdot\frac{\Delta^4 y_{-2}+\Delta^4 y_{-1}}{2}+\ldots\right]$$

We have $h = 0.01$, $x = x_0 = 7.5$. Employing $u=\dfrac{x-x_0}{h}$, we get $u = 0$. Substituting $u = 0$, we obtain

$$\left.\frac{dy}{dx}\right|_{x=x_0}=\frac{1}{h}\left[\Delta y_0-\frac{1}{2}\frac{\Delta^2 y_{-1}+\Delta^2 y_0}{2}+\frac{1}{12}\Delta^3 y_{-1}+\frac{1}{24}(\Delta^4 y_{-2}+\Delta^4 y_{-1})+\cdots\right] \quad \ldots(1)$$

The central difference table is given below:

Table 5.40

u	x	$f(x)$	Δf	$\Delta^2 f$	$\Delta^3 f$	$\Delta^4 f$	$\Delta^5 f$
–3	7.47	0.193					
			0.002				
–2	7.48	0.195		– 0.001			
			0.003		– 0.001		
–1	7.49	0.198		0.000		0.000	
			0.003		– 0.001		0.003
0	7.50	0.201		– 0.001		0.003	
			0.002		0.002		– 0.007
1	7.51	0.203		0.001		– 0.004	
			0.003		– 0.002		
2	7.52	0.206		– 0.001			
			0.002				
3	7.53	0.208					

Central line (through the row $u = 0$)

Take $\Delta y_0 = 0.002$; $\Delta^2 y_0 = 0.001$; $\Delta^2 y_{-1} = -0.001$; $\Delta^3 y_{-1} = 0.002$; $\Delta^4 y_{-1} = -0.004$; $\Delta^4 y_{-2} = 0.003$.
Substituting these values in (1), we get

$$f'(7.5)=\left.\frac{dy}{dx}\right|_{x=7.5}=100\left[0.002-\left(\frac{-0.001+0.001}{4}\right)+\frac{1}{12}(0.002)+\frac{1}{24}\{0.003+(-0.004)\}\right]$$

$$=100[0.002+0.000167-0.000042]=100(0.002125)=0.2125.$$

Problems based on Non-uniform Nodal Points

EXAMPLE 5.55 Find $f'(x)$ and $f''(x)$ given f_0, f_1, f_2 at x_0, x_1, x_2 respectively, using the Lagrange interpolation.

Solution By Lagrange's interpolation formula

$$f(x) = P_2(x) = L_0(x)\,f_0 + L_1(x)\,f_1 + L_2(x)\,f_2$$

where,

$$L_0(x)=\frac{(x-x_1)(x-x_2)}{(x_0-x_1)(x_0-x_2)},\qquad L_0'(x)=\frac{2x-x_1-x_2}{(x_0-x_1)(x_0-x_2)},$$

$$L_1(x)=\frac{(x-x_0)(x-x_2)}{(x_1-x_0)(x_1-x_2)},\qquad L_1'(x)=\frac{2x-x_0-x_2}{(x_1-x_0)(x_1-x_2)},$$

$$L_2(x)=\frac{(x-x_0)(x-x_1)}{(x_2-x_0)(x_2-x_1)},\qquad L_2'(x)=\frac{2x-x_0-x_1}{(x_2-x_0)(x_2-x_1)}.$$

Hence, $f'(x) = P_2'(x) = L_0'(x)\, f_0 + L_1'(x) f_1 + L_2'(x) f_2$

$$= \frac{2x - x_1 - x_2}{(x_0 - x_1)(x_0 - x_2)} f_0 + \frac{2x - x_0 - x_2}{(x_1 - x_0)(x_1 - x_2)} f_1 + \frac{2x - x_0 - x_1}{(x_2 - x_0)(x_2 - x_1)} f_2,$$

and $f''(x) = P_2''(x) = L_0''(x)\, f_0 + L_1''(x) f_1 + L_2''(x) f_2$

$$= \frac{2f_0}{(x_0 - x_1)(x_0 - x_2)} + \frac{2f_1}{(x_1 - x_0)(x_1 - x_2)} + \frac{2f_2}{(x_2 - x_0)(x_2 - x_1)}.$$

EXAMPLE 5.56 Given the following values of $f(x) = \log x$, find the approximate value of $f'(2.0)$ and $f''(2.0)$.

x:	2.0	2.2	2.6
$f(x)$:	0.69315	0.78846	0.95551

Solution Using the Lagrange's interpolation formula, we have

$$f'(x_0) = P_2'(x_0) = \frac{2x_0 - x_1 - x_2}{(x_0 - x_1)(x_0 - x_2)} f_0 + \frac{x_0 - x_2}{(x_1 - x_0)(x_1 - x_2)} f_1 + \frac{x_0 - x_1}{(x_2 - x_0)(x_2 - x_1)} f_2.$$

Therefore, $$f'(2.0) = \frac{4.0 - 2.2 - 2.6}{(2.0 - 2.2)(2.0 - 2.6)}(0.69315) + \frac{2.0 - 2.6}{(2.2 - 2.0)(2.2 - 2.6)}(0.78846) + \frac{2.0 - 2.2}{(2.6 - 2.0)(2.6 - 2.2)}(0.95551)$$

$$= -\frac{0.8(0.69315)}{0.12} + \frac{0.6(0.78846)}{0.08} - \frac{0.2(0.95551)}{0.24}$$

$$= -4.621 + 5.91345 - 0.79626 = 0.49619.$$

The exact value of $f'(2.0) = 0.50$.

$$f''(x_0) = 2\left[\frac{f_0}{(x_0 - x_1)(x_0 - x_2)} + \frac{f_1}{(x_1 - x_0)(x_1 - x_2)} + \frac{f_2}{(x_2 - x_0)(x_2 - x_1)}\right].$$

Therefore, $$f''(2.0) = 2\left[\frac{0.69315}{(2 - 2.2)(2 - 2.6)} + \frac{0.78846}{(2.2 - 2.0)(2.2 - 2.6)} + \frac{0.95551}{(2.6 - 2.0)(2.6 - 2.2)}\right]$$

$$= 2[5.77625 - 9.85575 + 3.98130] = -0.19642.$$

The exact value of $f''(2.0) = -0.25$.

EXAMPLE 5.57 The following data gives the values of x and $f(x)$. Find the first derivative of $f(x)$ at $x = 5$.

x:	0	2	3	4	7	9
$f(x)$:	4	26	58	112	466	922

Solution Since the values of x are not equi-spaced, we solved this problem by using Newton's divided difference formula. The divided difference table is as follows:

Table 5.41

x	$f(x)$	$\Delta f(x)$	$\Delta^2 f(x)$	$\Delta^3 f(x)$
0	4			
		$\frac{26-4}{2-0}=11$		
2	26		$\frac{32-11}{3-0}=7$	
		$\frac{58-26}{3-2}=32$		$\frac{11-7}{4-0}=1$
3	58		$\frac{54-32}{4-2}=11$	
		$\frac{112-58}{4-3}=54$		$\frac{16-11}{7-2}=1$
4	112		$\frac{118-54}{7-3}=16$	
		$\frac{466-112}{7-4}=118$		$\frac{22-16}{9-3}=1$
7	466		$\frac{228-118}{9-4}=22$	
		$\frac{922-466}{9-7}=228$		
9	922			

From the above table it is obvious that third differences are constant.

Newton's divided difference interpolation formula is given by

$$f(x) = f(x_0) + (x - x_0) f[x_0, x_1] + (x - x_0)(x - x_1) f[x_0, x_1, x_2] + (x - x_0)(x - x_1)(x - x_2) f[x_0, x_1, x_2, x_3]. \quad ...(1)$$

Here $x_0 = 0$, $x_1 = 2$, $x_2 = 3$, $f(x_0) = 4$, $f[x_0, x_1] = 11$, $f[x_0, x_1, x_2] = 7$ and $f[x_0, x_1, x_2, x_3] = 1$.

Substituting the appropriate values in (1), we obtain

$$\begin{aligned} f(x) &= 4 + x(11) + x(x-2)(7) + x(x-2)(x-3)(1) \\ &= 4 + 11x + 7x^2 - 14x + x^3 - 5x^2 + 6x \\ &= x^3 + 2x^2 + 3x + 4. \end{aligned} \quad ...(2)$$

This is a third degree polynomial interpolating the above data. Differentiating (2) with respect to x, we obtain

$f'(x) = 3x^2 + 4x + 3$. Therefore, $f'(x)|_{x=5} = 3(5)^2 + 4(5) + 3 = 98$.

EXAMPLE 5.58 The following table gives the values of x and $f(x)$.

x:	3	5	11	27	34
$f(x)$:	–13	23	899	17315	35606

Find $f'(10)$.

Solution Since the values of x are not equally-spaced, we use Newton's divided difference formula. The divided difference table is given below :

Table 5.42

x	$f(x)$	$\Delta f(x)$	$\Delta^2 f(x)$	$\Delta^3 f(x)$
3	−13			
		$\frac{23+13}{5-3}=18$		
5	23		$\frac{146-18}{11-3}=16$	
		$\frac{899-23}{11-5}=146$		$\frac{40-16}{27-3}=1$
11	899		$\frac{1026-146}{27-5}=40$	
		$\frac{17315-899}{27-11}=1026$		$\frac{69-40}{34-5}=1$
27	17315		$\frac{2613-1026}{34-11}=69$	
		$\frac{35606-17315}{34-27}=2613$		
34	35606			

The third differences are constant. Newton's divided difference interpolation formula is given by

$$f(x) = f(x_0) + (x - x_0) f[x_0, x_1] + (x - x_0)(x - x_1) f[x_0, x_1, x_2] + (x - x_0)(x - x_1)(x - x_2) f[x_0, x_1, x_2, x_3]. \quad ...(1)$$

Substituting $x_0 = 3$, $x_1 = 5$, $x_2 = 11$, $f(x_0) = -13$, $f[x_0, x_1] = 18$, $f[x_0, x_1, x_2] = 16$ and $f[x_0, x_1, x_2, x_3] = 1$ in relation (1), we get

$$f(x) = -13 + (x-3)(18) + (x-3)(x-5)16 + (x-3)(x-5)(x-11)1$$

$$= -13 + 18x - 54 + 16x^2 - 128x + 240 + x^3 - 19x^2 + 103x - 165$$

$$= x^3 - 3x^2 - 7x + 8. \quad ...(2)$$

This is the third degree polynomial interpolating the above data.

Differentiating (2) with respect to x, we get

$f'(x) = 3x^2 - 6x - 7$. Therefore, $f'(10) = 3(10)^2 - 6(10) - 7 = 233$.

Maximum and Minimum Value of a Tabulated Function

To find the maximum or minimum value of a tabulated function, we compute the various differences from the given table, insert them in the appropriate interpolation formula, find the first derivative of this formula and put it equal to zero and solve for u. The value of x then is found from the relation $x = x_0 + uh$.

EXAMPLE 5.59 From the following table, find the value of x for which y is minimum and find this value of y.

x:	10	15	20	25
y:	62.21	61.55	61.38	61.70

Solution The forward difference table is as under:

Table 5.43

x	y	Δy	$\Delta^2 y$	$\Delta^3 y$
10	**62.21**			
		– 0.66		
15	61.55		**0.49**	
		– 0.17		0
20	61.38		0.49	
		0.32		
25	61.70			

Taking $x_0 = 10$, $h = 5$, $y_0 = 62.21$, $\Delta y_0 = -0.66$, $\Delta^2 y_0 = 0.49$ and $\Delta^3 y_0 = 0$, and substituting these values in Newton's forward difference formula, we obtain

$$y = 62.21 + u(-0.66) + \frac{u(u-1)}{2}(0.49). \qquad ...(i)$$

Differentiating (*i*) with respect of u, we obtain

$$\frac{dy}{du} = (-0.66) + \frac{(2u-1)}{2}(0.49).$$

For y to have the minimum value, we put $\frac{dy}{du} = 0$. This gives $u = \frac{1.32 + 0.49}{0.98} = 1.847$.

Now $x = x_0 + uh = 10 + 1.847(5) = 19.235$.

Hence y will have minimum value for value of $x = 19.235$.

Putting the value of $u = 1.847$ in Newton's formula, the minimum value of y is given by

$$y_{\min} = 62.21 + (1.847)(-0.66) + \frac{1.847(1.847-1)}{2}(0.49) = 62.21 - 1.219 + 0.383 = 61.374$$

EXAMPLE 5.60 A function $f(x)$ representing the following data values has a minimum in the interval (0.5, 0.8). Find the point of minimum and the minimum value.

x:	0.5	0.6	0.7	0.8
$f(x)$:	2.3256	1.4632	0.9842	1.3282

Solution We construct the forward difference interpolating polynomial and differentiate it to find the point of minimum.

Table 5.44

x	$f(x)$	Δf	$\Delta^2 f$	$\Delta^3 f$
0.5	**2.3256**			
		– 0.8624		
0.6	1.4632		**0.3834**	
		– 0.4790		**0.4396**
0.7	0.9842		0.8230	
		0.3440		
0.8	1.3282			

We have the Newton's interpolating polynomial as

$$f(x) = 2.3256 + (x-0.5)\left(\frac{-0.8624}{0.1}\right) + (x-0.5)(x-0.6)\left(\frac{0.3834}{0.02}\right) + (x-0.5)(x-0.6)(x-0.7)\left(\frac{0.4396}{0.006}\right).$$

Differentiating and setting it to zero, we obtain

$$f'(x) = 0 = -8.624 + (2x - 1.1)\,19.17 + (3x^2 - 3.6x + 1.07)\left(\frac{1099}{15}\right) = 3297x^2 - 3381.3x + 730.265.$$

The roots of this equation are $x = 0.7164, 0.3092$ of which the first root lies in the interval (0.5, 0.8). The point of minimum is $x \approx 0.7164$ and the minimum value is

$$f(0.7164) = 2.3256 - (0.2164)(8.624) + (0.2164)(0.1164)(19.17) + (0.2164)(0.1164)(0.0164)\,(73.2667)$$

$$= 2.3256 - 1.86623 + 0.48287 + 0.03027 = 0.9725$$

[Note that $f''(x) = 6594x - 3381.3$. Therefore, $f''(0.7164) = 4723.94 - 3381.3 = 1342.64 > 0$].

EXERCISE 5.7

1. From the data in the following table, compute $f'(x)$ at $x = 0$.

x:	0.00	0.05	0.10	0.15	0.20	0.25
y:	0.00000	0.10017	0.20134	0.30452	0.41075	0.52110

Ans. 2.0034.

2. Find the first and second derivatives of $\sqrt{x}$ at $x = 2.5$ from the table:

x:	2.50	2.55	2.60	2.65	2.70	2.75
$\sqrt{x}$:	1.58114	1.59687	1.61245	1.62788	1.64317	1.65831

Ans. 0.3160; – 0.0600. The correct values are 0.3162, – 0.0632.

3. From the following table, obtain the value of d^2y/dx^2 at the point $x = 0.96$:

x:	0.96	0.98	1.00	1.02	1.04
y:	0.7825	0.7739	0.7651	0.7563	0.7473

Ans. –1.91675.

4. Find the first and second derivatives of the function $y = f(x)$ tabulated below at the point $x = 1.2$

x:	1.0	1.2	1.4	1.6	1.8	2.00	2.2
$f(x)$:	2.7183	3.3201	4.0552	4.9530	6.0496	7.3891	9.0250

Ans. 3.3205; 3.32.

5. Find the first derivative of $f(x)$ at $x = 1.1$ from the table given below:

x:	1.1	1.2	1.3	1.4	1.5	1.6
$f(x)$:	8.403	8.781	9.129	9.451	9.750	10.031

(AMIE, W-2010) **Ans.** 3.9515.

6. Find $f'(x)$ at $x = 0.4$ from the following table of values given below:

x:	0.1	0.2	0.3	0.4
$f(x)$:	1.10517	1.22140	1.34986	1.49182

Ans. 1.49133.

7. From the following table of values of x and y, find dy/dx and d^2y/dx^2 for $x = 1.05$

x:	1.00	1.05	1.10	1.15	1.20	1.25	1.30
y:	1.00000	1.02470	1.04881	1.07238	1.09544	1.11803	1.14017

(AMIE, W-2000) **Ans.** 0.48763; – 0.2144.

8 From the following table, find $f'(1.4)$:

x:	1.2	1.3	1.4	1.5	1.6
$f(x)$:	1.5095	1.6984	1.9043	2.1293	2.3756

Ans. 1.81075.

9. The following values of x and y are given. Find the value of dy/dx and also of d^2y/dx^2 when $x = 6.5$.

x:	5.0	5.5	6.0	6.5	7.0	7.5	8.0
$f(x)$:	3.2188	3.4096	3.5836	3.7436	3.8918	4.0298	4.1588

Ans. 0.3076; – 0.04698.

10. Find the first and the second derivatives of $f(x)$ at $x = 0.04$ from the table given below:

x:	0.01	0.02	0.03	0.04	0.05	0.06
$f(x)$:	0.1023	0.1047	0.1071	0.1096	0.1122	0.1148

Ans. 0.256; 1.1.

11. A rod is rotating in a plane. The following table gives the angle θ (radians) through which the rod has turned for various values of the time t seconds.

t seconds:	0.0	0.2	0.4	0.6	0.8	1.0	1.2
θ radians:	0.00	0.12	0.49	1.12	2.02	3.20	4.67

Calculate the angular velocity and angular acceleration of the rod, when $t = 0.7$ second.

Ans. 4.402 radian/sec; 6.875 rad / sec^2.

12. From the following table, calculate $f'(x)$ and $f''(x)$ at $x = 1.35$.

x:	1.1	1.2	1.3	1.4	1.5	1.6
$f(x)$:	–1.62628	0.15584	2.45256	5.39168	9.12500	13.83072

Ans. 29.32795; 71.33.

13. The population of a certain town (as obtained from census data) is shown in the following table

Year:	1976	1986	1996	2006	2016
Population: (in thousand)	19.96	39.65	58.81	77.21	94.61

Estimate the rate of growth of population in 2006. **Ans.** 1.80 thous./year.

14. For the set of points (0, 2), (2, –2), (3, –1) evaluate $\left.\frac{dy}{dx}\right|_{x=2}$. **Ans.** 0.

15. A function $f(x)$ representing the following data has a minimum in the interval (0.5, 0.8). Find the point of minimum and the minimum value.

x:	0.5	0.6	0.7	0.8
$f(x)$:	1.3254	1.1532	0.9432	1.0514

Ans. 0.7297; 0.9260.

NUMERICAL INTEGRATION

5.11 INTRODUCTION

It is sometimes difficult or even impossible to evaluate a definite integral $\int_a^b f(x)dx$ by method of direct integration. In other cases the integral cannot be expressed in terms of elementary functions or the integrand is defined by a set of numerical values given in the tabular form. In such case we may obtain approximate numerical values of the definite integral by a numerical integration method.

Thus numerical integration is the process of computing the value of a definite integral from a set of numerical values of the integrand.

The problem of numerical integration is solved by representing the integrand by an interpolation formula and then integrating it, within the given limits to get an approximate value of the definite integral. We shall now derive the most general quadrature formula using Newton's forward interpolation formula as the integrand.

5.12 NEWTON-COTES QUADRATURE OR INTEGRATION FORMULA

Consider an integral

$$I = \int_a^b f(x)\,dx. \qquad ...(5.99)$$

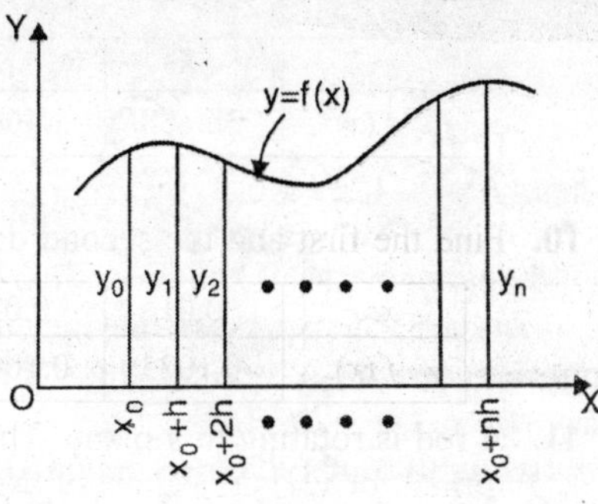

Fig. 5.1

Let $f(x)$ take the values $y_0, y_1, y_2, \ldots, y_n$ when $x = x_0, x_0 + h, x_0 + 2h, \ldots, x_0 + nh$, respectively (Refer Fig. 5.1). To evaluate the integral I, we replace $f(x)$ by an interpolating formula. It will be obvious that we will obtain different integration formulae depending upon the type of interpolation formula used.

Let the interval of integration (a, b) be divided into n sub-intervals of equal length h with $(n + 1)$ ordinates corresponding to $(n + 1)$ arguments. Let $a = x_0 < x_0 + h < x_0 + 2h \ldots < x_0 + nh = b$. Approximating $f(x)$ by Newton's forward interpolation formula, we can write the integral (5.99) as

$$I = \int_{x_0}^{x_0+nh} f(x)\,dx = \int_{x_0}^{x_0+nh}\left[y_0 + u\Delta y_0 + \frac{u(u-1)}{2!}\Delta^2 y_0 + \frac{u(u-1)(u-2)}{3!}\Delta^3 y_0 + \ldots\right]dx.$$

Since $x = x_0 + uh$, $dx = h\,du$, and the above integral is transformed to

$$I = h\int_0^n\left[y_0 + u\Delta y_0 + \frac{u(u-1)}{2!}\Delta^2 y_0 + \frac{u(u-1)(u-2)}{3!}\Delta^3 y_0 + \ldots\right]du.$$

Performing the desired integration and substituting the limits, we obtain

$$I = \int_{x_0}^{x_0+nh} f(x)\,dx = nh\left[y_0 + \frac{n}{2}\Delta y_0 + \frac{n(2n-3)}{12}\Delta^2 y_0 + \frac{n(n-2)^2}{24}\Delta^3 y_0 + \ldots\right] \qquad ...(5.100)$$

Relation (5.100) is known as **Newton-Cote's quadrature formula.**

Being a very general formula, we can obtain a variety of important quadrature formulae by setting $n = 1, 2, 3 \ldots$, etc.

Setting $n = 1, 2, 3$ and 6 in relation (5.100), we obtain four quadrature formulae which are named respectively as Trapezoidal rule, Simpson's 1/3 rule, Simpson's 3/8th rule and Weddle's rule.

5.12.1 Trapezoidal Rule (Newton-cote's Formula for $n = 1$)

Putting $n = 1$ in relation (5.100) above, and neglecting all differences above the first, we get

$$I = \int_{x_0}^{x_0+h} f(x)\,dx = h\left(y_0 + \frac{1}{2}\Delta y_0\right) = \frac{h}{2}(y_0 + y_1). \qquad ...(5.101)$$

For the next interval $(x_0 + h, x_0 + 2h)$, we obtain

$$\int_{x_0+h}^{x_0+2h} f(x)\,dx = h\left(y_1 + \frac{1}{2}\Delta y_1\right) = h\left(y_1 + \frac{y_2 - y_1}{2}\right) = \frac{h}{2}(y_1 + y_2). \qquad ...(5.102)$$

Proceeding in a similar manner, the value of the integral for the last interval $[x_0 + (n-1)h, x_0 + nh]$ will be given by

$$\int_{x_0+(n-1)h}^{x_0+nh} f(x)\,dx = \frac{h}{2}(y_{n-1} + y_n). \qquad ...(5.103)$$

Combining all the expressions like (5.101), (5.102) and (5.103), we obtain

$$\int_a^b f(x)\,dx = \int_{x_0}^{x_0+nh} f(x)dx = \frac{h}{2}[y_0 + 2(y_1 + y_2 + \ldots + y_{n-1}) + y_n] \quad \ldots(5.104)$$

where $h = \frac{b-a}{n}$.

Relation (5.104) is known as *Composite Trapezoidal rule* and is of order 1. It expresses the fact that the curve $y = f(x)$ is approximated by n straight lines through the points (x_0, y_0), $(x_0 + h, y_1)$, $(x_0 + 2h, y_2)$, ..., $[x_0 + (n-1)h, y_{n-1}]$, $(x_0 + nh, y_n)$. It also signifies that the area bounded by the part of the curve $y = f(x)$, the two ordinates $x = x_0$ and $x = x_0 + nh$ and the x-axis is approximately equal to the sum of the areas of the n trapeziums (see Fig. 5.2)

It is clear that the greater the number of intervals (that is, the smaller h is), the closer will be sum of the areas of the trapezoids approach the area under the curve.

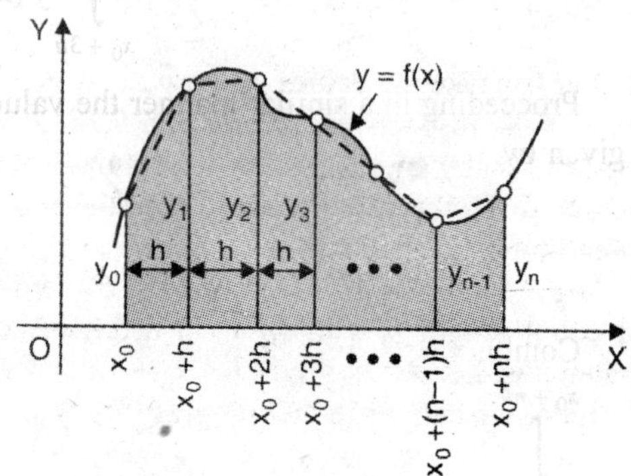

Fig. 5.2: Trapezoidal rule

5.12.2 Simpson's 1/3 Rule (Newton-Cote's Formula for $n = 2$)

Putting $n = 2$ in relation (5.100) above and neglecting all differences above the second, we obtain

$$\int_{x_0}^{x_0+2h} f(x)\,dx = 2h\left[y_0 + \Delta y_0 + \frac{1}{6}\Delta^2 y_0\right] = 2h\left(y_0 + y_1 - y_0 + \frac{y_2 - 2y_1 + y_0}{6}\right) = \frac{h}{3}(y_0 + 4y_1 + y_2) \ldots(5.105)$$

For the next two intervals, we get

$$\int_{x_0+2h}^{x_0+4h} f(x)\,dx = \frac{h}{3}(y_2 + 4y_3 + y_4) \quad \ldots(5.106)$$

and finally for the last interval of three ordinates, namely y_{n-2}, y_{n-1}, y_n, we obtain

$$\int_{x_0+(n-2)h}^{x_0+nh} f(x)\,dx = \frac{h}{3}(y_{n-2} + 4y_{n-1} + y_n). \quad \ldots(5.107)$$

Adding all such expressions from x_0 to x_n, where n is even, we obtain

$$\int_{x_0}^{x_0+nh} f(x)\,dx \approx \frac{h}{3}[y_0 + 4(y_1 + y_3 + \ldots + y_{n-1}) + 2(y_2 + y_4 + \ldots + y_{n-2}) + y_n] \quad \ldots(5.108)$$

This is known as *composite Simpson's 1/3 Rule* or *simply Simpson rule of integration.*

As in the case of the trapezoidal rule, the greater the number of parts into which range of integration is divided, the closer will the result be the area under the curve.

5.12.3 Simpson's 3/8th Rule: (Newton cote's Formula for $n = 3$)

Putting $n = 3$ in relation (5.100) above and neglecting all differences above the three, we have

$$\int_{x_0}^{x_0+3h} f(x)\,dx = 3h\left(y_0 + \frac{3}{2}\Delta y_0 + \frac{3}{4}\Delta^2 y_0 + \frac{1}{8}\Delta^3 y_0\right)$$

$$= \frac{3h}{8}[8y_0 + 12(y_1 - y_0) + 6(y_2 - 2y_1 + y_0) + (y_3 - 3y_2 + 3y_1 - y_0)]$$

$$= \frac{3h}{8}(y_0 + 3y_1 + 3y_2 + y_3). \quad ...(5.109)$$

For the next interval $(x_0 + 3h, x_0 + 6h)$, we obtain

$$\int_{x_0+3h}^{x_0+6h} f(x)\,dx = \frac{3h}{8}(y_3 + 3y_4 + 3y_5 + y_6). \quad ...(5.110)$$

Proceeding in a similar manner the value of the integral for the last interval $(x_0 + \overline{n-3}\,h, x_0 + nh)$ will be given by

$$\int_{x_0+(n-3)h}^{x_0+nh} f(x)\,dx = \frac{3h}{8}(y_{n-3} + 3y_{n-2} + 3y_{n-1} + y_n) \quad ...(5.111)$$

Combining all the expressions (5.109), (5.110) and (5.111), we obtain

$$\int_{x_0}^{x_0+nh} f(x)\,dx = \frac{3h}{8}[(y_0 + y_n) + 3(y_1 + y_2 + y_4 + y_5 + ... + y_{n-2} + y_{n-1}) + 2(y_3 + y_6 + ... + y_{n-3}) \quad ...(5.112)$$

which is known as Simpson's three-eigth rule. While using this formula, the given interval of integration must be divided into sub-intervals whose number n is a multiple of 3.

5.12.4 Weddle's Rule (Newton Cote's Formula for $n = 6$)

Putting $n = 6$ in relation (5.100) above and neglecting all differences above the sixth, we have

$$\int_{x_0}^{x_0+6h} f(x)\,dx = h\left[6y_0 + 18\Delta y_0 + 27\Delta^2 y_0 + 24\Delta^3 y_0 + \frac{123}{10}\Delta^4 y_0 + \frac{33}{10}\Delta^5 y_0 + \frac{41}{140}\Delta^6 y_0\right]$$

Here the coefficient of $\Delta^6 y_0$ that is, 41/140 differs from 3/10 by a small fraction 1/140. Hence, if we replace 41/140 by 3/10, we are committing an error of $\frac{h}{140}\Delta^6 y_0$. If the value of h is such that the sixth differences are small, the error committed will be negligible and can be ignored. We therefore, replace 41/140 by (3/10) as the coefficient of $\Delta^6 y_0$.

The result reduces to

$$\int_{x_0}^{x_0+6h} f(x)\,dx = \frac{3h}{10}[y_0 + 5y_1 + y_2 + 6y_3 + y_4 + 5y_5 + y_6].$$

For the next set of six intervals from $x_0 + 6h$ to $x_0 + 12h$, we get in the same way

$$\int_{x_0+6h}^{x_0+12h} f(x)\,dx = \frac{3h}{10}[y_6 + 5y_7 + y_8 + 6y_9 + y_{10} + 5y_{11} + y_{12}].$$

Adding all such expressions, we obtain the value of the integral from x_0 to x_n $(= x_0 + nh)$

$$\int_{x_0}^{x_0+nh} f(x)\,dx = \frac{3h}{10}[(y_0 + 5y_1 + y_2 + 6y_3 + y_4 + 5y_5 + y_6) +$$

$$(y_6 + 5y_7 + y_8 + 6y_9 + y_{10} + 5y_{11} + y_{12}) + ...] \quad ...(5.113)$$

This formula is known as Weddle's Rule. It is more accurate, in general, than Simpson' Rule, but it requires at least seven consecutive values of the function, that is, n should be taken as a multiple of 6.

Remark 12

(a) Simpson's 1/3 rule for approximating $\int_a^b f(x)\,dx$ is based on approximating the graph of $f(x)$ with quadratic polynomials that is, parabolic arcs instead of linear polynomials that is, line segments. We approximate the graph of $f(x)$ with parabolic arcs instead of line segments. In simpson's 1/3 rule, we divide the interval of integration $a \le x \le b$ into an **even number** of equal sub-intervals, say into $n = 2m$ sub-intervals of length $h = (b-a)/2m$, with end points $x_0 (= a)$, $x_1, ..., x_{2m-1}, x_{2m} (= b)$. We have odd number of ordinates.

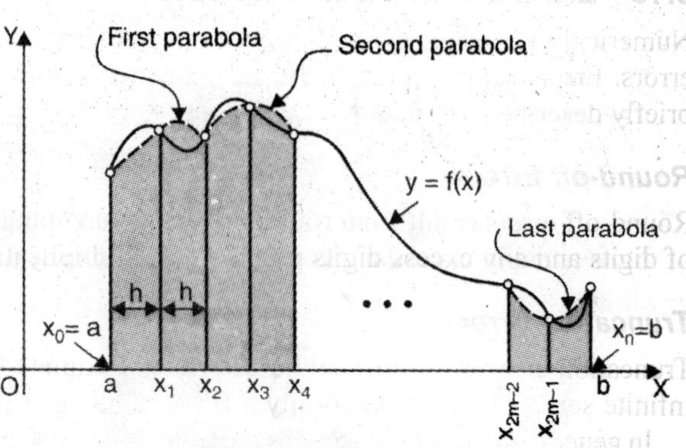

Figure 5.3: Simpson's rule.

Simpson's rule is usually much more accurate than the midpoint or trapezoidal rule.

(b) *Applications of Simpson's rule:* Simpson's rule has a number of applications in practical problems.

(*i*) If the *y*-values, that is, the ordinates denote the velocities of the body at equal interval of time, then Simpson's rule gives the distance covered during the time.

(*ii*) *Evaluation of Volumes.* If the ordinate values pertain to equi-spaced cross-sectional areas, then this rule gives the volume of the solid. Let $A_0, A_1, A_2, ..., A_{2m}$ are the areas of the sections of a solid by parallel planes at equal distances h apart. The volume of the solid

$$= \int_a^b A\,dx \approx \frac{h}{3}[A_0 + 4(A_1 + A_3 + ... + A_{2m-1}) + 2(A_2 + A_4 + ... + A_{2m-2}) + A_{2m}] \qquad ...(5.114)$$

(*iii*) *Evaluation of volume of revolution of areas.* Let the area under the curve $y = f(x)$, above the x-axis, between the lines $x = a$ and $x = b$ revolves about x-axis, the volume of the solid thus formed

$$= \int_a^b \pi[f(x)]^2\,dx \approx \frac{\pi h}{3}[\{f(a)\}^2 + 4\{[f(x_1)]^2 + [f(x_3)]^2 + ... + [f(x_{2m-1})]^2\}$$

$$+ 2[\{[f(x_2)]^2 + [f(x_4)]^2 + ... + [f(x_{2m-2})]^2\} + [f(b)]^2. \qquad ...(5.115)$$

Problems of computing the moment of inertia of the areas can also be solved with the help of this rule. Further this rule finds use to a number of civil engineering problems.

(c) Weddle's Rule: The interval is to be divided in $6n$ (where n is at least 1) equal parts so that $\frac{b-a}{6} = h.$

Then, $\int_a^b f(x)\,dx \approx \frac{3h}{10}(y_0 + 5y_1 + y_2 + 6y_3 + y_4 + 5y_5 + y_6).$

In case the interval is divided in 12 equal parts then

$$\int_a^b f(x)\,dx \approx \frac{3h}{10}[(y_0 + 5y_1 + y_2 + 6y_3 + y_4 + 5y_5 + y_6) + (y_6 + 5y_7 + y_8 + 6y_9 + y_{10} + 5y_{11} + y_{12})]$$

$$= \frac{3h}{10}[(y_0 + y_{12}) + 5(y_1 + y_5 + y_7 + y_{11}) + (y_2 + y_4 + y_8 + y_{10}) + 2y_6 + 6(y_3 + y_9)]$$

Similarly, for 18, 24, etc. parts.

5.13 ERRORS IN COMPUTATION

Numerical methods give approximate values of quantities, that is, they are not exact and contains a number of errors. Errors arise from rounding off, inaccuracy of measured values, truncation and so on. These errors are briefly described as under:

Round-off Errors

Round-off errors result from rounding, that is, computation is carried out to only with a specified/fixed number of digits and any excess digits produced in multiplications or divisions are lost.

Truncation Error

Truncation error arises due to the use of approximate formulas which are generally obtained by truncating an infinite series, that is, taking only a finite number of terms in an infinite series.

In general, increasing *n* reduces the truncation error but increases the round off error, since more computations are required for large *n*. In practical applications, it is important to know how large n must be taken to ensure that a specified degree of accuracy is obtained.

5.13.1 Error Bounds and Estimate for the Trapezoidal Rule

If f'' is continuous and M_2 is any upper bound for the values of $|f''|$ on $[a, b]$, then the error estimate expression is as follows

$$|E_T| \le \frac{(b-a)}{12} h^2 M_2, \qquad ...(5.116)$$

where $M_2 = \max\limits_{a \le x \le b} |f''(x)|$.

If an error tolerance ε is prescribed, it is possible of find n, the number of sub-intervals required to achieve that accuracy.

Since $h = (b-a)/n$, we get from Eq. (5.116)

$$\frac{(b-a)^3}{12n^2} M_2 \le \varepsilon \text{ or } n^2 \ge \frac{(b-a)^3 M_2}{12\varepsilon}. \qquad ...(5.117)$$

5.13.2 The Error of Estimate for Simpson's Rule

If the fourth derivative $f^{(iv)}(x)$ exists and is continuous on $a \le x \le b$, the *error* of relation (5.108) is

$$|E_s| \le \frac{(b-a)}{180} h^4 M_4, \qquad ...(5.118)$$

where $M_4 = \max\limits_{(a,b)} |f^{(iv)}(x)|$.

If the error tolerance ε is prescribed, it is possible to find $n = 2m$, the number of sub-intervals required to achieve that accuracy. Since $h = (b-a)/2m$, where m is an integer, we get from (5.118)

$$\frac{(b-a)^5 M_4}{2880m^4} \le \varepsilon \text{ or } m^4 \ge \frac{(b-a)^5 M_4}{2880\,\varepsilon} \qquad ...(5.119)$$

Which Rule Gives Exact Results?

Simpson's rule gives exact results for polynomials of degree 3 or less. We have $f^{(4)}(x) = 0$ for all x, so $M_4 = 0$ and consequently $|E_s| = 0$. The trapezoidal approximations give exact results for polynomials of degree 1 or less.

Remark 13

As *n* increases, the accuracy of Simpson's rule improves much more rapidly than that of other approximations.

EXAMPLE 5.61 Using the following data:

x:	1	2	3	4
$f(x)$:	0.3679	0.1353	0.0498	0.0183

and the trapezoidal rule with $n = 3$, determine the approximate value of $\int_1^4 f(x)\,dx$.

Solution We have $a = 1$, $b = 4$. For $n = 3$, we get $h = (4-1)/3 = 1$. The abscissas are $x_0 = 1$, $x_1 = 2$, $x_2 = 3$, and $x_3 = 4$.

Hence $\int_1^4 f(x)\,dx \approx \frac{h}{2}[f(1) + 2\{f(2) + f(3)\} + f(4)] = \frac{1}{2}[0.3679 + 2(0.1353 + 0.0498) + 0.0183] = 0.3782.$

EXAMPLE 5.62 Obtain the approximate value of $\int_{-1}^{1} \sqrt{1-x^2}\cos x\,dx$, using the trapezoidal rule with $n = 8$.

Solution We have $a = -1$, $b = 1$ and $f(x) = \sqrt{1-x^2}\cos x$. For $n = 8$, we get $h = (b-a)/n = \{1-(-1)\}/8 = 0.25$. The abscissas are $x_0 = -1$, $x_1 = -0.75$, $x_2 = -0.50$, $x_3 = -0.25$, $x_4 = 0$, $x_5 = 0.25$, $x_6 = 0.50$, $x_7 = 0.75$ and $x_8 = 1$.

The following table is prepared with increasing value of x by 0.25.

x:	–1	–0.75	–0.50	–0.25	0	0.25	0.50	0.75	1
$f(x)$:	0	0.48401*	0.76000	0.93818	1	0.93818	0.76000	0.48401	0

By trapezoidal rule

$$\int_{-1}^{1} \sqrt{1-x^2}\cos x\,dx \approx \frac{0.25}{2}[0 + 2\,(0.48401 + 0.76 + 0.93818 + 1 + 0.93818 + 0.76 + 0.48401) + 0]$$

$$= 0.125\,(10.72876) = 1.3411$$

EXAMPLE 5.63 Evaluate $\int_1^2 \cos x\,dx$ using the trapezoidal rule with $n = 2$. Compare with the exact solution. Find the maximum error.

Solution We have $a = 1$, $b = 2$, and $f(x) = \cos x$. For $n = 2$, we get $h = \frac{(b-a)}{n} = \frac{(2-1)}{2} = 0.5$. The abscissas are $x_0 = 1$, $x_1 = 1.5$ and $x_2 = 2$.

The following table is prepared with increasing value of x by 0.5.

x:	1	1.5	2.0
$f(x) = \cos x$:	0.54024**	0.07063	–0.41628

Hence, $\int_1^2 \cos x\,dx = \frac{h}{2}[f(1) + 2f(1.5) + f(2.0)] = \frac{1}{4}[0.54024 + 0.14126 - 0.41628] = 0.0663$

The exact solution is $[\sin x]_1^2 = \sin 2 - \sin 1 = 0.90924 - 0.84151 = 0.0677.$

---

* For $x = -0.75$, $f(x) = \sqrt{1-x^2}\cos x = \sqrt{1-(-0.75)^2}\cos(-0.75) = \sqrt{1-0.5625}\cos(0.75)$

$= 0.66144\cos 42.972°$ ($\because$ 1radian $= 57.296°$) $= 0.66144\cos 42°58' = 0.66144 \times 0.73175 = 0.48401.$

** $f(1) = \cos 1 = \cos 57.296° = \cos 57°18' = 0.54024$

The actual error in magnitude is (0.0677 – 0.0663) = 0.0014.
We have $f(x) = \cos x, f'(x) = -\sin x, f''(x) = -\cos x \Rightarrow |f''(1)| = |-\cos 1| = 0.54024$.

$$M_2 = \max_{(1,2)} |\cos x| = 0.54024.$$

Hence, $\quad |\text{Error}| \le \frac{(b-a)h^2}{12} M_2 = \frac{1}{12}(0.54024)h^2.$

For $h = 0.5$ $\;|\text{Error}, \varepsilon| \le \frac{1}{12}(0.54024) \times \frac{1}{4}$, that is, $|E| \le 0.0113$. The maximum magnitude of error = 0.0113.

EXAMPLE 5.64 Evaluate the following integrals using trapezoidal rule with $n = 4$. Compare with the exact solution.

(*i*) $\int_0^1 \frac{dx}{3+2x}$, $\qquad$ (*ii*) $\int_0^2 \frac{dx}{x^2+2x+10}$.

Find the bound on the error. Also, find the number of sub-intervals required if the error is to be less than 5×10^{-4}.

Solution (*i*) We have $a = 0$, $b = 1$ and $f(x) = 1/(3+2x)$. For $n = 4$, we get $h = (b-a)/n = (1-0)/4 = 1/4$. The abscissas are $x_0 = 0, x_1 = 1/4, x_2 = 1/2, x_3 = 3/4$ and $x_4 = 1.0$. The following table in prepared with increasing value of x by 1/4.

x:	0	1/4	1/2	3/4	1
$f(x)$:	1/3	2/7	1/4	2/9	1/5

Hence,
$$\int_0^1 \frac{dx}{3+2x} \approx \frac{h}{2}\left[f(0) + 2\left\{f\left(\frac{1}{4}\right) + f\left(\frac{1}{2}\right) + f\left(\frac{3}{4}\right)\right\} + f(1)\right]$$
$$= \frac{1}{8}\left[\frac{1}{3} + 2\left\{\frac{2}{7} + \frac{1}{4} + \frac{2}{9}\right\} + \frac{1}{5}\right] = 0.25615.$$

The exact solution is $\frac{1}{2}[\log_e(3+2x)]_0^1 = 0.5 \log_e\left(\frac{5}{3}\right) = 0.5\,(0.69897 - 0.47712)\,2.30258 = 0.25541.$

The actual error in magnitude is 0.00074.

We have $f(x) = \frac{1}{3+2x}$, $f'(x) = -\frac{2}{(3+2x)^2}$, $f''(x) = \frac{8}{(3+2x)^3}$ and $M_2 = \max_{[0,1]}\left[\frac{8}{(3+2x)^3}\right] = \frac{8}{27}$.

Hence, $|\text{Error}| \le \frac{(b-a)\,h^2}{12} M_2$. For $h = 1/4$ $|\text{Error}| \le \frac{1}{12}\left(\frac{8}{27}\right)\left(\frac{1}{16}\right) = 0.00154.$

The actual error is smaller than the maximum magnitude error.
If $\varepsilon = 5 \times 10^{-4}$, than the number of sub-intervals required is given by

$$n^2 \ge \frac{(b-a)^3 M_2}{12\varepsilon} = \frac{1^3 \times 8 \times 10^4}{27 \times 12 \times 5} = 49.38 \text{ or } n \ge 7.03.$$

Since n is an integer, we require $n = 8$.

(*ii*) We have $a = 0$, $b = 2$ and $f(x) = \frac{1}{x^2+2x+10}$. For $n = 4$, we get $h = \frac{(2-0)}{4} = \frac{1}{2}$. The abscissas are $x_0 = 0, x_1 = 1/2, x_2 = 1, x_3 = 3/2$ and $x_4 = 2$.

The following table is prepare with increasing value of x by 1/2.

x:	0	1/2	1	3/2	2
$f(x)$:	0.1	0.08889	0.07692	0.06557	0.05556

Hence, $$\int_0^2 \frac{dx}{x^2+2x+10} \approx \frac{h}{2}\left[f(0)+2\left\{f\left(\frac{1}{2}\right)+f(1)+f\left(\frac{3}{2}\right)\right\}+f(2)\right]$$

$$=\frac{1}{4}[0.1+2(0.08889+0.07692+0.06557)+0.05556]=0.15458.$$

The exact solution is as under:

$$\int_0^2 \frac{dx}{x^2+2x+10}=\int_0^2 \frac{dx}{(x+1)^2+(3)^2}=\frac{1}{3}\left|\tan^{-1}\left(\frac{x+1}{3}\right)\right|_0^2=\frac{1}{3}\left[\frac{\pi}{4}-\tan^{-1}\left(\frac{1}{3}\right)\right]=0.15455.$$

The actual error in magnitude is 0.15458 – 0.15455 = 0.00003. We have

$$f(x)=\frac{1}{x^2+2x+10},\quad f'(x)=-\frac{2(x+1)}{(x^2+2x+10)^2},$$

$$f''(x)=-\frac{[2(x^2+2x+10)^2-8(x+1)^2(x^2+2x+10)]}{(x^2+2x+10)^4}=\frac{6(x^2+2x-2)}{(x^2+2x+10)^3}.$$

For determining $\max\limits_{(0,2)} 6\dfrac{(x^2+2x-2)}{(x^2+2x-10)^3}$. We have $\max\limits_{(0,2)}(x^2+2x-2)=6$ and $\min\limits_{(0,2)}(x^2+2x+10)^3=1000$.

Hence, $M_2 \le \dfrac{6\times 6}{1000}$. Therefore, $|\text{Error}| \le \dfrac{(b-a)\,h^2}{12}M_2=\dfrac{2}{12}\times\dfrac{36}{1000}h^2=\dfrac{3}{500}h^2.$

For $h = 1/2$, $\quad |\text{Error}| \le 0.0015.$

If $\varepsilon = 5\times 10^{-4}$, then the number of sub-intervals required is given by

$$n^2 \ge \frac{(b-a)^3 M_2}{12\varepsilon}=\frac{8\times 36\times 10^4}{1000\times 12\times 5}=48,\ \text{Thus number of sub-intervals, } n=7.$$

EXAMPLE 5.65 (*i*)The cross-section of a tree is A sq. cm at a distance x cm from one end. Corresponding values of A and x are given

x:	10	30	50	70	90	110	130	150	170
A:	120	123	129	131	131	135	142	156	177

Find by Simpson's rule, the volume of the tree in cubic cm between x = 10 and x = 170.

(*ii*) A rocket is launched from the ground. Its acceleration is registered during the first 80 seconds and is given in the table below. Using Simpson's 1/3-rule, find the velocity of the rocket at t = 80 seconds.

t(sec):	0	10	20	30	40	50	60	70	80
f(cm/sec^2):	30.0	31.63	33.34	35.47	37.75	40.33	43.25	46.69	50.67

Solution (*i*) Volume, $V=\displaystyle\int_{10}^{170} A\,dx$. Use Simpson's 1/3-rule with n = 8 and h = 20.

$$\text{Required volume} = \int_{10}^{170} A\,dx \approx \frac{20}{3}[120+4(123+131+135+156)+2(129+131+142)+177]$$

$$=\frac{20}{3}(297+2180+804)=\frac{20}{3}(3281)=21873.33\text{ cm}^3.$$

(*ii*) The velocity of the rocket is given by, $v=\displaystyle\int_0^{80} f\,dt$. Using Simpson's 1/3-rule. Here h = 10 sec, y_0 = 30.0,

y_1 = 31.63, y_2 = 33.34, y_3 = 35.47, y_4 = 37.75, y_5 = 40.33, y_6 = 43.25, y_7 = 46.69, y_8 = 50.67.

Therefore, $v = \int_0^{80} f\,dt \approx \frac{h}{3}[(y_0 + y_8) + 4(y_1 + y_3 + y_5 + y_7) + 2(y_2 + y_4 + y_6)]$

$$= \frac{10}{3}[(30.0 + 50.67) + 4(31.63 + 35.47 + 40.33 + 46.69) + 2(33.34 + 37.75 + 43.25)]$$

$$= \frac{10}{3}[80.67 + 4(154.12) + 2(114.34)] = \frac{10}{3}[80.67 + 616.48 + 228.68]$$

$$= \frac{10}{3}(925.83) = 3086.1 \text{ cm/sec}$$

EXAMPLE 5.66 (*i*) The velocity v (km/min) of a moped which starts from rest, is given at fixed intervals of time t(min) as follows:

t:	2	4	6	8	10	12	14	16	18	20
v:	10	18	25	29	32	20	11	5	2	0

Estimate approximately the distance covered by the moped in 20 minutes, using Simpson's 1/3-rule.

(*ii*) The velocity v(*m/sec*) of particle at a distance s(metre) from a point on its path is given by the table

s:	0	10	20	30	40	50	60
v:	47	58	64	65	61	52	38

Estimate the time taken to travel 60 metre.

(*iii*) A variable force F kgf acts in the direction of motion of a moving body. The following table gives values of F at distances x metres along the path of the body.

x:	2	3	4	5	6	7	8	9	10	11	12
F:	122	159	188	210	231	246	259	268	277	282	287

Find the work done by the force on the body as its moves from $x = 2$ to $x = 12$.

Solution (*i*) If s (km) be the distance covered in time t (min), then $v = ds/dt$ or $ds = v.dt.$ It follows on integration that $s = \int v \cdot dt.$ Distance covered in 20 minutes is, $s = \int_0^{20} v\,dt.$ Since the train starts from rest, $v = 0$, when $t = 0$. Thus there are 11 values of v. Use Simpson's 1/3 rule, with $h = 2$.

Therefore,

$$\int_0^{20} v\,dt \approx \frac{2}{3}[0 + 4(10 + 25 + 32 + 11 + 2) + 2(18 + 29 + 20 + 5) + 0] = \frac{2[320 + 144]}{3} = \frac{2(464)}{3} = 309.33 \text{ kms.}$$

(*ii*) The time taken to travel 60 metre is $t = \int_0^{60} \frac{1}{v} ds$ (sec). We form a table of $1/v$ vers s

s:	0	10	20	30	40	50	60
$1/v$:	0.02127	0.01724	0.01563	0.01538	0.0164 0	0.01923	0.02631

Hence, $t = \int_0^{60} \frac{1}{v} ds \approx \frac{10}{3}[0.02127 + 4(0.01724 + 0.01538 + 0.01923) + 2(0.01563 + 0.01640) + 0.02631]$

$$= \frac{10}{3}[0.04758 + 0.20740 + 0.06406] = \frac{10}{3}(0.31904) = 1.06 \text{ seconds.}$$

(*iii*) Work done by the force, $= \int_2^{12} F \cdot dx$. Use Simpson's 1/3 rule, with $h = 1$.

$$\text{Hence, } \int_2^{12} F \cdot dx \approx \frac{1}{3}[122 + 4(159 + 210 + 246 + 268 + 282) + 2(188 + 231 + 259 + 277) + 287]$$

$$= \frac{1}{3}[409 + 4(1165) + 2(955)] = \frac{1}{3}(6979) = 2326.33 \text{ kgf-m.}$$

EXAMPLE 5.67 (*i*) Use Simpson's rule to determine (*a*) the area under the curve (*b*) the volume generated by revolving the area about the *x*-axis determined by the given data

x:	1	2	3	4	5
y:	1.8	4.2	7.8	9.2	12.3

(*ii*) A body is in the form of a solid of revolution and is 15 cm long. The following table gives the diameter *D* in cm of the section at distance *x* cm from one end. Find the volume of the solid generated.

x:	0	2.5	5.0	7.5	10.0	12.5	15.0
D:	5	5.5	6.0	6.75	6.25	5.5	4.0

Solution (*i*) Here $h = 1$. (*a*) Area under the curve $= \int_1^5 y\,dx \approx \frac{1}{3}[1.8 + 4(4.2 + 9.2) + 2(7.8) + 12.3]$

$$= \frac{1}{3}[14.1 + 4(13.4) + 2(7.81)] = 27.767 \text{ sq. units.}$$

(*b*) Volume of solid generated by revolving the area about the x-axis

$$\approx \frac{\pi \times 1}{3}[(1.8)^2 + 4\{(4.2)^2 + (9.2)^2\} + 2(7.8)^2 + (12.3)^2]$$

$$= \frac{\pi}{3}[3.24 + 4(17.64 + 84.64) + 2(60.84) + 151.29]$$

$$= \frac{\pi}{3}[154.53 + 409.12 + 121.68] = 228.44\pi \text{ cubic units.}$$

(*ii*) Here $h = 2.5$. Since the body is in the form of a solid of revolution, the volume of the solid generated is given by

$$\int_0^{15} \frac{\pi}{4} D^2 dx = \frac{\pi}{4}\int_0^{15} D^2 dx \approx \frac{\pi}{4}\frac{(2.5)}{3}[(5)^2 + 4\{(5.5)^2 + (6.75)^2 + (5.5)^2\} + 2\{(6.0)^2 + (6.25)^2\} + (4)^2]$$

$$= \frac{5\pi}{24}[25 + 4(30.25 + 45.5625 + 30.25) + 2(36 + 39.0625) + 16]$$

$$= \frac{5\pi}{24}[41 + 4(106.0625) + 2(75.0625)] = \frac{5\pi}{24}[41 + 424.25 + 150.125] = 402.9241 \text{ cm}^3$$

EXAMPLE 5.68 Using Simpson's rule, evaluate the following

(*i*) $\int_0^{\pi/2} e^{\sin x}\,dx$, with $n = 8$, (*ii*) $\int_0^{\pi} \frac{\sin x}{x}\,dx$, with seven ordinates, (*iii*) $\int_0^{\pi/2} \sqrt{\sin\theta}\,d\theta$ with $n = 6$.

Solution: (*i*) We have $a = 0$. $b = \pi/2$ and $f(x) = e^{\sin x}$. For $n = 8$, we get $h = (\pi/2 - 0)/8 = \pi/16$. The following table is prepared with increasing value of *x* by $\pi/16$.

x:	0	$\pi/16$	$\pi/8$	$3\pi/16$	$\pi/4$	$5\pi/16$	$3\pi/8$	$7\pi/16$	$\pi/2$
$\sin x$:	0	0.1951	0.3827	0.5555	0.7071	0.8315	0.9239	0.9808	1
$e^{\sin x}$:	1	1.215	1.467	1.743	2.028	2.297	2.519	2.667	2.718

Hence, $\int_0^{\pi/2} e^{\sin x} dx \approx \frac{\pi}{16 \times 3}[1 + 4(1.215 + 1.743 + 2.297 + 2.667) + 2(1.467 + 2.028 + 2.519) + 2.718]$

$$= \frac{\pi}{48}[3.718 + 4(7.922) + 2(6.014)] = \frac{\pi}{48}[3.718 + 31.688 + 12.028] = \frac{(\pi \times 47.434)}{48} = 3.1058.$$

(*ii*) We have $a = 0$, $b = \pi$ and $f(x) = (\sin x)/x$. For $n = 6$, we get $h = \frac{(\pi - 0)}{6} = \frac{\pi}{6}$.

We obtain the following table

x:	0	$\pi/6=30°$	$\pi/3=60°$	$\pi/2=90°$	$2\pi/3=120°$	$5\pi/6=150°$	$\pi=180°$
$\sin x$:	0.00	0.50	0.86603	1.00	0.86603	0.50	0.00
$\frac{(\sin x)}{x}$:	1	$\frac{3}{\pi}$	$\frac{(2.59809)}{\pi}$	$\frac{2}{\pi}$	$\frac{(1.29904)}{\pi}$	$\frac{(0.6)}{\pi}$	0

Hence, $\int_0^{\pi} \frac{\sin x}{x} dx \approx \frac{\pi}{18}\left[1 + 4\left(\frac{3}{\pi} + \frac{2}{\pi} + \frac{0.6}{\pi}\right) + 2\left(\frac{2.59809}{\pi} + \frac{1.29904}{\pi}\right) + 0\right]$

$$= \frac{\pi}{18}\left(1 + \frac{22.4}{\pi} + \frac{7.79426}{\pi}\right) = \frac{33.3371}{18} = 1.8521.$$

(*iii*) Here $h = \frac{(\pi/2 - 0)}{6} = \frac{\pi}{12}$.

The following table is prepared with increasing value of θ by $\pi/12$.

θ:	0	$\frac{\pi}{12} = 15°$	$\frac{\pi}{6} = 30°$	$\frac{\pi}{4} = 45°$	$\frac{\pi}{3} = 60°$	$\frac{5\pi}{12} = 75°$	$\frac{\pi}{2} = 90°$
$\sin\theta$:	0.0	0.2588	0.500	0.7071	0.8660	0.9659	1.000
$\sqrt{\sin\theta}$:	0	0.5087	0.7071	0.8409	0.9306	0.9828	1.000

Hence, $\int_0^{\pi/2} \sqrt{\sin\theta}\, d\theta \approx \frac{\pi}{36}[0 + 4(0.5087 + 0.8409 + 0.9828) + 2(0.7071 + 0.9306) + 1]$

$$= (\pi \times 13.605)/36 = 1.1877.$$

EXAMPLE 5.69 (*i*) By Simpson's 1/3 rule, find the value of $\log_e (3/2)$.

(*ii*) Using Simpson's rule calculate the approximate value of π from $\int_0^1 \frac{dx}{1+x^2}$. (*AMIE, Winter 2005*)

(*iii*) Calculate the value of $\log_e 2$ by finding $\int_0^1 \frac{x^2 dx}{1+x^3}$, using Simpson's rule, by dividing the interval into four equal parts.

(*iv*) Find an approximate value of $\log_e 5$, correct to 4 decimal places by Simpson's 1/3 rule the integral $\int_0^5 \frac{dx}{4x+5}$, divide the range into 10 equal parts. (*AMIETE, June 2013, Dec 2011*)

Solution (*i*) $\log_e \frac{3}{2}$ may be regarded as $\int_2^3 \frac{1}{x}\,dx$. Dividing the range into 10 equal intervals so that $h = (3-2)/10 = 0.1$. We obtain the following table:

x:	2.0	2.1	2.2	2.3	2.4	2.5	2.6	2.7	2.8	2.9	3.0
$y = 1/x$:	0.5000	0.4762	0.4545	0.4348	0.4167	0.4000	0.3846	0.3704	0.3571	0.3448	0.3333

By Simpson's 1/3-rule,

$$\int_2^3 \frac{dx}{x} \approx \frac{0.1}{3}[0.50 + 4(0.4762 + 0.4348 + 0.4000 + 0.3704 + 0.3448)$$

$$+ 2(0.4545 + 0.4167 + 0.3846 + 0.3571) + 0.3333]$$

$$= \frac{0.1}{3}[0.8333 + 4(2.0262) + 2(1.6129)] = 0.4055$$

(*ii*) Dividing the range into 10 equal intervals, so that $h = (1-0)/10 = 0.1$. The following table is prepared with increasing value of x by 0.1.

x:	0	0.1	0.2	0.3	0.4	0.5	0.6	0.7	0.8	0.9	1.0
$y = (1+x^2)^{-1}$:	1.0000	0.9901	0.9615	0.9174	0.8621	0.8000	0.7353	0.6711	0.6097	0.5525	0.5000

By Simpson's 1/3-rule,

$$\int_0^1 \frac{dx}{1+x^2} \approx \frac{0.1}{3}[1.00 + 4(0.9901 + 0.9174 + 0.8 + 0.6711 + 0.5525)$$

$$+ 2(0.9615 + 0.8621 + 0.7353 + 0.6097) + 0.50]$$

$$= \frac{0.1}{3}[1.5 + 4(3.9311) + 2(3.1686)] = \frac{1}{30}(1.5 + 15.7244 + 6.3372) = 0.78538.$$

We also know that

$$\int_0^1 (1+x^2)^{-1}\,dx = |\tan^{-1} x|_0^1 = \tan^{-1} 1 = \frac{\pi}{4}. \text{ Therefore, } \frac{1}{4}\pi \approx 0.78538 \text{ or } \pi \approx 3.1415.$$

(*iii*) First we shall determine the approximate value of $\int_0^1 \frac{x^2\,dx}{1+x^3}$ by employing Simpson's 1/3-rule. We have $a = 0, b = 1, f(x) = x^2/(1+x^3)$. For $n = 4$, we get $h = (1-0)/4 = 0.25$. The following table is prepared with increasing value of x by 0.25.

x:	0	0.25	0.50	0.75	1.00
$y = \frac{x^2}{(1+x^3)}$:	0	0.06154	0.22222	0.39560	0.5000

Hence, $$\int_0^1 \frac{x^2\,dx}{1+x^3} \approx \frac{0.25}{3}[0 + 4(0.06154 + 0.39560) + 2(0.22222) + 0.50]$$

$$= 0.25[0.5 + 1.82856 + 0.44444]/3 = 0.231083.$$

Also by definite integral $$\int_0^1 \frac{x^2}{1+x^3}\,dx = = \frac{1}{3}\int_0^1 \frac{3x^2}{1+x^3}\,dx = \frac{1}{3}[\log_e(1+x^3)]_0^1 = \frac{1}{3}(\log_e 2 - \log_e 1) = \frac{1}{3}\log_e 2$$

Therefore, $$\log_e 2 = 3\int_0^1 \frac{x^2\,dx}{1+x^3} \approx 3(0.231083) = 0.693249.$$

(*iv*) We have $h = (5-0)/10 = 0.5$. The following table is prepared with increasing value of x by 0.5.

x:	0.0	0.5	1.0	1.5	2.0	2.5	3.0	3.5	4.0	4.5	5.0
$y=f(x)=\frac{1}{4x+5}$:	0.20000	0.14286	0.11111	0.09091	0.07692	0.06667	0.05882	0.05263	0.04762	0.04348	0.04000

By Simpson's 1/3-rule,

$$\int_0^5 \frac{dx}{4x+5} \approx \frac{0.5}{3}[0.2+4(0.14286+0.09091+0.06667+0.05263+0.04348)$$

$$+2(0.11111+0.07692+0.05882+0.04762)+0.04]$$

$$=\frac{1}{6}[0.24+4(0.39655)+2(0.29447)]=\frac{1}{6}[0.24+1.5862+0.58894]=0.40252.$$

We also know that

$$\int_0^5 \frac{dx}{4x+5}=\frac{1}{4}[\log_e(4x+5)]_0^5=\frac{1}{4}[\log_e 25-\log_e 5]=\frac{1}{4}[2\log_e 5-\log_e 5]=\frac{1}{4}\log_e 5.$$

Therefore, $\frac{1}{4}\log_e 5 \approx 0.40252$ or $\log_e 5 = 1.6101$, correct to four decimal places.

EXAMPLE 5.70 A tank is discharging water through an orifice at a depth x metre below the surface of water whose area is A sq. metres. The following are the values of x for the corresponding values of A

A:	1.257	1.39	1.52	1.65	1.809	1.962	2.123	2.295	2.462	2.650	2.827
x:	1.50	1.65	1.80	1.95	2.10	2.25	2.40	2.55	2.70	2.85	3.00

Using the formula $(0.018)T=\int_{1.5}^{3.0}\frac{A}{\sqrt{x}}dx$, calculate T, the time in seconds for the level of water to drop from 3.0 m to 1.5 m above an orifice.

Solution Here $h = 0.15$ m. The following table is prepared.

x:	1.50	1.65	1.80	1.95	2.10	2.25	2.40	2.55	2.70	2.85	3.00
A:	1.257	1.39	1.52	1.65	1.809	1.962	2.123	2.295	2.462	2.650	2.827
$y=\frac{A}{\sqrt{x}}$	1.026	1.082	1.133	1.182	1.248	1.308	1.370	1.437	1.498	1.570	1.632

By Simpson's 1/3-rule,

$$\int_{1.5}^{3.0}\frac{A}{\sqrt{x}}dx \approx \frac{0.15}{3}[1.026+4(1.082+1.182+1.308+1.437+1.570)$$

$$+2(1.133+1.248+1.370+1.498)+1.632]$$

$$=\frac{0.15}{3}[2.658+26.316+10.498]=\frac{0.15}{3}\times 39.472=1.9736.$$

Therefore, $(0.018)T = 1.9736$, that is, $T=(1.9736)/0.018=110$ sec. (nearly).

EXAMPLE 5.71 Evaluate the following integrals using Simpson's 1/3-rule with $n = 4$. Compare with the exact solutions.

(*i*) $\int_0^1 \frac{dx}{3+2x}$, (*ii*) $\int_0^2 \frac{dx}{x^2+2x+10}$.

Solution (*i*) We have $a = 0$, $b = 1$ and $f(x) = 1/(3+2x)$. For $n = 4$, we obtain $h = (b-a)/n = 1/4$. The abscissas are $x_0 = 0, x_1 = 1/4, x_2 = 1/2, x_3 = 3/4$ and $x_4 = 1$. The following table is prepared with increasing values of x by 1/4.

x:	0	1/4	1/2	3/4	1
$f(x)$:	1/3	2/7	1/4	2/9	1/5

Hence, $$\int_0^1 \frac{dx}{3+2x} \approx \frac{h}{3}[f(0) + 4\{f(1/4) + f(3/4)\} + 2f(1/2) + f(1)]$$

$$= \frac{1}{12}\left[\frac{1}{3} + 4\left(\frac{2}{7} + \frac{2}{9}\right) + 2\left(\frac{1}{4}\right) + \frac{1}{5}\right] = 0.25542.$$

The exact solution is 0.5 log (5/3) = 0.25541. The error in magnitude is 0.00001.

(*ii*) We have $a = 0$, $b = 2$ and $f(x) = 1/(x^2 + 2x + 10)$. For $n = 4$, we obtain $h = (2-0)/4 = 1/2$. The abscissas are $x_0 = 0, x_1 = 1/2, x_2 = 1, x_3 = 3/2$ and $x_4 = 2$. The following table is prepared with increasing value of x by 1/2.

x:	0	1/2	1	3/2	2
$f(x)$:	1/10	4/45	1/13	4/61	1/18

Hence, $$\int_0^2 \frac{1}{x^2 + 2x + 10}\,dx \approx \frac{h}{3}[f(0) + 4\{f(1/2) + f(3/2)\} + 2f(1) + f(2)]$$

$$= \frac{1}{6}\left[\frac{1}{10} + 4\left(\frac{4}{45} + \frac{4}{61}\right) + 2\left(\frac{1}{13}\right) + \frac{1}{18}\right]$$

$$= \frac{1}{6}[0.1 + 4(0.0889 + 0.06557) + 2(0.07692) + 0.05556] = \frac{0.92728}{6} = 0.15454.$$

The exact solution is 0.15455. The error in magnitude is 0.00001.

EXAMPLE 5.72 Evaluate the integral $\int_{-1}^{1} x^2 e^{-x} dx$, by composite Simpson's 1/3-rule with spacing $h = 0.25$. You may use any other known numerical method to evaluate the given integral. *(AMIE, S-2008)*

Solution We have $a = -1$, $b = 1$, $h = 0.25$ and $f(x) = x^2 e^{-x}$. The abscissas are $x_0 = -1$, $x_1 = -0.75$, $x_2 = -0.50$, $x_3 = -0.25$, $x_4 = 0$, $x_5 = 0.25$, $x_6 = 0.50$, $x_7 = 0.75$ and $x_8 = 1$. The following table is prepared with increasing values of x by 0.25.

x:	−1.0	−0.75	−0.50	−0.25	0	0.25	0.50	0.75	1.0
$f(x)=x^2e^{-x}$	2.71828	1.19082	0.41218	0.08025	0	0.04868	0.15163	0.26571	0.36788

Hence, $$\int_{-1}^{1} x^2 e^{-x} dx \approx \frac{h}{3}[f(-1) + 4\{f(-0.75) + f(-0.25) + f(0.25) + f(0.75)\}$$

$$+ 2\{f(-0.50) + f(0) + f(0.50)\} + f(1.0)]$$

$$= \frac{0.25}{3}[2.71828 + 4(1.19082 + 0.08025 + 0.04868 + 0.26571) + 2(0.41218 + 0 + 0.15163) + 0.36788]$$

$$= \frac{1}{12}(3.08616 + 6.34184 + 1.12762) = 0.87964.$$

EXAMPLE 5.73 Apply the trapezoidal and Simpson's 1/3-rule to evaluate $\int_{1.00}^{1.30} \sqrt{x}\, dx$ using the data of following table. Compare with the correct value of the integral. Estimate the truncation error of the trapezoidal rule and Simpson's rule.

x	1.00	1.05	1.10	1.15	1.20	1.25	1.30
$y = f(x) = \sqrt{x}$	1.00000	1.02470	1.04881	1.07238	1.09544	1.11803	1.14017

Solution Here $h = 0.05$. Apply the trapezoidal rule, we obtain

$$\int_{1.00}^{1.30} \sqrt{x}\, dx \approx \frac{0.05}{2}[1.00000 + 2(1.02470 + 1.04881 + 1.07238 + 1.095544 + 1.11803 + 1.14017]$$

$$= \frac{0.05}{2}(12.85889) = 0.32147 \text{ (approx.)}$$

The correct value is $\frac{2}{3}[(1.3)^{3/2} - 1] = 0.32149$ to five places, making the actual error 0.00002.

By Simpson's 1/3-rule,

$$\int_{1.00}^{1.30} \sqrt{x}\, dx = \frac{0.05}{3}[1 + 4(1.02470 + 1.07238 + 1.11803) + 2(1.04881 + 1.09544) + 1.14017]$$

$$= \frac{0.05}{3}[2.14017 + 12.86044 + 4.2885] = \frac{0.05}{3}(19.28911) = 0.32149 \text{ (approx.)}$$

Truncation error of the trapezoidal rule

We have $f(x) = \sqrt{x},\ f'(x) = x^{-1/2}/2,\ f''(x) = -x^{-3/2}/4.\quad M_2 = \max\limits_{[1,\ 1.30]}\left|\frac{1}{4}x^{-3/2}\right| = \frac{1}{4}.$

Hence, $|\text{Error},\ \varepsilon_T| \le \frac{(b-a)h^2}{12} M_2 = \frac{0.30}{12}\left(\frac{1}{4}\right)h^2 = 0.00625\, h^2.$

For $h = 0.05$, $|\text{Error},\ \varepsilon_T| \le 0.000016$ which is slightly less than the actual error of 0.00002.

Truncation error of Simpson's rule

$$f'''(x) = \frac{3}{8}x^{-5/2},\ f^{iv}(x) = -\frac{15}{16}x^{-7/2}.\qquad M_4 = \max_{[1,\,1.30]}\left|\frac{15}{16}x^{-7/2}\right| = \frac{15}{16}.$$

Hence, $|\text{Error},\ \varepsilon_s| \le \frac{(b-a)}{180}h^4 M_4 = \frac{0.30}{180}(0.05)^4\left(\frac{15}{16}\right) \sim 0.00000001$ which is minute.

EXAMPLE 5.74 A curve passes through the points (1, 2), (1.5, 2.4), (2, 2.7), (2.5, 2.8), (3, 3), (3.5, 2.6), (4, 2.1) and (4.5, 1.9). Use Simpson's one third rule to obtain the volume of the solid of revolution got by revolving this area about x-axis. *(AMIE, S-2009)*

Solution Here the number of strips h are 7, that is, odd and number of values of y that is, ordinates are even. Therefore, Simpson rule cannot be applied directly to this problem. Refer Fig. 5.4(b), the volume of region $AA'BJ$ when revolved about x-axis is computed by using *Pappus theorem on volume of solid of revolution* and Simpson's 1/3-rule is applied to the remaining region JBCDEFGHI.

Refer Figure 5.4(b) which is a enlarged view of region $AA'BJ$ [shown in Fig. 5.4(a)] We have Area, $A_1 = 2 \times 0.5 = 1$, Area, $A_2 = 1/2 \times 0.5 \times 0.4 = 0.1$, $y_1 = 1$ unit from $0X$, $y_2 = 2 + \frac{1}{3}(0.4) = \frac{32}{15}$, unit from OX. Let $\bar{y}$ be the CG of the whole region, $AA'BJ$.

Taking moment about $0X$.

$$(A_1 + A_2)\,\bar{y} = A_1 y_1 + A_2 y_2 \Rightarrow (1.00 + 0\cdot 1)\bar{y} = 1 \times 1 + 0.1 \times \frac{32}{15} \Rightarrow \bar{y} = 1.103.$$

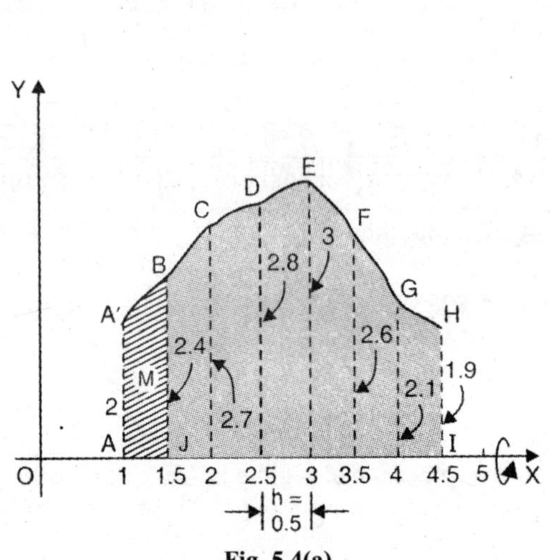

Fig. 5.4(a)

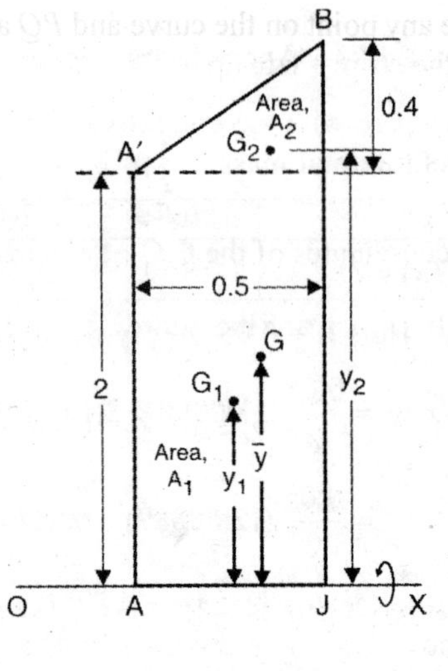

Fig. 5.4(b)

Therefore, volume of region $AA'BJ$ when revolved about OX = (1.1) $2\pi\bar{y}$ = (1.1) 2π (1.103) = 7.626 cubic units.

Now, we prepared the following table for computing volume of solid of revolution about X-axis, the portion JBCDEFGHI. [Fig. 5.4(a)].

x:	1.5	2.0	2.5	3.0	3.5	4.0	4.5
y:	2.4	2.7	2.8	3.0	2.6	2.1	1.9
y^2:	5.76	7.29	7.84	9	6.76	4.41	3.61

Therefore required volume of the solid of revolution about OX

$$= 7.626 + \frac{\pi(0.5)}{3}[(5.76 + 3.61) + 4(7.29 + 9 + 4.41) + 2(7.84 + 6.76)]$$

$$= 7.626^* + \frac{\pi}{6}(9.37 + 82.80 + 29.20) = 7.626 + 63.574 = 71.2 \text{ cubic units.}$$

Remark 14 The volume, V_1 of the solid of revolution of area $AA'BJ$ when revolved about OX may be calculated directly by employing the formula of volume of frustum of cone having left hand side area $\pi(2)^2$, right hand side area $\pi(2.4)^2$ and height of the frusturn of cone equal to 0.5 unit.

$$V_1 = \frac{h}{3}[A_1 + \sqrt{A_1A_2} + A_2] \text{ cubic units} = \frac{0.5}{3}\pi[4 + \sqrt{4 \times 5.76} + 5.76]$$

$$= \frac{\pi}{6}[4 + 4.8 + 5.76] = \frac{\pi}{6}(14.56) = 7.626^* \text{ cubic units}$$

EXAMPLE 5.75 A curve is drawn to pass through the points which are given in the following table:

x:	0	1	2	3	4	5	6
y:	0	2	2.5	2.3	2	1.7	1.5

Find by Simpson's 1/3-rule (*i*) the co-ordinates of the centroid of the area, (*ii*) the volume of revolution of the area about x-axis (*iii*) the moment of interia about the x-axis for the plane area, (*iv*) radius of gyration of the area about OX.

Solution Refer Fig. 5.5, The area $OABO = \int_0^6 y\,dx$. Volume of solid of revolution about $OX = \int_0^6 \pi y^2 dx$. Let

$P(x, y)$ be any point on the curve and PQ an elementary strip of width dx. If ρ be assumed to be the density, then mass of the strip $= ydx \cdot \rho$.

Therefore, total mass $= \int_0^6 \rho y \, dx.$

The coordinates of the C.G of an elementary strip are located at the mid-point of the strip and are given by $\left(x, \frac{y}{2}\right)$. If $(\bar{x}, \bar{y})$ are the coordinates of the C.G of the area, then

$$\bar{x} = \frac{\int_0^6 \rho x dA}{\int_0^6 \rho dA} = \frac{\int_0^6 (xy)\, dx}{\int_0^6 y dx}, \qquad \bar{y} = \frac{\int_0^6 \rho y dx (y/2)}{\int_0^6 \rho y dx} = \frac{\frac{1}{2}\int_0^6 y^2 \, dx}{\int_0^6 y dx}.$$

We know that, the moment of inertia of a rectangle about its base is $bh^3/3$. The moment of inertia of the elementary area $y \, dx$ about the x-axis is $y^3 \, dx/3$. Hence, for the whole area, we have

$$I_{xx} = (1/3)\int_0^6 y^3 dx \approx \left(\frac{1}{3}\right) \times \left(\frac{h}{3}\right)[y_0^3 + 4y_1^3 + 2y_2^3 + 4y_3^3 + ...]$$

To evaluate the various integrals, we prepare the following table. The area of the curve is determined using simpson's 1/3-rule.

x:	0	1	2	3	4	5	6
y:	0.0	2.0	2.5	2.3	2.0	1.7	1.5
xy:	0.0	2.0	5.0	6.9	8.0	8.5	9.0
y^2:	0.0	4.00	6.25	5.29	4.00	2.89	2.25
y^3:	0.0	8.0	15.625	12.167	8.00	4.913	3.375

(i) $$\bar{x} \approx \frac{(h/3)(x_0y_0 + 4x_1y_1 + 2x_2y_2 + 4x_3y_3 + ...)}{(h/3)(y_0 + 4y_1 + 2y_2 + 4y_3 + ...)}.$$

$$= \frac{(0+9) + 4(2+6.9+8.5) + 2(5.0+8.0)}{(0+1.5) + 4(2+2.3+1.7) + 2(2.5+2)}$$

$$= \frac{9 + 4(17.4) + 2(13)}{1.5 + 4(6) + 2(4.5)} = \frac{104.6}{34.5} = 3.03.$$

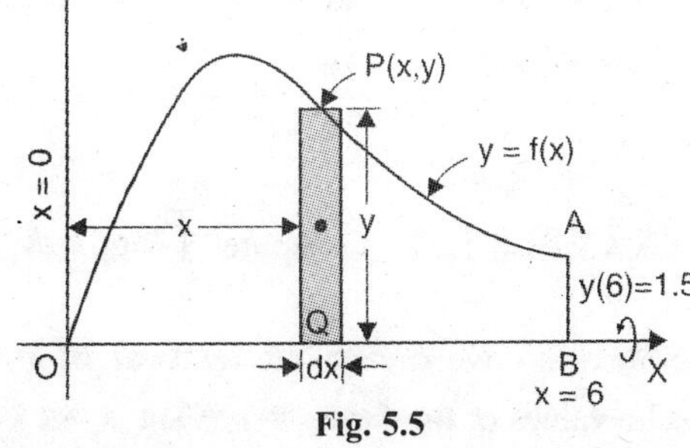

Fig. 5.5

$$\bar{y} \approx \frac{1}{2} \cdot \frac{(h/3)(y_0^2 + 4y_1^2 + 2y_2^2 + 4y_3^2 + ...)}{(h/3)(y_0 + 4y_1 + 2y_2 + 4y_3 + ...)}.$$

$$= \frac{1}{2} \cdot \frac{(0+2.25) + 4(4+5.29+2.89) + 2(6.25+4)}{(0+1.5) + 4(2+2.3+1.7) + 2(2.5+2)} = \frac{1}{2} \cdot \frac{71.47}{34.5} = 1.036.$$

Therefore, the coordinates of the centroid of the area are (3.03, 1.036).

(ii) Volume of revolution about x-axis

$$= \int_0^6 \pi y^2 \, dx \approx \frac{\pi}{3}[2.25 + 4(4 + 5.29 + 2.89) + 2(6.25 + 4)]$$

$$= \frac{\pi}{3}[2.25 + 48.72 + 20.50] = \frac{(71.47)(22)}{21} = 74.873 \text{ cubic units (approx.)}$$

(*iii*) $I_{xx} = \frac{1}{9}[3.375 + 4(8 + 12.167 + 4.913) + 2(15.625 + 8)]$

$= \frac{1}{9}[3.375 + 100.32 + 47.25] = 16.771$.

Area under the curve $= \frac{1}{3}[(0 + 1.5) + 4(2.0 + 2.3 + 1.7) + 2(2.5 + 2)] = \frac{1}{3}(1.5 + 24 + 9) = 11.5$.

(*iv*) If k is radius of gyration of the area A about OX, then

$$I_{xx} = Ak^2 \text{ or } k = \sqrt{\frac{I_{xx}}{A}} = \sqrt{\frac{16.771}{11.5}} = \sqrt{1.458} = 1.20.$$

EXAMPLE 5.76 A curve is drawn to pass through the following points:

x:	1	1.5	2	2.5	3	3.5	4
$y = f(x)$:	2	2.4	2.7	2.8	3	2.6	2.1

Using Simpson's 3/8-rule estimate the area bounded by the curve, x-axis and lines $x = 1$, $x = 4$. Also find the volume of solid generated by revolving this area . *(AMIETE, June, 2011, 2009)*

Solution Here $h = 0.5$. By Simpson's 3/8-rule

$$\text{Area} \approx \frac{3h}{8}[(y_0 + y_6) + 3(y_1 + y_2 + y_4 + y_5) + 2(y_3)].$$

$$= \frac{3}{16}[(2 + 2.1) + 3(2.4 + 2.7 + 3 + 2.6) + 2(2.8)] = \frac{3}{16}[4.1 + 32.1 + 5.6] = 7.838 \text{ square units.}$$

Volume of the solid generated by revolving the above area about x-axis

$$= \int_1^4 \pi y^2 dx \approx \pi \cdot \frac{3h}{8}[(y_0^2 + y_6^2) + 3(y_1^2 + y_2^2 + y_4^2 + y_5^2) + 2(y_3^2)]$$

$$= \frac{3\pi}{16}[(2^2 + 2.1^2) + 3(2.4^2 + 2.7^2 + 3^2 + 2.6^2) + 2(2.8)^2]$$

$$= \frac{3\pi}{16}[(4 + 4.41) + 3(5.76 + 7.29 + 9 + 6.76) + 2(7.84)]$$

$$= \frac{3\pi}{16}(8.41 + 86.43 + 15.68) = \frac{3\pi}{16} \times 110.52 = 65.13 \text{ cubic units.}$$

EXAMPLE 5.77 Compute $\int_4^{5.2} \log_e x\, dx$, using Weddles rule and compare the results, with its true value.

Solution We divide the interval of integration into six equal parts each of width $= 5.2 - 4.0/6 = 0.2$. The values of the function $y = \log_e x$ are computed for each point of subdivision. These vales are given in the table below:

x:	4.0	4.2	4.4	4.6	4.8	5.0	5.2
$y = \log_e x$:	1.38629	1.43508	1.48160	1.52606	1.56862	1.60944	1.64866

By Weddles rule, we have

$$I_w \approx \frac{3h}{10}[(y_0 + y_6) + 5(y_1 + y_5) + (y_2 + y_4) + 6(y_3)]$$

$$= \frac{3 \times (0.2)}{10}[(1.38629 + 1.64866) + 5(1.43508 + 1.60944) + (1.48160 + 1.56862) + 6 \times 1.52606]$$

$$= \frac{0.6}{10}[3.03495 + 5 \times 3.04452 + 3.05022 + 6 \times 1.52606]$$

$$= \frac{6}{100}[3.03495 + 15.2226 + 3.05022 + 9.15636) = \frac{6}{100} \times 30.46413 = 1.8278478$$

The true value of the integral is

$$I_t = \int_4^{5.2} \log_e x\, dx = [x(\log_e x - 1)]_4^{5.2} = 5.2(0.64866) - 4(0.38629) = 3.373032 - 1.54516 = 1.827872.$$

Remark 15 If the value of the function $y = \log_e x$ are computed for each point of sub-division upto 8th place of decimals, we have $I_w = 1.82784741$. The true value of the integral $I_t = 1.82784741$. Hence the error $E_w = 0$.

EXAMPLE 5.78 Compute the value of the integral $I = \int_{0.2}^{1.4} (\sin x - \log_e x + e^x)\, dx$ by (*i*) trapezoidal rule (*ii*) Simpson's 1/3-rule (*iii*) Simpson's 3/8-rule (*iv*) Weddle's rule.

Solution We shall divide the interval of integration into twelve equal parts by taking $h = (1.4 - 0.2)/12 = 0.1$.

The value of the function $y = \sin x - \log_e x + e^x$ are then computed for each point of subdivision. These values are given in the table below. The following table gives the values of y for different values of x.

x:	0.2	0.3	0.4	0.5	0.6	0.7	0.8	0.9	1.0	1.1	1.2	1.3	1.4
y:	3.02950	2.84935	2.79753	2.82130	2.89759	3.01464	3.16604	3.34830	3.55975	3.80008	4.06984	4.37050	4.70418

(*i*) By trapezoidal rule

$$I_T \approx \frac{h}{2}[y_0 + 2(y_1 + y_2 + \cdots + y_{n-1}) + y_n]$$

$$= \frac{0.1}{2}[(3.02950 + 4.70418) + 2(2.84935 + 2.79753 + 2.82130 + 2.89759 + 3.01464 + 3.16604 + 3.34830 + 3.55975 + 3.80008 + 4.06984 + 4.37050)]$$

$$= \frac{1}{20}[7.73368 + 2(36.69492)] = \frac{1}{20}(81.12352) = 4.05617$$

(*ii*) By Simpson's 1/3-rule,

$$I_s \approx \frac{h}{3}[(y_0 + y_n) + 4(y_1 + y_3 + \cdots + y_{n-1}) + 2(y_2 + y_4 + \cdots + y_{n-2})]$$

$$= \frac{0.1}{3}[(3.02950 + 4.70418) + 4(2.84935 + 2.82130 + 3.01464 + 3.34830 + 3.80008 + 4.37050) + 2(2.79753 + 2.89759 + 3.16604 + 3.55975 + 4.06984)]$$

$$= \frac{1}{30}[7.73368 + 4(20.20417) + 2(16.49075)]$$

$$= \frac{1}{30}[7.73368 + 80.81668 + 32.9815] = \frac{121.53186}{30} = 4.05106.$$

(*iii*) By Simpson's 3/8-rule

$$I_s \approx \frac{3h}{8}[(y_0 + y_n) + 3(y_1 + y_2 + y_4 + y_5 + \cdots y_{n-2} + y_{n-1}) + 2(y_3 + y_6 + \cdots + y_{n-3})].$$

$$= \frac{3(0.1)}{8}[(3.02950 + 4.70418) + 3(2.84935 + 2.79753 + 2.89759 + 3.01464 + 3.34830 + 3.55975 + 4.06984 + 4.37050) + 2(2.82130 + 3.16604 + 3.80008)$$

$$= \frac{0.3}{8}[7.73368 + 3(26.90750) + 2(9.78742)] = \frac{0.3}{8}(108.03102) = 4.05116$$

(*iv*) By Weddle's rule

$$I_w \approx \frac{3h}{10}[(y_0 + y_n) + 5(y_1 + y_5 + y_7 + y_{11} \cdots + y_{n-5} + y_{n-1}) + (y_2 + y_4 + y_8 + y_{10} + \cdots y_{n-4} + y_{n-2})$$
$$+ 2(y_6 + \cdots y_{n-6} + 6(y_3 + y_9 + \cdots y_{n-3})]$$

Therefore, $I_w = \dfrac{3(0.1)}{10}[7.73368 + 5(13.58879) + 13.32471 + 2(3.16604) + 6(6.62138)] = 4.05098$.

Remark 16 The true value of the integral is 4.05095. We see that trapezoidal rule is the least and Weddle's rule is the most accurate among all the above stated four rules.

EXAMPLE 5.79 Estimate the length of the arc of the curve $y = x^3/3$ from (0, 0) to $\left(1, \frac{1}{3}\right)$ using Simpson's 1/3-rule, taking 8 sub-interval.

Solution We have $a = 0$, $b = 1$. For $n = 8$, we get $h = (1 - 0)/8 = 0.125$. The abscissas arc $x_0 = 0$, $x_1 = 0.125$, $x_2 = 0.250$, $x_3 = 0.375$, $x_4 = 0.500$, $x_5 = 0.625$, $x_6 = 0.750$, $x_7 = 0.875$, $x_8 = 1.000$.

$$\text{Length of the arc} = \int_0^1 \sqrt{1 + \left(\frac{dy}{dx}\right)^2}\, dx.$$

We have $y = x^3/3$. Therefore, $\dfrac{dy}{dx} = x^2$, $ds = \sqrt{1 + \left(\dfrac{dy}{dx}\right)^2} \cdot dx$

$s = \displaystyle\int_0^1 \sqrt{1 + \left(\frac{dy}{dx}\right)^2}\, dx.$ The table of the values is given below:

x:	0	0.125	0.250	0.375	0.500	0.625	0.750	0.875	1.000
$\sqrt{1 + \left(\frac{dy}{dx}\right)^2} = \sqrt{1 + x^4}$:	1	1.000122	1.001951	1.009839	1.030776	1.073586	1.147347	1.259437	1.414214

$$\text{Therefore, } s = \int_0^1 \sqrt{1 + \left(\frac{dy}{dx}\right)^2}\, dx \approx \frac{h}{3}[(y_0 + y_8) + 4(y_1 + y_3 + y_5 + y_7) + 2(y_2 + y_4 + y_6)]$$

$$= \frac{0.125}{3}[(1 + 1.414214) + 4(1.000122 + 1.009839 + 1.073586 + 1.259437)$$
$$+ 2(1.001951 + 1.030776 + 1.147347)]$$

$$= \frac{1}{24}[2.414214 + 4(4.342984) + 2(3.180074)]$$

$$= \frac{1}{24}[2.414214 + 17.371936 + 6.360148] = \frac{1}{24}(26.1463) = 1.08943 \text{ units.}$$

EXERCISE 5.8

1. Evaluate the following definite integrals using trapezoidal rule.

(*a*) $\displaystyle\int_2^6 \frac{dx}{x}$ with $n = 4$, (*b*) $\displaystyle\int_1^3 \log_e x\, dx$ with $n = 5$, (*c*) $\displaystyle\int_0^1 \sqrt{1 + x^3}\, dx$ with $n = 5$,

(d) $\int_1^5 \sqrt{35+x}\,dx$ with $n = 4$, (e) $\int_0^6 \frac{dx}{1+x^2}$ with $n = 6$, (f) $\int_1^5 \log_{10} x\,dx$ with $n = 8$.
(AMIE-W-2010)

Ans. (a) 1.117, (b) 1.2870, (c) 1.115, (d) 24.654, (e) 1.4108.

2. (a) Evaluate $\int_0^1 \sqrt{1-x^3}\,dx$ by trapezoidal rule, taking six equal intervals correct to three decimal places. *(AMIE, W-2004)*

(b) Consider the function $f(x)$ given by

x:	0.1	0.2	0.3	0.4	0.5	0.6
f(x):	0.425	0.475	0.400	0.450	0.575	0.675

Use composite trapezoidal rule to evaluate the integral $\int_{0.1}^{0.6} [f(x)]dx$. *(AMIE, S-2008)* **Ans.** 0.245.

3. Use Simpson's 1/3-rule to approximate the integral $\int_0^2 x^3 dx$ with $n = 4$. **Ans.** 4.

4. Evaluate by Simpson's 1/3-rule, $\int_0^1 \frac{dx}{1+x}$ taking four number of sub-intervals and compare the result with the exact value. *(AMIE, W-2008; W-2006)* **Ans.** 0.69323; exact value = 0.693147.... .

5. Given that $e^{\circ} = 1$, $e^1 = 2.72$, $e^2 = 7.39$, $e^3 = 20.09$, $e^4 = 54.60$. Find an approximate value of $\int_0^4 e^x\,dx$ by (a) the trapezoidal rule (b) Simpson's 1/3-rule. Compare it with the actual value.

Ans. (a) 58.00 (b) 53.8733; Actual value = 53.5982

6. Using the following data and the Simpson's 1/3 -rule determine the approximate value of $\int_0^6 y\,dx$.

x:	0	1	2	3	4	5	6
y:	0.146	0.161	0.176	0.190	0.204	0.217	0.230

Ans. 1.136

7. (a) Using Simpson's 1/3-rule, calculate the value of the integral $\int_1^9 y\,dx$, for two, four and eight intervals.

x:	1	2	3	4	5	6	7	8	9
y:	2.061	2.312	2.819	3.106	3.670	4.721	6.103	7.95	9.942

(b) From the following data, estimate $\int_{1.1}^{1.9} f(x)\,dx$ as best as you can

x:	1.1	1.2	1.3	1.4	1.5	1.6	1.7	1.8	1.9
y:	3.95254	4.06587	4.18162	4.29982	4.42051	4.54372	4.66949	4.80135	5.02365

Ans. (a) 35.577; 36.687; 36.514, (b) 3.54541

8. (*a*) Approximate $\int_{-4}^{4} e^{-x^2} dx$ by Simpson's 1/3 -rule, using intervals with $h = 0.5$. (*AMIE, W-2001*)

(*b*) Calculate by Simpson's 1/3 -rule an approximate value of $\int_{-3}^{3} x^4 dx$ by taking seven equidistant ordinates. Compare it with exact value. **Ans.** (*a*) 1.77241, (*b*) 98; 97.2.

9. Using Simpson's 1/3 -rule with four sub-intervals, calculate the value of the integral $\int_{1.2}^{1.6} (x + x^{-1}) dx$.

(*AMIE, W-2002*) **Ans.** 0.84768.

10. (*a*) Using Simpson's 1/3 -rule, estimate the area bounded by the curve $y = f(x)$, x-axis and ordinates $x = 0$ and $x = 5$ from the following data:

x_i :	0	0.5	1	1.5	2	2.5	3	3.5	4	4.5	5
$f(x_i)$:	1	2.1	3.2	4	5.1	6.2	7.2	8.3	9	10.2	12

(*AMIE, W-2005*) **Ans.** 30.867.

(*b*) The following table gives the values of a function at equal intervals.

x:	0.0	0.5	1.0	1.5	2.0
$f(x)$:	0.3989	0.3521	0.2420	0.1295	0.0540

Evaluate $f(1.8)$, $f'(1.5)$ and $\int_{0}^{2} f(x) dx$ stating the formula used. **Ans.** 0.0782, –0.1868, 0.4772.

11. Evaluate $\int_{0}^{1} \frac{dx}{1+x^2}$ approximately, using (*a*) the trapezoidal rule, (*b*) Simpson's 1/3 -rule with five ordinates. (*AMIETE, June 2009*) **Ans.** (*a*) 0.7828; (*b*) 0.7854.

12. (*a*) Evaluate the value of $\log_e 2$ by finding $\int_{0}^{1} \frac{2x}{1+x^2} dx$, using Simpson's 1/3 rule, by taking $n = 4$.

(*b*) Find an approximate value of $\log_e 5$ by evaluating the integral $\int_{0}^{5} \frac{dx}{4x+5}$, by Simpson's 1/3 -rule, dividing the range into ten equal parts. (*AMIE, S-2007; Madras 2000*)

Ans. (*a*) 0.6933, (*b*) 1.61.

13. Find an approximate value of $\int_{0}^{\pi/2} \sqrt{\cos\theta}\, d\theta$, using Simpson's 1/3 -rule with six intervals. **Ans.** 1.1872

14. Using Simpson's 1/3 -rule evaluate the following:

(*a*) $\int_{0}^{4} \sqrt{x^3 + 3x}\, dx$, with 4 intervals, (*b*) $\int_{0}^{\pi/2} \sqrt{2 + \sin x}\, dx$, with 7 ordinates,

(*c*) $\int_{0}^{\pi} \frac{\sin^2\theta}{5 + 4\cos\theta} d\theta$, with 5 ordinates, (*d*) $\int_{200}^{1000} \frac{dx}{\log_{10} x}$, taking $n = 8$, (*e*) $\int_{4}^{5.2} \log_e x\, dx$.

(*f*) $\int_{0}^{1} e^{-x^2} dx$, by taking $h = 0.1$. (*AMIE - S, 2000*)

Ans. (*a*) 12.07, (*b*) 2.545, (*c*) 0.388, (*d*) 293.4, (*e*) 1.82785, (*f*) 0.746826.

15. A river is 80 unit wide. The depth d at a distance x unit from one bank is given by the following table:

x:	0	10	20	30	40	50	60	70	80
d:	0	4	9	7	12	15	14	8	3

Find approximate area of the cross-section of the river by Simpson's 1/3 -rule.

Ans. 696.67 sq. units.

16. The ordinates of a curve $y = f(x)$ for given values of x are tabulated below:

x:	1	2	3	4	5	6	7
y:	2.105	2.808	3.614	4.604	5.857	7.451	9.467

Using Simpson's 1/3 -rule evaluate the area bounded by the curve, the ordinates at $x = 1$ and $x = 7$ and the x-axis. **Ans.** 29.99 sq. units.

17. The following table gives the velocity v of a particle at time t

t(seconds):	0	2	4	6	8	10	12
v(m/sec):	4	6	16	34	60	94	136

Find the distance moved by the particle in 12 seconds and also the acceleration at $t = 2$ sec.

Ans. 552 m; 3 m/sec^2.

18. A solid of revolution is formed by rotating about the x-axis the area between the x-axis, the lines $x = 0$ and $x = 1$, and a curve through the points with the following coordinates:

x:	0.00	0.25	0.50	0.75	1.0
y:	1.000	0.9896	0.9589	0.9089	0.8415

Using Simpson's 1/3 -rule, estimate the volume of the solid formed, giving the answer to three decimal places. *(AMIE, S-2006)* **Ans.** 2.819 cubic units.

19. A curve is given by the following pairs of rectangular coordinates. Using Simpson's 1/3 -rule, find the volume generated by the revolution of the area under the curve from $x = 1$ to $x = 9$ about the x-axis.

x:	1	2	3	4	5	6	7	8	9
y:	0.0	0.6	0.9	1.2	1.5	1.4	1.7	1.8	2.0

Ans. 14.83 π cubic units.

20. A curve is given by the table:

x:	0	1	2	3	4	5	6
y:	0.0	2.0	2.5	2.3	2.0	1.7	1.5

Find by Simpson's 1/3 -rule, volume of solid of revolution about x-axis. **Ans.** 74.873 units^3.

21. Evaluate $\int_{0.2}^{1.4} (e^x + \sin x)\,dx$, using Simpson's 1/3 -rule of integration. Taking $n = 12$. **Ans.** 3.64389.

[Hint: Here $h = (1.4 - 0.2) / 12 = 0.1$. The following table as prepared with increasing value of x by 0.1.

x:	0.2	0.3	0.4	0.5	0.6	0.7	0.8	0.9	1.0	1.1	1.2	1.3	1.4
$y = f(x)$:	1.4201	1.64536	1.88122	2.12812	2.38672	2.65795	2.94294	3.2429	3.55978	3.89537	4.25212	4.6329	5.0406

Hence,

$$\int_{0.2}^{1.4}(e^x+\sin x)\,dx \approx \frac{0.1}{3}\{(1.4201+5.0406)+4(1.64536+2.12812+2.65795+3.2429+3.89537+4.6329)$$

$$+2(1.88122+2.38672+2.94294+3.55978+4.25212)\}$$

$$=\frac{0.1}{3}\{6.4607+4(18.2026)+2(15.02278)\}=3.64389.]$$

22. Using the data of the following table, compute the integrals

(*a*) $\int_{0.5}^{1.1} xy\,dx$, (*b*) $\int_{0.5}^{1.1} y^2 dx$, (*c*) $\int_{0.5}^{1.1} x^2 y\,dx$, (*d*) $\int_{0.5}^{1.1} y^3 dx$ by Simpson's 1/3 -rule.

x:	0.5	0.6	0.7	0.8	0.9	1.0	1.1
y:	0.4804	0.5669	0.6490	0.7262	0.7985	0.8658	0.9281

Ans. (*a*) 0.3585, (*b*) 0.3201, (*c*) 0.3104, (*d*) 0.2444

23. The following table gives the values of a function at equal intervals

x:	0.0	0.5	1.0	1.5	2.0
f(*x*):	0.3989	0.3521	0.2420	0.1295	0.0540

Evaluate $f(1.8), f'(1.5)$ and $\int_0^2 f(x)\,dx$. **Ans.** 0.0782; – 0.1868; 0.4772 (Simpson's 1/3 -rule)

24. The value of $\log_e 2$ can be calculated from the integral $\log_e 2=\int_1^2 \frac{1}{x}dx$. How many subintervals (steps) should be used in the trapezoidal rule to approximate the integral with an error of magnitude less than 10^{-4}? **Ans.** $n = 41$.

25. Use Simpson's 3/8 – rule to obtain an approximate value of the following definite integral:

(*a*) $\int_0^{0.3}(1-8x^3)^{1/2}\,dx$, (*b*) $\int_0^1 \frac{dx}{1+x^2}$.

Ans. (*a*) 0.29159, (*b*) 3.1312.

26. A curve is drawn to pass through the points given by the following table:

x:	1.0	1.5	2.0	2.5	3.0	3.5	4.0
y:	2.0	2.4	2.7	2.8	3.0	2.6	2.1

Calculate:

(*a*) the area bounded by the curve, the *x*-axis and the ordinates $x = 1$ and $x = 4$.

(*b*) the coordinates of the centroid of the area.

(*c*) the volume of solid of revolution of the area about *x*-axis.

(*d*) the radius of gyration of the area about *x*-axis.

Ans. (*a*) 7.78 sq. units; (*b*) (2.53, 1.31); (*c*) 64.07 cub. units; (*d*) $k = 1.1358$

CHAPTER 6

Statistical Methods

6.1 INTRODUCTION

Statistics may be defined as the scientific method that is employed for collecting, presenting, analysing and interpreting data leading finally to drawing statistical inferences about some important characteristics. The raw material of statistics is the data which is the quantitative/qualitative informations about some particular characteristic under consideration. The data may be the measurements of the dimensions of a mass produced article in certain industry, collected for the purpose of quality control, or it may be the marks of a number of students collected to determine their performance.

When the data consists of a few figures, it can be easily presented and understood. But when the number of figures is very large, a proper classification is essential for analysis and deriving valid inferences.

Variable: A variable may be defined as "A factor or character, which can take different values." If the variable can assume only one value it is called a *constant*. A *variable* is classified into two types:

(*a*) Discrete variable (*b*) Continuous variable

Discrete Variable

A variable which takes a whole number value, a finite number of times is called a *discrete variable*. Discrete values are generally whole numbers, for examples, test scores, family size (the number N of children in a family which can assume any of the values 0, 1, 2, 3, ..., but cannot be 2.5 or 3.74, number of employees, number of goals in fotball matches etc.

Continuous Variable

A variable which can take any numerical value between two given values is called a *continuous variable*. For example, height of persons (is given as 1.72, 1.84, 1.91 m etc,) or weight of persons (50.4, 58.3, 68.9 kgm etc,) distance, temperature, rainfall per day, yield of a crop per hectare etc.

Frequency

The number of times a variate value is repeated in a collection of data is known as *frequency* of that variate value.

6.2 FREQUENCY DISTRIBUTION

A frequency distribution may be defined as a tabular representation of statistical data with their corresponding frequencies. When the number of observations for a variable is small and we have repetitions of the values, we can arrange them in the form of a table according to their magnitudes with corresponding frequencies. For example if the variable values are:

18, 12, 9, 11, 12, 11, 6, 18, 19, 11, 12, 9, 3, 11, 8, 9, 11, 9, 8, 12,

we can write down the frequency distribution as:

Table 6.1: Frequency distribution

Variable value (x_i):	3	6	8	9	11	12	18	19
Frequency (f_i):	1	1	2	4	5	4	2	1

When the number of observations is large and the difference between the largest value and smallest value is large, we have classification according to class interval. Let us illustrate this with an example.

Let the marks of 50 students of a class out of 100 be:

46, 58, 54, 52, 55, 59, 52, 62, 65, 67, 64, 63, 77, 78, 92,
6, 7, 12, 18, 16, 3, 23, 25, 25, 27, 81, 88, 24, 29, 22,
34, 33, 30, 37, 36, 42, 48, 28, 22, 28 17, 13, 70, 37, 32,
36, 41, 40, 43, 44.

Here the largest mark is 92 and smallest is 3. We take 9 class intervals. Therefore, the width of the class interval can be calculated as (largest value – smallest value)/number of class intervals.

Thus, width of class interval = $(92-3)/9 \approx 10$. Therefore, the classes are 3 – 12, 13 – 22, 23 – 32, 33 – 42, 43 – 52, 53 – 62, 63 – 72, 73 – 82, 83 – 92.

Present the class interval in a table known as *frequency distribution table.*

Apply 'tally mark', that is, a stroke against the occurrence of a particular value in a class interval. After every four lines, we cross them for the 5th entry. Count the tally marks and present these numbers in the next column, known as frequency column, and finally check whether the total of all these frequencies tally with the total number of observations.

Table 6.2: Frequency distribution

Class	Tally marks	Frequency
3–12	𝍷𝍷𝍷𝍷	4
13–22	𝍸 𝍷	6
23–32	𝍸 𝍸	10
33–42	𝍸 𝍷𝍷𝍷𝍷	9
43–52	𝍸 𝍷	6
53–62	𝍸	5
63–72	𝍸	5
73–82	𝍷𝍷𝍷	3
83–92	𝍷𝍷	2
		Total= 50

6.2.1 Some important terms associated with a frequency distribution

Class Limit (CL)

Corresponding to a class interval, the class limits may be defined as the minimum value and the maximum value the class interval may contain. The minimum value is known as the lower class limit (LCL) and the maximum value is known as the upper class limit (UCL). For the above table of frequency distribution, the LCL and UCL of the fourth class interval are 33 and 42 respectively.

Class Boundary (CB)

Class boundaries may be defined as the actual class limit of a class interval. For *overlapping or mutually exclusive type* that excludes the upper class limits like 40–50, 50–60, 60–70, ..., etc. the class boundaries coincide with the class limits. However, for *non-overlapping* or mutually *inclusive type* that includes both the class limits like 16–25, 26–35, 36–45, ..., etc., we have LCB = LCL – $(D/2)$ and UCB = UCL + $(D/2)$, where D is the difference between the LCL of the next class interval and the UCL of the given class interval. For the data presented in the above table LCB of the first class interval = $3 - \{(13-12)/2\} = 2.5$ and the corresponding UCB = $12 + \{(13-12)/2\} = 12.5$.

Class Mark

The class mark is the mid point or mid value of the class limits or class boundaries and is obtained by adding the lower and upper limit or LCB and UCB and dividing by two.

Thus, we have class mark = $\dfrac{\text{LCL} + \text{UCL}}{2} = \dfrac{\text{LCB} + \text{UCB}}{2}$.

Refer Table 6.3, the class mark for the first two class intervals are $(3+12)/2$ and $(13+22)/2$ that is, 7.5 and 17.5 respectively.

Width or size of a class interval

The width of a class interval may be defined as the difference between the UCB and the LCB of that class interval. Refer Table 6.3, the class width for the third class interval is $C = 32.5 - 22.5 = 10$. For the other class intervals also, C remain same.

Remark 1

(a) Classes should be continuous that is, the upper limit of the preceding class should be the lower limit of the following class. If it is not then it should be made continuous, that is, inclusive type should be converted to exclusive type.

(b) Border line cases are taken in the higher class. For example a student getting 30 marks belongs to the class 30–40 and not 20–30.

Table 6.3

Mark (Class Interval)	Mark (Class Boundary)	Mid-value (Class Mark)	No. of students (Frequency)
3–12	2.5–12.5	7.5	4
13–22	12.5–22.5	17.5	6
23–32	22.5–32.5	27.5	10
33–42	32.5–42.5	37.5	9
43–52	42.5–52.5	47.5	6
53–62	52.5–62.5	57.5	5
63–72	62.5–72.5	67.5	5
73–82	72.5–82.5	77.5	3
83–92	82.5–92.5	87.5	2
			Total = 50

Cumulative Frequency

The total frequency of all values less than the upper class boundary of a given class interval is called the *cumulative frequency* upto and including the class interval. For example, the cumulative frequency upto and including the class interval 43–52 in Table 6.3 is $4 + 6 + 10 + 9 + 6 = 35$, signifying that 35 students have marks less than 52.5. This definition refers to the less that cumulative frequency. We can define more that cumulative frequency in a similar manner. Both types of cumulative frequencies are shown in Table 6.5. Whenever we refer to cumulative distribution or ogives without qualification, the "less than" type is implied.

EXAMPLE 6.1 Following are the weights in kgm of 36 students.

49, 61, 70, 73, 61, 47, 57, 59, 50, 68, 79, 55, 45, 65, 68, 56, 68, 70,
55, 70, 57, 69, 44, 64, 73, 49, 63, 70, 65, 62, 65, 64, 73, 67, 50, 60.

Construct a frequency distribution and cumulative frequency distribution of weights of 36 students, taking class length as 5.

Solution We have, Range = highest weight – lowest weight = (73 – 44) kgm = 29 kgm.

Number of class interval × class lengths ≈ Range.

or Number of class interval = 29/5 ≈ 6 (we always take the next integer as the number of class interval).

Table 6.4: Frequency distribution of weight of 36 students

Weight in kgm (Class Interval)	Weight in kgm (Class Boundary)	Mid-value (Class mark)	Tally marks	No. of students (Frequency)
44–48	43.5–48.5	46	|||	3
49–53	48.5–53.5	51	||||	4
54–58	53.5–58.5	56	~~||||~~	5
59–63	58.5–63.5	61	~~||||~~ ||	7
64–68	63.5–68.5	66	~~||||~~ ||||	9
69–73	68.5–73.5	71	~~||||~~ |||	8
				Total = 36

Table 6.5: Cumulative Frequency Distribution of weights of 36 students

Weight in kgm (Class Boundary)	Cumulative Frequency	
	Less than	More than
43.50	0	36 – 0 = 36
48.50	0 + 3 = 3	36 – 3 = 33
53.50	3 + 4 = 7	36 – 7 = 29
58.50	7 + 5 = 12	36 – 12 = 24
63.50	12 + 7 = 19	36 – 19 = 17
68.50	19 + 9 = 28	36 – 28 = 8
73.50	28 + 8 = 36	36 – 36 = 0

6.3 GRAPHICAL REPRESENTATION OF A FREQUENCY DISTRIBUTION

We consider the following types of graphical representation of frequency distribution:

(*a*) Histogram or Area diagram; (*b*) Frequency Polygon; (*c*) Ogives or Cumulative Frequency graphs.

(*a*) **Histogram or Area diagram:** Histogram helps us to get an idea of the frequency curve of the variable under study. In order to draw a histogram, the class limits are first converted to the corresponding class boundaries and a series of adjacent rectangles, one against each class interval, with the class interval as base on a horizontal axis (the x-axis) and the frequency (when the classes are of equal width, as is generally the case) as height on a vertical axis (the y-axis) is errected. We now construct histogram for the weight distribution of 36 students. Refer Ex. 6.1 and Fig. 6.1. The histogram for the distribution of weight of 36 students is shown in Fig. 6.1, using Table 6.4. The mode of the weights has also been determined using the histogram. Mode = 66.50 kgms.

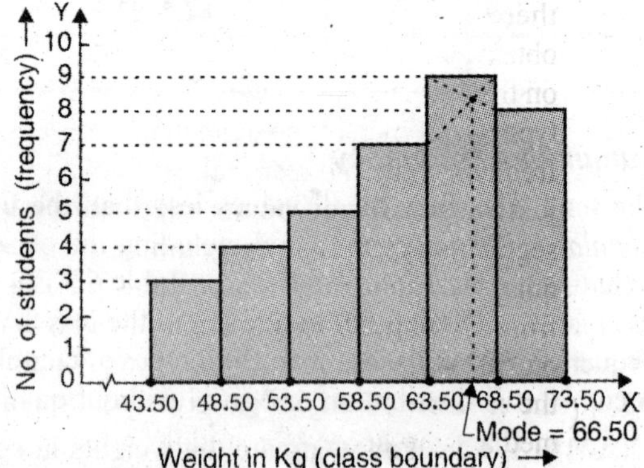

Fig. 6.1: Showing histogram for the distribuiton of weight of 36 students

(b) **Frequency Polygon:** A frequency polygon is a line graph of class frequency plotted against class mark. It is customary to add the extensions AF and EG to the next lower and higher class marks which have corresponding class frequency of zero. The frequency polygon for the distribution of weights of 36 students is shown in Fig. 6.2. We can also obtain a frequency polygon starting with a histogram by connecting mid-points of the tops of the rectangles successively and then completing the figure by joining the two ends AF and EG as before. The histogram and frequency polygon corresponding to the above frequency distribution of weights of 36 students are shown on the same set of axes in Fig 6.3.

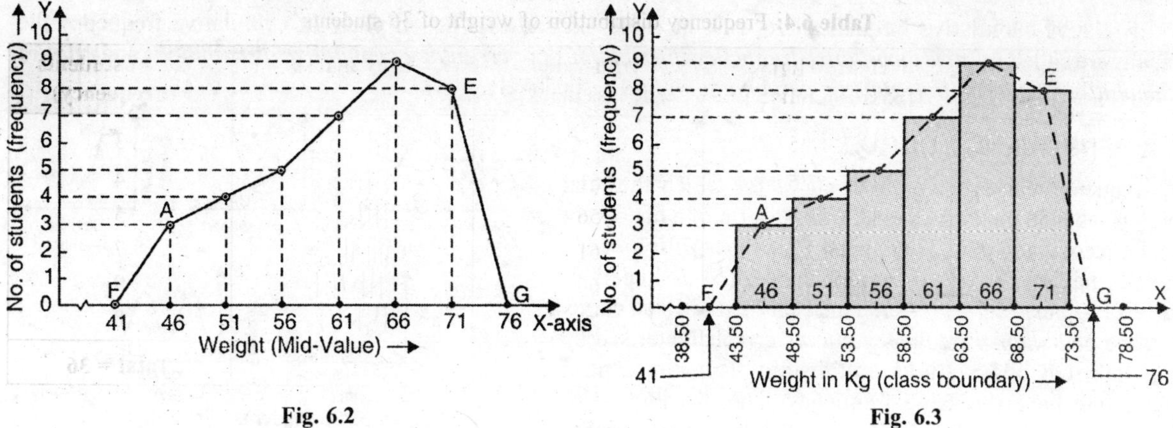

Fig. 6.2

Fig. 6.3

Relative Frequency Distributions

The *relative frequency* of a class interval may be defined as the ratio of the class frequency to the total frequency. Percentage frequency of a class interval may be defined as the ratio of class frequency to the total frequency, expressed as a percentage. For example, the percentage frequency of the class 59–63 in Table 6.4 is $(7/36) \times 100 = 19.44\%$. The sum of the relative frequencies of all classes is clearly 1 and percentage frequencies add up to or 100. If frequencies in the above frequency table are replaced by corresponding relative frequencies, the resulting table is called a *relative frequency distribution*, or *relative frequency table*. Graphical representations of relative frequency distributions can be obtained from the histogram or frequency polygon by simply changing the vertical scale from frequency to relative frequency, keeping exactly the same diagram. The resulting graphs are called *relative frequency histograms* or *percentage histograms* and *relative frequency polygons* or *percentage polygons*, respectively.

(c) Ogives or Cumulative Frequency Graph: By plotting cumulative frequency against the respective class boundary, we get ogives. As such there are two ogives–less than type ogives, obtained by taking less than cumulative frequency on the vertical axis (that is y-axis) and more than type ogives by plotting more than type cumulative frequency on the vertical axis and thereafter joining the plotted points successively by line segments. From ogives graph, we may obtain quartiles graphically. If a perpendicular is drawn from the point of intersection of the two ogives on the horizontal axis (that is, on x-axis), then the x-value of this point gives us the value of median, the second or middle quartile. Ogives graph further give us the following information

Ist quartile or lower quartile (Q_1) = 55 kgm

2nd quartile or median (Q_2 or Me) = 62.5 kgm

3rd quartile or upper quartile (Q_3) = 68 kgm.

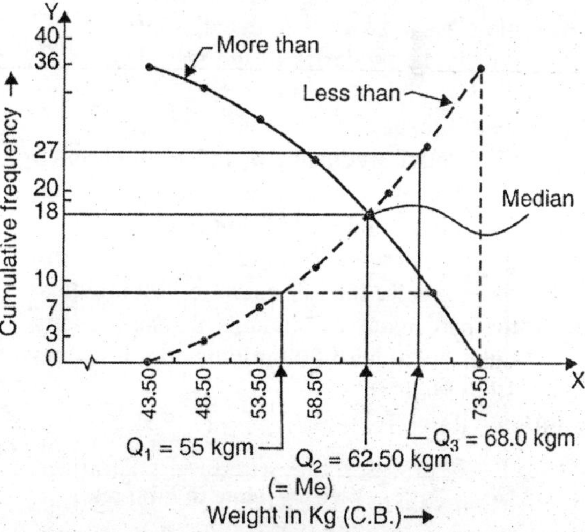

Fig. 6.4: Showing the ogives for the distribution of weights of 36 students. Refer Ex. 6.1 and Table 6.5

Relative Cumulative Frequency and Percentage Cumulative Frequency

The *relative cumulative frequency* is the cumulative frequency divided by the total frequency. For example, the relative cumulative frequency of weight less than 63.50 kgm is $(19/36) = 0.5278$. Percentage cumulative frequency is the ratio of cumulative frequency divided by total frequency, expressed as a percentage.

In the above case, the percentage cumulative frequency is 52.78% signifying that 52.78% of the students have weights less than 63.50 kgm.

If relative cumulative frequency are used in Table 6.5 and Fig. 6.4 in place of cumulative frequency the results are called *relative cumulative frequency distributions* or *percentage cumulative distributions* and *relative cumulative frequency polygons* or *percentage ogives* respectively.

6.4 FREQUENCY CURVE

A frequency curve is a smooth curve for which the total area is taken to be unity. It is a limiting form of a histogram or frequency polygon. There exist four types of frequency curves namely: (*a*) Bell-shaped curve, (*b*) *J*-shaped or reverse J-shaped curve, (*c*) *U*-shaped curve, (*d*) Mixed curve.

The *bell-shaped* frequency curves are characterised by the fact that observations equidistant from the central maximum have the same frequency. The distribution of height, weight, mark, profit, etc. usually belong to the category. On a bell-shaped curve, the frequency, starting from a low value, gradually reaches the maximum value, somewhere near the central part and then gradually decreases to reach its lowest value at the other extremity.

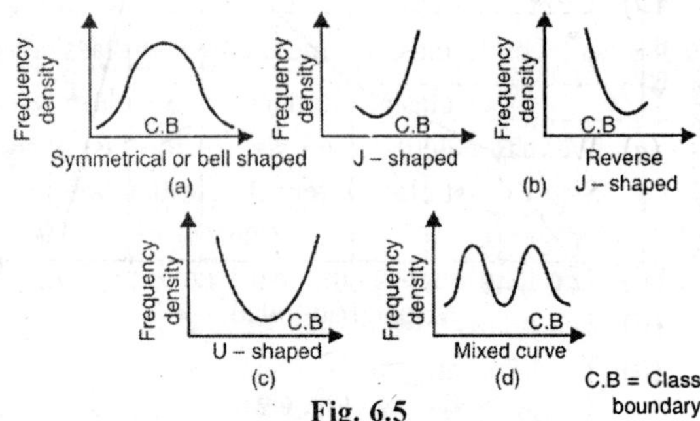

Fig. 6.5

The *J-shaped* or *reverse J-shaped curve* starts with a minimum frequency and then gradually reaches its maximum frequency at the other extremity. The distribution of commuters coming to metro station from the early morning hour to peak morning hours follows such a distribution. For a *U-shaped curve*, the frequency is minimum near the central part and the frequency slowly but steadily reaches its maximum at the two extremities. The traffic rush during the peak hours in the morning and in the evening belongs to this type of curve.

Lastly, we may have a combination of these frequency curves, known as mixed curve. These frequency curves are exhibited in the Fig 6.5.

EXAMPLE 6.2 With references to the Table 6.6 determine:

(*a*) the upper limit of the fifth class,
(*b*) the lower limit of the eighth class,
(*c*) the class mark of the seventh class,
(*d*) the class boundaries of the last class,
(*e*) the class interval size,
(*f*) the frequency of the fourth class,
(*g*) the relative frequency of the sixth class
(*h*) the percentage of students whose viewing time does not exceed 600 minutes,
(*i*) the percentage of students with viewing times greater than or equal to 900 minutes,
(*j*) the percentage of students whose viewing times are at least 500 minutes but less than 1000 minutes.

Table 6.6

S.No.	Viewing Time (in minutes) per week	Number of students
1	300–399	14
2	400–499	46
3	500–599	58
4	600–699	76
5	700–799	68
6	800–899	62
7	900–999	48
8	1000–1099	22
9	1100–1199	6
		Total = 400

Solution Refer Table 6.7.

Table 6.7

S.No.	Viewing Time in minutes (Class Interval)	Viewing Time in minutes (Class Boundary)	Number of students (Frequency)
1	300–399	299.5–399.5	14
2	400–499	399.5–499.5	46
3	500–599	499.5–599.5	58
4	600–699	599.5–699.5	76
5	700–799	699.5–799.5	68
6	800–899	799.5–899.5	62
7	900–999	899.5–999.5	48
8	1000–1099	999.5–1099.5	22
9	1100–1199	1099.5–1199.5	6
			Total = 400

(*a*) The upper limit of the fifth class = 799 minutes.

(*b*) The lower limit of the eighth class = 1000 minutes.

(*c*) The class mark of the seventh class = $(900+999)/2 = 949.5$ minutes.

(*d*) The class boundaries of the last class.

Lower class boundary of last class = $(1100+1099)/2 = 1099.5$ minutes.

Upper class boundary of last class = $(1199+1200)/2 = 1199.5$ minutes.

(*e*) We may calculate the size of the last class interval.

Size of last class interval = upper boundary of last class – lower boundary of last class
= 1199.5 – 1099.5 = 100 minutes.

(*f*) The frequency of the fourth class = 76.

(*g*) The relative frequency of the sixth class = $62/400 = 0.155$ or 15.5%

(*h*) The percentage of students whose viewing time does not exceed 600 minutes.

Total number of students whose viewing time less than 600 minutes = 58 + 46 + 14 = 118.

Percentage of students whose viewing time does not exceed 600 minutes = $(118/400)\times 100 = 29.5\%$.

(*i*) Percentage of students with viewing times greater than or equal to 900 minutes.

Total number of students whose viewing time greater than or equal to 900 minutes = 48 + 22 + 6 = 76.

Percentage of students whose viewing time greater than or equal to 900 minutes = $(76/400)\times 100 = 19\%$.

(*j*) The percentage of students with viewing times are at least 500 minutes but less than 1000 minutes.

Number of students with viewing times less than 1000 minutes but at least 500 minutes = 48 + 62 + 68 + 76 + 58 = 312.

Percentage of students with viewing times less than 1000 minutes but at least 500 minutes = $(312/400)\times 100 = 78\%$.

6.5 MEASURES OF CENTRAL TENDENCY

With a few exceptions all frequency distributions show a clustering of data around some central value. This central value gives a representative value of the whole distribution and is described as a measure of central tendency or average. A good average should be rigidly defined, based on all the observations taken, simple to compute and it should be least affected by the presence of extreme observations. Following are the different measures of central tendency: (*i*) Arithmetic mean (A.M.) (*ii*) Median (Me) (*iii*) Mode (Mo) (*iv*) Geometric mean (G.M.) (*v*) Harmonic mean (H.M.). We shall briefly discuss them one by one.

6.6 ARITHMETIC MEAN (AM)

We are all familiar with this measure. It is so popular that whenever we talk of average we mean arithmetic average or arithmetic mean. For example if the daily wages of 5 labourers are ₹ 100, 120, 125, 130, 135 respectively then the average wage = ₹ (100 + 120 + 125 + 130 + 135)/5 = ₹ 122 per head.

Arithmetic mean is defined as the sum of the observations (Σx_i) divided by the number of observations (N) that is, $\bar{x}$ (read as x bar) = $\Sigma x_i / N$.

The symbol x_i denotes the observation corresponding to the i-th individual in the sample. If there are N individual, we write $i = 1, 2, 3, \ldots, N$.

(*i*) **Arithmetic Mean of ungrouped data:** The arithmetic mean of a set of N numbers $x_1, x_2, \ldots, x_N$ is denoted by $\bar{x}$ and is defined as $\bar{x} = \dfrac{x_1 + x_2 + \cdots + x_N}{N} = \sum_{i=1}^{N} x_i \Big/ N$ or $\dfrac{\Sigma x}{N}$. ...(6.1)

(*ii*) **Arithmetic Mean of grouped data:** When the values of the variable are given in the form of a frequency table then we multiply each value by the corresponding frequency and find their sum. This sum divided by the total of frequencies is the arithmetic mean, that is, $$\bar{x} = \frac{f_1x_1 + f_2x_2 + \cdots + f_kx_k}{f_1 + f_2 + \cdots + f_k}. \quad ...(6.2)$$

We can thus write Eq. (6.2) in the compact form $\bar{x} = \sum_{i=1}^{k} f_i x_i \Big/ \sum_{i=1}^{k} f_i = \frac{\Sigma fx}{\Sigma f} = \sum f x / N$

where $N = \Sigma f$ is the total frequency, that is, the total number of cases.

In the case of grouped frequency distribution that is when the values are given in the form of class intervals, x is the mid-value of the class interval and f the corresponding frequency.

Weighted Arithmetic Mean

Sometime we associate with the numbers $x_1, x_2, ..., x_n$ certain weighting factors or weights $w_1, w_2, ..., w_n$, depending on the significance or importance attached to the numbers. In this case

$$\bar{x}_w = \frac{w_1x_1 + w_2x_2 + \cdots + w_nx_n}{w_1 + w_2 + \cdots + w_n} = \frac{\sum_{i=1}^{n} w_i x_i}{\sum_{i=1}^{n} w_i} \quad ...(6.3)$$

is called the *weighted arithmetic mean.*

EXAMPLE 6.3 Find the average earning per labourer from the following data:

Daily earning (Rs.):	100	110	120	150	160
No. of labourers:	12	9	8	2	1

Solution Denoting the daily wages by x_i and the corresponding number of labourers by f_i, we construct the following table:

x_i	f_i	f_ix_i
100	12	1200
110	9	990
120	8	960
150	2	300
160	1	160
Total	$N = \Sigma f_i = 32$	$\Sigma f_i x_i = 3610$

Therefore, Arithmetic mean $= \frac{\Sigma f_i x_i}{\Sigma f_i} = \frac{3610}{32} \approx$ ₹113.

EXAMPLE 6.4 Find the arithmetic mean for the following frequency distribution:

Class interval:	0 to under 4	4 to under 8	8 to under 12	12 to under 16	16 to under 20
Frequency:	6	11	20	7	4

Solution We make the table as follows:

Class interval	Mid-values x_i	Frequency f_i	f_ix_i
0–4	(0 + 4)/2 = 2	6	12
4–8	(4 + 8)/2 = 6	11	66
8–12	(8 + 12)/2 = 10	20	200
12–16	(12 + 16)/2 = 14	7	98
16–20	(16 + 20)/2 = 18	4	72
		$\Sigma f_i = 48$	$\Sigma f_i x_i = 448$

Therefore, $\bar{x} = \Sigma f_i x_i / \Sigma f_i = 448/48 \approx 9.33.$

6.7 SHORT METHODS FOR COMPUTING THE ARITHMETIC MEAN

When the frequencies and the values of the variables in the series are given to be large quantities, their products necessarily come out to be large and consequently mathematical labour in the computation of the arithmetic average becomes enormous. To minimize this labour, the following short cut methods prove very handy.

(*a*) **Assumed mean method.**

$$\bar{x} = a + \frac{\sum_{i=1}^{k} f_i d_x}{N} \qquad \text{...(6.4)}$$

- 'a' is called the assumed mean or provisional mean. It is usually a value of the variable x which lies somewhere in the middle of the highest and the lowest value in the given data, preferably the value having the maximum frequency. The symbol d_x is called the deviation of x from a, that is, $d_x = x_i - a$.

(*b*) **Step deviation method**: When the values of the variables are given at equal spacing or when the length of the class intervals are equal, a further reduction in labour is achieved by using the step deviation method. This is well expressed mathematically by the formula.

$$\bar{x} = a + \left(\frac{\Sigma f_i u_x}{N}\right) h, \text{ where } u_x = \frac{x_i - a}{h} = \frac{d_x}{h} \qquad \text{...(6.5)}$$

is the step deviation and h, is the length of the class interval.

EXAMPLE 6.5 Trying to determine the calories per serving of diet in a laboratory assignment in nutrition, fifteen students obtained 329, 335, 347, 318, 322, 330, 351, 362, 315, 342, 346, 353, 316, 327 and 333. Find the mean:

Solution We prepare the following table by taking assumed mean or provisional mean, a = 340 calories. Deviation from a is x_i – 340. Here N = 15.

x_i	$(x_i - 340)$	x_i	$(x_i - 340)$
315	–25	342	+2
316	–24	346	+6
318	–22	347	+7
322	–18	351	+11
327	–13	353	+13
329	–11	362	+22
330	–10		
333	–7		
335	–5		
	Total: –135		**Total: 61**

$\Sigma(x_i - 340) = -135 + 61 = -74.$ Therefore, $\bar{x} = a + \frac{[\Sigma(x_i - a)]}{N} = 340 - \frac{74}{15} = 340 - 4.933 \approx 335.1$ calories.

EXAMPLE 6.6 Compute the arithmetic mean for the following distribution:

Class interval:	1 to under 4	4 to under 9	9 to under 16	16 to under 27
Frequency:	6	12	26	20

Solution We prepare the following table and apply formula (6.4). Taking assume mean, a = 12.5.

Size of items	Mid values x_i	f_i	$d_x = x_i - a$	$f_i d_x$
1–4	2.5	6	–10	–60
4–9	6.5	12	–6	–72
9–16	a = 12.5	26	0	0
16–27	21.5	20	9	180
Total		$N = \Sigma f_i = 64$		$\Sigma f_i d_x = 48$

The required arithmetic mean is given by $\bar{x} = a + \frac{\Sigma f_i d_x}{N} = 12.5 + \frac{48}{64} = 12.5 + 0.75 = 13.25.$

EXAMPLE 6.7 The marks (out of 100) obtained by different students in an examination are shown in the following frequency table. Compute the mean.

Marks:	17.5 to 22.5	22.5 to 27.5	27.5 to 32.5	32.5 to 37.5	37.5 to 42.5	42.5 to 47.5	47.5 to 52.5	52.5 to 57.5	57.5 to 62.5	62.5 to 67.5	67.5 to 72.5
No. of students:	2	8	33	80	170	243	213	145	67	35	4

Solution We shall solve this problem by step deviation method taking $a = 45$ and $h = 5$ (the size of class intervals).

Class interval	(Mid values) x_i	(Frequency) f_i	$d_x = x_i - 45$	$u_i = d_x/5$	$f_i u_i$
17.5 – 22.5	20	2	–25	–5	–10
22.5 – 27.5	25	8	–20	–4	–32
27.5 – 32.5	30	33	–15	–3	–99
32.5 – 37.5	35	80	–10	–2	–160
37.5 – 42.5	40	170	–5	–1	–170
42.5 – 47.5	$a = 45$	243	0	0	0
47.5 – 52.5	50	213	5	1	213
52.5 – 57.5	55	145	10	2	290
57.5 – 62.5	60	67	15	3	201
62.5 – 67.5	65	35	20	4	140
67.5 – 72.5	70	4	25	5	20
		$N = \Sigma f_i = 1000$			$\Sigma f_i u_i = 393$

Therefore, $$\text{Mean} = a + \left(\frac{\sum_{i=1}^{11} f_i u_i}{N}\right) \times h = 45 + \left(\frac{393}{1000}\right) \times 5 = 45 + 1.965 \approx 46.965.$$

EXAMPLE 6.8 Find the mean from the following data:

Marks	No. of students
Below 10	5
Below 20	9
Below 30	17
Below 40	29
Below 50	45

Contd.

Marks	No. of students
Below 60	60
Below 70	70
Below 80	78
Below 90	83
Below 100	85

Solution Converting the given cumulative frequency distribution into continuous frequency distribution and using step deviation method. Taking $a = 55$ and $h = 10$.

Class intervals	(Mid value) x_i	f_i	$d_x = x_i - 55$	$u_i = d_x/10$	$f_i u_i$
0–10	5	5	–50	–5	–25
10–20	15	4	–40	–4	–16
20–30	25	8	–30	–3	–24
30–40	35	12	–20	–2	–24
40–50	45	16	–10	–1	–16
50–60	$a = 55$	15	0	0	0
60–70	65	10	10	1	10
70–80	75	8	20	2	16
80–90	85	5	30	3	15
90–100	95	2	40	4	8
Total		$N = \Sigma f_i = 85$			$\Sigma f_i u_i = -56$

Therefore, $$\text{Mean} = a + \left(\sum_{i=1}^{10} f_i u_i \Big/ N\right) \times h = 55 + \left(\frac{-56}{85}\right) \times 10 = 48.41.$$

EXAMPLE 6.9 Find the mean marks of students from the following cumulative frequency table:

Marks	No. of students
0 and above	80
10 and above	77
20 and above	72
30 and above	65
40 and above	55
50 and above	43
60 and above	28
70 and above	16
80 and above	10
90 and above	8
100 and above	0

Solution First convert the cumulative frequency distribution into an ordinary frequency distribution. Arrange the marks in groups which may be 0–10, 10–20, 20–30, etc. The number of students getting marks 0 and above 0 is 80, and the number of students getting mark 10 and above 10 is 77. Hence, the number of students getting marks between 0–10 is 80–77 = 3. Similarly, the number of students getting marks between 10–20 is 77 – 72 = 5 and so on. We take 55 as assumed mean and $h = 10$.

The work table for computing the mean is given below:

Marks	(Mid values) x_i	(Frequency) f_i	$d_x = x_i - 55$	$u_i = d_x/10$	$f_i u_i$
0 – 10	5	3	–50	–5	–15
10 – 20	15	5	–40	–4	–20
20 – 30	25	7	–30	–3	–21
30 – 40	35	10	–20	–2	–20
40 – 50	45	12	–10	–1	–12
50 – 60	$a = 55$	15	0	0	0
60 – 70	65	12	10	1	12
70 – 80	75	6	20	2	12
80 – 90	85	2	30	3	6
90 – 100	95	8	40	4	32
		$\Sigma f_i = 80$			$\Sigma f_i u_i = -26$

Hence, by step-deviation method, Eq. (6.5), we have

$$\bar{x} = a + h\frac{\sum_{i=1}^{10} f_i u_i}{N} = 55 + \frac{10}{80}(-26) = 55 - 3.25 = 51.75 \text{ marks.}$$

6.8 TWO IMPORTANT PROPERTIES OF ARITHMETIC MEAN

1. *The algebraic sum of deviations of a set of observations from their arithmetic mean is zero.*

   Proof: Let $\bar{x}$ be the A.M. and n the number of variates. Then $d = x_i - \bar{x}$.

   $$\therefore\ \Sigma d = \Sigma(x_i - \bar{x}) = \Sigma x_i - \Sigma \bar{x} = n\bar{x} - n\bar{x} = 0. \qquad \left(\because\ \bar{x} = \frac{\Sigma x_i}{n} \text{ and } \Sigma x_i = n\bar{x}\right)$$

2. *If $\bar{x}$ is the mean of a set $x_1, x_2 \ldots x_n$ of n observations and $\bar{y}$ is the mean of another set $y_1, y_2, \ldots, y_m$ of m observations, then the arithmetic mean of combined set is given by*

   $$\text{A.M.} = \frac{n\bar{x} + m\bar{y}}{n+m}.$$

   Proof: $\bar{x} = \dfrac{\Sigma x_i}{n}, i = 1, 2, \ldots n$ and $\bar{y} = \dfrac{\Sigma y_j}{m}, j = 1, 2, \ldots m.$

   Therefore, $\text{A.M. of combined set} = \dfrac{\Sigma x_i + \Sigma y_j}{n+m} = \dfrac{n\bar{x} + m\bar{y}}{n+m}.$

   This property could be extended to k sets.

EXAMPLE 6.10 There are 250 observations in a group. The average of first 100 observations is 5 and the average of remaining 150 is $8\frac{1}{3}$. Find the average of the whole group.

Solution We have $\frac{x_1 + x_2 + \cdots + x_{100}}{100} = 5$. Therefore $\sum_{i=1}^{100} x_i = 500$. ...(*a*)

Also $\frac{x_{101} + x_{102} + \cdots + x_{250}}{150} = \frac{25}{3}$. Therefore $\sum_{j=101}^{250} x_j = 1250$. ...(*b*)

Adding (*a*) and (*b*), we obtain $x_1 + x_2 + \ldots x_{100} + x_{101} + \ldots + x_{250} = 1750$.

Thus, A.M. of whole group $= \frac{1750}{100 + 150} = \frac{1750}{250} = 7$.

EXAMPLE 6.11 The mean weight of 98 students is found to be 50 kg. It is later discovered that the frequency of class-interval (30 – 40) was wrongly taken as 8 instead of 10, calculate the correct mean.

Solution Incorrect $\Sigma f_i = 98$. Incorrect $\bar{x} = 50$.

Therefore, incorrect $\Sigma f_i x_i = 98 \times 50 = 4900$. $\left(\because \bar{x} = \frac{\Sigma f_i x_i}{\Sigma f_i}\right)$

Now, correct $\Sigma f_i x_i = 4900 - \left(\frac{30+40}{2}\right) \times 8 + \left(\frac{30+40}{2}\right) \times 10 = 4970$.

correct $\Sigma f_i = 98 - 8 + 10 = 100$.

Therefore, correct mean, $\bar{x}_c = \frac{\text{Correct } \Sigma f_i x_i}{\text{Correct } \Sigma f_i} = \frac{4970}{100} = 49.70$ kgm.

6.9 MEDIAN (MED.)

The median of a series of data is defined as that value which divides the whole series into two equal parts.

Computation of Median for Raw or ungrouped data

For finding the median of raw or ungrouped data, we go through the following steps:

(*a*) Arrange the data in ascending or descending order of magnitude.

(*b*) If the number of observations N in the data is odd, then $[(N+1)/2]$th observation is the *median.*

(*c*) If N is even, there will be two middle terms. Then the average of $\left(\frac{N}{2}\right)$th and $\left(\frac{N}{2}+1\right)$th observation is the *median.*

EXAMPLE 6.12 Find the median of the following data:

(*i*) 35, 31, 12, 16, 29, 74, 45, 40, 49, 57, 62. (*ii*) 29, 34, 11, 17, 36, 21, 46, 23, 39, 41.

Solution (*i*) Arranging the data in ascending order, we obtain 12, 16, 29, 31, 35, 40, 45, 49, 57, 62, 74.

Here $N = 11$ (odd). Therefore the median is $\frac{(11+1)}{2}$th, that is, 6th observation which is 40 and so median = 40.

(*ii*) Arranging the observations in ascending order, we obtain 11, 17, 21, 23, 29, 34, 36, 39, 41, 46.

Here $N = 10$ (even). Therefore the median is the average of $\left(\frac{10}{2}\right)$th and $\left(\frac{10}{2}+1\right)$th observations, that is, 5th and 6th observations, which is $\frac{29+34}{2} = 31.5$. Therefore median = 31.5.

Computation of median for Grouped data

There are two cases:

Case I When the data is discrete. To determine the median, we follow the following steps.

(*a*) Arrange the data in ascending or descending order of magnitude along with their frequencies.

(*b*) Append a column of cumulative frequencies.

(*c*) If the total frequency N is odd, then $\frac{(N+1)}{2}$th observation is the *median.*

If the total frequency N is even, then the mean of $\frac{N}{2}$th and $\left(\frac{N}{2}+1\right)$th observations is the *median.*

EXAMPLE 6.13 Find the median for the following data:

Marks (out of 20):	5	9	10	12	13	16	18	20
No. of students:	4	5	6	12	11	6	4	2

Solution. We form the table as given below:

Marks (x_i)	No. of Students (f_i)	Cumulative Frequency (c_f)
5	4	4
9	5	9
10	6	15
12	12	27
13	11	38
16	6	44
18	4	48
20	2	50
	$\Sigma f_i = 50$	

Here $N = 50$ (even). Therefore, the median is the mean of $\left(\frac{50}{2}\right)$th and $\left(\frac{50}{2}+1\right)$th observations, that is, 25th and 26th observations, which are 12 and 12. Thus median $= \frac{12+12}{2} = 12.$

Case II When the data is continuous and in the form of a frequency distribution.

The median, is given by the following formula:

$$\text{Median} = l_1 + \left(\frac{\frac{N}{2} - f_c}{f}\right) \times h \qquad \text{...(6.6)}$$

where l_1 is the lower limit of the class in which median lies, f is the frequency of the median class.

f_c is the cumulative frequency of the class preceding the median class.

N is the total of frequency and

h is the length of class-interval (or class size).

Geometrically the median in the value of x (abscissa) corresponding to that vertical line which divides a histogram into two parts having equal areas.

Remark 2 For a grouped frequency distribution, first the class limits are made continuous (if not already continuous) and then the above formula is used.

EXAMPLE 6.14 Find the median for the following data:

Class intervals:	110–119	120–129	130–139	140–149	150–159	160–169	170–179
Frequency:	5	25	40	60	40	25	5

Solution The frequency table is given in inclusive form here. The exclusive form of the cumulative frequency table is given below:

Class boundary	Frequency	Cumulative Frequency (f_c)
109.5–119.5	5	5
119.5–129.5	25	30
129.5–139.5	40	f_c = 70
l_1= 139.5–149.5	f = 60	130 ← Median Class
149.5–159.5	40	170
159.5–169.5	25	195
169.5–179.5	5	200
	$N = \Sigma f = 200$	

We have $N/2 = 100$. Therefore, median class = 139.5 – 149.5. $l_l = 139.5$, $f_c = 70$, $f = 60$, $h = 10$.

$$\therefore \quad \text{Median } = 139.5 + \left(\frac{100-70}{60}\right) \times 10 = 144.5.$$

EXAMPLE. 6.15 An incomplete frequency function values distribution is given as follows:

Variable:	10–20	20–30	30–40	40–50	50–60	60–70	70–80	**Total**
Frequency:	12	30	?	65	?	25	18	**229**

Given that the median value is 46, determine the missing frequencies using the median formula. (*AMIE, S-2006*)

Solution Let the frequency of the class 30–40 be f_1 and that of 50–60 be f_2. It is given that $N = 229$. Making the table for the given data, we have

Class intervals	Frequency (f_i)	Cumulative Frequency (f_c)
10–20	12	12
20–30	30	42
30–40	f_1	$42 + f_1$
l_1 = 40–50	65	$107 + f_1$
50–60	f_2	$107 + f_1 + f_2$
60–70	25	$132 + f_1 + f_2$
70–80	18	$150 + f_1 + f_2$

Therefore, $150 + f_1 + f_2 = 229$, which gives, $f_1 + f_2 = 79$. The median, 46 lies in the class interval 40–50. Therefore, $l_1 = 40$, $h = 10$, $f = 65$, $f_c = 42 + f_1$ and $N = 229$.

Employing the formula, median $= l_1 + \left(\dfrac{\frac{N}{2} - f_c}{f}\right) \times h$, we obtain

$$46 = 40 + \left(\frac{(229/2) - (42 + f_1)}{65}\right) \times 10 \text{ or } 46 = 40 + \frac{145 - 2f_1}{13} \text{ which gives } f_1 = 34.$$

Therefore, $f_2 = 79 - f_1 = 79 - 34 = 45$.

6.10 MODE

Mode or the modal value is *defined as that value of a variate that occurs maximum number of times in a distribution.* It is also characterised as the point having maximum frequency or greatest density. The mode is of the greatest practical utility to large scale manufacturer of consumer goods.

Computation of Mode

The following methods prove helpful in the computation of the mode.

Determination of mode for an individual series.

If the measure of each item in a series is known, then the mode is the size of the item which occurs most frequently.

Illustration 1. The set 7, 4, 10, 9, 15, 12, 7, 9, 7 has mode 7, since this number occurring most frequently.

Illustration 2. The set 85, 76, 93, 82 and 96 has no mode.

Illustration 3. The set 2, 3, 4, 4, 4, 5, 5 7, 7, 7, 9 has two modes, 4 and 7, and is called *bimodal.*

Remark 3 A distribution having only one mode is called *unimodal.*

EXAMPLE. 6.16 Find the mean, median and mode for the sets:

(*i*) 3, 5, 2, 6, 5, 9, 5, 2, 8, 6 and (*ii*) 51.6, 48.7, 50.3, 49.5, 48.9

Solution (*i*) Arranging the data in an ascending order, we have 2, 2, 3, 5, 5, 5, 6,6, 8, 9.

$$\text{Mean} = \frac{1}{10}(2\times 2+3\times 1+5\times 3+6\times 2+8\times 1+9\times 1)=5.1.$$

$$\text{Median} = \text{arithmetic mean of two middle numbers} = \frac{1}{2}(5+5)=5.$$

Mode = 5, as this number occurring most frequently.

(*ii*) Arrange the data in an ascending order, we have 48.7, 48.9, 49.5, 50.3, 51.6.

$$\text{Mean} = \frac{1}{5}(48.7+48.9+49.5+50.3+51.6)=\frac{1}{5}(249.0)=49.8.$$

Median = middle number = 49.5.

Mode = It does not exist here.

Determination of mode for a discrete frequency distribution

In a frequency distribution sometimes the item with the highest frequency is not considered as mode because the mode has highest density or frequency in its neighbourhood. So we employ grouping method to find such value.

Method of Preparing grouping table

In **column 1**, we write the frequencies corresponding to the values of the variate.

In **column 2**, add frequencies in pairs starting from the first and place them in a position in the middle of the two frequencies.

In **column 3**, add the frequencies pairs starting from the second and place the sum in the middle of the two frequencies.

In **column 4**, add the frequencies of groups of three values of the variate starting with first and place the sum against the middle value of the group.

In **column 5**, skip the first frequency of column 1 and then add the frequencies in group of three.

In **column 6**, skip the first two frequencies and then add the frequencies in groups of three. Mark the highest frequency in each column 1 to 6 in bold black type.

We then prepare the **analysis table.**

For each column tick against the values having the maximum frequency. Record the number of ticks for each value. The value with maximum number of ticks is the **mode.**

EXAMPLE 6.17 Compute the mode of the following distribution:

Size (x):	10	11	12	13	14	15	16	17	18	19
Number of items (f):	8	15	25	100	98	95	70	55	35	30

Solution To determine the modal size, we use the method of grouping. The Grouping table is given below:

Frequency (f)

x_i	1	2	3	4	5	6
10	8	23				
11	15		40	48		
12	25	125			140	
13	100		198			223
14	98	193		293		
15	95		165		263	
16	70	125				220
17	55		90	160		
18	35	65			120	
19	30					

Analysis Table

↓ Column no. / $x_i \rightarrow$	10	11	12	13	14	15	16	17	18	19
1				✓						
2					✓	✓				
3				✓	✓					
4				✓	✓	✓				
5					✓	✓	✓			
6			✓	✓	✓					
Maximum number of times-size repeated		–	1	4	5	3	1	–	–	–

Since the value of 14 has been repeated maximum number of times, so 14 is the modal size.

Calculation of Mode for Grouped Frequency Distribution (continuous)

Here we assume that all classes are of the same size. First of all we mark the class with highest frequency, called the **modal class**. The mode is computed from the following formula:

$$\text{Mode} (= m_0) = l_1 + \left(\frac{f_m - f_1}{2f_m - f_1 - f_2} \right) h, \qquad ...(6.7)$$

where l_1 = lower limit of the class in which mode lies, f_m = frequency in the modal class, f_1 = frequency of the class just preceeding the modal class, f_2 = frequency of the class just succeeding the modal class, h = size of class interval.

EXAMPLE 6.18 Calculate the mean and mode for the following frequency distribution:

Class-interval:	0–8	8–16	16–24	24–32	32–40	40–48
Frequency:	8	7	16	24	15	7

Solution The computation of the mean is performed in the following table:

Class-interval	Mid values x_i	$u_i = (x_i - 28) / 8$	f_i	$f_i u_i$
0–8	14	–3	8	–24
8–16	12	–2	7	–14
16–24	20	–1	16	–16
i_1 = 24–32	a = 28	0	24	0
32–40	36	1	15	15
40–48	44	2	7	14
Total			$N = \Sigma f = 77$	$\Sigma f_i u_i = -25$

(*i*) $\text{Mean} = 28 + \left(\frac{-25}{77}\right) \times 8 = 28 - 2.6 = 25.4$

(*ii*) Class interval with maximum frequency is 24–32. Here $f_m = 24, f_1 = 16, f_2 = 15, l_1 = 24$ and $h = 8$.

$$\text{Mode} = l_1 + \left(\frac{f_m - f_1}{2f_m - f_1 - f_2}\right)h = 24 + \left(\frac{24 - 16}{2 \times 24 - 16 - 15}\right) \times 8 = 24 + \frac{8}{17} \times 8 = 27.76.$$

Hence, modal marks are 27.76.

EXAMPLE 6.19 The following table gives the length of life of 150 lamps.

Life (100 hours):	0–4	4–8	8–12	12–16	16–20	20–24	24–28	28–32
No. of lamps:	4	12	40	41	27	13	9	4

Calculate the median and modal life of the lamp.

Solution We prepare the following table which shows the relevant computation.

Class-interval	Mid values x_i	f_i	Cumulative frequency f_c
0–4	2	4	4
4–8	6	12	16
8–12	10	40	56
l_1 = 12–16	14	41	97 ← Median class
16–20	18	27	124
20–24	22	13	137
24–28	26	9	146
28–32	30	4	150
Total		$N = \Sigma f = 150$	

$$\text{Median} = l_1 + \left(\frac{\frac{N}{2} - f_c}{f}\right) \times h. \qquad \text{...(}i\text{)}$$

Here $N/2 = 75$, 12–16 is the median class. Thus $l_1 = 12, f_c = 56, f = 41$ and $h = 4$. Using the formula (*i*), and substituting values, we obtain.

$$\text{Median} = 12 + \left(\frac{75 - 56}{41}\right) \times 4 = 13.8537.$$

$$\text{Mode} = l_1 + \left(\frac{f_m - f_1}{2f_m - f_1 - f_2}\right) \times h = 12 + \left(\frac{41 - 40}{2 \times 41 - 40 - 27}\right) \times 4 = 12 + \frac{4}{15} = 12.2667.$$

Since units in class intervals is 100 hours.

∴ Median life of the lamp = 1385.37 hours, and modal life of the lamp = 1226.67 hours.

6.11 EMPIRICAL RELATION BETWEEN MEAN, MEDIAN AND MODE

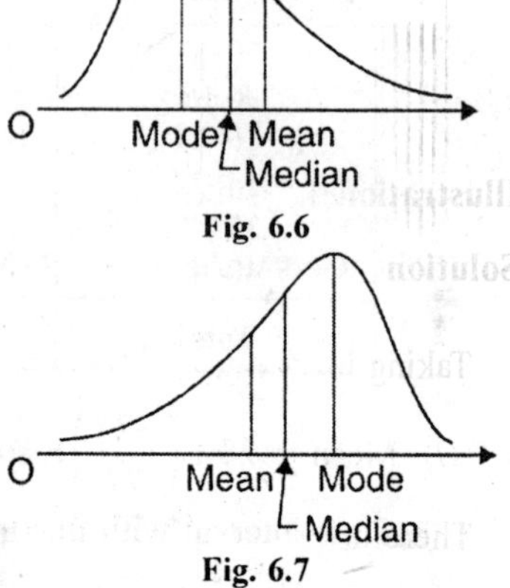

Fig. 6.6

Fig. 6.7

For unimodal frequency curves which are moderately skewed (asymmetrical), we have the empirical relation

$$\text{Mean} - \text{Mode} = 3\,(\text{Mean} - \text{Median}),$$

that is,

$$\text{Mode} = 3\,(\text{Median}) - 2\,(\text{Mean}). \qquad \text{...(6.8)}$$

In Figs. 6.6 and 6.7 are shown the relative positions of the mean, median and mode for frequency curves which are skewed to the right and left respectively. For symmetrical curves the mean, mode and median all coincides.

EXAMPLE 6.20 Use the empirical formula: Mean – Mode = 3 (Mean – Median) to find the modal life of the lamp. The following table gives the length of 150 lamps.

Life (100 hours):	0–4	4–8	8–12	12–16	16–20	20–24	24–28	28–32
No. of lamps:	4	12	40	41	27	13	9	4

Compare your result with the mode obtain in Example 6.19.

Solution We calculate the arithmetic mean by step-deviation method, taking $a = 18$, and $h = 4$. The following table is prepared.

Class-interval	Mid values x_i	f_i	$d_x = x_i - 18$	$u_i = d_x / 4$	$f_i u_i$
0–4	2	4	–16	–4	–16
4–8	6	12	–12	–3	–36
8–12	10	40	–8	–2	–80
12–16	14	41	–4	–1	–41
16–20	$a = 18$	27	0	0	0
20–24	22	13	4	1	13
24–28	26	9	8	2	18
28–32	30	4	12	3	12
		$N = \Sigma f_i = 150$			$\Sigma f_i u_i = -130$

$$\text{Mean} = a + \left(\frac{\Sigma f_i u_i}{N}\right) h = 18 - \left(\frac{130}{150}\right) 4 = 14.5333.$$

Since unit in class intervals is 100 hours.

Therefore, mean life of the lamp = 1453.33 hours.

Refer Example 6.19 Median life of the lamp = 1385.37 hours.

Employing the empirical formula. Mode = 3 Median – 2 Mean, we obtain

$$\text{Mode} = 3(1385.37) - 2\,(1453.33) = 1249.45.$$

From Example 6.19 the modal life of the lamp = 1226.67 hours, so there in good agreement with the empirical result in this case.

6.12 THE GEOMETRIC MEAN (G.M.) AND HARMONIC MEAN (H.M.)

The geometric mean (G.M.) of a set of N positive observations $x_1, x_2, x_3, ..., x_N$ is defined as the Nth root of the product of the observation.

$$\text{G.M.} = \sqrt[N]{x_1 x_2 x_3 \ldots\ldots x_N} \qquad \text{...(6.9)}$$

when the number of items is three or more the task of multiplying the numbers and of extracting the root becomes excessively difficult. To simplify calculations logarithms are used. Geometric mean, G is calculated as follows:

$$\log \text{G.M.} = \frac{1}{N}[\log x_1 + \log x_2 + \log x_3 + ... + \log x_N]$$

$$= \frac{1}{N}\sum_{i=1}^{N} \log x_i. \quad \text{Therefore, G.M.} = \text{Antilog}\left\{\frac{\Sigma(\log x_i)}{N}\right\}. \qquad ...(6.10)$$

Illustration 4. Find the geometric mean of the numbers 3, 5, 6, 6,7, 10, 12.

Solution Geometric mean, (G.M.) = $\sqrt[7]{(3)(5)(6)(6)(7)(10)(12)}$.

Taking logarithms on both sides, we obtain $\log \text{G.M.} = \frac{1}{7}(\log 3 + \log 5 + 2\log 6 + \log 7 + \log 10 + \log 12)$.

$$= \frac{1}{7}(0.4771 + 0.6990 + 2 \times 0.7782 + 0.8451 + 1.0000 + 1.0792) = 0.8081.$$

Therefore G..M. = Antilog (0.8081) = 6.43.

6.12.1 Calculation of Geometric Mean—Discrete Series

$$\text{G.M.} = \text{Antilog}\left[\frac{\Sigma f_i(\log x_i)}{N}\right]. \qquad ...(6.11)$$

Find the logarithms of the variable x_i, multiply these logarithms with the respective frequencies and obtain the total *i.e.* $\Sigma f_i(\log x_i)$. Divide $\Sigma f_i(\log x_i)$ by the total frequency and take the antilog of the value so obtained.

6.12.2 Calculation of Geometric Mean—Continuous Series

$$\text{G.M} = \text{Antilog}\left[\frac{\Sigma(f \log X)}{N}\right], \qquad ...(6.12)$$

where X is the mid value of each class. Geometric mean in always smaller than arithmetic mean.

In most of the cases arithmetic mean is a desirable central value. But in cases where the observations are given in ratios, percentages or proportions, geometric mean is more appropriate than other means.

6.12.3 The Harmonic Mean (H.M.)

For a given set of N non-zero observations, harmonic mean H is defined as the reciprocal of the AM of the reciprocals of the observations. So, if a variable x assumes N non-zero values x_1, x_2, x_3, ..., x_N then the harmonic mean of x is given by

$$\text{H.M.} = \frac{N}{\Sigma 1/x_i}. \qquad ...(6.13)$$

Harmonic mean is always less than the geometric mean. Harmonic mean is suitable for the data pertaining to time, rates, speed, velocity, etc.

Illustration 5. Find the harmonic mean (H.M.) of the numbers 6, 14, 21, 30.

Solution Harmonic mean or H.M. = Reciprocal $\left(\frac{\text{Reci } x_1 + \text{Reci } x_2 + \cdots \cdots + \text{Reci } x_N}{N}\right)$

or $$\text{H.M.} = \left(\frac{N}{\text{Reci } x_1 + \text{Reci } x_2 + \cdots + \text{Reci } x_N}\right) = \frac{4}{\frac{1}{6}+\frac{1}{14}+\frac{1}{21}+\frac{1}{30}} = \frac{4 \times 210}{67} = 12.54.$$

6.12.4 Calculation of Harmonic Mean—Discrete Series

In discrete series, harmonic mean (H.M.) is computed by applying the following formula:

$$\text{H.M.} = \frac{\Sigma f}{\sum_{i=1}^{N}(f_i/x_i)} = \frac{N}{\sum_{i=1}^{N}(f_i/x_i)}. \quad ...(6.14)$$

6.12.5 Calculation of Harmonic Mean—Continuous Series

For calculating harmonic mean in continuous series the procedure is the same as applied to discrete series. The only difference is that here we take the reciprocal of the mid-points. The formula is as under:

$$\text{H.M.} = \frac{N}{\Sigma(f_i/X)} \quad \text{or} \quad \text{H.M.} = \frac{\Sigma f_i}{\Sigma(f_i/X)}, \quad ...(6.15)$$

where X is the mid value of each class.

EXAMPLE 6.21 The marks obtained by 25 students in Civil Engineering are as given below. Find their Geometric and Harmonic means.

Marks (x):	11	12	13	14	15
No. of students (f):	3	7	8	5	2

Solution We prepare the following table for calculation of G.M. and H.M.

x_i	f_i	$1/x_i$	f_i/x_i	$\log x_i$	$f_i \log x_i$
11	3	0.0909	0.2727	1.0414	3.1242
12	7	0.0833	0.5831	1.0792	7.5544
13	8	0.0769	0.6152	1.1139	8.9112
14	5	0.0714	0.3570	1.1461	5.7305
15	2	0.0666	0.1332	1.1761	2.3522
Total	$N = \Sigma f = 25$		$\Sigma f_i/x_i = 1.9612$		$\Sigma f_i \log x_i = 27.6725$

$$\text{G.M.} = \text{Antilog}\left[\frac{\Sigma(f_i \log x_i)}{N}\right] = \text{Antilog}\left(\frac{27.6725}{25}\right) = \text{Antilog}\ (1.1069) \approx 12.8.$$

$$\text{H.M.} = \frac{N}{\Sigma(f_i/x_i)} = \frac{25}{1.9612} = 12.747.$$

EXAMPLE 6.22 A man travels from X to Y at an average speed of 30 km/hour and returns from Y to X along the same route at an average speed of 45 km/hour. Find the average speed for the entire trip.

Solution Let the distance from X to Y is a km. Then time taken to move from X to $Y = a/30$ hour. Time taken from Y to $X = a/45$ hour.

$$\therefore \text{ Average speed for entire trip} = \frac{\text{Total distance moved}}{\text{Total time taken}} = \frac{2a}{\frac{a}{30} + \frac{a}{45}} = 36 \text{ km/hour.}$$

Note The above average is the harmonic mean of 30 and 45, that is $\frac{2}{(1/30) + (1/45)} = 36$ km/hour.

6.13 RELATION BETWEEN ARITHMETIC, GEOMETRIC AND HARMONIC MEANS

The geometric mean of a set of positive numbers $x_1, x_2, ..., x_N$ is less than or equal to their arithmetic mean but is greater than or equal to their harmonic means. We have the following inequality:

$$\text{H.M.} \leq \text{G.M.} \leq \text{A.M.} \quad ...(6.16)$$

and that

$$(\text{A.M.}) \times (\text{H.M.}) = (\text{G.M.})^2.$$

The equality sign hold only if all the numbers $x_1, x_2, ..., x_N$ are equal.

6.14 PARTITION VALUES — QUARTILES, DECILES AND PERCENTILES

Partition values divide the given set of data into a number of equal parts. If a set of data is arranged in order of magnitude, the middle value (or arithmetic mean of the two middle values) which divides the set into two equal parts is the median. We consider some partition values here.

Quartiles: The values which divide the data into four equal parts are called *quartiles* and are denoted by Q_1, Q_2 and Q_3.

Q_1 = *First quartile* or the *lower* quartile is the value below which the first 25% observations lie = $\frac{(n+1)}{4}$th item.

Q_2 = *Second quartile* is the value below which 50% of the observations are present. The value Q_2 being equal to the median = $2\left(\frac{n+1}{4}\right)$th item.

Q_3 = *Third quartile* or the *upper* quartile is the value below which 75% of the observations lie = $3\left(\frac{n+1}{4}\right)$th item.

For grouped frequency distribution, the formulae for Q_1, Q_2, Q_3 are as follows:

$$Q_1 = l_1 + \left(\frac{\frac{N}{4} - f_c}{f}\right)h; \qquad Q_2 = l_1 + \left(\frac{\frac{2N}{4} - f_c}{f}\right)h; \qquad Q_3 = l_1 + \left(\frac{\frac{3N}{4} - f_c}{f}\right)\times h.$$

In general,
$$Q_k = l_1 + \left(\frac{k\cdot\frac{N}{4} - f_c}{f}\right)h$$

where Q_k = kth quartile, kth quartile class is the class interval in which $k\cdot\frac{N}{4}$ th term lies, N = total frequency, l_1 = lower limit of the kth quartile class, f_c = cumulative frequency of the class preceding this quartile class, f = frequency of the quartile class, h = size of the quartile class.

Deciles and Percentiles

Similarly, the values which divide the data into ten equal parts are called deciles and are denoted by D_1, D_2, D_3, ..., D_9, while the values dividing the data into one hundred equal parts are called percentiles and are denoted by P_1, P_2, P_3, ..., P_{99}. The 5th decile and the 50th percentile correspond to the median. The 25th and the 75th percentile correspond to the first and third quartiles respectively. We may note here that in case of partition values [Q_r, D_r and P_r]'s the data are to be arranged in ascending order but in case of median, they can be arranged in ascending or descending order of magnitude.

EXAMPLE 6.23 Find Q_1, D_6 and P_{82} for the following numbers: 82, 56, 90, 50, 120, 75, 75, 80, 130, 65.

Solution The given data is arranged in ascending order 50, 56, 65, 75, 75, 80, 82, 90, 120, 130.

Here n = 10. Therefore $Q_1 = \frac{(n+1)\text{th}}{4}$ observation = $\frac{(10+1)}{4}$th observation = (2.75)th observation.

= 2nd + 0.75 (3rd – 2nd) observation = 56 + 0.75 (65 – 56) = 62.75.

$D_6 = \left(\frac{6(n+1)}{10}\right)$th observation = (6.6)th observation

= 6th + 0.6 (7th – 6th) observation = 80 + 0.6 (82 – 80) = 81.20.

$P_{82} = \left(\frac{82(n+1)}{100}\right)$th observation = 9.02 observation

= 9th + 0.02 (10th – 9th) observation = 120 + 0.02 (130 – 120) = 120.2.

EXAMPLE 6.24 Calculate the lower and the upper quartiles, fourth decile and the 60th percentile of the following distribution.

Marks group:	5–10	10–15	15–20	20–25	25–30	30–35	35–40	40–45
Numbers of students:	5	6	15	10	5	4	2	2

Solution We construct the cumulative frequency table:

Class-interval	Frequency (f_i)	Cumulative frequency (f_c)
5–10	5	5
10–15	6	**11**
15–20	**15**	26 ← Q_1, D_4
20–25	10	36 ← P_{60}
25–30	5	41 ← Q_3
30–35	4	45
35–40	2	47
40–45	2	49
	$N = 49$	

Here $N = 49$, $N/4 = 12.25$, (12.25)th term lies in class 15–20.

Thus, $l_1 = 15, f_c = 11, f = 15, h = 20 - 15 = 5$.

(*i*) $$Q_1 = l_1 + \left(\frac{(N/4) - fc}{f}\right)h = 15 + \left(\frac{12.25 - 11}{15}\right)5 = 15.41.$$

(*ii*) For Q_3: (3/4) $N = 36.75$. (36.75)th term lies in the class 25–30. Therefore, $l_1 = 25, f_c = 36, f = 5, h = 5$.

$$Q_3 = l_1 + \left(\frac{(3/4)N - fc}{f}\right)h = 25 + \left(\frac{36.75 - 36}{5}\right)5 = 25.75.$$

(*iii*) For D_4: (4/10)N = (4/10)49 = 19.6. (19.6)th term lies in the class 15–20.
Thus, $l_1 = 15, f_c = 11, f = 15, h = 5$.

Therefore, $$D_4 = l_1 + \left(\frac{(4/10)N - fc}{f}\right)h = 15 + \left(\frac{19.6 - 11}{15}\right)5 = 17.87.$$

(*iv*) For P_{60}: (60/100)N = (3/5)49 = 29.4. (29.4)th term lies in the class 20–25.
Thus, $l_1 = 20$, $f_c = 26$, $f = 10$, $h = 5$.

Therefore, $$P_{60} = l_1 + \left(\frac{(60/100)N - fc}{f}\right)h = 20 + \left(\frac{29.4 - 26}{10}\right)5 = 21.7.$$

EXAMPLE 6.25 Find Q_1, Q_3, D_6 and P_{45} for the following data:

x_i:	18	19	20	21	22	23	24	25	26	27
f_i:	15	18	25	27	40	25	19	16	8	7

Solution We form the following table:

x_i	f_i	Cumulative frequency
18	15	15
19	18	33
20	25	58 ← Q_1
21	27	85
22	40	125 ← D_6, P_{45}
23	25	150
24	19	169 ← Q_3
25	16	185
26	8	193
27	7	200
	$N = \Sigma f_i = 200$	

Here $N = 200$, $\frac{N+1}{4} = 50.25$. Therefore, $Q_1 = 20$.

Again $\frac{3}{4}(N+1) = \frac{3}{4} \times 201 = 150.75$. Therefore, $Q_3 = 24$.

Also $\frac{6}{10}(N+1) = \frac{6}{10} \times 201 = 120.6$. Therefore, $D_6 = 22$.

Finally $\frac{45}{100}(N+1) = \frac{45}{100} \times 201 = 90.45$. Therefore, $P_{45} = 22$.

EXERCISE 6.1

1. Draw a histogram for the following frequency distribution and estimate the number of persons under the age of 55 years.

Age in years:	10–19	20–39	40–59	60–69	70–79
No. of persons:	20	80	120	10	70

2. Calculate arithmetic mean of the following data:
   2.4, 2.8, 3.2, 1.9, 2.7, 4.2, 3.8, 2.2. **Ans.** 2.9.
3. A set of numbers consists of six 6's, seven 7's, eight 8's, nine 9's and ten 10's. What is the arithmetic mean of the numbers? **Ans.** 8.25.

$$\left[\textbf{Hint: } \bar{x} = \frac{6\times6+7\times7+8\times8+9\times9+10\times10}{6+7+8+9+10} = \frac{330}{40} = 8.25\right]$$

4. The mean weight of 8 employees in a company is 70 kg and another 7 employees have an average of 55 kg and remaining 5 employees have an average of 71 kg. Find the mean weight of all 20 employees. **Ans.** 65 kg.

$$\left[\textbf{Hint: } \text{Mean weight of all 20 employees} = \frac{8\times70+7\times55+5\times71}{8+7+5} = \frac{1300}{20} = 65 \text{ kg}\right]$$

5. Find the arithmetic mean of (*a*) the first *n* natural numbers (*b*) weighted arithmetic mean of first n natural numbers when weights are equal to the corresponding numbers. **Ans:** (*a*) $(n+1)/2$, (*b*) $(2n+1)/3$.

$$\left[\textbf{Hint: } (b)\ \bar{x} = \frac{1.1+2.2+3.3+\cdots+n.n}{1+2+3+\cdots+n} = \frac{\Sigma n^2}{\Sigma n} = \frac{n(n+1)(2n+1)/6}{n(n+1)/2} = (2n+1)/3.\right]$$

6. A candidate obtain the following percentage of marks in an examination: English 60, Drawing 75, Mathematics 63, Physics 59 and Chemistry 55. Find the weighted mean if weights 2, 1, 5, 5, 3 are allotted to the subjects. **Ans.** 60.63% marks.
7. Compute the arithmetic mean from the following frequency table:

Height in inches:	58	60	62	64	66	68
No. of persons:	12	14	20	13	8	5

Ans. 67.12 inches approx.

8. Find the weighted Arithmetic mean of first n natural numbers, the weights being equal to the squares of the corresponding numbers. **Ans.** $\frac{3n(n+1)}{2(2n+1)}$.

[**Hint:**

x :	1	2	3	...	$n-1$,	n
w :	1^2	2^2	3^2	...	$(n-1)^2$,	n^2

$$\text{Weighted A.M.} = \frac{\Sigma w_i x_i}{\Sigma w_i} = \frac{1\times1^2+2\times2^2+3\times3^2\cdots+n\times n^2}{1^2+2^2+3^2+\cdots+n^2} = \frac{1^3+2^3+3^3+\cdots+n^3}{1^2+2^2+3^2+\cdots+n^2}$$

$$= \left[\frac{n(n+1)}{2}\right]^2 \cdot \frac{6}{n(n+1)(2n+1)} = \frac{3n(n+1)}{2(2n+1)}.\Big]$$

9. Find the mean for the following frequency distribution:

Class-interval:	0–8	8–16	16–24	24–32	32–40	40–48
Frequency:	8	7	16	24	15	7

Ans. 25.4.

10. The mean of 200 items was 50. It was later discovered that two items were misread as 92 and 8 instead of correct values 192 and 88. Calculate the correct mean. **Ans.** 50.9.

[**Hint:** Incorrect mean $\bar{x} = \dfrac{(\Sigma x_i)}{n}$. Therefore, incorrect $\Sigma x_i = n\bar{x} = 200 \times 50 = 10{,}000$.

Correct $\Sigma x = 10{,}000 - (92 + 8) + (192 + 88) = 10180$

Therefore, correct mean $= 10180/200 = 50.9$].

11. Find the missing frequency from the following data:

Marks:	0–10	10–20	20–30	30–40	40–50	50–60
No. of students:	5	15	20	—	20	10

The arithmetic mean is 34 marks. **Ans.** 180.

12. Given that the mean height of a group of students is 67.45 inches. Find the missing frequencies for the following incomplete distribution of height of 100 students.

Height in inches:	60–62	63–65	66–68	69–71	72–74
No. of persons:	5	18	–	–	8

Ans. 42; 27.

13. The following table shows the age distribution of cases of a certain disease reported during a year:

Age:	5–15	15–20	20–25	25–30	30–35	35–40
No. of cases:	5	10	20	22	13	5

Calculate the mean for the above distribution. **Ans.** 25.2.

14. The mean age of 50 persons was found to be 30 years. Later it was detected that the age 57 was misread as 75, age 65 was misread as 47 and age 67 was misread as 17. Find the correct mean age. **Ans.** 31 years.

15. Find the median of the sets of numbers:

(*a*) 5, 4, 8, 3, 6, 2, 9 (*b*) 18.2, 20.6, 19.3, 22.3, 20.4, 18.7, 19.7, 20.0. **Ans.** (*a*) 5, (*b*) 19.85.

16. Compute the median from the following data:

x:	115	125	135	145	155	165	175	185	195
f:	6	25	48	72	116	60	38	22	3

Ans. 153.8.

17. Calculate the median for the following data:

Marks obtained	No. of students
Less than 20	0
Less than 30	4
Less than 40	16
Less than 50	30
Less than 60	46 *Contd.*
Less than 70	66
Less than 80	82
Less than 90	92
Less than 100	100

Ans. 62.

18. Obtain the missing frequency for the following frequency distribution when the value of median is 86.

Class interval:	40–50	50–60	60–70	70–80	80–90	90–100	100–110
Frequency:	2	1	6	6	—	12	5

Ans. 10.

19. Find the missing frequencies for the distribution of marks of 100 pupils, median mark is known to be 32.

Marks:	0–10	10–20	20–30	30–40	40–50	50–60
No. of pupils:	10	—	25	30	—	10

Ans. 9, 16.

20. Find the mean, median and mode for the set of numbers:
(*a*) 7, 4, 10, 9, 15, 12, 7, 9, 7 and (*b*) 85, 76, 93, 82, 96.
Ans. (*a*) 8.9; 9; 7, (*b*) 86; 85; it does not exist.

21. Calculate the mode from the following frequency distribution:

Size (*x*):	4	5	6	7	8	9	10	11	12	13
Frequency (*f*):	2	5	8	9	12	14	14	15	11	13

Ans. 10.

22. Find the mode for the following distribution:

Class interval:	0–10	10–20	20–30	30–40	40–50	50–60	60–70	70–80
Frequency:	5	8	7	12	28	20	10	10

(*AMIE, S-2006*) **Ans.** 47.67.

23. The following are the marks obtained by students in a class test. Find the modal mark.

Class interval:	32–35	36–39	40–43	44–47	48–51	52–55	56–59	60–63
No. of students:	10	37	65	80	51	35	18	4

[**Hint:** Take $l_1 = 43.5, h = 4$] **Ans.** 44.9.

24. Calculate the mode and the median for the following frequency distribution:

Marks:	1-5	6-10	11-15	16-20	21-25	26-30	31-35	36-40	41-45
No. of students:	7	10	16	32	24	18	10	5	1

Ans. 18.4, 19.56.

25. Find the geometric mean of the numbers:
(*a*) 3, 5, 8, 3, 7, 2 and (*b*) 28.5, 73.6, 47.2, 31.5, 64.8. **Ans.** (*a*) 4.14 (*b*) 45.8.

26. Find the harmonic mean of the numbers: 3.2, 5.2, 4.8, 6.1, 4.2. **Ans.** 4.48.

[**Hint:** $\text{H.M.} = \dfrac{5}{\text{Reci } 3.2 + \text{Reci } 5.2 + \text{Reci } 4.8 + \text{Reci } 6.1 + \text{Reci } 4.2}$

$$= \frac{5}{0.3125 + 0.1923 + 0.2083 + 0.1639 + 0.2381} = \frac{5}{1.1151} = 4.48$$]

27. Find the harmonic mean for the following data:

x:	2	4	4	16
f:	2	3	3	2

Ans. 4.44.

28. An aeroplane files from *A* to *B* at the speed of 500 km/hour and comes back from *B* to *A* at the speed of 700 km/hour. Find the average speed of the aeroplane. **Ans.** 583.33 km/hour.

29. Three cities *A*, *B* and *C* are equidistant from each other. A motorist travels from *A* to *B* at 30 km/hour, from *B* to *C* at 40 km/hour and from *C* to *A* at 50 km/hour. Determine his average speed for the entire trip. **Ans.** 38.3 km/hour.

30. Find Q_1, Q_3, D_3, P_{67} for the following numbers:
15, 10, 20, 25, 18, 11, 9, 12, 22, 28, 16, 13. **Ans.** 11.25, 21.5, 11.9, 19.42.

6.15 DISPERSION OF VARIATION

In the preceeding sections, we have discussed different measures of central tendency and with their aid, it was possible to represent a given distribution by a single number e.g., mean, mode, median, etc. Let us now consider two distributions representing the marks scored by two groups of 6 students each viz.

	Students						Total Marks
	1	2	3	4	5	6	
Group A marks:	20	22	23	27	28	30	150
Group B marks:	15	18	21	29	32	35	150

It is obvious from the given data that mean of each series is 25 and there is no difference between the two series so for as their means are concerned. However, if we examine the items of the two distributions, we find that the marks in one group vary between 20 and 30 while in the second the variation is between 15 and 35. This certainly indicates a difference between the two distributions. The difference arises due to the fact that the divergence from the mean is different in the two groups. In other words the scatter or dispersion from the mean is less in the first group than in the second. This scatter or lack of uniformity in the sizes of the items of a group is known as *dispersion*. Dispersion or variation for a given set of observations may be defined as the amount of deviation of the observations, usually, from an appropriate measure of central tendency.

The following are some measures of dispersion which are most commonly used.

(*i*) Range (*ii*) Inter quartile range or Quartile deviation
(*iii*) Mean deviation (*iv*) Standard deviation (*v*) Coefficient of variation.

We shall briefly discuss them one by one.

6.16 RANGE

It is the simplest measure of dispersion. It is given by the difference between the largest and the smallest values of the variable appearing in the distribution. It is a very crude measure of dispersion as it depends only on two extreme items. It is particularly used in quoting interest rates and security prices. For a grouped frequency distribution, *range* is defined as the difference between the two extreme class boundaries.

Symbolically, Range = L–S, (where L is the largest value and S is the smallest value)

An absolute measure of range is coefficient of range.

The formula for coeff. of range $= \dfrac{\text{L} - \text{S}}{\text{L} + \text{S}}$. ...(6.17)

6.17 QUARTILE DEVIATION OR SEMI-INTER-QUARTILE RANGE

Quartiles are the magnitudes of the items in a series which divide it into four equal parts. The difference between the upper and the lower quartiles that is, $Q_3 - Q_1$ is called the inter-quartile range (IQR) and half of this range is defined as quartile deviation. Mathematically, the Quartile deviation or semi-inter-quartile range is given by

$$Q_d = \frac{1}{2}(Q_3 - Q_1). \qquad ...(6.18)$$

Q_3, the upper quartile is given by

$$Q_3 = l_1 + \left(\frac{\frac{3N}{4} - f_c}{f}\right) \times h, \qquad ...(6.19)$$

where l_1 is the lower limit of the quartile class, f is the frequency of the quartile class, h is its class intervals, f_c is the cumulative frequency before the quartic class, and N is the total of all the frequencies Q_1. The lower quartile, is calculated by a similar formula except that in place of $3N/4$ we write $N/4$.

$$\text{Coefficient of Quartile Deviation} = \frac{Q_3 - Q_2}{Q_3 + Q_1}. \qquad ...(6.20)$$

A small value of coefficient of quartile deviation indicates less variation in the series.

EXAMPLE 6.26 Find the range of the sets of numbers:

(*i*) 5, 3, 8, 4, 7, 6, 12, 4, 3 and (*ii*) 8.772, 6.453, 10.624, 8.628, 9.434, 6.351.

Solution (*i*) Arrange the given data in ascending order, we obtain

3, 3, 4, 4, 5, 6, 7, 8, 12. Range = largest number – smallest number = 12 – 3 = 9.

(*ii*) Range = largest number – smallest number = 10.624 – 6.351 = 4.273.

EXAMPLE 6.27 Calculate the inter-quartile range, quartile deviation and coefficient of quartile deviation for the distribution of the weight of 150 students from the data given below:

Weights in Kgs:	30–39	40–49	50–59	60–69	70–79	80–89
Numbers of students:	18	37	45	27	15	8

Solution The work can be arranged in the following table:

Class-interval (C.I.)	Class boundary (C.B)	Frequency f	Cumulative frequency ($c.f$)
30–39	29.5–39.5	18	18
40–49	39.5–49.5	37	55
50–59	49.5–59.5	45	100
60–69	59.5–69.5	27	127
70–79	69.5–79.5	15	142
80–89	79.5–89.5	8	150
Total		$N = \Sigma f = 150$	

Q_1 = size of (N/4)th observation = (150/4)th observation = (37.5)th observation. Q_1 lies in the class 40–49.

$$Q_1 = l_1 + \left(\frac{(N/4) - f_c}{f}\right)h = 39.5 + \left(\frac{37.5 - 18}{37}\right)10 = 44.77.$$

Q_3 = size of 3(N/4)th observation = 3(37.5)th observation = (112.5)th observation. Q_3 lies in the class 60–69.

$$Q_3 = l_1 + \left(\frac{3(N/4) - f_c}{f}\right)h = 59.5 + \left(\frac{112.5 - 100}{27}\right)10 = 64.13.$$

Thus inter-quartile range = $Q_3 - Q_1$ = 64.13 – 44.77 = 19.36.

$$\text{Quartile deviation} = \frac{Q_3 - Q_1}{2} = 9.68.$$

$$\text{Coefficient of quartile deviation} = \frac{Q_3 - Q_1}{Q_3 + Q_1} = \frac{64.13 - 44.77}{64.13 + 44.77} = \frac{19.36}{108.90} = 0.177.$$

6.18 MEAN DEVIATION

It is defined as the arithmetic mean of all the absolute values of the deviations from the central value. Mean deviation may be obtained from any measure of central tendency. However, mean deviation from mean and median are commonly used in statistical studies.

(A) Mean deviation about the mean

(*i*) **For Ungrouped Data.** Let N observations be $x_1, x_2, x_3, ..., x_N$.

$$\text{M.D.}(\bar{x}) = \frac{\sum_{i=1}^{N} |x_i - \bar{x}|}{N}, \text{ where } \bar{x} = \sum x_i / N. \quad ...(6.21)$$

(*ii*) **For Grouped Data — Discrete Frequency Distribution**

$$\text{M.D.}(\bar{x}) = \frac{\sum_{i=1}^{K} f_i |x_i - \bar{x}|}{N} = \frac{\Sigma f\,|x - \bar{x}|}{N}, \text{ where } \bar{x} = \frac{\sum f_i x_i}{\sum f_i},\ N = \sum_{i=1}^{K} f_i = \Sigma f. \quad ...(6.22)$$

(B) Mean derivation about median

(i) For Ungrouped Data

$$\text{M.D.(Med.)} = \frac{\sum_{i=1}^{N} |x_i - \text{Median}|}{N}. \qquad ...(6.23)$$

(ii) For Grouped Data – Discrete Frequency Distribution

$$\text{M.D.(Med.)} = \frac{\sum_{i=1}^{K} f_i\,|x_i - \text{Median}|}{\sum f_i}. \qquad ...(6.24)$$

$$\text{Coefficient of mean deviation about its mean} = \frac{\text{M.D.}(\bar{x})}{\bar{x}}. \qquad ...(6.25)$$

Similarly, coefficient of mean deviation based on median can be found out. Since, the mean deviation is based on all the values of the variate it provides a better measure of dispersion than the range and the quartile deviation.

EXAMPLE 6.28 Calculate the mean deviation about the mean and also about the median for following data: Height in cm of 7 soldiers: 168, 164, 172, 169, 178, 173, 173.

Solution (i) Arithmetic mean of the given data =

$$\bar{x} = \frac{\Sigma x_i}{N} = \frac{164 + 168 + 169 + 172 + 173 + 173 + 178}{7} = \frac{1197}{7} = 171 \text{ cm}.$$

The deviations of the respective observations from the mean $\bar{x}$, that is, $x_i - \bar{x}$ are –7, –3, –2, 1, 2, 2, 7.

Therefore, $\sum_{i=1}^{7} |x_i - \bar{x}| = 7 + 3 + 2 + 1 + 2 + 2 + 7 = 24$ and $\text{M.D.}(\bar{x}) = \dfrac{\sum_{i=1}^{7} |x_i - \bar{x}|}{7} = \dfrac{24}{7} = 3.43$ cm.

(ii) Here the number of observations is odd ($n = 7$). Arranging the data in ascending order, we have 164, 168, 169, 172, 173, 173, 178. Now median is $\dfrac{(7+1)}{2}$th or 4th observation. Therefore median = 172 cm.

$$\text{Hence M.D. (Med)} = \frac{\sum_{i=1}^{7} |x_i - \text{Median}|}{N} = \frac{\{|164-172| + |168-172| + |169-172| + |172-172| + |173-172| + |173-172| + |178-172|\}}{7}$$

$$= \frac{8 + 4 + 3 + 0 + 1 + 1 + 6}{7} = \frac{23}{7} = 3.29 \text{ cm}.$$

EXAMPLE 6.29 Calculate the mean deviation about the mean for the following data:

Score (x_i):	9	11	3	5	13	7
Frequency (f_i):	9	5	2	7	1	10

Solution. We construct the following table taking the scores in ascending order and append other columns after calculations.

x_i	f_i	$f_i x_i$	$\lvert x_i - \bar{x}\rvert$	$f_i \lvert x_i - \bar{x}\rvert$
3	2	6	4.65	9.30
5	7	35	2.65	18.55
7	10	70	0.65	6.50
9	9	81	1.35	12.15
11	5	55	3.35	16.75
13	1	13	5.35	5.35
Total	**N = Σf_i = 34**	**$\Sigma f_i x_i$ = 260**		**$\Sigma f_i \lvert x_i - \bar{x}\rvert$ = 68.60**

Therefore, mean, $\bar{x} = \dfrac{\Sigma f_i x_i}{\Sigma f_i} = \dfrac{260}{34} = 7.65$ (approx.)

and M.D. $(\bar{x}) = \dfrac{\Sigma f_i |x - \bar{x}|}{\Sigma f_i} = \dfrac{68.60}{34} = 2.02$ (approx.)

EXAMPLE 6.30 Find the mean deviation about the median for the following data:

x_i:	3	6	9	12	13	15	21	22
f_i:	3	4	5	2	4	5	4	3

Solution The data is already arranged in ascending order. Adding a new row of cumulative frequency to the given table. We have

x_i:	3	6	9	12	13	15	21	22
f_i:	3	4	5	2	4	5	4	3
c.f:	3	7	12	14	18	23	27	30

Here $N = \Sigma f_i = 30$, which is even.

Median is the mean of 15th and 16th observations. Since 15th and 16th observations are both 13. Therefore,

median $= \dfrac{13+13}{2} = 13.$

$|x_i - \text{Median}|$ has respective values 10, 7, 4, 1, 0, 2, 8, 9.

We have $\sum_{i=1}^{8} f_i |x_i - \text{Median}| = 30 + 28 + 20 + 2 + 0 + 10 + 32 + 27 = 149.$

Therefore, M.D. (Median) $= \dfrac{\sum_{i=1}^{8} f_i |x_i - \text{Median}|}{\sum f_i} = \dfrac{149}{30} = 4.97.$

EXAMPLE 6.31 Find the mean deviation about the mean for the following data:

Class-interval:	95–105	105–115	115–125	125–135	135–145	145–155
Frequency:	9	13	26	30	12	10

Hence, find the coefficient of mean deviation about its mean.

Solution We make the following table from the given data. Let the assumed mean of the given data be 120.

Class-interval	Mid. value x_i	$u_i = \dfrac{x_i - 120}{10}$	f_i	$f_i u_i$	$\lvert x_i - \bar{x}\rvert$	$f_i \lvert x_i - \bar{x}\rvert$
95–105	100	−2	9	−18	25.30	227.70
105–115	110	−1	13	−13	15.30	198.90
115–125	$a \to$ 120	0	26	0	5.30	137.80
125–135	130	1	30	30	4.70	141.00
135–145	140	2	12	24	14.70	176.40
145–155	150	3	10	30	24.70	247.00
Total			$\Sigma f_i = 100$	$\Sigma f_i u_i = 53$		$\Sigma f_i \lvert x_i - \bar{x}\rvert = 1128.80$

Mean, $\bar{x} = 120 + \dfrac{\sum f_i u_i}{\sum f_i} \times h = 120 + \dfrac{53}{100} \times 10 = 125.30.$

$$\text{M.D.}(\bar{x}) = \frac{1}{N}\sum_{i=1}^{6} f_i\,|x_i - \bar{x}| = 1128.80/100 = 11.288.$$

Coefficient of mean deviation about its mean $= \dfrac{\text{Mean deviation about the mean}}{\text{Mean}} = \dfrac{11.288}{125.30} = 0.09.$

EXAMPLE 6.32 Compute the coefficient of mean deviation about median for the following distribution:

Wt. in kgs.:	40–50	50–60	60–70	70–80
No. of persons:	8	12	20	10

Solution First we compute the median. Form the following table from the given data:

Wt. in kg (Class interval (x_i))	No. of persons (frequency, (f_i))	Cumulative frequency ($C.f$)
40–50	8	8
50–60	12	f_c : **20**
$l_1 \rightarrow$ 60–70	**20**	40
70–80	10	50
Total	$N = \Sigma f_i = 50$	

We find from the table, $N/2 = 25$. The class interval containing 25th item is 60–70. Therefore, 60–70 is the median class. Thus, we have $l_1 = 60, f_c = 20, f = 20$.

$$\text{Median} = l_1 + \frac{\frac{N}{2} - f_c}{f_i} \times h = 60 + \left(\frac{25-20}{20}\right) \times 10 = 62.5 \text{ kg.}$$

Now we prepare the following table for computation of mean deviation about the median.

Class-interval (wt. in kgm)	mid-value (x_i) kgs.	No. of persons f_i	$\|x_i - Me\|$ kgs	$f_i\|x_i - Me\|$
40–50	45	8	17.50	140
50–60	55	12	7.50	90
60–70	65	20	2.50	50
70–80	75	10	12.50	125
Total		$N = \sum_{i=1}^{4} f_i = 50$		$\sum_{i=1}^{4} f_i\,\|x_i - Me\| = 405$

Thus mean deviation about median is given by M.D. (Median) $= \dfrac{\sum f_i |x_i - M_e|}{\sum f_i} = \dfrac{405}{50} = 8.10$ kg.

Hence coefficient of mean deviation about the median $= \dfrac{\text{M.D. (Med.)}}{\text{Med.}} = \dfrac{8.10}{62.5} = 0.1296$.

6.19 VARIANCE AND STANDARD DEVIATION

Variance: The mean of the squares of the deviations from mean is called the variance and is denoted by σ^2 (read as sigma square).

Standard Deviation: It is defined as the positive square root of the mean of squared deviations, all deviations taken from the actual mean. It is denoted by the Greek letter σ (sigma). Standard deviation is considered as the best measure of dispersion and is a most used measure.

(*a*) **Standard deviation in case of ungrouped data** that is, data having N individual observations $x_1, x_2, \ldots\ldots, x_N$ is given by

$$\sigma = \sqrt{\frac{\sum_{i=1}^{N}(x_i - \bar{x})^2}{N}}, \text{ where } \bar{x} = \frac{x_1 + x_2 + \ldots + x_N}{N} \qquad \ldots(6.26)$$

This formula is appropriate if $\bar{x}$ is a whole number.

(*b*) **Standard deviation in case of grouped data without using arithmetic mean** that is, if $\bar{x}$ is not a whole number.

(*i*) **Direct method:** The formula in this case, can be expressed as

$$\text{S.D.} = \sqrt{\frac{\sum x_i^2}{N} - \left(\frac{\sum x_i}{N}\right)^2}, \qquad \text{...(6.27)}$$

where Σx_i^2, Σx_i denotes respectively the sum of the squares and that of the terms of the given data.

(*ii*) **Deviation method:** By assuming an arbitrary mean a, we can find from the formula

$$\text{S.D.} = \sqrt{\frac{\sum_{i=1}^{n} d_i^2}{N} - \left(\frac{\sum_{i=1}^{n} d_i}{N}\right)^2}, \qquad \text{...(6.28)}$$

where $d_i = x_i - a$, $\sum d_i^2, \sum d_i$ and N denote the sum of squares of deviations, the sum of the deviations and the number of items respectively.

(*c*) **Standard deviation in case of grouped data — Discrete frequency distribution**

$$\text{S.D.} = \sqrt{\frac{\sum_{i=1}^{K} f_i (x_i - \bar{x})^2}{\sum_{i=1}^{K} f_i}} = \sqrt{\frac{\sum_{i=1}^{K} f_i (x_i - \bar{x})^2}{N}}. \qquad \text{...(6.29)}$$

(*d*) **Standard deviation in case of grouped data without using mean**

(*i*) **Direct method:** Find the mid-point of the class interval, square it, multiply it with the frequencies and find the sum. Similarly, find the sum of product of mid-point of the class with frequencies. The formula is

$$\text{S.D.} = \frac{1}{N}\sqrt{N \sum f_i x_i^2 - \left(\sum f_i x_i\right)^2}. \qquad \text{...(6.30)}$$

(*ii*) **Deviation Method:** This method can be used only when the class-intervals are of the same size. In this method, we choose a point as assumed mean say a. We have

$$\sigma = h\sqrt{\frac{\Sigma f u_x^2}{N} - \left(\frac{\Sigma f u_x}{N}\right)^2}, \quad \text{where } u_x = \frac{x-a}{h}. \qquad \text{..(6.31)}$$

EXAMPLE 6.33 Find the standard deviation, from the following ungrouped data 12, 6, 7, 3, 15, 10, 18, 5 by using mean as well as not using mean.

Solution (*i*) ***Using mean.*** The mean of the given set of observations is given by

$$\bar{x} = \frac{\sum_{i=1}^{8} x_i}{N} = \frac{12+6+7+3+15+10+18+5}{8} = \frac{76}{8} = 9.5.$$

The respective $(x_i - \bar{x})^2$ are $2.5^2, 3.5^2, 2.5^2, 6.5^2, 5.5^2, 0.5^2$, 8.5^2 and 4.5^2 so that

$$\sum_{i=1}^{8} (x_i - \bar{x})^2 = 6.25 + 12.25 + 6.25 + 42.25 + 30.25 + 0.25 + 72.25 + 20.25 = 190.$$

$$\text{Standard Deviation } \sigma = \sqrt{\frac{\sum_{i=1}^{8} (x_i - \bar{x})^2}{N}} = \sqrt{\frac{190}{8}} = \sqrt{23.75} = 4.873.$$

(*ii*) **Finding standard deviation without using Arithmetic mean.**

(*a*) *Direct method:*

x_i :	12	6	7	3	15	10	18	5	$\Sigma x_i = 76$
x_i^2 :	144	36	49	9	225	100	324	25	$\Sigma x_i^2 = 912$

$$\text{S.D.} = \sqrt{\frac{\sum x_i^2}{N} - \left(\frac{\sum x_i}{N}\right)^2} = \sqrt{\frac{912}{8} - \left(\frac{76}{8}\right)^2} = \sqrt{114 - 90.25} = 4.873.$$

(*b*) Deviation method:

Take Assume Mean, $a = 10$.

x_i	$d_x = x_i - a$	d_x^2
12	2	4
6	–4	16
7	–3	9
3	–7	49
15	5	25
$a \to 10$	0	0
18	8	64
5	–5	25
Total	$\Sigma d_x = -4$	$\Sigma d_x^2 = 192$

$$\text{S.D.} = \sqrt{\frac{\Sigma d_x^2}{N} - \left(\frac{\Sigma d_x}{N}\right)^2}$$

$$= \sqrt{\frac{192}{8} - \left(\frac{-4}{8}\right)^2} = \sqrt{24 - 0.25} = 4.873.$$

EXAMPLE 6.34 Calculate the mean, variance and standard deviation for the following data:

Size of item:	6	7	8	9	10	11	12
Frequency:	3	6	9	13	8	5	4

(*AMIE, W-2006; VTU, 2001*)

Solution Denoting the size of item by x_i, the calculations for finding mean and standard deviation are shown in the following table.

x_i	f_i	$f_i x_i$	$x_i - \bar{x}$	$(x_i - \bar{x})^2$	$f_i(x_i - \bar{x})^2$
6	3	18	–3	9	27
7	6	42	–2	4	24
8	9	72	–1	1	9
9	13	117	0	0	0
10	8	80	1	1	8
11	5	55	2	4	20
12	4	48	3	9	36
Total	$\Sigma f = 48$	$\Sigma f_i x_i = 432$	**0**	**28**	$\Sigma f_i(x_i - \bar{x})^2 = 124$

Here $N = \Sigma f_i = 48$, $\sum_{i=1}^{7} f_i x_i = 432$, $\sum_{i=1}^{7} f_i (x_i - \bar{x})^2 = 124$. Therefore, $\bar{x} = \frac{\Sigma f_i x_i}{N} = \frac{432}{48} = 9.$

$$\text{Variance } (\sigma^2) = \frac{\sum_{i=1}^{7} f_i (x_i - \bar{x})^2}{N} = \frac{124}{48} = 2.583 \text{ and standard deviation } (\sigma) = \sqrt{2.583} = 1.607.$$

EXAMPLE 6.35 Calculate the mean, variance and standard deviation for the following distribution:

Class interval:	30–40	40–50	50–60	60–70	70–80	80–90	90–100
Frequencies:	3	7	12	15	8	3	2

Solution: From the given data, we construct the following table.

Class interval	Frequency (f_i)	Mid-point (x_i)	f_ix_i	$d_x = (x_i - \bar{x})$	d_x^2	$f_id_x^2$
30–40	3	35	105	–27	729	2187
40–50	7	45	315	–17	289	2023
50–60	12	55	660	–7	49	588
60–70	15	65	975	3	9	135
70–80	8	75	600	13	169	1352
80–90	3	85	255	23	529	1587
90–100	2	95	190	33	1089	2178
Total	$N = \Sigma f_i = 50$		**3100**			**10050**

$$\text{Mean, } \bar{x} = \frac{\sum_{i=1}^{7} f_ix_i}{N} = \frac{3100}{50} = 62.$$

$$\text{Variance } (\sigma^2) = \frac{1}{N}\sum_{i=1}^{7} f_id_x^2 = \frac{1}{50} \times 10050 = 201 \text{ and standard deviation } (\sigma) = \sqrt{201} = 14.18.$$

EXAMPLE 6.36 Calculate the standard deviation for the following distribution:

Class-interval:	0–5	5–10	10–15	15–20	20–25	25–30	30–35	35–40
Frequency:	2	5	7	13	21	16	8	3

Solution Denoting the mid value by x_i and taking h=5.

Let the assumed mean $a = 22.5$. $u_x = (x_i - a)/h = (x_i - 22.5)/5$. The following table is prepared.

Class-interval	Frequency (f_i)	Mid-point (x_i)	$u_x = (x_i - a)/h$	u_x^2	f_iu_x	$f_iu_x^2$
0–5	2	2.5	–4	16	–8	32
5–10	5	7.5	–3	9	–15	45
10–15	7	12.5	–2	4	–14	28
15–20	13	17.5	–1	1	–13	13
20–25	21	$a \to 22.5$	0	0	0	0
25–30	16	27.5	1	1	16	16
30–35	8	32.5	2	4	16	32
35–40	3	37.5	3	9	9	27
Total	$N = \Sigma f_i = 75$				$\Sigma f_iu_x = -9$	$\Sigma f_iu_x^2 = 193$

We have, $$\text{standard deviation } (\sigma) = \frac{h}{N}\sqrt{N\Sigma f_iu_x^2 - (\Sigma f_iu_x)^2} = \frac{5}{75}\sqrt{(75)(193) - (-9)^2}$$

$$= \frac{1}{15}\sqrt{14475 - 81} = \frac{1}{15} \times 119.975 \approx 8.0.$$

EXAMPLE 6.37 For a group of 200 candidates, the mean and standard deviation were found to be 40 and 15 respectively. Later on it was discovered that the scores 43 and 35 were misread as 34 and 53 respectively. Find the corrected mean and standard deviation corresponding to the corrected figures.

Solution We have $n = 200$, incorrect mean $(\bar{x}) = 40$ and incorrect standard deviation $(\sigma) = 15$.

We know, that $\bar{x} = \frac{1}{N}\sum_{i=1}^{200} x_i \quad \Rightarrow \quad$ Incorrect $\sum_{i=1}^{200} x_i = N\bar{x} = 200\,(40) = 8000$

and $$\sigma^2 = \frac{\Sigma x_i^2}{N} - \left(\frac{\Sigma x_i}{N}\right)^2 = \frac{\Sigma x_i^2}{N} - \bar{x}^2$$

$\Rightarrow$ Incorrect $\sum x_i^2 = N(\sigma^2 + \bar{x}^2) = 200\ (225 + 1600) = 365000.$

Therefore, corrected $\sum x_i = 8000 - 34 - 53 + 43 + 35 = 7991,$

and corrected $\sum x_i^2 = 365000 - (34)^2 - (53)^2 + (43)^2 + (35)^2 = 364109.$

Thus, corrected mean $= \dfrac{7991}{200} = 39.955$

and corrected standard deviation $= \sqrt{\dfrac{364109}{200} - (39.955)^2} = \sqrt{1820.545 - 1596.402} = \sqrt{224.143} = 14.97.$

6.20 VARIANCE OF THE COMBINED SERIES

Let the standard deviations of two sets containing N_1 and N_2 members be σ_1 and σ_2 respectively, being measured from their respective mean $\bar{x}_1$ and $\bar{x}_2$. If the two sets are grouped together as one set of $(N_1 + N_2)$ members, then the S.D, σ of this set measured from its mean is given by

$$\sigma^2 = \frac{N_1\sigma_1^2 + N_2\sigma_2^2}{N_1 + N_2} + \frac{N_1N_2}{(N_1 + N_2)^2}(\bar{x}_1 - \bar{x}_2)^2. \qquad ...(6.32)$$

6.21 RELATION BETWEEN QUARTILE DEVIATION, MEAN DEVIATION AND STANDARD DEVIATION

(*a*) Quartile deviation $= \dfrac{2}{3}$ (standard deviation). (*b*) Mean deviation $= \dfrac{4}{5}$ (standard deviation).

This may be represented as 4 SD = 5 MD = 6 SD. ...(6.33)

6.22 COEFFICIENT OF VARIATION (C.V.)

The coefficient of variation is defined as

$$\text{C.V.} = \frac{S.D}{A.M} \times 100 = \frac{\sigma}{\bar{x}} \times 100 \qquad ...(6.34)$$

where σ and $\bar{x}$ are the standard deviation and mean of the data.

For comparing the variability or dispersion of two series, we calculate the coefficient of variance for each series. The series having greater C.V. is said to be more variable than the other. Whereas, the series having lesser C.V. is said to be more consistent than the other.

EXAMPLE 6.38 Fluctuations in the rates of two articles x and y are given below. Find out which of the two shows greater variability.

x:	618	619	616	623	620	624	622	625	622	625	626	625	
y:	2152.5	2132.5	2134.25	2132.5	2145	2142.5	2146.25	2130	2146.25	2142.2	2150	2135	2152.5

Solution Arranging the x series in tabular form. Take assumed mean $a = 622$, we obtain

x_i	f	$d_x = x_i - 622$	d_x^2	fd_x	fd_x^2
616	1	–6	36	–6	36
618	1	–4	16	–4	16
619	1	–3	9	–3	9
620	1	–2	4	–2	4
$a \rightarrow$ 622	2	0	0	0	0
623	1	1	1	1	1
624	1	2	4	2	4
625	3	3	9	9	27
626	1	4	16	4	16
Total	$\Sigma f_i = 12$			$\Sigma fd_x = 1$	$\Sigma fd_x^2 = 113$

Mean, $\bar{x} = 622 + (1/12) = 622.0833.$ $\sigma_x = \sqrt{\frac{\Sigma f d_x^2}{N} - \left(\frac{\Sigma f d_x}{N}\right)^2} = \sqrt{\frac{113}{12} - \left(\frac{1}{12}\right)^2} = \sqrt{\frac{1355}{144}} = 3.0675.$

Coefficient of variation of x series $= \frac{\sigma_x}{\bar{x}} \times 100 = \frac{3.0675}{622.0833} \times 100 = 0.4931.$

Arranging y series in tabular form. Take assumed mean, $b = 2141.5$, we obtain

y_i	f_i	$d_y = y_i - 2141.5$	d_y^2	$f_i d_y$	$f_i d_y^2$
2130.0	1	−11.5	132.25	−11.5	132.25
2132.5	2	−9.0	81.00	−18.0	162.00
2134.25	1	−7.25	52.5625	−7.25	52.5625
2135	1	−6.5	42.25	−6.5	42.25
$b \to$ 2142.5	2	1.0	1.00	2.0	2.00
2145	1	3.5	12.25	3.5	12.25
2146.25	2	4.75	22.5625	9.5	45.125
2150	1	8.5	72.25	8.5	72.25
2152.5	2	11.0	121.00	22.0	242.00
Total	$N = 13$			$\Sigma f_i d_y = 2.25$	$\Sigma f_i d_y^2 = 762.6875$

Mean, $\bar{y} = 2141.5 + \frac{[illegible]}{13} = 2141.5 + 0.1731 = 2141.6731.$

$$\sigma_y = \sqrt{\frac{762.6875}{13} - \left(\frac{2.25}{13}\right)^2} = \sqrt{58.6682 - 0.0299} = \sqrt{58.6383} = 7.6576.$$

Coefficient of variation of y series $= \frac{\sigma_y}{\bar{y}} \times 100 = \frac{7.6576}{2141.6731} \times 100 = 0.3575.$

Since C.V for x is greater than C.V. for y, therefore, x shows greater variability.

EXAMPLE 6.39 From the prices of shares x and y below, state which share is stable or more consistent in value:

x:	55	54	52	53	56	58	52	50	51	49
y:	108	107	105	105	106	107	104	103	104	101

Solution We have to calculate coefficient of variation (CV) for two series as follows:
Take assumed mean for series $x = 53$ and series $y = 104$.

x_i	$d_x = x_i - a$	d_x^2	y_i	$d_y = y_i - b$	d_y^2
55	2	4	108	4	16
54	1	1	107	3	9
52	−1	1	105	1	1
$a \to$ 53	0	0	105	1	1
56	3	9	106	2	4
58	5	25	107	3	9
52	−1	1	$b \to$ 104	0	0
50	−3	9	103	−1	1
51	−2	4	104	0	0
49	−4	16	101	−3	9
Total	$\Sigma d_x = 0$	$\Sigma d_x^2 = 70$		$\Sigma d_y = 10$	$\Sigma d_y^2 = 50,\ n_1 = n_2 = 10$

For x-series: Mean $\bar{x} = a + \frac{\Sigma d_x}{n} = 53 + \frac{0}{10} = 53.$

$$\sigma_x = \sqrt{\frac{\Sigma d_x^2}{n_1} - \left(\frac{\Sigma d_x}{n_1}\right)^2} = \sqrt{\frac{70}{10} - \left(\frac{0}{10}\right)^2} = \sqrt{7} \approx 2.65.$$

Therefore, coefficient of variation for $x = \frac{\sigma_x}{\overline{x}} \times 100 = \frac{2.65}{53} \times 100 = 5.$

For y-series: Mean $\overline{y} = b + \frac{\Sigma d_y}{n_2} = 104 + \left(\frac{10}{10}\right) = 105.$

$$\sigma_y = \sqrt{\frac{\Sigma d_y^2}{n_2} - \left(\frac{\Sigma d_y}{n_2}\right)^2} = \sqrt{\frac{50}{10} - \left(\frac{10}{10}\right)^2} = \sqrt{4} \approx 2.$$

Therefore, C.V for $y = \frac{\sigma_y}{\overline{y}} \times 100 = \frac{2}{105} \times 100 = 1.905.$

Since C.V for y is less than C.V for x, therefore the share y is more consistent or stable in value.

EXAMPLE 6.40 The means of two samples of size 50 and 100 respectively are 54.1 and 50.3 and the standard deviation 8 and 7. Obtain the mean and S.D of the sample of size 150 obtained by combining the two samples. Also find the coefficient of variations.

Solution We have $\overline{x} = \frac{n_1\overline{x}_1 + n_2\overline{x}_2}{n_1 + n_2} = \frac{50\,(54.1) + 100\,(50.3)}{50 + 100} = 51.567,$

and
$$\sigma^2 = \frac{n_1\sigma_1^2 + n_2\sigma_2^2}{n_1 + n_2} + \frac{n_1 n_2}{(n_1 + n_2)^2}(\overline{x}_1 - \overline{x}_2)^2$$

$$= \frac{50\,(8)^2 + 100\,(7)^2}{50 + 100} + \frac{50\,(100)}{(50 + 100)^2}(54.1 - 50.3)^2 = 54 + 0.096 = 54.096\,.$$

Therefore, $\sigma \approx 7.355.$

Coefficient of variation (C.V) $= \frac{\sigma}{\overline{x}} \times 100 = \frac{7.355}{51.567} \times 100 = 14.263.$

6.23 MOMENTS—INTRODUCTION

The term *moment* is widely used in the topics of General Mechanics of Physics. However, the term *moment* used in statistics has a completely different meaning. In Physics, the moment of a force acting on a rigid body about a point determines the tendency of the body to rotate about the point. In statistics also, the moment has some significance. It describes various characteristics of a frequency distribution like central tendency of a set of observations, their variation, skewness and Kurtosis.

6.24 MOMENTS ABOUT AN ARBITRARY POINT '*a*'

The rth moment or the moment of order r, where r is a non-negative integer, of a random variable x about any point $x = a$ is usually denoted by μ'_r

and is defined by
$$\mu'_r = \frac{1}{N}\sum_{i=1}^{n} f_i\,(x_i - a)^r \qquad \text{...(6.35)}$$

where $x_1, x_2, ..., x_n$ are the values of the random variable x_i and $f_1, f_2, ..., f_n$ are the corresponding frequencies and $N = \sum_{i=1}^{n} f_i$.

Thus the r th moment of a random variable can be regarded as the arithmetic mean of the variate $(x - a)^r$. Setting $r = 0$ and 1 in Eq. (6.35), we obtain

$$\mu_0' = \frac{1}{N}\sum_{i=1}^{n} f_i\,(x_i - a)^\circ = \frac{1}{N}\sum_{i=1}^{n} f_i = \frac{N}{N} = 1. \qquad ...(6.36)$$

$$\mu_1' = \frac{1}{N}\sum_{i=1}^{n} f_i\,(x_i - a) = \frac{1}{N}\sum_{i=1}^{n} f_i x_i - \frac{a}{N}\sum_{i=1}^{n} f_i = \bar{x} - a = d, \text{ (say)} \qquad ...(6.37)$$

where $\bar{x}$ is the mean of the given frequency distribution.

The 2nd, 3rd and 4th moments, about any arbitrary point $x = a$ can be easily derived from Eq. (6.35).

Thus $\mu_2' = \frac{1}{N}\sum_{i=1}^{n} f_i\,(x_i - a)^2, \quad \mu_3' = \frac{1}{N}\sum_{i=1}^{n} f_i\,(x_i - a)^3, \quad \mu_4' = \frac{1}{N}\sum_{i=1}^{n} f_i\,(x_i - a)^4.$

Remark 4

In particular, if we take $a = 0$, we get

$$\mu_1' = \bar{x} = \text{mean}, \quad \mu_2' = \frac{1}{N}\sum_{i=1}^{n} f_i x_i^2, \quad \mu_3' = \frac{1}{N}\sum_{i=1}^{n} f_i x_i^3 \text{ and } \mu_4' = \frac{1}{N}\sum_{i=1}^{n} f_i x_i^4,$$

that is, first, second, third and fourth moments about the origin.

6.25 MOMENTS ABOUT THE MEAN

The r th moment of x about the mean $\bar{x}$ is denoted by μ_r and is defined as

$$\mu_r = \frac{1}{N}\sum_{i=1}^{n} f_i\,(x_i - \bar{x})^r, \text{........} \qquad ...(6.38)$$

where $N = \sum_{i=1}^{n} f_i, \quad r = 0, 1, 2, 3.$

In particular,

$$\mu_0 = \frac{1}{N}\sum_{i=1}^{n} f_i = \frac{N}{N} = 1. \qquad ...(6.39)$$

$$\mu_1 = \frac{1}{N}\sum_{i=1}^{n} f_i\,(x_i - \bar{x}) = \frac{1}{N}\sum_{i=1}^{n} f_i x_i - \frac{\bar{x}}{N}\sum_{i=1}^{n} f_i = \bar{x} - \bar{x} = 0.$$

$$\mu_2 = \frac{1}{N}\sum_{i=1}^{n} f_i\,(x_i - \bar{x})^2 = \text{var}\,(x) \quad \text{or} \quad \sigma_x^2,$$

thus the second moment about the mean is equal to the variance.

$$\mu_3 = \frac{1}{N}\sum_{i=1}^{n} f_i\,(x_i - \bar{x})^3 \text{ and } \mu_4 = \frac{1}{N}\sum_{i=1}^{n} f_i\,(x_i - \bar{x})^4.$$

The moment about the mean are also called the central moments.

6.26 RELATION BETWEEN THE MOMENTS ABOUT THE MEAN AND MOMENTS ABOUT ANY POINT – RELATION BETWEEN μ_r AND μ_r'

We have by definition

$$\mu_r = \frac{1}{N}\sum_{i=1}^{n} f_i\,(x_i - \bar{x})^r = \frac{1}{N}\sum_{i=1}^{n} f_i\,\{(x_i - a) - (\bar{x} - a)\}^r$$

$$= \frac{1}{N}\sum_{i=1}^{n} f_i\,(x_i - d)^r, \text{........} \qquad ...(6.40)$$

where $X_i = x_i - a$ and $d = \bar{x} - a = \mu_1'$ [from Eq. (6.37)]

Therefore, $$\mu_r' = \frac{1}{N}\sum_{i=1}^{n} f_i (x_i - a)^r = \frac{1}{N}\sum_{i=1}^{n} f_i X_i^r, \ldots\ldots \qquad \ldots(6.41)$$

Using Binomial expansion, we obtain from Eq. (6.40)

$$\mu_r = \frac{1}{N}\sum_{i=1}^{n} f_i [X_i^r - {}^rC_1 X_i^{r-1} d + {}^rC_2 X_i^{r-2} d^2 - \cdots\cdots + {}^rC_{r-1} X_i (-1)^{r-1} d^{r-1} + (-1)^r d^r]$$

$$= \frac{1}{N}\left[\sum_{i=1}^{n} f_i X_i^r - {}^rC_1 d \sum_{i=1}^{n} f_i X_i^{r-1} + {}^rC_2 d^2 \sum_{i=1}^{n} f_i X_i^{r-2} + \cdots\cdots \right.$$

$$\left. \cdots + {}^rC_{r-1} (-1)^{r-1} d^{r-1} \sum_{i=1}^{n} f_i X_i + (-1)^r d^r \sum_{i=1}^{n} f_i \right]$$

$$= \mu_r' - {}^rC_1 d \mu_{r-1}' + {}^rC_2 d^2 \mu_{r-2}' - \cdots + (-1)^{r-1} r d^{r-1} \mu_1' + (-1)^r d^r \qquad \text{[using (6.41)]}$$

$$= \mu_r' - {}^rC_1 \mu_1' \mu_{r-1}' + {}^rC_2 \mu_1'^2 \mu_{r-2}' - \cdots + (-1)^{r-1} [r \mu_1'^r - \mu_1'^r]$$

$$= \mu_r' - {}^rC_1 \mu_1' \mu_{r-1}' + {}^rC_2 \mu_1'^2 \mu_{r-2}' - \cdots + (-1)^{r-1} (r-1) \mu_1'^r. \qquad \ldots(6.42)$$

In particular taking r = 2, 3, 4 in Eq. (6.42), we obtain

$$\mu_2 = \mu_2' - \mu_1'^2.$$

$$\mu_3 = \mu_3' - 3\mu_2' \mu_1' + 2\mu_1'^3.$$

$$\mu_4 = \mu_4' - 4\mu_3' \mu_1' + 6\mu_2' \mu_1'^2 - 3\mu_1'^4.$$

6.27 KARL PEARSON'S BETA (β) AND GAMMA (γ) COEFFICIENTS BASED ON MOMENTS

Karl Pearson defined the following four coefficients, based upon the first four moments of a frequency distribution about it mean:

$$\beta_1 = \frac{\mu_3^2}{\mu_2^3} = \frac{\mu_3^2}{\sigma^6}, \quad \ldots(6.43) \qquad \beta_2 = \frac{\mu_4}{\mu_2^2} = \frac{\mu_4}{\sigma^4}, \quad \ldots(6.44) \qquad (\beta\text{-coefficients})$$

$$\gamma_1 = +\sqrt{\beta_1} = \frac{\mu_3}{\sigma^3}, \quad \ldots(6.45) \qquad \gamma_2 = \beta_2 - 3 = \frac{\mu_4 - 3\sigma^4}{\sigma^4}. \quad \ldots(6.46) \qquad (\gamma\text{-coefficients})$$

The practical utility of these coefficient is to measure the skewness and kurtosis of a frequency distribution. It should be noted that these coefficients are all pure numbers.

It may be remarked that it is only the first four moments which are of practical importance in the study of frequency distribution. As remarked earlier, the mean is the first moment about origin and the variance the second moment about the mean. The third moment about the mean is the measure of skewness while the fourth moment defines kurtosis, that is, the flatness of the curve.

EXAMPLE 6.41 The first four moments of a distribution about the value 5 of the variable are 2, 20, 40 and 50. Obtain the moments about (*i*) mean and (*ii*) origin. Also find β_1 and β_2.

Solution Let μ_1', μ_2', μ_3' and μ_4' be the first four moments of the given distribution about the value 5. Then $a = 5$, $\mu_1' = 2$, $\mu_2' = 20$, $\mu_3' = 40$ and $\mu_4' = 50$.

(*i*) *Computation of moments about mean*

Let μ_1, μ_2, μ_3 and μ_4 be first four moments about the mean. We know that $\mu_1 = 0$ for every distribution.

Now, $\mu_2 = \mu_2' - \mu_1'^2 = 20 - (2)^2 = 16.$

$$\mu_3 = \mu_3' - 3\mu_2' \mu_1' + 2\mu_1'^3 = 40 - 3\,(20)\,(2) + 2(2)^3 = -64.$$

$$\mu_4 = \mu_4' - 4\mu_3'\mu_1' + 6\mu_2'\mu_1'^2 - 3\mu_1'^4$$

$$= 50 - 4(40)(2) + 6(20)(2)^2 - 3(2)^4 = 50 - 320 + 480 - 48 = 162.$$

Note We know $\mu_1' = \bar{x} - a \Rightarrow 2 = \bar{x} - 5$ or $\bar{x} = 7$ that is, mean is 7 and μ_2 = variance = 16.

(ii) Computation of Moments about the Origin

Let the first four moments about zero be denoted by ν_1, ν_2, ν_3 and ν_4.

To find the moments about the origin, that is, when $a = 0$, we have

$$\nu_1 = \frac{1}{N}\Sigma f(x-0) = \frac{1}{N}\Sigma f\overline{(x-7+7)} = \mu_1 + 7 = 7. \qquad (\because \mu_1 = 0)$$

$$\nu_2 = \frac{1}{N}\Sigma f(x-0)^2 = \frac{1}{N}\Sigma f\overline{(x-7+7)}^2 = \frac{1}{N}\Sigma f(x-7)^2 + (2)(7)\frac{1}{N}\Sigma f(x-7) + 7^2$$

$$= \mu_2 + 14\mu_1 + 49 = 16 + 0 + 49 = 65. \qquad (\because \mu_1 = 0)$$

$$\nu_3 = \frac{1}{N}\Sigma f(x-0)^3 = \frac{1}{N}\Sigma f\overline{(x-7+7)}^3$$

$$= \frac{1}{N}\Sigma f(x-7)^3 + (3)(7)\frac{1}{N}\Sigma f(x-7)^2 + 3(7)^2\frac{1}{N}\Sigma f(x-7) + (7)^3$$

$$= \mu_3 + 21\mu_2 + 147\mu_1 + 343 = -64 + 21(16) + 0 + 343 = 615.$$

and
$$\nu_4 = \frac{1}{N}\Sigma f(x-0)^4 = \frac{1}{N}\Sigma f\overline{(x-7+7)}^4$$

$$= \mu_4 + {}^4C_1(7)\mu_3 + {}^4C_2(7)^2\mu_2 + {}^4C_3(7)^3\mu_1 + (7)^4$$

$$= 162 + (4)(7)(-64) + 6(49)(16) + 4(343)(0) + 2401$$

$$= 162 - 1792 + 4704 + 2401 = 5475.$$

Values of β_1 and β_2

We know that $\beta_1 = \dfrac{\mu_3^2}{\mu_2^3}$ and $\beta_2 = \dfrac{\mu_4}{\mu_2^2} \Rightarrow \beta_1 = \dfrac{(-64)^2}{(16)^3} = 1$ and $\beta_2 = \dfrac{162}{(16)^2} = 0.6328.$

EXAMPLE 6.42 Calculate the first four moments about the mean from the following data:

Marks:	0–10	10–20	20–30	30–40	40–50	50–60	60–70
No. of students:	8	12	20	30	15	10	5

Solution We shall first compute moments about an arbitrary point $a = 35$ (say).

Marks	Mid-point x_i	Frequency f	$u_x = \dfrac{x_i - 35}{10}$	$f_i u_i$	$f_i u_i^2$	$f_i u_i^3$	$f_i u_i^4$
0–10	5	8	–3	–24	72	–216	648
10–20	15	12	–2	–24	48	–96	192
20–30	25	20	–1	–20	20	–20	20
30–40	$a \to 35$	30	0	0	0	0	0
40–50	45	15	1	15	15	15	15
50–60	55	10	2	20	40	80	160
60–70	65	5	3	15	45	135	405
		N = Σf_i = 100		**$\Sigma f_i u_i = -18$**	**$\Sigma f_i u_i^2 = 240$**	**$\Sigma f_i u_i^3 = -102$**	**$\Sigma f_i u_i^4 = 1440$**

We have $h = 10$, $N = 100$, $\Sigma f_i u_i = -18$, $\Sigma f_i u_i^2 = 240$, $\Sigma f_i u_i^3 = -102$, $\Sigma f_i u_i^4 = 1440$.

$$\therefore \quad \mu_1' = h\left(\frac{1}{N}\sum f_i u_i\right) = 10\left(\frac{-18}{100}\right) = -1.8, \quad \mu_2' = h^2\left(\frac{1}{N}\sum f_i u_i^2\right) = 100\left(\frac{240}{100}\right) = 240,$$

$$\mu_3' = h^3\left(\frac{1}{N}\sum f_i u_i^3\right) = 1000\left(\frac{-102}{100}\right) = -1020 \text{ and } \mu_4' = h^4\left(\frac{1}{N}\sum f_i u_i^4\right) = 10000\left(\frac{1440}{100}\right) = 144000.$$

Computation of Moments about Mean

We know that $\mu_1 = 0$

$$\mu_2 = \mu_2' - \mu_1'^2 = 240 - (-1.8)^2 = 236.76.$$

$$\mu_3 = \mu_3' - 3\mu_2'\mu_1' + 2\mu_1'^3 = -1020 - 3\,(240)\,(-1.8) + 2\,(-1.8)^3 = -1020 + 1296 - 11.664 = 264.336.$$

$$\mu_4 = \mu_4' - 4\mu_3'\mu_1' + 6\mu_2'\mu_1'^2 - 3\mu_1'^4 = 144000 - 4\,(-1020)\,(-1.8) + 6\,(240)\,(-1.8)^2 - 3\,(-1.8)^4$$

$$= 144000 - 7344 + 4665.6 - 31.4928 = 141290.11.$$

EXAMPLE 6.43 For a distribution the mean is 10, variance is 16, γ_1 is 1 and β_2 is 4. Find the first four moments about the origin.

Solution Let $\bar{x}$ be the mean and μ_1, μ_2, μ_3 and μ_4 be first four moments about the mean.

It is given that $\bar{x} = 10$, variance $= 16$, $\gamma_1 = 1$ and $\beta_2 = 4$.

We know that $\mu_2 =$ variance $= 16$.

$$\gamma_1 = 1 \Rightarrow \sqrt{\beta_1} = 1 \Rightarrow \beta_1 = 1 \Rightarrow \frac{\mu_3^2}{\mu_2^3} = 1 \Rightarrow \mu_3^2 = \mu_2^3 = 16^3 \Rightarrow \mu_3 = 64.$$

$$\beta_2 = 4 = \frac{\mu_4}{\mu_2^2}. \quad \text{Therefore } \mu_4 = 4\mu_2^2 = 4\,(16)^2 = 1024.$$

Moments about the origin

$$v_1 = \frac{1}{N}\Sigma f(x-0) = \frac{1}{N}\Sigma f(\overline{x-10}+10) = \mu_1 + 10 = 10. \qquad (\because \mu_1 = 0)$$

$$v_2 = \frac{1}{N}\Sigma f(x-0)^2 = \frac{1}{N}\Sigma f(\overline{x-10}+10)^2 = \mu_2 + 20\mu_1 + 100 = 16 + 0 + 100 = 116.$$

$$v_3 = \frac{1}{N}\Sigma f(x-0)^3 = \frac{1}{N}\Sigma f(\overline{x-10}+10)^3 = \mu_3 + 3(10)\mu_2 + 3(10)^2\mu_1 + 10^3 = 64 + 30\,(16) + 0 + 1000 = 1544.$$

$$v_4 = \frac{1}{N}\Sigma f(x-0)^4 = \frac{1}{N}\Sigma f(\overline{x-10}+10)^4 = \mu_4 + 40\,\mu_3 + 600\,\mu_2 + 4000\,\mu_1 + 10000.$$

$$= 1024 + 40\,(64) + 600\,(16) + 0 + 10000 = 23184.$$

6.28 SKEWNESS

The term '*Skewness*' refer to lack of symmetry, or departure from symmetry of a distribution. For a symmetrical distribution, mean = mode = median, skewness = 0. In Fig. 6.8, of a symmetrical distribution the regions L and M are mirror images of each other. In an asymmetrical or skewed distribution the mean, mode and the median don't coincide and therefore, the range is usually greater on one side of the mode than on the other. The curve has a longer tail on one side. The behaviour of the distribution exhibits skewness.

Skewness can be positive as well as negative. If the mean is greater than either the mode or the median, the curve will have the longer tail on the right and skewness will be positive. On the other hand, if the curve has a longer tail on the left, the mean will be less than the mode and median and the skewness will be negative as is exhibited by the following figures. The median lies between the mode and the mean in both the cases.

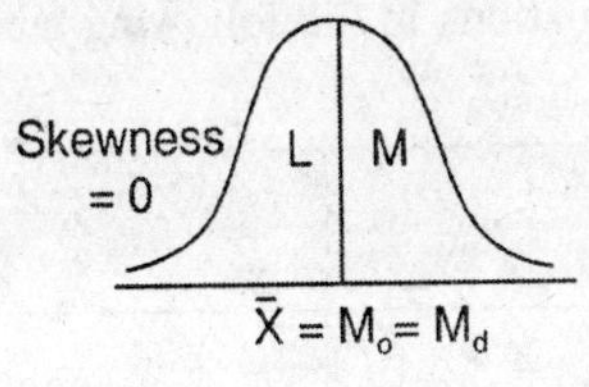

Fig. 6.8.

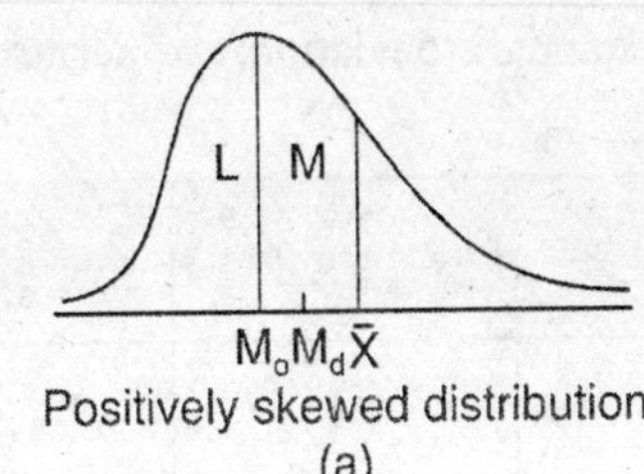

Positively skewed distribution
(a)

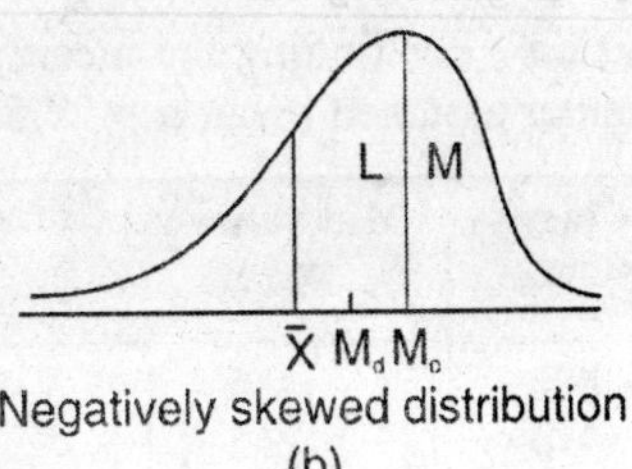

Negatively skewed distribution
(b)

Fig. 6.9.

6.29 MEASURES OF SKEWNESS

The aim of measuring skewness is to estimate the direction in which the distribution is distorted from symmetrical distribution.

We compute the relative measure of skewnes, which is called the coefficient of skewness. This coefficient of skewness is given by the following formulae:

(A) Karl Pearson's Coefficient of Skewness

Karl Pearson's coefficient of skewness $$S_{k_p} = \frac{\text{Mean} - \text{Mode}}{\text{Standard Deviation}} = \frac{\bar{x} - \text{Mode}}{\sigma}. \qquad ...(6.47)$$

This formula cannot, however, be used when the mode is ill-defined. In such cases, we can employ the empirical formula:

$$\text{Mean} - \text{Mode} = 3\ (\text{Mean} - \text{Median}) \text{ and define}$$

$$\text{coefficient of skewness} = \frac{3(\text{Mean} - \text{Median})}{\text{Standard Deviation}} = \frac{3(\bar{x} - \text{Median})}{\sigma}. \qquad ...(6.48)$$

If $S_{k_p} = 0$, the curve is symmetrical. If $S_{k_p} < 0$, the curve is negative skew and if $S_{k_p} > 0$, the curve is positive skew.

(B) Bowley's Coefficient of Skewness

Bowley's proposed the following measure of skewness based on the quartiles and is defined as

$$\text{Quartile coefficient of skewness, } S_{k_B} = \frac{Q_3 + Q_1 - 2Q_2}{Q_3 - Q_1} = \frac{(Q_3 + Q_1) - 2\,(\text{Median})}{Q_3 - Q_1}. \qquad ...(6.49)$$

(C) Coefficient of Skewness based on Third Moment

Karl Pearson, an English statistician, defined skewness based on moments as

$$\beta_1 = \frac{\mu_3^2}{\mu_2^3}, \qquad ...(6.50)$$

where μ_2 and μ_3 are second and third moments about mean respectively. β_1 measures skewness.

From Eq. (6.50), we have $\sqrt{\beta_1} = \dfrac{\mu_3}{\sqrt{\mu_2^3}}$. We denotes $\sqrt{\beta_1}$ by γ_1, so that $\gamma_1 = \sqrt{\beta_1} = \dfrac{\mu_3}{\sqrt{\mu_2^3}}$. ...(6.51)

A distribution is symmetric if $\gamma_1 = 0 \Rightarrow \mu_3 = 0$.

The greater the value of β_1, the more skewed is the distribution that is, it enable us to know the extent of symmetry in a series or distribution.

However, the direction of skewness (positive or negative) is determined by γ_1. If $\gamma_1 > 0$, that is, $\mu_3 > 0$, the skewness is positive and if $\gamma_1 < 0$, that is, $\mu_3 < 0$ than the skewness is negative.

EXAMPLE 6.44 Find out standard deviation and the coefficient of skewness for the following distribution:

Variable:	0–5	5–10	10–15	15–20	20–25	25–30	30–35	35–40
Frequency:	2	5	7	13	21	16	8	3

Solution For finding the mean and standard deviation, we perform the calculations in the following table. Consider assumed mean $a = 22.5$.

Class interval	Mid-value x_i	Frequency f_i	$u_x = \frac{x - 22.5}{5}$	u_x^2	$f_i u_x$	$f_i u_x^2$
0–5	2.5	2	–4	16	–8	32
5–10	7.5	5	–3	9	–15	45
10–15	12.5	7	–2	4	–14	28
15–20	17.5	13	–1	1	–13	13
20–25	$a \to 22.5$	21	0	0	0	0
25–30	27.5	16	1	1	16	16
30–35	32.5	8	2	4	16	32
35–40	37.5	3	3	9	9	27
		$N = \Sigma f_i = 75$			$\Sigma f_i u_x = -9$	$\Sigma f_i u_x^2 = 193$

$$\text{Mean } = a + \left(\frac{\Sigma f_i u_x}{\Sigma f_i}\right) \times h = 22.5 - \frac{9 \times 5}{75} = 21.9.$$

$$\text{S.D} (= \sigma) = h\sqrt{\frac{f_i u_x^2}{N} - \left(\frac{f_i u_x}{N}\right)^2} = \sqrt{\left[\frac{193}{75} - \left(\frac{-9}{75}\right)^2\right]} \times 5 \approx 8.$$

The mode is given by

$$\text{Mode } = l_1 + \left(\frac{f_m - f_1}{2f_m - f_1 - f_2}\right) \times h = 20 + \frac{(21 - 13)}{(42 - 13 - 16)} \times 5 = 20 + 3.076 = 23.076.$$

$$\therefore \quad \text{Skewness } = \frac{\text{Mean} - \text{Mode}}{\text{S.D.}} = \frac{(21.9 - 23.076)}{8} = -0.147.$$

EXAMPLE 6.45 From the following table of marks obtained by 10 candidates, find the coefficient of variation and skewness:

Statistics:	25	50	45	30	70	42	36	38	34	60
Mathematics:	10	70	50	20	95	55	42	60	48	80

Solution The calculation work is tabulated as under.

We have $\bar{x} = \Sigma X/N = 430/10 = 43$ and $\bar{y} = \Sigma Y/N = 530/10 = 53$.

Marks in Statistics X	Marks in Maths Y	Dev. of X from $\bar{x} = 43$, $d_x = X - 43$	Dev. of Y from $\bar{y} = 53$, $d_y = Y - 53$	d_x^2	d_y^2
25	10	–18	–43	324	1849
50	70	7	17	49	289
45	50	2	–3	4	9
30	20	–13	–33	169	1089
70	95	27	42	729	1764
42	55	–1	2	1	4
36	42	–7	–11	49	121
38	60	–5	7	25	49
34	48	–9	–5	81	25
60	80	17	27	289	729
$\Sigma X = 430$	$\Sigma Y = 530$			$\Sigma d_x^2 = 1720$	$\Sigma d_y^2 = 5928$

$$\sigma_x = \sqrt{\frac{1}{n}\Sigma d_x^2} = \sqrt{\frac{1720}{10}} = 13.11487, \text{ and } \sigma_y = \sqrt{\frac{1}{n}\Sigma d_y^2} = \sqrt{\frac{1}{10}(5928)} = 24.34748.$$

Therefore, coefficient of variation for $X = \frac{\sigma_x}{\bar{x}} \times 100 = \frac{13.11487}{43} \times 100 = 30.50\,\%$,

and coefficient of variation for $Y = \frac{\sigma_y}{\bar{y}} \times 100 = \frac{24.34748}{53} \times 100 = 45.94\,\%$.

To find the coefficient of skewness, proceed as under

Arrange X items to ascending order of magnitude such as 25, 30, 34, 36, 38, 42, 45, 50, 60, 70

$$\text{Median for } X = \frac{38+42}{2} = 40.$$

Similarly. median for $Y = \frac{50+55}{2} = 52.5$.

Therefore, coefficient of skewness for X $= \frac{3(\text{Mean} - \text{Median})}{\sigma_x} = \frac{3(43-40)}{13.11487} \approx 0.686,$

and coefficient of skewness for Y $= \frac{3(53-52.5)}{24.34748} \approx 0.062.$

EXAMPLE 6.46 Find the mean coefficients of dispersion and coefficient of skewness from the following table:

Wages in Rs.:	70–80	80–90	90–100	100–110	110–120	120–130	130–140	140–150
No. of persons:	12	18	35	42	50	45	20	8

Solution The computation work is shown in the table. Consider assumed mean $a = 115$.

Wages Class-interval	Mid-value (x_i)	No. of persons (Frequency, f_i)	$d_x = x_i - 115$	$u_x = d_x/10$	$f_i u_x$	$f_i u_x^2$
70–80	75	12	–40	–4	–48	192
80–90	85	18	–30	–3	–54	162
90–100	95	35	–20	–2	–70	140
100–110	105	42	–10	–1	–42	42
110–120	$a \to$ 115	50	0	0	0	0
120–130	125	45	10	1	45	45
130–140	135	20	20	2	40	80
140–150	145	8	30	3	24	72
		$N = \Sigma f_i = 230$			$\Sigma f_i u_x = -105$	$\Sigma f_i u_x^2 = 733$

Mean, $\bar{x} = a + \left(\frac{\Sigma f_i u_x}{N}\right) \times h$, (where $N = \Sigma f_i$) $= 115 + (-105/230)10 = 115 - 4.565 = 110.435$.

$$\text{Standard Deviation} = h\sqrt{\frac{\Sigma f u_x^2}{N} - \left(\frac{\Sigma f u_x}{N}\right)^2} = 10\sqrt{\frac{733}{230} - \left(\frac{-105}{230}\right)^2} = 10\sqrt{3.187 - 0.208} = 17.26.$$

Since the maximum frequency is 50, the mode lies in class 110–120. Employing the mode formula,

$$\text{mode} = l_1 + \left(\frac{f_m - f_1}{2f_m - f_1 - f_2}\right) \times h.$$

Substituting these values $l_1 = 110, f_m = 50, f_1 = 42, f_2 = 45$, and $h = 10$ in the above formula.

Therefore, $\quad \text{mode} = 110 + \left(\dfrac{50-42}{100-42-45}\right) \times 10 = 110 + 6.154 = 116.154.$

$$\text{Coefficient of skewness} = \frac{\text{Mean} - \text{Mode}}{\text{S.D}} = \frac{110.435 - 116.154}{17.26} = -0.331.$$

Now, $\Sigma f_i\,|x_i - \bar{x}| = 12\,|75 - 110.435| + 18\,|85 - 110.435| + 35\,|95 - 110.435| + 42\,|105 - 110.435|$
$+ 50\,|115 - 110.435| + 45\,|125 - 110.435| + 20\,|135 - 110.435| + 8\,|145 - 110.435|$
$= 12\,(35.435) + 18\,(25.435) + 35\,(15.435) + 42\,(5.435) + 50\,(4.565) + 45\,(14.565)$
$+ 20\,(24.565) + 8\,(34.565)$
$= 425.22 + 457.83 + 540.225 + 228.27 + 228.25 + 655.425 + 491.3 + 276.52 = 3303.04$

$$\text{M.D (Mean)} = \frac{\Sigma f_i\,|x_i - \bar{x}|}{\Sigma f_i} = \frac{3303.04}{230} = 14.361.$$

$$\text{Therefore, mean coefficient of dispersion} = \frac{\text{Mean deviation from the mean}}{\text{Mean}} = \frac{14.361}{110.435} = 0.13.$$

EXAMPLE 6.47 The sum of 50 observations is 500, its sum of square is 6000 and median 12. Find the coefficient of variation and coefficients of skewness.

Solution We have $n = 50$, $\Sigma x_i = 500$, $\Sigma x_i^2 = 6000$, median = 12. $\bar{x} = \Sigma x_i / N = 500/50 = 10.$

$$\text{S.D} = \sqrt{\frac{\Sigma x_i^2}{N} - (\bar{x})^2} = \sqrt{\frac{6000}{50} - (10)^2} = 4.472.$$

$$\text{Therefore, C.V.} = \frac{\sigma}{\bar{x}} \times 100 = \frac{4.472}{10} \times 100 = 44.72\%.$$

$$\text{Coefficient of skewness, } {}^{S}k_{P} = \frac{3(\bar{x} - \text{Median})}{\sigma} = \frac{3(10-12)}{4.472} = \frac{-6}{4.472} = -1.34.$$

EXAMPLE 6.48 For a moderately skewed distribution, the arithmetic mean is 100 and coefficient of variation is 35% and Karl Pearson's coefficient of skewness is 0.2. Find the mode and median

Solution We have $\bar{x} = 100$, C.V. = 35%, ${}^{S}k_{P} = 0.2$.

Employing $\text{C.V.} = \dfrac{\sigma}{\bar{x}} \times 100$, we get $\sigma = \dfrac{(\text{C.V.})\,\bar{x}}{100} = \dfrac{35 \times 100}{100} = 35.$

We know ${}^{S}k_{P} = \dfrac{\bar{x} - \text{Mode}}{\sigma}$. Therefore, mode $= \bar{x} - \sigma\,({}^{S}k_{P}) = 100 - 35\,(0.2) = 93.$

Employing, Mean – Mode = 3 (Mean – Median) to get

$$\text{Median} = \frac{2\,\text{Mean} + \text{Mode}}{3} = \frac{2(100) + 93}{3} = 97.67.$$

6.30 KURTOSIS

Meaning of Kurtosis

Kurtosis refers to the degree of flatness or peakedness in the region about the mode of a frequency curve relative to a normal distribution. Measures of kurtosis tell us the extent to which a distribution is more peaked or flat topped than the normal curve. If a curve is more peaked than the normal curve, it is known as *leptokurtic* and if it is move flat-topped than the normal curve, it is called *platy kurtic*. The normal curve has been named by the Greek work *mesokurtic*.

6.31 MEASURE OF KURTOSIS

Karl Pearson defined a coefficient $\beta_2 = \dfrac{\mu_4}{\mu_2^2}$, ...(6.52)

where μ_2 is the second moment and μ_4 is the fourth moment about the mean. The value of β_2 measures the kurtosis. The greater the value of β_2, the more peaked is the distribution.

For a normal or mesokurtic curve $\beta_2 = 3$. When $\beta_2 > 3$, the curve is more peaked than the normal curve. Hence the curve is leptokurtic curve. When $\beta_2 < 3$, then the curve is less peaked than the normal curve. Hence, the curve is platykurtic curve.

Another measure of Kurtosis is γ_2 which is given by

$$\gamma_2 = \beta_2 - 3. \quad ...(6.53)$$

For a normal curve, $\gamma_2 = 0$ and so the curve is mesokurtic. If $\gamma_2 > 0$, then the curve is leptokurtic and if $\gamma_2 < 0$, then the curve is platykurtic.

The three types of curves with different shapes are shown in Fig. 6.10.

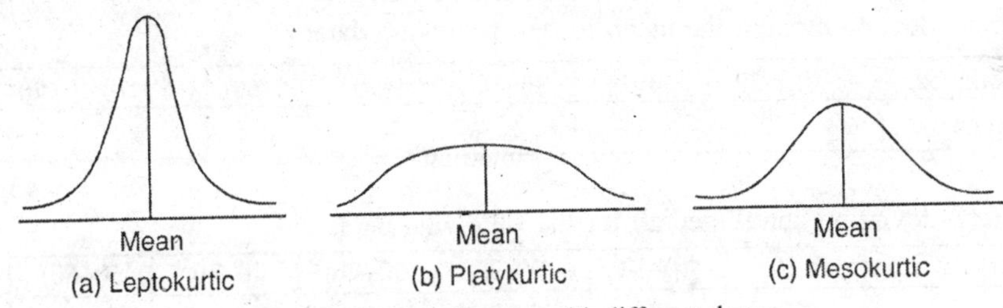

Fig. 6.10: Types of curves with different shapes

EXAMPLE 6.49 The standard deviation of a symmetric distribution is 5. What must be the value of the fourth moment about the mean in order that the distribution be (*i*) leptokurtic, (*ii*) mesokurtic, (*iii*) platykurtic.

Solution We have $\sigma = 5 \Rightarrow \mu_2 = \sigma^2 = 25$. $\beta_2 = \mu_4/\mu_2^2 \Rightarrow \beta_2 = \mu_4/(25)^2 = \mu_4/625$.

The distribution will be (*i*) leptokurtic if $\beta_2 > 3$ that is, if $\dfrac{\mu_4}{625} > 3$ or if $\mu_4 > 1875$.

(*ii*) Mesokurtic if $\beta_2 = 3$, that is, if $\dfrac{\mu_4}{625} = 3$ or if $\mu_4 = 1875$.

(*iii*) Platykurtic if $\beta_2 < 3$, that is, if $\dfrac{\mu_4}{625} < 3$ or if $\mu_4 < 1875$.

EXERCISE 6.2

1. Find the range of each set of numbers: (*a*) 12, 6, 7, 3, 15, 10, 18, 5; (*b*) 9, 3, 8, 8, 9, 8, 9, 18.
   Ans. (*a*) 15, (*b*) 15.
2. Following are the marks of the 10 students: 56, 48, 65, 35, 42 75, 82, 60, 55, 50. Find quartile deviation and also its coefficient.
   Ans. $Q_1 = 46.50$, $Q_3 = 67.50$, $Q.D = 10.50$, coefficient of quartile deviation = 0.1842.
3. Calculate the inter-quartile range, quartile deviation and coefficient of quartile deviation for the following data:

Class interval:	0–4	4–6	6–8	8–12	12–18	18–20
Frequency:	4	6	8	12	7	2

   Ans. 5.83, 2.915, 0.33.
4. What is the range and its coefficient for the following distribution of weights?

Weight in kgs:	50–54	55–59	60–64	65–69	70–74
No. of students:	12	18	23	10	3

Ans. 25 kg, 0.202.

[**Hint:** Lowest class boundary = 49.5 kg, Highest class boundary = 74.50 kgs]

5. Find the mean deviation about the mean for the following data:
12, 18, 17, 3, 4, 9, 17, 19, 20, 15, 17, 8, 2, 3, 16, 11, 3, 1, 0, 5 **Ans.** 6.2.

6. Find the mean deviation about the median for the following data:
3, 9, 5, 3, 12, 10, 18, 4, 7, 19, 21. **Ans.** Median = 9, M.D (Me) = 5.27.

7. Find the mean deviation about the mean for the following data:

x_i:	5	7	9	10	12	15
f_i:	8	6	2	2	2	6

Ans. 3.39.

8. Find the mean deviation about the median for the following data:

x_i:	15	21	27	30	35
f_i:	3	5	6	7	8

Ans. 5.1.

9. Find the mean deviation about the mean for the following data:

Marks obtained:	10–20	20–30	30–40	40–50	50–60	60–70	70–80
No. of students:	2	3	8	14	8	3	2

Ans. $\bar{x} = 45$, M.D. $(\bar{x}) = 10$.

10. Find the mean deviation about median for the following data:

Class interval:	0–10	10–20	20–30	30–40	40–50	50–60
Frequency:	6	8	14	16	14	2

Ans. 10.34.

11. Calculate the mean deviation about median for the age distribution of 100 persons given below:

Age:	16–20	21–25	26–30	31–35	36–40	41–45	46–50	51–55
Number:	5	6	12	14	26	12	16	9

Ans. 7.35.

12. The following are the haemoglobin values of ten children receiving treatment for the hemolytic anemia 9.1, 10.0, 11.4, 12.4, 9.8, 8.3, 9.9, 9.1, 7.5, 6.7. Compute the mean and the variance. **Ans.** 9.42, 2.6056.

13. Show that for any two numbers a and b, the standard deviation is given by $\frac{|a-b|}{2}$.

[**Hint:** $\bar{x} = (a+b)/2$, $S.D = \sqrt{\frac{\Sigma(x-\bar{x})^2}{n}} = \sqrt{\frac{\left(a - \frac{a+b}{2}\right)^2 + \left(b - \frac{a+b}{2}\right)^2}{2}} = \sqrt{\frac{(a-b)^2}{4}} = \frac{|a-b|}{2}$.]

14. The mean of two numbers is 20 and their standard deviation is 5. Find the two numbers. **Ans.** 25, 15.

15. The mean of 5 observations is 4.4 and their variance is 8.24. If three of the observations are 1, 2, and 6, find the other two observations. **Ans.** 9 and 4.

[**Hint:** Let the two observations be p and q.
Therefore, $1 + 2 + 6 + p + q = 5\bar{x} = 5(4.4) = 22 \Rightarrow p + q = 13$. ...(*i*)

Employing, variance $= \frac{\Sigma x_i^2}{n} - (\bar{x})^2$ to get $\Sigma x_i^2 = n[\sigma^2 + (\bar{x})^2] \Rightarrow \Sigma x_i^2 = 5(8.24 + 19.36) = 138$.

Therefore, $1^2 + 2^2 + 6^2 + p^2 + q^2 = 138 \Rightarrow p^2 + q^2 = 97$. ....(*ii*). On solving (*i*) and (*ii*) simultaneously, we get $p = 9$, $q = 4$.]

16. Show that the variance of the first n natural numbers is $(n^2 - 1)/12$. (*VTU, 2003*)

[**Hint:** The first n natural numbers are 1, 2, 3, ..., n.

$$\text{Variance } (\sigma^2) = \frac{\Sigma x^2}{n} - \left(\frac{\Sigma x}{n}\right)^2 = \frac{1^2 + 2^2 + 3^2 + \cdots + n^2}{n} - \left(\frac{1 + 2 + 3 + \cdots + n}{n}\right)^2$$

$$= \frac{n(n+1)(2n+1)}{6n} - \left[\frac{n(n+1)}{2n}\right]^2 = \frac{n+1}{2}\left(\frac{2n+1}{3} - \frac{n+1}{2}\right) = \frac{n+1}{2}\left(\frac{n-1}{6}\right) = \frac{(n^2-1)}{12}.\Bigg]$$

17. The crushing strength of 8 cement concrete experimental blocks, in tonnes per sq. cm was 4.8, 4.2, 5.1, 3.8, 4.4, 4.7, 4.1, 4.5. Find the mean crushing strength and the standard deviation.

Ans. 4.45, 0.36.

18. Calculate the standard deviation of the following data:

x_i:	10	11	12	13	14	15	16	17	18
f_i:	2	7	10	12	15	11	10	6	3

Ans. 1.99.

19. Calculate the mean, variance and standard deviation for the following frequency distribution:

Class interval:	30–40	40–50	50–60	60–70	70–80	80–90	90–100
Frequencies:	3	7	12	15	8	3	2

Ans. 62; 201; 14.17.

20. Compute the standard deviation from the following distribution of marks obtained by 90 students:

Marks:	20–29	30–39	40–49	50–59	60–69	70–79	80–89	90–99
No. of students:	5	12	15	20	18	10	6	4

Ans. 17.65 approximate.

21. Use the frequency distribution of heights in the following table to find the mean height and standard deviation of the 100 male students at *XYZ* university.

Height in inch:	60–62	63–65	66–68	69–71	72–74
Frequency:	5	18	42	27	8

Ans. 67.45; 2.92 approx.

22. The following table shows the marks obtained by 100 candidates in an examination. Calculate the mean, median, and standard deviation.

Marks obtained:	1–10	11–20	21–30	31–40	41–50	51–60
No. of candidates:	3	16	26	31	16	8

(*Osmania, 2003 S; VTU, 2003 S*) **Ans.** 32, 32.6, 12.4.

23. The mean and standard deviation of 200 items are found to be 60 and 20 respectively. While calculating these measures, two items were wrongly taken as 3 and 67 instead of 13 and 17, find the correct mean and standard deviation. Also determine the correct coefficient of variation.

[**Hint:** We have incorrect $\bar{x} = 60$, incorrect $\sigma = 20$, $n = 200$. Therefore, incorrect $\sum x_i = n\bar{x} = 200 \times 60 = 12{,}000$. Correct $\Sigma x_i = 12000 - (3 + 67) + (13 + 17) = 11960$.

$\therefore$ Correct $\bar{x} = 11960/200 = 59.8$.

Employing, $\sigma = \sqrt{\frac{\Sigma x_i^2}{n} - (\bar{x})^2}$ to get incorrect $\sum x_i^2 = \{20^2 + (60)^2\}\, 200 = 8{,}00{,}000$.

Correct $\sum x_i^2 = 8{,}00{,}000 - (3)^2 - (67)^2 + (13)^2 + (17)^2 = 795960$.

$\therefore$ Correct $\sigma = \sqrt{\frac{795960}{200} - (59.8)^2} = 20.094$. Correct C.V. $= \frac{\sigma}{\bar{x}} \times 100 = \frac{20.094}{59.8} \times 100 = 33.6\%$.]

24. The following are scores of two batsmen A and B in a series of innings:

A:	12	115	6	73	7	19	119	36	84	29
B:	47	12	16	42	4	51	37	48	13	0

Who is the better score getter and who is more consistent? (*VTU, 2004*)

Ans. *A* is better score getter; *B* is more consistent than *A*.
[**Hint:** Arithmetic mean of *A* is greater than arithmetic mean of *B*. Coefficient of variation *B* is less than the coefficient of variation of *A*.]

25. The first four moments of a distribution about the value 5 are –4, 22, –117 and 560. Calculate the moments about (*a*) mean and (*b*) origin.
Ans. (*a*) 0, 6, 19, 32; (*b*) 1, 7, 38, 145.

26. For a frequency distribution, the mean = 1.5, variance = 0.64, $\beta_2 = 2.5$ and $\gamma_i = 0.3$. Find μ_3, μ_4.
Ans. 0.1536, 1.024.

27. From a moderately skewed data, the arithmetic mean is 100, the variance is 35 and Karl Pearson's coefficient of skewness is 0.2. Find its mode and median. **Ans.** 98.92 (nearly); 99.61 (nearly).

28. In a discrete series of 20 observation, the sum of the observations is 200, the sum of the square is 5000 and median is 15. Find the coefficient of skewness. **Ans.** –1.22.

29. From the following data, calculate Karl Pearson's coefficient of skewness:

Class-interval:	0–10	10–20	20–30	30–40	40–50	50–60	60–70	70–80
No. of students:	5	6	11	21	35	30	22	18

Ans. Mean = 48.176, Mode = 47.37, S.D = 17.71, $S_{k_P} = 0.046$.

30. In a frequency distribution, coefficient of skewness based on quartiles is 0.6. If the sum of upper and lower quartiles is 100 and median is 38, find the values of lower and upper quartiles.
Ans. $Q_1 = 30$, $Q_3 = 70$.

31. The marks obtained by ten students in a test are: 43, 12, 31, 20, 17, 26, 35, 40, 5, 37. Compute the Bowley's coefficient of skewness for the above data.
Ans. $Q_1 = 15.75$, $Q_3 = 37.35$, Median = 28.5, $S_{k_B} = -0.158$.

32. The first four central moments of a distribution are 0, 2.5, 7 and 18.75. Comment on the Kurtosis of the distribution.
Ans. $\beta_2 = 3$, the distribution is mesokurtic.

6.32 CURVE FITTING

Suppose *x* and *y* are two related variables and corresponding to a set of values of *x*, we have measured a set of values of *y*. Let the pairs of values thus obtained be (x_1, y_1), (x_2, y_2), ..., (x_n, y_n). If these points are plotted on a graph paper, we get a set of points called *scatter diagram*. From this diagram we can find whether the points have the trend of a curve (this can be done by visualising a smooth curve about which these points cluster).

The general problem of finding approximating curves which fit a given data, is called *curve fitting*.

Y
y = a + bx
O
X

Fig. 6.11

The scatter diagram of Fig. 6.11 indicates a linear trend and the data can be approximated by a straight line. Obviously no straight line passes exactly through all the points although a number of lines can be drawn which nearly do so.

In the present text our study is restricted to the fitting of a straight line only. In the next section we give the procedure to find the line of best fit for a given data. It is based on the method of least squares.

6.33 THE LINES OF BEST FIT

The line of best fit depends upon whether *x* is independent variable and *y* the dependent variable; or *y* independent and *x* dependent variable. There are two lines of best fit, one corresponding to each of these cases.

(a) *x* independent and *y* dependent variable

Procedure to get line of best fit

(*i*) Assume that the line approximating the set of points (x_1, y_1) (x_2, y_2), ..., (x_n, y_n) has the equation

$$y = a + bx. \qquad ...(6.54)$$

(*ii*) Form the equations

$$\left.\begin{aligned}\Sigma y &= na + b\Sigma x,\\ \Sigma xy &= a\Sigma x + b\Sigma x^2.\end{aligned}\right\} \qquad ...(6.55)$$

These equations are called normal equations for this line.

(*iii*) Solve the normal equations (6.55) simultaneously for a and b.

(*iv*) Substitute these values of a and b in Eq. (6.54). The line thus obtained is the line required.

(*b*) *y* independent and *x* dependent variable.

Procedure to get line of best fit

(*i*) *Assume the equation to the line to be* $x = a + by$. ...(6.56)

(*ii*) *Form the normal equations*

$$\left.\begin{aligned}\Sigma x &= na + b\Sigma y,\\ \Sigma xy &= a\Sigma y + b\Sigma y^2.\end{aligned}\right\} \qquad ...(6.57)$$

(*iii*) *Solve the equations* (6.57) *for a and b.*

(*iv*) *Put these values of a and b in Eq.* (6.56). *The resulting line is the line required.*

The following solved examples will make the procedure clear.

EXAMPLE 6.50 Fit a straight line to the following data taking y as the dependent variable.

x :	1	2	3	4	6	8
y :	2.4	3	3.6	4	5	6

Solution Let the required line is $y = a + bx$. The normal equations to determine a and b are:

$$\left.\begin{aligned}\Sigma y &= na + b\Sigma x,\\ \Sigma xy &= a\Sigma x + b\Sigma x^2.\end{aligned}\right\} \qquad ...(i)$$

x	y	xy	x^2
1	2.4	2.4	1
2	3.0	6.0	4
3	3.6	10.8	9
4	4.0	16.0	16
6	5.0	30.0	36
8	6.0	48.0	64
$\Sigma x = 24$	$\Sigma y = 24$	$\Sigma xy = 113.2$	$\Sigma x^2 = 130$

Here $n = 6$, $\Sigma x = \Sigma y = 24$, $\Sigma xy = 113.2$ and $\Sigma x^2 = 130$.

Putting these values in (*i*), the normal equations become:

$$24 = 6a + 24b, \quad \text{and} \quad 113.2 = 24a + 130b$$

or

$$a + 4b = 4, \text{ and } 12a + 65b = 56.6.$$

On solving these equations, we get $a = 2$, $b = 0.5$.

Hence, the equation of the line of best fit is $y = 2 + 0.5x$.

EXAMPLE 6.51 Fit a straight line to the following data taking y as the independent variable:

x :	1	1.8	3.3	4.5	6.3
y :	0	1	2	3	4

Solution Let the required line is $x = a + by$. The normal equations to determine a and b are:

$$\left.\begin{aligned}\Sigma x &= na + b\Sigma y,\\ \Sigma xy &= a\Sigma y + b\Sigma y^2.\end{aligned}\right\} \qquad ...(i)$$

x	y	xy	y^2
1.0	0	0	0
1.8	1	1.8	1
3.3	2	6.6	4
4.5	3	13.5	9
6.3	4	25.2	16
$\Sigma x = 16.9$	$\Sigma y = 10$	$\Sigma xy = 47.1$	$\Sigma y^2 = 30$

Here $n = 5$, $\Sigma x = 16.9$, $\Sigma y = 10$, $\Sigma xy = 47.1$, $\Sigma y^2 = 30$.
Putting these values in (*i*) we get $16.9 = 5a + 10b$, and $47.1 = 10a + 30b$.
On solving these equations, we obtain $a = 0.72$, $b = 1.33$.
Therefore, the required straight line is $x = 0.72 + 1.33y$.

Note When values of variables are large we can reduce simplification work by using suitable transformations (which implies shifting the origin to suitable points). The advantage of this will be clear from the following example.

EXAMPLE 6.52 Fit a straight line to the following data:

x:	1	2	3	4	5	6	7	8	9
y:	9	8	10	12	11	13	14	16	15

Solution Let $X = x - 5$ and $Y = y - 12$.

x	y	$X = x - 5$	$Y = y - 12$	X^2	XY
1	9	–4	–3	16	12
2	8	–3	–4	9	12
3	10	–2	–2	4	4
4	12	–1	0	1	0
5	11	0	–1	0	0
6	13	1	1	1	1
7	14	2	2	4	4
8	16	3	4	9	12
9	15	4	3	16	12
$\Sigma x = 45$	$\Sigma y = 108$	$\Sigma X = 0$	$\Sigma Y = 0$	$\Sigma X^2 = 60$	$\Sigma XY = 57$

Let the line of best fit be $Y = a + bX$.

The normal equations are: $$\begin{cases} \Sigma Y = na + b\Sigma X, \\ \Sigma XY = a\Sigma X + b\Sigma X^2. \end{cases}$$

Substituting $\Sigma X = 0, \Sigma Y = 0, n = 9, \Sigma X^2 = 60, \Sigma XY = 57$, we get

$$0 = 9a + 0 \Rightarrow a = 0, \quad 57 = 0 + 60b \Rightarrow b = 0.95.$$

Therefore, the equation of the line of best fit is

$$Y = 0 + 0.95X \quad \text{or} \quad y - 12 = 0 + 0.95(x - 5) \quad \text{or} \quad y = 7.25 + 0.95x.$$

EXAMPLE 6.53 Fit a straight line to the following data showing production of a commodity in different years in certain state and find the trend value for 2001.

Year:	2001	2002	2003	2004	2005
Production (in 1000 tonnes):	10	12	8	10	14

Solution Let the year be denoted by x and production by y. Since the trend value for a given year is to be found, x is independent and y dependent variable. Now if we keep the values of x and y as they are, the computation work will be quite laborious. To avoid this we put $X = x - 2003$ and $Y = y - 10$.

Now X is the independent variable and Y dependent variable. Let the line of best fit be $Y = a + bX$. The normal equations are

$$\begin{cases} \Sigma Y = na + b\Sigma X \\ \Sigma XY = a\Sigma X + b\Sigma X^2 \end{cases}$$

x	y	$X = x - 2003$	$Y = y - 10$	XY	X^2
2001	10	−2	0	0	4
2002	12	−1	2	−2	1
2003	8	0	−2	0	0
2004	10	1	0	0	1
2005	14	2	4	8	4
		$\Sigma X = 0$	$\Sigma Y = 4$	$\Sigma XY = 6$	$\Sigma X^2 = 10$

Here $n = 5$, $\Sigma X = 0$, $\Sigma Y = 4$, $\Sigma XY = 6$, $\Sigma X^2 = 10$.

Substituting the values in the normal equations, we get

$$4 = 5a + 0 \Rightarrow a = 0.8 \quad \text{and} \quad 6 = 0 + 10b \Rightarrow b = 0.6.$$

Therefore, the equation of the line of best fit is $Y = 0.8 + 0.6X$

or $\quad y - 10 = 0.8 + 0.6(x - 2003)$ or $y = 10.8 + 0.6(x - 2003)$,

which is the line of best fit required.

The trend value for $x = 2001$, that is, the value of y when $x = 2001$ is

$$y = 10.8 + 0.6(2001\text{–}2003) \quad \text{or} \quad y = 10.8 + (0.6)(-2) = 9.6 \text{ thousand tonnes.}$$

6.34 CORRELATION

So far our data for statistical analysis consisted of observations concerning single variable. This data could be heights of students in a particular class, yield of grain per acre in a state etc. There are instances in which two variables are such that an increase or decrease in the value of one variable is accompanied by increase or decrease in the value of the other. Two such variables are said to be correlated. If two variables move in the same direction, that is, an increase (or decrease) on the part of one variable introduces an increase (or decrease) on the part of the other variable, then the two variables are known to be *positively* or *direct correlated.*

On the other hand, if the two variables move in the opposite directions that is, an increase (or a decrease) on the part of one variable result a decrease (or an increase) on the part of the other variable, then the two variables are known to have a *negative* or *inverse* correlation. For example, the price and demand of an item, the profits of Insurance Company and the number of claims it has to meet etc. are examples of variables having a negative correlation. The degree of relationship existing between three or more variables is called multiple correlation.

If there is no relationship indicated between the variables, we say that there is no correlation between them, that is, they are uncorrelated. As for examples, shoe-size and intelligence are uncorrelated. On the basis of what has been stated above, correlation may be defined as "*a measure of extent or degree of mutual dependence between two variables*". Here we study only simple correlation.

6.35 SCATTER DIAGRAM OR DOT DIAGRAM

This is a simple diagrammatic method to represent the correlation between a pair of variables. Under this method each data point, which in this case a pair of values (x_i, y_i) is represented by a dot on the rectangular coordinate system. The totality of all the plotted dots forms the scatter diagram.

The pattern of the plotted points reveals the nature of correlation. Scatter diagram can distinguish between different types of correlation although it fails to measure the extent of relationship between the variables. The following figures show different types of correlation.

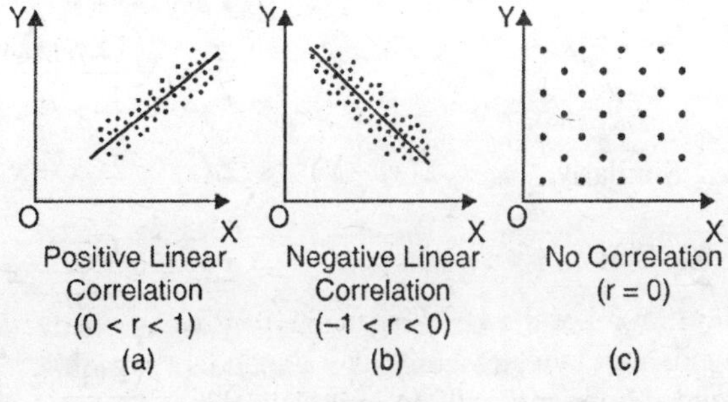

Fig. 6.12

6.36 KARL PEARSON'S (PRODUCT MOMENT) OF CORRELATION COEFFICIENT

For measuring the degree of closeness of linear relationship between the two variables when there is reasonable ground to believe they are correlated. Karl Pearson has developed a formula, called coefficient of correlation. This formula gives the numerical measure of the amount of correlation. This numerical measure is denoted by r. It is defined as the ratio of the covariance between the two variables to the product of the standard deviations of the two variables.

If the two variables are denoted by x and y and if the corresponding bivariate data are (x_i, y_i) for $i = 1, 2, 3, ..., n$ then the coefficient of correlation between x and y due to Karl-Pearson's is given by

$$r = r_{xy} = \frac{\text{Cov}(x, y)}{\sigma_x \sigma_y}, \qquad ...(6.58)$$

where $\text{Cov}(x, y) = \dfrac{\Sigma(x_i - \bar{x})(y_i - \bar{y})}{n} = \dfrac{\Sigma x_i y_i}{n} - \bar{x}\,\bar{y}$, n is the number of pairs.

The above formula may be written as $r = \dfrac{\Sigma(x_i - \bar{x})(y_i - \bar{y})}{n\sigma_x \sigma_y}$. ...(6.59)

If we put $x_i - \bar{x} = X$ and $y_i - \bar{y} = Y$, then $r = \dfrac{\Sigma XY}{n\sigma_x\sigma_y}$. ...(6.60)

Since standard deviation of $x = \sigma_x = \sqrt{\dfrac{\Sigma(x - \bar{x})^2}{n}} = \sqrt{\dfrac{\Sigma X^2}{n}}$.

S.D. of $y = \sigma_y = \sqrt{\dfrac{\Sigma(y - \bar{y})^2}{n}} = \sqrt{\dfrac{\Sigma Y^2}{n}}$.

Eq. (6.60) can be written in the form, $r = \dfrac{\Sigma XY}{\sqrt{\Sigma X^2}\sqrt{\Sigma Y^2}}$. ...(6.61)

This formula, which automatically gives the proper sign of r, is called the *product-moment formula.* Another form of the formula which is quite handy for computing correlation coefficient is given below:

$$r = \frac{n\,\Sigma x_i y_i - (\Sigma x_i)(\Sigma y_i)}{\sqrt{n\,\Sigma x_i^2 - (\Sigma x_i)^2}\,\sqrt{n\,\Sigma y_i^2 - (\Sigma y_i)^2}}. \qquad ...(6.62)$$

Proof:

Refer Eq. (6.61), we have

$$r = \frac{\Sigma XY}{\sqrt{\Sigma X^2}\sqrt{\Sigma Y^2}} = \frac{\Sigma(x_i - \bar{x})(y_i - \bar{y})}{\sqrt{\Sigma(x_i - \bar{x})^2}\,\sqrt{\Sigma(y_i - \bar{y})^2}}. \qquad ...(i)$$

But $\Sigma(x_i - \bar{x})(y_i - \bar{y}) = \Sigma(x_i y_i - \bar{x}y_i - x_i\bar{y} + \bar{x}\,\bar{y}) = \Sigma x_i y_i - \bar{x}\Sigma y_i - \bar{y}\,\Sigma x_i + n\bar{x}\,\bar{y}$

$= \Sigma x_i y_i - n\bar{x}\,\bar{y} - n\bar{y}\bar{x} + n\bar{x}\,\bar{y} = \Sigma x_i y_i - n\bar{x}\,\bar{y}$

$= \Sigma x_i y_i - \dfrac{(\Sigma x_i)(\Sigma y_i)}{n}$, since $\bar{x} = (\Sigma x_i)/n$ and $\bar{y} = (\Sigma y_i)/n$.

Similarly, $\Sigma(x_i - \bar{x})^2 = \Sigma(x_i^2 - 2x_i\bar{x} + \bar{x}^2) = \Sigma x_i^2 - 2\bar{x}\,\Sigma x_i + n\bar{x}^2$

$= \Sigma x_i^2 - 2\dfrac{(\Sigma x_i)^2}{n} + \dfrac{(\Sigma x_i)^2}{n} = \Sigma x_i^2 - \dfrac{(\Sigma x_i)^2}{n}$,

and $\Sigma(y_i - \bar{y})^2 = \Sigma y_i^2 - \dfrac{(\Sigma y)^2}{n}$. Then (i) becomes

$$r = \frac{\Sigma x_i y_i - \dfrac{(\Sigma x_i)(\Sigma y_i)}{n}}{\sqrt{\Sigma x_i^2 - \dfrac{(\Sigma x_i)^2}{n}}\sqrt{\Sigma y_i^2 - \dfrac{(\Sigma y_i)^2}{n}}} = \frac{n\,\Sigma x_i y_i - (\Sigma x_i)(\Sigma y_i)}{\sqrt{n\,\Sigma x_i^2 - (\Sigma x_i)^2}\,\sqrt{n\,\Sigma y_i^2 - (\Sigma y_i)^2}}.$$

Properties of Correlation Coefficient

1. The coefficient of correlation always lies between –1 and +1, that is, $-1 \le r \le 1$.
2. The coefficient of correlation is independent of the units measurement. It is a pure number having no units. The means that if x denotes height of a group of students expressed in cm and y denotes their weight expressed in kgm, then the correlation coefficient between height and weight would be free from any unit.
3. The coefficient of correlation is not affected by the choice of x or y, that is $r_{xy} = r_{yx}$.
4. The coefficient of correlation is independent of the choice of both origin and the scale of the variables under consideration.
5. If x and y are independent than $r_{xy} = 0$. But the converse is not true.

This property states that if the original pair of variables x and y is changed to a new pair of variables u and v by effecting a change of origin and scale for both x and y, that is

$u = \dfrac{x-a}{b}$ and $v = \dfrac{y-c}{d}$, where a and c are the origins of x and y and b and d are the respective scales.

Then, we have
$$r_{xy} = \frac{bd}{|b||d|} r_{uv} \qquad \text{...(6.63)}$$

r_{xy} and r_{uv} being the coefficient of correlation between x and y and u and v respectively. Eq. (6.63) represents the two correlation coefficient remain numerically equal, and they would have opposite signs only when b and d, the two scales, differ in sign.

EXAMPLE 6.54 (*i*) Given that the correlation coefficient between x and y is 0.8, write down the correlation coefficient between u and v where $2u - 3x + 4 = 0$ and $4v - 16y + 11 = 0$.

(*ii*) If $u + 3x = 5$, $2y - v = 7$ and correlation coefficient of x and y is 0.12, then find the correlation coefficient of u and v. (*AMIE, Dec. 2011*)

Solution Employing Eq. (6.63), we find that $r_{xy} = \dfrac{bd}{|b||d|} r_{uv}$, that is $r_{xy} = r_{uv}$ if b and d are of the same sign and $r_{uv} = -r_{xy}$ when b and d are of opposite signs, b and d being the scales of x and y respectively.

(*i*) We have $u = \dfrac{x-(4/3)}{2/3}$ and $v = \dfrac{y-(11/16)}{1/4}$.

Since $b = 2/3$ and $d = 1/4$ are of same sign, the correlation coefficient between u and v would be the same as that between x and y, that is $r_{xy} = 0.8 = r_{uv}$.

(*ii*) We have $u = \dfrac{x-(5/3)}{-1/3}$ and $v = \dfrac{y-(7/2)}{1/2}$.

Here $b = -1/3$ and $d = 1/2$ are of opposite signs and we have $r_{uv} = -r_{xy} = -0.12$.

EXAMPLE 6.55 Compute the coefficient of correlation between the heights of fathers and sons from the following table:

Heights of fathers (inches):	65	66	67	68	69	70	71
Heights of sons (inches):	67	68	66	69	72	72	69

Solution Denoting the heights of fathers by x and those of sons by y, the calculations for correlation are given in the following table:

x_i	y_i	Deviations from the Average $(X = x_i - \bar{x}) = x_i - 68$	$(Y = y_i - \bar{y}) = y_i - 69$	XY	X^2	Y^2
65	67	–3	–2	6	9	4
66	68	–2	–1	2	4	1
67	66	–1	–3	3	1	9
68	69	0	0	0	0	0
69	72	1	3	3	1	9
70	72	2	3	6	4	9
71	69	3	0	0	9	0
$\Sigma x_i = 476$	$\Sigma y_i = 483$	0	0	$\Sigma XY = 20$	$\Sigma X^2 = 28$	$\Sigma Y^2 = 32$

From this table, $\bar{x}$ = the mean height of fathers $= \frac{\Sigma x_i}{N} = \frac{476}{7} = 68$ inches.

$\bar{y}$ = the mean height of sons $= \frac{\Sigma y_i}{N} = \frac{483}{7} = 69$ inches.

From the above table $\Sigma XY = 20$. $\Sigma X^2 = 28$, $\Sigma Y^2 = 32$. As per Karl Pearson, coefficient of correlation,

$$r = \frac{\Sigma XY}{\sqrt{\Sigma X^2 \; \Sigma Y^2}} = \frac{20}{\sqrt{28 \times 32}} \approx +0.67.$$ The correlation is positive.

EXAMPLE 6.56 The following table gives marks obtained by two students in 10 tests during the year. Find the coefficient of correlation.

Test No.:	1	2	3	4	5	6	7	8	9	10
Marks in Math.:	77	54	27	52	14	35	90	25	56	60
Marks in English:	35	58	60	40	50	40	35	56	34	42

Solution Let x denote marks in Maths and y marks in English. The calculations for r are as follows:

x_i	y_i	$X = x_i - \bar{x}$	$Y = y_i - \bar{y}$	XY	X^2	Y^2
77	35	28	–10	–280	784	100
54	58	5	13	65	25	169
27	60	–22	15	–330	484	225
52	40	3	–5	–15	9	25
14	50	–35	5	–175	1225	25
35	40	–14	–5	70	196	25
90	35	41	–10	–410	1681	100
25	56	–24	11	–264	576	121
56	34	7	–11	–77	49	121
60	42	11	–3	–33	121	9
$\Sigma x_i = 490$	$\Sigma y_i = 450$	0	0	$\Sigma XY = -1449$	$\Sigma X^2 = 5150$	$\Sigma Y^2 = 920$

$$\bar{x} = \frac{490}{10} = 49, \quad \bar{y} = \frac{450}{10} = 45.$$

Also $\Sigma XY = 1449$, $\Sigma X^2 = 5150$, $\Sigma Y^2 = 920$.

Therefore, $$r = \frac{\Sigma XY}{\sqrt{\Sigma X^2 \Sigma Y^2}} = \frac{-1449}{\sqrt{5150 \times 920}} = \frac{-1449}{2176.6947} = -0.66.$$

EXAMPLE 6.57 Compute the coefficient of correlation between x and y from the following data:

$n = 10$, $\Sigma x_i y_i = 220$, $\Sigma x_i^2 = 200$, $\Sigma y_i^2 = 262$, $\Sigma x_i = 40$, and $\Sigma y_i = 50$.

Solution Applying the following formula:

$$r = \frac{n\Sigma x_i y_i - \Sigma x_i \Sigma y_i}{\sqrt{n\Sigma x_i^2 - (\Sigma x_i)^2} \times \sqrt{n\Sigma y_i^2 - (\Sigma y_i)^2}}.$$

We have $$r = \frac{10 \times 220 - 40 \times 50}{\sqrt{10 \times 200 - (40)^2} \times \sqrt{10 \times 262 - (50)^2}}$$

$$= \frac{2200 - 2000}{\sqrt{2000 - 1600} \times \sqrt{2620 - 2500}} = \frac{200}{20 \times 10.9545} = 0.913.$$

Thus, there is a good amount of positive correlation between the two variables x and y.

6.37 SHORT-COMPUTATIONAL FORMULAE

A. Assumed mean method. When the arithmetic means are not whole numbers, it is more convenient to compute the value of r by taking assumed mean a say, for x and b for y. d_x is deviation from assumed average a, that is, $d_x = x_i - a$ and d_y is deviation from assumed average b, that is, $d_y = y_i - b$. When deviations are taken from the assumed average and not the actual average, the formula is as follows:

$$r = \frac{n\Sigma d_x d_y - (\Sigma d_x)(\Sigma d_y)}{\sqrt{\{n\Sigma d_x^2 - (\Sigma d_x)^2\}}\ \sqrt{\{n\Sigma d_y^2 - (\Sigma d_y)^2\}}}. \quad ...(6.64)$$

B. Change of origin and change of scale. Sometimes, *change of origins* and *change of scales* reduces the computational labour to a great extent. When the deviations d_x and d_y have common factors say h and k respectively. Then putting $d_x/h = u_x$ and $d_y/k = v_y$. The correlated coefficient formula is as follows:

$$r = \frac{n\Sigma u_x u_y - (\Sigma u_x)(\Sigma v_y)}{\sqrt{\{n\Sigma u_x^2 - (\Sigma u_x)^2\}}\ \sqrt{\{n\Sigma v_y^2 - (\Sigma v_y)^2\}}}. \quad ...(6.65)$$

EXAMPLE 6.58 Find the coefficient of correlation between x and y.

x :	1	3	5	7	8	10
y :	8	12	15	17	18	20

Solution Here the actual average ($\bar{x}$) of the variable data x_i is 34/6 or 5.66 which is not an integral. To reduce the computational labour to a great extent, we make use of an assumed average. Refer the following table, d_x is deviation from assumed average 7 and d_y is deviation from assumed average 15.

x_i	y_i	$d_x = x_i - 7$	$d_y = y_i - 15$	$d_x d_y$	d_x^2	d_y^2
1	8	–6	–7	42	36	49
3	12	–4	–3	12	16	9
5	$b = 15$	–2	0	0	4	0
$a = 7$	17	0	2	0	0	4
8	18	1	3	3	1	9
10	20	3	5	15	9	25
$\Sigma x_i = 34$	$\Sigma y_i = 90$ $N = 6$	$\Sigma d_x = -8$	$\Sigma d_y = 0$	$\Sigma d_x d_y = 72$	$\Sigma d_x^2 = 66$	$\Sigma d_y^2 = 96$

Correlation coefficient is not affected by shift of origin and change of scale. We use the following formula.

$$r = \frac{n\Sigma d_x d_y - (\Sigma d_x)(\Sigma d_y)}{\sqrt{\{n\Sigma d_x^2 - (\Sigma d_x)^2\}}\ \sqrt{\{n\Sigma d_y^2 - (\Sigma d_y)^2\}}} \text{ and obtain}$$

$$r = \frac{6(72) - (-8)(0)}{\sqrt{6(66) - (-8)^2}\ \sqrt{6(96) - 0}} = \frac{432}{\sqrt{332}\ \sqrt{576}} \approx 0.988.$$

Remark 5

In the above example $\bar{x} = \Sigma x_i / n = 34/6 = 5.66$ and deviations from $\bar{x}$ involve fractions which make computations for r (using the formula $r = \Sigma XY / \sqrt{\Sigma X^2 \Sigma Y^2}$) quite inconvenient. This shows the advantage of using assumed average and not the actual average.

EXAMPLE 6.59 Calculate the coefficient of correlation for the following paired data:

x :	10	20	30	40	50	60	70	80
y :	32	20	24	36	40	28	48	44

Solution The correlation coefficient is not affected by change of origin and change of scale. To reduce the computational labour to a great extent we take origin of x and y as 40 and 36 and 10 and 4 are the respective change of scales. We prepare the following table:

x_i	y_i	$u_x = \frac{x_i - 40}{10}$	$v_y = \frac{y_i - 36}{4}$	$u_x v_y$	u_x^2	v_y^2
10	32	–3	–1	3	9	1
20	20	–2	–4	8	4	16
30	24	–1	–3	3	1	9
40	36	0	0	0	0	0
50	40	1	1	1	1	1
60	28	2	–2	–4	4	4
70	48	3	3	9	9	9
80	44	4	2	8	16	4
$\Sigma x_i = 360$	$\Sigma y_i = 272$	$\Sigma u_x = 4$	$\Sigma v_y = -4$	$\Sigma u_x v_y = 28$	$\Sigma u_x^2 = 44$	$\Sigma v_y^2 = 44$

Using the formula:

$$r = \frac{n\Sigma u_x v_y - (\Sigma u_x)(\Sigma v_y)}{\sqrt{\{n\Sigma u_x^2 - (\Sigma u_x)^2\}}\ \sqrt{\{n\Sigma v_y^2 - (\Sigma v_y)^2\}}}, \text{ we obtain}$$

$$r = \frac{8(28) - (4)(-4)}{\{\sqrt{8(44) - 4^2}\}\ \{\sqrt{8(44) - (-4)^2}\}} = \frac{224 + 16}{\sqrt{336}\ \sqrt{336}} \approx 0.714.$$

EXAMPLE 6.60 Calculate correlation coefficient from the following results:

$n = 10;\ \Sigma x_i = 140;\ \Sigma y_i = 150;\ \Sigma(x_i - 10)^2 = 180;\ \Sigma(y_i - 15)^2 = 215;\ \Sigma(x_i - 10)(y_i - 15) = 60.$

Solution Since the correlation coefficient r is not affected by change of origin of reference, $r_{xy} = r_{uv}$, where $u = x_i - 10$ and $v = y_i - 15$. From the given data, we find

$$\Sigma u = \Sigma(x_i - 10) = \Sigma x_i - \Sigma 10 = \Sigma x_i - 10n = 140 - 10(10) = 40,$$

$$\Sigma v = \Sigma(y_i - 15) = \Sigma y_i - \Sigma 15 = \Sigma y_i - 15n = 150 - 15(10) = 0,$$

$$\Sigma u^2 = \Sigma(x_i - 10)^2 = 180,\ \Sigma v^2 = \Sigma(y_i - 15)^2 = 215,$$

and $\Sigma uv = \Sigma(x_i - 10)(y_i - 15) = 60.$

Therefore,

$$r_{uv} = \frac{n\Sigma uv - (\Sigma u)(\Sigma v)}{\sqrt{n\Sigma u^2 - (\Sigma u)^2}\ \sqrt{n\Sigma v^2 - (\Sigma v)^2}} = \frac{10(60) - 40(0)}{\sqrt{10(180) - (40)^2}\ \sqrt{10(215) - 0}}$$

$$= \frac{600 - 0}{\sqrt{1800 - 1600}\ \sqrt{2150}} = \frac{600}{\sqrt{200 \times 2150}} = \frac{6}{\sqrt{2 \times 21.5}} = \frac{6}{\sqrt{43}} \approx 0.915.$$

EXAMPLE 6.61 While calculating correlation coefficient between two variables x and y from 25 pairs of observations, the following results were obtained:

$$n = 25;\ \Sigma x_i = 125;\ \Sigma y_i = 100;\ \Sigma x_i^2 = 650,\ \Sigma y_i^2 = 460,\ \Sigma x_i y_i = 508.$$

It was however discovered at the time of checking that two pairs of observations were not correctly copied.

They were taken as

x_i	6	8
y_i	14	6

while the correct values were

x_i	8	6
y_i	12	8

. Obtain the correct value of correlation coefficient.

Solution To get the correct results, we subtract the incorrect values and add the corresponding correct values.

∴ The correct result would be

$$\Sigma x_i = 125 - 6 - 8 + 8 + 6 = 125, \qquad \Sigma y_i = 100 - 14 - 6 + 12 + 8 = 100,$$

$$\Sigma x_i^2 = 650 - 6^2 - 8^2 + 8^2 + 6^2 = 650, \qquad \Sigma y_i^2 = 460 - 14^2 - 6^2 + 12^2 + 8^2 = 436,$$

and $\Sigma x_i y_i = 508 - 6(14) - 8(6) + 8(12) + 6(8) = 520.$

$$\therefore \text{ Corrected } r(x, y) = \frac{n\Sigma x_i y_i - (\Sigma x_i)(\Sigma y_i)}{\sqrt{n\Sigma x_i^2 - (\Sigma x_i)^2} \times \sqrt{n\Sigma y_i^2 - (\Sigma y_i)^2}}$$

$$= \frac{25(520) - 125(100)}{\sqrt{[25\,(650) - (125)^2]}\ \sqrt{[25\,(436) - (100)^2]}} = \frac{20}{\sqrt{25 \times 36}} \approx 0.667.$$

EXAMPLE 6.62 Coefficient of correlation between x and y for 50 observations is 0.3. The arithmetic means and standard deviations of x and y are known to be 10 and 6 and 3 and 2 respectively. However, on subsequent verification it was found that one value of x (= 10) and one value of y (= 6) were inaccurate and hence weeded out. With the remaining 49 pairs of values, how is the original value of correlation coefficient affected?

Solution We know that correlation coefficient

$$r = \frac{\Sigma(x_i - \bar{x})(y_i - \bar{y})}{n\sigma_x\,\sigma_y} \quad \text{or} \quad \Sigma(x_i - \bar{x})(y_i - \bar{y}) = nr\sigma_x\sigma_y = 50 \times 0.3 \times 3 \times 2 = 90;$$

or $\Sigma x_i y_i - \bar{x}\Sigma y_i - \bar{y}\Sigma x_i + \Sigma\bar{x}\,\bar{y} = 90$ or $\Sigma x_i y_i = 90 + n\bar{x}\,\bar{y}$ $[\because \Sigma y = n\bar{y},\ \Sigma x_i = n\bar{x}$ and $\Sigma\bar{x}\,\bar{y} = n\bar{x}\,\bar{y}]$

$$= 90 + 50 \times 10 \times 6 = 3090.$$

We also know that $\sigma_x^2 = \dfrac{\Sigma x_i^2}{n} - \left(\dfrac{\Sigma x_i}{n}\right)^2 = \dfrac{1}{n}\Sigma x_i^2 - \bar{x}^2$ or $\Sigma x_i^2 = n(\sigma_x^2 + \bar{x}^2) = 50(9 + 100) = 5450.$

Similarly, $\Sigma y_i^2 = n(\sigma_y^2 + \bar{y}^2) = 50(4 + 36) = 2000.$

Corrected value

$$\Sigma x_i y_i = 3090 - 10 \times 6 = 3030, \quad \Sigma x_i = n\,\bar{x} - 10 = 50 \times 10 - 10 = 490,$$

$$\Sigma y_i = n\,\bar{y} - 6 = 50 \times 6 - 6 = 294, \quad \Sigma x_i^2 = 5450 - (10)^2 = 5350, \quad \text{and} \quad \Sigma y_i^2 = 2000 - (6)^2 = 1964.$$

With the actual values of the variables x, y, the correlation coefficient is found from

$$r = \frac{n\Sigma x_i y_i - \Sigma x_i\,\Sigma y_i}{\sqrt{n\Sigma x_i^2 - (\Sigma x_i)^2}\ \sqrt{n\Sigma y_i^2 - (\Sigma y_i)^2}}.$$

Corrected value of the correlation coefficient with the remaining 49 pairs of values is therefore given by

$$r = \frac{49 \times 3030 - 490 \times 294}{\sqrt{49 \times 5350 - (490)^2}\ \sqrt{49 \times 1964 - (294)^2}} = \frac{49 \times 90}{\sqrt{49 \times 450 \times 49 \times 200}} = \frac{49 \times 90}{49 \times 300} = 0.3.$$

Therefore, original value of r remains unaffected.

6.38 CORRELATION TABLE

When the number of pairs of values is large they are grouped into a table of double frequency known as *correlation table*. An example of this type of table is given below:

Age of wives	Age of husbands					Total
	20-30	30-40	40-50	50-60	60-70	
15–25	5	9	3	0	0	17
25–35	0	10	**(25)**	2	0	37
35–45	0	1	12	2	0	15
45–55	0	0	4	16	5	25
55–65	0	0	0	4	2	6
Total	5	20	44	24	7	**100**

The bold numbers in the body of the table indicate frequencies. For example the number 25 shown bold in the body of the table indicates that there are 25 couples with ages of husbands between 40-50 and ages of wives between 25-35. Such frequency distribution is known as *bivariate frequency distribution*.

The formula for r for such frequency distribution is

$$r = \frac{n\Sigma fd_xd_y - (\Sigma fd_x)n(\Sigma fd_y)n}{\sqrt{n(\Sigma fd_x^2) - (\Sigma fd_x)^2}\ \sqrt{[n(\Sigma fd_y^2) - (\Sigma fd_y)^2]}} \quad ...(6.66)$$

where f represents frequencies.

In case deviations for x and y, that is, d_x, d_y have common factors then

$$r = \frac{n\Sigma fuv - (\Sigma f_xu)\,(\Sigma f_yv)}{\sqrt{n(\Sigma f_xu^2) - (\Sigma f_xu)^2}\ \sqrt{n(\Sigma f_yv^2) - (\Sigma f_yv)^2}} \quad ...(6.67)$$

where $u = d_x/h$ and $v = d_y/k$, h and k having their usual meanings. In case of grouped frequency distribution h and k are magnitudes of classes. The evaluation of r in this case will be clear from the following example.

EXAMPLE 6.63 **Calculate the coefficient of correlation for the correlation table given in article 6.38.**

Solution **Let x denote the central values of ages of husbands and y those of wives. The calculations for the value of r are given in the following table.**

Age of wives in years		Age of husbands in years									
	$y\downarrow$ \ $x\rightarrow$	25	35	a = 45	55	65	f_y	$v = \frac{y-40}{10}$	f_yv	f_yv^2
	20	5 **20**	9 **18**	3 **0**	0	0	17	−2	−34	68
	30	0	10 **10**	25 **0**	2 **−2**	0	37	−1	−37	37
	$b = 40$	0	1 **0**	12 **0**	2 **0**	0	15	0	0	0
	50	0	0	4 **0**	16 **16**	5 **10**	25	1	25	25
	60	0	0	0	4 **8**	2 **8**	6	2	12	24
	f_x	5	20	44	24	7	$\Sigma f_x = \Sigma f_y = n = 100$	Total	$-34 = \Sigma f_yv$	$154 = \Sigma f_yv^2$
	$u = \frac{x-45}{10}$	−2	−1	0	1	2	Total			
	f_xu	−10	−20	0	24	14	$8 = \Sigma f_xu$			
	f_xu^2	20	20	0	24	28	$92 = \Sigma f_xu^2$			

As is clear from the table, we have taken $a = 45$ as assumed mean for x and $b = 40$, as assumed mean for y.

Here $\Sigma f_x u = 8$, $\Sigma f_x u^2 = 92$, $\Sigma f_y v = -34$, $\Sigma f_y v^2 = 154$ and $n = 100$.

Thick figures in small squares represent the product fuv.

$\therefore \quad \Sigma fuv = 20 + 18 + 10 - 2 + 16 + 10 + 8 + 8 = 88.$

Thus,
$$r = \frac{n\Sigma fuv - (\Sigma f_x u)(\Sigma f_y v)}{\sqrt{n(\Sigma f_x u^2) - (\Sigma f_x u)^2}\ \sqrt{n(\Sigma f_y v^2) - (\Sigma f_y v)^2}}$$

$$= \frac{(100)\,88 - (8)(-34)}{\sqrt{\{(100)92 - 8^2\}\{100(154) - (-34)^2\}}} = \frac{8800 + 272}{\sqrt{9136 \times 14244}} = \frac{9072}{11407.6} = 0.795 \text{ approx.}$$

EXAMPLE 6.64 From the following table of bivariate frequency distribution, calculate the coefficient of correlation between heights and weights of children:

Weight (in pounds)	Height (in inches)						Total
	40–44	44–48	48–52	52–56	56–60	60–64	
35–55	4	40	60	—	—	—	104
55–75	—	—	24	88	12	—	124
75–95	—	—	—	8	32	8	48
95–115	—	—	—	—	4	8	12
115–135	—	—	—	4	—	—	4
135–155	—	—	—	—	4	4	8
Total	**4**	**40**	**84**	**100**	**52**	**20**	**300**

Solution Let x denote the central values of height and y-those of weight of children. The calculations for the value of r are given in the following table.

Height of children in inches (rows: Weight of children in pounds)

$x\rightarrow$ / $y\downarrow$	42	46	$a = 50$	54	58	62	f_y	$v = \dfrac{y-85}{20}$	$f_y v$	$f_y v^2$
45	4 **16**	40 **80**	60 **0**	0	0	0	104	−2	−208	416
65	0	0	24 **0**	88 **−88**	12 **−24**	0	124	−1	−124	124
$b = 85$	0	0	0	8 **0**	32 **0**	8 **0**	48	0	0	0
105	0	0	0	0	4 **8**	8 **24**	12	1	12	12
125	0	0	0	4 **8**	0	0	4	2	8	16
145	0	0	0	0	4 **24**	4 **36**	8	3	24	72
f_x	4	40	84	100	52	20	$\Sigma f_x = \Sigma f_y = n = 300$	Total	$-288 = \Sigma f_y v$	$640 = \Sigma f_y v^2$
$u = \dfrac{x-50}{40}$	−2	−1	0	1	2	3	Total			
$f_x u$	−8	−40	0	100	104	60	$216 = \Sigma f_x u$			
$f_x u^2$	16	40	0	100	208	180	$544 = \Sigma f_x u^2$			

We have taken $a = 50$ as assumed mean for x and $b = 85$ as assumed mean for y.

Here $\Sigma f_x u = 216$, $\Sigma f_x u^2 = 544$, $\Sigma f_y v = -288$, $\Sigma f_y v^2 = 640$ and $n = 300$.

Thick figures in small squares represent the product fuv.

Therefore, $\Sigma fuv = 16 + 80 - 88 - 24 + 8 + 24 + 8 + 24 + 36 = 84.$

Thus, $$r = \frac{n\Sigma fuv - (\Sigma f_x u)(\Sigma f_y v)}{\sqrt{n(\Sigma f_x u^2) - (\Sigma f_x u)^2}\sqrt{n(\Sigma f_y v^2) - (\Sigma f_y v)^2}} = \frac{300(84) - (216)(-288)}{\sqrt{300(544) - (216)^2}\sqrt{300(640) - (-288)^2}}$$

$$= \frac{25200 + 62208}{\sqrt{(163200 - 46656) \times (192000 - 82944)}} = \frac{87408}{\sqrt{116544 \times 109056}} = \frac{87408}{341.385 \times 330.236} \approx 0.775.$$

EXAMPLE 6.65 Show that the coefficient of correlation r between two variable x and y is given by

$r = \dfrac{\sigma_x^2 + \sigma_y^2 - \sigma_{x-y}^2}{2\sigma_x\sigma_y}$, where σ_x^2, σ_y^2 and σ_{x-y}^2 are the variance of x, y and $x-y$ respectively

Solution We have

$$\sigma_{x-y}^2 = \text{var.}(x - y) = \text{var.}(x) + \text{var.}(y) - 2\,\text{cov.}(x, y)$$

$$= \sigma_x^2 + \sigma_y^2 - 2\,\text{cov.}(x, y) \text{ giving}$$

$$\text{cov}(x, y) = \frac{\sigma_x^2 + \sigma_y^2 - \sigma_{x-y}^2}{2}, \text{ we know } r = \frac{\text{cov}(x, y)}{\sigma_x\,\sigma_y}.$$

Therefore, $$r = \frac{\sigma_x^2 + \sigma_y^2 - \sigma_{x-y}^2}{2\,\sigma_x\sigma_y}.$$

EXAMPLE 6.66 Using the result of Example 6.65 that, is $r = (\sigma_x^2 + \sigma_y^2 - \sigma_{x-y}^2)/2\,\sigma_x\sigma_y$, to compute the coefficient of correlation between x and y from the following data:

Price (x):	3	5	6	9	10	12	15	20	22	28
Supply (y):	10	12	15	18	20	22	27	30	32	34

Solution To find r, we have to calculate σ_x, σ_y and σ_{x-y}. We make the following table:

x	$X = x - \bar{x}$ $= x - 13$	X^2	y	$Y = y - \bar{y}$ $= y - 22$	Y^2	$x-y$	$(x-y)^2$
3	–10	100	10	–12	144	–7	49
5	–8	64	12	–10	100	–7	49
6	–7	49	15	–7	49	–9	81
9	–4	16	18	–4	16	–9	81
10	–3	9	20	–2	4	–10	100
12	–1	1	22	0	0	–10	100
15	2	4	27	5	25	–12	144
20	7	49	30	8	64	–10	100
22	9	81	32	10	100	–10	100
28	15	225	34	12	144	–6	36
$\Sigma x = 130$	$\Sigma X = 0$	$\Sigma X^2 = 598$	$\Sigma y = 220$	$\Sigma Y = 0$	$\Sigma Y^2 = 646$	$\Sigma(x - y) = -90$	$\Sigma(x - y)^2 = 840$

$$\bar{x} = \frac{\Sigma x}{N} = \frac{130}{10} = 13, \quad \bar{y} = \frac{\Sigma y}{N} = \frac{220}{10} = 22.$$

$$\therefore \quad \sigma_x^2 = \frac{\Sigma X^2}{N} - \left(\frac{\Sigma X}{N}\right)^2 = \frac{598}{10} - 0 = 59.8, \quad \sigma_y^2 = \frac{\Sigma Y^2}{N} - \left(\frac{\Sigma Y}{N}\right)^2 = \frac{646}{10} - 0 = 64.6.$$

$$\sigma_{x-y}^2 = \frac{\Sigma(x-y)^2}{N} - \left\{\frac{\Sigma(x-y)}{N}\right\}^2 = \frac{840}{10} - \frac{8100}{100} = 3.$$

From the above formula, $r = \dfrac{\sigma_x^2 + \sigma_y^2 - \sigma_{x-y}^2}{2\sigma_x\sigma_y} = \dfrac{59.8 + 64.6 - 3}{2\sqrt{(59.8)(64.6)}} = \dfrac{121.4}{2(62.154)} \approx 0.976.$

6.39 RANK CORRELATION

The coefficient of correlation (Pearson's coefficient of correlation) between the variables as have already been discussed is based on the values of the variables. Sometimes these values are not available or it may not be possible to find the magnitudes of the variables. For example we cannot measure intelligence or beauty quantitatively. In such cases it is possible to rank the individuals in some order on the basis of these characteristics. If the ranks assigned are 1 to n, then correlation coefficient between two series of ranks is called *rank correlation coefficient* and is given by

$$R = 1 - \frac{6\Sigma d_i^2}{n(n^2-1)} \qquad \text{...(6.68)}$$

where d_i is the difference between corresponding ranks of the two series and n is total number of paired items.

Formula (6.68) is known as *Edward Spearman's Formula for rank correlation coefficient.*

If all the d's are zero, then correlation is perfect and $R = 1$ otherwise $-1 \le R < +1$.

EXAMPLE 6.67 Ten students get the following ranks in two subjects A and B.

A:	3	5	8	4	7	10	2	1	6	9
B:	6	4	9	8	1	2	3	10	5	7

Compute the rank correlation coefficient

Solution Computation of Rank correlation coefficient is given in the following table:

Rank in A (x_i)	**Rank in B** (y_i)	**Rank Difference** $d_i = x_i - y_i$	d_i^2
3	6	–3	9
5	4	1	1
8	9	–1	1
4	8	–4	16
7	1	6	36
10	2	8	64
2	3	–1	1
1	10	–9	81
6	5	1	1
9	7	2	4
Total			$\Sigma d_i^2 = 214$

Here $n = 10, \Sigma d_i^2 = 214$.

$\therefore$ Rank correlation coefficient, $R = 1 - \left[\dfrac{6\Sigma d_i^2}{n(n^2-1)}\right] = 1 - \dfrac{6(214)}{10(10^2-1)} = 1 - \dfrac{1284}{990} = -0.3$ approx.

EXAMPLE 6.68 Calculate the coefficient of correlation from the data given below by the method of differences:

Series A :	78	89	97	69	59	79	68	57
Series B :	125	137	156	112	107	136	123	108

(*JNTU, 2005*)

Solution Computation of Rank correlation is given in the following table:

Series A	Series B	Rank in A (x_i)	Rank in B (y_i)	Rank Diff. $d_i = x_i - y_i$	d_i^2
78	125	4	4	0	0
89	137	2	2	0	0
97	156	1	1	0	0
69	112	5	6	–1	1
59	107	7	8	–1	1
79	136	3	3	0	0
68	123	6	5	1	1
57	108	8	7	1	1
Total					$\Sigma d_i^2 = 4$

Here $n = 8$ and $\Sigma d_i^2 = 4$.

Therefore, Rank correlation coefficient,

$$R = 1 - \left[\frac{6\Sigma d_i^2}{n(n^2-1)}\right] = 1 - \frac{6(4)}{8(64-1)} = 1 - \frac{24}{8\,(63)} = \frac{20}{21} = 0.95.$$

There is a high degree of closeness between the series A and B.

6.40 SPEARMAN'S MODIFIED FORMULA FOR TIED RANKS

When the values of the variables are known, we assign rank 1 to the largest value. The next largest is assigned rank 2 and so on. But if two values are equal then there is a tie. In such a situation, these values are assigned the average of the ranks. For example if two items are tied for rank 4, then each is ranked $(4+5)/2 = 4.5$ and next item will be given the rank 6.

Edward spearman's formula for tied rank, the correlation coefficient takes the following form:

$$R = 1 - \frac{6\left\{\Sigma_i d_i^2 + \dfrac{\Sigma_j (t_j^3 - t_j)}{12}\right\}}{n(n^2-1)}. \qquad ...(6.69)$$

In this formula, t_j represents the jth tie length and the summation $\Sigma_j (t_j^3 - t_j)$ extends over the lengths of all the ties for both the series The following example will make the procedure clear.

EXAMPLE 6.69 Find the rank correlation coefficient of the following data:

Series A :	114	108	110	88	97	119	97	100	97	118
Series B :	74	72	83	70	75	82	64	72	68	80

Solution In series A, the highest score is 119 and, therefore, its rank is 1. The next highest score is 118 and its rank is 2. Proceeding in this way we note that rank of 97 is 7. Since it occur 3 times, there is a tie for 7th place. We find the average rank $(7+8+9)/3 = 8$ and assign this rank to each of 97.

The score next to 97 is 88 and its rank is 10. Similarly the ranks for the second series are determined. There 72 occurs twice and there is a tie for 6th place. The average rank is $(6+7)/2 = 6.5$ and this rank is assigned to 72 etc.

Computation of rank correlation coefficient

Series A		Series B		Rank Difference	d_i^2
Score	Rank (x_i)	Score	Rank (y_i)	$d_i = x_i - y_i$	
114	3	74	5.0	–2.0	4.0
108	5	72	6.5	–1.5	2.25
110	4	83	1.0	3.0	9.0
88	10	70	8.0	2.0	4.0
97	8	75	4.0	4.0	16.0
119	1	82	2.0	–1.0	1.0
97	8	64	10.0	–2.0	4.0
100	6	72	6.5	–0.5	0.25
97	8	68	9.0	–1.0	1.0
118	2	80	3.0	–1.0	1.0
				0	$\Sigma d_i^2 = 42.50$

By Spearman's modified formula for tied rank, the correlation coefficient is given by

$$R = 1 - \frac{6\left[\Sigma d_i^2 + \dfrac{\Sigma(t_j^3 - t_j)}{12}\right]}{n(n^2-1)}. \text{ Here } \Sigma d_i^2 = 42.50, \quad n = 10, \quad t_j = 3, 2.$$

$$\therefore \frac{\Sigma(t_j^3 - t_j)}{12} = \frac{3^3-3}{12} + \frac{2^3-2}{12} = 2 + 0.5 = 2.5. \text{ Hence } R = 1 - \frac{6\,\{42.50 + 2.5\}}{10\,(10^2-1)} = 1 - \frac{6\times 45}{990} = 1 - \frac{3}{11} \approx 0.73.$$

EXAMPLE 6.70 A group of 7 students get the following percentage of marks in test in Applied Mechanics and Mathematics:

Roll no. of students :	01	02	03	04	05	06	07
% of marks in Applied Mechanics :	80	56	50	48	50	62	60
% of marks in Mathematics :	90	75	75	65	65	50	65

Compute the rank correlation coefficient.

Solution This is a case of tied ranks as more than one student share the same mark both for Applied Mechanics and Mathematics. For Applied mechanics subject the student receiving 80 marks gets rank 1, one getting 62 marks receives rank 2, the student with 60 marks receives rank 3, student with 56 marks gets rank 4 and since there are two students, each getting 50 marks, each would be receiving a common rank, the average of the next two ranks 5 and 6 *i.e.* $(5+6)/2$ *i.e.* 5.5 and lastly the last rank 7 goes to the student getting the lowest marks in Applied Mechanics. In a similar manner, we award ranks to the students with marks in Mathematics.

Computation of Rank Correlation Coefficient between marks in Applied Mechanics and marks in Maths with tied marks.

Marks in Applied Mechanics	Marks in. Maths	Ranks for Applied Mechanics (x_i)	Ranks for Maths (y_i)	Difference $d_i = x_i - y_i$	d_i^2
80	90	1.0	1.0	0.0	0.0
56	75	4.0	2.5	1.5	2.25
50	75	5.5	2.5	3.0	9.0
48	65	7.0	5.0	2.0	4.0
50	65	5.5	5.0	0.5	0.25
62	50	2.0	7.0	–5.0	25.0
60	65	3.0	5.0	–2.0	4.0
Total				0	$\Sigma d_i^2 = 44.5$

For marks in Applied Mechanics there is one tie of length 2 and for marks in Maths, there are two ties of lengths 2 and 3 respectively, $n = 7$.

Thus $$\frac{\sum(t_j^3 - t_j)}{12} = \frac{(2^3 - 2) + (2^3 - 2) + (3^3 - 3)}{12} = 3.$$

Therefore, $$R = 1 - \frac{6\left\{\sum d_i^2 + \frac{\sum(t_j^3 - t_j)}{12}\right\}}{n(n^2 - 1)} = 1 - \frac{6(44.5 + 3)}{7(7^2 - 1)} = 1 - \frac{6\,(47.5)}{336} = 1 - 0.848 \approx 0.15.$$

EXAMPLE 7.71 Ten competitors in a beauty contest are ranked by three judges in the following order:

Ist judge :	1	6	5	10	3	2	4	9	7	8
2nd judge :	3	5	8	4	7	10	2	1	6	9
3rd judge :	6	4	9	8	1	2	3	10	5	7

Use the rank correlation coefficient to determine which pair of judges has the nearest approach to common tastes in beauty.

Solution Taking x, y, z as the ranks in the opinion of first, second and third judges respectively. We prepare the following table for computation of rank correlation coefficients between the ranks given by three judges.

Ranks by x	Ranks by y	Ranks by z	$d_{xy} = x - y$	$d_{xz} = x - z$	$d_{yz} = y - z$	d_{xy}^2	d_{xz}^2	d_{yz}^2
1	3	6	–2	–5	–3	4	25	9
6	5	4	1	2	1	1	4	1
5	8	9	–3	–4	–1	9	16	1
10	4	8	6	2	–4	36	4	16
3	7	1	–4	2	6	16	4	36
2	10	2	–8	0	8	64	0	64
4	2	3	2	1	–1	4	1	1
9	1	10	8	–1	–9	64	1	81
7	6	5	1	2	1	1	4	1
8	9	7	–1	1	2	1	1	4
						$\Sigma d_{xy}^2 =$ **200**	$\Sigma d_{xz}^2 =$ **60**	$\Sigma d_{yz}^2 =$ **214**

If R_{12}, R_{13} and R_{23} be the rank correlation coefficients between first-second, first-third and second-third judges respectively, then we have

$$R_{12} = 1 - \frac{6\Sigma d_{xy}^2}{n(n^2 - 1)} = 1 - \frac{6\,(200)}{10\,(99)} = -0.21,$$

$$R_{13} = 1 - \frac{6\Sigma d_{xz}^2}{n(n^2 - 1)} = 1 - \frac{6\,(60)}{10\,(99)} = 0.64, \qquad R_{23} = 1 - \frac{6\Sigma d_{yz}^2}{n(n^2 - 1)} = 1 - \frac{6\,(214)}{10\,(99)} = -0.29.$$

R_{12} and R_{23} are negative which implies that the opinions of first-second and second-third judges oppose each other. R_{13} is positive implies that the opinions of first and third judge is similar. Thus, the pair of first and third judge have the nearest approach to common tastes in beauty.

6.41 LINES OF REGRESSION

The lines of best fit of Art 6.33 are also known as *lines of regression*.

When x is treated as independent variable and y as dependent then the line is called ***the line of regression of y on x***.

When y is treated as independent variable and x as dependent variable, the line is known as **the line of regression of x on y.**

From the solved examples of Art. 6.33, we know how to find the line of regression of y on x and also the line of regression of x on y.

In Example 6.50, we found the line of regression of y on x as $y = 2 + 0.5x$.

In Example 6.51, the equation of the line of regression of x on y is $x = 0.72 + 1.33y$.

6.42 CORRELATION AND REGRESSION

We have seen that the line of regression of y on x (which is the line of best fit when x is treated as independent variable and y as dependent variable) is

$$y = a + bx \quad \text{...(6.70)}$$

where a and b are given by the normal equations

$$\Sigma y = na + b\Sigma x \quad \text{...(6.71)}$$

$$\Sigma xy = a\Sigma x + b\Sigma x^2 \quad \text{...(6.72)}$$

and n is the number of pairs of values of x and y.

Eq. (6.71) can be written in the form

$$\frac{\Sigma y}{n} = a + b\frac{\Sigma x}{n} \quad \Rightarrow \quad \bar{y} = a + b\bar{x}. \quad \text{...(6.73)}$$

Eq. (6.73) shows that the point $(\bar{x}, \bar{y})$ lies on (6.70).

Thus the line of regression passes through $(\bar{x}, \bar{y})$, where $\bar{x}$ is the mean of x's and $\bar{y}$ the mean of y's.

Shifting the origin to the point $(\bar{x}, \bar{y})$ we find that (6.70) becomes

$$y - \bar{y} = a + b(x - \bar{x}) \quad \text{...(6.74)}$$

and Eq. (6.71) reduces to the form $\quad \Sigma(y - \bar{y}) = na + b\Sigma(x - \bar{x})$.

Since $\Sigma(y - \bar{y}) = 0$ and $\Sigma(x - \bar{x}) = 0$ by Art. 6.8, therefore, $a = 0$.

Also Eq. (6.72) becomes (as a result of shifting the origin) $\Sigma(x - \bar{x})(y - \bar{y}) = b\Sigma(x - \bar{x})^2 \quad (\because a = 0)$

or

$$\Sigma XY = b\Sigma X^2,$$

where X, Y are deviations of x and y from their means.

$$\therefore \quad b = \frac{\Sigma XY}{\Sigma X^2} = \frac{\Sigma XY}{\sqrt{\Sigma X^2 \Sigma Y^2}} \times \sqrt{\frac{\Sigma Y^2}{\Sigma X^2}} = r \cdot \frac{\sigma_y}{\sigma_x}.$$

Putting the values of a and b in Eq. (6.74), the line of regression of y on x becomes

$$y - \bar{y} = r\frac{\sigma_y}{\sigma_x}(x - \bar{x}).$$

$r\frac{\sigma_y}{\sigma_x}$ is called the *regression coefficient of y on x* and is denoted by b_{yx}. It is, obviously, the slope of this line.

Interchanging x and y we find that the line of regression of x on y is

$$x - \bar{x} = r\frac{\sigma_x}{\sigma_y}(y - \bar{y}). \quad \text{...(6.75)}$$

Here $r\frac{\sigma_x}{\sigma_y}$ is the *regression coefficient of x on y* and is denoted b_{xy}.

Regression coefficient b_{yx} and b_{xy} can also be calculated using following formulae:

$$b_{yx} = \frac{n\Sigma xy - \Sigma x\Sigma y}{n\Sigma x^2 - (\Sigma x)^2}, \quad b_{xy} = \frac{n\Sigma xy - \Sigma x\Sigma y}{n\Sigma y^2 - (\Sigma y)^2}.$$

From Eq. (6.75) it is clear that this line also passes through $(\bar{x}, \bar{y})$.

Properties of Regression Coefficients

1. Regression coefficient of y on x, measures the change in y corresponding to a unit change in x and regression coefficient of x on y measures the change in x corresponding to unit change in y.
2. The coefficient of correlation is the geometric mean of the two regression coefficients, that is

$$r = \pm\sqrt{b_{yx} \cdot b_{xy}}. \qquad \left[\because\ b_{yx} \cdot b_{xy} = r\frac{\sigma_y}{\sigma_x} \times r\frac{\sigma_x}{\sigma_y} = r^2\right]$$

3. Both regression coefficient must have the same sign as that of r, that is, all three b_{yx}, b_{xy} and r are of same sign.
4. If one of the regression coefficient is less than unity, then the other must be greater than unity, so that $b_{yx} \times b_{xy} \leq 1$.
5. Regression coefficients remain unchanged due to a shift of origin but change due to a shift of scale.

This property states that if the original pair of variables is (x, y) and if they are changed to the pair (u, v) where $u = (x - a)/p$ and $v = (y - c)/q$.

$$b_{yx} = \frac{q}{p} \times b_{vu}, \qquad \text{...(6.76)} \qquad \text{and} \qquad b_{xy} = \frac{p}{q} \times b_{uv}. \qquad \text{...(6.77)}$$

EXAMPLE 6.72 If the relationship between two variables x and u is $u + 5x = 6$ and between two other variables y and v is $2y + 3v = 21$, and the regression coefficient of y on x is known as 0.80. What would be the regression coefficient of v on u?

Solution We have

$$u + 5x = 6 \quad \Rightarrow \quad u = \frac{x - (6/5)}{-1/5}$$

$$2y + 3v = 21 \quad \Rightarrow \quad v = \frac{y - (21/2)}{-3/2}.$$

Employing $b_{yx} = \frac{q}{p} \times b_{vu}$ to get $b_{vu} = \frac{p}{q} \times b_{yx} = \frac{(-1/5)}{(-3/2)} \times 0.8 = \frac{8}{75}$.

Properties of Regression Lines

1. The two lines of regression intersect at the point $(\bar{x}, \bar{y})$. Hence, when the equations of two regression lines are given, $\bar{x}, \bar{y}$ can be found by solving them.
2. If $r = \pm 1$, that is, in case of perfect correlation, the two regression lines coincide.
3. If $r = 0$, that is, in case of no correlation, the two regression lines are perpendicular to each other.

Remark 6

When the line of regression of y on x is written in the form $y = a + bx$, then $b = r\frac{\sigma_y}{\sigma_x}$ which is called the coefficient of regression of y on x. Obviously it is the slope of this line.

Similarly, when the line of regression of x on y is written in the form $x = c + dy$ then $d = r\frac{\sigma_x}{\sigma_y}$ which is called the regression coefficient of x on y. It is not the slope of this line.

EXAMPLE 6.73 Using the following data, obtain the two regression equations:

x :	14	19	24	21	26	22	15	20	19
y :	31	36	48	37	50	45	33	41	39

Solution Presenting the data in tabular form, we get

x	y	$X = x - \bar{x}$	$Y = y - \bar{y}$	XY	X^2	Y^2
14	31	–6	–9	54	36	81
19	36	–1	–4	4	1	16
24	48	4	8	32	16	64
21	37	1	–3	–3	1	9
26	50	6	10	60	36	100
22	45	2	5	10	4	25
15	33	–5	–7	35	25	49
20	41	0	1	0	0	1
19	39	–1	–1	1	1	1
$\Sigma x = 180$	**$\Sigma y = 360$**	**0**	**0**	**$\Sigma XY = 193$**	**$\Sigma X^2 = 120$**	**$\Sigma Y^2 = 346$**

$$\bar{x} = \frac{180}{9} = 20, \quad \bar{y} = \frac{360}{9} = 40.$$

$$b_{xy} = \frac{\Sigma XY}{\Sigma Y^2} = \frac{193}{346} = 0.558 \quad \text{and} \quad b_{yx} = \frac{\Sigma XY}{\Sigma X^2} = \frac{193}{120} = 1.608.$$

Regression equation of x on y is given by

$$x - \bar{x} = b_{xy}(y - \bar{y}) \quad \text{or} \quad x - 20 = 0.558\,(y - 40) \quad \Rightarrow \quad x = 0.558y - 2.32.$$

Regression equation of y on x is given by

$$y - \bar{y} = b_{yx}\,(x - \bar{x}) \quad \text{or} \quad y - 40 = 1.608\,(x - 20) \quad \Rightarrow \quad y = 1.608x + 7.84.$$

EXAMPLE 6.74 From the following data find the regression equations, and estimate the likely value of y when $x = 100$.

x :	78	89	97	69	59	79	68	61
y :	125	137	156	112	107	136	123	108

Solution We construct the following table:

x	y	$d_x = x - 78$	$d_y = y - 125$	$d_x d_y$	d_X^2	d_Y^2
$a = 78$	$b = 125$	0	0	0	0	0
89	137	11	12	132	121	144
97	156	19	31	589	361	961
69	112	–9	–13	117	81	169
59	107	–19	–18	342	361	324
79	136	1	11	11	1	121
68	123	–10	–2	20	100	4
61	108	–17	–17	289	289	289
$\Sigma x = 600$	**$\Sigma y = 1004$**	**$\Sigma d_x = -24$**	**$\Sigma d_y = 4$**	**$\Sigma d_x d_y = 1500$**	**$\Sigma d_X^2 = 1314$**	**$\Sigma d_Y^2 = 2012$**

$\bar{x} = \dfrac{600}{8} = 75$ and $\bar{y} = \dfrac{1004}{8} = 125.5$. Since $\bar{y}$ is in fraction and not exact value.

It will be appropriate to apply the following formulae to evaluate b_{xy} and b_{yx}.

$$b_{xy} = \frac{n\Sigma d_X d_Y - \Sigma d_X \Sigma d_Y}{n\Sigma d_Y^2 - (\Sigma d_y)^2} = \frac{8(1500) - (-24)(4)}{8(2012) - (4)^2} = \frac{12096}{16080} = 0.752.$$

$$b_{yx} = \frac{n\Sigma d_X d_Y - \Sigma d_X \Sigma d_Y}{n\Sigma d_X^2 - (\Sigma d_x)^2} = \frac{8(1500) - (-24)(4)}{8(1314) - (-24)^2} = \frac{12096}{9936} = 1.217.$$

Regression equation of x on y is given by

$$x-\bar{x}=b_{xy}(y-\bar{y}) \quad \text{or} \quad x-75=0.752\,(y-125.5) \;\Rightarrow\; x=0.752y-19.376.$$

Regression equation of y on x is given by

$$y-\bar{y}=b_{yx}(x-\bar{x}) \quad \text{or} \quad y-125.5=1.217\,(x-75) \;\Rightarrow\; y=1.217x+34.225.$$

When $x = 100$, $y = 1.217\,(100) + 34.225 = 155.93$.

EXAMPLE 6.75 Find the regression equation from the following data; $\Sigma x = 15$, $\Sigma y = 25$, $\Sigma xy = 83$, $\Sigma x^2 = 55$, $\Sigma y^2 = 135$, $n = 5$. Estimate the values of x and y, if $y = 12$, $x = 8$.

Solution $\bar{x}=\sum x/n=15/5=3, \quad \bar{y}=\sum y/n=25/5=5.$

$$b_{xy}=r\cdot\frac{\sigma_x}{\sigma_y}=\frac{n\Sigma xy-\Sigma x\,\Sigma y}{n\Sigma y^2-(\Sigma y)^2}=\frac{5(83)-(15)(25)}{5(135)-(25)^2}=\frac{83-75}{135-125}=\frac{8}{10}=0.8.$$

$$b_{yx}=r\cdot\frac{\sigma_y}{\sigma_x}=\frac{n\Sigma xy-\Sigma x\,\Sigma y}{n\Sigma x^2-(\Sigma x)^2}=\frac{5(83)-(15)(25)}{5(55)-(15)^2}=\frac{83-75}{55-45}=\frac{8}{10}=0.8.$$

Regression equation of x on y is given by

$$x-\bar{x}=b_{xy}(y-\bar{y}) \quad \text{or} \quad x-3=0.8\,(y-5) \;\Rightarrow\; x=0.8y-1. \qquad ...(i)$$

Regression equation of y on x is given by

$$y-\bar{y}=b_{yx}(x-\bar{x}) \quad \text{or} \quad y-5=0.8\,(x-3) \;\Rightarrow\; y=0.8x+2.6. \qquad ...(ii)$$

The line of regression of x on y gives the best estimated value of x for given y. On putting $y = 12$, in (*i*), we obtain $x = 8.6$.

The line of regression of y on x gives the best estimated value of y on x. We put $x = 8$ in (*ii*), and obtain $y = 9$.

EXAMPLE 6.76 The two regression equations of the variables x and y are:

$$100x + 87y - 1913 = 0 \quad \text{and} \quad 50x + 100y - 1164 = 0.$$

Find (*i*) the mean of x's, mean of y's, (*ii*) the regression coefficients and (*iii*) the correlation coefficient between x and y.

Solution (*i*) Since the lines of regression pass through the point $(\bar{x}, \bar{y})$, therefore,

$$100\,\bar{x}+87\,\bar{y}-1913=0, \quad 50\,\bar{x}+100\,\bar{y}-1164=0.$$

Solving these equations, we find $\bar{x}=15.94$ and $\bar{y}=3.67$.

(*ii*) Let us assume that $100x + 87y - 1913 = 0$ is the regression equation of x on y. This may be written as

$x = 19.13 - 0.87y$. Therefore, $b_{xy}=r\cdot\dfrac{\sigma_x}{\sigma_y}=-0.87.$

The other equation $50x + 100y - 1164 = 0$ will then be the regression equation of y on x, which may be written as $y = 11.64 - 0.50x$. Therefore, $b_{yx}=r\cdot\dfrac{\sigma_y}{\sigma_x}=-0.50.$

Since $b_{xy}\cdot b_{yx} = (-0.87)\,(-0.50) < 1$, therefore our supposition is correct.

(*iii*) $r=\sqrt{b_{xy}\bullet b_{yx}}=\sqrt{(-0.87)\,(-0.50)}=\pm 0.66.$

As the regression coefficients have negative signs, we have to consider negative value for r. Thus $r = -0.66$.

EXAMPLE 6.77 In a partially destroyed laboratory record of an analysis of a correlation data only, the following two regression lines are available:

$$8x - 10y + 66 = 0 \quad \text{and} \quad 40x - 18y = 214.$$

Find (*i*) the average values of x and y, (*ii*) the correlation coefficient (*iii*) variance of y if var $(x) = 9$ (*iv*) the covariance of x and y.

Solution (*i*) Since the lines of regression pass through the point $(\bar{x}, \bar{y})$. Therefore,

$$8\,\bar{x} - 10\,\bar{y} + 66 = 0 \quad \text{and} \quad 40\,\bar{x} - 18\,\bar{y} - 214 = 0..$$

Solving these equations, we get $\bar{x} = 13$ and $\bar{y} = 17$.

(*ii*) Since we donot know from the two regression equations which is the regression equation of x on y. We make an assumption: let us take equation $8x - 10y + 66 = 0$ as the regression equation of x on y.

This may be written as $x = \frac{5}{4}y - \frac{33}{4}$. $\therefore\ b_{xy} = r \cdot \frac{\sigma_x}{\sigma_y} = \frac{5}{4}$.

The other equation $40x - 18y - 214 = 0$ will then be the regression equation of y on x, which may be re-written as

$$y = \frac{20}{9}x - \frac{107}{9}. \quad \therefore\ b_{yx} = r\frac{\sigma_y}{\sigma_x} = \frac{20}{9}.$$

Since both the regression coefficients are exceeding 1, our assumption is wrong. Hence $40x - 18y - 214 = 0$ is the regression equation of x on y and $8x - 10y + 66 = 0$ is the regression equation of y on x.

$\therefore\ b_{xy} = \frac{9}{20}$ and $b_{yx} = \frac{4}{5}$. Thus $r = \sqrt{b_{xy} \cdot b_{yx}} = \sqrt{\left(\frac{9}{20}\right) \times \left(\frac{4}{5}\right)} = \pm 0.6$. As the regression coefficients have positive signs, we have to-consider positive value of r. Hence $r = 0.6$.

(*iii*) $\sigma_x = \sqrt{9} = 3$, $r\frac{\sigma_x}{\sigma_y} = b_{xy} = \frac{9}{20}$. Therefore $\sigma_y = 20(0.6)(3)/9 = 4$.

(*iv*) We know $r = \frac{\text{Cov}(x, y)}{\sigma_x \bullet \sigma_y}$. Therefore, $\text{Cov}(x, y) = r\,\sigma_x \sigma_y = 0.6\,(3)\,(4) = 7.2$

EXAMPLE 6.78 You are given the following data:

Series	x	y
Arithmetic mean	36	85
Standard deviation	11	8

Coefficient of correlation between x and $y = 0.66$. Calculate the regression equations and estimate the value of x when $y = 75$.

Solution We have $\bar{x} = 36$, $\bar{y} = 85$, $\sigma_x = 11$, $\sigma_y = 8$ and $r = 0.66$. The regression line of x on y is given by:

$$x - \bar{x} = r \cdot \frac{\sigma_x}{\sigma_y}(y - \bar{y}).$$

Substituting the given values, we obtain the regression line equation of x on y as

$$x - 36 = 0.66 \times \frac{11}{8}(y - 85) = 0.9075\,(y - 85) = 0.9075y - 77.1375 \quad \Rightarrow \quad x = 0.9075y - 41.1375.$$

Substituting $y = 75$ in the above equation, we get $x = 0.9075(75) - 41.1375 = 68.0625 - 41.1375 = 26.925$. Therefore, when $y = 75$, $x = 26.93$. (approx.)

The regression line of y on x is given by $y - \bar{y} = r \cdot \frac{\sigma_y}{\sigma_x}(x - \bar{x})$. Substituting the given values, we obtain

$$y - 85 = 0.66 \times \frac{8}{11}(x - 36) = 0.48\,(x - 36) = 0.48x - 17.28 \quad \Rightarrow \quad y = 0.48x + 67.72.$$

EXAMPLE 6.79 The data given below show the test scores made by 10 salesmen on an intelligence test and their weekly sales :

Sales man :	1	2	3	4	5	6	7	8	9	10
Test score (x) :	40	70	50	60	80	50	90	40	60	60
Sales (in '000) (y) :	2.5	6.0	4.5	5.0	4.5	2.0	5.5	3.0	4.5	3.0

Calculate the regression line of sales on test scores and estimate the most probable weekly sales volume if a salesman makes a sure of 70. *(AMIE-W-2010)*

Solution We prepare the following table:

Total scores (x_i)	Sales (y_i)	$d_x = x_i - a$	$d_y = y_i - b$	$(d_x)(d_y)$	d_x^2	d_y^2
40	2.5	–20	–2.0	40	400	4.0
70	6.0	10	1.5	15	100	2.25
50	$b = 4.5$	–10	0	0	100	0
$a = 60$	5.0	0	0.5	0	0	0.25
80	4.5	20	0	0	400	0
50	2.0	–10	–2.5	25	100	6.25
90	5.5	30	1.0	30	900	1.00
40	3.0	–20	–1.5	30	400	2.25
60	4.5	0	0	0	0	0
60	3.0	0	–1.5	0	0	2.25
		$\Sigma d_x = 0$	$\Sigma d_y = -4.5$	$\Sigma d_x d_y = 140$	$\Sigma d_x^2 = 2400$	$\Sigma d_y^2 = 18.25$

From the table, we have

Mean of test score (x), $$\bar{x} = a + \frac{\Sigma d_x}{n} = 60 + \frac{0}{10} = 60.$$

Mean of sales (y), $$\bar{y} = b + \frac{\Sigma d_y}{n} = 4.5 - \frac{4.5}{10} = 4.05.$$

Regression line of sales (y) on score (x) is given by

$$y - \bar{y} = r\frac{\sigma_y}{\sigma_x}(x - \bar{x}) = b_{yx}(x - \bar{x}),$$ where b_{yx} is the regression coefficient of y on x.

or $$y - \bar{y} = \frac{n\Sigma d_x d_y - \Sigma d_x \Sigma d_y}{n\Sigma d_x^2 - (\Sigma d_x)^2}(x - \bar{x}).$$ Substituting the values, we obtain

$$y - 4.05 = \frac{10(140) - 0}{10(2400) - 0}(x - 60) = 0.06(x - 60)$$

or $y = 0.06x + 0.45$, the required regression line.

For $x = 70$, $y = 0.06(70) + 0.45 = 4.65$.

Thus the most probable weekly sales volume for a score of 70 is 4.65.

EXAMPLE 6.80 If θ is the angle between the two regression lines, show that $\tan\theta = \frac{1 - r^2}{r} \cdot \frac{\sigma_x \sigma_y}{\sigma_x^2 + \sigma_y^2}$.

Explain the significance when $r = 0$ and $r = \pm 1$. *(VTU, 2007; Andhra, 2000)*

Solution The equations to the line of regression of y on x and x on y are:

$$y - \bar{y} = r\frac{\sigma_y}{\sigma_x}(x - \bar{x}) \quad \text{and} \quad x - \bar{x} = r\frac{\sigma_x}{\sigma_y}(y - \bar{y}) \text{ respectively.}$$

Therefore, their slopes are $m_1 = r(\sigma_y/\sigma_x)$ and $m_2 = \sigma_y/r\,(\sigma_x)$.

Then $$\tan\theta = \frac{m_2 - m_1}{1 + m_1 m_2} = \frac{\dfrac{\sigma_y}{r\sigma_x} - r\dfrac{\sigma_y}{\sigma_x}}{1 + \left(\dfrac{\sigma_y}{r\sigma_x}\right)\left(\dfrac{r\sigma_y}{\sigma_x}\right)} = \frac{\dfrac{\sigma_y}{\sigma_x}\left(\dfrac{1}{r} - r\right)}{1 + \dfrac{\sigma_y^2}{\sigma_x^2}} = \frac{1-r^2}{r} \cdot \frac{\sigma_x \sigma_y}{\sigma_x^2 + \sigma_y^2}.$$

When $r = 0$, that is, there is no correlation between the variables, then $\tan\theta \to \infty$ or $\theta = \pi/2$. So, the two lines of regression are perpendicular to each other.

When $r = \pm 1$, $\tan\theta = 0$ that is, $\theta = 0$ or π. Thus, the lines of regression coincide, that is, there is perfect correlation between the two variables.

EXERCISE 6.3

1. Fit a straight line to the following data, taking y as dependent variable:

x:	1	2	3	4	5
y:	5	7	9	10	11

Ans. $y = 3.9 + 1.5\,x$.

2. Show that the line of best fit to the following data is given by $y = 8 - 0.5x$:

x:	6	7	7	8	8	8	9	9	10
y:	5	5	4	5	4	3	4	3	3

3. The weight of a calf taken at weekly intervals are given below:

Age in weeks (x):	1	2	3	4	5
Wt in kg (y):	30	34	37	43	45

Fit a straight line to the above data and then give an estimate of weight for the calf when it becomes 7 weeks old. **Ans.** $y = 26.1 + 3.9x$, 53.4 kgm.

4. Calculate the coefficient of correlation from the following data:

x:	9	8	7	6	5	4	3	2	1
y:	15	16	14	13	11	12	10	8	9

Ans. 0.95.

5. Calculate the value of r between X and Y for the values given below:

X:	2	5	7	9	19	17
Y:	25	27	26	29	34	35

Ans. 0.97.

6. Find the coefficient of correlation for the following table:

x:	10	14	18	22	26	30
y:	18	12	24	6	30	35

Ans. 0.6.

7. Find the coefficient of correlation from the following data:

x:	65	63	61	64	68	62	70	66
y:	68	66	68	65	69	66	68	65

Ans. 0.359.

8. From the following data, obtain the value of the correlation coefficient:

$n = 10$, $\Sigma x = 55$, $\Sigma y = 40$, $\Sigma x^2 = 385$, $\Sigma y^2 = 192$ and $\Sigma(x + y)^2 = 947$. **Ans.** -0.6812.

9. From the following data, compute the coefficient of correlation between x and y.

	x-series	y-series
No. of items:	15	15
Arithmetic mean:	25	18
Sum of squares of deviations from mean	136	138

Summation of product of deviations of x and y series from the respective arithmetic means = 122.

Ans. 0.891.

10. Calculate Karl Pearson's coefficient of correlation between x and y from the following information:

$$n = 12,\ \Sigma x_i = 120,\ \Sigma y_i = 130,\ \Sigma(x_i - 8)^2 = 150,\ \Sigma(y_i - 10)^2 = 200,\ \Sigma(x_i - 8)(y_i - 10) = 50.$$

[**Hint:** Let $u = x_i - 8$ and $v = y_i - 10$.

Therefore, $\Sigma u = \Sigma(x_i - 8) = \Sigma x_i - \Sigma 8 = \Sigma x_i - 8n = 120 - 8(12) = 24.$

$\Sigma v = \Sigma(y_i - 10) = \Sigma y_i - \Sigma 10 = \Sigma y_i - 10n = 130 - 10(12) = 10.$

$\Sigma uv = \Sigma(x_i - 8)(y_i - 10) = 50,\quad \Sigma u^2 = \Sigma(x_i - 8)^2 = 150,\quad \Sigma v^2 = \Sigma(y_i - 10)^2 = 200.$

Employing $$r_{uv} = \frac{n\Sigma uv - \Sigma u \Sigma v}{\sqrt{n\Sigma u^2 - (\Sigma u)^2}\ \sqrt{n\Sigma v^2 - (\Sigma v)^2}} = \frac{12(50) - 24(10)}{\sqrt{12\,(150) - (24)^2}\ \sqrt{12\,(200) - (10)^2}}$$

$$= \frac{360}{\sqrt{1224 \times 2300}} = \frac{360}{1677.85} = 0.214 \Bigg].$$

11. Examine whether there is any correlation between age and blindness on the basis of the following data:

Age in yrs:	0-10	10-20	20-30	30-40	40-50	50-60	60-70	70-80
No. of persons: (in thousands)	90	120	140	100	80	60	40	20
No. of blind persons:	10	15	18	20	15	12	10	6

[**Hint:** Let us denote the mid-value of age in years as x and the no. of blind persons per lakh as y. Computation of correlation between age and blindness is given in the following table:

Age in years	Mid-value x	No. of Persons, P (in thousands)	No. of blind B	No. of blind per lakh $y = (B/P) \times 1$ lakh	xy	x^2	y^2
0–10	5	90	10	11	55	25	121
10–20	15	120	15	12	180	225	144
20–30	25	140	18	13	325	625	169
30–40	35	100	20	20	700	1225	400
40–50	45	80	15	19	855	2025	361
50–60	55	60	12	20	1100	3025	400
60–70	65	40	10	25	1625	4225	625
70–80	75	20	6	30	2250	5625	900
Total	**320**	—	—	**150**	**7090**	**17000**	**3120**

Substituting the values in the formula $r = \dfrac{n\Sigma xy - (\Sigma x)(\Sigma y)}{\sqrt{n\Sigma x^2 - (\Sigma x)^2}\ \sqrt{n\Sigma y^2 - (\Sigma y)^2}}$.

The correlation coefficient between age and blindness is given by

$$r = \frac{8\times7090 - 320\times150}{\sqrt{8\times17000-(320)^2}\,\sqrt{8\times3120-(150)^2}} = \frac{8720}{183.303\times49.5984} = 0.96,$$

which exhibits a very high degree of positive correlation between age and blindness.]

12. From the data given below find the number of items n: $r_{xy} = 0.5$, $\Sigma XY = 120$, $\sigma_y = 8$, $\Sigma X^2 = 90$, where X and Y are deviations from arithmetic mean.

[Hint: Given that $r_{xy} = 0.5,\ \Sigma XY = \Sigma(x-\bar{x})(y-\bar{y}) = 120,\ \Sigma X^2 = \Sigma(x-\bar{x})^2 = 90$

$$\sigma_y = \sqrt{\frac{\Sigma(y-\bar{y})^2}{n}} = 8 \Rightarrow \Sigma(y-\bar{y})^2 = 64n.$$

Employing $r_{xy} = \dfrac{\Sigma(x-\bar{x})(y-\bar{y})}{\sqrt{\Sigma(x-\bar{x})^2}\,\sqrt{\Sigma(y-\bar{y})^2}} \Rightarrow 0.5 = \dfrac{120}{\sqrt{90}\,\sqrt{64N}}$ or $N = \dfrac{(120)(120)}{(90)(64)(0.25)} = 10.$

13. In two sets of variables x and y with 20 observations each, the following data were observed:

$$\bar{x} = 12,\ \bar{y} = 15,\ \text{S.D. of } x = 3,\ \text{S.D. of } y = 4.$$

Coefficient of correlation between x and y is 0.4. Later on, it was found that the pair (20, 15) was wrongly taken as (15, 20). Find the correct value of the correlation coefficient. **Ans.** 0.314.

14. Calculate the coefficient of correlation for the following table:

y↓ \ x→	0–4	4–8	8–12	12–16	Total
0–5	7	0	0	0	7
5–10	6	8	0	0	14
10–15	0	5	3	0	8
15–20	0	7	2	0	9
20–25	0	0	0	9	9
Total	**13**	**20**	**5**	**9**	**47**

Ans. 0.8.

15. If $u + 5x = 6$ and $3y - 7v = 20$ and the correlation coefficient between x and y is 0.58, then what would be the correlation coefficient between u and v? **Ans.** $r_{uv} = -0.58$.

[Hint: $u = \dfrac{x-(6/5)}{-1/5}$, $v = \dfrac{y-(20/3)}{7/3}$.]

16. If the relation between x and u is $3x + 2u + 4 = 0$ and y and v is $4v + 16y + 11 = 0$ and the correlation coefficient between x and y is 0.8, then what is the correlation coefficient between u and v? **Ans.** $r_{uv} = 0.8$.

[Hint: Here $u = \dfrac{x-(-7/3)}{-4/3}$, $v = \dfrac{y-(-11/16)}{-1/4}$.]

17. For a group of 8 students, the sum of squares of differences in ranks for Maths and Hindi marks was found to be 50. What is the value of rank correlation coefficient? **Ans.** 0.40.

18. Marks of 12 students in Arithmetic and Algebra are given below:

Arithmetic :	60	34	40	50	45	40	22	43	42	66	64	46
Algebra :	75	32	33	40	45	33	12	30	34	72	41	57

Calculate the rank correlation coefficient. **Ans.** 0.84.

19. Find the rank correlation between marks obtained in Mathematics and statistics.

Students :	A	B	C	D	E	F	G	H	I	J
Marks in Maths :	30	20	40	50	30	20	30	50	10	0
Marks in Statistics :	15	40	40	45	20	30	15	50	20	10

Ans. 0.63.

20. Two judges in a beauty contest rank the ten competitors in the following order:

I :	6	4	3	1	2	7	9	8	10	5
II :	4	1	6	7	5	8	10	9	3	2

Do the judges appear to agree in their standard.

Ans. $R = 0.224$. Since R is +ve, the judges appear to agree. The agreement is complete only if $R = 1$.

21. Establish the formula $\sigma_{x+y}^2 = \sigma_x^2 + \sigma_y^2 + 2r\,\sigma_x\sigma_y$, where r is the correlation coeff. between two variables x and y, σ_x^2, σ_y^2 and σ_{x+y}^2 are the variance of x, y and $x + y$ respectively.

22. If the regression coefficient of x on y is $-1/6$ and that of y on x is $-3/2$, what is the correlation coefficient between x and y?

Ans. -0.5.

23. Find the equations of regression lines for the following values of x and y.

x :	1	2	3	4	5
y :	2	5	3	8	7

Hence find the value of r.

Ans. $x = 0.5y + 0.5$, $y = 1.3x + 1.1$, $r = 0.81$.

24. The observed values of a function are respectively 168, 120, 72 and 63 at the four positions 3, 7, 9 and 10 of the independent variable. What is the best estimate you can give for the value of the function at the position 6 of the independent variable?

[**Hint**: The line of regression of y on x is $y = 217.91 - 15.47x$. When $x = 6$, $y = 125.09$, the required best estimate].

25. Heights of fathers and sons are given in inches:

Height of father :	65	66	67	67	68	69	71	73
Height of son :	67	68	64	68	72	70	69	70

Find the two lines of regression and find the expected average height of son when the height of father is 67.5 inches.

Ans. $y = 0.421x + 39.77$, $x = 0.52y + 32.63$; 68.19.

26. The following regression equations were obtained from a correlation table:

$$y = 0.516x + 33.73,\ x = 0.512y + 33.52.$$

Find the value of (*a*) r (*b*) the mean of x's (*c*) the mean of y's:

Ans. $r = 0.514$, $\bar{x} = 69.0268$, $\bar{y} = 69.3476$.

27. From a partially destroyed laboratory only following records could be available: $x = 4y + 5$ and $y = mx + 4$ are the regression lines of x on y and y on x respectively. Show that $0 < m < 1/4$. If $m = 1/16$, find the means of two variables and the coefficient of correlation between them. Also if var. $x = 9$, find var. y when $m = 1/16$.

Ans. 28, 23/4; 1/2; $\sigma_y = 3/8$.

28. The following data are given for marks in English and Mathematics in an examination:

Mean marks in English = 39.5,
Mean marks in Maths = 47.6,
r between marks in English and Maths = 0.42,
S.D. of marks in English = 10.8,
S.D. of marks in Maths = 16.9.

Form the two lines of regression, calculate the expected average marks in Maths of a candidate who secured 50 marks in English.

Ans. $y - 47.6 = 0.657\,(x - 39.5)$, $x - 39.5 = 0.268\,(y - 47.6)$, $y = 54.5$.

where x denotes marks in English and y those in Maths.

29. In a partially destroyed laboratory record of an analysis of correlation data, the following results are legible.

Variance of $x = 9$.

Regression equations: $4x - 5y + 33 = 0$ and $20x - 9y = 107$.

What were (*a*) the mean values of x and y (*b*) the standard deviation of y (*c*) the coeff. of correlation between x and y? (*VTU, 2005; Anna, 2003S, UPTU, 2003*) **Ans.** $\bar{x} = 13$, $\bar{y} = 17$, $\sigma_y = 4$, $r = 0.6$

30. Following are the marks in Hindi & English in an annual examination:

	Hindi (x)	English (y)	
Mean	40	50	
S.D.	10	16	$r = +0.5$

Estimate the score of English when score in Hindi is 50 and score in Hindi when score in English is 30.

Ans. 58; 34.

31. The equation of two lines of regression obtained in a correlation analysis are the followings:

$$2x + 3y - 8 = 0 \quad \text{and} \quad x + 2y - 5 = 0.$$

Obtain the value of the correlation coefficient and the variance of y, given that the variance of x is 12.

Ans. –0.866; 2.

32. The correlation coefficient between two variables x and y is 0.66. If $\sigma_x = 11.0$, $\sigma_y = 8$, $\bar{x} = 36$, $\bar{y} = 85$. Find the two regression lines.

Ans. Regression line of y on x is $y = 0.48x + 67.72$, regression line of x on y is $x = 0.9075y - 41.137$.

33. If $y = kx + 4$ and $x = 4y + 5$ are the regression lines of y on x and x on y respectively, then find the interval in which k lies.

[**Hint:** We have $b_{yx} = k$, $b_{xy} = 4$. Now $r^2 = b_{yx} \cdot b_{xy} = 4k$. $0 \le r^2 \le 1 \Rightarrow 0 \le 4k \le 1$ or $0 \le k \le 1/4$.]

34. If $u = 2x + 5$ and $v = -3y - 6$ and regression coefficient of y on x is 2.4, what is the regression coefficient of v on u? **Ans.** $b_{vu} = -3.6$.

[**Hint:** We write $u = \dfrac{x - (-5/2)}{1/2}$, and $v = \dfrac{y - (-2)}{-1/3}$. Here $p = 1/2$, $q = -1/3$.

Therefore, $b_{vu} = \dfrac{p}{q} \times b_{yx} = \dfrac{1/2}{-1/3}\,(2.4) = -3.6.$]

35. If the regression coefficient of y on x, the coefficient of correlation between x and y and variance of y are $-3/4$, $-\sqrt{3}/2$ and 4 respectively. What is the variance of x? **Ans.** $\sigma_x^2 = 16/3$.

36. If $x - (a - 2)y + 9 = 0$ and $ax - 3y + 9 = 0$ are the lines of regression of x on y and y on x respectively. Show that $-1 < a < 3$.

7.1 IN

The word
head up
say with
situation

Suppos
cards etc
a 5 will
cards. W
pack of
of this exp
above sta
uncertain

7.2 RA

Before de
Experime
outcomes
experimen
outcomes
a fair di
In all the
in advan

Sampl
associate
is called

Event
the samp
tail appe
events m
to be s
two or m

For e
getting
HT and

Sure
by S is c
one of t

Impos
considere
belong to

CHAPTER 7

Theory of Probability and Random Variables

7.1 INTRODUCTION

The word "*probable*" means "likely, though not certain to occur". Thus when a coin is tossed, it will show a head up or it will show a tail up, that is, there are two possible outcomes "head up" or "tail up" but we cannot say with certainty which of these two possible outcomes will materialise. In other words, we have an uncertain situation where we cannot state with certainty what outcome will really be.

Suppose, we conduct a certain experiment like throwing a die, a card is drawn from a pack of 52 playing cards etc. If a die is tossed in the air, then it is certain that the die will come down, but it is not certain that, say, a 5 will appear. Now consider another experiment that a card is drawn from a well-shuffled pack of playing cards. When a card is drawn from a well-shuffled pack of playing cards, we know that there are 52 cards in a pack of playing cards and the card drawn may be any one of these 52 cards. so there are 52 possible outcomes of this experiment. But we cannot say with certainty which of these 52 outcomes will materialise. In fact, the above statement involves an element of uncertainty. Probability theory is designed to estimate the degree of uncertainty regarding the happening of a given phenomenon.

7.2 RANDOM EXPERIMENTS AND EVENTS

Before defining probability we shall explain the following basic, widely used concepts and terms.

Experiment: Any operation that results in two or more outcomes is called an experiment. When all the outcomes of an experiment can be enumerated but which particular outcome will result is not known, the experiment is called a random experiment. For example if a fair coin is tossed, then it has two possible outcomes–"Head up or Tail up" but which face will actually turn up is not known. Similarly, result of rolling a fair die, result of drawing a card from a pack of 52 playing cards–are random events or random experiments. In all these experiments, there are more than one possible outcomes but it is not possible to predict the outcome in advance.

Sample Space: The set of all possible outcomes of a random experiment is called the the *sample space* associated with the experiment. Sample space is denoted by the symbol S. Each element of the sample space is called a *sample point.*

Events: One or more possible outcomes of an experiment are said to form an event, that is, any subset of the sample space is an event. For example when two fair coins are tossed, the events are two tails appear, one tail appears, no tail appear, when a card is drawn from a well shuffled pack of playing cards, the following events may be considered: The card is spade. The card is an ace. The card is heart or diamond. An event is said to be *simple* if it corresponds to a single possible outcome of an experiment, whereas the joint occurrence of two or more simple events is called a *compound* event.

For example, tossing a coin once provides us two simple events namely head and tail. On the other hand, getting a head when a coin is tossed twice is an example of composite event as it can be split into the events HT and TH which are both elementary events.

Sure Event: The sample space S of an experiment is the set of all possible outcomes. The event represented by S is called a *sure event.* In the case of throw of a die, the event $S = \{1, 2, 3, 4, 5, 6\}$ is a sure event because one of the outcomes from S must occur.

Impossible event: An empty set ϕ is always a subset of a set S. Hence, the emptyset ϕ can always be considered as representing an event of an experiment. But, there is no outcome of the experiment which can belong to ϕ. Hence, the event corresponding to the null set ϕ is known as an *impossible event.* In the case of a throw of a die 'getting a number more than 6', getting a number less than 1 or in the case of a throw of two dice together, getting a sum more than 12 etc., are examples of impossible events.

Union of two events: The *union* of two events A and B denoted by $A \cup B$ is defined as a set of events containing all the sample points of *event A or event B or both* the events A and B. This is shown in Fig. 7.1.

Intersection of two events: The *intersection* of two events A and B denoted by $A \cap B = B \cap A$ or simply AB or BA may be defined as the set containing all the sample points that are common to both the events A and B. This is shown in Fig. 7.2.

Complement of an event: The *complement* of an event A with respect to the sample space S is the set of all elements of S which are **not** in A. We denote the complement of A by A^c or A' or $\bar{A}$. Thus $\bar{A} = S - A$. Obviously, $A \cup \bar{A} = S$ and $A \cap \bar{A} = \phi$. In Fig. 7.3, the shaded part of the sample space S denotes $\bar{A}$.

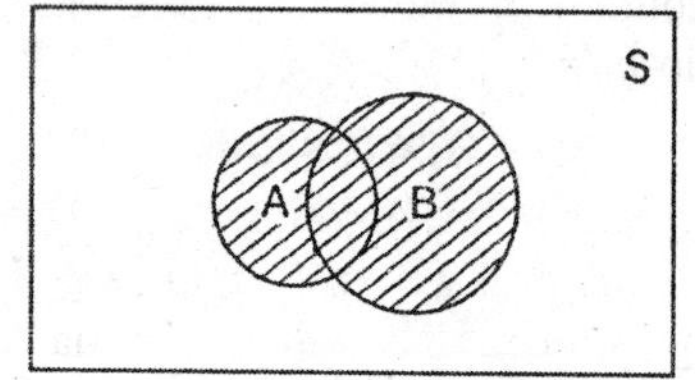

Fig. 7.1: $A \cup B$ is shaded.

The **difference** of two events A and B be denoted by $A - B$, may be defined as the set of sample points present in set A but **not** in B, that is, $A - B = A \cap \bar{B}$ and $B - A = B \cap \bar{A}$ (Refer Fig. 7.4).

Equally likely events or Equi-probable events: Two or more outcomes of an experiment are said to be equally likely if any one of them cannot be expected to occur in preference to the others. Thus when a fair coin is tossed the coming up of head or tail is equally likely.

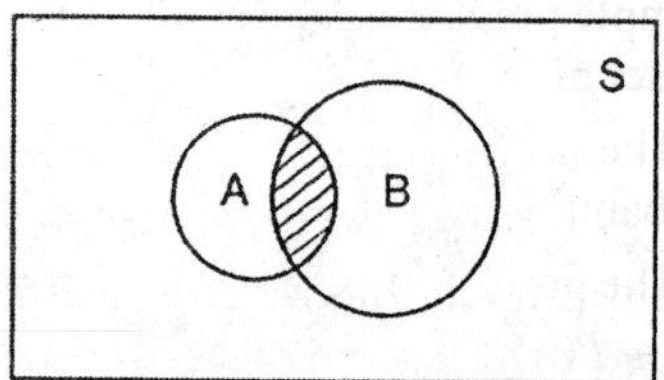

Fig. 7.2: $A \cap B$ is shaded.

Mutually exclusive events: Two or more events are said to be *mutually exclusive events* if the occurrence of any one of them excludes the occurrence of the other events. For example if a coin is tossed, either the head or tail comes up. Both can't occur simultaneously. Such events are called mutually exclusive events. In terms of sets, two events A and B are said to be mutually exclusive events if A and B are disjoint sets, that is, if $A \cap B = \phi$. (Fig. 7.5)

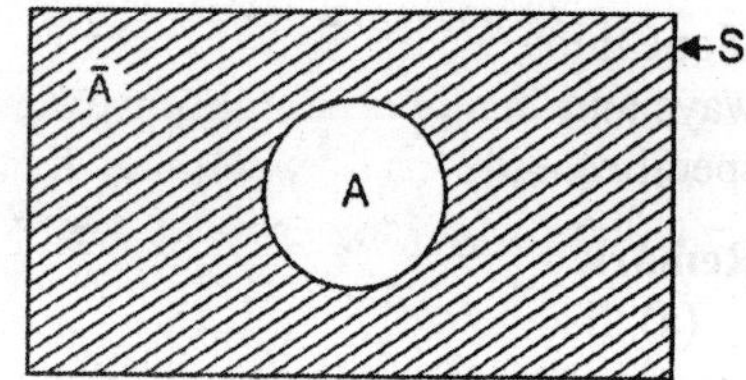

Fig. 7.3: $\bar{A}$ is shaded.

Exhausitive set of events: A set of events is said to be *exhaustive* if it includes all the possible events. If we consider the tossing of a coin, head and tail are exhaustive events that is, there is no other third possibility.

Now it is clear to the readers that if we draw a card from a well shuffled pack of playing card, we may get any one of the 52 cards. The 52 cases are mutually exclusive, equally likely and exhaustive. Similarly if a cubical uniform die is thrown, the turning up of each of six faces is equally likely, mutually exclusive and these 6 cases are exhaustive.

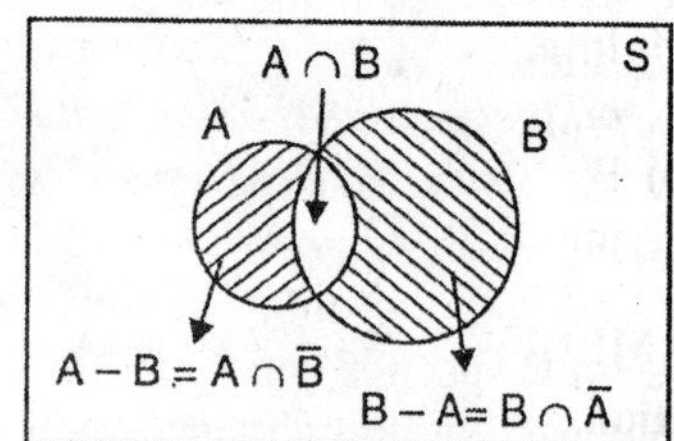

Fig. 7.4: Showing $(A - B)$ and $(B - A)$.

Combination of events: This can be done by using the operations "or", "and", "not".

Algebra of events: Let A, B and C be the events of a sample space S:

1. $(A \cup B)$ or (A or B) is the event "either A or B or both
2. $(A \cap B)$ or (A and B) is the event "both A and B"
3. A' is the event "not A"
4. $A - B$ is the event "A but not B"
5. $A \cup B = B \cup A, \quad A \cap B = B \cap A$
6. $A \cup \phi = A, \quad A \cap \phi = \phi$
7. $(A \cup B)' = A' \cap B'$ (De Morgan's Laws)
8. $(A \cap B)' = A' \cup B'$ (De Morgan's Laws)
9. $A = (A \cap B) \cup (A \cap B')$
10. $A \cup B = (A \cap B) \cup (A \cap B') \cup (B \cap A')$

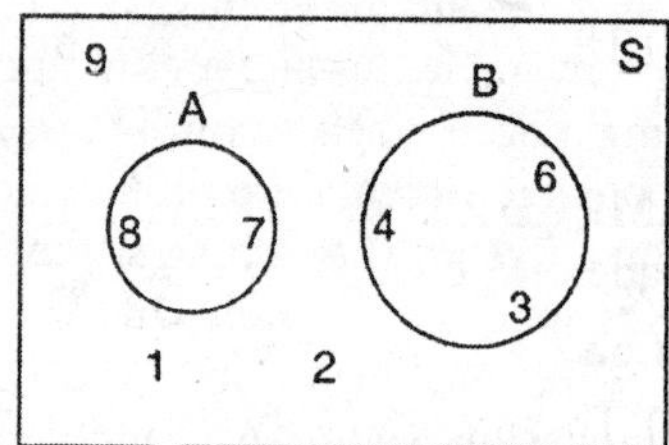

Fig. 7.5: Two disjoint sets A and B.

11. $A \cup (B \cup C) = (A \cup B) \cup C$
12. $A \cap (B \cap C) = (A \cap B) \cap C$
13. $A \cup (B \cap C) = (A \cup B) \cap (A \cup C)$
14. $A \cap (B \cup C) = (A \cap B) \cup (A \cap C)$

We may use these results in the study of probability of events.

7.3 DEFINITION OF PROBABILITY

If an experiment can result in n exhaustive, mutualy exclusive and equally likely cases and m of those are favourable to the event A, then the probability of occurrence of the event A, written as $P(A)$, is defined as the ratio m/n.

$\therefore$ The probability of occurrence of $A = P(A)$

$$= \frac{\text{The number of outcomes favourable to event } A}{\text{The total number of exhaustive mutually exclusive and equally likely cases which the experiment can result}} = \frac{m}{n}.$$

Using set notation, we can write $P(A) = n(A)/n(S)$, where $n(S)$ denotes the number of elements (sample points) in the sample space S and likewise $n(A)$ denotes the number of sample points in A. Event A is a subset of S.

The probability of occurrence of an event is a number lies between 0 and 1. If the event cannot occur, its probability is 0. If it must occur, that is, its occurrence is certain, its probability is 1.

The probability of non-occurrence of the event (called its *failure*) is denoted by $P(\bar{A}) = (n-m)/n = 1-(m/n)$ $= 1 - P(A)$.

Fundamental Principle of Counting

If one thing can be performed in n_1 different ways and after this a second thing can be performed in n_2 different ways and finally a kth thing can be performed in n_k different ways, then all k things can be performed in the specified order in $n_1\, n_2 \ldots n_k$ different ways.

Remark 1

(a) If an event A can happen in m ways and fail in n ways, all these (that is, $m + n$) ways being equally likely, mutually exclusive, and exhaustive then the probability of the happening of A is $\mathbf{m/(m+n)}$ and that of its failing is $\mathbf{n/(m+n)}$.

In place of saying that probability of happening of an event is $m/(m+n)$, we sometimes say that *the odds are m to n in favour of the event or n to m against the event.*

(b) By total number of ways' or possible number of ways' we shall mean "ways which are equally likely, mutually exclusive and exhaustive".

EXAMPLE 7.1 Find the probability of throwing (*i*) 5 (*ii*) an even number, with an ordinary six faced die.

Solution (*i*) In this case there are 6 possible ways out of which only one (that is, appearing of 5) is favourable.

Therefore, $$p = \frac{\text{number of favourable ways}}{\text{total number of ways}} = \frac{1}{6}.$$

(*ii*) Here number of favourable ways (that is, appearing of faces with numbers 2, 4, 6) = 3. Therefore $p = 3/6 = 1/2$.

EXAMPLE 7.2 If a card is drawn from a well shuffled pack, find the probability that the card is (*i*) a queen of spades (*ii*) a king (*iii*) a diamond.

Solution (*i*) There are 52 cards in a pack and there is only one queen of spades. Therefore, $p = 1/152$.

(*ii*) There are four Kings in a pack. Hence $p = 4/52 = 1/13$.

(*iii*) There are 13 cards of diamonds. Hence $p = 13/52 = 1/4$.

EXAMPLE 7.3 A card is drawn from a pack of 52 cards and a gambler bets that it is a spade or an ace. What are the odds against his winning the bet?

Solution Number of favourable ways, that is, the card is a spade or an ace = 13 + 3 = 16.

The probability that the gambler wins $= 16/52 = 4/13 = 4/(4+9)$.

Hence odds against his winning are 9 to 4.

EXAMPLE 7.4 Four cards are drawn at random from a pack of well shuffled cards. Find the probability that they are all cards of spades.

Solution Four cards can be choosen from 13 cards of spades in ${}^{13}C_4 = \dfrac{13\cdot 12\cdot 11\cdot 10}{4!}$ ways.

Total number of possibilities $= {}^{52}C_4 = \dfrac{52\cdot 51\cdot 50\cdot 49}{4!}$.

Hence, required probability $= \dfrac{{}^{13}C_4}{{}^{52}C_4} = \dfrac{13\cdot 12\cdot 11\cdot 10}{52\cdot 51\cdot 50\cdot 49} = \dfrac{11}{4165} = 0.0026.$

EXAMPLE 7.5 Two balls are drawn from a bag containing 5 white and 7 red balls. Find the probability that both are red.

Solution Two red balls can be drawn from 7 red balls in ${}^{7}C_2 = \dfrac{7.6}{2!} = 21$ ways.

Also there are 12 balls in the bag and 2 can be drawn in ${}^{12}C_2 = \dfrac{12\times 11}{2!} = 66$ ways.

Hence, required probability $= {}^{7}C_2/{}^{12}C_2 = 21/66 = 7/22.$

EXAMPLE 7.6 Five balls are drawn from a bag containing 6 white and 4 black balls. What is the probability that 3 are white and 2 black?

Solution There are in all 10 balls in the bag and total number of ways of drawing 5 from them $= {}^{10}C_5$.

Now 3 white balls can be drawn from 6 in ${}^{6}C_3$ ways and two black balls can be drawn from 4 in ${}^{4}C_2$ ways and any group of 3 white balls can be associated with 2 black balls in ${}^{6}C_3 \times {}^{4}C_2$ ways (Fundamental Theorm).

Hence $$\text{Prob.} = \frac{{}^{6}C_3 \times {}^{4}C_2}{{}^{10}C_5} = \frac{6\times 5\times 4}{3!} \times \frac{4\times 3}{2!} \times \frac{5!}{10\cdot 9\cdot 8\cdot 7\cdot 6} = \frac{10}{21}.$$

EXAMPLE 7.7 In a single throw of two dice, what is the probability of getting a sum of 7?

Solution The sample space consisting of $6 \times 6 = 36$ sample point.

Favourable ways for getting a sum of 7 are: (1, 6), (6, 1), (2, 5), (5, 2), (3, 4), (4, 3) that is, 6 ways.

$$\therefore \qquad \text{Prob.} = \frac{6}{36} = \frac{1}{6}.$$

Theorems on Probability

Theorem 7.1 (Addition rule for probabilities of mutually exclusive events)

Statement If $P(A)$ and $P(B)$ are the probabilities for success of two mutually exclusive events A and B then the probability that one or the other event happening is given by $P(A \cup B) = P(A \text{ or } B) = P(A) + P(B)$.

Proof Let n be the total number of *elementary events in the sample space S. Let m_1 be the number of ways favourable to the first event A and m_2, the number of ways favourable to the event B. Thus the number of ways favourable to either A or B is $m_1 + m_2$. The probability P, that either A or B happening is given by

$$P(A \cup B) = P(A \text{ or } B) = \frac{m_1 + m_2}{n} = \frac{m_1}{n} + \frac{m_2}{n} = P(A) + P(B) \qquad \text{...(7.1)}$$

---

* Suppose that we have performed a random experiment. It is completely defined when we know all the possible outcomes. Each outcome of this experiment is called an elementary event.

Remark 2

(a) If $A_1, A_2, ..., A_n$ are n mutually exclusive events in a sample space S, then probability of the occurence of any one of them is given by

$$P(A_1 \cup A_2 \cup ... \cup A_n) = P(A_1) + P(A_2) + \cdots + P(A_n). \quad ...(7.2)$$

(b) If $A_1, A_2, ..., A_n$ are n mutually exclusive and exhaustive events, that is, $A_1 \cup A_2 \cup ... \cup A_n = S$, then $P(A_1 \cup A_2 \cup ... \cup A_n) = P(S) = 1$ or $P(A_1) + P(A_2) + \cdots + P(A_n) = 1$.

Theorem 7.2 (Law of addition of probabilities)

If A and B are any two events of a sample space, then the probability that atleast one of the events A or B will occur is given by

$$P(A \cup B) = P(A) + P(B) - P(A \cap B). \quad ...(7.3)$$

Proof Consider the venn diagram (see Figure 7.6). Portion I, II, III make up $A \cup B$ and are mutually exclusive. Hence by Theorem I, we get

$$P(A \cup B) = P(I) + P(II) + P(III) = \{P(I) + P(II)\} + P(III)$$
$$= P(A) + [P(B) - P(II)] = P(A) + P(B) - P(A \cap B). \quad ...(7.4)$$

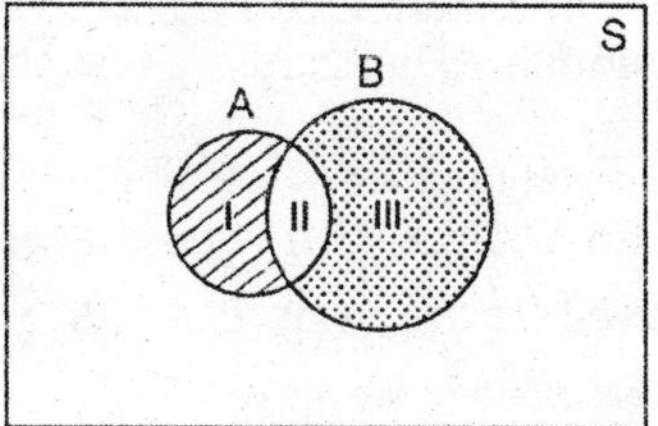

Fig. 7.6: Venn diagram

Remark 3

(a) If A and B are mutually exclusive events, that is, if the events A and B do not occur simultaneously we have $A \cap B = \phi$. Hence $P(A \cap B) = P(\phi) = 0$.

Then (7.4) reduces to $P(A \cup B) = P(A) + P(B)$ which is same as (7.1).

(b) If A, B, C are three events associated with a random experiment, then

$$P(A \cup B \cup C) = P(A) + P(B) + P(C) - P(A \cap B) - P(B \cap C) - P(C \cap A) + P(A \cap B \cap C).$$

(c) Portion I, II and III may be rewritten as $A \cap \bar{B}$, $A \cap B$ and $\bar{A} \cap B$ respectively.

Thus, $P(A) = P(A \cap \bar{B}) + P(A \cap B)$ and $P(B) = P(\bar{A} \cap B) + P(A \cap B)$.

Theorem 7.3 If $\bar{A}$ is any event associated with an experiment, then $P(\text{not } A) = P(\bar{A}) = 1 - P(A)$.

Proof The sample space S can be decomposed into mutually exclusive events A and $\bar{A}$, that is, $A \cup \bar{A} = S$ and $A \cap \bar{A} = \phi$.

Hence $P(A \cup \bar{A}) = P(A) + P(\bar{A}) = P(S) = 1$, thus $P(\bar{A}) = 1 - P(A)$.

Theorem 7.4 If A and B are two events such that $A \subset B$, then $P(A) \leq P(B)$.

Proof We are given that $A \subset B$ (See the venn diagram, Fig. 7.7). We write $B = A + (B - A)$, where $B - A$ is the shaded portion in Fig. 7.7

The events A and $B - A$ are mutually exclusive. Hence

$$P(B) = P[A + (B - A)] = P(A) + P(B - A).$$

Since, $P(B - A) \geq 0$, we obtain $P(B) \geq P(A) \Rightarrow P(A) \leq P(B)$.

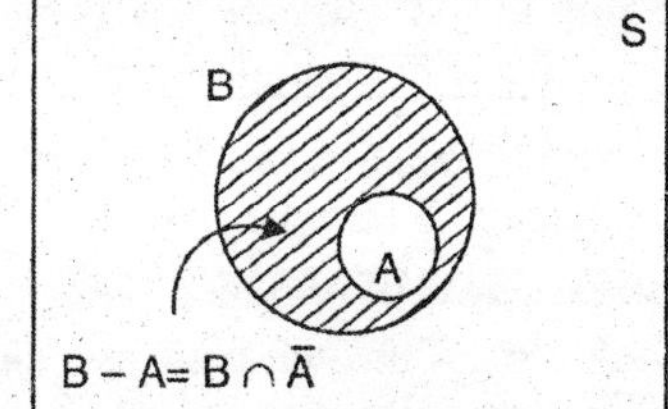

Fig. 7.7: Venn diagram

EXAMPLE 7.8 If A, B, C are three events associated with a random experiment, prove that $P(A \cup B \cup C) = P(A) + P(B) + P(C) - P(A \cap B) - P(B \cap C) - P(A \cap C) + P(A \cap B \cap C)$. *(AMIETE, June 2010)*

Solution Let $E = B \cup C$ so that $P(A \cup B \cup C) = P(A \cup E) = P(A) + P(E) - P(A \cap E)$. ...(*i*)

Now, $P(E) = P(B \cup C) = P(B) + P(C) - P(B \cap C)$. ...(*ii*)

Also $A \cap E = A \cap (B \cup C) = (A \cap B) \cup (A \cap C)$ [using distribution property of intersection of sets over the union]. Thus

$$P(A \cap E) = P(A \cap B) + P(A \cap C) - P[(A \cap B) \cap (A \cap C)] = P(A \cap B) + P(A \cap C) - P(A \cap B \cap C). \quad ...(iii)$$

Using (*ii*) and (*iii*) in (*i*), we obtain

$$P(A \cup B \cup C) = P(A) + P(B) + P(C) - P(A \cap B) - P(B \cap C) - P(A \cap C) + P(A \cap B \cap C).$$

EXAMPLE 7.9 Two dice are thrown simultaneously. Find the probability of getting "an odd number on the first dice" or "a total of 7".

Solution Let the event that getting an odd number on the first dice be denoted by A and that getting a total of 7 be denoted by B. We note that $S = 36$.

A: {(1, 1), (1, 2)... (1, 6), (3, 1), (3, 2)... (3, 6), (5, 1), (5, 2)... (5, 6)}.

B: Total of 7 = {(1, 6), (2, 5), (3, 4), (4, 3), (5, 2), (6, 1)} and $A \cap B: \{(1, 6), (3, 4), (5, 2)\}$.

Hence $P(A) = 18/36 = 1/2$, $P(B) = 6/36 = 1/6$ and $P(A \cap B) = 3/36 = 1/12$.

Therefore P(odd number on the first dice or a total of 7) = $P(A \cup B)$

$$= P(A) + P(B) - P(A \cap B) = \frac{1}{2} + \frac{1}{6} - \frac{1}{12} = \frac{7}{12}.$$

EXAMPLE 7.10 A and B are two mutually exclusive events. If $P(A) = 0.25$, $P(B) = 0.40$ and $P(A \cup B) = 0.50$, find the values of $P(A \cap B)$ and $P(A \cap \bar{B})$.

Solution We have

$$P(A \cup B) = P(A) + P(B) - P(A \cap B) \Rightarrow 0.50 = 0.25 + 0.40 - P(A \cap B) \text{ or } P(A \cap B) = 0.15.$$

Now $P(A \cap \bar{B}) = P(A) - P(A \cap B) = 0.25 - 0.15 = 0.10$.

EXAMPLE 7.11 A number is selected at random from the first 1000 natural numbers. What is the probability that it would be a multiple of 5 or 9?

Solution We have $S = \{1, 2, 3, ..., 1000\}$. Let A denote the event the number is divisible by 5. Then A consists of 200 sample points 5, 10, 15, ..., 995, 1000 and $P(A) = 200/1000$. Let B denotes the event that the number is divisible by 9. Then B consists of 111 sample points 9, 18, ..., 990, 999 and $P(B) = 111/1000$.

Then event $A \cap B$ consists of sample points 45, 90, ..., 945, 990 and $P(A \cap B) = 22/1000$.

Hence the probability that the selected number would be a multiple of 5 or 9 is given by

$$P(A \cup B) = P(A) + P(B) - P(A \cap B) = \frac{200}{1000} + \frac{111}{1000} - \frac{22}{1000} = 0.29.$$

EXAMPLE 7.12 Two dice are thrown simulaneously. Find the probability that the sum of points on the two dice would be 9 or more.

Solution If two dice are thrown then total number of elementary events is 6^2 or 36. Let A denote the event of throwing 9 in a single throw with two dice. Therefore, A consists of 4 sample points: (3, 6), (4, 5), (5, 4), (6, 3) and hence $P(A) = 4/36$. Let B denote the event of throwing 10 in single throw of two dice. Therefore, B consists of 3 sample points: (4, 6), (5, 5), (6, 4) and $P(B) = 3/36$. Let C denote the event of throwing 11 in single throw of two dice. Therefore, C consists of sample points: (5, 6), (6, 5), and $P(C) = 2/36$. Let D denote the event of throwing 12 in a single throw of two dice. Therefore, D consists of only one sample point: (6, 6) and $P(D) = 1/36$.

Now events A, B, C, D are mutually exclusive and hence we get

$$P(A \cup B \cup C \cup D) = P(A) + P(B) + P(C) + P(D) = 4/36 + 3/36 + 2/36 + 1/36 = 10/36 = 5/18.$$

EXAMPLE 7.13 If A, B, C are mutually exclusive and exhaustive events associated with a random experiment and $P(B) = 1.5\ P(A)$ and $P(C) = 0.5\ P(B)$, then find $P(A)$.

Solution Since, A, B, C are mutually exclusive and exhaustive events, we have

$$P(A) + P(B) + P(C) = 1 \text{ or } P(A) + 1.5\ P(A) + (0.5)(1.5)\ P(A) = 1 \text{ or } P(A) = 4/13.$$

EXAMPLE 7.14 The probability that atleast one of the events A and B occurs is 0.75 and the probability that both the events occur simultaneously is 0.30. Find the probability $P(\bar{A}) + P(\bar{B})$.

Solution We are given that $P(A \cup B) = 0.75$ and $P(A \cap B) = 0.30$.

Since, $$P(A \cup B) = P(A) + P(B) - P(A \cap B).$$

We have $$P(A) + P(B) = P(A \cup B) + P(A \cap B) = 0.75 + 0.30 = 1.05. \quad ...(i)$$

Also, $$P(A)+P(B)=1-P(\bar{A})+1-P(\bar{B})=2-P(\bar{A})-P(\bar{B})$$

or $$P(\bar{A})+P(\bar{B})=2-\{P(A)+P(B)\}=2-(1.05)=0.95, \text{ from } (i).$$

EXAMPLE 7.15 If $P(A-B)=0.20$, $P(A)=0.33$ and $P(B)=0.50$, what is the probability that out of the two events A and B, only B would occur?

Solution Refer Fig. 7.8 $P(A-B)=P(A\cap\bar{B})=P(A)-P(A\cap B)$ ...(*i*)

and $P(B-A)=P(B\cap\bar{A})=P(B)-P(A\cap B)$. ...(*ii*)

Equations (*i*) and (*ii*) describe the probabilities of occurrence of the event only A and only B respectively.

We have from (*i*), $P(A\cap B)=P(A)-P(A-B)=0.33-0.20=0.13$.

The probability that the event B only would occur $= P(B-A)$
$= P(B)-P(A\cap B)=0.50-0.13=0.37$.

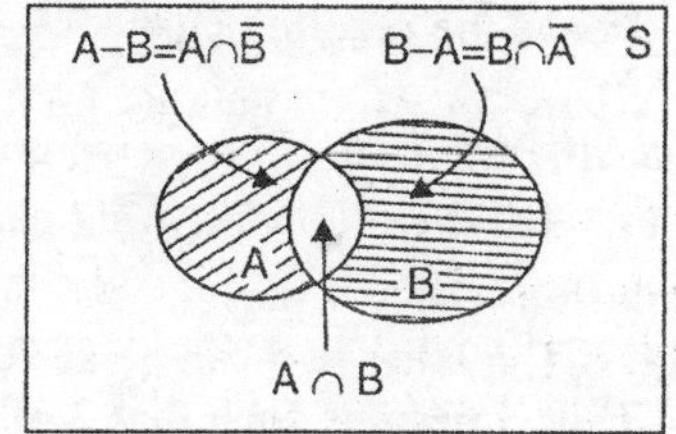

Fig. 7.8: Showing $(A-B)$ and $(B-A)$.

EXAMPLE 7.16 Given that $P(A\cap\bar{B})=0.25$, $P(\bar{A}\cap B)=0.33$ and $P(A\cup B)=0.75$, find (*i*) $P(B)$, (*ii*) $P(A)$.

Solution (*i*) We know $P(A\cup B)=P(A)+P(B)-P(A\cap B)=P(A)-P(A\cap B)+P(B)=P(A\cap\bar{B})+P(B)$

$[\because P(A\cap\bar{B})=P(A)-P(A\cap B)]$

or $P(B)=P(A\cup B)-P(A\cap\bar{B})$. Substituting the given values, we get $P(B)=0.75-0.25=0.50$.

(*ii*) $P(A\cup B)=P(A)+P(B)-P(A\cap B)=P(A)+[P(B)-P(A\cap B)]=P(A)+P(\bar{A}\cap B)$.

$$\therefore \quad P(A)=P(A\cup B)-P(\bar{A}\cap B)=0.75-0.33=0.42.$$

EXAMPLE 7.17: Two students Sita and Geeta appeared in an examination The probability that Sita will qualify the examination is 0.05 and that Geeta will qualify the examination is 0.10. The probability that both will qualify the examination is 0.02. Find the probability that (*i*) both Sita and Geeta will not qualify the examination, (*ii*) atleast one of them will not qualify the examination and (*iii*) only one of them will qualify the examination.

Solution Let A and B denote the events that Sita and Geeta will qualify the examination respectively. Given that $P(A)=0.05$, $P(B)=0.10$ and $P(A\cap B)=0.02$.

(*i*) The event both Sita and Geeta will not qualify the examination may be expressed as $A'\cap B'$.

Also $\quad A'\cap B'=(A\cup B)'$ (by Demorgan Law)

Now $\quad P(A\cup B)=P(A)+P(B)-P(A\cap B)=0.05+0.10-0.02=0.13$.

Therefore, $\quad P(A'\cap B')=P(A\cup B)'=1-P(A\cup B)=1-0.13=0.87$.

(*ii*) P (atleast one of them will not qualify) $= 1 - P$ (both of them will qualify) $= 1 - 0.02 = 0.98$.

(*iii*) The event only one of them will qualify the examination is same as the event either (Sita will qualify and Geeta will not qualify) or (Sita will not qualify and Geeta will qualify), that is, $A\cap B'$ or $A'\cap B$, where $A\cap B'$ and $A'\cap B$ are mutually exclusive.

Therefore, P (only one of them will qualify)

$= P(A\cap B' \text{ or } A'\cap B)=P(A\cap B')+P(A'\cap B)=P(A)-P(A\cap B)+P(B)-P(A\cap B)$

$= 0.05-0.02+0.10-0.02=0.11$.

EXAMPLE 7.18 From a pack of 52 cards three are drawn at random. Find the chance that they are a king, a queen and a knave.

Solution From a pack of 52 cards, 3 can be drawn in ${}^{52}C_3$ ways. Now there are 4 kings, 4 queens and 4 knaves. A king can be drawn in ${}^4C_1$ ways, a queen in ${}^4C_1$ ways and a knave also is ${}^4C_1$ ways. Since each can be associated with each of the other, a king, a queen, and a knave can be drawn in ${}^4C_1\times{}^4C_1\times{}^4C_1 = 4\times4\times4=64$ ways.

$$\therefore \text{ Required probability } = 64/{}^{52}C_3=\frac{64\times3!}{52\times51\times50}=\frac{16}{5525}.$$

EXAMPLE 7.19 If from a lottery of 30 tickets marked 1, 2, 3, ..., 30, four are drawn, find the chance that those marked 1 and 2 are among them.

Solution Any four tickets can be drawn in ${}^{30}C_4$ ways.

When two tickets marked 1 and 2 are always there, the remaining two can be drawn from remaining 28 in ${}^{28}C_2$ ways.

Hence the required chance $= {}^{28}C_2 \big/ {}^{30}C_4 = \dfrac{28\times 27\times 4\times 3\times 2}{2\times 30\times 29\times 28\times 27} = \dfrac{2}{145}.$

EXAMPLE 7.20 Four persons are chosen at randam from a group containing 3 men, 2 women and 4 children. Show that the probability that exactly two of them are children is 10/21.

Solution Now 4 persons can be chosen out of 9 in ${}^{9}C_4$ ways. Two children can be selected out of 4 in ${}^{4}C_2$ ways. The other two can be selected from the remaining (3 men and 2 women) 5 persons in ${}^{5}C_2$ ways.

Thus 4 persons such that 2 of them are children can be selected in ${}^{4}C_2 \times {}^{5}C_2$ ways.

Hence the required probability $= \dfrac{\text{No. of favourable cases}}{\text{Total possible cases}} = \dfrac{{}^{4}C_2\times {}^{5}C_2}{{}^{9}C_4} = \dfrac{10}{21}.$

EXAMPLE 7.21 One bag contains 4 white and 2 black beads and another contains 3 of each colour. A bead is drawn from each bag. What is the probability that one is white and one is black.

Solution One white and one black bead can be drawn in the following two mutually exclusive ways:

(*i*) white from first bag and black from second

(*ii*) black from first and white from second.

p_1, the probability of the first event $= (4/6)\times(3/6) = 1/3.$

p_2, the probability of the second event $= (2/6)\times(3/6) = 1/6.$

$\therefore$ Required probability $= p_1 + p_2 = 1/3 + 1/6 = 1/2.$

EXAMPLE 7.22 A bag contains 4 white and 2 black balls, and a second bag contains 3 of each colour. A bag is drawn at random and a ball is then selected at random from the bag chosen. What is the probability that the ball selected is white.

Solution We define the following events:

A: the ball is drawn from the first bag,

B: the ball is drawn from the second bag.

Since, both the bags are equally likely to be selected, we have $P(A) = P(B) = 1/2.$

Now, probability of drawing a white ball from first bag is $p_1 = 4/6 = 2/3$, and the probability of drawing a white ball from second bag is $p_2 = 3/6 = 1/2.$

Hence the required probability $= P(A)p_1 + P(B)p_2 = \dfrac{1}{2}\cdot\dfrac{2}{3} + \dfrac{1}{2}\dfrac{1}{2} = \dfrac{7}{12}.$

EXAMPLE 7.23 There are 3 boxes with the following composition:

Box I: 8 Red + 5 White + 3 Blue balls

Box II: 6 Red + 5 White + 3 Blue balls

Box III: 4 Red + 3 White + 2 Blue balls.

One of the boxes is selected at random and a ball is drawn from it. What is the probability that the drawn ball is red?

Solution Let A denote the event that the drawn ball is red. Since any of the 3 boxes may be drawn, we have $P(\text{Box I}) = P(\text{Box II}) = P(\text{Box III}) = 1/3.$

Also $P(R_1/B_I)$ = probability of drawing a red ball from the first box $= 8/16 = 1/2.$

$P(R_2/B_{II}) = 6/14 = 3/7$ and $P(R_3/B_{III}) = 4/9.$

Thus we have, $P(A)\ P(R_1 \cap B_I) + P(R_2 \cap B_{II}) + P(R_3 \cap B_{III})$

$= P(R_1/B_I)\times P(\text{Box I}) + P(R_2/B_{II})\times P(\text{Box II}) + P(R_3/B_{III})\times P(\text{Box III})$

$= \dfrac{1}{2}\times\dfrac{1}{3} + \dfrac{3}{7}\times\dfrac{1}{3} + \dfrac{4}{9}\times\dfrac{1}{3} = \dfrac{1}{6} + \dfrac{1}{7} + \dfrac{4}{27} = \dfrac{173}{378}.$

EXAMPLE 7.24 Three groups of children contain respectively 3 girls and 1 boy; 2 girls and 2 boys; 1 girl and 3 boys. One child is selected at random from each group. Find the chance that the three selected consist of 1 girl and 2 boys.

Solution One girl and two boys can be selected in the following mutually exclusive ways :

(*i*) Girl from first group, boy from second and boy from third. The probability, p_1, of selecting 1 girl and 2 boys from three groups as per arrangement (*i*) is given by $=\frac{3}{4}\times\frac{2}{4}\times\frac{3}{4}=\frac{9}{32}$.

(*ii*) Boy from first, girl from second and boy from third group. The probability, p_2, of selecting 1 girl and 2 boys from three groups as per arrangement (*ii*) is given by $=\frac{1}{4}\times\frac{2}{4}\times\frac{3}{4}=\frac{3}{32}$.

(*iii*) Boy from first group, boy from second and girl from third. The probability, p_3, of selecting 1 girl and 2 boys from three groups as per arrangement (*iii*) is given by $=\frac{1}{4}\times\frac{2}{4}\times\frac{1}{4}=\frac{1}{32}$.

Since these are mutually exclusive ways, the required probability $= p_1+p_2+p_3=\frac{9}{32}+\frac{3}{32}+\frac{1}{32}=\frac{13}{32}$.

EXAMPLE 7.25 An urn *A* contains 2 white and 4 black balls. Another urn *B* contains 5 white and 7 black balls. A ball is transferred from the urn *A* to the urn *B* then a ball is drawn from the urn *B*. Find the probability that it is white. (*AMIETE, June 2012*)

Solution There are two mutually exclusive cases: (*i*) when a white ball is transferred from *A* to *B* (*ii*) when a black ball is transferred from *A* to *B*.

Case 1 The probability of drawing white ball from *A* is $p_1=\frac{2}{6}=\frac{1}{3}$. When a white ball is transferred to *B*, it contains 6 white and 7 black balls. The probability of drawing white ball from *B* is $p_2=\frac{6}{13}$.

$\therefore$ Probability of transferring white ball to *B* and then drawing a white ball from *B* that is, these two events occuring simultaneously is $p_3=\frac{1}{3}\cdot\frac{6}{13}=\frac{2}{13}$.

Case 2 The probability of drawing a black ball from *A* is $p_4=4/6=2/3$ and then the probability of drawing a white ball from *B* (which now contains 5 white and 8 black balls) is $p_5=5/13$.

Probability of these two events occuring simultaneously is $p_6=\frac{2}{3}\cdot\frac{5}{13}=\frac{10}{39}$.

$\therefore$ The required probability is $p=p_3+p_6=\frac{2}{13}+\frac{10}{39}=\frac{16}{39}$.

EXAMPLE 7.26 A batch of 100 lamps is known to have 5% defectives. If 10 lamps are chosen at random, what is the probability that (*i*) none is defective (*ii*) two are defective.

Solution (*i*) There are 95 good and 5 defective lamps.

Therefore, the probability *p*, that all the lamps are good $=\frac{{}^{95}C_{10}}{{}^{100}C_{10}}=\frac{95\times94\times\cdots\times86}{100\times99\times\cdots\times91}=0.584.$

(*ii*) Probability that 2 are defective and 8 good $=\frac{{}^{95}C_8\times{}^5C_2}{{}^{100}C_{10}}=0.07.$

EXAMPLE 7.27 Three light bulbs are chosen at random from fifteen bulbs of which five are defective. Find the probability that (*i*) none is defective (*ii*) exactly one is defective (*iii*) at least one is defective.

Solution There are 5 defective and 10 non-defective or good bulbs.

(*i*) The probability that all three bulbs chosen are good $=\frac{{}^{10}C_3}{{}^{15}C_3}=\frac{10\times9\times8}{15\times14\times13}=\frac{24}{91}$.

(*ii*) The probability that of 3 bulbs chosen, 1 is defective and 2 non-defective

$$=\frac{{}^5C_1\times{}^{10}C_2}{{}^{15}C_3}=\frac{5\times10\times9}{2!}\times\frac{3!}{15\times14\times13}=\frac{45}{91}.$$

(*iii*) Here the mutually exclusive cases are

(*i*) one defective, two non-defective, (*ii*) two defective, one non-defective, (*iii*) all three defective.

In case (*i*) probability $p_1=\frac{{}^5C_1\times{}^{10}C_2}{{}^{15}C_3}=\frac{45}{91}.$

In case (*ii*) probability $p_2=\frac{{}^5C_2\times{}^{10}C_1}{{}^{15}C_3}=\frac{10\times10}{455}=\frac{20}{91}.$

In case (*iii*) probability $p_3=\frac{{}^5C_3}{{}^{15}C_3}=\frac{10}{455}=\frac{2}{91}.$

$\therefore$ p, the probability of atleast one defective $=\frac{45}{91}+\frac{20}{91}+\frac{2}{91}=\frac{67}{91}.$

(*d*) ***Alternative Solution:*** It is easier to find the probability of none defective which is 24/91.

$\therefore$ Probability of at least one defective = 1 – (Probability of none defective) $=1-\frac{24}{91}=\frac{67}{91}.$

EXAMPLE 7.28 Two flower seeds are randomly selected from a bag containing 10 seeds for red flowers and 5 seeds for white flowers. What is the probability that (*i*) Both result in white flowers, (*ii*) one of each colour is selected.

Solution Number of seeds for red flowers = 10.

Number of seeds for whilte flowers = 5. Total number of seeds = 15.

Therefore, number of ways of slected two flower seeds = ${}^{15}C_2=\frac{15!}{2!13!}=\frac{15.14}{2}=105.$

(*i*) Probability of selecting both white flowers $=\frac{{}^5C_2}{{}^{15}C_2}=\frac{(5)(4)(2)}{(2)(15)(14)}=\frac{2}{21}.$

(*ii*) Probability of selecting one of each colour $=\frac{{}^{10}C_1\times{}^5C_1}{{}^{15}C_2}=\frac{10\times5\times2}{15\times14}=\frac{10}{21}.$

7.4 CONDITIONAL PROBABILITY

If A and B are two events, the probability that B occurs under the condition that an event A has occurred is called the *conditional probability* of B given that A has occurred and is denoted by $P(B/A)$. In this case A serve as a new (reduced) sample space, and that probability is the fraction of $P(A)$ which corresponds to $A\cap B$.

Thus $$P(B/A)=\frac{P(B\text{ and }A)}{P(A)}=\frac{P(B\cap A)}{P(A)}=\frac{P(A\cap B)}{P(A)},\text{ where }P(A)\neq0. \quad ...(7.5)$$

Similarly, the *conditional probability* of A given that B has occurred is

$$P(A/B)=\frac{P(A\text{ and }B)}{P(B)}=\frac{P(A\cap B)}{P(B)},\text{ where }P(B)\neq0. \quad ...(7.6)$$

From (7.5) and (7.6), we have the following result.

Theorem 7.5 (*Multiplication law of probability*). If A and B are events in a sample space S and $P(A)\neq0$, $P(B)\neq0$. Let $P(A/B)$ denote the conditional probability of A when B has occurred. Then,

$$P(A\cap B)=P(A)\,P(B/A)=P(B)\,P(A/B). \quad ...(7.7)$$

EXAMPLE 7.29 For two events A and B, $P(A)=3/8$, $P(B)=5/8$ and $P(A\cup B)=3/4$. Find (*i*) $P(A/B)$, (*ii*) $P(B/A)$, (*iii*) $P(\bar{A}/\bar{B})$, (*iv*) $P(\bar{A}/B)$, (*v*) $P(A/\bar{B})$.

Solution We have $P(A \cup B) = P(A) + P(B) - P(A \cap B)$.

Therefore, $P(A \cap B) = P(A) + P(B) - P(A \cup B) = 3/8 + 5/8 - 3/4 = 1/4$. Hence,

(*i*) $P(A/B) = \dfrac{P(A \cap B)}{P(B)} = \dfrac{1/4}{5/8} = \dfrac{2}{5}$. (*ii*) $P(B/A) = \dfrac{P(A \cap B)}{P(A)} = \dfrac{1/4}{3/8} = \dfrac{2}{3}$.

(*iii*) First compute $P(\bar{B})$ and $P(\bar{A} \cap \bar{B})$. Now $P(\bar{B}) = 1 - P(B) = 1 - (5/8) = 3/8$.

By De Morgan's law, $\overline{A \cup B} = \bar{A} \cap \bar{B}$; hence $P(\bar{A} \cap \bar{B}) = P(\overline{A \cup B}) = 1 - P(A \cup B) = 1 - (3/4) = 1/4$.

Thus
$$P(\bar{A}/\bar{B}) = \frac{P(\bar{A} \cap \bar{B})}{P(\bar{B})} = \frac{1/4}{3/8} = \frac{2}{3}.$$

(*iv*) $P(\bar{A}/B) = \dfrac{P(\bar{A} \cap B)}{P(B)} = \dfrac{P(B) - P(A \cap B)}{P(B)} = \dfrac{(5/8) - (1/4)}{(5/8)} = \dfrac{3/8}{5/8} = 3/5.$

(*v*) $P(A/\bar{B}) = \dfrac{P(A \cap \bar{B})}{P(\bar{B})} = \dfrac{P(A) - P(A \cap B)}{1 - P(B)} = \dfrac{(3/8) - (1/4)}{1 - (5/8)} = \dfrac{1/8}{3/8} = \dfrac{1}{3}.$

EXAMPLE 7.30 What is the probability that in a single throw of two dice of getting a total of 7 or 11?

Solution Define the events *A*: getting a total of 7, *B*: getting a total of 11.

Now the first die may appear in 6 different ways and according to any one way in which the first appears, the second can appear in 6 ways. The two dice, therefore, may appear in 6 × 6 = 36 ways.

The sum of 7 may appear in 6 different ways. *A*: (1, 6), (2, 5), (3, 4), (4, 3), (5, 2), (6, 1).

∴ $P(A)$ (the probability of getting a sum of 7) = $6/36 = 1/6$.

The sum of 11 may appear in two different ways. *B*: (5, 6), (6, 5).

∴ $P(B)$ (the probability of getting a sum of 11) = $2/36 = 1/18$.

$A \cap B = \phi$. Therefore $P(A \cap B) = 0$.

Therefore, *P* (probability of getting a total of 7 or 11) = $P(A \cup B) = P(A) + P(B) = (1/6) + 1/18 = 2/9$.

EXAMPLE 7.31 Two dice are thrown twice. What is the probability that the first throw is 7 and the second throw is 11?

Solution The probability of the first throw of two dice to 7 has been shown (Refer Example 7.30) to be 1/6. The probability for 11 is 1/18.

Hence the probability for first throw to be 7 and second to be 11 is $(1/6)(1/18) = 1/108$.

EXAMPLE 7.32 Two cards are drawn one after the other from a well shuffled deck of 52 cards. Find the probability that both are ace cards, if the first card is (*i*) not replaced, (*ii*) replaced.

Also find this probability if the first card is replaced before the second is drawn.

Solution Let A_1 = ace card on first draw, A_2 = ace card on second draw. We have to find $P(A_1 \cap A_2)$.

(*i*) Let the first card is replaced. Then, we have $P(A_1 \cap A_2) = P(A_1)\,P(A_2/A_1) = \left(\dfrac{4}{52}\right)\left(\dfrac{4}{52}\right) = \dfrac{1}{169}$.

(*ii*) Consider the case when the drawn first card is not replaced

$$P(A_1 \cap A_2) = P(A_1)\,P(A_2/A_1) = \left(\frac{4}{52}\right)\left(\frac{3}{51}\right) = \frac{1}{221}.$$

EXAMPLE 7.33 A die is thrown twice and the sum of numbers appearing is noted to be 8. What is the conditional probability that the number 2 has appeared atleast once?

Solution Let the event *A*: define the number 2 has appeared at least once.
B: define the sum of the numbers appearing is 8.

The sets of elementary events are therefore

$A = \{(2, 1), (2, 2), (2, 3), (2, 4), (2, 5), (2, 6), (1, 2), (3, 2), (4, 2), (5, 2), (6, 2)\}$.

$B = \{(2, 6), (3, 5), (4, 4), (5, 3), (6, 2)$ and $A \cap B = (2, 6), (6, 2)\}$.

The total outcomes of throwing two die is $6 \times 6 = 36$.

Hence, $P(A) = 11/36, P(B) = 5/36, P(A \cap B) = 2/36$.

$$\therefore \quad P(A/B) = \frac{P(A \cap B)}{P(B)} = \frac{2/36}{5/36} = \frac{2}{5}.$$

EXAMPLE 7.34 Given $P(\bar{A}) = 0.2, P(B) = 0.5$ and $P(A \cap \bar{B}) = 0.4$; find $P(B/A \cup \bar{B})$. *(AMIE, S-2003)*

Solution We have $P(\bar{A}) = 0.2$, therefore $P(A) = 1 - P(\bar{A}) = 0.8$. $P(B) = 0.5$, therefore, $P(\bar{B}) = 1 - 05 = 0.5$. $P(A \cap \bar{B}) = 0.4$.

We know $P(A \cap \bar{B}) = P(A) - P(A \cap B)$, therefore $P(A \cap B) = P(A) - P(A \cap \bar{B}) = 0.8 - 0.4 = 0.4$.

Now $B \cap (A \cup \bar{B}) = A \cap B$.

$$\therefore \quad P[B \cap (A \cup \bar{B})] = P(A \cap B) = 0.4.$$

We also know that $P(A \cup \bar{B}) = P(A) + P(\bar{B}) - P(A \cap \bar{B}) = 0.8 + 0.5 - 0.4 = 0.9$.

$$\therefore \quad P(B/A \cup \bar{B}) = \frac{P[B \cap (A \cup \bar{B})]}{P(A \cup \bar{B})} = \frac{0.4}{0.9} = \frac{4}{9}.$$

EXAMPLE 7.35 There are two bags. The first bag contains 10 white and 3 black balls, and second bag contains 3 white and 5 black balls. Two balls are drawn at random from the first bag and are put into the second bag and then a ball is drawn at random from the latter. What is the probability that it is a white ball?

Solution The two balls drawn from the first bag may be (*i*) both white, (*ii*) one white and one black, (*iii*) both black. Denote these possibilities as events B_1, B_2, B_3 respectively. We have $P(B_1) = {}^{10}C_2/{}^{13}C_2 = 15/26$; $P(B_2) = ({}^{10}C_1 \times {}^{3}C_1)/{}^{13}C_2 = 5/13$, $P(B_3) = {}^{3}C_2/{}^{13}C_2 = 1/26$.

When two balls are transferred from first bag to second bag, the second bag will contain:

(*i*) 5 white and 5 black balls, (*ii*) 4 white and 6 black balls, (*iii*) 3 white and 7 black balls.

Denote E: event of drawing a white ball from the second bag after transfer.

Now $P(E/B_1) = 5/10, P(E/B_2) = 4/10, P(E/B_3) = 3/10$.

$$\therefore \quad P(E) = P(B_1) \cdot P(E/B_1) + P(B_2) \cdot P(E/B_2) + P(B_3) \cdot P(E/B_3)$$

$$= \frac{15}{26} \cdot \frac{5}{10} + \frac{5}{13} \cdot \frac{4}{10} + \frac{1}{26} \cdot \frac{3}{10} = \frac{15}{52} + \frac{2}{13} + \frac{3}{260} = \frac{59}{130} = 0.45.$$

7.5 INDEPENDENT EVENTS

Two events A and B are said to be independent, if the occurrence (or non-occurrence) of A does not affect the probability of occurrence of the B event. This means that the probability of event A does not depend on the occurrence (or non-occurance) of event B and conversely. For example if two coins are tossed, the probability for head on one coin is not affected by the result of the toss on the second coin.

Hence, in this case the conditional probability $P(A/B)$ is same as $P(A)$, that is $P(A/B) = P(A)$ and $P(B/A) = P(B)$. Substituting in (7.7), we obtain $P(A \cap B) = P(A/B)\, P(B) = P(A) \cdot P(B)$. ...(7.8)

Thus, if two events A and B are such that $P(A \cap B) = P(A)\, P(B)$, they are called independent events otherwise they are dependent.

Remark 4

(a) Three events A, B and C are known as independent if the following conditions hold:

$$P(A \cap B) = P(A) \times P(B)$$
$$P(B \cap C) = P(B) \times P(C)$$

If A, B, C are three independent events having respective probabilities p_1, p_2, p_3 then the probability of occurrence of A and B and C is $p_1 p_2 p_3$.

(b) If two events A and B are independent, then the following pairs of events are also independent:

(*a*) A and $\bar{B}$ (*ii*) $\bar{A}$ and B, (*iii*) $\bar{A}$ and $\bar{B}$.

EXAMPLE 7.36 Prove that if the events A and B are independent, then $\bar{A}$ and $\bar{B}$ are also independent.

Solution We have $P(A \cap B) = P(A)\, P(B)$.

$$\begin{aligned} P(\bar{A} \cap \bar{B}) &= P(\overline{A \cup B}) = 1 - P(A \cup B) \\ &= 1 - [P(A) + P(B) - P(A \cap B)] = 1 - P(A) - P(B) + P(A \cap B) \\ &= 1 - P(A) - P(B) + P(A)\, P(B) = [1 - P(A)] - P(B)\, [1 - P(A)] \\ &= [1 - P(A)]\, [1 - P(B)] = P(\bar{A})\, P(\bar{B}). \end{aligned}$$

Hence the events $\bar{A}$ and $\bar{B}$ are also independent.

EXAMPLE 7.37 Prove that if events A and B are independent, then $\bar{A}$ and B are also independent.

Solution Since the events A and B are independent, we have

$$P(A \cap B) = P(A)\, P(B).$$

The events $A \cap B$ and $\bar{A} \cap B$ are mutually exclusive and $(A \cap B) \cup (\bar{A} \cap B) = B$. (Refer Fig. 7.9)

From addition theorem, we obtain

$$P(B) = P(A \cap B) + P(\bar{A} \cap B)$$

or $$P(\bar{A} \cap B) = P(B) - P(A \cap B) = P(B) - P(A)\, P(B) = P(B)\, [1 - P(A)] = P(B)\, P(\bar{A}).$$

Therefore, the events $\bar{A}$ and B are independent.

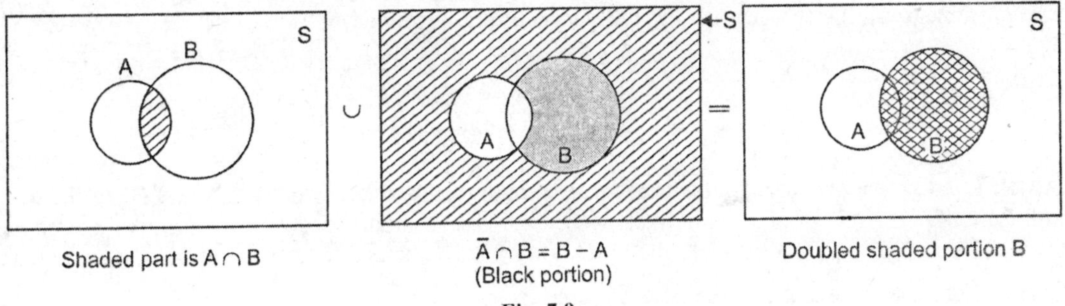

Shaded part is A ∩ B

Ā ∩ B = B − A
(Black portion)

Doubled shaded portion B

Fig. 7.9

EXAMPLE 7.38 A and B are two events such that $P(A) = 2/5$ and $P(A \cup \bar{B}) = 7/10$. If A and B are independent events, then find $P(B)$.

Solution Since, A and B are indepencent, A and $\bar{B}$ are also independent. We have

$$P(A \cup \bar{B}) = P(A) + P(\bar{B}) - P(A \cap \bar{B}) = P(A) + P(\bar{B}) - P(A)\, P(\bar{B}).$$

Substituting the values, we obtain

$$7/10 = (2/5) + P(\bar{B})\, [1 - 2/5] \Rightarrow P(\bar{B}) = \frac{[(7/10) - (2/5)]}{3/5} = \frac{1}{2}.$$

Hence $P(B) = 1 - 1/2 = 1/2$.

EXAMPLE 7.39 A and B are two independent events such that $P(A) = 0.6$ and $P(A \cup B) = 0.8$. Find $P(B)$.

Solution Since A and B are independent, we have $P(A \cap B) = P(A)\, P(B)$. Hence $P(A \cup B) = P(A) + P(B) - P(A)\, P(B) = 0.8$ or $P(B)\, [1 - P(A)] = 0.8 - P(A)$ or $P(B)\, [1 - 0.6] = 0.8 - 0.6 = 0.2$ or $P(B) = 0.2/0.4 = 0.5$.

EXAMPLE 7.40 A student A can solve 85% of the problems in a question paper and student B can solve 65% of the problems from the same question paper. If a problem from the question paper is selected at random, what is the probability that atleast one of them will solve the problem?

Solution The probability that A can solve the problems is $P(A) = 17/20$, and the probability that B can solve the problem is $P(B) = 13/20$. Since the events are independent. The probability that atleast one of them will solve the problem is given by

$$P(A \cup B) = P(A) + P(B) - P(A \cap B) = P(A) + P(B) - P(A)\,P(B)$$

$$= \frac{17}{20} + \frac{13}{20} - \left(\frac{17}{20}\right)\left(\frac{13}{20}\right) = \frac{379}{400} = 0.948.$$

EXAMPLE 7.41 The odds in favour of an event is 2:3 and the odds against another event is 3:7. Final the probability that only one of the two events occurs.

Solution We denote the two events by A and B repectively. We have

$$P(A) = 2/(2+3) = 2/5 \text{ and } P(B) = 7/(7+3) = 7/10.$$

As A and B are independent, $P(A \cap B) = P(A) \times P(B) = \frac{2}{5} \times \frac{7}{10} = \frac{7}{25}$.

Probability that either only A occurs or only B occurs $= P(A - B) + P(B - A)$

$$= [P(A) - P(A \cap B)] + [P(B) - P(A \cap B)] = P(A) + P(B) - 2P(A \cap B) = \frac{2}{5} + \frac{7}{10} - 2\cdot\left(\frac{7}{25}\right) = \frac{27}{50}.$$

EXAMPLE 7.42 A problem in mechanics is given to three students A, B, C in a class to be solved. The probabilities of their solving the problem are $1/2, 1/3, 1/4$ respectively. What is the probability that the problem will be solved? *(AMIETE, June 2011; VTU 2000)*

Solution. We have

$$P(\text{problem will be solved}) = 1 - P(\text{problem will not be solved})$$

$$= 1 - P(\bar{A})\,P(\bar{B})\,P(\bar{C}) = 1 - [1 - P(A)][1 - P(B)][1 - P(C)]$$

$$= 1 - \left(1 - \frac{1}{2}\right)\left(1 - \frac{1}{3}\right)\left(1 - \frac{1}{4}\right) = 1 - \left(\frac{1}{2}\right)\left(\frac{2}{3}\right)\left(\frac{3}{4}\right) = 1 - \frac{1}{4} = \frac{3}{4}.$$

EXAMPLE 7.43 An article manufactured by a company consists of two parts A and B. In the process of manufacture of part A, 9 out of 100 are likely to be defective. Similarly 5 out of 100 are likely to be defective in the manufacture of part B. Calculate the probability that the assembled parts will not have any defect.

Solution Let X be the event that the part A is free from defect

$$P(X) = 1 - P(\text{part A is defective}) = 1 - \frac{9}{100} = \frac{91}{100} = 0.91.$$

Let Y be the event that part B is free from defect

$$P(Y) = 1 - P(\text{part B is defective}) = 1 - \frac{5}{100} = \frac{95}{100} = 0.95.$$

The two events X and Y are independent. Let E be the event that the assembled product will not have any defect. It will be so if both A and B are free from defects.

Therefore, $E = X \cap Y$. $P(E) = P(X \cap Y) = P(X) \cdot P(Y)$ (Independent events) $= (0.91)\,(0.95) = 0.8645$.

EXAMPLE 7.44 The odds in favour of one student passing a test are 3 : 7. The odds against another student passing at are 3 : 5. What are the probabilities that (*i*) both pass, (*ii*) both fail?

Solution Let A be the event that the first student passes the test.

Therefore, $P(A) = \frac{3}{3+7} = \frac{3}{10} = 0.3$. $P(\bar{A}) = 1 - P(A) = 0.7$.

Let B be the event that the second student passes at the test.

Therefore, $P(B) = \frac{5}{3+5} = \frac{5}{8}$. $P(\bar{B}) = 1 = P(B) = \frac{3}{8}$.

(*i*) Probability both pass = $P(A) \cdot P(B)$, as te events are independent = $(0.3)\left(\frac{5}{8}\right) = \frac{3}{16}$

(*ii*) Probability both fail = $P(\bar{A}) \cdot P(\bar{B})$, the events are independent = $(0.7)\left(\frac{3}{8}\right) = \frac{21}{80}$.

EXAMPLE 7.45 The odds that a Ph.D thesis will be favourably reviewed by three independent examiners are 5 to 2, 4 to 3 and 3 to 4 respectively. What is the probability that a majority approve the thesis?

(*AMIETE, June 2004; VTU 2003 S*)

Solution The probabilities that the book shall be favourably reviewed by first second and third examiners are $5/7, 4/7, 3/7$ respectively. A majority approve the thesis if *atleast* two examiners are favourable.

∴ Probability that the first two examiners favourable the thesis and the third unfavour the thesis

$$= \frac{5}{7} \times \frac{4}{7} \times \left(1 - \frac{3}{7}\right) = \frac{80}{343}.$$

Probability that the first and third examiners favourable the thesis and second examiner unfavour the thesis

$$= \frac{5}{7} \times \frac{3}{7} \times \left(1 - \frac{4}{7}\right) = \frac{45}{343}.$$

Probability that the second and third examiners favourable the thesis and the first examiner unfavourable the thesis

$$= \frac{4}{7} \times \frac{3}{7} \times \left(1 - \frac{5}{7}\right) = \frac{24}{343}.$$

Finally, probability that all the three examiners favourable the thesis

$$= \frac{5}{7} \times \frac{4}{7} \times \frac{3}{7} = \frac{60}{343}.$$

Hence, the required probability that a majority favourable the thesis

$$= \frac{80}{343} + \frac{45}{343} + \frac{24}{343} + \frac{60}{343} = \frac{209}{343}.$$

EXERCISE 7.1

1. A coin is tosses thrice. What is the probabaility of getting (a) 2 heads (b) atleast 2 heads?
   [**Hint:** The number of elementary events is 8, that is HHH, HHT, HTH, HTT, THH, THT, TTH, TTT.]
   Ans. 3/8; 1/2.
2. A bag contains 5 green and 7 red balls. Two balls are drawn. What is the probability that one is green and other is red? **Ans.** 35/66
3. From a set of 17 cards, numbered 1, 2, 3, ..., 17, one is drawn at random. Show that the chance that its number is divisible by 3 or 7 is 7/17.
4. Out of 50 rare books, 3 of which are especially valuable, 5 are stolen at random by a thief. What is the probability that
   (*a*) none of the three is included? (*b*) two of the three are included?
   Ans. (*a*) ${}^{47}C_5/{}^{50}C_5$, (*b*) $({}^{3}C_2 \times {}^{47}C_3)/{}^{50}C_5$.
5. A bag contains 10 white and 15 black balls. Two balls are drawn in succession. What is the probability that
   (*a*) first is white and second is black? (*b*) one is white and one is black?
   Ans. (*a*) 1/4, (*b*) 1/2.
6. If 20% of the bulbs produced by a factory are defective, find the probability that out of 4 bulbs chosen at random atmost 2 bulbs will be defective. **Ans.** 0.9728.
7. An urn contains 10 black and 10 white balls. Find the probability of drawing two balls of the same colour.
   $\left[\textbf{Hint:} (a) \frac{{}^{10}C_2}{{}^{20}C_2} + \frac{{}^{10}C_2}{{}^{20}C_2} = \frac{9}{19}\right]$

8. Find the probabiltiy that a five digit number comparing the digits 1, 2, 3, 4 and 5 would be divisible by 4.
[**Hint:** The total number of five digit numbers that can be formed without any restriction is 5! = 120. Now a five digit number would be divisible by 4 if the number formed by the last two digits is divisible by 4. This could happen when the five digit number ends with 12 or 24 or 32 or 52. If we fix the last two digits by 12, then the first three places of five digit number can be filled up using the remaining 3 digits in 3! or 6 ways. Therefore, the total number of ways a five digit number which is divisible by 4 can be formed is 6 × 4 = 24. Probability of such an event = $\frac{24}{120} = \frac{1}{5}$.]

9. Find the probability of a 4 turning up at least once in two tosses of a fair die. **Ans.** 11/36.

10. A box contains 100 transistors, 20 of which are defective. 10 are selected for inspection. Indicate what is the probability that (*a*) All 10 are defective, (*b*) All 10 are good, (*c*) At least one is defective.

Ans. (*a*) $\frac{{}^{20}C_{10}}{{}^{100}C_{10}}$, (*b*) $\frac{{}^{80}C_{10}}{{}^{100}C_{10}}$, (*c*) $\frac{1-{}^{80}C_{10}}{{}^{100}C_{10}}$.

11. A box contains 40 good and 10 defective fuses. If 10 fuses are selected at random, what is the probability that they will all be good. **Ans.** $({}^{40}C_{10}/{}^{50}C_{10}) = 0.0825$.

12. A purse contains 2 silver and 4 copper coins and a second purse contains 4 silver and 4 copper coins. If a coin is selected at randon from one of the two purses, what is the probability that it is a silver coin?
[**Hint:** The chance of choosing 'first purse is 1/2 and if the first purse is chosen, the chance of drawing a silver coin is 2/6 = 1/3. Hence the chance of drawing a silver coin from first purse = (1/2) (1/3) = 1/6. Similarly the chance of drawing a silver coin from the second purse = (1/2) (4/8) = 1/4. Since the events are mutualy exclusive, the probability required, by addition law = (1/6) + (1/4) = 5/12].

13. A husband and wife appear in an interview for two vacancies in the same post. The probability of husband's selection is 1/7 and that of wife's selection is 1/5. What is the probability that
(*a*) both of them will be selected,
(*b*) only one of them will be selected, (*AMIETE, June 2012*)
(*c*) None of them will be selected? **Ans.** 1/35; 2/7; 24/35.
[**Hint:** Let *A* be the event: Husband's is selected and *B* the event: wife's is selected.
Clearly *A* and *B* are independent.

(*a*) P (both of them will be selected) $= P(A \cap B) = P(A)\ P(B) = \frac{1}{7} \times \frac{1}{5} = \frac{1}{35}$.

(*b*) P (only one of them will be selected) $= P(A$ and not B or B and not $A)$
$= P(A$ and not $B) + P(B$ and not $A) = P(A)\ P(\text{not } B) + P(B)\ P\ (\text{not } A)$
$= P(A)[1 - P(B)] + P(B)[1 - P(A)] = \frac{1}{7} \times \frac{4}{5} + \frac{1}{5} \times \frac{6}{7} = \frac{2}{7}$.

(*c*) P (none of them will be selected) $= P(\text{not } A \text{ and not } B) = P(\text{not } A) \cdot P(\text{not } B)$
$= [1 - P(A)][1 - P(B)] = \left(1 - \frac{1}{7}\right)\left(1 - \frac{1}{5}\right) = \frac{6}{7} \times \frac{4}{5} = \frac{24}{35}$.

14. What is the prabability that a leap year selected at random will contain 53 Sundays?
(*AMIETE, Dec. 2009, June 2002*) **Ans.** 2/7.
[**Hint:** A leap year contains 366 days, that is 52 weeks and 2 days. These 2 days may be one of the following: Sun-Mon, Mon-Tue, Tue-Wed, Wed-Thu, Thu-Fri, Fri-Sat, Sat-Sun.
For 53 Sundays the 2 days must include a Sunday. Therefore, 2 outcomes are favourable to the event of 53 sundays in leap year.
P(53 Sundays) = No. of favourable outcomes/Total number of outcomes = 2/7].

15. If the probabilities of happening of some events are $p_1, p_2, ..., p_n$, then find the probability of happening of at least one of them. (*AMIETE, S 2009*)
Use this result to find the chance of getting at least one six in a throw of 4 dice.

[**Hint:** The probability that none of the events happens = $(1 - p_1)(1 - p_2) ...(1 - p_n)$.
Hence the probability that at least one event happens = $1 - (1 - p_1)(1 - p_2) ...(1 - p_n)$.
The probability of getting six in a throw of one die = 1/6.

Probability of not getting six = 5/6. Probability of not getting six in a throw of 4 dice = $(5/6)^4$.

Probability of getting at least one six = $1 - (5/6)^4$].

16. A candidate is selected for interview for three posts. For the first post there are 3 candidates, for the second there are 4, and for the third there are 2. What is the chance of his getting at least one post?
(*AMIETE, S 2001*) **Ans.** 3/4.

17. A manufacturer supplies cheap quarter horse-power motors in lots of 25. A buyer, before taking a lot, tests at random sample of 5 motors and accepts the lot if they are all good; otherwise he rejects the lot. Find the probability that
(*a*) he will accept a lot containing 5 defective motors,
(*b*) he will reject a lot containing only one defective motor.
Ans. (*a*) ${}^{20}C_5/{}^{25}C_5 = 0.292$, (*b*) ${}^{24}C_4/{}^{25}C_5 = 0.20$.

18. A can hit a target 4 times in 5 shots, B : 3 times in 4 shots, C : twice in 3 shots. Find the probability that
(*a*) the target will not be hit,
(*b*) the target will be hit by at least one of them.

[**Hint:** Probibility that target will not be hit $= \left(1 - \frac{4}{5}\right)\left(1 - \frac{3}{4}\right)\left(1 - \frac{2}{3}\right) = \frac{1}{60}$.

Probability that target will be hit (*i.e.*, by at least one of them) $= 1 - (1/60) = 59/60$].

19. If it is 9 to 7 against a person A who is now 35 years of age living till he is 65 and 3 to 2 against a person B now 45 living till he is 75; find the chance that one at least of these person will be alive 30 years hence.
[**Hint:** Probability that A will die within 30 years = 9/16.
Probability that B will die within 30 years = 3/5.
The chance that both will die = $(9/16) \times (3/5) = 27/80$.
$\therefore$ Probability that both will not be dead *i.e.*, at least one will be alive = $1 - (27/80) = 53/80$].

20. The probability that a student passes in English is 0.75 and the probability that he passes in Hindi is 0.65. If 0.6 is the probability that he will pass in both the examination then what is the probability that he will pass in atleast one of them.
[**Hint:** P (passing at least one of English or Hindi) = P(English $\cup$ Hindi)
= P(English) + P(Hindi) – P(English and Hindi) = 0.75 + 0.65 – 0.6 = 0.8].

21. Each of the five rats are injected with the certain poison and the number of rats dying within 24 hours is observed. Suppose that the probability of one or more rats dying is 0.7 and the probability of four or less rats dying is 0.3. Find the probability that (*a*) 5 rats die, (*b*) No rat die.
[**Hint:** Let X be the number of rats dying within 24 hours.
We have $P(X \geq 1) = 0.7$ and $P(X \leq 4) = 0.3$.
(*a*) $P(X = 5) = 1 - P(X = 0) + P(X = 1) + P(X = 2) + P(X = 3) + P(X = 4)]$
$= 1 - P(X \leq 4) = 1 - 0.3 = 0.7$.
(*b*) $P(\text{No rat die}) = P(X = 0) = 1 - [P(X = 1) + P(X = 2) + \cdots + P(X = 5)]$
$= 1 - P(X \geq 1) = 1 - 0.70 = 0.3$].

22. The probability that a 50 year old man will be alive at 60 is 0.83 and the probability that a 45 year old woman will be alive at 55 in 0.87. What is the probability that a man who is 50 and his wife who is 45 will both be alive 10 years hence? Also find the probability that atleast one of them will be alive 10 years hence.
Ans. 0.722, 0.978.
[**Hint :** $P(A \cap B) = P(A)\, P(B)$; $P(A \cup B) = P(A) + P(B) - P(A)\, P(B)$].

23. A speaks truth in 75 per cent cases and B in 80 per cent cases. In what percentage of cases are they likely to contradict each other in stating the same fact? (*AMIETE, 2013*) **Ans.** 46%

[**Hint:** They contradict if any one of them speaks the truth and other not. Therefore, probability that A speaks the truth but B does not
$= (75/100)\,[1-(80/100)] = 3/20$. Probability that B speaks the truth but A does not
$= (80/100)\,[1-(75/100)] = 1/5$. The two events being mutually exclusive. The required probability
$= (3/20) + (1/5) = 7/20$].

24. Of the cigarette smoking population, 70% are men and 30% women. 10% of these men and 20% of these women smoke "WILLS". What is the probability that a person seen smoking "WILLS" will be a man.
[**Hint:** In smoking population: Men smokers = 70%, women smokers = 30%.
Out of 100 men, WILLS smokers $= 70 \times (1/10) = 7$.
Out of 100 women, WILLS smokers $= 30 \times (1/5) = 6$.
Probability that a person smoking WILLS is a man $= \dfrac{7}{7+6} = \dfrac{7}{13}$].

25. An urn contains nine balls, two of which are red, three blue and four black. Three balls are drawn from the urn at random. What is the probability that (*a*) the three balls are of different colours?
(*b*) the three balls are of the same colour? (*AMIETE, S 2000*) **Ans.** (*a*) $2/7$, (*b*) $5/84$.

26. A card is drawn from a well-shuffled deck of 52 cards. What is the probability of this card being a heart or a king. **Ans.** $4/13$.
[**Hint:** $P(A) = 13/52$, $P(B) = 4/52$, $P(A \cup B) = 1/52$, $P(A \cap B) = 4/13$].

27. If A and B be events with $P(A \cup B) = 7/8$, $P(A \cap B) = 1/4$ and $P(\bar{A}) = 5/8$, find $P(A)$, $P(B)$ and $P(A \cap \bar{B})$. **Ans.** $3/8, 3/4, 1/8$.
[**Hint:** $P(A \cap \bar{B}) = P(A - B) = P(A) - P(A \cap B) = (3/8) - (1/4) = 1/8$].

28. The probability that a boy will pass an examination is $3/5$ and the probability that the girl pass in the same examination is $2/5$. Find the probability that either boy or girl pass in the examination.
[**Hint:** Event A: Boy pass in the examination. Event B: girl pass in the examination. Both these events are independent, we have
$P(A \text{ or } B) = P(A \cup B) = P(A) + P(B) - P(A \cap B)$
$= P(A) + P(B) - P(A) \cdot P(B) = (3/5) + (2/5) - (3/5)(2/5) = 19/25$].

29. Let A and B be events with $P(A) = 1/3$, $P(B) = 1/4$ and $P(A \cup B) = 1/2$. Find
(*a*) $P(A/B)$, (*b*) $P(B/A)$, (*c*) $P(A \cap \bar{B})$, (*d*) $P(A/\bar{B})$. **Ans.** (*a*) $1/3$, (*b*) $1/4$, (*c*) $1/4$, (*d*) $1/3$.
[**Hint:** (*c*) $P(\bar{B}) = 1 - P(B) = 3/4$. Since A and B are independent, A and $\bar{B}$ are also independent. We obtain $P(A \cap \bar{B}) = P(A)\,P(\bar{B}) = (1/3)(3/4) = 1/4$.
(*d*) $P(A/\bar{B}) = P(A \cap \bar{B})/P(\bar{B})\, 1/4 \div 3/4 = 1/3$.]

30. Let A and B be two events with $P(A) = 1/2$, $P(B) = 1/3$ and $P(A \cap B) = 1/4$. Find $P(A/B)$, $P(A \cup B)$, $P(\bar{A}/\bar{B})$. **Ans.** $3/4, 7/12, 3/8$.

31. If A and B are two events such that $P(A) = 3/8$, $P(B) = 5/8$ and $P(A \cup B) = 3/4$, find $P(A/B)$ and $P(B/A)$. Also find $P(A \cap \bar{B})$ and $P(A/\bar{B})$. **Ans.** $2/5, 2/3, 1/8, 1/3$.

32. A can hit a target 3 times in 5 shots, B 2 times in 5 shots and C 3 times in 4 shots. They fire a volley. What is the probability that (*a*) two shots hit, (*b*) atleast two shots hit? **Ans.** (a) 0.45, (b) 0.63.

33. For two events A and B, if $P(A) = 0.5$, $P(B) = 0.6$ and $P(A \cap B) = 0.25$, then find the conditional probability $P(A/B)$ and $P(B/A)$.
Ans. $P(A/B) = \dfrac{P(A \cap B)}{P(B)} = \dfrac{0.25}{0.60} = \dfrac{5}{12}$, $P(B/A) = \dfrac{P(B \cap A)}{P(A)} = \dfrac{0.25}{0.50} = \dfrac{1}{2}$.

THEOREM OF TOTAL PROBABILITY

Theorem 7.5 Let $B_1, B_2, ..., B_n$ be a set of exhaustive and mutually exclusive events of the sample space S with $P(B_i) \neq 0$, $i = 1, 2, ..., n$ and let A be any event of S. Then

$$P(A) = \sum_{i=1}^{n} P(B_i \cap A) = \sum_{i=1}^{n} P(B_i)\, P(A/B_i). \qquad ...(7.9)$$

Proof: Since $B_1, B_2, ..., B_n$ are exhaustive and mutually exclusive events of S, we have

$$S = B_1 \cup B_2 \cup ... \cup B_n$$

and $$B_1 \cap B_j = \phi,\ i \neq j,\ i, j = 1, 2, ..., n.$$

We know that for any event A,

$$A = A \cap S = A \cap (B_1 \cup B_2 \cup ... \cup B_n)$$
$$= (A \cap B_1) \cup (A \cap B_2) \cup ... \cup (A \cap B_n).$$

Hence, by additive rule for mutually exclusive events (see Eq. 7.2) we obtain

$$P(A) = P[(A \cap B_1) \cup (A \cap B_2) \cup ... \cup (A \cap B_n)]$$
$$= P(A \cap B_1) + P(A \cap B_2) + ... + P(A \cap B_n)].$$

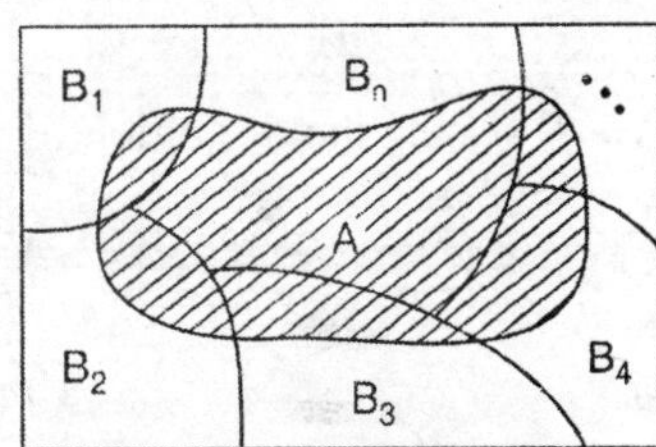

Fig. 7.10: Total Probability

Using the multiplicative rule (see Eq. 7.6), we obtain

$$P(A) = \sum_{i=1}^{n} P(B_i \cap A) = \sum_{i=1}^{n} P(B_i)\, P(A/B_i).$$

EXAMPLE 7.46 A and B throw alternately with a pair of dice. The one who throws 9 first wins. If A starts the game, compare their chances of winning. *(AMIETE, Dec. 2012)*

Solution A sum of 9 is obtained if the numbers appearing on the dice are (3, 6) (4, 5) (5, 4), (6, 3).

Hence, $$P\text{ (sum is 9)} = \frac{4}{36} = \frac{1}{9} = p.$$

$$P\text{ (not getting sum as 9)} = 1 - \frac{1}{9} = \frac{8}{9} = q.$$

Players A wins, if he gets a sum of 9 in the first or third or fifth, ..., throw. Player B wins, if he gets a sum of 9 in the second or fourth or sixth, ..., throw. Player A starts the game.

$$P\text{ (sum of 9 in first throw)} = p = \frac{1}{9}.$$

$$P\text{ (sum of 9 in third throw)} = q^2 p = \left(\frac{8}{9}\right)^2 \left(\frac{1}{9}\right) = \frac{64}{729}.$$

$$P\text{ (sum of 9 in fifth throw)} = q^4 p = \left(\frac{8}{9}\right)^4 \left(\frac{1}{9}\right) = \frac{4096}{59049}.$$

$$P(A\text{ wins}) = p + q^2 p + q^4 p + ... = p(1 + q^2 + q^4 + ...)$$
$$= \frac{p}{1 - q^2} = \frac{1}{9\left(1 - \frac{64}{81}\right)} = \frac{9}{17}.$$

$$P(B\text{ wins}) = 1 - P(A\text{ wins}) = 1 - \frac{9}{17} = \frac{8}{17}.$$

Therefore. $$\frac{P(A\text{ wins})}{P(B\text{ wins})} = \frac{9}{8},\ \text{that is,}\ P(A\text{ wins}) : P(B\text{ wins}) = 9 : 8.$$

7.6 INVERSE PROBABILITY—BAYE'S THEOREM

If an event has occurred through one of many mutually exclusive and exhaustive events then the conditional probability that it has occurred due to a particular event, is called its *inverse probability* (also known as posteriori probability). A theorem (or rule) named after the English mathematician Thomas Bayes (1702–1761) is used to determine inverse probability.

Bayes' Theorem: *If an event A of sample space S with $P(A) \neq 0$ can occur in conjunction with one of the n mutually exclusive and exhaustive events $B_1, B_2, ..., B_n$ of the sample space S with $P(B_i) \neq 0$, and if A has actually occurred then the probability that it was preceded by a particular event B_i ($i = 1, 2, ..., n$) is given by*

$$P(B_i/A) = \frac{P(B_i)\,P(A/B_i)}{\sum_{i=1}^{n} P(B_i)\,P(A/B_i)}, \quad i = 1, 2, ..., n \qquad ...(7.10)$$

Proof Using the definition of conditional probability, we have

$$P(B_i/A) = \frac{P(B_i \cap A)}{P(A)} = \frac{P(B_i)\,P(A/B_i)}{P(A)} \qquad \text{(by multiplication rule of probability)}$$

$$= \frac{P(B_i)\,P(A/B_i)}{\sum_{i=1}^{n} P(B_i)\,P(A/B_i)} \qquad \text{(by the result of theorem of total probability)}$$

Remark 5

(a) The denominator in Equation (7.10) gives the total probability.

(b) The probability $P(B_i)$, $i = 1, 2, ..., n$, which was already known before performing an experiment is called *priori* probability.

(c) $P(B_i/A)$, $i = 1, 2, ..., n$ is called *posteriori* probability. It gives the probability of occurrence of the event B_i with respect to the occurance of the event A.

(d) Baye's theorem is also called the theorem of *probability of causes*.

EXAMPLE 7.47 The chances that doctor A will diagnose a disease X correctly is 60%. The chance that a patient will die by his treatment after correct diagnosis is 40% and the chance of death by wrong diagnosis is 70%. A patient of doctor A who had disease X died. What is the chance that his disease was diagnosed correctly.

Solution Denote the events as

A_1: The disease is diagnosed correctly.

A_2: The disease is diagnosed incorrectly. E: The patient dies.

Then

$$P(A_1) = 60/100 = 0.6; \quad P(A_2) = 1 - 0.6 = 0.40$$

$$P(E/A_1) = 40/100 = 0.4; \quad P(E/A_2) = 70/100 = 0.7.$$

By Bayes'theorem

$$P(A_1/E) = \frac{P(A_1)\,P(E/A_1)}{P(A_1)\,P(E/A_1) + P(A_2)\,P(E/A_2)} = \frac{(0.6)\,(0.4)}{(0.6)\,(0.4) + (0.4)\,(0.7)} = \frac{6}{13}.$$

EXAMPLE 7.48 In a bolt factory, machines A, B and C manufacture respectively 25%, 35% and 40% of the total of the output. The defective bolts percentage for these machines are 5, 4, 2. A bolt is drawn at random from the output and is found to be defective. What are the probabilities that it was manufactured by machine A, B or C? *(VTU, 2006; Rohtak 2005; Madras 2000)*

Solution Denote the event as

B_1: the bolt is manufactured by machine A,

B_2: the bolt is manufactured by machine B,

B_3: the bolt is manufactured by machine C.

Note that B_1, B_2, B_3 are mutually exclusive and exhaustive events.

Let E = event, 'the bolt is defective'.

The event E occurs with B_1 or with B_2 or with B_3. Given that

$$P(B_1)=\frac{25}{100}=0.25,\ P(B_2)=\frac{35}{100}=0.35 \text{ and } P(B_3)=\frac{40}{100}=0.40.$$

Again $P(E/B_1)$ = Probability that the bolt drawn is defective given the condition that it is manufactured by machine $A=\frac{5}{100}=0.05$. Similarly, $P(E/B_2)=0.04,\ P(E/B_3)=0.02$.

Hence, by Bayes' Theorem, we have

$$P(B_1/E) = \frac{P(B_1)\,P(E/B_1)}{P(B_1)\,P(E/B_1)+P(B_2)\,P(E/B_2)+P(B_3)\,P(E/B_3)}$$

$$= \frac{0.25\times 0.05}{0.25\times 0.05+0.35\times 0.04+0.40\times 0.02}=\frac{0.0125}{0.0125+0.014+0.008}=\frac{0.0125}{0.0345}=0.36.$$

Similarly $$P(B_2/E) = \frac{0.35\times 0.04}{0.25\times 0.05+0.35\times 0.04+0.40\times 0.02}=\frac{0.014}{0.0345}=0.41,$$

and $$P(B_3/E) = \frac{0.40\times 0.02}{0.0345}=\frac{0.008}{0.0345}=0.23.$$

EXAMPLE 7.49 An insurance company insured 2000 scooter drivers, 4000 car drivers and 6000 truck drivers. The probability of an accident is 0.01, 0.03, 0.15 respectively. One of the insured persons meets an accident. What is the probability that he is a scooter driver? *(AMIETE, June 2013, Dec. 2011)*

Solution Let E_1, E_2, E_3 be the event that person chosen is a scooter driver, car driver and truck driver respectively. A be the event that person meets with an accident.

Now $P(E_1)=\frac{2000}{2000+4000+6000}=\frac{1}{6},\ P(E_2)=\frac{4000}{12000}=\frac{1}{3}$ and $P(E_3)=\frac{6000}{12000}=\frac{1}{2}$.

It is given that $P(A/E_1)=0.01,\ P(A/E_2)=0.03$ and $P(A/E_3)=0.15$.

It is required to find $P(E_1/A)$.

By Bayer's rule, we have

$$P(E_1/A)=\frac{P(E_1)\,P(A/E_1)}{P(E_1)\,P(A/E_1)+P(E_2)\,P(A/E_2)+P(E_3)\,P(A/E_3)}$$

$$=\frac{(1/6)\times 0.01}{(1/6)\times 0.01+(1/3)\times 0.03+(1/2)\times 0.15}=\frac{1}{1+6+45}=\frac{1}{52}.$$

EXAMPLE 7.50 The contents of three urns are as follows:

Urn I: 1 white, 2 black and 3 red balls,

Urn II: 2 white, 1 black and 1 red balls,

Urn III: 4 white, 5 black and 3 red balls.

One urn is chosen at random and two balls are drawn. These happen to be white and red. What is the probability that they come from Urn I?

Solution Denoting the following events.

A: two balls drawn are white and red. U_1, U_2 and U_3 be the events that the balls drawn are from the Urn I, Urn II, Urn III respectively. Then

$$P(U_1) = P(U_2) = P(U_3) = 1/3.$$

Also $P(A/U_1)$ = Probability of drawing a white ball and a red ball from the Urn I

$$=({}^1C_1\times {}^3C_1)/{}^6C_2=\frac{1\times 3}{15}=\frac{1}{5}.$$

$P(A/U_2)$ = Probability of drawing a white ball and a red ball from the Urn II

$$({}^2C_1 \times {}^1C_1)/{}^4C_2 = \frac{2\times 1}{6} = \frac{1}{3}.$$

Similarly
$$P(A/U_3) = \frac{{}^4C_1 \times {}^3C_1}{{}^{12}C_2} = \frac{4\times 3}{66} = \frac{2}{11}.$$

Hence, by Bayes' Theorem, we have

$P(U_1/A)$ = Probability that the balls drawn are from the Urn I

$$= \frac{P(U_1)\,P(A/U_1)}{P(U_1)\,P(A/U_1) + P(U_2)\,P(A/U_2) + P(U_3)\,P(A/U_3)}$$

$$= \frac{(1/3)\,(1/5)}{(1/3)\,(1/5) + (1/3)\,(1/3) + (1/3)\,(2/11)} = \frac{33}{33+55+30} = \frac{33}{118}.$$

Thus, the required probability is 33/118.

EXAMPLE 7.51 There are three bags: first containing 1 white, 2 red, 3 green balls; second 2 white, 3 red, 1 green balls and third 3 white, 1 red, 2 green balls. Two balls are drawn from a bag chosen at random. These are found to be one white and one red. Find the probability that the balls so drawn came from the second bag. *(AMIETE, June 2011; JNTU 2003)*

Solution Let E be the events that two balls drawn are white and red. Let B_1, B_2 and B_3 be the events that the balls drawn are from the Bag I, Bag II and Bag III respectively.

Now $P(B_1) = P(B_2) = P(B_3) = 1/3$.

$P(E/B_1)$ = Probability of drawing a white ball and red ball from the bag I

$$= \frac{({}^1C_1 \times {}^2C_1)}{{}^6C_2} = \frac{2}{15}.$$

Similarly,
$$P(E/B_2) = \frac{{}^2C_1 \times {}^3C_1}{{}^6C_2} = \frac{2}{5}, \qquad P(E/B_3) = \frac{{}^3C_1 \times {}^1C_1}{{}^6C_2} = \frac{1}{5}.$$

Hence, by Bayes' theorem

$$P(B_2/E) = \frac{P(B_2)\,P(E/B_2)}{P(B_1)\,P(E/B_1) + P(B_2)\,P(E/B_2) + P(B_3)\,P(E/B_3)} = \frac{(1/3)\,(2/5)}{(1/3)\,(2/15) + (1/3)\,(2/5) + (1/3)\,(1/5)} = \frac{6}{11}.$$

EXAMPLE 7.52 In a class of 75 students, 15 were considered to be intelligent, 45 as medium and the rest below average. The probability that a very intelligent student fails in a viva-voce examination is 0.005, the medium student failing has a probability of 0.05, and the correspoding probability for a below average student is 0.15. If a student is known to have passed the viva-voce examination, what is the probability that he is below average?

Solution Denoting A_1, A_2 and A_3 the events that a student is very intelligent, medium and below average respectively and E be the event that a student passes in a viva-voce examination. Given that $P(A_1) = 15/75 = 0.20$,

$$P(A_2) = 45/75 = 0.6,\ P(A_3) = 15/75 = 0.20.$$

$$P(E/A_1) = 1 - 0.005 = 0.995,$$

$$P(E/A_2) = 1 - 0.05 = 0.95,$$

$$P(E/A_3) = 1 - 0.15 = 0.85.$$

The probability that a student is below average level has passed the viva-voce examination = $P(A_3/E)$

$$= \frac{P(A_3)\,P(E/A_3)}{P(A_1)\,P(E/A_1)+P(A_2)\,P(E/A_2)+P(A_3)\,P(E/A_3)}$$

$$= \frac{0.20\times 0.85}{0.20\times 0.995+0.6\times 0.95+0.20\times 0.85} = \frac{0.17}{0.199+0.57+0.17} = \frac{0.17}{0.939} = 0.181.$$

EXERCISE 7.2

1. Machines I and II produce 30% and 70% respectively of a factory's output. Machine I produces 3% defectives and machine II produces 4% defectives item.
   (*a*) If an item is defective what is the chance it was produced by machine I?
   (*b*) If an item is not defective, what is the chance it was produced by machine I?
   Ans. 0.24 (approx.), 0.30 (approx).
2. There are two bags I and II. Bag I contains 2 white and 3 red balls, bag II contains 4 white and 5 red balls. One ball is drawn at random from one of the bags and is found to be red. Find the probability that it was drawn from bag II. **Ans.** 25/32.
3. There are three boxes *A*, *B*, *C* containing screws, some of which are defective. The proportion of defective screws in the three boxes are $1/4, 1/8, 1/3$ respectively. A box is selected at random and a screw drawn from it at random is found to be defective. What is the probability that the box *A* was selected? **Ans.** 6/17.
4. A doctor is to visit a patient. From the past experience, it is known that the probabilities that he will come by train, bus, scooter or by other means of transport are respectively 30%, 20%, 10% and 40%. The probabilities that he will be late are $1/4, 1/3$ and $1/12$, if he comes by train, bus and scooter respectively, but if he comes by other means of transport, then he will not be late. When he arrives, he is late. What is the probability that comes by train? **Ans.** 1/2.
5. In a test, an examinee either guesses or copies or knows the answer to a multiple choice question with four choices. The probability that he makes a guess is $1/3$ and the probability that he copies the anser is $1/6$. The probability that his answer is correct, given that he copied it, is $1/8$. Find the probability that he knew the answer to the question, given that he correctly answered it. **Ans.** 24/29.
6. In the coming year, there will be three candidates for the position of Principal—Dr. Sinha, Mr. Veenu Gopal and Dr. Chatterji whose chances of getting the appointment are in the proportion 4 : 2 : 3 respectively. The probability that Dr. Sinha if selected will abolish co-education in the college is 30%. The probability of Mr. Veenu Gopal and Dr. Chatterji doing the same are respectively 50% and 80%. What is the probability that co-education will be abolished in the college next year. **Ans.** 23/45.
7. A man is known to speak truth 3 out of 4 times. He throws a die and reports that it is a six. Find the probability that it is actually a six. **Ans.** 3/8.

7.7 RANDOM VARIABLES

In many experiments we are interested in *numbers* associated with the outcomes of the experiment. For example, when *n* coins are tossed simultaneously, one may be interested in knowing the number of heads. When a pair of dice is rolled and we ask for the sum of the points on the upturned faces. Whenever we associate a real number with each outcome of an experiment we are considering a *function* whose domain is the set of possible outcomes and whose range is the set of real numbers in question. Such a function is called a *random variable.*

Definition of a Random Variable (RV): Let *S* denote a sample space corresponding to a random experiment *E*. A *random variable X* on a sample space *S* is a function from *S* into the set *R* of real numbers such that the pre-image of every interval of *R* is an event of *S*.

Notations: Capital letters such as X, Y, Z are ordinarily used to denote one-dimensional random variables. A typical outcome of the experiment is usually denoted by the Greek letter ω (omega). Thus $X(\omega)$ denotes that real number which the random variable X associates with the outcome ω.

The following are some simple illustrations of random variables.

Illustration 7.1 If the random experiment E consists of two tosses of a coin, the random variable X defining the number of heads will assume the values 0, 1, 2 represented by

Outcome (ω)	TT	TH	HT	HH
Value of $X(\omega)$	0	1	1	2

Thus to each outcome say ω of E, there corresponds a real number $X(\omega)$. The experiment E contains four outcomes TT, TH, HT and HH. These are also called sample points and they constitute the entire sample space S of E. Further, since the sample points correspond to outcomes, it is obvious that the random variable $X(\omega)$ is defined for each ω belonging to S and thus a random variable corresponds to a real function on S.

Illustration 7.2 An experiment consists of rolling a die and reading the number of points on the upturned face. The most natural random variable X to consider is $X(\omega) = \omega$ for $\omega = 1, 2, 3, 4, 5, 6$.

If we are interested in whether the number of points is even or odd, then we can consider a random variable Y, which is defined as follows:

$$Y(\omega) = \begin{cases} 0, & \text{if } \omega \text{ is even,} \\ 1, & \text{if } \omega \text{ is odd.} \end{cases}$$

The values 0 and 1 are not essential—any two distinct real numbers could be used instead. However, 0 and 1 suggest "even" and "odd" respectively, because they represent the remainder obtained when the outcome ω is divided by 2.

There are two types of random variables *discrete* and *continuous*.

Discrete Random Variable

A random variable X defined on a discrete sample space is known as a *discrete random variable* if X assume either only a finite number or a countably infinite number of values $x_1, x_2, x_3, \ldots, x_n$. called the possible values of X, with positive probabilities. Examples of discrete random variables are marks obtained in a test, number of accidents per month, deaths by cancer, tosses until a die shows the first six etc.

Probability Function of a Discrete Random Variable

Let S be a discrete sample space associated with an experiment and X be discrete random variable which takes the values $x_1, x_2, x_3, \ldots, x_n$. Let each x_i be associated with a number p_i called the probability of x_i and denoted by $p_i = P(X = x_i) = p(x_i)$. The number $p(x_i)$ must satisfy the following conditions:

(*a*) $p(x_i) \geq 0$ for all values of i, and (*b*) $\sum_{i=1}^{n} p(x_i) = 1$.

The set of values x_i with their probabilities p_i constitutes the *probability distribution* or the *discrete probability distribution* of the discrete random variable X. The probability function is also called *probability mass function* or *point probability function*.

EXAMPLE 7.53 A random variable X has the following probability function:

x :	0	1	2	3	4	5	6	7	8
$p(x)$:	k	$3k$	$5k$	$7k$	$9k$	$11k$	$13k$	$15k$	$17k$

(*i*) Determine the value of k,

(*ii*) Find $P(X < 4)$, $P(X \geq 6)$, $P(3 < X \leq 5)$,

(*iii*) Find the minimum value of λ such that $P(X \leq \lambda) > 1/2$.

Solution (*i*) If X is a random variable, then $\sum_{i=0}^{8} p(x_i) = 1.$

This gives $k + 3k + 5k + 7k + 9k + 11k + 13k + 15k + 17k = 1$ or $k = 1/81$.

(*ii*) $P(X < 4) = P(X = 0) + P(X = 1) + P(X = 2) + P(X = 3)$

$= k + 3k + 5k + 7k = 16k = 16/81.$

$P(X \geq 6) = P(X = 6) + P(X = 7) + P(X = 8)$

$= 13k + 15k + 17k = 45k = 45/81 = 5/9.$

$P(3 < X \leq 5) = P(X = 4) + P(X = 5) = 9k + 11k = 20k = 20/81.$

(*iii*) $P(X \leq 1) = 4k = \frac{4}{81} < \frac{1}{2}$; $P(x \leq 2) = k + 3k + 5k = 9k = \frac{9}{81} < \frac{1}{2}$;

$P(X \leq 3) = k + 3k + 5k + 7k = 16k = \frac{16}{81} < \frac{1}{2}$;

$P(X \leq 4) = k + 3k + 5k + 7k + 9k = 25k = \frac{25}{81} < \frac{1}{2}$;

$P(X \leq 5) = k + 3k + 5k + 7k + 9k + 11k = 36k = \frac{36}{81} < \frac{1}{2}$;

$P(X \leq 6) = k + 3k + 5k + 7k + 9k + 11k + 13k = 49k = \frac{49}{81} > \frac{1}{2}$.

Therefore, the minimum value of λ so that $P(X \leq \lambda) > 1/2$ is 6.

EXAMPLE 7.54 From a lot of 10 items containing 3 defectives items, a sample of 4 items are drawn at random without replacement. Let a random variable X denote the number of defective items in the sample. Find the probability distribution of X. Calculate $P(X \leq 1)$, $P(X < 1)$, $P(0 < X < 2)$.

Solution The lot containing 7 non-defective and 3 defective items. Since X denotes the number of defective item, x can take the values 0, 1, 2, 3 such that

For $x = 0$, $p(x)$ = Probability of getting no defective item $= \frac{^7C_4}{^{10}C_4} = \frac{1}{6}$.

For $x = 1$, $p(x)$ = Probability of getting one defective item $= \frac{^7C_3 \times {^3C_1}}{^{10}C_4} = \frac{1}{2}$.

For $x = 2$, $p(x)$ = Probability of getting two defective item $= \frac{^7C_2 \times {^3C_2}}{^{10}C_4} = \frac{3}{10}$.

For $x = 3$, $p(x)$ = Probability of getting three defective item $= \frac{^7C_1 \times {^3C_3}}{^{10}C_4} = \frac{1}{30}$.

We have the following probability distribution.

x:	0	1	2	3
$p(x)$:	5/30	15/30	9/30	1/30

Now $P(X \leq 1) = P(X = 0) + P(X = 1) = (1/6) + (1/2) = 2/3.$

$P(X < 1) = P(X = 0) = 1/6.$

$P(0 < X < 2) = P(X = 1) = 1/2.$

Distribution Function

The distribution function $F(X)$ of random variable X is defined by

$F(x) = P(X \le x)$. It has the following properties.

(*a*) $0 < F(X) \le 1$,

(*b*) If $x_1 < x_2$, then $F(x_1) \le F(x_2)$,

(*c*) $P(a \le X \le b) = F(b) - F(a)$.

In the case of a discrete RV, the discrete distribution function is defined by $F(X) = \Sigma_{x_i \le x} p(x_i)$.

If X takes only a finite set of values $x_1, x_2, ..., x_n$ then the distribution function is given by

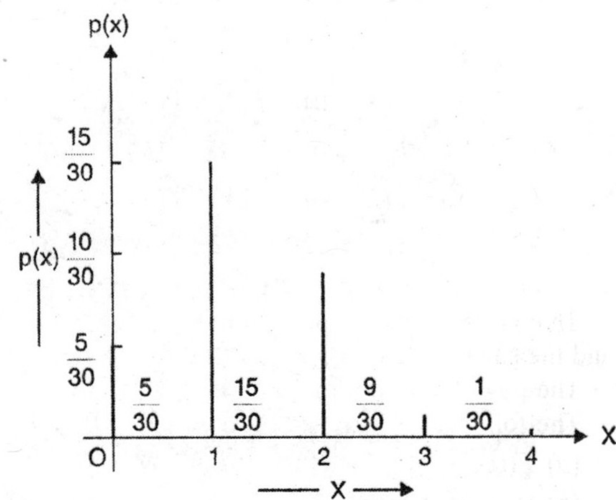

Fig. 7.11: Shows the probability function $p(x)$ of the discrete random variable X.

$$F(x) = \begin{cases} 0 & -\infty < x < x_1 \\ p(x_1) & x_1 \le x < x_2 \\ p(x_1) + p(x_2) & x_2 \le x < x_3 \\ \vdots & \vdots \\ p(x_1) + p(x_2) + \cdots + p(x_n) & \cdots x_n \le x < \infty. \end{cases}$$

From this definition, we have

$$p(x_i) = P(X = x_i) = F(x_i) - F(x_{i-1}). \quad ...(7.11)$$

The distribution function $F(x)$ is also called the *cumulative distribution function* (*cdf*) of X.

EXAMPLE 7.55 The probability function of the random variable X denoting the number of heads in two tosses of a coin is represented by the following Table 7.1:

Table 7.1

X:	0	1	2
$p(x)$:	1/4	1/2	1/4

Find the distribution function of X.

Solution The distribution function of $F(x)$ is given by

$$F(x) = \begin{cases} 0 & -\infty < x < 0 \\ 1/4 & 0 \le x < 1 \\ \frac{1}{4} + \frac{1}{2} = \frac{3}{4} & 1 \le x < 2 \\ \frac{3}{4} + \frac{1}{4} = 1 & 2 \le x < \infty. \end{cases}$$

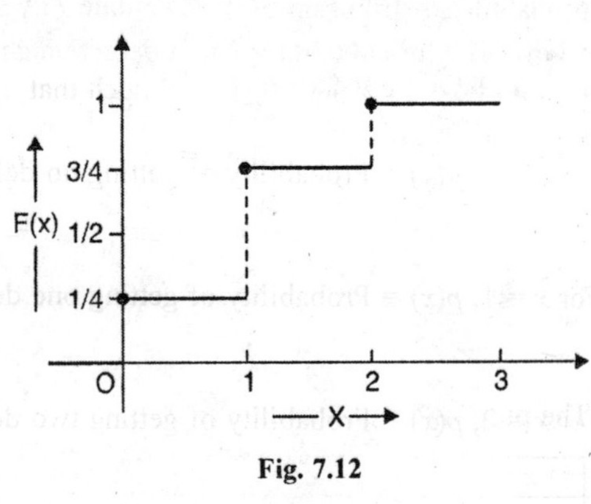

Fig. 7.12

Expectation and Variance of a Discrete Random Variable

If X denotes a discrete random variable which can assume the values $x_1, x_2, x_3, ..., x_n$ with respective probabilities $p_1, p_2, p_3, ..., p_n$ where $p_1 + p_2 + p_3, ..., p_n = 1$, then the *mean* or *expectation* or *expected value* of X, denoted by $E(X)$ or μ_x or simply E or μ_x is defined by $E(X) = \mu_x = p_1x_1 + p_2x_2 + p_3x_3 + \cdots + p_nx_n = \sum_{i=1}^{n} p_ix_i$. ...(7.12)

The *variance* of the random variable X, denoted by Var (X), is defined by

$$\text{Var (X)} = \sigma_X^2 = E[(X - \mu_x)^2] = \sum_{i=1}^{n} (x_i - \mu_x)^2 p_i$$

$$= \sum_{i=1}^{n}(x_i^2 - 2\mu_x x_i + \mu_x^2)p_i = \sum_{i=1}^{n} x_i^2 p_i - 2\mu_x \sum_{i=1}^{n} x_i p_i + \mu_x^2 \sum_{i=1}^{n} p_i$$

$$= \sum_{i=1}^{n} x_i^2 p_i - 2\mu_x^2 + \mu_x^2 \quad \left(\text{Using } \sum_i x_i p_i = \mu_x \text{ and } \sum_i p_i = 1\right)$$

$$= \sum_{i=1}^{n} x_i^2 p_i - \mu_x^2 = E(X^2) - \mu_x^2 = E(X^2) - [E(X)]^2. \qquad ...(7.13)$$

That is, the variance is equal to the difference between the expectation of the square of the random variable and the square of the expectation of the random variable.

The positive square root of variance is known as standard deviation and is denoted by σ.

The following relations hold good.

(*a*) $E(\alpha X + \beta Y) = \alpha E(X) + \beta E(Y)$.

(*b*) $\text{Var}(cX) = c^2 \text{Var}(X)$.

(*c*) $\text{Var}(aX + b) = a^2 \text{Var}(X)$.

Illustration 7.3 If $E(X) = 2$, then $(2X + 3) = 2E(X) + 3 = 2(2) + 3 = 7$.

Illustration 7.4 If $\text{Var}(X) = 2$, then $\text{Var}(2X + 3) = 2^2 \text{Var}(X) = 4(2) = 8$.

EXAMPLE 7.56 If four coins are tossed. What is the expectation of the number of heads?

(AMIETE, Dec. 2011)

Solution Let X denote the number of heads obtained. Then X can take the values of 0, 1, 2, 3, 4.

$$P(X = 0) = P(4 \text{ tails}) = \frac{1}{2^4} = \frac{1}{16}.$$

$$P(X = 1) = P(1 \text{ head and } 3 \text{ tails}) = \frac{{}^4C_1}{16} = \frac{1}{4}.$$

$$P(X = 2) = P(2 \text{ heads and } 2 \text{ tails}) = \frac{{}^4C_2}{16} = \frac{(4)(3)}{(2)(16)} = \frac{3}{8}.$$

$$P(X = 3) = P(3 \text{ heads and } 1 \text{ tail}) = \frac{{}^4C_3}{16} = \frac{4}{16} = \frac{1}{16}.$$

$$P(X = 4) = P(4 \text{ heads}) = \frac{1}{16}.$$

The probability distribution of X is shown below:

x :	0	1	2	3	4
$p(x)$:	1/16	1/4	3/8	1/4	1/16

$$E(X) = \frac{1}{16}(0) + \frac{1}{4}(1) + \frac{3}{8}(2) + \frac{1}{4}(3) + \frac{1}{16}(4) = 2.$$

EXAMPLE 7.57 A coin is tosses three times. A man gets Rs. 25 for every head and losses Rs 10 for every tail. Find his expectation.

Solution Let X be the money received by the man.

A man gets either 3H or 2HIT or 1H2T or 3T.

$X(3H) = 25 + 25 + 25 = 75$, $X(2HIT) = 25 + 25 - 10 = 40$,

$X(1H2T) = 25 - 10 - 10 = 5$, $X(3T) = -10 - 10 - 10 = -30$.

$$P(3H) = \frac{1}{8}, \quad P(2HIT) = \frac{{}^3C_2}{8} = \frac{3}{8}, \quad P(1H2T) = \frac{{}^3C_1}{8} = \frac{3}{8}, \quad P(3T) = \frac{1}{8}.$$

Therefore, X can takes values 75, 40, 5, – 30.

$P(X = 75) = P(3H) = \frac{1}{8}$, $P(X = 40) = P(2H1T) = \frac{3}{8}$,

$P(X = 5) = P(1H2T) = \frac{3}{8}$, $P(X = -30) = P(3T) = \frac{1}{8}$.

The probability distribution of X is shown below

x :	75	40	5	−30
$p(x)$:	1/8	3/8	3/8	1/8

$$E(X) = 75(1/8) + 40(3/8) + 5(3/8) - 30(1/8) = \frac{1}{8}(75 + 120 + 15 - 30) = 22.5.$$

Thus, expectation of the money is ₹ 22.50.

EXAMPLE 7.58 A random variable X has the following probability disribution.

x :	1	2	3	4	5
$p(x)$:	k	k	$3k$	$k^2 + k$	$6k^2$

Find the value of k. Evaluate $P(X < 3)$ and $P(1 < X < 4)$. Determine the distribution function of X. Find the mean, variance of X and standard deviation. Plot the probability function $p(x)$ and distribution function $F(x)$ of X.

Solution Since $\sum_{x=1}^{5} p(x) = 1$, we obtain

$$k + k + 3k + (k^2 + k) + 6k^2 = 1, \text{ or } 7k^2 + 6k - 1 = 0$$

or $$(7k - 1)(k + 1) = 0 \text{ or } k = 1/7, -1.$$

Since $p(x) \geq 0$, the possible value of $k = 1/7$.

Now, $P(X < 3) = P(X = 1) + P(X = 2) = k + k = 2k = 2/7.$

$P(1 < X < 4) = P(X = 2) + P(X = 3) = k + 3k = 4k = 4/7.$

We have the following probability distribution.

x :	1	2	3	4	5
$p(x)$:	1/7	1/7	3/7	8/49	6/49

Table 7.2: The results for the probability distribution and distribution function are given below

x :	1	2	3	4	5
$p(x)$:	1/7	1/7	3/7	8/49	6/49
$F(x)$:	7/49	14/49	35/49	43/49	1

$$\text{Mean} = E(X) = \sum_{i=1}^{5} x_i p_i = (1)(1/7) + (2)(1/7) + (3)(3/7) + (4)(8/49) + (5)(6/49)$$

$$= 1/7 + 2/7 + 9/7 + 32/49 + 30/49 = 146/49.$$

$$\text{Variance} = E(x^2) - [E(x)]^2 = \left[1^2\left(\frac{1}{7}\right) + 2^2\left(\frac{1}{7}\right) + 3^2\left(\frac{3}{7}\right) + 4^2\left(\frac{8}{49}\right) + 5^2\left(\frac{6}{49}\right)\right] - \left(\frac{146}{49}\right)^2$$

$$= \left(\frac{1}{7} + \frac{4}{7} + \frac{27}{7} + \frac{128}{49} + \frac{150}{49}\right) - \left(\frac{146}{49}\right)^2 = \frac{502}{49} - \left(\frac{146}{49}\right)^2 = 10.245 - 8.878 = 1.367$$

$$\text{Standard deviation} = \sqrt{\text{Var}} = \sqrt{1.367} = 1.169.$$

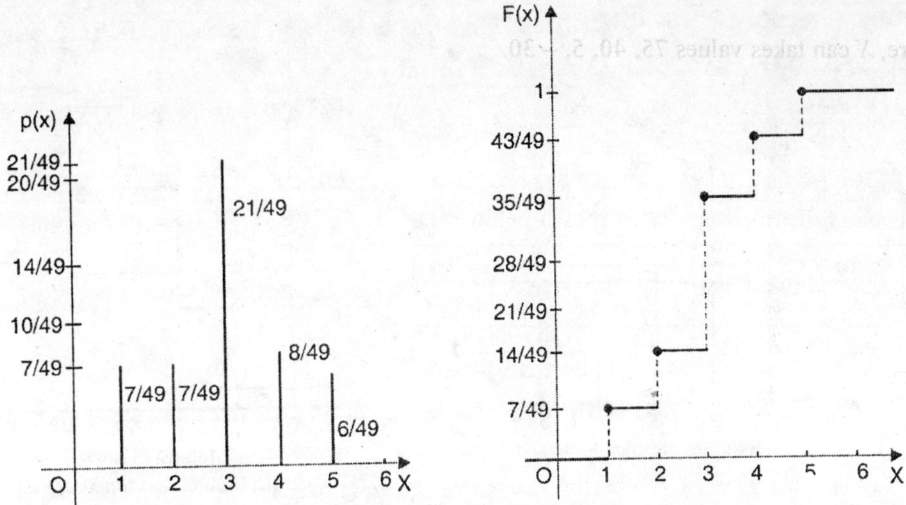

Fig. 7.13: Shows the probability function $p(x)$ and distribution function $F(x)$ of the discrete random variable X having the values x given in the table.

EXAMPLE 7.59 Find the probability density function of boys and girls in families with 4 children, assuming equal probability for boys and girls. Represent graphically the probability distribution. Find the distribution function of the random variable X of the number of boys.

Solution We have

$$P(0 \text{ boys and } 4 \text{ girls}) = \left(\frac{1}{2}\right)^4 = \frac{1}{16};$$

$$P(1 \text{ boy and } 3 \text{ girls}) = {}^4C_1\left(\frac{1}{2}\right)\left(\frac{1}{2}\right)^3 = 4\left(\frac{1}{2}\right)^4 = \frac{4}{16};$$

$$P(2 \text{ boys and } 2 \text{ girls}) = {}^4C_2\left(\frac{1}{2}\right)^2\left(\frac{1}{2}\right)^2 = 6\left(\frac{1}{2}\right)^4 = \frac{6}{16};$$

$$P(3 \text{ boys and } 1 \text{ girl}) = {}^4C_3\left(\frac{1}{2}\right)^3\left(\frac{1}{2}\right) = 4\left(\frac{1}{2}\right)^4 = \frac{4}{16};$$

$$P(4 \text{ boys and } 0 \text{ girl}) = \left(\frac{1}{2}\right)^4 = \frac{1}{16}.$$

The probability distribution is as indicated in the Table 7.3.

Table 7.3

Number of boys x :	0	1	2	3	4
Probability $P(x)$:	1/16	4/16	6/16	4/16	1/16

The graph is represented in Fig. 7.14(a).
The distribution function $F(x)$ is given by Table 7.4 and graph is represented in Fig. 7.14(b).

Table 7.4

x:	0	1	2	3	4
$F(x)$:	1/16	5/16	11/16	15/16	1

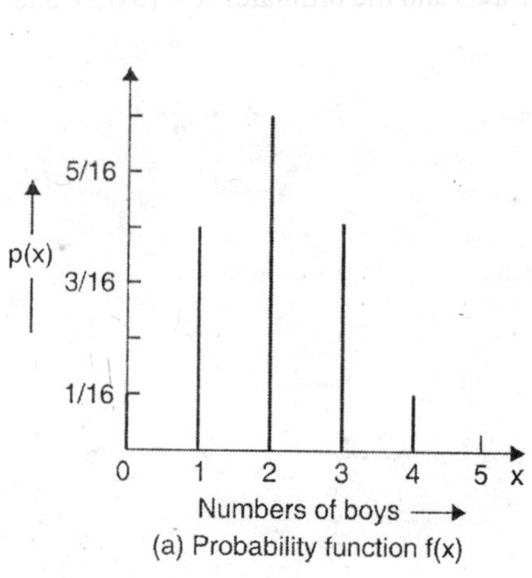

(a) Probability function f(x)

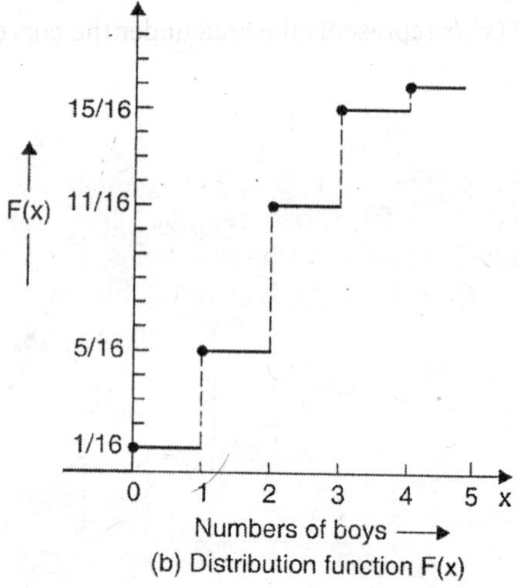

(b) Distribution function F(x)

Fig. 7.14

EXAMPLE 7.60 If X be a random variate with the following probability distribution.

X :	8	12	16	20	24
$p(x)$:	1/8	1/6	3/8	1/4	1/12

Find (*i*) $E(X)$, (*ii*) $E(X^2)$, (*iii*) $E(2X+1)^2$, (*iv*) $E[(X-\bar{X})^2]$, and (*v*) Var $(3X-4)$.

Solution (*i*) $E(X) = \Sigma X \cdot p(x) = (8)(1/8) + (12)(1/16) + (16)(3/8) + (20)(1/14) + (24)(1/12)$

$= 1 + 2 + 6 + 5 + 2 = 16$. This represents the *mean* of the distribution.

(*ii*) $E(X^2) = \Sigma X^2 \cdot p(x) = (8)^2 (1/8) + (12)^2 (1/16) + (16)^2 (3/8) + (20)^2 (1/14) + (24)^2 (1/12)$

$= 8 + 24 + 96 + 100 + 48 = 276.$

(*iii*) $E(2X+1)^2 = E(4X^2 + 4X + 1) = 4E(X^2) + 4E(X) + 1 = 4(276) + 4(16) + 1 = 1169.$

(*iv*) $E[(X-\bar{X})^2] = \Sigma(X-\bar{X})^2 \cdot p(X)$

$= (8-16)^2 (1/8) + (12-16)^2 (1/6) + (16-16)^2 (3/8) + (20-16)^2 (1/4) + (24-16)^2 (1/12)$

$= 8 + 8/3 + 4 + 16/3 = 20$. This represents the *variance* of the distribution.

(*v*) var $(3X-4) = 3^2$ Var$(X) = 9(20) = 180.$

Continuous Random Variables and Distributions

A random variable X is said to be a continuous random variable if it takes all possible values in a given interval I_x. The distribution defined by the variates like ages, heights, weights and temperature are examples of continuous distributions.

Unlike the discrete probability distribution in which a variate can be associated with a definite probability; it is not possible in the case of continuous distribution to assign a set probability to each variate. In such cases it is more useful and practicable to assign probabilities with intervals. Thus if $f(x)$ be a continuous function of the random variable X, the probability that the variable X falls in the infinitesimal small interval $x - \frac{dx}{2}$ to $x + \frac{dx}{2}$ can be expressed as $f(x)\,dx$.

Symbolically it can be expressed as,

$$P\left(x - \frac{dx}{2} \le X \le x + \frac{dx}{2}\right) = f(x)\,dx. \qquad ...(7.14)$$

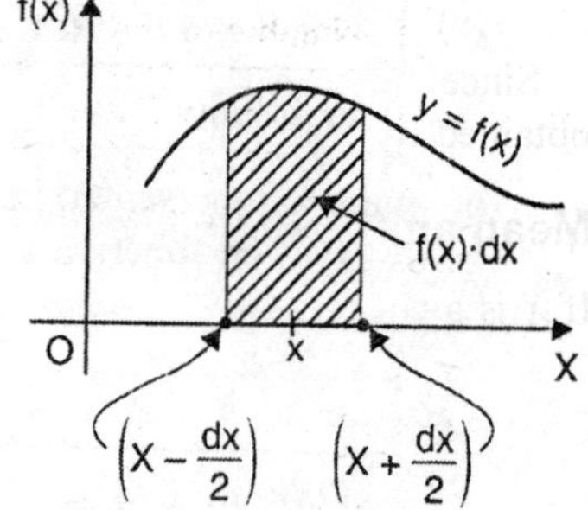

Fig. 7.15: Probability Curve.

Now, $f(x)dx$ represents the area under the curve $y = f(x)$, x-axis and the ordinates $x - (dx/2)$ and $x + (dx/2)$ (Refer Fig. 7.15)

The function $f(x)$ is called the *probability density function (pdf)* of X or simply the *density function.* The expression $f(x)\,dx$, usually written as $dF(x)$ is known as the *probablity differential.* The continuous curve represented by $y = f(x)$ is called the *probability density curve* or briefly the *probability curve.*

The interval of the variate may be finite or infinite but even when the interval is finite it is advisable to consider this as infinite by simple assumption that the density function is zero outside the specified range. Thus, if $f(x) = \phi(x)$ is the density function defined for the variate values of X in the interval $a \le x \le b$, it can be put in the form $f(x) = \begin{cases} 0 & \text{if } -\infty \le x < a \quad \text{or} \quad x < a, \\ \phi(x) & \text{if } a \le x \le b, \\ 0 & \text{if } b < x \le \infty \quad \text{or} \quad x > b. \end{cases}$

The Probability density function $f(x)$ possesses the following properties.

(*i*) $f(x) \ge 0$ for all $x \in I_x$, (*ii*) $\int_{I_x} f(x)\,dx = 1.$

The property (*i*) expresses the fact that probabilities can never be negative. Property (*ii*) stipulates the requirement that the probability of an event that is certain to happen must always be unity.

For a probability density function $f(x)$ the probability that the variate X lies in the interval (a, b) is given by

$$P(a \le X \le b) = \int_a^b f(x)\,dx. \qquad ...(7.15)$$

The probability defined by the above integral geometrically denotes the area bounded by the part of the curve $y = f(x)$, the two ordinates $x = a$ and $x = b$ and the x-axis while $\int_{-\infty}^{\infty} f(x)\,dx = 1 = P(-\infty < X < \infty)$ signifies that the total area under the curve is unity.

Cumulative Distribution Function (cdf)

If X is a continuous random variable, then $F(x) = P(X \le x)$ is called the *cdf* as in the case of a discrete random variable, that is,

$$F(x) = P(-\infty < X \le x) = \int_{-\infty}^{x} f(x)\,dx. \qquad ...(7.16)$$

The probability distribution of a continuous random variable is known if either its *pdf* or *cdf* is given.

The cumulative distribution function $F(x)$ has the following important properties

(*i*) $0 \le F(x) \le 1$, $-\infty < x < \infty$.

(*ii*) $F(x)$ is a non-decreasing function, that is, if $x_1 < x_2$, then $F(x_1) \le F(x_2)$.

(*iii*) $F(-\infty) = 0$ and $F(\infty) = 1$.

(*iv*) $F(x)$ is continuous function of x on the right side.

(*v*) $f(x) = F'(x)$ at all point where $F(x)$ is differentiable.

Since $f(x)$ is always positive, property (*i*) gives that $F(x)$ is a non-decreasing function. Also $f(x)$ can be obtained by differentiating $F(x)$.

Mean and Variance

If X is a continuous random variable and $f(x)$ is the *pdf* of X, then we define

$$\text{Mean} = \mu_x = \int_R x\,f(x)\,dx,$$

$$\text{Variance} = \sigma_x^2 = \int_R (x - \mu_x)^2\,f(x)\,dx.$$

Since $\int_R (x-\mu_x)^2 f(x)\,dx = \int_R (x^2 - 2\mu_x x + \mu_x^2) f(x)\,dx$

$$= \int_R x^2 f(x)\,dx - 2\mu_x \int_R x f(x)\,dx + \mu_x^2 \int_R f(x)\,dx$$

$$= \int_R x^2 f(x)\,dx - 2\mu_x\mu_x + \mu_x^2 \cdot 1 = \int_R x^2 f(x)\,dx - \mu_x^2.$$

The mean and variance of the distribution defined on $R = [a, b]$, are defined as

$$\text{Mean} = \mu = \int_a^b x f(x)\,dx. \qquad ...(7.17)$$

$$\text{Variance} = \sigma^2 = \int_a^b x^2 f(x)\,dx - \mu^2 \qquad ....(7.18)$$

$$\Rightarrow \quad \sigma^2 + \mu^2 = \int_a^b x^2 f(x)\,dx$$

EXAMPLE 7.61 Find the constant d so that the function $f(x)$ defined as below may be a density function:

$$f(x) = \begin{cases} 1/d & \text{if} \quad \alpha \le x \le \beta \\ 0 & \text{elsewhere.} \end{cases}$$

Find also the cumulative distribution of the random variable X when d satisfies the requirements for $f(x)$ to be a density function.

Solution If $d > 0$, $f(x) \ge 0$ for every x. Hence for $f(x)$ to be a density function, we have

$$\int_\alpha^\beta f(x)\,dx = 1 \text{ or } \int_\alpha^b \frac{1}{d}\,dx = 1 \text{ or } \frac{1}{d}(\beta - \alpha) = 1 \text{ which gives } d = \beta - \alpha.$$

For the distribution function, we have

$$F(x) = P(X \le x) = P(-\infty < X \le x) = \int_{-\infty}^x f(x)\,dx.$$

If $x < \alpha$, then $F(x) = 0$. If $\alpha \le x < \beta$, then $F(x) = \int_\alpha^x \frac{1}{d}\,dx = \frac{1}{\beta - \alpha}(x - \alpha)$.

If $x \ge \beta$, then $F(x) = \int_\alpha^\beta f(x)\,dx + \int_\beta^x f(x)\,dx = \frac{1}{\beta - \alpha}\int_\alpha^\beta dx + \int_\beta^x 0 \cdot dx = 1..$

Thus the required distribution function is $F(x) = \begin{cases} 0, & x < \alpha \\ \dfrac{x - \alpha}{\beta - \alpha}, & \alpha \le x < \beta. \\ 1, & x \ge \beta \end{cases}$

EXAMPLE 7.62 Let X be a continuous random variable with distribution

$$f(x) = \begin{cases} \frac{1}{6}x + k, & \text{if} \quad 0 \le x \le 3, \\ 0 & \text{elsewhere} \end{cases}$$

Evaluate (*i*) the value of k. (*ii*) $P\{1 \le X \le 2\}$.

Solution (*i*) The graph of f is shown in Fig. 7.16. Since f is a continuous probability function, the shaded region A must have area equal to 1. Trapezoidal area A (having parallel sides k, $k + (1/2)$ and perpendicular distance between parallel sides equal to 3) = $\frac{1}{2}\left(k + k + \frac{1}{2}\right)\cdot 3 = 1$ or $k + \frac{1}{4} = \frac{1}{3}$ or $k = \frac{1}{12}$.

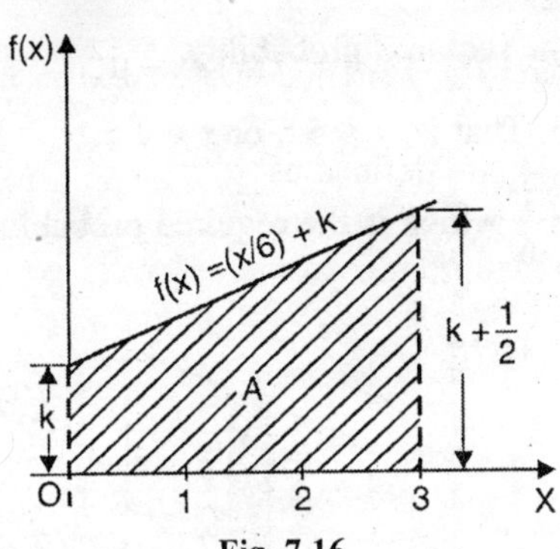

Fig. 7.16

Fig. 7.17

(*ii*) $P\{1 \le X \le 2\}$ is equal to the area of portion B which is under the graph of f and between ordinates $x = 1$ and $x = 2$ as shown in Fig. 7.17. We have

$f(x) = \frac{1}{6}x + \frac{1}{12}$. Therefore, $f(1) = \frac{1}{6} + \frac{1}{12} = \frac{3}{12}$ and $f(2) = \frac{2}{6} + \frac{1}{12} = \frac{5}{12}$.

Trapezoidal area $B = \frac{1}{2}\left(\frac{3}{12} + \frac{5}{12}\right)\cdot 1 = \frac{1}{3}$, which is the required probability.

EXAMPLE 7.63 A continuous random variable X which can assume only values between $x = 2$ and 8 inclusive, has a density function given by $k(x + 3)$ where 'k' is a constant. Calculate (*i*) the value of k (*ii*) $P\{3 < X < 5\}$, (*iii*) $P\{X \ge 4\}$, (*iv*) $P\{|X - 5| < 0.5\}$.

Solution (*i*) The graph of $f(x) = k(x + 3)$ is a straight line shown in Fig. 7.18. To find k, we proceed as under: The total area under the line between $x = 2$ and $x = 8$ and above the x-axis must be 1.
At $x = 2$, $f(2) = 5k$ and at $x = 8$, $f(8) = 11k$.

Trapezoidal area = $\frac{1}{2}$ (sum of parallel sides) × perpendicular distance between them

$= \{(5k + 11k)\cdot(6)/2\} = 48k = 1$ or $k = 1/48$.

The correct graphical representation is given in Fig. 7.19.

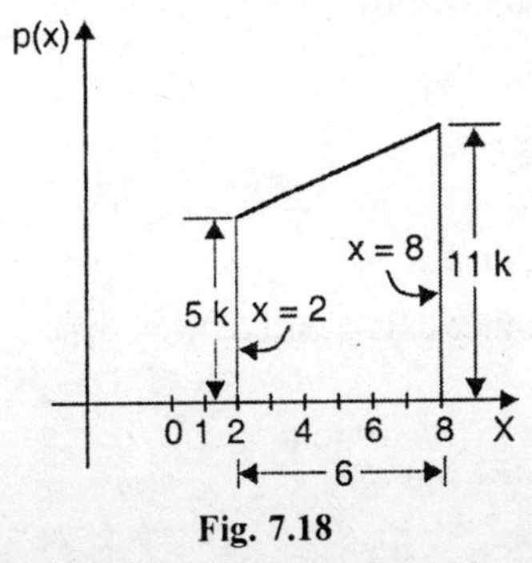

Fig. 7.18

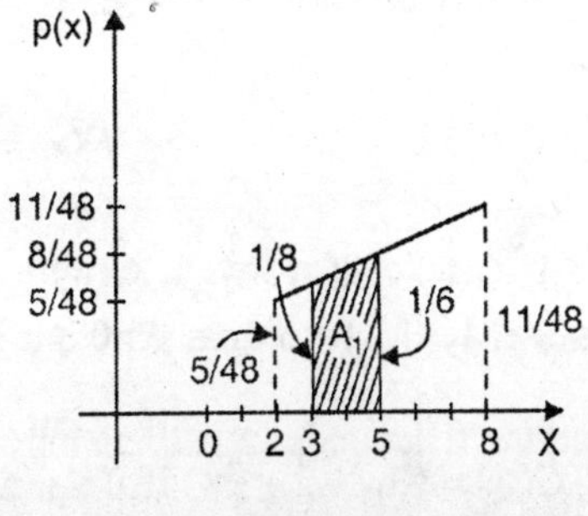

Fig. 7.19

(*ii*) The required probability is the area between $x = 3$ and $x = 5$ shown shaded in Fig. 7.19.

$p(3) = \frac{6}{48} = \frac{1}{8}$ and $p(5) = \frac{8}{48} = \frac{1}{6}$ are the ordinates at $x = 3$ and $x = 5$ respectively.

Shaded trapezoidal area $A_1 = \frac{1}{2}\left(\frac{1}{8}+\frac{1}{6}\right)\cdot 2 = \frac{7}{24}$, which is the required probability.

(*iii*) Refer Fig. 7.20. Now $p(4) = 7/48$, $p(8) = 11/48$.

$\therefore$ Trapezoidal area $B = \frac{1}{2}\left(\frac{7}{48}+\frac{11}{48}\right)\cdot 4 = \frac{3}{4}$, which is the required probability.

(*iv*) $|x-5| < 0.5$ may be written as $x - 5 < 0.5$ or $-x + 5 < 0.5$. That is, $x < 5.5$ or $x > 4.5$.

Refer Fig. 7.21. Trapezoidal area, $C = \frac{1}{2}\left(\frac{7.5}{48}+\frac{8.5}{48}\right)(1) = \frac{1}{6}$ which is the required probability.

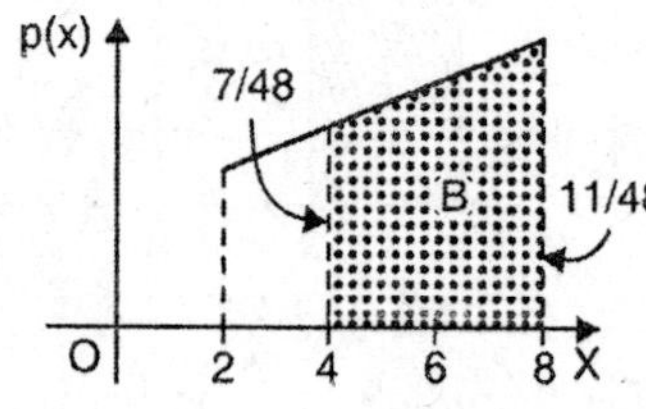

Fig. 7.20

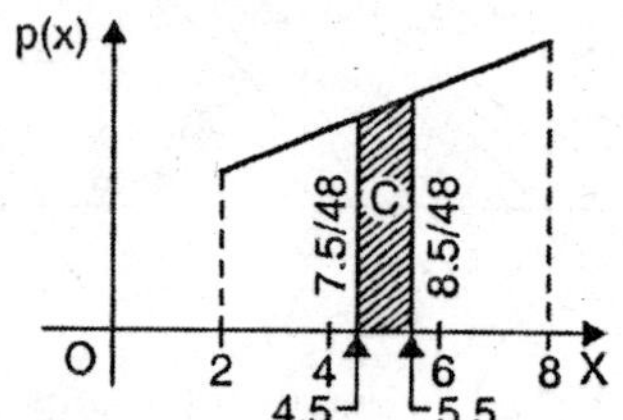

Fig. 7.21

EXAMPLE 7.64 The density function, f, is defined by

$$f(x) = \begin{cases} 0 & \text{if } x < 0 \\ kx(1-x) & \text{if } 0 \le x \le 1. \\ 0 & \text{if } x > 1 \end{cases}$$

Find (*i*) the value of k; (*ii*) the distribution function F. (*iii*) the density and distribution functions, g and G, of the random variable W where $w = x^2$. Sketch the graphs of f and F.

Solution (*i*) Since $\int_{-\infty}^{\infty} f(x)\,dx = 1$, we have $\int_0^1 kx(1-x)\,dx = 1$ or $k\left[\frac{x^2}{2} - \frac{x^3}{3}\right]_0^1 = 1$ or $k = 6$.

(*ii*) If $x < 0$, $F(x) = 0$. If $x > 1$, $F(x) = 1$.

If $0 \le x \le 1$, $F(x) = \int_0^x 6x(1-x)\,dx = [3x^2 - 2x^3]_0^x = x^2(3-2x)$.

Thus F is the function F: Range, $R \to \{y \in R: 0 \le y \le 1\}$ defined by

$$F(x) = \begin{cases} 0, & \text{if } x < 0, \\ x^2(3-2x), & \text{if } 0 \le x \le 1, \\ 1, & \text{if } x > 1. \end{cases}$$

(*iii*) If $w < 0$, $G(w) = 0$; if $w > 1$, $G(w) = 1$.

If $0 \le w \le 1$, $G(w) = P\{z \in R: 0 \le z \le w\} = w(3 - 2\sqrt{w})$ (Since x is positive, $x = \sqrt{w}$).

Thus $G(w) = \begin{cases} 0 & \text{if } w < 0 \\ 3w - 2w^{3/2} & \text{if } 0 \le w \le 1 \\ 1 & \text{if } w > 1. \end{cases}$

Also, $g = dG/dw$, giving

$$g(w) = \begin{cases} 0 & \text{if } w < 0 \\ 3(1-\sqrt{w}) & \text{if } 0 \le w \le 1 \\ 0 & \text{if } w > 1. \end{cases}$$

The graph of f, F are shown in Figs. 7.22 and 7.23.

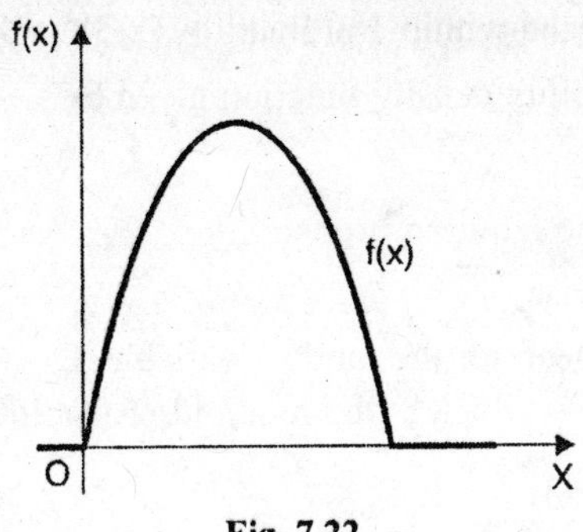

Fig. 7.22

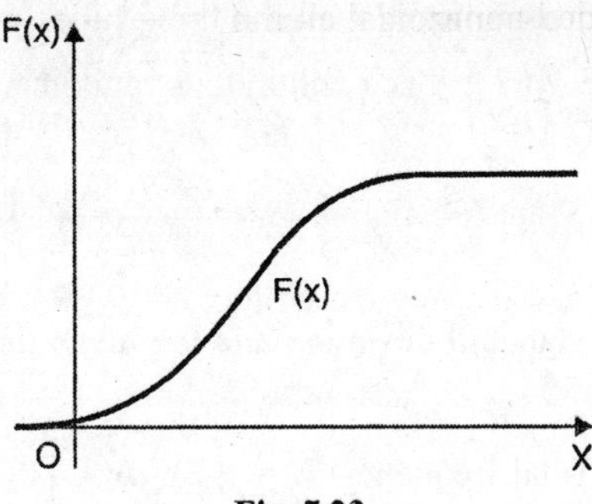

Fig. 7.23

EXAMPLE 7.65 Find the mean and variance of the random variable X whose density function, f, is defined by

$$f(x)=\begin{cases}0 & \text{if } x<0\\ 4x(1-x^2) & \text{if } 0\le x\le 1\\ 0 & \text{if } x>1.\end{cases}$$

Solution We have mean $= \mu_x = E(x) = \int_{I_x} x\, f(x)\, dx = \int_0^1 4x^2(1-x^2)\, dx = \left[\frac{4x^3}{3}-\frac{4x^5}{5}\right]_0^1 = \frac{8}{15}.$

$$\text{Variance} = \sigma_x^2 = \int_{I_x}(x-\mu_x)^2\, f(x)\, dx = \int_{I_x} x^2 f(x)\, dx - (\mu_x)^2 = \int_0^1 x^2[4x(1-x^2)]\, dx - \left(\frac{8}{15}\right)^2$$

$$= \int_0^1 4x^3(1-x^2)\, dx - \frac{64}{225} = \left[x^4 - \frac{2}{3}x^6\right]_0^1 - \frac{64}{225} = \frac{1}{3}-\frac{64}{225} = \frac{11}{225}.$$

EXAMPLE 7.66 Suppose the life in hours of a certain kind of radio tube has the probability density function:

$$f(x)=\begin{cases}100/x^2, & \text{for } x\ge 100\\ 0, & \text{for } x<100.\end{cases}$$

Find the distribution function. What is the probability that none of the 3 tubes in a given radio set will have to be replaced during the first 150 hours of operation? What is the probability that all three of the original tubes will be replaced during the first 150 hours? *(AMIETE, Dec. 2007)*

Solution Let $F(x)$ be the distribution function, then

$$F(x) = \int_{-\infty}^{x} f(x)\, dx = \int_{-\infty}^{100} f(x)\, dx + \int_{100}^{x} f(x)\, dx$$

$$= 0 + \int_{100}^{x} \frac{100}{x^2}\, dx = 100\left[-\frac{1}{x}\right]_{100}^{x} = 1-\frac{100}{x}.$$

∴ Probability that a tube will last for 150 hours,

$$F(150) = P(X \le 150) = 1-\left(\frac{100}{150}\right) = \frac{1}{3}.$$

Hence the probability that all the three tubes will last for 150 hours each $= (1/3)^3 = 1/27$.

Since the probability that a tube will have to be replaced within 150 hours of its life $= 1-(1/3) = 2/3$. Hence the probability that all the three tubes will have to be replaced within 150 hours $= (2/3)^3 = 8/27$.

EXAMPLE 7.67 X is a continuous random variable with probability density function given by

$$f(x) = \begin{cases} 2x^3, & 0 \le x \le 1 \\ 2(2-x)^3, & 1 \le x \le 2 \\ 0, & \text{otherwise.} \end{cases}$$

Find the standard deviation and the mean deviation about the mean for the random variable X.

(AMIETE, Dec. 2008)

Solution Total frequency, $N = \int_0^1 2x^3 dx + \int_1^2 2(2-x)^3\, dx = \frac{1}{2} + \frac{1}{2} = 1.$

We have mean, μ_1' (about the origin) $= \frac{1}{N}\left[\int_0^1 x(2x^3)\, dx + \int_1^2 2x(2-x)^3\, dx\right]$

$$= \left[\frac{2x^5}{5}\right]_0^1 + \left[\frac{-2x^5}{5} + 3x^4 - 8x^3 + 8x^2\right]_1^2 = 1.$$

$$\mu_2' \text{ (about the origin)} = \frac{1}{N}\left[\int_0^1 x^2(2x^3) + \int_1^2 2x^2(2-x)^3\, dx\right] = \left[\frac{x^6}{3}\right]_0^1 + \left[\frac{-x^6}{3} + 12\frac{x^5}{5} - 6x^4 + 16\frac{x^3}{3}\right]_1^2 = \frac{16}{15}.$$

Therefore, variance $= \mu_2' - \mu_1'^2 = \frac{16}{15} - 1 = \frac{1}{15} \Rightarrow \text{S.D.} = \frac{1}{\sqrt{15}}.$

Mean Deviation about the mean $= \frac{1}{N}\left[\int_0^1 |x-1|\, 2x^3 dx + \int_1^2 |x-1|\, \{2(2-x)^3\}\, dx\right]$

$$= 2\int_0^1 (1-x)x^3 dx + 2\int_1^2 (x-1)(2-x)^3 dx$$

$$= 2\int_0^1 (x^3 - x^4) dx + 2\int_1^2 x(2-x)^3 dx - 2\int_1^2 (2-x)^3\, dx$$

$$= 2\left[\frac{x^4}{4} - \frac{x^5}{5}\right]_0^1 + 2\left[\frac{-x^5}{5} + \frac{3}{2}x^4 - 4x^3 + 4x^2\right]_1^2 - 2\left[\frac{-x^4}{4} + 2x^3 - 6x^2 + 8x\right]_1^2$$

$$= \frac{1}{10} + 2\left[\frac{-x^5}{5} + \frac{7x^4}{4} - 6x^3 + 10x^2 - 8x\right]_1^2 = \frac{1}{10} + \frac{1}{10} = \frac{1}{5}.$$

Hence mean deviation about the mean standard deviation are $1/5$ and $1/\sqrt{5}$ respectively.

EXERCISE 7.3

1. 3 rotten eggs are accidentally mixed with 7 good ones. Find the probability distribution of the number of rotten eggs, if 3 eggs are drawn at random.

Ans.

X :	0	1	2	3
$p(X)$:	35/120	63/120	21/120	1/120

2. A random variable X has the following probability function:

Ans.

X :	0	1	2	3	4	5	6	7
$p(X)$:	0	k	$2k$	$2k$	$3k$	k^2	$2k^2$	$7k^2 + k$

(*a*) Find k, (*b*) Evaluate $P(X < 6)$, $P(X \geq 6)$, $P(3 < X \leq 6)$ and $P(0 < X < 5)$
(*c*) Find the minimum value of α so that $P(X \leq \alpha) > 1/2$. (*WBTU 2005, JNTU, 2003*)

Ans. $1/10, 81/100, 19/100, 33/100, 4/5; 4.$

3. Plot the graph of the commulative distribution function F of the random variable X with distribution.

x_i :	−3	2	6
$p(x_i)$:	1/4	1/2	1/4

Ans.

x_i :	−3	2	6
$p(x)$:	1/4	1/2	1/4
$F(x)$:	1/4	3/4	$4/4 = 1$

(See Fig. 7.24)

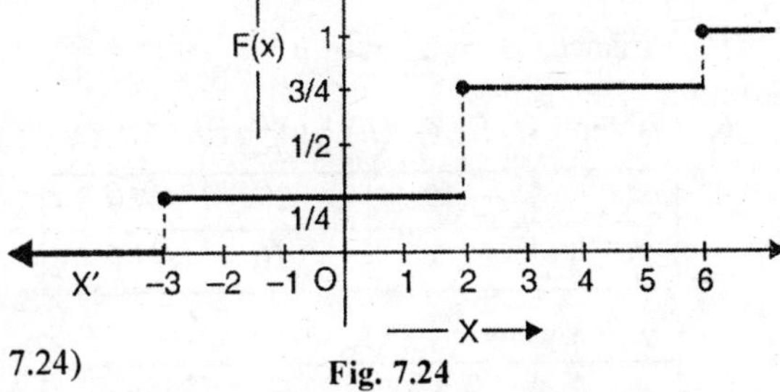

Fig. 7.24

4. Plot the graph of the cumulative distribution function F of the random variable X with the following distribution

x_i:	−2	1	2	4
$p(x_i)$:	1/4	1/8	1/2	1/8

Ans. The graph of the cumulative distribution of function F of X follows:

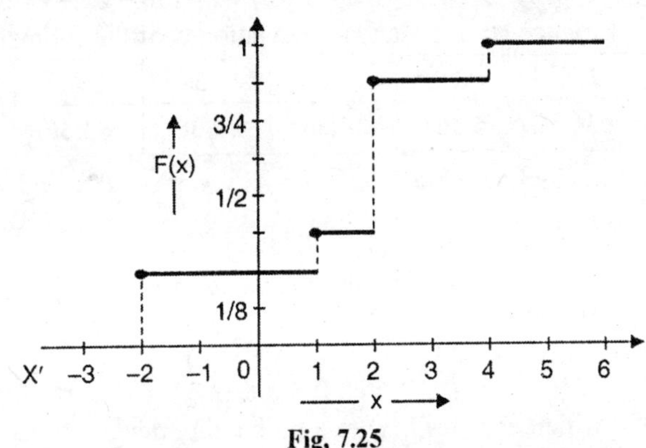

Fig, 7.25

5. A die is tossed thrice. A success is 'getting 1 or 6' on a toss. Find the mean and variance of the number of success. (*AMIETE, Dec. 2010*)

[**Hint:** Let X denote the number of success. Clearly X can take the values 0, 1, 2, or 3. Probability of a success $= 2/6 = 1/3$. Probability of failure $= 2/3$.

$P(X = 0)$ = Probability of no success, i.e. all 3 failure $= \frac{2}{3} \times \frac{2}{3} \times \frac{2}{3} = \frac{8}{27}$.

$P(X = 1)$ = Probability of one success and 2 failure $= {}^3C_1 \times \frac{1}{3} \times \frac{2}{3} \times \frac{2}{3} = \frac{4}{9}$.

$P(X = 2)$ = Probability of two success and one failure $= {}^3C_2 \times \frac{1}{3} \times \frac{1}{3} \times \frac{2}{3} = \frac{2}{9}$.

$P(X = 3)$ = Probability of three successes $= \frac{1}{3} \times \frac{1}{3} \times \frac{1}{3} = \frac{1}{27}$.

The probability distribution of the random variable X is

x_i :	0	1	2	3
p_i :	8/27	4/9	2/9	1/27

Mean $= \mu = \Sigma p_i x_i = \frac{4}{9} + \frac{4}{9} + \frac{1}{9} = 1$. Also $\Sigma p_i x_i^2 = \frac{4}{9} + \frac{8}{9} + \frac{9}{27} = \frac{5}{3}$.

Variance, $\sigma^2 = \Sigma p_i x_i^2 - \mu^2 = \frac{5}{3} - 1^2 = \frac{2}{3}$].

6. (*a*) Find (*i*) $E(X)$, (*ii*) $E(X^2)$, (*iii*) $E[(X - \bar{X})^2]$, and (*iv*) $E(X^3)$ for the following probability distribution.

X :	–10	–20	30
$p(X)$:	1/5	3/10	1/2

Ans. (*i*) 7, (*ii*) 590, (*iii*) 541, (*iv*) 10900.

(*b*) A variate X has p.d.f.

x :	–3	6	9
$p(x)$:	1/6	1/2	1/3

Find $E(X)$, $E(X^2)$ and $E(2X + 1)^2$. (*AMIETE, Dec. 2007*) **Ans.** 11/2; 93/2; 209.

[**Hint:** $E(2X + 1)^2 = E(4X^2 + 4X + 1) = 4E(X^2) + 4E(X) + 1$

$= 4[(-3)^2(1/6) + (6)^2(1/2) + (9)^2(1/3)] + 4[(-3)(1/6) + (6)(1/2) + (9)(1/3)] + 1$

$= 4[3/2 + 18 + 27] + 4(-1/2 + 3 + 3) + 1 = 186 + 22 + 1 = 209$].

7. Find the expectation μ, variance σ^2 and standard deviation σ of the following distribution:

X :	1	2	3	4	5	6
$p(X)$:	1/36	3/36	5/36	7/36	9/36	11/36

Ans. 4.47, 1.99, 1.4

8. The expected value of a random variable X is 2 and its variance is 1, find var$(3X + 4)$.

(*AMIETE, Dec. 2004*) **Ans.** 9.

9. (*a*) If $p(x) = \begin{cases} kx^2, & -3 \le x \le 3 \\ 0, & \text{elsewhere} \end{cases}$ is a pdf. Find the value of k. (*AMIETE, June 2010*) **Ans.** 1/8.

(*b*) A function $f(x)$ is defined by

$$f(x) = ce^{-x}, \ 0 \le x \le \infty.$$

Find the value of c which changes $f(x)$ to a probability density function. **Ans.** $c = 1$.

10. A function $f(x)$ is defind as follows:

$$f(x) = \begin{cases} kx^{-2}, & \text{if } x \ge 1 \\ 0, & \text{if } x < 1. \end{cases}$$

Find the value of k and hence find $P(x < 2)$. **Ans.** 1, 1/2.

11. $f(x) = ke^{-2x}$ if $x \ge 0$, $f(x) = 0$ if $x < 0$; find k and hence find $P(1 < x < 2)$. **Ans.** 2, 117/1000.

12. A function $f(x)$ is defined as follows:

$$f(x) = \begin{cases} 0, & \text{if } x < 2 \\ \frac{1}{18}(2x + 3) & \text{if } 2 \le x \le 4 \\ 0 & \text{if } x > 4. \end{cases}$$

Show that it is a density function. Find the probability that a variate having this density will fall in the interval $2 \le x \le 3$.

Ans. 4/9.

13. A continuous random variable X having values only between 0 and 4 has a density function given by $p(X) = (1/2) - kX$, where k is a constant. Calculate: (*a*) the value of k (*b*) $p\{1 < x < 2\}$. **Ans.** 1/8; 5/16.

14. State the condition under which a function is said to be a probability density function. Examine whether the function f defined by

$$f(x) = \begin{cases} \frac{3}{8}(4x - 2x^2), & 0 \le x \le 2 \\ 0, & \text{otherwise} \end{cases}$$

is a probability density function.

15. The probability distribution function of a continuous random variable is given by $f(x) = k(3x - x^2 - 2)$, $1 \le x \le 2$. Calculate

(*a*) the value of k, (*b*) $p(x > 3/2)$, (*c*) $p(5/4 \le x \le 3/2)$. **Ans.** (*a*) $k = 6$, (*b*) 1/2, (*c*) 1/4.

$$\left[\textbf{Hint: } (b)\ p(x > 3/2) = 1 - p(x \le 3/2) = 1 - \int_1^{3/2} 6(3x - x^2 - 2)\,dx\right].$$

16. Is the following a probability distribution function?

$$F(x) = \begin{cases} 2x, & \text{when } 0 \le x \le 1 \\ 3 - 2x, & \text{when } 1 < x \le 3 \\ 0, & \text{elsewhere}. \end{cases}$$

[**Hint:** A probability distribution function must satisfy the following conditions:

(*i*) $f(x) \le 0$ and (*ii*) $\int_{-\infty}^{\infty} f(x)\,dx = 1$]. **Ans.** No.

17. Let X be a random variable, a possible choice of $f(x)$ would be

$$f(x) = \begin{cases} 0 & \text{if } x < 0 \\ kx & \text{if } 0 \le x < 50 \\ k(100 - x) & \text{if } 50 \le x < 100 \\ 0 & \text{if } x \ge 100, \text{ where } k \text{ is some number.} \end{cases}$$

The graph of $y = f(x)$ is shown in Fig. 7.26. Calculate

(*i*) k, (*ii*) probability that P lies between 30 and 60 from A.

Ans. (*a*) 1/2500; (*b*) 0.5.

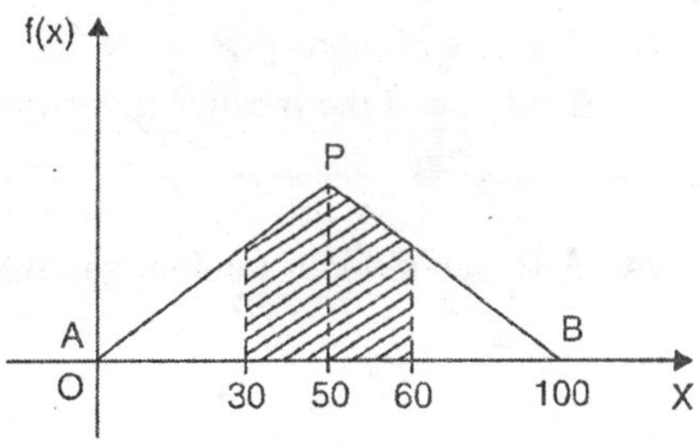

f(x)
P
A
B
O
30 50 60
100
X

Fig. 7.26

$$\left[\textbf{Hint: } (a) \int_0^{50} kx\,dx + \int_{50}^{100} k(100 - x)\,dx = 1;\quad (b)\ P = \int_{30}^{50} \frac{1}{2500} x\,dx + \int_{50}^{60} \frac{1}{2500}(100 - x)\,dx\right].$$

18. A continuous random variable X has a *pdf* $f(x) = 3x^2$, $0 \le x \le 1$. Find a and b when

(*i*) $p(X \le a) = p(X > a)$ and (*ii*) $p(X > b) = 0.05$. **Ans.** $(1/2)^{1/3}, (19/20)^{1/3}$.

[**Hint:** $p(X \le a) = 0.5 = p(X > a)$].

19. Verify that the following is a distribution function.

$$F(x) = \begin{cases} 0, & \text{for } x < 0 \\ x/3, & \text{for } 0 \le x < 1 \\ 1/3, & \text{for } 1 \le x < 2 \\ x/6, & \text{for } 2 \le x < 6 \\ 1, & \text{for } x \ge 6. \end{cases}$$

20. Define distribution function. Verify that the following is a distribution function.

$$F(x)=\begin{cases}0, & \text{when } x<-a\\ \frac{1}{2}\left(\frac{x}{a}+1\right), & \text{when } -a\le x\le a\\ 1, & \text{when } x>a.\end{cases}$$

21. If the density function, *f*, of a distribution is given by

$$f(x)=\begin{cases}0, & \text{if } x<0\\ \frac{1}{2}x, & \text{if } 0\le x\le 1\\ \frac{1}{2}, & \text{if } 1<x\le 2\\ \frac{1}{2}(3-x), & \text{if } 2<x\le 3\\ 0, & \text{if } x>3.\end{cases}$$

find the distribution function *F*. **Ans.** $F(x)=0,\, x^2/4,\, (2x-1)/4,\, (6x-2x^2-5)/4,\, 1.$

22. Let X be a continuous random variable with the following distribution:

$$f(x)=\begin{cases}x/2, & \text{if } 0\le x\le 2\\ 0, & \text{elsewhere}.\end{cases}$$

Calculate $p(1\le x\le 3/2)$; the expectation, variance and standard deviation of *X*.

Ans. $5/16; 4/3; 2/9; \sqrt{2}/3.$

23. For a continuous probability distribution $f(x)=y_0\,x\,(2-x),\ 0\le x<2.$
Find y_0 and the mean and the variance of the distribution.

(*AMIETE, Dec. 2009; Kerala 2005; JNTU 2003*)

Ans. $3/4, 1, 1/5.$

24. *X* is a continuous random variable with probability density function given by

$$f(x)=\begin{cases}kx, & (0\le x<2)\\ 2k, & (2\le x<4)\\ -kx+6k, & (4\le x\le 6).\end{cases}$$

Find *k* and mean value of *X*. (*AMIETE, June 2012, 2009, 2008; JNTU 2003*)

[**Hint:** Since the total probability is unity.

Therefore, $\int_0^6 f(x)\,dx=1$, that is, $\int_0^2 kx\,dx+\int_2^4 2k\,dx+\int_4^6 (6k-kx)\,dx=1\Rightarrow k=\frac{1}{8}.$

Mean, $\mu=\int_0^6 x\,f(x)\,dx=\int_0^2 kx^2 dx+\int_2^4 2kx\,dx+\int_4^6 x(6k-kx)\,dx=3.$]

25. If *X* be a continuous random variate with *pdf*

$$f(x)=\begin{cases}ax, & 0\le x<1\\ a, & 1\le x\le 2\\ -ax+3a, & 2\le x\le 3\\ 0, & \text{elsewhere}.\end{cases}$$

Find (*i*) the value of *a*, (*ii*) the *cdf* of *X*, (*iii*) $P(X\le 1.5)$.

[**Hint :** (*i*) $\int_{-\infty}^{+\infty} f(x)\,dx = 1$, that is $\int_{-\infty}^{0} f(x)\,dx + \int_{0}^{1} f(x)\,dx + \int_{1}^{2} f(x)\,dx + \int_{2}^{3} f(x)\,dx + \int_{3}^{\infty} f(x)\,dx = 1$, that is,

$\int_0^1 ax\,dx + \int_1^2 a\,dx + \int_2^3 (3-x)\,a\,dx = 1$. This gives $a = 1/2$.

(*ii*) $F(x) = P(-\infty < x \le x) = \int_{-\infty}^{x} f(x)\,dx$. We have $x < 0 : F(x) = 0$. $0 \le x < 1 : F(x) = \int_0^x \frac{1}{2} x\,dx = \frac{x^2}{4}$.

$$1 \le x \le 2 : F(x) = \int_0^1 \frac{x}{2}\,dx + \int_1^x \frac{1}{2}\,dx = \frac{1}{4} + \frac{1}{2}(x-1) = \frac{1}{2}x - \frac{1}{4}.$$

$$2 \le x \le 3 : F(x) = \int_0^1 \frac{x}{2}\,dx + \int_1^2 \frac{1}{2}\,dx + \int_2^x \frac{1}{2}(3-x)\,dx = \frac{1}{4} + \frac{1}{2} + \frac{3}{2}(x-2) - \frac{1}{4}(x^2-4) = \frac{3}{2}x - \frac{x^2}{4} - \frac{5}{4}.$$

$x > 4$: $F(x) = 1$.

(*iii*) Now $P(X \le 1.5) = \int_{-\infty}^{1.5} f(x)\,dx = \int_{-\infty}^{0} f(x)\,dx + \int_0^1 f(x)\,dx + \int_1^{1.5} f(x)\,dx$

$$= 0 + \int_0^1 \frac{1}{2}x\,dx + \int_1^{1.5} \frac{1}{2}\,dx = \frac{1}{4}(1-0) + \frac{1}{2}(1.5-1) = \frac{1}{2}\Bigg].$$

26. A continuous probability distribution has density function *f*, where $f(x) = a + bx + cx^2$ for $0 \le x \le 1$ and $f(x) = 0$ outside this range. The mean is $2/3$ and the variance is $4/45$. Find the values of *a*, *b* and *c*.

Ans. 1, –4, 6.

27. A random variable *X* is distributed at random between the values 0 and 1 so that its probability density function is: $f(x) = kx^2(1-x^3)$, where *k* is a constant. Find the value of *k*. Using this value of *k*, find its mean and standard deviation.

[**Hint:** Since $\int_{-\infty}^{\infty} f(x)dx = 1$, $k\int_0^1 (x^2 - x^5)\,dx = 1 \Rightarrow k = 6$.

Mean $= \mu_1' = \int_{-\infty}^{\infty} x\,f(x)\,dx = 6\int_0^1 (x^3 - x^6)\,dx = \frac{9}{14}$. $\mu_2' = \int_{-\infty}^{\infty} x^2 f(x)\,dx = 6\int_0^1 (x^4 - x^7)\,dx = \frac{9}{20}$.

Variance $= \mu_2 = \mu_2' - \mu_1'^2 = \left(\frac{9}{20}\right) - \left(\frac{9}{14}\right)^2 = \frac{9}{245}$. Therefore, S.D. $= \sqrt{\text{Variance}} = \frac{3}{7\sqrt{5}}$].

28. *X* is a continuous random variable with probability density function given by

$$f(x) = \begin{cases} x^3, & 0 \le x \le 1 \\ (2-x)^3 & 1 \le x \le 2. \end{cases}$$

Find the standard deviation and also the mean deviation about the mean.

(*AMIETE, June 2011, 2010, 2009, Dec 2009*) **Ans.** $\frac{1}{\sqrt{15}}, \frac{1}{5}$.

29. A continuous random variable *X* has a *pdf* $f(x) = kx^2 e^{-x}, x \ge 0$ find *k*. Also find mean and variance of this random variable. (*AMIETE, June 2009*) **Ans.** $k = 1/3$; mean = 3, variance = 3.

30. A random variable X have the density function $f(x) = \frac{3}{4}(1 - x^2)$ if $-1 \le x \le 1$ and zero otherwise. Find

(*a*) its distribution function, (*b*) probabilities $P\left(\frac{-1}{2} \le X \le \frac{1}{2}\right)$ and $P\left(\frac{1}{4} \le X \le 2\right)$.

[**Hint:** $F(x) = 0$ if $x \le -1$. $F(x) = 0.75 \int_{-1}^{x} (1 - v^2)\, dv = 0.5 + 0.75\, x - 0.25\, x^3$ if $-1 < x \le 1$,

and $F(x) = 1$ if $x > 1$. Also $P\left(\frac{-1}{2} \le x \le \frac{1}{2}\right) = F\left(\frac{1}{2}\right) - F\left(-\frac{1}{2}\right)$

$$= 0.75 \int_{-1/2}^{1/2} (1 - v^2)\, dv = 1.5 \int_{0}^{1/2} (1 - v^2)\, dv = \frac{3}{2}\left[v - \frac{v^3}{3}\right]_0^{1/2} = \frac{11}{16} = 0.6875 \text{ and}$$

$$P\left(\frac{1}{4} \le X \le 2\right) = F(2) - F\left(\frac{1}{4}\right) = 0.75 \int_{1/4}^{1} (1 - v^2)\, dv = \frac{3}{4}\left[v - \frac{v^3}{3}\right]_{1/4}^{1} = 0.3164.$$

(Note that the upper limit of integration is 1 not 2)].

7.8 THEORETICAL PROBABILITY DISTRIBUTIONS

The probability distributions which are not obtained by actual experiments or observations but are mathematically deduced on certain assumptions are called theoretical probability distributions. Here total probability of 1 is distributed over different values of the random variable X according to some definite law or pattern. We shall discuss the following discrete and contnuous probability distributions.

Discrete probability distributions: (*i*) Binomial distribution, (*ii*) Poisson distribution.

Continuous probability distributions: (*i*) Normal distribution, (*ii*) Uniform (or rectangular) distribution, (*iii*) Exponential distribution and, (*iv*) Beta distribution.

7.8.1 *Binomial Distribution

The Binomial distribution is a discrete probability distribution. It is derived from a random experiment known as Bernoulli process. Let an event or a trial associated with the experiment. Noting that a trial is an atempt to produce a particular outcome which is neither certain nor impossible. The characteristics of Bernoulli trial are stated below:

(*i*) Each trial has exactly two outcomes called success (occurrence of the event) and failure (non-occurrence of the event).

(*ii*) The trials are independent, that is, the outcomes of the one trial does not affect the probability of an outcome of the other trial.

(*iii*) The probability of a success usually denoted by p and that of a failure denoted by $q = 1 - p$, remain unchanged throughout the process.

(*iv*) The number of trials is a finite.

Let the probability of the success of an event in a single trial be p and probability of failure be q $(= 1 - p)$. It is assumed that p is the same from trial to trial.

We first find the probability that the event succeeds twice in three trials. The different possible ways of 2 successes in 3 trials are ${}^3C_2$, *i.e.*, 3 may be shown as follows:

SSF; SFS; FSS, where S stands for success and F for failure.

Now the probability that it succeeds in first two trials and fails in third, (*i.e.*, SSF) is $pp\,q = p^2q$.

---

* Binomial distribution is also known as *Bernoulli distribution* after James Bernoulli who discovered it at the end of 17th century.

Similarly the probability of SFS and FSS is also p^2q.

($\because$ in each case, number of p's is equal to number of times S appears and number of q's is equal to number of F's).

Since ${}^3C_2$ ways are mutually exclusive, the probability of 2 successes in 3 trials is by addition law

$$^3C_2p^2q = {}^3C_2p^2q^{3-2}.$$

In general the probability of r successes in n trial with probability of success p is

$$f(r) = P(X = r) = P(r) = {}^nC_rp^rq^{n-r}, \ r = 0, 1, 2, ..., n. \quad ...(7.19)$$

where $q = 1 - p$. X is called a binomial random variable with parameters n and p. Denote the binomial distribution by $X \sim B(n, p)$.

For $r = 0$, the probability $= {}^nC_0p^0q^n = q^n$ which is the probability that the event fails in all trials.

The probability distribution (7.19) is called *Binomial distribution* because for $r = 0, 1, 2..., n$ it gives probabilities of 0, 1, 2, ..., n successes which correspond to successive terms in the binomial expansion

$$(q + p)^n = q^n + {}^nC_1pq^{n-1} + {}^nC_2p^2q^{n-2} + ... + p^n.$$

Recurrence Formula for the Binomial Distribution

We have $P(r) = {}^nC_rp^rq^{n-r}$ and $P(r+1) = {}^nC_{r+1}p^{r+1}q^{n-r-1}$.

Therefore $$\frac{P(r+1)}{P(r)} = \frac{{}^nC_{r+1}p^{r+1}q^{n-r-1}}{{}^nC_rp^rq^{n-r}} = \frac{n!\,p^{r+1}q^{n-r-1}}{(r+1)!(n-r-1)!}\cdot\frac{r!(n-r)!}{n!\,p^rq^{n-r}} = \frac{n-r}{r+1}\cdot\frac{p}{q}.$$

Thus $$P(r+1) = \left(\frac{n-r}{r+1}\right)\cdot\frac{p}{q}P(r); \ r = 0, 1, 2,...., (n-1). \quad ...(7.20)$$

This is the required *recurrence* formula. Applying this formula successively, we can find the probabilities $P(1)$, $P(2)$, $P(3)$, etc. if $P(0)$ is known.

Thus, when $r = 0$, we have from (7.20), $P(1) = n(p/q)\,P(0)$.

But $P(0) = p^\circ q^n = q^n$. Therefore $P(1) = npq^{n-1}$.

Again, putting $r = 1$ in (7.20), we get $P(2) = \frac{n-1}{2}\frac{p}{q}P(1) = \frac{n-1}{2}\cdot\frac{p}{q}\cdot npq^{n-1} = \frac{n(n-1)}{2}p^2q^{n-2}$ and so on.

Mean (or Expectation) and Variance of Binomial Distribution

(AMIETE, June 2013, Dec. 2006; AMIE, W-2002)

Before we proceed to find the mean and S.D., we must bear in mind that in binomial distribution the variate values are 0, 1, 2,...., n and corresponding to frequencies, we have the probabilities of these values given by ${}^nC_rp^rq^{n-r}$. Also sum of all these probabilities $= (p + q)^n = 1^n = 1$.

Now, probabilities of getting 0, 1, 2, ..., n successes in n trials are ${}^nC_0p^0q^n, {}^nC_1p^1q^{n-1}, \cdots, {}^nC_np^nq^0$

or $$q^n, {}^nC_1pq^{n-1}, \cdots, p^n \quad (\because {}^nC_0 = {}^nC_n = 1, p^0 = q^0 = 1)$$

$\therefore$ The mean, $\mu = E(X) = \sum_{r=0}^{n} x_rp_r = \sum_{r=0}^{n} r\cdot P(r) = \sum_{r=0}^{n} (r)\,{}^nC_rp^rq^{n-r}$

$$= 0(q^n) + 1\cdot{}^nC_1pq^{n-1} + 2\cdot{}^nC_2p^2q^{n-2} + \cdots + np^nq^{n-n}$$

$$= np[q^{n-1} + {}^{n-1}C_1pq^{n-2} + \cdots + p^{n-1}] = np(q+p)^{n-1} = np. \quad ...(7.21)$$

$$\left(\text{as } p + q = 1, {}^nC_1 = n, \ 2\cdot{}^nC_2 = 2\cdot\frac{n(n-1)}{2} = n(n-1) = n\cdot{}^{n-1}C_1, \ 3\cdot{}^nC_3 = 3\cdot\frac{n(n-1)(n-2)}{3\cdot2\cdot1} = n\cdot{}^{n-1}C_2, \text{etc.}\right)$$

Thus, the mean of the binomial distribution is np.

The Variance, σ^2. For the binomial distribution, Var(X) is given by

$$\text{Var}(X) = E(X^2) - [E(X)]^2$$

$$E(X^2) = \sum_{r=0}^{n} x_r^2 P_r = \sum_{r=0}^{n} r^2 P(r).$$

Now $\sum_{r=0}^{n} r^2 P(r) = 0^2 \cdot q^n + 1^2 \cdot {}^nC_1 pq^{n-1} + 2^2 \cdot {}^nC_2 p^2 q^{n-2} + \cdots + n^2 p^n$

$$= npq^{n-1} + 2n(n-1)p^2 q^{n-2} + 3\frac{n(n-1)(n-2)}{2!} p^3 q^{n-3} + \cdots + n^2 p^n$$

$$= np\left[q^{n-1} + 2(n-1)q^{n-2}p + 3 \cdot \frac{(n-1)(n-2)}{2!} q^{n-3}p^2 + \cdots + np^{n-1}\right]$$

$$= np\left[q^{n-1} + (n-1)q^{n-2}p + \frac{(n-1)(n-2)}{2!} q^{n-3}p^2 + \cdots + p^{n-1}\right]$$

$$+ np\left[(n-1)q^{n-2}p + 2 \cdot \frac{(n-1)(n-2)}{2!} q^{n-3}p^2 + \ldots + (n-1)p^{n-1}\right]$$

$$= np(q+p)^{n-1} + np(n-1)p\left[q^{n-2} + {}^{n-2}C_1 q^{n-3} p + \cdots + p^{n-2}\right]$$

$$= np + n(n-1)p^2 (q+p)^{n-2} = np + n(n-1)p^2.$$

Now Var(X) = $\{n(n-1)p^2 + np\} - (np)^2 = n(n-1)p^2 + np - n^2p^2 = np - np^2 = np(1-p) = npq.$

Thus the variance of the binomial distribution is $\sigma^2 = npq$ and standard deviation = $\sqrt{\sigma^2} = \sqrt{npq}$. ...(7.22)

Since p and q are numerically less than or equal to 1, $npq < np$, that is, variance of a binomial variable is always less than its mean.

Applications of Binomial Distribution

Binomial distribution is applicable when the trials are independent and each trial has just two outcomes success and failure. It is applied in coin tossing experiments, sampling inspection plan etc.

EXAMPLE 7.68 If on an average, 1 ship in every 10 is wrecked, find the probability that out of 5 ships expected to arrive the port, four atleast will arrive safely. *(PTU. 2006)*

Solution The probability p that a ship is wrecked $= 1/10$.

$\therefore$ Probability of its arriving safely $= 1 - (1/10) = 9/10$.

The chance of atleast 4 arriving safely = chance of 4 arriving safely + chance of 5 arriving safely

$$= {}^5C_4\left(\frac{9}{10}\right)^4\left(\frac{1}{10}\right) + {}^5C_5\left(\frac{9}{10}\right)^5 = 5\frac{9^4}{10^5} + \frac{9^5}{10^5} = \frac{9^4}{10^5}(5+9) = 0.91854.$$

EXAMPLE 7.69 The probability that a man aged 60 will live to be 70 is 0.65. What is the probability that out of 10 men now 60, atleast 7 will live to be 70? *(AMIETE, June 2009, June 2008)*

Solution The probability of success p, (*i.e.*, a man living upto 70) is 0.65 and probability of failure, $q = 1 - 0.65 = 0.35$. Therefore, the probability of atleast 7 successes in 10 trials

$$= {}^{10}C_7(0.65)^7(0.35)^3 + {}^{10}C_8(0.65)^8(0.35)^2 + {}^{10}C_9(0.65)^9(0.35) + (0.65)^{10}$$

$$= 0.2522 + 0.1756 + 0.0725 + 0.0135 = 0.5138.$$

EXAMPLE 7.70 If 10% bolts produced by a machine are defective, determine the probability that out of 10 bolts chosen at random (*i*) 1, (*ii*) none, (*iii*) at least 1, (*iv*) atmost two bolts, will be defective.

Solution (*i*) Here the probability p, of a bolt being defective $=10/100=0.1$. Therefore, $q=1-p=1-0.1=0.9$.

By Binomial distribution, the probability of 1 defective bolt out of $10={}^{10}C_1(0.1)^1\,(0.9)^9=0.3874$.

(*ii*) The probability that none is defective $={}^{10}C_0\,(0.1)^0\,(0.9)^{10}=(0.9)^{10}=0.3487$.

(*iii*) Probability of at least one defective bolt = 1 – (Probability of none defective) = 1 – 0.3487 = 0.6513.

(*iv*) Probability of 2 defective bolts $={}^{10}C_2\,(0.1)^2\,(0.9)^8=0.1937$.

∴ Probability of at most 2 defective = Probability of none defective + Probability of 1 defective + Probability of two defective

= 0.3487 + 0.3874 + 0.1937 = 0.9298.

EXAMPLE 7.71 The incidence of occupational disease in an industry is such that the workmen have a 20% chance of suffering from it. What is the probability that out of seven workmen 5 or more will contact the disease?

Solution Let p denotes chance of suffering and q chance of not suffering.

$$\therefore\quad p=20\%=\frac{1}{5}\quad\text{and}\quad q=1-p=1-\left(\frac{1}{5}\right)=\frac{4}{5}.\quad\text{Here } n=7.$$

The binomial expression is $(q+p)^7=q^7+7q^6p+21q^5p^2+35q^4p^3+35q^3p^4+21q^2p^5+7qp^6+p^7$.

The probability of 5 or more (*i.e.*, 5, 6, 7) successes is $=21q^2p^5+7qp^6+p^7$. ...(*i*)

Substituting the values of p and q in (*i*) we obtain,

$$\text{required probability}=21\left(\frac{4}{5}\right)^2\left(\frac{1}{5}\right)^5+7\left(\frac{4}{5}\right)\left(\frac{1}{5}\right)^6+\left(\frac{1}{5}\right)^7=\frac{336}{78125}+\frac{28}{78125}+\frac{1}{78125}=\frac{73}{15625}.$$

EXAMPLE 7.72 Out of 320 families with 5 children each how many families would be expected to have (*i*) 2 boys and 3 girls (*ii*) atleast one boy, (*iii*) no girl and (*iv*) at most 2 boys? Assume equal probability for boys and girls.

Solution Let X denote the number of boys, then X is a binomial variate with parameters $p=1/2$, $n=5$, that is, $X\sim B(5, 1/2)$.

(*i*) Probability of 2 boys and 3 girls

$r=2$, $n-r=5-2=3$. Employing the formula: $P(x=r)={}^nC_r\,p^r q^{n-r}$.

$$P(r=2)={}^5C_2\left(\frac{1}{2}\right)^2\left(\frac{1}{2}\right)^3=\frac{5.4}{2}\cdot\frac{1}{4}\cdot\frac{1}{8}=\frac{5}{16}.$$

Expected number of families with 2 boys and 3 girls $=320\times(5/16)=100$.

(*ii*) Probability of at least one boy = 1 – probability of no boy

$$=1-P(r=0)=1-{}^5C_0\left(\frac{1}{2}\right)^0\left(\frac{1}{2}\right)^5=1-\frac{1}{32}=\frac{31}{32}.$$

∴ Expected number of families with atleast one boy $=320\times\dfrac{31}{32}=310$.

(*iii*) Prob. (no girl) = Prob. (all 5 boys) $={}^5C_5\left(\dfrac{1}{2}\right)^5\left(\dfrac{1}{2}\right)^0=\dfrac{1}{32}$.

∴ Number of families with no girl $=320\times\dfrac{1}{32}=10$.

(*iv*) Prob. (at most 2 boys) = Prob. (no boy) + Prob. (one boy) + Prob. (two boy)

$$= {}^5C_0\left(\frac{1}{2}\right)^0\left(\frac{1}{2}\right)^5 + {}^5C_1\left(\frac{1}{2}\right)^1\left(\frac{1}{2}\right)^4 + {}^5C_2\left(\frac{1}{2}\right)^2\left(\frac{1}{2}\right)^3 = \frac{1}{32}+\frac{5}{32}+\frac{10}{32}=\frac{1}{2}.$$

$\therefore$ Number of families with at most 2 boys $= 320 \times 1/2 = 160$.

EXAMPLE 7.73 A binomial distribution has mean 20 and standard deviation 4. Find the biomial distribution.

Solution Let $X \sim B(n, p)$. We know that mean = np and standard deviation = $\sqrt{npq}$.

Therefore $npq/np = 16/20$, $\Rightarrow$ $q = 0.8$, and $p = 1 - q = 1 - 0.8 = 0.2$. $n = 20/p = 20/0.2 = 100$.

Parameters n and p are 100 and 1/5 respectively.

Thus the probability mass function of x is given by

$$f(x) = {}^nC_x p^x q^{n-x} = {}^{100}C_x \left(\frac{1}{5}\right)^x \left(\frac{4}{5}\right)^{100-x} \qquad \text{for } x = 0, 1, 2, ..., 100.$$

EXAMPLE 7.74 The following data are the numer of seeds germinating out of 10 on damp filter paper for 80 sets of seeds. Fit a binomial distriution of this data. (*AMIETE, Dec. 2009*)

x :	0	1	2	3	4	5	6	7	8	9	10	
f:	6	20	28	12	8	6	0	0	0	0	0	**Total 80**

Solution Mean $= \dfrac{6(0)+20(1)+28(2)+12(3)+8(4)+6(5)}{6+20+28+12+8+6} = \dfrac{174}{80} = 2.175.$

Therefore, mean = np = $10p$ = 2.175 $\Rightarrow$ p = 0.2175, and $q = 1 - p = 0.7825$.

Binomial probabilities are given by $f(x) = P(X = x) = \sum_{x=0}^{10} {}^{10}C_x p^x q^{10-x}$.

Therefore, Binomial expected frequencies are the terms of

$$f(x) = Nf(x) = 80\left[\sum_{x=0}^{10} {}^{10}C_x\,(0.2175)^x\,(0.7825)^{10-x}\right].$$

Table 7.5: Fitting Binomial Distribution to an observed distribution.

x	$f(x) = {}^{10}C_x\,(0.2175)^x\,(0.7825)^{10-x}$	Expected frequency = $Nf(x) = 80\,f(x)$	**Observed frequency**
0	0.0862	6.9 or 7	6
1	0.2392	19.1 or 19	20
2	0.2993	24 24	28
3	0.2226	17.8 or 17	12
4	0.1075	8.6 or 9	8
5	0.0363	2.9 or 3	6
6	0.0089	0.7, or 1	0
7	0	0	0
8	0	0	0
9	0	0	0
10	0	0	0
	Total = 1.0000	**80**	**80**

EXAMPLE 7.75 Fit a binomial distribution to the following frequency distribution:

x:	0	1	2	3	4	5	6	
f:	13	25	52	58	32	16	4	**Total 200**

(*Madras; 2000s*)

Solution The fitted binomial distribution is given by $f(x) = {}^nC_x p^x q^{n-x}$ on the basis of the given data, we have $n = 6$ and $N = \Sigma f = 200$.

$$\bar{x} = \frac{\sum f_i x_i}{\sum f_i} = \frac{13(0)+25(1)+52(2)+58(3)+32(4)+16(5)+4(6)}{200} = \frac{535}{200} = 2.675.$$

Thus $p = \frac{\bar{x}}{n} = \frac{2.675}{6} = 0.446$ and $q = 1 - p = 0.554.$

The fitted binomial distriution is $f(x) = {}^6C_x\ (0.446)^x\ (0.554)^{6-x}$ for $x = 0, 1, 2, ..., 6.$

Table 7.6: Fitting Binomial Distriution to an observed distribution

x	$f(x) = {}^6C_x\ (0.446)^x\ (0.554)^{6-x}$	Expected or theoretical frequency $= Nf(x) = 200 f(x)$	Observed frequency
0	0.02891	5.782 or 6	13
1	0.13965	27.93 or 28	25
2	0.28106	56.212 56	52
3	0.30169	60.338 or 60	58
4	0.18216	36.432 or 36	32
5	0.05866	11.732 or 12	16
6	0.00787	1.574 or 2	4
	Total = 1.00000	**200**	**200**

EXAMPLE 7.76 A target is to be destroyed in a bombing exercise. There is 75% chance that any one bomb will strike the target. Assume that two direct hits are required to destroy the target completely. How many bombs must be dropped in order that the chance of destroying the target is > 99%? *(AMIETE, June 2011)*

Solution Let p = probability the bombs hits the target = 3/4, and q the failure not to hit the target $= 1 - p = 1/4$. Let n numbers of bombs required to destroy the target. If X be the random variate representing the number of bombs striking the target, then

$$P(X = r) = {}^nC_r (3/4)^r\ (1/4)^{n-r}, \quad r = 0, 1, 2, ..., n.$$

Therefore, $P(X \geq 2) \geq 0.99$, or $[1 - P(X \leq 1)] \geq 0.99.$

or $1 - [P(X = 0) + P(X = 1)] \geq 0.99$ or $1 - (1/4)^n - n(3/4)\,(1/4)^{n-1} \geq 0.99$

or $(1/4)^n + n(3/4)\,(1/4)^{n-1} \leq 0.01$ or $1 + 3n \leq (0.01)4^n$ or $100 + 300n \leq 4^n.$

By trial, $n = 6$ (since $1900 \leq 4^6$). Thus we obtain $n = 6$.

Moment Generating Function of Binomial Distribution

Let X be a binomial variate. Then, the moment generating function (MGF) of X is defined by

$$MX(t) = E(e^{tx}) = \sum_{r=0}^{n} e^{tr}\ {}^nC_r p^r q^{n-r} = \sum_{r=0}^{n} {}^nC_r (e^t p)^r q^{n-r} = (q + pe^t)^n$$

$$= \left[q + p\left(1 + t + \frac{t^2}{2!} + \frac{t^3}{3!} + ...\right)\right]^n = \left[q + p + pt + \frac{pt^2}{2!} + \frac{pt^3}{3!} + ...\right]^n = \left(1 + pt + \frac{pt^2}{2!} + \frac{pt^3}{3!} + ...\right)^n \text{ as } p + q = 1. \quad ...(7.23)$$

7.8.2 Poisson Distribution

Poisson distribution is a discrete probability distribution related to the probabilities of events which are extremely rare, but which have a large number of independent opporturnities for occurrence. Poisson distribution is a limiting case of the binomial distribution. We state below the conditions when Poisson distribution can be used:

(i) Each trial results in two mutually exclusive outcomes that can be classified as success or failure.
(ii) The number of trials, n is very large, that is, $n \to \infty$.
(iii) The probability of success p, is very small, that is, $p \to 0$.
(iv) $np = m$, where m is a positive real number, is finite.

Application of Poisson Distribution

(i) The number of printing mistakes per page of a large book.
(ii) The number of road accidents on a busy road per minute.
(iii) The emission of radioactive particles per minute in a fusion process.
(iv) The numer of defective items out of lots produced in a large manfacturing factory.

If X denotes a discrete reandom variable following Poisson distribution with parameter m, then the probability function of X is given by

$$P(X=r)=\frac{e^{-m}m^r}{r!},\quad r=0,1,2,...,m>0 \qquad ...(7.24)$$

where e is a constant with value $e = 2.71828$. We denote the Poisson distribution by $P(m)$.

From (7.24) the probability of 0, 1, 2, ...r,... successes is

$$e^{-m}, e^{-m}m, e^{-m}\frac{m^2}{2!}, ..., e^{-m}\cdot\frac{m^r}{r!}, ... \qquad ...(7.25)$$

It can be easily seen that sum to infinity of probabilities (7.25) is 1.

$$\left[\because \sum_{r=0}^{\infty} P(X=r) = e^{-m}\left(1+\frac{m}{1!}+\frac{m^2}{2!}+\frac{m^3}{3!}+\ldots \text{ to } \infty\right) = e^{-m}\cdot e^{m} = 1\right]$$

Poisson Distribution as a Limiting Case of Binomial Distribution

We shall now deduce the Poisson distribution from the binomial distribution by assuming that $n \to \infty$ and $p \to 0$ such that the product np remains finite, say m.

Let the random variable X be binomially distributed with parameters n and p.

As $n \to \infty$, $p \to 0$ and $np = m$, we obtain $P(X=r)={}^nC_r p^r q^{n-r}$

$$=\frac{n(n-1)(n-2)\ldots(n-r+1)}{r!}\cdot\left(\frac{m}{n}\right)^r\left(1-\frac{m}{n}\right)^{n-r}$$

$$=\frac{m^r}{r!}\left[1\left(1-\frac{1}{n}\right)\left(1-\frac{2}{n}\right)\ldots\left(1-\frac{r-1}{n}\right)\right]\left(1-\frac{m}{n}\right)^n\cdot\left(1-\frac{m}{n}\right)^{-r}. \qquad ...(7.26)$$

Now $\lim_{n\to\infty}\left(1-\frac{m}{n}\right)^n = e^{-m}$, $\lim_{n\to\infty}\left(1-\frac{m}{n}\right)^{-r} = 1$, $\lim_{n\to\infty}\left(1-\frac{a}{n}\right) = 1$, where 'a' is a finite number. Therefore, taking limits as $n \to \infty$ in (7.26), we obtain

$$\lim_{\substack{n\to\infty \\ np=m=\text{finite}}} P(X=r)=\frac{m^r}{r!}(1)\cdot e^{-m}=\frac{e^{-m}m^r}{r!},$$ which is the probability function of the Poisson variate X.

Mean and Variance of the Poisson Distribution *(AMIE, S-2000)*

From the definition, we have

$$E(X)=\sum_{r=0}^{\infty} x_r p_r = \sum_{r=0}^{\infty} r\, e^{-m}\frac{m^r}{r!} = me^{-m}\sum_{r=1}^{\infty}\frac{m^{r-1}}{(r-1)!} = me^{-m}e^{m} = m. \qquad ...(7.27)$$

Thus, the mean of the Poisson distribution is equal to the parameter m.

Variance σ^2

The *variance of a probability distribution* is given by variance $(X) = E(X^2) - [E(X)]^2$.

$$E(X^2)=\sum_{r=0}^{\infty}x_r^2\,p_r=\sum_{r=0}^{\infty}r^2\frac{e^{-m}m^r}{r!}=\sum_{r=0}^{\infty}[r(r-1)+r]\frac{e^{-m}m^r}{r!}=\sum_{r=0}^{\infty}r(r-1)\frac{e^{-m}m^r}{r!}+\sum_{r=0}^{\infty}r\frac{e^{-m}m^r}{r!}$$

$$=\sum_{r=0}^{\infty}r(r-1)\frac{e^{-m}m^r}{r!}+m=m^2e^{-m}\sum_{r=2}^{\infty}\frac{m^{r-2}}{(r-2)!}+m=m^2e^{-m}\left[1+\frac{m}{1!}+\frac{m^2}{2!}+\ldots\right]+m$$

$$=m^2\cdot e^{-m}\cdot e^m+m=m^2+m.$$

Therefore, variance $(X)=E(X^2)-[E(X)]^2=m^2+m-m^2=m.$ ...(7.28)

Therefore, for Poission distribution, mean = variance = m.

Recurrence Formula for the Probabilities of Poisson Distribution

In poisson distribution, the probability of r successes is given by

$$P(r)=\frac{e^{-m}\cdot m^r}{r!},\quad r=0,1,2,\ldots$$

Setting $r = r + 1$, we obtain

$$P(r+1)=e^{-m}\cdot\frac{m^{r+1}}{(r+1)!}=\frac{m}{r+1}\cdot\frac{e^{-m}\cdot m^r}{r!}=\frac{m}{r+1}P(r),\quad r=0,1,2,3,\ldots \qquad \text{...(7.29)}$$

This formula is called the *recurrence formula* for the poission distribution. If we know $P(0)=e^{-m}$, then with the formula (7.29), we can calculate all the other probabilities, $P(1)$, $P(2)$, $P(3)$ etc.

EXAMPLE 7.77 Suppose that on an average 1 house in 1000 houses get fire in a year in a district. If there are 2000 houses in that district, find the probability that exactly 5 houses will have fire during the year. (given $e^{-2}=0.1352$). *(AMIETE, Dec. 2006)*

Solution Applying the poisson distribution $P(X=r)=e^{-m}\cdot\frac{m^r}{r!}$

Mean, $m=np=(2000)\,(1/1000)=2,\ r=5.$

Therefore, $P(X=5)=e^{-2}\cdot\frac{2^5}{5!}=0.1352\times\frac{32}{120}=0.036.$

EXAMPLE 7.78 Ten per cent of tools produced in a certain manufacturing process turn out to be defective. Find the probability that in a sample of 10 tools chosen at random. (*i*) exactly two will be defective, (*ii*) more than one will be the defective.

Solution Probability to defective tools, $p = 10/100 = 0.1$. Therefore, $m = np = 10\ (0.1) = 1$.

(*i*) Here $r = 2$, $m = 1$.

Probability of 2 defective tools $=\frac{e^{-1}\cdot(1)^2}{2!}=\frac{e^{-1}}{2}=\frac{1}{2e}=0.18$ (approx).

(*ii*) P(more than 1 defective) = 1 – P (zero defective) – P (one defective)

$$=1-e^{-1}-e^{-1}=1-\frac{2}{e}=\frac{e-2}{e}=0.26\ \text{(approx)}.$$

EXAMPLE 7.79 The mortality rate for a certain disease is 7 per 1000. What is the probability for just 5 deaths from this disease in a group of 400?

Solution Here $m=np=400\times(7/1000)=2.8$ and $r=5$.

Therefore, probability required $=\frac{e^{-2.8}(2.8)^5}{5!}=0.087.$

EXAMPLE 7.80 If the probability that an individual suffers a bad reaction from a certain injection is 0.001; determine the probability that out of 2000 individuals (*i*) exactly 3, (*ii*) more than 2 individuals will get a bad reaction. It is given that $e^{-2} = 0.13534$. *(AMIETE, June 2012, Dec 2010, AMIE, W-2004)*

Solution Here $p = 0.001$, $n = 2000$. $\therefore\ m = np = 2000 \times 0.001 = 2$.

(*i*) Probability that 3 individuals suffer from bad reaction $= \dfrac{e^{-2}2^3}{3!} = \dfrac{4}{3e^2} = 0.18$.

(*ii*) Probability that none suffer $= \dfrac{e^{-2}2^0}{0!} = \dfrac{1}{e^2}$. Probability that 1 suffer $= \dfrac{e^{-2}2^1}{1!} = \dfrac{2}{e^2}$.

Probability that 2 suffers $= \dfrac{e^{-2}2^2}{2!} = \dfrac{2}{e^2}$.

$\therefore$ Prob. of more than 2 suffer from bad reaction

$= 1 - $ [Probability $(r = 0)$ + Probability $(r = 1)$ + Probability $(r = 2)$]

$$= 1 - \left(\frac{1}{e^2} + \frac{2}{e^2} + \frac{2}{e^2}\right) = 1 - \frac{5}{e^2} = 0.323.$$

EXAMPLE 7.81 A factory producing cycle tyres, there is a small chance of 1 in 500 tyres to be defective. The tyres are supplied in lots of 20. Using Poisson distribution, calculate the approximate number of lots containing no defective, one defective and two defective tyres in a consignment of 20,000 tyres. Given $e^{-0.04} = 0.9608$.

Solution Here $n = 20$, $p = 1/500$. Therefore, $m = np = 20/500 = 0.04$.

Also $e^{-m} = e^{-0.04} = 0.9608$. Now $N = 20{,}000/20 = 1000$ lots of 20 each.

The number of lots containing no defective, one defective, two defective tyres are given by

$$Ne^{-m},\quad Ne^{-m}\cdot m,\quad Ne^{-m}(m^2/2) \text{ that is,}$$

$$1000(0.9608),\ 1000(0.9608) \times 0.04,\ 1000(0.9608) \times \frac{(0.04)^2}{2}, \text{ that is, } 961, 38, 1 \text{ respectively.}$$

EXAMPLE 7.82 Out of 1000 balls 50 are red and the rest white. If 60 balls are picked at random, what is the probability of picking up (*i*) 3 red balls (*ii*) not more than 3 red balls in the sample? Assume Poisson distribution for the number of red balls picked up in the sample. Given $e^{-3} = 0.0498$.

Solution Since out of 1000 balls, 50 are red, the probability of drawing a red ball is given by $p = 50/1000 = 1/20$.

(*i*) As $n = 60$, $m = np = 60 \times 1/20 = 3$. $\therefore$ $P(X = 3)$ *i.e.,* probability of drawing 3 red balls

$$= \frac{3^3\, e^{-3}}{3!} = \frac{9}{2}(0.0498) = 0.2241.$$

(*ii*) P (X not more than 3 red balls) $= p(0 \text{ red}) + p(1 \text{ red}) + p(2 \text{ red}) + p(3 \text{ red})$

$$= \frac{3^0\, e^{-3}}{0!} + \frac{3^1\, e^{-3}}{1!} + \frac{3^2\, e^{-3}}{2!} + \frac{3^3 e^{-3}}{3!} = e^{-3}\left(1 + 3 + \frac{3^2}{2!} + \frac{3^3}{3!}\right)$$

$$= 0.0498(1 + 3 + 4.5 + 4.5) = 0.6474.$$

EXAMPLE 7.83 A car-hire firm has two cars, which it hires out day by day. The number of demands for a car on each day is distributed as a Poisson distribution with mean 1.5. Calculate the proportion of days (*i*) on which there is no demand, (*ii*) on which a demand is refused. (given that $e^{-1.5} = 0.2231$)

(AMIETE, Dec 2011, June 2011, June 2009, Dec. 2008; AMIE-S 2006; JNTU-2003)

Solution (*i*) Here the random variable X, which denotes the demands for a car on any day follows Poisson distribution with mean, $m = 1.5$. The proportion of days on which there are r demand for a car is given by:

$$P(X = r) = \frac{e^{-1.5}(1.5)^r}{r!};\ r = 0, 1, 2, \ldots..$$

$\therefore$ P (no demand) $= P(X=0) = \dfrac{e^{-1.5}(1.5)^0}{0!} = e^{-1.5} = 0.2231 = 22.31\% \approx 22\%$ days neither car is used.

(*ii*) P (a demand is refused): Prob. the demand to be more than 2 car that is, $P(X > 2) = 1 - P(X \le 2)$

$= 1 - [P(X = 0) + P(X = 1) + P(X = 2)] = 1 - e^{-1.5}\left[1 + 1.5 + \dfrac{(1.5)^2}{2!}\right] = 1 - 0.2231\,(3.625) = 0.1913.$

Thus 19% days some demand is refused.

EXAMPLE 7.84 The following table shows the number of days *f* in a 320 day period during which *X* accident occurred in a factory. Fit a Poisson distribution to the data.

No. of accidents, x_i :	0	1	2	3	4	
No. of days, f_i :	192	100	24	3	1	**Total 320**

Given that $e^{-0.5} = 0.6065$.

Solution Mean number of accidents is

$$m = \frac{\Sigma f_i x_i}{\Sigma f_i} = \frac{(192)(0) + (100)(1) + (24)(2) + (3)(3) + (1)(4)}{192 + 100 + 24 + 3 + 1} = \frac{161}{320} = 0.503 \approx 0.5.$$

Then, according to the Poisson distribution, $P\,(X \text{ accidents}) = \dfrac{e^{-0.5}\,(0.5)^X}{x!}$.

In the following table, we have listed the probabilities for 0, 1, 2, 3 and 4 accidents as obtained from this Poisson distribution, as well as the expected or theoretical number of days during which X accidents take place. This has been obtained by multiplying the respective probabilities by 320. For convenience of comparison, the fourth column giving the actual number of days has been repeated.

$P(X = 0) = e^{-m} = e^{-0.5} = 0.6065, \qquad P(X = 1) = 0.6065 \times 0.5 = 0.30325,$

$P(X = 2) = 0.6065 \times \dfrac{(0.5)^2}{2} = 0.07581, \qquad P(X = 3) = 0.6065 \times \dfrac{(0.5)^3}{3} = 0.01263.$

Lastly $P(X \ge 4) = 1 - P(X < 4) = 0.00181$ (by subtraction).

Table 7.7

X (No. of accidents)	P_r (X accidents)	Expected no. of days: $N \times P_r$	Actual no. of days
0	0.60650	194.08 or 194	192
1	0.30325	97.04 or 97	100
2	0.07581	24.259 or 24	24
3	0.01263	4.042 or 4	3
4	0.00181	0.5792 or 1	1
	Total = 1.00	**320**	**320**

For a true Poisson distribution, the variance σ^2 = mean (= *m*). Now we compute variance of the given distribution. Refer the following table:

Table 7.8

X	X^2	Frequency, *f*	fX^2
0	0	192	0
1	1	100	100
2	4	24	96
3	9	3	27
4	16	1	16
		$N = \Sigma f = 320$	$\Sigma f X^2 = 239$

$$\text{Variance} = \frac{\Sigma fX^2}{N} - \left(\frac{\Sigma fX}{N}\right)^2 = \frac{239}{320} - (0.503)^2 = 0.7469 - 0.253 = 0.4939.$$

Computation of the variance of the given distribution gives 0.4939. This compares favourably with the value 0.503 for m. Thus, we conclude that fit of the Poisson distribution to the given data is good.

EXAMPLE 7.85 Show that in a Poisson distribution with mean 1, then its mean deviation about mean is (2/e) times the standard deviation. *(AMIETE, Dec. 2005)*

Solution If $p(X = r)$ denotes probability of r successes, then $p(X = r) = \dfrac{e^{-m} m^r}{r!} = \dfrac{e^{-1}}{r!}$.

($\because$ $m = 1$, given); $r = 0, 1, 2, \ldots$

Mean deviation about mean 1 in a Poisson distribution

$$= \sum_{r=0}^{\infty} p(r)\,|r-1| = \sum_{r=0}^{\infty} |r-1| \frac{e^{-1}}{r!} = e^{-1} \sum_{r=0}^{\infty} \frac{|r-1|}{r!} = e^{-1}\left[\frac{|0-1|}{0!} + \frac{|1-1|}{1!} + \frac{|2-1|}{2!} + \frac{|3-1|}{3!} + \cdots \infty\right]$$

$$= e^{-1}\left[1 + 0 + \frac{2-1}{2!} + \frac{3-1}{3!} + \cdots\cdots\right] = e^{-1}\left[1 + \left(1 - \frac{1}{2!}\right) + \left(\frac{1}{2!} - \frac{1}{3!}\right) + \cdots \infty\right]$$

$$= e^{-1}\left[\left(1 + 1 + \frac{1}{2!} + \frac{1}{3!} + \cdots \infty\right) - \left(\frac{1}{2!} + \frac{1}{3!} + \cdots \infty\right)\right] = e^{-1}[e - (e-2)] = 2/e = (2/e) \text{ times the S.D.}$$

[Since $\text{S.D.} = \sqrt{m} = 1 \text{ (given)}$.]

Thus, Mean deviation about mean = (2/e) times the standard deviation.

Moment Generating Function (MGF) of Poisson Distribution

Let X be a Poisson variate. Then, the moment generating function of X is defined by

$$M_X(t) = E(e^{tX}) = \sum_{r=0}^{\infty} e^{tr} \cdot \left(\frac{e^{-m} m^r}{r!}\right) = e^{-m} \sum_{r=0}^{\infty} \frac{(me^t)^r}{r!}$$

$$= e^{-m}\left[1 + me^t + \frac{(me^t)^2}{2!} + \ldots\right] = e^{-m} \cdot e^{me^t} = e^{m(e^t - 1)}. \qquad \ldots(7.30)$$

7.8.3 Normal Distribution

The normal distribution can be regarded as the limiting form of the Binomial distribution when n is very large and p and q are not very small. It is one of the most important continuous probability distributions because a large number of frequency distributions approximate this distribution. Normal distribution is also called *Gaussian distribution*.

A continuous random variable X is said to have a normal distribution with parameters μ (mean) and σ^2 (variance) if its probability density function $f(x)$ is given by the equation

$$f(x) = \frac{1}{\sigma\sqrt{2\pi}} e^{-(x-\mu)^2/(2\sigma^2)} \qquad -\infty < x < \infty, \quad -\infty < \mu < \infty, \quad \sigma > 0. \qquad \ldots(7.31)$$

The graph of $y = f(x)$ is a bell shaped curve called the normal probability curve. This is shown in the Fig. 7.27. The curve is symmetric about the mean. The maximum occurs at $x = \mu$ and the maximum value is $1/(\sigma\sqrt{2\pi})$. The ordinate $f(x)$ decreases rapidly as $x \to \pm\infty$ and x-axis is an asymptote of the curve. If σ is large, the curve tends to be flat and if σ is small, the curve tends to be more peaked at $x = \mu$. We denote the above normal distribution by $X \sim N(\mu, \sigma^2)$.

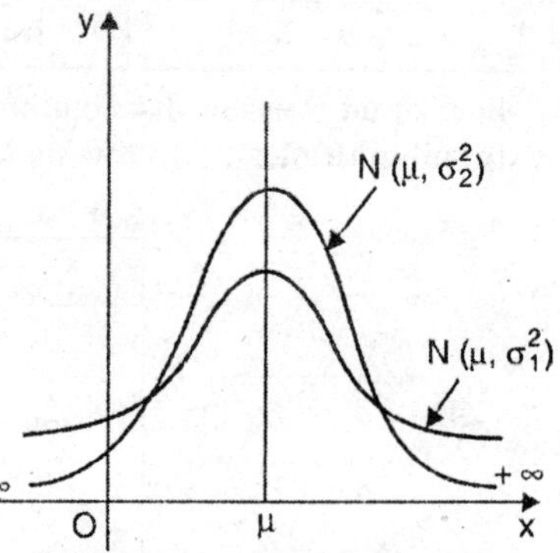

Fig. 7.27: Normal probability curve, $\sigma_1 > \sigma_2$

The total area bounded by the curve (7.31) and the x-axis is 1. Hence the area bounded by this curve and the two ordinates $x = a$ and $x = b$ $(a < b)$ gives the probability that x lies between a and b. In brief the probability between a and b is denoted by $P(a < X < b)$ and is given by

$$P(a < X < b) = \frac{1}{\sigma\sqrt{2\pi}} \int_a^b e^{-\frac{(x-\mu)^2}{2\sigma^2}} dx.$$

Let $z = (x - \mu)/\sigma$. Then $dx = \sigma\, dz$ and the limits $(a, b) \to \left(\frac{a-\mu}{\sigma}, \frac{b-\mu}{\sigma}\right)$. Therefore,

$$P(a < X < b) = P\left(\frac{a-\mu}{\sigma} < Z < \frac{b-\mu}{\sigma}\right) = \frac{1}{\sqrt{2\pi}} \int_{\frac{a-\mu}{\sigma}}^{\frac{b-\mu}{\sigma}} e^{-z^2/2} dz = \int_{\frac{a-\mu}{\sigma}}^{\frac{b-\mu}{\sigma}} \phi(z)\, dz \qquad ...(7.32)$$

where $\phi(z) = \dfrac{e^{-z^2/2}}{\sqrt{2\pi}}$ is the probability density function of the standard normal variate. We write standard normal variate $z \sim N(0, 1)$. The definite integral $\int_{\frac{a-\mu}{\sigma}}^{\frac{b-\mu}{\sigma}} \phi(z)\, dz$ is called the *normal probability integral.* Its gives the area under the standard normal curve $y = \phi(z)$, between the ordinates at $Z = \dfrac{a-\mu}{\sigma}$ and $Z = \dfrac{b-\mu}{\sigma}$. These areas are tabulated for different values of Z at intervals of 0.01. (See Table -1, Appendix-4).

The symmetry of the curve about $z = 0$ permits us to obtain the area between any two values of z.

The graph of the standard normal curve is shown in Fig. 7.28.

The area under the normal standardised curve is 1 and area between $z = -1$ and $z = +1$ is 0.6826 (or 68.26% of the total area which is 1).

Area between $z = -2$ and $z = +2$ is 0.9544 (or 95.44% the total area).

Area between $z = -3$ and $z = +3$ is 0.9973 (or 99.73% of the total area).

Sailent Features of Normal Probability Distribution

(*i*) The normal curve is symmetrical about a vertical line through mean, that is, about $x = \mu$. In the standardized form it is symmetrical about the ordinate at $z = 0$.

(*ii*) The curve is known when mean and standard deviation are given. We have verified that if $X \sim N(\mu, \sigma^2)$, then $E(X) = \mu$ and var . $(X) = \sigma^2$.

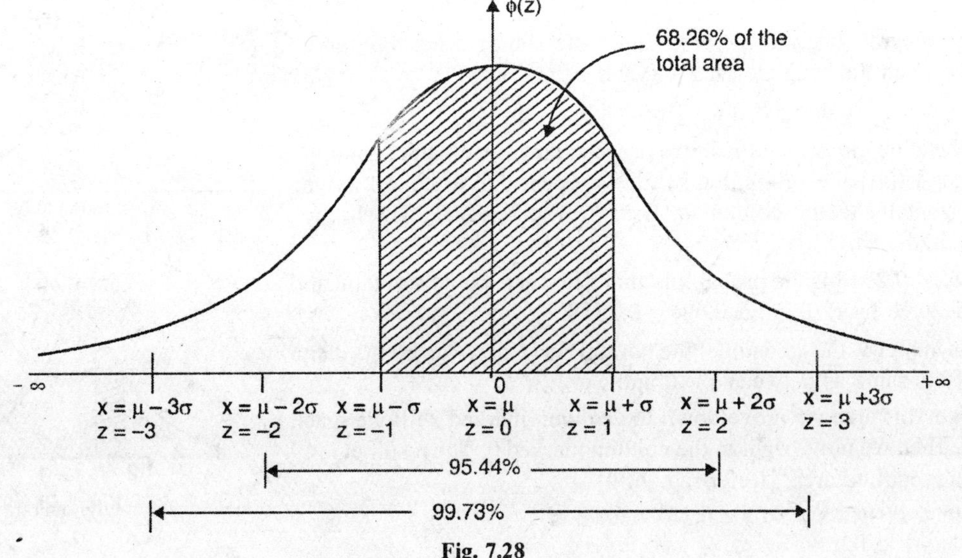

Fig. 7.28

(*iii*) The curve of normal distribution has a single peak, that is, it is unimodal (i.e., has one mode at x = mean). The value of y decreases rapidly as x increases numerically. The two tails of the curve extend infinitely and never touch the horizontal axis. The x-axis is its asymptote.

(*iv*) The area under the normal curve is unity, that is, $\int_{-\infty}^{\infty} f(x)\,dx = 1$. As a result, 0.5 area lies to the right of mean and balance 0.5 area to the left of mean. The probability for the variable to lie in any interval (a, b) in the range of the variable is given by the area under the normal curve, the ordinates $x = a$ and $x = b$.

(*v*) About 68.26% the area fall between $\mu - \sigma$ and $\mu + \sigma$;
About 95.44% the area fall between $\mu - 2\sigma$ and $\mu + 2\sigma$;
About 99.73% the area fall between $\mu - 3\sigma$ and $\mu + 3\sigma$.

(*vi*) $\mu \pm \sigma$ are the points of inflexion.

EXAMPLE 7.86 If $\phi(z) = \frac{1}{\sqrt{2\pi}} e^{-z^2/2}$ is the probability density function of the standard normal variate, evaluate normal probability integral (*i*) to (*ix*).

(*i*) $\int_0^{0.68} \phi(z)dz$, (*ii*) $\int_{-1.2}^{0} \phi(z)dz$, (*iii*) $\int_{-0.80}^{1.53} \phi(z)dz$,

(*iv*) $\int_{1.23}^{1.87} \phi(z)dz$, (*v*) $\int_{-2.35}^{-0.50} \phi(z)dz$, (*vi*) $\int_{1}^{\infty} \phi(z)dx$,

(*vii*) $\int_{-\infty}^{-0.6} \phi(z)dz$. ***Alternately:*** Find the area under the normal curve to the left of $z = -0.6$.

(*viii*) $\int_{-1.28}^{\infty} \phi(z)dz$. ***Alternately:*** Find the area under the normal curve to the right of $z = -1.28$.

(*ix*) $\int_{-\infty}^{1.09} \phi(z)dz$. ***Alternately:*** Find the area under the normal curve to the left of $z = 1.09$.

(*x*) Find the area under the normal curve to the right of $z = 2.05$ and to the left of $z = -1.44$.

Solution The definite integral $\int_0^{z_1} \phi(z)dz$ is called the normal probability integral. It gives the area under the standard normal curve $y = \phi(z)$ between the ordinates at $z = 0$ and $z = z_1$.

(*i*) Between $z = 0$ and $z = 0.68$. [Refer Fig. 7.29]

In the table I in Appendix 4, we proceed downward in the column under z (*i.e.*, the first column) till we reach 0.6. Then we move horizontally to the column marked 8. The result 0.2518 is the required area.

Note: 0.2518 is the probability that z lies between 0 and 0.68 and is denoted by $P(0 \le z \le 0.68) = 0.2518$.

(*ii*) By symmetry, the area under the normal curve between $z = -1.2$ and $z = 0$ is same as between $z = 0$ and $z = 1.2$.

To get this area we move down the column marked z till we reach 1.2. Then we move right to the column marked 0. The result 0.3849 is the required area. [Refer Fig. 7.30]

Note: $P(-1.2 \le z \le 0) = 0.3849$.

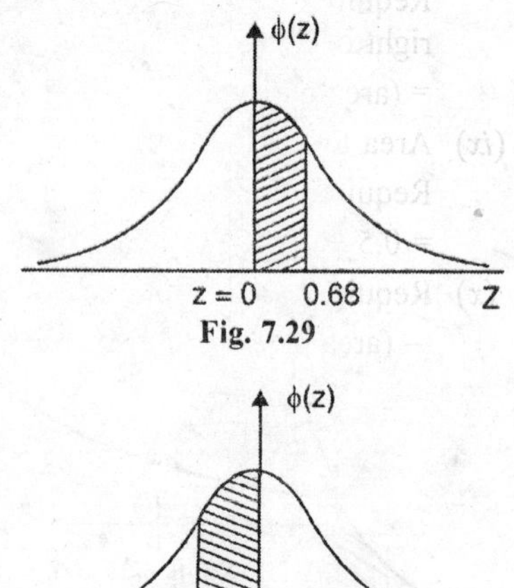

Fig. 7.29

Fig. 7.30

(*iii*) Required area = (area between $z = -0.80$ and $z = 0$) + (Area between $z = 0$ and $z = 1.53$). [Refer Fig. 7.31]
= (Area between $z = 0$ and $z = 0.80$) + (Area between $z = 0$ and $z = 1.53$) = 0.2881 + 0.4370 = 0.7251.

(*iv*) Required area = (area between $z = 0$ and $z = 1.87$) – (area between $z = 0$ and $z = 1.23$) = 0.4693 – 0.3907 = 0.0786. (Refer Fig. 7.32)

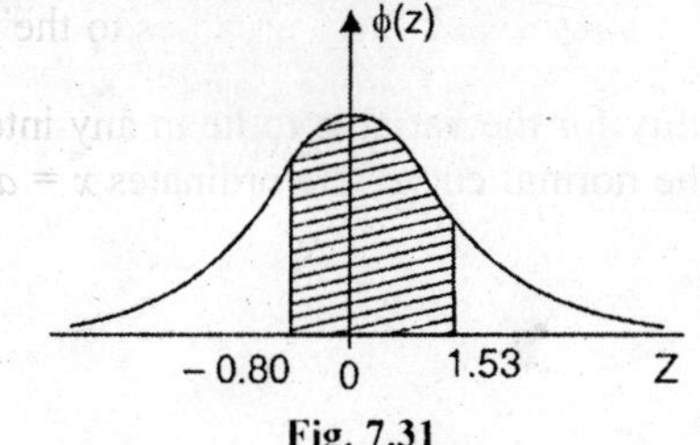

Fig. 7.31

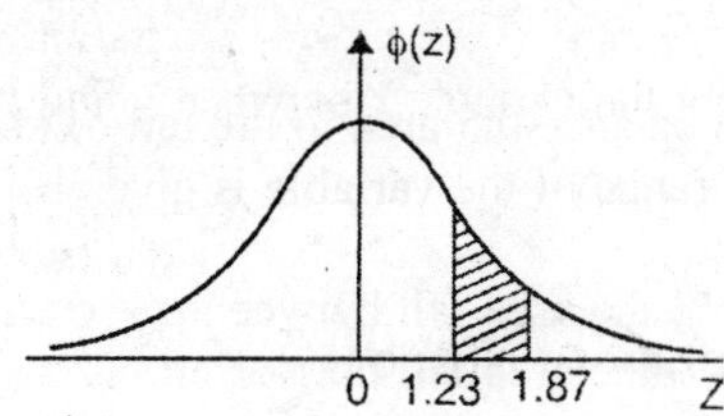

Fig. 7.32

(*v*) Required area = (area between $z = -2.35$ and $z = 0$) – (area between $z = -0.5$ and $z = 0$) [Refer Fig. 7.33]
= (area between $z = 0$ and $z = 2.35$) – (area between $z = 0$ and $z = 0.5$) = 0.4906 – 0.1915 = 0.2991.

(*vi*) Required area = area to the right of $z = 1$. [Refer Fig. 7.34]
= (area to the right of $z = 0$) – (area between $z = 0$ and $z = 1$) = 0.5 – 0.3413 = 0.1587.

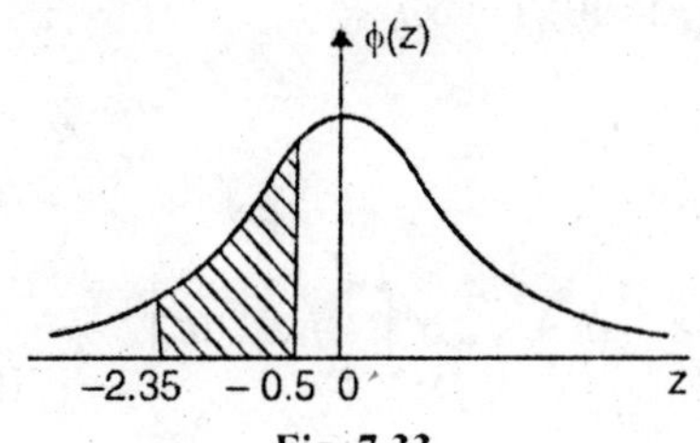

Fig. 7.33

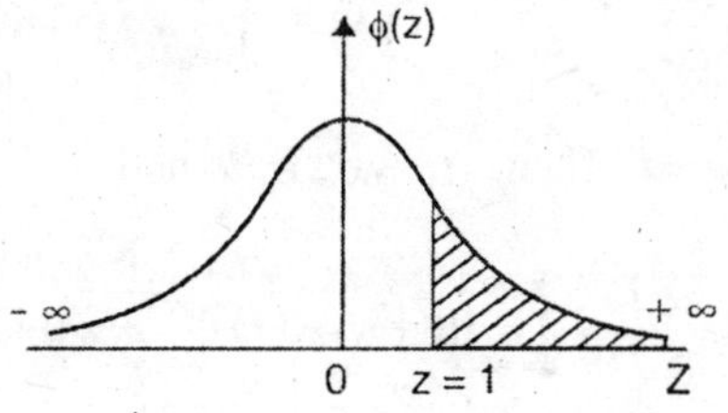

Fig. 7.34

(*vii*) Area to the left of $z = -0.6$.
Required area = (area to the left of $z = 0$) – (area between $z = -0.6$ and $z = 0$). [Refer Fig. 7.35].
= (area to the left of $z = 0$) – (area between $z = 0$ and $z = 0.6$)
= 0.5 – 0.2258 = 0.2742.

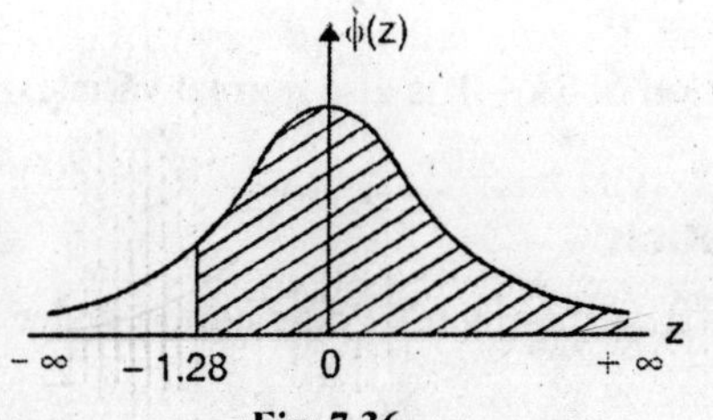

(*viii*) Area to the right of $z = -1.28$.
Required area = (area between $z = -1.28$ and $z = 0$) + (area to right of $z = 0$). [Refer Fig. 7.36].
= (area between $z = 0$ and $z = 1.28$) + (area to right of $z = 0$) = 0.3997 + 0.5 = 0.8997.

Fig. 7.35

(*ix*) Area to the left of $z = 1.09$.
Required area = (area to the left of $z = 0$) + (area between $z = 0$ and $z = 1.09$) [Refer Fig. 7.37].
= 0.5 + 0.3621 = 0.8621.

(*x*) Required area = total area – (area between $z = -1.44$ and $z = 0$)
– (area between $z = 0$ and $z = 2.05$) = 1– 0.4251 – 0.4798 = 0.0951. (Refer Fig. 7.38)

Fig. 7.36

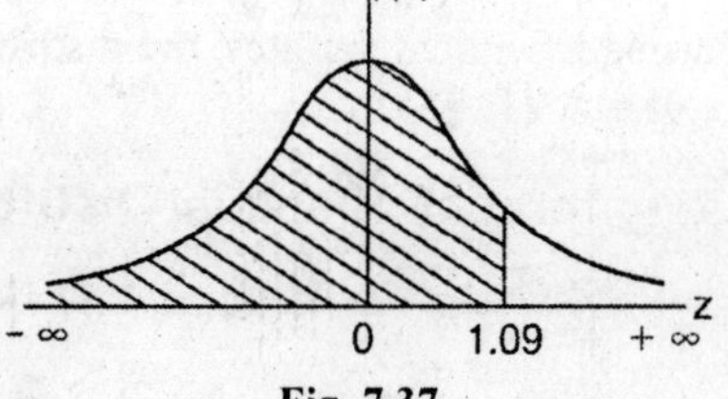

Fig. 7.37

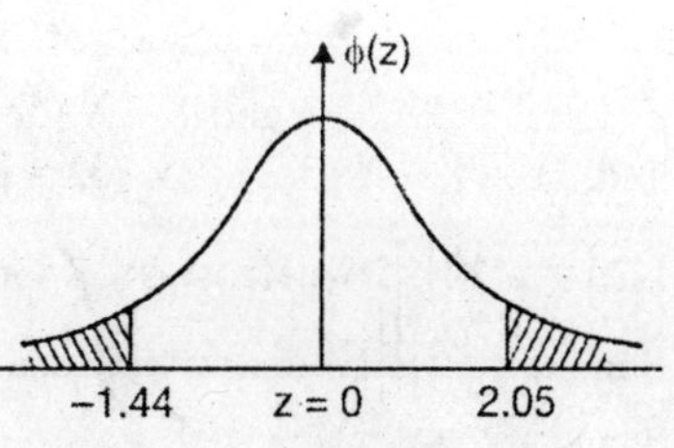

Fig. 7.38

Remark 6

The normal curve can be fitted to any given distribution. If μ and σ are the mean and standard deviation of the frequency distribution and N, the total frequency, then the normal frequency distribution is given by

$$\frac{N}{\sigma\sqrt{2\pi}}\, e^{-(x-\mu)^2/(2\sigma^2)}. \qquad ...(7.33)$$

The frequency of the variable x between a and b is given by

$$N\cdot\frac{1}{\sigma\sqrt{2\pi}}\int_a^b e^{-(x-\mu)^2/(2\sigma^2)}dx.$$

Mean and Variance of $N(\mu, \sigma^2)$

The density function for the normal distribution is given by

$$f(x)=\frac{1}{\sigma\sqrt{2\pi}}\, e^{-(x-\mu)^2/2\sigma^2}, \qquad -\infty<x<\infty$$

where μ and σ^2 are parameters of the distribution.

We can verify that if $X \sim N(\mu, \sigma^2)$, then $E(X) = \mu$ and $\text{Var}(X) = \sigma^2$.

We have
$$E(X) = \int_{-\infty}^{\infty} x f(x)\,dx = \frac{1}{\sigma\sqrt{2\pi}}\int_{-\infty}^{\infty} xe^{-(x-\mu)^2/(2\sigma^2)}dx.$$

Setting $\frac{x-\mu}{\sigma\sqrt{2}}=t$. Then, $dx=\sqrt{2}\,\sigma\, dt$ and

$$E(X) = \frac{1}{\sigma\sqrt{2\pi}}\int_{-\infty}^{\infty}(\mu+\sqrt{2}\,\sigma\, t)\,e^{-t^2}\sqrt{2}\,\sigma\, dt = \frac{1}{\sqrt{\pi}}\left[\mu\int_{-\infty}^{\infty}e^{-t^2}dt+\sqrt{2}\,\sigma\int_{-\infty}^{\infty}t\,e^{-t^2}dt\right]$$

$$= \frac{2\mu}{\sqrt{\pi}}\int_0^{\infty}e^{-t^2}dt+0 \quad \text{[the second expression has an odd integrand]} = \frac{2\mu}{\sqrt{\pi}}\cdot\frac{\sqrt{\pi}}{2}=\mu.$$

$$\text{Variance }(X) = E(X-\mu)^2 = \frac{1}{\sigma\sqrt{2\pi}}\int_{-\infty}^{\infty}(x-\mu)^2 e^{-(x-\mu)^2/2\sigma^2}dx$$

$$= \frac{1}{\sigma\sqrt{2\pi}}\int_{-\infty}^{\infty}\sigma^2 2t^2e^{-t^2}\sqrt{2}\,\sigma\, dt, \qquad \text{(using the same substitution as above)}$$

$$= \frac{4\sigma^2}{\sqrt{\pi}}\int_0^{\infty}\underset{\text{I}}{t^2}\,\underset{\text{II}}{e^{-t^2}}\,dt = \frac{4\sigma^2}{\sqrt{\pi}}\left[\frac{-1}{2}[te^{-t^2}]_0^{\infty}+\frac{1}{2}\int_0^{\infty}e^{-t^2}dt\right] = \frac{2\sigma^2}{\sqrt{\pi}}\int_0^{\infty}e^{-t^2}dt = \frac{2\sigma^2}{\sqrt{\pi}}\cdot\frac{\sqrt{\pi}}{2}=\sigma^2.$$

Remark 7

Let X and Y be two independent normal variates with means μ_1, μ_2 and variances $\sigma_1{}^2$, $\sigma_2{}^2$ respectively, that is, $X \sim N(\mu_1, \sigma_1^2)$, $Y \sim N(\mu_2, \sigma_2^2)$. Then the sum and the differences of two independent normal variates is also a normal variate, that is,

$$X+Y \sim N(\mu_1+\mu_2,\ \sigma_1^2+\sigma_2^2),$$
$$X-Y \sim N(\mu_1-\mu_2,\ \sigma_1^2+\sigma_2^2).$$

Let $X \sim N(2, 4)$ and $Y \sim N(5, 9)$ are independent variates, the distribution of $3X - Y$ is also normal variates and $3X - Y \sim N(3(2) - 5, (3)^2 (4) + 9) = N(1, 45)$.

Normal Distribution as a Limiting form of Binomial Distribution

Standard normal distribution can be considered as a limiting form of the standard binomial distribution under the following conditions

(*a*) n is very large, $n \to \infty$, (*b*) neither p nor q is very small, that is, too close to zero.

Let $X \sim B(n, p)$. Then, the standard binomial variate is defined as $Z = \dfrac{X - np}{\sqrt{npq}}$, $X = 0, 1, 2, 3, \ldots n$.

Now, when $X = 0$, $Z = \dfrac{-np}{\sqrt{npq}} = -\sqrt{\dfrac{np}{q}}$, and when $X = n$, $Z = \dfrac{n - np}{\sqrt{npq}} = \dfrac{nq}{\sqrt{npq}} = \sqrt{\dfrac{nq}{p}}$. $(\because p + q = 1)$

Therefore, as $n \to \infty$, Z varies from $-\infty$ to $+\infty$ with increment of $1/\sqrt{npq}$.

EXAMPLE 7.87 Assuming that the diameters of 1000 brass plugs taken consecutively from a machine, form a normal distribution with mean 1.9088 cm and standard deviation 0.0050 cm, how many of the plugs are likely to be rejected if the approved diameter is 1.910 ± 0.010 cm?

Solution The approved range of diameters is 1.90 to 1.92. The corresponding values of the normal variates are

$$z_1 = \frac{1.90 - 1.9088}{0.005} = -1.76, \quad z_2 = \frac{1.92 - 1.9088}{0.005} = 2.24.$$

Therefore, the number of plugs with approved diameters

$$= 1000\left[\int_{-1.76}^{2.24} \frac{1}{\sqrt{2\pi}} e^{-z^2/2} dz\right] = 1000\left[\frac{1}{\sqrt{2\pi}}\int_0^{1.76} e^{-z^2/2} dz + \frac{1}{\sqrt{2\pi}}\int_0^{2.24} e^{-z^2/2}\, dz\right]$$

$= 1000\,[0.4608 + 0.4875] = 1000[0.9483] = 948$ approx.

Therefore, the number of plugs likely to be rejected $= 1000 - 948 = 52$.

EXAMPLE 7.88 The marks of 500 candidates appearing in an English paper have a mean 41.6 and standard deviation 11.2. Assumming a normal distribution, find how many students obtain first class marks (for which a minimum of 60 marks is necessary).

Solution The number of students getting first class marks

$$= \frac{500}{11.2\sqrt{2\pi}} \int_{60}^{\infty} e^{-(x-41.6)^2/[2(11.2)^2]} dx. \qquad [i.e., 500 \times P(x \geq 60)].$$

Setting $(x - 41.6)/11.2 = z$, so that when $x = 60$, $z = 1.643$, the above integral

$$= \frac{500}{\sqrt{2\pi}} \int_{1.643}^{\infty} e^{-z^2/2} dz = 500\left[0.5 - \frac{1}{\sqrt{2\pi}}\int_0^{1.643} e^{-z^2/2} dz\right]$$

$= 500[0.5 - 0.45] = 500\,[0.05] = 25$ nearly. [using Table I in Appendix 4]

EXAMPLE 7.89 If (*i*) $X \sim N(75, 25)$, find $P(X > 80/X > 77)$. (*ii*) $X \sim N(10, 4)$, find $P(|X| \leq 5)$.

Solution (*i*) Since $z_1 = (80 - 75)/5 = 1$. Therefore, $P(X > 80) = \int_1^{\infty} \phi(z)\, dz$,

where $\phi(z) = \dfrac{1}{\sqrt{2\pi}} e^{-z^2}/2 = \int_0^{\infty} \phi(z) dz - \int_0^1 \phi(z) dz = 0.5 - 0.3413 = 0.1587$.

As $z_2 = (77 - 75)/5 = 0.4$, $P(X > 77) = \int_{0.4}^{\infty} \phi(z) dz = \int_0^{\infty} \phi(z) dz - \int_0^{0.4} \phi(z) dz = 0.5 - 0.1554 = 0.3446$.

Now, $P(X > 80/X > 77)$ means probability that $X > 80$ relative to the hypothesis that $X > 77$.

$\therefore \quad P(X > 80/X > 77) = P(X > 80)/P(X > 77) = (0.1587/0.3446) = 0.461$.

[**Note:** $X \sim N(75, 25)$ means that the random variable X is normally distributed with mean = 75 and variance = 25.]

In this case the p.d.f. of X is $f(x) = \dfrac{1}{5\sqrt{2\pi}} \exp\cdot\left(-\dfrac{(x - 75)^2}{2(25)}\right)\Big]$.

(*ii*) Here $z_1 = (5-10)/2 = -2.5$ and $z_2 = (-5-10)/2 = -7.5$.

$$\therefore \quad P(|X| \le 5) = \int_{z_2}^{z_1} \phi(z)dz = \int_{-7.5}^{-2.5} \phi(z)dz = \int_{2.5}^{7.5} \phi(z)dz = \int_0^{7.5} \phi(z)dz - \int_0^{2.5} \phi(z)dz$$

$= 0.5 - 0.4938 = 0.0062.$

EXAMPLE 7.90 (*i*) If $\log_e x$ is normally distributed with mean 1 and variance 4, find $P\left(\frac{1}{2} < x < 2\right)$ given that $\log_e 2 = 0.693$.

(*ii*) In a normal distribution, 31% of the items are under 45 and 8% are over 64. Find the mean and the standard deviation of the distribution. [Given: For a normal distribution the area between $z = 0$ and $z = 0.5$ is 0.19 and that between $z = 0$ and $z = 1.4$ is 0.42]

(AMIETE, W-2010; PTU 2006, JNTU 2003; AMIETE June 2003; Madras, 2000S; VTU, 2000)

Solution (*i*) Let $L = \log_e x$, Here $\mu = 1$, $\sigma^2 = 4$.

$$z = \frac{L-1}{2} = \frac{\log_e x - 1}{2}. \text{ When } x = \frac{1}{2},$$

$$z = \frac{\log(1/2) - 1}{2} = \frac{-1 - \log 2}{2} = -0.8465.$$

When $x = 2$,

$$z = \frac{\log 2 - 1}{2} = \frac{0.693 - 1}{2} = -0.1535.$$

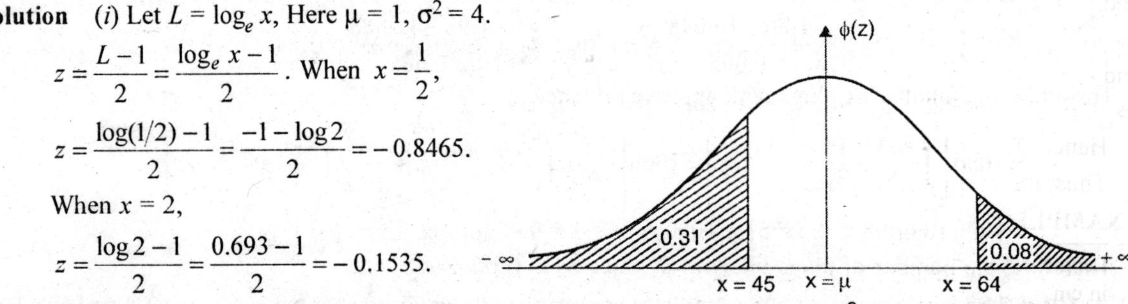

Fig. 7.39

Therefore,

$$P\left(\frac{1}{2} < x < 2\right) = \int_{-0.8465}^{-0.1535} \frac{1}{\sqrt{2\pi}} e^{-z^2/2} dz = \int_{0.1535}^{0.8465} \frac{1}{\sqrt{2\pi}} e^{-z^2/2} dz$$

= (area between $z = 0$ and $z = 0.8465$) – (area between $z = 0$ and $z = 0.1535$)
$= 0.2996 - 0.0596 = 0.24$ approx.

(*ii*) Let $X \sim N(\mu, \sigma^2)$. Then $P(X > 64) = 0.08$ and $P(X < 45) = 0.31$.

When $X = 45$, $Z = (45 - \mu)/\sigma = -z_1$. When $X = 64$, $Z = (64 - \mu)/\sigma = z_2$.

Therefore, $P(0 < Z < z_1) = 0.19$ and $P(0 < Z < z_2) = 0.42$.

From the table, we obtain $z_1 \approx 0.5$, and $z_2 \approx 1.4$.

Hence, $45 - \mu = -\sigma z_1 = -0.5\sigma$ and $64 - \mu = \sigma z_2 = 1.4\sigma$.

Solving these equations, we obtain $\mu = 50$ and $\sigma = 10$.

Thus, mean of the distribution is 50 and the standard deviation is 10.

EXAMPLE 7.91 The length of a certain species of fish are normally distributed with mean 60 cm, If 1.22% of the fish are less than 51, determine the standard deviation.

Solution Let x denotes the length of fish. Given that mean, $\mu = 60$.

Also, $P(x < 51) = 0.0122$.

Therefore, $P(x \ge 51) = 1 - 0.0122 = 0.9878$.

$$P\left[\frac{x-\mu}{\sigma} \ge \frac{51-60}{\sigma}\right] = 0.9878 \text{ or } P\left(Z > \frac{-9}{\sigma}\right) = 0.9878. \quad ...(i)$$

From table: $P(0 < Z < 2.25) = 0.4875$.

Therefore, $P(Z > -2.25) = 0.5 + 0.4878 = 0.9878$. ...(*ii*)

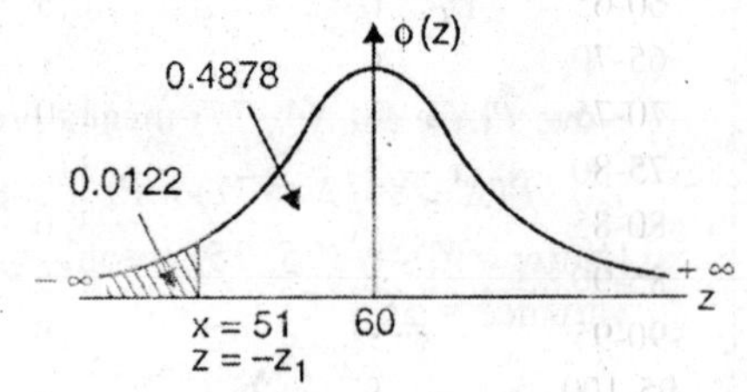

Fig. 7.40

From (*i*) and (*ii*), we have $\frac{-9}{\sigma} = -2.25 \Rightarrow \sigma = \frac{9}{2.25} = 4.$

EXAMPLE 7.92 If the skulls are classified as A, B and C according as the length breadth index is under 75, between 75 and 80 and over 80 respectively. Find approximately the mean and standard deviation of a series in which A are 58%, B are 38% and C are 4%. Assume that the distribution is normal.

Solution Let the length-breadth index is denoted by X, then $P(X < 75) = 0.58$ and $P(X > 80) = 0.04$.

$$\Rightarrow \quad P\left(\frac{X-\mu}{\sigma} < \frac{75-\mu}{\sigma}\right) = 0.58 \quad \text{and} \quad P\left(\frac{X-\mu}{\sigma} > \frac{80-\mu}{\sigma}\right) = 0.04$$

$\Rightarrow \quad P(Z < z_1) = 0.58 \quad$ and $\quad P(Z > z_2) = 0.04.$

where $\quad z_1 = (75-\mu)/\sigma \quad$ and $\quad z_2 = (80-\mu)/\sigma.$

$\Rightarrow \quad [0.5 + P(0 < Z < z_1)] = 0.58$

and $\quad [0.5 - P(0 < Z < z_2)] = 0.04$

$\Rightarrow \quad P(0 < Z < z_1)] = 0.08$

and $\quad P(0 < Z < z_2)] = 0.46$

$\Rightarrow \quad z_1 \approx 0.20 \quad$ and $\quad z_2 \approx 1.75 \quad$ [From the table].

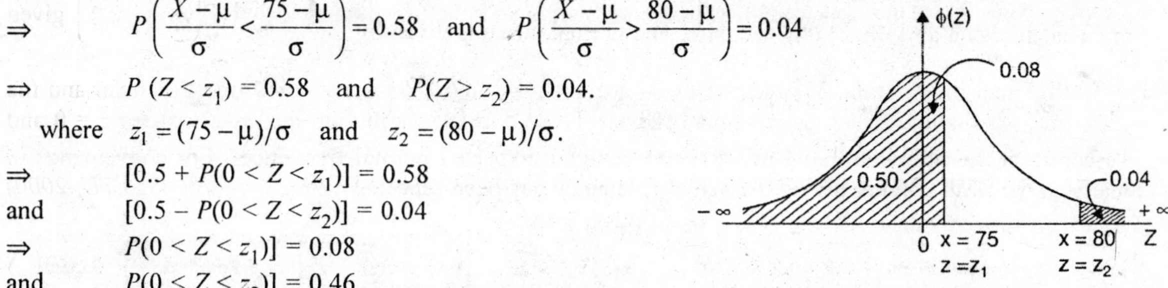

Fig. 7.41

Hence, $75 - \mu = 0.20\sigma$ and $80 - \mu = 1.75\sigma$. Solving these equations, we obtain $\mu = 74.4$ and $\sigma = 3.2$.

Thus, the mean of the series is 74.4 and the standard deviation is 3.2.

EXAMPLE 7.93 Fit a normal curve to the following data:

Length of line in cm,	8.60	8.59	8.58	8.57	8.56	8.55	8.54	8.53	8.52
Frequency	2	3	4	9	10	8	4	1	1

(PTU 2005)

Solution Let x denote the length of line and f, the corresponding frequency. We can easily find that $\mu = 8.563$, $\sigma_x = 0.0173$. and N (total frequency) $= \Sigma f_i = 42$.

Therefore, the equation of the normal curve fitted to this data is

$$y = \frac{N}{\sigma\sqrt{2\pi}} e^{-(x-\mu)^2/(2\sigma^2)} \quad -\infty \le x \le \infty = \frac{42}{0.0173\sqrt{2\pi}} e^{-(x-8.563)^2/0.0006}.$$

EXAMPLE 7.94 Fit a normal curve to the following data:

Class	60-65	65-70	70-75	75-80	80-85	85-90	90-95	95-100
Frequency	3	21	150	335	326	135	26	4

Compute the expected normal frequencies.

Solution First, we compute the mean and standard deviation of the distribution.

Table 7.9

Class	**Mid value x_i**	f_i	$d_x = \frac{x_i - 77.5}{5}$	d_x^2	$f_i d_x$	$f_i d_x^2$
60-65	62.5	3	–3	9	–9	27
65-70	67.5	21	–2	4	–42	84
70-75	72.5	150	–1	1	–150	150
75-80	$a \rightarrow$ **77.5**	335	0	0	0	0
80-85	82.5	326	1	1	326	326
85-90	87.5	135	2	4	270	540
90-95	92.5	26	3	9	78	234
95-100	97.5	4	4	16	16	64
		$N = \Sigma f_i = 1000$			$\Sigma f_i d_x = 489$	$\Sigma f_i d_x^2 = 1425$

$$\therefore \quad \bar{x} = a + h\frac{\Sigma f_i d_x}{N} = 77.5 + \frac{(489)5}{1000} = 79.945 (= \mu).$$

$$\sigma^2 = h^2\left[\frac{1}{N}\Sigma f_i d_x^2 - \left(\frac{1}{N}\Sigma f_i d_x\right)^2\right] = 25\left[\frac{1425}{1000} - \left(\frac{489}{1000}\right)^2\right] = 29.647.$$

σ (standard deviation) $= \sqrt{29.647} = 5.445$. The normal curve is given by

$$f(x) = \frac{1}{5.445\sqrt{2\pi}} \exp\left\{-\frac{1}{2}\left(\frac{x - 79.945}{5.445}\right)^2\right\}.$$

We construct the following table of values to compute expected normal frequences. For convenience of comparison, the forth column giving the actual frequency has been repeated.

Table 7.10

Class	Lower class boundary	$Z = (X - \mu)/\sigma$	$\phi(z) = 1/\sqrt{2\pi} \times \int_{-\infty}^{z} e^{-\mu^2/2} du$	$\Delta\phi(z) = \phi_{z+1} - \phi_z$	Expected frequency $= N\Delta\phi(z)$	Actual frequency
Below 60	$-\infty$	$-\infty$	0			
				0.0001	0.1	
60-65	60	–3.663	0.0001			
				0.0030	3.0} = 3.0	3
65-70	65	–2.745	0.0031		31 = 31	21
				0.0310		
70-75	70	–1.826	0.0341		148 = 148	150
				0.1480		
75-80	75	–0.908	0.1821		321.9 = 322	335
				0.3219		
80-85	80	0.0101	0.5040		319.2 = 319	326
				0.3192		
85-90	85	0.928	0.8232		144.40 = 144	135
				0.1444		
90-95	90	1.847	0.9676		29.60 = 30	26
				0.0296		
95-100	95	2.765	0.9972		2.7 = 3	4
				0.0027		
Over 100	100	3.683	0.9999			

EXAMPLE 7.95 It is given that X and Y are independent normal variates with mean 6, 7 and standard deviations 3, 4 respectively. Find the value of K such that $P(2X + Y \le K) = P(4X - 3Y \ge 4K)$.

Solution We have $X \sim N(6, 9)$, $Y \sim N(7, 16)$ and X, Y are independent. Define $U = 2X + Y$ and $V = 4X - 3Y$. Then U and V are also normal variates and

$$U = 2X + Y \sim N(2(6) + 7,\ 4(9) + 16) = N(19, 52).$$

$$V = 4X - 3Y \sim N(4(6) - 3(7),\ 16(9) + 9(16)) = N(3, 288).$$

Define, $Z = (U - 19)/\sqrt{52}$. Now $Z \sim N(0, 1)$ and for $U = K$, $Z = [(K - 19)\sqrt{52}] = z_1$.

Define, $Z = (V - 3)/\sqrt{288}$. Again $Z \sim N(0, 1)$ and for $V = 4K$, $Z = (4K - 3)/\sqrt{288} = z_2$.

We require $\quad P(2X + Y \le K) = P(4X - 3Y \ge 4K) \quad$ or $\quad P(Z \le z_1) = P(Z \ge z_2)$.

Since the normal probability curve is symmetric about $Z = 0$, this implies that $z_1 = -z_2$. Hence,

$$\frac{K - 19}{\sqrt{52}} = -\left[\frac{4K - 3}{\sqrt{288}}\right] \Rightarrow 6\sqrt{2}\,(K - 19) = \sqrt{13}\,(3 - 4K) \quad \text{or} \quad K = \frac{3\sqrt{13} + 114\sqrt{2}}{6\sqrt{2} + 4\sqrt{13}} \approx 7.51.$$

Moment Generating Function (MGF) of the Distribution

(a) MGF about origin

The MGF about origin is given by

$$M_X(t) = \int_{-\infty}^{\infty} e^{tx} \cdot f(x)\, dx = \frac{1}{\sigma\sqrt{2\pi}} \int_{-\infty}^{\infty} e^{tx} \cdot e^{-(x-\mu)^2/(2\sigma^2)} dx.$$

Putting $(x - \mu)/\sigma = z$. Then, $dx = \sigma\, dz$, we have

$$M_X(t) = \frac{1}{\sqrt{2\pi}} \int_{-\infty}^{\infty} e^{t(\mu + \sigma z)} \cdot e^{-z^2/2} dz = \frac{e^{\mu t}}{\sqrt{2\pi}} \int_{-\infty}^{\infty} e^{-(z^2 - 2\sigma t z)/2} dz$$

$$= \frac{e^{\mu t}}{\sqrt{2\pi}} \int_{-\infty}^{\infty} e^{-[(z-\sigma t)^2 - \sigma^2 t^2]/2} dz = \frac{e^{[\mu t + (\sigma^2 t^2/2)]}}{\sqrt{2\pi}} \int_{-\infty}^{\infty} e^{-(z-\sigma t)^2/2} dz$$

$$= \frac{e^{[\mu t + (\sigma^2 t^2/2)]}}{\sqrt{\pi}} \int_{-\infty}^{\infty} e^{-u^2} du \quad \left[\text{Set } \frac{z - \sigma t}{\sqrt{2}} = u\right] = \frac{e^{[\mu t + (\sigma^2 t^2/2)]}}{\sqrt{\pi}} \cdot \sqrt{\pi} = e^{[\mu t + (\sigma^2 t^2/2)]}. \quad \text{...(7.34)}$$

(b) MGF of the standard normal distribution N (0, 1).

The standard normal variate is defined by $z = (x - \mu)/\sigma$ and

$$M_z(t) = \frac{1}{\sigma\sqrt{2\pi}} \int_{-\infty}^{\infty} e^{[(x-\mu)t/\sigma]} \cdot e^{-(x-\mu)^2/(2\sigma^2)} dx = e^{-\mu t/\sigma} \left[\frac{1}{\sigma\sqrt{2\pi}} \int_{-\infty}^{\infty} e^{xt/\sigma} e^{-(x-\mu)^2/(2\sigma^2)} dx\right]$$

$$= e^{-\mu t/\sigma} M_X\left(\frac{t}{\sigma}\right) = e^{-\mu t/\sigma} \cdot e^{\left[\frac{\mu t}{\sigma} + \frac{\sigma^2}{2}\left(\frac{t^2}{\sigma^2}\right)\right]} = e^{t^2/2}. \quad \text{...(7.35)}$$

7.9 SOME OTHER DISTRIBUTIONS

Now, we shall discuss some other distributions of continuous random variables:

(i) Uniform (or rectangular) distribution,
(ii) Exponential distribution,
(iii) Beta distribution.

7.9.1 Uniform (or Rectangular) Distribution

If a continuous random variable can assume all real values between $x = a$ and $x = b$ and if the density function f is independent of position within this interval, then the random variable is said to possess a *uniform* distribution. The probaility density function $f(x)$ and the distribution function $F(x)$ are thus given, for the *uniform* distribution over the interval $(a \le x \le b)$ by

$$f(x) = \begin{cases} 0 & \text{if } x < a \\ \dfrac{1}{b-a} & \text{if } a \le x \le b \\ 0, & \text{if } x > a; \end{cases} \qquad F(x) = \begin{cases} 0 & \text{if } -\infty < x < a \\ \dfrac{x-a}{b-a} & \text{if } a \le x \le b \\ 1 & \text{if } b < x < \infty. \end{cases}$$

The mean and variance of the uniform distribution defined over the interval (a, b) are given by

$$\text{Mean} = \mu = \frac{1}{2}(a + b), \qquad \text{Variance} = \sigma^2 = \frac{1}{12}(b - a)^2.$$

7.9.2 Exponential Distribution

Def. The *exponential distribution* is defined by the density function

$$f(x)=\begin{cases}0 & \text{if } x<0,\\ \lambda e^{-\lambda x} & \text{if } x\geq 0, \text{ where } \lambda \text{ is a positive constant.}\end{cases}$$

Exponential distributions have a characteristic property which suggests their use in certain problems involving radioactive decay, traffic accidents, and failure of electronic equipment such as vacuum tubes.

For the exponential distribution, $E(X) = 1/\lambda$, variance $(X) = 1/\lambda^2$.

Mean and Standard deviation of exponential distribution

$$\text{Mean } =\lambda\int_0^\infty xe^{-\lambda x}dx=\lambda\left\{\left[x\frac{e^{-\lambda x}}{-\lambda}\right]_0^\infty+\frac{1}{\lambda}\int_0^\infty e^{-\lambda x}dx\right\}a=\lambda\left[0+\frac{1}{\lambda}\left[\frac{e^{-\lambda x}}{-\lambda}\right]_0^\infty\right]=\frac{1}{\lambda}.$$

Now $$\mu_2'=\lambda\int_0^\infty x^2e^{-\lambda x}dx=\lambda\left[x^2\left(\frac{e^{-\lambda x}}{-\lambda}\right)-(2x)\left(\frac{e^{-\lambda x}}{\lambda^2}\right)+2\left(\frac{e^{-\lambda x}}{-\lambda^2}\right)\right]_0^\infty=\frac{2}{\lambda^2}.$$

Variance, $\sigma^2=\mu_2'-\mu_1'^2=\mu_2'-(\text{mean})^2=\dfrac{2}{\lambda^2}-\dfrac{1}{\lambda^2}=\dfrac{1}{\lambda^2}.$ $(\because \mu_1'=\text{mean})$

$\therefore$ Standard deviation, $\sigma = 1/\lambda$. *[AMIE W-2005; MDU 2004; PTU-2003]*

EXAMPLE 7.96 The p.d.f. of X is given by $f(x)=\lambda e^{-\lambda x}$, $x\geq 0$, $\lambda>0$. Calculate $P[X > E(X)]$.

Solution $$E(X)=\lambda\int_0^\infty x\lambda e^{-\lambda x}dx=\lambda\left[x\left(\frac{e^{-\lambda x}}{-\lambda}\right)-\left(\frac{e^{-\lambda x}}{\lambda^2}\right)\right]_0^\infty=\lambda\left[\frac{1}{\lambda^2}\right]=\frac{1}{\lambda}.$$

$\therefore$ $$P[X>E(X)]=\int_{1/\lambda}^\infty \lambda e^{-\lambda x}dx=\lambda\left[\frac{e^{-\lambda x}}{-\lambda}\right]_{1/\lambda}^\infty=-(0-e^{-1})=\frac{1}{e}.$$

7.9.3 Beta Distribution

To understand *Beta distribution,* the readers should be familiar with the following definitions.

Gamma Function. Def. The gamma function denoted by $\Gamma(n)$ is defined by the improper integral

$$\Gamma(n)=\int_0^\infty x^{n-1}e^{-x}dx,\ n>0.$$

We note the following values of the Gamma function.

(*a*) $\Gamma(n+1)=n\Gamma(n)$, (*b*) $\Gamma(1)=1$, (*c*) $\Gamma(n+1)=n!$ for any positive integer m,

(*d*) $\Gamma(1/2)=\sqrt{\pi}$, (*e*) $\Gamma(n)=\dfrac{\Gamma(n+1)}{n}$ $(0>n\neq -1,-2,-3,\cdots)$.

The Beta Function. Def. The Beta function denoted by $\beta(m, n)$ or $\beta(n, m)$ is defined by the integral.

$$\beta(m,n)=\int_0^1 x^{m-1}(1-x)^{n-1}dx. \qquad (m>0,\ n>0)$$

We know that $\beta(m, n) = \beta(n, m)$.

In terms of the Γ function, $\beta(m,n)=\dfrac{\Gamma(m)\,\Gamma(n)}{\Gamma(m+n)}$ $m, n > 0$. ...(7.36)

Beta Distribution

A random variable X has a *beta distribution* or is a *beta random variable* if X has the interval (0, 1) as range and there are positive numbers m and n for which the probability density function f satisfies.

$$f(x) = \begin{cases} \dfrac{1}{\beta(m,n)} x^{m-1}(1-x)^{n-1}, \ 0 \le x \le 1, & n, m > 0 \quad = \dfrac{\Gamma(m+n)}{\Gamma(m)\,\Gamma(n)} x^{m-1}(1-x)^{n-1}, \\ 0 \text{ elsewhere}, \end{cases}$$

where Γ is the *gamma function* and β is the *beta function*.

Mean and Standard Deviation of Beta Distribution

(*a*) Mean $= \int_0^1 xf(x)\,dx = \dfrac{1}{\beta(m,n)} \int_0^1 x \cdot x^{m-1}(1-x)^{n-1}\,dx = \dfrac{1}{\beta(m,n)} \int_0^1 x^m (1-x)^{n-1}\,dx = \dfrac{\beta(m+1,n)}{\beta(m,n)}$

$$= \frac{\Gamma(m+1)\,\Gamma(n)}{\Gamma(m+n+1)} \div \frac{\Gamma(m)\,\Gamma(n)}{\Gamma(m+n)} = \frac{m}{m+n}. \qquad ...(7.37) \qquad [\because \Gamma(m+1) = m\Gamma(m)]$$

(*b*) Now $\mu_2' = \dfrac{1}{\beta(m,n)} \int_0^1 x^2 \cdot x^{m-1}(1-x)^{n-1}\,dx = \dfrac{1}{\beta(m,n)} \int_0^1 x^{m+1}(1-x)^{n-1}\,dx = \dfrac{1}{\beta(m,n)} \beta(m+2,n)$

$$= \frac{\Gamma(m+2)\Gamma(n)}{\Gamma(m+n+2)} \div \frac{\Gamma(m)\Gamma(n)}{\Gamma(m+n)} = \frac{m(m+1)}{(m+n+1)(m+n)}.$$

Now Variance, $\sigma^2 = \mu_2' - \mu^2 = \mu_2' - (\text{mean})^2 = \dfrac{m(m+1)}{(m+n+1)(m+n)} - \left(\dfrac{m}{m+n}\right)^2 = \dfrac{mn}{(m+n)^2(m+n+1)}.$

Standard deviation, $\sigma = \sqrt{\dfrac{mn}{(m+n)^2(m+n+1)}} = \dfrac{1}{m+n}\sqrt{\dfrac{mn}{m+n+1}}.$ ...(7.38)

EXERCISE 7.5

The Binomial Distribution

1. The incidence of occupational disease in an industry is such that the workers have a 20% chance of suffering from it. What is probability that out of 6 workers chosen at random 4 or more will suffer from the disease? **Ans.** 53/3125.
2. The probability that a pen manufactured by a company will be defective is 0.1. If 12 such pens are manufactured, find the probability that
   (*a*) exactly two will be defective, (*b*) at least two will be defective, (*c*) none will be defective.
   (*VTU, 2003, PTU 2002*) **Ans.** (*a*) 0.2301; (*b*) 0.3412; (*c*) 0.2833.
3. A group of 20 aeroplanes are sent on an operational flight. The chance that an aeroplane fails to return from the flight is 5%. Determine the probability that (*a*) one plane does not return (*b*) at the most 5 planes do not return.
   Ans. (*a*) ${}^{20}C_1(0.05)(0.95)^{19}$ (*b*) $\sum_{r=0}^{5} {}^{20}C_r (0.05)^r (0.95)^{20-r}$.
4. A factory finds that on the average 20% of the bolts produced by a given machine will be defective for certain specified requirements. If 10 bolts are selected at random from the day production of this machine, find the probability that (*a*) exactly 2 will be defective (*b*) 2 or more will be defective (*c*) more than 5 will be defective?
   Ans. (*a*) ${}^{20}C_2(0.2)^2(0.8)^8 = 0.302$, (*b*) $1 - (0.8)^{10} - 10(0.2)(0.8)^9 = 0.6242$,
   (*c*) $\sum_{r=6}^{10} {}^{10}C_r (0.2)^r (0.8)^{10-r} = 0.00637$.
5. (*a*) If X is a binomial variate with $p = 1/5$, for the experiment of 50 trials. Find the S.D.
   (*AMIETE, June 2011, 2010, 2009, 2008*) **Ans.** $2\sqrt{2}$.

(*b*) The mean and variance of the binomial distribution are 4 and 4/3 respectively. Find $P(X \geq 1)$.

Ans. 0.9986 (approx.)

[**Hint:** (*b*) $np = 4$, $npq = 4/3$, $\therefore$ $q = 1/3, p = 1 - q = 2/3$. $\therefore$ $n = 4/p = 6$.

$P(X \geq 1) = 1 - P(X = 0) = 1 - q^n = 1 - (1/3)^6 = 0.9986$.]

(*c*) With the usual notations, find the probability of a success for a binomial variate X, if $n = 6$ and $9P(X = 4) = P(X = 2)$.

Ans. 0.25.

[**Hint:** According to the given condition, $9p\ (x = 4) = p(x = 2)$.

or $9 \times {}^6C_4\ (p)^4\ (1 - p)^2 = {}^6C_2 p^2\ (1 - p)^4 \Rightarrow 9p^2 = (1 - p)^2$

$\Rightarrow 8p^2 + 2p - 1 = 0 \Rightarrow (4p - 1)\ (2p + 1) = 0$.

(*d*) Find the binomial distribution for which mean and standard deviation are 6 and 2 respectively.

Ans. $p(X = x) = {}^{18}C_x (1/3)^x (2/3)^{18-x}$ $\quad x = 0, 1, 2, ..., 18$.

6. A man takes a step forward with probability 0.4 and backward with probability 0.6. Find the probability that at the end of eleven steps, he is just one step away from the starting point.

(*AMIETE, Dec. 2003*) **Ans.** 3679/10000.

7. Assuming that half the population of a town consumes chocolates and that 100 investigators each take 10 individuals to see whether they are consumers, how many investigators would you expect to report that three people or less were consumers?

Ans. 17.

8. Out of 800 families with 4 children each, how many familes would you expect to have (*a*) 2 boys and 2 girls, (*b*) at least one boy, (*c*) no girl, (*d*) at most two girls assuming equal probabilities for boys and girls? (*VTU, 2004; AMIETE June 2002*) **Ans.** (*a*) 300; (*b*) 750; (*c*) 50; (*d*) 550.

9. Six dice are thrown together 729 times. How many times do you expect at least three dice to show a five or six.

Ans. 233.

10. Two players A and B play tennis games. Their chances of winning a game are in the ratio 3:2 respectively. Find *A*'s chance of winning at least two games out of four games played. **Ans.** 513/625.

[**Hint:** $P(X \geq 2) = 1 - (P(X = 0) + P(X = 1))$

$$= 1 - {}^4C_0 \left(\frac{3}{5}\right)^0 \left(\frac{2}{5}\right)^4 - {}^4C_1 \left(\frac{3}{5}\right)^1 \left(\frac{2}{5}\right)^3 = 1 - \frac{1}{5^4}[16 + 4(3)(8)] = \frac{513}{625}.\Bigg]$$

11. The probability that a bomb dropped from a plane will strike the target is 20%. If six bombs are dropped, find the probability that (*a*) exactly 2 will strike the target, (*b*) at least 2 will strike the target.

(*MDU 2007*) **Ans.** 0.246; 0.345.

12. Fit a binomial distribution for the following data and compare the theoretical frequencies with the actual one:

x:	0	1	2	3	4	5
f:	2	14	20	34	22	8

(*AMIETE, Dec 2011*) **Ans.** $100(0.432 + 0.568)^5$.

[**Hint:** $$\bar{x} = \frac{2(0) + 14(1) + 20(2) + 34(3) + 22(4) + 8(5)}{2 + 14 + 20 + 34 + 22 + 8} = \frac{14 + 40 + 102 + 88 + 40}{100} = 2.84.$$

$p = \dfrac{2.84}{5} = 0.57$, $q = 1 - p = 0.43$. The fitted binomial distribution is

$f(x) = {}^5C_x\ (0.57)^x\ (0.43)^{5-x}$ for $x = 0, 1, 2, 3, 4, 5$.

Table 7.11: Fitting Binomial Distribution to an observed distribution

x	$f(x) = {}^5C_x (0.57)^x (0.43)^{5-x}$	Expected frequency $Nf(x) = 100 f(x)$			Observed frequency
0	0.0147	1.47	or	1	2
1	0.0974	9.74	or	10	14
2	0.2583	25.83	or	26	20
3	0.3424	34.24	or	34	34
4	0.2270	22.70	or	23	22
5	0.0602	6.02	or	6	8

The Poisson Distribution

13. Six rats are administered a certain does of poison and the number of rats dying within 24 hours is observed. Suppose that each rat has a probability 1/4 of dying and that the survival of each rat is independent of the survival of the other rats. What is the probability that (a) exactly 4 rats die (b) all the rats die.

[**Hint:** That is a binomial experiment, number of trials = 6, $p = 1/4$, $q = 3/4$.

$$P(\text{exactly 4 rats die}) = {}^6C_4 p^4 q^2 = \frac{6!}{4!2!}\left(\frac{1}{4}\right)^4\left(\frac{3}{4}\right)^2 = \frac{15\times 9}{4^6} = \frac{135}{4^6}.$$

$$P(\text{all rats die}) = P(X=6) = p^6 = \left(\frac{1}{4}\right)^6 = \frac{1}{4^6}.\Big]$$

14. Find the probability of a success for the binomial distribution satisfying the following relation $4P(X=4) = P(X=2)$ and having the other parameter as six.

[**Hint:** We have $4P(X=4) = P(X=2) \Rightarrow 4\cdot {}^6C_4 p^4 q^2 = {}^6C_2 p^2 q^4$

$$\Rightarrow \quad 60p^4q^2 - 15p^2q^4 = 0 \quad \Rightarrow \quad 15p^2q^2(4p^2 - q^2) = 0$$

$$\Rightarrow \quad 4p^2 - q^2 = 0 \ (\text{as } p \neq 0, q \neq 0) \quad \Rightarrow \quad 4p^2 - (1-p)^2 = 0$$

$$\Rightarrow \quad (2p+1-p)(2p-1+p) = 0 \quad \Rightarrow \quad p = -1 \text{ or } p = 1/3.$$

Thus, $p = 1/3$ (as $p \neq -1$)].

15. Find the probability that at most 5 defective fuses will be found in a box of 200 fuses if experience shows that 2% of such fuses are defective ($e^{-4} = 0.0183$). **Ans.** 0.784.

[**Hint:** $m = np = 200 \times \dfrac{2}{100} = 4$. Probability of at most 5 successes = $P(X \le 5)$]

$$= p(0) + p(1) + p(2) + \cdots + p(5) = e^{-m}\frac{m^0}{0!} + e^{-m}\frac{m^1}{1!} + \cdots + e^{-m}\frac{m^5}{5!}$$

$$= e^{-4}\left[1 + 4 + 8 + \frac{32}{3} + \frac{32}{3} + \frac{128}{15}\right] = 0.0183 \times \frac{643}{15} = 0.784\Big].$$

16. Using Poisson distribution, find the probability that the ace of spades will be drawn from a pack of well shuffled cards at least once in 104 consecutive trials.

[**Hint:** $p = 1/52$, $n = 104$, $\therefore m = np = (104)(1/52) = 2$.

$$\text{Reqd. prob} = 1 - (\text{Prob. of not drawing an ace}) = 1 - \frac{e^{-2}2^0}{0!} = 1 - \frac{1}{e^2} = 0.864\Big].$$

17. If 2% of electric bulbs manufactured by a company are defective, find the probability that in a sample of 100 bulbs (*a*) 0 (*b*) 1 (*c*) 2 (*d*) 3 bulbs will be defective (use Poisson distribution).

Ans. (*a*) 0.135 (*b*) 0.271 (*c*) 0.271 (*d*) 0.180

18. A hospital administrator, who has been studying daily emergency admissions over a period of several years, has concluded that they are distributed according to the Poission Law. Hospital records reveal that emergency admissions have averaged 3 per day during this period. Find the proability that

(*a*) exactly three emergency admission wil occur on a given day.

(*b*) Either three or four emergency cases will be admited. Use $e^{-3} = 0.050$.

[**Hint:** (*a*) Here $m = 3$. Prob. of 3 admission per day $= P(X=3) = e^{-m}\dfrac{m^3}{3!} = 0.050 \times \dfrac{3^3}{3\times 2} = 0.225$.

(*b*) Prob. of either 3 or 4 admissions $= P(X = 3 \text{ or } 4) = P(X=3) + P(X=4)$

$$= e^{-3}\frac{(3)^3}{3!} + e^{-3}\frac{(3)^4}{4!} = e^{-3}\left(\frac{9}{2}+\frac{27}{8}\right) = 0.05 \times \frac{63}{8} = 0.394\Bigg].$$

19. Suppose that over a period of several years the average number of deaths from a certain non-contagious disease has been 10. If the number of deaths from this disease follow poisson distribution. What is the probability that during the current year.

(*a*) exactly seven people will die from te disease.

(*b*) ten or more people will die from the disease (given $e^{-10} = 0.000045$).

[**Hint:** Here $m = 10$. Let x = numer of deaths in the current year.

The Poisson distribution: $P(X = x) = e^{-m}\dfrac{m^x}{x!}$.

(*a*) $P(X=7) = e^{-10}\dfrac{(10)^7}{7!}$, (*b*) $P(x \geq 10) = 1 - p\,(x < 10) = 1 - \sum_{x=0}^{9} P(x) = 1 - \sum_{x=0}^{9}\left[e^{-10}\dfrac{(10)^x}{x!}\right]$.

20. Wireless sets are manufactured with 25 soldered joints each on the average 1 joint in 500 is defective. How many sets can be expected to be free from defective joints in a consignment of 10,000 sets.

Ans. 9512.

21. The number of telephone calls arriving on an internal switch board of an office is 90 per hour. Find the probability that at the most 1 to 3 calls in a minutes on the board arrive. (use $e^{-1.5} = 0.2231$).

Ans. 0.711.

22. Suppose a book of 585 pages contains 43 printing errors. If these errors are randomly distributed throughout the book, find the probability that 10 pages selected at random, will be free of errors? (Use $e^{-0.735} = 0.4795$).

[**Hint:** Here $p = 43/585 = 0.0735$, $n = 10$, $m = np = 0.735$.

$P(X=r) = \dfrac{e^{-m}\cdot m^r}{r!} = \dfrac{e^{-0.735}\cdot(0.735)^r}{r!}$. Therefore, $P(\text{no error}) = P(X=0) = e^{-0.735} = 0.4795$]

23. (*a*) Using Poisson's distribution, find the probability that the aces of spades will be drawn from a pack of well-suffled cards at least once in 104 times consecutive trials (Use $e^{-2} = 0.1353$).

(*b*) Six coins are tossed 6400 times. Using the Poisson distribution determine the approximate probability of getting six heads x times.

Ans. $[e^{-100}(100)^x]/x!$

[**Hint:** (*a*) $p(r) = e^{-m}\dfrac{m^r}{r!}$. Here $p = \dfrac{1}{52}$, $n = 104$. Therefore, $m = np = 2$.

Probability of drawing ace of spades at least once

$= 1 - p$ (drawing no ace of spades) $= 1 - \dfrac{e^{-2}\cdot 2^0}{0!} = 1 - 0.1353 = 0.8647$].

24. A discrete random variable X follows Poisson distribution. Find the values of $P(X \leq 2/X \geq 1)$. Given $E(X) = 2.20$ and $e^{-2.20} = 0.1108$.

[**Hint:** $P(x \le 2/x \ge 1) = \dfrac{P[(x \le 2) \cap (X \ge 1)]}{P(X \ge 1)}$, as $P(A/B) = \dfrac{P(A \cap B)}{P(B)}$

$$= \frac{P(X=1)+P(X=2)}{1-P(X<1)} = \frac{f(1)+f(2)}{1-f(0)}$$

$$= \left(e^{-m}\cdot m + e^{-m}\cdot \frac{m^2}{2!}\right)\Big/1-e^{-m} = \left(e^{-2.2}\cdot(2.2)+e^{-2.2}\cdot\frac{(2.2)^2}{2}\right)\Big/1-e^{-2.2}$$

$$= \frac{0.1108\,(4.62)}{0.8892} = \frac{0.5119}{0.8892} = 0.58.]$$

25. In a certain factory turning out razor blades, there is a small chance of 0.002 for any blade to be defective. The blades are supplied in packets of 10. Use Poisson distribution to calculate the approximate number of packets containing no defective, one defective and two defective blades respectively in a consignment of 10,000 packets. Use $e^{-0.02} = 0.9802$. *(VTU 2004; UPTU 2002, DU, 2000)*

[**Hint:** Here $N = 10{,}000$, $n = 10$ and $p = 0.002$, $m = np = 0.02$. Required frequencies are given by Ne^{-m}, $Ne^{-m}\cdot m$, $Ne^{-m}\cdot m^2/2!$, that is, 10000 (0.9802), 10000 (0.9802) (0.02), 10000 (0.9802)$(0.02^2)/2$, that is, 9802, 196, 2 packets respectively].

26. Data was collected over a period of 10 years, showing number of deaths from horse kicks in each of the 200 army corps. The distribution of deaths was as following:

No. of deaths, X:	0	1	2	3	4	
Frequency f:	109	65	22	3	1	**Total 200**

Fit a Poisson distribution to the above data. *(PU, 2005)*

[**Hint:** $m = 0.61$, Poisson distribution $= \dfrac{e^{-0.61}(0.61)^X}{X!}$. Expected or theoretical frequencies are given by $\dfrac{200e^{-0.61}(0.61)^X}{X!}$ for $X = 0, 1, 2, 3, 4$. From the table $e^{-0.61} = 0.5434$ (approx.). The expected frequencies are 108.7, 66.3, 20.2, 4.1 and 0.7 *i.e.*, 109, 66, 20, 4 and 1 respectively.]

27. Fit a Poisson distribution to the following data:

No. of deaths, x_i:	0	1	2	3	4
Frequency f_i:	122	46	23	8	1

Given that $e^{-0.6} = 0.5488$.

[**Hint:** $m = \dfrac{122(0)+46(1)+23(2)+8(3)+1(4)}{122+46+23+8+1} = \dfrac{120}{200} = 0.6.$

$$P(X=0) = e^{-m} = e^{-0.6} = 0.5488, \qquad P(X=1) = e^{-0.6}\times(0.6) = 0.3293,$$

$$P(X=2) = 0.5488\times\frac{(0.6)^2}{2!} = 0.0988,\ P(X=3) = 0.5488\times\frac{(0.6)^3}{3!} = 0.0198,$$

lastly $P(X \ge 4) = 1 - P(X < 4) = 0.0033$.

Table 7.12: Fitting Binomial Distribution to an Observed distribution

x	$P(X = r)$	Expected or Theoretical frequency $N \times P(X = r)$			Observed frequency
0	0.5488	109.76	or	110	122
1	0.3293	65.86	or	65	46
2	0.0988	19.76	or	20	23
3	0.0198	3.96	or	4	8
4 or more	0.0033 (By subtraction)	0.66	or	1	1
Total	**1**			**200**	**200**

28. **Fit a poisson distribution to the following data and calculate theoretical frequencies:**

Deaths:	0	1	2	3	4	
Frequency:	122	60	15	2	1	**Total = 200.**

(*PTU 2006; MDU 2006; VTU 2004; UPTU 2003*) **Ans.** Theoretical frequencies are: 121, 61, 15, 2, 1.

29. **The distribution of typing mistakes per page committed by a typist is given below. Assuming a Poisson model, find out the expected frequencies:**

No. of mistakes /page (x):	0	1	2	3	4	5	
No. of pages (f) :	142	156	69	27	5	1	**Total = 400.**

Ans. Expected frequencies of Poisson distribution are:

x :	0	1	2	3	4	5	
f:	147	147	74	25	6	1	**Total = 400.**

30. **Fit a Poisson distribution for the following data:**

x :	0	1	2	3	4	5	
$f(x)$:	110	170	130	60	23	7	**Total = 500.**

[Hint: $P(X = r) = \dfrac{e^{-m} m^r}{r!}, r = 0, 1, 2, ...$ $\quad \bar{x} = \dfrac{\Sigma f_i x_i}{\Sigma f_i} = \dfrac{737}{500} = 1.474 = m$

The expected frequencies are given by $500 \dfrac{e^{-m} m^r}{r!}$, $r = 0, 1, 2, ...$

Expected or theoretical frequencies are 115, 169, 124, 61, 22, 9].

31. (*a*) If the variance of the Poisson distribution is 2, find the probabilities for $r = 1, 2, 3, 4$ from the recurrence relation of the poisson distribution. Use $e^{-2} = 0.1353$.

Ans. 0.2706, 0.2706, 0.1804, 0.0902.

(*b*) In a Poisson distribution if $2P(X = 1) = P(X = 2)$. Find the variance.

(*AMIETE, June 2011, Dec 2008*) **Ans.** 4.

32. (*a*) Let X be a Poisson random variable such that $2P(X = 0) = P(X = 2)$. Find standard deviation of X.

Ans. $\sqrt{2}$.

(*b*) If a random variable X has a Poisson distribution such that $p(X = 1) = P(X = 2)$, then find $P(X = 4)$.

Ans. $2/3e^2$.

33. If X is a poisson variate such that $P(X = 2) = 9P(X = 4) + 90P(X = 6)$. Calculate (*a*) mean (*b*) standard deviation.

Ans. 1, 1.

[**Hint:** We have $\dfrac{e^{-m}m^2}{2!} = \dfrac{9e^{-m}m^4}{4!} + 90\dfrac{e^{-m}m^6}{6!} \Rightarrow m^4 + 3m^2 - 4 = 0$

$\Rightarrow (m^2 + 4)(m^2 - 1) = 0 \Rightarrow m^2 = 1$ or $m = 1$ ($\because m > 0$).

Hence, mean = variance = 1].

The Normal Distribution

34. A manufacturer known from experience that the resistance of resistors he produces is normal with mean = 100Ω and standard deviation = 2Ω. What percentage of resistors will have resistance between 98 and 102Ω? *(AMIETE W-2002)* **Ans.** 68.26%.

35. The average height of soliders of a country is given as 68.22 inches with variance 10.8 sq inch. How many soliders out of 1000 would you expect to be over 72 inches tall? Given that the area under the standard normal curve between $z = 0$ to $z = 0.35$ is 0.1368 and between $z = 0$ and $z = 1.15$ is 0.3749. *(UPTU 2002)* **Ans.** 125.

36. The mean inside diameters of a sample of 200 washers produced by a machine is 5.02 mm and the standard deviation is 0.05 mm. The purpose for which these washers are intended allows a maximum tolerance in the diameter of 4.96 to 5.08 mm; otherwise, the washers are considered defective. Determine percentage of defective washers produced by the machine assuming the diameters are normally distributed. [Area under the standard normal curve between $z = 0$ and $z = 1.2$ is 0.3849] *(AMIE, W-2005)*
Ans. 23%; Number of defective washers = 47.

37. The length of an item manufactured on an automatic machine tool is a normally distributed random variable with parameters: mean = 10 and variance = 1/200. Find the probability of defective production, if the tolerance is 10 ± 0.05. *(AMIE, W-2001)* **Ans.** 0.48.

38. In sampling a large number of parts manufactured by a machine, the mean number of defectives in a sample of 20 is 2. Out of 1000 such samples, how many would be expected to contain at least 3 defective parts? **Ans.** 324.

39. In a test on 2000 electric bulbs, it was found that the life of a particular make, was normally distributed with an average life of 2400 hours and standard deviation of 60 hours. Estimate the number of bulbs likely to burn for (*a*) more than 2150 hours, (*b*) less than 1950 hours, and (*c*) more than 1920 hours but less than 2160 hours. *(AMIETE, Dec 2010, June 2010; VTU, 2005)* **Ans.** (*a*) 68, (*b*) 184, (*c*) 1908.

40. A sample of 100 dry battery cells tested to find the length of life produced the following result:
Mean = 12 hours, standard deviation = 3 hours. Assuming the data to be normally distributed, what percentage of battery cells are expected to have life (*a*) more than 15 hours, (*b*) less than 6 hours (*c*) between 10 and 14 hours. *(UPTU 2004)* **Ans.** (*a*) 15.87% (*b*) 2.28% (*c*) 49.74%.

41. A manufacturer produces air mail envelopes whose weight is normally distributed with mean 1.95 gms and standard deviation 0.05 gm. The envelops are sold in lots of 1000. How many envelops in a lot will be weighing between 1.95 gm to 2.00gm each. Given that $\dfrac{1}{\sqrt{2\pi}}\int_{-\infty}^{1} e^{-z^2/2}\, dz = 0.8413$.
(AMIE, W-2000) **Ans.** 341.

42. If $\log_{10}X$ is normally distributed with mean 4 and variance 4, find the probability of $1.202 < X < 8318 \times 10^4$, given that $\log_{10} 1202 = 3.08$ and $\log_{10} 8318 = 3.92$. **Ans.** 0.95.

43. In a normal distribution, 7% of the items are under 35 and 89% are under 63. Determine mean and standard deviations of the distribution. *(AMIETE Dec. 2001)* **Ans.** 50.3; 10.36.

44. The mean height of 500 student's is 151 cm and the standard deviation is 15 cm. Assuming that the heights are normally distributed, find how many students heights lie between 120 and 155 cm.
Ans. 294.

45. The following table gives a set of observations x of frequency f:

x :	8.53	8.54	8.55	8.56	8.57	8.58	8.59	8.60	
f:	1	4	8	10	9	4	3	1	**Total =40.**

Determine the mean and the standard deviation and fit a normal curve to this distribution.

Ans. $\mu = 8.5628;\ \sigma = 0.0156;\ y = \dfrac{40}{\sigma\sqrt{2\pi}} \exp\left[-\dfrac{(x-\mu)^2}{2\sigma^2}\right] = 1023\ \text{exp.}\{-2055\,(x-8.5628)^2\},$

since $\dfrac{40}{\sigma\sqrt{2\pi}} = \dfrac{40}{0.0156 \times 2.50713} = 1023$, and $\dfrac{1}{2\sigma^2} = \dfrac{0.5}{(0.0156)^2} = 2055.$

46. Fit a normal distribution to the following data:

Variable:	60-62	63-65	66-68	69-71	72-74
Frequency:	5	18	42	27	8

Total = 100.

Ans. $y = \dfrac{100}{2.92\sqrt{2\pi}} e^{-(x-67.45)^2/17.06}$

47. Five thousand candidates appeared in a certain examination carrying a maximum of 100 marks. It was found that the marks were normally distributed with mean $\mu = 39.5$ and with standard deviation $\sigma = 12.5$. Determine approximately the number of candidates who secured a first class for which a minimum of 60 marks is necessary. You may use the table given below (d_x denotes the deviation from the mean). The proportion A of the whole area of the normal curve lying to the left of the ordinate at the deviation d_x/σ is

d_x/σ	1.5	1.6	1.7	1.8
A	0.93319	0.94520	0.95543	0.96407

[**Hint:** We have $\dfrac{d_x}{\sigma} = \dfrac{X-\mu}{\sigma} = \dfrac{60-39.5}{12.5} = 1.64$. Interpolation of the value of A for $d_x/\sigma = 1.64$ can be carried out as below:

Table 7.13

d_x/σ	A	ΔA	Δ^2A	Δ^3A
1.5	**0.93319**			
		0.01201		
1.6	0.94520		**–0.00178**	
		0.01023		**0.00019**
1.7	0.95543		–0.00159	
		0.00864		
1.8	0.96407			

Here $r = (1.64-1.5)/(0.1) = 1.4$. Use Newton's interpolation formula: we have A for d_x/σ (= 1.64)

$= 0.93319 + 1.4 \times 0.01201 + \dfrac{1.4(1.4-1)}{2!}(-0.00178) + \dfrac{1.4(0.4)(-0.6)}{3!} \times 0.00019 + \ldots\ldots$

$= 0.93319 + 0.016814 - 0.0004984 - \ldots = 0.95$ (approx).

Hence number of students getting 60 or more than 60 marks is 5%.

$\therefore$ Required number of students = 5000 × 0.05 = 250.]

Exponential Distribution

48. Find the points of inflexion of the normal curve $f(x) = \dfrac{1}{4\sqrt{2\pi}} e^{-(x-10)^2/32}$ for $-\infty < x < \infty$.

Ans. 6 and 14.

[**Hint:** The points of inflexion are given by $\mu - \sigma$ and $\mu + \sigma$. Comparing $f(x)$ to the p.d.f. of a normal variable x, we find that $\mu = 10$ and $\sigma = 4$.]

49. It is given that X and Y are independent normal variates with mean 2, 5 and variances 4, 9 respectively. Find the value of K such that $P(X+Y \le K) = P(3X - Y \ge 3K)$. **Ans.** $K \approx 2.885$.

50. It is given that X and Y are independent normal variates and $X \sim N(1, 4)$, $Y \sim N(3, 16)$. Find the value of K such that $P(2X+Y \le K) = P(4X - Y \ge 2K)$. **Ans.** $K \approx 2.487$.

51. The probability density function for the life of a motor car tyre is given by $f(x) = \lambda e^{-\lambda x}$ for $x \ge 0$, where $\lambda = 0.04$ and x is measured in units of 1500 km. What is the probability that a single tyre will last more than 30 units? **Ans.** 0.301.

Appendices

APPENDIX-1 SOME CONSTANTS, SOME IMPORTANT CONVERSION FACTORS, GREEK ALPHABET

SOME CONSTANTS

$e = 2.71828$	$\log_e 2 = 0.69315$
$\pi = 3.14159$	$\log_e 3 = 1.09861$
$\sqrt{2} = 1.41421$	$\log_e 10 = 2.30259$
$\sqrt{3} = 1.73205$	$\log_{10} e = 0.43429$
$\sqrt{10} = 3.16228$	$\log_e \pi = 1.14473$
$1° = 60' = 3600'' = 0.01745$ rad	$\log_{10} \pi = 0.49715$
1 radian $= 57° \ 17' \ 44.81''$ $= 57.29578°$	π radians = 180 degrees

SOME IMPORTANT CONVERSION FACTORS

1 inch = 2.54000 cm	1 acre = 4840 yd^2 = 4046.8564 m^2
100 sq. metres = 1 are	100 ares = 1 hectare = 10,000 m^2
	1 lb = 453.59 gm
100 hectares = 1 sq. kilometre	1 Kilogm.weight = 2.20462 lb.
	1 gallon of water = 4.5460 litres
1 Quintal = 100 kgm	1 metric Ton (Tonne) = 1000 kgm
1 H.P = 178.298 cal/sec = 0.74570 kW	1 kW = 1000 watts = 238.662 cal/sec.
°F = °C (1.8) + 32	

GREEK ALPHABET

α	Alpha	ι	Iota	ρ	Rho
β	Beta	κ	Kappa	σ, Σ	Sigma
γ, Γ	Gamma	λ, Λ	Lambda	τ	Tau
δ, Δ	Delta	μ	Mu	υ, Y	Upsilon
$\varepsilon, \in$	Epsilon	ν	Nu	$\varphi, \phi \ \Phi$	Phi
ζ	Zeta	ξ	Xi	χ	Chi
η	Eta	o	Omicron	ψ, Ψ	Psi
$\theta, \vartheta, \Theta$	Theta	π	Pi	ω, Ω	Omega

APPENDIX-2 MATHEMATICAL FORMULAS

ALGEBRA

1. Laws of Exponents

$$a^m . a^n = a^{m+n}, \quad (ab)^m = a^m b^m, \quad (a^m)^n = a^{mn}, \quad a^{m/n} = \sqrt[n]{a^m}$$

If $\quad a \neq 0, \quad a^m / a^n = a^{m-n}, \quad a^0 = 1, \quad a^{-m} = 1/a^m.$

2. Cross-multiplication

If $\quad a_1x + b_1y + c_1 = 0$ and $a_2x + b_2y + c_2 = 0,$

then $$\frac{x}{b_1c_2 - b_2c_1} = \frac{y}{c_1a_2 - c_2a_1} = \frac{1}{a_1b_2 - a_2b_1}.$$

$$\therefore \quad x = \frac{b_1c_2 - b_2c_1}{a_1b_2 - a_2b_1}, \; y = \frac{c_1a_2 - c_2a_1}{a_1b_2 - a_2b_1}.$$

3. Quadratic Equation

The solutions of the quadratic equation $ax^2 + bx + c = 0$, $a \neq 0$ are given by $x = \dfrac{-b \pm \sqrt{b^2 - 4ac}}{2a}$.

If α and β are the roots of this quardratic equation with $\alpha > \beta$, then $\alpha + \beta = -b/a$, $\alpha\beta = c/a$. The roots are (*i*) real and different, (*ii*) equal (*iii*) imaginary according as the discriminant $b^2 - 4ac$ is (*i*) positive, (*ii*) zero (*iii*) negative.

4. The Cubic equation

If $\quad \alpha, \beta, \gamma$ are the roots of cubic equation $ax^3 + bx^2 + cx + d = 0$, then
$$\alpha + \beta + \gamma = -b/a, \quad \alpha\beta + \beta\gamma + \gamma\alpha = c/a, \quad \alpha\beta\gamma = -d/a.$$

5. The Factor therorem

If for $x = a$, $f(x) = 0$, then $(x - a)$ is a factor of $f(x)$ or $x = a$ is a root of $f(x) = 0$.

6. Series

If the series in arithmetical progression (A.P.) is $a + (a + d) + (a + 2d) + \ldots$, that is, first term a and the common difference d. Then nth term of A.P. is $a + (n - 1)\,d$.

S_n = Sum of n terms of the A.P. $= \dfrac{n}{2}(a + l)$, where l is the last term $= \dfrac{n}{2}[2a + (n-1)d]$.

If the series in geometrical progression (G.P.) is $a + ar + ar^2 + \ldots$, a is called the first term and the constant number $r \neq 0$ is called the common ratio of the G.P. Then nth term of G.P. $= ar^{n-1}$.

S_n = Sum of n terms of G.P. $= \dfrac{a(1 - r^n)}{1 - r}$, if $|r| < 1 = \dfrac{a(r^n - 1)}{r - 1}$, if $|r| > 1$.

If $|r| < 1$, $r^n \to 0$ as $n \to \infty$. Therefore S_n or S, sum to infinity $= \dfrac{a}{1 - r}$.

Sum of first n natural numbers, that is, $1 + 2 + 3 + \ldots + n$ is $S_1 = \dfrac{n(n+1)}{2}$.

Sum of squares of the first n natural numbers, that is, $1^2 + 2^2 + 3^2 + \ldots + n^2$ is $S_2 = \dfrac{n(n+1)(2n+1)}{6}$.

Sum of cubes of the first n natural numbers, that is, $1^3 + 2^3 + 3^3 + ... + n^3$ is $S_3 = \left[\frac{n(n+1)}{2}\right]^2$.

7. Permutations and Combinations

$$n! = n(n-1)(n-2)...4 \cdot 3 \cdot 2 \cdot 1. \quad n \geq 1, \quad 0! = 1, \quad {}^nP_r = \frac{n!}{(n-r)!}$$

$${}^nC_r = \frac{n!}{r!(n-r)!}, 0 \leq r \leq n \text{ and } n \text{ is a non-negative integer } = \frac{{}^nP_r}{r!}.$$

$${}^nC_0 = 1 = {}^nC_n, \quad {}^nC_{n-r} = {}^nC_r, \quad {}^nC_r + {}^nC_{r-1} = {}^{n+1}C_r.$$

8. Binomial Theorem

(*i*) For any positive integer n,

$$(a+b)^n = a^n + na^{n-1}b + \frac{n(n-1)}{1.2}a^{n-2}b^2 + \frac{n(n-1)(n-2)}{1.2.3}a^{n-3}b^3 + \cdots\cdots + nab^{n-1} + b^n.$$

(*ii*) Certain particular cases of Binomial Formula, where we assume $|x| < 1$

$$(1+x)^{-1} = 1 - x + x^2 - x^3 + \cdots\cdots + (-1)^r x^r + \cdots\cdots$$

$$(1+x)^{-2} = 1 - 2x + 3x^2 - \cdots\cdots + (-1)^r (r+1)x^r + \cdots\cdots$$

$$(1+x)^{-3} = 1 - 3x + 6x^2 - \cdots\cdots + (-1)^r \frac{(r+1)(r+2)}{2}x^r + \cdots\cdots$$

$$(1-x)^{-1} = 1 + x + x^2 + x^3 + \cdots\cdots + x^r + \cdots\cdots$$

$$(1-x)^{-2} = 1 + 2x + 3x^2 + 4x^3 + \cdots\cdots + (r+1)x^r + \cdots\cdots$$

$$(1-x)^{-3} = 1 + 3x + 6x^2 + \cdots\cdots + \frac{(r+1)(r+2)}{2}x^r + \cdots\cdots$$

9. Determinants

(*i*) Det. of order 2: $\begin{vmatrix} a_1 & b_1 \\ a_2 & b_2 \end{vmatrix} = a_1b_2 - a_2b_1.$

(*ii*) Det. of order 3: $\begin{vmatrix} a_1 & b_1 & c_1 \\ a_2 & b_2 & c_2 \\ a_3 & b_3 & c_3 \end{vmatrix} = a_1\begin{vmatrix} b_2 & c_2 \\ b_3 & c_3 \end{vmatrix} - b_1\begin{vmatrix} a_2 & c_2 \\ a_3 & c_3 \end{vmatrix} + c_1\begin{vmatrix} a_2 & b_2 \\ a_3 & b_3 \end{vmatrix}.$

$$= a_1(b_2c_3 - b_3c_2) + b_1(a_3c_2 - a_2c_3) + c_1(a_2b_3 - a_3b_2)$$

LOGARITHMS

Laws of logarithms

(*i*) $\log_a(mn) = \log_a m + \log_a n$

(*ii*) $\log_a\left(\frac{m}{n}\right) = \log_a m - \log_a n$

(*iii*) $\log_a m^n = n\log_a m$

(*iv*) $\log_b r = \frac{\log_a r}{\log_a b}$ (Base-changing formula)

(Note that we can use any base in place of (a))

Remark $\log_a 1 = 0, \quad \log_a a = 1, \quad \log_a a^x = x, \quad e^{\log x} = x. \quad a^x = e^{x \log a}$, for $a > 0$.

MENSURATION

(A) Formulas from Plane Geometry:

Notations

a, b and c are sides of a triangle; $s=(a+b+c)/2$ = semi-perimeter; h = altitude; p = perimeter; $\angle A$, $\angle B$ and $\angle C$ are angles of a triangle, r = radius of a circle, C = circumference, R = radius of a sphere.

Triangle:

1. Area of a triangle $=\dfrac{1}{2}bh.$

$$=\sqrt{s(s-a)(s-b)(s-c)} \text{ (Hero's formula)}$$

$$=\frac{1}{2}ab\sin C=\frac{1}{2}ca\sin B=\frac{1}{2}bc\sin A$$

$$=a^2\cdot\frac{\sqrt{3}}{4} \text{ (for an equilateral triangle)}$$

$$=\frac{a^2\sin B\sin C}{2\sin A}=\frac{b^2\sin C\sin A}{2\sin B}=\frac{c^2\sin A\sin B}{2\sin C}$$

$$=\frac{1}{2}ab \text{ (right-angled triangle, hypotenuse } c\text{)}$$

2. Area of a parallelogram = bh

3. Area of a trapezoid $=\dfrac{1}{2}(b_1+b_2)\,h$

4. Area of a circle $=\pi r^2, C=2\pi r$

5. Area of a sector of a circle with central angle θ radians $=\dfrac{1}{2}r^2\theta$, Arc length = $r\theta$, (θ in radians)

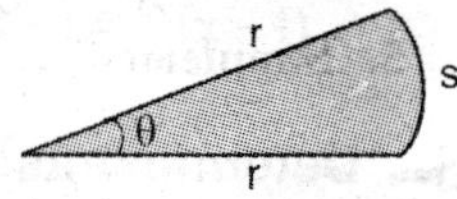

6. Area of a cyclic quadrilateral $=\sqrt{(s-a)(s-b)(s-c)(s-d)}$, where a, b, c, d are sides, and $s=(a+b+c+d)/2.$

7. Area of an ellipse = πab; (semi-major axis a and semi-minor axis b)

(B) Formulas from Solid Geometry:

1. *Right Circular Cylinder:*
   Volume = $\pi r^2 h$; Curved surface, S = $2\pi rh$.

2. *Right Circular Cone:*
   Volume $=\pi r^2 h/3$; Curved surface, $S=\pi rl=\pi r\sqrt{r^2+h^2}$.

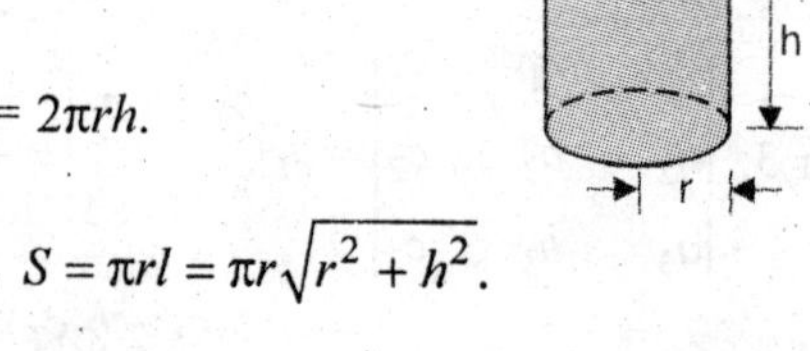

3. *Any Cylinder or Prism with Parallel Bases*

Volume = (Base area) × height = (B)(h)

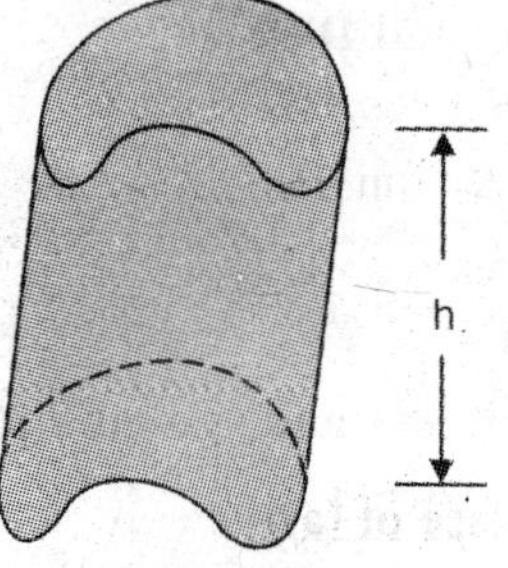

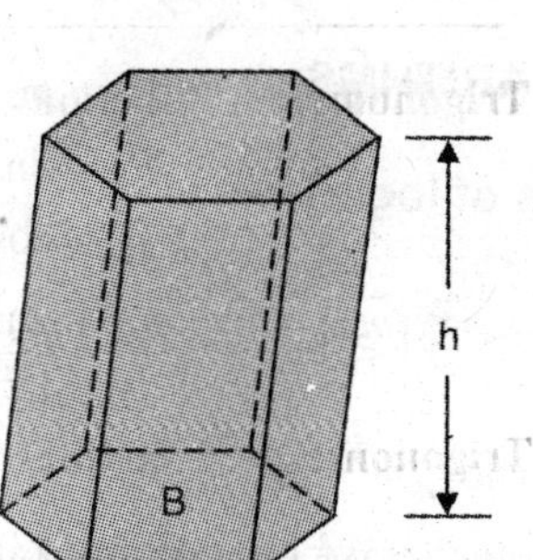

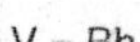

V = Bh

4. **Frustum of Pyramids and Cones:**

$$\text{Volume} = \frac{h}{3}\left[A_1 + A_2 + \sqrt{A_1 A_2}\right]; \text{ where } h = \text{height} \quad A_1 \text{ and } A_2 \text{ are the areas of ends.}$$

5. **Frustum of Right Circular Cone**

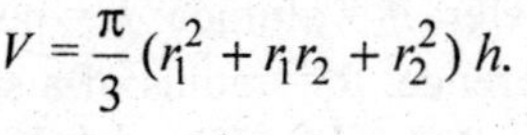

$$V = \frac{\pi}{3}(r_1^2 + r_1 r_2 + r_2^2)\, h.$$

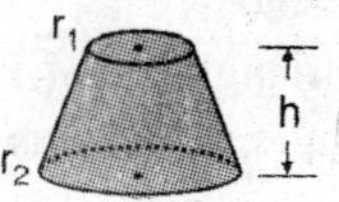

6. **Sphere:**

$$\text{Volume} = 4\pi R^3/3, \text{ Surface, } S = 4\,\pi R^2.$$

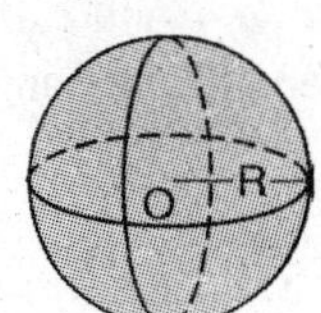

TRIGONOMETRY

1. **Radian:** An angle subtended at the centre of a circle by an arc equal in length to the radius of a circle is said to have a measure of 1 radian.
   $\theta = l/r$, where l is the length of arc which subtends angle θ (in radians) at the centre of the circle with radius r.
   π radians = 180°.

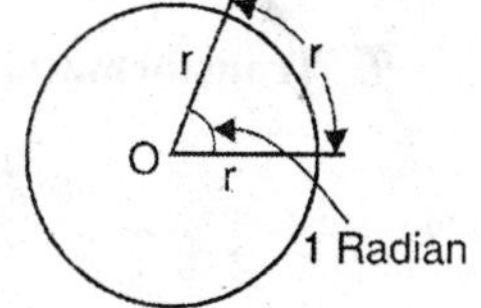

2. **Pythagorean Identities**

$$\sin^2\theta + \cos^2\theta = 1, \qquad \sec^2\theta = 1 + \tan^2\theta, \qquad \text{cosec}^2\,\theta = 1 + \cot^2\theta.$$

3. **Sign Identities**

$$\sin(-\theta) = -\sin\theta, \qquad \cos(-\theta) = \cos\theta, \qquad \tan(-\theta) = -\tan\theta.$$

II sin | I all
tan III | cos IV

4. **Complement Identities**

$$\sin\left(\frac{\pi}{2} - \theta\right) = \cos\theta, \qquad \cos\left(\frac{\pi}{2} - \theta\right) = \sin\theta, \qquad \tan\left(\frac{\pi}{2} - \theta\right) = \cot\theta.$$

$$\sin\left(\theta - \frac{\pi}{2}\right) = -\cos\theta, \qquad \cos\left(\theta - \frac{\pi}{2}\right) = \sin\theta, \qquad \sin\left(\frac{\pi}{2} + \theta\right) = \cos\theta, \qquad \cos\left(\frac{\pi}{2} + \theta\right) = -\sin\theta.$$

5. **Supplement Identities**

$$\sin(\pi - \theta) = \sin\theta, \quad \cos(\pi - \theta) = -\cos\theta, \quad \tan(\pi - \theta) = -\tan\theta.$$

$$\sin(\pi + \theta) = -\sin\theta, \quad \cos(\pi + \theta) = -\cos\theta, \quad \tan(\pi + \theta) = \tan\theta.$$

6. **Values of trigonometric functions**

Angle	0°	$\pi/6$	$\pi/4$	$\pi/3$	$\pi/2$	π	$3\pi/2$	2π
sin	0	1/2	$1/\sqrt{2}$	$\sqrt{3}/2$	1	0	−1	0
cos	1	$\sqrt{3}/2$	$1/\sqrt{2}$	1/2	0	−1	0	1
tan	0	$1/\sqrt{3}$	1	$\sqrt{3}$	not defined	0	not defined	0

7. **Trigonometric Functions of Sum and Difference**

$$\sin(A \pm B) = \sin A \cos B \pm \cos A \sin B.$$

$$\cos(A \pm B) = \cos A \cos B \mp \sin A \sin B.$$

$$\tan(A \pm B) = \frac{\tan A \pm \tan B}{1 \mp \tan A \tan B}.$$

8. **Trigonometric Functions of Multiples and sub-multiples Angles**

$$\sin 2A = 2\sin A \cos A = \frac{2\tan A}{1 + \tan^2 A}.$$

$$\cos 2A = \cos^2 A - \sin^2 A = 2\cos^2 A - 1 = 1 - 2\sin^2 A = \frac{1-\tan^2 A}{1+\tan^2 A}.$$

$$\sin^2 \frac{A}{2} = \frac{1-\cos A}{2}, \qquad \cos^2 \frac{A}{2} = \frac{1+\cos A}{2}.$$

$$\tan 2A = \frac{2\tan A}{1-\tan^2 A}, \qquad \sin 3A = 3\sin A - 4\sin^3 A, \qquad \cos 3A = 4\cos^3 A - 3\cos A,$$

$$\tan 3A = \frac{3\tan A - \tan^3 A}{1-3\tan^2 A}.$$

9. Transformation of a Product into a Sum and vice versa

$$\sin A \cos B = \frac{1}{2}\sin(A+B) + \frac{1}{2}\sin(A-B),$$

$$\cos A \sin B = \frac{1}{2}\sin(A+B) - \frac{1}{2}\sin(A-B),$$

$$\cos A \cos B = \frac{1}{2}\cos(A+B) + \frac{1}{2}\cos(A-B),$$

$$\sin A \sin B = \frac{1}{2}\cos(A-B) - \frac{1}{2}\cos(A+B).$$

$$\sin C + \sin D = 2\sin\frac{C+D}{2}\cos\frac{C-D}{2}, \qquad \sin C - \sin D = 2\cos\frac{C+D}{2}\sin\frac{C-D}{2},$$

$$\cos C + \cos D = 2\cos\frac{C+D}{2}\cos\frac{C-D}{2}, \qquad \cos C - \cos D = -2\sin\frac{C+D}{2}\sin\frac{C-D}{2},$$

10. Law of Sines or the Sine Formula

If A, B and C are the angles of a triangle and if a, b and c are the lengths of the sides opposite to A, B and C, respectively, then

$$\frac{\sin A}{a} = \frac{\sin B}{b} = \frac{\sin C}{c}.$$

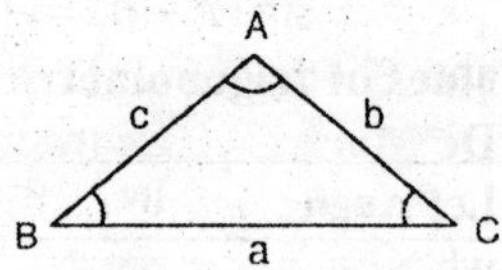

11. Law of Cosines or the Cosine Formulae

$$a^2 = b^2 + c^2 - 2bc\cos A,$$

$$b^2 = c^2 + a^2 - 2ca\cos B,$$

$$c^2 = a^2 + b^2 - 2ab\cos C.$$

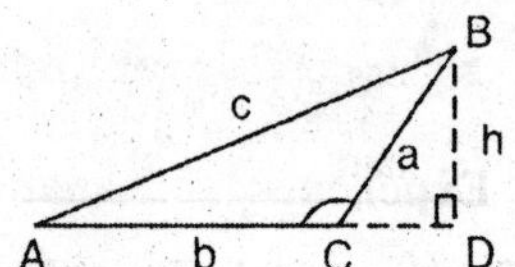

12. Projection Formulae

$$a = b\cos C + c\cos B, \qquad b = c\cos A + a\cos C, \qquad c = a\cos B + b\cos A.$$

13. Circumradius of the triangle, when sides are known:

$$R = \frac{a}{2\sin A} = \frac{b}{2\sin B} = \frac{c}{2\sin C} = \frac{abc}{4\Delta}.$$

14. Inradius of the triangle

$$r = \frac{\Delta}{s} = \frac{\text{area of the triangle}}{\text{semi-perimeter}}.$$

15. Inverse Circular Functions

(a) (*i*) $\sin^{-1}\dfrac{1}{x} = \operatorname{cosec}^{-1}x, x \geq 1$, (*ii*) $\cos^{-1}\dfrac{1}{x} = \sec^{-1}x, x \geq 1$, (*iii*) $\tan^{-1}\dfrac{1}{x} = \cot^{-1}x, x > 0$.

(b) (*i*) $\sin^{-1}(-x) = -\sin^{-1}x; x \in [-1, 1]$, (*ii*) $\tan^{-1}(-x) = -\tan^{-1}x, x \in R$,

(*iii*) $\operatorname{cosec}^{-1}(-x) = -\operatorname{cosec}^{-1}x, \quad |x| \geq 1$.

(c) (*i*) $\cos^{-1}(-x) = \pi - \cos^{-1}x, x \in [-1, 1]$, (*ii*) $\sec^{-1}(-x) = \pi - \sec^{-1}x, |x| \geq 1$,

(*iii*) $\cot^{-1}(-x) = \pi - \cot^{-1}x, x \in R$.

(d) (*i*) $\sin^{-1}x + \cos^{-1}x = \dfrac{\pi}{2}, x \in [-1, 1]$, (*ii*) $\tan^{-1}x + \cot^{-1}x = \dfrac{\pi}{2}, x \in R$,

(*iii*) $\operatorname{cosec}^{-1}x + \sec^{-1}x = \dfrac{\pi}{2}, |x| \geq 1$, (*iv*) $\tan^{-1}x = \dfrac{\pi}{2} - \tan^{-1}\left(\dfrac{1}{x}\right), \ x > 0$.

(e) $\sin^{-1}x + \sin^{-1}y = \sin^{-1}(x\sqrt{1-y^2} + y\sqrt{1-x^2})$.

(f) $\cos^{-1}x + \cos^{-1}y = \cos^{-1}(xy - \sqrt{1-x^2}\sqrt{1-y^2})$.

(g) (*i*) $\tan^{-1}x + \tan^{-1}y = \tan^{-1}\left(\dfrac{x+y}{1-xy}\right)$, if $xy < 1$,

(*ii*) $\tan^{-1}x - \tan^{-1}y = \tan^{-1}\left(\dfrac{x-y}{1+xy}\right)$, if $xy > -1$.

(h) (*i*) $2\tan^{-1}x = \sin^{-1}\left(\dfrac{2x}{1+x^2}\right), |x| \leq 1$, (*ii*) $2\tan^{-1}x = \cos^{-1}\left(\dfrac{1-x^2}{1+x^2}\right), x \geq 0$.

(*iii*) $2\tan^{-1}x = \tan^{-1}\left(\dfrac{2x}{1-x^2}\right), -1 < x < 1$.

COMPLEX NUMBERS AND FUNCTIONS

1. Every complex number $x + iy$ can always be expressed in the form $r(\cos\theta + i\sin\theta)$, where $r = \sqrt{x^2 + y^2}$ and $\theta = \tan^{-1}(y/x)$.

2. De'Moivre's Theorem

Let $z = r(\cos\theta + i\sin\theta)$. Then $z^n = r^n[\cos n\theta + i\sin n\theta]$,
where n is any positive or negative integer or a real rational number.

$$z^{-n} = r^{-n}[\cos n\theta - i\sin n\theta] = \frac{1}{r^n[\cos n\theta + i\sin n\theta]} = \frac{1}{z^n}.$$

3. Exponential, circular functions of a complex variable, z

(*i*) e^z or $\exp(z) = e^{x+iy} = e^x \cdot e^{iy} = e^x(\cos y + i\sin y)$,

(*ii*) Exponential form of $z\,(= x + iy) = re^{i\theta}$ (*iii*) $\sin z = \dfrac{e^{iz} - e^{-iz}}{2i}$,

(*iv*) $\cos z = \dfrac{e^{iz} + e^{-iz}}{2}$, (*v*) $\tan z = \dfrac{\sin z}{\cos z} = \dfrac{e^{iz} - e^{-iz}}{i(e^{iz} + e^{-iz})}$.

4. Euler's Theorem:

$$e^{i\theta} = \cos\theta + i\sin\theta, \text{ where } \theta \text{ is real or complex.}$$

5. Hyperbolic Functions: If x be real or complex,

(a) $\sinh x = \dfrac{e^x - e^{-x}}{2}$, $\cosh x = \dfrac{e^x + e^{-x}}{2}$, $\tanh x = \dfrac{\sinh x}{\cosh x} = \dfrac{e^x - e^{-x}}{e^x + e^{-x}}$,

$\coth x = 1/\tanh x = \cosh x/\sinh x$, $\quad \text{sech}\, x = 1/\cosh x$, $\quad \text{cosech}\, x = 1/\sinh x$.

(b) *Fundamental formulae*

$$\cosh^2 x - \sinh^2 x = 1, \quad \text{sech}^2 x + \tanh^2 x = 1, \quad \coth^2 x - \text{cosech}^2 x = 1.$$

6. Relations between hyperbolic and circular functions

$\sin ix = i \sinh x$; $\quad \cos ix = \cosh x$; $\quad \tan ix = i \tanh x$.

7. Inverse Hyperbolic function

$$\sinh^{-1} x = \log_e (x + \sqrt{x^2 + 1}); \quad \cosh^{-1} x = \log_e (x + \sqrt{x^2 - 1}); \quad \tanh^{-1} x = \frac{1}{2} \log \frac{1+x}{1-x}.$$

EXPONENTIAL AND LOGARITHMIC SERIES

1. Exponential Series

$$e^x = 1 + \frac{x}{1!} + \frac{x^2}{2!} + \frac{x^3}{3!} + \cdots + \frac{x^n}{n!} + \cdots \infty, \quad e = 1 + \frac{1}{1!} + \frac{1}{2!} + \frac{1}{3!} + \cdots \infty$$

2. Logarithmic Series: If $|x| < 1$, then $\log (1 + x) = x - \dfrac{x^2}{2} + \dfrac{x^3}{3} - \dfrac{x^4}{4} + \cdots$

Some Important Series

$$\sin x = x - \frac{x^3}{3!} + \frac{x^5}{5!} - \frac{x^7}{7!} + \cdots \qquad \log (1 - x) = -\left(x + \frac{x^2}{2} + \frac{x^3}{3} + \cdots \infty \right)$$

$$\cos x = 1 - \frac{x^2}{2!} + \frac{x^4}{4!} - \frac{x^6}{6!} + \cdots \qquad \sinh x = x + \frac{x^3}{3!} + \frac{x^5}{5!} + \frac{x^7}{7!} + \cdots \infty$$

$$a^x = e^{x \log a} = 1 + \frac{\log a}{1!} x + \frac{(\log a)^2}{2!} x^2 + \frac{(\log a)^3}{3!} x^3 + \cdots \infty$$

Polar Coordinates

$x = r\cos\theta,\ y = r\sin\theta, \qquad r = \sqrt{x^2 + y^2},\ \theta = \tan^{-1}(y/x), \qquad dxdy = x\, dr\, d\theta.$

APPENDIX-3 SOME DERIVATIVE AND INTEGRAL FORMULAS

Derivative Formulas

***u* and *v* are functions of *x*, c_1, and c_2 are constants.**

1. $\frac{d}{dx}(c_1u + c_2v) = c_1u' + c_1v'$ (Sum Rule)

2. $\frac{d}{dx}(uv) = uv' + vu'$ (Product Rule)

3. $\frac{d}{dx}\left(\frac{u}{v}\right) = \frac{vu' - uv'}{v^2}$ (Quotient Rule)

4. $\frac{dy}{dx} = \frac{dy}{du} \cdot \frac{du}{dx}$ (Chain Rule)

5. $\frac{d}{dx}(x^n) = nx^{n-1}$

6. $\frac{d}{dx}(\log_e x) = \frac{1}{x}$

7. $\frac{d}{dx}(e^{ax}) = ae^{ax}$

8. $\frac{d}{dx}(a^x) = a^x \log_e a$

9. $\frac{d}{dx}(\log_a x) = \frac{\log_a e}{x} = \frac{1}{x \log_e a}$

10. $\frac{d}{dx}(\sin x) = \cos x$

11. $\frac{d}{dx}(\cos x) = -\sin x$

12. $\frac{d}{dx}(\tan x) = \sec^2 x$

13. $\frac{d}{dx}(\cot x) = -\operatorname{cosec}^2 x$

14. $\frac{d}{dx}(\sec x) = \sec x \tan x$

15. $\frac{d}{dx}(\operatorname{cosec} x) = -\operatorname{cosec} x \cot x$

16. $\frac{d}{dx}(\sinh x) = \cosh x$

17. $\frac{d}{dx}(\cosh x) = \sinh x$

18. $\frac{d}{dx}(\tanh x) = \operatorname{sec}h^2 x$

19. $\frac{d}{dx}(\coth x) = -\operatorname{cosech}^2 x$

20. $\frac{d}{dx}(\operatorname{sech} x) = -\operatorname{sech} x \tanh x$

21. $\frac{d}{dx}(\operatorname{cosech} x) = -\operatorname{cosech} x \coth x$

22. $\frac{d}{dx}(\sin^{-1} x) = \frac{1}{\sqrt{1-x^2}}$

23. $\frac{d}{dx}(\cos^{-1} x) = \frac{-1}{\sqrt{1-x^2}}$

24. $\frac{d}{dx}(\tan^{-1} x) = \frac{1}{1+x^2}$

25. $\frac{d}{dx}(\cot^{-1} x) = \frac{-1}{1+x^2}$

26. $\frac{d}{dx}(\sec^{-1} x) = \frac{1}{x\sqrt{x^2 - 1}}, x > 1$

27. $\frac{d}{dx}(\operatorname{cosec}^{-1} x) = \frac{-1}{x\sqrt{x^2 - 1}}, x > 1$

28. $\frac{d}{dx}(\sinh^{-1} x) = \frac{1}{\sqrt{x^2 + 1}}$

29. $\frac{d}{dx}(\cosh^{-1} x) = \frac{\pm 1}{\sqrt{x^2 - 1}}, \ (x^2 > 1)$

30. $\frac{d}{dx}(\tanh^{-1} x) = \frac{1}{1 - x^2}, \ (x^2 < 1)$

31. $\frac{d}{dx}(\coth^{-1} x) = \frac{-1}{x^2 - 1}, \ (x^2 > 1)$

32. $\frac{d}{dx}(\operatorname{sech}^{-1} x) = \frac{\pm 1}{x\sqrt{1 - x^2}}, \ (x^2 < 1)$

33. $\frac{d}{dx}(\operatorname{cosech}^{-1} x) = \frac{\pm 1}{x\sqrt{x^2 + 1}}.$

Integral Formulas

u and v are functions of x, c_1, and c_2 are constants.

1. $\int (c_1 u + c_2 v)\, dx = c_1 \int u dx + c_2 \int v dx + c$

2. $\int uv'\, dx = uv - \int u' v dx + c,$ $\left(\begin{array}{c}\text{integration}\\ \text{by parts}\end{array}\right)$

3. $\int x^n dx = \dfrac{x^{n+1}}{n+1} + c \quad (n \neq -1)$

4. $\int \dfrac{1}{x}\, dx = \log_e |x| + c$

5. $\int e^{ax} dx = \dfrac{1}{a} e^{ax} + c$

6. $\int a^x dx = \dfrac{a^x}{\log_e a} + c$

7. $\int \sin ax\, dx = -\dfrac{1}{a} \cos ax + c$

8. $\int \cos ax\, dx = \dfrac{1}{a} \sin ax + c$

9. $\int \sec^2 x\, dx = \tan x + c$

10. $\int \operatorname{cosec}^2 x\, dx = -\cot x + c$

11. $\int \sec x \tan x\, dx = \sec x + c$

12. $\int \operatorname{cosec} x \cot x\, dx = -\operatorname{cosec} x + c$

13. $\int \tan ax\, dx = -\dfrac{1}{a} \log|\cos ax| + c = \dfrac{1}{a} \log|\sec ax| + c$

14. $\int \cot ax\, dx = \dfrac{1}{a} \log|\sin ax| + c$

15. $\int \sec x\, dx = \begin{cases} \log|\sec x + \tan x| + c \\ \log\left|\tan\left(\dfrac{\pi}{4} + \dfrac{x}{2}\right)\right| + c \end{cases}$

16. $\int \operatorname{cosec} x\, dx = \begin{cases} \log|\operatorname{cosec} x - \cot x| + c \\ \log\left|\tan \dfrac{x}{2}\right| + c \end{cases}$

17. $\int \dfrac{dx}{\sqrt{a^2 - x^2}} = \sin^{-1}\left(\dfrac{x}{a}\right) + c$

18. $\int \dfrac{dx}{x^2 + a^2} = \dfrac{1}{a} \tan^{-1}\left(\dfrac{x}{a}\right) + c$

19. $\int \sinh x\, dx = \cosh x + c$

20. $\int \cosh x\, dx = \sinh x + c$

21. $\int \dfrac{dx}{\sqrt{x^2 + a^2}} = \begin{cases} \sinh^{-1}\left(\dfrac{x}{a}\right) + c \\ \log_e (x + \sqrt{x^2 + a^2}) + c \end{cases}$

22. $\int \dfrac{dx}{\sqrt{x^2 - a^2}} = \begin{cases} \cosh^{-1}\left(\dfrac{x}{a}\right) + c \\ \log_e \left|x + \sqrt{x^2 - a^2}\right| + c \end{cases}$

23. $\int \dfrac{1}{a^2 - x^2}\, dx = \dfrac{1}{2a} \log\left|\dfrac{x+a}{x-a}\right| + c$

24. $\int \dfrac{dx}{x^2 - a^2} = \dfrac{1}{2a} \log\left|\dfrac{x-a}{x+a}\right| + c$

25. $\int \sqrt{a^2 - x^2}\, dx = \dfrac{x}{2}\sqrt{a^2 - x^2} + \dfrac{a^2}{2} \sin^{-1}\left(\dfrac{x}{a}\right) + c$

26. $\int \sqrt{x^2 + a^2}\, dx = \dfrac{x}{2}\sqrt{x^2 + a^2} + \dfrac{a^2}{2} \log\left|x + \sqrt{x^2 + a^2}\right| + c$

27. $\int \sqrt{x^2 - a^2}\, dx = \dfrac{x}{2}\sqrt{x^2 - a^2} - \dfrac{a^2}{2} \log\left|x + \sqrt{x^2 - a^2}\right| + c$

28. $\int \log x\, dx = x \log x - x + c$

29. $\int e^{ax} \sin bx\, dx = \dfrac{e^{ax}}{a^2 + b^2} (a \sin bx - b \cos bx) + c$

30. $\int e^{ax} \cos bx\, dx = \dfrac{e^{ax}}{a^2 + b^2} (a \cos bx + b \sin bx) + c.$

APPENDIX-4 TABLE I: NORMAL DISTRIBUTION

Areas under the Standard Normal Curve from 0 to Z

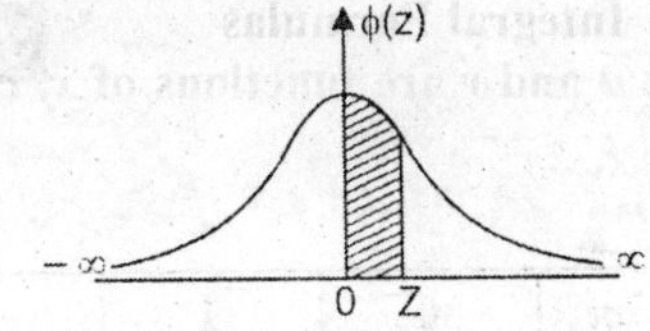

z	0	1	2	3	4	5	6	7	8	9
0.0	0.0000	0.0040	0.0080	0.0120	0.0160	0.0199	0.0239	0.0279	0.0319	0.0359
0.1	0.0398	0.0438	0.0478	0.0517	0.0557	0.0596	0.0636	0.0675	0.0714	0.0754
0.2	0.0793	0.0832	0.0871	0.0910	0.0948	0.0987	0.1026	0.1064	0.1103	0.1141
0.3	0.1179	0.1217	0.1255	0.1293	0.1331	0.1368	0.1406	0.1443	0.1480	0.1517
0.4	0.1554	0.1591	0.1628	0.1664	0.1700	0.1736	0.1772	0.1808	0.1844	0.1879
0.5	0.1915	0.1950	0.1985	0.2019	0.2054	0.2088	0.2123	0.2157	0.2190	0.2224
0.6	0.2258	0.2291	0.2324	0.2357	0.2389	0.2422	0.2454	0.2486	0.2518	0.2549
0.7	0.2580	0.2612	0.2642	0.2673	0.2704	0.2734	0.2764	0.2794	0.2823	0.2852
0.8	0.2881	0.2910	0.2939	0.2967	0.2995	0.3023	0.3051	0.3078	0.3106	0.3133
0.9	0.3159	0.3186	0.3212	0.3238	0.3264	0.3289	0.3315	0.3340	0.3365	0.3389
1.0	0.3413	0.3438	0.3461	0.3485	0.3508	0.3531	0.3554	0.3577	0.3599	0.3621
1.1	0.3643	0.3665	0.3686	0.3708	0.3729	0.3749	0.3770	0.3790	0.3810	0.3830
1.2	0.3849	0.3869	0.3888	0.3907	0.3925	0.3944	0.3962	0.3980	0.3997	0.4015
1.3	0.4032	0.4049	0.4066	0.4082	0.4099	0.4115	0.4131	0.4147	0.4162	0.4177
1.4	0.4192	0.4207	0.4222	0.4236	0.4251	0.4265	0.4279	0.4292	0.4306	0.4319
1.5	0.4332	0.4345	0.4357	0.4370	0.4382	0.4394	0.4406	0.4418	0.4429	0.4441
1.6	0.4452	0.4463	0.4474	0.4484	0.4495	0.4505	0.4515	0.4525	0.4535	0.4545
1.7	0.4554	0.4564	0.4573	0.4582	0.4591	0.4599	0.4608	0.4616	0.4625	0.4633
1.8	0.4641	0.4649	0.4656	0.4664	0.4671	0.4678	0.4686	0.4693	0.4699	0.4706
1.9	0.4713	0.4719	0.4726	0.4732	0.4738	0.4744	0.4750	0.4756	0.4761	0.4767
2.0	0.4772	0.4778	0.4783	0.4788	0.4793	0.4798	0.4803	0.4808	0.4812	0.4817
2.1	0.4821	0.4826	0.4830	0.4834	0.4838	0.4842	0.4846	0.4850	0.4854	0.4857
2.2	0.4861	0.4864	0.4868	0.4871	0.4875	0.4878	0.4881	0.4884	0.4887	0.4890
2.3	0.4893	0.4896	0.4898	0.4901	0.4904	0.4906	0.4909	0.4911	0.4913	0.4916
2.4	0.4918	0.4920	0.4922	0.4925	0.4927	0.4929	0.4913	0.4932	0.4934	0.4936
2.5	0.4938	0.4940	0.4941	0.4943	0.4945	0.4946	0.4948	0.4949	0.4951	0.4952
2.6	0.4953	0.4955	0.4956	0.4957	0.4959	0.4960	0.4961	0.4962	0.4963	0.4964
2.7	0.4965	0.4966	0.4967	0.4968	0.4969	0.4970	0.4971	0.4972	0.4973	0.4974
2.8	0.4974	0.4975	0.4976	0.4977	0.4977	0.4978	0.4979	0.4979	0.4980	0.4981
2.9	0.4981	0.4982	0.4982	0.4983	0.4984	0.4984	0.4985	0.4985	0.4986	0.4986
3.0	0.4987	0.4987	0.4987	0.4988	0.4988	0.4989	0.4989	0.4989	0.4990	0.4990
3.1	0.4990	0.4991	0.4991	0.4991	0.4992	0.4992	0.4992	0.4992	0.4993	0.4993
3.2	0.4993	0.4993	0.4994	0.4994	0.4994	0.4994	0.4994	0.4995	0.4995	0.4995
3.3	0.4995	0.4995	0.4995	0.4996	0.4996	0.4996	0.4996	0.4996	0.4996	0.4997
3.4	0.4997	0.4997	0.4997	0.4997	0.4997	0.4997	0.4997	0.4997	0.4997	0.4998
3.5	0.4998	0.4998	0.4998	0.4998	0.4998	0.4998	0.4998	0.4998	0.4998	0.4998
3.6	0.4998	0.4998	0.4999	0.4999	0.4999	0.4999	0.4999	0.4999	0.4999	0.4999
3.7	0.4999	0.4999	0.4999	0.4999	0.4999	0.4999	0.4999	0.4999	0.4999	0.4999
3.8	0.4999	0.4999	0.4999	0.4999	0.4999	0.4999	0.4999	0.4999	0.4999	0.4999
3.9	0.5000	0.5000	0.5000	0.5000	0.5000	0.5000	0.5000	0.5000	0.5000	0.5000

APPENDIX-4 TABLE II: VALUES OF e^{-m}

$(0 < m < 1)$

m	0	1	2	3	4	5	6	7	8	9
0.0	1.0000	0.9900	0.9802	0.9704	0.9608	0.9512	0.9418	0.9324	0.9231	0.9139
0.1	0.9048	0.8958	0.8869	0.8781	0.8694	0.8607	0.8521	0.8437	0.8353	0.8270
0.2	0.8187	0.8106	0.8025	0.7945	0.7866	0.7788	0.7711	0.7634	0.7558	0.7483
0.3	0.7408	0.7334	0.7261	0.7189	0.7118	0.7047	0.6977	0.6907	0.6839	0.6771
0.4	0.6703	0.6636	0.6570	0.6505	0.6440	0.6376	0.6313	0.6250	0.6188	0.6126
0.5	0.6065	0.6005	0.5945	0.5886	0.5827	0.5770	0.5712	0.5655	0.5599	0.5543
0.6	0.5488	0.5434	0.5379	0.5326	0.5273	0.5220	0.5169	0.5117	0.5066	0.5016
0.7	0.4966	0.4916	0.4868	0.4819	0.4771	0.4724	0.4677	0.4630	0.4584	0.4538
0.8	0.4493	0.4449	0.4404	0.4360	0.4317	0.4274	0.4232	0.4190	0.4148	0.4107
0.9	0.4066	0.4025	0.3985	0.3946	0.3906	0.3867	0.3829	0.3791	0.3753	0.3716

$(m = 1, 2, 3, ..., 10)$

m	1	2	3	4	5	6	7	8	9	10
e^{-m}	0.36788	0.13534	0.04979	0.01832	0.006738	0.002479	0.000912	0.000335	0.000123	0.000045

Note: To obtain value of e^{-m} for other values of m, use the laws of exponents.

Examples: $e^{-3.48} = (e^{-3.00})\,(e^{-0.48}) = (0.04979)\,(0.6188) = 0.03081.$

$e^{-1.3} = (e^{-1})\,(e^{-0.3}) = (0.36788)\,(0.7408) = 0.27253.$

$e^{-2.5} = (e^{-2})\,(e^{-0.5}) = (0.13534)\,(0.6065) = 0.08208.$

APPENDIX-5 SOLVED EXAMINATION PAPER AMIETE (NEW SCHEME)

Code: AE/56/AC56/AT56 Subject: Engineering Mathematics-II

AMIETE-ET/CS/IT

Time: 3 Hours Max. Marks: 100

June 2013

Note: There are 9 Questions in all.

- Question 1 is compulsory and carries 20 marks. Answer to Q.1 must be written in the space provided for it in the answer book supplied and nowhere else.
- The answer sheet for the Q.1 will be collected by the invigilator after 45 minutes of the commencement of the examination.
- Out of the remaining **EIGHT** Questions answer any **FIVE** Questions. Each question carries 16 marks.
- Any required data not explicitly given, may be suitably assumed and stated.

Q.1 Choose the correct or the best alternative in the following: **(2 × 10)**

a. The invariant points of the bilinear transformation $w = \dfrac{az+b}{cz+d}$

(A) $az^2 + (b-c)z - d = 0$ (B) $az^2 + (b+c) + d = 0$

✓(C) $cz^2 + (d-a)z - b = 0$ (D) $cz^2 + (d+a)z + b = 0$

b. If $f(A) \int_c \dfrac{2z^2 - z - 2}{z - a} dz$ where C is the circle $|z| = 2.5$, the value $f(2)$ is

(A) 0 (B) $2\pi i$

(C) $4\pi i$ ✓(D) $8\pi i$

c. Residue of $\dfrac{1-e^{2z}}{z^4}$ at its pole is

(A) $\dfrac{4}{3}$ ✓(B) $-\dfrac{4}{3}$

(C) $\dfrac{3}{4}$ (D) $-\dfrac{3}{4}$

d. A unit vector normal to the surface $x^2y + 2xz = 4$ at the point $(2, -2, 3)$ is

✓(A) $\dfrac{1}{3}(-i + 2j + 2k)$ *(see Example 1.23 on Page 1.16)*

(B) $\dfrac{1}{3}(i - 2j + 2k)$

(C) $\dfrac{1}{3}(i - 2j - 2k)$ (D) $\dfrac{1}{3}(i + 2j + 2k)$

e. If $\vec{A}$ is a constant vector and $\vec{R} = xi + yj + zk$, then $\text{div}(\vec{A} \times \vec{R})$ is

(A) $\vec{A}$ (B) $\vec{R}$

(C) $\vec{A} + \vec{R}$ ✓(D) 0 *(see Example 1.39(ii) on Page 1.27)*

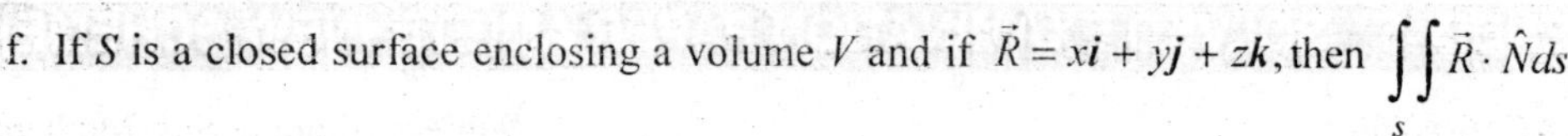

f. If S is a closed surface enclosing a volume V and if $\vec{R} = xi + yj + zk$, then $\iint_S \vec{R} \cdot \hat{N} ds$

(A) V (B) $2V$

✓(C) $3V$ (D) $4V$

g. If Δ is forward difference operator and interval of differencing is one; $\Delta^3[(1 - x)(1 - 2x)(1 - 3x)]$ is equal to

✓(A) –36 *(see Example 5.3 on Page 5.7)* (B) –18

(C) –12 (D) –6

h. The solution of partial differential equation $(y - z)p + (z - x)q = (x - y)$ is

(A) $x^2 + y^2 + z^2 = f(xyz)$ ✓(B) $x^2 + y^2 + z^2 = f(x + y + z)$

(C) $x + y + z = f(xyz)$ (D) None of these

i. If the mean and variance of a binomial distribution are 12 and 4 respectively, then the binomial distribution is

(A) $\left(\frac{1}{3} + \frac{2}{3}\right)^6$ (B) $\left(\frac{2}{3} + \frac{1}{3}\right)^{12}$

✓(C) $\left(\frac{1}{3} + \frac{2}{3}\right)^{18}$ (D) $\left(\frac{1}{3} + \frac{2}{3}\right)^{24}$

j. 'A' speaks the truth in 75% cases and 'B' in 80% of the cases. The percentage of cases in which they are likely to contradict each other in stating the same fact is

✓(A) 45% (B) 35%

(C) 30% (D) 25%

Answer any FIVE questions out of EIGHT Questions.

Each question carries 16 marks.

Q.2 a. Show that every analytic function $f(z) = u(x, y) + iv(x, y)$ defines two families of curves $u(x, y) = C_1$ and $v(x, y) = C_2$ which form an orthogonal system. **(8)**

(See Article 4.9, Property 2 of Analytic Function on Page 4.14.)

b. Find the bilinear transformation which maps the points 1, i, –1 into the points 0, 1, ∞.

Ans. $w = \frac{i(1 - z)}{1 + z}$.

Q.3 a. State and prove Cauchy's integral formula.

(See Article 4.14 on Page 4.40.)

b. Find the Laurent's series expansion of $\frac{Z^2 - 6Z - 1}{(Z - 1)(Z - 3)(Z + 2)}$ in the region $3 < |Z + 2| < 5$. **(8)**

Ans. $\frac{2}{z+2} + \frac{3}{(z+2)^2} + \frac{3^2}{(z+2)^3} + \cdots + \frac{1}{5}\left[1 + \frac{z+2}{5} + \frac{(z+2)^2}{5^2} + \frac{(z+2)^3}{5^3} + \cdots\right]$

Q.4 a. Find the angle between the surfaces $x^2 + y^2 + z^2 = 9$ and $Z = x^2 + y^2 - 3$ at the point (2, –1, 2).

(See Example 1.25 on Page 1.17.)

b. Show $\text{div}(\vec{A} \times \vec{B}) = \vec{B} \cdot \text{curl } \vec{A} - \vec{A} \cdot \text{curl } \vec{B}$.

(See Article 1.17 vector identity 3 on Page 1.31.)

Q.5 a. Apply Green's theorem to evaluate $\int_C [(y - \sin x)\,dx + \cos x\,dy]$ where C is the plane triangle enclosed by the lines $y = 0$, $y = \frac{2}{\pi}x$ and $x = \frac{\pi}{2}$. **(8)**

(See Example 1.72 on Page 1.56.)

b. For any closed surface S, use divergence theorem to evaluate

$$\int_S [x(y-z)\boldsymbol{i} + y(z-x)\boldsymbol{j} + z(x-y)\boldsymbol{k}]\cdot ds$$ **Ans.** Zero. **(8)**

Q.6 a. Find an approximate value of $\log_e 5$ by calculating to 4 decimal places by Simpson's $\frac{1}{3^{rd}}$ rule. $\int_0^5 \frac{dx}{4x+5}$ dividing the range into ten equal parts. **(8)**

(See Example 5.69(iv) on Page 5.81.)

b. Given the values

X	5	7	11	13	17
$f(x)$	150	392	1452	2366	5202

Evaluate $f(9)$, using
(i) Lagrange's formula (ii) Newton' divided difference formula

(See Example 5.39 on Page 5.42.)

Q.7 a. Use Charpit's method to solve $pxy + pq + qy = yz$. **(8)**

(See Example 2.32 on Page 2.21)

b. Use method of separation of variables to solve $3\frac{\partial U}{\partial x} + 2\frac{\partial U}{\partial y} = 0$, given that $U(x, 0) = 4e^{-x}$. **(8)**

(See Example 3.1 on Page 3.2)

Q.8 a. Two persons 'A' and 'B' toss an unbiased coin alternately on the understanding that the first who gets the head wins. If 'A' starts the game, compare their chances of winning. **(8)**

(Ans. 4 : 1)

b. An insurance company insured 2000 scooter drivers, 4000 car drivers and 6000 truck drivers. The probability of accident is 0.01, 0.03 and 0.15 respectively. One of the insured person meets an accident. What is the probability that he is a scooter driver? **(8)**

(See Example 7.49 on Page 7.21.)

Q.9 a. The diameter of an electric cable is assumed to be a continuous variate with probability density function given by $f(x) = Kx(1 - x)$, $0 \le x \le 1$. Find the number K. Also find the mean and the variance.

(Refer Q.27 on Page 7.41) **(2 + 3 + 3)**

b. Find the mean and variance of a Binomial Distribution. **(4 + 4)**

(See Page 7.43.)